THE PICTURE OF THE TAOIST GENII PRINTED ON THE COVER
of this book is part of a painted temple scroll, recent but traditional, given to
Mr Brian Harland in Szechuan province (1946). Concerning these four divinities,
of respectable rank in the Taoist bureaucracy, the following particulars have been
handed down. The title of the first of the four signifies 'Heavenly Prince', that
of the other three 'Mysterious Commander'.

At the top, on the left, is Liu *Thien Chün*, Comptroller-General of Crops and
Weather. Before his deification (so it was said) he was a rain-making magician
and weather forecaster named Liu Chün, born in the Chin dynasty about + 340.
Among his attributes may be seen the sun and moon, and a measuring-rod or
carpenter's square. The two great luminaries imply the making of the calendar, so
important for a primarily agricultural society, the efforts, ever renewed, to reconcile
celestial periodicities. The carpenter's square is no ordinary tool, but the gnomon
for measuring the lengths of the sun's solstitial shadows. The Comptroller-General
also carries a bell because in ancient and medieval times there was thought to be
a close connection between calendrical calculations and the arithmetical acoustics
of bells and pitch-pipes.

At the top, on the right, is Wên *Yuan Shuai*, Intendant of the Spiritual Officials
of the Sacred Mountain, Thai Shan. He was taken to be an incarnation of one of
the Hour-Presidents (*Chia Shen*), i.e. tutelary deities of the twelve cyclical characters
(see Vol. 4, pt. 2, p. 440). During his earthly pilgrimage his name was Huan Tzu-Yü
and he was a scholar and astronomer in the Later Han (b. + 142). He is seen
holding an armillary ring.

Below, on the left, is Kou *Yuan Shuai*, Assistant Secretary of State in the Ministry
of Thunder. He is therefore a late emanation of a very ancient god, Lei Kung.
Before he became deified he was Hsin Hsing, a poor woodcutter, but no doubt an
incarnation of the spirit of the constellation Kou-Chhen (the Angular Arranger),
part of the group of stars which we know as Ursa Minor. He is equipped with
hammer and chisel.

Below, on the right, is Pi *Yuan Shuai*, Commander of the Lightning, with his
flashing sword, a deity with distinct alchemical and cosmological interests. According to tradition, in his early life he was a countryman whose name was Thien Hua.
Together with the colleague on his right, he controlled the Spirits of the Five
Directions.

Such is the legendary folklore of common men canonised by popular acclamation.
An interesting scroll, of no great artistic merit, destined to decorate a temple wall,
to be looked upon by humble people, it symbolises something which this book has
to say. Chinese art and literature have been so profuse, Chinese mythological
imagery so fertile, that the West has often missed other aspects, perhaps more
important, of Chinese civilisation. Here the graduated scale of Liu Chün, at first
sight unexpected in this setting, reminds us of the ever-present theme of quantitative measurement in Chinese culture; there were rain-gauges already in the Sung
(+ 12th century) and sliding calipers in the Han (+ 1st). The armillary ring of
Huan Tzu-Yü bears witness that Naburiannu and Hipparchus, al-Naqqāsh and
Tycho, had worthy counterparts in China. The tools of Hsin Hsing symbolise that
great empirical tradition which informed the work of Chinese artisans and technicians all through the ages.

SCIENCE AND CIVILISATION IN CHINA

When they strive only to 'understand the high' without 'studying the low', how can their understanding of the high be right?

> CHHÊNG MING-TAO (+1032 to +1085)
> *Honan Chhêng Shih I Shu*, ch. 13, p. 1b.

If you do not understand what is earthly, how then will you understand what is heavenly?

> *Aurea Catena Homeri* (+1723)
> attrib. Joseph Kirchweger.

If one disdains the low and the near, and restlessly seeks the high and the far, skips over the steps and crosses the limits, one will be drifting on emptiness and vacuity, without anything to rely on. Can that be called 'reflecting on things at hand'?

> From LÜ TSU-CHHIEN's preface to the
> *Chin Ssu Lu*, +1176

The student must first of all know how to doubt.

> CHU HSI (+1130 to +1200)
> *Chin Ssu Lu*, ch. 3, para. 15.

Heaven has its generative power, Earth its numinous efficacy,
If a man can achieve the mastery of these, he will be able to attain to life immortal.

> *Tho Yo Tzu*
> (The Bellows-and-Tuyère Master;
> Sung or Yuan, between the +10th
> and the +14th century)
> In *Tao Tsang Chi Yao*,
> *hsia mao chi*, 5.

中國科學技術史

李約瑟著

冀朝鼎

SCIENCE AND CIVILISATION IN CHINA

BY

JOSEPH NEEDHAM, F.R.S., F.B.A.

MASTER OF GONVILLE AND CAIUS COLLEGE, CAMBRIDGE
FOREIGN MEMBER OF ACADEMIA SINICA

With the collaboration of

HO PING-YÜ, PH.D.

PROFESSOR OF CHINESE AT GRIFFITH
UNIVERSITY, BRISBANE

and

LU GWEI-DJEN, PH.D.

FELLOW OF LUCY CAVENDISH COLLEGE, CAMBRIDGE

VOLUME 5

CHEMISTRY AND CHEMICAL TECHNOLOGY

PART III: SPAGYRICAL DISCOVERY AND
INVENTION: HISTORICAL SURVEY, FROM
CINNABAR ELIXIRS TO SYNTHETIC INSULIN

CAMBRIDGE
AT THE UNIVERSITY PRESS
1976

Published by the Syndics of the Cambridge University Press
The Pitt Building, Trumpington Street, Cambridge, CB2 1RP
Bentley House, 200 Euston Road, London NW1 2DB
32 East 57th Street, New York, NY 10022, USA
296 Beaconsfield Parade, Middle Park, Melbourne 3206, Australia

Library of Congress catalogue card number: 54-4723

ISBN: 0 521 21028 3

First published 1976

Printed in Great Britain
at the
University Printing House, Cambridge
(Euan Phillips, University Printer)

To
two welcoming friends
expositors of chemistry in China

TSÊNG CHAO-LUN
Professor of Organic Chemistry at Peking University

and

HUANG TZU-CHHING
Professor of Physical Chemistry at Chhinghua University
1942 to 1946

as also in memory of
a great historian of chemistry

JAMES RIDDICK PARTINGTON
Professor of Chemistry at Queen Mary College, University of London
1919 to 1951

unassuming
unaffected
unpretentious
in learning, in factual exposition, in kindness, untiring
convivial, unforgettable,

this volume is dedicated

CONTENTS

List of Illustrations . . . . *page* xi

List of Tables xiv

List of Abbreviations xv

Acknowledgements xviii

Author's Note xix

33 ALCHEMY AND CHEMISTRY (*continued*) . *page* 1

(*e*) The historical development of alchemy and early chemistry, *p.* 1

   (1) The origins of alchemy in Chou, Chhin and Early Han;
      its relation with Taoism, *p.* 1

      (i) The School of Naturalists and the First Emperor, *p.* 12

      (ii) Aurifiction and aurifaction in the Han, *p.* 20

      (iii) The three roots of elixir alchemy, *p.* 44

   (2) Wei Po-Yang; the beginnings of alchemical literature in the
      Later Han (+2nd cent.), *p.* 50

   (3) Ko Hung, systematiser of Chinese alchemy (*c.* +300), and his
      times, *p.* 75

      (i) Fathers and masters, *p.* 75

      (ii) The *Pao Phu Tzu* book and its elixirs, *p.* 81

      (iii) Character and contemporaries, *p.* 106

   (4) Alchemy in the Taoist Patrology (*Tao Tsang*), *p.* 113

   (5) The golden age of alchemy; from the end of Chin (+400)
      to late Thang (+800), *p.* 117

      (i) The Imperial Elaboratory of the Northern Wei and the
         Taoist Church at Mao Shan, *p.* 117

      (ii) Alchemy in the Sui re-unification, *p.* 132

      (iii) Chemical theory and spagyrical poetry under the
         Thang, *p.* 141

      (iv) Chemical lexicography and classification in the Thang,
         p. 151

      (v) Buddhist echoes of Indian alchemy, *p.* 160

   (6) The silver age of alchemy; from the late Thang (+800) to the
      end of the Sung (+1300), *p.* 167

      (i) The first scientific printed book, and the court alchemist
         Mistress Kêng, *p.* 167

      (ii) From proto-chemistry to proto-physiology, *p.* 171

      (iii) Alchemy in Japan, *p.* 174

(iv) Handbooks of the Wu Tai, *p.* 180

(v) Theocratic mystification, and the laboratory in the National Academy, *p.* 182

(vi) The emperor's artificial gold factory under Metallurgist Wang Chieh, *p.* 186

(vii) Social aspects, conventional attitudes and gnomic inscriptions, *p.* 190

(viii) Alchemical compendia and books with illustrations, *p.* 196

(ix) The Northern and Southern Schools of Taoism, *p.* 200

(7) Alchemy in its decline; Yuan, Ming and Chhing, *p.* 208

(i) The Emaciated Immortal, Prince of the Ming, *p.* 209

(ii) Ben Jonson in China, *p.* 212

(iii) Chinese alchemy in the age of Libavius and Becher, *p.* 216

(iv) The legacy of the Chinese alchemical tradition, *p.* 219

(8) The coming of modern chemistry, *p.* 220

(i) The failure of the Jesuit mission, *p.* 221

(ii) Mineral acids and gunpowder, *p.* 237

(iii) A Chinese puzzle—eighth or eighteenth?, *p.* 242

(iv) The Kiangnan Arsenal and the sinisation of modern chemistry, *p.* 250

BIBLIOGRAPHIES   263

Abbreviations, *p.* 264

A. Chinese and Japanese books before +1800, *p.* 272

Concordance for *Tao Tsang* books and tractates, *p.* 323

B. Chinese and Japanese books and journal articles since +1800, *p.* 326

C. Books and journal articles in Western languages, *p.* 345

GENERAL INDEX   433

Table of Chinese Dynasties   479

Summary of the Contents of Volume 5   480

LIST OF ILLUSTRATIONS

Plates CDLXII to CDLXXVI will be found between pages 240 and 241.

1342 Chhih Sung Tzu, the Red Pine Master (from *Lieh Hsien Chhüan Chuan*, c. + 1580) . . . . . *page* 10

1343 Tomb of Hsü Fu (Jofuku) of the Chhin dynasty, in the coastal town of Shingū, south of Kyoto (orig. photo., 1971) (Pl. CDLXII) . . . . .

1344 The votive shrine of Hsü Fu at the Asuka Jinja, a Shinto temple on the outskirts of Shingū (orig. photo., 1971) (Pl. CDLXIII) . . . . . .

1345 Main hall of the Asuka Jinja (orig. photo., 1971) (Pl. CDLXIII) . . . . . .

1346 A bush of the drug-plant *tendai wuyaku* (*Lindera strychnifolia*) growing in the garden of the Asuka Jinja (orig. photo., 1971) (Pl. CDLXIV) . . . . .

1347 The representations on the painted silk banner buried with the Lady of Tai (–166) . . . . . *page* 22

1348 Li Shao-Chün praying in front of his alchemical furnace, which is visited by a crane (from *Shen Hsien Thung Chien*, + 1640) . . . . . *page* 30

1349 The ascension of the alchemical Taoist hierarch Mao Ying into the realm of the holy immortals (*Lieh Hsien Chhüan Chuan*) . . . . . . *page* 41

1350 A woman alchemist, Thai Hsüan Nü (*Lieh Hsien Chhüan Chuan*) . . . . . *page* 42

1351 Diagram of 'fire-times' (the programme for the heating and cooling of an elixir preparation in tune with the diurnal, lunar, monthly and annual cosmic cycles) to illustrate the *Tshan Thung Chhi* . . . . *page* 63

1352 Another diagram elucidating 'fire-times' . . . *page* 64

1353 A third diagram elucidating 'fire-times' . . . *page* 65

1353*a* Chhung-Hsü Kuan, a Taoist temple in the Lo-fou Shan mountains (Pl. CDLXV) . . . .

1354 The image and altar of Ko Hung in the votive temple dedicated to him at Hangchow (orig. photo., 1964) (Pl. CDLXVI) . . . . . .

1355 Graph of macrobiogens constructed from figures given in
      the *Pao Phu Tzu* book by Ko Hung, *c.* +320 .   .   *page* 86

1356 Thao Hung-Ching (+456 to +536), the great physician,
      alchemist and pharmaceutical botanist (*Lieh Hsien Chhüan
      Chuan*)   .   .   .   .   *page* 122

1357 Sublimation of calomel, mercurous chloride, *shui yin fên*;
      an illustration from a MS. copy of the *Pên Tshao Phin
      Hui Ching Yao*, +1505 (Pl. CDLXVII) .   .   .

1358 The eminent Sui and Thang physician and alchemist, Sun
      Ssu-Mo, d. +682 (*Lieh Hsien Chhüan Chuan*)   .   *page* 134

1359*a* Cross-section of an aludel or reaction-vessel drawn accord-
      ing to the specifications of Sun Ssu-Mo (Pl. CDLXVIII)

1359*b* Sealed two-part reaction-vessels after heating in an ex-
      perimental reconstruction of Sun Ssu-Mo's 'Scarlet Snow
      and Flowing Pearls Elixir' (Pl. CDLXVIII) .   .

1360 Bronze statue of the celebrated Taoist alchemist Lü
      Tung-Pin, at Thaiyuan (orig. photo., 1964) (Pl. CDLXIX)

1361 Lü Tung-Pin's image and altar at Chhing Shan (orig.
      photo., 1972) (Pl. CDLXX) .   .   .   .   .

1362 The development of symbolic notation in late Chinese
      alchemy (from the *Nei Chin Tan*, +1622)   .   *pages* 154–5

1362*a* Alchemical talismans, from the *Pao Phu Tzu* book
      (Pl. CDLXXI) .   .   .   .   .   .

1363 An indication from an Arabic source of Japan's richness
      in gold; the Queen of Wāqwāq and her court depicted
      in a coloured illustration from al-Qazwīnī's world
      geography '*Ajā'ib al-Makhlūqāt* (Pl. CDLXXII)   .

1364 Specimens of pharmaceutical chemicals from the +8th
      century preserved in the Shōsōin Treasury at Nara in
      Japan (Pl. CDLXXIII) .   .   .   .   .

1365 Passage from a manuscript of Hsü Kuang-Chhi, *c.* +1620, on
      the mercury-sulphur theory of metals (Pl. CDLXXIV)

1366 Another passage in the handwriting of Hsü Kuang-Chhi,
      on nitric acid (Pl. CDLXXV) .   .   .   .

1367 A corner of the Pai Yün Kuan (Taoist Temple of the White
      Clouds) at Peking (orig. photo., 1964) (Pl. CDLXXVI)

1368 An early nineteenth-century scene outside the Thung Jen
      Thang pharmacy at Peking .   .   .   .   .   *page* 251

1369   A page from the first book in Chinese dealing with modern
          chemistry, the *Po Wu Hsin Phien*, 1855 .   .   . *page* 253

1370   Illustration from the *Ko Chih Tshung Shu* (General Treatise
          on the Natural Sciences), ed. by Hsü Chien-Yin between
          1897 and 1901 .   .   .   . *page* 256

1371   Flasks and beakers from the *Ko Chih Tshung Shu*   . *page* 257

1372   Retorts, basins, crucibles and a wash-bottle from the *Ko Chih
          Tshung Shu* .   .   .   . *page* 258

1373   Some of the earliest illustrations of chemical apparatus in
          Japan, from the *Ranka Naigai Sanbō Hōten*, 1805 . *page* 260

LIST OF TABLES

106 Agents of immortality in the *Lieh Hsien Chuan* . . *page* 11

107 Symbolic correlations of the Five Elements in different interpretations of the *Tshan Thung Chhi* . . . 58

108 'Fire-times' in elixir preparation; the aegis of hexagrams during the day and night 61

109 'Fire-times' in elixir preparation; the aegis of trigrams during the lunation 62

110 Chart attempting a filiation of alchemists in Han and Chin (−2nd to +4th century) 77

111 Chemical preparations and elixirs described in the *Pao Phu Tzu* (*Nei Phien*), *c.* +320 . . . *pages* 90–7

112 The main alchemical tradition from Chin to Wu Tai (+4th to +10th century) . . . . . *page* 121

113 The Southern School of Taoism, in Sung and Yuan (+10th to +14th century) 202

LIST OF ABBREVIATIONS

The following abbreviations are used in the text and footnotes. For abbreviations used for journals and similar publications in the bibliographies, see pp. 264 ff.

B   Bretschneider, E. (1), *Botanicon Sinicum*.

CC   Chia Tsu-Chang & Chia Tsu-Shan (*1*), *Chung-Kuo Chih Wu Thu Chien* (Illustrated Dictionary of Chinese Flora), 1958.

CCIF   Sun Ssu-Mo, *Chhien Chin I Fang* (Supplement to the Thousand Golden Remedies), *c.* +660.

CHS   Pan Ku (and Pan Chao), *Chhien Han Shu* (History of the Former Han Dynasty), *c.* +100.

CJC   Juan Yuan, *Chhou Jen Chuan* (Biographies of Mathematicians and Astronomers), +1799. With continuations by Lo Shih-Lin, Chu Kho-Pao and Huang Chung-Chün. In *HCCC*, chs. 159 ff.

CLPT   Thang Shen-Wei *et al.* (ed.), *Chêng Lei Pên Tshao* (Reorganised Pharmacopoeia), ed. of +1249.

CSHK   Yen Kho-Chün (ed.), *Chhüan Shang-ku San-Tai Chhin Han San-Kuo Liu Chhao Wên* (Complete Collection of prose literature (including fragments) from remote antiquity through the Chhin and Han Dynasties, the Three Kingdoms, and the Six Dynasties), 1836.

CTPS   Fu Chin-Chhüan (ed.), *Chêng Tao Pi Shu Shih Chung* (Ten Types of Secret Books on the Verification of the Tao), early 19th cent.

HFT   Han Fei, *Han Fei Tzu* (Book of Master Han Fei), early −3rd cent.

HHS   Fan Yeh & Ssuma Piao, *Hou Han Shu* (History of the Later Han Dynasty), +450.

HNT   Liu An *et al.*, *Huai Nan Tzu* (Book of the Prince of Huai-Nan), −120.

ICK   Taki Mototane, *I Chi Khao* (*Iseki-kō*) (Comprehensive Annotated Bibliography of Chinese Medical Literature [Lost or Still Existing]), finished *c.* 1825, pr. 1831; repr. Tokyo 1933, Shanghai 1936.

K   Karlgren, *Grammata Serica* (dictionary giving the ancient forms and phonetic values of Chinese characters).

KHTT   Chang Yü-Shu (ed.), *Khang-Hsi Tzu Tien* (Imperial Dictionary of the Khang-Hsi reign-period), +1716.

Kr   Kraus, P. *Le Corpus des Écrits Jābiriens* (*Mémoires de l'Institut d'Égypte*, 1943, vol. 44, pp. 1–214).

LPC   Lung Po-Chien (*1*), *Hsien Tshun Pên Tshao Shu Lu* (Bibliographical Study of Extant Pharmacopoeias and Treatises on Natural History from all Periods).

MCPT   Shen Kua, *Mêng Chhi Pi Than* (Dream Pool Essays), +1089.

N   Nanjio, B., *A Catalogue of the Chinese Translations of the Buddhist Tripiṭaka*, with index by Ross (3).

NCCS   Hsü Kuang-Chhi, *Nung Chêng Chhüan Shu* (Complete Treatise on Agriculture), +1639.

PPT/NP   Ko Hung, *Pao Phu Tzu (Nei Phien)* (Book of the Preservation-of-Solidarity Master; Inner Chapters), c. +320.

PPT/WP *Idem (Wai Phien)*, the Outer Chapters.

PTKM   Li Shih-Chen, *Pên Tshao Kang Mu* (The Great Pharmacopoeia), +1596.

PWYF   Chang Yü-Shu (ed.), *Phei Wên Yün Fu* (encyclopaedia), +1711.

R   Read, Bernard E. *et al.*, Indexes, translations and précis of certain chapters of the *Pên Tshao Kang Mu* of Li Shih-Chen. If the reference is to a plant see Read (1); if to a mammal see Read (2); if to a bird see Read (3); if to a reptile see Read (4 or 5); if to a mollusc see Read (5); if to a fish see Read (6); if to an insect see Read (7).

RP   Read & Pak (1), Index, translation and précis of the mineralogical chapters in the *Pên Tshao Kang Mu*.

SC   Ssuma Chhien, *Shih Chi* (Historical Records), c. -90.

SF   Thao Tsung-I (ed.), *Shuo Fu* (Florilegium of (Unofficial) Literature), c. +1368.

SHC   *Shan Hai Ching* (Classic of the Mountains and Rivers), Chou and C/Han.

SIC   Okanishi Tameto, *Sung I-Chhien I Chi Khao* (Comprehensive Annotated Bibliography of Chinese Medical Literature in and before the Sung Period). Jen-min Wei-shêng, Peking, 1958.

SKCS   *Ssu Khu Chhüan Shu* (Complete Library of the Four Categories), +1782; here the reference is to the *tshung-shu* collection printed as a selection from one of the seven imperially commissioned MSS.

SKCS/TMTY Chi Yün (ed.), *Ssu Khu Chhüan Shu Tsung Mu Thi Yao* (Analytical Catalogue of the *Complete Library of the Four Categories*), +1782; the great bibliography of the imperial MS. collection ordered by the Chhien-Lung emperor in +1772.

SNPTC   *Shen Nung Pên Tshao Ching* (Classical Pharmacopoeia of the Heavenly Husbandman), C/Han.

SSIW   Toktaga (Tho-Tho) *et al.*; Huang Yü-Chi *et al.* & Hsü Sung *et al. Sung Shih I Wên Chih, Pu, Fu Phien* (A Conflation of the Bibliography and Appended Supplementary Bibliographies of the History of the Sung Dynasty). Com. Press, Shanghai, 1957.

SYEY   Mei Piao, *Shih Yao Erh Ya* (The Literary Expositor of Chemical Physic; or, Synonymic Dictionary of Minerals and Drugs), +806.

TKKW   Sung Ying-Hsing, *Thien Kung Khai Wu* (The Exploitation of the Works of Nature), +1637.

TPHMF   *Thai-Phing Hui Min Ho Chi Chü Fang* (Standard Formularies of the (Government) Great Peace People's Welfare Pharmacies), +1151.

TPYL Li Fang (ed.), *Thai-Phing Yü Lan* (the Thai-Phing reign-period (Sung) Imperial Encyclopaedia), +983.

TSCC Chhen Mêng-Lei *et al.* (ed.), *Thu Shu Chi Chhêng* (the Imperial Encyclopaedia of +1726). Index by Giles, L. (2).

TSCCIW Liu Hsü *et al.* & Ouyang Hsiu *et al.*; *Thang Shu Ching Chi I Wên Ho Chih*. A conflation of the Bibliographies of the *Chiu Thang Shu* by Liu Hsü (H/Chin, +945) and the *Hsin Thang Shu* by Ouyang Hsiu & Sung Chhi (Sung, +1061). Com. Press, Shanghai, 1956.

TT Wieger, L. (6), *Taoïsme*, vol. 1, Bibliographie Générale (catalogue of the works contained in the Taoist Patrology, *Tao Tsang*).

TTC *Tao Tê Ching* (Canon of the Tao and its Virtue).

TTCY Ho Lung-Hsiang & Phêng Han-Jan (ed.). *Tao Tsang Chi Yao* (Essentials of the Taoist Patrology), pr. 1906.

TW Takakusu, J. & Watanabe, K., *Tables du Taishō Issaikyō* (nouvelle édition (Japonaise) du Canon bouddhique chinoise), Index-catalogue of the Tripiṭaka.

V Verhaeren, H. (2) (ed.), Catalogue de la Bibliothèque du Pé-T'ang (the Pei Thang Jesuit Library in Peking).

WCTY/CC Tsêng Kung-Liang (ed.), *Wu Ching Tsung Yao* (*Chhien Chi*), military encyclopaedia, first section, +1044.

YCCC Chang Chün-Fang (ed.), *Yün Chi Chhi Chhien* (Seven Bamboo Tablets of the Cloudy Satchel), Taoist collection, +1022.

YHL Thao Hung-Ching (attrib.), *Yao Hsing Lun* (Discourse on the Natures and Properties of Drugs).

YHSF Ma Kuo-Han (ed.), *Yü Han Shan Fang Chi I Shu* (Jade-Box Mountain Studio collection of (reconstituted and sometimes fragmentary) Lost Books), 1853.

ACKNOWLEDGEMENTS

LIST OF THOSE WHO HAVE KINDLY READ THROUGH SECTIONS IN DRAFT

The following list, which applies only to Vol. 5, pts 2–5, brings up to date those printed in Vol. 1, pp. 15 ff., Vol. 2, p. xxiii, Vol. 3, pp. xxxix ff., Vol. 4, pt. 1, p. xxi, Vol. 4, pt. 2, p. xli and Vol. 4, pt. 3, pp. xliii ff.

Prof. Derk Bodde (Philadelphia)	Introductions.
Mr J. Charles (Cambridge)	Metallurgical chemistry.
Prof. A. G. Debus (Chicago)	Modern chemistry (Mao Hua).
Prof. A. F. P. Hulsewé (Leiden)	Theories.
Dr Edith Jachimowicz (London)	Comparative (Arabic).
Mr S. W. K. Morgan (Bristol)	Metallurgy (zinc and brass).
The late Prof. Ladislao Reti (Milan)	Apparatus (alcohol).
Dr Kristofer M. Schipper (Paris)	Theories.
Prof. R. B. Serjeant (Cambridge)	Comparative (Arabic).
Mr H. J. Sheppard (Warwick)	Introductions.
Prof. Cyril Stanley Smith (Cambridge, Mass.)	Metallurgy, and Theories.
Mr Robert Somers (New Haven, Conn.)	Theories.
Dr Michel Strickmann (Kyoto)	Theories.
Dr Mikuláš Teich (Cambridge)	Introductions.
Mr R. G. Wasson (Danbury, Conn.)	Introduction (ethno-mycology).
Mr James Zimmerman (New Haven, Conn.)	Theories.

AUTHOR'S NOTE

It is now nearly a dozen years since the preface for Vol. 4 of this series (Physics and Physical Technology) was written; since then much has been done towards the later volumes. We are now happy to be able to present a substantial part of Vol. 5 (Spagyrical Discovery and Invention), i.e. alchemy and early chemistry, which go together with the arts of peace and war, including military and textile technology, mining, metallurgy and ceramics. The point of this arrangement was explained in the preface of Vol. 4 (e.g. pt. 3, p. 1). Exigences not of logic but of collaboration are making it obligatory that these other topics should follow rather than precede the central theme of chemistry, which here is printed as Vol. 5, parts 2, 3, 4 and 5, leaving parts 1 and 6 to appear at a later date.

The number of physical volumes (parts) which we are now producing may give the impression that our work is enlarging according to some form of geometrical progression or along some exponential curve, but this would be largely an illusion, because in response to the reactions of many friends we are now making a real effort to publish in books of less thickness, more convenient for reading. At the same time it is true that over the years the space required for handling the history of the diverse sciences in Chinese culture has proved singularly unpredictable. One could (and did) at the outset arrange the sciences in a logical spectrum (mathematics—astronomy—geology and mineralogy—physics—chemistry—biology) leaving estimated room also for all the technologies associated with them; but to foresee exactly how much space each one would claim, that, in the words of the Jacobite blessing, was 'quite another thing'. We ourselves are aware that the disproportionate size of some of our Sections may give a mis-shapen impression to minds enamoured of classical uniformity, but our material is not easy to 'shape', perhaps not capable of it, and appropriately enough we are constrained to follow the Taoist natural irregularity and surprise of a romantic garden rather than to attempt any compression of our lush growths within the geometrical confines of a Cartesian parterre. The Taoists would have agreed with Richard Baxter that ''tis better to go to heaven disorderly than to be damned in due order'. By some strange chance our spectrum meant (though I thought at the time that the mathematics was particularly difficult) that the 'easier' sciences were going to come first, those where both the basic ideas and the available source-materials were relatively clear and precise. As we proceeded, two phenomena manifested themselves: first, the technological achievements and amplifications proved far more formidable than expected (as was the case in Vol. 4, pts. 2 and 3); and secondly, we found ourselves getting into ever deeper water, as the saying is, intellectually (as will fully appear in the Sections on medicine in Vol. 6).

Alchemy and early chemistry, the central subjects of the present volume, exemplified the second of these difficulties well enough, but they have had others of their own. At one time I almost despaired of ever finding our way successfully through the inchoate mass of ideas, and the facts so hard to establish, relating to alchemy, chemistry, metallurgy and chemical industry in ancient, medieval and traditional China. The facts indeed were much more difficult to ascertain, and also more perplexing to interpret, than anything encountered in subjects such as astronomy or civil engineering. And in the end, one must say, we did not get through without cutting great swathes of briars and bracken, as it were, through the muddled thinking and confused terminology of the traditional history of alchemy and early chemistry in the West. Here it was indispensable to distinguish alchemy from proto-chemistry and to introduce words of art such as aurifiction, aurifaction and macrobiotics. It is also fair to say that the present subject has been far less well studied and understood either by Westerners or Chinese scholars themselves than fields like astronomy and mathematics, where already in the eighteenth century a Gaubil could do outstanding work, and nearer our own time a Chhen Tsun-Kuei, a de Saussure, and a Mikami Yoshio could set them largely in order. If the study of alchemy and early chemistry had advanced anything like so far, it would be much easier today than it actually is to differentiate with clarity between the many divergent schools of alchemists at the many periods, from the −3rd century to the +17th, with which we have to deal. More adequate understanding would also have been achieved with regard to that crucial Chinese distinction between inorganic laboratory alchemy and physiological alchemy, the former concerned with elixir preparations of mineral origin, the latter rather with operations within the adept's own body; a distinction hardly realised to the full in the West before the just passed decade. As we shall show in these volumes, there was a synthesis of these two age-old trends when in iatro-chemistry from the Sung onwards laboratory methods were applied to physiological substances, producing what we can only call a proto-biochemistry. But this will be read in its place.

Now a few words on our group of collaborators. Dr Ho Ping-Yü,[1] since 1972 Professor of Chinese and Dean of the Faculty of Asian Studies at Griffith University, Brisbane, in Queensland, was introduced to readers in Vol. 4, pt. 3, p. lv; here he has been responsible for drafting the major part of the sub-section on the history of alchemy in China. Dr Lu Gwei-Djen,[2] my oldest collaborator, dating (in historian's terms) from 1937, has been involved at all stages of the present volumes, especially in that seemingly endless mental toil of ours which resulted in the introductory sub-sections on concepts, definitions and terminology, with all that that implies for theories of alchemy, ideas of immortality, and the physiological pathology of the elixir complex. But her particular domain has been that of physiological alchemy, and it was her discoveries, just at the right moment, of what was meant by the three primary vitalities, mutationist inversion, counter-current flow, and such abstruse matters, which alone permitted the unravelling, at least in the provisional form here presented (in the relevant sub-section j), of that strange and unfamiliar system, quasi-Yogistic perhaps, but full of interest for the pre-history of biochemical thought.[a] A third collaborator is now to be welcomed for the first time, Dr Nathan Sivin, Professor

[a] Some of her findings have appeared separately (Lu Gwei-Djen, 2).

[1] 何丙郁   [2] 魯桂珍

at the Massachusetts Institute of Technology, who has contributed the sub-section on the general theory of elixir alchemy.

Although Prof. Sivin has helped the whole group much by reading over and suggesting emendations for all the rest, it is needful to make at this point a proviso which has not been required in previous volumes. This is that my collaborators cannot take a collective responsibility for statements, translations or even general nuances, occurring in parts of the book other than that or those in which they each themselves directly collaborated. All incoherences and contradictions which remain after our long discussions must be laid at my door, in answer to which I can only say that the state of the art is as yet very imperfect, that it will certainly be improved by later scholars, and that in the meantime we have done the best we can. If fate had granted to the four of us the possibility of all working together in one place for half-a-dozen years, things could have been rather different, but in fact Prof. Ho and Prof. Sivin were never even in Cambridge at one and the same time. Thus these volumes have come into existence the hard way, drafted by different hands at fairly long intervals of time, and still no doubt containing traces of various levels of sophistication and understanding. Indeed it would have been reasonable to mark the elixir theory sub-section 'by Nathan Sivin', rather than 'with Nathan Sivin', if it had not been for the fact that some minor embroideries were offered by me, and that a certain part of it, not perhaps the least interesting, is a revised version of a memoir by Ho Ping-Yü and myself first published in 1959. Lacking the unities of time and place, complete credal unity, as it were, has been unattainable, but that does not mean that we are not broadly at one over the main facts and problems of the field as a whole; so that rightly we may be called co-workers.

Besides this I am eager to make certain further acknowledgements. During the second world war I was instrumental in securing for Cambridge copies of the *Tao Tsang* and the *Tao Tsang Chi Yao*. At a somewhat later time (1951–5) Dr Tshao Thien-Chhin,[1] then a Fellow of Caius, made a most valuable pioneer study of the alchemical books in the Taoist Patrology, using a microfilm set in our working collection (now the East Asian History of Science Library, an educational Trust). After his return to the Biochemical Institute of Academia Sinica, Shanghai, of which he has been in recent years Vice-Director, these notes were of great help to Dr Ho and myself, forming the ultimate basis for another sub-section, that on aqueous reactions. Secondly, when we were faced with the fascinating but difficult study of the evolution of chemical apparatus, especially for distillation, in East and West, Dr Dorothy Needham put in a considerable amount of work, including some drafting, in what happened to be a convenient interval in work on her own book on the history of muscle biochemistry, *Machina Carnis*. She has also read all our pages—perhaps the only person in the world who ever does so!

While readers of sub-sections in typescript and proof have not been as numerous, perhaps, as for previous volumes, a special debt of gratitude is due to Mr J. A. Charles of St John's College, chemist, metallurgist and archaeologist, whose advice to Prof. Ho

[1] 曹天欽

and myself from the earliest days has been extremely precious. Valuable consultations also took place with Mr H. J. Sheppard of Warwick, especially during his time in Cambridge as a Schoolmaster-Fellow of Churchill College. And a special debt is owing to the late Prof. Ladislao Reti of Milan, who advised us with a lifetime's experience of chemical industry on many subjects connected with distillation and other processes; no correspondence was ever too tedious for him. Few chemists in Cambridge, by some chance, happen to be interested at the present time in the history of their subject, but if Dr A. J. Berry and Prof. J. R. Partington had lived we could have profited greatly from their help. With the latter, indeed, we did have fruitful and most friendly contact, but it was in connection mainly with the gunpowder epic, Prof. Wang Ling[1] and I endeavouring, not unsuccessfully, to convince him of the real and major contribution of China in that field; those were days however before any word of the present volume had been written. In 1968, well after it had been started, there was convened the First International Conference of Taoist Studies at the Villa Serbelloni at Bellagio on Lake Como; Ho Ping-Yü, Nathan Sivin and myself were all of the party, and here much stimulus was obtained from that remarkable *Tao shih* Kristofer Schipper—hence the unexpected sub-section on liturgiology and alchemical origins in our introductory material. Four years later we met again for the Second Conference, at Tateshina near Chino in Japan, and some of the material in the present part was there contributed to a stimulating discussion of the social aspects of Taoist alchemy. In addition to the invaluable advice of many other colleagues in particular areas, we record especially the kindness of Professor Cyril Stanley Smith in commenting upon the whole sub-sections on metallurgy and on the theory of elixir alchemy. Dr N. Sivin also expresses his gratitude to Prof. A. F. P. Hulsewé and his staff for the open-hearted hospitality which they gave him during the gestation of the latter study, carried out almost entirely at the Sinologisch Instituut, Leiden.

It is right to record that certain parts of these volumes have been given as lectures to bodies honouring us by such invitations. Thus various excerpts from the introductory sub-sections on concepts, terminology and definitions were given for the Rapkine Lecture at the Pasteur Institute in Paris (1970) and the Bernal Lecture at Birkbeck College in London in the following year. Portions of the historical sub-sections, especially that on the coming of modern chemistry, were used for the Ballard Matthews Lectures of the University of Wales at Bangor. A considerable part of the physiological alchemy material formed the basis of the Fremantle Lectures at Balliol College, Oxford,[a] and had been given more briefly as the Harvey Lecture to the Harveian Society of London the year before. Lastly, four lectures covering the four present parts of this volume were given at the Collège de France in Paris at Easter, 1973, in fulfilment of my duties as Professeur Étranger of that noble institution.

If there is one question more than any other raised by this present Section 33 on alchemy and early chemistry, now offered to the republic of learning in these volumes,

[a] The relevant volume is therefore offered to the Trustees of the late Sir Francis Fremantle's benefaction in discharge of the duty of publication of his Lectures.

[1] 王鈴

it is that of human unity and continuity. In the light of what is here set forth, can we allow ourselves to visualise that some day before long we shall be able to write the history of man's enquiry into chemical phenomena as one single development throughout the Old World cultures? Granted that there were several different foci of ancient metallurgy and primitive chemical industry, how far was the gradual flowering of alchemy and chemistry a single endeavour, running contagiously from one civilisation to another?

It is a commonplace of thought that some forms of human experience seem to have progressed in a more obvious and palpable way than others. It might be difficult to say how Michael Angelo could be considered an improvement on Pheidias, or Dante on Homer, but it can hardly be questioned that Newton and Pasteur and Einstein did really know a great deal more about the natural universe than Aristotle or Chang Hêng. This must tell us something about the differences between art and religion on one side and science on the other, though no one seems able to explain quite what, but in any case within the field of natural knowledge we cannot but recognise an evolutionary development, a real progress, over the ages. The cultures might be many, the languages diverse, but they all partook of the same quest.

Throughout this series of volumes it has been assumed all along that there is only one unitary science of Nature, approached more or less closely, built up more or less successfully and continuously, by various groups of mankind from time to time. This means that one can expect to trace an absolute continuity between the first beginnings of astronomy and medicine in Ancient Babylonia, through the advancing natural knowledge of medieval China, India, Islam and the classical Western world, to the break-through of late Renaissance Europe when, as has been said, the most effective method of discovery was itself discovered. Many people probably share this point of view, but there is another one which I may associate with the name of Oswald Spengler, the German world-historian of the thirties whose works, especially *The Decline of the West*, achieved much popularity for a time. According to him, the sciences produced by different civilisations were like separate and irreconcilable works of art, valid only within their own frames of reference, and not subsumable into a single history and a single ever-growing structure.

Anyone who has felt the influence of Spengler retains, I think, some respect for the picture he drew of the rise and fall of particular civilisations and cultures, resembling the birth, flourishing and decay of individual biological organisms, in human or animal life-cycles. Certainly I could not refuse all sympathy for a point of view so like that of the Taoist philosophers, who always emphasised the cycles of life and death in Nature, a point of view that Chuang Chou himself might well have shared. Yet while one can easily see that artistic styles and expressions, religious ceremonies and doctrines, or different kinds of music, have tended to be incommensurable, for mathematics, science and technology the case is altered—man has always lived in an environment essentially constant in its properties, and his knowledge of it, if true, must therefore tend towards a constant structure.

This point would not perhaps need emphasis if certain scholars, in their anxiety to do justice to the differences between the ancient Egyptian or the medieval Chinese, Arabic or Indian world-views and our own, were not sometimes tempted to follow lines of thought which might lead to Spenglerian pessimism.[a] Pessimism I say, because of course he did prophesy the decline and fall of modern scientific civilisation. For example, our own collaborator, Nathan Sivin, has often pointed out, quite rightly, that for medieval and traditional China 'biology' was not a separated and defined science. One gets its ideas and facts from philosophical writings, books on pharmaceutical natural history, treatises on agriculture and horticulture, monographs on groups of natural objects, miscellaneous memoranda and so on. He urged that to speak without reservations of 'Chinese biology' would be to imply a structure which historically did not exist, disregarding mental patterns which did exist. Taking such artificial rubrics too seriously would also imply the natural but perhaps erroneous assumption that medieval Chinese scientists were asking the same questions about the living world as their modern counterparts in the West, and merely chanced, through some quirk of national character, language, economics, scientific method or social structure, to find different answers. On this approach it would not occur to one to investigate what questions the ancient and medieval Chinese scientists themselves were under the impression that they were asking. A fruitful comparative history of science would have to be founded not on the counting up of isolated discoveries, insights or skills meaningful for us now, but upon 'the confrontation of integral complexes of ideas with their interrelations and articulations intact'. These complexes could be kept in one piece only if the problems which they were meant to solve were understood. Chinese science must, in other words, be seen as developing out of one state of theoretical understanding into another, rather than as any kind of abortive development towards modern science.

All this was well put; of course one must not see in traditional Chinese science simply a 'failed prototype' of modern science, but the formulation here has surely to be extremely careful. There is a danger to be guarded against, the danger of falling into the other extreme, and of denying the fundamental continuity and universality of all science. This could be to resurrect the Spenglerian conception of the natural sciences of the various dead (or even worse, the living) non-European civilisations as totally separate, immiscible thought-patterns, more like distinct works of art than anything else, a series of different views of the natural world irreconcilable and un-

---

[a] Just recently a relevant polemical discussion has been going on among geologists. Harrington (1, 2), who had traced interesting geological insights in Herodotus and Isaiah, was taken to task by Gould (1), maintaining that 'science is no march to truth, but a series of conceptual schemes each adapted to a prevailing culture', and that progress consists in the mutation of these schemes, new concepts of creative thinkers resolving anomalies of old theories into new systems of belief. This was evidently a Kuhnian approach, but no such formulation will adequately account for the gradual percolation of true knowledge through the successive civilisations, and its general accumulation. Harrington himself, in his reply (3), maintained that 'there is a singular state of Nature towards which all estimates of reality converge', and therefore that we can and should judge the insights of the ancients on the basis of our own knowledge of Nature, while at the same time making every effort to understand their intellectual framework. In illustration he took the medieval Chinese appreciation of the meaning of fossil remains (cf. Vol. 3, pp. 611 ff.). We are indebted to Prof. Claude Albritton of Texas for bringing this discussion to our notice.

connected. Such a view might be used as the cloak of some historical racialist doctrine, the sciences of pre-modern times and the non-European cultures being thought of as wholly conditioned ethnically, and rigidly confined to their own spheres, not part of humanity's broad onward march. Moreover it would leave little room for those actions and reactions that we are constantly encountering, deep-seated influences which one civilisation had upon another.

In another place Nathan Sivin has written: 'The question of why China never spontaneously experienced the equivalent of our scientific revolution lies of course very close to the core of a comparative history of science. My point is that it is an utter waste of time, and distracting as well, to expect any answer until the Chinese tradition has been adequately comprehended from the inside.' The matter could not be better put; we must of course learn to see instinctively through the eyes of those who thought in terms of the Yin and Yang, the Five Elements, the symbolic correlations, and the trigrams and hexagrams of the *Book of Changes*. But here again this formulation might suggest a purely internalist or ideological explanation for the failure of modern natural science to arise in Chinese culture. I don't think that in the last resort we shall be able to appeal primarily to inhibiting factors inherent in the Chinese thought-world considered as an isolated Spenglerian cell. One must always expect that some of these intellectual limiting factors will be identifiable, but for my part I remain sceptical that there are many factors of this kind which could not have been overcome if the social and economic conditions had been favourable for the development of modern science in China. It may indeed be true that the modern forms of science which would then have developed would have been rather different from those which actually did develop in the West, or in a different order, that one cannot know. There was, for example, the lack of Euclidean geometry and Ptolemaic planetary astronomy in China, but China had done all the ground-work in the study of magnetic phenomena, an essential precursor of later electrical science;[a] and Chinese culture was permeated by conceptions much more organic, less mechanistic, than that of the West.[b] Moreover Chinese culture alone, as we shall see, perhaps, provided that materialist conception of the elixir of life which, passing to Europe through the Arabs, led to the macrobiotic optimism of Roger Bacon and the iatro-chemical revolution of Paracelsus, hardly less important in the origins of modern science than the work of Galileo and Newton. Whatever the ideological inhibiting factors in the Chinese thought-world may turn out to have been, the certainty always remains that the specific social and economic features of traditional China were connected with them. They were clearly part of that particular pattern, and in these matters one always has to think in terms of a 'package-deal'. In just the same way, of course, it is impossible to separate the scientific achievements of the ancient Greeks from the fact that they developed in mercantile, maritime, city-state democracies.

To sum it up, the failure of China to give rise to distinctively modern science while having been in many ways ahead of Europe for some fourteen previous centuries is

a See our discussions in Vols. 3 and 4, pt. 1.
b This was emphasised in Vol. 2, *passim*.

going to take some explaining.[a] Internalist historiography is likely to encounter grave difficulties here, in my opinion, because the intellectual, philosophical, theological and cultural systems of ideas of the Asian civilisations are not going to be able to take the causal stress and strain required. Some of these idea-systems, in fact, such as Taoism and Neo-Confucianism, would seem to have been much more congruent with modern science than any of the European ones were, including Christian theology. Very likely the ultimate explanations will turn out to be highly paradoxical—aristocratic military feudalism seeming to be much stronger than bureaucratic feudalism but actually weaker because less rational—the monotheism of a personal creator God being able to generate modern scientific thought (as the San Chiao could never do) but not to give it an inspiration enduring into modern times—and so on. We do not yet know.

A similar problem has of late been worrying Said Husain Nasr, the Persian scholar who is making valuable contributions to the history of science in Islam. He, for his part, faces the failure of Arabic civilisation to produce modern science. But far from regretting this he makes a positive virtue of it, rejecting belief in any integral, social-evolutionary development of science. Opening one of his recent books we read as follows:[b]

The history of science is often regarded today as the progressive accumulation of techniques and the refinement of quantitative methods in the study of Nature. Such a point of view considers the present conception of science to be the only valid one; it therefore judges the sciences of other civilisations in the light of modern science, and evaluates them primarily with respect to their 'development' with the passage of time. Our aim in this work, however, is not to examine the Islamic sciences from the point of view of modern science and of this 'evolutionist' conception of history; it is on the contrary to present certain aspects of the Islamic sciences as seen from the Islamic point of view.

Now Nasr considers that the Sufis and the universal philosophers of medieval Islam sought and found a kind of mystical *gnosis*, or cosmic *sapientia*, in which all the sciences 'knew their place', as it were (like servitors in some great house of old), and ministered to mystical theology as the highest form of human experience. In Islam, then, the philosophy of divinity was indeed the *regina scientiarum*. Anyone with some appreciation of theology as well as science cannot help sympathising to some extent with this point of view, but it does have two fatal drawbacks: it denies the equality of the forms of human experience, and it divorces Islamic natural science from the grand onward-going movement of the natural science of all humanity. Nasr objects to judging medieval science by its outward 'usefulness' alone. He writes:[c] 'However important its uses may have been in calendrical computation, in irrigation or in architecture, its ultimate aim always was to relate the corporeal world to its basic spiritual principle through the knowledge of those symbols which unite the various orders of reality. It can only be understood, and should only be judged, in

---

[a] We set forth in a preliminary way what is at issue here in Vol. 3, pp. 150ff. Some 'thinking aloud' done at various times has also been assembled in Needham (65).

[b] (1), p. 21.   [c] (1), pp. 39–40.

terms of its own aims and its own perspectives.' I would demur. It was part, I should want to maintain, of all human scientific enterprise, in which there is neither Greek nor Jew, neither Hindu nor Han. 'Parthians, Medes and Elamites, and the dwellers in Mesopotamia, and in Judaea and Cappadocia, in Pontus and Asia . . . and the parts of Libya about Cyrene. . .we do hear them speak in our tongues the marvellous works of God.'[a]

The denial of the equality of the forms of human experience comes out clearly in another work of Said Husain Nasr (2). Perhaps rather under-estimating the traditional high valuation placed within Christendom upon Nature—'that universal and publick manuscript', as Sir Thomas Browne said,[b] 'which lies expans'd unto the eyes of all'— he sees in the scientific revolution at the Renaissance a fundamental desacralisation of Nature, and urges that only by re-consecrating it, as it were, in the interests of an essentially religious world-view, will mankind be enabled to save itself from otherwise inevitable doom. If the rise of modern science within the bosom of Christendom alone had any causal connections with Christian thought that would give it a bad mark in his view. 'The main reason why modern science never arose in China or Islam', he says,[c]

is precisely because of the presence of a metaphysical doctrine and a traditional religious structure which refused to make a profane thing of Nature. . . .Neither in Islam, nor India nor the Far East, was the substance and the stuff of Nature so depleted of a sacramental and spiritual character, nor was the intellectual dimension of these traditions so enfeebled, as to enable a purely secular science of Nature and a secular philosophy to develop outside the matrix of the traditional intellectual orthodoxy. . . .The fact that modern science did not develop in Islam is not a sign of decadence [or incapacity] as some have claimed, but of the refusal of Islam to consider any form of knowledge as purely secular, and divorced from what it conceived to be the ultimate goal of human existence.

These are striking words,[d] but are they not tantamount to saying that only in Europe did the clear differentiation of the forms of experience arise? In other terms, Nasr looks for the synthesis of the forms of experience in the re-creation of a medieval world-view, dominated by religion,[e] not in the existential activity of individual human beings dominated by ethics. That would be going back, and there is no going back. The scientist must work *as if* Nature was 'profane'. As Giorgio di Santillana has said:[f]

Copernicus and Kepler believed in cosmic vision as much as any Muslim ever did, but when they had to face the 'moment of truth' they chose a road which was apparently not that of *sapientia*; they felt they had to state what appeared to be the case, and that on the whole it would be more respectful of divine wisdom to act thus.

[a] Acts, 2. 1.
[b] *Religio Medici*, I, xvi. 'Thus there are two Books from whence I collect my Divinity; besides that written one of God, another of his servant Nature . . . .'
[c] (2), p. 97.
[d] Views such as this are by no means restricted to Muslim scholars. From within the bosom of Christendom a very similar attitude is to be found in the book on alchemy by Titus Burckhardt (1), cf. esp. pp. 66, 203.
[e] It seems very strange to us that he should regard Chinese culture as having been dominated by religion at any time.   [f] In his preface to Said Husain Nasr (1), p. xii.

And perhaps it is a sign of the weakness of what can only be called so conservative a conception that Nasr is driven to reject the whole of evolutionary fact and theory, both cosmic, biological and sociological.

In meditating on the view of modern physical science as a 'desacralisation of Nature' many ideas and possibilities come to mind, but one very obvious cause for surprise is that it occurred in Christendom, the home of a religion in which an incarnation had sanctified the material world, while it did not occur in Islam, a culture which had never developed a soteriological doctrine.[a] This circumstance might offer an argument in favour of the primacy of social and economic factors in the break-through of the scientific revolution. It may be that while ideological, philosophical and theological differences are never to be undervalued, what mattered most of all were the facilitating pressures of the transition from feudalism to mercantile and then industrial capitalism, pressures which did not effectively operate in any culture other than that of Western, Frankish, Europe.

In another place Nasr wonders what Ibn al-Haitham or al-Bīrūnī or al-Khāzinī would have thought about modern science. He concludes that they would be amazed at the position which exact quantitative knowledge has come to occupy today. They would not understand it because for them all *scientia* was subordinated to *sapientia*. Their quantitative science was only one interpretation of a segment of Nature, not the means of understanding all of it. '"Progressive" science', he says,[b] 'which in the Islamic world always remained secondary, has now in the West become nearly everything, while the immutable and "non-progressive" science or wisdom which was then primary, has now been reduced to almost nothing.' It happened that I read these words at a terrible moment in history. If there were any weight in the criticism of the modern scientific world-view from the standpoint of Nasr's perennial Muslim *sapientia* it would surely be that modern science and the technology which it has generated have far outstripped morality in the Western and modern world, and we shudder to think that man may not be able to control it. Probably none of the human societies of the past ever were able to control technology, but they were not faced by the devastating possibilities of today, and the moment I read Nasr's words was just after the Jordanian civil war of September, 1970, that dreadful fratricidal catastrophe within the bosom of Islam itself. Since then we have had the further shocking example of Bengali Muslims being massacred by their brothers in religion from the Indus Valley. *Sapientia* did not prevent these things, nor would it seem, from the historical point of view, that wars and cruelties of all kinds have been much less within the realms of Islam or of East Asia than that of Christendom. Modern science, at all events, is not guilty as such of worsening men's lot, on the contrary it has immensely ameliorated it, and everything depends on what use humanity will

---

[a] This point was made by the Rev. D. Cupitt in discussion following a lecture for the Cambridge Divinity Faculty (1970) in which some of these paragraphs were used. It was afterwards published in part (Needham, 68). The contrast may be to some extent a matter of degree, since Islamic philosophy tended to recognise the material world as an emanation of the divine.

[b] (1), p. 145.

make of these unimaginable powers for good or evil. Something new is needed to make the world safe for mankind; and I believe it can and will be found.

In later discussions Nathan Sivin has made it clear that he is just as committed to a universal comparative history of science as any of the rest of us. That would be the ultimate justification of all our work. His point is not that the Chinese (or Indian, or Arabic) tradition should be evaluated only in the light of its own world-view, then being left as a kind of museum set-piece, but that it must be understood as fully as possible in the light of this as a prelude to the making of wide-ranging comparisons. The really informative contrasts, he suggests, are not those between isolated discoveries, but between those whole systems of thought which have served as the matrices of discovery.[a] One might therefore agree that not only particular individual anticipations of modern scientific discoveries are of interest as showing the slow development of human natural knowledge, but also that we need to work out exactly how the world-views and scientific philosophies of medieval China, Islam or India differed from those of modern science, and from each other. Each traditional system is clearly of great interest not only in itself but in relation to our present-day patterns of ideas. In this way we would not only salute the Chinese recording of sun-spots from the −1st century,[b] or the earliest mention of the flame test for potassium salts by Thao Hung-Ching in the +5th century, or the first correct explanation of the optics of the rainbow by Quṭb al-Dīn al-Shīrāzī in +1300,[c] as distinct steps on the way to modern science, but also take care to examine the integral systems of thought and practice which generated these innovations. Modern science was their common end, but their evolution can only be explained (that is to say, causally accounted for) in the context of the various possibilities opened and closed by the totality of ideas, values and social attitudes of their time.

Section 33(h), on the theoretical background of proto-chemical alchemy, may be taken as an exemplification and a test of this way of looking at early science.[d] Nathan Sivin's contribution deals with an abstract approach to Nature which has little to do with post-Galilean physical thought. Looking at the aims of the theoretically-minded alchemists as expressed in their own words, they turn out to be concerned with the design and construction of elaborate chemical models of the cyclic Tao of the cosmos which governs all natural change. A multitude of correspondences and resonances inspire the design of these models. One can distinguish as elements in their rationale the archaic belief in the maturation of minerals within the earth, the complex role of time, and the subtle interplay of quantity and numerology in ensuring that the elaboratory would be a microcosmos. Once we have reached at least a rough comprehension of the system which unites these elements, we can apprehend the remarkable culmination envisaged by the Chinese alchemists: to telescope time by reducing the grand overriding cycles of the universe to a compass which would allow of their contemplation by the adept—leading, as we have phrased it, to perfect freedom in perfect fusion with the cosmic order. But in the course of our reconnaissance we gather

[a] Cf. Sivin (10).   [b] Cf. Vol. 3, p. 435.   [c] Cf. Vol. 3, p. 474.
[d] Another attempt at this approach, applied to mathematical astronomy, will be found in Sivin (9).

a rich harvest of ideas worth exploring and comparing with those of other cultures, including those of the modern world—for instance, the notion of alchemy as a quint-essentially temporal science, springing from a unique concept of material immortality, a sublime conviction of the possibility of the control of change and decay. And we make a beginning towards understanding how the alchemist's concepts determined the details—the symmetries and innovations of materials, apparatus, and exquisitely phased combustions—of his Work, and how new results were reflected in new theoretical refinements as the centuries passed.

It is no less important to be aware that every anticipatory feature of a pre-modern system of science had its Yin as well as its Yang side, disadvantages as well as advantages. Thus the polar-equatorial system of Chinese astronomy delayed Yü Hsi's recognition of the precession of the equinoxes by six centuries after Hipparchus, but on the other hand it gave to Su Sung an equal priority of time over Robert Hooke in the first application of a clock-drive to an observational instrument; and the mechanisation of a demonstrational one by I-Hsing and Liang Ling-Tsan was no less than a thousand years ahead of George Graham and Thomas Tompion with their orrery of 1706.[a] In a similar way, perhaps, the conviction of the existence of material life-elixirs cost the lives of untold numbers of royal personages and high officials no less than of Taoist adepts, but it did lead to the accumulation of a great fund of knowledge about metals and their salts, in the pursuit of which such earth-shaking discoveries as that of gunpowder were incidentally made. So also the ancient idea of urine and other secretions as drugs might easily be written off as 'primitive superstition' if we did not know that it led, by rational if quasi-empirical trains of thought, combined with the use of chemical techniques originally developed for quite different purposes, to the preparation of steroid and protein hormones many centuries before the time of experimental endocrinology and biochemistry.

The only danger in the conception of human continuity and solidarity, as I have outlined it, is that it is very easy to take modern science as the last word, and to judge everything in the past solely in the light of it. This has been justly castigated by Joseph Agassi, who in his lively monograph on the historiography of science (1) satirises the mere 're-arranging of up-to-date science textbooks in chronological order', and the awarding of black and white marks to the scientific men of the past in accordance with the extent to which their discoveries still form part of the corpus of modern knowledge. Of course this Baconian or inductivist way of writing the history of science never did justice to the 'dark side' of Harvey and Newton, let alone Paracelsus, that realm of Hermetic inspirations and idea-sources which can only be regained by us with great difficulty, yet is so important for the history of thought, as the life-work of Walter Pagel has triumphantly shown. One can see immediately that this difficulty is even greater in the case of non-European civilisations, since their thought-world has been even more unfamiliar. Not only so, but the corpus of modern knowledge is changing and increasing every day, and we cannot foresee at all what its aspect will be a century from now. Fellows of the Royal Society like to speak of

[a] On all these subjects see Vol. 3 and Vol. 4, pt. 2.

the 'true knowledge of natural phenomena', but no one knows better than they do how provisional this knowledge is. It is neither independent of the accidents of Western European history, nor is it a final court of appeal for the eschatological judgment of the value of past scientific discoveries, either in West or East. It is a reliable measuring-stick so long as we never forget its transitory nature.

My collaborators and I have long been accustomed to use the image of the ancient and medieval sciences of all the peoples and cultures as rivers flowing into the ocean of modern science. In the words of the old Chinese saying: 'the Rivers pay court to the Sea'.[a] In the main this is indubitably right. But there is room for a great deal of difference of opinion on how the process has happened and how it will proceed. One might think of the Chinese and Western traditions travelling substantially the same path towards the science of today, that science against which, on the inductivist view, all ancient systems can be measured. But on the other hand, as Nathan Sivin maintains, they might have followed, and be following, rather separate paths, the true merging of which lies well in the future. Undoubtedly among the sciences the point of fusion varies, the bar where the river unites at last with the sea. In astronomy and mathematics it took but a short time, in the seventeenth century; in botany and chemistry the process was much slower, not being complete until now, and in medicine it has not happened yet.[b] Modern science is not standing still, and who can say how far the molecular biology, the chemistry or the physics of the future will have to adopt conceptions much more organicist than the atomic and the mechanistic which have so far prevailed? Who knows what further developments of the psychosomatic conception in medicine future advances may necessitate? In all such ways the thought-complex of traditional Chinese science may yet have a much greater part to play in the final state of all science than might be admitted if science today was all that science will ever be. Always we must remember that things are more complex than they seem, and that wisdom was not born with us. To write the history of science we have to take modern science as our yardstick—that is the only thing we can do—but modern science will change, and the end is not yet. Here as it turns out is yet another reason for viewing the whole march of humanity in the study of Nature as one single enterprise. But we must return to the volume now being introduced.

Although the other parts of Vol. 5 are not yet ready for press we should like to make mention of those who are collaborating with us in them. Much of the Section on martial technology for Vol. 5, pt. 1, has been in draft for many years now,[c] but it has been held up by delays in the preparation of the extremely important sub-section on the invention of the first chemical explosive known to man, gunpowder, even though all the notes and books and papers necessary for this have long been collected.

---

[a] *Chhao tsung yü hai*.[1] Cf. Vol. 3, p. 484.

[b] This picture has been elaborated elsewhere; Needham (59), reprinted in (64), pp. 396 ff.

[c] Including an introduction on the literature, a study of close-combat weapons, the sub-sections on archery and ballistic machines, and a full account of iron and steel technology as the background of armament. The first draft of this last has been published as a Newcomen Society monograph; Needham (32), (60).

[1] 朝宗于海

At present our collaborators Wang Ling (Wang Ching-Ning[1]), of the Institute of Advanced Studies at Canberra, and Ho Ping-Yü at Brisbane, Australia, are seeing what can be done about this.[a] Meanwhile Prof. Lo Jung-Pang,[2] of the University of California at Davis, spent the winter of 1969–70 in Cambridge, accomplishing not only the sub-section on the history of armour and caparison in China but also the draft of the whole of Section 37 on the salt industry, including the epic development of deep borehole drilling (Vol. 5, pt. 6). Other military sub-sections, such as those on poliorcetics, cavalry technology and signalling we have been able to place in the capable hands of Dr Corinna Hana of Göttingen. About the same time we persuaded Dr Tsien Tsuen-Hsuin (Chhien Tshun-Hsün[3]), the Regenstein Librarian at the University of Chicago, to undertake the writing of Section 32 on the great inventions of paper and printing and their development in China; this is now actively proceeding. For ceramic technology (Section 35) we have obtained the collaboration of Mr James Watt (Chhü Chih-Jen), Curator of the Art Gallery of the Institute of Chinese Studies in the Chinese University of Hongkong. The story of these marvellous applications of science will be anticipated by many with great interest. Finally non-ferrous metallurgy and textile technology, for which abundant notes and documentation have been collected, found their organising genii in two other widely separated places. For the former we have Prof. Ursula Martius Franklin and Dr Hsü Chin-Hsiung[4] at Toronto; for the latter Dr Ohta Eizō[5] and his colleagues at Kyoto. When their work becomes available, Volume 5 will be substantially complete.

As has so long been customary, we offer our grateful thanks to those who try to keep us 'on the rails' in territory which is not our own: Prof. D. M. Dunlop for Arabic, Dr Charles Sheldon for Japanese, Prof. G. Ledyard for Korean and Prof. Shackleton Bailey for Sanskrit.

Next comes our high secretariat—Miss Muriel Moyle, who continues to give us impeccable indexes; Mrs Liang Chung Lien-Chu[6] (wife of another Fellow of Caius, the physicist Dr Liang Wei-Yao[7]), who has inserted many a page of well-written characters and made out many a biographical reference-card; and Miss Philippa Hawking, who hewed away manfully at translations from the Japanese. We are also happy to acknowledge the skilled and accurate typing help of Mrs Diana Brodie and Mrs Evelyn Beebe, and the editorial work of Mrs Janin Hua Chhang-Ming.[8]

All that has been said in previous volumes (e.g. Vol. 4, pt. 3, p. lvi) about the University Press, our treasured medium of communication with the world, and Gonville and Caius College, that milieu in which we live and move and have our being, has become only truer as the years go by—their service and their encouragement continues unabated and so does our heartfelt gratitude. If it were not for the devotion of the typographical—and typocritical—masters, and if one could not count

---

[a] A preliminary treatment of the subject, still, we think, correct in outline, was given in an article in the *Legacy of China* eleven years ago; Needham (47). This has recently been re-issued in paper-back form.

[1] 王靜寧　　[2] 羅榮邦　　[3] 錢存訓　　[4] 許進雄　　[5] 太田英藏
[6] 梁鍾連杼　　[7] 梁維耀　　[8] 華昌明

on the understanding, kindness and appreciation of one's academic colleagues, nothing of what these volumes represent could ever have come into existence. We have taken pleasure on previous occasions of paying a tribute to our friend Mr Peter Burbidge of the University Press, and this time perhaps we may be allowed to add mention also of our gratitude to Miss Judith Butcher, the amiable Lucina who presided over the monstrous birth of Vol. 4, pt. 3.

As for finance, continuing gratitude is ever due to the Wellcome Trust of London, whose generous support has upheld us throughout the period of preparation of these chemical volumes. Since the history of medicine is touched upon at so many points in them we feel some sense of justification in accepting their unfailing aid. It can hardly be too much emphasised that in China proto-chemistry was elixir alchemy from the very beginning (as it was not in other civilisations of equal antiquity), and by the same token alchemists were very often physicians too (much more so than they tended to be in other civilisations). For the basic elixir notion was a pharmaceutical and therapeutic one, even though its optimism regarding the conquest of death reached a height which modern medical science dare not as yet contemplate. All this will be clarified in what follows. More recently our project received a notable benefaction from the Coca-Cola Company of Atlanta, Georgia, through the kind intermediation of Dr C. A. Shillinglaw, and for this also our grateful thanks are due. Meanwhile, and lastly, it should be added that Dr N. Sivin wishes to acknowledge financial assistance from the National Science Foundation (U.S.A.) and the Department of Humanities at the Massachusetts Institute of Technology.

Let us end with a few words of help to the prospective reader, as on previous occasions, offering some kind of waywiser to guide him through those pages of type not always possible to lighten by some memorable illustration. This is not intended as a substitute for the contents-table, the *mu lu*, or as any enlargement of it; but rather as some useful tips of 'inside information' to tell where the really important paragraphs are, and to distinguish them from the supporting detail secondary in significance though often fascinating in itself.

First, then, we would recommend a reader to study very carefully our introduction (Sect. 33*b*) on concepts, terminology and definitions, especially pp. 9–12; because once one has obtained a clear idea of the distinctions between aurification, aurifaction and macrobiotics (already referred to, p. xx above), everything that one encounters in the proto-chemistry and alchemy of all the Old World civilisations falls into place. There is a parallel here with the history of time-keeping, for the radical gap between the clepsydra and the mechanical clock was only filled by half-a-dozen centuries of Chinese hydro-mechanical clockwork. So in the same way the radical gap between Hellenistic aurifictive and aurifactive proto-chemistry at one end, and late Latin alchemy and iatro-chemistry at the other, could only be explained by a knowledge of Chinese chemical macrobiotics.

After that the argument develops in several directions, among which the reader can take his choice. How could belief in aurifaction ever have arisen when the cupellation test had been known almost since the dawn of the ancient empires? Look at

33 b, 1–2, and especially p. 44. What was the position of China in this respect, and what were the ancient Chinese alchemists probably doing experimentally? Read 33 b, 3–5; and c, 1–8. Why were they so much more occupied with the perpetuation of life on earth, even in ethereal forms, than with the faking or making of gold? We try to explain it in 33 b, 6. Such an induction of material immortality was indeed the specific characteristic of Chinese alchemy, and our conclusion is that the world-view of ancient China was the only milieu capable of crystallising belief in an elixir (*tan*[1]), good against death, as the supreme achievement of the chemist (see esp. pp. 71, 82, 114–15).

This is the nub of the argument, and in later parts (33 i, 2–3, in Vol. 5, pt. 4) we follow the progress of that great creative dream through Arabic culture into the Latin Baconian and Paracelsian West. Differences of religion, theology and cosmology did not stop its course, but there can be no doubt that it was born within the bosom of the Taoist religion, and hence the reader is invited to participate in a speculation that the alchemist's furnace derived from the liturgical incense-burner no less than from the metallurgical hearth (33 b, 7, see esp. pp. 127, 154). Finally something is said on the physiological background of the ingestion of elixirs (33 d, 1, see esp. p. 291); why were they so attractive to the consumer initially and why so lethal later? Here belongs also the conservation of the body of the adept after death, so important in the Taoist mind in connection with material immortality (33 d, 2, see esp. pp. 106, 297–8).

In the sub-section giving the straight historical account of Chinese alchemy from beginning to end, *chi shih pên mo*,[2] as the phrase was (33 e, 1–8, in Vol. 5, pt. 3), no part is really more significant than any other. Yet special interest does attach to the oldest firm records of aurifiction and macrobiotics expounded in (1), and to the study of the oldest alchemical books in (2) and (6, i). Now and then the narrative is interrupted by passages of detail, especially in (1), (2), (3, iii) and (6, vii) which readers not avid for minutiae may like to pass over (esp. pp. 42–4, 52–6, 76–8, 111–13, 201–5); such is the wealth of information not previously available in the West. The following parts on laboratory apparatus, aqueous reactions, and alchemical theory (all in Vol. 5, pt. 4) explain themselves from the contents table, and again no passage stands out as particularly crucial; unless it were the relation of the Chinese alchemist to time (33 h, 3–4). His was indeed the science (or proto-science) of the Change and Decay Control Department, as one might say, for he could (as he believed) accelerate enormously the natural change whereby gold was formed from other substances in the earth, and conversely he could decelerate asymptotically the rate of decay and dissolution that human bodies, each with their ten 'souls' (*hun*[3] and *pho*[4]), were normally subject to. Thus in the words of the ancient Chinese slogan (33 e, 1) 'gold *can* be made, and salvation *can* be attained'. And the macrobiogens were thus essentially time- and rate-controlling substances—a nobly optimistic concept for a nascent science of two thousand years ago.

Lastly we pass from the 'outer elixir' (*wai tan*[5]) to the 'inner elixir' (*nei tan*[6]), from proto-chemistry to proto-biochemistry, from reliance on mineral and inorganic

[1] 丹   [2] 紀事本末   [3] 魂   [4] 魄   [5] 外丹   [6] 內丹

remedies to a faith in the possibility of making a macrobiogen from the juices and substances of the living body. For this new concept we coin a fourth new word, the enchymoma; its synthesis was in practice the training of mortality itself to put on immortality. This 'physiological alchemy' occupies Vol. 5, pt. 5 (Sect. 33*j*, 1–8), and the basic ideas may be found in two places, (2) especially (i, ii), and (4). It was not primarily psychological, like the 'mystical alchemy' of the West, though it made much use of meditational techniques, as did the Indian *yogacāra* with which it certainly had connections. Our conclusion is, at the end of (4) and in (8), that most of its procedures were highly conducive to health, both mental and physical, even though its theories embodied much pseudo-science as well as proto-science.

In the end, the iatro-chemistry of the late Middle Ages in China began to apply *wai tan* laboratory procedures to *nei tan* materials, bodily secretions, excretions and tissues. Hence arose some extraordinary successes and anticipations (33*k*, 1–7), but we must not enlarge on them now. And this may suffice for a reader's guide, hoping only that he may fully share with us the excitement and satisfaction of many new insights and discoveries.

33. ALCHEMY AND CHEMISTRY

(e) THE HISTORICAL DEVELOPMENT OF ALCHEMY AND EARLY CHEMISTRY

(1) THE ORIGINS OF ALCHEMY IN CHOU, CHHIN AND EARLY HAN; ITS RELATION WITH TAOISM

In this sub-section we must try to lay bare the roots of chemistry in Chinese culture; a development inspired by the profound belief that longevity and material immortality were possible, and that these could be attained by a variety of practical techniques.[a] It was natural that dietary regimens should have been one form of these,[b] but from the 'nutritional' to the 'pharmaceutical' was only a short step, and before long the belief grew up that macrobiotic effects could be obtained by the ingestion of all kinds of strange substances, mineral, vegetable and even animal. Since metallic gold was the most beautiful and imperishable metal, it naturally came to be associated with the imperishability of the immortals,[c] and if the mortal was to put on immortality it must somehow associate itself with the metal or its inner principle or nature.[d] At first it seems to have been thought that the foods or potions of eternal life should be taken from vessels of gold, or that libations to the spirits should be poured from them, to induce these perhaps to appear and to confer immortality upon the devout invocator. Later it was felt that the human body itself must somehow be transformed to a gold-like state, and later again that this could be effected by drinking or absorbing preparations of some kind of 'potable gold'. The idea that gold and silver could be made from other substances arose in China at least as early,[e] and this aurifaction or argentifaction had two aspects, first a purely practical one in that the adepts in their mountain hermitages were not rich men or rulers well furnished with such specie,[f] but secondly also that the making of the noble metals yielded a better product and necessitated spiritual and bodily disciplines which would aid the attainment of perpetual life on earth.

There is an interesting discussion in the *Pao Phu Tzu* book on this. Writing about +300 Ko Hung says:[g]

I once enquired of my teacher Chêng (Yin),[1] saying: 'Lao Tzu tells us not to prize things that are hard to get,[h] and says that in a well-ordered society all the gold would be thrown

---

[a] Cf. Vol. 2, pp. 141 ff.   [b] Cf. Sect. 40 in Vol. 6.
[c] This link was established in China, as we shall see, by the middle of the −2nd century.
[d] The Chinese, like other people, accepted a hierarchy of metals among which gold was naturally supreme.
[e] Certainly by the end of the −3rd century, perhaps by the end of the −4th.
[f] As in the West, alchemical gold was sometimes made with the altruistic aim of relieving poverty (cf. pt. 2, p. 234).
[g] Ch. 16, pp. 5 a ff., tr. auct. adjuv. Ware (5), p. 267.
[h] *TTC*, ch. 64.

[1] 鄭隱

NEE

away in the mountains, and all the jade scattered in the wilderness.[a] Why then did the Ancients value gold and silver, and why did they hand down to us records of their processes?' To which he replied, 'What Lao Tzu had in mind was the intolerable effort of the people involved in sifting sand,[b] splitting rocks, overthrowing mountains, draining gulfs, and going heaven knows where to risk their lives by being crushed or drowned—all in the search for gems and jewels. This is what interferes with the proper use of the people's time. This is not knowing where to stop[c] in the pursuit of useless ornaments. If anyone wishes to take the Tao seriously, striving to attain the life of the immortals, and yet engages in mercantile affairs, he shows himself lacking in faith and humility. Tossing on the deep and courting dangers, he will take unfair advantages to make a profit; careless of his life, he will never discipline himself to minimise his covetousness. But when an adept (*chen jen*[1]) makes gold (by transformation) he does so, not in order to become rich, but to consume it himself so as to attain (the blessedness of) the holy immortals. This is why the Manuals say that gold can be made, and that men can attain salvation (*chin kho tso yeh, shih kho tu yeh*[2]).[d] Silver can also be ingested, but it is not the equal of gold.'

I then made difficulties, saying: 'Why not use ordinary gold and silver? Why (go to the trouble of) making it by transformation (*hua*[3])? The transformed product will not be genuine (*fei chen*[4]), and if not genuine it will be counterfeit (*cha wei*[5]).' Master Chêng replied, 'Mundane gold and silver are indeed good. But Taoists and their disciples are poor. As the proverb says: "Who ever heard of a fat immortal or a rich Taoist?" A teacher and his pupils may be five or ten persons—how could so many be supplied with gold and silver? Besides, they could not travel far and wide to collect them. Therefore it is fitting that they should make them. And finally the gold which is made by transformation embodies the essences of many different ingredients, so that it is superior to natural gold.'[e]

Here we have a clear distinction of motives. The exploiter of gem and gold mining for the luxury of rulers and officials is condemned, so also is the merchant who piles up profits of gold; but the adept who uses it for the techniques of immortality is praised, and most significantly, artificially made gold is considered positively better than natural gold.

All this was alchemy in its purest sense. Of the meaning and etymology of the word 'alchemy' something has already been said,[f] and we are now in a position therefore to trace the development of the concept of the elixir of immortality in early China. As we shall see, this concept became firmly crystallised by the −4th century in the time of Tsou Yen, and the belief in the feasibility of achieving physical immortality was so strong that the emperor Chhin Shih Huang Ti sent several expeditions to search for the elixir during the −3rd century.[g] By the time of the Han dynasty we have contemporary records of the activities of aurifactors and magicians at the imperial court, due to the great historian Ssuma Chhien.

As far back as prehistoric times people were wont to paint the human remains in burials with colours which would give the appearance or significance of life. Red was

---

[a] *TTC*, ch. 9.   [b] Probably a reference to placer mining.
[c] Cf. *Ta Hsüeh* I, 1, cf. Vol. 2, p. 566.   [d] Cf. p. 27 below.
[e] *Hua tso chê chin, nai shih chu yao chih ching, shêng yü tzu-jan chê yeh.*[6]
[f] Vol. 5, pt. 2, pp. 9 ff., 12 ff.
[g] Cf. Vol. 1, p. 240, Vol. 4, pt. 3, pp. 551 ff. Also pp. 13, 17–18 below.

[1] 眞人   [2] 金可作也世可度也   [3] 化   [4] 非眞   [5] 詐偽
[6] 化作之金乃是諸藥之精勝於自然者也

the colour of blood and its ceaseless movement, so it was a natural piece of sympathetic magic to use red pigments in symbolic revivification of the entombed dead. It has been reported that ornamental stone beads worn by the Upper Cave Man of Chou-khou-tien,[1] dating from the very end of the Pleistocene, were painted red with haematite, and that a large quantity of haematite powder was also found scattered around the body. This custom persisted through historical times.[a] There have been many reports of the use of red ochre in colouring skulls and skeletons in palaeolithic and neolithic graves.[b] But mixtures of iron compounds were not the only red substances used in this way.[c] Pigment on oracle-bones has been ascertained to be cinnabar by micro-chemical methods.[d] As we found in another connection,[e] amulets of jade, beads or cicadas, were placed in the mouth of the dead during the Chou period, and these were sometimes painted with the life-giving colour of red cinnabar or haematite.[f] The *I Chou Shu*[2] relates that cinnabar was presented to the Chou king Chhêng Wang[3] by the tribal people of Phu,[4] and according to this account, the event would have taken place in the late −11th or early −10th century.[g] Tombs of the −6th century have been found to contain masses of cinnabar, and vermilion paint on the remains and the bronzes.[h] It can hardly be without significance that this bright red substance, used in such ancient times as what might almost be called a strong magic of resurrection, should have turned out to give rise to the most living of the metals, quick- (or living) silver, metallic mercury.[i] Compounds of this with other elements, as also salts of lead, were in widespread common use in high antiquity in China (see pp. 15 ff., 123 ff. below). As we shall find, all these substances became common ingredients and reagents in the elixir-making of the Chinese alchemists.

[a] Chêng Tê-Khun (4), p. 32, (9), vol. 1, pp. 35, 37; Loewenstein (1). Oakley (2) has now been able to date the 'Red Lady of Paviland' skeleton by radiocarbon methods to −16,500. Okladnikov found red decoration on remains in the Shilka cave, Upper Amur Valley; and down to our own time the Ob Ugrians, in the Upper Konda Valley, clothe their dead in red wrappings (Mr R. G. Wasson, priv. comm.).

[b] Black (1); Vogt (1).

[c] Tombs in the Near and Middle East dating back to the −7th millennium have been found to contain bodies or skeletons painted with ochre or cinnabar (Widengren, 1). This was the finding of Mellaert, too at Çatal Hüyük (−6th millennium).

[d] Benedetti-Pichler (1).

[e] Vol. 4, pt. 3, pp. 544, 545. Cf. Laufer (8), pp. 294 ff., 301; Biot (1), vol. 1, pp. 40, 389; Wieger (2), p. 90. There are close Amerindian parallels.

[f] It is interesting that the magic vermilion was also used in European antiquity:

'Pan deus Arcadiae venit, quem vidimus ipsi
sanguineis ebuli bacis minioque rubentem' (Virgil, *Ecl.* x, 27)

Conington notes that the Greeks and Romans seem not infrequently to have painted their gods red (cf. Plutarch, *Quaestiones Romanae*, 98), especially country deities such as Bacchus and Priapus. In Tibullus (2, i, 55) the rustic worshipper of Bacchus paints himself with cinnabar. For minium, as in Virgil's line above, was, for Pliny and his contemporaries (cf. *Hist. Nat.* XXXIII, 111), cinnabar, usually from the Spanish mines of Almadén which are still working today. See Crosland (1), p. 105; Bailey (1), vol. 1, pp. 119, 217; and on the ancient religious practices Wunderlich (1). A Hebrew parallel can be seen in Wisdom 13.14.

[g] Ch. 59, p. 12 b; cf. *Yuan Chien Lei Han*, ch. 119, p. 5 b.

[h] See Bishop (12).

[i] Cf. the possible connection with liturgical practices, discussed in pt. 2, pp. 128 ff.

[1] 周口店   [2] 逸周書   [3] 成王   [4] 僕

The concept of material immortality went back to prehistorical days in Chinese legend. We have earlier referred to Yi[1] the Archer, who obtained the medicine of immortality from Hsi Wang Mu[2].[a] But it was his wife Chhang O[3] who stole and ate the elixir and subsequently became the Lady of the Moon.[b] An ageless story universal in Chinese folklore and art motifs has a jade rabbit (*yü thu*[4]) on the moon working untiringly mixing and pounding the drugs of immortality.[c] There are a thousand legends, which the Taoists perpetuated, of people attaining immortal life. For example, Lung-Yü,[5] daughter of Mu Kung,[6] Duke of Chhin in the −7th century, was courted by an alchemical immortal, Hsiao Shih,[7] who provided her with mercurous chloride face-powder[d] and also taught her to play the flute, after which music they both soared into the empyrean—he on a dragon and she on a phoenix.[e] In the fabulous account of the travels of the Chou High King Mu Wang[8] (−10th century) to visit the immortal goddess Hsi Wang Mu in the West we are told about a certain 'gold paste' (*huang chin chih kao*[9]),[f] which he was shown in the palace of the River Spirit (Ho Po[10]),[g] and which may already imply amalgamation, a favourite process of the Taoist alchemists, as we shall later see.

The question of the antiquity of cinnabar and mercury in China has been raised by Dubs (5), and it is one of importance because of the central position of these substances in all later alchemy. Perhaps the oldest Chinese reference to cinnabar (mercuric sulphide) occurs in the *Shu Ching* (Historical Classic), where the Yü Kung (Tribute of Yü) chapter[h] lists *tan*[11] among the tribute products of the region called Ching-chou.[12] The date of this text is still uncertain but it is truly ancient, perhaps of the early −5th century, perhaps based on documents or oral traditions going back to the −8th. When mercury was first distilled from cinnabar we do not know, but in the Chhin and Early Han (−3rd century onwards) references come thick and fast, and it is hard to say whether *hung*[13] or *shui yin*[14] (liquid silver) was the earlier term. As we saw on a previous occasion, the supply was sufficiently abundant at the time of the death of Chhin Shih Huang Ti (−210) to allow a relief map of China to be set up in the chamber of his mausoleum, having the great rivers and streams running with mercury circulated by a machine.[i] According to the *Huang Lan*[15] encyclopaedia, edited by Miu Shih-Têng[16] in +220, robbers from Kuantung later broke into the tomb and made off with the valuable metal.[j] But one cannot be sure that Chhin Shih

---

[a] Vol. 2, p. 71; cf. Granet (1), p. 376.
[b] Cf. *Wei Lüeh*, ch. 6, p. 3*a*, and many other texts.
[c] This is constantly represented in art—see, for example, *Chin Shih So*, Shih sect. ch. 1, (p. 101). Cf. Janse (5), pp. 48 ff., (6), p. 210.
[d] Cf. p. 125 below.   [e] See *Chung Hua Ku Chin Chu*,[17] p. 6*b*.
[f] See *Mu Thien Tzu Chuan*,[18] p. 1*b* (tr. Chêng Tê-Khun, 2), and *TPYL*, ch. 811, p. 1*a*.
[g] Cf. Vol. 2, p. 103.   [h] See on this Vol. 6; and Karlgren (12), pp. 13, 15.
[i] Vol. 3, p. 582, where the passage in *Shih Chi*, ch. 6, p. 31*a*, is given in translation. Cf. Chavannes (1), vol. 2, p. 194.
[j] Quoted in *TPYL*, ch. 812, p. 6*b*.

[1] 羿	[2] 西王母	[3] 嫦娥	[4] 玉兔	[5] 弄玉	[6] 穆公
[7] 蕭史	[8] 穆王	[9] 黃金之膏	[10] 河伯	[11] 丹	[12] 荊州
[13] 汞	[14] 水銀	[15] 黃覽	[16] 繆十等	[17] 中華古今注	
[18] 穆天子傳					

Huang Ti's tomb was the first to have mercury, for the *Wu Yüeh Chhun Chhiu*[1] says that the King of Wu, Ho Lu,[2] who died about −495, was buried with a whole pool of mercury in his tomb-chamber.[a] Although this is a Later Han book, written by Chao Yeh[3] in the +1st century, it may report an authentic tradition.

The term *hung*[4] is defined as *shui yin*[5] in the *Shuo Wên* of +121,[b] the *Kuang Ya* of +230 and many subsequent dictionaries,[c] and it occurs in the −2nd century both in the *Huai Nan Tzu* book[d] and in the *Huai Nan Wan Pi Shu*.[e] The oldest extant pharmaceutical work, however, also attributable (at least in part) to the −2nd century, the *Shen Nung Pên Tshao Ching*,[f] uses the term *shui yin*, and in general this was always later preferred in the pharmacological literature. Among the alchemical writers, on the other hand, a different way of writing *hung*[6] grew up, and this must have happened quite early because the *Huai Nan Tzu* book also has it;[g] after which (though not in the *Shuo Wên*) it became the commonest form in the spagyrical literature, sanctified both by the *Tshan Thung Chhi* (+142) and the *Pao Phu Tzu* (c. +300). Many dictionaries such as the *Kuang Yün* of +1011 define it as *shui yin*.[h] To sum up, therefore, the distillation of mercury, presumably *per descensum*, must have started in China sometime between the life of Confucius and the first unification of the empire. The School of Naturalists and Tsou Yen, whom we shall shortly discuss,[i] would fit the picture very well chronologically, though as yet we have no positive evidence to link them with this fundamental chemical discovery.

Further thoughts relevant at this point were voiced long ago in an interesting paper by Lao Kan (6), who drew attention to the social valuation of the colours in Chinese antiquity. In close connection with Five-Element theory there grew up gradually a body of lore about the colours which were chosen as ceremonially dominant in different dynasties. As we saw at an earlier stage,[j] this was much discussed in the school of Tsou Yen himself; what is important here is that throughout the Chou period and perhaps in late Shang, red was the imperial colour, used for robes and vestments, carriages and palace buildings, every kind of object indeed from banners to brush-pen holders.[k] The early Shang was supposed to have used white, and the Chhin and Han certainly went over to black,[l] but red had a long innings and

---

[a] Also quoted in *TPYL, loc. cit.* It gives a length of 60 ft, but one may accept a width of 6, and the idea was doubtless to arrest decomposition of the body. We have seen already (pt. 2, pp. 298, 304) how this use of mercury and its salts continued down throughout Chinese history.

[b] Ch. 11 A, (p. 237.2).

[c] E.g. *PWYF*, ch. 31, (p. 1500.23).

[d] Cf. p. 23.

[e] Cf. p. 25. Some texts write the related *hung*,[7] perhaps then interchangeable.

[f] Cf. Vol. 6, pt. 1.

[g] In the passage on the growth of metals in the earth; see Vol. 3, p. 640, and here in pt. 4.

[h] This was based on the *Chhieh Yün* of +601, now lost as such. Under *shui yin*, *PWYF*, ch. 11, (p. 463.2) relates an abridgment of the story of Chhêng Wei, on which see p. 38 below.

[i] Pp. 12ff. below.

[j] Vol. 2, p. 238 and Table 12 on p. 263.

[k] Also, significantly, coffins, as in the case of one of the princes of Chhu, excavated at Shouhsien.

[l] The imperial yellow, so evocative of Thang, Sung and Ming China, started only in the Sui time. We have seen already in Sect. 30 one of the beginnings of this colour preference change.

[1] 吳越春秋   [2] 闔廬   [3] 趙曄   [4] 汞   [5] 水銀   [6] 汞   [7] 鴻

continued in favour for centuries with the Taoists,[a] indeed becoming, as we know, the popular 'auspicious' colour in all the folk custom of traditional China.[b] Consequently during the early Chou and Warring States periods, formative as they were for Chinese proto-chemistry as for many other things, there was a special emphasis on the colour red, and this must have meant a wide employment of vermilion (cinnabar) as well as its cheaper substitute, red lead.

Correspondingly the mining of mercury ores becomes prominent quite early. The famous 'capitalists' chapter of the *Shih Chi* has an account of a Szechuanese widow of the −3rd century who managed mines like those of Almadén with great success. Her story cannot be better placed than here. We read:[c]

There was a widow of Pa (Szechuan) by the name of Chhing,[1] whose husband's ancestors had found mercury mines (lit. caves or pits of cinnabar, *tan hsüeh*[2]) and monopolised the profits for several generations. The family wealth was beyond counting. This woman had the ability to look after her enterprises, using much of her wealth as protection so that no one molested her or them. The first Chhin emperor considered her a virtuous widow, treated her as a protégée (Kho[3]), and caused to be built in her honour a monument called the Nü Huai Chhing Thai[4] (Tower of the Women's Remembrance of Mistress Chhing).

Particularly significant here, since Chhing's *floruit* must have been *c.* −245 to −210, is the statement that the mines had been worked by several generations of her husband's forebears. Thus the industry must have been flourishing in the −4th century, the time of Tsou Yen, if not already somewhat earlier, in Chi Ni Tzu's time. Another mention of 'caves of cinnabar' occurs in some texts of the *I Lin*[5] (Forest of Symbols of the *Book of Changes*) where it is said that three men went out to collect oranges but found a mercury mine instead, so that their wives attained dignity and wealth, each worth a hundred ingots of gold.[d] Although this book was traditionally attributed to Chiao Kan,[6] *c.* −40, it is more probably of the +2nd century or a little later, yet the incident itself may well be of the Early Han.[e]

Another point of Lao Kan's is the ancient use of vermilion for red ink on particularly important documents of state, hence also for magical exorcisms and the like. The *Tso Chuan* affords an immediate example, involving the burning of a feudal register kept in vermilion script (*tan shu*[7]);[f] this incident is datable at −549. From the mathematics Section it will be remembered that one of the two most ancient magic squares, the Lo Shu, was supposed to have been written in red characters on

---

[a] This is evident from the mid +2nd-century *Thai Phing Ching*,[8] which extols red as the colour of fire, the sun, and pure Yang (chs. 2, 4, 7, 119, cf. (p. 682) for example). The motif of blood seems not present here, but 'our Thai-Yang Tao is the Way of humane administration, which has no desire to harm any of the people'. It is interesting that a symbol still so powerful today as the Red Flag can be traced back as far as the Taoist social reformers of the Later Han.

[b] See, for example, Tiefensee (1).

[c] *Shih Chi*, ch. 129, pp. 6*b*, 7*a*; *Chhien Han Shu*, ch. 91, p. 6*a*; tr. auct., adjuv. Swann (1), p. 431.

[d] P. 2*b*, the combination of the *Kua* Chhien and Hsien.

[e] On the mercury deposits and resources of China see Tegengren (3); Wei Chou-Yuan (1) and Torgashev (1), pp. 243 ff.

[f] Duke Hsiang, 23rd year; tr. Couvreur (1), vol. 2, p. 391.

[1] 清   [2] 丹穴   [3] 客   [4] 女懷清臺   [5] 易林   [6] 焦贛
[7] 丹書   [8] 太平經

its first appearance.[a] By the time of the *Pao Phu Tzu* book (see pp. 81 ff. below), all the talismans and charms were necessarily to be written in red script.[b] Since we even have the names of some of these exorcistic scribes from the Hou Han period,[c] the practice must have arisen during the −1st century if not before.

The concept of a medicine of immortality was undoubtedly much occupying the minds of Chinese scholars and projectors by the time of the Warring States period. A good example is found in early −3rd-century texts such as the *Han Fei Tzu*[1] book. What they say is this:[d]

Once upon a time someone presented an elixir of life (*pu ssu chih yao*[2]) to the Prince of Ching.[3] As the chamberlain was taking it into the palace, the guard at the gate asked if it was edible, and when he answered yes, the guard took it from him and ate it. The prince was [extremely] annoyed and condemned the guard to death. But the latter sent a friend to persuade the prince on his behalf, saying: 'After all, the guard did ask the chamberlain whether the elixir could be eaten before he ate it. Hence the blame attaches to the chamberlain and not to him. Besides, what the guest presented was an elixir of life, but if you now execute your servant after eating it, it will be an elixir of death [and the guest will be a liar]. Now rather than killing an innocent officer in order to demonstrate a guest's false claim it would be better to release the guard.' So the prince let him off.

This event would have occurred in the State of Chhu during the reign of Prince Ching Hsiang[4] between −294 and −261. The passage was no doubt composed or preserved as an exercise in sophistic argument, but it takes its place among a whole series of texts of an alchemical or quasi-alchemical character which establish the origins of Chinese alchemy in the Warring States period (−480 to −221) from at least the time of Tsou Yen onwards (cf. p. 12).

An interesting story given in the *Chuang Tzu*[5] book of a man who acquired the art of butchering dragons goes as follows:[e]

Chu Phing-Man[6] learnt how to slay dragons from (his teacher) Chihli I,[7] expending (in doing so) all his wealth of 1000 ounces of gold. In three years he became perfect in the art, but he never used his skill.

The commentators say that Chu Phing-Man never used his art because there were no dragons to kill, and that the *chün-tzu* does not value such techniques but only the Tao of the Golden Mean (*chung yung chih tao*[8]); yet if we may assume that the story

---

[a] Vol. 3, p. 56. Cf. *Ku Wei Shu*, ch. 34, p. 1a.

[b] Ch. 17; Ware tr. p. 296, for example.

[c] E.g. Chhü Shêng-Chhing.[9]

[d] *Chan Kuo Tshê*, ch. 5, p. 33b, tr. Ho Ping-Yü & Needham (4), adjuv. Liao Wên-Kuei (1), p. 235, translating the parallel passage in *Han Fei Tzu*, ch. 22, pp. 5bff. Words within square brackets are in the latter text only. It is interesting that this story got into one of the earliest books of European sinology, the *De Re Literaria Sinensium* of Theophilus Spizel (+1660), where we may read a Latin version of it on pp. 262–3. It was later re-translated by Imbault-Huart (2), who added (1, p. 70) a very similar story with Tungfang Shuo as the hero, from the *Po Wu Chih*, ch. 8, p. 6a.

[e] Ch. 32 (*Pu Chêng*, ch. 10A, p. 17a), tr. auct. adjuv. Legge (5), vol. 2, p. 206. Attention to this passage was first drawn by Barnes (2). Dubs (5) is not disposed to accept an alchemical allusion, and in any case regards the chapter as a Han interpolation, but we differ on both points.

[1] 韓非子   [2] 不死之藥   [3] 荊   [4] 景襄   [5] 莊子   [6] 朱泙漫
[7] 支離益   [8] 中庸之道   [9] 麴聖卿

was based on something factual, it is easier to attach to it an alchemical meaning. Chu Phing-Man might well have learnt the art in secrecy but neither spoke about it nor used it. The metaphor of dragon-slaying is common to alchemy all over the world, connected, of course, with the formation of the calx or the sulphide of a metal.[a] In any case the story would be of the late −4th century.

The Taoist cult of the holy immortals, so closely associated with alchemy, does not loom very large in the *Chuang Tzu* book, but there are passages which speak about them clearly enough. For example, Kuang Chhêng Tzu[1] was already 1200 years old when Huang Ti went to call upon him;[b] and Nü Yü[2] transmitted his art of maintaining perpetual youth to Puliang I,[3] refusing to reveal it to Nanpo Tzu-Khuei,[4] whom he considered unsuitable as a disciple.[c] In another place a significant passage reads:[d]

Far away on the Ku-shê Shan[5] (mountains)[e] there live numinous men whose flesh and skin are smooth as ice and pale as snow. Their ways are (innocent) like those of young girls. They do not eat the Five Cereals but inhale the wind and drink of the dew. They ride on the *chhi* of the clouds, and drive flying dragons which carry them roaming beyond the Four Seas; yet when they concentrate their spirit-like powers, living things are not attacked by the corruption of disease, and every year brings plentiful harvests.

Here Maspero was surely right[f] in seeing a reference to the dietary regimen, the respiratory exercises, and the meditational techniques with a background of Shamanic flight ecstasy,[g] all characteristic of the early aspirants to Taoist material immortality. But of alchemy itself there is no further word from Chuang Chou.

The belief in material immortality during the time of the Warring States was sufficiently strong to draw the criticism of the philosopher Lieh Yü-Khou, who affirmed that 'he who hopes to perpetuate his life or to shut out death is deceived as to his destiny'.[h] The following remark also comes from him:

That skull and I both know that there is no such thing as absolute life or death. This knowledge is better than all your methods of prolonging life, a more potent source of happiness than any other.[i]

The dating of such passages is as usual very difficult, for some parts of the *Lieh Tzu* book may go back to the −5th century and others may be as late as the +4th.

---

[a] The difficulty about this interpretation is that in the later alchemical literature one can 'slaughter' metals (e.g. *ssu shui yin*[6] in the *Tshan Thung Chhi*) but where dragons are concerned one is generally marrying them off. In Thang texts *ssu* usually stands for *ssu tu*,[7] 'to kill the toxicity of' something, especially metals. It is not equivalent to *fu*[8] or *chih*[9] (cf. pt. 4). Also the phrase *thu lung*[10] does not recur in later alchemical writing.

[b] Ch. 11, Legge (5), vol. 1, p. 299.

[c] Ch. 6, Legge (5), vol. 1, p. 245.

[d] Ch. 1, tr. auct. adjuv. Legge (5), vol. 1, pp. 170ff.; Fêng Yu-Lan (5), p. 36; Maspero (13), pp. 205-6.

[e] Or Ku-i Shan.

[f] (13), pp. 205ff.

[g] Quite possibly induced by cryptogamic or other plant hallucinogens.

[h] *Lieh Tzu*, ch. 1, p. 10b, tr. L. Giles (4), p. 24.

[i] *Lieh Tzu*, ch. 1, p. 6b, tr. L. Giles (4), pp. 22-3.

| [1] 廣成子 | [2] 女偊 | [3] 卜梁倚 | [4] 南伯子葵 | [5] 故射山 |
| [6] 死水銀 | [7] 死毒 | [8] 伏 | [9] 制 | [10] 屠龍 |

A study of the hagiography of the immortals, the large literature about the lives, achievements and 'miracles' of famous *hsien*,[a] quickly gives an idea of the Chinese conception of the early adepts and how they achieved immortality. The *Lieh Hsien Chuan*[1] (Lives of Famous Immortals),[b] a work attributed to Liu Hsiang[2] (*c.* −50) but certainly by an unknown Taoist who lived between the +2nd and +4th centuries, gives accounts of 71 immortals, many of whom inhabited the legendary period of Chinese antiquity. Prominent among the means of attaining the state of *hsien*-ship was the ingestion of a number of unprocessed mineral and plant substances (see Table 106). It is hard to say how seriously any real adepts followed such procedures,[c] and doubtless what ethereality and longevity they attained was due to an ascetic dietary regimen, but at any rate these pharmacological-alchemical methods were part of the corpus of legend from early times, and the urge towards macrobiotic achievements is beyond doubt. Among the legendary immortals were Chhih Sung Tzu,[3] who ingested jade as a suspension in blood, Fang Hui[4] who treated mica and other minerals to render them potable, Wo Chhüan,[5] who presented pine-seed elixirs to the emperor Yao[6] (but the latter could not find time to try them out), Phêng Tsu,[7] who lived on cinnamon and magic mushrooms,[d] Jen Kuang,[8] who took an elixir, and Chhih Fu,[9] who prepared one from mercury and consumed it together with nitre (or saltpetre).[e] It is of interest to note that Fan Li,[10] the putative −5th-century author of

---

[a] This subject has been discussed at some length earlier; see Vol. 2, pp. 152ff.

[b] Tr. Kaltenmark (2). This book is not mentioned in the bibliographical chapters of the *Chhien Han Shu*, but Ko Hung discusses it in his *Pao Phu Tzu* (Nei Phien), ch. 2, p. 5*a*; tr. Ware (5), p. 41.

[c] Most probably many did. When one of us (N.S.) was in Thaiwan in 1962 he met a Taoist who told him that he used *shih chung huang tzu*[11] frequently (brown haematite, RP 81) together with certain berries which he showed him. Abstaining from cereals (*chüeh ku*[12]) was of course a widespread practice, discussed in many of the medical books.

[d] The identification of what was meant by this Taoist religious symbol is a matter of much difficulty (cf. pt. 2, pp. 116, 121 above). It certainly had a reference wider than anything we should call a fungus today, for it could include mineral excrescences recalling the shape of mushrooms, all kinds of cryptogams, and doubtless some fictitious plants. See, for example, *Pao Phu Tzu* (Nei Phien), ch. 11; tr. Feifel (3) and Ware (5), pp. 177ff.; cf. Chikashige (1), pp. 29ff. We shall discuss the *chih* at length in Section 45.

[e] One curious account is that of a man called 'Father Cinnamon' (Kuei Fu[13]), who surprised people by turning black, white, yellow or red in succession (Kaltenmark (2), no. 31, p. 118). The fact that these were emblematic colours of three of the four quarters and the centre is perhaps less interesting than the statement that he lived upon *kuei*[14] (cinnamon) and *khuei*[15] (mallow), together with the brains of tortoises. One wonders whether this idea of colour-changes was inspired by some ancient observations that red or blue extracts of certain plants would change colour according to acidity, salt content, etc. One might think of Kuei Fu therefore as the litmus-immortal. Litmus itself comes from a lichen *Lecanora tartarea*, but anthocyanin extracts of higher plants behave in the same way. The mallow of the *Shen Nung Pên Tshao Ching* is *Malva verticillata* (R280; CC747). On the general history of colour-tests see Greenaway (1). Colour-change indicators did not become really important until the rise of volumetric analysis in the +18th century, with its associated techniques of quantitative washing. In this connection some liturgiologist ought to write the history of the ablutions at the Christian eucharist, when the chalice and the fingers of the celebrant are washed, first with unconsecrated wine, then with this mixed with water, and finally with water alone. To a modern chemist, washing out a vessel with the solvent would seem the merest common sense, yet in earlier ages it would not have been so obvious, and one would like to know what branches of chemical technology first developed it, and whether it was in East or West. Quantitative analysis was of course only a special case of the conservation of what was precious.

[1] 列仙傳	[2] 劉向	[3] 赤松子	[4] 方回	[5] 偓佺	[6] 堯
[7] 彭祖	[8] 任光	[9] 赤斧	[10] 范蠡	[11] 石中黃子	
[12] 絕穀	[13] 桂父	[14] 桂	[15] 葵		

Fig. 1342. Chhih Sung Tzu, the Red Pine Master (from *Lieh Hsien Chhüan Chuan*, ch. 1, p. 8*a*).

Table 106. *Agents of immortality in the 'Lieh Hsien Chuan'*

		No. of adepts consuming
tan sha[1]	cinnabar (HgS)[a]	I
hung[2]	mercury (Hg)	I
yü[3]	jade	I
yün mu[4]	mica	I
shih sui[5]	stalactite (CaCO$_3$) RP 68	I
shih chih[6]	siliceous clays RP 57	3
hsiao shih[7]	nitre (NaNO$_3$) or saltpetre (KNO$_3$)	I
sung yeh[8]	pine tree leaves	I
sung tzu[9]	pine tree seeds	2
kuei[10]	cinnamon, *Cinnamomum Cassia* (= *aromaticum*) Lauraceae; CC 1318; R 494[b]	I
(tshang-)shu[11]	a composite, *Atractylis ovata*; CC 34; R 14	I
thien mên tung[12, 13, 14]	a liliaceous plant, *Asparagus lucidus*; CC 1830; R 676	I
phêng lei[15]	a bramble, *Rubus hirsutus* (= *Thunbergii*) Rosaceae; CC 1162; R 459; Anon. (109), vol. 2, p. 275	I
fu ling[16]	a fungus, *Poria* (*Pachyma*) *cocos*, 'Indian bread' or 'tuckahoe'; CC 2320; R 838	I
(ling-)chih[17]	perhaps a lichen *Gyrophora esculenta* (= *vellea*); CC 2327; R 818 or a fungus *Fomes japonicus*; CC 2301[c]	I
—	miscellaneous plants in combinations	3
tan[18]	unidentified elixir, perhaps cinnabar	I

the *Chi Ni Tzu*[19] book, which we shall quote presently, is listed among the immortals—he also is said to have been fond of eating cinnamon. Among the many immortals described by Huangfu Mi[20] in his *Kao Shih Chuan*[21] (Lives of Men of Lofty Attainments), *c.* +275, is one Ho Shang Chang Jen,[22] said to be the teacher of An Chhi shêng,[23] the famous immortal so much sought after by the emperors Chhin Shih Huang Ti and Han Wu Ti, as we shall see later.[d] It was said that An Chhi shêng had among his pupils one Mao Shih Kung,[24] who in turn imparted his knowledge to Lo Hsia Kung,[25] who had a disciple called Lo Chhen Kung[26] (*fl.* −3rd cent.). The adept Kai Kung,[27] the teacher of the Early Han minister Tshao Tshan[28]

[a] Cinnabar is still used as a drug in India today, *anuma-karadhwaja* (Mahdihassan (16), p. 23).

[b] The following five clearly identifiable plants all occur in the oldest extant pharmacopoeia *Shen Nung Pên Tshao Ching*, on which see Vol. 6, Sect. 38.

[c] *Amanita* spp. may have been included, significantly, here; see the discussion in pt. 2, pp. 116 ff. and Vol. 6.

[d] *Kao Shih Chuan*, (p. 62).

[1] 丹砂   [2] 澒   [3] 玉   [4] 雲母   [5] 石髓   [6] 石脂
[7] 硝石   [8] 松葉   [9] 松子   [10] 桂   [11] 蒼朮   [12] 天門多
[13] 天蘪多   [14] 天蘁多   [15] 蓬蘽   [16] 伏苓   [17] 靈芝   [18] 丹
[19] 計倪子   [20] 皇甫謐   [21] 高士傳   [22] 河上丈人   [23] 安期生   [24] 毛翕公
[25] 樂瑕公   [26] 樂臣公   [27] 蓋公   [28] 曹參

(d. −190), was said to have been his disciple.[a] Such is the sort of chain of transmission which we meet with constantly in early Chinese macrobiotics.[b]

One character who cannot be omitted here is Lingyang Tzu-Ming[1] (or Tou Tzu-Ming[2]) who must have lived in the Han before the end of the +1st century, and who has a place in the *Lieh Hsien Chuan*.[c] We have encountered him before in connection with what must be the earliest reference in any culture to the reel of the fishing-rod,[d] the which he used to capture a white dragon, ultimately his vehicle to the heavens, or at least to the top of a sacred mountain. He was also famous for consuming mineral substances, especially coloured clays, which were supposed to have elixir properties. Whether Lingyang was fully a historical person or not, he had the almost unique fate of giving his name in perpetuity to a chemical substance, metallic mercury, for which it became one of the synonyms (cf. Table 95, no. 125).[e] His fame grew as time went on, and the Sui bibliography lists in its medical section a *Ling Yang Tzu Shuo Huang Chin Pi Fa*[3] (Lecture on a Secret Method for Yellow Gold by Master Lingyang).[f] By the Sung he could be regarded as the first originator of aurifaction itself.[g] At the same time his name was also closely connected with the beginnings of physiological alchemy, especially respiratory techniques and some sort of heliotherapy (cf. pt. 5), on which a manual entitled *Lingyang Tzu-Ming Ching*[4] undoubtedly existed in the time of Wang I,[5] the commentator on the Odes of Chhu (*fl.* +115 to +135).

(i) *The School of Naturalists and the First Emperor*

The beginnings of alchemy in China are very much connected with the School of Naturalists. We have mentioned earlier that its founder Tsou Yen[6] (*c.* −350 to −270) made lists of natural products, describing minerals, chemical substances, plants and animals.[h] He and his followers were undoubtedly responsible for the first systematisation of the theories of Yin and Yang and of the Five Elements, so fundamental in all later natural philosophy.[i] We repeat here two important passages which point to the alchemical interest of his school. The first, from the *Shih Chi*, runs thus:[j]

---

[a] See *Shih Chi*, ch. 54 and *CHS*, ch. 39, p. 11a on Tshao Tshan. Cf. Dubs (2), vol. 1, p. 143.
[b] See Table 110 on p. 77. Cf. *TT*878, ch. 1, p. 1a, ch. 2, p. 5a, a Sung text.
[c] Tr. and annot. Kaltenmark (2), pp. 183 ff.
[d] Vol. 4, pt. 2, p. 100.
[e] Ko Hung mentions it in a list of cover-names, *PPT/NP*, ch. 16, p. 6b, cf. Ware (5), p. 271. A parallel is the case of the Sung or pre-Sung physician Kuo Shih-Chün[7] who gave his name to the Rangoon creeper *Quisqualis indica*, an anthelminthic plant (Sect. 38).
[f] Ch. 3, (p. 109), among books on elixirs and other macrobiotic techniques.
[g] As by Li Kuang-Hsüan[8] in *TT*263, p. 24a.   [h] Vol. 2, p. 233.
[i] Among the thaumaturgical legends that grew up about him was one according to which he had rendered a cold and barren valley suitable for raising crops, fruits and vegetables, by playing certain airs, like Orpheus, on his pipe. The *chhi* of the place responded (Vol. 4, pt. 1, pp. 29, 135). It is interesting to find a similar story told of another reputed alchemist, Albertus Magnus, who made a garden blossom out of season for the visit of Duke William II of Holland in +1249 (Partington (3), p. 7).
[j] Ch. 28, pp. 10b–11b, tr. Chavannes (1), vol. 2, p. 152, vol. 3, p. 435; Dubs (5), eng. et mod. auct. Parallel passage in *CHS*, ch. 25A, pp. 12a–13a.

[1] 陵陽子明   [2] 竇子明   [3] 陵陽子說黃金祕法   [4] 陵陽子明經
[5] 王逸   [6] 鄒衍   [7] 郭使君   [8] 李光玄

From the time of (Kings) Wei[1] and Hsüan[2] of the State of Chhi[a] the disciples of Master Tsou discussed and wrote about the cyclical succession of the Five Powers. When (the King of) Chhin became (the First) Emperor (in −221), people from Chhi sent in memorials (bringing these theories to his notice). And the First Emperor (Chhin Shih Huang Ti) chose them and gave them employment. Moreover from first to last Sung Wu-Chi,[3] Chêng Po-Chhiao,[4] Chhung Shang[5] and Hsienmên Kao[6] were all people[b] from (the State of) Yen who practised the method of (becoming) immortals by the use of magical techniques, so that their bodies would be etherealised and metamorphosed by some transmutation (*hsing chieh hsiao hua*[7]).[c] For this they relied upon their services to the gods and spirits.

Tsou Yen was famous among the feudal lords (for his doctrine) that the Yin and the Yang control the cyclical movements of destiny. The men who possessed magical techniques, and who lived along the sea-coast of Yen and Chhi, transmitted his arts, but without being able to understand them. From this time on one cannot count the constantly increasing number of those persons who performed deceptive wonders, flatteries, and illicit practices.[d]

Then beginning with (Kings) Wei and Hsüan (of Chhi) and (King) Chao of Yen,[e] people were sent out into the ocean to search for (the fairy isles of) Phêng-Lai,[8] Fang-Chang,[9] and Ying-Chou.[10] These three divine (island) mountains were reported to be in the Sea of Po,[11] not so distant from human (habitations), but the difficulty was that when they were almost reached, boats were blown away from them by the wind. Perhaps some succeeded in reaching (these islands). (At any rate, according to report) many immortals (*hsien*) live there, and the drug which will prevent death (*pu ssu chih yao*[12]) is found there. Their living creatures, both birds and beasts, are perfectly white, and their palaces and gate-towers are made of gold and silver. Before you have reached them, from a distance they look like clouds, but (it is said that) when you approach them, these three divine mountain-islands sink below the water, or else a wind suddenly drives the ship away from them. So no one can really reach them. Yet none of the lords of this age would not be delighted to go there.[f]

The transmission of secret writings or perhaps oral traditions of the school of Naturalists to the circle surrounding Liu An,[13] the Prince of Huai Nan,[14] during the −2nd century is revealed in a second passage from the *Chhien Han Shu*, which says,[g] referring to about −60:

At that time the emperor Hsüan[15] (r. −73 to −48) was desirous of following in the footsteps of Han Wu Ti.[h] He summoned to his side eminent scholars and able men. (Liu) Kêng-Shêng[16] (i.e. Liu Hsiang) was (a young man) of penetrating intellect and well versed in literature. Together with Wang Pao[17] and Chang Tzu-Chhiao[18] and others he was called to court, and presented a poetical writing in several dozen chapters. Now the emperor was

[a] The two reigns covered −377 to −312.
[b] These may or may not have been historical personages; Han writers refer to them as 'former immortals'. Yet they may well have been magician-Naturalists of the State of Yen contemporary with, or earlier than, Tsou Yen himself. It is suggested (cf. Vol. 2, p. 133) that *hsien-mên* transliterates shaman.
[c] Or, 'so that they might be released from the flesh by fusion and transformation'.
[d] A strong hint of alchemy.
[e] R. −311 to −278.
[f] The expeditions into the Eastern Ocean are quite historical and have been discussed at length in Vol. 4, pt. 3, pp. 551 ff.
[g] Ch. 36, p. 6b, 7a; tr. auct. adjuv. Dubs (5).   [h] Cf. p. 35.

[1] 威	[2] 宣	[3] 宋毋忌	[4] 正伯僑	[5] 充尚	[6] 羨門高
[7] 形解銷化	[8] 蓬萊	[9] 方丈	[10] 瀛洲	[11] 渤	[12] 不死之藥
[13] 劉安	[14] 淮南	[15] 宣帝	[16] 更生	[17] 王褒	[18] 張子僑

interested in the matter of reviving the arts and techniques of the holy immortals. (The Prince of) Huai-Nan had had in his pillow (for safe-keeping) certain writings entitled *Hung Pao Yuan Pi Shu*[1] (Book of the Infinite Treasure in the Secret Garden). These writings told about the holy immortals and their arts of conjuring spirits and making gold, together with Tsou Yen's technique for prolonging life by a method of repeated (transmutation) (*chhung tao*[2]).[a] Most people at that time had never seen these writings, but (Liu) Tê,[3] the father of (Liu) Kêng-Shêng, had, in the time of the emperor Wu (−123), investigated the case of the (Prince of) Huai Nan, and, (after his downfall) had secured his books. . . .[b]

This story remained famous for long afterwards.[c] It will be remembered from previous volumes that the group of proto-scientists and naturalist philosophers gathered around Liu An is regarded as responsible for that great compendium of natural philosophy, the *Huai Nan Tzu* book (*c*. −120). There is also the important collection of quasi-technological, quasi-magical procedures mentioned above and usually entitled *Huai Nan Wan Pi Shu* (The Ten Thousand Infallible Arts of the Prince of Huai-nan) which has come down to us from the same group. To this, as also to the government-supported alchemical trials of Liu Hsiang, we shall return presently (pp. 25, 35).

A list of chemical substances is included in the *Chi Ni Tzu*,[4] also known as *Fan Tzu Chi Jan*,[5] a book attributed to the −5th-century administrator, wealthy merchant and alchemist Fan Li, whom we first came across long ago.[d] It records conversations between a naturalist philosopher Chi Ni Tzu or Chi Yen[6] with Kou Chien,[7] King of Yüeh. Whether Chi Yen was a historical character or an invention of the writer of the book, who fathered it on Fan Li, we do not know, but it must belong to a southern school of natural philosophy connected somehow with Tsou Yen, and therefore datable in the late −4th or early −3rd century. Besides the theoretical discussions it contains elaborate lists of things, minerals, plants and animals, sometimes with prices and notes about provenance and quality, a very intriguing feature in view of the fact that Fan Li was a famous merchant as well as a princely adviser. Among these lists are found metallic lead (*hei chhien*[8]), lead monoxide, litharge (*huang tan*[9]), lead carbonate (*shui fên*[10]), sulphur (*shih liu huang*[11]), red bole clay (*chhih shih chih*[12]), calcareous spar (trigonal calcium carbonate), or perhaps gypsum, hydrated calcium sulphate (*ning shui shih*[13]), stalactitic calcium carbonate (*shih chung ju*[14]), brown haematite, hydrated ferric oxide (*yü yü liang*[15]), saltpetre or potassium nitrate

[a] This phrase is extremely significant because of the later obsession of the Taoist alchemists for repeatedly separating and combining mercury and sulphur in cyclical transformations. See pp. 86, 110, below.

[b] Already in the +11th century it was pointed out by Liu Fêng-Shih that Liu Tê cannot have been the investigator in the Huai Nan Tzu case, for he was only a few years old at the time, not born before −126. It was probably the grandfather, Liu Pi-Chiang[16] (−164 to −85), who made the investigation and obtained the books. Nevertheless, we have a statement in *CHS*, ch. 36, p. 5*a*, that Liu Tê, when young, was fond of Taoism and liked to cultivate the techniques of the immortals.

[c] Cf. *Po Wu Chih* (*c*. +270), ch. 5, p. 5*b*.

[d] Vol. 2, pp. 275, 554. The extant fragments have been collected in *YHSF*, ch. 69, pp. 19*a*ff.

[1] 鴻寶苑祕書	[2] 重道	[3] 劉德	[4] 計倪子	[5] 范子計然
[6] 計研	[7] 句踐	[8] 黑鉛	[9] 黃丹	[10] 水粉	[11] 石硫黃
[12] 赤石脂	[13] 凝水石	[14] 石鍾乳	[15] 禹餘糧	[16] 劉辟彊

(*hsiao shih*[1]),[a] talc or soapstone, hydrated magnesium silicate (*hua shih*[2]), potash alum, the double sulphate of potassium and aluminium (*fan shih*[3]), malachite (basic copper carbonate) in the nodular form with large holes (*khung chhing*[4]), and in the stratified form (*tshêng chhing*[5]), azurite (*pai chhing*[6]), blue azurite (*fu chhing*[7]),[b] both with less copper hydroxide than malachite, red haematite, specular iron ore, ferric oxide or ochre (*shih chê*[8]),[c] and a blue variety of siliceous clay (*chhing o*[9]).[d]

The *Chi Ni Tzu* book contains an entry of great chemical significance, but the only text now available to us, which was preserved in the *Thai-Phing Yü Lan*, includes the word *tsho*[10] (generally meaning 'a mistake') in the passage,[e] which reads: *hei chhien chih tsho hua chhêng huang tan, tan tsai hua chih chhêng shui fên.*[11] It would be difficult to make sense of this as it stands,[f] but the text presents a clear and definite meaning if we assume that it originally had the word *tshu*[12] instead of *tsho*, but that corruption had substituted the *chin*[13] radical for the *yu*[14] radical. *Hei chhien* is metallic lead;[g] *huang tan* normally refers to litharge, the red or yellow oxide of lead;[h] and *shui fên* is ceruse or basic lead carbonate, $Pb_3(OH)_2(CO_3)_2$.[i] The passage would thus read: 'When metallic lead meets with vinegar it turns into "litharge", which in turn changes into white lead.'[j] Here perhaps the basic lead acetate was mistaken by the early Chinese alchemists for litharge, especially if there was discolouring by hydrogen sulphide. The text must surely therefore have referred to the so-called 'Dutch vinegar process' of making white lead,[k] well known later in the Chinese pharmaceutical natural histories and in the traditional chemical technology of China and Japan.[l] There is a full account of it in the *Pên Tshao Kang Mu* (+1596)[m] under the name *fên hsi*,[15] 'powder of (black) tin (i.e. lead)'. Before the Ming the pharmacal naturalists mostly confined themselves to its medical uses, but they attest its antiquity, for the *Shen Nung Pên Tshao Ching* includes it, under the same name, with a synonym *chieh hsi*,[16] 'dissociated (black) tin (i.e. lead)'.[n] This is good evidence, in our view, for a date

---

[a] See the discussion in pt. 4 below.

[b] The text gives *lu chhing*,[17] where *lu* is probably a misprint for *fu*.

[c] *Shih chê* is probably the same as *tai chê shih*.[18]

[d] Not in RP. But since *pai o*[19] is synonymous with *pai shih chih*,[20] we may assume that *chhing o* may be synonymous with *chhing shih chih*,[21] the blue variety of siliceous clay according to RP 57*a*.

[e] *TPYL*, ch. 812, p. 7*a*, and *Chi Ni Tzu* (in *YHSF*), ch. 3, p. 1*b*.

[f] It is true that one of the ancient meanings of *tsho* was 'gilding' (*Shuo Wên*).

[g] RP 10; *TPYL* gives *hei kung*[22] which is synonymous with *hei chhien*.

[h] Though a sample of the traditional substance analysed at Peking in 1928 was found to contain lead peroxide and carbonate. RP 13 and 14.

[i] RP 12.

[j] Schafer took note of this passage, (9), p. 422, but did not hit upon the emendation which clarifies the matter chemically.

[k] See Mellor (1), pp. 664–5; Sherwood Taylor (4), pp. 83ff. Tan bark, wet hay or dung provides CO_2 and enough heat to volatilise the acetic acid, acetate being the intermediate product.

[l] Cf. Atkinson (2).

[m] Ch. 8, (pp. 15, 16), RP 12. Also a little later in *TKKW*, ch. 14, pp. 11*b*, 12*a*, tr. Sun & Sun (1), p. 256. Unfortunately Sung Ying-Hsing gave no illustration. We shall return to the process in Sect. 34.

[n] Mori ed., p. 85.

[1] 滑石	[2] 滑石	[3] 礬石	[4] 窒青	[5] 曾青	[6] 白青	[7] 膚青
[8] 石赭	[9] 青堊	[10] 錯	[11] 黑鉛之錯化成黃丹丹再化之成水粉			
[12] 醋	[13] 金	[14] 酉	[15] 粉錫	[16] 解錫	[17] 盧青	
[18] 代赭石	[19] 白堊	[20] 白石脂	[21] 青石脂	[22] 黑鈆		

in the −2nd century, not so long after the *Chi Ni Tzu* book. Since the provenance is said to be from valleys in the mountains it might be argued that naturally occurring cerussite was alone known, but the name reveals that this cannot be so, for the conscious connection with lead would imply a knowledge of the artificial product. In the works of the *pên tshao* tradition lead carbonate is recommended as antiparasitic, anthelminthic, emmenagogue and abortifacient, also in soothing antiseptic plasters for burns, wounds or ulcers; and many other uses.[a] The old looseness of the name, confusing lead with tin, was already corrected in the *Hsin Hsiu Pên Tshao* of +659.

The use of white lead for cosmetic purposes as 'face-powder' or 'face-cream' goes back a long way, it seems, both in East and West. It has been found in Indus Valley sites and in Greek graves of the −4th century,[b] as also in ancient Egyptian remains,[c] but one can never be sure that these samples were artificially made, since impure lead carbonate (cerussite) can occur, as just mentioned, in Nature. It is usually said that the vinegar process was described by Theophrastus (d. −287),[d] as well as by Pliny, Vitruvius and Dioscorides, but a close examination of the passages led Bailey to the view that their *psimithion* or *cerussa* was the acetate rather than the carbonate; though the latter was known as a natural earth.[e] In any case lead salts of white character in some form or other were being used for cosmetic purposes in Western antiquity already by Xenophon's time (d. *c.* −350).[f] The same would be true of China, where lead carbonate face-powder has been found in tombs of Chhin and Han date by archaeologists;[g] but again we do not know whether it was artificially made. Here it is worth pointing out, however, the great significance of the two distinct stages mentioned so clearly in the *Chi Ni Tzu* statement, for the acetate is indeed an intermediate product in the process.

If this text can be placed about −300, there is a still earlier reference in the *Mo Tzu* book, perhaps nearer −400, which says that 'Yü the Great invented (face-)powder'.[h] The attribution is of course legendary, and one cannot rule out natural ceruse, or even a simple paste of rice-flour and water, but the date is near enough to the *Chi Ni Tzu* to warrant a possible connection. Presumably the oldest name for lead carbonate was *shui fên*,[1] 'water powder', as in the *Chi Ni Tzu* list (p. 14), but its most common ancient name was *hu fên*,[2] not indicating any foreign origin but, as the *Shih Ming*[3] explains (+100),[i] a corruption of *hu fên*,[4] 'paste, or ointment, powder'. Later tradition ascribed lead carbonate cosmetic to the Shang period as the invention

---

ᵃ Cf. *CLPT* (+1249), ch. 5, (p. 127.1).   ᵇ Caley (5).   ᶜ Lucas (1), pp. 100ff., 276.
ᵈ *Peri Lithōn*, 101. Cf. Neuburger (1), p. 193; Smythe (1), pp. 16ff.; Pulsifer (1), pp. 205–7; Mellor (2), vol. 7, p. 828.
ᵉ (1), vol. 2, pp. 204, 213. Cerussite is quite a common mineral (Mellor (2), vol. 7, pp. 829–30). But the lead carbonate process must have been worked by Galen's time in the +2nd century since he mentions the use of dung in his *De Simplicium Medicamentorum Temperamentis*.
ᶠ *Oikonomikos*, x, 7.
ᵍ Harada & Tazawa (1), p. 33 for Lo-lang in Korea; Bergman (1), p. 125 for Lou-lan in Sinkiang.
ʰ Fide *TPYL*, ch. 719, p. 1a; *Shih Wu Chi Yuan*, ch. 3, p. 18a, b; *PTKM*, ch. 8, (p. 16). The passage seems not to be in the *Mo Tzu* now, but it may easily have dropped out since the Sung. It would be natural for Mo Ti to have inveighed against face-powder, though Ta Yü was a great hero of his.
ⁱ Ch. 15, (p. 240 in Wang Hsien-Chhien's ed.).

¹ 水粉   ² 胡粉   ³ 釋名   ⁴ 餬粉

of the last ruler, the voluptuary High King Chou,[1] but this legend hardly arose before the end of the Han.[a] The *Huai Nan Tzu* book (−120) also has a revealing saying: 'Lacquer does not refuse to be black, lead does not decline to be white.'[b] White lead paint for walls is recorded in the *Han Kuan I*,[c] and, at a more elegant level, white lead ink was used by scholars for commenting on textual passages written in black on wood or bamboo.[d] The former use at least is one very good reason for believing that the Chinese product was indeed lead carbonate from *Chi Ni Tzu* onwards, since the acetate has nothing like the same pigmentary density or 'covering power'.[e] Artificial lead carbonate face-powder must have been used during the Han because a poem of the Lady Pan chieh-yü[2] (*c.* −20) refers to it as 'blended lead' (*tiao chhien*[3]).[f] Lastly, Wei Po-Yang in the mid +2nd century remarks in the *Tshan Thung Chhi* that ceruse goes back to metallic lead when treated with charcoal, an early testimony to the reducing power of carbon.[g] All in all, it would seem sure that the Chinese were making basic lead carbonate by the 'Dutch process' by about −300, whatever may have been happening in other cultures at that time. We must now return to the more elated realms of alchemical aspiration.

The search for the medicine of immortality by the first emperor, Chhin Shih Huang Ti, in the −3rd century, and the activities of the shamans, *wu* and *fang-shih* around him have already been referred to in previous volumes of this work.[h] It remains only to quote the following passages from the *Shih Chi*:[i]

During the 28th year of his reign Shih Huang Ti (−219)[j] despatched Hsü Fu[4] with several thousand young men and maidens to go and look for (the abodes of) the immortals (hidden) in the Eastern Ocean. . . . During the 32nd year (−215) he went to Chieh-shih[5] and sent

[a] See *Po Wu Chih* (+290), in *TPYL*, ch. 719, p. 1*a* (not in all reconstructed versions); *Chung Hua Ku Chin Chu* (+925), ch. 2, p. 6*b*; *Hsü Shih Shih* (+960), in *SF*, ch. 10, p. 51*a, b*. Copied in *PTKM*, ch. 8, (p. 16) and elsewhere.

[b] Ch. 11, p. 3*a*, also cit. *TPYL*, ch. 719, p. 1*b*, which writes *chhien*,[6] lead, instead of *fên*,[7] powder. Wallacker in his translation, (1), pp. 31–2, prefers a different interpretation. The saying occurs in the context of Liu An's excellent contention that there is no nobility or baseness inherent in natural things.

[c] By Ying Shao (d. +195), cit. *TPYL*, ch. 719, p. 2*a*. Again in *Yeh Chung Chi*, cit. ibid. with reference to Shih Hu of the Later Chao (r. +334 to +349). Cf. Vol. 4, pt. 2, p. 287.

[d] See Yang Hsiung's letter to Liu Hsin (*c.* −20) in *CSHK* (Chhien Han sect.), ch. 52, p. 9*a*. Schafer, who noted this from another collection, shows that the practice can be attested also from the +5th century, when such white brushes were called *fên pi*,[8] 'powder-pens', (9), p. 437. We shall have more to say about white and red inks as valuable tools of scientific writers when describing the *pên tshao* tradition in Sect. 38.

[e] Furthermore, lead acetate was not listed and separately discussed, under the quite different name of *chhien shuang*[9] (frost of lead) until the *Jih Hua Chu Chia Pên Tshao* of +972; cf. *PTKM*, ch. 8, (pp. 14, 15), RP 11.

[f] This is the *Tao Su Fu*[10] (Ode on a Girl of Matchless Beauty), in *CSHK* (Chhien Han sect.), ch. 11, p. 7*b*, also noted by Schafer, from another source, (9), p. 435.

[g] Ch. 12 (ch. 1, p. 25*b*). Wu & Davis (1), p. 241.

[h] Vol. 2, pp. 132 ff., Vol. 4, pt. 3, pp. 551 ff.

[i] Ch. 6, pp. 18*a*; 20*b*, 21*a*; 21*a, b*; 24*b*, 25*a*; 26*a, b*; tr. auct. adjuv. Chavannes (1), vol. 2, pp. 139, 152, 164, 167, 176 ff., 180 ff.

[j] The count is from his accession to the throne of Chhin State, not that of the empire.

[1] 紂	[2] 班婕妤	[3] 調鉛	[4] 徐市	[5] 碣石	[6] 鉛
[7] 粉	[8] 粉筆	[9] 鉛霜	[10] 擣素賦		

2

Master Lu,[1] a man of Yen,[2] to sea to find (the immortal) Hsienmên Kao-Shih[3]. . . . In the same year he ordered Han Chung,[4] the Venerable Hou[5] and Master Shih[6] to (set sail to) search for the elixir of life of the immortals (*hsien jen pu ssu chih yao*[7]). . . . During the 35th year (−212) he styled himself 'Perfected Adept' (*chen jen*[8]—one who has attained the Tao) because of his desire to achieve immortality. . . . (In the same year, in a speech complaining that his experts had left him) he said that he had summoned all the scholars and naturalists (*wên hsüeh fang shu shih*[9]) (to his court) as (he) wished (the former) to help to bring peace and prosperity (to the empire) and the naturalists to search out, select and prepare wonderful medicines (*chhi yao*[10]). But now Han Chung had gone off without letting him know, and Hsü Fu, though supported at vast expense, had never brought back the drug (of immortality). . . .[a]

Hsü's name is also written Hsü Fu,[11] read in Japanese as Jofuku. Although his mission was to find the fabled isle of Phêng-Lai, it seems quite probable that he and his people settled in Japan; they certainly never returned to China. Today there is a tomb of Jofuku at Shingū, a coastal town on Honshu south of Kyoto.[b]

No one knows exactly what kind of medicine Chhin Shih Huang was looking for. We have seen reasons (pt. 2, pp. 116, 122) for surmising a mushroom. There were later suggestions that it might have been a plant, perhaps a kind of mulberry. The following account is given in the *Chin Lou Tzu*[12] (Book of the Golden Hall Master), compiled by Hsiao I[13] (the emperor Liang Yuan Ti) *c.* +550:

On the magical island of Shen-Chou[14] there grows the herb of immortality (*pu ssu chih tshao*[15])[c] new sprouts of which come up in great abundance. People who have been dead for some time rise again if this herb is strewn upon them. In the time of Chhin Shih Huang Ti many people in Ferghana (Ta-Yuan[16])[d] died unjustly, but certain birds looking like crows took this herb (in their bills) and dropped it on the ground so as to cover them, whereupon the dead immediately sat up. (Chhin) Shih Huang Ti sent (someone) to enquire of Master

---

[a] He apparently never caught any of the proto-scientific experts, but he put to death 460 literati at Hsienyang and exiled many more to the frontiers. On the expression *chen jen* cf. Porkert (2), pp. 444 ff.

[b] See Davis & Nakaseko (1, 2). The authenticity cannot of course be guaranteed because Japanese scholars were familiar from early times with the Chinese historical records. But pottery and other artifacts of the Yayoi culture in Japan (−3rd to +2nd centuries) strongly suggest Chinese influence (cf. Kidder (1), pp. 91 ff.), so Hsü Fu and many others may really have ended their days there. We had the pleasure of visiting this place in 1971 and following his traces with Dr Hashimoto Keizō; Fig. 1343 shows the tomb as it is today. The votive shrine, however (*tzhu*,[17] Fig. 1344), is some distance away across the town forming part of a Shinto temple, the Asuka Jinja[18] (Fig. 1345), which is protected on the sea side by a conical forested hill called Hōraisan[19] (Phêng-Lai Shan[19]), the name of the chief mountain-island sought by Chhin Shih Huang Ti. In the temple grounds, and elsewhere in the city, there grow bushes of a valued drug-plant, *tendai wuyaku*,[20] which was, according to Japanese tradition, what Hsü Fu was seeking. This has been identified as *Lindera strychnifolia* (Lauraceae, CC 1325), tonic and tranquillising, native to China also (especially Thien-thai Shan), and recorded first in the *Khai-Pao Pên Tshao* of +970, though doubtless known much earlier (Fig. 1346). The Japanese tradition is that the forested mountain country of Kumano[21] (now a national park), inland from Shingū, was the land of Phêng-Lai[19] coveted by Chhin Shih Huang Ti.

[c] This is the ancient phrase so often met with, cf. e.g. *Po Wu Chih* (*c.* +290), ch. 1, p. 7b.

[d] Reading *yuan*[22] for *yuan*[23]. The history had gone astray for there were no campaigns in Central Asia until Han times.

[1] 盧生	[2] 燕	[3] 羨門高誓	[4] 韓終	[5] 侯生	[6] 石生
[7] 仙人不死之藥		[8] 眞人	[9] 文學方術士	[10] 奇藥	[11] 徐福
[12] 金樓子	[13] 蕭繹	[14] 神州	[15] 不死之草	[16] 大苑	[17] 祠
[18] 阿須賀神社	[19] 蓬萊山	[20] 天台烏藥	[21] 熊野	[22] 宛	[23] 苑

Kuei Ku[1] (who dwelt) by the northern city wall,[a] and (the answer) came that on Tan-Chou[2] (another magical island) in the Eastern Sea the herb of immortality grew in beautiful fields. It was on hearing this that Chhin Shih Huang Ti sent Hsü Fu to sea to look for certain golden and jade-like vegetables and also a tree producing mulberries one inch across. Thus did the Chhin emperor send Hsü Fu to search for the mulberry (sang jen[3]) in the midst of the blue sea. There grew the fu-sang[4] mulberry tree several times ten thousand feet high. There was a pair of them supporting each other; hence the name 'mutually-supporting mulberry' (fu-sang). The immortals ate the mulberries, their bodies gave out a golden glow, and they flew in and out of the Palace of Primal Vitality (yuan kung[5]).[b]

The interest of this lies mainly in the fact which it evokes that in the most ancient times medicaments of longevity and immortality had been sought just as much in the realm of botany and pharmacy as in that of mineralogy and alchemy. This we have already seen in Table 106 from the Lieh Hsien Chuan. The fu-sang was an entirely legendary tree with a very long career in folk cosmology;[c] it was one of the 'Arbores Solis et Lunae', growing in the most easterly of all islands and serving as a perch for the ten suns before they took off on their regular flights throughout the ten-day week. Naturally any vegetable product connected with such an exalted tree could be expected to have magical macrobiotic powers. Ko Hung says that peach gum should be macerated in an extract of ashed mulberries; this if taken cures all illnesses, and if taken for a long time it will make the body light and glowing, so much so as to light up dark places at night.[d] Or, to take another example, the Shun-Yang Lü Chen-Jen Yao Shih Chih[6] (The Adept Lü Shun-Yang's (i.e. Lü Tung-Pin's) Book on Drugs and Minerals), a Thang work,[e] recommends an elixir called Pao sha lung ya[7] which seems to be nothing but mulberry leaves.

Besides the fang shih in the service of Chhin Shih Huang Ti many other contemporary figures are mentioned in the hagiography of the immortals.[f] For example, there were the two friends Chiang Shu-Mou[8] and Liu Thai-Pin,[9] both aspirants to hsien-ship; it is said that the former planted fruits and vegetables and sold them in exchange for cinnabar.[g] Chao Tao-I[10] tells us of Thang Kung-Fang,[11] who became an immortal after eating an elixir, that he had been a disciple of the adept Li Pa-Pai[12].[h]

---

[a] Again an anachronism, for Master Kuei Ku was a Warring States personality.

[b] Ch. 5, p. 16a, b, tr. auct. Evidently a re-telling of the legend in the earlier Hai Nei Shih Chou Chi (pt. 2, p. 122 above).

[c] See Vol. 3, p. 567; Vol. 4, pt. 3, pp. 540ff. Fu-sang was also the name of the Hibiscus genus; cf. Li Hui-Lin (1), pp. 138ff.

[d] Pao Phu Tzu, ch. 11, p. 11a; Ware (5), p. 190.

[e] TT 896.

[f] Much material about the magician-technicians of the Warring States, Chhin and Han periods has been collected by Chhen Phan (7).

[g] Hsüan Phin Lu[13] (TT773), ch. 1, p. 11a.

[h] See Li Shih Chen Hsien Thi Tao Thung Chien[14] (Comprehensive Mirror of the Embodiment of the Tao by Adepts and Immortals throughout History), TT293, ch. 10. Also pt. 2, pp. 124 ff.

[1] 鬼谷	[2] 亶州	[3] 桑椹	[4] 扶桑	[5] 元宮
[6] 純陽呂眞人藥石製	[7] 寶砂龍芽	[8] 姜叔茂	[9] 劉太賓	[10] 趙道一
[11] 唐公房	[12] 李八百	[13] 玄品錄	[14] 歷世眞仙體道通鑑	

(ii) *Aurifiction and aurifaction in the Han*

After the fall of the Chhin empire the belief in the feasibility of material immortality was so strong that a number of enthusiasts were found among the senior administrators serving the emperor Kao Tsu[1] of the Early Han dynasty. We have already mentioned (p. 11) Tshao Tshan[2] (d. −190), the minister who acquired the art from the adept Kai Kung.[3] There was also Chhen Phing[4] (d. −178), a trusted adviser of Kao Tsu. According to the *Chhien Han Shu* Chhen Phing lived in poverty in his early days, was fond of reading and had studied the art of immortality.[a] However, as aspirants to *hsien*-ship, they were far overshadowed by the reputation of their colleague Chang Liang[5] (d. −187), another trusted adviser of Kao Tsu. The story of the immortal Huang Shih Kung[6] (the Old Gentleman of the Yellow Stone) testing the patience of Chang Liang by thrice making him go and fetch the sandals which the sage had thrown down from a bridge is famous among the literati. He resigned from office in order that he might fulfil his ambition to become an immortal himself. According to the *Chhien Han Shu*:

He said to the emperor (Kao Tsu), 'My desire is only to shun all worldly affairs and become a follower of Chhih Sung Tzu[7] (the Red Pine Master).'[b] After which he retired to learn the Tao, hoping to achieve immortality.[c]

The *Shih Chi* version specifies that for this purpose he engaged in dietary and gymnastic exercises.[d] Chang Liang later came to be regarded as the ancestor of Chang Tao-Ling[8] (*fl.* +156), the first prominent figure in the development of Taoism into an organised religion.[e]

The hagiographies mention a number of other adepts who were active during the −2nd century. For example, Chao Tao-I says[f] that Huang Hua,[9] also known as Chiu Ling Tzu,[10] succeeded in preparing an elixir, that Yin Hêng,[11] otherwise called Pei Chi Tzu,[12] had ingested a 'magical (or spiritual) elixir' (*shen tan*[13]), that Liu Jung,[14] known also by the pseudonym Nan Chi Tzu,[15] had taken a 'cloud-and-frost elixir' (*yün shuang tan*[16]),[g] and that Li Hsiu,[17] also called Chüeh Tung Tzu,[18] had eaten a 'cyclically-transformed elixir' (*huan tan*[19]).[h]

---

[a] Ch. 40, p. 12*a*.

[b] Cf. p. 9 above.

[c] Ch. 40, p. 11*b*, tr. auct. At least this was the official excuse he gave for his resignation. He was aware of the fact that Kao Tsu, after becoming emperor, began to feel suspicious of some of his followers who had supported him to the throne, for example in the case of Han Hsin.[20]

[d] Ch. 55, p. 13*b*, tr. Watson (1), vol. 1, p. 150.

[e] Cf. Vol. 2, pp. 155ff.

[f] In *TT* 293, ch. 5. A late writer, probably of the Yuan period.

[g] These terms were often used, as we shall see, for purified white powdery or crystalline precipitates or sublimates.

[h] On cyclical transformation, see pp. 60, 83, 86, 90 below.

[1] 高祖	[2] 曹參	[3] 蓋公	[4] 陳平	[5] 張良	[6] 黃石公
[7] 赤松子	[8] 張道陵	[9] 皇化	[10] 九靈子	[11] 陰恒	[12] 北極子
[13] 神丹	[14] 柳融	[15] 南極子	[16] 雲霜丹	[17] 李修	[18] 絕洞子
[19] 還丹	[20] 韓信				

This was about the time of the Lady of Tai (*Tai Hou chhi tzu*[1]) who died in −186 or perhaps somewhat later,[a] and whose body, uncorrupt through more than two millennia, was found at Ma-wang Tui[2] near Chhangsha in 1972. This discovery, unique in modern times, though by no means without parallels in textual accounts from earlier ages,[b] formed, as will be remembered, a kind of climax to Vol. 5, pt. 2; because it demonstrated that the Taoist conception of *shih chieh hsien jen*[3] (corpse-free immortals), whose bodies would remain century after century like those of persons still living, was not entirely imaginary, and that the adepts knew ways of accomplishing this. Such methods were used, no doubt, for the bodies of fellow-adepts, but also, it seems, if a suitable fee was forthcoming, for the bodies of members of any patrician families rich enough to be able to afford it. We still do not know quite how it was done, but several facts are clear: the innermost of the four coffins of the Lady of Tai contained a certain amount of an aqueous solution or suspension of mercuric sulphide, the atmosphere within them was largely methane under some pressure, all were remarkably air-tight and water-tight, and the temperature (some fifty feet down) had been rather constant at 13–14°C.[c]

To throw a little more light on the Taoist ideas of the time, it is well worth while to take a look at the famous banner of painted silk (*po hua*[4]) which covered her coffin.[d] The T-shaped form of this *hua fan*[5] (painted standard), *ming ching*[6] (personal ensign) or *yin hun fan*[7] (psychopompic banner), is shown in the drawing of Fig. 1347.[e] Regarded as a temporary dwelling-place for the souls of the deceased,[f] it is divided into three levels, the heavenly world of the holy immortals at the top, the earthly world in the middle, and the underworld (*shui fu*[8] or *huang chhüan*[9])[g] at the bottom. High up on the right we see the sun with its crow, and on the left the moon with its toad and rabbit; just below them respectively the Fu-Sang[10] tree with its ten suns,[h] and a great dragon carrying up Chhang O with her elixir into the Palace of the Moon.[i] In the centre writhes Fu-Hsi with his serpent tail, the organiser god, surrounded by magical crane birds. Lower down at the gates of the heavens sit two guardian-immortals, while above them two strange animals (*shêng lung*[11] or *fei lien*[12]) are bearing successful candidates upwards into the empyrean. On the level next below, the Lady

---

[a] It depends on which of the four successive Lords of Tai was her husband. Her death can hardly have been earlier than −193 or later than −110, but the most probable estimate is about twenty years after the time when the first Lord died (−186), i.e. the neighbourhood of −166. See Anon. (*104, 105*); Anon. (*113, 114*); Miyagawa Torao et al. (*1*); Rudolph (*8*); Chhen Shun-Hua (*1*).

[b] See Yang Po-Chün (*1*).

[c] See the discussions in Anon. (*117*); Ku Thieh-Fu (*1*); Shih Wei (*1*); Chhen Shun-Hua (*1*).

[d] Full descriptions and elucidations will be found in An Chih-Min (*1*); Ma Yung (*1*) and Sun Tso-Yün (*3*). Cf. Bulling (*15*); Wên Pien (*1*).

[e] Detailed pictures have been published in colour as an album, Anon. (*118*). Such banners were previously known only from textual mentions, as in the *Shih Chi*.

[f] Cf. Vol. 5, pt. 2, p. 91.

[g] See also Vol. 5, pt. 2, pp. 84ff.

[h] See Vol. 4, pt. 3, pp. 540–1. Only eight are shown in the banner.

[i] Cf. p. 4 above.

[1] 軑侯妻子	[2] 馬王堆	[3] 尸解仙人	[4] 帛畫	[5] 畫幡
[6] 銘旌	[7] 引魂幡	[8] 水府	[9] 黃泉	[10] 扶桑
[11] 升龍	[12] 飛廉			

Fig. 1347. The representations on the painted silk banner buried with the Lady of Tai
(d. c. −166). Drawing from An Chih-Min (1). For description see pp. 21, 23.

of Tai, attended by three maids of honour, greets two envoy-immortals,[a] and lower still a family ancestral sacrifice is seen in progress beneath a huge chime-stone, either under her presidency or for her spirit.[b] Clearly she is destined to become a *hsien*.[1] Finally there are the dragons and strange beasts of the underworld, one of them, a dwarf-like creature, supporting the visible world above. These animals echo a veritable ballet of similar creatures very finely drawn, and human beings with thero-morphic heads, depicted on the lacquered sides of the Lady of Tai's coffins.[c] These are all forms of Thu Po[2] (the Earth Lord)[d] doing battle with evil in the form of snakes, birds and even cattle.[e] All in all, the Taoist myths and legends so richly portrayed on the banner and the coffins correspond closely with texts such as the *Shan Hai Ching*, the *Chhu Tzhu* and the *Huai Nan Tzu* book. To this last we must now turn.

It belongs to an important nodal point in the history of Chinese alchemy. Liu An[3] (−178 to −122), the Prince of Huai-Nan[4] and a grandson of the emperor Kao Tsu, was a great patron of magicians and alchemists.[f] He himself had acquired the art from the adepts 'Pa Kung'[5] and Wang Chung-Kao.[6] Hsü Ti-Shan[7] says that 'Pa Kung' is not a personal name but refers to eight alchemists among the many guests or retainers surrounding Liu An, and gives their names as Su Fei,[8] Li Shang,[9] Tso Wu,[10] Thien Yu,[11] Lei Pei,[12] Mao Pei,[13] Wu Pei,[14] and Chin Chhang[15].[g] The *Chhien Han Shu* has the following to say about Liu An:

He gathered guests and those versed in the art (*fang shu chih shih*[16]) to the number of several thousands, and wrote an 'Inner Book' (*Nei Shu*[17]) consisting of 21 chapters and an 'Outer Book' (*Wai Shu*[18]) of many (chapters). There was also written a 'Middle Volume' (*Chung Phien*[19]) in 8 chapters, dealing with the art of the immortals and the transmutation of gold and silver (*shen hsien huang pai chih shu*[20]).[h]

Liu An was an uncle of the emperor Wu Ti[21] and at first highly respected by him. However, he was later implicated in a conspiracy, and took his own life or disappeared before the commission sent by the emperor could lay hands on him.[i] But the Taoists claimed that Liu An had attained *hsien*-ship; accepting the situation as an

---

[a] One is reminded of the folk-song 'Diverus and Lazarus'. But here Dives could pay for the angel's knee, and succeeded in getting it, or something equivalent.

[b] One might like to think of this as a scene in an elixir laboratory, but speculation must really have its limits.   [c] A full description will be found in Sun Tso-Yün (2).

[d] Probably identical with the *Chou Li*'s *fang hsiang shih*,[22] mentioned already in previous volumes, and with the later tomb-guardian figures with long tongues (*chen mu shou*[23]).

[e] Auspicious cranes also occur in places, some shown breathing upwards, others downwards; an early reference to the respiratory exercises of physiological alchemy (cf. Vol. 5, pt. 5).

[f] There is now a biography of him by Wallacker (2), cf. Morgan (1), p. xliii.

[g] Hsü Ti-Shan (1), vol. 1, p. 119. For the story of how Liu An became an immortal with the help of the 'Pa Kung' see *Shen Hsien Thung Chien*, ch. 8, sect. 2, pp. 36 ff., and Davis (2, 4).

[h] Ch. 44, p. 8*b*, tr. auct.

[i] See *SC*, ch. 118, p. 19*a*; *CHS*, ch. 44, p. 14*a*. He may well have been a victim of one of Han Wu Ti's attempts to centralise power by 'framing-up' possible rivals.

[1] 仙	[2] 土伯	[3] 劉安	[4] 淮南	[5] 八公	[6] 王仲高
[7] 許地山	[8] 蘇飛	[9] 李尚	[10] 左吾	[11] 田由	[12] 雷被
[13] 毛被	[14] 吳被	[15] 晉昌	[16] 方術之士		[17] 內書
[18] 外書	[19] 中篇	[20] 神仙黃白之術		[21] 武帝	[22] 方相氏
[23] 鎭墓獸					

opportunity for ingesting a powerful elixir, he had only pretended to die as a mortal, thus escaping public humiliation and imperial ire.[a]

It has been thought that of all the writings of Liu An and his school only the *Huai Nan Tzu*,[1] otherwise called the *Huai Nan Hung Lieh Chieh*,[2] has come down to us. This important book, frequently quoted in our previous volumes, constitutes, with the *Lü shih Chhun Chhiu*, the greatest monument of Chhin and Han natural philosophy. But there is little of a strictly chemical nature in the *Huai Nan Tzu*, always excepting that fundamental passage on the natural growth and transmutation of metals and minerals in the bowels of the earth, which we studied at an earlier stage,[b] and shall have occasion to examine again in due course.[c] This was the doctrine of spontaneous very slow natural change which the alchemists might hope to accelerate in their laboratories.

One may say that there is no overt alchemy in the *Huai Nan Tzu* book. But there are passages verging on the alchemical, recognisable by their imagery of gold, cinnabar, herbs and immortality. In ch. 4 the writer describes a fabulous geography somewhat reminiscent of the *Shan Hai Ching*,[d] but shot through with many observations and statements, technical terms and natural history names, which show that a body of real knowledge about real things was being collected by Liu An's proto-scientists. Our point can be illustrated by one of the more fantastic passages, which runs as follows:

> The (mountains called) 'Hanging Gardens' (Hsüan-Phu[3]), 'Cool Breezes' (Liang-Fêng[4]) and 'Crystal Trees' (Fan-Thung[5]),[e] lie between the Khun-Lun[6] (massif)[f] and the 'Gate of Heaven' (Chhang-Ho[7]). They are (as it were) its vegetable gardens. The pools in these gardens are filled with yellow water, which flows thrice roundabout and returns to its source (continually). It is called 'cinnabar water' (*tan shui*[8]), and if a man drinks of it he will never die. . . .
>
> All the four streams are the magical fountains of the divinities. They are useful for mixing the hundred medicines which fertilise the ten thousand things. Twice the height of Khun-Lun towers the Cool-Breeze peak; if a man can get to the top of that he will never see death. Twice the height of the Cool-Breeze peak tower the Hanging-Gardens; if one can reach them one becomes mighty in magic and can control the wind and the rain. Twice the height of the Hanging-Gardens is Highest Heaven itself; if any mortal wings his way thither he becomes one of the spirits of the blest. For it is the dwelling-place of the highest divinity (Thai Ti[9]).[g]

Now we quoted earlier a passage from the *Chhien Han Shu* about certain writings entitled *Hung Pao Yuan Pi Shu*, which came down to Liu Hsiang through Liu Tê.[h] The *Sui Shu* bibliography mentions two different books by Liu An, namely a *Huai*

---

a See, for example, *TT* 293, ch. 5. Cf. also Fig. 1313 (pt. 2, p. 127) above.
b Vol. 3, p. 640. It is in ch. 4, p. 12*a*, tr. Erkes (1); Dubs (5); cf. Eliade (4, 5).
c See pt. 4 below.   d See Vol. 3, pp. 504 ff.
e Lit. 'alum *Paulownia*-trees', but surely the thought must have been of a dendritic crystallisation.
f Mythologically the Khun-Lun Shan was the Himalayan massif, corresponding to Mt Meru, the central mountain of Indian and Buddhist cosmology. See Vol. 3, pp. 565 ff., Vol. 4, pt. 2, pp. 529 ff.
g Ch. 4, pp. 2*b* ff., tr. auct. adjuv. Erkes (1).   h See p. 14 above.

1 淮南子   2 淮南鴻烈解   3 懸圃   4 涼風   5 礬桐
6 昆侖   7 閶闔   8 丹水   9 太帝

Nan Wan Pi Ching[1] and a *Huai Nan Pien Hua Shu*[2] which existed during the Sui period, but the *Hsin Thang Shu* gives only the name *Huai Nan Wan Pi Shu*[3].[a] Yeh Tê-Hui[4] thinks that the two titles in the *Sui Shu* were incorporated under a single title in the *Hsin Thang Shu*, and this may well have been the same text as one or other of the books referred to by the *Chhien Han Shu* under the names 'Inner Book', 'Outer Book', and 'Middle Chapter'. Perhaps Chen-Chung,[5] Hung-Pao,[6] Wan-Pi[7] and Yuan-Pi[8] were originally titles of parts of a Corpus of writings bearing the name of the Prince of Huai-Nan and including (possibly as the Nei Shu[9]) what we have now as the *Huai Nan Tzu* book.[b] Fragments of the *Huai Nan Wan Pi Shu* are quoted in several texts.[c] Much of the material deals with magic, as well as various rational techniques,[d] but we may quote a few statements of alchemical significance:

(a) An old basket (put into *chiang*[10] sauce) removes excess salt (by acting as a focus of crystallisation).[e]

(b) To make weak wine strong, put in (more) Ferghana grapes (i.e. add more sugar for fermentation).[f]

(c) When jade is made into a jade suspension it is called *yü chhüan*[11] and if this is consumed eternal youth may be attained.[g]

(d) Light-coloured azurite (*pai chhing*[12]) and iron instantly changes into copper.[h]

(e) Cinnabar can be made to turn into mercury (by distillation).[i]

(f) Mulberry wood is the *ching*[13] (essence) of *Chi hsiu*[14];[j] it is a numinous wood. When insects bore into it and eat it they form marks in patterns, and if a man eats this he will be rejuvenated.[k]

(g) The *Atractylis* plant (*shu*[15])[l] is the *ching*[16] (essence) of mountains; it gathers together the essential *chhi* of Yin and Yang. If a man eats it he will be able to abstain from cereals and become one of the holy immortals.[m]

[a] See *Sui Shu*, ch. 34, p. 27a and *Hsin Thang Shu*, ch. 49, p. 16b.

[b] Cf. p. 14 above. On the whole subject, cf. Kaltenmark (2), p. 32. Among interesting later references in Chinese literature cf. *Chin Lou Tzu*, ch. 3, p. 10b.

[c] Such as Ouyang Hsün's *I Wên Lei Chü*[17] (c. +622), Hsü Chien's *Chhu Hsüeh Chi*[18] (+700), Li Fang's *Thai-Phing Yü Lan*, and under various titles such as *Huai Nan Fang*[19] and *Huai Nan San-Shih-Liu Shui Fa*[20] in Chang Tshun-Hui's[21] *Chung-Hsiu Chêng-Ho Ching-Shih Chêng-Lei Pei-Yung Pên Tshao*[22] pharmacopoeia (+1249). The book has been reconstructed from fragments found in the above works independently by Yeh Tê-Hui (in the *Kuan Ku Thang So Chu Shu*[23] collection), Sun Fêng-I[24] (in the *Wên Ching Thang Tshung Shu*[25]), and Mao Phan-Lin[26] (in *Lung Chhi Ching Shê Tshung Shu*[27]).

[d] Cf. Vol. 4, pt. 2, p. 596.

[e] In Yeh Tê-Hui's recension, ch. 1, p. 6b, Sun Fêng-I's, p. 3a, and Mao Phan-Lin's, p. 7a.

[f] In Mao Phan-Lin, p. 11a; Yeh Tê-Hui (ch. 1, p. 10a) gives *huan*[28] instead of *yuan*.[29]

[g] In Yeh Tê-Hui, ch. 1, p. 13b. See pt. 4 below on aqueous solutions.

[h] In Yeh Tê-Hui, ch. 2, p. 1b, Sun Fêng-I, p. 6b, and Mao Phan-Lin, p. 5b. This is the wet precipitation of copper from solutions by iron. See pt. 4 below.

[i] In Yeh Tê-Hui, ch. 2, p. 1b, Sun Fêng-I, p. 6b, and Mao Phan-Lin, p. 6a; TPYL, ch. 988, p. 6a.

[j] One of the twenty-eight lunar mansions. See Vol. 3, p. 235, Table 24.

[k] In Yeh Tê-Hui, ch. 2, p. 5a. See p. 19 above. For the lunar mansion named see Vol. 3, p. 235.

[l] Cf. p. 11 above, and p. 177 below.   [m] Yeh Tê-Hui, from *I Wên Lei Chü*, ch. 81.

[1] 淮南萬畢經	[2] 淮南變化術	[3] 淮南萬畢術	[4] 葉德輝	
[5] 枕中	[6] 鴻寶	[7] 萬畢	[8] 苑祕	[9] 內書
[10] 醬	[11] 玉泉	[12] 白青	[13] 精	[14] 箕宿
[15] 朮	[16] 精	[17] 藝文類聚	[18] 初學記	
[19] 淮南方	[20] 淮南三十六水法	[23] 觀古堂所著書	[21] 張存惠	
[22] 重修政和經史証類備用本草			[24] 孫馮翼	
[25] 問經堂叢書	[26] 茆泮林	[27] 龍溪精舍叢書	[28] 莧	[29] 苑

(h) Thung-tree wood (*Aleurites*) forms clouds.[a]

(i) Water can be clarified by stirring with glue (clearing with isinglass).[b]

(j) Malachite (*tshêng chhing*[1]) used as a medicine stops a man from ageing.[c]

This is therefore one of the earliest texts in Chinese literature which gives evidence of alchemical and chemical procedures embodied among a great variety of techniques. It seems quite fair to date this material between −150 and −120.

It is usually supposed that the circle of alchemical adepts gathered by Liu An broke up at the time of his death or disappearance leaving no perceptible trace. But a curious tradition preserved in encyclopaedias and so far not much noticed suggests that it may have gone on working for some time afterwards. The *Hsin Lun*[2] of Huan Than,[3] written between +10 and +20 and presented to the throne in +25 or +26, is quoted in the *Thai-Phing Yü Lan* as follows:[d]

Huan Than, in his *Hsin Lun*, says that the son of the Prince of Huai-Nan, (Liu) Phing,[4] welcomed Taoists who could make gold and silver (from lead). He also talks about the character for gold (*chin*[5]), saying that lead (*chhien* or *yuan*[6]) is the grandfather (*kung*[7]) of gold, and that silver (*yin*[8]) is its younger brother (*ti*[9]).[e]

We need not dwell on Huan's punning etymologies, but Liu Phing is both interesting and obscure, for no such son is known to the historians,[f] and as the name is a girl's name some daughter may perhaps have been intended.[g]

By this time the making of artificial or counterfeit gold had apparently become so common in China that in the year −144 the emperor Ching Ti[10] (r. −156 to −141) issued an edict prohibiting it and punishing offenders by public execution.[h]

In the 6th year of the middle (part of the reign)[i] ... in the twelfth month ... a statute was established forbidding the (unauthorised private) minting (lit. casting) of coin, and the (making of) artificial or counterfeit gold (*wei huang chin*[11]) under penalty of death.

The early date of this evidence and its historical reliability make it so important for the history of alchemy or aurifaction in all cultures that any further light which can be thrown on it is welcome. Fortunately two commentators in the *Chhien Han Shu*

---

a Yeh Tê-Hui and Sun Fêng-I, from *TPYL*, chs. 736 and 956. Perhaps this is a reference to the distillation of the oil, or its volatilisation when the wood was heated.

b *TPYL*, ch. 736, p. 8a.

c In Yeh Tê-Hui, ch. 1, p. 13a, Sun Fêng-I, p. 6b, and Mao Phan-Lin, p. 5b.

d Ch. 812, p. 7a, tr. auct.

e Because Kên,[12] the phonetic in *yin*, is the trigram corresponding to the youngest son.

f The elder was Liu Chhien[13] and the younger, by a concubine, Liu Pu-Hai.[14]

g The quotation in *CSHK* (Hou Han sect.) writes Sou[15] instead of Phing, which could mean a sister-in-law of one of the sons.

h *CHS*, ch. 5, p. 7b, tr. Dubs (2), vol. 1, p. 323, (5), mod. auct.

i See Dubs (2), vol. 1, p. 316. Strictly speaking, reign-period names were not introduced until −114 or −113 under Han Wu Ti.

---

1 曾青	2 新論	3 桓譚	4 劉娉	5 金	6 鉛
7 公	8 銀	9 弟	10 景帝	11 僞黃金	12 艮
13 劉遷	14 劉不害	15 嫂			

undertook to explain the situation further according to the traditions and documents which had come down to them. Ying Shao[1] (c. +140 to +206) wrote:[a]

The emperor Wên Ti[2] (r. −179 to −156), in the fifth year of his reign,[b] had allowed the common people to mint (lit. cast) coin (without special authorisation), and this law had not yet been abrogated. During the intervening time, and earlier (hsien shih[3]), many people had made artificial gold. But in the end they did not succeed in fabricating gold, in spite of heavy expenses, mutual deceptions and magniloquent claims. Those who became impoverished took to banditry and robbery, hence this edict.

A second gloss was added by Mêng Khang[4] (c. +180 to +260):[a]

In earlier times many people had made artificial gold. This was why there was a saying of theirs: 'Gold can be made, and men can find salvation (Chin kho tso, shih kho tu[5])'.[c] But they incurred heavy expenses and in the end they did not succeed (in fabricating gold). Most of the people understood the difficulties involved, so there were few offenders, and the (new) ordinance was generally followed.

Finally Yen Shih-Ku[6] in +641 signified his agreement with the words of Ying Shao.

The negative form of the famous proverbial saying comes in another writing of Ying Shao, the Fêng Su Thung I[7] of +175, where he is talking about a scholar-official named Wang Yang.[8]

Although a Confucian scholar, Wang Yang [he says] was from a poor family, but he loved sumptuous equipages and fine food and raiment. Yet he had no gold or silver or precious stuffs of his own. When he moved (from one official position to another) he had no more than a sackful of clothes, and when he finally went home he wore hempen cloth and resumed a very ordinary diet. People admired his probity but were somewhat shocked at his extravagances, and put it about that he was able to make gold (tso huang chin[9]).[d]

Ying Shao, debunking, says that many had tried this and everyone had failed—Chhin Shih Huang Ti, Hsü Fu, Han Wu Ti, the Prince of Huai-Nan, and even Liu Hsiang under imperial auspices—so how could Wang Yang have succeeded? 'As the proverb says: Gold can not be made, and this world can not be transcended.' Wang Yang did it out of his official revenues, that was all. Pan Ku never ought to have repeated the silly story.[e]

[a] CHS, ch. 5, p. 7b, tr. Dubs (2), vol. 1, p. 323, (5), mod. auct.

[b] See CHS, ch. 4, p. 12a, referring to −175; tr. Dubs (2), vol. 1, p. 250.

[c] Dubs (5) took this as a sceptical or ironical deduction: 'If gold could be made, the world could be measured.' But obviously many people believed in it, so the saying must have had an affirmative character. His interpretation of the second half seems to us meaningless, though no doubt he avoided 'salvation' as too Buddhist for this early date. Our rendering assumes a parallel Taoist connotation which the date may permit. Certainly tu shih[10] and pu ssu[11] are frequent and interchangeable both in the Lun Hêng and the Thai Phing Ching. A third possibility is that the words meant: 'and the country can be saved', i.e. by the larger resources of steeds and weapons against the Huns which greater wealth would be able to purchase. But this would be very far-fetched. The saying must have been widely known, for we find it again in the Pao Phu Tzu book; see the translation given on p. 2 above.

[d] Ch. 2, pp. 16b to 17b, tr. auct.

[e] Ying Shao attributed it to the Chhien Han Shu. Two men of this name are indeed referred to in chs. 76, 81 and 89 of this history, but we are not sure which one was the person in question.

[1] 應劭　　　[2] 文帝　　　[3] 先時　　　[4] 孟康　　　[5] 金可作世可度
[6] 顏師古　　[7] 風俗通義　[8] 王陽　　　[9] 作黃金　　[10] 度世
[11] 不死

It is interesting that there had already been grave anxieties about the debasement of the coinage thirty or forty years earlier. The *Chhien Han Shu* records a speech by Chia I[1] in −175 directed against private minting and ingenious alloying. Although individual masters and craftsmen were allowed to make coins, he said,

those who dare to alloy the metal with lead and iron, in order cleverly to counterfeit it, are guilty of a crime, and liable to have their faces tattooed black as a punishment. It is true that the business of coin-moulding is such that unless the alloying is adroitly done one cannot obtain a material advantage, but the more the skill the more the profit. . . .

In the past, when the casting of coins was prohibited, the capital cases waiting for verdicts piled up continually; now, when it is free, the cases deserving face-tattooing are piling up in a similar way. . . .

Nowadays agriculture is being abandoned for copper-mining. People have dropped their ploughs and hoes in order to smelt (copper and other metals), make coin-moulds and blow up their charcoal fires without ceasing. Counterfeit coins increase daily.[a]

And he went on to recommend the nationalisation of copper and the prohibition of minting by individual persons. But the emperor did not listen to him and no action was taken. Counterfeiting was still going on in −120, and particularly after −119 when the government introduced a white metal (*pai chin*[2]) coinage of silver and tin; even the death penalty for debasement could not stop it.[b] Finally a government monopoly of minting was introduced in −112, and after that 'only the most expert artisans and criminal incorrigibles illicitly manufactured coins'.[c]

All this demonstrates that proto-chemical and metallurgical experimentation (whether aurifiction or aurifaction) was already being actively pursued from about −200 onwards during the first half of the −2nd century. It also strongly suggests that in order to become so widespread the attempts to make artificial or counterfeit gold must have started in the previous century at least, which takes us back to the time of Chhin Shih Huang Ti (−3rd cent.) or indeed to that of Tsou Yen (−4th). Furthermore two other important conclusions emerge, first that there was a group of metallurgical aurifictors in those times whose interest in gold was economic rather than macrobiotic, and secondly that analytical methods were available for proving that the artificial gold produced by them was not true gold (cf. pt. 2, p. 48 above), otherwise the failures could not have been recognised. Erratic or haphazard application of these methods would account for the self-contradictory phraseology of the commentators, who evidently found it hard to be sure whether artificial gold had really been successfully made or not. One is tempted to compare these Chinese aurifictors with the writers of the Graeco-Egyptian papyri, while the Taoist philosophical adepts rather

---

[a] Ch. 24B, pp. 3 *b* ff., tr. auct. adjuv. Swann (1), pp. 233 ff.

[b] *Ibid.* p. 10 *b*, Swann (1), p. 270. Silver-tin coinage ceased in −113 and returned in +148.

[c] *Ibid.* p. 14 *a*, Swann (1), pp. 292 ff. By way of reminder, there is an excellent book on the history of coinage in China by Yang Lien-Shêng (3), but it has relatively little on counterfeiting and other metallurgical aspects.

[1] 賈誼   [2] 白金

resemble the writers of the Hellenistic proto-chemical Corpus (cf. pt. 2, pp. 16–17).[a] We shall encounter presently other instances of the making of artificial gold during the − 1st century (p. 35) and in Wang Mang's[1] time at the turn of the era (p. 38).

The emperor Wu Ti (r. − 140 to − 86), who ordered the punishment of Liu An, surpassed even Chhin Shih Huang Ti in his efforts to achieve immortality. Ssuma Chhien, the Historiographer-Royal and chief author of the *Shih Chi*, served under him, so that his account of the magicians and alchemists thronging the court is of great interest and importance. The following passage refers to the year − 133. Ssuma Chhien says:[b]

The Empress Dowager Tou[2] was fond of the teachings of the adepts and had little liking for the scholars. . . . At that time Li Shao-Chün[3] was granted an audience by the Emperor because he possessed the art of making offerings to the (spirit of the) Furnace (*tzhu tshao*[4])[c] (i.e. carrying on alchemical practices), and knew how to live without (eating) cereals and without growing old. The emperor honoured him. (Li) Shao-Chün had formerly been a familiar of the Marquis of Shen-Tsê, for whom he took charge of magical techniques (and medical arts). He dissembled his age and place of birth and breeding, saying always that he was seventy years old. He had the power of using natural substances (*shih wu*[5])[d] to bring about perpetual youth. He said that he was travelling about in order to make his techniques known to the nobles, and that he had no wife or children.

People who heard that he had power over things and was not subject to death brought him eatables and other presents, so that he always had abundance of gold, money, clothes and food. As people saw that he was provided with everything yet followed no trade or calling, they all had faith in him and vied with one another to serve him. . . .[e] (Li) Shao-Chün said to the emperor, 'By making offerings to the Furnace (-Spirit) natural substances can be caused to change (*chih wu*[6]).[f] If one can cause substances to change, cinnabar can

[a] Later on (pt. 4), in comparing datings for the Chinese and the Mediterranean developments, we shall note a slight but regular priority in favour of the former. The same seems to hold true for the government prohibition of counterfeiting and aurification. For China we have − 175 and − 144, while in Rome there was the Cornelian Law of − 81, and eventually the Diocletian decree against 'alchemists' (specifically aurifictors) in + 296. On this see further in pt. 4.

[b] Ch. 12, pp. 1 *b* to 16 *a*, ch. 28, pp. 22 *a* to 32 *b*, tr. auct. adjuv. Chavannes (1), vol. 3, pp. 463–93; Johnson (1), pp. 76 ff.; Dubs (5); Watson (1), vol. 2, pp. 38 ff.

[c] Commentators say that it would appear in the form of a beautiful girl dressed in red robes. On sacrifices to stove and furnace spirits in general see de Visser (1), pp. 119 ff. The prominent role of the Kitchen God (Tsao Chün[7]) in Chinese folk custom (cf. Vol. 2, p. 159) is widely known. Holmes Welch (1), p. 100, believes that the Spirit of the Furnace was anciently identified with an even older divinity, Ssu Ming,[8] the Director of Destinies, whose position as the regulator of the lengths of human lives would link proto-chemistry with longevity–immortality in yet another way. Cf. *Li Chi*, ch. 23, p. 35 *b* (tr. Legge (7), vol. 2, pp. 206 ff.).

[d] Some commentators and most translators take *wu* here to mean *kuei wu*,[9] ghosts and spirits, but we follow those who interpret it as *yao wu*,[10] chemical substances and drugs.

[e] There follow two stories, here omitted, showing Li's extraordinary skill at giving the impression that he had lived through several centuries.

[f] Or, 'natural phenomena can be caused to happen'. Again we follow the interpretation of *wu* as natural substances or phenomena and not spiritual beings. It is true that the phrase *chih wu* was often afterwards used in the latter sense, as in *Pao Phu Tzu* (Ware tr. pp. 83, 84, 316), where the adept can summon nectar and ambrosia (*chih hsing chhu*[11])—the 'Travelling Canteen', in Ware's bizarre translation. On this see the special study of Stein (6). But it will be perceived that the interpretation of spiritual beings at this point spoils the sorites, which works up to a climax and almost ends with them.

[1] 王莽	[2] 竇	[3] 李少君	[4] 祠竈	[5] 使物	[6] 致物
[7] 竈君	[8] 司命	[9] 鬼物	[10] 藥物	[11] 致行厨	

Fig. 1348. Li Shao-Chün praying in front of his alchemical furnace, which is visited by a crane. Mr Dark-Valley and the Venerable Mr Quiescence pass by below. A drawing, admirably economical in its wood-cut line, from *Shen Hsien Thung Chien*, ch. Shou, p. 29*a*.

be transformed into gold. When this gold has been produced it can be made into vessels for eating and drinking, the use of which will prolong one's life.[a] If one's life is prolonged one will be able to meet the immortals (of the isle) of Phêng-Lai in the midst of the sea. When one has seen them one will be able to make the *feng*[1] and *shan*[2] sacrifices, and after that one will never die.[b] The Yellow Emperor did just this. Your subject formerly, when sailing on the sea, encountered Master An-Chhi, with whom he ate jujube-dates as large as melons. Master An-Chhi is in communication with the Isle of Phêng-Lai; when it pleases him to appear to men, he does so, otherwise he remains invisible.' Thereupon the emperor personally for the first time made offerings to the furnace-spirit, and despatched magicians and alchemists to the sea to search for (the isle of) Phêng-Lai and (immortals) like Master An-Chhi. He also occupied himself with the business of transforming cinnabar and other substances into gold.

Some time afterwards Li Shao-Chün fell ill and died. However, the emperor believed that he had undergone a transfiguration and disappeared, not having really died. He ordered (the alchemists) Huang Chui[3] and Shih Khuan-Shu[4] to learn his arts and to search for Phêng-Lai and An-Chhi, but nothing came of it. Nevertheless, many queer and devious adepts from the coastal districts of Yen and Chhi came (to the court) and spoke more and more about the affairs of the spirits. . . .

In the following year (−121) Shao Ong,[5] a man of Chhi, came to see the emperor with his art of (influencing) the ghosts and spirits. At that time the emperor had just lost a favourite consort, Wang fu-jen.[6] Shao Ong therefore used his arts to bring back her simulacrum and that of the Furnace-Spirit during the night, and the emperor was able to view them through a curtain.[c] Hence the emperor gave to Shao Ong (the title) of Perfected-Learning General (Wên Chhêng Chiang-Chün[7]), rewarded him handsomely and treated him as a

---

[a] As Waley (14) pointed out, the idea of the macrobiotic efficacy of eating and drinking from plate and vessels of gold still persisted in Ko Hung's time; see *PPT/NP*, ch. 4, p. 14b, tr. Ware (5), p. 90. Not only this; it can be found in texts presumably of Thang date, e.g. *TT*910, ch. 1, p. 6b. An associated belief was that certain metals would ward off evil. We noted earlier (pt. 2, p. 203) an instance of this in the case of 'yellow silver' (*huang yin*[8]), almost certainly brass, considered by the *Hsin Hsiu Pên Tshao* (+659) an auspicious thing (*jui wu*[9]), cit. *CLPT*, ch. 4, (p. 110.1). Chhen Tshang-Chhi, in his *Pên Tshao Shih I* of +739, still approved of this, but Thang Shen-Wei denied it (*CLPT*, ch. 3, p. 39a, b). He went on to say, however, that in his time (+1080) silver blackened by exposure to the vapours of sulphur for several days (*niello, wu yin*[10]) was used to make 'dew-mirrors' (cf. Vol. 4, pt. 1, p. 89). Exposed at the top of ten-foot columns, they collected the dew at night, and this was drunk to obtain longevity and immortality; *CLPT*, *loc. cit.* (p. 97.1).

[b] Chavannes annotated this fundamental sorites with a full recognition of what it implied for the origins of all alchemy; it shows, he said, that all the elements of this were present in the −2nd century in China, i.e. aurifaction and macrobiotics. And he quoted Berthelot's encyclopaedia article on alchemy which instanced the 'gilding' of copper by zinc-mercury amalgam. If one tries to think out exactly what technique could have been at the back of Li Shao-Chün's prescription, it would seem certain that mercury must have come into it somehow, otherwise he would not have started with cinnabar, and thick films of gold amalgam on bronze or copper would have been the essence of the technique. The really important thought-linkage was that between the gold and the immortality. But the association of gold and mercury also continued throughout Chinese history. Looking back now, and bearing in mind what has been said above about the striking colours of red and gold and the striking properties of quicksilver, it is strange to note how close the two metals are in the Periodic Table, with a difference of only one proton and one electron between them. My friend the late Professor Wacław Nowinski reminded us that Max Planck remarked on this in one of his lectures.

[c] This incident has been discussed already; Vol. 4, pt. 1, p. 122. It is curious that a similar exploit is recorded in European history concerning the alchemical Abbot Johannes Trithemius of Sponheim (+1462 to +1516) and the Emperor Maximilian (Partington (11), p. 56). Trithemius seems to have been interested in aurifaction but to have condemned the dream of aurifaction.

[1] 封 [2] 禪 [3] 黃錘 [4] 史寬舒 [5] 少翁
[6] 王夫人 [7] 文成將軍 [8] 黃銀 [9] 瑞物 [10] 烏銀

guest of honour. . . . Later Shao Ong made an ox swallow a piece of silk on which he had previously written something. He then told the emperor that strange phenomena would be discovered in the beast's stomach. On killing the animal a letter was indeed found, but the writing was very strange, and the emperor became suspicious. The handwriting was recognised after enquiries were made, and Shao Ong was executed (in −119), but the matter was hushed up. . . .

That spring (−113) the Marquis of Lo-Chhêng[1] presented a memorial recommending Luan Ta[2] to the emperor. Luan Ta had been one of the eunuchs of Prince Khang[3] of Chiao-Tung,[a] and had had the same teacher as the Perfected-Learning General. It was this fact which had made him Magician-Technician and Pharmacist-Royal (Shang Fang)[b] to the Prince of Chiao-tung. Now his elder sister had become the wife of Prince Khang, though childless, but upon his death, and the son of a secondary spouse succeeding, she gave her-self over to debauchery and quarrelled all the time with the new Prince. So hearing of the death of the Perfected-Learning (General), and wishing to ingratiate herself with the Emperor, she sent Luan Ta, through the introduction of the Marquis of Lo-Chhêng, to expound his arts and techniques in an audience. The Emperor was now regretting that he had put to death the Perfected-Learning (General), and that the fullness of his arts had not been experienced, so he welcomed Luan Ta warmly.

Luan Ta was tall and a brilliant talker, fertile in techniques, and daring in promises, never hesitating. He said to the Emperor: 'Your subject has often been overseas and seen (Master) An Chhi, Hsienmên (Kao) and other great magicians, but as I was an ordinary commoner they despised me and did not take me into their confidence. In fact they even considered Prince Khang and the other noble lords as unworthy of their secrets. I often used to speak with Prince Khang about these things, but he would never make use of me. My teacher maintained that gold can be (artificially) produced, that the breach in the Yellow River (dykes) can be (permanently) closed, that the herb of immortality can be found, and that the *hsien* can be made to appear. But all your subjects are afraid that they will meet with the same fate as the Perfected-Learning (General), so none of them dare to open their mouths. How then should I make bold to speak to you of my arts?' The emperor replied: 'The Perfected-Learning (General) died through eating horse liver. If you are capable of restoring his arts, there is no love and respect that I will not show you.'[c]

Accordingly Luan Ta asked for hitherto unheard-of honours, such that he should be treated as an imperial relative and be made ambassador to the world of the immortals. At the end of the interview he demonstrated one of his lesser arts by making mag-netised chessmen appear to move of themselves.[d] At that time, the text goes on, 'the Emperor was distressed that the Yellow River breaches could not be closed, and that no (artificial) gold had been made'; so he conferred the title of Five Boons General (Wu Li Chiang Chün[4]) on Luan Ta, and five other similar titles, besides giving him the eldest daughter of the empress *née* Wei in marriage, and a dowry of immense

---

[a] I.e. Liu Chi,[5] one of the fourteen sons of Han Ching Ti, and closest to his half-brother Han Wu Ti, because their mothers were sisters. See *CHS*, ch. 53, p. 17*a*.

[b] Yen Shih-Ku comments (*CHS*, ch. 25, p. 23*a*) that this was then a title implying superintendence of technical arts, drugs and chemical substances (*chu fang yao*[6]).

[c] The whole of this long passage appears almost identically in *Chhien Han Shu*, ch. 25, pp. 18*b* to 30*a*, tr. Wieger (1), vol. 1, pp. 443 ff.; and (on Li Shao-Chün only) Waley (14); Yetts (4).

[d] This affair has been fully discussed in Vol. 4, pt. 1, pp. 316 ff.

---

¹ 樂成   ² 欒大   ³ 康王   ⁴ 五利將軍   ⁵ 劉寄   ⁶ 主方藥

size. Luan Ta wore feather robes to show his kinship with the winged immortals, and conducted liturgical sacrifices to the gods and spirits.[a] He set out for the coast to lead an expedition to the magic islands in the Eastern Sea, but had not the courage to embark, so he went and sacrificed on Mt Thai Shan instead. The emperor now tiring of him had him watched without result, but 'his magical arts were failing, and were often contradicted by the facts', so in −112 the emperor had him executed. Thus ends the fascinating account of alchemy in Ssuma Chhien's chapter on the *fêng* and *shan* sacrifices.

Luan Ta and Shao Ong have their significance but the really important person here was Li Shao-Chün, for his were the words which transmitted to posterity the essential alchemical doctrine of the Chinese −2nd century.[b] Together with the edict of −144, they fix a historical period when the full combination of aurifaction and macrobiotics had come into being, with all that that implied for the inspiration and stimulus to chemical invention and discovery of succeeding generations of spagyrical seekers. Although at first the effect of the artificial gold was thought of as mediated through food and drink consumed from vessels made of it, no long time elapsed before the consumption of the actual gold itself in some form or other was regarded as essential. In 1927 so great an authority as Partington, writing on 'Li Siao Kiun', could say that 'if the dates are authentic, this is before the earliest alchemy otherwise known'.[c] His judgment has been fully confirmed.

It thus appears that the emperor Han Wu Ti followed the example of Chhin Shih Huang Ti in the search for medicines of immortality. This is amplified in a curious Taoist text, which has received little attention so far, entitled *Phêng-Lai Shan Hsi Tsao Huan Tan Ko*[1] (Mnemonic Rhymes of the Cyclically Transformed Elixir from the Western Furnace on Phêng-Lai Island),[d] written by Huang Hsüan-Chung,[2] according to his own statement an alchemical official during the time of Han Wu Ti. The preface of the book takes the form of a memorial to the emperor saying that during the Yuan-Fêng reign-period (−110 to −105) the author had been sent by Wu Ti to look for the medicine of immortality on Phêng-Lai Island, but when he returned with the drug the emperor did not believe him, and asked him to try it out on himself first to see if he would become an immortal. It seems that the author left in disgrace after testing the drug, and retired to the mountains, as we read from the preface that he wrote his book ten years after leaving the palace, and that with it he tried to convince the emperor that the elixir was achievable. The whole text deals essentially with plant drugs, about 172 of them, written in the form of 168 verses.[e]

---

[a] Here cf. Vol. 2, p. 141.

[b] Li Shao-Chün and Shao Ong were afterwards sometimes confused, as in the *Han Wu Ti Ku Shih*, which omits the statements about alchemy and gives only the evocation of the simulacrum of the dead girl, ascribing it to Li (cf. d'Hormon tr., pp. 39, 53; and also, on related subjects, pp. 45, 62, 87).

[c] Partington (8). Cf. Maspero (13), pp. 218–19.

[d] *TT*909. The bibliographical chapter of *Sung Shih* lists this title, only substituting *ao*[3] for *tsao*.

[e] A special study of the plants considered by Taoists to be favourable for longevity and immortality has been made by Roi & Wu Yün-Jui (1).

[1] 蓬萊山西竈還丹歌   [2] 黃玄鍾   [3] 鰲

The dating of a work such as this is always difficult, but the prosody gives evidence that as we have it now it is a text of the Thang period.[a]

Another book which has to be mentioned here is the *Han Wu Ti Nei Chuan*[1] (Inside Story of the Emperor Wu of the Han), a text of novelistic type datable about the middle of the +6th century when the Liang were in power, but doubtless based on older legendary traditions. As Schipper (1) points out in the introduction to his translation, the account concerns mainly a visit paid by the Goddess of the West, Hsi Wang Mu,[2] to Wu Ti in his court; and although Li Shao-Chün is occasionally referred to, alchemy is not prominent in spite of the many allusions to material immortality and the magical drugs which can confer it. A list of 100 such drugs is given, but there is no overlap at all with the list mentioned in the previous paragraph, and while there the names are genuinely botanical in character, here they are fanciful and poetical, so much so indeed that Schipper believes they were obtained by mediumistic revelation and destined for liturgical recitation.

That the tradition of the first Chhin emperor's search for elixirs of immortality connected in some way with chemical processes and gold was still vividly present in the minds of the men of the −1st century appears well from one of the discussions in the *Yen Thieh Lun*[3] (Discourses on Salt and Iron), written by Huan Khuan[4] in −80 and later.[b] As will be remembered,[c] he was recording the results of a conference on the principles of administration between the leading bureaucrats and the Confucian scholars. In the chapter entitled San Pu Tsu[5] (Extravagance leads to Want), the spokesman of the Confucian Worthies (*hsien liang*[6]) is addressing the meeting in the presence of the Literati (*wên hsüeh*[7]), the Lord Grand Secretary (Ta Fu[8]) and the Lord Chancellor (Chhêng Hsiang[9]). He says:[d]

The sages of old personally took part in labour and preserved the balance of their mind, restricting their desires and moderating their passions, honouring heaven and paying respect to earth, always treading in the way of virtue and love of humanity. Therefore Heaven accepted the fragrance of their sacrifices, lengthening their days and filling up their years with abundance. Hence Yao's ruddy face and bushy eyebrows, hence his happy reign of a hundred years.

But when it came to Chhin Shih Huang Ti, what a fancy he took for strange and impracticable enterprises! How he believed in liturgies and sacrifices! He sent Master Lu to look for Hsienmên Kao, and Hsü Fu and others out to sea to search for the medicine of immortality (*pu ssu chih yao*[10]). At that time, therefore, the gentlemen of Yen and Chhi laid aside their hoes and digging-sticks, and competed to make themselves heard on the subject of immortals and magical technicians.[e] Consequently those who headed for the capital at Hsienyang were to be numbered in thousands. They averred that the immortals had eaten

---

[a] Cf. Yuan Han-Chhing (1), pp. 197–8.

[b] The conference took place in −81 and it was on the basis of verbatim reports finished in the following year that Huan Khuan prepared his amplified and dramatised text some time before −50.

[c] Cf. Vol. 2, pp. 251 ff., Vol. 4, pt. 2, pp. 21 ff.

[d] Ch. 29, pp. 8 b ff. tr. auct. adjuv. Sivin (1) in part.

[e] On the association of Yen and Chhi with early science and technology see Vol. 2, pp. 133, 232 ff., 240 ff.

[1] 漢武帝內傳　　　　[2] 西王母　　　[3] 鹽鐵論　　　[4] 桓寬　　　[5] 散不足
[6] 賢良　　　[7] 文學　　　[8] 大夫　　　[9] 丞相　　　[10] 不死之藥

of gold and partaken of (the juice of) pearls; only after that had they achieved longevity as enduring as the sky and earth. This was why the emperor made several visits to the five sacred mountains and to the pavilions along the ocean shore, always earnestly looking for traces of the Isles of Phêng-Lai and the Holy Immortals. But no matter what commanderies and counties the emperor visited, the rich had to give contributions and the poor were conscripted to erect buildings along the roads. Soon many of the small people ran away, while men of consequence went into hiding; the officials made arrests and put those they found in bonds, all without any pretence of justice. Famous halls there might be, and humble cottages, but not a plant was coming up among them, and no young trees were growing. The masses withdrew their support, and out of every ten people six were disaffected.

This was the voice of traditional Confucianism, which always merits sympathy, but the beginnings of science were rather to be found among the 'magical technicians' of the former seaboard States (by no means all charlatans) who flocked to find support and protection under the imperial wing.

After a short intervening reign, the emperor Hsüan Ti[1] (r. −73 to −48) revived the interest in aurifaction. He entrusted this work to Liu Hsiang,[2] who (as we saw above, pp. 13–14) had acquired the art from books which his father or grandfather had obtained from among the possessions of Liu An.[a] Following on from the passage on p. 14 above, concerning the events of about −60, the *Chhien Han Shu* says:[b]

(Liu) Kêng-Shêng (i.e. Liu Hsiang) was young, but he had read and rehearsed (the writings of the group of Liu An on alchemy) and was quite thrilled by them, so he presented them (to the throne) saying that gold could indeed be made (artificially) (*huang chin kho chhêng*[3]). The emperor ordered that he should be put in charge of (some of the) Imperial Workshops (Shang Fang[4]) for the purpose of making (alchemical gold). He expended very much (money) but the procedures could not be verified.[c] The emperor therefore committed (Liu) Kêng-Shêng to the public prosecutor, who impeached him for (attempting to make and) cast false gold.[d] He was thus imprisoned and sentenced to death, but his elder brother, (Liu) An-Min,[5] the Marquis of Yang-chhêng, presented a memorial offering to surrender to the nation half his estate in order to ransom the fault of (Liu) Kêng-Shêng. Besides, the emperor also valued his ability. So he got through the winter in prison, and then his death-sentence was commuted.[e]

Thus ended a government-supported experimental programme most remarkable for the −1st century. A Confucian speech by Chang Chhang[6] adjured the emperor to pay more attention to the arts of peaceful government and less to military men and alchemists; whereupon all the laboratory technicians of the Imperial Workshops (Shang Fang Tai Chao[7]) were dismissed.[f]

[a] On the biography of Liu Hsiang see Forke (12), p. 65.
[b] Ch. 36, p. 7a, tr. auct. adjuv. Dubs (5).
[c] I.e. they did not yield the hoped-for results—*fei shen to, fang pu yen*.[8]
[d] Dubs (5) suggests that the basis for this was the law of −144 against making false (alchemical) gold.
[e] Executions were always carried out in autumn or winter, so this means that sentence was postponed in Liu Hsiang's case, and he could be liberated in the spring.
[f] This we know from the shorter parallel passage in *CHS*, ch. 25B, p. 7b. On the Tai Chao cf. Vol. 6, Sect. 38.

[1] 宣帝   [2] 劉向   [3] 黃金可成   [4] 尚方   [5] 安民   [6] 張敞
[7] 尚方待詔   [8] 費甚多方不驗

This whole episode has weighty significance for the ancient history of aurifiction and aurifaction. Why was Liu Hsiang[a] unable to achieve success in the Imperial Workshops, backed by all the resources of the State? Because that was precisely where the artisans were who understood the technique of cupellation, and they could disprove time after time whatever he did. Why did Ko Hung and so many other Taoists later on put so much emphasis on carrying out alchemical processes in the mountains far away from the turmoil of worldly life? Because that was precisely where they would not be disturbed by inconvenient metallurgical technicians suggesting that the results ought to be cupelled. We do not have to suppose that this consideration ever came into the consciousness of the Taoist alchemists, nor need we write them all off as charlatans, for we have clearly seen already (pt. 2, pp. 10, 19 ff.) that their definitions of gold (like those of the Hellenistic proto-chemical aurifactors) were different from that of the technical goldsmiths. What we gain from this analysis is a deeper insight into what was really implied by the antithesis of 'worldly' versus 'recluse' environments.

In spite of all these failures and disappointments, the Early Han emperors continued to follow the quest of the elixir of immortality. Regarding the time of Chhêng Ti[1] (r. −32 to −7) the *Chhien Han Shu* says:

Towards the end of the reign of Chhêng Ti, the emperor was rather interested in matters relating to the ghosts and spirits. As he had no descendants, many people presented memorials talking of sacrifices and magical arts, and they were all named Experts-in-Attendance (Tai Chao[2]).[b] Offerings were made in the Shang-lin[3] Park and outside the capital at Chhang-an, involving heavy expenditure but producing no notable result. Ku Yung[4] therefore spoke to the emperor, saying: 'Your servant has heard that he who understands the nature of heaven and earth cannot be deceived by strange spiritual occurrences, and he who knows the behaviour of the ten thousand things cannot be led astray by unnatural phenomena. (To follow) what goes against the Tao of benevolence and righteousness or contradicts the teachings of the Five Classics, emphasising mysterious affairs of ghosts and spirits, or propagating methods of rite and sacrifice, is to ask from the gods blessings which can never give happiness. Moreover those who say that there are immortals who consume elixirs (*yu hsien jen fu shih pu chung chih yao*[5]), so that they can fly far away with light body, or ascend into the heavens with inverted shadow,[c] or leisurely gaze down upon the Hanging-Gardens (peak, near the Khun-Lun Mountain),[d] or float on the sea to Phêng-Lai (Island), or cultivate the five-coloured cereals sown in the morning and harvested at dusk, with life as long-lasting as the mountains and rocks; or operate the transformations of the (alchemical) furnace (turning cinnabar into gold, *huang yeh pien hua*[6]), or draw a crystal elixir from seethed urine,[e] or practise the art of changing the colours of the five viscera (so that one knows hunger and thirst no more)—all, all is the babbling of meretricious charlatans, who employ sinister arts (*tso tao*[7])[f] and false pretensions to deceive the ruler of men. When one hears them talk, their

---

[a] And Shih Tzu-Hsin a little later, for whom see p. 82.   [b] Cf. pp. 35, and Sect. 38.
[c] I.e. in the regions above the sun, where the shadow is cast upwards, not downwards.
[d] Cf. p. 24 above.
[e] On this strange phrase, see pt. 5 below.
[f] Lit. 'left-hand arts', cf. Tantric *vamacāra*.

[1] 成帝   [2] 待詔   [3] 上林   [4] 谷永   [5] 有仙人服食不終之藥
[6] 黃冶變化   [7] 左道

copious words sound well in the ear and it seems as if their aims can be attained, but when one seeks to follow their arts, everything is vague and formless, like trying to tie up the wind or to capture a shadow, and no result is ever achieved. . . .' The emperor considered that he had spoken well.[a]

Here we see alchemy through the veil of the words of a typically sceptical Confucian scholar. The original language of his speech is highly elegant.

Wang Mang,[1] the only emperor of the intrusive Hsin dynasty, naturally had a share in the enterprise of aurifaction.[b] The *Chhien Han Shu* has this to say:

In the 2nd year after (Wang) Mang had usurped the throne (+10) he took in hand the business (of becoming) a holy immortal. There was a magician-technician (*fang shih*) called Su Lo[2] who advised him to construct an 'Eight Winds Tower' (Pa Fêng Thai[3]) within the palace, which when completed cost 10,000 pieces of gold.[c] (Su) Lo took his place upon it,[d] and according to the wind (he) prepared various potions (*i thang*[4]).[e] Also the five grains were planted within the palace in plots facing according to the colour of each one. The seeds had been soaked in (a liquid made from) the marrow of the bones of cranes, tortoise-shell (*tu mao*[5]),[f] rhinoceros (horn), and jade, in all more than twenty constituents. One bushel of this grain cost one piece of gold. This was called Huang Ti's cereal method for becoming a holy immortal. So (Su) Lo was made a Palace Officer (Huang-Mên Lang[6]) and put in charge of these affairs. Thus (Wang) Mang made extravagant worship of the ghosts and spirits. In his last year he built 1700 temples for the worship of different ranks of spirits, ranging from the six highest categories down to the smallest, using three kinds of sacrifices with more than 3000 kinds of animals and birds. As they could not get all the species complete they used cocks as substitute for wild ducks and wild geese, and dogs instead of deer. He often issued edicts saying that he must become a holy immortal—see his Biography.[g]

Chang Tzu-Kao (2) points out that there has never been any archaeological evidence to show that money used during Wang Mang's time was made of alloys looking like gold, though the gold and silver used have been found to vary in purity.[h] He has calculated that when Wang Mang died the Chinese treasury had some 6,333,000 ounces of gold; not far from the somewhat lower figure given in the study of Dubs (4).[i] Chang remarks on the approximate coincidence of this with the amount of gold held at its height by the Roman Empire, but Dubs had estimated that it was much higher than the total supply of medieval Europe.

[a] Ch. 25B, pp. 14b to 16b, tr. auct. I like to recall that this passage was copied out for me early in 1943 by colleagues at the Academia Sinica History Institute near Lichuang in Szechuan. On the reforms of this period see Loewe (6).

[b] Cf. Vol. 1, pp. 109 ff. where his proto-scientific predilections were pointed out.

[c] The *Chin Shih So* gives in Shih sect. ch. 6, (p. 83), a rubbing of a moulded tile end from the Pa Fêng Thai.

[d] We translate thus, but the text is ambiguous, having at least two other equally plausible interpretations, depending on how it is read. If read *tso yo chhi shang*[7] it means that 'music was played at the top', but if read *tso lo chhi shang* it means either that 'they made merry inside it', or that a Taoist hierogamy was performed. *Tso lo* as an expression for sexual intercourse occurs in ch. 30, p. 52a.

[e] The commentary says that the Bibliography has a *I Thang Ching*,[8] but actually it has a *Thang I Ching Fa*[9] in its medical section. See ch. 30, p. 51b.

[f] *Tu mao* is the same as *tai mei*[10] (R202).   [g] Ch. 25B, pp. 22b, 23a, tr. auct.

[h] P. 81.   [i] Cf. Vol. 1, p. 109; Vol. 4, pt. 3, pp. 518 ff.

[1] 王莽	[2] 蘇樂	[3] 八風臺	[4] 液湯	[5] 毒冒
[6] 黃門郎	[7] 作樂其上	[8] 液湯經	[9] 湯液經法	[10] 瑇瑁

Further evidence of practical alchemical experiments in Wang Mang's time comes from a source written about +10 to +20, the *Hsin Lun*[1] (New Discussions) of Huan Than,[2] a book not fully extant now, but quoted in many places.[a] The passage is particularly important because it attests the existence in China in the −1st century of the idea of 'projection',[b] so important in Western proto-chemistry and later alchemy often as an attribute of the 'philosophers' stone'. Ko Hung wrote:[c]

Also Huan Chün-Shan[3] (Huan Than) tells us that there was a gentleman of the Han Imperial Court (Huang-Mên Lang[4]) named Chhêng Wei[5] who had a liking for the Art of the Yellow and the White (*huang pai shu*,[6] aurifaction and argentifaction). His wife came from a family of technicians skilled in magical arts. Chhêng often had to take part in imperial processions but did not have the clothes appropriate to the season; seeing his distress, his wife said: 'I will ask for two lengths of silk', whereupon the silk suddenly appeared in front of them. Chhêng Wei tried to make gold following the directions in the *Chen Chung Hung Pao*[7] book,[d] but without success. One day his wife went to see him just as he was fanning the charcoal to increase the heating of a reaction-vessel in which there was mercury. She said 'Let me show you what I can do', and taking a small amount of some substance from her pouch she threw it into the vessel. After about the space of time in which a man could take a meal, she opened the vessel, and they saw that the contents had all turned to silver.

Chhêng Wei was astounded and said: 'How is it possible that you could possess this Tao and never let me know about it?' She answered 'It cannot be gained unless one has the right destiny.' But he kept on at her day and night for the secret. He sold lands and buildings to provide her with the best of food and clothing, but all to no effect. Then he schemed with a friend to beat her, but of course she got to know of this and told him that the process was to be transmitted to the right person alone. If she met such a one, even only by chance at the roadside, she would tell him, but to anyone who was not the right man—anyone whose words and secret thoughts were different—she would never divulge the art even though she were to be cut into pieces and dismembered. Chhêng Wei obstinately kept on pressing her, however, until finally she went mad, rushed out naked, smeared herself with mud and died.[e]

We shall return to the idea of projection later (p. 100), only drawing attention here to the emphasis placed upon a kind of vow of secrecy. One wonders whether the 'unworthiness' of Chhêng Wei was really his mercenary motive; we are to understand

[a] Cf. Vol. 3, p. 219 on cosmology, and Vol. 4, pt. 2, p. 392 for his important statement on water-mills.

[b] In later times the Chinese term for this was *tien*,[8] lit. 'adding a spot of . . .' some chemical. The general idea both in East and West was the transformation of a large amount of substance by the addition of a very small amount of something else.

It is interesting that a parallel idea ran through the eucharistic theology of Christendom from the +4th to the +16th century, namely consecration by contact (*immixtio*). As we know from the interesting monograph of Andrieu (1), it was believed that wine could be consecrated, without the repetition of the words of institution, by the mere addition of a drop of previously consecrated wine, or a particle of a previously consecrated host. The same principle applied in some traditions to the oil of chrism. These practices were evidently related not only to the idea of projection in alchemy but also to that of contagion in medicine; and both perhaps originated from early thought about fermentation. We shall return to these matters in pt. 4.

[c] *Pao Phu Tzu*, ch. 16, pp. 3 *a*; tr. auct. adjuv. Ware (5), p. 264; Dubs (5); Waley (14).

[d] Liu An's of course; see pp. 14, 23, 25 above.

[e] This text is also in *CSHK* (Hou Han sect.), ch. 15, p. 7 *a*. *TPYL*, ch. 812, p. 4*b*, has an abbreviated version. Huan Than wrote an ode on the immortals; see pt. 2, pp. 111–12 above, and Pokora (3).

| [1] 新論 | [2] 桓譚 | [3] 桓君山 | [4] 黃門郎 | [5] 程偉 |
| [6] 黃白術 | [7] 枕中鴻寶 | [8] 點 | | |

that he was not seeking for immortality or salvation but rather for funds to keep up his position at court. Indeed the story follows in *Pao Phu Tzu* immediately upon another one relating how Wu Ta-Wên,[1] a government secretary at Chhêngtu, studied under a master of the art, Li Kên,[2] yet was never able to succeed in transmutation by projection because he could never sufficiently withdraw from worldly business.[a]

It may well be that the idea or principle, or practice (whatever it was), of projection, began some time before Chhêng Wei, for there remain records of another alchemist or aurifactor, the great Taoist Mao Ying,[3] who lived in the time of Han Yuan Ti (r. −48 to −33). His biography, entitled *Mao Chün Nei Chuan*,[4] has long been lost,[b] but later encyclopaedias have quotations from it, such as the following matter-of-fact statement in *Thai-Phing Yü Lan*:[c]

Take ten catties of lead, put in an iron vessel, and heat intensely. After it has been brought to the boil three times, throw into the lead 1 *shu*[5] of the nine-times cyclically-transformed floriate essence (*chiu chuan chih hua*[6]). On stirring it will instantly turn to gold.

Or did Chhêng Wei precede Mao Ying? One usually tends to think of Chhêng's story as almost contemporary with Huan Than's account of it in the first couple of decades of the +1st century, i.e. to place it in Wang Mang's time; but according to one tradition it was a whole century earlier. This tradition is also interesting because of the leading role played by women, so consonant with Taoist ideas in general.[d] The *Shen Hsien Thung Chien* gives a specific date for the death of Chhêng Wei's wife, namely −95, and says that Han Wu Ti was so incensed about it that he had Chhêng executed.[e] According to this account her family name was Fang,[7] and she had studied alchemy with one of Han Wu Ti's favourite spouses, Court Beauties or Attendant Nymphs.[f] This lady, a specialist in Taoist sexual techniques (*Huang Ti Su Nü chih shu*[8]),[g] who was said to have acquired her knowledge of Taoism from the goddess Ma-Ku,[9] was named Chao Chieh-Hao[10];[h] she too came to a sad end, for some years later, falling from favour, she was involved in one of the periodical plots and witch-hunts which convulsed the court, and was also executed. How much to believe

---

[a] Or, as we might now secretly say, because he could never get far enough away from the artisans who understood the cupellation techniques.
[b] There is however an account of his life in *YCCC*, ch. 104, pp. 10a to 19a, by a disciple, Li Tao[11] (or Li Tsun[12]). Though rather hagiographical, it gives one exact date, −39, and this is confirmed by a second quotation in *TPYL*, ch. 661, p. 4b, which specifies the reign-period in which Mao was active. Mao Ying was the eldest of three brothers, the second being Mao Ku[13] and the third Mao Chung,[14] hence the expression San Mao Chün;[15] all three were very important in the founding of one of the greatest ancient centres of Taoism, at Mao Shan[16] near Nanking. This was the group with which Thao Hung-Ching was associated later on. We have already become acquainted with Mao Ying in connection with the making of cupro-nickel (pt. 2, pp. 234–5 above). Exactly how historical the three brothers are remains an open question. Cf. Fig. 1349.
[c] Ch. 812, p. 7a.
[d] Cf. pp. 169, 191, and one may remember that Ko Hung's wife, Pao Ku, was also an alchemist, p. 76.
[e] Ch. 8, p. 1a, b.   [f] Cf. Vol. 4, pt. 2, p. 477.
[g] Cf. Vol. 2, pp. 146 ff.   [h] Or Chieh-Yü,[17] which could have been a title. Cf. p. 17 above.

[1] 吳大文   [2] 李聡   [3] 茅盈   [4] 茅君內傳   [5] 銖
[6] 九轉之華   [7] 方   [8] 黃帝素女之術   [9] 麻姑   [10] 趙婕妤
[11] 李道   [12] 李遵   [13] 茅固   [14] 茅衷   [15] 三茅君
[16] 茅山   [17] 婕妤

of this tradition may rest with personal choice, but if it were true it would throw a welcome light on the activities of the generation just following Li Shao-Chün, and in its emphasis on the role of women Taoists with all that that implies for the deep and ancient connection of chemical theory with sexual phenomena and Yin-Yang philosophy, it bears a certain stamp of truth (cf. Fig. 1350).

The official dynastic histories also disclose the names of other alchemists besides those favoured at the imperial court. Kan Shih,[1] already mentioned as an expert in Taoist sexual techniques, is said to have joined his master Han Ya[2] in the South Seas in making gold, while Fêng Chün-Ta,[3] also known as the Blue Ox Master (Chhing Niu Shih[4]) and another expert on sexual techniques, is said to have taken mercury as elixir. The same source mentions that both of them flourished at the same time as a third adept, Tungkuo Yen-Nien,[5] namely during the time of Emperor Wu Ti;[a] but in so far as any of them are fully historical characters they lived more probably in the +1st or +2nd century. Later Taoist texts tell about other adepts of this time, such as Wang Hsing,[6] who ingested a 'potable gold elixir' (chin i chih tan[7]) and Wei Shu-Chhing,[8] who achieved immortality through eating mica.[b] Flourishing at about the same time as Liu Hsiang were Su Lin[9] (d. c. −60) and his pupil Chou Chi-Thung,[10] better known under the name Tzu-Yang Chen Jen,[11] to whom is attributed a book in the Tao Tsang entitled Thai-Shang Tung Fang Nei Ching Chu[12] (Esoteric Manual of the Innermost Chamber, with Commentary).[c]

During the time of the Later Han, the emperor Huan Ti[13] (r. +147 to +168) hoped, so we are informed, to attain eternal life by his frequent worship of Lao Tzu and the holy immortal Huang Shih Kung.[d] A distinguished scholar of the age, Tshai Yung,[14] also expressed interest in the art of the alchemists.[e]

The Tao Tsang contains accounts of many Later Han adepts. For example, the Hsüan Phin Lu[f] mentions Liu Khuan[15] (traditionally +121 to +186), who was a disciple of Chhing Ku hsien-sêng,[16] i.e. Han Chhung.[17] Han acquired the art of making the 'flowing-pearl elixir' (liu chu tan[18]) from Wang Wei-Hsüan[19] and found it effective. Similarly, Hsia Fu[20] studied alchemy from his youth upwards and ingested the thistle-like Atractylis (pai shu[21])[g] and mica.[h] Chao Tao-I tells us about Lung Shu[22] (also called Lung Po[23]) who ate magic mushrooms (shen chih[24]) in the early days of the

---

[a] HHS, ch. 112B, p. 18a. On all these cf. Vol. 2, p. 148, and in relation to iatrochemistry in pt. 5 below, where the full translation is given. Meanwhile it can be found in Lu Gwei-Djen & Needham (3).

[b] See TT293, ch. 7, pp. 1b ff., a Yuan text.

[c] TT130; cf. TT293, pp. 6b ff. On the term tung in Taoism cf. Porkert (2), pp. 447 ff.

[d] HHS, ch. 7, pp. 15aff.   [e] HHS, ch. 90B, p. 1a.

[f] Another Yuan text (TT773).

[g] Probably extracts of the roots of Atractylis ovata (=chinensis), R14; CC34; now ranged as species and varieties of Atractylodes, cf. Sato Junpei (1), p. 65; Kimura Koichi (2), p. 107. The roots give a very bitter aromatic extract with a resinous yellow pigment, and figured in many classical tonic prescriptions (Stuart (1), p. 57).   [h] Ch. 2, pp. 11a, 15b, and 16b.

[1] 甘始	[2] 韓雅	[3] 封君達	[4] 青牛師	[5] 東郭延年
[6] 王興	[7] 金液之丹	[8] 衛叔卿	[9] 蘇林	[10] 周季通
[11] 紫陽眞人	[12] 太上洞房內經注	[13] 桓帝	[14] 蔡邕	[15] 劉寬
[16] 青谷先生	[17] 韓崇	[18] 流珠丹	[19] 王緯玄	[20] 夏馥
[21] 白朮	[22] 龍述	[23] 龍伯	[24] 神芝	

Fig. 1349. The ascension of the alchemical Taoist hierarch Mao Ying into the realm of the holy immortals, admired by two disciples of the Mao Shan school (from *Lieh Hsien Chhüan Chuan*, ch. 2, p. 11 *a*).

Fig. 1350. A woman alchemist, Thai Hsüan Nü (or Thai Yang Nü, or Thai Yin Nü), compounding elixirs (from *Lieh Hsien Chhüan Chuan*, ch. 2, p. 7a).

Later Han;[a] and Chiu Yuan Tzu[1] who became an immortal after taking a 'magical elixir' (*shen tan*[2]) he had made himself. He also wrote a book called *Kêng Hsin Ching*[3] (Book of the Realm of Kêng and Hsin, i.e. the metals and minerals symbolised by these two cyclical characters).[b] Other experts mentioned by Chao Tao-I are Liu Chêng[4] and Chhang Shêng Tzu,[5] the former a consumer of a 'red flowery elixir' (*chu ying tan*[6]), and the latter using a suspension of jade to achieve immortality. There was also a certain Mao Po-Tao,[7] who compounded a 'magical elixir' (*shen tan*[2]) with three friends, Liu Tao-Kung,[8] Hsieh Chih-Chien[9] and Chang Chao-Chhi[10].[c] One must not forget to mention Yin Chhang-Shêng[11] (*fl.* +122), whose name appears frequently in many Taoist alchemical texts. He is said to have acquired the art of preparing the 'Grand Purity Golden Potion Magical Elixir' (*thai-chhing chin i shen tan*[12]) from the adept Ma Ming-Shêng,[13] and also to have known how to make gold. Ma Ming-Shêng[d] had another name, Ho Chün-Chih,[14] and was said to have learnt the art of elixir-making from Master An Chhi himself.[e] Although we cannot now reconstruct exactly what all these men did, or what knowledge they had, the great number of names of such people in Chinese literature is significant in itself.

The +2nd century was a milestone in the history of both Taoism and alchemy in China, for it was then that Taoism first became an organised religion, and that the earliest extant treatise on alchemical theory was written. The part played by Chang Ling[15] (*fl. c.* +156), afterwards called Chang Tao-Ling,[16] in establishing the Taoist religion has already been described.[f] However, very little is known about the connection of Chang Tao-Ling with alchemy, so that some do not believe he was ever an alchemist,[g] though later Taoists certainly claimed that he practised the art. For example, Chang Thien-Yü informs us that he learnt the *Huang Ti Chiu Ting Ching*[17] (the Yellow Emperor's Manual of the Nine Vessels) and prepared an elixir at Fan-yang[18] mountain;[h] while Chao Tao-I, after claiming that Chang Tao-Ling was born in the 14th year of the Chien-Wu reign-period (+34),[i] affirms that he compounded the 'great dragon-and-tiger elixir' (*lung hu ta tan*[19]) and the 'nine-vessel magical elixir' (*chiu ting shen tan*[20]).[j] Two of his disciples, Wang Chhang[21] and Chao Shêng[22] (otherwise known as Lu Thang Tzu[23]), were said to have also ingested elixirs. The

---

a Was this not a garbled echo of Indian Nāgārjuna? See p. 161 below.
b Cf. pp. 104, 210.
c All this is from *TT*293, chs. 5, 6 and 7.
d This might be a case similar to that of Lung-Shu and Nāgārjuna, for as Sivin has pointed out (1), p. 58, the Mahāyanist monk Aśvaghoṣa, active at about the same time, was named Ma Ming[24] in Chinese.   e See *TT*293, ch. 13, or better *Yün Chi Chhi Chhien*, ch. 106, pp. 20*b* to 23*a*.
f See Vol. 2, Sect. 10 (*j*).   g E.g. Spooner (1, 2).
h *Hsüan Phin Lu*, ch. 2, p. 4*b*. Part at least of the *Huang Ti Chiu Ting Ching* may be preserved in the *Huang Ti Chiu Ting Shen Tan Ching Chüeh* (*TT* 878), especially because of the *chiu ting shen tan*[20] elixir mentioned. Chhen Kuo-Fu believes that the first chapter of this work is Han material.
i This is obviously impossible, though he may have claimed it himself.
j *TT*293, ch. 18.

[1] 九元子	[2] 神丹	[3] 庚辛經	[4] 劉政	[5] 常生子
[6] 朱英丹	[7] 毛伯道	[8] 劉道恭	[9] 謝稚堅	[10] 張兆期
[11] 陰長生	[12] 太清金液神丹	[13] 馬鳴生	[14] 和君實	[15] 張陵
[16] 張道陵	[17] 黃帝九鼎經	[18] 繁陽	[19] 龍虎大丹	[20] 九鼎神丹
[21] 王長	[22] 趙昇	[23] 鹿堂子	[24] 馬鳴	

Yün Chi Chhi Chhien describes how he tested the sincerity of these two disciples before imparting his secrets to them.[a] Several writings in the *Tao Tsang* that we have are attributed to Chang Tao-Ling, but none of them contains anything of alchemical interest.

Chang Tao-Ling claimed that Chang Liang[1] had been one of his ancestors.[b] His descendants later acquired the title Chang Thien Shih[2] and became hereditary heads of the Taoist church called Thien Shih Tao,[3] the fame of which lay more in its magical and apotropaic rites than in alchemy.[c] Nevertheless we owe much to at least two members of this same Chang family for their preservation of the *Tao Tsang* including so many of the important Taoist alchemical texts we now have.[d] During the time of the Mongols this vast collection of Taoist books suffered destruction, first at the hands of Mangu Khan (r. +1251 to +1259), and then Khubilai Khan (r. +1280 to +1294), and it was Chang Yü-Chhu[4] who collected and re-edited the *Chêng-Thung Tao Tsang*[5] in the +15th century.[e] Then Chang Kuo-Hsiang,[6] another descendant of Chang Tao-Ling, compiled the *Wan-Li Hsü Tao Tsang*[7] in 1607 by order of the emperor Shen Tsung.[8] The Taoist patriarchate still exists (1974) in Thaiwan, the present Heavenly Master being the sixty-fourth of the line.[f]

(iii) *The three roots of elixir alchemy*

In a moment we shall be speaking of the oldest Chinese alchemical book, written in the +2nd century. But before going further it will be well to consider a point much emphasised by Lao Kan (6) and expressible as follows: during Chou, Chhin and Early Han the accent is on *chhiu*,[9] 'searching for', while in the Later Han it begins to be on *lien*,[10] 'effecting chemical transformation in', or 'preparing by heating'. It is true that the earliest texts do generally speak of searching for the drug, or herb, of immortality,[g] but at the same time the making of artificial gold is indisputably attested from the −2nd century and by implication the −3rd.[h] The systematic handling of chemical change to produce desired inorganic substances clearly goes back at least to the −5th.[i] The decisive point arises when someone begins to consume orally inorganic substances

---

[a] Ch. 109, p. 18*b* to p. 20*b*.

[b] His family tree is reconstructed in the *Han Thien Shih Shih Chia*[11] (Genealogy of the Family of the Han Heavenly Teacher), *TT* 142; and the *Pu Thien Shih Shih Chia*[12] (Supplement to the 'Genealogy of the Family of the Heavenly Teacher'). See also Chao Tao-I (*TT* 293) and Fu Chhin-Chia (*1*). Of course these Taoist genealogies have to be accepted with all reserve.

[c] A general history of this patriarchate was written long ago (1884) by Imbault-Huart (*1*).

[d] The first compiler, Chang Chün-Fang,[13] early in the +11th century, was of the same clan-name but not of the same family.

[e] See p. 116 below.

[f] Much information on the recent history of the patriarchate is contained in the book of Li Shu-Huan (*1*), pp. 144–55, and the account of a personal visit by Holmes Welch (*2*).

[g] Cf. pp. 13, 18, 31 above.

[h] See Li Shao-Chün and Luan Ta (pp. 29, 32 above), the Coining Edict (p. 26), Liu An (p. 24), Liu Hsiang (p. 35), Chhêng Wei and Mao Ying (p. 39), Kan Shih (p. 40).

[i] See Chi Ni Tzu (p. 14), Liu An (p. 25).

[1] 張良	[2] 張天師	[3] 天師道	[4] 張宇初	[5] 正統道藏
[6] 張國祥	[7] 萬曆續道藏	[8] 神宗	[9] 求	[10] 鍊
[11] 漢天師世家	[12] 補天師世家	[13] 張君房		

and elixir preparations made from them, in heavy dosage or long-continued administration. This we do not get for sure until the Later Han (+ 1st and + 2nd centuries),
some names backed with good evidence,[a] others more shadowy and traditional.[b]
Further evidence will accrue in the next sub-section. So perhaps what Liu I-Chhing[1]
said in the + 5th century was a fair statement of the case. We find it in the *Shih Shuo
Hsin Yü*,[2] often before quoted (New Discourses on the Talk of the Times), as follows:

Ho Phing-Shu[3][c] said that those who eat the medicinal powder of the Five Minerals[d]
(*wu shih san*[4]) do so not only for curing diseases but also to enhance their vitality.

[Comm.][e] (The physician) Chhin Chhêng-Tsu[5] wrote a tractate entitled *Han Shih San
Lun*[6] (Discourse on the Cooling Regimen Powder), but in Han times few people used it and
the method was hardly handed down.[f] However, in the Wei, Ho Yen[7] took it with marvellous
results, so its knowledge and use began to circulate more in the world, and people sought
eagerly to obtain (the formula) from one another.[g]

Here the reference is to the Three Kingdoms period in the early + 3rd century as
the date of popularisation, and to the Han in the + 2nd or + 1st century as the date
of first use. From this one can begin to see clearly that originally there were two
traditions (doubtless not far separated), the macrobiotic pharmaceutical-botanical one
and the chemical-metallurgical one, macrobiotic in a slightly different sense.[h] After
the beginning of the Later Han these two completely coalesced and the adepts and
their disciples set forth on the hard and painful process, centuries long, of testing the
effects of inorganic drugs upon the human body and mind.

The clue to the union of the two traditions lies, we think, in the history of medicine,
if cautious doses of inorganic substances were being given therapeutically some
considerable time before large amounts of metallic preparations began to be consumed habitually in the form of elixirs. The evidence for this is very strong. In a
striking paper already mentioned, Temkin (3) contrasted Hellenistic with Arabic
chemistry, pointing out that the former was much closer to metallurgy than to medicine,
while the latter, in such men as al-Razī and al-Kindī, was medically oriented from the

---

[a] E.g. Fêng Chün-Ta (p. 40 above and pt. 5 below), also in Lu Gwei-Djen & Needham (3). The
Lieh Hsien Chuan is evidence for the Later Han, but not good for earlier times.

[b] See Liu Khuan, Liu Chêng, Yin Chhang-Shêng and Ma Ming-Shêng (pp. 43 ff. above).

[c] I.e. Ho Yen,[7] + 190 to + 249, a scholar and courtier of renown.

[d] Not here Ko Hung's[8] (p. 96). Hg and As were not involved, only forms of $CaCO_3$ and silica, with
nine powders of plants containing alkaloids, most still used in traditional medicine; cf. the monograph
of Wagner (1). The effect was tonic and calefacient, hence a cooling regimen was needed.

[e] By Liu Hsün[9] in the + 6th century.

[f] The *hsiang*[10] of most editions has to be emended to *tsu*.[11] Chhin Chhêng-Tsu (*fl.* + 420 to + 470)
was Director-General of Medical Services (Thai I Ling[12]) and Professor of Medicine (I Hsüeh Po
Shih[13]). We shall discuss the early growth of State medicine and medical schools in Vol. 6; meanwhile
see Lu Gwei-Djen & Needham (2). Chhin wrote many other books, including a *Yen Tshê Jen Ching*[14]
(Manual of Clinical Diagnosis), a *Yao Fang*[15] (Collection of Prescriptions), a *Mo Ching*[16] (Pulse
Manual) and a *Pên Tshao*[17] (Pharmaceutical Natural History). All are lost. For this information we are
indebted to Prof. Richard Mather, whose annotated translation of the *Shih Shuo Hsin Yü* has now in
part appeared (1).   [g] Ch. 2, p. 23 *b*, tr. auct.   [h] Cf. p. 33 above.

[1] 劉義慶   [2] 世說新語   [3] 何平叔   [4] 五石散   [5] 秦丞祖
[6] 寒食散論   [7] 何晏   [8] 葛洪   [9] 劉峻   [10] 相
[11] 祖   [12] 太醫令   [13] 醫學博士   [14] 偃側人經   [15] 藥方
[16] 脈經   [17] 本草

beginning. For the Chinese scene this would be equally true, as the names of Ko Hung, Thao Hung-Ching and Sun Ssu-Mo alone suffice to bear witness; also the early date at which the chemical-metallurgical line joined with the pharmaceutical ones to form the classical elixir pattern of developed Chinese alchemy. Now it is important that in the oldest Chinese pharmacopoeia, the *Shen Nung Pên Tshao Ching*,[1] there are 41 entries for inorganic substances,[a] including both cinnabar and mercury. This text is generally considered to be of the Early Han (−1st or −2nd century), even containing Chou and Warring States material, though not finalised until the end of the Later Han (+2nd century).[b] Of cinnabar powder[c] it is said that it is found in mountain valleys, its sapidity is sweet and slightly cold, it controls the hundred diseases of the body and the five viscera, it nourishes the *ching*[2] and *shen*[3] vitalities, calms the *hun*[4] and *pho*[5] 'souls', benefits the *chhi*, clears the eyes, kills *mei*[6] parasites and destroys malignant humours (*hsieh o chhi*[7]).[d] It can change into mercury (*hung*[8]). If taken constantly it puts one in touch with spiritual beings and gives longevity (*thung shen ming pu lao*[9]). Significantly the *Pên Ching* never speaks of becoming an immortal (*chhêng hsien*[10]) or ascending to the heavens (*fei thien*[11]),[e] but always confines its remarks to longevity, lightness of body and length of years in such phrases as *chhing shen*,[12] *yen nien*,[13] *tsêng shou*,[14] etc. Significantly also the same terms are applied to many plant drugs.

For the first well-documented evidence of the taking of inorganic substances in this way, the pharmaceutical use of the Five Minerals, we can go back to a time much earlier than that of Ho Yen. The first half of the −2nd century, a period just before the Coining Edict (cf. p. 26 above), saw the activity of one of the greatest physicians in Chinese history, Shunyü I.[15] Born under the Chhin in −216 he lived till about −147, and practised extensively among the princely families of the Han. Our information about him is of high authenticity because Ssuma Chhien devoted half a chapter

---

[a] 18 in the highest category (the safest), 14 in the middle, and 9 in the lowest (the most active). For an account of the pharmacodynamic classification used, see Sects. 38 and 45 in Vol. 6.

[b] For example, there are Later Han place-names for provenances, but these could easily have been up-dated by revisers without changing anything else.

[c] Since in the following pages we shall find many references to the constant taking of mercuric sulphide *per os*, it is natural to inquire how much of it would have been absorbed into the system since the salt is so insoluble. Stimulated by the parallel use in Indian Ayurvedic medicine, Ghosh (1) made a study of the subject in 1931 using the sensitive Bardach's test. He found positive results under conditions *in vitro* simulating the action of the gastric juice at HCl concentrations from 0·05 to 0·2%, incubation at 37°C and continual stirring; in 2 hours some 0·02% of the sulphide present went into solution. Extracts of liver in experimental dogs, but not of other organs, also showed positive reactions. The general conclusion would be that cinnabar ingestion alone, unless habitually in large amounts, was a fairly safe way, because of its low solubility, of increasing the mercury content of intestinal contents, organs and body fluids, whether for disinfection, anti-luetic or anti-suppurative effects. Thus the ancient and medieval physicians were not wrong in using it.

[d] *Kuei*[16] in some versions, for *chhi*.

[e] It does use the expression *shen hsien*,[17] holy immortals, but apparently only as a comparative phrase: 'longevity (like that of) the immortals', or, 'so light that you can fly a thousand *li* (like the) immortals'.

[1] 神農本草經	[2] 精	[3] 神	[4] 魂	[5] 魄
[6] 魅	[7] 邪惡氣	[8] 汞	[9] 通神明不老	[10] 成仙
[11] 飛天	[12] 輕身	[13] 延年	[14] 增壽	[15] 淳于意
[16] 鬼	[17] 神仙			

to him in the *Shih Chi*, hence we can rely on what is told of the relations between Shunyü and another physician named Sui.[1] On two occasions the former got into trouble with the authorities (−167 and −154), and on the second of these he was commanded by imperial decree to declare his methods of practice and to reply to certain specific questions; this is the reason why we have an unexampled set of 25 clinical histories of the −2nd century as well as details of the titles of the medical books handed down to Shunyü by his teachers. Of these histories case no. 22 concerned Dr Sui,[1] an Archiater-Royal or Physician-in-attendance to the Prince of Chhi (Chhi Wang Shih I[2]),[a] and the account in the *Shih Chi* runs as follows:[b]

Sui, the Physician-in-waiting to the Prince of Chhi, fell ill. He had himself made preparations of the Five Minerals and had consumed them (*tzu lien wu shih fu chih*[3]). Your servant therefore went to visit him, and he said 'I am unworthy and have some disease; would you be so kind as to examine me?' So I did, and then addressed him in these terms: 'Your malady is a fever (*chung jê*[4]). The *Discussions*[c] say that when there is a central fever along with obstipation and suppression of urine one must not ingest the Five Minerals. Mineral substances considered as drugs are fierce and potent, and it is because of having taken them that you have several times failed to evacuate. They ought not to be hastily taken. Judging by the colour, an abscess is forming.'

And then the two physicians go off into a lengthy discussion of the principles of Yin and Yang in therapy where we need not follow them here, leaving it to the pharmacological Section later on. Shunyü I concluded his case-history thus:

I warned him that after a hundred days or more an abscess would gather above the pectoral region, that it would penetrate (the flesh by) the collar-bone, and that he would die. Such was the general tenor of our discussion. Of course the rules are laid down in the (medical) manuals and treatises, but if there should be a single one which a maladroit practitioner has failed to study, the explanations of the texts on the Yin and Yang may remain a dead letter.

Thus his opinion of Sui's skill was not high, and perhaps in the absence of antibiotics and sulpha-drugs there was little he could do against a staphylococcal infection. For Bridgman interpreted the case as one of an encysted pulmonary abscess opening through the skin. The medical details are unimportant here, what we must retain is the clear evidence of the therapeutic use of inorganic substances in the early −2nd century by a physician connected with the land of Chhi; for the events would have taken place about −160. And not only so, but among the books which Shunyü I received from his teacher there was a *Lun Shih*[5] (Discussion on (the Use of) Mineral Drugs),[d] perhaps one section of a *Mo Shu Shang Hsia Ching*[6] (Pulse Manual in Two Chapters) which he also had, and this itself perhaps an early version of the *Huang Ti Nei Ching*[7] of classical tradition. If he was studying medicine in −190, as is practically

---

[a] It can surely be no coincidence that Sui was connected with Chhi, the seaboard State whence came so many of the earliest alchemists and magician-technicians; cf. pp. 31, 34 above, and Vol. 2, p. 240.

[b] Ch. 105, pp. 21*b*, 22*a*, tr. Bridgman (2), p. 42, eng. auct.

[c] Not identifiable exactly, but doubtless one of the books which Shunyü had studied.

[d] The punctuation of the titles of Shunyü I's books is difficult, and we diverge in some ways from the conclusions of Bridgman (2), pp. 65 ff., who may however well be right.

[1] 遂   [2] 齊王侍醫   [3] 自煉五石服之   [4] 中熱   [5] 論石

[6] 脈書上下經   [7] 黃帝內經

certain, the *Lun Shih* itself must go back at least to the second half of the −3rd century, that is to say, before the time of Bolus of Mendes and long before that of Pseudo-Democritus (cf. pt. 2, pp. 17, 25). Of course the parallelism is not close, but it helps our sense of historical proportion to visualise the extremely early date of 'Paracelsian' mineral remedies in Chinese culture.

All this was also a long time before Fêng Chün-Ta and Ho Yen. So perhaps we should speak of three strands out of which the typical elixir alchemy of the Chinese middle ages was woven: (*a*) the pharmaceutical-botanical tradition of the herb or plant of immortality, (*b*) the metallurgical-chemical tradition of the making of artificial gold, and (*c*) the medical-mineralogical tradition of the use of inorganic and metallic substances in therapy. Though we cannot pin-point as yet the first origin of any of these they must all come down in embryo from the Warring States period, and their fusion was certainly completed during the +1st century. It is on the +2nd that we must now concentrate our attention.

Before proceeding to this, however, we may pause to look at a tradition, Indian seemingly rather than Chinese, which connects the ancient herbal complex with the probably somewhat later mineral-metallic complex, and might throw light on how the latter developed. Mahdihassan, in his studies of alchemical origins in the Asian cultures, noting that 'plant cults preceded metal cults',[a] points out that a mixed technique has continued down in Ayurvedic medicine and pharmacy until the present day.[b] 'Herbo-metallic preparations' are made by the repeated calcination of metals in the presence of vegetable extracts, often preceded by amalgamation and removal of the mercury by distillation, a process which leaves the metal finely comminuted or spongy and therefore more easily oxidised when roasted.[c] Although the end result must be nothing but the oxide of the metal together with the ash of the plant extract, all the rest of which will have steamed and burnt away, classical Indian theory considers that the metals have been 'activated therapeutically by means of the plant decoctions'.[d] This is expressed by saying that there has been a transfer of the plant 'soul' (originating from starry emanations) to the metal;[e] and here a sexual element also enters in because the plants were considered rich in *rūḥ* (somewhat analogous to the Chinese *hun*, or Yang 'soul', cf. pt. 2, pp. 85 ff.) while metals were rich in *nafs* (somewhat analogous to *pho*, or Yin 'soul'), hence a hermaphrodite thing would have been created.[f] The oxide was also regarded as a base or vehicle like sandalwood oil for essential perfume oils.[g] Hundreds of such herbo-metallic preparations are still used in Ayurvedic medicine under the names (Skr.) *bhasma* (i.e. ashed), or (Pers., Urdu)

---

[a] (13), p. 243.

[b] (12), pp. 88ff., 91; cf. also (15), p. 88, (17), p. 85, (18), pp. 43ff., (20), p. 39, (25), pp. 42, 46.

[c] Mahdihassan (12), cf. Neogi & Adhikari (1); Sengupta (1), vol. 2, p. 22. Mahdihassan (32), pp. 341–2, has described an interesting traditional indicator test for the degree of fineness of the calcined powder. A hand-spun cotton thread dyed yellow with turmeric is stretched horizontally between two poles on a humid windless day, then the alkaline powder is placed on it at one end, and the length to which the colour turns red is taken as a measure of the quality, i.e. the degree of comminution, of the oxide. This 'thread chromatography' has been experimentally verified (M. Sreenivasaya).

[d] Mahdihassan (9), p. 123.

[e] Mahdihassan (21), pp. 183ff., 203, (23); Bhagvat Singhji (1), p. 137.

[f] Mahdihassan (32), pp. 339ff.   [g] Mahdihassan (12), *loc. cit.*

kushta (i.e. killed), plant extracts being thus 'combined' with metals in a great variety of permutations and combinations.[a]

All this clearly constitutes in principle a kind of intermediate stage between the 'plant of immortality' that Chhin Shih Huang Ti was looking for, and the later mineral-metallic *tan*, though we do not recall any text which would show that the alchemists of ancient China actually sought to capture herbal virtues in this way. At all events the latter were evidently still in favour as late as Ko Hung's time, for he sometimes felt he had to defend the mineral-metallic preparations.

Ordinary people [he says in one place] do not compound the spiritual (metallic) elixirs, but they have great faith in vegetable medicines—though these decay when buried, soften when cooked, and scorch when roasted. Since these substances cannot even maintain life in themselves, how could they give life to others?[b]

Nevertheless, the idea lay near at hand that a herbal drug might heal a diseased metal and make it into a healthy one, that is to say, act as an elixir and turn it into gold.[c] Such a belief, indeed, might have been the origin of the *bhasmas* and *kushtas*. It is possible that a factual basis existed for it, namely the use of plant acids, such as oxalic, malic, citric, etc., in cementation process, especially leaching (cf. pt. 2, p. 250), for the surface enrichment of gold-containing alloys. Both in India[d] and in Cyprus[e] the popular belief has been recorded that thaumaturgical devotees could turn copper coins into gold ones by the use of certain plant juices, and there is at least one first-hand account of the procedure, in which the heat of glowing tobacco was used to convert a 'copper' coin to a gold mohur.[f] Surface enrichment is unquestionably a very ancient device, as witness the electrum spear-head of Chaldaean Ur, discovered by Woolley in 1926. Dating from *c.* −2700, this consists of some 30% Au, 60% Ag and 10% Cu, but there is so obvious a concentration of gold in the surface layers that the object appears to be electro-gilded.[g] If then the discovery was anciently made that plant juices or extracts could under some conditions accomplish an 'ennoblement' such as this, the Ayurvedic plant-ash metal-oxide combination drugs would have followed naturally enough, and some similar train of thought and practice may well have assisted in China the passage from the 'herb of deathlessness' to the mineral-metallic elixirs and the potable gold.

There was also the transition from the vessels of gold (Li Shao-Chün, p. 31 above) to the actual ingestion of gold (e.g. Wang Hsing or Yin Chhang-Shêng, pp. 40, 43 above). Some light on how this may have happened is thrown by Mahdihassan's study of preparations of 'colloidal gold' still being made by the traditional Ayurvedic and Unani physicians in India today.[h] Sometimes gold leaf is triturated with plant powders in pestle and mortar, sometimes gold filings are heated with plant powders

[a] Cf. Jamshed Bakht & Mahdihassan (1); Shastri (1).
[b] *PPT/NP*, ch. 4, p. 5*b*, tr. auct., adjuv. Ware (5), p. 76; Wu & Davis (2), p. 239.
[c] This would have been very close to the central concepts of Arabic alchemy (cf. pt. 4).
[d] Mahdihassan (12), p. 89, (15), pp. 83–4; cf. also (9, 18, 21, 24, 31). [e] Gildemeister (1).
[f] Mukand Singh (1), p. 19.
[g] Woolley (4), p. 24 and pl. VIII, 2; cf. Stapleton (4), p. 31.
[h] (55), esp. pp. 115ff. Electron microscope photographs of the finely divided gold are presented.

at about 400°C; the former labour may take three months at five hours a day, the latter some forty heatings of four hours each. The resulting brick-red powders (*suvarna bhasma*) are so finely divided as to form 'colloidal gold' on contact with liquids,[a] and they may well afford a clue as to the nature of some of the ancient Chinese *chin i*[1] elixirs.

(2) WEI PO-YANG; THE BEGINNINGS OF ALCHEMICAL LITERATURE IN THE LATER HAN (+2ND CENT.)

The earliest book on alchemical theory extant[b] was written in the +2nd century by Wei Po-Yang,[2] who has been called the 'father of alchemy' for this reason.[c] Very little is known about his life; his name is not mentioned in any of the official dynastic histories. It was Ko Hung[3] who first wrote in the +4th century that Wei Po-Yang came from the region of Wu[4] (roughly corresponding to parts of modern Chekiang and Kiangsu provinces) and that he was a son of the Kaomên[5] clan, probably one of the shaman or magician-technician families.[d] Wei Po-Yang states in his own work, the *Tshan Thung Chhi*[6] (Kinship of the Three), that he came from Kuei-chhi[7] and that he had kept himself away from government service.[e] According to Phêng Hsiao,[8] who wrote a commentary on the work in +947, Wei Po-Yang was a native of Shang-yü,[9] in modern Chekiang, and confided his art to one Hsü Tshung-Shih,[10] who, after adding his own commentary, passed the book in turn to Shunyü Shu-Thung[11] during the time of the emperor Huan Ti[12] (r. +147 to +167).[f] Others say that Wei Po-Yang transmitted his art directly to Shunyü Shu-Thung.[g] We know a little more about the latter as he was a senior administrative officer in Loyang and relinquished his post about the year +150 to seek the Tao. From this Yuan Han-Chhing (1) concludes that Wei Po-Yang flourished between +100 and +170. Later on Wei became known in Taoism under the names Thai-Su Chen Jen[13] (Highest Purity Adept) and Yün Ya Tzu[14] (the Cloudy-Banner Master), and his personal name was said to have been Wei Ao[15].[h] Of course one can read more about Wei Po-Yang in

a The nature of the suspension needs further study; it may or may not be like the stable colloidal gold formed by chemical precipitation in the absence of electrolytes (cf. the 'purple of Cassius', Vol. 5, pt. 2, p. 268). At any rate it can be ingested *per os*.

b It should be understood that this book is not necessarily the earliest book on alchemical practice. The *Huang Ti Chiu Ting Shen Tan Ching* (*Chüeh*), which we discuss elsewhere (pp. 84-5), vies with it in date, since the first chapter of that may well belong to the +2nd century, but it gives practical instructions for elixir-making and does not go into the theory.

c Wilson (2c), who puts his *floruit* between +120 and +150.

d *Shen Hsien Chuan*, ch. 1, p. 4a. Ko Hung's authorship is not certain.

e *Tshan Thung Chhi*, ch. 35, p. 12a. All references quoted in this form by us are to the *Tshan Thung Chhi Fên Chang Chu* (*Chieh*), see p. 54 below.

f See *TT* 993, cf. p. 53 below.

g See *Ssu Khu Chi Yao Pien Chêng*, ch. 19, (p. 1206).

h See *Tao Shu*[16] (*TT* 1005), ch. 34, p. 1a.

1 金液	2 魏伯陽	3 葛洪	4 吳	5 高門
6 參同契	7 會稽	8 彭曉	9 上虞	10 徐從事
11 淳于叔通	12 桓帝	13 太素眞人	14 雲牙子	15 魏翱
16 道樞				

the hagiography of the immortals.[a] We need not repeat the story often given about how he shared his elixir of immortality with one of his three disciples, Yü shêng,[1] and a dog.[b] The *Shen Hsien Thung Chien* tells us that he learnt the art of immortality from Yin Chhang-Shêng[2] and also gives the name of Yü shêng as Yü Hsün[3].[c] Master Yin is the putative author of a text called *Chin Pi Wu Hsiang Lei Tshan Thung Chhi*[4] (Golden Jade Treatise on the Similarities and Categories of the Five (Substances) and the *Kinship of the Three*)[d] but there is insufficient evidence to warrant its dating as early as this. Wei is said to have written, as we shall see, another alchemical text called the *Wu Hsiang Lei*[5] (Similarities and Categories of the Five (Substances)), and hence we find his name associated with the *Tshan Thung Chhi Wu Hsiang Lei Pi Yao*[6] (Arcane Essentials of the Similarities and Categories of the Five (Substances) in the *Kinship of the Three*), an important, but later, text on alchemical theory which we shall discuss presently in this Section.[e] Another text in the Taoist Patrology, entitled *Ta Tan Chi*[7] (Record of the Great Elixir),[f] also ascribed to Wei Po-Yang, is probably of Thang if not of Sung origin.

The text of the *Tshan Thung Chhi* (The Kinship of the Three) has been dated about the year +142.[g] The name of its author is concealed in a cryptogram in the last paragraph of the epilogue. Known also as *Chou I Tshan Thung Chhi*[8] (The Kinship of the Three; or, the Accordance of the *Book of Changes* with the Phenomena of Composite Things), its title has given rise to much speculation. Phêng Hsiao in the +10th century explains that *tshan* means 'to mix', *thung* is 'to communicate', and *chhi* denotes 'combination', i.e. substances are mixed and then allowed to communicate with one another to bring about alchemical changes.[h] Chu Hsi[9] adds that such operations correspond with the *Book of Changes*.[i] Chhen Chih-Hsü[10] in the +14th century elucidates by saying that *tshan* is to compare, i.e. to compare the Natural Order of Heaven and Earth, *thung* (meaning 'similar') refers to the assistance given to the production and formation of things of similar kinds, and *chhi*, which is 'agreement', like a tally, signifies the agreement with production and growth in the Order of Nature.[j] Yü Yen[11] in the late +13th century says that the word 'three' refers to the *Book of Changes*, immortality and alchemy.[k] A totally different interpretation is given by one of the anonymous commentators of the text, saying that *chou i* refers to the 'cyclical changes' of the elixir, and *tshan thung chhi* means the mixing of the three substances of the Water, Earth and Metal elements together so that they combine to

---

[a] For example, in *Shen Hsien Chuan*, ch. 1, pp. 4a,b, *YCCC*, ch. 109, pp. 5aff., and *Shen Hsien Thung Chien*, ch. 9, sect. 5, pp. 1bff. A translation from a Yuan work of the same kind is given in L. Giles (6), p. 67.

[b] See Wu & Davis (1); and pt. 2, p. 295 above.   [c] Ch. 9, sect. 5, pp. 1bff.

[d] *TT*897; we shall glance at it again in due course (p. 150).   [e] See p. 145 and pt. 4 below.

[f] *TT*892.   [g] Wu & Davis (1).

[h] *TT*994, p. 11a. His view is shared by the anonymous commentator of *TT*991.

[i] *TT*992, preface, p. 1a.   [j] *Tshan Thung Chhi*, ch. 15, p. 35a.

[k] *TT*996, ch. 9, pp. 13b and 14a. This view is shared by Wang Ming (3). Wu & Davis (1) and Yuan Han-Chhing (1) simply say the *Changes*, Taoism and alchemy.

[1] 虞生   [2] 陰長生   [3] 虞巡   [4] 金碧五相類參同契
[5] 五相類   [6] 參同契五相類秘要   [7] 大丹記   [8] 周易參同契
[9] 朱熹   [10] 陳致虛   [11] 兪琰

form a single entity.[a] And lastly there must have been some who interpreted the title as a reference to the tying together of the three primary vitalities in man, which, when regained, gave longevity and immortality.[b]

The *Tshan Thung Chhi* begins by quoting the *Book of Changes*, saying: '*Chhien* and *Khun* are the gateway to the Changes.'[c] Some of its quotations come also from the Taoist canon *Tao Tê Ching*,[1] for example in the following instances:

He who knows the white, yet cleaves to the black. . .

and

The man of highest virtue[d] never acts (contrary to Nature)
And so has no need to apply force to things,
The man of inferior virtue tries to bend things to his will
And so in the end there is just (unsuccessful) force.

Besides this one finds much in the text about the theory of Yin and Yang as well as the Five Elements in relation to the processes of alchemy. Towards the end of the book there is a remark that the *Book of Changes*, the study of Huang (Ti) and Lao (Tzu), which can mean either Taoism or the art of immortality, and the alchemical art, are like three roads leading up the same mountain.[e] It seems therefore that the treatise is speaking in terms of the three things, namely the theory of the *Book of Changes*, which in a broader sense includes those of the Yin and Yang as well as the Five Elements,[f] the philosophical teachings of Taoism, and the processes of alchemy.

Important for the dating of the book is the fact that a commentary on parts of it was written by Yü Fan[2] about +230.[g] Ko Hung late in the same century never mentions the *Tshan Thung Chhi*,[h] but lists in his bibliography a *Wei Po-Yang Nei Ching*[3] which may well have been the same work.[i] The poet Chiang Yen[4] (+444 to +505) refers to the usual title in a verse towards the end of the +5th century,[j] and it appears in both the Sui and Thang bibliographies. The *Shen Hsien Chuan* says:[k]

---

[a] See *TT* 995, p. 1 a. The ambiguity in the word *chou* has already been encountered in our translation of the mathematical classic *Chou Pei Suan Ching*, see Vol. 3, p. 19.

[b] See, on this, pt. 5 below.

[c] *Tshan Thung Chhi*, ch. 1, p. 1 a. From *I Ching*, ch. 3, p. 23 b; tr. auct., adjuv. Legge (9), p. 395 and Wilhelm (2), vol. 1, p. 369.

[d] Mana, charisma, spiritual power. On the idea of not acting contrary to Nature see Vol. 2, pp. 68 ff.

[e] *Tshan Thung Chhi*, ch. 34, p. 11 b.

[f] See Vol. 2, Sect. 13 for a detailed account of these fundamental ideas of Chinese science.

[g] Cited in *Ching Tien Shih Wen*[5] (Textual Criticism of the Classics) by Lu Tê-Ming,[6] c. +600, *I Ching* section. Hu Shih (7) has suggested that Yü Fan was none other than the disciple Yü sheng (or Hsün) himself, in which case the date of the *Tshan Thung Chhi* would be rather toward the end of the +2nd century than just before the middle of it.

[h] That is to say, in the *Pao Phu Tzu*. The *Shen Hsien Chuan* speaks of both the author and the book under its subsequent name, but Ko Hung's connection with it is uncertain.

[i] Perhaps he was not very interested in the theoretical linkage of alchemy with the *Book of Changes*. He was essentially a practitioner.

[j] *Chiang Wên-Thung Chi*,[7] ch. 3, p. 5 b. These early mentions were noted by Waley (13), p. 55, (14), pp. 7 ff. Waley agreed with the consensus of Chinese scholarly opinion that the +2nd-century date is justified by the internal evidence of the rhyme-system used.   [k] Ch. 1, p. 4 b.

[1] 道德經   [2] 虞翻   [3] 魏伯陽內經   [4] 江淹   [5] 經典釋文
[6] 陸德明   [7] 江文通集

Wei Po-Yang wrote the *Tshan Thung Chhi*, including the *Wu Hsing Hsiang Lei*[1] (Similarities and Categories of (Substances formed by) the Five Elements), in three chapters. This purports to be about the *Book of Changes* (*Chou I*) but actually it uses the symbols to discuss the concepts of alchemy (*tso tan*[2]). Thus ordinary Confucians, knowing nothing of alchemy, have commented on it as though it were a treatise on Yin and Yang, thus completely misunderstanding it.

The earliest edition now extant of the *Tshan Thung Chhi* text is said to be a Ming block-printed copy made by the Chao Fu Wei Ching Thang[3] during the Chia-Ching reign-period (+1522 to +1566) and now preserved in the Peking Library.[a]

Seldom has a book been so much commented on, and with so much criticism of their predecessors by commentators, as the *Tshan Thung Chhi*. The *Tao Tsang* alone includes the following ten commentaries each reproducing the original text:

(a) *Chou I Tshan Thung Chhi Chu*[4] (*The Kinship of the Three* with Commentary)[b] attributed to Yin Chhang-Shêng[5] (*fl.* +120 to +210), but probably of Sung origin.[c]

(b) *Chou I Tshan Thung Chhi Chu*,[4] by an anonymous commentator and believed to be of Sung date.[d]

(c) *Chou I Tshan Thung Chhi Chu*,[4] by Chu Hsi[6] (+1197) under the pseudonym Tsou Hsin.[7] This text[e] sometimes appears under the title *Chou I Tshan Thung Chhi Khao I*[8] (Investigations and Criticisms of the *Kinship of the Three*).[f]

(d) *Chou I Tshan Thung Chhi Fên Chang Thung Chen I*[9] (*The Kinship of the Three* divided into (short) chapters for the Understanding of its Real Meanings)[g] by Phêng Hsiao[10] in +947, also known as Chen I Tzu[11] (the Real-Unity Master).

(e) *Chou I Tshan Thung Chhi Ting Chhi Ko Ming Ching Thu*[12] (Bright Mirror Chart illuminating the Mnemonic Rhymes about Reaction-Vessels in the *Kinship of the Three*)[h] also by Phêng Hsiao (+947).[i]

[a] See Yuan Han-Chhing (*1*), p. 177. The *Han Wei Tshung Shu* gives the text with an epilogue by Wang Mu[13] and a preface by Chu Chhang-Chhun.[14] The text is also included in the *Shuo Fu*[15] and the *Pai Ling Hsüeh Shan*[16] collections. A commentary by Wang Wên-Lu[17] (+1534) is appended to the text in the latter edition, which is also reproduced in the *Tshung Shu Chi Chhêng*.[18] In the *An Chhi Chhüan Shu*[19] Li Kuang-Ti[20] (+1717) splits the text into two, one called the *Tshan Thung Chhi* and the other the *San Hsiang Lei*[21] (Similarities and Categories of the Three (Substances)), the two being combined under a single title *Tshan Thung Chhi Chang Chü*[22] (the *Kinship of the Three* (arranged in) Chapters and Sections).

[b] TT990.

[c] According to Yuan Han-Chhing (*1*), p. 171. If Yin was indeed the teacher of Wei, it would have been very unusual for him to comment upon the work of a disciple, yet possible, if he lived till about +160.　　　[d] See TT991. Yuan Han-Chhing, *ibid.*　　　[e] TT992.

[f] For example in the *Shou Shan Ko Tshung Shu*[23] and the *Ssu Khu Chhüan Shu*.[24] The version in the former is reproduced in the *Ssu Pu Pei Yao* and the *Tshung Shu Chi Chhêng*. We have already made extensive quotations from Chu Hsi's work (Vol. 2, pp. 330ff., 441ff.).

[g] TT993. This text with commentary also appears in the *Hsü Chin Hua Tshung Shu*.[25]

[h] TT994.

[i] We shall have a good deal more to say about this in the sub-section on physiological alchemy (pt. 5).

[1] 五行相類	[2] 作丹	[3] 趙府味經堂	[4] 周易參同契註	[5] 陰長生
[6] 朱熹	[7] 鄒訢	[8] 周易參同契考異	[9] 周易參同契分章通眞義	
[10] 彭曉	[11] 眞一子	[12] 周易參同契鼎器歌明鏡圖		
[13] 王謨	[14] 朱長春	[15] 說郛	[16] 百陵學山	[17] 王文祿
[18] 叢書集成	[19] 安溪全書	[20] 李光地	[21] 三相類	[22] 參同契章句
[23] 守山閣叢書	[24] 四庫全書	[25] 續金華叢書		

(f) *Chou I Tshan Thung Chhi Chu*, by an anonymous commentator, believed to be of Sung date.[a]

(g) *Chou I Tshan Thung Chhi Fa Hui*[1] (Elucidation of the *Kinship of the Three*),[b] by Yü Yen[2] in +1284 (also known by the names Yü Wu-Yü,[3] or Chhüan Yang Tzu,[4] the Complete-Yang Master). This is the text that was used by Wu & Davis (1) in translating the *Tshan Thung Chhi*.

(h) *Chou I Tshan Thung Chhi Shih I*[5] (Clarification of Doubtful Matters in the *Kinship of the Three*),[c] also by Yü Yen (+1284).

(i) *Chou I Tshan Thung Chhi Chieh*[6] (The *Kinship of the Three* with Explanations),[d] by Chhen Hsien-Wei[7] in +1234 (also known by the names Chhen Tsung-Tao,[8] or Pao I Tzu,[9] the Unity-Embracing Master).

(j) *Chou I Tshan Thung Chhi Chu*,[e] by Chhu Hua-Ku;[10] c. +1230.

Other important commentaries on the *Tshan Thung Chhi* appear in another version of the Patrology, the *Tao Tsang Chi Yao*.[11] There we find the *Tshan Thung Chhi Chhan Yu*[12] (Explanation of the Obscurities in the *Kinship of the Three*), by Chu Yuan-Yü[13] in +1669 (otherwise known as Yün Yang Tao Jen[14]); and the *Chou I Tshan Thung Chhi Fên Chang Chu*[15] (Commentary on the *Kinship of the Three* divided into (short) chapters), by Chhen Chih-Hsü[16] c. +1330 (known also as Shang Yang Tzu,[17] the Honouring-the-Yang Master). Chhen Hsien-Wei's *Chou I Tshan Thung Chhi Chieh*[f] is also included in the *Tao Tsang Chi Yao*.

Chhen Chih-Hsü's commentary seems to have become rather popular, partly perhaps because of his position as seventh Patriarch of the Northern School of Taoism.[g] Punctuations and further explanations were added by Fu Chin-Chhüan[18] (also known as Chi I Tzu,[19] the Complete-Unity Master) about 1820. The text we have contains a preface by Yü Mu-Shun[20] (1841), also known as Chhien Yang Tzu,[21] a pupil of Fu.[h] Wylie reports a prevailing view that Chhen Chih-Hsü gives the text in its purest state and provides one of the clearest of the later commentaries.[i]

There are yet other commentaries like the *Chou I Tshan Thung Chhi Shu Lüeh*[22] (Brief Explanation of the *Kinship of the Three*) by Wang Wên-Lu[23] in +1564.[j] Some of them, like Lu Hsi-Hsing's[24] *Chou I Tshan Thung Chhi Tshê Su*[25] (Penetrating

---

[a] *TT*995. See Yuan Han-Chhing (*1*), p. 171.

[b] *TT*996.   [c] *TT*997.   [d] *TT*998.   [e] *TT*999. See Yuan Han-Chhing, *ibid.*

[f] *TT*998.   [g] Chhen Kuo-Fu (*1*), 2nd ed. vol. 2, p. 442.

[h] We have two editions of this text, one printed from blocks preserved in the Lu Yeh Chai[26] and one published by the Chin Chang Thu Shu Chü[27] at Shanghai. All references to the *Tshan Thung Chhi*, unless otherwise stated, are to the former. Although the outlook of Chhen Chih-Hsü was more physiological-spiritual than chemical, he did take into account some of the previous commentaries, and his text contains far fewer errors than that used by Wu & Davis (1) for their translation.

[i] See Wylie (1), p. 175.

[j] Incorporated in the *Pai Ling Hsüeh Shan*[28] collection, and from which the edition in the *Tshung Shu Chi Chhêng* is derived.

[1] 周易參同契發揮	[2] 兪琰	[3] 兪吾玉	[4] 全陽子
[5] 周易參同契釋疑	[6] 周易參同契解	[7] 陳顯微	[8] 陳宗道
[9] 抱一子	[10] 儲華谷	[11] 道藏輯要	[12] 參同契闡幽
[13] 朱元育	[14] 雲陽道人	[15] 周易參同契分章註	[16] 陳致虛
[17] 上陽子	[18] 傅金銓	[19] 濟一子	[20] 兪慕純
[21] 乾陽子	[22] 周易參同契疏略	[23] 王文祿	[24] 陸西星
[25] 周易參同契測疏	[26] 綠野齋	[27] 錦章圖書局	[28] 百陵學山

Explanation of the *Kinship of the Three*),[a] or *Tshan Thung Chhi Khou I*[1] (Oral Explanation of the *Kinship of the Three*), or Wang Fu's[2] *Tu Tshan Thung Chhi*[3] (On reading the *Kinship of the Three*), are rare books which have not been available to us.[b] And there must be other commentaries that we have not heard of.[c]

During the +16th century a claim was made to the effect that the original text of the *Tshan Thung Chhi* had been recovered by excavation, found in a rock chamber. Thus came the so-called *ku wên*[4] (ancient script) text. The obvious motive of this operation was to enhance the prestige of the book in particular, and Taoism in general, by imitating the famous −2nd-century 'discovery' of ancient-script versions of the Confucian classics, which caused infinite controversy and division through the centuries.[d] The Ming scholar Yang Shen,[5] also called Yang Shêng-An,[6] was so much taken in that he wrote a preface to it in +1546.[e] It was claimed that the various other existing versions of the *Tshan Thung Chhi* were not arranged in the proper textual sequence, and that in some places commentaries had got mixed up with the text proper. Hence the *ku wên* comes in three parts, the first supposed to be written by Wei Po-Yang, the second by Hsü Tshung-Shih,[7] and the third by Shunyü Shu-Thung.[8] All this has little importance for the history of chemistry, but it does show the importance of the *Tshan Thung Chhi* during the Ming.[f]

To add to the confusion of the already numerous conflicting commentaries and textual condition of the *Tshan Thung Chhi*, we find a totally different text, yet known by the same title, incorporated in the *Tao Shu*[9] (Axis of the Tao), a compendium put together by one Tsêng Tshao[10] during the Sung.[g] The first chapter of this so-called

[a] Written in +1569 and +1573; included in his *Fang Hu Wai Shih*[11] (Unofficial History of the Land of the Immortals). We now have an interesting paper on this by Liu Tshun-Jen (1) who elucidates its strongly *nei tan* tendency, sexual rather than respiratory.

[b] We know them only from the catalogue of the Jimbun Kagaku Kenkyusō in Kyoto, which does not divulge all their dates.

[c] Dr Friedrich Litsch has kindly informed us that the Collection Billigé of the École des Langues Orientales Vivantes at Paris contains three Manchu alchemical books or MSS, all commentaries on the *Tshan Thung Chhi*, and one supposedly by the Khang-Hsi emperor himself. The subject of alchemy among the Manchus briefly recurs on p. 229 below, but no real study, so far as we know, has yet been made of it.

[d] See Vol. 2, p. 248 and Vol. 7 below (in the meantime Needham (56), p. 30).

[e] The text we have contains also another preface by Chiang I-Piao[12] (+1614).

[f] They are known as follows: *Ku Wên Tshan Thung Chhi Chi Chieh*[13] (Collected Explanations of the *Kinship of the Three* in Ancient Script), by Wei Po-Yang; *Ku Wên Tshan Thung Chhi Chien Chu Chi Chieh*[14] (Commentary and Collected Explanations of the *Kinship of the Three* in Ancient Script), by Hsü Tshung-Shih; and *Ku Wên Tshan Thung Chhi San Hsiang Lei Chi Chieh*[15] (Collected Explanations of the *Kinship of the Three*, and the Similarities and Categories of the Three (Substances), in Ancient Script), by Shunyü Shu-Thung. These are contained in the *Chin Tai Pi Shu*[16] and the *Hsüeh Chin Thao Yuan*[17] collections. The copies in the former are reproduced in the *Tshung Shu Chi Chhêng*. Another copy of the *ku wên* text with a commentary by Yuan Jên-Lin[18] and entitled *Ku Wên Chou I Tshan Thung Chhi Chu*[19] (Commentary on the *Kinship of the Three* in Ancient Script) is found in the *Hsi Yin Hsien Tshung Shu*.[20]

[g] *TT* 1005; see chs. 32, 33, 34.

[1] 參同契口義	[2] 汪紱	[3] 讚參同契	[4] 古文	[5] 楊愼
[6] 楊升菴	[7] 徐從事	[8] 淳于叔通	[9] 道樞	[10] 曾慥
[11] 方壺外史	[12] 蔣一彪	[13] 古文參同契集解		[16] 津逮祕書
[14] 古文參同契箋註集解		[15] 古文參同契三相類集解		
[17] 學津討原	[18] 袁仁林	[19] 古文周易參同契註		[20] 惜陰軒叢書

Tshan Thung Chhi bears no author's name at all. The second purports to have been written by Tshao I Tzu[1] (the Straw Coat Master), supposedly the style of a man named Lou Ching,[2] of the Former Han period.[a] The third chapter carries the name of Yün Ya Tzu[3] (cf. p. 50); and a commentary follows saying that Wei Ao,[4] whose courtesy name was Po-Yang,[5] was a man of the Han period whose style was Yün Ya Tzu.[b] All three chapters give a text quite unlike the *Tshan Thung Chhi* that we know. We need not deal with them here as they lack chemical interest, being primarily physiological in character. The last takes the form of a dialogue between Yün Ya Tzu and Yuan Yang Tzu[6] the commentator. Some of the terms in Wei Po-Yang's *Tshan Thung Chhi* are of course used, for example *chin kung*,[7] *chha nü*,[8] *ho chhê*,[9] etc.[c]

The *Tshan Thung Chhi* is considered exquisite in literary style by many Chinese scholars, but at the same time extremely obscure in meaning. To quote from Wu & Davis, 'the perfection of the imagery and the intricacy of the rhetoric give the treatise an atmosphere similar to that of a piece of old Chinese embroidery'.[d] It would rather have occurred to us to say that the text is composed of short sentences resembling gnomic or oracular utterances,[e] in which the allusive undertones, as well as the overt meanings, of each word, and its character-structure, have to be taken into account— as in some modern poetry. This makes it almost impossible to translate—unless one should annotate every other word.[f] The great +12th-century Neo-Confucian scholar Chu Hsi appreciated its style so much that (as we have seen) he did not disdain to write a commentary on it, though it might also be said that Taoism was much in vogue at his time, receiving the patronage of the Sung Emperors. On the other hand, the obscurity and the various possible interpretations of the text itself give every reason for the long list of commentaries. We can find much divergence of opinion among the different commentators, one generally criticising his predecessors. They are not even unanimous on the origins of the text; for example one anonymous commentator[g] says that the *Tshan Thung Chhi* was derived from an earlier work en- titled *Lung Hu Shang Ching*[10] (Exalted Manual of the Dragon and Tiger), but Yü Yen holds exactly the opposite view, admitting the existence of this text but saying that it was itself derivative from *Tshan Thung Chhi*.[h]

The commentators belong, broadly speaking, to two different schools of thought. The first[i] maintained that the *Tshan Thung Chhi* was basically a work of practical

---

a *TT* 1005, ch. 33, p. 1 a. On Lou Ching, see pt. 2, pp. 258 ff. above.   b *TT* 1005, ch. 34, p. 1 a.
c For the alchemical explanation of these terms see pp. 104, 127–8, 152 ff. below. Sexual *nei tan* meanings are always implicit in them.
d (1), p. 216. This judgment would be even more applicable, to the rhapsodical odes (*fu*[11]) of these centuries, however.
e Almost with a folk-proverb flavour.
f Truly translation should only be attempted by those who have saturated themselves in the Yin- Yang, Five Elements and *I Ching* philosophies, and can give the time required for the study of all the commentaries. A full presentation would be the work of years.
g *TT* 990.   h *TT* 997, ch. 3, p. 3 b.
i To which belong the commentators of *TT* 990, 993, 994, 995, 998, 999.

¹ 草衣子	² 婁敬	³ 雲牙子	⁴ 魏翱	⁵ 伯陽	⁶ 元陽子
⁷ 金公	⁸ 姹女	⁹ 河車	¹⁰ 龍虎上經	¹¹ 賦	

alchemy, *wai tan*[1] (external elixir); while the second,[a] to which belonged generally speaking most of the later commentators, regarded it as a book of *nei tan*[2] (internal elixir).[b] At this point we encounter for the first time that grave division, with its obvious (but, as we shall see, illusory) Western analogue of practical versus allegorical-mystical alchemy. The parallelism between *chrysopoia*, the making of the perfect or noble metal, gold (whether by aurifiction or aurifaction), and the refinement or ascent of the soul or body towards perfection or perpetuation is a pattern which goes back right to the very beginning of proto-chemical enterprise, and it is remarkable that it should have appeared concurrently in the seemingly separated civilisations of East and West.[c] Physiological alchemy involved sexual techniques, and there must also have been some who interpreted the *Tshan Thung Chhi* as primarily concerned with these, for this idea came under the criticism of Chhen Chih-Hsü in the +14th century.[d] The great argument is still pursued by modern writers, as witness Wu & Davis (1) and Wang Ming (3) on one side, and Chhen Kuo-Fu (1), Liu Tshun-Jen (1) and Fukui Kōjun (1) on the other, the former favouring the practical alchemical view and the latter the psycho-physiological one. It would serve no useful purpose to follow this further here in view of the many possible interpretations of the text. What is undeniable is that the *Tshan Thung Chhi* contains many chemical terms and much alchemical phraseology. These terms and phrases must imply a knowledge of practical operations on the part of the author of the book. We wish to deal only with such knowledge, not speculating on how it would have been put into use in ancient and medieval times by aspirants to immortality.

The diversity of opinion between different commentators and the self-contradictory explanations of the same commentator add to the immense difficulty of understanding the *Tshan Thung Chhi*. For example, an anonymous commentator explains *liu chu*[3] (lit. flowing pearls) as mercury (*hung*[4]) in one place and as cinnabar (*tan sha*[5]) in another,[e] while a *nei tan* commentator interprets it as something having to do with the lungs, calling it *fei i*[6] (lit. lungs fluid, i.e. saliva).[f] Table 107 on p. 58 illustrates further the diverse interpretations made by three different commentators. With such richness of symbolism one cannot help musing over Wei Po-Yang's statement that '... A thousand readings will bring out some points, and ten thousand perusals will enable a man to see others. At last revelation will come to bring him enlightenment. ...'[g] Indeed it was because he believed himself to have attained

---

[a] See *TT* 991, 996.　　　　　　　　　　[b] Cf. the discussion on pp. 182, 201, 208 below.

[c] Further on this subject see pt. 5. Already here, however, it may be well to point out that these ideas were independent of the macrobiotic element, for the Alexandrian proto-chemists lacked this and yet had elaborate ideas about the development of the practitioner.

[d] Cf. *Tshan Thung Chhi*, ch. 12, p. 26b. Naturally, because Chhen Chih-Hsü was of the ascetic celibate Northern School. Lu Hsi-Hsing's +16th-century commentary, on the other hand, emphasised such explanations; cf. Liu Tshun-Jen (1). On this subject see Vol. 2, pp. 146 ff., and below, pp. 205, 209.

[e] See *TT* 990, ch. 1, p. 33b and p. 35b.

[f] *TT* 991, ch. 2, p. 14a. This will become quite understandable later on (pt. 5).

[g] *Tshan Thung Chhi*, ch. 32, p. 5a, Wu & Davis tr., p. 260.

[1] 外丹　　　[2] 內丹　　　[3] 流珠　　　[4] 汞　　　[5] 丹砂　　　[6] 肺液

enlightenment that each commentator sat down to write and criticise his predecessors for not understanding the real meaning of the text.

Table 107. *Symbolic correlations of the Five Elements in different interpretations of the 'Tshan Thung Chhi'*

Elements	Metal	Water	Wood	Fire	Earth
trigrams	*Tui*[1]	*Khan*[2]	*Chen*[3]	*Li*[4]	
colours	white	black	caerulean	red	yellow
symbolic animals} {	White Tiger		Blue Dragon	Red Dragon	
wai tan interpretation TT 994[a]	metallic lead (*chin ching*[5])	lead ore (*chhien*[6])	mercury (*hung*[7])	cinnabar (*chu sha*[8])	
nei tan interpretation TT 991[b]	lungs	reins	liver	heart	
TT 996[c]	lungs mother	reins son	liver father	heart daughter	spleen grand-parent

Many of the obscure passages in the *Tshan Thung Chhi* seem to be explicable in terms of the fundamental ideas of Chinese science already discussed in Section 13. Take as a first example the following translation of Wu & Davis: '... Yellow earth is the father of gold, and flowing pearls (mercury?) the mother of water.'[d] This obscure passage makes good sense to a Chinese reader if the word *chin*[9] in this particular instance is interpreted as 'metal' and not 'gold', and both 'earth' and 'water' are regarded also as elements. The passage then reads as follows: '...The yellow Earth (element) is the father of the (element) Metal and mercury (*liu chu*[10]) is the mother of the (element) Water.' According to the Mutual Production Order of the Five Elements,[e] Earth produces Metal, which in turn produces Water. The colour yellow is associated with the Earth element and mercury belongs to the Metal element. Hence Wei Po-Yang says here that Earth is the father of Metal and mercury is the mother of Water.

Let us next take the following passage from the *Tshan Thung Chhi*, quoting again the translation by Wu & Davis (1) without the slightest change.[f] The meaning will be by no means obvious, but we shall then show that it can be intelligibly elucidated.

---

[a] Ch. 1, pp. 1*a*, *b* and 14*a*.　　　[b] Ch. 2, p. 19*a*.　　　[c] Ch. 5, p. 19*b*.
[d] *Tshan Thung Chhi*, ch. 5, p. 13*b*, Wu & Davis tr., p. 240. See also *TT* 990, ch. 1, p. 35*b*.
[e] See Vol. 2, p. 255ff.
[f] *Tshan Thung Chhi*, ch. 19 *in toto*, Wu & Davis tr., pp. 247ff. We deliberately refrain from making the many corrections and emendations which present-day knowledge would impose. The version of Liu Tshun-Jen (1), pp. 92ff., is scarcely better.

[1] 兌　　　[2] 坎　　　[3] 震　　　[4] 離　　　[5] 金精　　　[6] 鉛　　　[7] 汞
[8] 朱砂　　　[9] 金　　　[10] 流珠

At the first double-hour of the day,[a] which corresponds to the *Fu kua*,[b] the *Yang chhi* (positive ether) begins to operate and at once appears to be slightly strong. At this time when the *Huang-chung*[1][c] coincides with the ordinal *tzu*,[d] a promising beginning flourishes forth. Let there be warmness and all will be well.

When the furnace is worked with sticks,[e] room is made for the propagation of light. With the increase in brilliance the day becomes longer. This corresponds to the ordinal *chhou* and to the *Ta-lü*.[2] Appropriateness is now realised.

Face upward to attain the *Thai* (greatness). *Kang* and *jou* (hardness and softness) both come to have sway. Yin and Yang (negativeness and positiveness) are in contact with one another. Undesirable things give place to desirable ones. Activity centres at this, the ordinal *yin*, when fortune is at its high tide.

Gradually the rule of the *kua* of *Ta Chuang* (great brave) is passed. This corresponds to the *Chia-chung*[3] and the ordinal *mao*. The elm seeds fall, returning to their origin. Just as punishment and forgiveness are opposite to one another, even so is the day distinguished from the night.

At the *Kuai kua* when the Yin (negativeness) beats its retreat, the Yang (positiveness) rises forth, washing its feathers to rid them of accumulated dust.

The strong light of the powerful *Chhien* (male, positiveness) covers the neighbourhood on all sides. The rule of Yang (positiveness) comes to an end at the ordinal *ssu*, which occupies a central position with good connections.

A new period begins with the *Kou kua*. As this is the transition to coldness, it should be faced with perfect calmness. It is now the ordinal *wu*, corresponding to the *Sui-pin*.[4] Yin has come to be the mistress.

When the *Tun kua* is here, retirement is in order. With the retirement the unusual powers go into hiding, waiting to reappear at the propitious time.

The *Phi kua* brings with it an unpropitious time, when Yin (negativeness) gathers power at the expense of Yang (positiveness) and vegetation grows no more.

At the *Kuan kua* the powers and capacities of things are observed. In mid-autumn different things happen to the plants. Some plants ripen their flowers into fruits and seeds so as to enable the aged and decaying to flourish anew. Wheat and the shepherd's purse bud forth to thrive. And then comes the *Po kua*. The body is torn to pieces so that the form is no more. For, as the *chhi* (ethereal essence) of transformation is exhausted, the divinest is lost.

When the limit of the *Tao* (Way) is reached, a return to the primordial *Khun* (negativeness) is made. The lay of the land should always be given due consideration and the sky should be obeyed. The mystically obscure and distantly indistinct are separated yet related. To propagate according to the proper measures is the foundation of Yin and Yang (negativeness and positiveness). Everything is obscure and unknowable. Although at a loss at first, it finally becomes the ruler.

[a] On Chinese time reckoning see Vol. 4, pt. 2, pp. 439, 461, and more fully in Needham, Wang & Price (1).

[b] On the system of the *kua* in the *I Ching* (Book of Changes) see Vol. 2, Table 14, and its accompanying explanation.

[c] This like *Ta-lü*, *Chia-chung* and *Sui-pin* (properly *Jui-pin*), appearing in the following sentences, is one of the twelve notes of the classical Chinese gamut. See Vol. 4, pt. 1, Table 47, and its accompanying explanations.

[d] One of the cyclical characters, like *chhou*, *yin* and *mao* appearing in the following sentences. The *yin* here is not to be confused with the Yin (of Yin and Yang). We have often explained this system and its applications; see Vol. 1, p. 79, Vol. 2, pp. 357ff., Vol. 3, pp. 82, 396ff., Vol. 4, pt. 1, pp. 297ff.

[e] Here the *kua Lin* was not recognised as such.

[1] 黃鐘   [2] 大呂   [3] 夾鐘   [4] 蕤賓

Without the valley there would be no hill. That is in the nature of the *Tao* (Way). Similarly there exists a contrast between rise and fall, and between growth and degeneration. The *Khun kua* marks the end and the *Fu kua* marks the beginning: like a cycle they go. Throughout and forever the monarch lives to rule.

At the end of this the reader will be fully ready to agree that 'everything is obscure and unknowable'. Nevertheless the original made good sense.[a] Wu & Davis did at least succeed in giving the correct explanation that it referred to the cyclical heating and cooling of the chemical reactants. During the period 11 p.m. to 11 a.m. the substances are to be heated and kept hot, and during the period between 11 a.m. and 11 p.m. they are to be chilled and kept cool. These four operations correspond to the four seasons, and if the four seasons are thus followed the Five Elements will do their work properly. We shall refrain from giving a better translation of the same passage because even in its present form it is quite possible to see what Wei Po-Yang had in mind. He was simply composing a cryptogram to give the proper times for heating and cooling, just as he did subsequently to conceal his own name. To decipher this we shall first take note that he used only twelve of the sixty-four hexagrams, namely *Fu* (24), *Lin* (19), *Thai* (11), *Ta Chuang* (34), *Kuai* (43), *Chhien* (1), *Kou* (44), *Thun* (33), *Phi* (12), *Kuan* (20), *Po* (23), and *Khun* (2).[b] In each of the sentences following the mentions of the first four hexagrams, the four cyclical signs *tzu*, *chhou*, *yin*, and *mao* appear in that order. For the fifth hexagram *Kuai* the cyclical sign *chhen*[1] is concealed in the word *chen*.[2] Attached to the sixth and seventh hexagrams, i.e. *Chhien* and *Kou*, are the two cyclical signs *ssu* and *wu* respectively. For the eighth hexagram, *Thun*, the cyclical sign *wei*[3] is concealed in the word *mei*;[4] and similarly for *Phi* the cyclical sign *shen*[5] is found in the right-hand portion of *shen*[6].[c] For the tenth hexagram, *Kuan*, the cyclical sign *yu*[7] does not appear directly in the sentences that follow, unless by any chance *mao*[8] is a misprint for this word in the phrase *yin mao i shêng*.[9] However, in the sentence following this hexagram mid-autumn (*chung chhiu*[10]) is referred to, and in the *Huai Nan Tzu*[11] we find the sentence 'In the month of mid-autumn (the asterism) *Chao yao*[12] points towards (the azimuthal cyclical sign) *yu*'.[d] For the eleventh hexagram, *Po*, the cyclical sign *hsü*[13] is concealed in the word *mieh*,[14] and Wei Po-Yang even hints that something has to be 'torn to pieces', undoubtedly a reference to this word. Finally, for the twelfth hexagram, *Khun*, the cyclical sign *hai*[15] is found enclosed within the word *ai*.[16] All the twelve cyclical characters here refer of course to the time of the day.

Taking a step further, let us write down the six lines for each of the twelve hexa-

[a] The elucidation which now follows is also contained, with some amplifications, in Ho Ping-Yü (16).
[b] The numbers within brackets refer to the order of the hexagrams given in Table 14.
[c] Yü Yen's text gives *shen*,[17] which makes the sentence even more unfathomable because suppressing a natural antithesis, though the character still contains the cryptographically significant phonetic component.
[d] *Huai Nan Tzu*, ch. 5, p. 10a. Cf. Vol. 3, p. 250.

[1] 辰	[2] 振	[3] 未	[4] 眛	[5] 申	[6] 伸
[7] 酉	[8] 冒	[9] 因冒以生	[10] 中秋	[11] 淮南子	
[12] 招搖	[13] 戌	[14] 滅	[15] 亥	[16] 閡	[17] 神

grams (Table 108).[a] The reason why these were selected out of the sixty-four then becomes obvious.

Table 108. *'Fire-times' in elixir preparation; the aegis of hexagrams during the day and night*

kua	cyclical character and time (double-hours)
Fu (24) 復	*tzu* (23 hr to 01 hr) 子
Lin (19) 臨	*chhou* (01 hr to 03 hr) 丑
Thai (11) 泰	*yin* (03 hr to 05 hr) 寅
Ta Chuang (34) 大壯	*mao* (05 hr to 07 hr) 卯
Kuai (43) 夬	*chhen* (07 hr to 09 hr) 辰
Chhien (1) 乾	*ssu* (09 hr to 11 hr) 巳
Kou (44) 姤	*wu* (11 hr to 13 hr) 午
Thun (33) 遯	*wei* (13 hr to 15 hr) 未
Phi (12) 否	*shen* (15 hr to 17 hr) 申
Kuan (20) 觀	*yu* (17 hr to 19 hr) 酉
Po (23) 剝	*hsü* (19 hr to 21 hr) 戌
Khun (2) 坤	*hai* (21 hr to 23 hr) 亥

It is clear that Wei Po-Yang chose his twelve hexagrams so that beginning from the bottom line of the first hexagram *Fu* we have a single full Yang line————, indicating the starting of the fire at the *tzu* double-hour (23 hr to 01 hr). Then Yang lines are added on, one at a time, to the following hexagrams, until *Chhien*, which denotes full intensity, is reached at the *ssu* double-hour (09 hr to 11 hr). The single broken *Yin* line —— —— in the next hexagram, *Kou*, shows that the fire is to be decreased from midday, and each additional broken line in the hexagrams from *Thun* to *Khun* tells us that the cooling is to be continued gradually until complete, as shown by the six broken lines in *Khun*, the final diagram.

Such is the basic meaning of Wei Po-Yang's passage on what afterwards came to be called 'fire-times' (*huo hou*[1]), i.e. the proper times, durations and intensities of heating in the alchemical processes.[b] The theory of the *Book of Changes* is again

---

[a] *Tshan Thung Chhi*, ch. 19. Table 17 on p. 332 of Vol. 2 is to be amended by this list. See further pp. 63 ff. below.

[b] Something was already said about this at a much earlier stage; Vol. 2, pp. 330 ff., 332 ff. Cf. here p. 196.

[1] 火候

employed in explaining the monthly cycle of firing. The six *kua*: *Chen*, *Tui*, *Chhien*, *Sun*, *Kên*, and *Khun*, are used to indicate the 3rd, 8th, 15th, 16th, 23rd, and 30th day of the lunar month respectively. Although the text itself, which we shall translate immediately, speaks in terms of hexagrams, it is more illuminating to represent these in their trigram form[a] as follows (Table 109):

Table 109. *'Fire-times' in elixir preparation; the aegis of trigrams during the lunation*

Day of lunar month		Trigram
3rd	☳	*Chen* 震
8th	☱	*Tui* 兌
15th	☰	*Chhien* 乾
16th	☴	*Sun* 巽
23rd	☶	*Kên* 艮
30th	☷	*Khun* 坤

Looking at the lines of the trigrams one can guess that heating is to start on the 3rd day, to be increased on the 8th, and given maximum intensity on the 15th. The fire is to be decreased on the 16th, and on the 23rd day there should be further cooling until on the 30th day room temperature is reached.

This in fact is just what Wei Po-Yang describes, but in a more abstract way, as shown in the passage below, which may conveniently be illustrated by three cyclical charts (Figs. 1351, 1352, 1353), the work of commentators of the Wu Tai and Sung.[b] The *Tshan Thung Chhi* says:[c]

At the time of the new moon (start heating but) act with moderation. Even when yet in a state of chaos, the male and the female pair together.[d] Moist juices, nourishing and fertilising, flow and penetrate, inducing change (*hua*[1]). The works of Nature are immeasurably marvellous. One should utilise profitably that which will hide the substance and conceal the form.[e]

The cycle of heaven begins in the northeast, the position of (the lunar mansions) *Chi* and *Tou*.[f] Thence the turning of the heavenly bodies towards the right shows forth the (moon)

---

[a] Vol. 2, Table 13.

[b] *TT*996, ch. 5, p. 31*a* (Yü Yen, +1284), *TT*998, ch. 3, p. 2*b* (Chhen Hsien-Wei, +1234), and *TT*994, p. 8*a*,*b* (Phêng Hsiao, +947).

[c] *Tshan Thung Chhi*, ch. 18, pp. 5*a* to 6*a*, tr. auct., cf. Wu & Davis (1), pp. 246ff. We cannot make the translation as sophisticated as it ought to be, but we shall try to convey something of the real spirit of the text.

[d] He probably refers here to the onset of chemical reaction.

[e] This probably refers to the charging of the reaction-vessel with the chemical substances which are to undergo change. But it should be realised that (as the commentaries show) a wealth of Yin-Yang, Five-Element, and *I Ching* theory is concealed within even the simplest statement, such as those which follow.

[f] See Fig. 338 in Vol. 4, pt. 1. The writer is analogising the stages of the reaction with the passage of the moon through the lunar mansions.

[1] 化

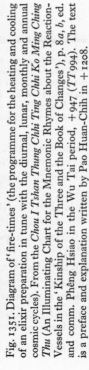

Fig. 1351. Diagram of 'fire-times' (the programme for the heating and cooling of an elixir preparation in tune with the diurnal, lunar, monthly and annual cosmic cycles). From the *Chou I Tshan Thung Chhi Ting Chhi Ko Ming Ching Thu* (An Illuminating Chart for the Mnemonic Rhymes about the Reaction-Vessels in the 'Kinship of the Three and the Book of Changes'), p. 8a, b, ed. and comm. Phêng Hsiao in the Wu Tai period, +947 (TT 994). The text is a preface and explanation written by Pao Huan-Chih in +1208. The circles (reading from inside outwards) are marked as follows:

1 the 5 elements,
2 the 4 seasons,
3 the 12 cyclical characters (*chih*), standing for double-hours, compass-points, etc.,
4a the 12 months of the year, numbered,
4b 12 hexagrams (*kua*) corresponding to the months (and double-hours),
4c the names of these hexagrams in writing. Here Chhien is in the SSE, Khun in the NNW (see Vol. 2, pp. 312ff.),
5 empty,
6 30 moon phases representing a lunation,
7 numbering for these, starting with the new moon at Khan,
8a the 28 lunar mansions (*hsiu*), cf. Vol. 3, pp. 234ff.,
8b *kua* names of 8 trigrams corresponding to the seasons (and spatial directions),
8c the trigrams, and the symbolic names of the 8 directions. Here Chhien is in the NW and Khun in the SW (see Vol. 2, pp. 312ff.).

Note that both Metal and Fire, the summer, noon, the hexagram *Kou* (copulation, reaction, fusion, no. 44), full moon, the trigram *Li* (no. 7), and the red bird of the South, are all at the top. Correspondingly, Water, winter, midnight, the hexagram *Fu* (return, no. 24), the darkest part of the lunation, the trigram *Khan* (no. 4), and the sombre warrior of the North, are all at the bottom.

It is also worthy of note that in circle 4c the arrangement of the hexagrams is like that of the Fu-Hsi system (see Vol. 4, pt. 1, p. 296) in so far as *Chhien* and *Khun* are arranged opposite one another; but in circle 8c the trigrams are arranged exactly as in the Wên Wang system. The opposition (and mutual interpenetration and transformation) of *Khan* and *Li* is a theme of great importance in physiological alchemy, as will appear in Vol. 5, pt. 5.

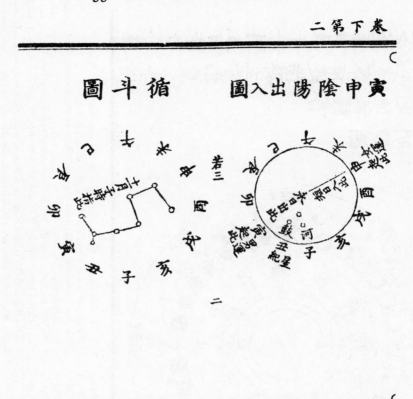

Fig. 1352. Another diagram elucidating 'fire-times'; from the *Chou I Tshan Thung Chhi Chieh* (The 'Kinship of the Three and the Book of Changes' with Explanations), ch. 3, p. 2b, comm. Chhen Hsien-Wei in the Sung period, +1234 (*TT*998). On the left the Great Bear (Pei Tou) surrounded by the 12 cyclical characters (*chih*). The legend says that at midnight in the 11th month the tail points at the character *mao*. On the right a similar diagram indicates the rise and fall of Yin and Yang; the male force beginning to arise at the character *yin* (cf. p. 59), the female force at *shen*.

disc and its shafts of light. Out of the obscure abyss comes this manifestation, the scattering of exquisite rays. When (the moon) comes to (the lunar mansions) *Mao* and *Pi*, *Chen kua* presides, and now the *Yang chhi* begins to be active. Nine–one, during this time the Dragon is in hiding.[a]

The Yang is based on Three and the Yin acts on Eight. Hence the *Chen kua* moves (into action) on the third day (of the moon) and the *Tui kua* operates on the eighth. Nine–two, the Dragon makes its appearance, and now there is a calm and peaceful balance (between Yin and Yang).[b]

On the fifteenth day the virtue (*tê*) of the work is achieved; and the *Chhien kua* presides over the substance formed. Nine–three, watchfulness is in order, for the working of the spiritual powers must now wane.[c]

---

[a] A typical *I Ching* phrase; see Vol. 2, pp. 304ff. and references there quoted. It will be seen in what follows that the *kua* are conceived of not only as symbolic diagrams but as actual forces working in the universe.

[b] This is the halfway point of the heating.   [c] He means that the heating is now to be reduced.

周易參同契發揮

用九翩翩為道規矩陽數已訖訖則復起推
情合性轉而相與循據璇璣昇降上下周流
六爻難以察觀故無常位為易宗祖
九陽數也易曰參天兩地而倚數蓋取五
行之生數天一天三天五參天相倚而成

止五
二十五

Fig. 1353. A third diagram elucidating the 'fire-times'; from the *Chou I Tshan Thung Chhi Fa Hui* (Elucidations of the 'Kinship of the Three and the Book of Changes'), ch. 5, p. 31a, ed. Yü Yen in the late Sung, +1284 (*TT* 996). Again we see the 30 moon phases, 5 of each corresponding to one of the 6 trigrams arranged in a mirror-image of the Fu-Hsi system (Vol. 4, pt. 1, p. 296). Here *Chhien*, corresponding to the south, is actually drawn in the SE, to the left at the top, much as in the hexagram series of Fig. 1351, circles *b, c*; with *Khun* opposite in the NW. In the innermost circle each *kua* carries its description in terms of 'nines' according to the definitions in the *I Ching* (Book of Changes), cf. p. 64. The text on the left proceeds to discuss these.

Growth and decay must gradually come and go; finally there must be a return to the beginning. Now the *Sun kua* takes command, consolidating and controlling. Nine–four, the Dragon may stir, and dangers are on each side of the road.

The *Kên kua* governs motion and rest. Action must be taken at the right time. On the twenty-third day of the month careful watch should be kept. Nine–five, the Dragon is flying, in this position Heaven sends success.

On the thirtieth day of the moon *Khun kua* comes into power to complete the cycle. As the mother of all, it shelters and brings up young things.[a] Nine at the top, the victorious dragon roams with embattled virtue in the wilderness.

The Nines,[b] elegant and exquisite in their workings, are the regulating forces of the Tao. When the Yang numbers are completed everything begins all over again. Knowing the nature of substances and how they combine, they may be submitted to cyclical change so as

[a] He means the products of the reaction.
[b] I.e. the symbols of the *I Ching* and the natural forces which they represent.

to join them in a better way. Following the rotation of the heavens (lit. of the circumpolar constellation template, *hsüan chi*[1]),[a] things rise and fall, ascend and descend. In ceaseless circulation the six lines (*hsiao*[2]) of the hexagrams perform their dance; difficult they are to observe and to investigate. They have no constant position, and this is just what the *Book of Changes* (*I Ching*) is based upon.

The use of the hexagrams and trigrams to explain the diurnal and lunar cycles of heating described above is another classical example of the function of the *Book of Changes* as a universal concept-repository which we have described at length in a previous volume.[b] There we also pointed out how the symbols were employed to denote alchemical substances and apparatus in the *Tshan Thung Chhi*. This we must not repeat here, only referring the reader to the appropriate Section.[c]

What the fundamental alchemical reactions mentioned in the *Tshan Thung Chhi* were is not at all certain, but a prominent place is taken in the text by the 'action between the Dragon and the Tiger (*lung hu*[3])'. The Dragon and the Tiger are generally taken by the *wai tan* school to mean mercury and lead.[d] This would mean amalgamation, but heating might not have been required. An anonymous commentator explains[e] that Dragon and Tiger have two different meanings depending on the context. The Dragon, or, more specifically, the Blue Dragon (*chhing lung*[4]), refers to mercury (*hung*[5]), and the Tiger, or the White Tiger (*pai hu*[6]), to lead (*chhien*[7]). On the other hand, the Dragon can also refer to cinnabar (*tan sha*[8]), and is then denoted, if desired, by the term Red Dragon (*chhih lung*[9]). If the Dragon refers to cinnabar then the White Tiger, or simply the Tiger, should mean mercury.[f] One fairly clear description in the text is that of the solution of lead in mercury, producing Hg (Pb) amalgam. The use of lead is particularly emphasised, for example, in the following passage:[g]

From the very beginning of Yin and Yang, *hsüan*[10] (lead ore)[h] encloses the 'yellow sprout' (*huang ya*[11]). The leader of the Five Metals is the 'river chariot' (*ho chhê*[12]) (lead) of the north. Hence lead is black outside but holds the 'golden flower' (*chin hua*[13]) in its bosom, like someone carrying a piece of jade but looking like a madman dressed in rags. Metal (element), being the mother of Water (element), conceals her offspring; while Water, being the offspring of Metal, is hidden in the womb of its mother. The adept can understand the mysterious (changes); he knows that things cannot always be said to be or not to be. The reaction is like a chaos, now sinking, now floating, advancing, retreating, dispersing, and concocting into bounded entities. First it seems white, treated, it turns red; but the heating affects only the external appearance and whiteness is still within it. Squareness, roundness, and dimensions matter nothing; if there is a mutual embrace within the chaos.

---

a See Vol. 3, pp. 334ff.
c Vol. 2, pp. 330 and 331.
e *TT*994.
g *Tshan Thung Chhi*, ch. 7, p. 16*a*, tr. auct., cf. Wu & Davis (*1*), p. 237.
b Vol. 2, pp. 322ff.
d See Tshao Yuan-Yü (*1*).
f *TT*994, preface, p. 1*a*, *b*.

h Non-magnetic iron oxide ore is the usual meaning of *hsüan shih* (RP 77) but a commentator explains (*TT*990, ch. 1, p. 28*b*) that here it means lead ore.

[1] 璇璣	[2] 爻	[3] 龍虎	[4] 靑龍	[5] 汞	[6] 白虎
[7] 鉛	[8] 丹砂	[9] 赤龍	[10] 玄	[11] 黃芽	[12] 河車
[13] 金華					

The liberal use of synonymic cloak-words is quite conspicuous here. Once again the commentators differ among themselves about interpretations of the terms. Most of them agree that the 'yellow sprout' (*huang ya*) refers to metallic lead smelted from its ore,[a] but Wu & Davis, on the basis of the words *Yin huo pai huang ya chhien*[1],[b] take it to be litharge. They interpret this phrase as meaning that the Yin fire (soft or gentle fire) produces *huang ya* from lead. And it is true that lead heated in air at a moderately elevated temperature, preferably slightly above its melting point, becomes coated with the yellow monoxide, litharge; this might easily be conceived to sprout from the shiny molten metal.[c] Perhaps Wu & Davis are right, but what the passage actually says is 'Yin fire, which is white, and yellow sprouts, which is lead, form the two sevens, and these come together to assist the operator'. One must remember that colour terms in this field are never necessarily the obvious colours of things seen in the laboratory, they may refer to the symbolic colours of the Five-Element system. If we think as the Chinese alchemist did, and consider metallic lead to be under Metal, which is generated by Earth according to the Order of Mutual Production, yellow being the colour of Earth, there is no reason why metallic lead should not have been given the name 'yellow sprout'. Lao Kan (6) agrees broadly with Wu & Davis, but goes further, taking the passage to refer to the making of minium as well as litharge from lead carbonate according to the reactions:

$$PbCO_3 \rightarrow PbO + CO_2 \qquad 6PbO + O_2 \rightarrow 2Pb_3O_4.$$

A third opinion is that of Tshao Yuan-Yü (1) who makes the 'yellow sprout' sulphur; and Yuan Han-Chhing (1) agrees with him, though the latter also recognises the references to lead amalgam in the text. The 'river chariot' (*ho chhê*) is, according to one commentator, lead,[d] while the 'golden flower' (*chin hua*) is said to be metallic lead by another.[e]

The two key passages referring to the production of amalgams are as follows:

The 'flowing pearls' (*liu chu*,[2] i.e. mercury), (which comes from the essence of) Thai Yang,[3] has a tendency to escape from man.[f] Eventually when (they) get the 'golden flower' (*chin hua*, metallic lead, or gold) (they) turn and react with it, melting into a white paste or solidifying into a mass.[g] It is the 'golden flower' that first undergoes change (lit. sings), (for) in a few moments it melts into a (viscous) liquid. (The two substances now fuse together and) assume a disorderly appearance like coral or horse-teeth.[h] The (essence of) Yang then comes forth to join it, and the nature of things is now working in harmony. Within a brief interval of time (the two substances) will be confined within a single gate.[i]

---

[a] For example in *TT* 996, ch. 3, p. 3 *a*.
[b] Found in ch. 33, p. 10 *a*.
[c] Wu & Davis (1), p. 283.
[d] *TT* 990, ch. 1, p. 29 *a*.
[e] *TT* 996, ch. 3, p. 3 *a*. However, Yoshida (5) interprets it as gold, see Yabuuchi (25), p. 206.
[f] This is mercury according to *TT* 990, ch. 1, p. 33 *b* and *TT* 996, ch. 3, p. 15 *a*.
[g] Depending on the proportions.
[h] See the commentary by Yuan Jen-Lin in *Ku Wên Chou I Tshan Thung Chhi Chu*, ch. 3, p. 4 *b*.
[i] *Tshan Thung Chhi*, ch. 24, p. 17 *a*, tr. auct.; cf. Wu & Davis (1), p. 252.

[1] 陰火白黃芽鉛   [2] 流珠   [3] 太陽

And again:

The 'elegant girl by the riverside' (*ho shang chha nü,*[1] mercury) is a numinous thing and profoundly mysterious.[a] In the fire she flies away and none can see the traces of her path.[b] Like the vanishing of a spirit or the concealment of a dragon her whereabouts cannot be known. In order to fix her (one may use) the 'yellow sprout' (*huang ya*) as the root.[c]

The kind of reaction described in the second paragraph depends on our interpretation of 'yellow sprout'. If we go by the words of the *Tshan Thung Chhi* commentators it is a case of amalgamation, but if, with Yuan Hang-Chhing (*1*), we take the 'yellow sprout' as sulphur, then we have the formation of mercuric sulphide. But Yoshida (*5*) thinks that the basic reaction in the whole of the *Tshan Thung Chhi* is the making of gold amalgam.

The *Tshan Thung Chhi* mentions one important fundamental concept of alchemy, namely that changes happen if there is similarity (*thung lei hsiang pien*[2]), but will not occur if there is dissimilarity (*i lei pu nêng hsiang chhêng*[3]). This is part of the most ancient history of the idea of chemical affinity, and we shall return to it in more detail at a later stage (pt. 4). The following passage is of interest not only because it explains this concept but also gives further evidence that the idea of the transmutation of gold was prevalent during the time of Wei Po-Yang:[d]

White lead (*hu fên*[4]), on being placed in the fire, becomes discoloured and changes back into lead.[e] In a hot liquid ice and snow turn into water (*thai hsüan*[5]). Cinnabar is the chief possessor of the Metal (element) for its natural endowment is to produce mercury. Transformations (*pien hua*[6]) depend on the true nature of the substances; their beginnings and endings have a mutual causation. The way to become an immortal through consuming (medicines) lies in the use of substances of similar category. Grains are used for raising crops, hen's eggs are used for hatching chickens. With substances of similar category to help Nature, the formation of things is easily moulded (or manipulated, *thao yeh*[7]). But fish eyes cannot replace pearls,[f] neither can wild raspberry or mugwort leaves be used for tea (*chia*[8]).[g] Things of the same category go together: precious substances cannot be made with incorrect procedures or wrong materials. This explains why swallows and sparrows do not give birth to the phoenix, and why foxes and rabbits do not produce horses. Flowing water does not heat what is above it, nor does a fire moisten what is underneath it.

---

[a] This is mercury according to the commentator of *TT* 993; see ch. 2, p. 25 *a*.

[b] A clear reference to distillation, or at least volatility.

[c] *Tshan Thung Chhi*, ch. 26, p. 20 *a*, tr. auct. Cf. Wu & Davis (*1*), p. 254; Liu Tshun-Jen (*1*), p. 83, seems even more off the rails.

[d] *Tshan Thung Chhi*, ch. 12, p. 25 *b*, tr. auct.; cf. Wu & Davis (*1*), p. 241; Liu Tshun-Jen (*1*), p. 82.

[e] *Hu fên* is lead carbonate, $PbCO_3$, white lead, according to RP 12 and general usage, but Yuan Han-Chhing (*1*), p. 176, regards it as minium, Pb_3O_4, red lead; and accordingly sees the effect as a reduction to the metal with release of CO_2. Yoshida even interprets *hu fên* as meaning a compound of mercury, (*5*), p. 205.

[f] See, however, Vol. 4, pt. 3, p. 677.

[g] The oldest references to tea-drinking (cf. Vol. 6) are of the +3rd century, but Kuo Pho, explaining this name in the *Erh Ya* (−4th cent.), says it was the same bush as *khu thu*,[9] and that a drink was made from its leaves (cf. B II, 292, 307).

| [1] 河上妊女 | [2] 同類相變 | [3] 異類不能相成 | [4] 胡粉 | [5] 太玄 |
| [6] 變化 | [7] 陶冶 | [8] 檟 | [9] 苦茶 | |

This important notion is emphasised again in another section of the *Tshan Thung Chhi*. The following passage also shows how much Wei Po-Yang appreciated natural regularities, convinced that the ultimate result of any operation depended chiefly on the use of proper materials and correct techniques, not on miracles.[a]

Let there be two very beautiful maidens living in the same house. Also let there be a Su Chhin[1] to start the match-making, and a Chang I[2] to act as the go-between.[b] Their persuasion and praise might succeed in bringing about a union, but the two could never be man and wife even if they lived till their hair fell off and their teeth dropped out. So also when substances of the wrong name, nature and category are used, or when the proper proportions and mixing have all gone astray (*shih chhi kang chi*[3]),[c] there can be no success. Under such conditions, even with Huang Ti to stoke the furnace, with Thai I to control the heating, with the Eight Venerable Masters (Pa Kung) to pound the materials and supervise the operation, with the Prince of Huai-Nan himself to do the mixing; failure will still be inevitable. Even if you build a laboratory with a high platform using jade for the steps, even if you sacrifice the fat of the unicorn and the phoenix, even if you make long prostrations after purificatory ablutions and fastings, praying to the spirits of Heaven and Earth, and imploring the aid of ghosts and demons, hoping to have some success to look forward to—all, all your efforts will be in vain. It is like trying to mend a metal reaction-vessel with glue, or to heal a boil by applying sal ammoniac (*nao*[4]),[d] or to get rid of cold with ice or heat with hot soup; these things are just as absurd and impossible as flying tortoises and dancing snakes.

This wonderful passage is interesting not only for the theory of action between substances of the same category, but also from two other points of view. First, it gives a glimpse of the ceremonies observed by the Taoist alchemists in the days of Wei Po-Yang before engaging in their experimental operations. Secondly, together with the previous passage, it brings out a striking feature in Wei Po-Yang's writing, namely his inclination to use one metaphorical phrase after another to illustrate his points and describe his operations. We find these particularly numerous in connection with the Dragon and the Tiger, for example, phrases like 'endowment by the Yang and acceptance by the Yin' (*Yang ping Yin shou*[5]),[e] and 'mutual need between the male and the female' (*hsiung tzhu hsiang hsü*[6]).[e] In +1284 Yü Yen collected the following list of such phrases saying that they all referred to the same thing.[f] This was the *conjunctio oppositorum*, whether proto-chemical or physiological.

'sexual union between the Dragon and the Tiger'   (*lung hu chiao kou*[7])
'combination of *Chhien* and *Khun*'   (*Chhien Khun phei ho*[8])
'combination of the Metal and Wood (elements)'   (*chin mu chiao ping*[9])

---

[a] *Tshan Thung Chhi*, ch. 30, p. 26*b*; cf. Wu & Davis (1), p. 256.

[b] A reference to the two great protagonists of the Diplomacy School in the Warring States period (cf. Vol. 2, p. 206; Vol. 4, pt. 3, p. 266).

[c] Recognition of the importance of careful weighing of the reactants goes back a long way in Chinese proto-chemistry. On the expression *kang-chi* or *chi-kang* cf. Vol. 2, pp. 554 ff.

[d] Pronounced *lu*, this character means simply sand. But some *Tshan Thung Chhi* texts write *nao*[10] for this, which is the more correct term for ammonium chloride (RP 126). We shall return presently to sal ammoniac, meanwhile see Stapleton (1) who thought it was first recognised and named in China, *vs.* Laufer (1), pp. 503 ff., who urged a Persian origin for the name.

[e] See *TT* 996, ch. 5, p. 3*a*.   [f] See *TT* 996, ch. 5, pp. 2*a* ff.

| [1] 蘇秦 | [2] 張儀 | [3] 失其綱紀 | [4] 硇 | [5] 陽稟陰受 |
| [6] 雌雄相須 | [7] 龍虎交媾 | [8] 乾坤配合 | [9] 金木交併 | [10] 砫 |

'entwining of the tortoise and the serpent'	(*kuei shê phan tou*[1])
'mutual attraction between the red and the black'	(*hung hei hsiang thou*[2])
'intermixture of heaven and earth'	(*thien ti chiao thai*[3])[a]
'mixture of the black and the yellow'	(*hsüan huang hsiang tsa*[4])
'mixing and fusion of Metal and Earth (elements)'	(*chin thu hun jung*[5])
'union of the red and the white'	(*chhih pai hsiang chiao*[6])[b]
'the crow and the hare in the same cave'	(*wu thu thung hsüeh*[7])
'conjugal felicity'	(*fu fu huan ho*[8])
'the moon and the sun in the same palace'	(*jih yüeh thung kung*[9])
'reunion of the Herd-boy and the Weaving Girl'	(*Niu Nü hsiang fêng*[10])[c]
'the male and female following one another'	(*phin mou hsiang tshung*[11])
'the harmony between *hun* and *pho*' (Yang and Yin 'souls')	(*hun pho hsiang thou*[12])[d]
'water and earth in the same village'	(*shui thu thung hsiang*[13])
'gold and mercury in the same reaction-vessel'	(*chin hung thung ting*[14])
'metal and fire sharing the same furnace'	(*chin huo thung lu*[15])

The numerous metaphors in Wei Po-Yang's writing cover a wide field of knowledge, ranging from some rudiments of embryology and physics to the sexual techniques practised in his day. The two passages below deal with embryology. For example:[e]

The body of man has a natural endowment, but it is a nothingness, and the primal essence gathers like a cloud. The causal *chhi* makes a beginning with the Yin and Yang for its measures. The *hun* and *pho* 'souls' begin to dwell in it; the former derived from the Yang spirit of the sun and the latter from the Yin spirit of the moon. . . . At first (the embryo) is like a hen's egg, black and white fitting together like a tally. It measures an inch in breadth and length. The four limbs, the five viscera, the muscles and bones, are then added. After ten months it comes out of the womb. Its bones are weak and flexible, and its flesh is as smooth as lead (*chhien*[16]).[f]

Or again:[g]

The mother harbours the nourishing fluids and the father is the donator (of what is nourished).[h]

    a When one of us (H. P. Y.) was working on this in 1958, he read in the press about the condition of 'white-out' encountered by the Fuchs Expedition in the Antarctic, when the horizon becomes indistinguishable.
    b This is a physiological, sexual and embryological metaphor. The ancients discussed which parts of the body were formed from red blood and which from white semen (cf. Needham, 2).
    c Two prominent asterisms (cf. Vol. 3, p. 282 and *passim*), Vega and Altair.
    d Or, more basically, the 'light' and 'heavy' souls; cf. Vol. 2, pp. 22, 490, and, more extensively, pt. 2, pp. 85ff. above.
    e *Tshan Thung Chhi*, ch. 20, p. 11*a*, tr. auct., cf. Wu & Davis (1), p. 249.
    f For this word *TT*996, ch. 6, p. 13*a* gives 'sweetmeats' (*i*[17]), hence Wu & Davis (1), p. 249 said 'soft candy'.
    g *Tshan Thung Chhi*, ch. 25, p. 19*a*, tr. auct., cf. Wu & Davis (1), p. 253.
    h This sounds very like the ancient Western theories which minimised the function of the maternal organism, not understanding the role of the egg-cell (see Needham, 2).

[1] 龜蛇蟠虯	[2] 紅黑相投	[3] 天地交泰	[4] 玄黃相雜
[5] 金土混融	[6] 赤白相交	[7] 烏兔同穴	[8] 夫婦歡合
[9] 日月同宮	[10] 牛女相逢	[11] 牝牡相從	[12] 魂魄相投
[13] 水土同鄉	[14] 金汞同鼎	[15] 金火同爐	[16] 鉛

[17] 飴

The burning-mirror and the dew-pan are mentioned in the following:[a]

The sun-mirror is used for making fire, but without the sun no light (-rays) would be produced. And the *fang chu*[1] mirror, how could it produce dew (by condensation) without the stars and the moon?

Of a prevailing technique of sexual intercourse Wei Po-Yang writes:[b]

In the natural workings of Heaven and Earth there is a spontaneity (*tzu-jan*[2]).[c] When fire burns heat rises up, when water moves it wets what is below. They need no teacher; it happens spontaneously. So it has been from the beginning; no one can change it. Look what happens when the moment comes for intercourse between a man and a woman. The hard and the soft inextricably intertwine and cannot be separated. They fit like a tally, and no special skill is involved in the management of it. Man is born to lie facing downwards and woman on her back. These modes of behaviour are derived from the very beginning of their existence in the womb, and are manifested not only during their lifetime but also after their death.[d] They have not been taught this by their parents: such things are already fixed at the time of their conception during sexual union.

Wei Po-Yang gives other descriptions of alchemical processes. At the end of the book we find an account of the reaction-vessel in the section entitled 'Mnemonic Rhyme of the Reaction-Vessel' (*ting chhi ko*[3]).[e] Of alchemical processes he writes:[f]

The 'Book of the Firing-Process' (*Huo Chi*[4]) was not written without a purpose, and can be understood in the light of the 'Book of Changes'. The reaction-vessel (*ting*[5]) and the furnace (*lu*[6]) resemble the crescent moon (half of which belongs to Yin and the other half to Yang) lying on its back. The 'white tiger' (possibly lead) forms the prime constituent in this heating (process). Mercury (*hung*[7]), (representing the) sun, is the 'flowing pearls' (*liu chu*[8]). The 'blue dragon' (*chhing lung*,[9] also mercury) goes together with (the 'white tiger') like the east merging with the west, or the *hun* and *pho* 'souls' capturing each other. At the first quarter of the moon (the trigram) *Tui*[10] numbers eight (i.e. take 8 oz. of mercury),[g] while at the last quarter of the moon (the trigram) *Kên*[11] also numbers eight (i.e. take 8 oz. of lead).[g] The essences of the two quarters combine (i.e. mercury and lead), so that the body of *Chhien* and *Khun* (i.e. the amalgam or elixir) is formed. Twice eight (i.e. 16 oz.) make one lb. The Tao of the 'Book of Changes' is always true and unmistakable.

The existence of the *Huo Chi* has long been a matter of controversy. Chu Yuan-Yü in +1669, the commentator of the *Tshan Thung Chhi Chhan Yu*, doubted that such a book ever existed. But there is no reason for thinking that a book of practical

[a] *Tshan Thung Chhi*, ch. 21, p. 13*b*, tr. auct., cf. Wu & Davis (1), p. 250. For both kinds of mirrors see Vol. 4, pt. 1, pp. 87, 89.

[b] *Tshan Thung Chhi*, ch. 26, p. 20*b*, tr. auct.; cf. Wu & Davis (1), p. 254; Liu Tshun-Jen (1), p. 84. Of course ancient Chinese sexology was much more complicated than this, as a Taoist like Wei Po-Yang would have known very well, but he was taking a simple example in illustration of the spontaneous. Cf. van Gulik (3, 8), and Vol. 2, pp. 146ff.

[c] See Vol. 2, pp. 50ff.

[d] Commentators say that people floated this way if drowned; cf. p. 21*b*.

[e] This is translated in connection with apparatus in pt. 4 below, from Ho Ping-Yü & Needham (3).

[f] *Tshan Thung Chhi*, ch. 9, p. 20*b*, tr. auct.; cf. Wu & Davis (1), p. 239.

[g] According to the commentator of *TT*995, ch. 2, p. 13*b*.

[1] 方諸	[2] 自然	[3] 鼎器歇	[4] 火記	[5] 鼎	[6] 爐
[7] 汞	[8] 流珠	[9] 青龍	[10] 兌	[11] 艮	

direction with this title did not circulate among the Han alchemists. The question has again been raised by Yoshida (5), who suggests it was a Zoroastrian text. He bases this hypothesis on a superficial similarity between the Yin and Yang system and Persian dualism, believing that Manichaeism came to the east about the +3rd century, and that its influence can be found reflected in Taoist ceremonies as well as in the traditional concept of Yin and Yang, the four seasons and the Five Elements.[a] However, we have already declared our agreement with Waley in his rejection of any direct influence.[b] Zoroastrianism and Manichaeism did not arrive in China from Persia until the +6th and the end of the +7th century respectively.[c] So until a Zoroastrian text with some similarity to the *Tshan Thung Chhi* can be found, the existence of any influence of this kind remains to be proved.

There is only one passage in which Wei Po-Yang waxes enthusiastic about the elixir of life as so many later writers did. In ch. 11 he included the following words which may be translated as a poem:[d]

If even the herb *chü-shêng*[1] can make one live longer[e]
Surely the elixir is worth taking into one's mouth,[f]
Prepared as it is by cyclical transformations?
Gold by its nature does not rot or decay
Therefore it is of all things the most precious.
If the chymic artist includes it in his diet
The duration of his life will become everlasting.
(Element) Earth endures through the Four Seasons
Keeping its bounds as if fixed by compass and square.
When gold and cinnabar permeate the Five Entrails
A fog is dispelled, like rain-clouds scattered by wind,
Fragrant exhalations pervade the four limbs,
The countenance beams with well-being and joy.
Hairs that were white all turn to black.
Teeth that had fallen grow in their former place.
The old dotard is again a lusty youth,
The decrepit crone is again a young girl.
He whose form has changed escapes the perils of life
And has for his title the name of True Man (*chen jen*[2]).

The next passage must have been regarded by aspirants to immortality as one of prime importance, as it describes the very process of how an elixir was made, but unfortunately Wei Po-Yang does not state explicitly what the two main ingredients

[a] In Yabuuchi (25), pp. 208 and 209.
[b] Waley (4), p. 112. See also Vol. 1, p. 154, Vol. 2, pp. 273 ff.   [c] See Vol. 1, p. 128.
[d] P. 14a, tr. Waley (14), mod. auct. Cf. Liu Tshun-Jen (1), p. 81.
[e] The identity of this plant is a difficult question (cf. p. 97); we discuss it in Vol. 6, Sect. 38.
[f] *Huan tan kho ju khou*.[3] Note the early appearance of this phrase so common in later times, as also elsewhere (e.g. ch. 14, p. 31a) many typical terms such as *tao kuei*,[4] 'a large knife-point', of some substance or other. We were still using this in my student days of practical biochemistry under Sydney W. Cole. Like so many other chemical terms, this also came to have an esoteric physiological significance (cf. pt. 5).

[1] 苣勝   [2] 眞人   [3] 還丹可入口   [4] 刀圭

really are. Instead he calls them by the names *chin*[1] and *shui*[2] respectively, both of which can be interpreted in several ways. Wu & Davis simply take their most common modern meanings and call them gold and water,[a] but these could not produce the phenomenon described in the text. It seems reasonably safe to translate these two words as the Metal element and the Water element respectively, but again one has to make a guess as to what these two actually represent. An anonymous commentator says that *chin* is the abbreviated form of *chin hua*,[3] which is metallic lead, and *shui* is the short form for *shui yin*,[4] mercury.[b] The text says:[c]

Chin (the metal) is used as an embankment (to prevent mercury from escaping), so that *shui* (mercury) can be put in and run about freely. The amount of *chin* is fifteen (oz.) and so is the amount of *shui* (mercury). Weighings should be made when the furnace is about to be heated. An excess amount of *shui* (mercury) by half should be used. These are the two genuine substances. The weight of *chin* will be the same as it was originally. A third (substance) therefore does not come in. But when fire (which is also represented by the number) Two is introduced these three will interpenetrate each other and marvellous changes (*pien hua*[5]) will take place. Below (the reaction-vessel) is the *chhi* of Thai-Yang (i.e. the fire). After a short time of heating (lit. steaming, *chêng*[6]) first liquefaction and then solidification take place. (The substance thus formed) is called the 'yellow carriage' (*huang yü*[7]). As the time (lit. month and year) draws to a close, the nature (of the original substances) is destroyed and their life shortened. (Eventually a transformation of) their form and matter comes about giving a sort of powdery ash, resembling 'bright window dust'.[d]

(The substance) is ground, mixed well and enclosed (in another reaction-vessel) before being introduced into the opening of a red(-hot furnace). Attention should be paid to the sealing of the edges of the container so as to keep the whole intact without leaking. The dazzling flame plays below, making a noise both day and night. At the start the flame should be gentle so as to be controllable, but eventually its strength should be increased until it reaches maximal intensity. The regulation of the temperature should be watched over with the greatest care. There are twelve periods in the (diurnal) cycle.[e] At the end of each period one should be particularly careful. When the *chhi* (i.e. the fire) is about to be let down, the (original) bodies have been killed, and the *hun* and *pho* 'souls' have disappeared (i.e. the substances have changed their nature). The colour has already turned purple, and thus is the 'cyclically-transformed elixir' (*huan tan*[8]) achieved. This is then made into pills which can be taken, and is magically effective even if (only) a knife-point of it is administered.

The reaction-vessels described in the *Tshan Thung Chhi* will be discussed in the subsection on laboratory equipment.[f] As for the explanation of what was actually going on, certainty will never be possible, but there are several statements which may be significant. Amalgams of mercury with gold or with lead may be referred to in the first paragraph, and there may also be some reference to the cupellation of gold.[g]

---

[a] (1), p. 243.   [b] See *TT*995.
[c] *Tshan Thung Chhi*, ch. 14, pp. 30*b*, 31*a*, tr. auct.; cf. Wu & Davis (1), p. 243; Liu Tshun-Jen (1), p. 87.
[d] The best interpretation of this is as a reference to the motes dancing in sunbeams from windows. We shall return to the subject on p. 149 below.
[e] The double-hours. See p. 61 above.   [f] See pt. 4 below.   [g] Cf. pt. 2, p. 57 above.

[1] 金   [2] 水   [3] 金花   [4] 水銀   [5] 變化   [6] 蒸
[7] 黃輿   [8] 還丹

The 'cyclically-transformed elixir' is a particularly important term, for in the light of subsequent Chinese alchemy it would refer to the synthesis of vermilion, mercuric sulphide, by the sublimation of mercury and sulphur. Indeed, one asks oneself whether any other substances were involved at all in the process described, so that the Metal Element would stand for sulphur and the Water Element for mercury. The elaboration would then have consisted mainly in different and varying conditions of the reaction-vessel and the heating. At any rate, this seems to be the first appearance of the classical term *huan tan*.[1]

To sum up, the *Tshan Thung Chhi* coined a style and terminology which dominated the earlier Chinese alchemists. Wei Po-Yang's application of the Yin and Yang concept, the theory of the Five Elements and that of the 'Book of Changes' to alchemy, influenced the philosophical thinking of his followers for centuries afterwards. He therefore opened a new field, but at the same time set a limitation to its development. From the large number of commentaries alone one can see that no other text on the same subject was so much studied both by the laboratory alchemists and the *nei tan* school. Secondly, the use of mercury and lead (or sulphur) as the prime sources of the elixir limited the scope and potentialities of later experimentation and gave rise to numerous cases of poisoning; indeed it is quite possible that many of the most brilliant and creative alchemists fell victim to their own experiments by taking dangerous elixirs.[a] Thirdly, Wei Po-Yang's concept of natural regularities was among the earliest beginnings of affinity theory, as is witnessed by his saying that 'changes happen in similarity but not in dissimilarity'. He also has another statement: 'it is easy to work on things of similar category, but with things not of similar categories it is difficult to display any subtle skill; and that is the whole secret of the art'.[b] Lastly, from the *Tshan Thung Chhi* we can understand one of the inhibiting factors for the spread of alchemical and chemical knowledge in ancient and medieval China. Books on alchemy were written in the most obscure manner, piling synonym on synonym and metaphor on metaphor, with cryptograms added in the case of the *Tshan Thung Chhi*, thus giving rise to scores of possible interpretations. Did their writers seek to protect against danger sometimes by purposive obfuscation? Moreover, the alchemist was warned to keep his knowledge to himself; the Venerable Masters did not encourage him to discuss the matter with others. A striking example is seen towards the end of the *Tshan Thung Chhi*, where we read:[c]

Those who love the Tao trace things to their roots. They carefully observe the Five Elements to determine the weights (of the materials used). Profound reflection should be made, but no discussion with others is necessary. The secrets should be carefully guarded, and the knowledge should not be handed down in writing.

Such precepts, discouraging free discussion and free circulation of knowledge, must obviously be considered an important factor hampering the progress of alchemy

---

a Elixir poisoning will be discussed in detail in Vol. 6, Section 45.
b *Tshan Thung Chhi*, ch. 32, p. 5a.
c *Tshan Thung Chhi*, ch. 33, p. 10b.
1 還丹

and chemistry in China. Of course their counterparts in the West were no more enlightened until the dawn of the scientific revolution.

Although a whole sub-section has been devoted to Wei Po-Yang, some contemporary evidence beyond that already given (pp. 40 ff. above) on elixir-making in the +2nd century does not come amiss. Therefore we may end with a note on the *Thai Phing Ching*[1] (Canon of the Great Peace and Equality), that fascinating book of Taoist religion and social philosophy which was being put together around +150. Near the beginning we find a list of twenty-four methods for attaining *hsien*-ship orally handed down,[a] and of these the tenth is 'consuming the Flowery Elixir (*fu hua tan*[2])',[b] and the twentieth 'making white silver and purple gold (*tso pai yin tzu chin*[3])'.[c] Coming as it did from a popular milieu, nothing could show better than this how widely macrobiotic alchemy and the consumption of mineral elixirs had become known by the middle of the +2nd century.

(3) KO HUNG, SYSTEMATISER OF CHINESE ALCHEMY (*c.* +300), AND HIS TIMES

(i) *Fathers and masters*

The Chinese alchemists of the Later Han period were so widely known in East Asia that at least one of them featured in the history of Annam. The *Đai-Viêt Sú-ký Toàn-thú*[4] (The Complete Book of the History of Annam), written by Ngô Si-Liên[5] about +1479, records the healing of the Annamese king Si-vúóng[6] during the +2nd century by the Chinese alchemist Tung Fêng,[7] who is mentioned in the hagiography of the immortals ascribed to Ko Hung,[8] the *Shen Hsien Chuan*[9].[d] This official Annamese history says:

In a *bính-ngo*[10] (*ping-wu*) year, his 40th (+187), the King passed away. Once before the king had been unconscious in an illness for three days. The (Chinese) immortal adept, Tung Fêng, administered to him a pill in some water, and supporting him by the head, shook him about to revive him. After a short while (the king) opened his eyes and moved his hands, while gradually his complexion returned to normal. The next day he presently sat up, and on the fourth day he could speak; thus eventually he recovered.[e]

Certain links between some of the alchemists are suggested by Taoist texts and even by the official histories. The adept Yin Chhang-Shêng[11] occupies a very important

[a] Ch. 1, (p. 8).
[b] As will be seen very shortly, this was precisely the name (inverted) of the first important elixir discussed in the *Pao Phu Tzu* book, ch. 4, a century and a half later (cf. p. 90). It was Lao Kan (6) who first called attention to this passage.
[c] Another of the twenty-four, the eighteenth, has *chen*[12] instead of *tso*, perhaps 'to make spells with' the artificial metals or the real; cf. pt. 2, pp. 204, 244. On purple gold, see pt. 2, pp. 257 ff. above.
[d] Ch. 5.
[e] See ch. 3, p. 7b. The same source tells us that Si-vúóng's grave was excavated more than 160 years later and little sign of putrefaction was found in the corpse. This could have been the result of having taken elixirs containing mercury or arsenic. Cf. pt. 2, pp. 298, 304. The reliability of the *Đai-Viêt Sú-ký Toàn-thú* is discussed in Ho Ping-Yü (7), whose translation this is.

[1] 太平經   [2] 服華丹   [3] 作白銀紫金   [4] 大越史記全書
[5] 吳士蓮   [6] 士王   [7] 董奉   [8] 葛洪   [9] 神仙傳
[10] 丙午   [11] 陰長生   [12] 鎮

position in perhaps the mainstream of the line of descent of Chinese alchemists. As we mentioned earlier (p. 43), he was said to have been a disciple of Ma Ming-Shêng,[1] who was supposed to have acquired the art of immortality from An Chhi shêng[2] (Master An Chhi) himself.[a] Among those to whom Yin Chhang-Shêng was believed to have imparted his knowledge were, besides Wei Po-Yang,[3] Lü Tzu-Hua[4] and a certain Chu hsien-sêng[5] (Mr Chu). Historians recorded that Lü Tzu-Hua eventually took a potion of a 'rainbow elixir' (hung tan[6]),[b] having had among his disciples Pao Ching,[7] a famous scholar of alchemical interests and the father-in-law of Ko Hung[8].[c] Pao Ching, who was for some time Governor of Nan-hai, also had two other distinguished disciples, Wu Mêng[9] and Hsü Mai.[10] The former went on to study under the adept Ting I,[11] and was later recognised as one of the leading figures of the Taoist Church; while the latter became a friend of the celebrated Chin calligrapher Wang Hsi-Chih[12] (+321 to +379), and the Taoist Yang Hsi[13] (+330 to +387), to whom is attributed the Shang-Chhing[14] Taoist scriptures.[d] The Chin Shu records that Hsü Mai and Wang Hsi-Chih used to go out together to search for plants and minerals for the making of elixirs with no regard for the distance they had to travel.[e]

Returning to the other disciple of Yin Chhang-Shêng, that is Chu hsien-sêng, we are told that he handed down his knowledge to Wang Ssu-Chen[15] (fl. +180), who in turn taught Ko Hsüan[16] (+164 to +244).[f] Now Ko Hsüan was Ko Hung's great-uncle, who had also learnt the art from Tso Tzhu[17] (or Tso Yuan-Fang[18]), the famous magician and alchemist of the Three Kingdoms period. Said to be a disciple of the adept Li Chung-Fu,[19] he roamed the Mao-Shan[20] mountains in search of cinnabar for making the 'nine flower elixir' (chiu hua tan[21]).[g]

Than Ssu-Hsien[22] afterwards wrote a biography of Ko Hsüan, entitled Thai-Chi Ko Hsien-Ong Chuan[23] (Biography of the Supreme-Pole Elder-Immortal Ko), which is in the Tao Tsang.[h] Ko Hsüan handed down his knowledge to Chêng Yin[24] (also known as Chêng Ssu-Yuan,[25] c. +220 to +300), and the latter became the teacher of Ko Hung. Table 110 shows at a glance the descent and connections of the alchemists so far mentioned. In Ko Hung's own book he tells the story of this transmission, giving

[a] Cf. pt. 2, p. 106 above.   [b] See TT 293, ch. 17, p. 4a.

[c] See Chin Shu, ch. 95, p. 10a; YCCC, ch. 106, p. 29a. It remains an open question whether or not Pao Ching (Pao Thai-Hsüan[26]) was identical with the outstanding radical, almost socialist, thinker Pao Ching-Yen,[27] with whom Ko Hung had discussions reported in PPT/WP, ch. 48. On these see Vol. 2, pp. 434ff. Pao Ching's daughter, Pao Ku,[28] was also skilled in alchemy, like her husband.

[d] See Wieger (6), p. 16.   [e] See Chin Shu, ch. 80, p. 5a.

[f] Also known by the names Ko Hsiao-Hsien[29] and Ko Hsien-Ong.[30]

[g] See TT 773, ch. 2, p. 19b. Both the names of Tso Tzhu and Ko Hsüan are mentioned in the Tunhuang MS. no. S2070 in the British Museum; a list of magician-technicians (fang shu[31] experts) with brief bibliographical details about them.   [h] TT 447.

[1] 馬鳴（明）生	[2] 安期生	[3] 魏伯陽	[4] 呂子華	[5] 朱先生
[6] 虹丹	[7] 鮑靚	[8] 葛洪	[9] 吳猛	[10] 許邁
[11] 丁義	[12] 王羲之	[13] 楊羲	[14] 上清	[15] 王思眞
[16] 葛玄	[17] 左慈	[18] 左元放	[19] 李仲甫	[20] 茅山
[21] 九華丹	[22] 譚嗣先	[23] 太極葛仙翁傳		[24] 鄭隱
[25] 鄭思遠	[26] 鮑太玄	[27] 鮑敬言	[28] 鮑姑	[29] 葛孝先
[30] 葛仙翁	[31] 方術			

Table 110. *Chart attempting a filiation of alchemists in Han and Chin*
(−2nd to +4th century)

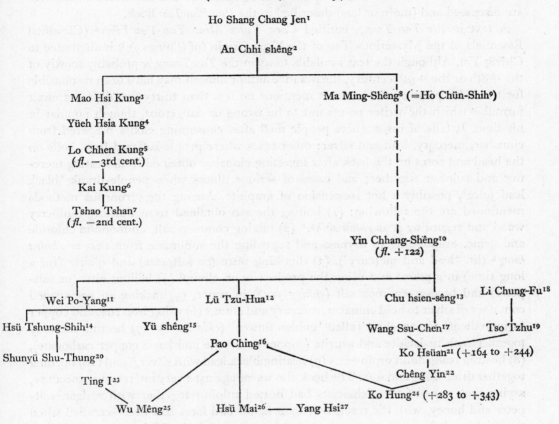

Ho Shang Chang Jen[1]

An Chhi shêng[2]

Mao Hsi Kung[3]

Yüeh Hsia Kung[4]

Lo Chhen Kung[5]
(*fl.* −3rd cent.)

Kai Kung[6]

Tshao Tshan[7]
(*fl.* −2nd cent.)

Ma Ming-Shêng[8] (=Ho Chün-Shih[9])

Yin Chhang-Shêng[10]
(*fl.* +122)

Wei Po-Yang[11]   Lü Tzu-Hua[12]   Chu hsien-sêng[13]   Li Chung-Fu[18]

Hsü Tshung-Shih[14]   Yü shêng[15]   Wang Ssu-Chen[17]   Tso Tzhu[19]

Shunyü Shu-Thung[20]   Pao Ching[16]   Ko Hsüan[21] (+164 to +244)

Ting I[23]   Chêng Yin[22]

Wu Mêng[25]   Hsü Mai[26] —— Yang Hsi[27]   Ko Hung[24] (+283 to +343)

N.B. This table of filiation is continued in Table 112 below (p. 121).

Entries earlier than the +1st century are mostly very legendary in character.

[1] 河上丈人	[2] 安期生	[3] 毛翕公	[4] 樂瑕公
[5] 樂臣公	[6] 蓋公	[7] 曹參	[8] 馬鳴生
[9] 和君實	[10] 陰長生	[11] 魏伯陽	[12] 呂子華
[13] 朱先生	[14] 徐從事	[15] 虞生	[16] 鮑靚
[17] 王思眞	[18] 李仲甫	[19] 左慈	[20] 淳于叔通
[21] 葛玄	[22] 鄭隱	[23] 丁義	[24] 葛洪
[25] 吳猛	[26] 許邁	[27] 楊羲	

the titles of certain precious texts that were handed down to him.[a] These were: *Thai-Chhing Tan Ching* [1] (Manual of the Grand-Purity Elixir), *Chiu Ting Tan Ching* [2] (Manual of the Nine Reaction-Vessels Elixir) and a *Chin I Tan Ching* [3] (Manual of the Potable Gold Elixir). These do not exist as such now but the elixirs themselves are discussed and (more or less) described in the *Pao Phu Tzu* book.

A text in the *Tao Tsang*, entitled *Chen Yuan Miao Tao Yao Lüeh* [4] (Classified Essentials of the Mysterious Tao of the True Origin (of Things)),[b] is attributed to Chêng Yin. Although the text available to us in the *Tao Tsang* is probably mostly of the +8th or the +9th century, the putative author himself may have been responsible for the older parts of the book. It mentions no less than thirty-five different elixir formulae which the writer points out to be wrong or dangerous, though popular in his time. It tells of cases where people died after consuming elixirs prepared from cinnabar, mercury, lead and silver; other cases where people suffered from boils on the head and sores on the back after ingesting cinnabar obtained from heating mercury and sulphur together; and cases of serious illness when people drank 'black lead juice', possibly a hot suspension of graphite. Among the erroneous methods mentioned are the following: (1) boiling the ash obtained from burning mulberry wood and regarding it as *chhiu shih* [5],[c] (2) mixing common salt, ammonium chloride and urine, evaporating to dryness and regarding the sublimate from that as *chhien hung* [6] (lit. 'lead and mercury'), (3) digesting nitre (or saltpetre) and quartz (for a long time) in a gourd and using the product as an elixir,[d] (4) boiling nitre (or saltpetre) and blue-green rock salt (*chhing yen* [7]) in water, (5) making an egg-shaped container of silver to hold cinnabar, mercury and alum,[e] (6) using iron rust and copper as ingredients for an elixir called 'golden flower' (*chin hua* [8]),[f] (7) heating mercury together with malachite and azurite (copper carbonate and basic copper carbonate), (8) heating realgar and orpiment,[g] (9) heating black lead with silver,[h] and (10) burning together dried dung and wax. The book also warns against a very interesting procedure, saying that some of the alchemists had heated sulphur together with realgar, saltpetre and honey, with the result that their hands and faces had been scorched when the mixture deflagrated, and even their houses burnt down. This passage is of outstanding importance because it is one of the first references to an explosive mixture,

---

[a] *PPT/NP*, ch. 4, p. 2*a*, tr. Ware (5), p. 70.   [b] *TT* 917.

[c] *Chhiu shih* ('autumn mineral') has been shown by Lu & Needham (3) to be a purified mixture of urinary hormones. It is discussed fully in pt. 5 below. Cf. p. 36. The recognisable methods of preparation date at least as far back as the early Sung, but the name may have been already current in the −2nd century, since it is connected at an early date with the Prince of Huai-Nan. There is a reference in the *Tshan Thung Chhi*, ch. 15, p. 34*a*; where the text says 'Huang Ti admired the "golden flower" (*chin hua* [8]), and (the Prince of) Huai-Nan prepared the "autumn mineral" (*chhiu shih* [5])'. Li Shih-Chen attributed the term to the *Huai Nan Tzu* book, but it is not in the text as we have it today; only a solitary reference to 'autumn drugs' can be found (ch. 19, p. 14*b*), but it has to do with something else.

[d] This is reminiscent of the digestion methods (*shui fa* [9]) which will be discussed below (pt. 4).

[e] Cf. the silver aludels of Thang date described in pt. 4.   [f] Cf. p. 73.

[g] This will produce the arsenical oxides, with evolution of SO_3.

[h] *Hei chhien* is a synonym for metallic lead, but sometimes, as above, may perhaps be interpreted as graphite.

---

[1] 太淸丹經   [2] 九鼎丹經   [3] 金液丹經   [4] 眞元妙道要略

[5] 秋石   [6] 鉛汞   [7] 靑鹽   [8] 金花   [9] 水法

proto-gunpowder, combining sulphur with nitrate and a source of carbon, in any civilisation.[a] The book also gives a test for saltpetre. Exactly how much of all this material goes back to the days of Chêng Yin himself is extremely difficult to determine, but future research may be expected to throw more light on the problem. In the meantime, having regard to the general pattern of development of chemical knowledge and use of explosives, we place the essential passages in the Thang period.

The *Tao Tsang* also includes two Taoist texts purporting to record dialogues between Chêng Yin and his celebrated disciple Ko Hung, namely the *Thai-Chhing Yü Pei Tzu*[1] (The Jade-Tablet Master, a Thai-Chhing Scripture)[b] and the *Ta Tan Wên Ta*[2] (Questions and Answers on the Great Elixir).[c] Neither book reveals its author's name, nor any indication of date. The *Thai-Chhing Yü Pei Tzu* is however mentioned in one of the bibliographical chapters of the *Sung Shih*,[d] so it cannot be later than the Sung. As the courtesy name Chih-Chhuan[3] appears in the text, it seems unlikely to have come from the pen of Ko Hung himself. In any case, it is not of great chemical interest, referring only to the combination of mercury and sulphur to form cinnabar, and the conversion of cinnabar back into mercury. The *Ta Tan Wên Ta* is quite a small text, again not of much alchemical interest, and the obscurity of its wording does not fit in well with Ko Hung's style.

Ko Hung, also known under other names as Ko Chih-Chhuan,[4] Chih-Chhuan Chen Jen,[5] Pao Phu Tzu[6] (the Preservation-of-Solidarity Master), and Hsiao Hsien-Ong,[7] to distinguish him from his great-uncle Ko Hsien-Ong,[8] was the greatest alchemist of his age, and the greatest Chinese alchemical writer of any age.[e] There has been some uncertainty about his exact dates. Yoshida saw that he was born between +280 and +286.[f] Feifel (1), following the *Chin Shu*, which says that Ko Hung lived to the age of 80 (or 81 by Chinese reckoning),[g] gives his dates as *c.* +253 to *c.* +333, while Chang Tzu-Kao (2, 3) makes them *c.* +281 to *c.* +361, also believing that Ko Hung lived to the age of 80. Yuan Han-Chhing (1) on the other hand, taking his age at death to be more probably 60, gives *c.* +281 to *c.* +340; Ware (5) chooses the same range (*c.* +280 to *c.* +340). Chhen Kuo-Fu, however, in the most elaborate study,[h] manages to arrive at definite years of birth and death as +283

---

[a] Attention was first drawn to it by Fêng Chia-Shêng (1), p. 42, (5), p. 38. We discuss it in its place when dealing with the history of gunpowder; Sect. 30.

[b] *TT*920.   [c] *TT*932.   [d] Cf. Wieger (6), p. 274.

[e] The second of these appellations may be truer than the first, for Ko Hung often deplores his inability to carry out all the processes which he describes (*Pao Phu Tzu* (*Nei Phien*), tr. Ware (5), pp. 70, 262, 269 ff.). The expenses are dauntingly heavy (*ibid.* pp. 92, 112), there is difficulty in getting reagents (*ibid.* pp. 91, 262, 269) and books (p. 92), while the actual procedures are taxing and complicated (pp. 91, 112). Nevertheless it seems certain from internal evidence that Ko Hung must have witnessed personally many operations, perhaps in the laboratories of friends and hosts, probably more than he could actually carry out himself, and in at least one case (see Table 111, no. 51) he appears to be describing a preparation discovered in his own laboratory. On the other hand the persuasiveness of his arguments and the beauty and elegance of his literary style (qualities not remarkable in the bulk of the alchemical literature) sufficed to enrol his book among the classics of Chinese literature, often to be read and savoured, no doubt, centuries after any widespread belief in the reality of alchemy had departed.

[f] (5), p. 210.   [g] Ch. 72, p. 7*b*.   [h] (1), vol. 1, pp. 95 ff.

[1] 太清玉碑子   [2] 大丹問答   [3] 稚川   [4] 葛稚川   [5] 稚川眞人
[6] 抱朴子   [7] 小仙翁   [8] 葛仙翁

and $+343$ respectively; and this seems the best estimate so far.[a] An autobiography of Ko Hung is included in the *Pao Phu Tzu* (*Wai Phien*[1]), (Book of the Preservation-of-Solidarity Master: Exoteric Chapters),[b] but since this contains surprisingly little of scientific interest, and has been fully translated into English by Ware,[c] we shall not quote at length from it here. It only remains for us to remind the reader that Ko Hung has featured regularly in the discussions of astronomy, meteorology and mineralogy in Volume 3 of this work, and that he was as great a physician as he was an alchemist.[d]

Let us sketch for a moment some rough picture of his life.[e] The autobiography is not very informative about Ko Hung's scientific work, emphasising rather the military exploits which he carried out as a young officer on the government side in the suppression of the rebellions of $+303$. But it does bring out, like all the other sources, his abstracted unworldliness, his devoted search for proto-scientific books and writings, his inability to engage in ordinary conversation though so eloquent with the brush, and his lack of any desire for a normal official career. His father, however, had risen to be Governor of Shao-ping. Before the troubles, besides his studies with Chêng Yin and Pao Ching, he had also engaged in the acquisition of medical knowledge from learned teachers. After his army service Ko Hung, seeking no reward, travelled to Loyang and other places collecting books; and then, when his friend Chi Han[2] was made Governor of Kuang-chou in $+306$, he accepted to go south with him as his military adviser, probably because he was attracted by the exotic plants and unusual mineral substances of the south. This friendship deserves note, for Chi Han was a scholar whose *Nan Fang Tshao Mu Chuang*[3] (Records of the Plants and Trees of the Southern Regions) makes him one of the greatest of all Chinese botanists. Unfortunately Chi was soon assassinated, but Ko remained in the south for many years, only returning for a brief visit to his birthplace, Nanyang in Chiangsu, after which he was belatedly given the title of Kuan Nei Marquis for his youthful military successes. About this time his friend Kan Pao[4] recommended him to the throne as a suitable member of the Bureau of Historiography. Again the connection is quite interesting, for Kan Pao was what might nowadays be called a psychical research specialist, the author of a famous book on all kinds of strange phenomena: *Sou Shen Chi*[5] (Reports on Spiritual Manifestations),[f] finished $c. +348$. Prominent

---

[a] It is indeed made almost certain by the parallel deductions and arguments of Hung Yeh (2), brought to the attention of Western scholars by Sivin (7).   [b] Ch. 50, the last in this part.

[c] (5), pp. 6ff. Partial translations are in L. Giles (6) and Davis & Wu Lu-Chhiang (2).

[d] The medical work of Ko Hung will be dealt with in Vol. 6.

[e] The standard biography is in *Chin Shu*, ch. 72, pp. 7bff., tr. Davis & Chhen Kuo-Fu (1), p. 299. Partial tr. L. Giles (6), p. 97. Later hagiographic embroideries enlarge it in works such as the *Lieh Hsien Chhüan Chuan*,[6] whence Davis (7), and *Li Tai Shen Hsien Thung Chien*,[7] ch. 12. Further biographical information is found in Chhen Kuo-Fu (1), vol. 1, pp. 95ff.; Hou Wai-Lu *et al.* (1), pp. 270ff. Forke (20), incorporated in (12), pp. 204ff., may also be consulted, especially as it analyses the *Wai Phien* in some detail. There is an interesting study by Ōbuchi Ninji (1).

[f] For a discussion and partial translation see Bodde (9, 10). A translation of the *Chin Shu* biography of Kan Pao is in L. Giles (14).

[1] 抱朴子外篇   [2] 稽含   [3] 南方草木狀   [4] 干寶   [5] 搜神記
[6] 列仙全傳   [7] 歷代神仙通鑑

appointments Ko Hung all declined on account of age, but hearing that Chiao-chih in the south produced much cinnabar, he asked for the post of Magistrate of Kou-lou; this the emperor hesitated to give, on account of Ko's superior talents, but when Ko said 'I am seeking not for a grand career but for the secrets of elixirs', the emperor consented. On the way, however, he was amicably detained by Têng Yo,[1] the Governor of Canton, another interesting man, noted in history as the great founder and patron of the iron and steel industry of Kuangtung province. It may well be significant that a technologist of this calibre was a friend of Ko Hung's. So he took up residence in the recesses of the Lo-fou Shan[2] mountains[a] and devoted himself to alchemical and proto-chemical experiments;[b] probably the *Pao Phu Tzu* book was written there. The year +317 is an acceptable date for its completion. And there he died.

(ii) *The* Pao Phu Tzu *book and its elixirs*

Ko Hung's contribution to alchemy and early chemistry is contained in his *Pao Phu Tzu* (*Nei Phien*),[3] (Book of the Preservation-of-Solidarity Master: Esoteric Chapters), consisting of twenty in all. Both the *Nei Phien* (Esoteric Chapters) and the *Wai Phien* (Exoteric Chapters) must have been written by Ko Hung during the first quarter of the +4th century. Ware gives the date of the *Nei Phien* as +320, while Chhen Kuo-Fu says it was completed by the year +317, but revised *c.* +323.[c] Among the 20 chapters in the *Nei Phien* three are of particular chemical interest: chapter 4 which deals exclusively with the preparations and types of elixirs; chapter 11 which describes the various natural substances that could bring about longevity; and chapter 16 devoted to the transmutation of base metals into gold or silver. Alchemical information is also found scattered among some of the other chapters.[d] Chapters 4 and 16 were put into English by Wu Lu-Chhiang & Tenney Davis (2), and chs. 8 and 11 by Davis & Chhen Kuo-Fu (1), but these translations were criticised for not being particular in matters of textual criticism by Feifel, who in turn translated chapters 1, 2, 3, 4 and 11 of the *Nei Phien*, besides giving a full discussion of its various editions.[e]

[a] There is a local topography, the *Lo-fou Shan Chih*,[4] from which Fig. 1353 *a* is taken. It shows one of the chief temples there associated with Ko Hung, the Chhung-Hsü Kuan[5] at Chu Ming Tung,[6] a description of which by a Western visitor in 1895 may be read in Bourne (2). One of the greatest benefactors to Ko Hung's memory at Lo-fou Shan was the emperor Hsüan Tsung of the Thang, who built a big temple there in his honour about +750. For full details on this and related subjects reference should be made to the elaborate monograph of Soymié (4), which draws on much earlier literature about the mountains, including a work by Tsou Shih-Chêng[7] written in the late +13th century.

[b] Other places are of course traditionally associated with his work, for example the Chhu-Yang Thai[8] temple at Ko Ling,[9] one of the highest points of the low ridge of hills protecting the West Lake at Hangchow from the north (see Fig. 1354). There is a well in the grounds from which he is supposed to have taken pure water for his experiments. I paid a memorable visit to this temple with Dr Lu Gwei-Djen and Dr Dorothy Needham in 1964.

[c] So also Soymié (6).

[d] Critical notes on the whole text have been published by Yang Ming-Chao (1).

[e] See Feifel (1, 2, 3). This was very well, but even today it is unwise to ignore the contributions of Tenney Davis and his collaborators, for he was a most distinguished chemist, and sometimes the modern scientific colleague of an ancient Chinese scientist can see further through a brick wall than any philologist.

[1] 鄧嶽   [2] 羅浮山   [3] 抱朴子內篇   [4] 羅浮山志   [5] 冲虛觀
[6] 朱明洞   [7] 鄒師正   [8] 初陽台   [9] 葛嶺

A recent translation of the complete text of the *Nei Phien*, with the omission of the talismanic diagrams, has been made by Ware (5).[a]

Ko Hung believed that the only way to achieve material immortality was to prepare and consume one of the major elixirs, such as the 'cyclically-transformed elixir' (*huan tan*[1]) and the 'potable gold (elixir)' (*chin i*[2]), which he described at length in chapter 4. Other methods, such as the ingestion of various natural products, respiratory exercises, the techniques of sex, and even the minor elixirs, were, according to him, only meant to prolong the human life-span for a couple of hundred years or so, in order to give the aspirant sufficient time to produce one of the major elixirs. Ko Hung informs us that he had to serve his teacher, Chêng Yin, as a domestic for a long time until the latter succeeded in preparing an elixir at the Ma-chi Shan[3] mountain. Even twenty years after acquiring the art Ko Hung himself could not proceed to make any of the more powerful elixirs on account of lack of funds, achieving only some minor ones and prolonging his life so that eventually he might have a chance to carry out the greater experiments. He stresses that even the least of the minor elixirs (*hsiao tan*[4]) was superior to the best things among the drugs.[b] In another place he warns his readers that the sexual techniques were not only incapable of leading to material immortality, but on the contrary could ruin one's health if not properly carried out.[c] Ko Hung affirms that it is never sufficient to study books when performing alchemical experiments, and that one must receive oral instructions from a teacher.[d] Indeed he gives this as the explanation of Liu Hsiang's[5] spectacular failure to accomplish the making of gold.[e]

Of course there were other explanations—the mercenary and non-spiritual motives of the operators in the Imperial Workshops, and also the impossibility of success in a crowded and worldly environment. Part of what that meant we have already guessed (p. 36). This comes out particularly well from another story which Ko Hung quotes from the *Hsin Lun* of Huan Than, written about +10 or +20. It reads:[f]

Huan Than tells us that 'after Shih Tzu-Hsin[6] was made a Secretary in the Prime Minister's office, he constructed an elaboratory and mobilised minor officials as well as government slaves to set on foot the making of gold, but it was not successful. The Prime Minister then decided that adequate resources were lacking, and brought the matter to the attention of the Dowager-Empress Fu. This lady was not interested in the enterprise of aurifaction as such, but when she heard that gold could be used for making medicines to lengthen one's life, she agreed to participate (with financial support), giving her blessing to the project.

---

[a] Note that the abbreviation *TT* in Ware refers to the catalogue of Ong Tu-Chien (*1*), while in the present work it refers to that of Wieger (6).

[b] *PPT/NP*, ch. 4, p. 3*b*; tr. Ware (5), p. 72.

[c] *PPT/NP*, ch. 6, p. 8*a, b*; tr. Ware (5), pp. 122–3; cf. pt. 5.

[d] The *locus classicus* for this is *PPT/NP*, ch. 16, pp. 6*a*ff., a wonderful passage on the multifarious synonyms and cover-names which need to be explained (Ware tr. pp. 270ff.). Cf. also pp. 71, 175, 319 of this translation. We shall return to this question presently (p. 153). The paramount necessity of personal discipleship and direct transmission of techniques by word of mouth is constantly emphasised in the literature, cf. the Thang text *TT*878, ch. 3, pp. 2*b*, 3*a*. Also p. 74 above.

[e] *PPT/NP*, ch. 2, p. 11*a*; tr. Ware (5), p. 51; Feifel (1), p. 177.

[f] *PPT/NP*, ch. 16, p. 4*a*, tr. auct. adjuv. Ware (5), p. 266.

[1] 還丹   [2] 金液   [3] 馬迹山   [4] 小丹   [5] 劉向   [6] 史子心

Shih Tzu-Hsin was then made a Court Gentleman, and transferred his activities to the Northern Palace, where he was waited upon by a whole retinue.' How could such a holy art succeed in a palace? How could such a sacred process be accomplished in the presence of a company of the profane? Everybody knows that even dyers of silks do not like to have miscellaneous people watching what they are doing, for fear that their work will be spoilt.[a] How much more so is this true for the changes and transformations effected in aurifaction and argentifaction (lit. the art of the yellow and the white).

Chapter 4 of the *Nei Phien* tells us about the various forms of 'cyclically transformed elixir' (*huan tan*), the 'magical elixir' (*shen tan*[1]), and thirdly how to make 'potable gold' (*chin i*). It is far easier, says the text, to prepare 'potable gold' than the other elixirs, but the difficulty lies in getting enough gold for the purpose. Two pounds of gold would be needed to make a single dose sufficient for eight aspirants. It quotes the names of nine different types of 'magical elixir' from the *Huang Ti Chiu Ting Shen Tan Ching*[2] (The Yellow Emperor's Manual of the Nine-Vessel Magical Elixir), namely:[b]

(1) 'elixir flower' (*tan hua*[3]);[c]
(2) 'magical elixir' (*shen tan*[1]) or 'magical amulet (elixir)' (*shen fu*[4]);
(3) 'magical elixir' (*shen tan*[1]);
(4) 'cyclically-transformed elixir' (*huan tan*);
(5) 'edible elixir' (*erh tan*[5]);
(6) 'refined elixir' (*lien tan*[6]);
(7) 'soft elixir' (*jou tan*[7]);
(8) 'fixed elixir' (*fu tan*[8]); and
(9) 'cold elixir' (*han tan*[9]).

In eight of these cases no information is vouchsafed apart from bare details of dosage and the wonderful results ensuing. In the first, however, something approximating to a preparative method is given, difficult though the interpretation of it is. It reads:[d]

The first elixir is called 'elixir flower'. One should first prepare the 'mysterious yellow (substance)' (*hsüan huang*,[10] perhaps lead-mercury amalgam, perhaps the mixed oxides). Add to it (lit. use) a solution of realgar (arsenic disulphide) and a solution of alum.[e] [One

---

[a] Cf. ch. 4, p. 16b (tr. Ware, p. 93). Physicians, says Ko Hung, preparing beneficial medicines or salves, dislike being watched by animals, children or married women. And he mentions again dyers afraid of the 'evil eye' (*o mu*[11]).

[b] *PPT/NP*, ch. 4, pp. 5b ff.; tr. Feifel (2); Ware (5), pp. 75 ff. These are listed with the same numbers in Table 111.

[c] Note the similarity of the name with that mentioned in the +2nd-century *Thai Phing Ching* (p. 75 above).

[d] *PPT/NP*, ch. 4, p. 6a, tr. auct., cf. Feifel (2); Ware (5), p. 76; Chikashige (1), pp. 41 ff.

[e] For the ancient methods of getting inorganic substances into solution with the aid of nitrate and acetic acid, see pt. 4 below. One such for realgar is detailed elsewhere in the book, ch. 16, p. 8b, tr. Ware (5), p. 274. Hsüeh Yü (1) took *fan shih* as copper sulphate here.

---

[1] 神丹   [2] 黃帝九鼎神丹經   [3] 丹華   [4] 神符   [5] 餌丹
[6] 鍊丹   [7] 柔丹   [8] 伏丹   [9] 寒丹   [10] 玄黃   [11] 惡目

text says 'alum and mercury'.][a] Take several dozen pounds each of rough Kansu salt (*jung yen*[1]), crude alkaline salt (*lu hsien*[2]),[b] alum,[c] (powdered) oyster-shells, red bole clay, (powdered) soapstone, and lead carbonate; and with these make the Six-One Lute [and seal (the reaction-vessel) with it].[d] After 36 days heating the elixir will be completed, and anyone who takes it continuously for 7 days will become an immortal. Now if this elixir is made into pills with 'mysterious fat' (*hsüan kao*[3]) and placed upon a fierce fire, it will very quickly turn into gold. Gold can also be made by taking 240 *chu*[4] (10 oz.) of this elixir and adding it to 100 catties (lb.) of mercury, then upon heating, it will all turn to gold. If this works we know that the elixir is right. If it does not, re-seal the constituents and heat for as long as before. This never fails.

The whole procedure is evidently interesting not only in itself but because of the mention of projection, and even a test by projection, at the end.

This was the passage which aroused the interest of one of the pioneers of the history of chemistry in China, Chikashige Masumi, some forty years ago.[e] He believed rightly that it was susceptible of a rational explanation, but fixed upon the 'red bole clay' (*chhih shih chih*[5]) as the most important thing, suggesting that the alchemists of Ko Hung's time used an auriferous variety. It is true that gold can occur in small quantities in alluvial clays, but perhaps it is hard to believe that we are dealing here only with a smelting process. Chikashige thought that the lime helped the silicate to slag, that the salt and soapstone acted as a flux, and that the lead from the carbonate dissolved the gold; then he daringly interpreted the two alternative 'projection' stages as amalgamation and cupellation.[f] It remains an open question whether his view of the whole process is still tenable. In one way research has failed to justify it. He himself drew attention to the existence of a specimen of *chhih shih chih*[5] in the Shōsōin Treasury at Nara, and hoped it might some day be analysed; twenty years later this was done, and no trace of gold was reported.[g] This in itself does not invalidate his argument, which will have to be judged on wider grounds of plausibility.[h]

Parallel passages occur in several other texts, notably the *Huang Ti Chiu Ting Shen Tan Ching Chüeh*[6] (Explanation of the Yellow Emperor's Manual of the Nine-Vessel Magical Elixir)[i], probably compiled by an early Thang or an early Sung writer, but incorporating some material even earlier than Ko Hung's time. The relevant passage here has been ably translated by Ware.[j] An important feature of it is that it distinctly

[a] An ancient gloss crept into the text here. *Shui*[7] and *hung*[8] were liable to be confused by copyists.
[b] Hsüeh Yü (*1*), giving no authority, interpreted this as ammonium chloride from urine, with organic impurities.
[c] Another gloss in some editions suggests a mis-reading for arsenolite (*yü*[9]) here.
[d] Some editions add these words.   [e] (*1*), pp. 49 ff.
[f] He applied (p. 51) the same hypothesis to another of the aurification procedures, the 'Child's-play Method of Making Gold' (*hsiao erh tso huang chin fa*,[10] ch. 16, p. 9*a*, *b*, Ware (*5*), p. 274). See Table 111, no. 54.   [g] Asahina (*1*), no. 21; Masutomi (*1*), p. 137.
[h] Chikashige's theory has found favour with others, e.g. Hsüeh Yü (*1*). And a parallel case in +1682 (Sir Kenelm Digby) is discussed by Dobbs (*3*), p. 8.   [i] *TT*878.
[j] (*5*), pp. 78, 79. Ware interpreted *hsüan huang*[11] as tin oxide, quoting *TT*878, ch. 1, p. 3*b*, as his authority, but the formula given in this source states the required ingredients as 10 lb. of mercury

[1] 戎鹽   [2] 鹵鹹   [3] 玄膏   [4] 銖   [5] 赤石脂
[6] 黃帝九鼎神丹經訣   [7] 水   [8] 汞   [9] 礜   [10] 小兒作黃金法
[11] 玄黃

starts with cinnabar (*chen sha*[1]) and only after a long heating and sublimation of this does the process with the *hsüan huang* begin; then the descriptions follow a similar course, ending with the projection test, which is given as the words of a spiritual being, the Mysterious Girl (Hsüan Nü[2]).[a] This book in fact relates the full details of all the nine elixir preparations listed by Ko Hung, and the study of what each of them implies in terms of modern chemistry will be a rewarding one for future investigators. They will also be able to draw on other parallel passages, such as those describing the techniques of preparation of the Nine Elixirs in the *Chiu Chuan Liu Chu Shen Hsien Chiu Tan Ching*[3] (Manual of the Nine Elixirs of the Holy Immortals and of the Ninefold Cyclically Transformed Mercury).[b] The first process here again starts with cinnabar, using its synonym 'the red boy' (*chu erh*[4]); the mercuric sulphide being decomposed and recombined in the form of a sparkling scarlet sublimate. The exact reactions would have depended on the size of the sublimatory container (two pots luted together) and the conditions of heating. In any case the meaning of the term *chiu chuan* is considered to have been the decomposition of mercuric sulphide and the re-sublimation of the cinnabar nine times repeated.[c]

Another important elixir which would transform a mortal into an immortal within three days was the 'Grand-Purity elixir' (*Thai-Chhing tan*[5]), derived by Ko Hung

(*shui yin*[6]) and 20 lb. of lead (*chhien*[7]) saying nothing about tin. Another alchemical text, the *Thai-Chhing Chin I Shen Tan Ching*[8] (Manual of the Potable Gold and Magical Elixir; a Thai-Chhing Scripture), says that *hsüan huang* is prepared by heating 9 lb. of mercury and 1 lb. of lead in an earthenware vessel over a strong fire from morning till dusk (*TT*873, ch. 1, p. 15*a*). Chhen Kuo-Fu (*1*), vol. 2, p. 379, suggests that in view of the strong heating prescribed, *hsüan huang* was a mixture of lead and mercuric oxides, the colour of which could well be a dirty yellow. Ware, making a guess as to the meaning of *hsüan kao*,[9] suggests that it might be human faeces but cites no authority. *TT*878 (ch. 1, p. 5*a*) calls the same substance *lung kao*,[10] so presumably these two terms refer to the same substance, unless one is a misprint. The *Yin Chen Chün Chin Shih Wu Hsiang Lei*[11] (The Adept Yin [Chhang-Shêng]'s (Classification of) Metals and Minerals (according to the) Similarities and Categories of the Five (Substances)), *TT*899, p. 6*b*, says that *hsüan ming lung kao*[12] is a synonym for mercury, and is abbreviated as *lung kao*. The *Shih Yao Erh Ya*[13] (Synonymic Dictionary of Minerals and Drugs), *TT*894, ch. 1, p. 1*b*, gives *hsüan shui lung kao*[14] as a synonym for mercury. If *hsüan kao* and *lung kao* are interchangeable then they would seem to be synonyms for mercury. It must be pointed out, however, that *lung kao* has also other meanings, for example a plant, the wild raspberry (*fu phên tzu*[15]). This is *Rubus coreanus* (R457, CC1153), well known anciently in China as it is in the *Shen Nung Pên Tshao Ching*. We shall meet it again before long (p. 98). Or alternatively, *lung kao* may mean the bile of a white dog (*pai kou tan*[16]), cf. R323. For these identifications see the *Shih Yao Erh Ya* (*TT*894), ch. 1, p. 4*a*, *b*.

---

[a] Cf. Vol. 2, pp. 147ff.

[b] *TT*945, pp. 4*a*ff. The identity of the compiler, Thai-Chhing Chen Jen,[17] is very hard to determine. Two semi-legendary alchemists of supposedly Chou date bore this title, Sung Lun[18] and Phêng Tsung[19] (*TT*293, ch. 9, pp. 2*b*, 4*a*). The material cannot be later than Sung and may be very much earlier.

[c] The idea of purification by repeated sublimation seems almost like a premonition of the practice of purification by repeated crystallisation, if indeed it did not derive from ancient processes used in the salt industry. The expression occurs at least as early as Ko Hung, see *PPT/NP*, ch. 4, pp. 10*a*ff. (Ware (5), p. 82). The efficacy of the substance as an elixir, as given by Ko Hung, can be represented graphically (p. 86). The *chiu chuan shen tan* is mentioned in other contemporary sources, e.g. *Shih I Chi*, ch. 4, p. 4*b*.

---

[1] 眞砂	[2] 玄女	[3] 九轉流珠神仙九丹經	[4] 朱兒
[5] 太淸丹	[6] 水銀	[7] 鉛 [8] 太淸金液神丹經	
[9] 玄膏	[10] 龍膏	[11] 陰眞君金石五相類	[12] 玄明龍膏
[13] 石藥爾雅	[14] 玄水龍膏	[15] 覆盆子	[16] 白狗膽 [17] 太淸眞人
[18] 宋倫	[19] 彭宗		

from the *Thai-Chhing Tan Ching*[1] documents which were in his possession.[a] This was supposed to be an elixir which had undergone nine cyclical changes, becoming more efficacious with each transformation (Fig. 1355). However, only the preliminary steps are given, not the exact composition of the elixir itself. Similarly, it is hinted that the 'Ninefold Radiance elixir' (*chiu kuang tan*[2]) could be made by transforming the

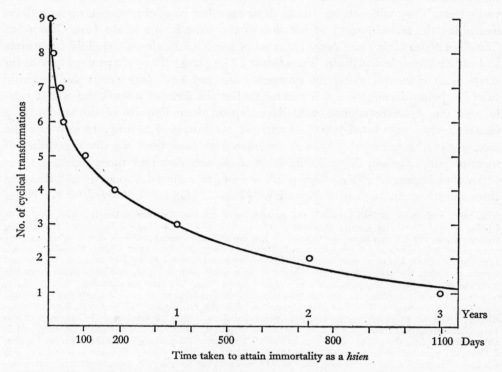

Fig. 1355. Graph constructed from figures given in the *Pao Phu Tzu* book by Ko Hung, *c.* +320. Time taken to attain material immortality as a *hsien* is plotted against the number of cyclical transformations to which the elixir has been subjected. Cf. the translation of Ware (5), p. 82.

'five minerals' (*wu shih*[3]), i.e. cinnabar (*tan sha*[4]), realgar (*hsiung huang*[5]), purified potash alum (*pai fan*[6]), stratified malachite (*tshêng chhing*[7]) and magnetite (*tzhu shih*[8]), by heating them together with certain unspecified ingredients.[b]

Next comes a series of twenty-seven elixirs, with different efficacies, though not quite comparable with those just mentioned.[c] Less complicated chemical operations are involved. In describing the constituents of these elixirs Ko Hung often uses the word *tan*.[9] This of course by itself can mean 'elixir' but makes no sense in some

[a] *PPT/NP*, ch. 4, p. 8*b*, tr. Ware (5), p. 81. Table 111, no. 10.
[b] See Table 111, no. 11, and the note there on the Five Minerals.
[c] Nos. 12 to 38 in Table 111. All are in *PPT/NP*, ch. 4.

[1] 太清丹經   [2] 九光丹   [3] 五石   [4] 丹砂   [5] 雄黃
[6] 白礬   [7] 曾青   [8] 慈石   [9] 丹

contexts unless taken as an abbreviated form of *tan (-sha)*,[1] i.e. cinnabar. In other cases *tan* has to mean 'elixir' because nothing is said of cinnabar; for example, in the making of the 'Instantly Successful elixir' (*li chhêng tan*[2]), no. 16, the only ingredients used are orpiment (*tzhu huang*[3]), realgar (*hsiung huang*[4]), copper and vinegar.[a] In at least one instance (no. 17) Ko Hung uses the word to denote a red colour, as when he speaks of 'vermilion goldfish' (*tan yü*[5]).[b] Hence like the word *chin*,[6] which can mean gold, or metal in a general sense, or even the element Metal, or just a golden colour (as mentioned in connection with the *Tshan Thung Chhi*), we can only hope to infer its exact meaning from the context. Unfortunately, it is not always easy to tell from the text what precisely the author had in mind. This is not the fault of the Chinese language itself, for there exist terms like *tan sha* and *huang chin*[7] which have been used exclusively for cinnabar and gold respectively. Perhaps this was another way of keeping alchemical secrets even though put down in writing, and doubtless this again is why Ko Hung stressed the importance of getting oral instructions from a teacher.[c]

Among these twenty-seven elixirs cinnabar is found as an ingredient in twenty-one of them, mercury in eleven or twelve, realgar and malachite are each used in eight cases, potash alum, sulphur[d] and magnetite in five, and mica in three. In five instances the reactions were carried out in copper vessels. Vinegar, wine, honey and blood are each used on three occasions. Orpiment is mentioned twice, and haematite, rock-salt, minium, lead, jade, and lacquer only once each. Two synonyms, namely the 'mysterious liquid' (*hsüan shui*,[8] nos. 22 and 24) and the 'dragon's fat' (*lung kao*,[9] no. 28), present a problem. The first is one of the many synonyms for mercury according to

[a] Ch. 4, p. 11a. This is a case where previous translators (Wu & Davis (2), p. 245; Ware (5), p. 84) went astray by not adhering stoutly enough to the simple principle that copper cannot be obtained from the sulphides of arsenic. Only Feifel (2), p. 20, saw that after the roasting of these to convert them partially to the oxides, copper ore is separately smelted and cast into a vessel, the arsenic salts then being placed in it and covered over with strong vinegar. 'After a hundred days the vessel will be covered with red exfoliations several tenths of an inch long, and five-coloured *lang-kan* masses like coral growths will have developed.' This material is the elixir. The description is evidently one of severe corrosion of copper with the formation of oxides and acetates of copper and arsenic with cupric arsenite and arsenates. Since these last are brilliantly coloured (Scheele's green, Paris green, etc.), the comparison with a bluish or greenish gem, *lang-kan*[10] (not so far definitely identifiable), was very appropriate; cf. Partington (10), pp. 627ff. The passage, it is true, is a little obscure, and perhaps two phrases should be interchanged.

[b] This may be the place to advert to the many meanings of the word *tan*. Besides 'cinnabar' and 'elixir', it also from ancient times meant simply 'red', as we see, for example, in the compass-points of sailing-directions (cf. Vol. 4, pt. 3, p. 564). Just as in ancient Rome (cf. p. 3 above), but less usually, there was confusion between cinnabar and red lead (minium, Pb_3O_4), and this use of *tan* became particularly common in Japan (cf. Asahina (1), no. 58). In medicine *tan* came traditionally to mean any compounded prescription for ingestion in solid form, generally as pills, not at all confined to inorganic drugs, though many of them naturally figured among these (cf. Liu Yu-Liang, 1). Okanishi Tameto (4) has made a valuable index of 2,405 *tan* prescriptions found in 322 medical books. In modern usage the term has acquired the special nuance of a proprietary remedy, and is used in the names of a number of Chinese and Japanese patent medicines.

[c] Cf. p. 82 above.

[d] In view of its reactivity it is perhaps surprising that sulphur was not more frequently used. Cf. p. 74 above.

[1] 丹砂   [2] 立成丹   [3] 雌黃   [4] 雄黃   [5] 丹魚   [6] 金
[7] 黃金   [8] 玄水   [9] 龍膏   [10] 琅玕

several sources, e.g. the *Yin Chen Chün Chen Shih Wu Hsiang Lei*, the *Lung Hu Huan Tan Chüeh*[1] (Explanation of the Dragon-and-Tiger Cyclically Transformed Elixir), and the *Shih Yao Erh Ya*;[a] but the third source also says that it is a synonym for vinegar.[b] As for *lung kao*, we have already suggested that it could be mercury, though we cannot rule out other possibilities, such as wild raspberry juice and dog's bile.

It is of interest that twenty of the elixirs are named after adepts, some, like Chhih Sung Tzu, Hsienmên Tzu and Wu Chhêng Tzu, mentioned in the *Pao Phu Tzu* (*Nei Phien*) itself, others, like Han Chung, Chi Chhiu Tzu, Tshui Wên Tzu (with some of the names already mentioned in Ko Hung's text) only found in the hagiographic literature (e.g. the *Lieh Hsien Chuan* and the *Li Shih Chen Hsien Thi Tao Thung Chien*). Not all the names have yet proved identifiable. The two elixirs 'Chhi-Li's' (no. 31) and the 'Handy' (no. 33) are of special significance from the point of view of projective aurifaction and argentifaction. A knife-point of the former, according to Ko Hung, turns a whole catty (lb.) of lead into silver when heated, and this silver becomes gold if heated with realgar solution for 100 days. Similarly a piece of the 'Handy' the size of a pea thrown into a quantity of copper from Tanyang, and heated, turns it all to gold.[c] Projection for gold and silver is also found in procedures described elsewhere in Ko Hung's book (see Table 111, nos. 39a, b, 51, 54). We shall return to this in a moment.

The list of ingredients for the making of 'potable gold' (no. 39 in Table 111)[d] has presented much difficulty, as can be seen by the translations of both Ware and Feifel. The method requires that 1 catty (lb.) of gold be placed with a number of substances in a container, which is then sealed and left over a period of time until a liquid is formed.[e] The question is, what the substances were; they are described in an unpunctuated succession, as usual, but the synonyms are not so usual. *Hsüan ming lung kao*,[2] 'mysterious bright dragon's fat', is, according to the *Shih Yao Erh Ya*,[f] a synonym for selenite (a naturally occurring form of calcium sulphate), but it is also said to be mercury by the *Yin Chen Chün Chin Shih Wu Hsiang Lei*.[g] On the other hand if one reads it as *hsüan ming* and *lung kao* then one gets vinegar and the plant *fu phên tzu*,[3] the wild raspberry (*Rubus coreanus*), the unripe fruits of which contain hydrocyanic acid.[h] Wang Khuei-Kho (2) suggests that it would have been possible to

---

[a] See *TT*899, p. 5*b*; *TT*902, ch. 1, p. 1*b*; *TT*894, ch. 1, p. 1*b*.

[b] See *TT*894, ch. 1, p. 5*b*.

[c] Ware (5), ignoring the grammar here, spoke of 'male copper', which he thought might be an alloy of arsenic and copper. But metals and alloys did have sexes; see no. 56 in Table 111.

[d] *PPT/NP*, ch. 4, p. 14*a*, cf. Ware (5), pp. 89ff.

[e] The formula is: *hsüan ming lung kao Thai-I hsün shou chung shih ping shih tzu yu nü hsüan shui i chin hua shih tan sha*.[4] A parallel formula occurs in *TT*910 (ch. 3, p. 11*a*), one of the texts attributed to Ko Hung (see p. 109 below). It says that 1 lb. of gold is to be mixed with the substances *hsüan ming lung kao ku Thai-I hsün shou chung shih shui tzhu yu nü hsüan shui i chin hua shih tan sha*.[5]

[f] *TT*894, ch. 1, p. 3*a*.   [g] *TT*899, p. 6*b*.

[h] Or of course cyanogenetic glucosides as in related species. On *R. coreanus* see R457, CC1153. Anon. (57), vol. 2, no. 123, describes and figures *R. Chingii*.

---

[1] 龍虎還丹訣   [2] 玄明龍膏   [3] 覆盆子
[4] 玄明龍膏太一旬首中石冰石紫遊女玄水液金化石丹砂
[5] 玄明龍膏骨太一旬首中石水紫遊女玄水液金化石丹砂

bring gold into solution by taking *hsüan ming lung kao* either as mercury alone or as the two separate substances. *Thai-I hsün shou chung shih*[1] is most probably realgar, since one of its synonyms given in the *Shih Yao Erh Ya* is *Thai hsün shou chung shih*, while *Thai-I hsün shou* appears in the *Yin Chen Chün Chin Shih Wu Hsiang Lei*.[a] Wang Khuei-Kho identifies *ping shih* as either common salt or *han shui shih*[2],[b] but thinks that in any case it played no part in bringing the noble metal into solution. We have not been able to find any other reference to the 'purple roving girl' (*tzu yu nü*), except that 'purple girl' (*tzu nü*[3]) is a synonym for amethyst.[c] Wang Khuei-Kho thought that it was perhaps sulphur, but Mêng Nai-Chhang (*1*) prefers to take it as *lü fan*,[4] i.e. green vitriol or copperas.[d] *Hsüan shui i*, the 'mysterious liquid', may be a synonym for mercury.[e] We have not been able to find the term *chin hua shih*[5] elsewhere, but *hua chin shih*,[6] the 'metal-dissolving mineral', is a synonym for saltpetre.[f] Since *chin hua shih* and *hua chin shih* have virtually the same meaning, we may infer that the former also means saltpetre. About *tan sha* there is no question. Hence the ingredients for making 'potable gold' could have been as follows: gold, mercury (or vinegar and wild raspberry juice), realgar, leonite (or common salt), iron alum (or copperas), mercury (or magnetite), saltpetre, and cinnabar. This 'potable gold' could be taken directly for the attainment of immortality, or submitted to further treatment in various ways and used for projective aurifaction or argentifaction. A knife-point added to a pound of mercury would convert the whole to silver, and a quantity mixed with yellow earth would turn to gold after strong heating.

According to Wang Khuei-Kho (*2*), if *hsüan ming lung kao* is mercury then gold amalgam will be formed. This would probably be in a solid state because of the large amount of gold. It would then have been put into vinegar together with all the other substances to form 'potable gold' on standing.[g] KNO_3 in vinegar would form an oxidising agent, dilute nitric acid, hence the Hg in the amalgam would gradually dissolve out. The reaction could have taken the following form:

$$3Hg + 2NO_3^- + 8H^+ + (2K^+) + (8C_2H_3O_2^-)$$
$$\rightarrow 3Hg^{++} + 2NO + (2K^+) + (8C_2H_3O_2^-) + 4H_2O$$

In this way the gold would gradually dissociate itself from the amalgam forming a

---

[a] *TT*894, ch. 1, p. 1*b* and *TT*899, p. 12*a*.

[b] This is usually considered one or another natural form of gypsum, calcium sulphate (RP 51, 119). The Shōsōin sample has been found to be calcite (calcium carbonate) but this was probably due to a confusion in the Thang (Asahina (*1*), no. 7). Wang Khuei-Kho suggests that it was leonite ($MgSO_4$. $K_2SO_4.4H_2O$), the double sulphate of magnesium and potassium.

[c] *TT*894, ch. 1, p. 3*b*.

[d] He gives the formula $Fe_2O(SO_4)_2$ rather than the usual $FeSO_4.7H_2O$ (RP 132). Of course the double sulphates of the iron-potash and chrome-potash alums are purple in colour.

[e] See p. 153 below. Wang Khuei-Kho interprets it as a suspension of magnetite (*hsüan shui shih*[7]).

[f] *TT*894, ch. 1, p. 3*b*.

[g] On this first interpretation the vinegar is an interpolated assumption. What follows will be better appreciated in the context of the discussion on medieval methods of bringing inorganic substances into aqueous solution; cf. pt. 4 below.

---

[1] 太一旬首中石　　[2] 寒水石　　[3] 紫女　　[4] 綠礬　　[5] 金化石
[6] 化金石　　[7] 玄水石

Table 111. Chemical preparations and elixirs described in the 'Pao Phu Tzu (Nei Phien)', c. +320

No.	Ware (5), tr. p.	Name	Presence of						Other constituents	Efficacies	Comments
			cinnabar	mercury	realgar	alum	malachite	magnetite			
Ch. 4[a]											
1	76	'elixir flower' (tan hua)[b]	·	+	+	+	·	·	lead, lead carbonate, salt, alkali, oyster-shells, red bole clay, soapstone, others uncertain	immortality, projection for Au	amalgamation; possibly a smelt of auriferous clay (Chikashige); use of aqueous solutions of minerals
2	77	'magical elixir' (shen tan) or 'magical amulet' (shen fu)[b]	no details given						—	immortality, invulnerability, walking on water, panacea	suggestions of Hg poisoning (visions)
3	77	'magical elixir' (shen tan)[b]	no details given						—	—do—	—
4	77	'cyclically-transformed elixir' (huan tan)[b]	no details given						—	immortality, projection for Au, exorcism	—do—
5	77	'edible elixir' (erh tan)[b]	no details given						—	immortality	—do—
6	77	'refined elixir' (lien tan)[b]	no details given						—		—
7	77	'soft elixir' (jou tan)[b]	no details given						—	immortality, projection for Au	—
8	77	'fixed elixir' (fu tan)[b]	no details given						—	rejuvenation, projection for Au	—
9	78	'cold elixir' (han tan)[b]	no details given						—	immortality, exorcism	—
10	81	'Grand-Purity elixir' (Thai-Chhing tan)[c]	+	details incomplete					vinegar, red salt (sulphates), calomel, lead, gold, others uncertain	immortality, flight immortality	visions; mineral solutions, Pb/Au amalgamation, cyclical transformation (see Fig. 1355)
11	82	'ninefold radiance elixir' (chiu kuang tan)[c]	+	+	+	+	+	+	uncertain	resuscitation of the dead, invisibility, foreknowledge, perpetual youth, longevity, telepathic knowledge longevity	visions, suggesting poisoning; use of the 'five minerals',[d] emphasis on colour changes
			details incomplete								
12	83	'five numinous elixirs' (wu ling tan)[e]	+	+	+	+	+	+	sulphur, orpiment, haematite, rock-salt[k]		—
13	83	'Min-shan (Mtn.) elixir methods' (Min-shan tan fa)[f]	·	+	+	+			yellow copper alloy (perhaps brass) made into a moon-mirror to collect dew, then heated with mercury by burning-glass, focusing sun's rays.[l] Subsequent heating with realgar	immortality, rejuvenation, cure of blindness	Hg amalgam, sun-heated

No.	Ref.	Elixir method				sulphur	immortality	
14	84	'Master Wu Chhêng's elixir method' (Wu Chhêng Tzu tan fa[5])g	·	+	·	sulphur	immortality	unsublimed mercuric sulphide. Cf. no. 55
15	84	'Master Hsienmên's elixir' (Hsienmên Tzu tan fa[6])h	·	+	·	wine	panacea, immortality, exorcism	direct consumption of cinnabar suspension; visions suggesting toxic effects
16	84	'instantly successful elixirs' (li chhêng tan[7])i	·	+	+	orpiment, vinegar, copper, dodder sap, others uncertain	transfiguration, flight, longevity	Cu and As acetates, Cu arsenite
17	85	'selected fixed elixir method' (chhü fu tan fa[8])	·	·	·	goldfish blood, obtained in specified places and at a special time	walking on water, life under water	use of animal material
18	85	'The Red Pine Master's elixir method' (Chhih Sung Tzu tan fa[9])j	·	+	·	sap from chhien sui lei[10] (Vitis flexuosa)m and from the 'alum-peach', mulberry juice	longevity	—
19	86	'Teacher Shih's elixir method' (Shih hsien-sêng tan fa[11])	·	+	·	fed to young birds, bodies dried and eaten	longevity to 500 years	animal feeding experiment
20	86	'Master Khang Fêng's elixir method' (Khang Fêng Tzu tan fa[12])	·	+	·	mica, lacquer, blood of the embryo of the black stork (Ciconia nigra),r aconite sap or extract	longevity from 100 to 1000 years depending on amount taken	—
21	86	'Master Tshui Wên's elixir method' (Tshui Wên Tzu tan fa[13])n	·	+	·	put in duck's stomach and steamed	longevity and immortality	—
22	86	'Liu Yuan's elixir method' (Liu Yuan tan fa[14])	·	+	(+)	mica ions in solution or suspension; mercury or vinegar (hsüan shui[15])	longevity	solutions of minerals
23	86	'Yüeh Tzu-Chhang's elixir method' (Yüeh Tzu-Chhang tan fa[16])	·	+	+	minium, heated in copper cylinder	immortality (gradual)t	—
24	86	'Li Wên's elixir method' (Li Wên tan fa[17])	·	+	(+)	haematite (pai su[18]),s bamboo sap or extract, mercury or vinegar (hsüan shui[19])	immortality (gradual)t	—
25	87	'Master Yin's elixir method' (Yin Tzu tan fa[20])	·	+	·	mica suspension, molten lead (chin hua chhih[21])	immortality	—

For Notes to Table 111 see pp. 96-7.

Table III (continued)

No. Ch. 4	Ware (5), tr. p.	Name	cinnabar	mercury	realgar	alum	malachite	magnetite	Other constituents	Efficacies	Comments
26	87	'The Thai-I Soul-Recalling elixir method' (Thai-I chao hun pho tan fa[22])o	+	.	+	+	+	+	concocted in aludel, taken with sulphur	will resuscitate one who has been dead for not more than 3 days	use of the 'five minerals', cf. nos. 11, 34
27	87	'The Chosen Girl's elixir method' (Tshai Nü tan fa[23])p	+	.	+	.	.	.	rabbit's blood, honey	appearance of heavenly hand-maidens	visions due to Hg poisoning?
28	87	'Master Chi Chhiu's elixir method' (Chi Chhiu Tzu tan fa[24])q	.	+	.	.	.	.	wine, hempseed oil, honey, others uncertain (lung kao[25])	longevity to 500 years	use of mainly organic substances
29	87	'Master Mo's elixir method' (Mo Tzu tan fa[26])u	+	.	+	+	+	.	heated in copper vessel	panacea, immortality	mineral solutions
30	87	'Chang Tzu-Ho's elixir method' (Chang Tzu-Ho tan fa[27])	+	.	+	.	+	.	lead, red panicled millet, with crushed jujube-dates as vehicle	longevity to 500 years	organic constituents
31	88	'Chhi-Li's elixir method' (Chhi-Li tan fa[28])	+	+	+	+	+	+	jade, heated in copper vessel, lead	immortality, projection for Ag and for Au	the 'gold' may be too hard or too soft; emphasis on colour changes
32	88	'Chu Chu's elixir method' (Chu Chu tan fa[29])v	+	.	+	.	+	.	sulphur, vinegar	appearance of attendant goddesses, omniscience	visions
33	88	'handy elixir method' (chou hou tan fa[30])	+	(+)x	.	.	.	.	lead (chin hua[31]),y copper or one of its alloys	longevity, projection for Au	—
34	88	'The Venerable Li's elixir method' (Li Kung tan fa[32])	.	+	+	+	.	+	concocted in aludel, sulphur	immortality (gradual)	the 'five minerals' in aqueous solution, cf. nos. 11, 26 etc.
35	88	'Mr Liu's elixir method' (Liu shêng tan fa[33])	+	.	.	.	.	.	saps or extracts of pai chü[34] (white variety of Chrysanthemum), ti chhu[35] (gromwell),z and chhu[36] (Ailanthus)aa	longevity up to 500 years (gradual), inhibition of ageing	link with plant pharmacology
36	89	'Wang Chün's elixir method' (Wang Chün tan fa[37])	+	+	.	.	.	.	ingredients placed inside an egg, lacquer-sealed and hen-incubated	inhibition of growth as well as of ageing	use of body temperature
37	89	'Mr Chhen's elixir method' (Chhen shêng tan fa[38])	+	.	.	.	.	.	honey, sealed in a copper container and sunk in a well	cessation of hunger, longevity for 100 years	—

No.	Page	Name	Markers	Constituents	Effects (classification)	Remarks
38	89	'Han Chung's elixir method' (Han Chung tan fa[39])w	+	honey, lacquer	longevity, casting no shadow, preservation of vision	—
39	89 (cf. also pp. 64, 68 ff.)	potable gold (chin i[40])bb	· + + + +	gold, saltpetre, leonite or common salt, perhaps vinegar and wild raspberry juice, perhaps copperas instead of alum, magnetite and selenite uncertain, possibly other constituents; see discussion on pp. 88 ff. no further heating	transfiguration, longevity, immortality, exorcism, fulfilment of all wishes (earthly or heavenly)	no heating, mineral substance in solution; probable eventual formation either of potassium auricyanide or of colloidal gold
39a	90	'jet(-black) chü-shêng seeds' (wei hsi chü-shêng[41])cc	· + ·	—	immortality, longevity, invulnerability, projection for Ag, conversion to Au ('red gold')	Hsüeh Yü (1) assumes amalgams. Noteworthy is the statement that bowls and plates made from this gold will confer immortality on those who eat from them (cf. pp. 31, 49 above)
39b	90	—	· ·	loess earth, strongly heated	conversion to real Au ('yellow gold'), immortality (earthly), double projection for Ag, immortality (earthly)	—
40	92	ingestion of small amounts of gold (erh huang chin[42])dd	(+) · · · (+)	gold beaten out to leaf, softened, dissolved, liquefied with wine, hog-fat, Ailanthus bark, or the constituents indicated, or orpiment	—	—
41	92, 189	silver and large pearls (yin chi pang chung ta chu[43])	· · ·	silver pearls, dissolved, or with wine, raspberry juice	very gradual longevity or immortality	silver cyanide ?
42	95, 198	lesser magical elixir formula (hsiao shen (tan) fang[44])	+	honey, sun-heated	rejuvenation, inhibition of ageing, longevity, immortality	—
43	95, 199	lesser elixir method (hsiao tan fa[45], erh tan sha fa[46])	+	strong vinegar, lacquer, heated	panacea, health, immortality, metamorphoses, casting no shadow	—
44	95, 198	lesser method for ingesting small amounts of gold (hsiao erh huang chin fa[47] or fang[48])	· · ·	gold or silver, wine	insensibility to heat and cold, visions, levitation, immortality (heavenly or earthly)	probably leaf

Table 111 (*continued*)

No.	Ware (5), tr. p.	Name	Presence of cin-nabar	mer-cury	real-gar	alum	mala-chite	mag-netite	Other constituents	Efficacies	Comments
Ch. 4 45	96, 198	'Master Liang I's method for ingesting small amounts of dispersed gold' (*Liang I Tzu erh hsiao huang chin fa*[49]ee)	·	·	·	·	·	·	gold, hog-fat, strong vinegar	immortality, longevity to 2000 years	probably leaf
46	179ff.	the five types of magic mushrooms (*wu chih*[50])	·	·	·	·	·	·	none	invulnerability, ff panacea, magical powers, longevity, immortality, sensations of multicoloured light, glowing body, life under water, invisibility	by no means all cryptogams, but a wide variety of excrescences, mineral, vegetable and animal, cf. Sect. 45 in Vol. 6. Among the fungi some may well have contained hallucinogenic substances
47	186	the five types of mica (*yün mu wu chung*[51])	·	·	·	·	·	·	plant saps or extracts, saltpetre and acetic acid, honey, black tea	employment of gods and spirits, invulnerability, panacea, rejuvenation	minerals in aqueous solution
48	187	realgar (*hsiung huang*)	·	·	+	·	·	·	wine, saltpetre and acetic acid, with animal material	panacea, longevity, rejuvenation, visions	suggestions of As poisoning
49	188	jade (*yü*)	·	·	·	·	·	·	wine, plant materials	invulnerability (but often causes fever)	intestinal inflammation?
50 gg	190	lacquer (*shun chhi*[52])	·	·	·	·	·	·	crab meat (to prevent setting, see pt. 4), mica or jade in solution	panacea, visions	minerals in aqueous solution
51	271	app. Ko Hung's own	+	+	+	·	·	·	ox bile, crude salt, copper sulphate, aqueous solution of cinnabar (cf. pt. 4)	projection for Au	preparation of a Cu/As alloy resembling gold
51a	272	method of making cinnabar solution (*tso tan sha shui fa*[53])	+	·	·	·	·	·	strong vinegar, saltpetre, copper sulphate	—	inorganic ions in aqueous solution (see pt. 4)

No.	page	method						materials	preparation	commentary
52	273	'method for making gold received by Teacher Chin Lou from Master Chhing Lin' (Chin Lou hsien-sêng so tshung Chhing Lin Tzu shou tso huang chin fa⁵⁴)	·				·	tin, 'red salt' (mixed sulphates), lime-water	preparation of an artificial Au	preparation of 'mosaic gold' or stannic sulphide (cf. pt. 2, pp. 69, 71, 271, and here, pp. 99, 103)
52a	273	method of making red salt (chih tso chhih yen fa⁵⁵)	+	+	+	+	·	gypsum, potash or iron alums, fused	—	mixing of inorganic sulphates for no. 52
53	273	'method for transmuting (into) gold received by Teacher Lu Li from Master Chi Chhiu' (Lu Li hsien-sêng tshung Chi Chhiu Tzu so shou hua huang chin fa⁵⁶)hh	+	+	+	+	·	none	conversion to Ag; conversion to 'purple sheen gold of the best colour' (shang sê tzu mo chin⁶⁰) (?Au). See pt. 2, pp. 70, 257 ff.	conversion of inorganic ions in aqueous solution, result not easily interpretable. Hsüeh Yü (1), taking fan shih shui abnormally as copper sulphate, sees first the formation of a low Cu amalgam and then of a high Cu one
53a	274	method of making realgar solution (chih tso hsiung huang shui fa⁵⁷)	·			+	·	strong vinegar, saltpetre	—	See pt. 4
53b	274	the same for azurite	+		·	+	·	—do—	—	—do—
53c	274,	the same for alum	·			·	·	—do—	—	—do—
54	274, 275	'child's-play method for making gold' (hsiao erh tso huang chin fa⁵⁸)ii	+		·	+	·	red bole clay, saltpetre, mica, haematite, sulphur, calcspar (gypsum), vinegar, Pb/Hg amalgam, lead	projection for Au from Pb, projection for Ag from Hg	formation of purple powder (tzu fên⁶¹), HgS, according to Hsüeh Yü (1), used for the projection. Cf. no. 1 for a possible explanation (Chikashige)
54a	274	to make lead-mercury amalgam (liang fei fa⁵⁹)	·			+	·	earthworm excreta (earth), cinnabar in solution, lead	—	—
55	275	'Master Wu Chhêng's (aurifaction) method' (Wu Chhêng Tzu fa⁶²)	+	+	+	+	+	—	conversion to Au, then softened and ingested; results: panacea, rejuvenation, invisibility, invulnerability, walking on water, protection by gods and spirits, and their employment, visions of ghosts and heavenly beings, with many other magical powers	formation of a Cu/As alloy; suggestions of As poisoning; cf. no. 14
Ch. 17 56	294	a formula of the Chin Chien Chi⁶³ (Gold Tablet Record)	+	+	+	+	+	orpiment, molten lead	used to make magic daggers, for invulnerability and safety in travel jj	formation of a Cu/Hg/As alloy

Notes to Table 111, pp. 90–5.

[a] With some duplicate descriptions in ch. 11.

[b] These nine, all taken by Ko Hung from a *Huang Ti Chiu Ting Shen Tan Ching*, have been discussed on p. 83 above, where the characters for their names will be found. *TT*878, a book with almost exactly the same title, seemingly compiled in the Thang or Sung, but probably including parts going back to the Later Han, gives the whole list with full details on the preparation of all of them.

[c] Both taken by Ko Hung from a *Thai-Chhing Tan Ching*. Ware (5), p. 367, has tentatively identified this book with the *Thai-Chhing Chin I Shen Tan Ching*[64] still extant (*TT*873), but although the substances used are similar, neither the names of the preparations nor the descriptions of the procedures coincide closely with the words of Ko Hung. This book is very hard to date, but it must have existed before +1022 because it is abridged in *YCCC*, ch. 65, pp. 5*a*ff. Whether or not it derives from an original text analogous to that which Ko Hung knew must be left for further research to determine. The preface bears the name of Chang Tao-Ling (see pp. 43–4) but its genuineness cannot be taken as established. Cf. Ho Ping-Yü (10).

[d] Cf. nos. 26 and 34 in this Table. The 'five minerals' are defined twice in the *Pao Phu Tzu* book, once here (ch. 4, p. 9*b*) and once in ch. 17, p. 10*b*, where magnetite is replaced by orpiment. The series cinnabar, realgar, alum, malachite and magnetite would be the most consonant with the colours (red, yellow, white, caerulean and black) required in the traditional five-element symbolic correlations (see Table 12 in Vol. 2), so it may have been one of the earliest. *Yü shih*[65] (arsenolite), also white, because of the similarity of its orthography, tended to get substituted for *fan shih*[66] (alum) but the latter is much more common in the alchemical texts—here perhaps was a real pitfall for the unwary experimentalist. If one takes the texts of other traditions, such as the pharmaceutical and medical, into account, one can compile at least half a dozen other lists, all quite different from this one, and all from authoritative sources. Cf. Sect. 45 in Vol. 6.

[e] From a *Wu Ling Tan Ching* which Ko Hung had at his disposal, not identifiable now. It contained five different methods, hence the name.

[f] From a *Min Shan Tan Fa* available to Ko Hung but no longer to us. It contained two methods, attributed to an alchemist named Chang Kai-Tha.[67]

[g] Legendary immortal, cf. *TT*293, ch. 2.

[h] Cf. Vol. 2, pp. 133ff.

[i] Nine methods.

[j] Legendary immortal, the first in the *Lieh Hsien Chuan*, cf. Kaltenmark tr. pp. 35ff.

[k] Some texts have arsenolite instead of alum.

[l] Cf. Vol. 4, pt. 1, pp. 87ff.

[m] R284 (= *V. parvifolia*), a wild vine. Also written *lei*.[68]

[n] Also in the *Lieh Hsien Chuan*.

[o] Thai-I was the name of an ancient god, and of one of the stars in the north polar region (cf. Vol. 3, p. 260).

[p] See Vol. 2, p. 148.

[q] In the *Lieh Hsien Chuan* (Kaltenmark tr. p. 132) and *TT*293, ch. 3, p. 21*b*, which says that Chi Chhiu Chün[69] flourished in the time of Han Wu Ti.

[r] R248, supposing *yang wu hao*[70] to be *yang wu hao*.[71]

[s] Here as elsewhere we diverge from Wu & Davis (2) and Ware (5). This synonym for haematite is given in the *Thai-Chhing Shih Pi Chi*, *TT*874, ch. 2, p. 9*a*, cf. p. 130 and Ho Ping-Yü (8); so no 'wrapping in plain silk' was involved.

[t] This means that the elixir has to be taken for a long time to produce its effect.

[u] Cf. Vol. 2, p. 202. The *San Kuo Chih* bibliography has a *Mo Tzu Tan Fa*,[72] but the exact connections between Mohists and alchemists are still highly obscure.

[v] Or Yü Chu's[73] in some texts; Yü Kuei's[74] in others. Perhaps all variants are mistakes for Chu Chu,[75] named as a (legendary) alchemist in *TT*293, ch. 3, p. 22*b*. Cf. p. 127 below.

[w] He occurs in *TT*293, ch. 4.

[x] In some texts.

[y] But Feifel (2) notes also that *chin hua*, according to *YCCC*, ch. 65, p. 14*b*, is something made from the three ingredients cinnabar, realgar and orpiment.

[z] This is *Lithospermum*. A gloss in Sun Hsing-Yen's edition says that *TPYL*, ch. 996, quotes it as *ti hsüeh*;[76] this fixes it as *L. officinale* (= *erythrorhizon*), R153, CC386.

[aa] *Ailanthus glandulosa* (= *altissima*), Simarubiaceae, the 'tree of heaven', R341, CC892.

[bb] Ko Hung was in possession of a *Chin I Ching*[77] (Manual of Potable Gold) on this process.

[cc] This is hard to translate, but the first two words must refer to the black colour of the pellets of elixir used, and the second two define their size. *Wei-hsi* is an old name for jet, i.e. a kind of hard and compact lignite (allied to coal), which can be easily cut and polished for ornamental use. In his entry

for *i*,[78] its more usual later name, Li Shih-Chen, about +1590, wrote as follows: '*I* is (a sort of) amber, but black. Some say it gets its colour because the earthy mould dyes it, others that it is a kind of wood which under the influence of water has concreted into this form. It is not at all necessarily produced from amber after a thousand years. The *Yü Tshê Ching*[79] (Questions about Jade) says that pine resin after a millennium turns into the *Pachyma* fungus, and this after another millennium into amber, and this again after a similar time into vitriol, and finally vitriol after yet another thousand years produces *wei-hsi* (i.e. *i*). All this is in general fabulous talk and cannot be taken seriously' (*PTKM*, ch. 37, (p. 8)). Li Shih-Chen's perspicacity was shown here in several ways: he included both jet and amber in a chapter on plants and plant materials, not as others had done, with minerals; he accepted that amber was formed from pine resin; and he recognised the plant origin of a kind of coal. Curiously enough, the book from which he quoted was listed in the *Chin Shu* bibliography as one of Ko Hung's own, though a *Chi*[80] rather than a *Ching*, but the text may have been earlier, as an anonymous one of identical title is mentioned in the *San Kuo Chih*.

The plant *chü-shêng* is hard to identify now, perhaps impossible, and the subject is discussed in Sect. 38 (Vol. 6, pt. 1). It was certainly not *Sesamum indicum* (Pedaliaceae) because it is referred to before the earliest time at which that could have been introduced. Composites such as *Mulgedium* or *Ixeris* have been suggested, but the name is at present untranslatable. Cf. p. 72.

dd *Erh* is generally a noun, meaning cakes, or meat dumplings such as the so familiar *pao tzu*,[81] but Ko Hung often uses it as a verb. Another meaning is 'bait', strangely significant in view of the argument in pt. 2, p. 283 above, but there could have been no thought of that here.

ee This person seems not to be otherwise known, but his philosophical name would refer to the Two Forces, Yin and Yang.

ff Here an animal experiment is even described, in which arrows are shot at a number of hens, only one of which is 'protected', the rest serving as controls and being killed.

gg The remainder of this chapter describes a number of plants which could be used as elixirs, but it would be out of place to list them here. Cf. Sect. 45 in Vol. 6.

hh Cf. no. 28. The name Lu Li is often found written Chio Li[82] but this is incorrect. Lu Li hsien-sêng, whose real name was Chou Shu,[83] was one of the four aged sages who visited the emperor in Chang Liang's time, *c.* −190, and advised against the supersession of the Crown Prince; cf. *Shih Chi*, ch. 55, p. 12*b*; tr. Watson (1), vol. 1, p. 148.

ii So we translate, but there may have been some concealed magical influence of Yang maleness if the presence of a young boy was really needed. Cf. no. 56.

jj The alloy is described as turning out either male or female (*mu*[84] or *phin*[85]) and these magic daggers, also male or female (*hsiung*[86] or *tzhu*[87]), are to be made from the two sorts. To test which is which, virgin boys and girls are asked to sprinkle water on the cooling metal, whereupon part of it rises to a male convexity and part sinks to a female concavity.

¹ 太清丹	² 九光丹	³ 五靈丹	⁴ 岷山丹法
⁵ 務成子丹法	⁶ 羨門子丹法	⁷ 立成丹	⁸ 取伏丹法
⁹ 赤松子丹法	¹⁰ 千歲藥	¹¹ 石先生丹法	¹² 康風子丹法
¹³ 崔文子丹法	¹⁴ 劉元丹法	¹⁵ 玄水	¹⁶ 樂子長丹法
¹⁷ 李文丹法	¹⁸ 白素	¹⁹ 玄水	²⁰ 尹子丹法
²¹ 金花池	²² 太乙招魂魄丹法	²³ 采女丹法	²⁴ 稷丘子丹法
²⁵ 龍膏	²⁶ 墨子丹法	²⁷ 張子和丹法	²⁸ 綺里丹法
²⁹ 主柱丹法	³⁰ 肘後丹法	³¹ 金華	³² 李公丹法
³³ 劉生丹法	³⁴ 白菊	³⁵ 地楮	³⁶ 樗
³⁷ 王君丹法	³⁸ 陳生丹法	³⁹ 韓終丹法	⁴⁰ 金液
⁴¹ 威喜亘勝	⁴² 餌黃金	⁴³ 銀及蚌中大珠	⁴⁴ 小神丹方
⁴⁵ 小丹法	⁴⁶ 餌丹砂法	⁴⁷ 小餌黃金法	⁴⁸ 方
⁴⁹ 兩儀子餌消黃金法	⁵⁰ 五芝	⁵¹ 雲母五種	⁵² 㴞漆
⁵³ 作丹砂水法	⁵⁴ 金樓先生所從青林子受作黃金法	⁵⁵ 治作赤鹽法	
⁵⁶ 角里先生從稷丘子所授化黃金法	⁵⁷ 治作雄黃水法	⁵⁸ 小兒作黃金法	
⁵⁹ 良非法	⁶⁰ 上色紫磨金	⁶¹ 紫粉	⁶² 務成子法
⁶³ 金簡記	⁶⁴ 太清金液神丹經	⁶⁵ 礜石	⁶⁶ 礬石
⁶⁷ 張蓋蹹	⁶⁸ 蘗	⁶⁹ 稷丘君	⁷⁰ 羊烏鶴
⁷¹ 陽烏鶴	⁷² 墨子丹法	⁷³ 玉柱	⁷⁴ 玉柱
⁷⁵ 主柱	⁷⁶ 地血	⁷⁷ 金液經	⁷⁸ 瑿
⁷⁹ 玉策經	⁸⁰ 記	⁸¹ 飽子	⁸² 角里
⁸³ 周術	⁸⁴ 牡	⁸⁵ 牝	⁸⁶ 雄 ⁸⁷ 雌

coloured suspension, probably purple-red. When the realgar was oxidised in the solvent the following reaction would have been possible:

$$3As_2S_2 + 22K^+NO_3^- + 4H_2O \rightarrow 6AsO_4^{--} + 6SO_4^{--} + 22NO + 8H^+ + 22K^+$$

With cinnabar the following would probably have taken place:

$$3HgS + 2K^+NO_3^- + 8HC_2H_3O_2 \rightarrow 3S + 3Hg^{++} + 2NO + 2K^+ + 4H_2O + 8C_2H_3O_2^-$$
$$3HgS + 8K^+NO_3^- + 8HC_2H_3O_2$$
$$\rightarrow 3SO_4^{--} + 8NO + 8K^+ + 3Hg^{++} + 4H_2O + 8C_2H_3O_2^-$$

Part of the sulphur could then have been oxidised to form sulphate:

$$S + 2K^+NO_3^- + 2HC_2H_3O_2 \rightarrow 2K^+ + 2H^+ + SO_4^{--} + 2C_2H_3O_2^- + 2NO$$

The sulphate would react with Hg forming the slightly water-soluble $HgSO_4$, and after hydrolysis becoming $2HgO.HgSO_4$ in the form of a yellow precipitate. Arsenious oxide and mercury together would form a lemon-coloured precipitate, $Hg_3(AsO_4)_2$. The ferric ions supplied by the magnetite, because of hydrolysis, would show a light brown colour in acid solution. In the presence of ferrous ions there could be other colours, but only in strong concentrations. $Fe(C_2H_3O_2)_3$ gives a deep red colour in solution. Whether *ping shih* was NaCl or $MgSO_4.K_2SO_4.4H_2O$ (*han shui shih*) makes no difference as both dissolve in water and the ions formed are colourless. The colour of the final state may well have been ruby red because of the presence of much colloidal gold, and presumably this was the condition in which the noble metal was 'liquefied'.[a]

On the other hand, if we take *hsüan ming* as vinegar and *lung kao* as wild raspberry juice (*fu phên tzu*), CN ions will be present as the unripe fruits of the latter contain hydrocyanic acid. Then saltpetre and *ping shih* (if sodium chloride) will provide K and Na ions in solution. In the presence of air gold will dissolve slowly as follows:

$$4Au + 8NaCN + O_2 + 2H_2O \rightarrow 4Na[Au(CN)_2] + 4NaOH \quad \text{and}$$
$$4Au + 8KCN + O_2 + 2H_2O \rightarrow 4K[Au(CN)_2] + 4KOH$$

The strong alkalis will be neutralised by the acetic acid and other newly formed acids. Since *fu phên tzu* also contains organic substances (especially glucose) there would be considerable reversibility; some of the gold dissolved would turn back into particles of colloidal gold in a suspension of purple-red or other colour. One difficulty is that KCN might be oxidised before it could have a chance to act on the gold and dissolve it.

Wang Khuei-Kho thinks that sulphur, magnetite, saltpetre and cinnabar were first put in vinegar to form the classical 'bath' or *hua chhih*.[1] If only a small amount of saltpetre was used, most of it would become K and NO ions in the solvent, and the slow rate of solubility might prevent the HCN ions from being oxidised after the *fu phên tzu* was introduced. Similarly, realgar would also remain unoxidised, and being in powder form would remain at the bottom of the solvent as a yellow precipitate. Since a solution of not more than 0·03 % CN is sufficient to dissolve gold, especially

---

[a] This assumes reducing conditions in a later stage of the process.

[1] 華池

over a long period of time, the soluble gold salt could have been formed.[a] Thus in this case Ko Hung ended with potassium auricyanide rather than with colloidal gold. As a means of attaining material immortality the former can hardly have been more salubrious (or pleasant to take) than the latter.

All this reasoning is of course hypothetical enough, but it constitutes only a preliminary attempt to explain what was happening in this *Pao Phu Tzu* process, assuming (*a*) the correct interpretation of the ingredients, and (*b*) that he did start from real gold and not from some artificial gold-like alloy. It is at any rate not invalidated by the argument of Mêng Nai-Chhang (*1*) that some descriptions speak only of gold and vinegar, e.g. that in the *Thai-Chhing Chin I Shen Tan Ching*,[b] a work of unknown date and authorship; for the writer might only too easily have been concealing the details of the process.[c] Dilute acetic acid would certainly do nothing to gold by itself, but one doubts whether anybody ever believed that it could.

Various other methods of ingesting gold are given by Ko Hung, who says that although these are inferior to the elixirs and the potable gold just mentioned they are still far better than alternative ways of achieving longevity and immortality.[d] According to him gold can be ingested if treated with the skin and fat of the hog in wine, or prepared with *Ailanthus* bark, or made into a suspension with wine from Ching[1] and magnetite, or drawn out into leaf (*chin*[2]),[e] or taken together with realgar and orpiment.[f] Similarly silver, and large pearls from oysters, could be made into a suspension or dissolved and so consumed.[g]

We find five further elixir formulae repeated by Ko Hung in different chapters.[h] These show how the same alchemical terms could be written in different ways even by the same author, always supposing that the divergences were not introduced by subsequent copyists. For example, in one place we find *tan*,[3] and in another *tan sha*,[4] for cinnabar; while the method *Liang I Tzu erh hsiao huang chin fa* (Master Liang I's method of ingesting small amounts of dispersed gold in suspension) in one place is called *liang (i) erh hsiao huang chin fa* (Method of using the Two Fundamental Forces (Yin and Yang) to bring gold into suspension) in another. This fluctuation is always happening in the old alchemical literature. For example, in the *Thai-Chhing Shih Pi Chi*,[i] the preparation called 'Soul-recalling elixir' (*chao hun tan*[5]) is given twice, but with distinct differences in each case. Presumably this reflects the cooperation of

[a] Nevertheless, Mêng Nai-Chhang (*1*) remains sceptical that even this low strength was reached.
[b] *TT*873. Cf. note c to Table 111.   [c] Cf. pp, 181, 195 below.
[d] *PPT/NP*, ch. 4, p. 15*b*; tr. Ware (5), p. 92. Nos. 40, 44, 45 in Table 111.
[e] This shows that the goldbeater's art was known. One of us (J. N.) vividly remembers the decoration of sweet dishes in India with edible gold leaf—perhaps an alchemical survival.
[f] As we have already suggested (pt. 2, pp. 69, 71, 271) many of these procedures would make sense if one supposes that the golden crystals of stannic sulphide (mosaic gold) were being taken up with various media, vehicles in pharmaceutical language, for ingestion. Cf. p. 103 below.
[g] Since the alchemists of Ko Hung's time made so much use of vinegar, they doubtless dissolved pearls (when they could get them) in the manner of the Egyptian queen, though the little extra dose of calcium carbonate would not have done them or their clients any particular good.
[h] Nos. 41, 42, 43, 44, 45 in Table 111; in ch. 4, pp. 15*b*, 17*b*, 18*a* and ch. 11, pp. 16*a, b*, 17*a*.
[i] *TT*874, mainly of the early +6th century; see pp. 130ff. below. Cf. no. 26 in Table 111.

[1] 荆   [2] 巾   [3] 丹   [4] 丹砂   [5] 召魂丹

numerous practitioners, each with his individual experiences, preferences, beliefs and modifications. The same thing may also come under two or more different names, and sometimes a single name covers a variety of constituents and processes.

The idea that a small amount of a certain potent substance could be used to transform a large amount of base metal into gold or silver is of course that of the 'philosopher's stone' later so prominent in Europe. The process was known as 'projection' (*tien*[1]).[a] Now in chapter 16 Ko Hung describes several instances where the noble metals were made by projection. First, Wu Ta-Wên[2] saw the Taoist Li Kên[3] putting a little of a chemical substance into boiling lead and tin, and stirring it with an iron spoon; then when allowed to cool it turned into silver. However, Wu Ta-Wên himself met with failure when he repeated the experiment after, so he thought, learning the art.[b] As we saw in the quotation given on p. 38 above, Chhêng Wei[4] tried to force his wife to divulge to him the secret of making gold and silver by projection until the latter had to run away from him.[c] Hua Ling-Ssu[5],[d] the Prefect of Lu-chiang,[6] did not believe in the possibility of making gold and silver, and asked a certain Taoist to demonstrate his skill before his eyes. Ko Hung relates that this Taoist melted some lead in an iron vessel, and silver was immediately formed when he put in some chemical substance. Then melting this 'silver' again he put in some other chemical substance and turned it all to gold.[e] Ko Hung himself undoubtedly believed in the feasibility of making gold and silver (as he defined them).[f] He even says that gold obtained by alchemical means is superior to gold found in nature as an ingredient of elixirs, because the former embodies the essence of the various chemical substances used for making it.[g] This is a very significant point, the point where the practical alchemy of China could conceivably have joined hands with a 'spiritual alchemy' such as grew up in the West. For if artificial gold was a finer thing than natural gold it was not a far cry to the idea that the process of making it was a parallel to, or an image of, the purification of oneself. Man in his perfected state, an artificial not a natural work, a gold produced by exhausting labour, not picked up as a spontaneous creation, would thus indeed be 'higher than the angels'. But there is no sign that Ko Hung himself, or the other Chinese alchemists, ever had any such ideas; they remained to the end believers in practical experimentation. And as we shall presently see (in pt. 5), the 'spiritual alchemy' of China was a far different thing from that of Europe.

Many accounts of alchemical transmutation by projection can be found at the time

---

[a] On the origins of this idea cf. pt. 2, p. 27 above. From p. 39 we know that it started in China at least as early as among the Hellenistic Egyptians. To date it in India is more difficult, but p. 161 below will presently demonstate that it can hardly be much younger there, for we find it in the first few centuries of the era.

[b] *PPT/NP*, ch. 16, p. 2b, tr. Ware (5), p. 264.

[c] *Ibid.* p. 3a, tr. pp. 264ff.

[d] I.e. Hua Than,[7] whose biography is in *Chin Shu*, ch. 52, p. 6b.

[e] *PPT/NP*, p. 3b, tr. pp. 265ff.

[f] See pt. 2, pp. 62ff., esp. p. 70.   [g] *PPT/NP*, p. 5a, tr. p. 268. See p. 2 above.

[1] 點   [2] 吳大文   [3] 李根   [4] 程偉   [5] 華令思
[6] 廬江   [7] 華譚

of Ko Hung. For example, one of his contemporaries was Yin Kuei[1] or Thai-Ho Chen Jen[2].[a] In the *Thai-Phing Yü Lan* we read:[b]

The *Shen Hsien Chuan* says: 'Yin Kuei's courtesy name was Kung-Tu.[3] There was a man whose father had died and was to be buried, but he was very poor (and had not the means). (Yin) Kung-Tu passing by asked him what was the matter and learnt of his plight. Moved by sympathy he said: "Can you borrow several tens of catties of lead?" The filial son said that he could, and soon came back with a hundred catties. (Yin) Kung-Tu took it up into the neighbouring mountains, where he built a small shelter and melted the lead in a furnace; then taking out from a tube which he had brought along with him a small pellet of chemical as big as a jujube-date, he threw it into the molten lead. On stirring, all became silver of good quality. This he gave to the poor man, saying that he had made it on account of his misfortune, and asking him not to talk much about it.'[c]

This story is curiously reminiscent of later European parallels where the good moral intentions and character of the alchemist tend often to be stressed. After Ko Hung's time there developed a special technical term for projection, *tien hua*,[4] change or transformation effected by a mere 'spot' of substance added (cf. pp. 193, 213 below).[d]

Several methods of aurifaction are described by Ko Hung. Of these we quote one as follows:[e]

First take any desired amount, but not less than five catties, of realgar obtained from Wu-tu,[5] vermilion in colour like a cock's-comb, lustrous and free from bits of rock. This is pounded to powder and mixed with ox bile and heated until dry. Take a red clay pot (*fu*[6]) with a capacity of one peck (as the lower part of a reaction-vessel), spreading crude Kansu salt (*jung yen*[7]) and blue vitriol (*shih tan*[8]) in powder form all over the inside to a thickness of three-tenths of an inch. Then put in the realgar powder, (spreading it) to a thickness of five-tenths of an inch, and placing more of the salt (mixture) over it until it is completely (covered). Next spread on top of this a layer of pieces of hot charcoal, about the size of jujube-date stones, two inches thick. The pot must be smeared all over outside with a lute made from the earth (excavated by) earth-worms and crude salt.[f] Another pot is then inverted over (the lower one) (to form the reaction-vessel), and all the outside smeared with lute to a thickness of three inches so that there can be no leaks. After allowing the whole to dry in the shade for a month it is heated in a fire of burning horse-dung for three days and three nights. When cool, remove the contents (and place it in a smelting furnace), then work the bellows to liquefy the copper (*ku hsia chhi thung*[9]), and it will flow like newly smelted copper or iron. This copper(-like) substance is then cast into the shape of a cylindrical container (*yung*[10]), and

---

[a] *TT*293, ch. 8, pp. 19a to 21b, places him in Chin Hui Ti's time, +290 to +307, and this is confirmed in a story similar to that quoted here, but in *TPYL*, ch. 812, p. 8a, where it is said that he went away into the mountains for good in +306.

[b] Ch. 812, p. 7a, tr. auct.

[c] The parallel story makes it tin, not lead.

[d] Literally identical with the current slang expression, 'a spot of' something or other.

[e] Table 111, no. 51. *PPT/NP*, ch. 16, p. 7a, b, tr. auct., cf. Sivin (1); Ware (5), pp. 271ff. This is the method which seems to be Ko Hung's own.

[f] This kind of instruction presumably indicates that glazed earthenware was not being used.

---

[1] 尹軌   [2] 太和眞人   [3] 公度   [4] 點化   [5] 武都
[6] 釜   [7] 戎鹽   [8] 石膽   [9] 鼓下其銅   [10] 筩

filled with an aqueous solution of cinnabar (*tan sha shui*[1]).[a] This is again to be heated in a horse-dung fire for thirty days and then (the contents) taken out, pounded, and smelted. Two parts of this with one part of crude cinnabar added to mercury will immediately solidify it into gold. It will be bright and shining with a beautiful colour, fit for making into *ting*[2].[b]

Whatever else this may mean it must involve the making of a copper-arsenic alloy. The arsenic and copper compounds would be reduced by the charcoal and the ox-bile, the salt acting as a flux and adding minor constituents. The initial heating in the aludel would not effect this, but the furnace would. Cinnabar solution however requires a little further explanation. As we shall see later, in the discussion of medieval methods for getting inorganic substances into solution, vinegar was constantly used with saltpetre, giving dilute nitric acid.[c] In the *Pao Phu Tzu* book the formula follows immediately on the passage just quoted;[d] the powdered cinnabar is placed with saltpetre and copper sulphate in a sealed bamboo container, the whole immersed in strong vinegar, and buried in the ground for a month. The resulting solution, red and bitter, must have contained mercuric, copper and potassium anions, and sulphate, nitrate, acetate and arsenate cations. Within the vessel of copper-arsenic alloy progressive corrosion will then take place during the slow heating, bringing more copper and arsenic into solution and finally into the dry contents. The eventual smelting presumably gave a copper with just sufficient arsenic to imitate more or less the colour of natural gold. The mercury and its sulphide were then unnecessary because volatile at the melting-point of the alloy, but this could not have been understood. As Sivin (1) rightly says, this rather involved procedure is quite as interesting in its deployment of techniques as any of those which formed the basis of Hellenistic 'alchemy'.

The theme of the dissolution of inorganic substances runs all through the writings of Ko Hung, and must go back well before his time.[e] For making aqueous solutions of realgar (As_2S_2), or azurite (*pai chhing*,[3] basic copper carbonate), or potash alum (*fan shih*[4]), the substance concerned was placed with saltpetre in a bamboo tube immersed in a bath of vinegar.[f] Cinnabar, as we have just seen, was treated in the same way.[g] In another section of the book,[h] Ko Hung mentions the feasibility of bringing all thirty-six minerals into aqueous solution,[i] of reducing jade to a potable form so that it has the appearance of a sweetmeat, and of breaking up gold into a paste.

[a] Sun Hsing-Yen has pointed out that twenty-seven mostly repetitive words after this sentence should be regarded as commentary and not the text itself. Hence we omit them in our translation.

[b] The obvious meaning here is nails, but it would be queer if anyone ever thought that real gold was hard enough for this. The *Shuo Wên*, however, defines *ting*[2] as melting gold and casting it into little ingots (*ping*[5]). Perhaps therefore *ting*,[6] ingot, was intended by Ko Hung. The same expression occurs elsewhere, e.g. ch. 16, p. 5a. On the other hand he may really have meant nails, feeling that his artificial gold was harder and better than natural gold. If our interpretation of the meaning of the process is right, it was indeed harder than natural gold. An emendation to needles, *chen*,[7] very similar in orthography, has even been suggested (Chang Tzu-Kao).

[c] Pt. 4 below.   [d] Ch. 16, p. 7b, Ware tr., p. 272.

[e] We discuss the probable date of origin in pt. 4 below.

[f] *PPT/NP*, ch. 16, p. 8b; Ware (5), p. 274. See nos. 53a, 53b, 53c in Table 111.

[g] No. 51a in Table 111.   [h] *PPT/NP*, ch. 3, p. 1b; Ware (5), p. 54.

[i] 36 here means 'many', not exactly 36.

[1] 丹砂水   [2] 釘   [3] 白青   [4] 礬石   [5] 餅   [6] 錠   [7] 針

One of the other gold-making processes described by Ko Hung has great chemical interest,[a] for it involved the earliest known preparation of stannic sulphide, a compound with properties just as interesting as those of the calcium polysulphides which were prominent in Hellenistic proto-chemistry (cf. pt. 2, pp. 252, 271). As will be seen, Ko Hung did not claim it as his own, but attributed it in his usual way to venerated predecessors, so that indeed it may well be older than his time. He wrote as follows:[b]

According to the method for making yellow gold received by the teacher Chin Lou (Chin Lou hsien-sêng[1]) from Master Chhing Lin (Chhing Lin Tzu[2]), one first casts cakes of tin six inches square and 1·2 inches thick. These little ingots are then covered all over to a thickness of a tenth of an inch with a paste of red salt[c] and lime-water, and packed one after another into a crucible of red (refractory) clay. For every ten catties of tin one uses four catties of red salt. Close the crucible and seal all cracks well; then heat in a fire of horse dung for thirty days. When opened and examined after removal from the furnace, it will be found that the tin has all gone to a kind of ash[d] in the midst of which there are lumps like clusters of beans—this is the gold. Or mix the substances together and put them into an earthenware pot for ten successive refinings over a charcoal fire blown by bellows; both (methods) will be successful.[e] The proportion is that for every ten catties of tin you get twenty ounces of gold. Only the clay crucibles made in Chhangsha, Kueiyang, Yüchang and Nanhai are suitable (for this work). Such pots are quite easy to get because the people of those places make them for cooking.

The convincing identification of the product here with stannic sulphide (SnS_2) or 'mosaic gold' was made by Wu & Davis.[f] It forms golden yellow glistening hexagonal scales or flakes which do not tarnish and are still used for gilding and bronzing. In this case therefore Ko Hung and his predecessors did really succeed in producing an artificial gold from substances manifesting none of the properties of the precious metal.

Other statements of proto-chemical interest made by Ko Hung are as follows:

(a) Ordinary people would not believe that minium (huang tan,[3] Pb_3O_4) and white lead (hu fên,[4] $PbCO_3$) are made from lead by chemical change.[g]

(b) I aver that the 'flowing pearls' (liu chu,[5] mercury) will 'fly' (fei,[6] i.e. vaporise or distil); and that gold and silver can be made.[h]

[a] No. 52 in Table 111.

[b] PPT/NP, ch. 16, p. 8a, b, tr. auct. adjuv. Ware (5), p. 273; Wu & Davis (2), pp. 264ff.

[c] The preparation of 'red salt' is described by Ko Hung in a paragraph immediately following the above passage. Some form of gypsum, i.e. calcium sulphate, was mixed with some variety or varieties of potash and iron alum and fused in an iron pot. This would have provided the sulphur for the reaction with the tin, but the yield of about 12% mentioned very candidly by Ko Hung was not high. The text here is somewhat uncertain (editions differ) and we have not followed the previous translators.

[d] Wu & Davis suggested that this may have been the powdery allotropic modification known as 'grey tin', but this usually appears only when tin is subjected to cold (Mellor (1), p. 789).

[e] Here we prefer the interpretation of Wu & Davis to that of Ware. The latter envisaged a second set of heatings, which would be pointless; surely Ko Hung was describing a slow way and a quick way of carrying out the preparation.

[f] (2), p. 232, accepted by Leicester (1), p. 57. See Mellor (1), p. 411; Durrant (1), p. 420. In Europe it was known in the +14th century and well described in +1679 by Johann Kunckel (Ars Vitraria Experimentalis); cf. p. 99 above, and Partington (7), vol. 2, p. 375; (10), p. 521.

[g] PPT/NP, ch. 2, p. 11b; Ware (5), p. 52.   [h] Ch. 3, p. 5b; Ware (5), p. 60.

[1] 金樓先生   [2] 青林子   [3] 黃丹   [4] 胡粉   [5] 流珠   [6] 飛

(c) Crabs affect the setting of (lit. change) lacquer, and hemp-seed oil spoils wine. We cannot give the explanation for such phenomena (*pu kho i li thui chê yeh*[1]). How can we hope to get to the bottom of the vast profusion of Nature's effects?[a]

(d) When salt and brine penetrate flesh and marrow, dried meat will not putrefy. Why be surprised then that longevity follows when people take substances that are beneficial for their lives and bodies?[b]

(e) When plants are burnt, they become ashes, but cinnabar when heated turns into mercury, and after many transformations changes back into cinnabar.[c]

(f) The Manuals of the Immortals say that the essence of cinnabar produces gold; this is another way of saying that gold can be made from cinnabar. This is why gold is generally found below deposits of cinnabar in the mountains.[d] Moreover when the process of making gold has been successful, it is the real thing, it will be homogeneous inside and out, and a hundred refinings will not change or diminish it. So when the formularies say that it can be made into ingots,[e] they mean that it is firm and stable. This is done by following the spontaneous Tao of Nature itself—how could such a substance be said to be counterfeit (*cha*[2])? Of course there are counterfeit things. For example, when iron is rubbed with stratified malachite (*tshêng chhing*,[3] basic $CuCO_3$), its colour changes to red like copper. Silver can be transformed by the white of an egg so that it looks yellow like gold. However, both have undergone changes outside, but not inside.[f]

Ko Hung often seems to try to explain some of the jargon used in alchemy and Taoism, although he does not always give the exact meanings of the secret cover-names. He emphasises the importance of getting oral instruction precisely because of the synonyms and metaphors used in alchemy. He illustrates this by saying that the 'elegant girl by the riverside' (*ho shang chha nü*[4]) does not refer to any woman, nor does Lingyang Tzu-Ming[5] (the name of an adept mentioned in the *Lieh Hsien Chuan*)[g] mean a particular person—but without further explanations.[h] However, in other places he explains that *hung*[6] means mercury (*shui yin*[7]), 'yellow' (*huang*[8]) and *kêng hsin*[9] both refer to gold, and 'white' (*pai*[10]) is silver.[i] The most peculiar synonyms given by Ko Hung relate to the animal kingdom; for example the 'elder' (*chang jên*[11]) is the rabbit, the 'rain master' (*yü shih*[12]) is the dragon, the 'Count of the River' (*ho po*[13]) mean fish (generally the carp), the 'gutless lordling' (*wu chhang*

---

[a] *PPT/NP*, ch. 3, p. 5*b*; Ware (5), p. 61. The lacquer story will be found fully discussed in pt. 4 below. Ko Hung's words here are a striking statement of his empirical attitude.

[b] *Ibid.* ch. 3, p. 6*b*; Ware (5), p. 62.

[c] *Ibid.* ch. 4, p. 3*b*; Ware (5), p. 72.

[d] On mineralogical prospecting cf. Vol. 3, pp. 673 ff.

[e] *Ting*[14] is used again; if he really meant nails, one should take the following word as 'hard'.

[f] *PPT/NP*, ch. 16, p. 5*a*; Ware (5), p. 268; Wu & Davis (2), pp. 263 ff. Cf. Vol. 5, pt. 2, pp. 67, 251 ff.

[g] Kaltenmark (2), no. 68, p. 183. Lingyang Tzu-Ming became the tutelary genius of the reel of the fishing rod, and his story may embody one of the oldest references to this invention, cf. Vol. 4, pt. 2, pp. 44, 100, and p. 12 above.

[h] In *PPT/NP*, ch. 16, p. 6*b*. Both these synonyms mean mercury, according to *TT*899, p. 6*b* and *TT*993, ch. 2, p. 25*a*.   [i] In *PPT/NP*, ch. 16, p. 7*b* and p. 1*a*.

[1] 不可以理推者也	[2] 詐	[3] 曾青	[4] 河上奼女	[5] 陵陽子明
[6] 汞	[7] 水銀	[8] 黃	[9] 庚辛	[10] 白	[11] 丈人
[12] 雨師	[13] 河伯	[14] 釘		

kung tzu[1]) refers to the crab, the 'scholar' (*shu shêng*[2]) the cow, the 'immortal' (*hsien jen*[3]) an old tree, and so on.[a]

Ko Hung mentions the names of many past adepts and alchemists, sometimes telling us how they achieved the state of immortality and sometimes describing their magical powers. For example, he says that when Kan Shih[4] put some elixir in the mouth of a fish it could swim about in boiling oil, when he fed it to a silkworm it stopped developing even after the tenth month, when he let chickens and puppies eat it they grew no more, and when he gave it to a white dog its hair turned black. Tso Tzhu[5] showed no sign of physical change when he ceased to eat for a month, and claimed that he could remain alive for fifty years without eating.[b]

The existence of a considerable number of alchemical texts before his time is indicated by Ko Hung at the beginning of chapter 19 of his book. In fact he gives us one of the oldest bibliographies of Taoist literature, amounting to 206 book-titles in all.[c] Unfortunately few or none of the alchemical texts are extant now, though one may wonder whether the *Wei Po-Yang Nei Ching*[6] (The Inner Book of Wei Po-Yang) did not mean the *Tshan Thung Chhi*; and it is also likely that some of the contents of another book, the *San-shih-liu Shui Ching*[7] (Manual of the Thirty-Six Aqueous Solutions), is quoted in the *San-shih-liu Shui Fa*[8] (Thirty-Six Methods for Bringing Solids into Aqueous Solution), a text in the present *Tao Tsang*.[d] Also at the beginning of chapter 11, Ko Hung quotes a book called the *Shen Nung Ssu Ching*[9] (Four Books of Shen Nung). Yoshida suggests[e] that this might be the same as the *Shên Nung Pên Tshao Ching*[10] (Pharmacopoeia of the Heavenly Husbandman); but there is considerable divergence between what Ko Hung quotes and the text of the latter that we have now.[f]

A number of instances of magic and strange arts are described in the *Pao Phu Tzu* book. The story of Luan Ta[11] making chessmen hit one another of their own accord is given.[g] Wei Shang,[12] we are told, was able to disappear from sight when sitting in calm meditation. Chang Khai[13] could at will produce cloud and mist.[h] Tso Tzhu and Chao Ming[14] had the power of making flowing water go backwards, lighting a fire on a thatched roof to cook their food without burning the hut, or sucking out

[a] All these are contained in *PPT/NP*, ch. 17, p. 7a, tr. Ware (5), p. 288.

[b] *Ibid.* ch. 2, pp. 5a, 6a; see Ware (5), pp. 40ff. It may be of interest that most of these examples come from the court of the Wei state in the Three Kingdoms period, where under the sons of Tshao Tshao there was a lively interest in natural wonders and strange phenomena (cf. Vol. 3, p. 659, Vol. 4, pt. 1, p. 39, and *Po Wu Chih*, ch. 5). Among the others mentioned by Ko Hung in his *Pao Phu Tzu* are Master An Chhi, Yin Chhang-Shêng, Li Shao-Chün, Liu Hsiang, Chang Liang, Huang Shih Kung, Ko Hsüan, Chêng Yin, and Chhih Sung Tzu, all of whom we have already had occasion to meet before.

[c] Pp. 3bff. See Ware (5), pp. 313, 379. There is a special study of the bibliography, identifying some other books quoted by Ko Hung; see Ishijima Yasutaka (1).

[d] *TT*923. This text has been translated in full by Tshao, Ho & Needham (1). For reactions in aqueous solution see pt. 4 below.

[e] (5), p. 217.   [f] Checked particularly by one of us (G. D. L.).

[g] Cf. p. 32 above, and for the magnetic explanation of this feat, Vol. 4, pt. 1, pp. 315ff.

[h] For all these three instances see *PPT/NP*, ch. 3, p. 6b; Ware (5), p. 63.

[1] 無腸公子   [2] 書生   [3] 仙人   [4] 甘始
[5] 左慈   [6] 魏伯陽內經   [7] 三十六水經   [8] 三十六水法
[9] 神農四經   [10] 神農本草經   [11] 欒大   [12] 魏尚
[13] 張楷   [14] 趙明

with their breath a nail driven deep into a wooden pillar, etc., etc.[a] Pu Chhêng[1] stepped up and up to the clouds until people lost sight of him.[b] A knowledge of magic was, Ko Hung believed, essential for aspirants to *hsien*-ship, so that they might protect themselves against calamities due to natural or human causes, and be enabled to carry out the Great Experiment. As the Subtle Work had to be performed in secrecy and seclusion among great mountains, Ko Hung gives detailed instructions on how to select by astrological calculation auspicious days on which to begin one's journey, what taboos one must guard against, and what proper charms or talismans one should bear or wear in order to keep wild animals, snakes, poisonous insects, and evil spirits away. There is a long and curious passage on mirrors as demonifuges.[c] Strong Taoist influence on alchemy is apparent from the use of charms and amulets, Taoist magic and ceremonies, and the frequent mention of terms like Taoist (*Tao jen*[2]) and Taoism (*Tao chia*[3]) coupled with the names of adepts and immortals with a Taoist flavour.

We have written already about the alchemists who thronged the vestibules of emperors and princes, like the *fang shih* of Han Wu Ti, Liu An, and Tshao Tshao. Ko Hung now tells us about certain charlatans who deceived their disciples by getting from them not only material support but also free labour in exchange for false promises.[d] Ko Hung himself had personally met a number of such false teachers and said that in the end they would be punished by the holy immortals for their wrong-doings.

(iii) *Character and contemporaries*

The moment has come to say a word about the personal character and cast of mind of the great alchemist, physician and natural philosopher whom we have been discussing.[e] A fascinating contrast is displayed if we compare him with another great scientific writer working fifteen hundred years before modern science was born. Much has been said in previous volumes of Wang Chhung[4] (+27 to +97), and in one place we quoted *in extenso* his demonstration that the tides of the sea depended on the moon's attraction and not on any fabled earthly influences.[f] It was typical of the way in which he would take a popular belief and tear it limb from limb, exposing its illogicalities to ridicule and ending with a rational theory or a determined suspension of judgment. Wang Chhung was the great representative in ancient China of the sceptical and rationalist frame of mind. He was a stout opponent of all forms of divination and derided those who believed that the lightning-stroke was a divine retribution, or that the unethical behaviour of rulers brought about natural calamities. He was unwavering in his attacks on the lore of ghosts and spirits, dismissing almost

---

[a] *Ibid.* ch. 5, p. 6a; Ware (5), p. 106.

[b] *Ibid.* ch. 5, p. 7a; Ware (5), p. 108. Sun Hsing-Yen thought that this name might be a mistake for one of the magician-technicians mentioned in the *Hou Han Shu*, Shangchhêng Kung.[5] We have already discussed this case (pt. 2, pp. 107, 109 above).

[c] *PPT/NP*, ch. 17, pp. 2aff. tr. Ware (5), pp. 281ff. Mirrors have been prominent in folk-lore all over the world, worn by ceremonial dancers, etc.

[d] Charlatans with hosts of disciples are mentioned in *PPT/NP*, ch. 14, pp. 7aff.; Ware (5), pp. 236ff.

[e] Discussions on the thought of Ko Hung will be found in Forke (12), pp. 204ff. and newly in Hou Wai-Lu *et al.* (1), pp. 263ff.; Murakami Yoshimi (3). Ko Hung's position within the schools of Taoism is discussed by Fukui Kōjun (1).   [f] *Lun Hêng*, ch. 16, tr. Vol. 3, pp. 485ff.

[1] 卜成   [2] 道人   [3] 道家   [4] 王充   [5] 上成公

contemptuously Taoist claims of longevity and material immortality, by whatever means to be attained. His scepticism, moderated by gentility, remained through the centuries the standard attitude of Confucian scholars.

How different a man was Ko Hung, the untiring experimentalist who worked with his own hands at bench and furnace;[a] the frequenter of the chemical technicians, smelters and metal-workers of his time. Typical of both was Wang Chhung's pessimism about natural knowledge, and Ko Hung's corresponding optimism. Wang's mind was occupied with Chance and Fate; Ko believed that men could change their fate. Wang would have written off Ko as hopelessly credulous, while Ko would have regarded Wang as uselessly sceptical.[b] The sceptical-rational typified by Wang Chhung was opposed by the mystical-empirical in Ko Hung, who was in effect constantly saying:

> There are more things in heaven and earth, Horatio,
> Than are dreamt of in your philosophy.

This can well be seen in the eloquent passage reproduced in full at an earlier stage, which begins: 'The rumbling thunder is inaudible to the deaf, and the three luminaries are invisible to the blind. . . .'[c] Interesting too is the fact that both their works, the *Lun Hêng* and the *Pao Phu Tzu*, are written in a clear, easy and discursive style. All this constitutes another example of the argument elaborated elsewhere[d] to show that in the opening phases of the development of natural science, mystical religion may be more valuable heuristically than rational philosophy. Wang Chhung and Ko Hung deeply represented two opposite poles in the psychology of all scientific endeavour to understand the world of Nature.[e] Could one not say that this antithesis had to be

---

[a] So we write, but there remains much diversity of opinion as to how far Ko Hung was really a practical experimenter, and to what extent he simply collected alchemical information from books and from adepts whom he knew. On the general relation between naturalism and alchemy, Yamada (2) has an interesting discussion.

[b] Ko Hung certainly knew the *Lun Hêng* well; indeed he recorded his great admiration for Wang Chhung's genius (*PPT/WP*, ch. 43, p. 1a). There is also a reference to Wang's autobiography (*Lun Hêng*, ch. 30) in Ko Hung's *PPT/WP*, ch. 50, p. 13a. However, it is clear that Ko Hung disagreed with him, as in the notable passage in *Chin Shu*, ch. 11, p. 3a, b, where Ko defended the Hun Thien theory against Wang Chhung's views (tr. Ho Ping-Yü (1), pp. 54ff.; cf. Vol. 3, p. 218). And many of the paragraphs in *Pao Phu Tzu* (*Nei Phien*) sound like defence speeches against the scepticism of Wang Chhung.

[c] *PPT/NP*, ch. 2, pp. 2aff., tr. Vol. 2, p. 438, cf. Ware (5), pp. 35ff. There is a parallel passage in ch. 8, pp. 7aff., tr. Ware (5), pp. 146ff.; Davis & Chhen Kuo-Fu (1), pp. 308ff. It is interesting to read the complementary passage in *Lun Hêng*, ch. 10 (tr. Forke (4), vol. 2, pp. 4ff., partly improved by Leslie (1), pp. 170ff.). Here all the emphasis is on pre-established harmony and on the constant universal cosmic rhythms, i.e. on regularity; and not on the strange exceptional things and occurrences which demonstrate the need for more subtle and further-reaching fundamental concepts, and which fascinated Ko Hung. This contrast between regularity and repetitiveness as against uniqueness and unpredictability crops up frequently in Chinese thought, cf. e.g. Vol. 3, p. 634 and Sivin (9).

[d] Vol. 2, pp. 89ff.

[e] This was well seen by Leslie (1), p. 165, in his study of Wang Chhung's biological philosophy. 'Wang Chhung's reasoning and metaphysics', he said, 'were eminently favourable to scientific research. A superior natural philosophy kept within bounds his speculations, the biological ones being of a very high order; but the crucial factors of systematic observation and experiment were lacking. Experimentation was found among the Taoists, with their alchemical and physiological search for immortality; but unfortunately Wang's restraining logic was not heeded. In the West, the scientific revolt from the +14th to the +16th centuries, though associated with mysticism and anti-authoritarianism (comparable with the Taoist revolt), managed to combine rationalism with its empiricism to produce modern science. In China this combination was never adequately made.'

reconciled in a synthesis when modern science was born? Did not Confucianism and Taoism, scepticism and enthusiasm, marry at last in the Renaissance among the men surrounding Paracelsus, Galileo and Francis Bacon, who had never heard of either of them? Alas that traditional Chinese society never permitted such a marriage to take place!

A lot more could be said of the psychology of Ko Hung and his contemporaries. How did they 'keep their heads' in the midst of so much religious-magical 'enthusiasm'? How did he manage to make so many true observations of chemical behaviour, and carry out so many interpretable experiments, even though he himself could never interpret them? Here a study of religious experience in other climes and contexts might be very revealing.[a] For example, take what Ronald Knox said about John Wesley: 'On these and many other occasions you feel that Wesley was in the position of the old prophet—"being in a trance, but having his eyes open". Wesley the enthusiast, rapt in the communal ecstasy of some consoling love-feast, is being watched all the time by Wesley the experimentalist in religion, who is taking notes, unobtrusively, for his "Journal".' So in some such way Ko Hung had one eye on the magic, the sacrifices, the Taoist temple liturgies, but he kept the other firmly fixed on the real changes and transformations which he observed at his bench and his furnaces.[b]

To take leave of Ko Hung it may be worth quoting his prose poem on the Tao with which his book opens.[c]

Pao Phu Tzu said: 'The Mysterious (Tao) is the first ancestor of the spontaneous Natural Order (*tzu-jan*[1]), and the oldest forefather of the ten thousand things. Boundless and impenetrable are its depths; we have to describe it as elusive. Endlessly continuous and prolonged is its length; we have to call it marvellous. Higher it is than the nine heavens, so vast that it envelops the eight corners of the universe. Brighter it is than the sun and moon, swifter than the lightning in the storm. Sometimes it flashes by and disappears like a spark in the air, sometimes it shoots like a meteor in the sky; sometimes it reveals itself in the rippling surface of deep water, sometimes in the drifting mists and floating clouds.

(The Tao) is the cause of the thousand categories of the things that exist, but it also hides itself in silence and emptiness. It reaches as far as the depth of the great abyss, and moves above the height of the pole of the heavens. Metal and stone are not comparable in hardness to the hardness (of the Tao), neither is its softness approached by that of a drop of clear dew. It has perfect rectangularity but never uses a carpenter's square, perfect roundness, yet knowing no compasses. Who can take note of its arrival, or follow it when it departs? *Chhien* relies upon it for its exaltation, and *Khun* relies upon it for its lowliness;[d] only because of its power do the clouds move, only because of it does the rain come down.

The (Tao) bore the primal unity (*yuan i*[2]) in its womb, and cast in their moulds the two fundamental forces (*liang i*[3]);[e] it breathed forth the great beginning (*ta shih*[4]), and blew the

[a] One thinks of course of James (1) and Knox (1).
[b] The same level-headedness was abundantly characteristic of Thao Hung-Ching also, two centuries later, deeply involved as he was with the 'enthusiasm' and *Schwärmerei* of the nascent Taoist Church at Mao Shan.
[c] *PPT/NP*, ch. 1, pp. 2b ff., tr. auct., cf. Ware (5), pp. 28 ff.
[d] The two chief of the *kua*, corresponding to Yang and Yin respectively (cf. Vol. 2 *passim*).
[e] Yin and Yang.

[1] 自然   [2] 元一   [3] 兩儀   [4] 大始

bellows for the smelting of the myriad categories of things. It set in motion the cycle of the twenty-eight (lunar mansions), and none other was the artisan (*chiang*[1]) of the completion of the early ages of the world. It stands at the controls of the numinous machine of the universe (*phei tshê ling chi*[2]), and its breath is the *chhi* of the four seasons. Abscondite it embraces all darkness and silence, invisibly it displays all things beautiful and fragrant. It makes turbidity settle, and brings forth clear waters, regulating the flow of the (silt-laden Yellow) River and the (transparent) Wei (River). Add what you will, it will never overflow, take away what you will, it will never be deficient; nothing you do will add to its glory or render it in any way impoverished. Therefore wherever the Mysterious (Tao) is there is infinite happiness, and wherever it withdraws the light of its countenance the vessels are broken and the spirits depart.

Such was the reverent mind in which Ko Hung manipulated his crucibles and recorded the changes of colour and substances during his operations. His contemporary Zosimos of Panopolis, though set in a world of creative divinity and personalised demiurges, would surely have appreciated it deeply.

Ko Hung's work remained so famous in subsequent periods that many passages from it were copied verbatim into other books, some of which consisted almost entirely of excerpts from him, while a legion of later writings were attributed to his pen. Almost the whole of chapter 4 of the *Pao Phu Tzu* (*Nei Phien*) is quoted in chapter 2 of the *Pao Phu Tzu Shen Hsien Chin Shuo Ching*[3] (The Preservation-of-Solidarity Master's Manual of the Bubbling Gold (Potion) of the Holy Immortals);[a] and a smaller part in the *Chin Mu Wan Ling Lun*[4] (Essay on the Tens of Thousands of Efficacious (Substances) among Metals and Plants).[b] Another text in the *Tao Tsang*, entitled *Pao Phu Tzu Yang Shêng Lun*[5] (The Preservation-of-Solidarity Master's Essay on Hygiene),[c] consisting of only about four printed pages, begins by quoting the last few sections of chapter 18 of the *Pao Phu Tzu* (*Nei Phien*). Chapter 1 of the *Pao Phu Tzu Shen Hsien Chin Shuo Ching* describes a method of making the 'cyclically-transformed elixir' (*huan tan*[6]), according to which 12 oz. of gold and 12 oz. of mercury are first mixed to form an amalgam, then washed several times with water and sealed in a bamboo tube after adding 2 oz. each of realgar and saltpetre, with some vinegar. After a hundred days a suspension will be formed. Then 2 lb. of mercury are introduced into this suspension and heated in the presence of vinegar for thirty days; after this time the mercury, having turned purple in colour, is taken out, sealed in an earthenware pot (and presumably heated) for a day and a night. Then the 'cyclically-transformed elixir' is completed.[d] Apart from the above procedure these three alchemical texts contain nothing which is not in the *Pao Phu Tzu* book itself.

Chapter 16 of the *Pao Phu Tzu* (*Nei Phien*) is also quoted in chapter 1 of the *Chu Chia Shen Phin Tan Fa*[7] (Methods of the Various Schools for Magical Elixir Prepara-

---

a *TT*910. *TT*933. c *TT*835.
d This text needs further study, but its processes may well have involved the production of colloidal gold.

1 匠 2 轡策靈機 3 抱朴子神仙金汋經 4 金木萬靈論
5 抱朴子養生論 6 還丹 7 諸家神品丹法

tions), with some minor textual variations.[a] Unlike the three previous texts, the dates and compilers of which are uncertain, the *Chu Chia Shen Phin Tan Fa* was, we know, compiled during the Sung period by Mêng Yao-Fu[1] (Hsüan Chen Tzu,[2] the Mysterious-Truth Master) and others. A further alchemical text in the *Tao Tsang* called *Chih-Chhuan Chen Jen Chiao Chêng Shu*[3] (Technical Methods of the Adept (Ko) Chih-Chhuan, with Critical Annotations)[b] was attributed to Ko Hung because of the inclusion of his name Chih-Chhuan Chen Jen in the title, but Ko Hung would not have used his own courtesy name in this way, and the text itself mentions a line of descent quite different from what one would expect from Ko Hung. Nevertheless this is an important text on alchemical laboratory equipment, and we shall have occasion to refer to it in the appropriate sub-section below (pt. 4). The *Huan Tan Chou Hou Chüeh*[4] (Handy Formulae for Cyclically-Transformed Elixirs)[c] has also been attributed to Ko Hung, but as it mentions Thao Chen Jen,[5] most probably Thao Hung-Ching,[6] who lived a century later, it must have been written by some other author whose name is unknown to us.[d] Two other texts in the *Tao Tsang*, namely the *Chen Chung Chi*[7] (Records of the Pillow-Book)[e] and the *Yuan Shih Shang Chen Chung Hsien Chi*[8] (Record of the Assemblies of Perfected Immortals; a Yuan-Shih Scripture),[f] have also been attributed to Ko Hung.[g] The former deals mainly with hygiene and has something to say about aqueous solutions, but although there is nothing in the text to indicate its authorship, it talks about the purchase of cinnabar in the early +7th century, thus ruling out Ko Hung.[h] The latter deals with the *hsien* and speaks of liturgy, visions and magic, but not alchemy.

Ko Hung's medical treatise, the *Ko Hsien Ong Chou Hou Pei Chi Fang*[9] (The Adept Ko (Hung)'s Prescriptions for Emergencies),[i] and his hagiography of the saints, the *Shen Hsien Chuan*[10] (Lives of the Holy Immortals), are believed by Yuan Han-Chhing to be the only two books besides the *Pao Phu Tzu* that were actually written by him.[j] There is no doubt about the medical treatise, but (as has been mentioned already) the present text of the book on the immortals is of somewhat uncertain authenticity.

There is little doubt that other alchemists were active during the time of Ko Hung, though we do not know as much about them. There lived for example Hsü Hsün[11]

---

[a] *TT*911, ch. 1, p. 1*a* to p. 12*a*.   [b] *TT*895.   [c] *TT*908.
[d] It incorporates a memorandum by Wu Ta-Ling[12] dated +875.
[e] *TT*830; the similarity to one of the titles of the Huai-Nan Corpus (cf. pp. 14, 25 above) may be noted.   [f] *TT*163.
[g] See, for example, Wieger (6), pp. 142, 163.
[h] In fact the bibliographical chapters of the *Thung Chih* (*Thung Chih Lüeh*) give Sun Ssu-Mo[13] as the author of the *Chen Chung Chi*; cf. Chhen Kuo-Fu (1), vol. 2, p. 415. It is also said to be identical with a book called *Ko Hung Chen Chung Shu*[14] (Pillow-Book of Ko Hung).
[i] *TT*1287.
[j] Of course one may except those later tractates which consist of almost nothing but what are clearly his own words.

¹ 孟要甫   ² 玄眞子   ³ 稚川眞人校証術   ⁴ 還丹肘後訣
⁵ 陶眞人   ⁶ 陶弘景   ⁷ 枕中記   ⁸ 元始上眞衆仙記
⁹ 葛仙翁肘後備急方   ¹⁰ 神仙傳   ¹¹ 許遜
¹² 仵達靈   ¹³ 孫思邈   ¹⁴ 葛洪枕中書

(c. +290 to +374)[a] to whom is attributed the *Hsü Chen Chün Shih Han Chi*[1] (The Adept Hsü's Treatise found in a Stone Coffer).[b] The text of the book is written in a rather obscure style, but it has the following to say about mercury and cinnabar:[c]

(Properly handled), mercury can be concreted to cinnabar, like little grains of golden sand. If one of these is ingested each day for a hundred days, changes in the body's form and transmutations within the bones (will be brought about). The bones will turn into metal and rock. When the form and the spirit fuse, they both become marvellously excellent and the extreme of limitless longevity is attained. It is the sagely property of these golden grains to produce matter from nothingness. Now matter comes into being from the void. The hexagrams are in harmony, so that there is agreement between the *Chen*[2] (and the *Li*[3]) *kua*, while the *Tui*[4] and *Khan*[5] *kua* can communicate. The emptiness and nothingness of the Four Symbols (*ssu hsiang*[6]) cannot be drawn in any diagram, neither can (a man's) original constitution (*yuan ching*[7]) be seen. When that which is without form combines with emptiness, it is the form (hidden in) formlessness combining with the change (hidden in) empty changelessness. The transformation of cinnabar is a spontaneous natural effect. The coming-into-being of cinnabar is a most mysterious operation of Nature.

Perhaps this shows that some of Ko Hung's colleagues could philosophise about chemical change quite in the manner of the Warring States scholars such as Chuang Chou.

Many other Taoist texts in the *Tao Tsang* are attributed to Hsü Hsün. It is said that he acquired the alchemical art from a certain adept named Lan Kung-Chhi[8].[d] Hsü Hsün himself had many disciples. One of them, Shih Ho,[9] knew how to transmute gold and jade, another, Kan Chan[10] (also called Kan Po-Wu[11]), learnt the secret of the 'gold elixir'. An account of his death was written by a third close disciple, Shih Tshên.[12] This is the *Hsü Chen Chün Pa-shih-wu Hua Lu*[13] (Record of the Transfiguration of the Adept Hsü (Hsün) at (the Age of) Eighty-Five).[e]

Another contemporary of Ko Hung was the Taoist Liang Shen[14] (Liang Khao-Chhêng,[15] d. +318), who sought immortality by ingesting cinnabar.[f] He was a disciple of the adept Thai-Ho Chen Jen[16] or Yin Kuei[17] (Yin Kung-Tu,[18] *fl.* late +3rd

---

[a] Also known by the names Hsü Ching-Chih,[19] Thai-Shih Chen Chün[20] and Hsü Chen Chün.[21]

[b] *TT*944.

[c] *TT*944, ch. 1, p. 11 *a, b*, tr. auct. We think the addition of the word *Li* after *Chen* is necessary not only for the antithetical style, but also to make it more easily understandable, since these two hexagrams were so often used by the alchemists to represent cinnabar and mercury respectively.

[d] See *TT*293, ch. 27, p. 10*b*.

[e] *TT*445. We know little about another disciple called Phêng Khang[22] (Phêng Wu-Yang,[23] d. +421). According to Chao Tao-I, Hsü Hsün also imparted his art to his two nephews Chungli Chia[24] and Hsü Lieh,[25] to his son-in-law Huang Jen-Lan,[26] also known as Huang Tzu-Thing,[27] to his own elaboratory assistant Chhen Hsün,[28] and to his two servants Chou Kuang[29] and Hsü Ta.[30] See *TT*293, ch. 27.

[f] All the following alchemists are mentioned in *TT*293, ch. 30.

[1] 許眞君石函記	[2] 震	[3] 離	[4] 兌	[5] 坎
[6] 四象	[7] 元精	[8] 蘭公期	[9] 時荷	[10] 甘戰
[11] 甘伯武	[12] 施岑	[13] 許眞君八十五化錄		[14] 梁諶
[15] 梁考成	[16] 太和眞人	[17] 尹軌	[18] 尹公度	[19] 許敬之
[20] 太史眞君	[21] 許眞君	[22] 彭抗	[23] 彭武陽	[24] 鍾離嘉
[25] 盱烈	[26] 黃仁覽	[27] 黃紫庭	[28] 陳勳	[29] 周廣
[30] 許大				

cent., *c.*+290 to +306), who possessed a 'magical elixir' (*shen tan*[1]).[a] Yin Thung[2] (Yin Ling-Chien,[3] d. +388), a descendant of Yin Kuei, took preparations of two liliaceous plants, *huang ching*[4] and *thien mên tung*[5] [b] with realgar, with what results is not recorded; he had two disciples, Niu Wên-Hou[6] and Wang Tao-I.[7] The latter became the teacher of Chhen Pao-Chih[8] (also known as Chêng-I hsien-sêng,[9] +472 to +549), who in turn taught Wang Yen[10] (Wang Tzu-Yuan,[11] d. +604). Wang Yen was ordered by the emperor of the Northern Chou, Wu Ti[12] (r. +561 to +578), to edit the Taoist literature, so he produced a book called *San Tung Chu Nang*[13] (A Sack of Pearls from the Three Heavens), which must have been one of the earliest collections or bibliographies. This text is no longer extant, but another treatise of exactly the same name by Wang Hsüan-Ho[14] of the Thang dynasty is included in the *Tao Tsang*.[c] It is said that Wang Yen also learnt from a certain adept Chiao-Kuang Chen Jen.[15] Chhen Pao-Chih had two other disciples, Li Shun-Hsing[16] and Hou Khai[17] (d. +573). The latter imparted his art to Yü Chang[18] (d. +614). In such ways as these was the alchemical art and hope of Ko Hung's time transmitted to the people of Sui and Thang.

Several alchemists are mentioned in official historiography of the time. The *Chin Shu*[19] (History of the Chin Dynasty) writes about Chang Chung[20] (also known as Chang Chü-Ho[21] and An-Tao hsien-sêng,[22] *fl.* +307) in the following words:[d] 'He practised the art of breathing, and how to ingest plants and minerals to nourish life, but when Fu Chien[23] invited him to Chhang-an[24] and offered him an appointment he declined, and died on the way home.' We also read in the same dynastic history about Thao Tan,[25] also called Thao Chhu-Ching,[26] who practised the art of immortality at the age of fifteen or sixteen, abstained from eating cereals and remained celibate.[e] Then there was Shan Tao-Khai,[27] a contemporary of the Central Asian missionary monk and thaumaturgist Fo-Thu-Têng[28] (*fl.* +310); he achieved a cicada-like metamorphosis by ingesting pills.[f] The Emporor Ai Ti[29] (r. +362 to +365) himself died of an overdose of elixir, but unfortunately we are not told what type he took.[g] A number of other contemporaries of Ko Hung are mentioned in the hagiography of the immortals, for example, Chao Kuang-Hsin[30] (d. *c.* +345), who bought cinnabar

---

[a] See *TT*293, ch. 8, pp. 19*a*–20*b*. We have given a passage about him in full on p. 101 above.

[b] The first was the 'deer-bamboo', not a bamboo at all, *Polygonatum falcatum* (R687), the second *Asparagus lucidus* (R676). The latter was an ancient drug-plant, described in the *Shen Nung Pên Tshao Ching*, the former was introduced only after Ko Hung's time; cf. *CLPT*, ch. 6, (pp. 142 ff.).

[c] *TT*1125. See Chhen Kuo-Fu (*1*), p. 115.

[d] Ch. 94, p. 15*a*. Fu Chien was the fourth and last ruler, Shih Tsu, of the Chhien Chhin dynasty, r. +357 to +385.

[e] Ch. 94, p. 19*b*.

[f] Ch. 95, p. 16*a*. Cf. *Lo-Fou Shan Chih*, ch. 3, p. 3*b*, ch. 4, p. 9*b*.

[g] Ch. 8, p. 8*a*; cf. Section 45 in Vol. 6.

[1] 神丹	[2] 尹通	[3] 尹靈鑒	[4] 黃精	[5] 天門多
[6] 牛文侯	[7] 王道義	[8] 陳寶熾	[9] 正懿先生	[10] 王延
[11] 王子元	[12] 周武帝	[13] 三洞珠囊	[14] 王懸河	[15] 焦曠真人
[16] 李順興	[17] 侯楷	[18] 于章	[19] 晉書	[20] 張忠
[21] 張亙和	[22] 安道先生	[23] 苻堅	[24] 長安	[25] 陶淡
[26] 陶處靜	[27] 單道開	[28] 佛圖澄	[29] 哀帝	[30] 趙廣信

for the making of the 'nine flower elixir' (*chiu hua tan*[1]); and Chu Ju-Tzu[2] (d. *c.* +345) who is said to have taken preparations of the chrysanthemum and the composite *Atractylis ovata* (*shu*[3]).[a]

(4) ALCHEMY IN THE TAOIST PATROLOGY (*Tao Tsang*)

Taoist texts were first listed in the bibliographical chapter of the *Chhien Han Shu*, but none of them seems to have been alchemical in nature. Many more came into existence during the period of the Three Kingdoms and the Chin Dynasty, as can be seen in the Bibliography of Ko Hung's *Pao Phu Tzu*,[b] which we have already noted. This includes the titles of many alchemical works now no longer extant. A catalogue of Taoist writings was then compiled in +471 by the Taoist Lu Hsiu-Ching,[4] under imperial order. At least so far as extant sources are concerned it was also in one of the writings of Lu Hsiu-Ching, dated +437, that the term 'Three Heavens' (*san tung*[5]) first made its appearance.[c] These were: the 'Heaven of Reality' (*tung chen*[6]), the 'Heaven of Mystery' (*tung hsüan*[7]) and 'Numinous Heaven' (*tung shen*[8]). These were the classical three divisions later employed in the classification of all Taoist canonical writings. According to Ōbuchi, by about the turn of the +6th century another four divisions, called the 'Four Ancillaries' (*ssu fu*[9]), were added to the original three. These were the 'Great Mystery' (*thai hsüan*[10]), the 'Great Peace' (*thai phing*[11]),[d] the 'Great Purity' (*thai chhing*[12]), and the 'Perfect Unity' (*chêng i*[13]).[e] The *san tung* and the *ssu fu* together form the 'seven divisions' (*pu*[14]). Without embarking on a lengthy description of these, we need only point out that the great majority of the alchemical writings can be found included in the *thai chhing* division.[f]

The catalogue of Lu Hsiu-Ching was soon followed by other catalogues due to the Taoist Mêng Fa Shih[15],[g] and to Thao Hung-Ching[16] (+456 to +536), the great naturalist, alchemist and physician.[h] In +523 Juan Hsiao-Hsü,[17] in his *Chhi Lu* (*Hsien Tao Lu*)[18] (Taoist Section of the Bibliography of the Seven Classes of Books), listed 425 Taoist works, amounting in all to 1138 rolls (*chuan*[19]) of manuscripts.[i] He

---

[a] *TT*293, ch. 17.　　　　　[b] Ch. 19; listed alphabetically in Ware (5), pp. 379 ff.

[c] Ōbuchi (1) thinks that the term *san tung* took shape somewhat earlier, about +400 or soon after.

[d] We shall return later to the social implications, some almost revolutionary, of this ancient phrase (Sect. 49 in Vol. 7), meanwhile see Needham (55, 56).

[e] The *thai hsüan* supplements the *tung chen*, the *thai phing* supplements the *tung hsüan*, the *thai chhing* supplements the *tung shen*, while the *chêng i* supplements all three of the *san tung*.

[f] Here I cannot refrain from referring to personal experiences recorded in Vol. 1, p. 12.

[g] The title was *Yü Wei Chhi Pu Ching Shu Mu*[20] (Catalogue of the Seven Divisions of the Jade Apocrypha). His own title was a borrowing from Buddhism.

[h] There were several titles of this, notably *Thai-Shang Chung Ching Mu*[21] (Catalogue of the Highly Exalted Assembly of Canonical Texts). See Chhen Kuo-Fu (*1*), 1st ed. p. 111, 2nd ed. vol. 1, p. 107.

[i] Anciently Chinese books were written as scrolls or rolls of silk, and then of paper, instead of flat bound volumes. These rolls were called *chuan*, and a single book title might include one or more rolls. In some cases there may have been more than one title in the same roll. When books were later pro-

[1] 九華丹	[2] 朱孺子	[3] 朮	[4] 陸修靜	[5] 三洞
[6] 洞眞	[7] 洞玄	[8] 洞神	[9] 四輔	[10] 太玄
[11] 太平	[12] 太清	[13] 正一	[14] 部	[15] 孟法師
[16] 陶弘景	[17] 阮孝緒	[18] 七錄（仙道錄）		[19] 卷
[20] 玉緯七部經書目		[21] 太上衆經目		

divided them into four sections, i.e. philosophy and precepts (*ching chieh pu*[1]), the nourishment of life by diet and medicines (*fu erh pu*[2]), sexual techniques (*fang chung shu*[3]), and talismanic magic and apotropaics (*fu thu pu*[4]). In +570 the Taoists of the Hsüan-Tu Kuan[5] abbey submitted to the emperor of the Northern Chou a list of Taoist books totalling 6363 rolls, of which 2040 were extant at that time. Within the next few years Wu Ti (p. 112) commissioned Wang Yen[6] to edit his collection of Taoist texts which had by that time increased to 8030 rolls. From these Wang Yen produced the catalogue entitled *San Tung Chu Nang*[7] (A Sack of Pearls from the Three Heavens).

A couple of other catalogues also appeared after Wang Yen before the year +712 when the Thang emperor Hsüan Tsung[8] formed a team consisting of scholars from the Imperial Academy and members of Taoist institutions to compile the catalogue *I Chhieh Tao Ching Yin I*[9] (Titles of all the Taoist Canons and their Meanings).[a] During the Khai-Yuan reign-period (+713 to +741) Hsüan Tsung followed up this project by issuing an order for the collection of all Taoist writings. This Corpus, consisting of some 3774 rolls (or, according to other versions, as many as 5700 rolls),[b] was given the name *San Tung Chhiung Kang*[10] (Essentials of the Magnificence of the Three Heavens), and from the year +748 scribes were set at work to make multiple copies of the texts included in it. However, most of these books were lost by fire during the rebellions of An Lu-Shan[11] and Shih Ssu-Ming.[12] Later efforts were made to restore the collection by another search throughout the empire for Taoist texts, and by the Hsien-Thung reign-period (+860 to +873) the number of Taoist writings again reached 5300 rolls.[c] Then during the reign of Hsi Tsung[13] (+874 to +888) the capital was seized by Huang Chhao[14] and most of the Taoist texts were burnt in the disturbances.[d] Remnants of the collection were later put together by the Taoist Shen Yin Tzu,[15] but again they met with the same fate during the upheavals accompanying the last days of the Thang.

duced in the form of stitched sheets the word *chuan* persisted, and because of the size of such logically convenient divisions, came to be equivalent to the term and conception of 'chapter'. However, there had always been smaller divisions, the *phien*,[16] a name which by its radical betrays its origin from the bamboo slips on which Chinese books had been written before the days of scrolls. We always represent the *chuan* by the abbreviated form for chapter (ch.) except (as stated in Vol. 1, p. 22) where the smaller divisions are available; these then take precedence as chapters (chs.). Besides all this, in Chinese novels, later terms such as *hui*[17] and *chang hui*,[18] with the idea of 'recitation', grew up to denote 'chapter'.

---

a This title was evidently borrowed from Buddhism, for in +649 the monk Hsüan-Ying[19] had written an *I Chhieh Ching Yin I* (Dictionary of Sounds and Meanings in the whole Tripiṭaka), but it was mainly concerned with the Vinaya portion. See Vol. 4, pt. 1, p. 105, pt. 3, p. 458. *I chhieh* was a transliteration of Skr. *sarva*, the whole, as in 'the Sarvāstivādin school' of Buddhism.
b I.e. chapters.
c It is memorable that just before this time one alchemical book at least had actually been printed. The text from which we know this constitutes the second oldest of all references to printed books, and we discuss it in its proper chronological place, p. 167 below.   d Cf. Vol. 1, pp. 215 ff.

¹ 經戒部	² 服餌部	³ 房中術	⁴ 符圖部	⁵ 玄都觀
⁶ 王延	⁷ 三洞珠囊	⁸ 玄宗	⁹ 一切道經音義	
¹⁰ 三洞瓊綱	¹¹ 安祿山	¹² 史思明	¹³ 僖宗	¹⁴ 黃巢
¹⁵ 神隱子	¹⁶ 篇	¹⁷ 回	¹⁸ 章回	¹⁹ 玄應

In the early days of the Sung the second emperor Thai Tsung (r. +976 to +997) instituted another search for Taoist writings among abbeys, temples and private libraries, obtaining some 7000 rolls. Hsü Hsüan[1] was asked to examine these, with the help of Wang Yü-Chhêng,[2] between the years +989 and +991. By removing all duplicates they reduced the number to 3737 rolls. In +1008 the emperor Chen Tsung ordered Wang Chhin-Jo[3] to edit and catalogue the resulting Taoist collection. Wang Chhin-Jo was assisted in this work by certain civil officials like Chhi Lun[4] and Chhen Yao-Tso,[5] and Taoists like Chu I-Chhien[6] and Fêng Tê-Chih.[7] By that time the number of rolls in the collection had again grown to 4359. The catalogue they compiled was given the title *Pao Wên Thung Lu*[8] (General Catalogue of Precious Writings) by the emperor. But the arrangement in this catalogue was later found unsatisfactory as there was much disagreement with the previous ones. Acting on the advice of Wang Chhin-Jo and Chhi Lun, the emperor commissioned Chang Chün-Fang[9] in the winter of +1012 to have the whole collection reclassified. The new Corpus thus established, augmented with new titles found by Chang Chün-Fang himself, consisted of 4565 rolls, and was completed in the year +1019 when the emperor bestowed upon it the title *Ta Sung Thien Kung Pao Tsang*[10] (Treasures of the Heavenly Palace; the Great Sung Patrology).[a] This has always been considered the definitive edition of what came to be generally referred to as the *Tao Tsang*[11] (Taoist Patrology).[b] Chang Chün-Fang also at this time selected the more important texts from his *Tao Tsang* to compile the *Yün Chi Chhi Chhien*[12] (Seven Bamboo Tablets of the Cloudy Satchel).[c]

Still more Taoist writings were found during the reign of the emperor Hui Tsung (+1101 to +1125) and the number of rolls in the *Tao Tsang* increased to 5387. During the Chêng-Ho reign-period (+1111 to +1117) blocks were made for its first printing. This was called the *Wan Shou Tao Tsang*[13] in honour of royal longevity.[d] Besides those in the Imperial Library (*pi ko*[14]), copies of this patrology were preserved in various temples and abbeys, such as the Chung Thai I Kung[15] temple and the Chien Lung Kuan,[16] both in Khaifêng, and at the Chhung Hsi Kuan[17] on Mao-shan[18] mountain. Already on several occasions we have had to emphasise the interest of Hui Tsung and his court in proto-science and technology, e.g. in relation to hydro-

[a] It is interesting that at least two Manichaean texts crept into this collection. The story is told by Chavannes & Pelliot (1), 2nd pt., pp. 327ff. Cf. p. 72 and pts. 4, 5.

[b] For a detailed account of the whole history of the *Tao Tsang*, see the excellent book of Chhen Kuo-Fu (1), on both editions of which the present résumé has been largely based. The Japanese counterpart to this is the important work of Yoshioka Yoshitoyo (1).

[c] *TT*1020. This remains to the present day an extremely important work because it includes a number of texts which are not in the Ming recension which constitutes the *Tao Tsang* of today.

[d] We may note here that of the writings of the Hellenistic proto-chemists no manuscript survives that is older than about +1000. The Chinese tradition, however, gives us texts which were already stabilised in printed form by about +1115.

[1] 徐鉉	[2] 王禹偁	[3] 王欽若	[4] 戚綸	[5] 陳堯佐
[6] 朱益謙	[7] 馮德之	[8] 寶文統錄	[9] 張君房	
[10] 大宋天宮寶藏		[11] 道藏	[12] 雲笈七籤	[13] 萬壽道藏
[14] 祕閣	[15] 中太一宮	[16] 建隆觀	[17] 崇禧觀	[18] 茅山

mechanical clockwork and rare drugs and minerals,[a] so that the printing of the Taoist literature at this time is not at all surprising.

The *Wan Shou Tao Tsang* unfortunately was not as immortal as its name suggested, for before long it became a prey to war and fire. Some of the blocks in the capital fell into the hands of the Jurchen Tartars. In +1164 the Jurchen emperor Shih Tsung[1] compiled the *Ta Chin Hsüan Tu Pao Tsang*[2] (Precious Patrology of the Mysterious Capital (i.e. the Taoist Church) collected in the Great Chin Dynasty) from these blocks, and from those made for the additional Taoist texts he managed to find. Completed within two years, this new Patrology consisted of 6455 chapters (*chuan*).[b] However, in +1202 the Thien-Chhang Kuan,[3] where the blocks were preserved, was burnt to the ground by fire. Meanwhile in +1175 copies of the *Wan Shou Tao Tsang* which had been treasured in Fukien province were brought to the Southern Sung capital at Hangchow, Lin-an-fu,[4] where several duplicates were copied and subsequently distributed to some Taoist abbeys.

In +1237 Sung Tê-Fang[5] embarked anew upon the Herculean task of recovering as many Taoist texts as possible, including those not then found in the Taoist Patrology, and with the help of his disciples he established twenty-seven centres in different places for this purpose. By +1244 he completed the *Hsüan Tu Pao Tsang*[6] (Precious Patrology of the Mysterious Capital), consisting of over 7800 chapters.[c] The blocks were first preserved in the Hsüan-Tu Kuan[7] at Phing-yang,[8] but were later (about the year +1247) transferred to the Shun-Yang Wan Shou Kung[9] temple in the same city.

Under the rule of the Mongols Taoism gradually fell out of favour. During the reign of Mangu Khan (Hsien Tsung, +1251 to +1259) an order was issued to burn all the books and blocks of the Taoist Patrology following a court debate between Taoists and Buddhists on the validity of the *Lao Tzu Hua Hu Ching*[10] (Book of Lao Tzu's Conversion of Foreigners).[d] Another edict in +1281 ordered the burning of the surviving Taoist books except the *Tao Tê Ching*,[11] but this was probably not carried out. Further destruction of Taoist books and blocks took place about +1294 by Khubilai Khan (Shih Tsu) when Taoism and Buddhism again confronted each other in an open debate at court. This inflicted the greatest blow of all on Taoist writings, many of which, including probably many alchemical texts, were lost for ever.

During the Yung-Lo reign-period (+1403 to +1424) the Ming emperor Chhêng Tsu ordered Chang Yü-Chhu[12] to edit and reprint the *Tao Tsang*, but printing did not begin until the year +1444. This resulted after a year in the *Chêng-Thung Tao Tsang*,[13] a collection of Taoist texts in 5305 chapters occupying 480 cases

---

[a] Vol. 4, pt. 2, pp. 501ff.

[b] Pelliot (58) believed that there was a further printing between +1186 and +1191.

[c] The edict of +1240 concerning this, in Mongolian as well as Chinese, has been translated by Cleaves (1).

[d] See Vol. 2, p. 159.

[1] 世宗	[2] 大金玄都寶藏	[3] 天長觀	[4] 臨安府
[5] 宋德方	[6] 玄都寶藏	[7] 玄都觀	[8] 平陽
[9] 純陽萬壽宮	[10] 老子化胡經	[11] 道德經	[12] 張宇初
[13] 正統道藏			

(*hsien*[1] or *han*[1]).[a] In +1607 the emperor Shen Tsung ordered Chang Kuo-Hsiang[2] to compile a supplement, and when this was completed it was given the title *Wan-Li Hsü Tao Tsang*[3] (Supplementary Taoist Patrology of the Wan-Li reign-period). The blocks were safely preserved till the end of the last century, when they were completely burnt during the Boxer Uprising. Between 1923 and 1926 the incomplete collections of the *Chêng-Thung Tao Tsang* and the *Wan-Li Tao Tsang* belonging to the Pai-Yün Kuan[4] abbey at Peking were borrowed by the Han Fên Lou[5] publishers at Shanghai and reprinted to give us our modern *Tao Tsang*. There is also available a 1906 printing of the *Tao Tsang Chi Yao*[6] (Selections from the *Tao Tsang*) from blocks preserved at Chhêng-tu in Szechuan.[b]

The *Tao Tsang* forms the main source of supply of our alchemical texts.[c] We shall now describe them according to their historical sequence as far as their dating permits, as also against the historical background of the development of alchemy and proto-chemistry in China. As we have already dealt with the *Tshan Thung Chhi* and the *Pao Phu Tzu* we shall take up the story from the time of the death of Ko Hung.

(5) THE GOLDEN AGE OF ALCHEMY; FROM THE END OF CHIN (+400) TO LATE THANG (+800)

(i) *The Imperial Elaboratory of the Northern Wei and the Taoist Church at Mao Shan*

The elixir of life continued to attract and fascinate many a Chinese emperor after the time of Ko Hung. Following in the steps of the Han emperors, Tao Wu Ti[7] (Thopa Kuei,[8] r. +386 to +409) of the Northern Wei Dynasty gave strong support to the study of alchemy by establishing a professional chair in the subject and arranging for the prosecution of large-scale experiments at the capital during the period +398 to +404. The *Wei Shu* says:[d]

In the 3rd year of the Thien-Hsing reign-period (+400), (the emperor) instituted a post of Hsien Jen Po Shih Kuan[9] (Professor of Macrobiotics, or Alchemist-Royal), to take charge of the preparation of drugs and elixirs (*chu lien pai yao*[10]).

The official history goes on to say:[e]

Thai Tsu (Tao Wu Ti) liked the words of Lao Tzu and never wearied of studying them. In the Thien-Hsing reign-period an official of the Board of Rites, Tung Mi,[11] accordingly

---

[a] There was a reprint in +1598 (Pelliot, 58).

[b] In the Erh Hsien Ssu[12] Taoist temple, during the second world war in November 1945, I had the pleasure of purchasing a complete set of it for the Cambridge University Library from the Taoists of this abbey.

[c] Reference is made elsewhere (p. xxi) to the most valuable work of our collaborator Dr Tshao Thien-Chhin, who during his time as a Fellow of Caius College carried out pioneer studies of these. Much of it is bearing fruit in the present volumes, but his notes are preserved intact in the archives of the East Asian History of Science Library at Cambridge for future use.

[d] Ch. 113, p. 3*a*, tr. auct.

[e] Ch. 114, pp. 32*b*, 33*a*, tr. auct., adjuv. Ware (1).

[1] 函　　[2] 張國祥　　[3] 萬曆續道藏　　[4] 白雲觀　　[5] 涵芬樓
[6] 道藏輯要　　[7] 道武帝　　[8] 拓跋珪　　[9] 仙人博士官
[10] 煮煉百藥　　[11] 董謐　　[12] 二仙寺

presented a *Fu Shih Hsien Ching*[1] (Manual of Longevity and Immortality produced by Diet and Drugs) in several dozen scrolls. Thereupon (the emperor) established a Chair of Macrobiotics (*hsien jen po shih*[2]) and built (at the capital, Phing-chhêng[3] in Shansi) an imperial elaboratory (*hsien fang*[4]) for the concocting of medicines and elixirs. He also reserved the Western Mountains for the supply of firewood (for the furnaces). Furthermore, he ordered that those who had been condemned for capital offences should test (*shih fu*[5]) (the preparations), but since it was not their original intention (to seek for immortality) many died. Thus (the experiments) gave no decisive result (*wu yen*[6]).[a]

Seeing that Thai Tsu continued to encourage these activities, the Imperial Physician Chou Tan[7] became much distressed at the labour involved in collecting and processing (the drug plants and minerals), so he desired to bring it to an end. Consequently he privily got his wife to bribe a concubine of the Professor of Macrobiotics, Chang Yao,[8] to reveal his secret misdoings. In this situation (Chang) Yao, fearing death, requested permission to abstain from cereals, and this Thai Tsu granted, building for him a Fasting Pavilion in the Imperial Park and giving him two families as domestics. But the concoction and preparation of drugs and elixirs continued without respite. Only after some time did Thai Tsu's interest wane and cease.

This revealing passage gives a vivid glimpse of an imperially maintained laboratory of alchemy at the beginning of the +5th century. Whether the objections of Chou Tan had a humanitarian ground, or whether the College of Physicians of the day was jealous of a competitor 'Society of Chymical Physitians', as exactly happened a thousand years later in England,[b] one cannot say.

The emperor Thai Wu Ti[9] (r. +424 to +452) of the Northern Wei was also a great patron of Taoism and alchemy. He showed special favour to the Taoist religious leader Khou Chhien-Chih[10] (d. +448), the great reformer of the teachings of Chang Tao-Ling[11] and the successor of Chang Yao.[c] Khou was supported by Tshui Hao[12] in the great laboratory, but no new addition was made to it. Khou Chhien-Chih himself was a pupil of two adepts, Chhêngkung Hsing[13] and Li Phu-Wên,[14] learning from the latter the methods of making 'gold elixir' and bringing mica and the eight minerals as well as jade into aqueous solution.[d] The emperor ordered his successor Wei Wên-Hsiu[15] and the secretary Tshui Tsê[16][e] to go to the Wang-wu[17] Mountains in Southern Shansi to compound an elixir, but the mission was not successfully accomp-

---

[a] This raises the whole question of the beginnings of systematic biological experimentation on man and animals. We shall return to it in Sect. 45 in Vol. 6. Meanwhile cf. Vol. 5, pt. 2, p. 295.

[b] See H. Thomas (1); Webster (1). The physicians were on the threshold of great things. In +466 provincial colleges were set up all over the empire, later to have medical departments, and in +493 the posts of Regius Professor and Regius Lecturer in Medicine are mentioned for the first time. These were the beginnings of the Imperial College of Medicine (see Lu & Needham, 2).

[c] It will be remembered that between +423 and +428 Khou had received the title of Thien Shih[18] or 'Pope' (Vol. 2, pp. 158, 441).

[d] *Wei Shu*, ch. 114, pp. 33 aff., 35 aff.; Ware (1), pp. 225 ff., 231 ff.

[e] Or perhaps better, Tshui I.[19]

[1] 服食仙經	[2] 仙人博士	[3] 平城	[4] 仙坊	[5] 試服
[6] 無驗	[7] 周澹	[8] 張曜	[9] 太武帝	[10] 寇謙之
[11] 張道陵	[12] 崔浩	[13] 成公興	[14] 李譜文	[15] 韋文秀
[16] 崔賾	[17] 王屋	[18] 天師	[19] 崔頤	

lished.[a] After this the self-professed adept Lo Chhung[1] was despatched to find the immortals. He returned empty-handed, but was spared by the emperor.[b]

The *Pei Shih*[2] tells us about another spagyrist Hsü Chien[3] (also called Hsü Chhêng-Po[4]) who tried to prepare an elixir for the emperor Hsiao Wên Ti[5] (r. +471 to +500) of Northern Wei. It says:[c]

He and his elder brother, Hsü Wên-Po,[6] were skilled in medicine. Intending to prepare a 'gold elixir' for the emperor Hsiao Wên Ti to make him an immortal, he went to live among the mountains of Sung-shan[7] in order to collect the (necessary) raw materials. Many years passed and yet he met with no success. At last he gave up the venture.

This was approximately the period of activity of the rather shadowy figure, Lei Hsiao,[8] whose book *Lei Kung Phao Chih Lun*[9] (The Venerable Master Lei's Treatise on the Decoction and Preparation of Drugs), now preserved only in quotations, takes a place of some importance in the history of pharmacy and medicine.[d] We mention it here only because it gives details of some of the methods of ingesting cinnabar as such. One procedure, preserved in the great pharmaceutical natural histories,[e] describes the repeated pounding and grinding of the sulphide and its successive boiling with the extracts of certain plants. There is an account of many natural varieties of the ore, including references to its crystalline structure, one with fourteen faces (*shih ssu mien*[10]) that shine like mirrors. Although the sulphide was to be taken in quiet retirement after fasting and bathing, with the accompaniment of incense, it was not necessarily part of a longevity–immortality exercise, for many pathological conditions are given in other books for which it was considered useful. It was no doubt a way of getting small doses of mercury into the system, and certainly the flora and fauna of the intestinal tract would not have been unaffected thereby.

The most celebrated alchemist in +5th- and +6th-century China was undoubtedly Thao Hung-Ching[11] (+456 to +536),[f] who was also a great physician and pharmaceutical naturalist like Ko Hung. In fact, he can be traced to the same alchemical tradition as Ko Hung himself (see Table 112). The fact that genealogies of this kind have come down to us with fair reliability is a remarkable witness to the tenacity of the Chinese alchemical transmissions, so much emphasis being placed on the personal and oral inheritance of exposition.[g] Living in the South, near Nanking, he prepared elixirs for the emperor Wu Ti[12] of the Liang dynasty (Hsiao Yen,[13] r. +502 to +549). The official 'History of the Southern Dynasties', *Nan Shih*,[14] says:[h]

---

[a] *Wei Shu*, ch. 114, pp. 39*b*, 40*a*; Ware (1), p. 239.
[b] *Wei Shu*, ch. 114, p. 40*b*; Ware (1), p. 240.   [c] *Pei Shih*, ch. 90, p. 2*b*, tr. auct.
[d] See further in Sect. 45.   [e] E.g. *CLPT*, ch 3, (p. 792).
[f] Also known under the names Thao Yin-Chü,[15] Thao Thung-Ming,[16] and Chen Pai hsien-sêng.[17]
[g] The table has been constructed from many sources, the dynastic histories, statements in the alchemical books themselves, general Taoist literature, and the hagiographic texts.
[h] *Nan Shih*, ch. 76, p. 9*a*, tr. auct.

[1] 羅崇	[2] 北史	[3] 徐謇	[4] 徐成伯	[5] 孝文帝
[6] 徐文伯	[7] 嵩山	[8] 雷斅	[9] 雷公炮炙論	[10] 十四面
[11] 陶弘景	[12] 武帝	[13] 蕭衍	[14] 南史	[15] 陶隱居
[16] 陶通明	[17] 貞白先生			

He was a friend of (the emperor) Liang Wu Ti. After (the latter) ascended the throne he treated (Thao Hung-Ching) with great respect and continued to correspond with him. After acquiring the secret art, Thao Hung-Ching thought that he could succeed in making elixirs, but was worried about the shortage of material. So the emperor supplied him with gold, cinnabar, copper sulphate, realgar, and so forth. When the process was accomplished the elixirs had the appearance of frost and snow[a] and really did make the body feel lighter (*thi chhing*[1]).[b] The emperor took an elixir and found it effective. His respect for (Thao Hung-Ching) grew so great that he burnt incense whenever he received a letter from him. During the Thien-Chien reign-period (+502 to +519) (Thao) presented another elixir to the emperor. At the beginning of the Chung-Ta-Thung reign-period (+529 to +534) he offered two more, one called 'Skilful Victory' (*shan shêng*[2]) and the other 'Accomplished Victory' (*chhêng shêng*[3]).

It is particularly interesting that Liang Wu Ti showed such keen interest in Thao Hung-Ching's preparations since he himself was a devout Buddhist. Other elixirs were also offered to him, as by Têng Yü[4] (d. +515), who claimed to have eaten no cereals for thirty years, but only pieces of mica with water from mountain streams.[c] The emperor declined to take it. Perhaps Liang Wu Ti knew of the reality of poisoning, and had confidence only in Thao Hung-Ching's work, because he admired his skill as a physician and naturalist.

The last half of Thao Hung-Ching's life, from +492, was spent in close association with the important group of Taoists which grew up at Mao Shan[5] near Nanking. There he was intimately concerned with the nascent phases of organisation of the Taoist Church, at the centre therefore of a wealth of magical, technical, proto-scientific and liturgical activities (cf. pt. 2, pp. 110, 152).

As the botanical, pharmaceutical and medical contributions of Thao Hung-Ching will call for extended discussion in the next volume of this work, we shall confine ourselves here to his alchemical and Taoist writings. No alchemical book now extant actually bears his name, but we suggested long ago that the important *San-shih-liu Shui Fa*[6] (Thirty-Six Methods for Bringing Solids into Aqueous Solution) may well be due to him.[d] Among writings attributed to him in the *Tao Tsang* are the *Yang Shêng Yen Ming Lu*[7] (Notes on the Nourishing and Prolonging of Life),[e] and the *Chen Kao*[8] (Declarations of Perfected Immortals) which he certainly edited by +499.[f] Biographical data on Thao Hung-Ching can be found in the *Hua-Yang Thao Yin-Chü Chuan*[9] (The Life of Thao Yin-Chü of Huayang), written by Chia Sung[10] in

---

[a] I.e. crystals grown from solutions or sublimed.
[b] I.e. more lively and active; probably a loss of weight is referred to.
[c] *Nan Shih*, ch. 76, p. 8*b*.
[d] *TT*923. Cf. p. 105. It is mentioned in his *Pên Tshao Ching Chi Chu* (+492), so it is not later. See Tshao, Ho & Needham (1) for full translation, and also pt. 4 below.   [e] *TT*831.
[f] *TT*1004. This book is one of the strange classics of religious Taoism, consisting of conversations with heavenly visitants from the Taoist pantheon. Such visions were basically a justification for the particular theological and ecclesiastical policies which the Taoist Church was adopting at the time. Cf. Ware (1), pp. 229ff., for a good example.

[1] 體輕	[2] 善勝	[3] 成勝	[4] 鄧郁	[5] 茅山
[6] 三十六水法	[7] 養生延命錄	[8] 眞誥	[9] 華陽陶隱居傳	
[10] 買嵩				

Table 112. *The main alchemical tradition from Chin to Wu Tai*
(*+4th to +10th century*)

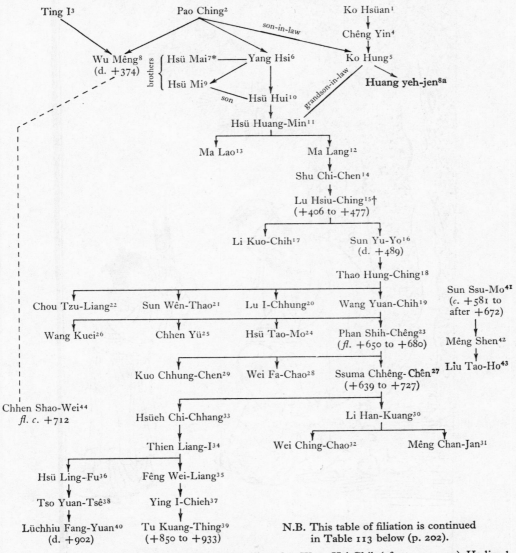

* *Fl.* +340 to +365, a friend of the famous calligrapher Wang Hsi-Chih (cf. pt. 2, p. 205). He lived in the hills west of Hangchow, and changed his name to Hsü Hsüan.[7a]   † The bibliographer, see p. 113 above.

1 葛玄	2 鮑靚	3 丁義	4 鄭隱	5 葛洪
6 楊羲	7 許邁	7a 許玄	8 吳猛	8a 黃野人
9 許謐	10 許翽	11 許黃民	12 馬朗	13 馬牢
14 殳季眞	15 陸修靜	16 孫遊嶽	17 李果之	18 陶弘景
19 王遠知	20 陸逸冲	21 孫文韜	22 周子良	23 潘師正
24 徐道邈	25 陳羽	26 王軌	27 司馬承貞	28 韋法昭
29 郭崇眞	30 李含光	31 孟湛然	32 韋景昭	33 薛季昌
34 田良逸	35 馮惟良	36 徐靈府	37 應夷節	38 左元澤
39 杜光庭	40 閭丘方遠	41 孫思邈	42 孟詵	43 劉道合
44 陳少微				

Fig. 1356. Thao Hung-Ching (+456 to +536), the great physician, alchemist and pharmaceutical botanist, listening to the music of a *shêng* played by a disciple (from *Lieh Hsien Chhüan Chuan*, ch. 5, p. 11*a*).

the Thang period,[a] and of course in the *Yün Chi Chhi Chhien*[1].[b] Biographical material in Western languages has been published by Barnes & Yuan (1); Davis & Wu Lu-Chhiang (3) and Strickmann (2); while Sanaka Sō (1) has essayed a scientific biography in Japanese.

Although it is true that no separate book on alchemy by Thao Hung-Ching has survived, a mass of his writings on minerals and chemical substances, as well as on plants, is contained in the pharmaceutical natural histories. As we shall explain later on,[c] the early history of his works is bibliographically complex, but most of his own contributions came into the codices through his *Pên Tshao Ching Chi Chu*,[2] now mostly lost in itself. The *Chêng Lei Pên Tshao*[3] of +1249, for example, preserves large amounts of his text, cut up under its many entries and headings. Just as one example, here is what he said on mercury:[d]

There are two sorts of mercury, crude and refined (*shêng shu*[4]). That which comes from the earthy plain of Fu-ling is obtained from cinnabar, but the ore is also found in pale sandy places. The best way is to powder it and roast it. The colour is (silvery) white, and not as impure as the crude product. Mercury is able to soften and change gold and silver with the formation of a paste (*ni*,[5] amalgam). People use it for the plating (*tu*,[6] gilding and silvering) of objects. It can be reclaimed and converted back into *tan*[7] (elixir, or cinnabar), so the Manuals of the Immortals (*Hsien Ching*[8]) say; and they add that if taken with wine warmed in the sun (*pao*[9]) it will give longevity and immortality. On being heated (with other substances) it volatilises (*fei*[10]) and a kind of ash sticks to the top of the reaction-vessel; this is called *hung fên*,[11] or popularly, *shui-yin hui*.[12] This (sublimate) is excellent for getting rid of fleas and lice (*sê*[13]).

The passage is interesting for its matter-of-fact statement about amalgams, and for Thao's characteristically cool and rational attitude to the 'Manuals of the Immortals'; he was a tireless experimentalist but never quite an 'enthusiast' in spite of his environment. The description at the end must be the making of either calomel (Hg_2Cl_2) or corrosive sublimate ($HgCl_2$), which would have depended on the relative amounts of mercury and the other constituents added to the sublimatory pot, salt, and sulphate in some form.[e]

This needs looking at a little closer, for the preparation of the chlorides of mercury is regarded as an achievement in Western chemical history, associated as it was with the Paracelsian revolution in medicine. 'Basil Valentine' (i.e. Thölde) at the end of the +16th century has often been considered the first to get mercuric chloride (corrosive sublimate),[f] but in fact a close study of Latin Geber shows that it was

---

[a] *TT* 297.
[b] *TT* 1020, ch. 107, p. 9*a*.
[c] See Sect. 38 in Vol. 6.
[d] *CLPT*, ch. 4, (p. 107.2).
[e] Further on this see pt. 4 below. Sanaka Sō (1) argues for calomel.
[f] Cf. Partington (7), vol. 2, p. 199.

[1] 雲笈七籤　　　[2] 本草經集注　　　[3] 證類本草　　　[4] 生熟　　　[5] 泥
[6] 鍍　　　　　　[7] 丹　　　　　　　[8] 仙經　　　　　[9] 暴　　　　[10] 飛
[11] 汞粉　　　　　[12] 水銀灰　　　　　[13] 蝨

certainly prepared already in the late +13th century.[a] Paracelsus undoubtedly knew mercurous chloride (calomel),[b] as also did his followers such as Quercetanus (du Chesne),[c] and it first became officinal in +1618 with the 'London Pharmaco-poeia'.[d] After Thao Hung-Ching, working about +490, there is a remark in the +8th-century *Tshan Thung Chhi Wu Hsiang Lei Pi Yao* that 'the white tiger can become as bright as frost', on which Lu Thien-Chi commented early in the +12th century (p. 145) that 'at certain times both lead and mercury will turn into the form of frost and snow'. As Ho Ping-Yü & Needham pointed out,[e] this must refer to the making of lead acetate and carbonate (by the vinegar method, cf. p. 15), and to the sublimation of the chlorides of mercury; processes clearly much older than the Thang.[f] As we have already seen (p. 16) the most ancient reference to white lead artificially made is of the early −4th century in China, with a clear though brief description about −300.[g] Calomel became officinal there from +972 in the *Jih Hua Chu Chia Pên Tshao*,[h] i.e. just about the period of the first description of its preparation in the Arabic-Syrian text translated by Berthelot & Duval.[i] The same text also describes the poisonous sublimate.[j] Calomel appears in India rather later, with the *Rasarṇava Tantra* of the +12th century.[k] An interesting account of the commercial preparation of 'muriate of quicksilver' in China 150 years ago can be found in Davis.[l] By then the name most standard for calomel was *shui yin fên*[1] (quicksilver powder), or, when purified, *fên shuang*[2] (frost powder); while that for corrosive sublimate was

[a] *Summa Perfectionis*, ch. 45, Darmstädter tr. (1), pp. 47, 150; Russell tr., pp. 86ff. The snow-white powder obtained must have been mostly corrosive sublimate because it melted, which would not be true of calomel. Also *De Inventione Veritatis*, ch. 8, Darmstädter tr. (1), pp. 108, 176; Russell tr., pp. 210–11. This was the passage cited in the interesting discussion of Ray (1), 1st ed., vol. 1, p. 256. The making of corrosive sublimate has now been traced back to the alchemists of Muslim Spain in the late +11th or early +12th centuries, since a couple of procedures for it have been found in the *De Aluminibus et Salibus*, an influential work believed to have been translated from the Arabic by Gerard of Cremona about +1160; cf. Multhauf (5), pp. 160ff. It was through this, rather than through his own writings, that the chemistry of al-Rāzī (cf. pt. 4) was transmitted to the medieval Latins. In any case, the preparation of the chlorides of mercury, or one of them, is in the *Kitāb al-Asrār* of al-Rāzī himself (i.e. *c.* +900); see Stapleton, Azo & Husain (1), pp. 386–7. It is also certainly in al-Khwārizmī al-Kathī (+1034; see Stapleton & Azo (1) and Ahmad & Datta (1), etc., discussed also in pt. 4); and may well be in the Jābirian Corpus (bk. 61 of the Seventy Books, Kr 183), which would mean a date about +870 if not a few decades earlier.
[b] Cf. Partington (7), vol. 2, p. 145.
[c] Partington, *ibid.* pp. 168–9.
[d] *Ibid.* p. 165. The story of its introduction in Europe has been well told by Urdang (1).
[e] (2), p. 186.
[f] As was realised by Divers (1), cit. Partington (10), p. 396, Chikashige (1), p. 10.
[g] On white lead see also Schafer (9) and further in Sect. 34.
[h] Confirmed in the *Chia-Yu Pu Chu Pên Tshao* (+1060), and never afterwards dropped.
[i] (1), pp. xviii, 143.
[j] *Ibid.* pp. 186–7. Cf. von Lippmann (1), pp. 388, 393. The late Prof. J. R. Partington, to whom we are indebted for emphasising to us in 1959 the historical importance of the chlorides of mercury, believed that this Arabic text written in Syriac derived from an earlier Greek predecessor, but there is no positive evidence for this. It is not even attributed to any ancient name like Pseudo-Democritus, as some other parts of the same Corpus are.
[k] Ray (1), 1st ed., vol. 1, pp. 73, 250, 255ff., 2nd ed., pp. 139, 206–7. Cf. von Lippmann (1), p. 437.
[l] (1), vol. 3, pp. 58ff.

[1] 水銀粉   粉霜

pai hsiang tan[1] (white quelled chemical).[a] Thus in general the Chinese alchemists seem to have been ahead of everyone else in preparing the chlorides of mercury (Fig. 1357).[b]

So much is this the case that we even find what can only be called a 'calomel legend'. Li Shih-Chen's entry for the purified salt contains two strange items, first a reference to the making of calomel at the −7th-century court of Chhin State, and secondly a quotation from Ko Hung the source of which is not easy to identify now, and which has its own difficulties.[c] We shall speak first of the former, together with a secondary legend, and return afterwards to the latter.

Let us give the story in the words of the oldest extant source, the *Chung Hua Ku Chin Chu*[2] (Commentary on Things Old and New in China), written about +925 by Ma Kao.[3]

Since the three dynasties of high antiquity, (beauty) powder has been made from lead. Now Lung-Yü,[4] the daughter of Duke Mu of Chhin (Chhin Mu Kung[5]),[d] was a girl both beautiful and virtuous. She was taken notice of by the Immortal, Hsiao Shih,[6] who made a powder by heating mercury, and gave it to her to paint (on her face). This was called 'flying cloud cinnabar' (*fei yün tan*[7]). It is said that it flew upwards (sublimed) at the ending of the music of his flute.[e]

One does not quite know how to take this, for the cosmetic could have been red, in which case it would have been a re-constituted vermilion, and some have so interpreted it,[f] yet Li Shih-Chen evidently thought of it as white, and so did many other later writers. For instance, Tung Yüeh[8] in his *Chhi Kuo Khao*[9] about +1660 put the story under *chhing fên*[10] ('light powder', one of the standard names for calomel), not, like Ma Kao, just under *fên* as such, and he also called the sublimate *fei hsüeh tan*,[11] 'flying snow cinnabar, or chemical'.[g] He further added that it was the same as what we now call *ni fên*[12] or 'glossy powder' (another calomel synonym). The

[a] *PTKM*, ch. 9, pp. 14*b*ff. (pp. 59 and 61), RP 45, 46. A Taoist name was *shêng tan*[13] (Hsüeh Yü (*1*), p. 31) and there were many other calomel synonyms (cf. Table 95 in pt. 2), some of which will crop up in a moment.
[b] The traditional Japanese process for calomel studied by Divers (1) at the end of the last century used the magnesium chloride of bittern and partially purified sea-salt as the source of HCl, together with metallic Hg and air, under conditions well below the volatilisation temperature of calomel itself (cf. pt. 4). In the Chinese process, on which Geerts (5) had also written, salt and alum or copperas gave the HCl, while by adding KNO_3 chlorine was formed instead, and the higher chloride, corrosive sublimate, was produced. In view of the extremely potent and widely different physiological properties of the two salts it would be of the greatest interest to know just how early the Chinese achieved full control of their products. The question is evidently related to the antiquity of knowledge of saltpetre among them, on which see pt. 4. In assessing, or repeating, the methods used in the different medieval cultures, account may also have to be taken of the possible catalytic effect of elements in other compounds present. Van Leersum (1) has given an interesting but inconclusive account of efforts he made to elucidate some of the Indian procedures, as described by Wise (1), pp. 119–20, (2), vol. 1, p. 215.
[c] *PTKM*, *loc. cit.* p. 61.   [d] R. −658 to −620.
[e] Ch. 2, p. 6*b*, tr. auct., adjuv. Kaltenmark (2), p. 147. Hsiao Shih was a well-known legendary piper and musician; he could summon auspicious phoenixes to Duke Mu's court, and finally took Lung-Yü away with him to the world of the immortals.   [f] E.g. Kaltenmark, *loc. cit.*
[g] Ch. 14, p. 3*a*. The 'snow' variant had been current long before, as in the mid +12th-century *Hsü Po Wu Chih*, ch. 10, p. 7*a*, which mentioned the term *ni fên*[12] too.

[1] 白降丹	[2] 中華古今注	[3] 馬縞	[4] 弄玉	[5] 秦穆公
[6] 蕭史	[7] 飛雲丹	[8] 董說	[9] 七國考	[10] 輕粉
[11] 飛雪丹	[12] 膩粉	[13] 昇丹		

Hsü Shih Shih[1] of +960 links the story directly with that of the legendary Shang origin of lead carbonate (ceruse) cosmetic,[a] thereby showing that it was thought to concern something white, and practically proving that in the Sung calomel was used as the basis of a face-cream.[b] This same text introduces us also to the oldest version of the legend (now presumably lost), a book called *Erh I Shih Lu*[2] (Veritable Records of Heaven and Earth) by Liu Hsiao-Sun[3] of the Sui (late +6th century) which sounds like a 'history' of the oldest legendary dynasties. An additional note is introduced in the *Shih Wu Chi Yuan* of +1085, which says that 'the product of the first transformation (*ti i chuan*[4]) was presented to Lung-Yü for use as a cosmetic, and was called *fên*. That was the beginning of our *chhing fên*.[5]'[c] This suggests vermilion rather than calomel, for we know from innumerable instances how the Chinese alchemists engaged in the repeated cyclical transformations of mercury, sulphur and mercuric sulphide (cf. pp. 14, 109, 131, 198, 250, 261). Moreover, the mid +12th-century *Hsü Po Wu Chih*, which uses the same words, goes on to say that later the pigments of two red dye-plants were used.[d] This is not evidence for vermilion, however, for there is much that goes to show that not only white, but also pink, 'rouge' powders and creams were aimed at, at least from the Han onwards. The *Shih Ming* tells us so in +100: 'the (lead) powder is dyed pink, and so applied to the cheeks',[e] while about +540 the *Chhi Min Yao Shu* gives a formula for *tzu fên*,[6] 'purple powder', fine rice flour being mixed with ceruse and coloured with 'mallow' anthocyanin.[f] All such preparations were called *yen chih*.[7] In the +17th century Sung Ying-Hsing says[g] that rouge was made by tinting white powder with the purple pigment of stick lac (*tzu kung*[8])[h] or, if of inferior quality, with safflower (*hung hua*[9])[i] or 'mountain pomegranate' (*shan shih liu*[10]).[j] During the Ming, calomel continued in use as well as ceruse for the white base.[k]

---

[a] Cf. p. 17 above.   [b] In *SF*, ch. 10, p. 51 *a*, *b*.

[c] Ch. 3, p. 18 *b*, purporting also to be quoting from Liu Hsiao-Sun's book.

[d] Ch. 10, p. 7 *a*. The two plants were *tzu tshao*[11] and *hung hua*.[9] 'Purple herb', i.e. alkanet or dyer's bugloss, is identified in China as *Lithospermum officinale = erythrorhizon* (R 153; CC 386). True alkanet in England is *Pentaglottis = Anchusa, sempervirens = officinalis*; and bastard alkanet is *Lithospermum arvense* (Bentham & Hooker (1), nos. 696, 704; Clapham, Tutin & Warburg (1), pp. 656, 663). All are borages. 'Red flower' is of course safflower, *Carthamus tinctorius* (R 21; CC 49). Cf. p. 96 above.

[e] Ch. 15, (p. 240), cf. p. 16 above.

[f] Ch. 52, p. 13 *a* (p. 72); comm. Shih Shêng-Han (3), vol. 2, p. 334. This was no true mallow but the *lo khuei tzu*[12] or Malabar nightshade, *Basella rubra* (R 553, cf. CC 1485) of the Basellaceae. One of its synonyms is *yen chih tshai*.[13]

[g] *TKKW*, ch. 3, p. 3 *b*, tr. Sun & Sun (1), p. 77.

[h] For this term Schafer (9), p. 436, guessed vermilion (mercuric sulphide), and the Suns guessed litmus, but neither hit the mark. *Tzu kung*, or *tzu kêng*,[14] the 'purple branches', is stick lac with its purple dye, produced by the coccids *Laccifer lacca*, *Lakshadia chinensis*, etc.; see *PTKM*, ch. 39, (p. 69), cf. R 12. Its first appearance in the pharmaceutical natural histories was in +659, but the oldest account of it in the literature of any culture was that by Chang Pho[15] in his *Wu Lu*[16] (+3rd century). See Mahdi-hassan (7, 8, 48).   [i] *Carthamus*, as just noted.

[j] Identification not quite clear. CC 530 has *Rhododendron Simsii*, but R 455 has *Rosa laevigata*. Perhaps the orange pigment of *Berberis Thunbergii* (R 518) is the most likely.

[k] Judging from an entry in the +15th-century *Wu Yuan*[17] (p. 32).

[1] 續事始	[2] 二儀實錄	[3] 劉孝孫	[4] 第一轉	[5] 輕粉
[6] 紫粉	[7] 燕脂	[8] 紫鉚	[9] 紅花	[10] 山石榴
[11] 紫草	[12] 落葵子	[13] 燕脂菜	[14] 紫梗	[15] 張勃
[16] 吳錄	[17] 物原			

Summing up, there is a legend here about the −7th century which seems to have started in the late +6th, not very long after Thao Hung-Ching's subliming of the chlorides of mercury at the end of the +5th. Since calomel was clearly sometimes used as well as ceruse for the white cosmetic colour,[a] and since both were mixed habitually with red, purple or orange pigments to give rouge, there seems to have been some confusion between the two, as well as with vermilion itself. The whole story looks like a bookish attempt to give a cachet of remote antiquity to a familiar commodity of the inner apartments—unless perchance the Immortal Hsiao Shih was working with lead, in which case, since Duke Mu of Chhin is a sufficiently historical character, the carbonate artificially made might go back several centuries beyond the *Mo Tzu* and *Chi Ni Tzu* texts. This one may doubt, but we shall see in a moment another uneasy case where the two metals may have been mixed up. Meanwhile there is a very different legend which could reasonably suggest, at least, that Thao Hung-Ching was not quite the first to make calomel.

The *Lieh Hsien Chuan* (Lives of Famous Immortals), considered to be of the +3rd or +4th century, with some parts going back to the +2nd, has a piece about Chu Chu.[1] With other Taoists, he found a deposit of cinnabar on Tang Shan;[2] this, when the civil authorities came to appropriate it, burst into flames and flew through the air, so they left him in peace to mine it himself. Having taken this cinnabar for three years, the local governor, Chang Chün-Ming,[3] got hold of some 'flying mercuric snow' (*shen sha fei hsüeh*[4]), presumably made by Chu Chu, and took that for another five, after which he could fly in the air. Eventually he went away with Master Chu and was no more seen.[b] One may of course insist that this passage was interpolated later than the time of Thao Hung-Ching, but there is no particular reason for supposing this, and the possibility should be left open that the sublimation of the chlorides of mercury was first accomplished in the time of Ko Hung rather than Thao.

Returning now to the entry for purified calomel in the *Pên Tshao Kang Mu*,[c] Li Shih-Chen gives a curious quotation apparently from the *Pao Phu Tzu* book. Ko Hung, it seems, wrote:

Pao Phu Tzu says that 'white snow' (*pai hsüeh*[5]) is the same thing as 'frost powder' (*fên shuang*,[6] calomel). You take sea salt to make a box, and cover it over inside an earthenware reaction vessel, which must be tight so as not to allow the essences and radiances to leak. It will be ready in seven days. It is important that the Yang *chhi* be sufficient, and that the Yin be not allowed to invade. (Add) only ginger, lotus root and fumitory, together with the *ho chhê*,[7] which can thus be transformed (*lien*[8]) and used in projection processes. Among the immortals this is called the 'mysterious pot', men call it 'the origin of seminal essence', for alchemists it is 'essence of the element Wood', in Nature it is 'white snow' and in Heaven it is 'sweet dew'.

a It may be well to recall that all through the ages paleness was regarded as an attribute of beautiful women in China, hence the use of artificial aids. Yet in the stage make-up of traditional opera, white has always symbolised resourcefulness and cunning (Chang Kuang-Yu & Chang Chêng-Yu, 1).

b No. 46, tr. Kaltenmark (2), p. 146.   c Ch. 9, (p. 61), under *fên shuang*.

1 主柱   2 宕山   3 章君明   4 神砂飛雪   5 白雪   6 粉霜
7 河車   8 鍊

Here the trouble is that *ho chhê*[1] seems invariably to mean lead, not mercury, yet Li filed it among his notes on calomel. In one case Ware rendered *ho chhê* as mercury,[a] but there seems no other authority for this. The passage does not occur in the *Pao Phu Tzu* as we now have it,[b] but it may be found some day in one of the other *Tao Tsang* texts attributed to Ko Hung. The inclusion of the plant material is puzzling, and although plenty of chloride is present, nothing is said of alum, vitriol or other materials. The possibility always remains, of course, that in fact they were there, and that *ho chhê* was a deliberate false appellation to put competing alchemists off the track. Li Shih-Chen indeed must himself have interpreted it in this way. So perhaps about +300 will turn out in the end to be a better date than +500 for the first preparation of the chlorides of mercury in China.

This suspicion is confirmed, transmuted indeed almost into a certainty, by a chain of inference which we can follow back from subsequent literature. The *Pao Phu Tzu* book has a bare reference to a preparation called 'hard snow' (*kên hsüeh*[2]), if we may venture thus to translate a name which may have had much deeper meaning, since *Kên* is one of the *I Ching* trigrams;[c] in any case it occurs in a list of things which the alchemist must learn to make before proceeding to higher elixirs.[d] Ware interpreted it as calomel,[e] and he was probably quite right. For in the *Tan Ching Yao Chüeh*, written by Sun Ssu-Mo about +640 (see p. 133), *liu kên hsüeh*[3] appears, used in a formula with the comment 'Take the sublimate obtained from (quick-)silver'.[f] So far corrosive sublimate could not be excluded, but evidence which rules this out is forthcoming from a book intermediate in date between Ko Hung and Sun Ssu-Mo, and not long after Thao Hung-Ching himself, namely the *Thai-Chhing Shih Pi Chi* (see p. 130), placeable in the early +6th century. Here is described an elaborate process for making *kên hsüeh tan*,[4] one of the synonyms of which is given as *shui-yin shuang tan*,[5] 'mercury frost elixir', i.e. one of the undoubted names of calomel throughout the centuries.[g] Elsewhere in the same work another method of making *shui-yin shuang* is given,[h] particularly interesting because it starts from tin amalgam, Sn presumably acting as a reducing agent to prevent the formation of corrosive sublimate.[i] That seven re-sublimations are specified as necessary to purify is also interesting in view of the finding that a medieval specimen of calomel in Japan has been found to be 99·55 % pure.[j] The term *kên hsüeh tan* for calomel was still familiar

[a] (5), pp. 269, 344. The *Shih Yao Erh Ya* says emphatically lead.
[b] Nor do the terms *pai hsüeh* and *fên shuang* either.
[c] See Vol. 2, Table 13.   [d] *PPT/NP*, ch. 4, p. 9a.   [e] (5), p. 82.
[f] *YCCC* text, ch. 71, p. 3a, tr. Sivin (1), p. 169, cf. p. 282. Soon afterwards, he uses the expression *shui-yin shuang*, ibid. p. 3b, see Sivin (1), p. 171, cf. pp. 288–9.
[g] Ch. 1, pp. 5b to 7a, tr. Ho Ping-Yü (8), pp. 22 ff.
[h] Ch. 2, p. 4a, b, tr. Ho Ping-Yü (8), pp. 51 ff.
[i] There is at least one other mention (Ho Ping-Yü tr., p. 19).
[j] By Masutomi Kazunosuke (1), in his work on the drugs of the Shōsōin treasury, p. 102. Calomel made by the traditional method in Japan was found by Divers (1) in 1894 to be 'of signal purity'. The process was not so much a sublimation of a compound already formed as a crystallisation at comparatively low temperature direct from three substances in the gaseous state, mercury, hydrochloric acid and oxygen. Hanbury (1), pp. 224–5, (2), found also that Chinese calomel in his time (1860) was exceedingly pure.

[1] 河車   [2] 艮雪   [3] 流艮雪   [4] 艮雪丹   [5] 水銀霜丹

at the beginning of the +9th century, when Mei Piao compiled his *Shih Yao Erh Ya* (see p. 152), for there we find among its synonyms such names as *fei hsien ying tan*[1] (glittering elixir of the flying immortals) and *chhing hsiang chu tan*[2] (minister-subverting pearly elixir).[a] Thus we have here a good example of the philological method by means of which one can go back from one text to another by little and little, elucidating in the end a phrase which would otherwise be incomprehensible, and establishing a terminus for the first appearance of a discovery or a technique. It does really seem then that the sublimation of the chlorides of mercury, and of calomel specifically, goes back to about +300 in China. But now we must return to the consideration of Thao Hung-Ching.

He is also worth reading on arsenic compounds. On *yü shih*,[3] arsenolite (As_4O_6), he wrote:[b]

Nowadays both Shu and Han have it, but the best comes from Nan-yeh-chhi near Nan-khang, from the borders of Phêng-chhêng and from south of Loyang. Most comes from the Hsiao-shih Mountains. If *yü*[3] is put in water it will stop it from freezing, which shows that it has an extreme internal heat. People wrap it in balls of yellow clay and heat it strongly for a day and a night, then it breaks up into small lumps and can be used. It is good for treating debility (*lêng*[4]),[c] and can be combined into valuable elixir recipes; besides this it is greatly used in aurifaction and argentifaction (*huang pai shu to yung chih*[5]). Hsin-ning east of the Hsiang River, and Ling-ling, all have white *yü* ore.[3] It can soften the metals (*nêng jou chih*[6]).

This last remark must refer to the greater malleability and workability without cracking of arsenic–copper alloys as opposed to bronze.[d]

If we add one further quotation it is because it points up one especially important thing. Throughout the early centuries of Chinese alchemy, proto-chemistry and metallurgy, there must have been, reading between the lines, great use of processes which made a veneer of one metal upon another, by enriching or impoverishing the surface layers. If there were those who could accomplish this there were also those who knew how to detect it when it had been done, and here Thao Hung-Ching states quite clearly that the inside of a vessel or an ingot was not always what it seemed to be from the outside. The entry is on alum (*fan shih*[7]):[e]

It comes now from Hsi-chhuan in the north of I-chou, crossing the (Yellow) River from the west. Its colour is greenish-white. The crude stuff is called 'horse-tooth alum', but after purifying it becomes very white. People in Szechuan often mistake it for nitre (*hsiao shih*[8]),[f]

[a] Ch. 2, p. 2a.
[b] *CLPT*, ch. 5, (p. 124.1), tr. auct.
[c] The aphrodisiac nuance here is important, see pt. 2, pp. 284ff. [d] Cf. pt. 2, pp. 223ff.
[e] Cit. *CLPT*, ch. 3, (p. 84.1), tr. auct. It probably came originally from the *Pên Tshao Ching Chi Chu*. On plating see pt. 2, pp. 67, 246ff., 255, and on the ionic exchange of Cu and Fe, pt. 4.
[f] This term is ambiguous, especially in pharmaceutical literature, meaning sodium sulphate and nitrate as well as potassium nitrate. In Europe nitre even included sodium carbonate, cf. Crosland (1), p. 106. For further discussion see pt. 4.

[1] 飛仙英丹   [2] 傾相珠丹   [3] 礜石   [4] 冷   [5] 黃白術多用之
[6] 能柔金   [7] 礬石   [8] 消石

but it is white alum. The dark yellow variety is called 'bird-droppings alum'; it is not suitable for medicinal use, but it can be employed for 'plating' (*tu*[1]). If it is combined with
vinegar (lit. bitter wine) and refined (or processed) copper, and the mixture rubbed over the
surface of a piece of iron, it will assume a completely copper colour; but although the outside
is copper-coloured the nature of the interior remains (*wai sui thung ssu, nei chih pu pien*[2]).
The Manuals of the Immortals say that it can be taken by itself, and it is used also in elixir
formulae. After being ground in water it can be combined with plant drugs, then boiled
and heated to dryness; it is good for toothache but spoils the teeth if used in excess. This
shows that it is injurious for bones, so I doubt the statement that it strengthens (lit. hardens)
bones and teeth.

An important alchemical and medical text, *Thai-Chhing Shih Pi Chi*[3] (The Records
in the Rock Chamber; a Thai-Chhing Scripture),[a] was edited about Thao Hung-
Ching's time by an alchemist called Chhu-Tsê hsien-sêng[4] from the original text of
Su Yuan-Ming.[5] Su is a rather shadowy figure who lived between the time of Ko
Hung and Thao Hung-Ching, or perhaps even a little earlier.[b] Known also under the
pseudonym Chhing Hsia Tzu[6] (the Caerulean-Clouds Master), Su Yuan-Ming seems
to have been a quite prolific writer, no less than nine alchemical treatises being listed
under his pseudonym in the bibliographical chapters of Chêng Chhiao's *Thung Chih
Lüeh*.[c] Unfortunately most of Su's writings did not survive, though some are quoted
from time to time in other alchemical texts and in the pharmaceutical natural histories.

There is a general impression, especially reinforced when one has read something
of the *Tshan Thung Chhi*, that the Chinese alchemists, like their counterparts in the
West, invariably attempted to conceal their secret art in highly obscure language.
Su Yuan-Ming gives us an example showing that this is by no means always true,
since he wrote in a simple and lucid style, not only refraining from using synonymic
cover-names in his instructions but also explaining many common synonyms which
the alchemist might encounter. The *Thai-Chhing Shih Pi Chi* is one of the very few
alchemical texts extant that reveals to us plainly the chemical substances and their
synonyms, with the procedures and apparatus employed in early medieval Chinese
alchemy and alchemical medicine. We may quote one of its procedural formulae for
the preparation of the 'Nine-vessel elixir of the Yellow Emperor'.

---

[a] *TT*874. A complete provisional translation has been made by one of us and will appear in due
course (Ho Ping-Yü, 8).

[b] His name is sometimes written Su Yuan-Lang;[7] probably the original form, changed for tabu
reasons.

[c] Unless of course someone else between the +6th and the +11th centuries appropriated the same
pseudonym. Among the books quoted is a *Pao Tsang Lun*[8] (Discourse on the Contents of the Precious
Treasury (of the Earth)), and if this was the same as that quite frequently quoted in the *Chêng Lei Pên
Tshao* and other pharmaceutical natural histories, it was a work of real importance for it has much to
tell us on metallurgical matters (cf. pt. 2, p. 273 above, and Sect. 30 on iron and steel technology). Sometimes it appears under the title *Hsien-Yuan Pao Tsang Lun*,[9] and possibly the *Huang Ti Pao Tsang Ching*,
also now lost as such, was identical with it. It is interesting that the *Lo-fou Shan Chih*, in its biography
of Su Yuan-Lang (-Ming), ch. 4, pp. 13 *a*ff., attributes the (or a) *Pao Tsang Lun* to him. But, as already
noted (pt. 2, p. 342 above), our present *Pao Tsang Lun* text dates from the close neighbourhood of +918
(cf. Tsêng Yuan-Jung, *1*).

[1] 鍍	[2] 外雖銅色內質不變	[3] 太清石壁記	[4] 楚澤先生
[5] 蘇元明	[6] 青霞子 [7] 蘇元朗	[8] 寶藏論	[9] 軒轅寶藏論

Method of making the 'Huang Ti chiu ting tan[1]*'.*

Realgar (*hsiung huang*[2]) and orpiment (*tzhu huang*[3])	½ lb each
Cinnabar (*chu sha*[4])	5 lb
Sulphur (*shih liu huang*[5]), quartz (*pai shih ying*[6]), stalactites (*chung ju*[7]), crude sodium sulphate (*phu hsiao*[8]), and arsenolite (*yü shih*[9]) .	3 ozs each
Stalagmites (*shih chhuang*[10]), calcium sulphate (*han shui shih*[11]), gypsum (*shih kao*[12]), brown haematite (*Yü yü liang*[13]), lapis lazuli (*chhing shih*[14]), selenite (*Thai Yin hsüan ching*[15]), red bole clay (*chhih shih chih*[16]), mica (*yün mu*[17]), and magnetite (*tzhu shih*[18])	5 ozs each

The above 17 ingredients are to be pounded together, mixed with vinegar (*tsho*[19]) until thoroughly soaked (before being placed in the lower bowl of the reaction-vessel). They are then covered with (a layer of) common salt (*Wu yen*[20]). (The upper bowl of the reaction-vessel is then placed over the lower mouth-to-mouth, and the vessel rendered airtight by applying a lute outside. It) is heated for three days and nights and then allowed to cool for half a day before being opened. This sublimation (*fei*[21]) process is to be repeated until seven cyclical changes have been performed. (The product) can be used to cure all illnesses, and once treated by this elixir the same illness will never recur.[a]

It is hardly possible to speculate about what the end-product would actually have been with so many constituents mixed together and containing impurities which we cannot now exactly know. Probably nothing very interesting would have taken place except for some mutual interchanges of anions and cations; while the sublimate would have been a mixture of sulphur and re-formed mercuric sulphide together with arsenic and its oxides and sulphides.[b] Most of the elixir formulae involve a large number of substances as in the above example, though there are some much simpler procedures. The 'Three Messengers elixir' (*san shih tan*[22]) required only calomel (*shui yin shuang*[23]), cinnabar (*chu sha*[24]), realgar (*hsiung huang*[25]) and common salt.[c] Yoshida has given a table showing the compositions of most of the elixirs described in the *Thai-Chhing Shih Pi Chi*.[d] Although it is not very accurate, since it omits some important ingredients and even some elixirs, the table serves its purpose by indicating the preponderance of arsenical and mercurial compounds in early medieval Chinese alchemy.

The Northern Chhi emperor Wên Hsüan Ti[26] (r. +550 to +560), Kao Yang,[27] while anxious to become an immortal, was rather cautious about the taking of elixirs. According to the *Pei Shih* (History of the Northern Dynasties):[e]

He ordered Chang Yuan-Yu[28] and other alchemists to prepare the 'Nine-fold Cyclically-transformed elixir' (*chiu chuan chin tan*[29]). When it was accomplished he put it in a jade box,

---

[a] *TT*874, ch. 1, p. 2*a*, tr. auct.
[b] Much would depend on the conditions, the temperature gradient, the size and shape of the vessel, etc.
[c] *TT*874, ch. 2, p. 1*a*.   [d] (5), pp. 220, 221.   [e] Ch. 88, p. 12*a*, tr. auct.

[1] 黃帝九鼎丹	[2] 雄黃	[3] 雌黃	[4] 朱砂	[5] 石硫黃
[6] 白石英	[7] 鐘乳	[8] 朴硝	[9] 礜石	[10] 石牀
[11] 寒水石	[12] 石膏	[13] 禹餘糧	[14] 青石	[15] 太陰玄精
[16] 赤石脂	[17] 雲母	[18] 礠石	[19] 酢	[20] 吳鹽
[21] 飛	[22] 三使丹	[23] 水銀霜	[24] 朱砂	[25] 雄黃
[26] 文宣帝	[27] 高洋	[28] 張遠遊	[29] 九轉金丹	

and said: 'I am still too fond of the pleasures of this world to wish to take flight to the heavens immediately. I shall keep (the elixir) and take it only when I am about to die.'

Where medieval Chinese elixirs were concerned, a death-bed conversion was very prudent.

(ii) *Alchemy in the Sui re-unification*

After the restoration of the empire under the Sui, the emperor Yang Ti[1] (Yang Kuang,[2] r. +605 to +617) patronised Wang Yuan-Chih,[3] one of Thao Hung-Ching's many disciples. When that short-lived dynasty had in turn come to an end Wang Yuan-Chih was visited personally by the Thang emperor Thai Tsung[4] during the Wu-Tê reign-period (+618 to +627), and was then given first the title Shêng-Chen hsien-sêng[5] and later Shêng-Hsüan hsien-sêng[6].[a]

Among the many disciples of Wang Yuan-Chih was Phan Shih-Chêng,[7] who was in his turn much respected by the Thang emperor Kao Tsung[8] (+650 to +684). He lived for twenty years in a valley among the Sung-shan[9] Mountains, eating only pine leaves and drinking nothing but water (according to the *Chiu Thang Shu*).[b] In this seclusion he was accompanied by another alchemist, Liu Tao-Ho.[10] This latter person draws our attention to a much greater alchemist of the early Thang, for Liu had been a disciple of Mêng Shên,[11] the well-known physician and alchemist,[c] and he when young had studied under Sun Ssu-Mo.[12] Now Sun ranks fully with Ko Hung and Thao Hung-Ching in the history of Chinese alchemy and medicine, but until recently we knew as little about his personal life as about that of Hippocrates of Cos. In fact his name was only perpetuated by his medical writings; for example, Okanishi (*1*) credits more than a dozen medical treatises to him.[d] The *Chiu Thang Shu* gives a short biography of Sun Ssu-Mo, saying that he died in the year +682, and claimed to have been born in a *hsin-yu* year of the Khai-Huang reign-period.[e] There has been doubt regarding his dates, and Sun himself undoubtedly believed in +673 that he had reached the age of ninety-two; Sivin, who has translated the biography and made a study of the subject, gives the best span as +581 to +682.[f]

We know of Sun Ssu-Mo's alchemical work chiefly through the *Thai-Chhing Tan Ching Yao Chüeh*[13][g] (Essentials of the Elixir Manuals for Oral Transmission, a Thai-Chhing Scripture),[h] which exists not in the present *Tao Tsang* as such but incorporated

---

a See *Hsin Thang Shu*, ch. 192, p. 6 *a*.   b Ch. 192, p. 7 *a*.   c See p. 140 below.
d The most important is of course the *Chhien Chin Yao Fang*[14] (Thousand Golden Remedies) written between +650 and +659, with its continuation *Chhien Chin I Fang*[15] finished between +660 and +675. A nutritional and hygiene section in this was originally a separate work, *Chhien Chin Shih Chih.*[16]
e Ch. 191, pp. 5 aff. There was no such year, but it could have meant +601.
f (*1*), pp. 120ff., 265–6.
g In the Ming the work was renamed *Thai-Chhing Chen Jen Ta Tan*[17] (The Great Elixirs of the Adepts; a Thai-Chhing Scripture).
h A full translation and critical study of this important text has been made in recent years by Sivin (*1*). Much of what follows here is based on the paper of Ho Ping-Yü (5).

1 煬帝	2 楊廣	3 王遠知	4 太宗	5 昇眞先生
6 昇玄先生	7 潘師正	8 高宗	9 嵩山	10 劉道合
11 孟詵	12 孫思邈	13 太清丹經要訣	14 千金要方	
15 千金翼方	16 千金食治	17 太清眞人大丹		

in the *Yün Chi Chhi Chhien* of +1022.[a] It would have been written about +640. Sun Ssu-Mo says that he describes only those procedures which he himself had successfully tried out. He lists the names of sixty-seven different elixirs, pointing out, however, that the methods of preparation were available only for a few of them. He then explains the making of the lute or sealing-compound, the 'six-and-one paste' (*liu i ni*[1]),[b] and the proper selection and preparation of raw materials, for example alum, for use in the Great Work. He goes on to describe the essential alchemical apparatus, including the furnace (*tsao*[2])[c] and the reaction-vessel (*fu*[3]), giving the necessary instructions for the hermetical sealing of the joints by the lute. The constituents of this included red bole clay, calcium carbonate from oyster-shells, kalinite (potash alum, potassium aluminium sulphate) from Tunhuang, talc (magnesium silicon oxide), 'Turkestan' salt from Shensi, lake salt from Shansi, and some arsenolite (arsenic trioxide).[d]

As a practical alchemist Sun Ssu-Mo described his procedures in notably simple language, avoiding the use of synonyms and obscure terms with concealed meanings. In fact, among the many ingredients recommended for the thirty-two formulae in the text,[e] he included only three such terms. In one instance he uses the two trigrams *Li*[4] and *Tui*[5] to denote two alchemical substances. According to some commentaries of the *Tshan Thung Chhi* these would mean cinnabar and lead respectively,[f] but Sivin has given a more plausible interpretation of them as cinnabar and white lead.[g] The other synonym appearing in the *Tan Ching Yao Chüeh* is 'flowing hard snow' (*liu kên hsüeh*[6]), which must mean calomel (p. 128).

Among the thirty-two formulae mentioned in the text, eighteen refer to the preparation of some fourteen elixirs. Most of them seem quite poisonous, containing mercury and lead, if not arsenic, as elements or compounds. For example, in the preparation of the 'Minor Cyclically-transformed elixir' (*hsiao huan tan*[7]) Sun Ssu-Mo recommends the use of 1 lb of mercury, 4 oz of sulphur, 3 oz of cinnabar, 4 oz of rhinoceros horn, and 2 oz of musk. These are to be ground or pounded separately into powder and then mixed together and made into pills.[h] For the 'Great Unity Three Messengers elixir' (*Thai-I san shih tan*[8]) the formula requires 1 lb of purified calomel, 10 oz of cinnabar, 10 oz of sulphur, and also 10 oz of realgar.[i] In the case of the 'Gold elixir' (*chin tan*[9]) 8 oz each of gold and mercury and 1 lb each of realgar

---

[a] *TT* 1020.   [b] Because seven constituents. An ancient term.

[c] He also mentions at least twice a forced-draught furnace (*fêng lu*[10]) which probably involved the use of the continuous-blast piston-bellows (*fêng hsiang*[11]), cf. Vol. 4, pt. 2, pp. 135 ff.

[d] Some added also the excreta of earthworms, but Sun considered this foolish, having found that it was just the same as any other earth, and no good as an ingredient anyway. The calcium carbonate was also useless, in his opinion. See Sivin (1), pp. 160 ff.

[e] Thirty-eight counting variants.

[f] See, for example, *TT* 991, ch. 2, p. 19*a*, and *TT* 994, ch. 1, p. 14*a*.

[g] (1), pp. 194 ff.

[h] *YCCC*, ch. 71, p. 5*a*, *b*; tr. Sivin, *op. cit.* pp. 174–5.

[i] *Ibid.* pp. 3*b*, 4*a*; tr. Sivin, *op. cit.* p. 171.

[1] 六一泥   [2] 竈   [3] 釜   [4] 離   [5] 兌
[6] 流艮雪   [7] 小還丹   [8] 太一三使丹   [9] 金丹   [10] 風爐
[11] 風箱

Fig. 1358. The eminent Sui and Thang physician and alchemist, Sun Ssu-Mo (d. +682). In the drawing, taken from *Lieh Hsien Chhüan Chuan*, ch. 5, p. 20*b*, he seems to be carrying scrolls of prescriptions or elixir recipes hanging from his rustic staff.

and orpiment are required, besides some vinegar.[a] The so-called 'Lead elixir' (*chhien tan*[1]) calls for as much as 4 lb of lead and 1 lb of mercury besides other ingredients.[b] It is not that Sun Ssu-Mo was unaware of the lethal effects of many of the chemical substances used. In his medical treatise, the *Chhien Chin I Fang*[2] (Supplement to the Thousand Golden Remedies), he states categorically that mercury, realgar, orpiment and sulphur are poisonous, as also are gold, silver and vitriol.[c] It is interesting that while in his medical prescriptions Sun Ssu-Mo recommended substances like mercury, it was in much more conservative doses than for the elixirs. While one must be mindful of the profound conviction of Paracelsus (+1493 to +1541) some eight centuries later that 'poisonous action and remedial virtue are intimately bound up with each other', as in the case of arsenic and especially mercury,[d] one cannot help being mystified by the fact that in spite of his knowledge of the toxic effect of certain inorganic substances Sun Ssu-Mo seems to have prescribed them in so much larger amounts for elixirs than for medicines.[e] Could the thought have been that when human beings were raised to a level approaching that of the immortals their bodies would no longer be susceptible to poison? Every student of pharmacology knows that although arsenic has a markedly cumulative action, astonishing degrees of tolerance can be achieved if it is taken by the mouth, as habitually among the peasants of Styrian Austria. The rationale appears to be a decreased absorptive power of the alimentary tract, as the effect is not shown when arsenic is taken in soluble form.[f]

We can adduce at least one piece of evidence to show that the Chinese alchemists of the Middle Ages were fully aware that these elixirs could not be taken in the ordinary way, and that one had to 'build up one's constitution' before the venture. Such a conviction may have come from Sun Ssu-Mo himself. The *Chen Chung Chi*[3] (Records of the Pillow-Book), which can be shown from internal evidence to have been written during or shortly after the Chen-Kuan reign-period (+627 to +649) and which is attributed to Sun Ssu-Mo in the *Thung Chih Lüeh* bibliography, gives the following warning: 'One must take vegetable and plant drugs at first until their beneficial effects are felt; then only can one take mineral drugs for the purpose of achieving longevity.'[g]

Many points of interest arise from a survey of the *Tan Ching Yao Chüeh*. Particu-

[a] *YCCC*, ch. 71, pp. 9b, 10a; tr. Sivin (1), pp. 185–6.

[b] *Ibid.* p. 10b; tr. Sivin, *op. cit.* pp. 187–8.

[c] Ch. 2, p. 2a to p. 6a. For elixir-poisoning see Ho Ping-Yü & Needham (4). The subject will be discussed in connection with industrial diseases in Vol. 6 (Sect. 45).

[d] Pagel (10), p. 145. We can almost call Paracelsus the father of modern pharmacology and chemotherapy because of his famous dictum 'Alein die Dosis macht daß ein Ding kein Gift ist' (cf. Lieben (1), pp. 13 ff.). This wonderful aphorism comes from the 'Sieben Defensiones' (+1537–8), Sudhoff ed., vol. 11, p. 138. It is a commonplace today that any powerful remedy will be, in certain conditions, a powerful poison, cf. Green (1). In classical Chinese the same word, *tu*,[4] means 'poison' and 'active principle'.

[e] This is only partly true, cf. Sivin (1), p. 143.

[f] Cf. Clark (1), p. 608, and the discussion in pt. 2, pp. 290 ff. above.

[g] *TT*830, p. 15b. This book has been mentioned in connection with Ko Hung (p. 110 above). It deals much with hygiene, but contains some methods for preparing aqueous solutions of inorganic substances, which will be quoted at a more appropriate place (see pt. 4 below).

[1] 鉛丹   [2] 千金翼方   [3] 枕中記   [4] 毒

larly striking in it is the Chinese form of that *diplōsis* or 'doubling' of valuable metals which (as we have seen above, pt. 2, pp. 18, 193) occurs so often in Alexandrian proto-chemistry. Here the technical term is *thien*,[1] to add to or to augment, and it is applied both to paktong (*pai thung*,[2] cupro-nickel)[a] and to brass (*thou*[3]).[b] The former is 'increased', i.e. diluted or adulterated, by heating with lead as basic lead carbonate (white lead), two similar processes with different auxiliary ingredients being given.[c] The latter is multiplied in a parallel way either by heating with *la*,[4] probably an alloy of zinc, tin and lead,[d] or with basic lead carbonate as for paktong.[e] It has naturally been assumed that these formulae are a link between the Western techniques and Thang China, presumably transmitted eastwards via Sassanian Persia, and it is true that the third method just mentioned is called the 'Persian method for using azedarach to augment brass' (*Po-Ssu yung khu-lien-tzu thien thou fa*[5]), this mysterious thing being the fruits of the so-called Persian lilac, *Melia Azedarach*, freely growing in China at least since the time of Thao Hung-Ching.[f] However, in view of what we have already seen concerning 'projection' (*tien*[6]), a term for another practice of the highest importance for ancient Western aurifaction, which seems to occur at least as early in the Chinese culture-area as in Hellenistic Egypt, we should perhaps be on our guard against the idea that 'doubling' must necessarily have started with the Alexandrians. And there is always the possibility that many of these ancient chemical ideas, terms and techniques radiated from the Iranian culture-area in both directions.[g]

Other sophistications also are to be found in the pages of Sun Ssu-Mo's book. There is a 'dyeing' of verdigris (copper acetate) with indigo (*chhing tai*[7])[h] to give it the bluer colour of the more valuable pigment malachite (basic copper carbonate), presumably useful in fresco and scroll painting.[i] There is also a formula for adding red sappan pigment to indigo[j] either to make it go further or to give it a purple tint;[k] moreover one can produce an artificial white jade from clamshell calcite or aragonite

---

[a] Cf. pt. 2, pp. 225 ff. above.   [b] Cf. pt. 2, pp. 195 ff. above.

[c] *YCCC*, ch. 71, pp. 13*a* to 14*b* and 15*b*, 16*a*. Because of certain difficult terms used, as well as misleading or partially misleading comments by later editors, the rationale took a good deal of unravelling; see Sivin (1), pp. 194 ff., 201 for translation.

[d] *YCCC*, ch. 71, pp. 16*b*, 17*a*, tr. Sivin (1), pp. 203–4, who took *la* to be pewter, i.e. an alloy of lead and tin, but see our discussion on pp. 211, 216 of pt. 2 above. Lead carbonate was also used, and in a variant recipe, tin.

[e] *YCCC*, ch. 71, p. 17*a, b*; tr. Sivin (1), p. 204.

[f] R335, also known as *chin ling tzu*.[8] Other vegetable material was also added, including dried plums and sparrow droppings, presumably as carbonaceous reducing agents.

[g] This will be better appreciated from pt. 4 below. But see also pt. 2, p. 220 above.

[h] Sun's text generally has *shih tai*,[9] which normally means graphite, as in *tai mei shih*[10] (black eye-brow mineral), but his meaning is unmistakable.

[i] *YCCC*, ch. 71, pp. 20*b*, 21*a*; tr. Sivin (1), p. 213. Cf. Yü Fei-An (1).

[j] *Lan chih*,[11] 'vegetable blue infusion', from *Polygonum tinctorium* or one of several species of *Indigofera*. See Sun & Sun (1), pp. 78–9, commenting on *TKKW*, ch. 3.

[k] *YCCC*, ch. 71, p. 21*a*; tr. Sivin (1), pp. 213–14. Water-soluble sappan red comes from the trunk wood of the tree *Caesalpinia sappan* (*su fang mu*,[12] Roi (1), pp. 173–4. Cf. Sun & Sun (1), p. 78.

---

[1] 添   [2] 白銅   [3] 鍮   [4] 鑞   [5] 波斯用苦楝子添鍮法
[6] 點   [7] 青黛   [8] 金鈴子   [9] 石黛   [10] 黛眉石   [11] 藍汁
[12] 蘇方木

fused with quartz,[a] as also artificial pearls from calcite and (probably) fish guanine.[b] In all these processes an 'ersatz' tradition very reminiscent of the Hellenistic papyri comes to the surface, but it can hardly be later than theirs since already nearly five hundred years earlier Li Shao-Chün had proposed to the emperor not natural but artificial gold.[c]

Quite apart from processes of this kind, however, the *Tan Ching Yao Chüeh* has many interesting things. There is, for example, an account of the calcination of mercury very like that of Maslama al-Majrītī (d. +1007), so much praised by Leicester[d] though more than three centuries later. Because of the remarkable reversibility of the Hg/HgO transformation at 630°, it was possible for Sun Ssu-Mo, by very careful attention to the temperatures used, to obtain a cyclical change from mercury to 'cinnabar' indefinitely, though with gradually decreasing yield as the mercury volatilised.[e] He did it in fact seven times, hence the name *Chhi fan tan sha fa*.[1] In another place where an elixir called *Chhih hsüeh liu chu tan*[2] (Scarlet snow and flowing pearls) is made by heating realgar, acetic acid and salt in an aludel (Fig. 1359, *a*, *b*), it has been found experimentally that the sublimate is pure metallic arsenic.[f] This then was an isolation long preceding the +16th-century experimentalists (Paracelsus, Libavius, 'Basil Valentine')[g] or even Albertus Magnus, if indeed he got it, as is said, by heating orpiment with soap.[h] It was used, says Sun Ssu-Mo, for reviving those who had fainted or were on the point of death.[i] There is also a careful, if complicated, sublimation of calomel.[j] Elsewhere again polymeric sulphur[k] is produced by boiling with sesame oil and caustic lye, then washed with wine and honey before being administered in the form of pills made with jujube-date pulp.[l] All in all, the *Tan Ching Yao Chüeh* is one of the most rewarding of Thang alchemical texts.

Some discoveries that may have been Sun Ssu-Mo's are embodied in short extracts quoted in other collections. For example, the *Chu Chia Shen Phin Tan Fa*[3] (see pp. 159, 197) appears to quote him as follows:[m]

Take of sulphur and saltpetre (*hsiao shih*[4]) 2 oz. each and grind them together, then put them in a silver-melting crucible or a refractory pot (*sha kuan*[5]). Dig a pit in the ground and

---

[a] YCCC, ch. 71, p. 20*a*; tr. Sivin (1), p. 211.

[b] *Ibid.* p. 20*a*, *b*; tr. Sivin, *op. cit.* pp. 212–13.   [c] Cf. p. 31 above.   [d] (1), p. 71.

[e] YCCC, ch. 71, p. 12*a*; tr. & comm. Sivin (1), p. 191.

[f] *Ibid.* p. 8*a*, *b*; tr. Sivin, *op. cit.* pp. 180 ff. There is an uncertainty at one point in the text, and the final product may have been crystalline or fibrous arsenic trioxide in purified form. Many of the formulae in Sun's book involve sublimation, but there seems to be no reference to distillation.

[g] See Partington (7), vol. 2, pp. 147, 199, 261.

[h] As is said by Weeks (1), p. 10, following Jagnaux (1), vol. 1, pp. 656 ff.

[i] Cf. pt. 2, p. 90 above.   [j] YCCC, ch. 71, pp. 6*a*ff.; tr. Sivin (1), pp. 176 ff.

[k] Durrant (1), pp. 492–3.   [l] YCCC, ch. 71, pp. 11*b*, 12*a*; tr. Sivin (1), pp. 189 ff.

[m] TT911, ch. 5, p. 11*a*, tr. auct. Attention was first drawn to this passage by Fêng Chia-Shêng (1), p. 41; cf. (4), p. 35, (6), p. 10. It is also quoted in Yen Tun-Chieh (20), p. 19. Its heading is *Tan Ching Nei Fu Liu Huang Fa*[6] (Process in the Elixir Manuals for the subduing of sulphur). Actually, the passage is anonymous. The one preceding it is attributed to a Huang San Kuan-Jen[7] (His Excellency Huang Tertius, as Sivin (1) renders it), otherwise unknown, while the one before that is given Sun Ssu-Mo's name—but neither concerns gunpowder. Nevertheless the process is archaic and may belong to his time, so that Fêng Chia-Shêng's attribution to Sun might be right.

[1] 七返丹砂法   [2] 赤雪流朱丹   [3] 諸家神品丹法   [4] 硝石

[5] 沙䥶   [6] 丹經內伏硫黃法   [7] 黃三官人

put the vessel inside it so that its top is level with the ground, and cover it all round with earth. Take three perfect pods of the soap-bean tree,[a] uneaten by insects, and char them so that they keep their shape, then put them into the pot (with the sulphur and saltpetre). After the flames have subsided close the mouth and place three catties (lb) of glowing charcoal (on the lid); when this has been about one third consumed remove all of it. The substance need not be cool before it is taken out—it has been 'subdued by fire' (fu huo[1]) (i.e chemical changes have taken place giving a new and stable product).

Someone seems to have been engaged here about +650 in an operation designed, as it were, to produce potassium sulphate, and therefore not very exciting, but on the way he stumbled upon the first preparation of a deflagrating (and later explosive) mixture in the history of all civilisation.[b] Exciting must have been the word for that.

The time between Ko Hung and Sun Ssu-Mo was an important period in alchemical development. Beginning from Ko Hung alchemical writings became less theoretical, developing finally into the lucid practical style of the *Thai-Chhing Shih Pi Chi* and the *Tan Ching Yao Chüeh*. However, after the time of Sun Ssu-Mo, many of the alchemical writers gradually returned to the fashion of using obscure synonyms, perhaps because of the alarm caused by many cases of elixir poisoning, and the desire that those without proper guidance should be dissuaded from trying out alchemical experiments by themselves.

In general, the number of elixir ingredients used in the *Tan Ching Yao Chüeh* is relatively fewer than those in the *Thai-Chhing Shih Pi Chi*. The maximum number in any one case occurs in the 'Purple travelling elixir' (tzu yu tan[2]),[c] requiring twelve different substances, while the minimum is found in the 'Seven-fold cyclically-transformed cinnabar' (chhi fan tan sha[3])[d] which needs only one, mercury itself. The average number of ingredients used is five. One may compare these with some of the recipes in the *Thai-Chhing Shih Pi Chi*, which require more than thirty ingredients each. After the time of Sun Ssu-Mo the Chinese alchemists never reverted to the complexity of the formulae which are found in this book.

Written by an anonymous author probably during the time of Sun Ssu-Mo (or soon after) is another important alchemical text entitled *Chin Shih Pu Wu Chiu Shu Chüeh*[4] (Explanation of the Inventory of Metals and Minerals according to the Numbers Five and Nine).[e] It is particularly interesting because it tells how substances can be identified, and says that their 'quality' must be known before they can be used for making elixirs, besides mentioning the occurrences and properties of some of them. But it contains notes on fewer chemical substances than the *Shih Yao Erh Ya*[5] (Synonymic Dictionary of Minerals and Drugs) by Mei Piao,[6] of which we

---

[a] *Gleditschia sinensis* (CC 587). Much will be said of this saponin source in Vol. 6, meanwhile cf. Needham & Lu Gwei-Djen (1).

[b] This subject is dealt with fully in Sect. 30.

[c] Or 'Empyrean-roving', *YCCC*, ch. 71, pp. 4a to 5a; tr. Sivin (1), pp. 171 ff.

[d] *Ibid.* p. 12a; Sivin, *op. cit.* p. 191. See p. 137 above.

[e] The numbers stand for the elements Earth and Metal.

[1] 伏火   [2] 紫遊丹   [3] 七返丹砂   [4] 金石簿五九數訣
[5] 石藥爾雅   [6] 梅彪

shall speak presently.[a] Of special interest are the names of foreign countries, such as Persia, Annam and Udyāna, and the names of Indian Buddhist monks mentioned in it.[b] The following passage illustrates this:[c]

Saltpetre (hsiao shih[1]).

Originally this was produced in I-chou[2] by the Chiang[3] tribes-people, Wu-tu[4] and Lung-hsi[5],[d] (but now) that which comes from the Wu-Chhang[6] country (Udyāna) is (also) of good quality. In recent times, during the Lin-Tê reign period of the Thang, in a *chia-tzu* year (+664) a certain Indian (lit. Brahmin) called Chih Fa-Lin[7][e] (came to China) bringing with him (some sūtras in) the Sanskrit (language) for translation. He asked if he might visit the Wu-thai Shan[8] mountains to study the (Buddhist) customs, (and was allowed to do so). When he reached the Ling-shih[9] district in Fen-chou[10] he said: 'This place abounds in saltpetre. Why is it not collected and put to use?' At that time this monk was in the company of twelve persons, among whom were Chao Ju-Kuei[11] and Tu Fa-Liang.[12] Together they collected some of the substance and put it to the test, but found it unsuitable (for use) and not comparable to that produced in Wu-Chhang. Later they came to Tsê-chou,[13] where they found a mountain covered with beautiful trees. (The monk) said once again: 'Saltpetre should also occur in this region. I wonder whether it will be as useless as (what we came across) before?' Whereupon together with the Chinese monk Ling-Wu[14] they collected the substance, and found that upon burning it emitted copious purple flames (lit. smoke).[f] The Indian monk said: 'This marvellous substance can produce changes in the Five Metals, and when the various minerals are brought into contact with it they are completely transmuted into liquid form (*chin pien chhêng shui*[15]).'[g] And the fact that its properties were indeed the same as those of the material from Wu-Chhang was confirmed by testing it several times on different metals. Compared to that from Wu-Chhang this from Tsê-chou was a little softer.

Here we have mention of the potassium flame[h] and of the use of saltpetre as a flux in smelting. This passage raises several important questions, notably the appearance of close Indian–Chinese chemical contacts during the Thang period, and the exact time when potassium nitrate was reliably discovered, identified and used. To all these we shall return in pt. 4, mindful of their great significance for the first knowledge of mineral acids, the invention of gunpowder, and the mutual indebtedness of China and

---

a Forty-one as against some 180, but the *SYEY* lists include plant and animal substances, apparatus, and the names of compounded elixirs.

b Cf. pp. 160, 174, 193–4.   c *TT* 900, pp. 5 b, 6 a, tr. auct.

d Mod. Kansu.   e This ethnikon shows that he was of Yüeh-chih or Śaka stock.

f The same statement is made in the immediately following entry (p. 6 b) for a substance called *thien ming sha*,[16] evidently another potassium salt. We have not come across this term anywhere else (it is not in *SYEY*); here the text says that it was imported from Persia. We suspect, therefore, that it was potash alum in powder form (*alumen minutum*, flour- or sugar-alum); cf. Singer (8), pp. 113, 117, 218 and de Mély (1), p. 145. Laufer (1), pp. 474–5, discusses this importation, but he believed of course that the 'Persia' in question was actually somewhere in Malaya (cf. pt. 2, p. 143).

g This may refer both to its use as a metallurgical flux and to the aqueous solution of inorganic substances by nitric acid (pt. 4).

h As we shall see in pt. 4, the flame test for K goes back to Thao Hung-Ching, c. +480.

¹ 硝石	² 益州	³ 羌	⁴ 武都	⁵ 隴西
⁶ 烏長	⁷ 支法林	⁸ 五臺山	⁹ 靈石	¹⁰ 汾州
¹¹ 趙如珪	¹² 杜法亮	¹³ 澤州	¹⁴ 靈悟	¹⁵ 盡變成水
¹⁶ 天明砂				

India.[a] It is also from the above passage that the date of the *Chin Shih Pu Wu Chiu Shu Chüeh* can be deduced.

Mêng Shen[1] (+621 to +718), the disciple of Sun Ssu-Mo, became an outstanding pharmaceutical naturalist.[b] He had occasion to demonstrate his skill as an alchemist by detecting the counterfeit gold which the empress Wu Tsê Thien[2] used to give away to her civil officials as awards. The *Chiu Thang Shu* has the following to say about him:[c]

> From early youth onwards (Mêng Shen) was fond of alchemy. On one occasion, he visited the home of Liu Wei-Chih,[3] Vice-President (Shih Lang[4]) of the Department of the Imperial Grand Secretariat (Fêng Ko[5]), and saw some gold (objects) given (by the empress) as rewards. Whereupon he said: 'This is alchemical gold (*yao chin*[6]), and if you submit it to the fire, coloured vapours will be seen above it.' Afterwards the matter was put to the test, and he was proved right, (but) when (the empress) Tsê Thien heard about it she was not amused. So she found a pretext for having Mêng re-posted (away from the capital) to Thaichow.

This must surely be a reference to the appearance of the green flame of copper at the flue of the cupellation furnace. Mêng Shen's disciple, Liu Tao-Ho,[7] enjoyed such a good reputation as an alchemist that the emperor Kao Tsung (r. +650 to +684) built for him a temple called the Thai-I Kuan[8] at the place where he lived, and asked him to prepare a cyclically-transformed elixir. In this he succeeded and eventually presented it to the emperor.[d]

Followers of Thao Hung-Ching continued to enjoy the confidence of the imperial court for several generations. Among the many disciples of Phan Shih-Chêng[9] were Wu Yün[10] and Ssuma Chhêng-Chên[11] (perhaps +639 to +727). Wu Yün was much respected, and even consulted, by the emperor Hsüan Tsung. When asked by him to reveal the secret of immortality Wu replied that such arts should neither be sought for nor practised by an emperor.[e] Wu was a friend of the great poet Li Pai[12] (+701 to +762), who during his later years became interested in the techniques of immortality.[f] Ssuma Chhêng-Chên[g] was much favoured by the empress Wu Tsê Thien (r. +685 to +705) and the two emperors Jui Tsung (r. +710 to +712) and Hsüan Tsung (r. +713 to +756).[h] Although famous as an alchemist, he unfortunately left behind very

---

[a] See pp. 160 ff. and pt. 5 below.

[b] He was the writer of the *Shih Liao Pên Tshao*[13] (Nutritional Therapy; a Pharmaceutical Natural History) *c.* +670, and one of the first to study deficiency diseases such as beri-beri. Mêng Shen had also a successful civil service career, becoming Governor of Thungchow.

[c] *Chiu Thang Shu*, ch. 191, p. 9a, tr. auct., cf. Sivin (1), p. 45. That the significance of this incident was widely appreciated may be seen perhaps from the fact that it was often quoted, as e.g. in *Li Tai Ming I Mêng Chhiu*[14] (Brief Lives of the Famous Physicians in all Ages), written by Chou Shou-Chung[15] in +1040 (ch. 1, p. 13a).   [d] See *Chiu Thang Shu*, ch. 192, p. 7a. Cf. p. 132 above.

[e] *Ibid.* p. 8b.   [f] See *Hsin Thang Shu*, ch. 202, p. 10a.

[g] Also known as Ssuma Tzu-Wei[16] and Chên-I hsien-sêng.[17]

[h] His period of eminence marks the ascendancy of the Mao Shan school (cf. p. 120 above, and pt. 2, pp. 110, 152, 235) at court.

[1] 孟詵	[2] 武則天	[3] 劉禕之	[4] 侍郎	[5] 鳳閣
[6] 藥金	[7] 劉道合	[8] 太一觀	[9] 潘師正	[10] 吳筠
[11] 司馬承貞	[12] 李白	[13] 食療本草	[14] 歷代名醫蒙求	
[15] 周守忠	[16] 司馬子微	[17] 貞一先生		

little in writing to enable us to judge of him. We only know a work of his called the *Shang-Chhing Han Hsiang Chien Chien Thu*[1] (The Sword and Mirror Diagram embodying the Image; a Shang-Chhing Scripture),[a] but it has little of alchemical interest except a process of making an amalgam from silver, cinnabar, lead and mercury.

(iii) *Chemical theory and spagyrical poetry under the Thang*

Among the Thang emperors Hsüan Tsung is most noted for his association with alchemists and Taoists. The *Hsin Thang Shu* has the following two passages about him:

During the tenth month of the 9th year of the Thien-Pao reign period (+750) he visited (the Taoist abbey) of Hua-Chhing Kung.[2] Wang Hsüan-I,[3] the adept of the Thai-pai Shan[4] mountain, announced that (the 'immortal emperor') Hsüan-Yuan Huang Ti[5] (i.e. Lao Tzu) would descend and visit (the cave) Pao Hsien Tung.[6] Accordingly, on an *i-hai* day in the twelfth month (+751), he proceeded there from Hua-Chhing Kung.[b]

On another occasion, suspecting him of unfaithfulness, he said to Kao Li-Shih[7]:[c]

It is already ten years since I last left Chhang-an. Now that there is peace in the empire I am thinking of pursuing the technique of breathing to nourish my own life. Should I not leave the government of the empire in the hands of (Li) Lin-Fu[8]?[d]

Another alchemist in Hsüan Tsung's entourage was Chang Kuo[9] (perhaps +685 to +756), destined to receive the title Thung-Hsüan hsien-sêng[10] (Mr See-Through) and later to be canonised in folklore as one of the Eight Immortals (Pa Hsien[11]). A book on alchemy entitled *Yü Tung Ta Shen Tan-Sha Chen Yao Chüeh*[12] (True and Essential Teachings about the Great Magical Cinnabar of the Jade Cavern),[e] compiled by Chang Kuo, the Inhabitant of Ku-shê Mountain (Ku-Shê Shan Jen[13]), is probably due to him. This book closely resembles two other alchemical treatises, of which indeed it was a condensed paraphrase, so it will be convenient to mention them together. The two texts in question are (*a*) the *Ta-Tung Lien Chen Pao Ching, Hsiu Fu Ling-Sha Miao Chüeh*[14] (Mysterious Teachings on the Processing of Numinous Cinnabar, according to the Precious Manual of the Re-casting of the Primary (Vitalities); a Ta-Tung Scripture)[f] and (*b*) the *Ta-Tung Lien Chen Pao Ching, Chiu Huan Chin Tan Miao Chüeh*[15] (Mysterious Teachings on the Ninefold Cyclically-Transformed Gold Elixir, according to the Precious Manual of the Re-casting of the

---

[a] *TT*428.   [b] Ch. 5, p. 14*b*, tr. auct.
[c] D. +763, a favourite eunuch and faithful servant of Hsüan Tsung, famous for the numerous water-mills which he owned.
[d] Ch. 207, p. 2*b*. Li Lin-Fu (d. +752) was one of the most eminent ministers of the Thang period, also famous as a proprietor of water-mills and patron of mill-wrights. The leader of the aristocratic party, he was in full power from +736 onwards, but his policies led to the disastrous rebellion of An Lu-Shan. He was the editor of the *Thang Liu Tien* (Institutes of the Thang Dynasty).
[e] *TT*889.   [f] *TT*883.

[1] 上清含象劍鑑圖   [2] 華清宮   [3] 王玄翼   [4] 太白山
[5] 玄元皇帝   [6] 寶仙洞   [7] 高力士   [8] 李林甫   [9] 張果
[10] 通玄先生   [11] 八仙   [12] 玉洞大神丹砂眞要訣
[13] 姑射山人   [14] 大洞鍊眞寶經修伏靈砂妙訣   [15] 大洞鍊眞寶經九還金丹妙訣

Primary (Vitalities); a Ta-Tung Scripture).[a] Both of these were originally intended by their author, Chhen Shao-Wei[1],[b] to form one single treatise, but have somehow come to be regarded as two separate texts in our present *Tao Tsang*. Circumstantial evidence suggests that Chhen Shao-Wei's version is older and more original than that of Chang Kuo. Although Chang was more widely known, Chang Chün-Fang in the early +11th century included part of Chhen Shao-Wei's version in his *Yün Chi Chhi Chhien* under another title, *Chhi Fan Ling Sha Lun*[2] (Discourse on the Seven-fold Cyclically-Transformed Cinnabar).[c] Chang Kuo's version is more concise, and after his name is found the word *chhi*,[3] which could be a printer's error for the word *tsuan*,[4] to compile. One can only conjecture because we do not have the exact date of Chhen Shao-Wei. All that we know is that he flourished during the Thang period, some time between the early +7th century and the early +10th century;[d] both his name and book appear in the bibliographical chapters of the Thang histories and in Chang Chün-Fang's collection. If Chang Kuo (d. +756) did actually derive his material from Chhen Shao-Wei then the latter must have flourished between the early +7th and the middle of the +8th.[e] We know nothing else about him except his claim to have been a descendant of the alchemical school of Wu Mêng,[5] the disciple of Ko Hung's father-in-law Pao Ching[6] as well as of the alchemist Ting I.[7]

The main theme of these treatises is cinnabar. Various types of cinnabar are identified according to their quality. The superior grade when ingested by itself could transform one into the state of immortality on the very same day, because cinnabar is the *natural* cyclically-transformed elixir (*tan sha chê tzu-jan chih huan tan yeh*[8]).[f] However, those of lower quality have to be treated each according to its type before they can be consumed.[g] The classification of the ores seems to employ the presence of impurities as a criterion. It is said that the best variety, called *kuang ming sha*,[9] would yield 14 oz of mercury for every 1 lb of cinnabar, the next grade, *ma ya sha*,[10] 12 oz, the next lower grade, *tzu ling sha*,[11] 10 oz, and the lowest grades of all, such as *chhi sha*,[12] *tsa sha*,[13] and *thu sha*,[14] only 6 to 7 oz.[h] The amount of the other ingredients used together with cinnabar in an elixir recipe would vary according to the type of cinnabar used. Chang Kuo's *Yü Tung Ta Shen Tan-Sha Chen Yao Chüeh* says, for example, that for 1 lb of the best-quality type of *kuang ming sha* obtained from Chhen(-chou)[15] or Chin(-chou),[16] 6 oz each of rock-salt (*shih yen*[17]) and purified

---

[a] *TT*884.
[b] Also known under the names Chhen Tzu-Ming[18] and Hêng-Yo Chen Jen.[19]
[c] Ch. 69, pp. 1 a ff.
[d] See further in pt. 4.
[e] On independent grounds, Chang Tzu-Kao (2), p. 116, gives the tentative date +712 or +713 for Chhen's *TT*883 and 884.
[f] Cf. Vol. 3, p. 640, and Vol. 5, pt. 4 below, on the transformation of metals and minerals in the earth.
[g] *TT*883, p. 3 a.
[h] *TT*883, preface p. 2 b. See further on this pt. 4 below, where the quantitative measurements of the Chinese alchemists are considered.

[1] 陳少微	[2] 七返靈砂論	[3] 縶	[4] 篡	[5] 吳猛
[6] 鮑靚	[7] 丁義	[8] 丹砂者自然之還丹也		[9] 光明砂
[10] 馬牙砂	[11] 紫靈砂	[12] 溪砂	[13] 雜砂	[14] 土砂
[15] 辰州	[16] 錦州	[17] 石鹽	[18] 陳子明	[19] 衡嶽眞人

sodium sulphate (*ma ya hsiao*[1]) should be used, for 1 lb of the second-grade *kuang ming sha* 4 oz each of the salt and the sulphate, for 1 lb of *ma ya sha* 3 oz each, for 1 lb of *tzu ling sha* 3 oz each, and for the lowest grade of cinnabar like *chhi sha*, *thu sha* and *tsa sha* rock-salt only should be used.[a]

Some technical terms in connection with the heating processes are also explained in these texts. For example, the trigram *Khan*[2] used in this context means boiling with water (*shui chu*[3]), while the trigram *Li*[4] would mean direct heating over fire (*Yang huo*[5]).[b] *Hou*[6] is defined as a period of five days, making three *hou* in every fortnightly period (*chhi*).[c] In the *Ta-Tung Lien Chen Pao Ching, Chiu Huan Chin Tan Miao Chüeh* Chhen Shao-Wei gives the *destillatio per descensum* method of extracting mercury from cinnabar using a bamboo tube.[d] Hints for identifying certain minerals are also given in the second part of this book. The following passage gives an example of how copper sulphate can be tested:

If you wish to identify genuine vitriol rub it over copper or iron, and on heating in the fire the colour of the metal will look red. Vitriol is effective either for stopping (substances) from changing or for making them change. If a little vitriol is put in a copper basin containing water, the water becomes jade-blue and will not change (its colour) after several days. What is genuine is indicated by the colour remaining unchanged.[e]

Rudiments of alchemical theory[f] appear during the time of Chhen Shao-Wei (*c.* +700), or at the latest by the early +8th century, to explain why certain ingredients are necessary to go with others in alchemical formulae. These were again based on the proto-scientific system of Yin and Yang, the Five Elements and the 'Book of Changes'. Chhen Shao-Wei has the following to say in his *Ta-Tung Lien Chen Pao Ching, Hsiu Fu Ling-Sha Miao Chüeh*:[g]

The Yang essence is Fire, while the Yin essence is Water. Yin and Yang subdue and control each other, while Water and Fire are in mutual opposition. Hence ice and (burning) charcoal do not get along together. (However,) thriving and declining must eventually come to a balance. Cinnabar is of a Yang essence, and so has to come under the control of Yin. Such control comes from Water. That is to say, stratified malachite (*tshêng chhing*[7]), hollow nodular malachite (*khung chhing*[8]), rock-salt (*shih yen*[9]), purified sodium sulphate (*ma ya hsiao*), and saltpetre (*hua shih*[10]) should be used (in connection with cinnabar).

This concept was further developed by a contemporary of Chang Kuo, Chang Yin-Chü,[11] in his *Chang Chen-Jen Chin Shih Ling-Sha Lun*[12] (A Discourse on Metals, Minerals and Cinnabar by the Adept Chang) written during or shortly after the Khai-Yuan reign-period (+713 to +742).[h] He says that cinnabar, including the best

---

[a] *TT*889, p. 2*b* and p. 3*a*. The number and quantity of ingredients mentioned in *TT*883 are somewhat different, see pp. 4*a*ff.

[b] *TT*883, p. 5*a* or *TT*889, p. 6*b*.   [c] *TT*883, p. 5*a* or *TT*889, p. 3*b*.

[d] See the sub-section on laboratory equipment in pt. 4 below.

[e] *TT*884, p. 6*b* tr. auct.   [f] Discussed at length in pt. 4.   [g] P. 4*a* to p. 4*b*, tr. auct.

[h] *TT*880, Chang Tzu-Kao (*1*), p. 117, suggests that Chang Yin-Chü and Chang Kuo may perhaps have been the same person.

[1] 馬牙硝	[2] 坎	[3] 水煮	[4] 離	[5] 陽火
[6] 候	[7] 曾青	[8] 空青	[9] 石鹽	[10] 化石
[11] 張隱居	[12] 張眞人金石靈砂論			

varieties like *kuang ming sha* and *tzu* (*ling*) *sha*, would not bring about immortality when taken alone, as he knew of many instances where people had died as a result of eating it.[a] He states very explicitly that substances like mercury, lead, realgar, and arsenious oxide (*phi huang*[1]) are toxic. Important though it certainly was as an elixir ingredient, even gold itself, according to Chang Yin-Chü, is poisonous. The following extract from his book shows how substances could, he thought, be employed together to overcome the poisonous effects:[b]

> Gold is the seminal essence of the sun, corresponding to the sovereign (*chün*[2]), and the principal *chhi* of Thai Yang. Mercury is the *pho* soul of the moon, and the principal *chhi* of Thai Yin. When they are combined and absorbed into a man's body he cannot die. . . . The ancients said 'If one ingests gold one will be like gold; if one eats jade one will become like jade.' The nature of gold is endurance and resilience. When heated it does not crack or soften, when buried it does not rust (lit. rot), when placed in the fire it will not burn. Hence it is a medicine which can make man live (for ever). After taking gold the skin will not wrinkle, the hair will not go white, and one will neither be affected by the lapse of time nor disturbed by ghosts and spirits. Hence there will be longevity without end. . . . Gold is the essence of the sun. It is the prince (*chün*[2]) among the substances (used in the elixir). After taking gold one can communicate with the immortals, and enjoy a lightness of the body. . . . Nevertheless gold by itself is poisonous, because of its accumulation of the *chhi* of Thai Yang; if native gold is made into a powder and consumed it will have deleterious effect on bones and marrow, and will cause death.[c] Gold has to be combined with mercury before it can be taken, in order to achieve immortality, because mercury is the essence of Thai Yin and the *pho* soul of the moon.

It looks as if elixir formulae were drawn up along the same lines as medical prescriptions when Chang Yin-Chü speaks about adjuvant and complementary ingredients in the following tone:[d]

> The *Lung Hu Ching*[3] (Dragon-and-Tiger Manual)[e] says: 'Gold is the princely (*chün*[2]) ingredient, silver is the minister (*chhen*[4]), malachite (*tshêng chhing*) is the adjutant (*shih*[5]) and realgar the commander (*chiang-chün*[6]). When all are assembled the elixir is formed.'[f]

He further mentions that there are seven Yang elixir ingredients, two of them metals and the other five minerals, i.e. gold, silver, realgar, orpiment, arsenious oxide, malachite, and sulphur. He also says there are seven Yin ingredients, three metals and four minerals, but he only gives the names of four in all, i.e. mercury, lead, saltpetre,

---

  [a] P. 4*b*.   [b] P. 1*b*, tr. auct.
  [c] On the danger of unmixed Yang *chhi* we shall remember the disquisitions of Wang Chhung in the +1st century (cf. Vol. 2, p. 369, Vol. 3, p. 481). On the toxicology of gold see Clark (1), pp. 463, 618. It does indeed poison the bone-marrow and injure haemopoiesis, so Chang Yin-Chü's warning was a particularly lucky shot.
  [d] P. 2*b*, tr. auct.   [e] Unidentifiable now.
  [f] This terminology, characteristic of Chinese pharmacy and prescribing, will be more fully explained in Vol. 6, Sect. 38 onwards. On the find of labelled inorganic medicines at Sian dating from +756 cf. pt. 2, p. 161, as also Kêng Chien-Thing (1) and Anon. (124).

  [1] 砒黄   [2] 君   [3] 龍虎經   [4] 臣   [5] 使   [6] 將軍

and crude sodium sulphate (*phu hsiao*[1]). Then he once again emphasises the importance of bringing Yin and Yang into proper combination and quotes the famous
dictum from the 'Book of Changes': 'One Yin and one Yang; that is the Tao' (*i
Yin i Yang chih wei Tao*[2]).[a] A parallel development of alchemical and medical principles is not at all unexpected since the most renowned alchemists from the time of
Ko Hung onwards, such as Thao Hung-Ching, Sun Ssu-Mo and Mêng Shen, not to
mention Su Yuan-Ming, were all themselves eminent physicians.

So also by the +8th century in the time of the Thang emperor Hsüan Tsung
(+713 to +756) a body of alchemical theory had already grown up in China.
Ho Ping-Yü & Needham (2) were the first to make a study, some years ago, of the
theories of categories which developed about this time or earlier. These first steps
towards ideas of chemical affinity were embodied, for example, in the *Tshan Thung
Chhi Wu Hsiang Lei*[3] (The Similarities and Categories of the Five (Substances) in
the *Kinship of the Three*). Although this book is no longer extant under its original
title we do have a version of the text, together with an early +12th-century commentary, under a slightly different title.[b]

During the reign of Sung Hui Tsung, between +1111 and +1117, Lu Thien-
Chi, an Education Commissioner, wrote a commentary on the *Tshan Thung Chhi
Wu Hsiang Lei*, which he called the *Tshan Thung Chhi Wu Hsiang Lei Pi Yao*[4]
(Arcane Essentials of the Similarities and Categories of the Five (Substances) in the
Kinship of the Three)[c] and presented it to the throne. This is probably the most
important book on Chinese alchemical theory which has so far come to light. Here we
find a combination of the principle of the mating of contraries with that of *similia
cum similibus agunt*, for this medieval category theory says that substances of opposite
sign (Yin or Yang) will react only if they belong to the same category (*thung lei*[5]).
Thus the *lei* or category was a new classification of substances quite distinct from
their Yin or Yang nature. Moreover, their behaviour might depend on what they were
reacting with, so that mercury might behave as a Yin substance to sulphur, but as a
Yang substance to silver. Bodies thus ascended for the first time to a quantitative
plane, at least in theory, taking precedence of one another in accordance with their
Yin–Yang *mistio*, so that a given substance might act as Yin in one relation and Yang
in another. Thus was formed a kind of hierarchy analogous to the electro-chemical
series of the elements, the order in which elements displace one another from their
salts.

There are several other texts in the *Tao Tsang* which touch on alchemical theory,
but we do not know even the approximate dates at which they were written. One of
them is the *Yin Chen-Chün Chin Shih Wu Hsiang Lei*[6] (The Deified Adept Yin
(Chhang-Shêng's Book) on the Similarities and Categories of the Five (Substances)

---

[a] P. 6a. Cf. Vol. 2, p. 274, for further explanation of this statement; also pp. 276 ff. there.
[b] See p. 51 above.
[c] *TT*898.

[1] 朴硝   [2] 一陰一陽之謂道   [3] 參同契五相類   [4] 參同契五相類秘要
[5] 同類   [6] 陰眞君金石五相類

among Metals and Minerals). It has, for example, the following to say about saltpetre:[a]

Saltpetre (*hsiao shih*[1]) is the *chhi* of all Yin minerals just as sal ammoniac (*nao sha*[2]) is the
chhi of all Yang ones. . . . The immortals make use of it to control the great toxicity of
(substances that are) strongly Yang (in nature).

And about sulphur:[b]

Sulphur also leads to the goal of becoming an immortal, but not a very distinguished one.
When using it one has to follow the principle of the similarities of categories of substances
(*hsiang lei*[3])[c] whether working externally or internally (*wei piao li*[4]),[d] or using it as a ministerial or an adjutant medicine (*wei chhen tso*[5]),[e] or for energy and strength (*wei chhi li*[6]). The
user knowing this secret will become an ever-living immortal of the middling sort, but he
who ignores it will be a dead alchemist.

It should not be thought that the books in the *Tao Tsang* and the records in the
dynastic histories are the only sources from which we can gain information about
alchemical ideas and processes in the Thang period. Some manuscript material, not
hitherto much investigated, remains at our disposal. For example, one of the Tunhuang MSS in the British Museum, probably of the early +8th century, gives a
recipe for the prevention of hunger, using 8 oz of realgar, 6 oz of brown haematite and
four-tenths of an oz of saltpetre.[f] Two words are missing in the recipe, but the procedure consists essentially of making pills out of these three ingredients, and it is
said that taking four of these will inhibit appetite while maintaining strength. Tonic
medication with arsenic can hardly ever have been more heroically applied.

The emperor Hsüan Tsung also made himself the patron of some other Taoists who
practised the art of attaining immortality by consuming elixirs of vegetable origin,
either as a preliminary measure following the advice given in the *Chen Chung Chi*[7],[g]
or because they were too fearful of poisoning by metallic and mineral elixirs. The *Hsin
Thang Shu* tells us about the adept Wang Hsi-I,[8] a recluse among the Sung-shan
Mountains, who managed to live to old age by feeding on pine leaves and flower
pollen. Hsüan Tsung gave him presents and asked him to visit his palace.[h] The same

---

 [a] *TT*899, p. 20*b*, tr. auct. It will be remembered from pp. 51, 77 above that Yin Chhang-Shêng was
the fabled teacher of Wei Po-Yang. His name here is certainly only putative, but we cannot fix the date of
the text nearer than the Thang period as a whole (*c.* +620 to +900).
 [b] P. 25*b*, tr. auct.
 [c] The enormous importance of this term and its synonym *thung lei*[9] will appear in the sequel; see
Vol. 5, pt. 4, and Vol. 6.
 [d] This antithesis is classical in medical language; see Vol. 6 and meanwhile Needham (64), p. 404.
 [e] Technical terms here from the classical pharmaceutics of the *Pên Ching*; see Vol. 6, Sect. 38.
 [f] S 5795. *Hsiung huang pa liang, yü yü liang liu liang, hsiao shih ssu fên, yu san fên . . . wei wan; thun
ssu wan chieh pu chi.*[10]
 [g] See p. 135 above.
 [h] *Hsin Thang Shu*, ch. 204, p. 6*b* and *Chiu Thang Shu*, ch. 192, p. 4*a*.

[1] 硝石   [2] 硇砂   [3] 相類   [4] 爲表裏   [5] 爲臣佐
[6] 爲炁力   [7] 枕中記   [8] 王希夷   [9] 同類
[10] 雄黃八兩禹餘粮六兩消石四分右三分□□爲丸吞四丸即不飢

official history also mentions another adept Chiang Fu,[1] who was rusticated in disgrace when he could not substantiate his claims. It says:[a]

He purported to know the art of avoiding death as practised by the immortals, saying that by taking a tincture of ivy (*chhang chhun thêng*[2])[b] white hair would turn black and one would gain longevity. . . . Many people, however, died a violent death after drinking a wine in which the plant had previously been steeped. (On hearing this) the emperor stopped taking Chiang's preparation. Much ashamed, he asked leave to return to the mountains to search for (the right) plant. (This was granted, but) he never made his appearance again.

Increased attention to the use of vegetable material and plants in alchemical processes can be seen in the treatise *Shun-Yang Lü Chen-Jen Yao Shih Chih*[3] (The Adept Lü Shun-Yang's (Book) on Preparations of Drugs and Minerals).[c] It describes in verse form some sixty-six different plants, named as though they were all species of a single genus, 'dragon sprout' (*lung ya*[4]), and each described with details on its alchemical properties. For example, mulberry leaves (*sang yeh*[5])[d] are called 'precious cinnabar dragon sprout' (*pao sha lung ya*[6]), and it is said that they can produce changes in copper. Similarly, the plant *Portulaca oleracea* (*ma chhih* (*hsien*)[7]),[e] called here the 'five-leaves dragon sprout' (*wu yeh lung ya*[8]), could produce gold when used with cinnabar, and would also bring longevity. Exactly what was going on in these experiments and preparations is rather difficult to say now, without laboratory investigation, but plants do contain chemical substances which can have striking effects on metals in different conditions, particularly (*a*) sulphydryl groups, (*b*) organic acids, and (*c*) cyanides or their precursors. The first could contribute to the formation of sulphide films (cf. pt. 2, pp. 251 ff.), the second could play a part as fluxes, in surface enrichment (pt. 2, p. 250), and in 'bronzing dips' (pt. 2, p. 253),[f] while the third may well have been involved in elixir recipes (cf. pp. 88, 98 above). The work as we have it now must on internal evidence date from a time much later than the +8th, but there is no reason for doubting an association between this kind of vegetable metallurgical biochemistry and the alchemists of the middle Thang. This text would certainly repay further study, interesting botanists and chemists alike.[g]

Unfortunately nothing can yet be said for certain about the exact date of Lü Shun-Yang, better known as Lü Tung-Pin[9] (Fig. 1360).[h] His year of birth has been variously given as +755 and +796, but Liu Tshao[10] in the early +11th century, and

[a] *Hsin Thang Shu*, ch. 204, p. 6 *b*, tr. auct.

[b] Identifiable as common ivy *Hedera helix* (R239; CC593).   [c] TT896.   [d] *Morus alba*, R605.

[e] A kind of purslane (R554; CC1487), of the Portulacaceae. Cf. Vol. 3, p. 679, where it was observed that in Chinese tradition metallic mercury could be obtained from this plant. Such plant accumulators are certainly known, and this one was noted already by Henckel in the +18th century. The plant was also used in medieval China as an important source of vitamins; cf. Vol. 6, Sect. 40, and meanwhile Lu Gwei-Djen & Needham (1).

[f] There is reason to think that fairly strong solutions of organic acids were prepared by the freezing-out process (cf. pt. 4).

[g] There is now an interesting translation into English verse by Ho Ping-Yü, Lim & Morsingh (1).

[h] Or as Lü Yen,[11] Shun Yang Tzu[12] or Shun Yang Ti Chün.[13] Cf. Fig. 1361.

[1] 姜撫	[2] 常春籐	[3] 純陽呂眞人藥石製	[4] 龍芽
[5] 桑葉	[6] 寶砂龍芽	[7] 馬齒莧	[8] 五葉龍芽	[9] 呂洞賓
[10] 劉操	[11] 呂嵒	[12] 純陽子	[13] 純陽帝君

Wang Chung-Fu[1] in the early +12th, both claimed to have had Lü Tung-Pin as their teacher. The name of Lü Tung-Pin enjoyed for centuries afterwards such great prestige as a numinous adept and alchemist that he was canonised in popular folklore as a leading member of the Eight Immortals (*pa hsien*[2]).[a] Votive temples and shrines in his honour are found all over China.[b] Now a close study of the rhymes, the prosody and the technical terminology, by Ho Ping-Yü & Chhen Thieh-Fan (1), has shown that the *Shun-Yang Lü Chen Jen Yao Shih Chih* must have been composed in the later +14th or early +15th century. Nevertheless, it may well have been in a tradition which took its origin from Lü Tung-Pin, and adherence to this may have been all that Liu and Wang meant by their claims.

Here it is worth emphasising that throughout the +8th century the poetry and literature of the Thang were saturated with the ideas of alchemy and immortality. Apart from anything else, imperial patronage conferred upon them a certain elegance and refinement as subjects of discussion. Of the two great poets of the age, Li Pai[3] (+701 to +762) and Pai Chü-I[4] (+722 to +846), both among the greatest Chinese poets of any age, both were concerned in one way or another with experimental alchemy. Each of them had a particular friend among the alchemists, and the names of these men have by good fortune come down to us.

In Li Pai's poetical works there are many references to alchemy and even whole poems on the subject, including one which essentially summarises the *Tshan Thung Chhi*.[c] He was a great admirer of the alchemist Sun Thai-Chhung,[5] who acquired fame by making a glittering elixir of some kind for the emperor in +744,[d] and he eulogised him in a still extant inscription dated +749.

Pai Chü-I also knew many Taoist alchemists at Lu Shan and elsewhere,[e] but there was one in particular, Kuo Hsü-Chou,[6] who lent him a copy of the *Tshan Thung Chhi* in +818. This led him once again to set up a private laboratory, and try out experiments in elixir-making with mineral and metallic substances.[f] But in the end he felt he had failed, and wrote the following poem:[g]

> I read it, and day by day the meaning grew clearer
> Till no doubt was left in my mind at all.
> The Yellow Sprout, yes, and the Purple Carriage
> Seemed to be perfectly easy things to produce. . . .

[a] Stories about him are innumerable. We may mention only sources such as the +12th-century *Nêng Kai Chai Man Lu*,[7] ch. 18, pp. 1 b ff.

[b] I myself particularly recall the temple of Lü Tung-Pin in the centre of Thai-yuan (Shansi), now the provincial museum, and his shrine at the great Taoist complex at Chin Tzhu in the same province.

[c] See Waley (13), pp. 53 ff.

[d] The account of this is in the *Wên Yuan Ying Hua*[8] (+987), ch. 562, p. 9 a. Waley thought it must have been phosphorescent, as some sulphides are, but this idea needs further examination.

[e] We shall see later what he had to say of the iatro-chemical activities of his scholarly friends Han Yü and Yuan Chen (pt. 5).

[f] There is now a special study of Pai Chü-I's poems on immortality and alchemy by Ho Ping-Yü, Ko Thien-Chi & Parker (1). Many of Lu Kuei-Mêng's have been translated by Yates (1).

[g] Tr. Waley (12), mod. auct. On the cover-names see pp. 67, 153. The last line obviously refers to the volatility of mercury. On the *mysterium conjunctionis* cf. Vol. 2, p. 333, and above, pp. 69 ff.

[1] 王中孚   [2] 八仙   [3] 李白   [4] 白居易   [5] 孫太冲
[6] 郭盧舟   [7] 能改齋漫錄   [8] 文苑英華

I bade a lofty farewell to the world of men;
All my hopes were set on the silence of the hills.
My platform of brick was accurately squared,
Compasses showed that my aludel was round.
At the very first motion of the furnace-bellows
A red glow augured that all was well;
I purified my heart and sat in solitary awe.
In the middle of the night I stole a furtive glance,
The Yin and Yang ingredients were in conjunction
Manifesting an aspect I had not foreseen,
Locked together in the posture of man and wife
Intertwined like dragons coil upon coil. . . .
The bell sounded from the Chien-Chi Kuan,
Dawn was breaking on the Peak of Purple Mist.
It seems that the dust was not yet washed from my heart;
The stages of the firing had gone all astray.
A pinch of elixir would have meant eternal life;
A hair's-breadth wrong, and all my labours lost!
The Master snapped his fingers and rose to go;
The Elegant Girl flew up with the smoke to the sky. . . .

Then 'I knew at last', Pai continues, in Buddhist phraseology, 'that on the plane of Assembled Occasions one cannot escape from the secret laws of predestination.' That night he dismantled his furnace, and on the next day, he tells us, he heard that he had been made Governor of Chung-chou.[a]

Some of the images of these poets have aspects of interest for the history of scientific philosophy. The idea of a solid substance so finely comminuted as to become an impalpable dust able to penetrate everywhere, even through apparently impenetrable solids, caught their imagination strongly. Hence the expression 'bright window dust (*ming chhuang chen*[1])'—assuredly a reference to the motes which can be seen dancing in sunbeams and shafts of light. It was perhaps rather characteristically Chinese that these observations did not arouse (in spite of Buddhist philosophers) any ideas of an atomist nature.[b] On the contrary, the poets laid their emphasis on permeation, penetration and rest as opposed to the ceaseless motion. They felt that the elixirs, if made correctly from cinnabar and other inorganic substances, must consist of such subtle matter, able to pass like incense smoke, as it were, through the minutest pores of bodies until they reached the most recondite and essential places. Here we touch upon something very deep-seated in Chinese medieval natural philosophy, the aversion from atomism,[c] and the assimilation of matter, almost infinitely divided, to *chhi*, *pneuma*, vapour or emanation.[d] In China the line between matter and spirit was almost infinitely thin;[e] thought-emanations wind like incense-wreaths through the

---

[a] See on all this Waley (12), pp. 127ff.
[b] Nor, in spite of Buddhist moralists, any idea of repulsion; *hung chhen*,[2] the 'red dust', was one of the most widely known expressions for 'this world' and all its emptiness.
[c] Cf. Vol. 4, pt. 1, pp. 3ff.
[d] Cf. Vol. 2, pp. 472ff.   [e] Cf. pt. 2, pp. 86, 92 above.

[1] 明窗塵   [2] 紅塵

frescoes of the Tunhuang cave-temples, and no flesh was so solid that the 'bright window dust' elixir could not penetrate it. Again, the Thang poets were deeply impressed by the ceaseless whirling motion of the motes, and took it as a symbol of life, contrasting it naturally with the quiescent ashes of the funeral pyre.

About +1110 Hsü Yen-Chou wrote:[a]

Li Thai-Pai (Li Pai) once made a poem on coming-into-being and passing away, which included the words:

> 'Restless is life like the dust in the window's brightness,
> Yet dust and ashes of death settle to quiet at the last.'[b]

For a long time I could not understand the meaning of these words, but then I got a book on alchemy by Mr Li (*Li shih Lien Tan Fa*[1])[c] in which 'bright window dust' is identified as the wonderful medicine cinnabar (*tan sha miao yao*[2]).

And indeed it is true that the phrase *ming chhuang chhen* appears, perhaps for the first time, in the +2nd-century *Tshan Thung Chhi*, where it is used as a poetical term for one of the phases in the making of an elixir.[d] Later in the +12th century Wu Tshêng[3] returned to the subject, saying, in a passage redolent of physiological alchemy:[e]

Li Thai-Pai once replied to Liu Kuan-Ti[4] with a poem in which he used the phrase 'it is like bright window dust, and death is the settling of it'. Ku Sung Tzu[5] has a 'Song of the Potable Gold' (Chin I Ko[6]), in which he says that the two *chhi* of the sun's *hun*[7] (soul) and the moon's *flos* (*hua*[8])[f] (uniting), foster and give birth to a mysterious spirit; this, after many natural changes and transformations have been wrought upon it, can clearly be seen as the dust in the rays from bright windows. The commentary says that this means it is an impalpable powder. Now the 'Song of the Potable Gold' was based on the *Chin Pi Ching*[9] (Gold and Caerulean Jade Manual)[g] which says that the 'House of the Spirit' (*shen shih*[10]) is the chief pivotal node of the elixir (*tan chih shu niu*[11])—not at all the sort of thing that ordinary people mean by metals—and that if the elixir succeeds it will appear as an impalpable powder

---

a *Hsü Yen-Chou Shih Hua*,[12] ch. 1, p. 1 *b*, tr. auct.
b *Fang fu ming chhuang chhen, ssu hui thung chih chi.*[13]
c Not now easily identifiable.
d Ch. 14, p. 30*b*; cf. Wu & Davis (1), p. 243. Cf. *PWYF*, ch. 11B, (p. 456.2); p. 73 above.
e *Nêng Kai Chai Man Lu*, ch. 7, pp. 31*b*, 32*a*, tr. auct. Cf. *Piao I Lu*, a Ming book, ch. 16, p. 2*a*.
f Surely here a synonym or scribal error for *pho*.[14]
g This book is not, as might be thought, the *Chin Pi Wu Hsiang Lei Tshan Thung Chhi*[15] (Gold and Caerulean Jade Treatise on the Similarities and Categories of the Five (Substances) and the *Kinship of the Three*), which has been analysed by Ho Ping-Yü (12). The date of this text (*TT*897) is extremely difficult to determine; Tshao Thien-Chhin was inclined to think that the verses might be as old as the +2nd century, with the prose much later. There is nothing about category-theory in it now. Cf. p. 51 above. No, the words quoted come from a work entitled *Chin Tan Chin Pi Chhien Thung Chüeh*[16] (Oral Instructions explaining the Abscondite Truths of the Gold and Caerulean Jade (Components of) the Metallous Enchymoma), now found only as a torso in *YCCC*, ch. 73, pp. 7*a*ff. It is likely therefore that this *nei tan* text may have had something to do with the elusive though often quoted *Chin Pi Ching*, one of the unsolved mysteries of Chinese alchemical literature. What we have now shows no signs of being older than the Wu Tai period, and lacks any attribution.

---

[1] 李氏鍊丹法	[2] 丹砂妙藥	[3] 吳曾	[4] 柳官廸	[5] 古嵩子
[6] 金液歊	[7] 魂	[8] 華	[9] 金碧經	[10] 神室
[11] 丹之樞紐	[12] 許彥周詩話	[13] 髮鬢明窗塵死灰同至寂	[14] 魄
[15] 金碧五相類參同契		[16] 金丹金碧潛通訣	

like bright window dust. If such an elixir (so full of motion, energy and vitality) is ingested, it will irrigate the three Red Regions (*tan thien*[1]) of the body of man (with a life-giving water).[a]

How delighted these medieval Chinese naturalists would have been by the phenomenon of Brownian motion, and indeed with the kinetic theory of matter in general, on which all life as well as non-life depends. But it would take us too far to follow further the dancing of the sunbeam motes and the thoughts of the poets and alchemists about them.

Of the other Thang emperors obsessed with the attainment of immortality after Hsüan Tsung, we can name Hsien Tsung[2] (r. +806 to +820) and Wu Tsung[3] (r. +841 to +847). Of the former the *Hsü Thung Chih* says:[b]

Deluded by the claims of alchemists, Thang Hsien Tsung consumed a 'gold elixir' and fell into a grave distemper. He daily became furious with those officials whom he had to meet, and as a result the prisons were over-crowded. At midnight on a *kêng-tzu* day in the first month of the fifteenth year (of the Yuan-Ho reign-period)[c] Wang Shou-Chhêng[4] and a Palace Attendant (*nei chhang shih*[5]), Chhen Hung-Chih,[6] assassinated the emperor at the Chung-Ho[7] Palace Hall.

And the *Hsin Thang Shu* tells us of Wu Tsung that:

on a *chia-shen* day in the sixth month of the fifth year of the Hui-Chhang reign-period[d] (the emperor) built a 'Tower of Waiting and Watching for the Immortals' (*wang hsien lou*[8]) at Shen-tshê[9].[e]

Hsien Tsung and Wu Tsung, together with two of the other three intervening Thang emperors, i.e. Mu Tsung[10] (r. +821 to +824) and Ching Tsung[11] (r. +825 to +826), all suffered from the ill-effects of elixir poisoning,[f] but the interest of the court in alchemy was apparently undiminished thereby.

(iv) *Chemical lexicography and classification in the Thang*

The beginning of the +9th century saw the advent of two very important alchemical texts, the first a lexicographic handbook called *Shih Yao Erh Ya*[12] (Synonymic

---

[a] On this subject see Sect. 43, and meanwhile Maspero (7), pp. 192ff., (13), pp. 92ff. There were three Red Regions, each consisting of nine vesicles or cavities, one in the head (*ni wan*[13]), one in the thorax (*chiang kung*[14]), and one in the lower abdomen (*ming mên*[15]). The system was not at all without anatomical basis, for the head vesicles probably represented the ventricles of the brain, those of the thorax represented the auricles and ventricles of the heart, while the viscera have many cavities which ancient anatomists would have known. No doubt in the minds of Thang literary men the Red Regions had something of the mystical quality associated with the solar plexus by D. H. Lawrence and his readers.

[b] Ch. 575, (p. 6495.1), tr. auct.   [c] 14 Feb. 820.
[d] 16 July 845.   [e] Ch. 8, p. 10*b*, tr. auct.
[f] Against which many poets at this time warned and inveighed, e.g. Li Ho about +810 (tr. Frodsham (1), pp. xlvi, 65–6, 170–1, 193).

[1] 丹田	[2] 憲宗	[3] 武宗	[4] 王守澄	[5] 內常侍
[6] 陳弘志	[7] 中和	[8] 望仙樓	[9] 神策	[10] 穆宗
[11] 敬宗	[12] 石藥爾雅	[13] 泥丸	[14] 絳宮	[15] 命門

Dictionary of Minerals and Drugs),[a] written by Mei Piao[1] in the year +806; and the other a compendium called *Chhien Hung Chia Kêng Chih Pao Chi Chhêng*[2] (Complete Compendium on the Perfect Treasure of Lead, Mercury, Wood and Metal),[b] compiled by Chao Nai-An[3] (otherwise known by the pseudonyms Chih I Tzu[4] and Chhing Hsü Tzu[5]) about the year +808.[c]

The *Shih Yao Erh Ya* gives a list of no less than 163 chemical substances together with their numerous synonyms, the names of 69 different elixirs (twenty-five with synonyms), and a bibliography of the alchemical texts which the author had seen. Since we have a firm date for this remarkable book we are provided with some clue to aid us in the dating of pre-ninth-century alchemical texts. Moreover, it is only with the help of Mei Piao's work that we are able today to identify many of the alchemical terms encountered in this literature. Even now it can be regarded as the most comprehensive dictionary of Chinese alchemical synonyms that we possess. It reminds one of nothing so much as the *Lexicon Alchemiae* of Martin Ruhland, published in +1612,[d] yet eight centuries had elapsed before the necessity for the same kind of thing was felt in Europe. This circumstance must give pause for serious thought to all those who accept without qualification the usual belief that alchemy in all ages and nations was essentially vowed to secrecy, and that the supersession of this by the 'plain naked natural manner of speaking' of the early Royal Society was one of the greatest steps in the beginnings of modern science.[e] Ruhland's work was doubtless connected with this movement, but what then was the environment in which Mei Piao sought for clarification and systematisation of 'chymical' experiments? Besides, as we have pointed out above, some of the alchemical books before his time were written in remarkably lucid language (pp. 107, 130). If, as we shall see, there was a return to obscurity and abstruse theorising after his time, then perhaps we are witnessing yet another instance of a truly scientific development which the social conditions of Chinese medieval culture did not permit to come to fruition.[f]

In saying that medieval Europeans felt no necessity for dictionaries like those of Mei and Ruhland we must not forget that there are extensive lists of technical terms in some of the Greek alchemical manuscripts.[g] But they often do no more than define the shorthand symbols used. It is rather interesting to reflect that the Chinese alchemists never developed a special system of symbols for chemical substances and operations. They probably had no need to do so, because the ideographic language itself provided plenty of symbolic patterns, stylised drawings of high antiquity. A

---

    [a] *TT*894.   [b] *TT*912.

    [c] The date of this text is not yet conclusively established, for there is a discrepancy in the cyclical characters given and the reign-period year. One of us therefore (N. S.) would prefer to place it in the Wu Tai or Sung. See further on this subject p. 158 below.

    [d] Second edition, Frankfurt, +1661. Cf. Vol. 3, p. 645.

    [e] One must not attribute too much to Mei Piao's book. It gives only synonyms, not definitions, and was probably meant to assist memorisation. There may also be more in it than meets the eye, as some of his terms seem to belong to physiological rather than proto-chemical alchemy.

    [f] For a glossary of cover-names used in Arabic alchemy, see Siggel (3).

    [g] E.g. Marcianus 299 of the +11th century, and Paris, Bib. Nat. 2327 of +1478. See Berthelot (2), pp. 92ff.; and especially Ruska (11), pp. 380ff.

    [1] 梅彪   [2] 鉛汞甲庚至寶集成   [3] 趙耐菴   [4] 知一子   [5] 清虛子

single character can carry great weight of meaning, signifying an element such as *hung*,[1] mercury, or a compound such as *chien*,[2] soda, or an alloy of a particular composition such as a kind of brass, *thou*.[3] At the same time it is true that the great majority of chemical terms throughout the centuries were of multiple-character nature, generally doublets, sometimes triplets and occasionally quadruplets. Nevertheless the pictorial and symbolic quality was always there;[a] presumably the Chinese alchemists had plenty of time to write two or three symbols instead of only one, but as we show elsewhere (p. 157) multiple-character synonymic names were constantly abbreviated. Only in the latest stages of Chinese alchemy do we find any tendency to elaborate symbols of a form quite different from the classical spirit and line of development of the script itself, and paradoxically this occurred in the realm of the physiological-psychological *nei tan* school rather than in that of the chemical *wai tan* experimentalists. Fig. 1362 shows two pages from the *Nei Chin Tan*[4] (Golden Elixir of the Inner World), written in +1622 by an unknown Taoist, and included in the nineteenth-century collection edited by Fu Chin-Chhüan[5] (Chi I Tzu). We shall try to explain these symbols in due course (Vol. 5, pt. 5 below).

Let us now reproduce two of Mei Piao's entries.[b]

Quicksilver (*shui yin*[6]) is also called mercury (*hung*[7]), 'essence of lead' (*chhien ching*[8]), 'magical glue' (*shen chiao*[9]), the 'elegant girl' (*chha nü*[10]), the 'mysterious liquid' (*hsüan shui*[11]), '(Master) Tzu-Ming' (*tzu ming*[12]),[c] the 'flowing pearls' (*liu chu*[13]), the 'mysterious pearl' (*hsüan chu*[14]), the 'flowing pearls of Thai Yin' (*Thai Yin liu chu*[15]), 'white tiger's brain' (*pai hu nao*[16]), the 'long-lived Master' (*chhang shêng tzu*[17]), the 'dragon fat of the mysterious liquid' (*hsüan shui lung kao*[18]), 'Master Yang-Ming' (*Yang Ming tzu*[19]), the 'elegant girl by the riverside' (*ho shang chha nü*[20]), the 'heaven-born' (*thien shêng*[21]), the 'mysterious girl' (*hsüan nü*[22]),[d] the 'Caerulean Dragon' (*chhing lung*[23]),[e] 'divine liquid' (*shen shui*[24]), the 'Great Yang' (*Thai Yang*[25]), 'red mercury' (*chhih hung*[26]), and 'granular mercury' (*sha hung*[27])....

Cinnabar (*tan sha*[28]) is also called 'essence of the Sun' (*jih ching*[29]), the 'real pearl' (*chen chu*[30]), 'sand of the Immortals' (*hsien sha*[31]), 'mercury sand' (*hung sha*[32]), the 'Red Emperor' (*chhih ti*[33]), the 'Great Yang' (*Thai Yang*[34]), 'vermilion sand' (*chu sha*[35]), the 'Red Bird' (*chu niao*[36]),[f] the 'red boy descending on the tumulus' (*chiang ling chu erh*[37]),

---

[a] It was greatly elaborated in the Taoist talismans (*fu*[38]) such as we find in ch. 17 of the *Pao Phu Tzu* book (Fig. 1362a). These were used (until very recently) in all sorts of ways—fixed to doorways like the Jewish *mezuzah*, worn on the body, burnt to ashes and the ashes taken with water, etc., etc.

[b] Ch. 1, p. 1b, tr. auct.

[c] Cf. p. 12 above, on Lingyang (or Tou) Tzu-Ming, and immediately below.

[d] A famous legendary character or goddess in Taoist sexology; cf. Vol. 2, p. 147.

[e] Presumably one of the symbolic animals of the four quarters, the East, more usually named *tshang lung*;[39] see Vol. 3, p. 242.

[f] Again one of the symbolic animals, representative of the red South; see Vol. 3, p. 242.

[1] 汞	[2] 鹼	[3] 鍮	[4] 內金丹	[5] 傅金銓
[6] 水銀	[7] 汞	[8] 鉛精	[9] 神膠	[10] 姹女
[11] 玄水	[12] 子明	[13] 流珠	[14] 女珠	[15] 太陰流珠
[16] 白虎腦	[17] 長生子	[18] 玄水龍膏	[19] 陽明子	[20] 河上姹女
[21] 天生	[22] 玄女	[23] 青龍	[24] 神水	[25] 太陽
[26] 赤汞	[27] 沙汞	[28] 丹砂	[29] 日精	[30] 眞珠
[31] 仙砂	[32] 汞砂	[33] 赤帝	[34] 太陽	[35] 朱砂
[36] 朱鳥	[37] 降陵朱兒	[38] 符	[39] 蒼龍	

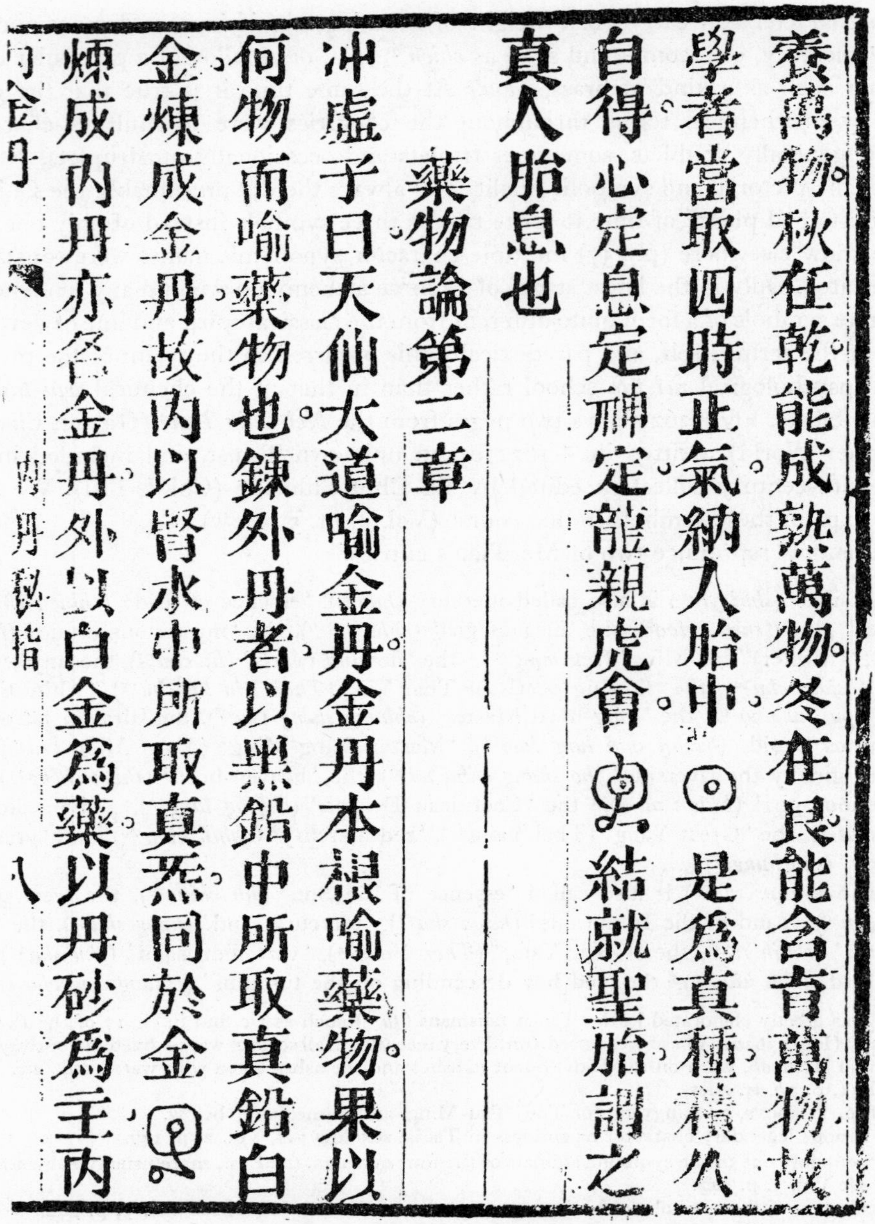

養萬物。秋在乾能成熟萬物。冬在艮能含育萬物。故

學者當取四時正氣納入胎中。☉ 是為真種積久

自得心定息定神定龍親虎會 ⊕ 結就聖胎謂之

真人胎息也

藥物論第二章

冲虛子曰天仙大道喻金丹金丹本根喻藥物果以

何物而喻藥物也錬外丹者以黑鉛中所取真鉛白

金錬成金丹故內以腎水中所取真炁同於金 ⊕

煉成內丹亦冬金丹外以白金為藥以丹砂為主內

內金丹

掛丹秘指

Fig. 1362. The development of symbolic notation in late Chinese alchemy, two pages (7b, 8a) from the *Nei Chin Tan* (Golden Elixir of the Inner World), written by an anonymous Taoist in the years just before +1622. Strangely, these symbols relate, not to laboratory alchemy or proto-chemistry, but to physiological alchemy (see Vol. 5, pt. 5); the recycling and imagined re-combination of bodily juices and *pneumata*.

 The text on the right is discussing the skill of recognising and collecting the *chhi* of primary vitality as it circulates, rising and descending in the machinery of the body, so that the preparation of the inner elixir (the enchymoma) can be achieved. The symbols have to do with the circulation.

 The centre paragraph is an instruction on the principle of 'embryonic respiration'; the formation of the 'divine embryo' (i.e. the enchymoma) by the embodiment of the *chhi* of the four seasons, and the

時之氣者戊已也春在巽能發生萬物夏在坤能長
萬物在人爲呼吸之氣在天爲寒暑之氣能改後四
夫元炁者大道之根天地之母 ◉ 一陰一陽生育
袁天剛胎息訣云
用火符亦是水火煮空鐺而已又何言伏氣也哉
降之機 ◉ 得理則能探取真炁不然不得真氣縱
陰符周天畢有分餘象潤等用 ◉ 探取之功由升
第者知藥生之眞時 ◉ 探取烹煉封固進陽火退
動動既不真則無眞氣者不知次第者亦不成丹次

encouragement of the play of dragon and tiger (Yang and Yin) in stillness and perfect psychosomatic calm. The symbols indicate the two forces.

　　Chapter 2, beginning on the left-hand page, explains what the physiological alchemists meant by 'chemical substances'. Just as the *wai tan* adepts extract 'true lead', i.e. silver, from (argentiferous) lead, for making outer elixirs, so the *nei tan* adepts extract the *chhi* of primary vitality from the juices of the reins (semen) for making inner elixirs. A symbol illustrates this. Lastly, the text goes on to say that while the *wai tan* adepts treat silver with cinnabar, the *nei tan* adepts treat the *chhi* of primary vitality and the metallous juice (saliva) with the spirit (*shen*) of primary vitality and original nature.

　　Since books of this kind have no keys for the symbols, they must have been explained by oral tradition.

the 'red boy of the crimson palace' (*chiang kung chu erh*[1]), 'seminal essence of the Red Emperor' (*chhih ti ching*[2]), 'marrow of the Red Emperor' (*chhih ti sui*[3]), and the 'red sparrow' (*chu chhiao*[4]).[a]

Mei Piao thus gives us twenty-two different names for mercury and fourteen for cinnabar. However, his list cannot be considered by any means complete, though the most comprehensive among extant Taoist alchemical texts. From the latter we can easily collect another sixteen synonyms for mercury to supplement Mei's list as follows:

'leavings of the Count of the River'[b]	(*Ho Po yü*[5])[c]
'son of the Chu Clan'	(*Chu shih tzu*[6])[d]
the 'Great Yin'	(*Thai Yin*[7])
the 'Red Emperor's flowing mercury'	(*chhih ti liu hung*[8])
the 'red-blooded General'	(*chhih hsüeh chiang-chün*[9])
'golden juice of the divine liquid'	(*hsüan shui chin i*[10])
'elixir powder summoned from the vasty deep'	(*chhou yün tan sha*[11])
'mysterious bright dragon fat'	(*hsüan ming lung kao*[12])
'mercurial tiger alum'	(*shui yin hu fan*[13])
'bone of the Sombre Warrior'	(*hsüan wu ku*[14])[e]
'granular mercury'	(*sha hung*[15])
'golden mercury'	(*chin hung*[16])[f]
the 'flowing pearls of Thai Yang'	(*Thai Yang liu chu*[17])
'Master Ling-Yang'	(*Ling Yang Tzu*[18])[g]
the 'mysterious bright dragon'	(*hsüan ming lung*[19])
'gold-and-silver mat'	(*chin yin hsi*[20])[h]

It would easily be possible to enlarge the list of synonyms for cinnabar, but there is no necessity to do so here.[i] For example, apart from alchemical writings,[j] the above lists do not include some of the synonyms given in the pharmaceutical literature. Moreover there was 'coding redundancy' because some of the names were applied to two or more different substances.[k] For instance, the 'Great Yang' in the passages just quoted from the *Shih Yao Erh Ya* is common to both mercury and cinnabar. The two synonyms 'mysterious liquid' (*hsüan shui*) and 'divine liquid' (*shen shui*) for

[a] Again one of the symbolic animals, representative of the red South; see Vol. 3, p. 242.
[b] Cf. Vol. 2, pp. 103, 137 above.   [c] *TT*945, ch. 1, p. 7*b*.   [d] *TT*945, ch. 2, p. 1*a*.
[e] The symbolic tortoise of the black North; cf. Vol. 3, p. 242.
[f] For all these ten synonyms see *TT*899, p. 6*a* to p. 7*a*.
[g] Again cf. p. 12 above.
[h] For these four synonyms see *TT*902.
[i] A valuable service to the history of science would be rendered by a complete and annotated translation of the *Shih Yao Erh Ya*.
[j] An explanation of very obscure synonyms for twenty-four substances is given in *YCCC*, ch. 68, pp. 1*a*ff.
[k] This may have been because some names were functional categories—like our oxidising or reducing agents.

[1] 絳宮朱兒	[2] 赤帝精	[3] 赤帝髓	[4] 朱雀	[5] 河伯餘
[6] 朱氏子	[7] 太陰	[8] 赤帝流汞	[9] 赤血將軍	[10] 玄水金液
[11] 抽暈丹砂	[12] 玄明龍膏	[13] 水銀虎礬	[14] 玄武骨	[15] 砂澒
[16] 金澒	[17] 太陽流珠	[18] 陵陽子	[19] 玄明龍	[20] 金銀席

mercury were also terms for vinegar, according to the same writer.[a] 'Essence of lead' (*chhien ching*) is also the proper name for metallic lead,[b] and according to another book the synonym given by Mei Piao for mercury, 'elegant girl by the riverside' (*ho shang chha nü*), is yet another name for metallic lead.[c] 'Real pearl' (*chen chu*) for cinnabar conflicts with its use as the common name for ordinary pearls. It is interesting, but at the same time confusing, to note that two synonyms with opposite meanings can refer to the same thing; for example the 'Great Yin' (*Thai Yin*) and the 'Great Yang' or 'flowing pearls of Thai Yin' and 'flowing pearls of Thai Yang' all refer to the same substance, mercury.

Apparently some of the synonyms were just abbreviated forms of others. One of the names for mercury not given above is 'Mr Lingyang Tzu-Ming' (*ling yang tzu ming*[1]),[d] i.e. the name of an ancient adept, who according to Taoist hagiography became an immortal by drinking water boiled with the five siliceous clays (*wu shih chih*[2]).[e] By omitting the given name one gets *Ling Yang Tzu*, by omitting the family name *tzu ming*. The *Yin Chen-Chün Chin Shih Wu Hsiang Lei* says that the name *hsüan ming lung kao* for mercury may be abbreviated to *lung kao*,[3] 'dragon fat', giving us yet another term.[f] Presumably then *chha nü* was derived from *ho shang chha nü*, *liu chu* from either *Thai Yang liu chu* or *Thai Yin liu chu*, and *hsüan shui* from *hsüan shui lung kao* in the same way. However, things may not have been quite so straightforward as this, for occasionally the alchemists did try to expound their allegorical terms. Here is an explanation of the synonym *tzu ming*, not as an abbreviation of the personal name Lingyang Tzu-Ming, but in punning derivation from the Yin and Yang and the Five Elements:[g] 'The essence of cinnabar is born from the Sun. Essence is bright; and the offspring of the sun is Fire.[h] Hence Fire is called the "son's brightness" (*tzu ming*). Mercury then is also known as the "son's brightness".'[i]

Even after understanding all the synonyms in an alchemical procedure one has to face difficulties of perhaps yet greater magnitude in identifying correctly the nature of the chemical substances used. Of course they usually contained various amounts of impurities depending on the places they came from, so that the same name cannot always have meant exactly the same thing to two men working in different periods and at different localities. An excellent illustration of this would be the case of 'nitre', *hsiao shih*,[4,5] *phu hsiao*[6,7] and *mang hsiao*,[8,9] to which we shall return in detail at a somewhat later stage;[j] it is particularly important because of its bearing on the

---

[a] These are given on p. 5*b* of *TT*894.   [b] *TT*894, p. 1*a*.
[c] *TT*945, ch. 1, p. 8*b*.   [d] This is given in *TT*899, p. 7*a*. Cf. p. 12 above.
[e] *TT*293, ch. 2, p. 29*b*, and of course the *Lieh Hsien Chuan* (tr. and annot. Kaltenmark (2), no. 67, pp. 183 ff.).
[f] *TT*899, p. 6*b*.
[g] *TT*945, ch. 1, p. 8*b*, tr. auct.   [h] Cf. p. 216 below.
[i] Because it belongs to the (hot and sunny) south, as indicated by the *Li* trigram (cf. Vol. 2, p. 333 and Table 14), and the compass-points *ping* and *ting* (cf. Vol. 4, pt. 1, p. 298, Table 51). It is therefore instinct with the element Fire.
[j] Pt. 4 below.

---

[1] 陵陽子明   [2] 五石脂   [3] 龍膏   [4] 硝石   [5] 滑石   [6] 朴硝
[7] 朴消   [8] 芒硝   [9] 芒消

discovery of the first mineral acid and the first explosive mixture.[a] There we shall have to deal with the nitrates of sodium and potassium, sodium carbonate, magnesium sulphate (Epsom salt, the heptahydrate) and sodium sulphate (Glauber's salt or mirabilite, the decahydrate). Since we have to do with days before modern chemistry there is only one way of identifying what people were dealing with, and that is the systematic study of exactly what they said about their products. In any medieval literature we have to expect much confusion, and there were sometimes differences of opinion and usage between the alchemists and the pharmaceutical naturalists; in such circumstances we have to fall back on our own judgment.

Chao Nai-An's *Chhien Hung Chia Kêng Chih Pao Chi Chhêng*, whether of +808 or later, is a florilegium of alchemical writings in five chapters. In ch. 4 it quotes from a treatise called *Tan Fang Ching Yuan*[1] (The Mirror of the Alchemical Elaboratory; a Source-book), long confused with the +10th-century book of Tuku Thao,[2] the *Tan Fang Chien Yuan*[3] (The Mirror of Alchemical Processes (and Reagents); a Source-book).[b] It can be shown that the *Chêng Lei Pên Tshao* pharmacopoeia in +1249 quoted directly from Chao Nai-An's compendium, not Tuku Thao's book.[c]

Chêng Chhiao, in the bibliographical section of his *Thung Chih*, about +1150, included Tuku Thao's *Tan Fang Chien Yuan*, but wrongly used the word *fang*[4] (chamber or laboratory) instead of *fang*[5] (procedure or formula) given in the *Tao Tsang*. The *Sung Shih* bibliography, about +1345, followed suit. Since *chien*[6] and *ching*[7] have much the same meaning (mirror), one could easily regard the two words as interchangeable, so it was not unnatural for Li Shih-Chen in his *Pên Tshao Kang Mu* about +1590 to alter the former to the latter. Almost certainly his only access was through the *Chêng Lei Pên Tshao*, which never quotes Tuku Thao's book, but instead of attributing the quotations to Chao Nai-An's *Tan Fang Ching Yuan*, as the *Chêng Lei* correctly does, he simply gave the name of Tuku Thao as the source. Thus

---

[a] Cf. pp. 78–9, 137–8 above, and p. 159 opposite.

[b] *TT*918. See, for example, Yoshida (5), p. 223, which still says that the *Tan Fang Ching Yuan* was written by Tuku Thao.

[c] This can be proved by examples such as the three following:

(a) '*Mang hsiao* is capable of subduing (*fu*[8]) orpiment' (i.e. reacting with it chemically and preventing it from subliming as usual). For example, when treated with sodium carbonate, CO_2 is evolved, arsenic disulphide precipitated and thioarsenite formed. This sentence comes from the *Tan Fang Ching Yuan* and the same words are quoted in the *Chêng Lei Pên Tshao* (*TT*912, ch. 4, p. 3*b* and *CLPT*, ch. 3, p. 18*b*). However, the *Tan Fang Chien Yuan* makes a different statement altogether, saying that '*mang hsiao* subdues realgar' (*TT*918, ch. 2, p. 4*a*).

(b) 'Mica powder controls (*chih*[9]) mercury and subdues (*fu*[8]) cinnabar, and it can also be ingested.' This comes from the *Tan Fang Ching Yuan* and is repeated word for word in the *Chêng Lei Pên Tshao* (*TT*912, ch. 4, p. 3*a* and *CLPT*, ch. 3, p. 7*b*). The *Tan Fang Chien Yuan* has a similar entry, but without the two words *i*[10] and *chih*[11] (*TT*918, ch. 2, p. 4*a*).

(c) 'Sulphur is capable of drying up (*kan*[12]) mercury (i.e. the product of their reaction is no longer a liquid). It is noted that sulphur gives a black colour with the Five Metals (forms sulphides), but turns red with mercury.' This passage from the *Tan Fang Ching Yuan* is repeated fully in the *Chêng Lei Pên Tshao* (*TT*912, ch. 4, p. 4*b* and *CLPT*, ch. 4, p. 8*a*), whereas the *Tan Fang Chien Yuan* merely says, 'Sulphur is capable of controlling (*chih*[9]) mercury' (*TT*918, ch. 1, p. 4*a*).

The whole case is demonstrated in the paper of Ho Ping-Yü & Su Ying-Hui (*1*).

---

[1] 丹房鏡源	[2] 獨孤滔	[3] 丹方鑑源	[4] 房	[5] 方
[6] 鑑	[7] 鏡	[8] 伏	[9] 制	
[10] 亦	[11] 之	[12] 乾		

two men and two centuries were confused, and statements of +808 or a good deal before were all attached to a writer of *c.* +938.[a]

In his collection Chao Nai-An furnished us with a list of twenty types of gold, saying that the majority of these were alchemical and not genuine. The list appears twice, once in ch. 1 and once in ch. 4 with minor textual variations, indicating that they came from different sources. The second one[b] turns up in other texts also, for example the book entitled *Pao Tsang Lun*,[1] datable at +918 and attributed to Chhing Hsia Tzu[2].[c] We have dealt with these lists already, when considering what gold-like alloys could have been produced in medieval China.[d] Chao's book also describes the properties of various substances such as malachite, realgar, orpiment, sal ammoniac, alum, vitriol, sulphur and mercury; and it contains three illustrations of alchemical apparatus. It is full of interesting things; it uses an empty hen's egg suitably supported as an aludel or 'chaos vessel' (*hun tun*[3]), it preserves an alchemical *mantram* in an Indian language, and most of its formulae include vegetable ingredients. For this reason it takes its place naturally as another of the earliest known records of a proto-gunpowder mixture, describing, under the heading *Fu huo fan fa*[4] (Method of Subduing Alum (or Vitriol) by Fire), a composition of sulphur, saltpetre and dried aristolochia (*ma tou ling*[5]) as the carbon source.[e] This would have ignited suddenly, bursting into flames, without actually exploding.[f] The exact sequence of these first accounts has yet to be determined, but if Sun Ssu-Mo was really the experimenter of the *Chu Chia Shen Phin Tan Fa* (pp. 137, 197) the middle of the +7th century would have seen that first beginning; and it does look like the most archaic procedure, for the carbon source in the shape of the soap-bean pods was doubtless added with far different intention. The *Chen Yuan Miao Tao Yao Lüeh* (p. 78), with its use of dried honey, is dated plausibly by Fêng Chia-Shêng between the mid +8th century and the end of the +9th. If our present text, which uses another kind of plant material for the carbon, is rightly placed at the beginning of the +9th, it could be the second oldest reference, but if it should turn out to be rather of Wu Tai or early Sung it could belong to the first or second half of the +10th or even the first half of the +11th. Yet in any case it must surely precede by some time the first regular gunpowder formulae in the military encyclopaedia *Wu Ching Tsung Yao* of +1044.[g] And most probably it will also be older than +919, the first appearance of gunpowder (*huo yao*[6]) in a military context.[h] To speak further of all this would too much trespass on the survey in Sect. 30, yet here we could hardly dispense with some reference to the earliest

[a] On the dating of Tuku Thao, and Li Shih-Chen's quotations, correct Vol. 3, pp. 671, 716.
[b] Ch. 1, p. 18*a, b*.
[c] See pt. 2, p. 273 above. The list is given from this book in *CLPT*, ch. 3, (p. 109.2).
[d] Cf. pt. 2, p. 276 above.
[e] *Aristolochia* sp. prob. *debilis* (R585; CC1559). This was recognised by Fêng Chia-Shêng, e.g. in (6), p. 10.
[f] As Fêng Chia-Shêng was able to prove by a personal experiment, (1), p. 41.
[g] Cf. Needham (47).
[h] By implication. Wang Ling (1).

[1] 寶藏論   [2] 青霞子   [3] 混沌   [4] 伏火礬法   [5] 馬兜鈴
[6] 火藥

phases of so cardinal a chemical discovery. All in all, the *Chhien Hung Chia Kêng Chih Pao Chi Chhêng* is one of the alchemical texts that deserve careful study.[a]

Looking back at the four centuries between Ko Hung and Mei Piao one feels that they saw a steady advance in chemical and metallurgical experience and technique, not unaccompanied by promising beginnings of theory and hypothesis. There were many examples of clear exposition, hampered though understanding was by grave deficiencies of laboratory equipment on the one hand and the continuing mystifications of professional thaumaturgists on the other. The +9th century, full blossoming time of Thang culture, was also perhaps the apogee of Chinese proto-chemistry, for though there were many notable figures yet to come, they tended to be either commentators and disputants, or poetical theoreticians of the old school, or else—a growing development—advocates of the Nei Tan psycho-physiological sect, who made use of alchemical terms and symbols for their very different conceptions of an inner elixir prepared by control of the adept's own mind and body.

(v) *Buddhist echoes of Indian alchemy*

The close relation of Taoism with alchemy and proto-chemistry in China needs no further emphasis, yet occasionally there were Buddhist monks who cultivated the Great Work, contrary though it might have seemed to their ethos of world-renouncing poverty.[b] Moreover there are a number of references to alchemy in the *Ta Tsang*[1] or *Tripiṭaka*, the Buddhist Patrology,[c] and these are of considerable interest because they mark the entry into China of Indian ideas, not of course by any means necessarily new to the Chinese.[d] Some seem to belong to a common fund of early belief about alchemy which expressed itself in a dozen languages all the way from Alexandria via Taxila and Nālanda to Chhang-an, but we also know of special cases where Indian chemical practitioners with particular skills were welcomed at the Chinese imperial court—e.g. Nārāyaṇasvāmin about +649, who seems to have had some process for preparing strong mineral acids or alkalis.[e] Moreover, in this or the previous century a number of 'Brahmin' books were translated into Chinese, as is shown by the *Sui Shu* bibliography (+636).[f] The *Po-lo-mên Yao Fang*[2] (Brahmin Pharmaceutics), for example, may well have contained some alchemical material bordering on medicine, but all such texts were lost centuries ago, and we shall probably never have a sight of their contents.

Waley (24), browsing through the *Tripiṭaka*, found several references to alchemy, and recorded them for the benefit of his grateful successors. Most of them belong to

---

[a] Among other interesting things to be found in it are a solubilisation process using vinegar and an extract of sour white plums (malic acid, cf. pt. 2, pp. 250, 265), a mention of the use of bits of tile or lampwick to stop 'bumping' during boiling, and what seems to be a reference to metacinnabarite, green or black (cf. Gowland (9), p. 348).

[b] Cf. pp. 139, 193. On Buddhism in general see Sect. 15.   [c] Cf. Vol. 2, p. 419.

[d] Nor always readily accepted by them either. Indian atomism is a case in point (cf. Vol. 4, pt. 1, pp. 5, 13), though its relations with Indian alchemy remain very obscure. On Indian five-element theories, and concepts of chemical combination, see P. Ray (1) and Subbarayappa (1).

[e] Vol. 1, p. 212.   [f] Vol. 1, p. 128.

[1] 大藏   [2] 婆羅門藥方

the +6th, 7th or 8th centuries, the period which we have just been discussing, but there is one which comes from a distinctly earlier time, raising questions about the comparative development of alchemical thought and practice, so we may deal with it first.

The *Mahā-prajñāpāramito-padeśa Śāstra*[a] is known to have been translated into Chinese by Kumārajīva[b] in +406, and this text, entitled *Ta Chih Tu Lun*,[1] is now the only extant version. The putative author was Nāgārjuna (a name to conjure with), the great Buddhist philosopher and patriarch in South India who is usually placed in the +2nd century,[c] the founder too of the very sophisticated dialectical Mādhyamika logic,[d] but also associated with the beginnings of Tantrism.[e] Hence we are not altogether surprised to hear him say:[f] 'By drugs and incantations (*chou shu*[2]) one can change bronze into gold.' Or again:[g] 'By the skilful use of chemical substances, silver can be changed into gold and gold into silver.' And elsewhere:[h] 'By spiritual power a man can change even pottery or stone into gold.' And finally:[i] 'One measure of a (certain) liquid (prepared) from minerals (*shih chih*[3]) can change a thousand measures of bronze into gold.' Here we have a clear reference to projection, but nothing in any of these passages would have surprised a Chinese scholar of the early +5th century interested in such matters.

All this raises the vexed question of 'Nāgārjuna' and alchemy, for there may have been half-a-dozen people of the same name during the centuries in India, and one at least was an outstanding alchemist. The question is, which one? We are certainly not in a position to exclude the Buddhist patriarch of the +2nd century, for a large body of semi-legendary tradition attributes just this quality to him, and he may well have been a polymath concerned with the study of Nature and 'natural magick' as much as with Buddhist philosophy and logic—besides, we remember how closely the family of Ko Hung was involved in the development of Taoism as an ecclesiastical institution. At the other end of the story is the alchemist Nāgārjuna whom al-Bīrūnī, writing on India in the +11th century, placed in the century immediately before his own.[j] In between come more than one hypothetical Nāgārjuna. What is probably the oldest surviving Sanskrit alchemical book, the *Rasaratnākara*, bears this name as author, but on internal evidence (one is not too sure how well founded) Ray (1) places it in the +7th or the +8th century.[k] Others assume an alchemist Nāgārjuna almost as old as the patriarch, perhaps of the +3rd or +4th century, to account for the legendary corpus without involving him.[l] The question is quite unsolved, but more light is to be expected, e.g. from further researches in Tibetan texts, which sometimes preserve crucial information lost from both Sanskrit and Chinese.

---

[a] 'Commentary on the Great Sūtra of the Perfection of Wisdom'; N 1169, TW 1509.
[b] Cf. Vol. 2, p. 424.
[c] Though a remarkable and important work, few believe that more than a small part of it originated with him, and it may have been mainly composed somewhere in Central Asia; cf. Zürcher (1), vol. 1, p. 212.
[d] Vol. 2, pp. 404, 423.
[e] Vol. 2, pp. 425 ff.
[f] Taishō Trip. vol. 25, p. 178.1.
[g] Taishō Trip. vol. 25, p. 195.3.
[h] Taishō Trip. vol. 25, p. 298.2.
[i] Taishō Trip. vol. 25, p. 401.1.
[j] Cf. Filliozat (10).
[k] 2nd ed. pp. 61, 116 ff.
[l] E.g. Tucci (4).

[1] 大智度論   [2] 咒術   [3] 石汁

Stein (1), commenting on Waley, wanted to bring down the date of the *Ta Chih Tu Lun* to the +8th century to make the alchemical references agree with the probable date of the author of the *Rasaratnākara*, but this is absolutely impossible because of Kumārajīva, as was shown by the translator of the French version, Lamotte (1).[a] We have to do unmistakably with Chin and pre-Chin times. This leads one to have another look at the traditions that Nāgārjuna the +2nd-century patriarch and Bodhisattva was an alchemist. One person who certainly thought so was the pilgrim Hsüan-Chuang,[1] for there are two significant references in his *Ta Thang Hsi Yü Chi*[2] (Records of the Western Countries in the Time of the Great Thang Dynasty) written in the decade before +646. The first concerns longevity, the second aurifaction. Concerning the former he says:[b]

The Bodhisattva Nāgārjuna (Lung-Mêng Phu-Sa[3])[c] was deeply versed in the techniques of pharmacy, and by eating certain preparations he had attained a longevity of several hundreds of years, without any decay either in mind or body. Sadvaha Raja (Yin Chêng Wang[4]) had also partaken of these mysterious medicines and had likewise reached an age of several centuries.

And a few pages further on he wrote:[d]

The Bodhisattva Nāgārjuna (Lung-Mêng Phu-Sa) scattered some drops of a numinous and wonderful liquid pharmakon over certain large stones, whereupon they all turned into gold,

much to the pleasure of the benevolent king when he returned the next day. Projection again.

It is convenient that King 'Sadvaha' introduced himself in the above passage because the corpus of traditions about the first Nāgārjuna has him intimately related to this (also somewhat shadowy) ruler. A King Sātavāhana of the +2nd century, heading a South Indian dynasty rivalling the Kushān empire of Kanishka in the North,[e] does seem to be a historical figure, however,[f] and the curious thing is that the *Rasaratnākara* contains dialogues of the alchemist with a king named Salivahana or Sadvahana who might be the person.[g] This is why historians of Indian chemistry such as Ray are unable to exclude completely the identification of the alchemical writer with the Buddhist patriarch. No one indeed can as yet do this, and it is certainly interesting that many books on science and proto-science bore Nāgārjuna's name. Parallel passages to those of Hsüan-Chuang on alchemy occur in a text entitled *Lung-Shu Phu-Sa Chuan*,[5] but although this is hagiographical rather than biographical,[h]

---

[a] See pp. xi, 383.

[b] Ch. 10, p. 11a, tr. auct., adjuv. Beal (2), vol. 4, p. 416, St Julien (1), vol. 2, p. 98.

[c] The most usual name for Nāgārjuna in Chinese was Lung-Shu[6] (cf. Vol. 2, p. 404), but Lung-Shêng[7] is also found.

[d] Ch. 10, p. 12a, tr. auct., adjuv. Beal (2), vol. 4, p. 419, St Julien (1), vol. 2, p. 103.

[e] Cf. Vol. 1, pp. 175, 182, 207.   [f] Renou & Filliozat (1), vol. 1, pp. 244, 377.

[g] Interestingly, it also contains at least two descriptions of projection (Ray (1), 2nd ed., pp. 129, 133).

[h] The best account of what is known of Nāgārjuna's life is the paper of Walleser (3).

[1] 玄奘   [2] 大唐西域記   [3] 龍猛菩薩   [4] 引正王
[5] 龍樹菩薩傳   [6] 龍樹   [7] 龍勝

never gained a listing in the dynastic history bibliographies, and was assuredly written much later than the time of Kumārajīva to whom it was wrongly attributed, it may not be much later than the time of Hsüan-Chuang.[a]

The *Sui Shu* bibliography lists three books by Nāgārjuna, a *Lung-Shu Phu-Sa Yao Fang*[1] (Pharmaceutics of the Bodhisattva Nāgārjuna),[b] a *Lung-Shu Phu-Sa Yang Shêng Fang*[2] (Macrobiotic Prescriptions of the Bodhisattva Nāgārjuna),[c] and a *Lung-Shu Phu-Sa Ho Hsiang Fa*[3] (Methods of the Bodhisattva Nāgārjuna for Compounding Perfumes, or Incense).[d] As all are lost, we shall never know much of their content, nor who was really responsible for them. Interesting also is his connection with ophthalmology. The *Sung Shih* bibliography names a *Lung-Shu Yen Lun*[4] (Discourse of Nāgārjuna on Eye (Diseases)),[e] and though it is lost, several derivatives of it from later dates are to be found. Since Wang Thao[5] does not quote it in his *Wai Thai Pi Yao*[6] of +752 it may have been either translated or compiled between then and the beginning of the +9th century, when Pai Lo-Thien[7] (Pai Chü-I) mentions the title in a poem. Early in the +12th the syllable Shu collided with a word in one of the imperial titles of Ying Tsung so that Lung-Shu became thereafter Lung-Mu,[8] and the ancient Bodhisattva was transformed into a spiritual prince as Lung-Mu Wang.[9] From the beginning of the Sung dynasty this treatise was the chief textbook of those students in the Imperial Medical College who followed the ophthalmic speciality.[f] Among later redactions and expansions one could instance the +16th-century *Yen Kho Lung-Mu Lun*[10] (Nāgārjuna's Discussions on Ophthalmology),[g] with a chapter by the Taoist physician Pao-Kuang Tao-Jen[11] (probably of the Yuan), and prefaced by Wang Wên.[12] What is certain is that someone called Nāgārjuna was regarded in post-Sung China as one of the founding fathers of the treatment of eye diseases. A connection with mineral remedies such as the salts of copper and silver would provide the obvious link. Finally, on the Indian side, there is a particularly interesting text attributed to Nāgārjuna, the *Rasa-vaiśeshika Sūtra*, apparently a treatise on theoretical alchemy seeking to demonstrate the unity of all tastes or essences, and thus suggesting a parallel with the *prima materia* of the West, which could be deprived of certain forms and endowed with others by means of alchemical procedures.[h]

On the entire subject, opinions differ widely; Lamotte would not wish wholly to reject the traditions that make the +2nd-century Nāgārjuna an alchemist, Renou & Filliozat doubt very much whether any of the Indian alchemical texts go back to him,[i] Filliozat (10) admits at least two persons of the name, Tucci (4) inclines to believe in a +3rd- or +4th-century alchemist, and Ray is prepared to identify his +8th-century

---

[a] TW 2047; see Zürcher (1), vol. 2, p. 340. [b] Cf. *SIC*, p. 771.
[c] Cf. *SIC*, p. 460. MSS of a Sanskrit work on the same subject attributed to Nāgārjuna and entitled *Yogasara* still exist in India (Ray (1), 2nd ed., p. 118).
[d] Cf. *SIC*, p. 793. [e] Cf. *SIC*, p. 448.
[f] Its content is excellently summarised by Pi Hua-Tê & Li Thao (1), who also unravelled the history given above.
[g] Cf. *SIC*, pp. 449, 450. [h] Renou & Filliozat (1), vol. 2, p. 168. [i] (1), vol. 2, p. 169.

[1] 龍樹菩薩藥方 [2] 龍樹菩薩養生方 [3] 龍樹菩薩和香法 [4] 龍樹眼論
[5] 王燾 [6] 外臺祕要 [7] 白樂天 [8] 龍木
[9] 龍木王 [10] 眼科龍木論 [11] 葆光道人 [12] 王問

one with the man referred to by al-Bīrūnī, assuming that the Arabic writer somewhat underestimated the time which had elapsed since his death.[a] Meanwhile evidence continues to accumulate associating the Buddhist patriarch with gold and gold-making, e.g. a text discovered by Lévi (8). And Indian alchemy may yet be traced back into the +1st or even the −1st century if anything historical can ever be established for names such as Vyāḍi, who is supposed to have antedated Nāgārjuna.[b] What we, at any rate, are left with, is the conclusion that clear statements of alchemical import came into China from India by +406. Since similar ideas had been current there, as we have seen, for some six centuries previously, the statements would presumably have passed as commonplaces.

Let us now look at some of the other passages that Waley found. The next oldest would be that in the *Mahāyāna-saṃgraha-bhāshya*, translated into Chinese about +650 by Hsüan-Chuang himself with the title *Shê Ta Chhêng Lun Shih*[1] (Explanatory Discourse to assist the Understanding of the Great Vehicle).[c] Referring to bodhisattvas this text says:[d] 'They can turn earth into gold or other precious substances just as they please.' As it is a commentary on a work of Asaṅga, whose date is unsure, one cannot fix it very well in time, but it can hardly be older than +300 nor later than +500.

The remaining two are of the +6th or +7th century. There is an interesting passage in the *Avataṃsaka Sūtra*,[e] which Śikshānanda translated into Chinese in +699 with the title *Ta Fang Kuang Fo Hua Yen Ching*.[2] It says:[f] 'The good man is like the chymical liquid called *hataka*,[3] a single ounce of which, for whoever is so lucky as to obtain it, will turn one thousand ounces of bronze completely into genuine gold (*chen chin*[4]). But a thousand ounces of bronze are not able to effect any change whatever in this chymical liquid.' Since the chapter in which this occurs is absent from the earlier recension of *c.* +420, it was presumably composed at a later date than that. Lastly there is an evocative story in the *Abhidharma Mahāvibhāshā* (A-Phi-Than-Phi Po-Sha Lun[5]),[g] another of Hsüan-Chuang's translations (+659), to the following effect:[h] 'It took Śāṇaka and his minister Moon-lover (Huai-Yüeh[6]) twelve years to learn how to make gold. At last they were able to produce a button of it, not larger than a grain of corn; but at once they said: "There is nothing now to prevent us from making a mountain of gold!".' The modern reader may be gratified to recognise the principle of the pilot plant in this early medieval account, but the identity of Śāṇaka is even more intriguing. There was no doubt a disciple of Ānanda called Śāṇaka-vāsin, but Stein (1) realised that 'Moon-lover' must be none other than King Chandragupta Maurya himself[i] and that Śāṇaka, by an inversion of rôles, is Cāṇakya or Chanakya, the hakim Sānāq of the Arabs, i.e. Kautilya by another name; famous prime minister of Chandragupta *c.* −300 and putative author of the

[a] (1), 2nd ed., p. 118.
[b] See Filliozat (10); Renou & Filliozat (1), vol. 2, p. 168.   [c] N1171 (4); TW1597.
[d] Taishō Trip. vol. 31, p. 358.2.   [e] N88; TW279.
[f] Taishō Trip. vol. 10, p. 432.2.   [g] N1263; TW1546.
[h] Taishō Trip. vol. 28.   [i] Cf. Vol. 1, pp. 102, 172, 177.

[1] 攝大乘論釋   [2] 大方廣佛華嚴經   [3] 訶宅迦   [4] 眞金
[5] 阿毘曇毘婆沙論   [6] 懷月

Arthaśāstra, that wonderful account of ancient Indian government administration.[a] It is certainly fascinating to visualise Chandragupta and Kautilya hard at it in their alchemical elaboratory, but whether legend or no, the story was probably recent when Hsüan-Chuang incorporated it, for his recension is more than three times as long as the similar work translated in the +5th century.

The general upshot is, then, that proto-chemical aurifaction may well go back in India to the turn of the era, even perhaps to the −1st century, making it contemporaneous or almost so with the same complex of ideas and practices in Hellenistic Egypt. But by the time its echoes reached China through the Buddhist literature, Chinese alchemy, macrobiotic aurifaction, was already six centuries old there.

In our chronological survey we are now standing at the beginning of the +9th century, so this may be the place to look at certain curious facts recorded by the Japanese monk Ennin,[1] who was in China between +838 and +847. He was unlucky in that his stay coincided with the purgation and even persecution of the Buddhist clergy under Thang Wu Tsung in +842, though he himself and those monks of well-known good life and learning were not greatly affected. In his diary entitled *Nittō-Guhō Junrei Giyōki*[2] (Record of a Pilgrimage to China in Search of the Buddhist Law), he wrote:[b]

On the 9th day of the tenth month of the 2nd year of the Hui-Chhang reign-period, an imperial edict was issued (to the effect that) all (Buddhist) monks and nuns throughout the empire who understood the forbidden arts of alchemy, incantations and anathemas, talismanic exorcisms and the like (*shao lien chou shu chin chhi*[3]); all monks who had fled from the army, or bore on their bodies scars or tattoo marks from former judicial punishment; those also who were avoiding terms of hard labour, or who had formerly committed sexual offences or maintained paramours; all moreover who were not observing the Buddhist rules (of the *vinaya*); all should be unfrocked and obliged to return to lay life.

So far so good, but the interpretation remains a little uncertain. *Shao lien*[4] is at first sight alchemy, for *lien*[5] can often stand for *lien*,[6] chemical transformation by heating, but as Okada Masayuki, Ennin's editor, has pointed out, the two words may be a reference to the Buddhist practice of self-mutilation, so abhorred by Confucians. An edict of the Northern Chou emperor Shih Tsung, *c.* +559, had spoken of *shao pei*,[7] burning off one's arm, and *lien chih*,[8] incinerating one's finger,[c] so far had the ideas of self-sacrifice and altruistic compassion run mad by this time.[d] Moreover, the

[a] Tr. Shamasastry (1). It is usually regarded as dating from the −1st century, but Kalyanov (1) shows that it cannot have reached its present form until the +3rd.
[b] Tr. Reischauer (2), p. 321, (3), p. 238, mod. auct.   [c] The text is in *Chhüan Thang Wên*, ch. 125, p. 10 a.
[d] Cf. Vol. 2, p. 413. And I myself, when visiting Buddhist monasteries during the second world war, saw photographs of monks and abbots who had mutilated themselves in just this way. The Tunhuang frescoes often represent the story of the compassionate king cutting off pieces of his own flesh to feed hungry animals, but that a symbolic legend should have turned into a quasi-liturgical practice was truly an aberration of religion. And still the fire-suicides go on, as in martyred Vietnam, political protest being one of the forms compassion takes today. On the general background of Buddhist suicide and self-mutilation in a non-persecuting society, see Welch (4), pp. 324 ff., Needham (43), pp. 293 ff.; and in relation to anthropophagic practices in famine times the elaborate monograph of des Rotours (3).

[1] 圓仁   [2] 入唐求法巡禮行記   [3] 燒練呪術禁氣   [4] 燒練
[5] 練   [6] 鍊   [7] 燒臂   [8] 鍊指

chin chhi[1] of Ennin's edict echoes the *fu chin tso tao*[2] of the earlier one, i.e. 'exorcisms and Tantric *vamacāra* practices'.[a]

Another reason for doubting whether alchemy was intended arises from the fact that the arch-opponent of the Buddhists at this time was himself closely connected with elixir-making.[b] This was the Taoist Chao Kuei-Chen,[3] executed in +846 after the death of Wu Tsung. His address to the throne in +844 mentioned Lao Tzu's *fei lien hsien tan*[4] (elixir prepared by heating and sublimation or distillation), and asked for the erection of a 'Terrace of the Immortals' (Hsien Thai[5]) in the palace, for Taoist liturgies. This was done. In the following year the Taoists were ordered to compound elixirs, and Chao asked to be despatched to Tibet to find some of the ingredients, but he was not allowed to go. Thus the fact that the chief enemies of the Buddhists at this time were (at least in the broad sense) alchemists, makes it perhaps unlikely that alchemy was prominent in the misdeeds of which they were accused. There was plenty of scope for other criticisms.

While we are on this subject we may glance for a moment at the little which is known in the West about alchemy in Burma, a country where it was closely associated with Buddhism. If only from the brief account of Htin Aung,[c] it is quite clear that the Burmese tradition of 'work with fire' (*aggiya*) was a very syncretistic one, as might perhaps be expected from its geographical position. There was indeed aurifaction, with references to the 'stone of live mercury'; but the main current was elixir-oriented, aiming at the attainment of the state of a *zawgyi* (=*siddhi*?), i.e. a *hsien*, enjoying longevity, immortality, invulnerability, impassibility, magical powers and eternal youth. Chinese influence may also be indicated by the division of chemical substances into male and female,[d] and a Tantric feature appears in the story of the trees bearing pure fruit-maidens with whom the *zawgyi* could couple. An Indian yogic trait occurs in the temporary death which the *zawgyi* must pass through, burial for seven days necessarily preceding permanent resurrection. Hellenistic influence might be traced in the *prima materia* beneath the four elements, an essence common to all metals from which the elixir could be formed. Alchemy seems to have been first cultivated by the Ari monks,[e] but its great period was between the +5th and the +11th centuries, ending with the imposition of Theravadin Buddhism by King Anawrahta in +1065. Finally, Arabic connections may be signified in the prominence of magic squares (cf. pt. 4), though this mathematical ploy reached its

---

[a] Including the worship of the feminine principles (*śaktis*), with sexual intercourse (*maithuna*), ceremonies of feeding hungry ghosts in burial grounds (*ko shen tan mo*[6]), necromancy and the conquest of natural aversions, etc., etc.

[b] See Reischauer (2), pp. 351, 354ff., (3), pp. 243ff., 248ff.

[c] (1), pp. 41ff.

[d] Nine female metals and twelve male minerals; hence the 'marriage of contraries' which was required.

[e] Believed to have been Mahāyana Buddhists.

[1] 禁氣   [2] 符禁左道   [3] 趙歸眞   [4] 飛鍊仙丹   [5] 仙臺
[6] 割身啖魔

full development a good deal later, in the time of King Dhammazedi (r. +1460 onwards).[a] Thus Burmese alchemy was a coat of many colours, linked with all the surrounding civilisations.[b]

(6) THE SILVER AGE OF ALCHEMY; FROM THE LATE THANG (+800) TO THE END OF THE SUNG (+1300)

(i) *The first scientific printed book, and the court alchemist Mistress Kêng*

On a previous page we had occasion to note the penetration of Taoist alchemy into Vietnam from the Han period onwards (p. 75); here we may begin by recording the similar spread of the Art of the Yellow and the White in Liu Chhao times into Korea.[c] The principles of Pao Phu Tzu were propagated with royal support under the King of Silla, Chinhǔng[1] (r. +540 to +575), while the well-developed state of precious-metal metallurgy in Koguryǒ was mentioned in the *Pên Tshao Ching Chi Chu*. Frescoes of Taoist immortals and Jade Girls gathering magic mushrooms occur in +6th- and +7th-century Koguryǒ tombs. Many Koreans went to China to study under the adepts, as for example Kim Kagi[2] in the middle of the +9th century, and the *Ishinhō* later recorded special pharmaco-sexual techniques of the Silla masters.[d] Thus there was a very long background of alchemical medicine behind Hǒ Chun's[3] *Tongǔi Pogam*[4] (Precious Mirror of Eastern Medicine, +1610) which we quote elsewhere in this Section.[e] In the +16th century physiological alchemy was particularly studied in Korea, as a life like that of Yun Kunphyǒng[5] shows. Only slowly did the belief in the holy immortals die out, upheld by Yi Su-gwang[6] in the +14th century but criticised by King Sǒnjo[7] (r. +1568 to +1608) in spite of the advocacy of Yi Chunmin.[8] But we are still in the +9th century, and must attend to events of that time in China.

Towards the middle of it we come upon an event which must always remain a landmark in the history of chemistry, indeed of science as a whole, the first printing of a book on a scientific subject; as it happened, the life and work of a particular alchemist. We know about it mainly from a book written not long afterwards, the *Yün Chhi Yu I*[9] (Discussions with Friends at Cloudy Pool). In this, about +870, the scholar and poet Fan Shu[10] wrote:[f]

A certain minister named Hokan Chi[11] had given more than fifteen years' arduous study to the preparation of the dragon-and-tiger elixir. When he was in charge to the right of the

[a] Htin Aung (1), pp. 54ff.

[b] And it is still very much alive. Dr Laurence Picken writes to us from Rangoon (April 1972) of his visit to the laboratory of a Buddhist adept (one of many hundreds). Transmutation (gold amalgams), elixirs (effective against diseases as well as for longevity, and even rejuvenating on plants), and ascetic self-purification of the operator — all are prominent.

[c] See further in the book of Chǒn Sangun (1), pp. 257ff. On alchemy in Tibet cf. Beyer (1), pp. 252ff., 261ff.

[d] Ch. 28, (p. 655).   [e] Vol. 5, pt. 5.

[f] Ch. 10, p. 6a, tr. auct. adjuv. Carter (1), 2nd edn ed. Goodrich, p. 59; Pelliot (41), p. 35.

[1] 眞興王   [2] 金可紀   [3] 許俊   [4] 東醫寶鑑   [5] 尹君平
[6] 李晬光   [7] 宣祖   [8] 李俊民   [9] 雲溪友議   [10] 范攄
[11] 紇干臬

River (i.e. Governor of Chiangsi),[a] he sent invitations to a large number of magician-technicians (*fang shu chih shih*[1]) (to join his entourage). He also composed a 'Biography of Liu Hung'[2] and had several thousand copies of it printed (*tiao yin*,[3] with blocks), which he sent out to all those who were devoting themselves to alchemy (*ching hsin shao lien*[4]), whether at court or scattered in other places within the Four Seas.

This passage is important not only for the history of chemistry but for that of printing, since it still constitutes the second oldest clear reference to the typographical art in any civilisation. It is gratifying to know that the printer and the practical experimentalist joined hands at such an early date. Fan Shu goes on immediately afterwards, however, to a general statement against projective aurifaction and argentifaction (*tien hua chin yin*[5]),[b] not on *nei tan* grounds but rather on the basis of Confucian morality, opposed to the acquisition of inordinate private wealth and estates.[c] More interesting is the question of the identity of Hokan Chi and Liu Hung. The Hokan family[d] was of Eastern Mongolian origin; sinified in Northern Wei, it contributed a number of quite eminent officials during the Thang period. Hokan Chi[e] was Governor of Chiangsi between +847 and +850, so the printing of the 'Life of Liu Hung' must have occurred between those years. But who was Liu Hung? Even Pelliot could only suggest a Liu Sung prince, Liu Hung,[6] but neither he nor others of the same or similar names were known to have had any interest in alchemy. One there was, nevertheless, who fills the bill, namely Liu Hung,[7] an alchemist ascribed to the Later Han period (*fl.* +122) and appearing in a Thang alchemical dialogue prefaced anonymously in +855, the *Hsüan Chieh Lu*[8] (Mysterious Antidotarium).[f] The special relevance of this is that it gives warnings against elixir poisoning and recommends the use of plant drugs as well as minerals and metals.[g] Thus the object of the use of the new method of mass documentary reproduction may have been a humanitarian one, to

---

[a] This curious phrase described the country south of the Yangtze as the emperor would see it looking south from his throne. As Chiangsi was beyond and to the right, so Chiangsu was beyond and to the left.

[b] Cf. on this term, pp. 38, 100.

[c] Others about the same time had no confidence that alchemical processes could ever bring such results. Li Shang-Yin[9] (+813 to *c.* +858) left an amusing collection of epigrams entitled *I-Shan Tsa Tsuan*[10] which is available in translation by Bonmarchand (1). For example, among 'things which it is better not to know' (ix, 5)—alchemy, for it will bring young men to ruin; and 'deceptive ideas' (xv, 2)—that one can get rich by alchemy.

[d] Pelliot conjectured Organ as the original form. *Kan* was sometimes mis-written as *yü*.[11]

[e] There was exceptional confusion about the Governor's name, which appears in various sources as Chhüan,[12] Chung[13] and even Hsiang.[14]

[f] *TT*921, also in *YCCC*, ch. 64, pp. 5 a ff. Authorship unknown. Conceivably Hokan Chi himself?

[g] One of Liu Hung's elixirs consisted of five plants, including the wild raspberry and the dodder, all their names ending with the character *tzu*.[15] Hence the name Shou Hsien Wu Tzu Wan.[16] This was essentially a detoxicant for avoiding the dangers of certain elixirs containing the heavy metals. According to the *Hsüan Chieh Lu*, Liu Hung inscribed the formula on a stone stele which was found again before the Khai-Yuan reign-period of Thang (+713 to +741) when Chang Kuo (cf. p. 141) presented it to the throne. See *YCCC*, ch. 64, pp. 11 b, 12 a. All plants rich in oxalic acid (e.g. spinach, *PTKM*, ch. 27, (p. 92), Schafer (13), p. 147) or selenium (e.g. milk-vetch, Benfey (3), cf. Vol. 3, p. 678) would detoxicate mercury.

[1] 方術之士	[2] 劉弘	[3] 雕印	[4] 精心燒煉	[5] 點化金銀
[6] 劉宏	[7] 劉泓	[8] 玄解錄	[9] 李商隱	[10] 義山雜纂
[11] 于	[12] 泉	[13] 㴂	[14] 象	[15] 子
[16] 守仙五子丸				

alert as many adepts as possible about the dangers of some of the paths they were pursuing.

It may indeed be that the *Hsüan Chieh Lu* itself was the book that Hokan Chi printed. The bibliography of the *Chiu Thang Shu* has a *Thung Chieh Lu*[1] prefaced by him, including his correct date and title.[a] The *Thung Chih* bibliography lists a *Hsien Chieh Lu*,[2] again with the preface by Hokan Chi.[b] So quite probably the 'Life of Liu Hung' (*Liu Hung Chuan*[3]) was only Fan Shu's casual way of referring to the book properly entitled *Hsüan Chieh Lu*.

Just while Hokan Chi was publishing in Chiangsi the first of all printed books on a scientific subject, another curious and interesting character was occupying the imperial attention far away in the north-west at the capital Chhang-an. Her story is worth reading in full, for it illustrates a notable feature in Chinese alchemy, the number of remarkable women who became adepts in it. From what we know of Taoist philosophy[c] this is not surprising theoretically, but the fact that it was possible within Chinese society is what is striking. In his *Chiang Huai I Jen Lu*[4] (Records of (Twenty-five) Strange Magician-Technicians between the Yangtze and the Huai River, during the Thang, Wu and Southern Thang Dynasties) Wu Shu[5] wrote, about +975:[d]

Teacher Kêng (Kêng hsien-sêng[6]) was the daughter of Kêng Chhien.[7] When she was young she was intelligent and beautiful and liked reading books. Fond of writing, she sometimes composed praiseworthy poetry, but she was also acquainted with Taoist techniques, and could control the spirits. She mastered the 'art of the yellow and the white' (alchemy), with many other strange transformations, mysterious and incomprehensible. No one knew how she acquired all this knowledge.

In the Ta-Chung reign-period (+847 to +859) the emperor, being fond of the elegant and fascinated by the strange, summoned her to the palace to observe her techniques. She was not added to the pool of palace ladies but was given special lodgings and called 'Teacher'. When having an audience she always wore green robes and held a tablet and when she spoke it was with brilliant eloquence and confident bearing. Her hands were so small that she relied upon others at meals, and she walked in the palace grounds very little, being rather carried about by attendants. Sometimes she would write poems on the walls, calling herself the 'Great Teacher of the North', but nobody knew what this signified. The Teacher did not often demonstrate her (divination) techniques, but when she did so the results always came out as she predicted. For this the emperor valued her all the more.

When she first entered the palace the emperor enquired about her alchemical procedures; and upon tests being made, all her experiments proved successful. When she repeated (some of) them, it was seen that they were simple and not difficult. Once in a leisure hour the emperor said to her: 'All these processes have been accomplished by the use of fire. Is it possible that something might be done without heating?' She answered, 'Let me try. It might be.' So the emperor took some mercury and enveloped it in several layers of beaten bark-cloth, closing it with the (imperial) seal; this she placed forthwith in her bosom. After a long time there suddenly came a sound like the tearing of a piece of silk. The Teacher

---

[a] Cf. Chhen Kuo-Fu (*1*), vol. 2, p. 406, and *Hsin Thang Shu*, ch. 59, p. 4a. *Thung* can be a tabu replacement for *hsüan*.   [b] Ch. 67, (p. 793.1).
[c] Cf. Vol. 2, pp. 57 ff.   [d] Pp. 7b ff., tr. auct.

[1] 通解錄   [2] 賢解錄   [3] 劉弘傳   [4] 江淮異人錄   [5] 吳淑
[6] 耿先生   [7] 耿謙

smiled and said: 'Your Majesty did not believe in my methods, but now you will see for yourself. Ought you not to trust me ever hereafter?' Then she handed the packet back to the emperor, who saw that the seal was unbroken, and upon opening it found that the mercury had all turned to silver.

On another occasion during heavy snow the emperor asked her if she could turn that into silver also. Saying that she could, she cut up some hard snow into the form of an ingot and threw it into a blazing charcoal fire. The ashes rose up and all was covered over with charcoal; then after about the time required to take a meal she said it was done, and taking out the red-hot substance, put it on the ground to cool, whereupon it brightened and took the form of a silver ingot, with the same knife-marks still on it. The under-surface, which the fire had reached first, looked like stalactites. Later the Teacher made a lot of 'snow-silver', so much that she presented vessels made of it to the emperor on his birthday.

She also had many other ingenious ideas, and what she made always exceeded other products in quality. (People from the) South Sea regions once paid tribute to the court of things rare and strange, such as distilled attar of roses (*chhiang wei shui*[1]) and Borneo camphor (*lung nao chiang*[2]). The former had a fragrance most pure and fresh, while the latter had aphrodisiac properties. Thus the emperor valued it very much, and often drank wine flavoured with it, finding that the fragrance lasted in his mouth for several days. He also bestowed (samples of) it on the officials about him. However, the Teacher said that there was something much better than this, and when the emperor asked if she could make it she replied that she would try, and that it should be possible. So she wrapped the camphor in a silk bag and hung it in a glass vessel (*liu-li phing*[3]). The emperor personally sealed the vessel, and then watched it (the process), with wine beside him. After a space of time sufficient for a meal, the Teacher said that the camphor had already been (transformed into) a liquid (*chiang*[4]). The emperor rose from his seat, put his ear to the vessel and listened; he heard, indeed, the sound of dripping (*ti li shêng*[5]), then he returned to his seat and went on drinking. After another interval, he observed some liquid like water within the glass vessel. On the following day he opened it and it was half full of liquid with a fragrance far stronger than before (i.e. than the original flavoured wine).

Later on the Teacher appeared to be pregnant, and told the emperor that she would give birth that night to a baby who would be a great sage. The emperor caused everything to be prepared, but towards midnight there came a violent thunderstorm, and the next morning she had regained her normal size, but no baby was to be seen. She said that the gods had taken it away and that it could not be got back.

The Teacher was fond of wine, and just like other people in love affairs and sexual relations. Moreover in the course of time she fell ill and died. Formerly the holy immortals often mixed with human beings—perhaps she was one of them. I myself often heard talk about her, but the rumours of the palace were various. Finally I was told the true story by Hsü Shuai,[6] the grandson of Hsü I-Tsu,[7] who had himself been (on duty) in the palace, and knew all about it.[a]

From this vivid picture of a woman adept at the Thang court the chemist will retain with interest the impression of a Chinese emperor watching what was essentially a Soxhlet continuous extraction process, the formation of a strong solution of camphor in hot alcohol. The connection of this with the distilling of essential vegetable oils, which Teacher Kêng certainly knew all about, is also evident enough. But of course

---

[a] The last two paragraphs are here slightly abridged.

[1] 薔薇水   [2] 龍腦漿   [3] 琉璃瓶   [4] 漿   [5] 滴瀝聲
[6] 徐率   [7] 徐義祖

she was obviously skilled, like other Taoist alchemists encountered in these pages, in what we should now call conjuring tricks; besides possessed, no doubt, of a hypnotic personality (cf. p. 186). At any rate she got away with it, and died in her bed, which was more than every one of her colleagues was able to achieve.

(ii) *From proto-chemistry to proto-physiology*

During and after the +9th century there seems to have been a general trend in alchemical writings from originality to compilation, from clarity in style to obscurity, and from proto-chemical techniques (*wai tan*[1]) to psycho-physiological exercises (*nei tan*[2]). We have just seen that both Mei Piao and Chao Nai-An were essentially compilers, one producing a dictionary and the other a florilegium. But there are three interesting alchemical texts definitely traceable to the second half of the +9th century. The first is the *Hsüan Chieh Lu*[3] (Mysterious Antidotarium) already mentioned,[a] the work on elixir poisoning by the anonymous writer of +855. A version with another title, *Yen Mên Kung Miao Chieh Lu*[4] (The Venerable Yen Mên's Explanations of the Mysteries),[b] is essentially the same text as this, but without the formula for the 'Five-Herbs Immortality-Safeguarding Pills' (*shou hsien wu tzu wan*[5]). We shall have occasion to refer to this book again in the discussion of elixir poisoning.[c]

The second book is the *Thung Hsüan Pi Shu*[6] (Secret Art of Penetrating the Mystery),[d] written by Shen Chih-Yen[7] soon after +864. This is mainly a treatise on the medical prescriptions used by the alchemists,[e] but it contains an interesting method for the preparation of lead acetate, with which the 'Imperial Baldachin elixir' (*hua kai tan*[8]) was made.[f] This method will be described in Section 34 on chemical technology.

The third book is the *Huan Tan Chou Hou Chüeh*[9] (Oral Instructions on Handy Formulae for Cyclically-Transformed Elixirs),[g] compiled by writers mostly anonymous probably during the period +874 to +879 or not much later. This book has been attributed to Ko Hung, but wrongly.[h] It deals mainly with cinnabar, mercury and lead, employing many synonyms and using the proto-scientific theories in its explanations; it mentions heating times, and gives one alchemical diagram. The following passage may serve to illustrate the difficulties of Five-Element theory in coping with the facts:[i]

The Oral Instructions say: 'red lead' (*hung chhien*[10]) is cinnabar. The liquid extracted from cinnabar indicates the presence of the (element) Water. This is the Yin within the Yang,

---

[a] P. 168. The *Hsüan* certainly stands for *Hsüan*;[11] cf. Chhen Yuan (4), p. 153.
[b] *TT*937.   [c] See Vol. 6 below.   [d] *TT*935.
[e] It is interesting to learn that elixirs containing heavy metals were never to be given to pregnant women, as the foetus could be harmed.
[f] On this constellation-name see Vol. 4, pt. 3, pp. 567, 571, 583. It was highly important in navigation. *TT*908.   [h] Cf. p. 110 above.
[i] Ch. 1, p. 5*a,b*, tr. auct. In understanding it, Vol. 2, p. 257, will be useful.

[1] 外丹   [2] 內丹   [3] 懸解錄   [4] 鴈門公妙解錄
[5] 守仙五子丸   [6] 通玄秘術   [7] 沈知言   [8] 華蓋丹
[9] 還丹肘後訣   [10] 紅鉛   [11] 玄

that is to say, mercury. What comes out of cinnabar indicates the presence of Yin. Yet when the Yang combines with the (element) Metal, it gives mercury. Hence this cannot be endured by the (element) Wood, which is controlled and conquered by the (element) Metal. Now the (element) Metal holds the (element) Water within its womb. To keep the (element) Water at ease one has to maintain the (element) Metal. The (element) Wood holds the (element) Fire within its womb. The (element) Water is needed to control the (element) Fire. Hence the (element) Water within lead answers to the same (element) Water within mercury, while the (element) Fire within lead also answers to the same (element) Fire within mercury. (Therefore amalgamation occurs.) The (element) Wood is embodied in their nature and the (element) Fire combines their form.

The Oral Instructions say: The (element) Water within lead (*chhien shui*[1]) implies (an affinity for) mercury. What is found in nature (disseminated as ores), cinnabar, contains Yang mercury (*Yang hung*[2]). Yet mercury belongs to the (element) Water, so mercury extracted from cinnabar is Yin mercury (*Yin hung*[3])....

Anyone who feels that this kind of thing is less illuminating than it might be is invited to reflect that exactly the same sort of argumentation was going on among the Aristotelian proto-chemists of late Renaissance Europe, and that Robert Boyle's 'Sceptical Chymist' of +1661 was written primarily to show that neither the Four Elements of scholasticism nor the Tria Prima of Western alchemy could ever hope to do justice to the volume of new chemical knowledge. But here we are in the +9th century, not the +17th, and the writer of the *Huan Tan Chou Hou Chüeh* was doing his best for a rational explanation.

Three other Thang alchemical texts in the *Tao Tsang* are worthy of mention here; (*a*) the *Ta Tan Chhien Hung Lun*[4] (Discourse on the Lead and Mercury of the Great Elixir)[a] by Chin Chu-Pho,[5] (*b*) the *Huan Chin Shu*[6] (Account of the Cyclically (-Transformed) Metallous Enchymoma)[b] by Thao Chih,[7][c] and (*c*) the *Tan Lun Chüeh Chih Hsin Chien*[8] (Elucidation of Secret Teachings concerning Elixirs)[d] written by Chang Hsüan-Tê.[9] It is not possible to give any of them an exact date. All three are more concerned with physiological than with proto-chemical alchemy, and urge great caution about the danger of poisoning by elixir preparations.

The *Ta Tan Chhien Hung Lun* is a small tractate purporting to deal with lead amalgamation, as its title implies. The following passage goes to show the mutationist style of Chin Chu-Pho's writing:

Lead is a Yin (substance), black in colour and corresponding to the Sombre Warrior (*hsüan wu*[10]).[e] Its trigram is *Khan*,[11] which belongs to the North, and its cyclical characters are *jen* and *kuei*.[f] The (element) Water is born from the (element) Metal, so that there is

---

a *TT*916.
b *TT*915. The text is also excerpted in *YCCC*, ch. 70, pp. 13*a*ff.
c See further on him in pt. 5 below.
d *TT*928. The text is also to be found in *YCCC*, ch. 66. Tr. Sivin (5).
e Symbol of the North (cf. Vol. 3, p. 242).
f Compass-points adjacent to due north (Vol. 4, pt. 1, Table 51).

---

¹ 鉛水　　² 陽汞　　³ 陰汞　　⁴ 大丹鉛汞論　　⁵ 金竹坡
⁶ 還金述　　⁷ 陶植　　⁸ 丹論訣指心鑑　　⁹ 張玄德　　¹⁰ 玄武
¹¹ 坎

Water within Metal.[a] The colour of Metal (as such) is white, corresponding to the White Tiger.[b] Its trigram is *Tui*,[1] which belongs to the West, and *kêng* and *hsin* are its cyclical characters.[c] This is gold.[d]

Mercury is a Yang (substance), blue-green in colour and corresponding to the Caerulean Dragon.[e] Its trigram is *Chen*,[2] which belongs to the (element) Wood in the East, and has *chia* and *i* as its cyclical characters.[f] Now the (element) Wood can give birth to the (element) Fire.[g] Hence mercury can give rise to cinnabar. The colour (of cinnabar) is red. It is the Red Bird,[h] and its trigram is *Li*,[3] which belongs to the South. Its cyclical characters are *ping* and *ting*,[i] and it is associated with the (element) Fire.

Hence *Khan* corresponds to the (element) Water and to the moon, (denoting) lead, while *Li* corresponds to the (element) Fire and to the sun, (denoting what can be made from) mercury.[j]

The *Huan Chin Shu* makes frequent reference to Wei Po-Yang's *Tshan Thung Chhi*, emphasising however that the mercury and lead mentioned therein must not be regarded as real metals. To quote from Thao Chih:[k] 'Those who say that mercury can be made into a "gold elixir" are deceiving others, (and those who) claim that cinnabar can arrest the process of ageing do not understand the Tao.'

In Chang Hsüan-Tê's *Tan Lun Chüeh Chih Hsin Chien* the number of substances used for elixir-making (if that was what it was) is much reduced. It talks of lead and mercury only, again in a way reminiscent of the time of Wei Po-Yang. It also warns of the danger of poisoning if such things as alum, sulphur and sal ammoniac are used. All these marks are characteristic of the *nei tan* school.

From the late +9th and early +10th centuries come many stories of projective aurifaction and argentifaction (often involving conscious deception),[l] as well as of successful longevity or preservation of youth by means of elixirs. Chhen Yün-Shêng[4] (*fl.* +904 to +943) was one who achieved all these.[m] Sometimes the production of 'purple sheen gold' (*tzu mo chin*[5]) is mentioned, curious in view of Hellenistic *iōsis*.[n] Tshai Thien[6] could make this, about +875, though later exposed as 'incompetent' and executed.[o] The same source tells us of another alchemist, Tsung Hsiao-Tzu,[7] who used to foretell the success or failure of his operations by the use of a diviner's board when in exile in Szechuan about +880.[p] Ho Fa-Chhêng[8] got artificial silver

---

[a] See Vol. 2, p. 257. [b] Symbol of the West.
[c] Compass-points adjacent to due west.
[d] One would have expected yellow gold to belong to yellow Centre.
[e] Symbol of the East.
[f] Compass-points adjacent to due east.
[g] See Vol. 2, p. 257. [h] Symbol of the South.
[i] Compass-points adjacent to due south.
[j] *TT*916, p. 1*a*, *b*, tr. auct. Slight emendations to the text as it has come down to us have been necessary in order to make sense of it.
[k] *TT*915, p. 5*a*, tr. auct.
[l] See, e.g., *Tu Hsing Tsa Chih* (+1176), ch. 6, p. 11*b*; *Kuei Yuan Tshung Than*,[9] in *Lei Shuo*, ch. 52, p. 5*a*.
[m] *Chiang Huai I Yen Lu* (*c.* +975), p. 15*a*. [n] Cf. the discussion in pt. 2, pp. 23, 253 ff. above.
[o] *Pei Mêng So Yen* (*c.* +950), ch. 11, p. 4*a*. Cf. Miyakawa Hisayuki (1).
[p] *Ibid.* p. 5*b*; we have mentioned him before in Vol. 4, pt. 1, p. 269.

[1] 兌   [2] 震   [3] 離   [4] 陳允升   [5] 紫磨金
[6] 蔡畋   [7] 宗小子   [8] 何法成   [9] 桂苑叢談

(*yao yin*[1]) from a Buddhist monk who eventually under duress taught him a misleading method, with which he nearly came to grief at the Szechuanese court of Wang Yen[2] (r. +918 to +925).[a] A general, Chang Chih-Fang,[3] successfully demonstrated a 'method of concreting mercury by projection' (*kan shui yin tien chih*[4])[b] before an assembly of scholars, saying that he himself did not pretend to understand it, but had forced a Taoist to give him the materials and instructions.[c] Again, about +875 Shenthu Pieh-Chia[5] was practising 'the gilding of tiles by a projection method' (*tien chuan wa pan yeh*[6]) but became involved with greedy and unscrupulous officials and so lost his life, though his son Shenthu Ssu-Ma[7] kept some of the mercury-projection chemicals (*tien hung yao*[8]).[d]

(iii) *Alchemy in Japan*

Standing now at the very end of that Thang period during which Japanese culture crystallised into its permanent form, one is naturally moved to ask questions about the role of alchemy there. On previous pages we have said something on the penetration of Chinese spagyrical art and knowledge southwards to Vietnam and Burma, northwards to Korea—what of its spread eastwards to Japan?

When one surveys medieval Japanese alchemy and early chemistry one notes first of all a general similarity to China in simple chemical and metallurgical technology.[e] But then there appears a distinct difference in that elixir alchemy was just as prominent as in China though aurifiction and aurifaction were hardly known or practised at all. It seems that ancient Japan was relatively rich in natural gold, both mineral[f] and alluvial,[g] as well as being able to get still more of it from the Siberian tribes;[h] for this

---

[a] *Ibid.* p. 4a.   [b] On the term *tien* see pp. 38, 100.
[c] *Ibid.* p. 3b.   [d] *Ibid.* p. 5a. Cf. Miyakawa Hisayuki (1).

[e] This can readily be seen by leafing through the inorganic chapters (59–61) of Terashima Ryōan's[9] *Wakan Sanzai Zue*[10] of +1712, translated by de Mély (1). Names of Japanese localities where mines and chemical industries existed are often given.

[f] The *Shoku-Nihongi*[11] records an edict by the emperor Mommu[12] about gold-mines in +698. One famous medieval source was the island of Kinkazan,[13] off the coast north-east of Sendai,[14] where the works must go back at least to +725, for in that year the emperor Shōmu,[15] pleased at their rich tribute, changed the reign-period name to Tempyō-Kampō[16] in its honour. I visited the beautiful Shinto temple on Kinkazan in 1971 with Dr Yoshida Tadashi, Dr Lu Gwei-Djen and Dr Dorothy Needham.

[g] Cf. de Mély (1), text, pp. 12–13, tr., pp. 14–15; Tsunoda & Goodrich (1), pp. 50, 83, 120.

[h] Evidence for the richness of medieval Japan in gold can be adduced from the *Kitāb al-Masālik wa'l-Mamālik* of Ibn Khurdādhbih, c. +885. 'In the eastern parts of al-Ṣīn (China)', he wrote, 'is the country of al-Wāqwāq, which is so rich in gold that its people make the chains of their dogs and the neck-rings of their apes from gold, and they bring to market shirts woven with gold (threads) for sale' (from *Bibl. Geogr. Arabicorum*, ed. de Goeje, vol. 6, p. 69, tr. Dunlop (6), p. 158). Marco Polo four centuries later emphasised the same thing (ch. 159 on Çipingu, Moule & Pelliot tr., vol. 1, pp. 357–8). And after yet another four centuries the readers of the *Philosophical Transactions of the Royal Society* learnt from 'an ingenious Person that hath many years resided in that country' how Japan was 'exceedingly stored with gold-mines' (Anon. (105), p. 985).

We are able to illustrate the Arabic accounts by a picture (Fig. 1363) from an illustrated manuscript (Roy. Asiat. Soc. 178) of Zakarīyā ibn Muḥammad ibn Maḥmūd al-Qazwīnī's world geography, '*Ajā'ib*

---

[1] 藥銀	[2] 王衍	[3] 張直方	[4] 乾水銀點制
[5] 申屠別駕	[6] 點甋瓦半葉	[7] 申屠司馬	[8] 點汞藥
[9] 寺島良安	[10] 和漢三才圖會	[11] 續日本記	[12] 文武
[13] 金華山	[14] 仙台	[15] 聖武	[16] 天平感寶

reason artificial 'gold' was not particularly interesting.[a] In the great collection of poetry assembled in +759, the *Manyōshū*[1] (Anthology of a Myriad Leaves), there are many references to gold and to gilt objects; moreover we know that such works as the gilding of the colossal Buddha image in the Tōdaiji[2] temple at Nara about +770 were done on a lavish scale. Thus when the novelist Takizawa Bakin[3] about 1820 wrote a story of aurifactive alchemy in his *Kinsei-setsu Bishōnen-roku*,[4] based on the early +17th-century Chinese collection *Chin Ku Chhi Kuan* (see p. 213 below), the idea was quite strange to Japanese readers, some of whom afterwards supposed that aurifaction was a modern art which had originated as late as that time.

But with the 'chemo-therapeutic' element of Chinese alchemy the case was quite otherwise; drugs of longevity and immortality, whether plant or mineral by origin, were accepted with eagerness. How far Taoist ideas penetrated into Japan, influencing the beginnings of Shintoism, is a very moot point—certainly no recognisable Taoist Church ever developed there, though similarities between the two religions and their history strike every observer.[b] In any case the conception of the holy immortals (*hsien*[5]), men persisting eternally deep in the forests or among the clouds, is to be found in Japanese culture fairly early, even though mixed already with Buddhism as soon as we can discern it. The *Nihon Ryo-iki*[6] (Record of Strange and Mysterious Things in Japan), written in +823, has several accounts of thaumaturgists and apotropaists who were believed to be able to fly through the air like *hsien* and to exist without food, or at the very least without cereal food. One of these we have already met with (pt. 2, p. 299), En-no-Shōkaku[7] (or En-no-Gyōja[8]), a quite historical character who was born in +634 and died after +701, undoubtedly a mountain magician but whether really the founder of the *shugendō*[9] cult or not remains still uncertain.[c] Another recluse, Kume no sennin,[10] in the +9th century, besides communicating with spirits,

al-Makhlūqāt (Marvels of Creation). The book was written about +1275, and the MS dates from some two centuries later (cf. B. W. Robinson (1), p. 206, who identifies the painter; and Massé (1) illustrating Bib. Nat. Suppl. Pers. 332). Al-Qazwīnī echoes Ibn Khurdādhbih verbally, attributing the account however to Muḥammad ibn-Zakarīyā al-Rāzī, but adds that all Japan's seventeen hundred islands are ruled by a queen (tr. in Ethé (1), pp. 217, 221–2, cf. 420; Ferrand (1), vol. 2, pp. 300–1). One Mūsa ibn al-Mubarraq al-Shīrafī says that he went there and saw her sitting on her throne in naked perfection, with a great golden crown on her head, and four thousand beautiful serving-maids around her similarly attired. This was either a reminiscence of +4th- or +5th-century Japan before the time of Chinese and Buddhist influence, or a confusion with some hotter country, possibly inspired by paintings or carvings of Indian apsaras or yoginis—but the gold is what is relevant here. Clearly the Japanese never lacked it. As for all the girls, there might be some connection with the voluminous legends of 'Kingdoms of Women', on which see Pelliot (47), vol. 2, pp. 671 ff., 681 ff.

[a] Such at any rate is the view of Yoshida Mitsukuni (6), pp. 206 ff., one of the few scholars who have investigated medieval Japanese alchemy. What follows in the next few paragraphs is based to a large extent on his researches.

[b] Cf. Kubo Noritada (1); Tamburello (1). Of the legendary (or semi-legendary) coming of Hsü Fu, Chhin Shih Huang Ti's −3rd-century macrobiotic explorer, to Shingū in Japan, we have said something at the appropriate place (p. 18 above).

[c] Biography in *Nihon Ryo-iki*, ch. 1, no. 28. The *Shoku-Nihongi* (+797) says that he had contact in +699 with a physician Karakuni Hirotari[11] who from his name could have been a Korean. Similar significant contacts earlier in En-no-Shōkaku's life would have been possible, even likely.

[1] 萬葉集	[2] 東大寺	[3] 瀧澤馬琴	[4] 近世說美少年錄
[5] 仙	[6] 日本靈異記	[7] 役小角	[8] 役行者
[9] 修驗道	[10] 久米仙人	[11] 韓國廣足	

was adept at flying, raising and quelling storms, etc.;[a] and similar powers were ascribed to Yōshyō sennin[1] (*fl.* +901).[b] Self-mummification is reported of a foreign magician in +854.[c] The Taoism latent in all these affairs is evidenced rather clearly by the broad spectrum of *yamabushi*[2] ('mountain monk') practices which have continued down to our own time, on the one hand asceticisms going much beyond the normal celibate community life of mainstream Buddhism, and on the other the almost Tantric valuation of sex which made it right for *yamabushi* to marry (often shaman-esses) and to be Shinto priests as well.[d]

Again, the literature of the +9th century contains some markedly Taoist elements. The oldest text written in the *katakana* syllabary, the *Taketori Monogatari*[3] (Tale of the Bamboo-Gatherer), probably dates from about +865.[e] Set in an atmosphere of *sennin*[4] (immortals) and *tennin*[5] (deities, jade girls, etc.), it describes the finding of a baby in a thicket, and her growth into an outstanding beauty. Kaguyahime,[6] however, though very dear to the old man and his wife, is really an immortal exiled from the moon.[f] She is courted by five distinguished suitors,[g] and eventually by the emperor himself, but when the time is ripe, upon a night of full moon, a heavenly host descends to conduct her home. She leaves with them, but not before sending a letter and a poem to the mikado, together with a vessel containing some of the elixir of immortality. These daring not to keep he later causes to be burned upon Mt Fuji no Yama. Thus in +9th-century Japan we find a legend profoundly Taoist in character witnessing to the great influence of Chinese culture.[h]

Many other signs of it can be found. The lore of lucky and unlucky days, for example (*shou kêng shen*[7]), and a great use of talismans (*fu lu*[8]) in the style of Ko Hung, written e.g. with cinnabar ink on white paper, permeated the Nara and Heian courts.[i] At the same time the Taoist theory of the 'three corpses' (*san shih*[9]) in the body,[j] and the Taoist respiratory exercises (*fu chhi fa*[10]),[k] figure prominently enough not

---

[a] His deeds are recorded in the *Konjaku Monogatari*,[11] a collection of traditions compiled in +1107, and in the later miscellany *Tsurezuregusa*[12] (+1338), Porter tr., p. 14.

[b] See the *Nihongi Ryaku*,[13] and the *Fusō Ryakuki*[14] of +1198.

[c] *Montoku-Jitsuroku*[15] (+879), one of the six classical histories of Japan. On self-mummification cf. Ando Kosei (2); Anon. (103); and pt. 2, pp. 299ff. above.

[d] Cf. Hori Ichiro (2), p. 78.

[e] The outside limits are +810 and +955. There is a noteworthy German translation by Matsubara Hisako (1). Cf. Yasuda Yuri (1), an abridged version in English.

[f] On the relation of the moon to alchemy and elixirs cf. p. 63 above.

[g] These, it is thought, probably symbolise the five elements.

[h] On the book and its background see the monograph by Matsubara Hisako (2), which includes a less literary translation more suitable for scholars. The *Taketori Monogatari* is full of alchemical echoes —one of the suitors has to bring asbestos cloth (see Vol. 3, pp. 655ff.), another is sent to the Isle of Phêng-Lai for a branch of the tree made of precious metals and jewels (cf. p. 19), a third must fetch the night-shining jewel (Vol. 1, p. 199), and finally Kaguyahime dons a cloak of feathers before taking off (cf. Vol. 2, p. 141).

[i] Yoshida Mitsukuni (6), pp. 189ff.

[j] Cf. Kubo Noritada (2).

[k] See our account of physiological alchemy in Vol. 5, pt. 5.

---

[1] 陽勝仙人	[2] 山伏	[3] 竹取物語	[4] 仙人	[5] 天人
[6] かじや姫	[7] 守庚申	[8] 符籙	[9] 三尸	[10] 服氣法
[11] 今昔物語	[12] 徒然草	[13] 日本記畧	[14] 桑扶畧記	[15] 文德實錄

only in 'esoteric' or Tantric *shingon*[1] (*mikkyō*[2]) Buddhism,[a] but also in the great medical work *I Hsin Fang*[3] (*Ishinhō*[3]) compiled by Tamba no Yasuyori[4] in +982.[b]

The fact is that almost from the beginning of the +6th century Chinese proto-scientific influence had been pouring into Japan, both directly and through Korea. The despatch of the first Japanese ambassador, Ono Imoko,[5] to the Sui court in +607,[c] is often taken as the focal point of such intercourse, but it had begun long before. Already in +554 a Korean Professor of Medicine, Wangyu Rungtha,[6] accompanied by two Pharmacognostic Masters, Pan Yungphung[7] and Chŏng Yutha,[8] had brought Chinese medicine (and naturally the iatro-chemistry of the time) from Paekche to Japan.[d] In +562 the Chinese monk Chih-Tshung[9] (Chisō[9]) brought many books on pharmaceutical natural history (*yao tien*[10]), anatomy and acupuncture (*ming thang thu*[11]);[e] while in +602 the Korean monk Kwŏllŭk[12] (Kwanroku[12]) introduced from Paekche not only the first of the learned calendrical systems,[f] together with geography, divination and natural magic, but also further medical knowledge and practice (including apotropaics), which he taught to one of the pioneer Japanese students, Hinamitachi[13].[g] Similarly, just after the first China embassy there came from Koguryŏ in Korea in +610 the priest Tamjing,[14] skilled in all kinds of technology, among which some chemical industry must certainly have figured as well as the millwright's art which the chronicle records.[h] Again, in +685 the Paekche priest Pŏpchang[15] appeared presenting tribute of drugs, notably *pai shu*,[16] that characteristically Taoist longevity medicine (cf. pp. 11, 40 above).[i] In the following century the flow continued, for the great Buddhist Chien-Chen[17] (Kanshin,[17] +688 to +763),

---

[a] Cf. the monograph of Sawa Ryūken (*1*).

[b] This collection preserved excerpts from many Chinese texts which would otherwise have become completely lost. The *Yü Fang Pi Chüeh*[18] and the *Hsüan Nü Ching*[19] (cf. Vol. 5, pt. 5, and meanwhile Vol. 2, p. 147) are well-known examples, but there is also the *Fu Shih Lun*[20] (Treatise on the Consumption of Mineral Drugs) listed in the Sui dynastic bibliography, and the *Yen Shou Chhih Shu*[21] (Red Book on the Promotion of Longevity) by Phei Yü[22] (or Hsüan[23]). Both these titles are indicative of elixir alchemy.

[c] *Nihongi*,[24] ch. 22 (vol. 3, pp. 130ff.), tr. Aston (*1*), pt. 2, pp. 136ff. A return embassy of twelve persons, headed by Phei Shih-Chhing,[25] accompanied him home in the following year, and when Phei went back to China after some months' stay eight Japanese students (of whom four were priests) went with him. Cf. Wang Chi-Wu (*1*), pp. 52ff.

[d] See Miki Sakae (*1*), p. 26; Kimiya Yasuhiko (*1*).

[e] Miki Sakae, *loc. cit.*; Shirai Mitsutaro (*1*).

[f] Cf. Vol. 3, p. 391, as also Rufus (*2*). See *Nihongi*, ch. 22 (vol. 3, p. 120), tr. Aston (*1*), pt. 2, p. 126.

[g] Also called Yamashiro no Omi.[26] Miki Sakae (*1*), p. 27.

[h] *Nihongi*, ch. 22 (vol. 3, p. 135), tr. Aston (*1*), pt. 2, p. 140. Tamjing's companion Pŏpchŏng[27] seems also to have been a technician.

[i] *Nihongi*, ch. 29 (vol. 3, p. 394), tr. Aston (*1*), pt. 2, pp. 371–2. The plant is a Composite, *Atractylis ovata* (R14). A lay Japanese collaborator Kinshō[28] (Masuda no Atahe[29]) was trained in making the preparation.

---

[1] 眞言	[2] 密敎	[3] 醫心方	[4] 丹波の康頼	[5] 小野妹子
[6] 王有陵陀	[7] 潘量豐	[8] 丁有陀	[9] 知聰	[10] 藥典
[11] 明堂圖	[12] 勸勒	[13] 日並立	[14] 曇徵	[15] 法藏
[16] 白朮	[17] 鑑眞	[18] 玉房秘訣	[19] 女女經	[20] 服石論
[21] 延壽赤書	[22] 裴煜	[23] 玄	[24] 日本記	[25] 裴世淸
[26] 山背臣	[27] 法定	[28] 金鐘	[29] 益田直	

builder of the Tōshōdaiji[1] temple at Nara,[a] had a reputation for pharmaceutical learning which earned him the appellation of 'the Shen Nung of Japan'.[b] And elsewhere (pt. 2, p. 161) we have seen something of the Chinese drug specimens deposited in the Shōsōin Treasury in +756 and preserved there to this day (cf. Fig. 1364 *a–d*). As for books, we are fortunate that there still exists a Japanese bibliography drawn up about +895 which notes many works of Chinese origin extant at that time, the *Nihon-koku Ganzai-sho Mokuroku*[2] compiled by Fujiwara no Sukeyo.[3] Among sixty-three Taoist titles and sixty-eight on medicine there are at least six clearly concerned with alchemy in one form or another.[c]

It is thus not surprising that throughout the whole of the Heian period, from about +795 to +1185, the imperial court and the nobility were consistently devoted to longevity medicines, at first primarily of plant origin but later more and more mineral and metallic. In the +8th century general *chhang shêng*[4] tonics, powders compounded from plant materials alone, were widely used;[d] but from the reign of the emperor Saga[5] (+809 to +823) more daring elixirs supervened. He himself took, contrary to the advice of his physicians, powdered quartz (*pai shih ying*[6]) and a 'potable gold

[a] This still exists in full use and preservation. The name commemorates the Chinese origin, and the words *chao-thi* transliterate Skr. *caturdiśa*, the four directions of space, whence would congregate the monks of the temple. Grateful thanks are due to the Rev. Endo Shōen, Sub-Dean, for his hospitality and tireless explanations on the occasion of our visit in the summer of 1971, accompanied by Prof. Yabuuchi Kiyoshi, Prof. Shimao Eikoh, Dr Hashimoto Keizō and other friends.

[b] See the biography by Ando Kosei (1).

[c] These are as follows. (1) A *Thai-Chhing Chu Tshao Mu Fang Chi Yao*[7] (Collection of all the Important Plant Drug Prescriptions; a Thai-Chhing Scripture); this is in both Thang dynastic bibliographies but was afterwards lost. (2) A *Thai-Chhing Shen Tan Ching*,[8] and (3) a *Thai-Chhing Chin I Tan Ching*.[9] These may have been parts of *TT*873, the still existing *Thai-Chhing Chin I Shen Tan Ching*[10] (on which see pp. 96, 99); though the Sui and both the Thang bibliographies list also a *Thai-Chhing Shen Tan Chung Ching*,[11] which does not seem to have been preserved, at least under that title. (4) A *Shen Hsien Chih Tshao Thu*[12] (Illustrations of the Mushrooms of the Holy Immortals), and (5) a *Hsien Tshao Thu*[13] (Illustrations of the Drug-Plants of the Immortals). Both these probably had some connection with the *Shen Hsien Yü Chih Jui Tshao Thu*[14] ascribed to Thao Hung-Ching[15] in the Sung bibliography but lost long since. There still exists, however, as *TT*837, the *Shen Hsien Fu Shih Ling Chih Chhang Phu Wan Fang*[16] (Prescriptions for making Pills from Numinous Mushrooms and Sweet Flag, as taken by the Holy Immortals), which may be a related text. Lastly (6), the *Chih Tshao Thu*[17] (Illustrations of Mushrooms and Plants) may have been identical with the treatise of the same title in the Sui and both Thang bibliographies, and attributed to Sun Ssu-Mo[18] in the Sung one. A presumably related work still exists, the *Thai-Shang Ling-Pao Chih Tshao Phin*[19] (*TT*1387). We shall return to these interesting texts in Section 38, when dealing with botanical iconography and mycology. In the meantime the Taoist flavour of all this is so obvious that it needs no emphasis.

[d] E.g. *tosusan*,[20] *shinmeihokusan*[21] and *doshiyōsan*;[22] the composition of all of which can be found in the *Chhien Chin Yao Fang* (c. +655), and similarly in the *Ssu Shih Tsuan Yao*,[23] a technical encyclopaedia for farming families put together by Han O[24] about a century later (pp. 47, 82, 168, 169).

[1] 唐招提寺	[2] 日本國現在書目錄	[3] 藤原佐世	[4] 長生
[5] 嵯峨	[6] 白石英	[7] 太清諸草朩方集要	[8] 太清神丹經
[9] 太清金液丹經	[10] 太清金液神丹經	[11] 太清神丹中經	
[12] 神仙芝草圖	[13] 仙草圖	[14] 神仙玉芝瑞草圖	[15] 陶弘景
[16] 神仙服食靈芝菖蒲丸方	[17] 芝草圖	[18] 孫思邈	
[19] 太上靈寶芝草品	[20] 屠蘇散	[21] 神明白散	
[22] 度嶂散 [23] 四時纂要	[24] 韓鄂		

elixir' (*chin i tan*[1]).[a] Towards the end of his reign the imperial physician Mononobe Kōsen[2] produced a treatise with the significant title *Shê Yang Yao Chüeh*[3] (*Setsuyō Yoketsu*[3]), i.e. 'Important Instructions for the Preservation of Health conducive to Longevity'. The court was always ready to listen to herbalists as well as to metallurgical alchemists, and under Montoku[4] (r. +851 to +858), a pharmaceutical adept Takada Chitsugi[5] was authorised to plant special physic gardens full of *kou chhi*[6].[b] So also, during the later +10th century there was a great vogue for black myrobalans.[c] The admiration of the Heian court for everything foreign, and its propensity for continual medication, can be seen in the life of the Regent Fujiwara no Tadahira,[7] who died in +949 after chronic illness of long duration. In his diary he reports the taking of elixirs also by the emperor Daigo[8] (r. +897 to +930). Besides *chin i tan*, there were 'stalactite pills' (*chung ju wan*[9]), 'red snow', i.e. powder (*hung hsüeh*[10]) and 'purple snow' (*tzu hsüeh*[11]); both these last complex mixtures containing mercuric sulphide.[d]

During the centuries of upheaval and civil strife after the Heian period interest in elixirs lessened, though it would be possible to follow much further the fortunes of mineral therapy and alchemical medicine in Japan. But enough has been said to suggest that of all the countries within China's intellectual field of force Japan was the

[a] This was almost certainly the *chin i hua shen tan*[12] described in slightly later books. It appears almost simultaneously in the *Ishinhō* (+982) and in the *Thai-Phing Shêng Hui Fang*[13] of +992 (now rare), by Wang Huai-Yin,[14] Chêng Yen[15] et al. A complex preparation, it started with magnetite and sulphur, and involved heating such inorganic substances with the urine of youths—if the total solids of this remained, the result could perhaps have contained some of the urinary steroid hormones. As we shall see in Vol. 5, pt. 5, procedures of this kind appear to go back at least as far as the latter part of the +8th century; meanwhile see Lu Gwei-Djen & Needham (3). *Thai-Phing Hui Min Ho Chi Chü Fang*[16] (p. 104) adds red bole clay but does not mention the urine (+1151).

The 'potable gold' preparation was favoured also by one of Saga's successors, Nimmyō[17] (r. +833 to +850), who supplemented it by 'pills of the seven chhi' (*chhi chhi wan*[18]). Full details are in the *Shoku-Nihonkoki*,[19] a history concerned only with this reign.

[b] I.e. *Lycium chinense* (R115, Stuart (1), p. 250). The account is in the *Montoku-Jitsuroku*.

[c] These are fruits of the Combretaceous vine *Terminalia chebula* (R247, Khung et al. (1), p.1120.1; Anon. (109), vol. 2, p. 988.1), essentially an Indian medicinal product. Jap. *karirokugan* corresponds to Chinese *ho-li-lê*[20] (cf. Skr. *haritaki*, Tam. *kadukai*); cf. Burkill (1), vol. 2, p. 2134; Ainslie (1), vol. 2, p. 128.

[d] The formulae are in the *Ishinhō*, the *Wai Thai Pi Yao*, and the *Ssu Shih Tsuan Yao* (pp. 161–2). Red snow contained nine plant drugs, two kinds of animal horn, Glauber's salt (sodium sulphate) and powdered cinnabar. Purple snow was still more complicated, comprising besides powdered metallic gold five other minerals (among which potassium nitrate), five plant drugs (including three perfume aromatics) and three animal substances (including one perfume, musk). These were veritable East Asian theriacs.

According to Yoshida Mitsukuni (6), p. 203, there are descriptions of six types of pharmaceutical preparations in the *Wamyō Ruijūshō*,[21] an encyclopaedia produced by Minamoto no Shitagau[22] in +934. One of the types is *tan yao*,[23] and under this head sixteen elixir preparations are described, including 'potable gold elixir', 'jade juice' (*yü i*[24]), red snow and purple snow. Minamoto's source for this seems to have been a *Ta Thang Yen Nien Ching*[25] (Great Thang Dynasty Manual of the Promotion of Longevity). No such title exists in the Chinese dynastic bibliographies themselves, but a *Yen Nien Pi Lu*,[26] though now lost, is in both the Thang lists and the Sung one.

[1] 金液丹	[2] 物部廣泉	[3] 攝養要訣	[4] 文德	[5] 竹田干繼
[6] 枸杞	[7] 藤原忠平	[8] 醍醐	[9] 鐘乳丸	[10] 紅雪
[11] 紫雪	[12] 金液華神丹	[13] 太平聖惠方	[14] 王懷隱	[15] 鄭彥
[16] 太平惠民和劑局方	[17] 仁明	[18] 七氣丸	[19] 續日本後記	
[20] 訶黎勒	[21] 倭名類聚抄	[22] 源順	[23] 丹藥	[24] 玉液
[25] 大唐延年經	[26] 延年秘錄			

one where the elixir idea developed in purest culture, almost free from the preoccupations of aurifiction and aurifaction so often elsewhere associated with it.

(iv) *Handbooks of the Wu Tai*

Some of the rulers of the Five Dynasties also took an interest in elixirs. The emperor
Thai Tsu of Later Liang (r. +907 to +914) became seriously ill as a result of elixir
poisoning, and fell a ready victim to a plot of assassination. Li Shêng,[1] the founder of
the Southern Thang kingdom, also died from elixir poisoning. The official history
Wu Tai Shih Chi tells us about the alchemists Wang Jung[2] (+873 to +921) and
Chêng Ao[3] (+866 to +939). It says:[a]

> Wang Jung was fond of unorthodox arts (*tso tao*[4]) and prepared elixirs for the achieve-
> ment of immortality. Together with the Taoist Wang Jo-Na,[5] he dwelt at Hsi-shan[6] and
> roamed about among the mountains. . . .
> Chêng Ao (Chêng Yün-Sou[7]) failed in the imperial examinations, which he took during
> the time of the Thang emperor Chao Tsung (+889 to +903). . . . He went to live in Hua-
> yin[8] to search for an elixir said to have been formed by five pine-seeds buried in the ground
> among the Hua-shan[9] Mountains for a thousand years. . . . One of his friends, Li Tao-Yin,[10]
> could fish successfully without using bait, and also knew how to turn minerals into gold
> (*nêng hua shih wei chin*[11]).

Three notable works, including two handbooks of considerable importance, were produced during the Five Dynasties period. These were (*a*) the *Pao Tsang Lun*[12] (Discourse on the Precious Treasury of the Earth),[b] a treatise on minerals and metals
produced in the close neighbourhood of +918 by an alchemist who adopted the
pseudonym Chhing Hsia Tzu,[13] (*b*) the *Ta Huan Tan Chao Chien*[14] (An Elucidation
of the Great Cyclically-Transformed Elixir),[c] written by an anonymous author in
+962, and (*c*) the *Tan Fang Chien Yuan*[15] (Mirror of Alchemical Processes and
Reagents)[d] written by Tuku Thao,[16] who seems to have flourished during the time of
the Later Shu emperor Mêng Chhang[17] (+938 to +965).[e]

[a] Ch. 39, p. 1 *a* and ch. 34, p. 2 *a*.
[b] Often quoted as *Hsien-Yuan Pao Tsang Lun*.[18] We no longer possess the full work, but there are a
fair number of quotations in the pharmaceutical natural histories and related literature.
[c] *TT*919.   [d] *TT*918.
[e] The dating of Tuku Thao to the time of Mêng Chhang derives from an opinion of our friend Dr
Li Hsiang-Chieh (cf. Vol. 1, p. 12), privately communicated. In any case he must have lived before
+1150, when Chêng Chhiao put him in the *Thung Chih Lüeh* bibliography. If he lived in Szechuan under
the Later Shu, he must surely have known the Li family of Earlier Shu, since both dynasties were very
short. In Vol. 6 we consider at some length Li Hsün[19] and his brother and sister, prominent between
+919 and +925. Of Persian origin, they were all excellent Chinese scholars, and Li Hsün produced
in +923 his *Hai Yao Pên Tshao*[20] (Natural History of Drugs from Overseas), now lost as such but
quoted frequently enough in later books. The brothers were merchants of drugs and perfumes, local
and exotic, and skilled in the distillation of the essential oils, Arabic, Persian, Indian and Malayan.
  Alchemical interest in exotic products is shown again by the *Hai Kho Lun*[21] (Guests from Overseas),
*TT*1033, written by Li Kuang-Hsüan[22] a century or more later. This short book deserves study.

[1] 李昇	[2] 王鎔	[3] 鄭遨	[4] 左道	[5] 王若納	[6] 西山
[7] 鄭雲叟	[8] 華陰	[9] 華山	[10] 李道殷	[11] 能化石爲金	[12] 寶藏論
[13] 青霞子	[14] 大還丹照鑑	[15] 丹方鑑源	[16] 獨孤滔	[17] 孟昶	
[18] 軒轅寶藏論	[19] 李珣	[20] 海藥本草	[21] 海客論	[22] 李光玄	

Of the value of the *Pao Tsang Lun* we have occasion to speak in other contexts.[a] The *Ta Huan Tan Chao Chien* is a tractate, rather short and seemingly obscure unless one is familiar with the theories and terminology of the *nei tan* (physiological alchemy) school.

Take the following example:

Oral instructions of (the Adept) Yen Chün-Phing.[1]

Lead is lead, mercury is mercury. Only these two are lead and mercury. The Dragon is the Dragon, the Tiger is the Tiger, and only these two share the same ancestors. If one can understand this (one can next proceed to consider) the positions of the trigram *Khan* and (the trigram) *Li*. Under *Li* is found 'yellow earth' (*huang thu*[2]), and from 'yellow earth' comes forth the 'yellow sprout' (*huang ya*[3]), which bears the 'golden flower' (*chin hua*[4]). When the 'golden flower' is formed under (*Li*) the 'purple essence' (*tzu ching*[5]) will congeal. When the golden flower comes into being the 'white metal' (*pai chin*[6]) will form below it, for white metal sinks, while 'yellow metal' (*huang chin*[7]) floats. The part that sinks is taken as the most important constituent of the elixir (*tan thou*[8]), (but) a search has to be made for the 'red marrow' (*chhih sui*[9]). After mixing they have to be sown. There is a proper time for sowing as well as a proper time for reaping. The muscles of the aspirant are transformed and his bones are strengthened so that he may live as long as the immortals of Heaven and Earth. Do not contemplate the transmutation of lead, iron and fragments of broken tile. One, Two, Three, Four, Five; Water, Fire, Wood, Metal, and Earth.[b] To understand just this one secret essential is the (sovereign) Way to the achievement of immortality.[c]

Here we seem to approach rather closely the style of writing of late medieval European alchemists, with their elliptical allusiveness and oracular pronouncements not meant to be understood too readily. At the same time one may note the role of the symbolic correlations supplementing the Five Elements and the Book of Changes, features of course exclusively Chinese.

The *Tan Fang Chien Yuan* details the properties of many alchemical reagents, including metals, minerals, plants and other organic substances such as urine.[d] It describes only one alchemical process, in connection with a formula for preparing the 'Five yellow (substances) pill' (*wu huang wan*[10]), in which equal weights of realgar (*hsiung huang*[11]), orpiment (*tzhu huang*[12]), arsenious oxide (*phi huang*[13]), sulphur (*liu huang*[14]), and yellow alum (*huang fan*[15]) are fused together, mixed with vinegar and common salt, and heated in a closed reaction-vessel until a sublimate is formed under the cover.[e] The book consists of three short chapters, the first of which has been translated into English in a pioneer work by Fêng Chia-Lo & Collier (1).

---

[a] See pt. 2, p. 273 and here, pp. 130, 211.

[b] Here the cosmogonic enumeration order will be recognised (cf. Vol. 2, pp. 253 ff.). It must have been prevalent among the alchemists, as we find it clearly set forth in *Tshan Thung Chhi*, ch. 34, p. 11 a.

[c] P. 23 a, b, tr. auct.

[d] This is particularly interesting in view of the work of the iatro-chemists which will be considered in Vol. 6, Sect. 45; meanwhile see Lu Gwei-Djen & Needham (3).

[e] In ch. 3, pp. 6b, 7a.

[1] 嚴君平	[2] 黃土	[3] 黃芽	[4] 金花	[5] 紫精
[6] 白金	[7] 黃金	[8] 丹頭	[9] 赤髓	[10] 五黃丸
[11] 雄黃	[12] 雌黃	[13] 砒黃	[14] 硫黃	[15] 黃礬

Chinese alchemy seems in a way to have reached its peak of development between the time of Ko Hung early in the +4th century and that of Mei Piao at the beginning of the +9th. This can be seen in the lucid style of many of the alchemical writings of this period (a great departure from the abstruse language used by Wei Po-Yang), in the adventurous experimentation with ever greater numbers of inorganic substances for elixir recipes, and in the development of alchemical theory. The majority of the most important proto-chemical writings we now possess belong to this period. From the +9th century onwards there was a tendency to revert to the gnomic and theoretical style, a marked decrease in the number of substances used in elixirs, and a growing interest in substances of plant and animal origin. Alchemical writings took the form of compendia like the *Yün Chi Chhi Chhien* collection[a] and the *Kêng Tao Chi*[1] (Collection of Procedures of the Golden Art);[b] or else of commentaries, such as the many annotations of the *Tshan Thung Chhi* already mentioned (p. 53). It now becomes difficult to decide whether a writer belonged to the *wai tan* or the *nei tan* school. Such changes may have been in part a direct consequence of elixir poisoning, for from the beginning of the +9th century onwards one Thang emperor after another, Hsien Tsung (r. +806 to +820), Mu Tsung (r. +821 to +824), Ching Tsung (r. +825 to +826), Wu Tsung (r. +841 to +846), and Hsüan Tsung (r. +847 to +859), died as a result of taking elixirs in the hope of becoming immortal. Many alchemists themselves, especially the more brilliant ones who had the most faith in their own preparations, must also have perished in the same way. Other alchemists must have been much alarmed at the numerous cases of elixir-poisoning, so they changed their approach; giving their allegiance rather to the relatively new school of *nei tan* (physiological) alchemy, subjecting the ancient texts to a 'modernist' exegesis, and turning away from minerals towards plant and animal products. Another factor for the decline of alchemy may be sought in the loss of Taoist writings during the successive upheavals of the last two centuries of the Thang, e.g. the uprisings of An Lu-Shan[2] and Shih Ssu-Ming[3] in the +8th century and that of Huang Chhao[4] in the +9th; then the great chaos during the transitional period of the Five Dynasties in the +10th century.

(v) *Theocratic mystification and the laboratory in the National Academy*

We shall now find that the continuation of the gradual decline of alchemy during the Sung was due neither to lack of enthusiasm on the part of the emperors nor to opposition from the literati. On the contrary, the Sung emperors were great supporters of Taoism.[c] We have already mentioned the effort made by the second Sung ruler Thai Tsung to search for Taoist writings (p. 115), and we need only say something of some of his successors. We are also told that certain scholars and high officials, like Fan Chung-Yen[5] (+989 to +1052) and Su Chhê,[6] were acquainted with the alchemi-

---

[a] *TT* 1020.   [b] *TT* 946. See p. 197 below.
[c] This period has already been discussed by Ho Ping-Yü (14).

[1] 庚道集   [2] 安祿山   [3] 史思明   [4] 黃巢   [5] 范仲淹
[6] 蘇轍

cal art, and we have already seen that the great neo-Confucian scholar, Chu Hsi himself, did not disdain to write a commentary on the *Tshan Thung Chhi*.

Earlier on the Thang emperors had found it convenient to adopt as their patronal ancestor Lao Tzu, the putative founder of Taoism, because of the common surname Li.[1] In the year +667 the Thang emperor Kao Tsung (r. +650 to +682) conferred upon Lao Tzu the posthumous title of a past emperor, Thai-Shang Hsüan-Yuan Huang Ti,[2] thus claiming him as the forbear of the reigning family.

During the Sung period it was thought expedient to select another figure from among the imperial clan to take the place of Lao Tzu. Hence the third emperor, Chao Hêng,[3] whose temple name is Chen Tsung (r. +997 to +1022), created a new putative founder of Taoism and a new ancestral patron in the person of one Chao Hsüan-Lang,[4] conferring upon him the brief title Thai-Shang Khai-Thien Chih-Fu Yü Li Han-Chen Thi-Tao Yü-Huang Ta-Thien Ti.[5] Chen Tsung was either a real lay Taoist in the Mao Shan revelatory tradition (pt. 2, p. 110) or a man prepared to use it for his own purposes—perhaps both. The *Sung Shih* reports many revelations vouchsafed to him in dreams,[a] and describes how he first received a letter from Heaven in the year +1008. This missive was supposed to have been fastened to a piece of yellow silk suspended from a corner of the roof of a palace-gate. Another message arrived engraved on a piece of jade and the Prime Minister was requested to read it aloud at the palace audience. The following extracts from the *Sung Shih* regale us with more information about Chen Tsung's activities:[b]

On a *jen-shen* day in the ninth month of the 1st year of the Hsien-Phing reign-period (+998) he gave (the adept) Chhung Fang[6] gifts of grain, cloth and money. . . . On a *ting-chou* day in the third month of the 4th year (+1001) he sent for Chhung Fang, (but the latter) could not come, saying that he was ill. . . . On an *i-ssu* day in the seventh month of the 5th year (+1002) he again sent for Chhung Fang, the adept of the Chung-nan[7] Mountain, and received a memorial (from him). . . . During the ninth month of the same year he met Chhung Fang at the palace, and after making him one of his advisers, offered him living quarters there. . . . On a *ting-yu* day in the second month of the 4th year of the Ching-Tê reign-period (+1007) he gave a gift of silk to the recluse Yang Phu.[8] . . . On an *i-chhou* day in the first month of the 1st year of the Ta-Chung Hsiang-Fu reign-period (+1008) a piece of yellow silk (with a letter) was found hanging from a fish-tail-shaped roof-ornament (*chhih wei*[9]) at the south corner of the left palace gate. This was reported to the authorities by Thu Jung[10] the gate-keeper. The emperor asked some of the officials to bring it back from there with great reverence to the Chao-Yuan[11] palace-hall, where it was opened (and read). He then pronounced it a 'Letter from Heaven' (*thien shu*[12]). . . . On an *i-wei* day in the sixth month another 'Letter from Heaven' appeared north of Li-chhüan[13] in Thai-shan Mountain. . . . On a *chi-yu* day in the eighth month Wang Chhin-Jo[14] presented to the emperor over eighty thousand magic mushroom plants (*chih tshao*[15]). . . . On a *hsin-mao* day in the tenth month the emperor set out from the capital, asking those who accompanied

---

[a] Cf. Vol. 2, p. 159; Eichhorn (12).   [b] *Sung Shih*, ch. 6, pp. 5aff., tr. auct.

[1] 李   [2] 太上玄元皇帝   [3] 趙恆   [4] 趙玄朗
[5] 太上開天執符御歷含眞體道玉皇大天帝   [6] 種放   [7] 終南
[8] 楊璞   [9] 鴟尾   [10] 塗榮   [11] 朝元   [12] 天書
[13] 醴泉   [14] 王欽若   [15] 芝草

him to lead the procession displaying the 'Letter from Heaven'. . . . On a *ting-chhou* day in the eleventh month the emperor returned to his palace from Thai-shan Mountain bringing back with him the 'Letter from Heaven'. . . . On a *ting-ssu* day in the twelfth month of the 3rd year (+1011) Li Tsung-O,[1] a Fellow of the Han-lin Academy (Han Lin Hsüeh Shih[2]), and others, presented to the throne an illustrated treatise on the geography of the empire.[a] . . . On a *ting-yu* day in the first month of the 4th year (+1011) (the emperor) left the capital, taking along with him the 'Letter from Heaven'. . . . On a *ting-ssu* day in the second month, a yellow cloud was seen following the 'Letter from Heaven'. . . . On a *kêng-wu* day in the second month of the 4th year (+1011) the emperor gave audience to (the adepts) Chêng Yin[3][b] and Li Ning.[4] . . . On a *hsin-wei* day he gave audience to the Taoist Chhai Yu-Hsüan.[5] . . . On a *hsin-chhou* day in the first month (+1012) he visited the Thai-Chhing Kung[6] temple. . . . On an *i-wei* day in the fifth month another 'Letter from Heaven' was found engraved on a piece of jade. . . . On an *i-mao* day in the first month of the 1st year of the Thien-Hsi reign-period (+1017) the Prime Minister read out publicly the new 'Letter from Heaven' at (an audience in) the Thien-An[7] palace-hall.

All this nonsense may seem as meaningless as it is amusing, but one's estimate of Chen Tsung and his time is somewhat altered if one realises that it was exactly at this juncture that a number of important military inventions were made, notably the development of gunpowder bombs thrown by trebuchets, and the first appearance of rocket weapons.[c] If Chen Tsung preferred to govern by means of theocratic mystification rather than by military might that should be considered civilised of him rather than the reverse. Moreover, it looks as though his Taoists, when consulting with their military colleagues, were capable of suggesting something more serious than magic mushrooms. It is interesting too to note that Chen Tsung had earlier enquired of the Taoist Holan Hsi-Chen[8] (d. +1010) about the transmutation of base metals into gold, but received the reply that the art of transmutation practised by an emperor should be that of transforming the chaotic world into one of peace and prosperity.[d] The emperor conferred upon this adept the title Thung-Hsüan Ta Shih[9] (Great Mystery-Penetrating Teacher). Chen Tsung died in +1022, and his son, who succeeded him as the emperor Jen Tsung, buried the 'letter from Heaven' with him.[e] Only twenty years afterwards Tsêng Kung-Liang[10] was to produce his military encyclopaedia, the *Wu Ching Tsung Yao*,[11] with its practical formulae for gunpowder, the first published in any civilisation.

It is interesting indeed to find, as a background for such discoveries, that under Chen Tsung and later an alchemical elaboratory was actually maintained for seventy

---

[a] We know that this is what it was, because his *Chu Tao Thu Ching*[12] (the same words) appears as a book in ninety-eight chapters, in the *Sung Shih* bibliographical register under geography. It has long been lost. We are grateful to Prof. Yabuuchi Kiyoshi for drawing our attention to this.

[b] This Chêng Yin must not be confused with Chêng Ssu-Yüan, Ko Hung's teacher, who was also known by the same name.

[c] See Sect. 30.

[d] *Sung Shih*, ch. 462, p. 1a and *Lu Huo Chien Chieh Lu*, p. 1b.

[e] *Sung Shih*, ch. 9, p. 2b.

[1] 李宗諤	[2] 翰林學士	[3] 鄭隱	[4] 李寧	[5] 柴又玄
[6] 太清宮	[7] 天安	[8] 賀蘭棲眞	[9] 通玄大師	[10] 曾公亮
[11] 武經總要	[12] 諸道圖經			

years in the Imperial Academy itself.[a] In his *Mo Chuang Man Lu*[1] (Recollections from the Estate of Literary Learning), written in +1131, Chang Pang-Chi[2] tells us that:

in the time of Chang-Shêng (part of one of the memorial titles of the emperor Chen Tsung) a furnace laboratory for making elixirs was installed in the Han-Lin Academy, under the name of the Hall of the Golden Elixir (Chin-Tan Ko[3]). It was supplied daily with five loads of charcoal, and down to the 1st year of the Hsi-Ning reign-period (+1068) the fire was continually kept going. But then the father of Liu Jou[4] (Liu Yen-Chung[5]) received an imperial order which cut down the size of the civil service establishment, so that the upkeep expenses (of the laboratory) were cancelled. The elixir which had been produced looked dark like iron. The emperor ordered that it should be kept in the Thien-Chang Ko[6].[b]

Chang Pang-Chi follows this up by relating a story intended to warn people of the danger of ingesting metallic and mineral elixirs. This is a matter to which we have to refer again and again.[c]

When Chang An-Tao[7] (Chung Ting Kung[8]) was living at Nantu he also kept a furnace laboratory for transmuting elixirs, but it was only after several dozen years of maintaining the fire that any were achieved. He himself, however, did not dare to eat any of them. At that time Chang Chhu[9] (Chang Shêng-Min[10]), the prefect of Nantu, was very thin and debilitated, so when he heard about this he importuned (Chang) An-Tao to be allowed to take some. But the latter said: 'I do not want to be miserly, but this elixir has been heated in the fire for a very long time, so it is either very efficacious or extremely poisonous. It must not be taken rashly.' But as (Chang) Shêng-Min kept on begging for it, he finally gave him a piece the size of a millet-grain, warning him once more to treasure it and not to take it lightly. He swallowed it as soon as he got it, however, and before a few days were out, he went down with an effusion of blood, all his viscera becoming rotten, and death shortly ensuing. These two affairs I heard about from Liu Yen-Chung himself.[b]

The most important circumstance about the reign of Chen Tsung was that it saw the definitive redaction of the Taoist Patrology (the *Tao Tsang*) in +1019, and the preparation of the *Yün Chi Chhi Chhien* a few years later.[d] Not only did these great collections preserve many books on alchemy, but just at this time several important individual treatises on the subject were written.[e] For example, this was the period of activity of Li Pi[11] and his great pupil Tshui Fang,[12] the author of the *Wai Tan Pên Tshao*[13] (Iatro-chemical Natural History),[f] *c.* +1045, men who may perhaps be regarded as the initiators, five centuries before Paracelsus, of the long movement

---

[a] This is reminiscent of the laboratory maintained at the Northern Wei court between +386 and +409; see p. 118 above. It also calls to mind the activities of the College of All Sages in horological engineering between +720 and +750 (cf. Vol. 4, pt. 2, pp. 471 ff.).

[b] Ch. 3, p. 11 *b*, tr. auct. [c] Pp. 74, 182 above, 194, 212 below.

[d] See p. 115 above. [e] See p. 197 below.

[f] We possess the remains of this book only in the form of quotations in later works. Those in the *Pên Tshao Kang Mu* (probably transmitted through the *Kêng Hsin Yü Tshê*, cf. p. 210) have been collected by Ho Ping-Yü & B. Lim (1). The postscript of the *Wai Tan Pên Tshao*, quoted in *Kêng Tao Chi* (cf. p. 197 below), gives +1043 as a firm date when Tshui Fang (Tshui Hui-Shu,[14] or Wên Chen Tzu[15]) took up an official appointment in Hunan.

[1] 墨莊漫錄	[2] 張邦基	[3] 金丹閣	[4] 劉宷	[5] 劉延仲
[6] 天章閣	[7] 張安道	[8] 忠定公	[9] 張錫	[10] 張聖民
[11] 李弼	[12] 崔昉	[13] 外丹本草	[14] 晦叔	[15] 文眞子

away from elixir-making towards the preparation of substances organic as well as inorganic genuinely useful in medicine. But they were also skilled in metallurgy, as we
have noted already in connection with the different kinds of artificial gold (pt. 2, p. 281).

(vi) *The emperor's artificial gold factory under Metallurgist Wang Chieh*

This is the place at which we may pause to consider the career of an eminent alchemist
who had, like Teacher Kêng, a peaceful and successful career at court, and died in
the odour of sanctity. From the quotations which follow it is possible to build up a
rather clear picture of what a man could accomplish who performed what he promised,
i.e. the making of reasonably gold-like alloys—surrounded though this had to be in
those times with strange behaviour, hypnotic powers, mystifications and abracadabra.
The activities of Wang Chieh[1] ranged from about +980 to +1020, precisely under
the patronage of Chen Tsung, and from the number of accounts of him which have
been preserved in the writings of notable scholars he must have impressed his personality very thoroughly on his contemporaries. Beginning as a merchant, he was
taught the chymical arts by an un-named Taoist, acquiring other uncanny techniques
into the bargain; then feigning madness, he secured protectors of steadily increasing
official rank until he finally won the confidence of the emperor himself. The gold-like
alloys were a matter of hard fact, which nobody could deny, and Wang Chieh went on
producing them until he died, after which he received high posthumous honours.

One of our oldest sources, the *Chhing Hsiang Tsa Chi*[2] (Miscellanous Records on
Green Bamboo Tablets), written about +1070 by Wu Chhu-Hou,[3] reads as follows:[a]

In the time of the emperor Chen Tsung (+997 to +1022) Wang Chieh of Tingchow
(Chhangting in Fukien), when travelling in his younger days near the Yangtze (as a merchant)
and staying overnight at Hsing-tzu Hsien, met a Taoist who taught him the 'art of the yellow
and the white' (alchemy), though he could not comprehend quite all of it. He met the Taoist
again in Mao Shan[4],[b] and when they both went to Liyang the Taoist showed him numinous
herbs (*ling tshao*[5]) and taught him secret methods of mixture and combination (*ho ho mi
chüeh*[6]). These, when tested, all proved to be effective (*shih chieh yu yen*[7]). The Taoist also
gave him numinous recipes (*ling fang*[8]) which were sealed and closed with the warning that
he should not tell anyone about them unless it were the emperor personally. Afterwards
Wang Chieh was exiled because of (assumed) madness to Ling-nan, where he managed to
see the official Hsieh Tê-Chhüan[9] who appreciated his extraordinary knowledge and looked
after him in his own home. There he made artificial silver (*yao yin*[10]) and artificial gold (*yao
chin*[11]), and presented it to the emperor, who gave him an audience and asked a high official
named Liu Chhêng-Kuei[12] to see about his affairs. Wang said that his teacher had strictly
enjoined him never to reveal his art except to the emperor, but Liu (believed in him and)
changed Wang's given name to Chung-Chêng.[13] Then from first to last Wang produced and
presented to the throne artificial gold and silver amounting to many tens of thousands (of
cash), brilliant and glittering beyond all ordinary treasures. Later he manufactured the

---

[a] Ch. 10, p. 1 a, tr. auct.   [b] The famous centre of Taoist techniques (cf. pt. 2, *sub voce*).

[1] 王捷	[2] 青箱雜記	[3] 吳處厚	[4] 茅山	[5] 靈草
[6] 和合密訣	[7] 試皆有驗	[8] 靈方	[9] 謝得權	[10] 藥銀
[11] 藥金	[12] 劉承珪	[13] 中正		

Golden Badge decorations. He was never spendthrift, but laid out his riches on the poor and needy, and for worshipping the Taoist immortals and Buddhas. He it was who built the Khai-Yuan temple at Tingchow. When he died he was given the posthumous title of Legate of Chen-nan[a]—such a thing had never been heard of before.

This account can be supplemented by another written a little later, *c.* +1090, in Wang Phi-Chih's[1] *Shêng Shui Yen Than Lu*[2] (Fleeting Gossip beside the River Shêng), from which we learn that a votive temple was built to Wang Chieh after his death.[b] 'Even now', wrote Wang Phi-Chih, 'there is still in the Treasury some of the (artificial) gold that Wang Chieh made and presented, and they also have some of his remaining chemicals, together with his crucibles, tongs and furnace.'[c]

The next spotlight on him brings up a point of particular interest, namely that he seems to have been assisted by an artisan destined to be much more illustrious than Wang Chieh himself, Pi Shêng,[3] the inventor of all movable-type printing. This we learn from the book so often quoted in these pages, the *Mêng Chhi Pi Than* of Shen Kua, finished about +1086. Here we read:[d]

In the (Ta-Chung) Hsiang-Fu reign-period (+1008 to +1016) there was an alchemist (lit. magician-technician, *fang shih*[4]) who was able to make gold. Originally he had been a branded convict banished to Monks' Island (Sha-mên Tao[5]).[e] An old craftsman named Pi Shêng[3f] who had formerly worked for (Wang) Chieh, casting gold in the imperial palace, said: 'His method was to station the men who blew the bellows in a separate room from the furnace (with the tuyère coming through a hole in the wall), because he did not want them to see what was going on, especially during the filling and emptying of the crucibles. His gold was made from iron. When it came out of the furnace it was still black.[g] More than a hundred ounces went to make an ingot, and each one was cut into eight pieces. This was called

[a] Other sources say Ling-nan.
[b] Ch. 10, p. 6a, tr. auct.
[c] These accounts seem to have been conflated in very full form in the +13th-century *Sung Chhao Shih Shih*[6] of Li Yu.[7]
[d] Ch. 20, p. 12a (para. 19), tr. auct., adjuv. Vacca (2) who first drew attention to the passage more than fifty years ago. Cf. Hu Tao-Ching (1), vol. 2, pp. 667ff.
[e] Doubtless at the time of his assumed madness.
[f] The element of doubt here is that the given name of Pi Shêng[8] the printer is written in a slightly different way (*MCPT*, ch. 18, pp. 7aff., para. 10, cf. Hu Tao-Ching (1), vol. 2, pp. 597ff.; and our Vol. 4, pt. 2, p. 33). But as the dates agree so well (the invention of movable-type printing would have taken place about +1045), it is almost certain that we are dealing with the same person. This was also the opinion of Vacca (2).
[g] These statements are very curious. A black metal such as lead might have been mistaken for iron, but certainly not by a craftsman such as Pi Shêng. One might conceive that zinc was an important constituent and that Wang Chieh's gold was some bright form of brass. On the other hand, although iron does not amalgamate, you can plate it with amalgams, though generally they do not stick very well unless copper has been deposited first, so possibly all his products were mercury-gilded (cf. C. S. Smith (7), p. 122). Alternatively, he was using some ion exchange method (cf. pp. 104, 129) such as the coating of iron surfaces by the sulphates of copper or silver, or the cyanide of gold—if he could have made it. Haschmi (5) quotes al-Bīrūnī as referring to the deposition of copper on silver by rubbing with copper acetate; this will not in fact work with silver, but it does with lead. Finally it is curious that one of the books of the Jābirian Corpus (cf. pt. 4 below) speaks of the transformation or iron into gold and silver, without making very clear, however, how to do it. This is the *Kitāb al-Naqd* (Book of Testing, or, of Coinage), Kr156; cf. Kraus (2), p. 53. Since this would date from the second half of the +9th century or the first half of the +10th, it might have been transmitted eastwards—assuming that the process was really a workable one.

[1] 王闢之 [2] 澠水燕談錄 [3] 方士 [4] 沙門島 [5] 畢升
[6] 宋朝事實 [7] 李攸 [8] 畢昇

"crow's-beak gold (*ya tsui chin*[1])".' Nowadays there are still people who have some of this. The emperor ordered the imperial workshops to cast it into gold tortoises and gold medals, several hundred of each. First he distributed the former to the high officials who were closest to him, one to each, and then, apart from the imperial family itself, seventeen other court personalities received them. The remainder were buried under the foundations of the Yü-Chhing Kung and the Chao-Ying Kung (palaces) and also the Pao-Fu Hall, as a talismanic protection. As for the golden medals he despatched one each to all the district prefects and military inspectors throughout the empire. This was called the Precious Golden Badge (Chin Pao Phai[2]). At the home of Li Chien-Fu[3] at Hungchow one of these gold tortoises is still preserved, because his great-uncle (Li) Hsü-Chi[4] was one of the seventeen recipients. (They say that) this tortoise goes out and wanders about at night, giving off a phosphorescent glow, but if you touch it there is nothing peculiar about it. Anyway, it is still kept as a precious treasure in the (Li) family.[a]

In a different place we come across another assistant of Wang Chieh's. In the entry for gold dust (*chin hsieh*[5]) in his *Pên Tshao Yen I*[6] of +1116, Khou Tsung-Shih[7] says:[b]

Chang Yung-Tê[8] of the present dynasty, whose *tzu* name was Pao-I,[9] was a man from Pingchow. In the Five Dynasties period he was a local commander in Shansi, but in the 2nd year of the Shun-Hua reign-period (+991) he was posted (back) to Pingchow, and lived at Suiyang; there he had occasion to cure successfully a neighbour who was bedridden and in pain. One day (long afterwards) this scholar came and begged Chang to provide him with five ounces of mercury, which being done the visitor put it into a reaction-vessel and heated the contents until it all became silver (*chung chin*[10]).[c] Chang Yung-Tê then earnestly asked him for the chemical technique, to which the scholar replied: 'You are one fated to be wealthy, so I do not mind giving you the secret. Yet I fear it may injure your blessings.' After the work of transformation (of the precious metal) was quite finished, he rose and explained that in the Hsiang-Fu reign-period (+1008 to +1016) he had been in the palace workshops making gold by transformation for the magician-technician (*fang shih*[11]) Wang Chieh. In his process iron was used, and they obtained large round flat pieces like cakes each weighing 100 ounces; these were cut like the spokes of a wheel into eight segments. Since in the early stages of the process the material was black in colour, the result was named 'crow's-beak gold'. Thus this kind (of metal) was made from mercury and iron and various chemicals, not being the product of Nature herself (*fei tsao-hua so chhêng*[12]); how could its curative powers not be different (from those of natural gold)? So, for example, when the National Dispensaries (Hui Min Chü,[13] of this dynasty)[d] compound their 'purple powder' (*tzu hsüeh*[14])[e] they

---

a Shen Kua's usual note of scepticism comes in here.
b Ch. 5, p. 1*b*, cit. *CLPT*, ch. 4, (p. 110.1), tr. auct. Parallel passage in *Lung Chhuan Pieh Chih*, ch. 1, p. 2*b*.   c Cf. p. 38 above.
d See Sect. 44 in Vol. 6, meanwhile Lu Gwei-Djen & Needham (2).
e What 'purple powder' was is known from the surviving formulary of *c.* +1106, *Thai-Phing Hui Min Ho Chi Chü Fang*,[15] ch. 6, (p. 113). It was a pale purple powder resulting from a complicated preparation involving eight mineral substances (including gypsum, cinnabar, saltpetre and gold), four plant extracts, three aromatic ingredients and two sorts of powdered animal horn. Conceivably some gold chloride or cyanide was present in the final product. For a Plinian parallel see *Nat. Hist.* XXXIII, xxv, 84, on which Berthelot (2), pp. 14, 15; Bailey (1), vol. 1, pp. 105, 204.

1 鴉嘴金	2 金寶牌	3 李簡夫	4 李虛己	5 金屑
6 本草衍義	7 寇宗奭	8 張永德	9 抱一	10 中金
11 方士	12 非造化所成	13 惠民局	14 紫雪	
15 太平惠民和劑局方				

start with some gold (dust or leaf) in order to take advantage of the *chhi* of natural gold; they do not like (components such as) tin (or iron). Moreover, gold from the South-east is deep in colour, while that from the South-west is pale,[a] so that different regions vary in what is proper to them. When (natural gold) is incorporated into medicines there is nothing so good as the dark-coloured variety. Besides, when (artificial 'golds') come into contact with (the juice of the) *yü-kan-tzu*[1] (fruit)[b] they turn soft by a reaction of mutual response.

Here the interesting thing is Khou Tsung-Shih's clear recognition of the artificial or counterfeit nature of the 'gold' and 'silver' made by Wang Chieh and his assistants. Writing as a pharmaceutical naturalist he was convinced that such things could not possibly have the medicinal value of natural gold.[c]

In all this Wang Chieh's secretive and yet successful ways are well portrayed. Other aspects of his personality were noted and preserved independently, as we see from an entry in the *Tu Hsing Tsa Chih*[2] (Miscellaneous Records of the Lone Watcher), written by Tsêng Min-Hsing[3] in +1176. He says:[d]

In the Hsiang-Fu reign-period (+1008 to +1016) Wang Chieh possessed alchemical techniques for producing artificial gold (*yu shao chin chih shu*[4]). Through Tsêng Hui[5] he met Liu Chhêng-Kuei[6] who introduced him to the Rt. Honourable Wang Chi,[7] through whom he got an audience with the emperor. His contemporaries called him 'Mr Wang the Alchemist' (Shao Chin Wang hsien-sêng). He could make people see anything that he chose to think about, so that many were astounded and perplexed. This is in general one of the magical arts of the South. . . .

And he goes on about ways of collecting the saliva of foxes, which was really something to conjure with. But the reference to hypnotic powers seems clear.

We can thus build up a rather clear picture of the kind of man that Wang Chieh was. He must have been essentially a level-headed metallurgical chemist, keenly observant and well acquainted with the properties of many minerals and inorganic substances. How much of medieval theory he used for guidance is perhaps what we should most like to know, for he certainly 'delivered the goods' in producing alloys, perhaps like pinchbeck, or gold-plated iron, and continued to do so till the end. Part of his success probably lay in making no claim to produce real gold, i.e. gold that could resist cupellation in the Imperial Workshops; as long as one had an emperor who was willing to accept as 'gold' anything that looked rather like it, one had nothing to fear. At the same time it is interesting that although Wang Chieh kept well clear of all entanglement with Taoist elixirs, he had to use much mystification and charisma in order to gain and keep his high position; such were the necessities of the age.

---

[a] Perhaps because alloyed naturally with silver, as electrum, cf. pt. 2, p. 18 *et passim*.

[b] This is *Phyllanthus Emblica = E. officinalis*, of the Euphorbiaceae (R330; CC875; Anon. (*109*), vol. 2, p. 587.2), another kind of myrobalan, the Indian gooseberry. The fruit is medicinal but used also for dyeing and tanning (Burkill (1), vol. 1, p. 920); possibly its organic acids could have had some corrosive action on metals such as tin, copper or iron in the artificial 'golds' (pt. 2, p. 250). Cf. da Orta (1), Markham ed., p. 320.

[c] Note the radical opposition of this doctrine to that of Ko Hung (p. 2 above).

[d] Ch. 7, p. 3b, tr. auct.

[1] 餘甘子　　　[2] 獨醒雜志　　　[3] 曾敏行　　　[4] 有燒金之術　　　[5] 曾繪
[6] 劉承珪　　　[7] 王冀

As a pendant to the story of Wang Chieh, it may be interesting to add (though it interrupts momentarily our chronological thread) some mention of a successor a century later who must have been a very similar sort of man. In his *Lao Hsüeh An Pi Chi*[1] (Notes from the Hall of Learned Old Age), written about +1190, Lu Yu[2] said:[a]

> In the Thien-Hsi reign-period (+1017 to +1021) the emperor bestowed as decorations all over the empire the Precious Golden Badges (Chin Pao Phai[3]) made by Wang Chieh. Towards the end of the Hsüan-Ho reign-period (+1119 to +1125) the reigning emperor similarly bestowed Golden Mandalas (Chin Lun[4]) transmuted by the magician-technician (*fang shih*[5]) Liu Chih-Chhang[6] upon the Numinous Empyrean Temples (Shen Hsiao Kung[7]) all over the country. They were called the Precious Mandalas of the Numinous Empyrean (Shen Hsiao Pao Lun[8]). Liu Chih-Chhang said that in his method the gold was formed by the transformation of mercury (*shui* (*yin*) *lien*[9]).[b] These discs could be used as a talismanic protection in the various astrological provinces of the empire[c] against war, famine and calamities. This was in the autumn of the 7th year of the Hsüan-Ho reign-period (+1125). But no sooner had the emperor sent out his commissioners to deliver the discs all over the empire, and just as the Court of Imperial Sacrifices was about to draft the edict on the ceremonies to be observed at their reception—the (barbarian) enemies attacked across the (Yellow) River.

This, as we know, was the death knell of the Northern Sung dynasty.[d] We are back again at the old Confucian criticism of all magic and nonsense, all trusting in 'golden *maṇḍalas*' and Taoist liturgies rather than arming the troops and ensuring their loyalty. The Confucians had right on their side, but so in a way had Wang Chieh and Liu Chih-Chhang, for if we now have powder metallurgy, beryllium alloys and liquid oxygen steel, it is due to them and their successors, not to the Confucian apostles of common-sense.

(vii) *Social aspects, conventional attitudes and gnomic inscriptions*

During the reign of the emperor Hui Tsung (Chao Chi,[10] r. +1101 to +1126) the *Tao Tsang*, as we have seen, was printed for the first time.[e] Many Taoist texts were written under imperial aegis, for example the *Yü Chieh Tao Tê Ching*[11] (Imperial Commentary on the Canon of the Virtue of the Tao). Hui Tsung even enthroned himself in the Apostolic See of the Taoist religion, assuming the title Chiao-Chu Tao-Chün Huang Ti.[12] Official ranks were created for the Taoist hierarchy to give them status equal with civil officials. Taoists and magicians thronged the imperial vestibules. In +1105 Chang Chi-Hsien,[13] claiming to be the twenty-sixth descendant of Chang Tao-Ling, received the title Hsü-Ching hsien-sêng,[14] in +1111 Wang Tzu-Hsi[15]

---

a Ch. 9, p. 5*b*, tr. auct.
b Or perhaps (without interpolation), 'by a wet method'.
c On the *fên yeh*[16] system, see Vol. 3, p. 545.
d Cf. Vol. 4, pt. 2, pp. 497ff.   e See p. 115 above.

[1] 老學庵筆記   [2] 陸游   [3] 金寶牌   [4] 金輪   [5] 方士
[6] 劉知常   [7] 神霄宮   [8] 神霄寶輪   [9] 水（銀）錬
[10] 趙佶   [11] 御解道德經   [12] 教主道君皇帝   [13] 張繼先
[14] 虛靜先生   [15] 王仔昔   [16] 分野

was created Chhung-Yin Chhu Shih,[1] and in +1118 Lin Ling-Su[2] obtained the title Thung-Chen Ta-Ling hsien-sêng[3] while Chang Hsü-Pai[4] became Thung-Yuan Chhung-Miao hsien-sêng.[5] By +1119 the Taoists were even undertaking the absorption of Buddhism, the patriarchs being turned into Taoist immortals by suitable changes of title, and many temples with their monks and nuns receiving new names of Taoist flavour.[a] All this should be viewed in conjunction with the background of interest in natural phenomena and mechanical invention which led us at an earlier stage to call Hui Tsung's court an 'entourage of virtuosi'.[b]

At this time we meet once again with a woman alchemist, whose story is worth giving as a picture from the life of medieval China. Though not successful like Teacher Kêng (p. 169 above), great strength of will evidently lay behind her lonely life, and the mention of quantitative measurements hints at a beginning of true science.[c] It is inspiring to realise that women have participated with men in the growth of the sciences from the beginning, from Mary the Jewess onwards, as they do in experimental research today, though as yet we do not know all the details of what in the past they contributed. The following account is a first-hand one from the pen of Hsü Yen-Chou[6] whose book on poetry was written about +1111. In this he says:[d]

There was a certain woman named Li Shao-Yün[7] who came of a scholarly family (and was happily married) but her husband died, and having no children she left home, put on Taoist robes, and wandered about (from one temple to another) in the region between the Huai and the Yangtze. I often used to meet her in (and around) Chinling (Nanking). She (was fond of) writing poems, one of which, (I remember) ran:

'The wind turns over the snow among the willow catkins
And there are rosy clouds of peach petals on the water below.'

She used no feminine arts of paint and powder—what really interested her was the trans-ᶜormation of cinnabar. (At one time) I obtained (some of) her procedures from her, and found that in general they were like those of Wei Po-Yang, only with much more detail and with precise statements of weights and measures. Once she said to me, 'My destiny is insufficient, I fear I shall never be able to bring this elixir to completion.' Two years later I saw her again, and she had become so thin that the bones were showing; indeed Shao-Yün was ill, and yet the elixir was not achieved. I said, jokingly, 'Surely your elixir is done; are you not about to become an immortal? You are so thin that a crane bird could come and

---

[a] Late Taoist hagiography includes accounts not only of Buddha, but also of Jesus and Muḥammad, telling how they each in their several ways attained to the Tao. See, for example, *Shen Hsien Thung Chien*, ch. 5, section 1, pp. 7*b* to 9*b*, and ch. 9, section 2, pp. 4*b* to 5*b*.
[b] Vol. 4, pt. 2, pp. 501ff.; cf. Needham, Wang & Price (1). It invites comparison with the courts of Alfonso X of Castile (+1252 to +1280), and of Rudolf II at Prague (+1576 to +1612), on which we now have the illuminating book of Evans (1). Our own Charles II, founder of the Royal Society, could certainly also be thought of in this connection.
[c] Cf. Vol. 3, pp. 150ff. above.
[d] *Hsü Yen-Chou Shih Hua*, ch. 1, p. 4*b*, tr. auct.

[1] 冲隱處士　　　　　[2] 林靈素　　　[3] 通眞達靈先生　　　[4] 張虛白
[5] 通元冲妙先生　　　[6] 許彥周　　　李少雲

carry you away.' She smiled and said, 'How can you have the heart to make fun of me?' While she was ill she had made a poem on the winter apricot, saying,

> 'Purity, beauty, brightness in the snow,
> Purity, fragrance, in the morning breeze,
> But pitiful too, the girl who wanders alone,
> Lost in the mazes of the mountain mists.'

Not long after, she died. Later I carefully studied her alchemical books. I read the accounts of the procedures, and I found this poem, therefore I record it here.

Some of the great Sung scholars and civil servants were also acquainted with the art of alchemy. Fan Chung-Yen[1] (+989 to +1052) had a book of secret formulae for converting mercury into silver which had been given to him by a dying classmate at the Imperial University who had at the time a very young son. However, Fan never made any use of it, and returned it to the boy when the latter became old enough.[a] Hu Su[2] acquired from a certain Buddhist monk the art of turning tiles and stones into gold, yet never put it into practice.[b] A striking story is told of the austere Chang Yung,[3] who bore rule in Szechuan at the time when Wang Chieh was so prominent at court.[c] In his *Tung Hsien Pi Lu*[4] (Jottings from the Eastern Side-Hall), written towards the end of the +11th century, Wei Thai[5] says:[d]

After the rebellion of Wang (Hsiao-Po[6]) and Li Shun[7] (+995),[e] most of the officials posted to Szechuan did not bring along their families, and even at the present time this regulation still holds good. When Chang Yung was made Governor of I-chou he went there on his own. At that time everyone was afraid of Chang's severity, so nobody dared to keep slave-girls or other women servants. However, not wishing to strain human nature too far, he himself (set the example by) buying a slave-girl to look after him; consequently all the other officials gradually acquired maids and concubines. After Chang had been there four years he was recalled to the capital, so he summoned the parents of the slave-girl and bestowing money and property on her arranged a suitable marriage. It was then found that she was still a virgin.

One day when Chang was in Szechuan he was visited by an alchemist (*shu shih*[8]) who said that he was able to transmute mercury into white metal (i.e. silver). Chang asked him whether he could transform 100 oz at one time, and he replied that he could. So Chang sent out to buy this amount, and the alchemist duly transformed it in a single heating without any loss. Chang Yung sighed and said: 'Indeed you have reached the pinnacle of art. However such things should not be used in private families.' So he immediately called in metal-workers, instructing them to cast a large (incense-)burner with it, having the inscription on its belly 'Presented to the Temple of Great Loving-Kindness for public use.' After the vessel had been sent to the temple, Chang entertained the alchemist with wine and sent him on his way. Thus Chang's upright character was admired by everyone.

---

[a] The story is found in many places, e.g. *Lu Huo Chien Chieh Lu*, p. 3*b*; *Hou Tê Lu*, ch. 2, p. 16*b*; *Yün Chai Kuang Lu*, cit. in *Lei Shuo*, ch. 18, p. 26*b*; and *Sun Kung Than Phu*, ch. 2, p. 11*b*. The last of these is the oldest, dating from not much more than thirty years after Fan Chung-Yen's death.
[b] *Lu Huo Chien Chieh Lu*, p. 3*b*.
[c] Chang Yung's biography will be found in *Sung Shih*, ch. 307, pp. 3*b*ff.
[d] Ch. 10, p. 2*b*, tr. auct.   [e] Cf. Vol. 4, pt. 2, p. 23.

[1] 范仲淹   [2] 胡宿   [3] 張雍   [4] 東軒筆錄   [5] 魏泰
[6] 王小波   [7] 李順   [8] 術士

The great poet Su Shih[1] (Su Tung-Pho,[2] +1036 to +1101) undoubtedly had a considerable knowledge of alchemy. This is seen in the book of his friend Ho Wei[3] written about +1095, *Chhun Chu Chi Wên*[4] (Record of Things Heard at Spring Island).[a] A whole chapter (ch. 10) of this is devoted to accounts of successful chemical and metallurgical experiments which Ho Wei and Su Tung-Pho had seen or heard of, mostly concerned with aurifaction, however, rather than the making of elixirs. That some of these were very valid processes for gold- and silver-like alloys we have seen in a quotation already given (pt. 2, p. 233) where cupro-nickel was almost certainly produced. Evidence of Su Tung-Pho's interest in iatro-chemical operations is to be found in the book of pharmacal procedures which bears the name *Su Shen Liang Fang*[5] (Beneficial Prescriptions collected by Su Tung-Pho and Shen Kua), published about +1120. This was not the result of direct collaboration between the two great men, but a conflation of their writings under the supervision of the Taoist Lin Ling-Su.[6]

In his *Lu Huo Chien Chieh Lu*,[7] Yü Yen[8] (c. +1285) says:[b]

Su Tung-Pho was conversant with aurifaction and argentifaction, and demonstrated his skill to his (elderly) friend Chhen Hsi-Liang,[9] a great enthusiast for the art.[c] This was expressly recorded by his brother, Su Chhê,[10] in the book *Lung Chhuan Lüeh Chih*.[11] He said in a letter to a friend, Wang Ting-Kuo: 'I have recently been given some cinnabar elixir which shows a very remarkable colour, but I do not dare to eat it. I only admire its brilliance.'

When Su Tung-Pho first took up appointment in the civil service at the age of 26 at Chhi-hsia[12] a strange monk obliged him to accept a formula for the transmutation of gold. After receiving it he sealed it (in a container). Later he passed it to his younger brother (Su) Ying-Pin,[13] who kept it (likewise). However, when Ying-Pin later went to live at Wu-chhang,[14] some old friend or relative came to know of this and called upon him to enquire about the art. Ying-Pin said 'I have, it is true, kept the formula carefully for many years since my late brother Tung-Pho gave it to me, and I will try to find it as soon as I have time.' After a long while he sent for the enquirer and showed him the container with the seal made by Su Tung-Pho himself still intact. Then he immediately threw it into the fire in a stove, saying: 'Poverty is something which man must learn to endure. Why go so far beyond the bounds as to want to try to make (artificial gold)?' So the enquirer left in shame, knowing not where to hide himself.

Another story, from a contemporary source, again shows Su Tung-Pho in relation with a Buddhist monk who practised alchemy. In his *Sun Kung Than Phu*[15] (The Venerable Mr Sun's Conversation Garden), written about +1085, Sun Shêng[16] says:[d]

When (Su) Tzu-Chan[17] (Su Tung-Pho) was an official in Fêng-hsiang the local Prefect, Chhen Chung-Liang,[18] was an admirer of the 'art of the yellow and the white'.[e] Now in the district there was a spagyrical monk who gave the impression of being quite out of the ordinary. (Chhen) Chung-Liang many times pressed him to tell his secrets, but he always

[a] Cf. Lin Yu-Thang (5), pp. 212ff., 353, 359.   [b] Pp. 7b, 8a, 9a, tr. auct.
[c] D. c. +1067; admired for his uprightness, however.   [d] Ch. 2, p. 1a, tr. auct.
[e] Possibly a younger brother or cousin of the Chhen Hsi-Liang just mentioned. But there are versions of the story which name the latter instead (e.g. *Wu Li Hsiao Shih*, ch. 7, p. 2a, b).

[1] 蘇軾   [2] 蘇東坡   [3] 何薳   [4] 春渚紀聞   [5] 蘇沈良方
[6] 林靈素   [7] 爐火監戒錄   [8] 俞琰   [9] 陳希亮   [10] 蘇轍
[11] 龍川略志   [12] 岐下   [13] 蘇顥（潁）濱   [14] 武昌   [15] 孫公談圃
[16] 孫升   [17] 蘇子瞻   [18] 陳仲亮

made excuses, avoiding the Prefect or refusing to come out. But (Su) Tzu-Chan found an opportunity to visit the temple, and opening a door, found the monk inside, so he asked him what it was all about. The monk replied: '(Chhen) Chung-Liang is a covetous man and therefore is not worthy to be taught the art', but he was willing to impart his knowledge to (Su) Tzu-Chan. His procedure was to take 1 oz of gold and 1/10th of an oz of cinnabar and heat them together, then very soon the mixture turned into purple gold worth many times the original (ingredients) in value. Afterwards (Su) Tzu-Chan went away and told (Chhen) Chung-Liang, who called the monk before him and had the matter verified; it turned out just as he said. Then (Chhen) Chung-Liang made a lot of (the purple gold) and built himself a residence (out of the proceeds). But soon his official career came to a disastrous end, and it was not long before he died.

Once again, then, we have the curious parallel to *iōsis*, and a tale of the retribution that overtook a man whose attitude to alchemy was dictated by the desire for wealth and riches. Yet the overt content was only a gold amalgam with mercury in low proportion.[a] Su Chhê (Su Ying-Pin, +1039 to +1112), the younger brother of Su Tung-Pho, and author of the *Lung Chhuan Lüeh Chih* as well as a commentary on the *Tao Tê Ching*, also knew something of alchemy, though declining, as we have seen, to use it. According to the *Lu Huo Chien Chieh Lu*:[b]

Su Tzu-Yu[1] (i.e. Su Chhê) once intended to practise alchemy. He set aside a room closed to the outside world, and installed a stove in it. When he was about to light the fire for the first time he saw a large cat approach the stove and piss against it; then the animal disappeared. Tzu-Yu gave up. He said: 'The art of the adepts is meant by Heaven for the relief of the poor. This art should be passed on to the right persons, but I am not one of them.' After that he never spoke about the art again.

Evidently this was a bad omen, but in general stories of the kind we have given, of which there are many more, illustrate typical Confucian moralistic attitudes towards alchemy.

Attitudes of caution and restraint were by no means confined to the Confucians, as we can see from the study of the Taoist adepts of the +10th and +11th centuries. The danger of elixir poisoning may well have been the main reason why the adept Chhen Thuan[2] gave negative advice to two emperors on two different occasions, first Hsi Tsung of the Later Chou dynasty in +956, and then Thai Tsung of the Sung between the years +976 and +984. Like Holan Hsi-Chen, Chhen Thuan said that they should not worry about elixirs but direct their minds to improving the administration of the State. The *Sung Shih* says that Chhen Thuan wrote a book called *Chih Hsüan Phien*[3] (On the Demonstration of the Mystery), which dealt with the techniques of nourishing life and of the cyclically-transformed elixir.[c] It was said that before

---

[a] On purple gold cf. pt. 2, pp. 257ff. above.

[b] Pp. 8b, 9a. The same story appears in other, more contemporary, sources, notably *Sun Kung Than Phu*, ch. 3, p. 1a.

[c] *Sung Shih*, ch. 457, p. 3b. We have already had a good deal to say about Chhen Thuan's role in the history of Chinese philosophy, especially in the interpretation of the Book of Changes, the fixation of the Ho Thu and Lo Shu diagrams, and the origins of the Thai Chi Thu (Vol. 2, pp. 442ff., 467, Vol. 4, pt. 1, p. 296). See on him *I Thu Ming Pien*,[4] ch. 1, pp. 2bff.

¹ 蘇子由   ² 陳摶   ³ 指玄篇   ⁴ 易圖明辨

Chhen Thuan died in +989 he passed his arts to Chang Wu-Mêng[1] (also called Chang Ling-Yin[2] and Hung Mêng Tzu[3]), who practised the art of immortality but never spoke about aurification and argentifaction though he knew how to do it.[a]

This may be the place to record that the adepts of medieval Chinese alchemy had a penchant, like their Arabic and European counterparts, for enigmatical epigrams analogous to the *Tabula Smaragdina*[b] and the gnomic sentences of natural philosophy embedded in the Hellenistic Corpus.[c] One can find a good example in the book of Hsü Yen-Chou[4] on poetry, written about +1111. There he says:[d]

In a Taoist temple in Szechuan province (not long ago) a stele was discovered in the ground when a well was being dug. There was an inscription on it resembling a rhapsodic ode (*fu*[5]) or an eulogy (*tsan*[6]) which read as follows:

> 'There is a thing, which contains another thing,
> It can be augmented, it can be prolonged,
> It must be plucked before it is gnawed by silk-worms
> And used after being transformed by fire.
> Thang the Completer showers down from above,
> Khua-Fu[7] being empty can receive his fill.
> The *chhi* responds to the light of the morning.
> The process accords with the night clepsydra.
> White flowers accumulate, putting the snow to shame,
> Yellow flakes solidify, surpassing gold in beauty.
> The cyclical process continuously goes on,
> Now there is rapid steaming and gassing,
> Now there is drastic solidifying shrinkage.
> What is it that appears, gold or jade?
> It brings longevity, eternal as the heavens.
> All this must never be recorded in writing,
> Only oral instruction can transmit it.'

Afterwards a certain recluse averred that this was what Yin Chen Jen[8] had written in the Han period on the alchemical process.[e] Subsequently it was copied on the Stele of Tzu-Yü[9].[f] I myself regret that I cannot understand these words, so I simply record them here.

Perhaps Hsü Yen-Chou was right to be a bit puzzled. By way of elucidation it is clear that the first two lines might refer to the preparation of chemical substances from minerals (e.g. mercury from cinnabar), or sal ammoniac from animal substances;[g] and to the artificial prolongation of the human life-span. Perhaps the third line

---

[a] See *Thien-thai Shan Fang Wai Chih*,[10] ch. 9, p. 15a.
[b] Cf. Steele & Singer (1); Holmyard (1), p. 95; and pt. 4 below.
[c] Cf. pt. 4 also.   [d] *Hsü Yen-Chou Shih Hua*, ch. 1, p. 34a, tr. auct.
[e] Presumably Yin Chhang-Shêng,[11] the teacher of Wei Po-Yang; see p. 77 above.
[f] Perhaps this has something to do with Yü Pei Tzu[12] (the Jade-Stele Master), whose name appears in the title of one of the books in the *Tao Tsang* (*TT*920), unknown in date and authorship. It consists of dialogues between Chêng Yin and Ko Hung, so it must be later than the Chin period.
[g] Cf. pt. 4 below.

[1] 張無夢   [2] 張靈隱   [3] 鴻濛子   [4] 許彦周   [5] 賦
[6] 讚   [7] 夸父   [8] 陰眞人   [9] 子玉碑   [10] 天台山方外志
[11] 陰長生   [12] 玉碑子

has to do with catching the right point in the natural development of minerals in the earth,[a] and the fourth could allude to the artificial acceleration of such processes by the Chinese *philosophi per ignem*. Next comes the legendary emperor Chhêng Thang,[1] who established a bond with his great minister, I Yin[2],[b] in a ceremony like that of marriage which included a lustration or aspersion (*fei*,[3] *fo*[3]),[c] hence a symbol here for the descending streams of reflux condensation or *chhi* circulation.[d] He is followed immediately by a mythological being, Khua-Fu the Boaster, who wanted to eat the sun and for whom all the rivers did not suffice for his drink;[e] this spherical 'belly-image' must thus symbolise the round reaction-vessel, aludel or still, whether real or figurative. The next lines (seven and eight) appear to concern the 'fire-times' (*huo hou*[4]) in a cycle of heating which runs continuously through many days and nights.[f] Lines eight and nine hint at the colours appearing in chemical operations, the white of sublimates or crystals deposited from solutions, the yellow of gold-like alloys or mosaic gold.[g] There follows a reference to the cyclical processes so often described in *huan tan*[5] methods; and then at lines twelve and thirteen we can recognise under the old terms *shen*[6] and *kuei*[7] a Neo-Confucian contrast between expansion-dispersion and condensation-aggregation.[h] At the end the goal of the entire process is plainly revealed, and the usual adjuration to oral transmission (*khou chüeh*[8]) completes the epigram. And yet the entire inscription would be interpretable purely in terms of physiological alchemy, without any reference to laboratory operations at all. Let us now turn to examine some of the most important books on alchemy which were produced during the Sung period.

(viii) *Alchemical compendia and books with illustrations*

Chang Wu-Mêng, mentioned a page or two above, was the teacher of the voluminous Taoist writer Chhen Ching-Yuan[9] (*fl.* early +11th century, also called Chhen Ta Shih[10] and Pi Hsü Tzu[11]).[i] The *Tao Tsang* contains seven treatises by this Chhen, among them one on alchemy called *Pi Yü Chu Sha Han Lin Yü Shu Kuei*[12] (Cold Forest Jade-Cinnabar Casing Process).[j] This book refers to the use of a number of chemicals such as borax, ammonium chloride, copper sulphate, saltpetre, sodium chloride, sulphur, etc., in elixir formulae, with lead and mercury or cinnabar alone as the principal ingredients. The uses of gold, organic substances like honey, and various plants are also mentioned.

The *Yün Chi Chhi Chhien*[13] (Seven Tablets of the Cloudy Satchel)[k] compiled under Chen Tsung by Chang Chün-Fang[14] between +998 and +1022 is a miniature *Tao Tsang* in itself; we have had occasion to quote it often. It reproduces a number of

---

[a] Cf. Vol. 3, p. 640, and Vol. 5, pt. 4 below.　[b] Mayers (1), no. 233.　[c] See Granet (1), pp. 418ff.
[d] Alternatively this line could refer simply to the placing of the substances or solutions in the reaction-vessel. But the comparison of reflux condensation within it is as old as the *Tshan Thung Chhi* itself, cf. ch. 32, p. 5a, tr. Wu & Davis (1), p. 259.
[e] See Granet (1), pp. 361ff.　[f] Cf. pp. 60ff. above, and pt. 4 below.　[g] Cf. p. 103 above.
[h] Vol. 2, p. 490.　[i] See Wieger (6), p. 328.　[j] *TT*891.　[k] *TT*1020.

[1] 成湯　　[2] 伊尹　　[3] 祓　　　　[4] 火候　　　　[5] 還丹
[6] 神　　　[7] 鬼　　　[8] 口訣　　　[9] 陳景元　　　[10] 陳大師
[11] 碧虛子　　　[12] 碧玉朱砂寒林玉樹匱　　[13] 雲笈七籤　　[14] 張君房

alchemical texts that are now no longer extant elsewhere. Among these are Sun Ssu-Mo's *Thai-Chhing Tan Ching Yao Chüeh*, and the *Chin Hua Yü Nü Shuo Tan Ching*[1] (Elixir Manual according to the Teaching of the Jade Girl on the Golden Flower). This consists of dialogues between Huang Ti and Yü Nü or Hsüan Nü on the principles of chemical reactions and changes.[a] While Chang Chün-Fang was doing his compiling work there appeared the *Tan Fang Ao Lun*[2] (Subtle Discourse on the Alchemical Elaboratory),[b] written by Chhêng Liao-I[3] *c.* +1020. It seems to speak about amalgamation, stressing the importance of distinguishing genuine materials from false in elixir-making, and the proper control of temperature; but it is really a treatise on physiological alchemy, hiding the interactions of organs and the regaining of the primary vitalities of youth under the terminology of proto-chemistry. The *Kêng Tao Chi*[4] (Collection of Procedures of the Golden Art),[c] consisting of nine chapters and over 60,000 words, is the most voluminous Sung (or even post-Sung) alchemical treatise in the *Tao Tsang*. As for its date we only know that it was compiled after the year +1144. Yuan Han-Chhing rightly suggests that such detailed descriptions as are given in the text could only have been written by one who had carried out the experiments himself, and that from the style of writing the book must have been written by more than one such person.[d] It will be remembered that this was just after the time (+1117) when Lu Thien-Chi wrote his commentary on the theories of the *Tshan Thung Chhi Wu Hsiang Lei* (cf. p. 145).[e]

Another important compendium probably put together during the Sung is the *Chu Chia Shen Phin Tan Fa*[5] (Methods of the Various Schools for Magical Elixir Preparations).[f] There is nothing in the book to indicate the name of its author or compiler, nor the date when it was written. In the second of its three chapters, however, one finds the name Mêng Yao-Fu,[6] also called Hsüan Chen Tzu.[7] Yuan Han-Chhing suggests that Mêng wrote one of the constituent parts of the book, and that the date of the compilation could not have been earlier than the beginning of the Sung (+960) nor later than its end (+1280).[g] The book contains many formulae for the transmutation of base metals and minerals into gold, and claims that most of the products when taken orally would enable one to attain *hsien*-ship. The collections mentioned in these two paragraphs are likely to yield important results in future research. As the proverb says: 'He who enters the mountain of treasures will never come back empty-handed.'[h]

---

[a] On the Jade Girl and the Mysterious Girl as interlocutors see Vol. 2, p. 147 above, and pt. 5 below.
[b] *TT*913.   [c] *TT*946.
[d] Yuan Han-Chhing (1), p. 199.
[e] It may be interjected here that interest in alchemy, proto-chemical as well as physiological, continued to flow strongly among scholarly amateurs during this period of the Southern Sung. For example, many of the poems of the eminent writer Lu Yu[8] (+1125 to +1209) were devoted to alchemy, as may be seen in the recent study with translations by Ho Ping-Yü, Ko Thien-Chi & Lim (1).
[f] *TT*911.
[g] Yuan Han-Chhing (1), p. 199.   [h] *Ju pao shan shou pu khung hui.*[9]

[1] 金華玉女說丹經   [2] 丹房奧論   [3] 程了一   [4] 庚道集
[5] 諸家神品丹法   [6] 孟要甫   [7] 女眞子   [8] 陸游
[9] 入寶山手不空回

The *Ling Sha Ta Tan Pi Chüeh*[1] (Secret Doctrine of the Numinous Cinnabar and the Great Elixir),[a] by an anonymous writer, is regarded by Chhen Kuo-Fu as also of Sung origin.[b] The preface says that the secret of success was to practise simultaneously elixir-making (*wai tan*) and respiratory exercises (*nei tan*).[c] Besides dealing with the cyclical transformation of sulphur and mercury, the author gives a method of manufacturing 'gold' from substances like realgar, orpiment and cinnabar. It is interesting to observe that the weight of 'gold' derived in the latter procedure far exceeds the total weight of the raw material used, contrary to the fundamental laws of chemistry (unless he was bringing in other metals or minerals which he forgot to mention).

The *Hsiu Lien Ta Tan Yao Chih*[2] (Essential Hints on the Preparation of Powerful Elixirs),[d] again by an anonymous author, was written before the +14th century but after the time of Lü Tung-Pin,[3] so it is reasonably considered by Yuan Han-Chhing as of Sung origin.[e] A special feature of this book lies in the conservative amounts of ingredients in its alchemical formulae as compared to the lavish quantities employed in well-known texts like the *Pao Phu Tzu*, the *Thai-Chhing Tan Ching Yao Chüeh* and the *Thai-Chhing Shih Pi Chi*. One of its elixir recipes, for example, requires the use of 2 oz of purified calomel, 1 oz of arsenious oxide, two-tenths of an oz of borax, one-tenth of an oz of sal ammoniac, three-tenths of an oz of *ju hsiang*[4],[f] and one-tenth of an oz of saltpetre. This is much more plausible than the almost industrial specifications of the Chin and Thang periods; one wonders whether it is not a significant index of increasing refinement of the available apparatus. Yet another text by an anonymous writer probably of the Sung period is the *Kan Chhi Shih-liu Chuan Chin Tan*[5] (The Sixteen-fold Cyclically-Transformed Gold Elixir prepared by the 'Responding to the Chhi' Method).[g] This is a rather small tractate dealing with the transformation of mercury into 'gold'. It illustrates several pieces of alchemical laboratory equipment, such as the aludel (or closed reaction-vessel) and the platform for the stove.

One of the great characteristics of Sung alchemical texts is the presence of many illustrations of alchemical apparatus. A number of such pictures and diagrams can be found in Wu Wu's[6] *Tan Fang Hsü Chih*[7] (Chymical Elaboratory Practice)[h] dated +1163. They include a still for distilling mercury, pestles and mortars, and the stove platform.[i] Wu Wu also wrote another text called *Chih Kuei Chi*[8] (Pointing the Way

---

[a] *TT*890.

[b] See Chhen Kuo-Fu (*1*), p. 391. The text was received by Chang Shih-Chung[9] in +1101.

[c] Cf. Vol. 2, pp. 143 ff.; Vol. 4, pt. 3, p. 674, and further in pt. 5 below.

[d] *TT*905.　　　　　　　　　　　[e] (*1*), p. 199.

[f] This is a tree of the *Pistacia* genus (Anacardiaceae, like the mango), either the Bombay mastic or terebinth tree *P. Khinjuk* (= *Terebinthus*), R313, or the Mediterranean species *P. lentiscus*, CC838. See Burkill (*1*), vol. 2, p. 1756. All the species are very resinous and aromatic, containing also, of course, tannins, essential oils and steroids. Presumably the nuts were used in the formula given.

[g] *TT*904.　　　　　　　　　　　[h] *TT*893.

[i] This book also gives much curious information on the choice of companions, the proper location of a laboratory, water-sources, furnace clays, and exorcisms against ghosts and malevolent spirits.

---

[1] 靈砂大丹祕訣　　　[2] 修鍊大丹要旨　　　　　　　[3] 呂洞賓　　　[4] 乳香
[5] 感氣十六轉金丹　　[6] 吳悮　　　[7] 丹房須知　　　[8] 指歸集　　　[9] 張侍中

Home to Life Eternal)[a] in which he expressly advocated the elixir, and said that respiratory exercises (*nei tan*) only served the purpose of maintaining one's health to enable the experiments to be accomplished. The most interesting Sung alchemical text from the point of view of equipment is the *Chin Hua Chhung-Pi Tan Ching Pi Chih*[1] (Confidential Hints on the Manual of the Heaven-Piercing Golden Flower Elixir)[b] written in +1225. The book consists of two parts, the first containing material transmitted to Phêng Ssu[2] by his master Pai Yü-Chhan,[3] while that in the second came from Lan Yuan-Lao[4] to Mêng Hsü[5].[c] Illustrations found in this book include reaction-vessels, stoves, an ambix called the 'pomegranate' vessel (*tzhu shih liu kuan*[6]) used in descensory distillation, and an elaborate series of different types of cooling system or water-jackets.

All this will of course be discussed in its place (pt. 4 below). Unfortunately some of the best illustrated texts present difficulties in dating. An example of this is the *Thai Chi Chen-Jen Tsa Tan Yao Fang*[7] (Tractate of the Supreme-Pole Adept on Miscellaneous Elixir Recipes),[d] which has diagrams of furnaces and aludels. We have placed this in the Sung mainly on account of the philosophical significance of the pseudonym,[e] though it was given (at dates uncertain) to at least three adepts of various historical periods, one in semi-legendary antiquity,[f] and one at Han Wu Ti's time;[g] but in content the material could easily be Thang or even Chin. The same applies to another book, the *Thai Chi Chen-Jen Chiu Chuan Huan Tan Ching Yao Chüeh*[8] (Essential Teachings of the Manual of the Supreme-Pole Adept on the Ninefold Cyclically-Transformed Elixir).[h] Without the last two characters, the Manual itself must be pre-Sui because it is listed in the *Sui Shu* bibliography,[i] and again the contents could reasonably be of Chin or Liang date. Minute directions are given for the firing cycle of an aludel containing all the five minerals (cf. p. 96) together with quartz and mercury (cf. Fig. 1359); this recalls the firing cycles adumbrated in the *Tshan Thung Chhi* (pp. 60 ff.). The process ends with projections for gold and silver which seem to imply the formation of alloys of copper, arsenic, mercury and lead. In its combination of projection with elixir-making this text has distinct resemblances to what was done in Ko Hung's time, and might in essence really be as old as that.[j]

a *TT*914.
b *TT*907.
c Mêng Hsü also edited and prefaced the whole. The second part is much longer than the first, and more clearly concerns practical laboratory alchemy. Lan Yuan-Lao was thought to be an avatar of Pai Yü-Chhan, and the title of the whole was taken from the name of his alchemical elaboratory.
d *TT*939, writer unknown.
e Cf. Vol. 2, pp. 459 ff.
f Tu Chhung,[9] ascribed to the −10th century, according to *TT*293, ch. 9, pp. 1aff. One wonders if there could have been any connection with Tu Chung,[10] also of semi-legendary times, who gave his name to the tree *Eucommia ulmoides* and will be met in the botany Section in Vol. 6.
g Wang Than,[11] −198 to −123, according to *TT*293, ch. 9, pp. 11bff. Little else is known of him.
h *TT*882, writer unknown. There is a preliminary study by Ho Ping-Yü (9).
i *Sui Shu*, ch. 34, p. 33b.
j A very similar process is described in *TT*873.

¹ 金華沖碧丹經祕旨   ² 彭耜   ³ 白玉蟾   ⁴ 闚元老   ⁵ 孟煦
⁶ 磁石榴罐   ⁷ 太極眞人雜丹藥方   ⁸ 太極眞人九轉還丹經要訣
⁹ 杜沖   ¹⁰ 杜仲   ¹¹ 王探

(ix) *The Northern and Southern Schools of Taoism*

While we can leave the laboratory equipment of the alchemist for the moment, we can hardly proceed further without noting the development of two important Taoist groups during the +11th and the +12th centuries, the Southern School (Nan Tsung[1]) and the Northern School (Pei Tsung[2]). Many of the later Taoist alchemical writings can be traced to the first of these,[a] and the *Chin Hua Chhung-Pi Tan Ching Pi Chih* is one of them. The founder of the Southern School, Liu Tshao[3],[b] flourished early in the +11th century, claiming himself a friend of the adepts Chhen Thuan,[c] Chhung Fang and Chang Wu-Mêng, and a disciple in the art of immortality of the two teachers Chêng Yang Tzu[4] and Lü Tung-Pin.[d] The reputation of Liu Tshao was surpassed by his disciple Chang Po-Tuan[5] (d. +1082),[e] the author of the *Wu Chen Phien*[6] (Poetical Essay on Realising the Necessity of Regenerating the Primary Vitalities),[f] the *Chin Tan Ssu-Pai Tzu*[7] (Four Hundred Word Epitome of the Metallous Enchymoma)[g] and the *Yü-Chhing Chin-Ssu Chhing-Hua Pi-Wên Chin-Pao Nei-Lien Tan Chüeh*[8] (The Green-and-Elegant Secret Papers in the Jade-Purity Golden Box on the Essentials of the Internal Refining of the Enchymoma, the Golden Treasure).[h] All these three texts have been translated into English and studied by Davis & Chao Yün-Tshung.[i]

Chang Po-Tuan's poetical style of writing long concealed from Western scholars the fact that he was talking almost exclusively about physiological alchemy, not proto-chemical operations at all. Nevertheless the *Wu Chen Phien* had been considered by Taoists in later times almost as important as the *Tshan Thung Chhi* itself. The book is cast in poetic verse form and divided into three chapters, which are again sub-divided into short paragraphs, or stanzas, ninety-nine in all. It opens with a call to the unworldly life:[j]

If (one) does not seek for the great Tao, leaving the paths of error, how can (one be called) wise even though highly talented? A hundred years of living is (as transient as) a spark (struck) from a stone, and the course of life (is like) a bubble floating on water. (Those who) only think of profits and emoluments, seeking worldly prosperity and glory, will soon find their faces turning pale and their bodies withering. Even if they have piled up riches mountain-high, may I be allowed to ask whether they can buy off the Messenger of Death (*wu chhang*[9]), (with their wealth)?

a On this school see Liu Tshun-Jen (1, 2). It was devoted to 'dual-cultivation' (cf. pt. 5).
b Also called Liu Tsung-Chhêng,[10] Hai Chhan Tzu,[11] Hai-Chhan Ti Chün[12] and Hsüan-Ying Yen Shan Jen.[13]
c Cf. p. 194 above.   d Cf. p. 147 above.
e Also called Chang Phing-Shu,[14] Chang Tzu-Yang[15] and Tzu-Yang Chen Jen.[16]
f *TT*138. See pt. 5 below.
g *TT*1067. See pt. 5 below.   h *TT*237.
i (2, 5, 7). They also translated three alchemical poems by this writer (3). Unfortunately, they were under the misapprehension that these texts concerned proto-chemical alchemy.
j *Wu Chen Phien*, stanza 1 (ch. 1, p. 5b), tr. auct., adjuv. Davis & Chao Yün-Tshung (7).

¹ 南宗	² 北宗	³ 劉操	⁴ 正陽子	⁵ 張伯端
⁶ 悟眞篇	⁷ 金丹四百字	⁸ 玉清金笥青華祕文金寶內鍊丹訣		⁹ 無常
¹⁰ 劉宗成	¹¹ 海蟾子	¹² 海蟾帝君	¹³ 玄英燕山人	¹⁴ 張平叔
¹⁵ 張紫陽	¹⁶ 紫陽眞人			

A typical passage clearly shows that Chang Po-Tuan's conception of alchemy was that of the psycho-physiological exercises rather than the work of the laboratory with its minerals, metals and drugs.[a]

Compound not the three yellow (substances, *san huang*[1]), (i.e. sulphur, orpiment and realgar), neither the four magical (things, *ssu shen*[2]), (alum, cinnabar, lead, and mercury); and make no search for particular plants (as elixir ingredients), for they are still further from the genuine (medicine). When Yin and Yang things are of the same category, they will resonate and fall spontaneously into conjunction (*Yin Yang tê lei kuei chiao kan*[3]). . . . People in this world should understand that true lead and true mercury are not at all the same things as common cinnabar and common quicksilver.[b]

Here Chang Po-Tuan evidently has partly in mind the category theory which developed so much during the Thang period (cf. p. 145 above and pt. 4 below), emphasising that reactions occur only between substances having properties of similitude, yet differing in Yin–Yang status. This idea is again reflected in another passage:[c]

When a bamboo (utensil) is broken it has to be repaired by using bamboo. For a hen to hatch chickens, eggs are required. You may bring together ten thousand substances, but if they are of disparate categories (*fei lei*[4]) all your labour will be wasted. One must strive for that union (within oneself) which, by a marvellous natural mechanism, will produce true lead.[d]

One remembers that eloquent passage of Wei Po-Yang (p. 69 above).

Although translators have so far numbered Chang Po-Tuan's works among those on experimental (or at least theoretical) elixir alchemy, they are in fact far more concerned with the enchymoma. Of course, writings of this kind were meant, we suspect, to be susceptible of several parallel and distinct interpretations—proto-chemical, mutational, meditational, respiratory, sexual, etc., in accordance with the interests, knowledge and mood of the reader; who was probably expected to practise them all. For example, a reference to the numbers 2 and 8 in one of the stanzas might be taken to be concerned either with the *kua* of the 'Book of Changes' or with the timing of a heating cycle, or with the rhythm of respiratory techniques, or (when multiplied) to the proper age of the girl ideally fitted to be the adept's partner in the practice of the sexual arts leading to immortality.

The descent of the Southern School is shown in Table 113.[e] Chang Po-Tuan's

---

[a] *Ibid.*, stanza 8 (ch. 1, p. 9*b*), tr. auct., adjuv. Davis & Chao Yün-Tshung (7).

[b] At first sight this sentence calls to mind the 'sophic mercury' and 'sophic sulphur' of Western alchemy, but in fact there is no parallel at all, as we shall explain in the sub-section on physiological alchemy, where we shall return to this same text and give a fuller exegesis of it.

[c] *Wu Chen Phien*, stanza 25 (ch. 2, p. 2*b*), tr. auct.; adjuv. Davis & Chao Yün-Tshung (7).

[d] The Adept Thao comments that the 'mother'-*chhi* of true lead and the 'child'-*chhi* of the semen are of the same category, hence the natural mechanism or reaction can go on. It is not possible to elucidate this further here, but our sub-section on physiological alchemy (pt. 5 below) will explain it at length. *Wu Chen Phien San Chu*, stanza 8 (ch. 2, p. 6*b*).

[e] After Chhen Kuo-Fu (*1*).

[1] 三黃   [2] 四神   [3] 陰陽得類歸交感   [4] 非類

Table 113. *The Southern School of Taoism, in Sung and Yuan* (+*10th to* +*14th century*)

FOUNDER Liu Tshao 劉操 (d. bef. +1050) —————Chhen Thuan 陳摶 (d. +989)
(personal name later changed to Liu Hsüan-Ying 劉玄英)
courtesy name: Liu Chao-Yuan 劉昭遠
(courtesy name later changed to Liu Tsung-Chhêng 劉宗成)
philosophical name: Hai Chhan Tzu 海蟾子

Chang Po-Tuan 張伯端 (d. +1082)
(personal name later changed to Chang Yüng-Chhêng 張用成)
courtesy name: Chang Phing-Shu 張平叔
temple name: Tzu-Yang Chen Jen 紫陽眞人

Shih Thai 石泰 (d. +1158)
courtesy name: Shih Tê-Chih 石得之
literary name: Shih Hsing-Lin 石杏林
philosophical name: Tshui Hsüan Tzu 翠玄子

Hsüeh Tzu-Hsien 薛紫賢 (d. +1191)
other personal names: Hsüeh Shih 薛式
                      Hsüeh Tao-Kuang 薛道光
                      Hsüeh Tao-Yuan 薛道源

Chhen Nan 陳楠 (d. +1213)
courtesy name: Chhen Nan-Mu 陳南木
literary name: Chhen Tshui-Hsü 陳翠虛
also called Chhen Ni-Wan 陳泥丸

Sha Chê-Hsü 沙蟄虛   Chü Chiu-Ssu 鞠九思   Pai Yü-Chhan 白玉蟾 (*fl.* +1209 to +1224) [Lan Yuan-Lao 蘭元老]
                                                    other name: Ko Chhang-Kêng 葛長庚
                        Chu Chü 朱橋 (d. +1242)   philosophical name: Hai Chhiung Tzu 海瓊子
                                                    temple name: Tzu-Chhing Chen Jen 紫清眞人
                                                    other appellations : Chhiung-Shan tao jen 瓊山道人
                                                                         Pin-An 蠙庵
                                                                         I san jen 夷散人
                                                                         Shen-Hsiao san li 紳宵散吏

Wang Chin-Chhan 王金蟾   Phêng Ssu 彭耜   Mêng Hsü 孟煦
                                      courtesy name: Phêng Chi-I 彭季益
Li Tao-Shun 李道純   appellation : Hao-Lin yin shih 鶴林隱士
  other name: Li Chhing-An 李清庵
  philosophical name: Ying Chhan Tzu 瑩蟾子   Hsiao Liao-Chen 蕭了眞

Miao Thai-Su 苗太素
  other name: Miao Shih-An 苗實庵

Wang Chih-Tao 王志道
  other name: Wang Chhêng-An 王誠庵

disciple, Shih Thai[1] (d. +1158),[a] has been thought responsible for two treatises in the *Tao Tsang*. These are the *Hsiu Chen Shih Shu*[2] (A Collection of Ten Books on the Regeneration of the Primary Vitalities)[b] and the *Huan Yuan Phien*[3] (Book of the Return to the Source),[c] but Shih Thai was not the editor of the first, and neither is concerned with laboratory alchemy. Shih Thai then imparted his knowledge to Hsüeh Tzu-Hsien[4],[d] who added a commentary to Chang Po-Tuan's *Wu Chen Phien* and wrote also the *Huan Tan Fu Ming Phien*[5] (Book on the Restoration of Life by the Cyclically-Transformed Elixir),[e] which again is not very alchemical in nature.[f] Hsüeh Tzu-Hsien became the teacher of Chhen Nan[6],[g] the author of the *Tshui Hsü Phien*[7] (Book of the Emerald Heaven),[h] purely concerned with Taoist religion. Chhen Ni-Wan then taught the famous adept Pai Yü-Chhan[8] (*fl.* +1209 to +1224),[i] a prolific writer of many Taoist canons. Thus all the writings of the Southern School after the time of Chang Po-Tuan seem to have had very little bearing on proto-chemical alchemy, the *Wu Chen Phien* being perhaps a partial exception. But then we suddenly come across the *Chin Hua Chhung-Pi Tan Ching Pi Chih*, that very important text on alchemical procedures and apparatus by two of Pai Yü-Chhan's disciples, Phêng Ssu[9] & Mêng Hsü[10] (see p. 199 above). This suggests that we can never say for certain that a particular adept was not a practical alchemist when all we know is that he only wrote on religious Taoism or physiological alchemy (*nei tan*). Such men were probably versed in all these fields, and probably practised both proto-chemical and physiological alchemy even though they might have some preference for one or the other.[j]

We must now turn our attention to the other important Taoist school, that of the North, founded by Wang Chung-Fu[11] (+1113 to +1170) who claimed that he had himself been a disciple of the famous Thang adept Lü Tung-Pin (p. 147 above). Like Liu Tshao, his origins are obscure. Known under a number of different names,[k] this mage contributed several tractates to the *Tao Tsang*, including the *Chhung-Yang Chhüan Chen Chi*[12] (Wang Chhung-Yang's Records of the Perfect-Truth School),[l]

[a] Also called Shih Hsing-Lin[13] and Tshui Hsüan Tzu.[14]
[b] *TT*260. We treat of this extensively in pt. 5 below.
[c] *TT*1077. It occurs also as ch. 2 of the *Hsiu Chen Shih Shu*, hence the mis-attribution.
[d] Also called Hsüeh Tao-Kuang.[15]   [e] *TT*1074.
[f] Accounts of Shih Hsing-Lin and Hsüeh Tao-Kuang are given in Davis & Chao Yün-Tshung (4). The latter has a special interest because he had previously been a Buddhist philosopher of note, with the monastic name of Tzu-Hsien[16] and the title Phi-Ling Chhan Shih.[17]
[g] Also known as Chhen Ni-Wan[18] or Ni-Wan Chen Jen.[19]   [h] *TT*1076.
[i] Otherwise called Hai Chhiung Tzu[20] or Tzu-Chhing Chen Jen.[21]
[j] In this particular case it looks as if Phêng Ssu was more interested in *nei tan* and Mêng Hsü in *wai tan*.
[k] Such as Wang Chê,[22] Wang Yün-Chhing,[23] Wang Shih-Hsiung,[24] Chhung Yang Tzu,[25] Wang Chhung-Yang,[26] and Chhung-Yang Ti Chün.[27]
[l] *TT*1139.

[1] 石泰	[2] 修眞十書	[3] 還源篇	[4] 薛紫賢	[5] 還丹復命篇
[6] 陳楠	[7] 翠虛篇	[8] 白玉蟾	[9] 彭耜	[10] 孟煦
[11] 王中孚	[12] 重陽全眞集	[13] 石杏林	[14] 翠玄子	[15] 薛道光
[16] 紫賢	[17] 毗陵禪師	[18] 陳泥丸	[19] 泥丸眞人	[20] 海瓊子
[21] 紫清眞人	[22] 王嚞	[23] 王允卿	[24] 王世雄	[25] 重陽子
[26] 王重陽	[27] 重陽帝君			

the *Chhung-Yang Chiao Hua Chi*[1] (Memorials of Wang Chhung-Yang's Preaching),[a] the *Chhung-Yang Fên-Li Shih-Hua Chi*[2] (Writings of Wang Chhung-Yang (to commemorate the time when he received a daily) Ration of Pears, and the Ten Precepts of his Teacher),[b] the *Chhung-Yang Chin-Kuan Yü-So Chüeh*[3] (Wang Chhung-Yang's Instructions on the Golden Gate and the Lock of Jade),[c] and the *Chhung-Yang Li-Chiao Shih-Wu Lun*[4] (Fifteen Discourses of Wang Chhung-Yang on the Establishment of his School).[d]

Wang Chung-Fu's disciples, the 'Seven Perfect-Truth Masters' (*chhüan chen chhi tzu*[5]), were all well known. Ma Yü[6] (+1123 to +1183),[e] the most senior member, was responsible for the *Tan-Yang Chen Jen Yü Lu*[7] (Precious Records of the Adept Tan-Yang),[f] the *Chien Wu Chi*[8] (On the Gradual Understanding (of the Tao)),[g] the *Tzu-Jan Chi*[9] (Collected (Poems) on the Spontaneity of Nature),[h] the *Tung-Hsüan Chin Yü Chi*[10] (Collections of Gold and Jade; a Tung-Hsüan Scripture),[i] and the *Tan-Yang Shen Kuang Tshan*[11] (Tan Yang (Tzu's Book) on the Resplendent Glow of the Numinous Light);[j] all in the *Tao Tsang*. The second master was Than Chhu-Tuan[12] (+1123 to +1185).[k] He wrote a *Than hsien-sêng Shui Yün Chi*[13] (Mr Than's Records of Life among the Mountain Clouds and Waterfalls).[l] Liu Chhu-Hsüan[14] (+1147 to +1203),[m] the third of the Seven Masters, wrote the *Hsien Lo Chi*[15] (Collected (Poems) on the Happiness of the Holy Immortals)[n] and a commentary on the *Huang Ti Yin Fu Ching*[16] (The Yellow Emperor's Book on the Harmony of the Seen and the Unseen).[o] The fourth and most celebrated master was Chhiu Chhu-Chi[17] (+1148 to +1227),[p] widely known as Chhang-Chhun Chen Jen[18] (the Adept of Eternal Spring). He wrote the *Chhiu Chhang-Chhun Chhing Thien Ko*[19] (Chhiu Chhang-Chhun's Song of the Blue Heavens),[q] and a *Hsüan Fêng Chhing Hui Lu*[20] (Record of the Auspicious Meeting of the Mysterious Winds),[r] which contained the answers

---

[a] *TT*1140.   [b] *TT*1141. This was a fasting diet during his first initiation.
[c] *TT*1142.   [d] *TT*1216.
[e] Known variously as Ma Chhu-Yü,[21] Ma Hsüan-Pao,[22] Ma Thung-Pao,[23] Ma Tshung-I,[24] Ma I-Fu,[25] Tan Yang Tzu[26] and Tan-Yang Ti Chün.[27]
[f] *TT*1044.   [g] *TT*1128.   [h] *TT*1130.   [i] *TT*1135.   [j] *TT*1136.
[k] He was known by different names, Than Thung-Chêng,[28] Than Yü,[29] Than Po-Yü,[30] Chhang Chen Tzu,[31] and Chhang-Chen Chen Chün.[32]
[l] *TT*1146. The book describes the ascetic life of the Taoist hermit. The similarity of its title to that of Yeh Mêng-Tê's[33] book, *Shui Yün Lu*,[34] which gives us one of the oldest accounts of the sublimation of steroid hormones, may be noted. Yeh (+1077 to +1148) was of a rather older generation than that of the Seven Masters. See Lu Gwei-Djen & Needham (3).
[m] Also called Liu Thung-Miao,[35] Chhang Shêng Tzu[36] and Chhang-Shêng Chen Chün.[37]
[n] *TT*1127.   [o] *TT*119. See Vol. 2, p. 447.
[p] Also called Chhiu Thung-Mi,[38] Chhiu Chhang-Chhun.[39]   [q] *TT*134.   [r] *TT*173.

[1] 重陽敎化集	[2] 重陽分梨十化集	[3] 重陽金關玉鎖訣	[4] 重陽立敎十五論
[5] 全眞七子	[6] 馬鈺	[7] 丹陽眞人玉錄	[8] 漸悟集 [9] 自然集
[10] 洞玄金玉集	[11] 丹陽神光燦	[12] 譚處端	[13] 譚先生水雲集
[14] 劉處玄	[15] 仙樂集 [16] 黃帝陰符經	[17] 邱處機	[18] 長春眞人
[19] 邱長春靑天歌	[20] 玄風慶會錄 [21] 馬處鈺		[22] 馬玄寶
[23] 馬通寶	[24] 馬從義 [25] 馬宜甫	[26] 丹陽子	[27] 丹陽帝君
[28] 譚通正	[29] 譚玉 [30] 譚伯玉	[31] 長眞子	[32] 長眞眞君
[33] 葉夢得	[34] 水雲錄 [35] 劉通妙	[36] 長生子	[37] 長生眞君
[38] 邱通密	[39] 邱長春		

given by Chhiu Chhu-Chi to Chingiz Khan when consulted by him about the way of immortality.[a] Chhiu also wrote a *Ta Tan Chih Chih*[1] (Direct Hints on the Great Elixir),[b] and the *Chhang-Chhun Tzu Phan-hsi Chi*[2] (Chhiu Chhang-Chhun's Collected (Poems) at Phan-hsi).[c] The fifth of the Seven Masters, Wang Chhu-I[3] (+1142 to +1198),[d] was the author of the *Hsi Yo Hua-shan Chih*[4] (Records of Huashan, the Great Western Mountain)[e] and the *Yün Kuang Chi*[5] (Collected (Poems) of Light (through the Clouds)).[f] The sixth master, Ho Ta-Thung[6] (+1140 to +1212),[g] contributed the *Thai-Ku Chi*[7] (Collected Works of (Ho) Thai-Ku)[h] to the *Tao Tsang*. Finally it is pleasant to record that the seventh and last divine, Sun Pu-Erh[8] (+1119 to +1183),[i] was a woman, and in fact the wife of Ma Yü.

The Northern School laid emphasis on self-cultivation (*hsiu ming*[9]) by the leading of an ascetic life and the suppression of desire for earthly things, so much so that its priests took to celibacy.[j] Practical alchemy seemed to be outside its concern, but the physiological alchemy of respiratory and other exercises was very much cultivated. On being given an audience by Chingiz Khan in +1222 and consulted about the way of achieving immortality, Chhiu Chhu-Chi told him that as an emperor he would automatically become an immortal when his reign ended if he had accomplished the mission of bringing peace and happiness to the world. In order to achieve this, however, it was necessary that he should conserve his health by exercising more restraint in his sexual life, especially as he was over forty and had more than one consort.[k] The greatest single contribution made by the Northern School to alchemy was no doubt the collection of the Taoist Patrology made by some of the disciples of Chhiu Chhu-Chi, under the leadership of Sung Tê-Fang,[10] from +1237 onwards, culminating in the printing of the *Hsüan Tu Pao Tsang*[11] collection in +1244.[l] Among Sung's disciples was Li Chüeh[12],[m] the teacher of Chang Mu[13],[n] who in turn taught

---

[a] Chhiu Chhu-Chi was summoned to the court of the Mongolian conqueror, then in Afghanistan, and travelled to Samarqand and back from Shantung between +1219 and +1224. Cf. Vol. 3, pp. 522 ff. A disciple's account of the journey has been translated by Waley (10) under the title 'Travels of an Alchemist'. This was *Chhang-Chhun Chen Jen Hsi Yu Chi*[14] (Journey of the Eternal-Spring Adept into the West), written by Li Chih-Chhang,[15] also called Thung-Hsüan Ta Shih.[16]

[b] *TT*241.　　[c] *TT*1145.　　[d] Also called Yü Yang Tzu[17] or Yü-Yang Chen Chün.[18]

[e] *TT*304.　　　　　　[f] *TT*1138.

[g] Also called Ho Ta-Ku,[19] Ho Thai-Ku,[20] Ho Lin,[21] Thien Jan Tzu,[22] Kuang Ning Tzu,[23] and Kuang-Ning Chen Chün.[24]　　　　　[h] *TT*1147.

[i] Also known as Chhing-Ching San Jen[25] and Chhing-Ching Yuan Chün.[26]

[j] This was such a departure from the general tendencies of early Taoism (cf. Vol. 2, pp. 57, 137, 146) that one can hardly fail to suspect the influence of Buddhist asceticism.

[k] The whole of *TT*173 is devoted to the advice given by Chhiu Chhu-Chi on this occasion.

[l] See p. 116 above.

[m] Also called Li Shuang-Yü,[27] Li Hsi-Chen,[28] and Thai-Hsü Chen Jen.[29]

[n] Also known by the names Chang Chün-Fan,[30] Chang Tao-Hsin[31] and Tzu-Chhiung Chen Jen.[32]

---

[1] 大丹直指	[2] 長春子磻溪集	[3] 王處一	[4] 西嶽華山誌
[5] 雲光集	[6] 郝大通	[7] 太古集	[8] 孫不二	[9] 修命
[10] 宋德方	[11] 玄都寶藏	[12] 李珏	[13] 張模	
[14] 長春眞人西遊記		[15] 李志常	[16] 通玄大師	[17] 玉陽子
[18] 玉陽眞君	[19] 郝大古	[20] 郝太古	[21] 郝璘	[22] 恬然子
[23] 廣寧子	[24] 廣寧眞君	[25] 清靜散人	[26] 清靜元君	[27] 李雙玉
[28] 李栖眞	[29] 太虛眞人	[30] 張君範	[31] 張道心	[32] 紫瓊眞人

Chao Yu-Chhin[1].[a] Chao Yu-Chhin then became the teacher of the greatest Taoist writer of the Yuan period, Chhen Kuan-Wu,[2] also called Chhen Chih-Hsü[3] and Shang Yang Tzu.[4] About +1331 Chhen Chih-Hsü wrote the *Chin Tan Ta Yao*[5] (Main Essentials of the Metallous Enchymoma; the true Gold Elixir), also called *Shang Yang Tzu Chin Tan Ta Yao*[6];[b] compiling for it also a *Chin Tan Ta Yao Thu*[7] (Illustrations for the 'Main Essentials...').[c] These are important books on physio-logical alchemy (cf. pt. 5 below) using many technical terms in common with the laboratory alchemists. This is why one cannot but feel that practical experimentation had remained a continuing, if esoteric, activity of Taoists who seemed primarily interested in the psycho-physiological techniques. Chhen Chih-Hsü also wrote commentaries on the *Tshan Thung Chhi* and the *Wu Chen Phien*, called respectively *Chou I Tshan Thung Chhi Fên Chang Chu*[8] and *Wu Chen Phien San Chu*[9] (Three Commentaries on the 'Essay on Regenerating the Primary Vitalities').[d] Although we shall have occasion in due course to quote from the text and diagrams of Chhen Chih-Hsü, who certainly knew the practical techniques of chemical operations, his own interests lay mainly in the field of psycho-physiological alchemy.

This sub-section may suitably be ended by a quotation from a scholar of about +1230 who knew both *wai tan* and *nei tan* but greatly preferred the latter. Chhu Yung[10] (Chhu Hua-Ku[11]) was, as we noted at an earlier stage, one of the commen-tators on the *Tshan Thung Chhi*, but he also produced an interesting work entitled *Chhü I Shuo Tsuan*[12] (Discussions on the Dispersal of Doubts) which we have often quoted already in these volumes.[e] In the course of this he wrote in the following vein:[f]

Those who practise the 'Art of the Yellow and the White' are deceitful and treacherous; they are called 'sooty empiricks' (*jê kho*[13]) or 'furnace firemen' (*lu huo*[14]). The lesser (artists) refine and thin down gold and silver to form cupel cakes (*san chih*[15]); the greater (ones) work to compose and perfect an 'elixir-mother' (*tan mu*[16]).[g] Both results are called 'casing-process ingredients' (*kuei thou*[17]).[h] (According to the old saying): 'swallows and sparrows do not lay phoenix eggs, nor are horses suckled by foxes and rabbits';[i] this, they say, supports the theory of (elixir-)mothers. Sometimes they steal the genuine (elixir-)mother and substitute another substance for it; sometimes they make (the metal) absorb (mercury), and expect to be profoundly thanked (and rewarded). When the mercury enters the absorbent (metal) (*ju kuei*[18]) it necessarily eats up the (elixir-)mother, thus forming the precious substance.

[a] Otherwise known as Yuan-Tu Chen Jen.[19]
[b] *TT* 1053.   [c] *TT* 1054.   [d] *TT* 139.
[e] Cf. e.g. Vol. 2, p. 387, Vol. 3, p. 323, Vol. 4, pt. 1, p. 307.
[f] Ch. 1, pp. 11*b*, 12*a*, tr. auct.
[g] This seems to be a distinction between the humble assayers and the more prestigious Taoist alchemists, but the products of both could be amalgamated. A technical term for just this follows in the next sentence. The 'elixir-mother' was evidently some precursor-substance of the product desired.
[h] Because they would form amalgams with mercury?
[i] A quotation from the *Tshan Thung Chhi*, ch. 12, p. 25*b*.

[1] 趙友欽	[2] 陳觀吾	[3] 陳致虛	[4] 上陽子	[5] 金丹大要
[6] 上陽子金丹大要		[7] 金丹大要圖	[8] 周易參同契分章註	
[9] 悟眞篇三註	[10] 儒泳	[11] 儒華谷	[12] 袪疑說纂	[13] 爇客
[14] 爐火	[15] 鏒制	[16] 丹母	[17] 匱頭	[18] 入匱
[19] 緣督眞人				

When this has been repeated several times the *chhi* of the (elixir-)mother becomes exhausted. When the gold and silver is quite finished, all the mercury goes off in fume and flame.[a] Others again use mercury to collect the body of the silver, and various chemicals to eat up the colour of the gold, then on long heating the precious product is seen; this process is called 'collecting the (elixir-)mother through a window in the wall (*ko chhuang chhü mu*[1])'.[b] Others yet again use reaction-vessels (*ting chhi*[2]) actually made of gold and silver, putting mercury in them with various chemicals derived from plants, then after transformation by heating another precious material is obtained; this process is called 'inverting the body of the Jade Girl (*Yü Nü fan shen*[3])'. Still others heat mercury with vitriol in iron vessels, gaining the result in a very short time. A product may look like silver but more yellow in colour and of a harder substance, or it may look like gold but also harder and paler, or it may look like copper though smoother in texture and brighter in hue.[c] This is because mercury eats up the formative essence (*ying hua*[4])[d] of the iron to form a (new) substance, and vitriol changes the colour of iron to yellow.[e] This is what they call 'instant transformation into a true precious substance (lit. before you can turn round, *chuan shen pien chhêng chen pao*[5])'. There is hardly anyone in the world who is not deceived by these things.

(On the other hand) there is, for example, the getting of mercury from lotus plants (*ho yeh*[6]), and the preparation of lead and tin from the ash of the purslane (*hui hsien*[7]);[f] all this is described in (the texts) concerning the 72 kinds of 'dragon sprouts' (*lung ya*[8]). These are the kinds of things that are worth looking into in the chymic art (*lu huo chung chih kho kuan chih*[9]). All the things more vulgar than these are really not worth talking about, and I do not intend to discuss them. I sincerely hope that scholars who wish to understand the Tao will not pay any attention to them.[g]

One may feel that Chhu Yung showed some discrimination in rejecting as vulgar both the making of metallic and mineral elixirs (now no longer convincing), and the alleged triumphs of aurifaction, while pointing to the presence of metals in plant tissues as a real and elegant natural wonder. If only he and his contemporaries could have found some way of breaking out of the charmed circle of practical alchemical procedures in which the 'sooty empiricks' had revolved for so long, and finding the road to a developed chemical science, we should be ready to give him far higher praise.

[a] This would seem to be a reference to gilding and silvering by the use of amalgams.

[b] This could well be a reference to surface enrichment processes.

[c] Doubtless these were all alloys resembling gold and silver in appearance, like those produced by Wang Chieh (p. 186 above).

[d] See the remarks in pt. 4.

[e] This must be an allusion to the precipitation of copper from copper sulphate solutions in the presence of iron scrap (the 'wet copper method') on which see p. 129, and pt. 4 below.

[f] Here Chhu Yung is clearly talking about the finding of metals in plants, a subject we have already discussed under the head of bio-geochemical prospecting (Vol. 3, pp. 675 ff.). Plant accumulators, which store astonishingly large amounts of metal elements in their bodies, are now well known, and as early as the +18th century it was appreciated in Europe that the Chinese had discovered this phenomenon long before. What evidence Chhu Yung had for his statement about mercury in the lotus (*Nelumbo*) we do not know, but it certainly accumulates in the purslane (*Portulaca oleracea*, R554, more often called *ma chhih hsien*[10]), which he mentions immediately afterwards (cf. Vol. 3, p. 678).

[g] The whole passage seems to be an inventory of 'casing' and 'irrigation' processes. For an attempt to elucidate what these were, see pt. 4 below.

[1] 隔窗取母   [2] 鼎器   [3] 玉女翻身   [4] 英華   [5] 轉身便成眞寶
[6] 荷葉   [7] 灰莧   [8] 龍牙   [9] 爐火中之可觀者
[10] 馬齒莧

But neither the needs of society nor the nature of the traditional natural philosophy permitted, and instead Chhu Yung followed the path of gentlemanly meditational and discreet physiological *nei tan* practices. One can sense in him the class-distinction between the clerkly administrator familiar with polite literature on the one hand and the quasi-artisanal manual operator on the other. The writ of Taoist religious sanction for the latter way of life was now fast running out. 'In general', Chhu Yung wrote,[a] 'those who are interested in the "art of the yellow and the white" are not really scholars of refinement and distinction (*fei chhing kao chih shih*[1]). How can they qualify as learned students of the Tao?'

(7) ALCHEMY IN ITS DECLINE; YUAN, MING AND CHHING

After its heyday in the Thang dynasty alchemy continued to flourish, as we have seen, during the Sung period. However, the dangers of elixir-poisoning had clearly manifested themselves since no less than six Thang emperors and a number of court officials died from this cause. By the time of the Sung more caution was exercised in the general approach to elixir-making, not only in the composition of the elixirs themselves, but also in attempts to elaborate pharmaceutical ways and means of counteracting the toxic effects. The number of ingredients used in elixir formulae was reduced and there was a tendency to return to the ancient and difficult theorising of the *Tshan Thung Chhi*, perhaps to conceal the processes from rash and ignorant operators. Psycho-physiological alchemy (*nei tan*) became steadily more popular than laboratory alchemy (*wai tan*). This can be seen from the works of Chang Po-Tuan and the numerous commentaries on the *Tshan Thung Chhi*. The Northern School founded by Wang Chê seems to have paid (at least on the surface) very little attention to alchemy, while in the Southern School we find only one treatise that can be said to be truly concerned with chemical operations: the *Chin Hua Chhung-Pi Tan Ching Pi Chih* of Phêng Ssu & Mêng Hsü.

When the empire fell into the hands of the Mongols towards the end of the +13th century Taoism lost favour, since it was suspected of subversive (and nationalist) political ideas.[b] As a result of the confrontation with Buddhism at this time there was a great loss of alchemical texts when the Taoist patrology was committed to the flames. Taoist alchemy was now down to its lowest ebb. The most important text outwardly dealing with alchemy, namely the *Chin Tan Ta Yao* (with its associated *Chin Tan Ta Yao Thu*), by Chhen Chih-Hsü, *c.* +1331, must be considered primarily a work on psycho-physiological alchemy rather than practical proto-chemistry.[c] On the other hand certain popular cyclopaedias continued to give details about alchemical operations and apparatus, e.g. the *Mo O Hsiao Lu*[2] (Secretary's Commonplace-Book) put together by an anonymous compiler about this time. Equally, certain emperors had

---

a Ch. 1, p. 12*b*, tr. auct.   b Cf. Vol. 2, pp. 100, 115, 138.
   c At this time a great number of writers were composing tractates and poems on physiological alchemy—to take one example Kao Hsiang-Hsien,[3] whose verses induced Davis & Chao Yün-Tshung (1), cf. (6), p. 397, to attempt a translation.

   1 非清高之士   2 墨娥小錄   3 高象先

dealings with alchemists, and certain sceptical Confucian scholars condemned them as of old. An example of the former might be found in Shih Tsu[1] (Khubilai Khan) himself, who ruled China from +1280 to +1294, and to illustrate the latter we could refer to the *Pien Huo Phien*[2] (Disputations on Doubtful Matters),[a] written by Hsieh Ying-Fang[3] in +1348. Here the writer runs through the whole story of alchemy and elixir-poisoning (cf. pp. 135, 185) through the ages, consigning it entire to the dustbin of charlatanry and religious superstition.[b]

The idea of the elixir of life lingered on in China for a few more centuries, though there was never any sign of a general revival of alchemy. Indeed it caught the fancy of some of the Ming emperors.[c] The official history (*Ming Shih*) informs us that the emperor Ming Thai Tsu granted audience to the adept Liu Yuan,[4] and sent messengers in +1391 to search for an alchemist called Chang San-Fêng.[5] He also graciously greeted Chang Chêng-Chhang,[6] descendant of Chang Tao-Ling in the 24th generation.[d] The emperor Chhêng Tsu was still looking for Chang San-Fêng during the Yung-Lo reign-period (+1403 to +1424), and in +1459 Ying Tsung at last honoured Chang by giving him the title Thung-Wei Hsien-Hua Chen Jen[7].[e] The name of Chang San-Fêng is nowadays generally associated with one of the schools of Chinese boxing called *thai chi chhüan*[8],[f] and we know very little about his history.[g] A number of books are however attributed to him and one of them, the *San-Fêng Tan Chüeh*[9] (Chang San-Fêng's Instructions about Enchymomas), talks of psycho-physiological alchemy in terms of lead amalgamation. While his books do not suggest that Chang was practising any alchemy, we are told that one of his disciples, Shen Wan-San,[10] was an experimental alchemist, familiar with the properties of lead and mercury, and able to transmute copper and iron into 'gold' and 'silver' by the application of mercury which had been previously treated. We are also told that Shen Wan-San's daughter, Shen Yü-Hsia,[11] was an alchemist too.[h]

(i) *The Emaciated Immortal, Prince of the Ming*

During the first half of the +15th century the imperial house of the Ming itself contributed several outstanding names to the history of Chinese science. Prince Chu Hsiao[12] (Chou Ting Wang,[13] c. +1380 to +1425) gained imperishable renown for his *Chiu Huang Pên Tshao*[14] (Natural History of Emergency Food Plants) printed in

---

[a] This work has already been described in Vol. 2, p. 389.

[b] Ch. 4, pp. 4b to 10b.

[c] A brief survey of alchemy during the Ming period has been given by Ho Ping-Yü (13).

[d] *Ming Shih*, ch. 202, p. 23a, and ch. 299, p. 8a.

[e] *Ming Shih*, ch. 202, p. 23a.

[f] A kind of physical exercise, part self-defence, part medical eurhythmics. We shall say more about it in connection with physiological alchemy in Vol. 5, pt. 5, and also in the sub-section on physiotherapy and remedial gymnastic techniques in Vol. 6, Sect. 44. It goes back at least as far as Hua Tho.

[g] An analytical study of his hagiography has been made by Seidel (1).

[h] See *Chang San-Fêng Chuan*, p. 4b (in Fu Chin-Chhuan's *Chêng Tao Pi Shu Shih Chung*[15]).

[1] 世祖	[2] 辯惑編	[3] 謝應芳	[4] 劉淵	[5] 張三峯
[6] 張正常	[7] 通微顯化眞人		[8] 太極拳	[9] 三峯丹訣
[10] 沈萬三	[11] 沈玉霞	[12] 朱橚	[13] 周定王	[14] 救荒本草
[15] 證道秘書十種				

+ 1406, a work which, together with others of the same kind, we shall discuss in detail in the botanical Section of Vol. 6.ᵃ But here it is his younger brother, also a prince, Chu Chhüan¹ (Ning Hsien Wang,² + 1390 to + 1448), who attracts us more, for he was greatly interested in the proto-chemistry of the alchemists, chiefly from the medical and technological angle.ᵇ Just as his brother maintained a great botanical garden at Khaifêng with many gardeners cultivating the plants which were safe to use for food in times of famine or scarcity, and some assistants studying methods of detoxicating the dangerous ones, so Chu Chhüan must have had an elaboratory in which he carried out or supervised experiments on iatro-chemistry, pharmacy and metallurgy. About + 1421 he produced a book which we should dearly like to see today, the *Kêng Hsin Yü Tshê*³ (Precious Secrets (lit. Jade Pages) of the Realm of Kêng and Hsin),ᶜ i.e. all things connected with metals and minerals, symbolised by these two cyclical characters, which also constitute an alchemical synonym for gold.ᵈ It enumerated under 541 entries the substances employed in chemical preparations. According to Li Shih-Chen,ᵉ it was divided into the following sections: (*a*) metals and inorganic materials, (*b*) 'subtle sprouts' (*ling miao*⁴), i.e. ore indications, or perhaps products of chemical reactions, (*c*) plants with active principles of remarkable properties, (*d*) feathers and hair,ᶠ (*e*) carapaces and hides,ᵍ (*f*) substances and liquids which may be eaten and drunk, (*g*) iatro-chemical and alchemical apparatus. Chu Chhüan drew largely upon a literary tradition which is tantalisingly obscure, not exactly alchemical, not exactly pharmaceutical, connected rather with mining and metallurgy, and perhaps because of its specialised interest not widely copied, hence now lost, or extant only in quotations. Among such sources Li mentioned the *Wai Tan Pên Tshao*⁵ (Iatro-chemical Natural History) of Tshui Fang⁶ in the early Sung,ʰ and the *Thu Hsiu Chen Chün Tsao-Hua Chih Nan*⁷ (Guide to (lit. South-pointing Compass for) the Creation,ⁱ by the Earth's Mansions Immortal), a valuable work on mining,

---

ᵃ Sect. 38.

ᵇ His philosophical name was Han Hsü Tzu,⁸ the Full-of-Emptiness Master, but he also called himself Chhü Hsien,⁹ the Emaciated Immortal, and others named him the Teacher of the Elixir Mound, Tan Chhiu hsien-sêng.¹⁰ His biography is found in *Ming Shih*, ch. 117, p. 4*a*.

ᶜ Already referred to in Vol. 1, p. 147, Vol. 3, p. 678.

ᵈ The book was certainly available to Li Shih-Chen. Bretschneider (1), vol. 1, p. 53, spoke as if he had seen it, but in our own time Ching Li-Pin (1) characterised it as 'introuvable'. If it still exists at all it must be extremely rare, but China has so many provincial libraries that one should not give up hope of encountering it one day.

ᵉ *PTKM*, ch. 1A, p. 12*b*.

ᶠ Important in relation to sal ammoniac, a key substance in chemical history. It is formed by the destructive distillation of keratin. Cf. pt. 4.

ᵍ No doubt Chu Chhüan was interested in the process of tanning. Cf. Vol. 6.

ʰ A very important book (cf. Vol. 5, pt. 2, pp. 201, 209, and p. 185 above), written about + 1045.

ⁱ On this phrase, *tsao-hua*, the Author or Foundation of Change, and its parallel locution, *tsao-wu chê*,¹¹ the Author of Things or Nature, see Vol. 2, pp. 564, 581 and Vol. 3, p. 599. More recently Schafer (17) has argued that these phrases did imply some kind of personal deity in ancient and medieval China, but since the concept of creation *ex nihilo* was so foreign to Chinese thought, we adhere to our view that the Author of Change was not a person, but rather a numinous poetical allegory of the Tao of all things.

---

¹ 朱權   ² 寧獻王   ³ 庚辛玉册   ⁴ 靈苗   ⁵ 外丹本草
⁶ 崔昉   ⁷ 土宿眞君造化指南   ⁸ 涵虛子   ⁹ 臞仙
¹⁰ 丹丘先生   ¹¹ 造物者

mineralogy, assaying, mineral remedies,[a] and probably alchemy, so far quite impossible to date, but hardly earlier than the Thang or later than the beginning of the Ming. One at any rate of the books prized and quoted by Chu Chhüan we still have available, namely the *Tan Fang Chien Yuan*[1] (Mirror of Alchemical Processes and Reagents) written about +950 by Tuku Thao;[2] on this see p. 180 above. Two others we know only by quotations, the *Hsien-Yuan Pao Tsang Lun*[3] (Discourse of the Yellow Emperor on the Contents of the Precious Treasury of the Earth), obscure in date and authorship but probably finished by +918;[b] and the *Tan Thai Hsin Lu*[4] (New Discourse on the Alchemical Laboratory), which must be early Sung or pre-Sung because mentioned in Sung bibliographies,[c] but attributed to an even earlier date because attributed to Chhing Hsia Tzu[5] (see pp. 159, 180). All in all we have here a fascinating Paracelsian realm where alchemy joined hands with practical mining and metallurgy as well as with medicine; and it is exasperating that fate has deprived us of so many of the documents concerned.

The *Kêng Hsin Yü Tshê* was far from being Chu Chhüan's only book. He wrote one on geriatric medicine, which hardly concerns us here, though mineral remedies were assuredly involved,[d] and another which must have been chemical or metallurgical since it was entitled *Tsao-Hua Chhien Chhui*[6] (The Hammer and Tongs of Creation).[e] So also some theoretical proto-chemistry was assuredly contained in two other books, *Chhien Khun Pi Yün*[7] (The Hidden Casket of Yin and Yang Opened) and *Chhien Khun Shêng I*[8] (Principles of the Coming into Being of Yin and Yang), now almost wholly lost. These were never very well known, and rarely quoted, but much greater prominence was given by the Chhing bibliographers to another work, the *Chhü Hsien Shen Yin Shu*[9] (Book of Daily Occupations for Scholars in Rural Retirement, by the Emaciated Immortal); this dealt with botanical and horticultural subjects, fermentation technology, fruit preservation and veterinary medicine, again doubtless emphasis-

[a] Hence it was sometimes known as *Thu Hsiu Pên Tshao*[10] (The Earth's Mansions Pharmaceutical Natural History).

[b] Chang Tzu-Kao (2), p. 118, considers that this book reached its final form in the Wu Tai period (+10th cent.), and speaks as if he thought that Hsien-Yuan was a name or pseudonym of a person living at that time. But, as we saw on p. 130 above, the book seems to have a connection with Su Yuan-Ming in the Chin (+4th cent.), who perhaps wrote the first recension of it. It may be of interest that there seems to have been in the late Thang an adept with a double family name, Hsienyuan Chi;[11] he was active under Wu Tsung and Hsüan Tsung in the middle of the +9th century. So far the *Lo-fou Shan Chih*, ch. 4, p. 17b, but the subject of this biography was more of a magician than a metallurgical chemist and mining expert.

[c] *Sung Shih*, ch. 205, p. 18a, *Thung Chih Lüeh*, ch. 43, p. 24a. The latter gives the author as Hsia Yu-Chang,[12] a person not otherwise known. This may be a convenient place to refer to the value of the *Thung Chih* bibliographies, compiled about +1150, much earlier than those in the dynastic history. Of *wai tan* books it lists 203 and of *nei tan* books 40; not to speak of 31 on metals and minerals, 56 on Taoist dietary regimen, 9 on sexual techniques, 107 on respiratory exercises, 20 on gymnastics and 74 on macrobiotics in general. These totals by themselves may afford some estimate of the extent to which the several specialities had been cultivated during the preceding centuries.

[d] *Shou Yü Shen Fang*[13] (Magical Prescriptions of the Realm of the Old).

[e] Cf. Vol. 3, p. 599.

[1] 丹方鑑源	[2] 獨孤滔	[3] 軒轅寶藏論	[4] 丹臺新錄	[5] 青霞子
[6] 造化鉗鎚	[7] 乾坤秘韞	[8] 乾坤生意	[9] 臞仙神隱書	
[10] 土宿本草	[11] 軒轅集	[12] 夏有章	[13] 壽域神方	

ing mineral remedies.[a] Besides these it is thought that Chu Chhüan's interests in-cluded the knowledge of distant lands, e.g. those of the Arabs, and their products, so that he seems to have been associated, perhaps as patron, with the *I Yü Thu Chih*[1] (Illustrated Record of Strange Countries), a geographical encyclopaedia compiled just before +1430. This we have already described in Sect. 22.[b]

The Ming emperor Shih Tsung (r. +1522 to +1567) showed great interest in the art of the immortals, and Chang Yen-Phien,[2] a descendant of Chang Tao-Ling, took advantage of this to gain his favour.[c] In fact it brought Shih Tsung to his death, for in his last years he put much confidence in Taoist physicians, magicians and alchem-ists. There was one especially, named Wang Chin,[3] who had been a protégé of the high official and favourite Thao Chung-Wên,[4] and the amateur alchemist Yin Ying-Lin,[5] a country magistrate.[d] In spite of various frauds Wang Chin was appointed a Physician-in-Attendance in the Imperial Academy of Medicine (Thai I Yuan[6]), but his ministrations brought about the emperor's last illness. Eventually Wang's over-bold iatro-chemical therapy caught up with him and he was exiled to the frontiers, lucky to escape execution, in +1570. The whole story, which involved a number of other dubious characters, is particularly interesting because of the persistence of the belief still recorded by the historians here that eating and drinking from vessels made of alchemical gold and silver would bring about immortality (cf. pp. 31, 49 above). For another protégé of Thao Chung-Wên, Tuan Chhao-Yung,[7] averred that 'the product of his transmutation was all silver of the immortals, and that anyone who supped off it would never die'.[e] He actually presented ten thousand pieces of gold to the throne, but later his art 'could not be verified (*shu pu yen*[8])', and a disciple revealed his secrets, so they both came to grief. But the idea had by then had a run of some sixteen centuries.

(ii) *Ben Jonson in China*

Meanwhile throughout this time the usual stories about alchemical charlatans and gullible scholar-officials continued to proliferate,[f] paralleling closely enough indeed the oft-repeated tales in European culture.[g] Since the Ming was a famous period for novels and plays, the elaboration of this particular theme is not at all surprising. It was frequently worked out in different forms by the prolific writer Fêng Mêng-Lung[9] (d. +1646),[h] partly in his *Chin Ku Chhi Kuan*[10] (Strange Tales New and Old),

[a] See *SKCS/TMTY*, ch. 147, p. 9a.
[b] Vol. 3, pp. 512 ff.
[c] *Ming Shih*, ch. 299, p. 22a.
[d] See *Ming Shih*, ch. 307, pp. 25a to 31a. Thao was a magician-technician himself.
[e] *So hua yin chieh hsien wu, yung wei yin shih chhi tang pu ssu.*[11]
[f] Cf. pp. 168, 170, 194, 206 above.
[g] See, for instance, Read (1); Holmyard (1) and many other histories of Western alchemy.
[h] A good account of him will be found in Chhen Shou-Yi (3), pp. 480 ff., and there are several references in Lu Hsün (1), pp. 257, 267, 418 for example.

| [1] 異域圖志 | [2] 張彥頨 | [3] 王金 | [4] 陶仲文 | [5] 陰應麟 |
| [6] 太醫院 | [7] 段朝用 | [8] 術不驗 | [9] 馮夢龍 | [10] 今古奇觀 |

[11] 所化銀皆仙物用爲飮食器當不死

partly in his *Tsêng Kuang Chih Nang Pu*[1] (Additions to the Enlarged Bag of Wisdom). Fêng was also responsible for three widely read collections called the *San Yen*,[2] in which again plausible projectors and holy deceivers prominently figure.[a] Another writer of the same period who liked this pseudo-scientific gambit was Chang Hsüan,[3] the author of the *Hsi Yuan Wên Chien Lu*[4] (Things seen and heard in the Western Garden).

The story in the *Chin Ku Chhi Kuan* about the alchemists is entitled 'Khua Miao Shu Tan Kho Thi Chin',[5] a phrase best translated perhaps in couplet form:

> How Spagyrists with vaunted occult Art
> Cozen the Host and with his Gold depart.

Comparison with the well-known and almost exactly contemporary English play, 'The Alchemist' by Ben Jonson (+1572 to +1637), acted in +1610 and printed two years later, springs to the mind. We cannot forbear from comparing the plots as a momentary recreation from the business of this book. In Jonson's play a London gentleman, Mr Love-Wit, leaves his town house on account of the plague, with his servant, Face, in charge, whereupon Face and the alchemist Subtle make use of the premises for cheating a number of characters. These include a wealthy citizen, Sir Epicure Mammon, a clerk and a shopkeeper, also two puritans, Mr Tribulation-Wholesome and Ananias, together with a quarrelsome young man, Kastril, and his sister Dame Pliant—only Surly, a gamester, disbelieves in the proceedings. In his text Jonson shows considerable knowledge of the technical terms and concepts of alchemy.[b] Then Love-Wit returns unexpectedly, Subtle decamps with his consort Doll Common, and Face arranges a marriage for Love-Wit with Dame Pliant. The 'Khua Miao Shu Tan Kho Thi Chin' is quite comparable. A learned and wealthy scholar named Phan, living at Sungchiang, becomes so infatuated with alchemy that he travels to Hangchow to seek out adepts. There he falls in with a grave stranger who consents to perform projection in Phan Fu-Ong's home, bringing his beautiful wife with him. The process is always a *diplōsis* (cf. pt. 2, p. 193) or multiplication, meaning that much silver or gold has to be provided to start with, but needless to say Phan never gets any of it back; a failure which the alchemist is able to attribute to a love affair between Phan and his wife, conducted in the laboratory during a temporary absence of the alchemist. On another occasion Phan is persuaded by a group of alchemists to shave his head and pose as their monastic teacher, but this plan also goes awry and the others all decamping Phan is left to bear the brunt of the wrath of a rich merchant, also their dupe. Finally while begging his way back to Sungchiang he passes a beautiful courtesan sitting in a boat—none other than the 'wife' of the first alchemist who had defrauded him. With the usual angelic kindness of such characters in Chinese novels

---

[a] These were *Hsing Shih Hêng Yen*[6] (Stories to Awaken Men), *Yü Shih Ming Yen*[7] (Stories to Enlighten Men) and *Ching Shih Thung Yen*[8] (Stories to Warn Men).

[b] Cf. pp. 24, 166.

[1] 增廣智囊補   [2] 三言   [3] 張萱   [4] 西園聞見錄
[5] 誇妙術丹客提金   [6] 醒世恆言   [7] 喻世明言   [8] 警世通言

she gives him money with which to get home, and after that he meddles with alchemy no more.[a]

Elsewhere we note several striking parallels between Ben Jonson's synopsis of European alchemical practice and features which were characteristic of the Chinese alchemical tradition,[b] but here it may be added that the similarities between the plots of Jonson and Fêng are especially striking where the relationship between alchemy and sex is concerned. There must be a certain ascesis on the part of the operator, yet the consumer of the elixir is expected to achieve sexual hyperactivity. In a long passage Mammon describes the paradise of concubines he will create, enjoying luxurious baths and feeding on nectar and ambrosia.

> Surly : And do you think to have the *Stone*, with this?
> Mam : No, I do think t' have all this, with the *Stone*.
> Surly : Why, I have heard, he must be *homo frugi*,
>   A Pious, Holy, and Religious Man.
>   One free from mortal sin, a very Virgin.
> Mam : That makes it, Sir, he is so. But I buy it.
>   My venture brings it me. . . .[c]

Here the language of nascent capitalism joins with the most ancient hopes of Chinese emperors and high officials. Later on Mammon becomes enamoured of Doll, who feigns to be attractively mad, and just as Phan Fu-Ong's dalliance with the 'wife' from Hangchow is made the excuse for the total failure of the Great Work, so Subtle, discovering Mammon and Doll together, arranges an explosion in the laboratory, so that equally all is lost, and for the same reason.[d]

> Sub : How! What sight is here!
>   Close deeds of darkness, and that shun the light!
>   Bring him again. Who is he? What, my Son!
>   O, I have liv'd too long.   Mam : Nay, good dear Father,
>   There was no unchaste purpose.   Sub : Not? And flee me,
>   When I come in?   Mam : That was my error.   Sub : Error?
>   Guilt, guilt, my Son. Give it the right name. No marvel,
>   If I found check in our *great work* within,
>   When such affairs as these were managing!
> Mam : Why, have you so?
> Sub : It has stood still this half hour
>   And all the rest of our *less works* gone back. . . .
>   This'll retard the *work*, a Month at least. . . .
> Mam : Our purposes were honest.   Sub : As they were,
>   So the reward will prove. How now! Aye me.
>   God and all Saints be good to us.

> [*A great Crack and Noise within*]

---

[a] The story has many times been translated into Western languages; see in French d'Hervey St Denys (3); in English Douglas (2); and in German Strzoda (1), Rudelsberger (1), Kühnel (1), Kühn (3).
[b] In pt. 4.   [c] P. 377.   [d] P. 430.

Sub :   What's that?

Face :   O Sir, we are defeated, all the *works*
        Are flown in fumo; every Glass is burst.
        Fornace and all rent down! As if a bolt
        Of Thunder had been driven through the House.
        Retorts, Receivers, Pellicanes, Bolt-heads,
        All struck in shivers! Help, good Sir! Alas. . . .

Mam :   O my voluptuous mind! I am justly punish'd. . . .

Sub :   O the curst fruits of Vice and Lust! . . .

Etc., etc.

It will hardly be believed that a third similar play exists, though it is much later—'Mullah Ibrahim Khalil the Alchemist', written in Azeri Turkish by Mirza Fath-ali Akhunzadé in 1851. From the translation of Barbier de Meynard (3) we find that it also ends with an explosion, and that the attempt at argentifaction by the projection of an *iksir* fails. Neither author nor translator seems to have known anything about Ben Jonson or Fêng Mêng-Lung.

Alchemy, or at least aurifiction, continued to figure in novels of the +18th century. For example, the *Ju Lin Wai Shih*[1] (Unofficial History of the World of Learning), that famous satire on the life of the literati in the Ming period written by Wu Ching-Tzu[2] between +1736 and +1749, has an incident of this kind.[a] An impecunious scholar, Ma Shun-Shang,[3] meets a strange adept in a temple and goes with him to his dwelling. The charlatan, Hung Han-Hsien,[4] who is posing as a Taoist immortal, gives Ma some black powder (*yin mu*[5]) which later on proves indeed to yield silver when heated according to the directions of the method (*shao yin chih fa*[6]). But eventually Ma finds that he is only a pawn in the early stages of an elaborate hoax on a wealthy patrician family, and Hung suddenly dying, it transpires that the powder was but chemically disguised silver all the time.

Besides the stories of ordinary life in the late Ming period, alchemy figured also, as would be expected, in the novels which concerned themselves with gods and spirits. Here the obvious example is the *Hsi Yu Chi*[7] (Pilgrimage to the West) written by Wu Chhêng-Ên[8] about +1570.[b] As readers of Waley's celebrated translation, 'Monkey', will remember, this contains a diverting scene in the alchemical laboratory of Lao Tzu in the Thirty-third Heaven.[c]

---

[a] Ch. 15, (pp. 154ff.); tr. Yang & Yang (1); Tomkinson (2), pp. 130ff.

[b] Based on the historical pilgrimage of Hsüan-Chuang to India in search of the sūtras (+629 to +645, cf. Vol. 1, p. 207), but embellished with a wealth of mythological and allegorical characters, including a monkey-spirit perhaps derived from Hanuman and symbolising the restless instability of genius. There had been an earlier version by Yang Chih-Ho;[9] cf. Lu Hsün (1), pp. 199, 203, 209.

[c] (17), pp. 196ff.

[1] 儒林外史	[2] 吳敬梓	[3] 馬純上	[4] 洪憨仙	[5] 銀母
[6] 燒銀之法	[7] 西遊記	[8] 吳承恩	[9] 楊志和	

(iii) *Chinese alchemy in the age of Libavius and Becher*

Perhaps the greatest contribution to alchemy during the Ming period was the printing of the Taoist patrology, first the *Chêng-Thung Tao Tsang*[1] (Taoist Patrology of the Chêng-Thung reign-period) in +1444, and then the *Wan-Li Hsü Tao Tsang*[2] (Supplementary Taoist Patrology of the Wan-Li reign-period) in +1607. These we have already mentioned (pp. 116–7). Almost all the Taoist alchemical texts described so far in this sub-section come from the Ming Taoist patrology. A few further alchemical texts have however emerged since then, and it may not be amiss to say a few words about them, though they are of no great significance for the history of chemistry.

Before doing this, however, some reference must be made to the great compendia of pharmaceutical natural history produced during the Ming period. We must not anticipate what will have to be said about them in Vol. 6, but since they incorporated a considerable amount of material chemical or proto-chemical in nature derived from the alchemical tradition they cannot be overlooked here. The *Pên Tshao Phin Hui Ching Yao*[3] (Essentials of the Pharmacopoeia Ranked according to Nature and Efficacity), imperially commissioned, was produced in +1505 by Liu Wên-Thai,[4] Wang Phan[5] and the physician Kao Thing-Ho.[6] It was not printed, however, until 1937, nor was its supplement of +1701, and even now the illustrations have never been made available, more's the pity, since from existing MSS they are known to be excellent. We shall draw further from this work in the Section on chemical technology. Also of the Ming, it may go without saying, so often have we referred to it in these volumes, was the *Pên Tshao Kang Mu*[7] (Great Pharmacopoeia) of Li Shih-Chen,[8] the 'uncrowned king' of Chinese naturalists. To the present day this great work remains an inexhaustible quarry for the history of chemistry, as of other scientific disciplines, in Chinese culture.

Many of the later books and tractates which seem at first sight to be concerned with chemical or elixir alchemy are really expositions of psycho-physiological alchemy. This can be illustrated by a couple of quotations from the *Huang Pai Ching*[9] (Mirror of Alchemy), written by Li Wên-Chu[10] in +1598. He says:[a]

(The Mirror) Reflecting Lead and Mercury. Within lead there is a small amount of the (element) Water, associated with the (cyclical character) *jen*.[11] The nature (of lead) belongs to Yang, but among the five elements its Water matches only with the (element) Fire associated with the (cyclical character) *ting*.[12] Within cinnabar is enclosed a little speck of the (element) Fire associated with the (cyclical character) *ting*. The nature (of cinnabar) belongs to Yin, but among the five elements its Fire matches only with the (element) Water associated with the (cyclical character) *jen*. At the time when cinnabar and lead come into conjunction, if the (element) Water (*jen*) arrives first and the (element) Fire (*ting*) comes later, then the Yang will surround the Yin and (the trigram) *Li* (☲) will be formed. The (broken) line in

---

[a] Ch. 13, in *Wai Chin Tan*,[13] ch. 2, p. 50a, b, tr. auct. This is part of the *Chêng Tao Pi Shu Shih Chung* just mentioned.

[1] 正統道藏	[2] 萬曆續道藏	[3] 本草品彙精要	[4] 劉文泰
[5] 王槃	[6] 高廷和	[7] 本草綱目	[8] 李時珍
[9] 黃白鏡	[10] 李文燭	[11] 壬	[12] 丁 [13] 外金丹

the middle of the *Li* (trigram) is the underworld of the 'prior to Heaven' (*hsien thien*,[1] system of arrangement of the *kua*).[a] But if by chance the (element) Fire (*ting*) arrives first and the (element) Water (*jen*) follows later, then Yin will enclose Yang, forming (the trigram) *Khan* (☵). The single solid line within the *Khan* (trigram) is the heaven of the 'prior to Heaven' (arrangement system). Just these two lines, one solid (denoting Yang or male) and one broken (denoting Yin or female) form the roots of Heaven and Earth in the 'prior to Heaven' (system), and the Gateway of the Mysterious Feminine (*hsüan phin chih mên*[2]).[b] Hence they are called real lead and real mercury. Other things like ordinary cinnabar and ordinary mercury, and the five metals together with the eight minerals, are all mere residues belonging to the 'posterior to Heaven' (system). How could they ever be called real? Hence the Sung people used to say: 'Those of our age who wish to understand (the meaning of) real lead and mercury (should know that) these are not common cinnabar and mercury.'[c]

Elsewhere he says:[d]

(*The Mirror*) *Reflecting the Mysterious Feminine*.[e] The two things within the Mysterious Feminine occupy the main thrones of Heaven and Earth, and hide within the middle line of the *Khan* and *Li* (trigrams). The line in the middle of the *Khan* (trigram) is a solid one, therefore the (element) Earth with the (cyclical character) *wu*,[3] the metal lead, the male Yang, and the real father. Hence it is said, 'The *Khan* (trigram's) *wu*, being male (element) Earth, is the father of (element) Metal.' The broken line in the middle of the *Li* (trigram) is the (element) Earth with the (cyclical character) *chi*,[4] mercury of the (element) Wood, the female Yin, and the real mother. Hence the saying, 'The *Li* (trigram's) *chi*, being the female (element) Wood and mercury, is the mother.'[f]

To the uninitiated this is likely to seem incomprehensible gibberish, 'rhapsodical' writing inspired by the *I Ching* and the *Tshan Thung Chhi*—unworthy of the +16th century, when modern science was already being born. In actual fact it makes very good sense, but only if one understands the principles of physiological alchemy, and these take quite a deal of explaining. The reader may like to return to Li Wên-Chu's words after following our account of these principles in Sect. 33 (*h*). We give them here only as a good example of what can make nonsense if taken unwittingly to be late chemical alchemy.

Since the terminology of physiological alchemy employed only a few names of metals and minerals, and so long as its different system remained unrecognised, historians were naturally struck by a decline in the number of substances mentioned in Ming alchemical works. Thus the tale of those mentioned in the *Huang Pai Ching* is even less than those in the *Tshan Thung Chhi*, being limited only to lead, mercury, cinnabar, and silver. The trend of Ming and post-Ming alchemical writing of both kinds may well have been to explain the real meaning of the words lead, mercury and

[a] See Vol. 4, pt. 1, p. 296. The real meaning of these mysterious expressions will be explained in pt. 5 below.
[b] See Vol. 2, p. 58. The classic phrase of the *Tao Tê Ching*, ch. 6.
[c] He is quoting the actual words of Chang Po-Tuan on 'sophic lead and mercury', in *Wu Chen Phien*; see p. 201 above.
[d] Ch. 14, in *Wai Chin Tan*, ch. 2, p. 50*b*, tr. auct.
[e] The commentary says that this includes the true or real Yang and Yin.
[f] This sentence seems to lack the final words 'of Fire (element)'.

[1] 先天　　[2] 玄牝之門　　[3] 戊　　[4] 己

cinnabar in an abstruse manner, leaving the teacher to give further details or make his own interpretations when oral instructions were given. An example of this might be found in the tractate entitled *Hsüan Shuang Chang Shang Lu*[1] (Mysterious Frost on the Palm of the Hand)[a] of unknown authorship and date, but roughly assignable to late Yuan or early Ming. This gives only one elixir formula using nothing but lead, mercury and vinegar,[b] and the 'mysterious frost' was presumably nothing but the mixed nitrates and acetates of the two metals, crystallised, precipitated or thrown down on evaporation. The tendency to simplify at this late date might reasonably be interpreted as a sign of the bankruptcy of traditional chemistry, lacking the infinite *élan* which the distinctively modern conceptions and methods soon to arise in the West would bring. For we have now reached approximately the time of Libavius[c] and Becher,[d] and the stage was set for Boyle, Priestley, Lavoisier and Dalton. In the next sub-section something will be said of the coming of modern chemistry to China.

Just as in the +14th century, when the *Mo O Hsiao Lu* (p. 208) was published, so at the beginning of the +17th alchemy still counted among the things which writers of encyclopaedic handbooks expected to discuss. The *Ko Chih Tshao*[2] (Scientific Sketches) of Hsiung Ming-Yü[3] would be a case in point; it deals with astronomy, astronomical instruments, the sphericity of the earth, the magnetic pole and the compass, and many other curious natural phenomena not excluding those seen in alchemical laboratories. This was first finished by +1620, but by the time it was printed in +1648 it had acquired a certain amount of wider information through Jesuit channels, reproducing for instance one version of Ricci's world-map.[e] The same applies to another excellent book, the *Wu Li Hsiao Shih*[4] (Small Encyclopaedia of the Principles of Things) completed in +1664 by Fang I-Chih.[5] This contains quite a lot of chemical knowledge, now rather free from alchemical preoccupations, and we have quoted it often in these volumes.[f]

In the early 19th century Min I-Tê[6] edited the *Tao Tsang Hsü Phien Chhu Chi*[7] (First Series of a Supplement to the Taoist Patrology), a collection of some thirty small Taoist works composed largely between the end of the +17th and the early 19th centuries. Min I-Tê was a pupil of Shên I-Ping[8] (+1708 to +1786), who claimed to be a descendant of the Southern School of Taoism founded in early Sung, but none of the books included in the series is truly chemical in nature. An early 19th-century authority on Taoist physiological alchemy appeared in the person of Fu Chin-Chhüan,[9] also known as Chi I Tzu,[10] who punctuated both the *Tshan Thung*

a *TT* 938.
b On the uses of strong acetic acid see pt. 4.
c +1540 to +1616, cf. Partington (7), vol. 2, p. 244.
d +1635 to +1682, cf. Partington (7), vol. 2, p. 637; Leicester (1), p. 121.
e See Hummel (6).
f E.g. Vol. 4, pt. 2, pp. 432, 524. On Fang I-Chih, a particularly interesting character, see Hou Wai-Lu (3, 4).

¹ 玄霜掌上錄   ² 格致草   ³ 熊明遇   ⁴ 物理小識   ⁵ 方以智
⁶ 閔一得   ⁷ 道藏續篇初集   ⁸ 沈一炳   ⁹ 傅金銓
¹⁰ 濟一子

Chhi and the *Wu Chen Phien*, together with Chhen Chih-Hsü's commentaries on these works, and also produced the *Chêng Tao Pi Shu Shih Chung*[1] (Ten Types of Secret Books on the Verification of the Tao). Most of the tractates included by him deal with physiological alchemy and its techniques, but a few of the others have some connection with proto-chemical alchemy. Once again the substances involved in these texts hardly go beyond lead, mercury, cinnabar, gold, and silver, and one of these books is Li Wên-Chu's *Huang Pai Ching*, which we have just discussed.

(iv) *The legacy of the Chinese alchemical tradition*

Standing now at the conclusion of the story, before modern science came surging in, one is moved to wonder who exactly inherited the wealth of practical technical knowledge accumulated by men such as Wang Chieh and Mêng Hsü, and secondly how it was that the adherents of psycho-physiological alchemy came to preponderate so vastly over the 'sooty empiricks' of the Taoist elaboratories. The latter question is the easier to answer. Surely the trend towards physiological manipulations was connected with the age-old Confucian disdain for artisanal technology,[a] one more indication of the preference of the literati for book-learning and self-cultivation rather than doing anything with their hands.[b] Taoism and all its works naturally suffered when Buddhism was in the ascendant under the Yuan, and could hope for no advancement once a Neo-Confucian orthodoxy had established itself under the Ming; but perhaps there was also an instrinsic core of frustration since Chinese society could not permit the appearance of a radically original scientific revolution. By the end of the Sung period a mass of facts, accumulated since the time of Ko Hung, Thao Hung-Ching and Sun Ssu-Mo, had become known and was recorded, but there arrives at length a limit to what can usefully be done without the illumination of a coordinating theory of modern type, and Mêng Hsü's successors could not go on indefinitely repeating with variations the experiments which had been started in the Han. But the liberating theory never came.

In these circumstances what then happened to the chymical lore accumulated in this tradition? At first sight it seems to disappear, but if we look more closely we shall see that it flowed strongly into other channels, the metallurgical, the industrial and (with particular finesse) the pharmaceutical, all of which benefited by the solid technical work of the alchemical ages. Since alchemy had started with a kind of metallurgy, aurifaction and argentifaction, it was natural for the knowledge of alloys and amalgams to pass over to the professional metal-workers, and indeed one particular reign-period in the Ming became famous for the subtlety of the special bronzes which it produced.[c] So also, as we shall see in another Section, the production of many industrial chemicals reached a high degree of excellence.[d] But it was pharmaceutics which drew most,

---

a Cf. Vol. 2, p. 9.   b Cf. Vol. 2, p. 122, and especially the discussion on pp. 89ff.
c Hsüan-Tê, +1426 to +1435; we shall return to this in Sect. 36.
d Here is the place to recall the *Thien Kung Khai Wu*[2] of +1637 (The Exploitation of the Works of Nature), by China's Diderot, Sung Ying-Hsing,[3] so often quoted in these volumes.

1 證道秘書十種   2 天工開物   3 宋應星

perhaps, from alchemy, generating what at a later stage we shall not hesitate to call a school of iatro-chemists, utilising the techniques of the alchemists to prepare medicines in the spirit of Paracelsus (though without any one particular leading figure), and scoring such extraordinary successes as the preparation of crystalline mixtures of steroid sex-hormones long before the end of the Ming.[a] The attitude of the greatest of the pharmaceutical naturalists, Li Shih-Chen,[1] towards the close of the +16th century, was perhaps typical; he regarded the aurifaction and the macrobiotics of the alchemical tradition as nonsense but was prepared to use to the full, with the greatest interest, all the techniques which in the course of their odyssey they had developed.[b] After all, many of the greatest names in the past, such as Thao Hung-Ching and Sun Ssu-Mo, had belonged fully to both traditions, and in the early Sung work *Wai Tan Pên Tshao*[2] of Tshui Fang,[3] about +1045, we have them combined in a single book-title. Revealing, too, is his other production, the *Lu Huo Pên Tshao*[4] (Pharmaceutical Natural History of the Stove and the Furnace). Thus when the psycho-physiological meditations and exercises of the Confucian-Buddhist gentlemen took possession of the field, Taoist practical alchemical technique, not yet ready to join the universal ocean of modern science, flowed underground in a number of powerful currents giving new life to the master-craftsmen, the *ta chiang*[5] and *chhiao kung*,[6] in many different directions, the metal-workers, the distillers and chemical artists, and the learned pharmacists. Such was one of the ways in which the Chinese technicians earned the reputation which they had among Arabs and Westerners alike in the late Middle Ages as being the greatest master-craftsmen in the world.

(8) THE COMING OF MODERN CHEMISTRY

This last phase of the history of Chinese chemistry has to begin with the time of the early Jesuit mission (+1580 onwards), but modern chemistry did not come to China through that intermediation because modern chemistry had not yet been born. The services of the Jesuits in bringing modern science to China almost as soon as it arose—with Kepler and Galileo—in Europe have often been hymned; and in these volumes we have devoted a good deal of space to their activities in the Sections on mathematics and astronomy.[c] These activities were indeed highly important, and deserve much of the praise which they have received. But astronomy, mathematics and physics are far from being the whole of natural philosophy, and now for the first time we reach a science which the Jesuits could not transmit, because it had not as yet 'found itself'. The fundaments of modern theoretical chemistry, with its atomic and molecular sub-structure, and its rational nomenclature (as opposed to empirical practice and chaotic terminology), were laid, as everyone knows, during the later

---

[a] See Section 45 in Vol. 6, and meanwhile Lu Gwei-Djen & Needham (3).
[b] Cf. his biography by Lu Gwei-Djen (1), and Sect. 38 in Vol. 6.
[c] Vol. 3, pp. 114, 437ff.

[1] 李時珍   [2] 外丹本草   [3] 崔昉   [4] 爐火本草   [5] 大匠
[6] 巧工

+18th and early 19th centuries; with the exploration of the nature of gases by Priestley and others (+1760 to +1780), the 'revolution in chemistry' effected by Lavoisier (+1789),[a] and then the atomic theory of Dalton (1810). The second wave was introduced by the far-reaching insights of Justus von Liebig (1830 to 1840), founder of organic chemistry.[b] In the days of Ricci and Adam Schall von Bell, all this still lay in the womb of time, and even Collas and Cibot in the following century could not do much more. By the time that modern chemistry could have been transported quickly to China the Jesuit mission had been dismantled, and a new generation of missionaries was no longer interested in contacts of so high an intellectual level. Moreover, what is true of chemistry is equally true of botany and zoology. There was a Jesuit transmission in all these sciences, but it was westwards, not eastwards, as we shall see.

(i) *The failure of the Jesuit mission*

During the second half of the +18th century there were still Jesuits in China, and one may ask why they could not have transmitted the discoveries of Black, Priestley and Cavendish; Scheele, Fourcroy and Vauquelin; Lavoisier and even Dalton; just as their predecessors brought joyously news of Kepler and Galileo, Vieta and Boyle. In the early +17th century mathematics and astronomy got through, in the late +18th century chemistry did not. The reason was that the Jesuit mission was now in complete disarray. What does it mean to say that it was being dismantled, and that there were to be no more contacts at high intellectual level until the Protestant period in the second half of the 19th? The role of the Jesuits as transmitters of scientific knowledge had been so outstanding that it is worth while to pause for a moment to consider the tragic story of their collapse. We saw their rise and now must see their fall.

The Rites Controversy was the heart of the matter, nothing to do (at first sight) with science and chemistry, but hardly to be ignored by historians of science.[c] The Jesuits (or the great majority of them) took the view that the veneration of Confucius as a sage and the honours universally paid to family ancestors throughout Chinese culture were both perfectly compatible with Catholic Christianity, and that the rites which expressed these principles were theologically unexceptionable. But this was fiercely contested by other parties in the Church, notably the Franciscans and Dominicans in China and the Jansenists in France and Italy;[d] eventually they gained the ear of the Papacy and events took an inevitable course. Two papal legates were sent to China in +1705 and +1720, and both had audiences with the Khang-Hsi emperor, who issued several edicts defining the Chinese position (which was just

a For reasons which will shortly become clear Lavoisier's work reached China only after nearly a century's delay. On its reception in Japan a little sooner see Shimao Eikoh (1).

b On the whole movement reference may be made to the usual accounts: Thorpe (1); Lowry (1); Partington (4, 7). On Dalton see the new study of Greenaway (4), and the lecture of Hartley (1). There are a number of good accounts of the history of modern chemistry in Chinese, e.g. Chang Tzu-Kao (1); Tsêng Chao-Lun (2).

c Cf. Vol. 2, pp. 498ff.

d The machinations of the Jansenists at Rome have been brilliantly revealed in the monograph of Hay (1), and much other information can be gathered by burrowing through the gingerly-written biographies of Pfister (1). A luminous résumé is to be found in Cordier (1), vol 3, pp. 318ff.

what the Jesuits said it was), all to no effect.[a] A Bull of +1707 against the rites was followed by a stronger one, *Ex illa die*, in +1715, and the final decree, which condemned the Jesuit position, was the Bull *Ex quo singulari*, of +1742. Complex political causes now added to the growing storm, and a Brief, *Dominus ac redemptor*, suppressing the Society completely, was promulgated in +1773, though it did not reach China till a year or more later.[b] This was the time when Lavoisier's quantitative chemical experiments were at their very height, but it is hardly surprising that the Jesuits had no chance to assimilate and expound them.[c]

It is interesting to see who was there when the blow fell in +1774. Out of about fifty Jesuits the majority were engaged in purely pastoral or religious work, but there was a strong minority of scholars and scientific men, if few as distinguished as the leading lights of the +17th century. J. J. M. Amiot (Chhien Tê-Ming[1]) was probably the most remarkable among them, a philologist, historian, geographer, anthropologist, physicist and meteorologist.[d] Most relevant to our present interests were J. P. L. Collas (Chin Chi-Shih[2]) and P. M. Cibot (Han Kuo-Ying[3]), writers of many papers on subjects in Chinese chemistry and natural history. Besides these there were five capable representatives of the old astronomical expertise[e] and four engineering specialists,[f] besides at least one physician[g] and three painters employed at court.[h] The last Jesuit to arrive in China (before modern times) was the greatest botanist of them all, João Loureiro, a man who had spent most of his life in Indo-China (Cochin-China[i] or South Vietnam) and did not reach Canton till +1779. His *Flora Cochin-*

[a] Both the legates were bishops who had obtained Latin patriarchal titles. Charles Maillard de Tournon, who died in Macao in +1710, was Latin Patriarch of Antioch, but for the Chinese he had the much less high-sounding name of To Lo.[4] He was very anti-Jesuit and anti-Chinese so that his visit had no conciliatory value whatever. Carlo Mezzabarba, Latin Patriarch of Alexandria, in China from +1720 to +1722, was a much more understanding man, and did his best to explain matters at Rome when he returned there, but fruitlessly.

[b] The best account of the Suppression in China is still that of Cordier (13). There is also the interesting monograph of de Rochemonteix (1), not so much a life of Amiot, as its title might lead one to believe, but a detailed description rather of these unhappy events.

[c] Later on, however (p. 242 below), we shall describe a fragment of what may have been a Jesuit attempt to explain the discovery of oxygen in Chinese.

[d] He gained permanent renown for his elaborate monographs on Chinese music and Chinese military technology (see Sects. 26 (*h*) and 30). The other philologist was J. F. M. D. Ollières (Fang Shou-I[5]), outstanding as linguist and translator in Manchu as well as Chinese.

[e] Felix da Rocha (Fu Tso-Lin[6]), the eighth Director of the Astronomical Bureau; José d'Espinha (Kao Shen-Ssu[7]), the ninth; J. B. d'Almeida (So Tê-Chhao[8]), the tenth and last Jesuit Director (d. 1805); André Rodrigues (An Kuo-Ning[9]), a Vice-Director and perhaps for a short interregnum Director; and J. J. de Grammont, whose Chinese name we do not know. It was Grammont who transmitted a copy of the *Tao Tê Ching* to the Royal Society (cf. Vol. 2, p. 163).

[f] Michel Benoist (Chiang Yu-Jen[10]); the two horologists J. M. de Ventavon (Wang Ta-Hung[11]) and Hubert de Méricourt (Li Chün-Hsien[12]); and an Italian gem- and glass-worker Luigi Cipolla.

[g] Louis Bazin (Pa Hsin-Mou[13]), a lay brother.

[h] Giuseppe Panzi (Phan Thing-Chang[14]), Louis de Poirot (Ho Chhing-Thai[15]), and the Czech Ignatius Sichelbarth (Ai Chhi-Mêng[16]).

[i] A name derived from the ancient Chinese appellation Chiao-chih[17] for Annam.

[1] 錢德明	[2] 金濟時	[3] 韓國英	[4] 鐸羅	[5] 方守義
[6] 傅作霖	[7] 高愼思	[8] 索德超	[9] 安國寧	[10] 蔣有仁
[11] 王達洪	[12] 李俊賢	[13] 巴新懋	[14] 潘廷璋	[15] 賀清泰
[16] 艾啓蒙	[17] 交趾			

chinensis of +1790, published after his return to Portugal in +1782, remains to this day one of the greatest landmarks in the taxonomy of the area. On the whole however it is clear that the mission had no one qualified to bring modern chemistry to China in the way in which Renaissance mathematics and astronomy had been brought in earlier days.

In many great human social organisations there occur from time to time processes which recall the autotomy seen in comparative physiology, occasions when the main body turns against a militant part of itself and amputates it to the accompaniment of untold anguish both physical and mental suffered by its most faithful supporters. The suppression of the Society can be compared only with the gruesome story of the extinction of the Knights Templars earlier and the liquidation of the Old Bolsheviks later.[a] True, the Chinese Jesuits were so fortunate as not to have to undergo the long years of imprisonment in dungeons which befell many of their brothers; they just lived on in Peking and a few provincial cities as a disorganised remnant, trying to hold together what remained of the past.[b] When in 1814 the Society was fully restored there was only one left to see it, Louis de Poirot, then in his last year aged eighty,[c] but a few years earlier five Peking fathers had re-joined as members of the West Russian province, which had been able to persist throughout the persecution. But when in 1939 the Rites judgment itself was at last revoked by the Papacy and the compatibility of Confucianism with Christianity recognised, those most concerned had been in their graves for two centuries or more.

Let us turn now to the minor part played by the later Jesuits in transmitting to the West knowledge of proto-chemistry in China. To begin with, the Chinese ideas about elixirs of immortality gradually became known.[d] In their influential *Confucius Sinarum Philosophus* of +1687, Prosper Intorcetta,[e] Philippe Couplet[f] and two other Jesuits published the first translation of the Confucian classics into a Western language, together with some of the commentaries, and further explanations of their own.[g] Here reference was made to the elixirs in the course of a discussion of ancient and traditional Chinese ideas on ancestors and immortality, ideas which Couplet and his colleagues were anxious to assimilate as much as possible to those held in the Christian West. Speaking therefore of the thirst for an after-life, they wrote:

Qui appetitus quam vehemens in quibusdam illorum fuerit, argumentum possunt esse pretiosae potiones illae seu Ambrosiae vitae immortalis quas Imperatores quoque fabulis

---

[a] One might instance also the dismantling of the Jesuit communities in Paraguay in the interests of Spanish colonial exploitation of the Indians (see Gide, 1).

[b] There were some painful internal disputes about the mission's property, until the Lazarists were designated as the inheritors of it in +1781.

[c] De Poirot had been a considerable linguist as well as a painter, and had translated much of the New Testament into Manchu as well as Chinese.

[d] They had of course exercised, as we emphasise elsewhere (pt. 2, pp. 14–15, and pt. 4), an overwhelmingly important influence on European alchemy and chemistry from the +12th century onwards, but their original provenance was not then, or for long afterwards, understood or recognised.

[e] Yin To-Tsê[1] (+1625 to +1696).

[f] Po Ying-Li[2] (+1624 to 1692).   [g] Cf. Vol. 2, pp. 163,

[1] 殷鐸澤   [2] 柏應理

impostorum, aliquot post Christum saeculis, delusi spe, immortalitatis hauriendae, sume-bant.[a]

Among later readers of this passage was Nicolas Fréret, whose notes on the whole work we still possess.[b] Writing about +1737, he said:[c]

The desire for immortality has been shared by the Chinese with other men, and history tells us of the ardour with which they sought for the Potion which was supposed to be able to give it to them. Those who searched the most assiduously for this Potion, however, were the Taoists or sectaries of Laokiun (Lao Chün), who maintain that the soul is mortal. With this in mind they are right to seek to prolong a life beyond which there is nothing in store for them. But the Fohists (the Buddhists), who do believe in another life, do not so greatly fear death.

He then went on to examine all the other Jesuit arguments for ancient Chinese belief in the immortality of the soul, and naturally found them not convincing.[d] But of course at this early stage the question was confused by the European inability to conceive of the Taoist idea of material immortality, a wraith-like duration not in heaven but on or above the earth.[e]

It is rather remarkable that the Jesuits themselves were at the beginning very often thought to be alchemists.[f] The fine scientific scholar Chhü Thai-Su,[1] who definitively accepted Christianity in +1605, had originally made friends with Matteo Ricci because he believed that he could get help from him in his alchemical work.[g] At an earlier stage we have related, in the original words, an account of the lamentable occasion when at this time a large library of valuable books on the sciences and proto-sciences was sent by Chhü Thai-Su to the Jesuits to be burned.[h] A good many writings on alchemy and early chemistry must have been among them. It was not an isolated example,[i] for there had been an even greater holocaust three years earlier, and for every similar *auto-da-fé* which was written about, when the convert was sufficiently distinguished, there must have been several others, records of which have not come down to us. Our judgment on these aberrations was rightly severe, yet they were strangely paradoxical in view of the exceptionally scientific character of the Jesuit

[a] P. 95. 'How vehement was this desire (for immortality) among certain of them can be seen from those potions or precious Ambrosias of immortal life which the emperors took in several centuries of the Christian era, thirsting for immortality but deluded in their hopes by the fables of charlatans.'

[b] In Pinot (2), p. 115. We have met this polymath before (Vol. 3, p. 182) in connection with the clash between Chinese and biblical chronology, in which he was much interested. Fréret (+1688 to +1749) is regarded as the founder of French sinology; a friend of Arcadius Huang,[2] who came to Paris in +1703, he explained the radical system, and composed a Chinese grammar while imprisoned in the Bastille in +1714. On Huang see Cordier (10), p. 15; Huard & Huang Kuang-Ming (5), p. 138.

[c] Pinot (2), p. 122, eng. auct.   [d] Cf. Vol. 2, pp. 490 ff.

[e] A long account of the Taoists by Amiot (9) appeared in +1791, and this had more to say about their ideas on the prolongation of individual existence, yet there is little concerning alchemy in it.

[f] See, for example, Ricci (in d'Elia (2), vol. 1, pp. 107, 240ff., 278, 313, 347, 359, 375; vol. 2, pp. 29, 108, 117 etc.). Cf. Mish (1), p. 76.

[g] Ricci (in d'Elia (2), vol. 1, pp. 296ff.).

[h] Vol. 4, pt. 1, p. 244, based on Ricci himself (in d'Elia (2), vol. 2, p. 342). Cf. Trigault (Gallagher tr.), p. 468. Parallel activities on the part of the Jesuits relative to Sanskrit writings took place in India; cf. Lach (5), p. 438.   [i] Cf. d'Entrecolles (1), p. 120.

[1] 瞿太素   [2] 黃

mission, and doubtless if chemistry at that time had been as advanced a science as astronomy the Jesuits might not have acted as they did. It was a strange fate for a science to suffer on account of its backwardness, but nothing can modify our condemnation of book-burning in whatever cause. It will of course have been obvious from the preceding sub-sections that alchemy had for many centuries been closely related to the 'illicit' magical arts of enchanters and wonder-workers—heterodox pseudo-science for the Confucians as much as for the Jesuits. But the Confucians would never have burnt the rules of the art, for in traditional China the written word had a sacredness of its own, and the supposed example of Chhin Shih Huang Ti was execrated throughout Chinese history.[a]

If the Jesuits were assumed now and then to be alchemists and magicians there is some reason for believing that they had only themselves to thank. This is illustrated in a fascinating letter addressed from Peking in the autumn of +1735 by Dominique Parennin (Pa To-Ming,[1] +1665 to +1741) to one of the Academicians in Paris, the physical chemist J. J. Dortous de Mairan.[b] After thanking him for a further sending of copies of his publications to the Jesuit library in Peking, Parennin recalls an earlier dissertation of his in +1716 on the freezing of water and salt solutions which had won a prize at Bordeaux and by which the Jesuit had been much impressed. Parennin then goes on to say that in the same year, being in attendance on the Khang-Hsi emperor during a winter hunting expedition, he had found himself drawn on to convince a group of scholars, including two Ministers of State and ten Han-Lin Academicians, that one could make water freeze near a hot brazier. Parennin's discourse on congelation in the President's tent, if we may judge from his account of it, was, though corpuscular, hardly less vague than that of the Chinese, and equally erroneous.[c] But when the experiment was tried with a bowl of snow standing in a dish of water, he managed to slip some saltpetre surreptitiously into the cup, so that the temperature of the melting snow dropped to $-20°$ C or so and induced the water in the saucer to freeze solid although quite close to the fire.[d] The platter then remained suspended by

[a] Cf. Vol. 1, p. 101. It is interesting that even in the Chhin 'burning of the books', all science and technology, including pseudo-sciences such as divination, were exempted.

[b] +1678 to +1771; cf. Partington (7), vol. 3, p. 153. Continuing to work on freezing mixtures, he published his most important paper on them in +1749.

[c] They spoke of course of the 'occult qualities' of Yin and Yang. He said that the liquid state of water was due to air mixed among the particles, and that in order to freeze it 'il ne s'agissoit que de la déranger', extracting the most subtle particles which prevented the others from mutual attachment, and introducing new ones capable of bringing about fixation and immobility.

[d] Knowledge of freezing mixtures goes back at least to the beginning of the +16th century, as may be seen from the special study of the subject by von Lippmann (8). One of the best-known mentions was that of della Porta in +1589, who used, in fact, ice and crude nitre, but it was not by any means the earliest. About +1550 a Spanish physician at Rome, Blasius Villafranca, published a tractate entitled *Methodus Refrigerandi ex vocato Salenitro Vinum Aquamque ac Potus quodvis aliud Genus*; in this he described the cooling of cold water by the addition of saltpetre. The endothermic effect of dissolving certain salts and crystalline hydrates had been mentioned even earlier (c. +1530) by the Italian physician Zimara in his *Problemata*. At this time the procedure seems to have been quite widely used for cooling wine in summer.

The depression of the freezing-point by salt and saltpetre added to ice and snow aroused the keen attention of Francis Bacon in +1623, and Robert Boyle wrote a special communication (4) on freezing mixtures for the *Philosophical Transactions* in +1666 (cf. Partington (7), vol. 2,

[1] 巴多明

itself, and after the contents of the bowl had been thrown into the glowing charcoal the Academicians were able to verify that the ice was pure and normal. The success of his demonstration was complete, but he neither told nor explained what he had done.[a]

Next day they fell to talking of hailstorms and thunder and lightning, which they said often exerted its effects downwards, rather than upwards like gunpowder; Parennin accordingly promised to show them a powder which would likewise blow a hole downwards through an iron spoon. Having the materials for making some fulminate[b] he did just this the following evening, awakening still further the admiration of the company, one of whom remarked that thenceforward he would feel compelled to believe anything Parennin said, and feared that he might therefore be deceived. 'I am incapable of deceiving anyone,' said Parennin, 'and on the contrary would be only too happy to be able to undeceive you from the religious errors which you are in, and which are of far greater consequence for your happiness than the ignorance of a few natural phenomena.' Another day the conversation turned to the formation of minerals in the bowels of the earth, and the Jesuit promised to show them, when they should all have returned to Peking, a solid white mass formed by the combination of two liquids—first there would be a violent effervescence or combat of the two, ending only with their mutual destruction and the formation of a white 'stone' at the bottom of the vessel.[c] 'But', he went on, 'you must remember that

pp. 21, 400, 539; vol. 3, pp. 64, etc.). Aggiunti before him (+1635) had been perhaps the first to try the effects of a variety of different salts, and this work was much followed up once the +18th century had begun. Dalencé's attempt to fix a thermometric zero by the aid of freezing mixtures in 1688 led later to the well-known experiments of Fahrenheit in +1724 and Reaumur in +1734. The invention of 'ice cream' is also part of the story, beginning with Procope Couteaux in Paris in +1660. Thus Parennin's parlour-trick had a considerable background.

But it seems unlikely that the refrigeration effects of salt solutions were really a European discovery. The great historian of medicine Ibn Abū Uṣaybi'a (cf. Hitti (1), 2nd ed. p. 686; Mieli (1), p. 168) has a passage in his Kitāb 'Uyūn al-Anbā' fī Ṭabaqāt al-Aṭibbā (Book of Sources of Information on the Classes of Physicians), in which he speaks of the making of artificial ice or snow by the addition of saltpetre to cold water. As he lived from +1203 to +1270 he long preceded the +16th-century Europeans, but even he ascribes the process to an older author, Ibn Bakhtawayhī, of whom nothing is known. The Muslims of Spain could have been the transmitters, but possibly the first observations were Indian, for the +4th-century Pancatantra has a verse saying that water can become really cold only if it contains salt (Fritze tr., p. 160). The evaporation of water at night in flat porous earthen vessels set on layers of straw in shallow pits is an ancient Indian custom, and might have facilitated the discovery. In pt. 4 below we shall return to these cryological topics in connection with alcohol in East Asia.

[a] Unless it was to the Han-Lin President, a particular friend, who divulged it no further at that time.

[b] Fulminating gold or aurum volatile was, it seems, the earliest of these compounds to be discovered. A complex formed from ammonia and gold chloride or hydroxide, the exact nature of which is still a little obscure (cf. Mellor (1), p. 390, Partington (10), p. 358), it is mentioned already in the late +16th-century 'Basil Valentine' Corpus, c. +1604, then by Croll in +1609. This was doubtless what Parennin made, for the silver compound (probably the nitride, Ag_3N, Durrant (1), p. 594), and the much more important mercury one (apparently $Hg(ONC)_2$), were not described until the work of Kunckel in +1716. On these developments see Partington (7), vol. 2, pp. 176, 197, 246, 377. It was in 1867 that Alfred Nobel introduced the use of mercuric fulminate in detonators for high explosives (Sherwood Taylor (4), p. 254), and I have vivid memories of visiting the fulminate plants in Chinese arsenals during the second world war. It is a tricky substance to work with.

[c] He did not say what he intended to use, but one could easily imagine a saturated solution of calcium hydroxide mixed with strong sulphuric acid. Some further light on the background of this demonstration may be gained from Debus (16, 17); attempts to reproduce geological and pedological phenomena in the laboratory had been a feature of the iatro-chemical movement of the preceding century.

you are giving me your word that you will hereafter listen to me with greater docility when I speak to you of a subject far more elevated and of infinite advantage to you, since it will gain for you eternal felicity.' And he concluded his account to de Mairan by saying that in dealing with Chinese scholars 'one must gain their esteem by a knowledge of natural things, which mostly they lack, but about which they are curious to learn—nothing better disposes them to attend to our preaching of the sacred truths of Christianity'.

Dominique Parennin was an outstanding linguist, an estimable man and a devoted priest, yet his action in using physico-chemical knowledge to perform tricks with the object of inducing belief in the body of religious dogma which he represented, and without explaining the meaning of his demonstrations to his listeners, was something we may find it hard to forgive today. Missionaries of all religions everywhere have sought to accredit themselves by signs and wonders, whenever they could contrive to do so, but seen in the light of the ethic of the scientific world community of today there was something almost sacrilegious in Parennin's attitude. Thaumaturgy was in bad taste, to say the least, when what was called for at that particular time and place, in eighteenth-century China, was the transmission of true chemical knowledge and understanding.

Ricci himself had a good deal to say on the Taoist alchemists.[a] In the chapter of his diary on superstitions and bad customs he wrote as follows:[b]

We may conclude all these distressful things by speaking of two rather fantastic obsessions common in all the fifteen provinces of China, where there are many who give themselves over to them. One is the claim that from mercury and other substances it is possible to make true silver. The other is the belief that by means of various medicines and exercises it is possible to attain perpetual life without ever dying. Ancient tradition holds that these two systems have been handed down by those who have been considered saints [i.e. the holy immortals], men who having, they say, accomplished many good works, flew up body and soul together at last into the heavens. At the present time the books on these two sciences circulate in ever-increasing and enormous numbers, some printed, and others, more greatly prized, in manuscript form.

He then continues with two long paragraphs very much as in the later recension of Trigault,[c] ending with a somewhat garbled version of the story in the *Han Fei Tzu* book already quoted (p. 7 above). His testimony must be of some value in any estimate of the prevalence of the different forms of alchemy in the late Ming (cf. pp. 208–9 above).

In the following century an interesting exchange on Chinese alchemy took place between a Dutch philosopher de Pauw and a Chinese scholar who had become a

---

[a] And on the Taoists themselves, cf. d'Elia (2), vol. 1, pp. 129, 131.

[b] D'Elia (2), vol. 1, pp. 104ff.; cf. Trigault (Gallagher tr.), pp. 90ff. eng. auct.

[c] The first on the *wai tan*, the second on the *nei tan*, schools. Of the latter (in Trigault's words): 'Here in the province of Peking, in which we are living, there are few if any of the magistrates, of the eunuchs, or of others of high station, who are not addicted to this foolish pursuit.'

Jesuit, Aloysius Ko (Kao Lei-Ssu,[1] +1734 to *c.* +1790).[a] De Pauw we have come across before;[b] he was one of the leaders of the movement against the sinophilism which the Jesuits had inspired. In his *Recherches Philosophiques sur les Egyptiens et les Chinois* of +1774 he set out to deny all connections between them, and to 'debunk' the claims which had been made on behalf of Chinese culture. Almost the only writing of Kao Lei-Ssu which ever saw print was a long and detailed reply to de Pauw's criticisms, partially published in the Jesuit *Mémoires* for +1777. Here are two of the remarks of the Netherlander to which the Chinese took exception. In each case (nos. 70 and 71) the Chinese Jesuit first quotes de Pauw's words, and then gives his opinions on them.

No. 70. 'The Jesuits have tried to depict the Chinese for us as determined chymists' (p. 356).

If the Missionaries have written only a portion of what he attributes to them, or even all of it, it is because—apart from the fact that our Annals speak in many places of the ridiculous mania of some emperors and scholars who believed they could obtain by means of chymistry a universal panacea—they (the Missionaries) had a thousand questions of the Chinese on this matter to answer, questions so full of conviction that they were obliged to reply to them in many of their books, and to state quite clearly that people in the West did not know the secret of the transformation of metals, nor any universal panacea either. If there is a copy of the *Tay-y-pien*[c] in the Royal Library (at Paris), as we think there is, one can see that the Author, writing more than a hundred and fifty years ago, took occasion of the superiority of Europeans in chymical knowledge to disabuse his compatriots of a chimaera with which they continued to be infatuated. If anything more were needed to offer to those who should think as our Author (de Pauw) does, we can quote to them from the Sung Annals the article on Yang-kiai[d] and that on Tchang-yong.[e] Of the first it says that in the belief that it was possible to change tiles and stones into gold (*Hoa-oua-che-ouei-hoang-kin*),[f] he left his

---

[a] He had been educated in France, and was, with Yang Tê-Wang,[2] one of the two Chinese commissioned under the minister Bertin to study physics, chemistry and industrial technology there before returning home with a view to subsequent liaison in these subjects between China and the West. They both took orders in +1763 and travelled widely throughout France, but once back in China no opportunity offered for using their knowledge, and they spent their lives in pastoral work in the provinces, not much affected, apparently, by the Suppression.

[b] Vol. 1, p. 38.

[c] Undoubtedly a reference to a small work entitled *Tai I Phien*[3] (On Replacing Doubts by Certainties), written and published in +1621 by a Christian apologist, Yang Thing-Yün[4] (Dr Michael), with a preface by Wang Chêng,[5] the engineering friend of Johann Schreck (Terrentius) whom we met so often in Vol. 4, pt. 2, e.g. pp. 170, 171. Our present text of it, however, says little to support Kao Lei-Ssu's interpretation. The only reference to chemistry occurs in ch. 1, p. 36*a*, where in a moralistic context we read: 'Brass looks like gold, but if it be tried in the fire it will produce nothing but ashes.' Since cupellation had been known and practised for many centuries in China, this did not say much for the superiority of European chemistry. On 'Dr Michael' and his book see Pfister (1), p. 109 and app. p. 15.

[d] Clearly identifiable as Yang Chiai,[6] whose biography is in *Sung Shih*, ch. 300, p. 1*a*. This says, however, that after an alchemist had demonstrated to him the art of turning tiles into gold, and offered to reveal the technique to him, he declined to accept it. He is also mentioned in *Wu Li Hsiao Shih*, ch. 7, p. 12*a*.

[e] Clearly Chang Yung,[7] whose story we have told on p. 192 above. His biography is in *Sung Shih*, ch. 307, p. 3*b*.

[f] *Hua wa shih wei huang chin.*[8]

[1] 高類思   [2] 楊德望   [3] 代疑篇   [4] 楊廷筠   [5] 王徵
[6] 楊偕   [7] 張雍   [8] 化瓦石爲黃金

official employments to go and carry on the Great Work. And the second believed that he had seen silver changed into gold by means of the vapour of some composition. It is remarkable that Tchang-yong was impressed by the antiquity of this secret. And indeed one finds in the ancient book *Tsay-y-chi*[a] that a certain person of olden times had changed roots and earth into gold by calcining them in a vessel shaped like the head of a bird.[b] Let us add one interesting fact. The lodging which the Kang-hi (Khang-Hsi) emperor gave to the French Jesuits, within the first (outer) enclosure of the Palace, had been, under the preceding Dynasty, a house set apart for the prosecution of the Great Work, and what doubled the value of the gift was that we found among these old walls and subterranean vaults much excellent material which was made use of to build a church. . . .[c]

The passage throws several interesting sidelights upon alchemy on the threshold of the chemical age. Kao Lei-Ssu is concerned to refute de Pauw's scepticism about Chinese chemical knowledge while at the same time not defending the belief in aurifaction which had previously been so widespread in China and which the Jesuits regarded as a superstition. At the same time he reveals the existence of what looks like an imperial alchemical elaboratory attached to the Palace in Ricci's own time, the late +16th century. In the second passage he points out against de Pauw that there is nothing about alchemy in the Confucian classics:

No. 71. 'This superstitious craze came to them from their Tartar ancestors' (p. 357).
In this case, the further one went back into antiquity, or, if you will, the nearer the time of our supposed descent from the Scyths, the more one would find this 'superstitious craze' evident, general and clearly stated. Unfortunately there is not a single word about it in our ancient King (*Ching*, classics) and suchlike books. True, the Chan-hai-king (*Shan Hai Ching*), Kouan-tsée (*Kuan Tzu*), Lié-tsée (*Lieh Tzu*) and other books and Authors of later antiquity, speak touching this matter of a garden of delights in which the herb of immortality grew, though some thought it was a tree, others a fountain, some a mushroom and others again a plant. Tsin-chi-hoang (Chhin Shih Huang Ti, the first emperor), who would very much have liked to escape death, consulted all his Geographers but no one could tell where this garden of delights was to be found; it was only then that recourse was had to alchemy with the object of preparing a potion that might substitute for the herb in the garden of delights. For the rest, if our Author wishes to think in this way of the garden of delights, we shall readily come to agreement with him, averring there can be no doubt but that the Tartars of the plain of Senaar,[d] our first ancestors, handed down this tradition to us, distorted and disfigured though it afterwards was by the Tao-sée (Taoists).[e]

Thus Kao explained, with considerable perspicacity, that the idea of the plant of immortality was assuredly a good deal older than that of a chemical elixir. But all these exchanges gave Europeans no idea of the real state of empirical chemistry and chemical technology after two millennia of development in China.

[a] This book we have so far been unable to identify. The last two characters are surely *I Chi*,[1] but there are a great many books the titles of which end in this way.
[b] This description evokes an apparatus for distillation, of which we would like to know more.
[c] Kao (1), p. 492.
[d] Surely the Sunggari Ula (Milk River), Sung-hua Chiang,[2] the principal river of Manchuria, cf. Gibert (1), p. 820. Kao speaks as if himself a Manchu, as perhaps he was.
[e] Kao (1), p. 494.

[1] 異記   [2] 松花江

Very little more was made available in the essay 'Des Sciences Chimiques des Chinois' which formed part of Grosier's seven-volume book on China completed by 1820. In this somewhat light-weight section[a] Grosier made rather generous remarks on traditional Chinese chemical technology but went widely astray in supposing that mineral remedies had been absent from Chinese pharmacy; the case was precisely the opposite, for there had been no Galenic herbalism in China. In the usual Jesuit style he was hostile to alchemy (not unreasonably, at this date), mentioning Yang Chiai and Chang Yung again, and reporting information he had had from F. X. d'Entre-colles[b] that some Chinese had adopted Christianity because they thought it would help them in their alchemical experiments. He ended, entertainingly, by repeating as a 'fait historique' obtained from d'Entrecolles the novelette from the late Ming *Chin Ku Chhi Kuan* which we have sketched on p. 213 above.[c]

D'Entrecolles' letter, however, had shown the enlightened spirit in which some of the eighteenth-century Jesuits aimed at transmitting to Europe knowledge of Chinese discoveries and inventions.[d]

I have been hesitating for some time [he wrote] as to whether or no I ought to acquaint you with certain secrets and other rather curious observations which I have found in Chinese books, because I have had neither the leisure nor the facilities for making any tests which could verify them; but I have been reassured by an ingenious reflection made by a celebrated Academician in a similar situation. This is how he expresses himself in the volume of the History of the Academy for the year 1722.

'The physicists, who ought naturally to be the most incredulous about these sorts of marvels, are nevertheless those who show the least contempt in rejecting them, and have the most favourable inclinations for examining them. For they know better than the majority of men how vast is the extent of what is still unknown to us in Nature.'

This encourages me much in venturing to entertain you with some of the Chinese discoveries, based upon the accounts only of their own writers. Even if these should serve but to exercise the sagacity of our learned artists, they will not be altogether useless.

And he went on to describe the production of artificial pearls,[e] several porcelain techniques, the surface coloration of metals by inorganic and organic substances, and the thermo-remanent magnetisation of steel needles heated (with a complex mixture of chemicals) and then cooled in the earth's magnetic field.[f]

During the sixties and seventies of the +18th century Cibot and Collas, armed with Macquer's *Élémens de Chymie*, made a real effort to find out something about the

---

[a] Grosier (1), vol. 6, pp. 94 ff.

[b] Yin Hung-Hsü[1] (+1662 to +1741). D'Entrecolles was the Jesuit who settled at the great porcelain centre of Ching-tê-chen in Chiangsi, and played an important part in transmitting knowledge of the technology to Europe (cf. Sect. 35).

[c] All this had come through Duhalde, in a letter from d'Entrecolles to him dated 4 November 1734, the original of which can be read in *Lettres Édifiantes et Curieuses*, vol. 22, pp. 120 ff.

[d] (1), pp. 91 ff.

[e] Cf. Vol. 4, pt. 3, p. 675.

[f] Cf. Vol. 4, pt. 1, pp. 252 ff. and W. A. Harland (1) who observed the same process in 1816. D'Entrecolles was particularly puzzled by this process because the ingredients were all unnecessary and what mattered was the orientation during the cooling, a point no doubt omitted in the texts.

[1] 殷弘緒

Chinese chemical manufactures and the knowledge implicit in them. The results were reported in the *Mémoires concernant les Chinois* . . . during the eighties, after the deaths of the two naturalists in +1780 and +1781 respectively; writings neither extensive nor very professional, yet a unique attempt to convey to Europeans that in China also an empirical chemistry had equally developed all ready to be incorporated into the new synthesis. Judging from the dates, Cibot and Collas probably had only one of the earlier editions of Macquer's book (+1749 to +1756), which did not contain the new discoveries, but the revised edition of +1775, which did, could have reached them just in time.[a] In reading their accounts one is impressed by the difficulties under which they laboured, they had no proper laboratory at their disposal, they had no means of travelling to the localities where chemical industries were at work, and apparently they could not even go out to chat with the apothecaries and ransack the drugstores in person.

Nevertheless, Cibot (5) led off with an interesting account of the fermentation industries, the distillation of spirits from the cereal wines characteristic of China, and the making of vinegar.[b] His next paper (11) was devoted to that ancient central substance, cinnabar. He described the various colours of the natural compound in commerce, told how the pharmacists made preparations of it evaporated to dryness with plant extracts, and expounded its use (as well as that of metallic mercury) in intestinal and external affections, not forgetting to say something of the danger of poisoning. He thought that the synthesis by subliming mercury and sulphur must be as old as the Thang period, but doubted whether it went back (as we think now that it did, cf. pp. 67–8) as far as the Han. He struck one note of particular interest in connection with what has been said on pp. 208, 217 above, namely the recognition of a progressive diminution during late times of the number of constituents of elixir-like medicines.

Since, according to Chinese philosophy, [he wrote] the treatment of diseases is a need common to all men, the case must be as it is with food and clothing, universal needs which it should be very easy to satisfy. . . . Imbued with this great principle, and fortified by the beliefs and practice of the ancients, it has succeeded in convincing the public mind that only few remedies suffice, and that a multitude of them would be a luxuriousness as absurd, unjust and ill-omened as that of banquets and rich apparel. . . . And indeed it is a fact that Chinese culture, which in the past revelled in multitudes of materia medica, and even the most subtle and learned of chymical remedies, has in the end come back to reliance upon very few, almost as if by a return to the primitive ages.

Hence the importance which cinnabar and mercurial preparations retained. Another point in Cibot's paper concerns the report of the isolation of mercury from plant tissues—well justified in fact.[c]

We are aware [he went on] that certain Chymists have sought to divert the public at the expense of the Missionaries who sent to Europe the news of the extraction of mercury from

[a] The point is important, as we shall shortly see (p. 247). On Macquer and his books see Partington (7), vol. 3, pp. 80 ff.; Coleby (1). [b] See further in Sect. 40.
[c] See Vol. 3, p. 679. D'Entrecolles (1), pp. 111 ff., had reported it in +1734.

plants, but that has not prevented us from mentioning it.[a] The Chinese books from different centuries are so unanimous in their accounts of this kind of mercury that we have preferred to rely upon their testimony in a factual matter, rather than to believe in the infallibility of a science which, after all, may well be backward on many points of the technical arts and natural history of East Asia. Since mercury is so common in many places among the different Provinces, why should it be impossible to find it in the water-lentils and other aquatic plants of the marshes?

He was quite right, for we know today that metal elements can be obtained in surprising amounts from particular plant species which act as accumulators for them, often serving as signposts in bio-geochemical prospecting. In a third paper Cibot (12) took up the question of borax in China. He knew that most of the sodium borate came from the Tibetan border (the regions around Lake Kokonor), and some from Hainan (but it was more probably Kuangsi). Although the information which he drew from them was somewhat vague—'nitre and arsenic can substitute for borax in solders, smelting fluxes and the purification of metals', or, 'certain plants (named) can dissolve and decompose borax'—his words show that he read the relevant authors quoted in the Pên Tshao literature quite carefully, and his datings of the books in question, e.g. the +10th and the +13th centuries, were perfectly correct.[b] He also dwelt briefly on the possible significance of the different ways of writing the characters of the Chinese name, *phêng sha*.

About the same time Collas (3) was devoting himself to the naturally-occurring 'soda', nitron or natron, of China, a mixture of salts called *chien*.[1] This soil-surface incrustation is, as Schlegel (11) reported, a mixture of some 60% sodium sulphate, 30% sodium chloride, and 10% sodium carbonate, analogous to the *tequesquite* of Chile.[c] The carbonate and sulphate were traditionally separated in China by fractional crystallisation, filtration and decantation, the former being used as a bleach and mordant, and for baking-powder, the latter for medicinal and industrial uses. The carbonate tends to be accompanied by organic pigments, hence its name *tzu chien*[2] or 'purple' soda; this was described by Collas, who noted its employment in soap-making, especially in the north, where the soap-bean saponin detergent was less common.[d] Collas found that the natural alkali tended to lose its water of crystallisation rather than show any deliquescence, and he tried to analyse it with the inadequate means at his disposal, obtaining 'quadrangular nitre' (sodium nitrate)[e] after treatment with nitric acid, and Glauber's salt (sodium sulphate) using a rough titration

[a] He must have been thinking of some more recent writers, but the preparation of mercury from plants had been strongly denied as early as +1661, when Werner Rolfinck listed it among the 'nonentia' of chemistry in his *Chimia in Artis Formam Redacta* (cf. Partington (7), vol. 2, p. 313). J. F. Henckel in +1760 recognised the reasonableness of the Chinese reports (*Flora Saturnisans*).

[b] See below, *sub voce*, in Sect. 34. From an aside in Cibot (11) we learn that the Jesuits had access to the great *Pên Tshao Phin Hui Ching Yao*, a pharmaceutical natural history which was still at that time in MS. form, and only to be consulted in the Imperial Library. On this, see further in Sect. 38.

[c] The old term *chien* finds use in contemporary chemical nomenclature as a general appellation (*chien shih*,[3] *chien hsing*[4]) for basic salts, slags, etc. See further, pt. 4 below.

[d] See Sect. 44, and in the meantime Needham & Lu Gwei-Djen (1).

[e] Cf. Crosland (1), p. 76.

[1] 鹼   [2] 紫鹼   [3] 鹼式   [4] 鹼性

with ferrous sulphate by taste. He concluded therefore that it contained the basic principle of marine salt, or, as we should say, sodium.

The other papers of Cibot and Collas were somewhat more metallurgical. Collas (5), however, discussed *huang fan*[1] and *nao sha*[2];[a] the former, he thought, resembled white vitriol (zinc sulphate),[b] though it must have been either yellow iron alum or ferric sulphate; as for the latter (sal ammoniac), several different varieties were circulating in commerce, some of which appeared to be forms of rock-salt and not ammonium chloride at all. Then an interesting paper on the coal industry around Peking was written by Collas (6). In another (4) he described a kind of 'black chalk' which was used for cement, sketched the making of incense, and expatiated on the beautiful coloured glaze frits used for the tiles of the great buildings in Peking—a subject also dealt with by Cibot (14). The characteristic Chinese alloy, paktong (*pai thung*[3]), was the subject of a note of Collas (7). As we have seen already in pt. 2, pp. 225 ff., this attractive metal, consisting of copper, zinc and nickel in the proportions 50%, 30%, and 20% approximately, with small amounts of tin, lead, iron and cobalt on occasion, is essentially what the West has called argentan or 'German silver',[c] but the Chinese manufacture long antedated the European. What chiefly puzzled Collas was whether the alloy, which he knew came largely from Yunnan, was the product of a mixed ore or not; in fact the Chinese used both pentlandite and nickel arsenides.[d] Of course Collas could not have guessed the most interesting constituent, nickel, for although the metal had been isolated first by A. F. Cronstedt in +1751 and named by him three years later, the full investigation of it was only now proceeding, by Collas' contemporary T. Bergman (+1775).[e] Finally Cibot (13) wrote well on the iron and steel industry,[f] while Collas (8) contributed an account of a kind of gilt paper much used in the Peking of his time, as also of the gold paint used on furniture and buildings, apparently some golden substance finely divided and suspended in tung oil, certainly not gold leaf, as he satisfied himself in trials by fire.[g] As a pendant to all this, one can read a curious paper by Collas (9), 'Sur la Quintessence Minérale de M. le Comte de la Garaye'. There was question of trying out this newly-found remedy in China as well as France. De la Garaye, Partington tells us, was a philanthropist and fanatic who converted his castle in Brittany into a hospital, using there curious preparations of his own, especially medicines made by long maceration of minerals with neutral salt solutions, and a mercurial tincture which took months to mature. In +1750 Macquer was charged by the court to investigate, and found that the tincture was only corrosive sublimate ($HgCl_2$) dissolved in alcohol, but he reported favourably and the king bought the secrets for a high price.[h] This, alas, was no 'Jesuits' bark', and if it had been taken to China would have been no great advertisement for European chemistry.

[a] See pt. 4.   [b] Cf. Crosland (1), p. 84.   [c] Cf. Hiscox (1), p. 70.
[d] Pt. 2, p. 232.   [e] Cf. Partington (7), vol. 3, pp. 173, 190.
[f] See further in Needham (31, 32) and Sects. 30, 36 below.
[g] Very probably it was stannic sulphide (cf. pp. 99, 103 above).
[h] Cf. Partington (7), vol. 3, pp. 88, 215; Coleby (1), pp. 59ff. De la Garaye seems not, however, to have been entirely useless as an experimenter, for he was one of the first to use hydrofluoric acid for etching graduations on glass.

[1] 黃礬   [2] 硇沙   [3] 白銅

Thus during the latter part of the +18th century we find a halting westward transmission of news about the chemical substances and processes of China, carried on under great difficulties by a handful of devoted amateurs. It was totally insufficient to give any idea of what the history of chemical discovery in China had been, but it can at least be set beside the more successful efforts of the botanists Michel Boym and João Loureiro. The only observations comparable, perhaps, to those of Cibot and Collas were those made a little later (+1793) by Hugh Gillan, the physician attached to the Macartney Embassy. His MS. 'Observations on the State of Medicine, Surgery and Chemistry in China' were written down at the request of Lord Macartney and attached to the MS. of the latter's 'Journal', so that they did not see the light until the publication of Cranmer-Byng's edition in 1962.[a] Gillan's studies were rather superficial, as they could hardly fail to be seeing that he was only six months in the country, yet they were not as supercilious as what he wrote on medicine.[b] He never came into any contact with the complexities of the millennial alchemical tradition, and must have depended for the most part on what could be seen in the shops, but internal evidence shows that he visited workshops and manufactories whenever he could. He naturally had a good deal to say about paktong, and gave a description of the preparation of zinc metal from natural calamine or zinc-bloom (zinc carbonate), a product in his time preferred by Europeans to their own because of its superior purity. He noticed the greatness of the iron and steel industry, and the special art the Chinese had for casting exceedingly thin-walled iron vessels; he noted also the relatively slight use of lead, and the abundant production of tin and tin-foil. Another thing which interested Gillan was the coal industry and the manufacture of coke. The porcelain industry he had no opportunity of seeing, but he gave an interesting description of the preparation of camphor at Hangchow. On the other hand he made many mistakes; he thought that *chien* alkali, saltpetre and common salt were the only salts known in China, that all Chinese wine whether distilled or not was fermented from some kind of fruit, and that all the paper was made from the 'thin filmy membranes that line the interior cavities of the bamboo'. In any case it did not matter, for what Gillan said was known only to the Embassy's staff and a few friends later on, not in any way enlightening contemporary Westerners in general. The next step will be to see what happened in the nineteenth century, but before looking at this we must return to the beginning of the seventeenth in order to see what fragments of European chemistry or chemical technology were conveyed to China through Jesuit channels.[c]

One could start, perhaps, with the introduction of the Greek Four-Element theory. The first indication of it at this time[d] seems to be in the *Thai Hsi Shui Fa*[1] (Hydraulic

---

[a] See Cranmer-Byng (2), pp. 291 ff., 311.
[b] What justification for this there was may be appreciated by the discussion in Needham (59).
[c] Cf. Ho Ping-Yü (13).
[d] We say advisedly at this time, because there is always speculation that the Empedoclean-Aristotelian quadruple elements or humours were introduced to China in various obscure forms at a much earlier date. They would have come then by means of Buddhist intermediation, as in the medical sūtras translated by Chih-Chhien[2] between +225 and +250 which explain disease as caused by imbalance

¹ 泰西水法   ² 支謙

Machinery of the West),[a] the book on the Archimedean screw and other water-raising devices produced in +1612 by Sabatino de Ursis (Hsiung San-Pa[1]) and Hsü Kuang-Chhi.[2] Not much more than a bare reference, it occurs in connection with water-tanks.[b] Much more elaborate explanations were given by another Jesuit, Alfonso Vagnoni (Kao I-Chih[3]).[c] His *Khung Chi Ko Chih*[4] (Treatise on the Material Composition of the Universe)[d] was printed in +1633, and four years later he produced a second book, *Huan Yu Shih Mo*[5] (On the Beginning and End of the World), which gave the Hebrew–Christian account of creation and explained the four Aristotelian causes as well as the four elements.[e] None of this had, so far as we know, any perceptible influence on indigenous Chinese thought, and perhaps it was just as well, for only three decades were to elapse before Robert Boyle was to publish his *Sceptical Chymist*, the doubts and paradoxes of which were a very land-mine that blew the four-element doctrine sky high.[f]

A more solid contribution was a copy of the +1556 edition of Agricola's *De Re Metallica*, which is still to be found in the Jesuit library at Peking.[g] This had almost certainly been brought out by Johann Schreck (Têng Yü-Han[6]) and Nicholas Trigault (Chin Ni-Ko[7]) who travelled together and arrived in +1621. Among the intense scientific activities of the former during the ensuing decade was, it seems, a Chinese translation of several chapters of Agricola,[h] never printed, though presented to the throne by Li Thien-Ching[8] for Adam Schall von Bell in +1639 or +1640. This was surely the *Khun Yü Ko Chih*[9] (Investigation of the Earth), a treatise on

---

of the elements earth, water, fire and wind (cf. Satiranjan Sen, 1). Many medical historians have suspected Indian influences about this time (e.g. Chhen Pang-Hsien (1), p. 98; Jen Ying-Chhiu (1), pp. 42ff.); and long ago Hsü Ta-Chhun[10] in his *I Hsüeh Yuan Liu Lun*[11] (+1757) remarked that he thought Thao Hung-Ching's ideas on the *ssu ta*[12] were of foreign origin. This reference is due to our colleague Nathan Sivin. Cf. Chhen Kuan-Shêng (6), p. 482. On Chih-Chhien see Zürcher (1), pp. 23, 36, 48ff. We return to the subject in pt. 4 below.

---

a Cf. Vol. 4, pt. 2, pp. 170, 211ff.

b In *NCCS*, ch. 20, p. 2a.

c +1556 to +1640. His original Chinese name had been Wang Fêng-Su,[13] but he changed it to avoid recognition when he re-entered China from Macao in +1624. A Chinese account of him is in *CJC*, pt. 2, app. no. 9.

d Cf. Wylie (1), p. 140; Pfister (1), p. 94; no. 227 in Bernard-Maître (18), who conjectures that it was an adaptation of the Coimbra work *In Libros Meteorum* of Aristotle (Lisbon, +1593, Lyon, +1594). It was prefaced by Han Lin and Chhen So-Hsing. Note the wording of the title (cf. Vol. 1, pp. 48–9). Cf. Chang Tzu-Kao (2), p. 183.

e Pfister (1), p. 93; no. 283 in Bernard-Maître (18). A third work of Vagnoni's, printed in +1636, the *Fei Lu Hui Ta*[14] (Questions and Answers on Things Material and Moral), no. 272, probably also dealt with the same subjects.

f The parallel of the crystalline celestial spheres (Vol. 3, p. 439) will be remembered.

g V730; Bernard-Maître (4). On the vicissitudes of the four Jesuit libraries there, now reduced to one, see Verhaeren (1)

h So Bernard-Maître (4), pp. 226, 230. Pfister (1) did not notice this, but we know of Schreck's chemical interests from him, since he tells us that the library at Montpellier possesses a MS. of Paracelsian commentary and glossary written by Schreck before his departure for China.

---

[1] 熊三拔	[2] 徐光啓	[3] 高一志	[4] 空際格致
[5] 寰宇始末	[6] 鄧玉函	[7] 金尼閣	[8] 李天經
[9] 坤輿格致	[10] 徐大椿	[11] 醫學源流論	[12] 四大
[13] 王豐肅	[14] 斐錄彙答		

mining with the methods of the West.[a] Fang I-Chih,[1] who knew Schall personally,[b] tells us that Ni Hung-Pao[2] was Minister of Agriculture at the time and held several discussions with him about it, but nothing was done, doubtless owing to the imminence of the Manchu invasion.[c]

In any case this must have been to some extent a work of supererogation, for the Chinese had perfectly adequate smelting methods of their own, their metallurgy was in some respects in advance of Europe (notably as concerned with zinc and nickel), the European ores were different in many cases, and although Agricola belonged to the old assay tradition rather than the alchemical, there is not a word of the new chemistry in him. It might have been better to start with Andreas Libavius' *Alchemia* of +1597, a copy of which had also been brought by Schreck and Trigault.[d] Libavius has been called the first modern chemical writer; his *Alchemia* is an excellent practical textbook, clear, concise and sensible, entirely different from the rambling, bombastic and obscure verbosity of Paracelsus and the earlier alchemists.[e] Libavius put new wine into old bottles by defining alchemy as 'the art of perfecting magisteries (i.e. technical methods and reagents), and of extracting pure essences by separating bodies from mixtures'. To have translated and expounded Libavius systematically would have been a real contribution to traditional Chinese chemistry and chemical technology, analogous to the transmigration of Kepler and Galileo, though in its way of course less immediately revolutionary than they—but alas, the only man who could have done it, Johann Schreck, was cut off after barely ten years of work in China, a time in any case already filled with scientific achievement.[f] A third book in this select library dealing with chemistry was Oswald Croll's *Basilica Chymica* of +1609, but this, though truly scientific in its practical part, describing silver chloride, antimonic acid, tin and calcium acetates, succinic acid and fulminating gold,[g] was impregnated with Paracelsian theosophy which the Chinese might well have been spared, as indeed they were, for Croll lay ignored among the Jesuit books through the following three centuries. It was a tantalising twist of history that such works could get as far as Peking yet fail to penetrate the language barrier.[h]

---

[a] Cf. Hummel (2), vol. 1, p. 489; Chang Tzu-Kao (2), pp. 188ff.

[b] There is now a new biography of Fang I-Chih by Yü Ying-Shih (1), though dealing only with his later years.   [c] *Wu Li Hsiao Shih*,[3] ch. 7, p. 10a.

[d] V2043. It is, or was, bound up with two tractates of Agricola; Bernard-Maître (4), p. 228. Libavius' dates are c. +1540 to +1616.   [e] See Leicester (1), pp. 98ff.; Partington (7), vol. 2, pp. 247ff.

[f] Apart from initiating the calendar reform which led to the institution of the Jesuit Directorate in the Bureau of Astronomy (cf. Vol. 3, pp. 444ff.), Schreck collected before his death in +1630 a great number of plants new to Western science and described them in a two-volume *Plinius Indicus* (long since unhappily lost); he produced the remarkable work with Wang Chêng on mechanics and mechanical engineering, *Chhi Chhi Thu Shuo* (cf. Vol. 4, pt. 2, pp. 170ff.); and in +1626 wrote a book of modern anatomy (pr. +1643), the *Jen Shen Shuo Kai*,[4] based on Caspar Bauhin's *Theatrum Anatomicum*, also among the books he himself had brought out.

[g] Cf. Partington (7), vol. 2, pp. 175ff. Croll died in the year of publication of his book, but it was highly regarded for long afterwards, and went into many editions. V1403.

[h] Nor were they the only ones. There were some of the books of the Lullian Corpus (cf. pt. 4) dating from the +14th century but printed in the +16th, e.g. *Mercuriorum Liber* (Cologne, +1567) and *Libelli Aliquot Chemici* (Basel, +1572); V2537-8; Ferguson (1), vol. 2, p. 54. On the Lullian

---

¹ 方以智   ² 倪鴻寶   ³ 物理小識   ⁴ 人身說概

(ii) *Mineral acids and gunpowder*

During this period there are two examples of some chemical contact between Europeans and Chinese. The first concerns the mineral acids, and the second that curious incident in the forties when Adam Schall von Bell was 'drafted' to cast cannon and organise the making of gunpowder for the Ming dynasty against the invading Manchus. Both were essentially practical matters with no lasting effects. The first concerns Ricci's friend Hsü Kuang-Chhi. As we shall explain presently (in pt. 4 below), weak nitric acid had been widely used in early medieval Chinese alchemical practice for getting inorganic substances into solution, though no one then recognised it for what it was; and some kind of mineral acid may have been known and used in the +7th century by an Indian alchemist at the Chinese court.[a] Furthermore there is evidence (cf. Sect. 34) for the employment of an acid which was probably hydrochloric in the +15th-century ceramics industry. In the West, nitric acid, *aqua fortis*, made by distilling saltpetre with alum and ferrous sulphate (copperas, green vitriol), dates from Vital du Four, *c.* +1295, not earlier,[b] and *aqua regia*, the mixture with hydrochloric acid, used for dissolving gold, first appears in the early +14th-century Geberian Corpus.[c] Pure hydrochloric acid (*aqua caustica*), made by distilling common salt with ferrous sulphate, comes in the 'Basil Valentine' writings (datable to the last years of the +16th century).[d] Sulphuric acid was first known at an intermediate time,

Corpus in general see Sherwood Taylor (3), pp. 109ff.; Leicester (1), p. 87; Thorndike (1), vol. 4, pp. 35ff. In the context of +17th-century China these were deservedly neglected. So also the Geber, *De Alchemia* (Strasburg, +1598), V1669; Ferguson (1), vol. 1, p. 302. Francis Kieser's *Cabala Chymica* (Mulhouse, +1606), V3935, judging from Ferguson (1), vol. 1, p. 464, was Paracelsian and therefore not very promising. There was also Joachim Tancke's *Promptuarium Alchemiae* (Leipzig, +1610), a rather traditional work with much symbolism which would have been no use at all in China (cf. the descriptions in Ferguson (1), vol. 2, p. 427 and Partington (7), vol. 2, p. 189). Tancke is one of the alchemists suspected of the authorship of 'Basil Valentine's' *Triumphal Chariot of Antimony*. V3988.

But besides these there were some other books which could have been very useful if there had been someone to expound them and take them through the language barrier. We find listed the *De Destillatione* of Geronimo Rossi, a papal physician (see Partington (7), vol. 2, p. 87), V2623, which described the best methods of preparing vegetable essential oils. It is interesting to note another work of Oswald Croll's, the *Tractatus de Signaturis Rerum Internis* (cf. Partington (7), vol. 2, p. 176; Ferguson (1), vol. 1, p. 185), marked in the catalogue 'prohibetur donec emendatur'. Much more important was Biringuccio's *Pirotechnia*, V3201, the Bologna edition of +1678, though again like Agricola its use would have been limited in Schreck's China. But one book about which no such reservations need be entertained was Antonio Neri's *l'Arte Vetraria* (Florence, +1612), V3375; cf. Partington (7), vol. 2, p. 368; Ferguson (1), vol. 2, p. 135. An up-to-date treatise on glass-making would have interested the Chinese.

[a] Cf. Vol. 1, p. 212; and pt. 4 below.

[b] Partington (4), p. 40; Sarton (1), vol. 3, p. 531; Multhauf (5), p. 207. Another name was *scheidewasser*, 'parting acid', because it dissolved silver but not gold; Sherwood Taylor (4), p. 92.

[c] Sal ammoniac (ammonium chloride) was simply added to the mixture in the retort; *De Inventione Veritatis*, ch. 23 (Russell tr., p. 223).

[d] Partington (4), p. 56, (7), vol. 2, p. 200. But della Porta in +1589 described an oily water produced by strongly heating bricks, quenching them in concentrated brine, and then distilling very hot from a retort (Partington (7), vol. 2, p. 23; Multhauf (5), p. 208). Jerome Cardan in the *De Rerum Varietate* of +1557 has something very similar (Partington, *op. cit.* p. 15). Now 'oil of bricks' (*oleum de lateribus*) goes a good deal further back. It seems to have developed from the distillation of essential oils mixed with brick fragments and salt (cf. Multhauf (5), pp. 204–5, 207–8, 210). It was mentioned, not very clearly, by Abū al-Qāsim ibn-'Abbas al-Zahrāwī of Cordoba (Abulcasis, d. +1013; Mieli (1), 2nd ed., p. 181), whose *Liber Salvatoris*, the Latin translation of +1471, was a very popular pharmaceutical book throughout the +16th century. It also comes in the materia medica of Mesue, the so-called

early in the +16th century; it was obtained in two ways, by the distillation of green vitriol (ferrous sulphate) alone, and by the collection of the condensate from a bell suspended over burning sulphur. The former was called *oleum vitrioli*, the latter *oleum sulphuris*, and their identity was not recognised until the following century.[a] Mention of the first occurs for the first time in Antonio Brasavola's *Examen Omnium Simplicium* . . . (+1534), and of the second in P. A. Mattioli's commentary on Dioscorides (+1535).[b] After that there are soon frequent mentions and descriptions of the technique.[c] Elsewhere we remark on the central importance of potassium nitrate both for the oldest mineral acid and for gunpowder (pt. 4); here one cannot help noting the key historical position of a seemingly most commonplace salt, ferrous sulphate, which figured in the preparation of all the three chief mineral acids. And further noteworthy, whatever it may mean, is the fact that the oldest procedure involved the most complicated mixture of substances in the retort. Perhaps this mirrors that other slow trend away from 'theriac' or 'polypharmacal' alchemy in China (cf. pp. 208, 217).

Now it is clear that Ricci and his colleagues did sometimes discuss chemical matters with their Chinese friends. An extant MS. of Hsü Kuang-Chhi shows that he must have heard about the sulphur-mercury theory of the metals (cf. pt. 4), and also the making of nitric acid, from the Jesuits. For example, he wrote:

Among the Five Metals those that are too brittle are so because of (excess of) the *chhi* of sulphur. When this sulphur is removed they become malleable. Those that are too soft are so because of (excess of) the *chhi* of mercury, and they harden when this mercury is taken away.[d]

Grabbadin, also of +1471 and also popular during the same period. Here the evidence is more significant, for Giovanni Costeo (d. +1603), one of Mesue's commentators, says that he himself saw the acid distillate instantly penetrate a marble or metal plate. John of Rupescissa (*fl. c.* +1345) had said something very similar (Multhauf (5), p. 213). The identity of Mesue (Māsūya) is not known; he may have been Māsawayah al-Mardīnī of Baghdad (d. +1015; Mieli (1), 2nd ed., p. 121), but recent opinion is more inclined to the view that the book was a European work of the +13th century with no Arabic original. Graphidion (γραφίδιον) means tractate, so a Byzantine origin might be suspected.

Since bricks are made of aluminous clays they could have substituted to some extent for the ferrous sulphate used in the perfected processes. In later times oil of bricks came to mean simply the evil-smelling 'empyreumatic oils' prepared by the partial destructive distillation of vegetable oils such as olive oil.

Of course, by the time of Oswald Croll (+1609) and J. B. van Helmont (+1648), all the chief mineral acids were clearly known (Partington (7), vol. 2, p. 225; Multhauf (5), p. 225). What appears from the curious story of 'oil of bricks' is that hydrochloric acid in half-recognised form may go back as far as the +10th century, in which case it would have anticipated nitric. It may well be earlier, for al-Zahrāwī said that this acid oil was known to the alchemists. We should bear this in mind when considering the case of the Indian (p. 237).

Finally, Reti (11) has adduced fairly strong evidence that the distillation of hydrochloric acid did not wait for the end of the +16th century, but was practised already in the +15th. This appears from a technical MS. preserved at Bologna. In that case hydrochloric would have preceded sulphuric but not nitric.

a Certainly by the time of Glauber, *c.* +1650.
b See Partington (7), vol. 2, p. 96; Sherwood Taylor (4), p. 96.
c E.g. Biringuccio (+1540), Valerius Cordus before +1544, Jerome Cardan in +1550, Conrad Gesner (+1552), J. B. della Porta (+1556), and in Gerard Dorn's *Clavis Totius Philosophiae Chymisticae* (+1567). See on these Partington (7), vol. 2, pp. 14, 23, 35, 37, 160, 166; Sherwood Taylor (4), p. 96.   d In *Hsü Kuang-Chhi Shou Chi*.

This is shown in Fig. 1365.[a] Another manuscript note, giving the method for making nitric acid by distilling green vitriol and alum together with saltpetre, is shown in Fig. 1366.

Fang I-Chih, in his *Wu Li Hsiao Shih* (Small Encyclopaedia of the Principles of Things) of +1664, mentioned that he had heard personally from Tao Wei Kung[1] (Adam Schall von Bell) about a strong acid, which he called *nao shui*[2,3] and that this had the property of etching metal plates.[b] It was distilled, he said, from a retort of earthenware or glass, and collected in a receiver at the end of a long tube. His actual words are as follows:

There is a remarkable liquid known as *nao shui*, which will quickly dissolve silver foil if this is put into it. When the precipitate has been formed (silver chloride) it can be poured out into a container and will take whatever shape the latter has. The *nao shui* can be decanted and returned to its bottle. The method of preparing *nao shui* is to obtain a long tube (retort) from the glass-blowers and heat *nao sha* (sal ammoniac) in it to collect its vapour.[c] It was the Venerable Tao Wei himself who informed me about this.

Also, in another place,[d] Fang I-Chih acutely noted that the fumes produced in the neighbourhood of places where copperas (ferrous sulphate, *lü fan*,[4] *chhing fan*,[5] cf. pt. 4) was being manufactured, destroyed the clothing of the workers and killed plants and bushes. He suspected no doubt that this effect was due to some other kind of vitriolic acid. The name *nao shui*, derived from sal ammoniac, suggests that it was *aqua regia*, the mixture of nitric and hydrochloric acids.[e] Hsü Kuang-Chhi had called nitric acid *chhiang shui*[6] (lit. strong liquid), and so did Chao Hsüeh-Min[7] later on in his *Pên Tshao Kang Mu Shih I*[8] (Supplementary Amplifications for the Great Pharmacopoeia) begun about +1760 and finished in the last years of the +18th century.[f] From the fumes observed during the process of distillation Chang Tzu-Kao concludes that the term *chhiang shui* found in Chinese writing between the 1620s and the 1780s meant only nitric acid,[g] though in present everyday usage it refers to *aqua regia*. Forty years ago Mikami Yoshio (*16*) considered all these points in an interesting study of mineral acids in China, and added that the production of a weak sulphuric

[a] It is ironical that he did not recognise for what it was a theory which may well have been developed by Chinese medieval alchemists in the first place (cf. pt. 4).

[b] Ch. 7, p. 10a. There is a new study of Fang I-Chih by Yü Ying-Shih (*1*).

[c] As Wang Chin (*2*) points out, it was a matter of common knowledge that sal ammoniac easily decomposed into ammonia and an acid substance. By +1590 Li Shih-Chen was noting that sal ammoniac had a corrosive or sometimes a brightening effect upon the metals gold, silver, copper and tin (*shan lan chin yin thung hsi*[9]). PTKM, ch. 11, (p. 59).   [d] Ch. 7, p. 20a.

[e] He also said that Schreck & Schall von Bell's version of Agricola (+1640) showed how the winning and isolation of the five metals could be facilitated by the strong acids, but the government did not take it up. With the Manchus at the door, it could hardly have been expected to do so.

[f] Ch. 1, p. 6a. It is worth noticing that this admirable pharmaceutical natural history (cf. Vol. 6, Sect. 38), which contains much chemical knowledge, was being produced just during the couple of decades when revolutionary discoveries in chemistry were being made in Europe. How much information about these reached the Jesuits in Peking is not really clear, but they were certainly too distracted to have been able to pass it on; moreover Chao Hsüeh-Min worked at Hangchow and so far as we know had no Jesuit contacts. How sad such barriers were.   [g] (*5*), pp. 190ff.

[1] 道未公   [2] 礠水   [3] 硇水   [4] 綠礬   [5] 青礬
[6] 强水   [7] 趙學敏   [8] 本草綱目拾遺   [9] 善爛金銀銅錫

acid may have been current much earlier in the Ming. For Li Shih-Chen wrote about +1580 that a 'sulphur juice' or 'liquor' (*liu huang i*[1]) was made by placing saltpetre with sulphur in a closed bamboo tube and heating it at the temperature of burning horse dung[a] for a month, after which a liquid was obtained.[b] He does not however mention the uses of it; they were probably medicinal.

An exchange about mineral acids took place in the polemic between Kao Lei-Ssu and de Pauw (cf. p. 228 above). It runs as follows:

No. 76. 'They know neither *aqua fortis* nor *aqua regia*' (p. 357).

If by the term 'they' one is supposed to understand the peasants, the artisans and the merchants, nothing could be more true. One could find only a couple out of ten thousand who have heard about them, but in fact *aqua fortis* and *aqua regia* are so well known here that people do not much like making them. If the pharmacists of Europe had been willing to manufacture all that which is needed by the Imperial Palace, we should have been most willing to yield them that honour, and indeed have often proposed it.

This was the eighteenth-century equivalent of our oft-repeated remark about fish-wives and morris-dancers.[c]

Now a word about the military chemistry and metallurgy of Joh. Adam Schall von Bell (Thang Jo-Wang[2]). In view of the threat of Manchu invasion, he and other Jesuits were called upon by the emperor in +1636 to advise the general staff about the state of the fortifications of the capital.[d] Then in +1642 the emperor sent the Minister of War, Chhen Hsin-Chia,[3] to discuss the art of gun-founding with Schall, who, showing some theoretical knowledge of the subject,[e] was surprised to be presented suddenly with the imperial order that he should set up a bronze cannon foundry. There was no getting out of it, and all that Schall could obtain was a reduction of the specification from iron-ball 75-pounders to 40-pounders. The work of casting twenty such guns was accordingly carried out in the same year, not without a thousand vexations from the jealousies and peculations of eunuchs, and in the midst of a severe epidemic of plague.[f] In the following year Schall superintended (much against his

[a] Or fermenting? Less likely.

[b] *PTKM*, ch. 11, (p. 63). The presence of the nitrate is significant, because this was the ingredient added by Joshua Ward in +1749 to increase greatly the yield of sulphuric acid 'by the bell' (Sherwood Taylor (4), p. 97), and so presaged the oxides of nitrogen essential in the classical chamber process (Mellor (1), pp. 430ff.). The similarity with the *shui fa* methods (pt. 4) will be noted.

[c] Cf. Vol. 4, pt. 1, p. 309.

[d] Fuller accounts of what follows here will be found in Bornet (2); Väth (1), pp. 111ff. and Duhr (Attwater adapt.), pp. 60ff., but it is best to have recourse to Schall von Bell's own words as given in his *Historica Relatio*, written before +1661 and published with a French translation in 1942 by Bornet (3). On the fortifications see this work, p. 34.

[e] Probably no more than would have been the common property of educated men interested in the scientific and technical culture of the period. But there was in one of the Jesuit libraries at Peking an up-to-date work by Luigi Colliado, *Prattica Manuale dell' Artigliere* (Milan, +1641), which no doubt proved useful to Schall. See Bornet (1), p. 82; V3249.

[f] We pass over the details, but one scene is unforgettable, Schall inducing the foundrymen, with the emperor's approbation, to worship at a Christian altar instead of making their customary sacrifice to the Spirit of the Furnace (*huo shen*[4]); cf. pp. 29, 31. Constrained by circumstances he was, but no one with a spark of Christian pacifist feeling can fail to regret the symbolism of this particular manifestation of Jesuit missionary zeal in Asia. Bornet (1), p. 84.

[1] 硫黄液   [2] 湯若望   [3] 陳新甲   [4] 火神

Fig. 1343. Tomb of Hsü Fu (Jofuku) of the Chhin dynasty, in the coastal town of Shingū south of Kyoto (orig. photo., 1971). Hsü Fu was the adept sent to sea by the First Emperor in −219 with a host of young people to search for the isles of the immortals where the plant of immortality grew (see p. 18). Japanese tradition preserves the belief that he landed at Shingū with a remnant of followers, and died there.

Fig. 1344. The votive shrine of Hsü Fu at the Asuka Jinja, a Shinto temple on the outskirts of Shingū (orig. photo., 1971). The conical forested hill behind it is known as Hōraisan (Phêng-lai Shan).

Fig. 1345. Main hall of the Asuka Jinja itself, with the usual offerings of *sake* piled up outside (orig. photo., 1971).

Fig. 1346. A bush of the drug-plant *tendai wuyaku* (*Lindera strychnifolia*) growing in the garden of the Asuka Jinja. According to Japanese tradition, this plant, with its tonic and tranquillising properties, was what Hsü Fu came to seek (orig. photo., 1971).

Fig. 1353a. On the tracks of Ko Hung; the Chhung-Hsü Kuan (Temple of the Vast Inane), a Taoist sacred place in the Lo-fou Shan mountains east of Canton (from *Lo-fou Shan Chih*, +1716, ch. *shou*, pp. 10b, 11a). The site is that of Ko Hung's southern hermitage-laboratory, and we are looking south-eastwards. The central building is the Hall of the Three Pure Ones (San Chhing, the Taoist Trinity, cf. Vol. 2, pp. 154, 158, 160), and just in front of it stands a pavilion honouring an imperial tablet of dedication dated +1419. On the left of the pavilion is the dwelling of the hermit Huang (one of Ko's disciples, cf. Table 112), and to its right a large alchemical furnace (*tan tsao*). At the back on the left is the votive chapel of Ko Hung himself, and in the centre the Phêng-Lai Hall, flanked by the Gallery of the Forgotten Shoe (left behind by one of the holy immortals). Here there were iron statues of Yü Huang Ti (the second, incarnate, person of the Taoist Trinity), and two attendants cast at some time during the S/Han dynasty (+917 to +971). In the foreground, outside the gate in the boundary wall, is the Bridge of Meeting with the Immortals, and to the left the Tomb of the Empty Clothes and Cap — i.e. Ko Hung's because he achieved *shih chieh* (cf. Vol. 5, pt. 2, pp. 297, 301). Here a materia medica market was held from time to time. Further details are given in ch. 3, pp. 2b ff., and biographies of Ko, Huang and other alchemists in ch. 4, pp. 7a ff., 9a ff.

Fig. 1354. The image and altar of Ko Hung in the votive temple
dedicated to him, the Chhu-Yang Thai (Terrace of the Arising
of the Yang) on the Ko Ling, one of the highest points of the
low ridge of hills protecting the West Lake at Hangchow from
the north (orig. photo., 1964). The significance of the name of
the temple can be appreciated from Fig. 1352 above. There is
a well in its grounds from which Ko Hung is supposed to have
drawn pure water for the experiments which he carried out at
this place.

Fig. 1357. Sublimation of calomel, mercurous chloride, *shui yin fên*; an illustration from a MS. copy of the *Pên Tshao Phin Hui Ching Yao* (+1505). Photo. Dr S. D. Sturton (1959), cf. Bertuccioli (1). This work (Essentials of the Pharmacopoeia ranked according to Nature and Efficacity) was edited by Liu Wên-Thai for the Imperial Library, and not printed at all until our own time. The picture comes from one of a family of MSS all having fine illustrations in ·olour, and preserved in various parts of the world.

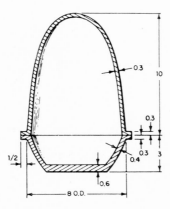

Fig. 1359a. Cross section of an aludel or reaction-vessel drawn according to the specifications of Sun Ssu-Mo (see Sivin (1), p. 167). Dimensions in Chinese inches (*tshun*), 10 *fên* going to the inch and 10 inches to the foot (*chhih*).

Fig. 1359b. Sealed two-part reaction-vessels after heating in the experimental reconstruction of Sun Ssu-Mo's 'Scarlet Snow and Flowing Pearls Elixir' (after Sivin (1), p. 182). Purified realgar and rice wine vinegar were heated to 900° C. within an aludel formed of two nickel crucibles luted together. It was not possible to make the system air-tight, so that water-vapour could have left and air entered later. The sublimate turned out to be pure metallic arsenic, which badly corroded the crucibles. These experiments were performed by Dr Nathan Sivin at the University of Singapore in 1962; and one may hope that his example will be more widely followed, for without attempts at reconstruction the procedures of the medieval alchemists must necessarily remain obscure.

Fig. 1360. Bronze statue of the celebrated Taoist alchemist Lü
Tung-Pin (*fl.* +8th and +9th centuries) in the central shrine of
his votive temple at Thaiyuan in Shansi (orig. photo., 1964).
The statue is of late Ming date, probably cast on the occasion
of the foundation of the Lü Thien Hsien Tzhu (Oratory of the
Heavenly Immortal Lü) in +1575. The temple (now the
Provincial Museum) has also been called Lü Tsu Miao and,
significantly from the alchemical viewpoint, Shun-Yang Kung
(Palace of the Pure Yang). At the time of the Boxer Uprising the
fleeing Empress Dowager Tzu-Hsi was lodged in this temple
late in 1900, and wrote an epigram still preserved there: 'Like
endlessness of days is the world of the Tao.'

Fig. 1361. Lü Tung-Pin's image and altar, the central one of three, in the Chhing-Sung Kuan (Taoist Temple of the Green Pine), near Castle Peak in the New Territories, Hongkong (orig. photo., 1972). Here Lü Tsu is venerated as Shun-Yang Ti, Emperor of the Pure Yang, the title of his deification.

Fig. 1362 a. Four of the eighteen alchemical talismans (*fu*) in the *Pao Phu Tzu* book, ch. 17. These magical devices of the early +4th century were believed to ensure safety for the Taoist alchemists in the depths of the mountains and forests, and success in preparing their elixirs there. They felt great need for protection, psychologically against the gods and spirits of the ravines and wildernesses, physically against noxious animals and plants, falling trees, landslides and the like. They entered the mountains (*ju shan*) only on propitious days, they danced a special step from time to time, the famous 'Pace of Yü' (*Yü pu*, cf. *PPT/NP*, ch. 17, p. 5 a, tr. Ware (5), p. 286, cf. Granet (1), vol. 2, pp. 549 ff.), they wore mirrors on their backs to ward off evil spirits, and carried with them diagrams like the *Wu Yo Chen Hsing Thu* (cf. Vol. 3, pp. 546, 566 ff.) or bunches of plants which gave invisibility or dispelled apparitions. For an insight into the 'other world' as it appeared to the ancient Taoist believers nothing serves better than the descriptions of Castaneda (1, 2, 3) of the ideas and powers of the Yaqui Indian 'sorcerers' of Mexico today.

(*a*) No. 1, a design to be carved on peach-wood plaques and fixed over the door, on the four sides, and the four corners, of the hermitage-laboratory (p. 14 a). The lower part of it resembles the standard way of drawing constellations in star-maps (cf. Vol. 3, p. 276).

(*b*) No. 6, also for the pillars and beams of the house (p. 17 a). One can make out the characters *wang* (king), *po* (count) and part of *tien* (lightning), doubtless all apotropaic. The constellation might be the Great Bear (cf. Vol. 3, p. 241).

(*c*) No. 10, the perfected immortal Chhen An-Shih's amulet for carving on a small plaque of jujube-tree pith, and wearing on the body; good against gods, ghosts, tigers, wolves and serpents (p. 19b). The character for woman is clear, perhaps one of the Taoist goddesses like Hsüan Nü (cf. Vol. 2, pp. 147–8).

(*d*) No. 11, anthropoid in character, for wearing at one's belt when travelling, or for protecting the stables of the hermitage against wolves and tigers (p. 20b).

From the Ssu Pu Tshung Khan edition. Cf. pp. 152–3 above.

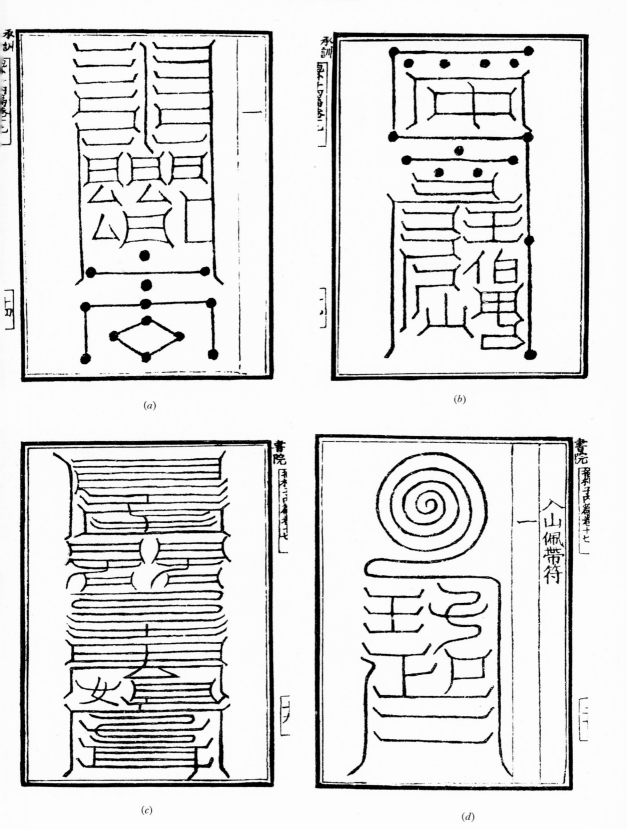

(a)

(b)

(c)

(d)

Fig. 1363. An indication from an Arabic source of Japan's richness in gold; the Queen of Wāqwāq and her court depicted in a coloured illustration from al-Qazwīnī's world geography 'Ajā'ib al-Makhlūqāt (Roy. As. Soc. MS. 178). Apart from earlier accounts of Japanese gold abundance, one Mūsa al-Shīrafī reported that he had seen the queen and her four thousand ladies wearing gold crowns and necklaces above their sarongs, otherwise their only dress. This was perhaps a confusion with some hotter country, but the queen's crown in the picture is curiously Taoist. Also it is interesting that among the names for Japan and her scattered islands given in the *San Kuo Chih* (*Wei Shu*), ch. 30, pp. 26a, b, 28b, 29a, are Queensland (Nü Wang Kuo) and Nakedland (Lo Kuo); cf. Pearson (1). Al-Qazwīnī was writing about +1275, and the date of the MS., copied by Muḥammad al-Baqqāl, is about +1475. But the story of the sorceress queen and her regiment of serving-maids was a very old one, going back to the Han (c. +160), as can be seen in the *Hou Han Shu* (ch. 115, pp. 12a ff.); and it was repeated in the *Sui Shu* (ch. 81, pp. 13a ff.) as well as the *San Kuo Chih*. Her name, Pimihu (Pimiko), is thought to conceal the Japanese term *himeko*, princess. In some accounts she was succeeded by a girl named I-Yü (Iyo). Cf. Tsunoda & Goodrich (1).

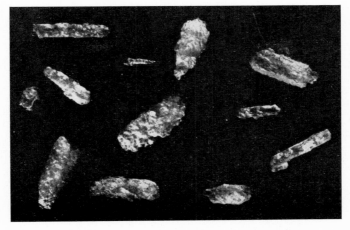

(a)

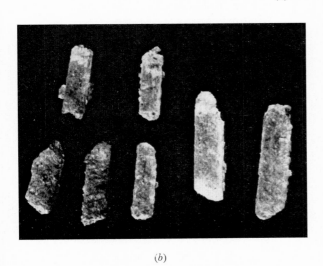

(b)

(c)

(d)

Fig. 1364. Specimens of pharmaceutical chemicals from the +8th century preserved in the Shōsōin Treasury at Nara in Japan

a, b Crystals of *mang hsiao* (here magnesium sulphate, Epsom salt). Asahina (*1*), pl. 27 B; p. 290, fig. 1.

c Specimen of *hua shih* (talc, soapstone), magnesium silicate, halloysite. Asahina (*1*), pl. 31 A.

d Bag containing red lead oxide (minium). Though labelled *shang tan*, 'superior quality red', it has only 26·2% of this oxide. Asahina (*1*), pl. 37 B.

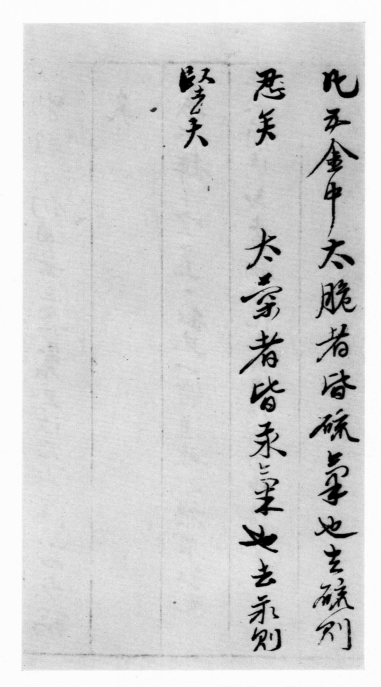

Fig. 1365. Passage from a manuscript of Hsü Kuang-Chhi (+1562 to +1634), eminent scholar and friend of Matteo Ricci and other Jesuits, on the mercury-sulphur theory of metals. See p. 238.

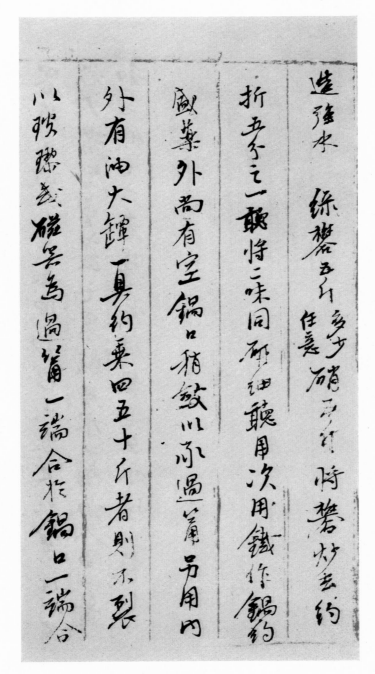

Fig. 1366. Another passage in the handwriting of Hsü Kuang-Chhi, describing (about +1620) the preparation of nitric acid (*chhiang shui*) by the distillation of green vitriol and alum together with saltpetre. See p. 239.

Fig. 1367. A corner of the Pai Yün Kuan (Temple of the White Clouds), a Taoist religious centre of great renown at Peking (orig. photo., 1964). Cf. p. 250.

will) the casting of some 500 hand-guns or portable culverins weighing 60 lb each;[a] and before they were finished he was called upon to demonstrate better designs of redoubts or ravelins, which he did with the aid of a triangular model, but this was rejected for astrological reasons.[b] In the following year the capital fell to the Manchus. Of great chemical interest is the fact that in +1643 Schall collaborated with a Chinese gunner, Chiao Hsü[1],[c] in producing a book on gunpowder manufacture, the casting and boring of cannon, and artillery tactics. This was entitled *Huo Kung Chhieh Yao*[2] (Essentials of Gunnery).[d] It was not the first Chinese book on modern artillery, but it became the most widely known and used.[e] Schall's gunpowder formulae[f] were of course not essentially different from those which had been given in the *Wu Ching Tsung Yao*[3] of +1044, the oldest in any civilisation, but they certainly contained a higher proportion of nitrate, as was necessary for a propellant charge of modern type rather than rocket compositions, carton bomb fillings, and quasi-incendiary mixtures.[g] But nothing in all this involved any serious transmission of nascent modern chemistry.

A short while ago we mentioned Robert Boyle, and it has to be recorded that he also got through to Peking (in the letter) towards the end of the century, though not in the spirit, for no one introduced his work into the realm of Chinese scientific discourse.[h] In +1685 five Jesuit Mathematicians-Royal of the King of France, led by the Breton missionary Jean de Fontaney (Hung Jo-Han[4]), set out for China, bringing with them large acquisitions of books for the libraries at the capital.[i] These included four works of Boyle, all in Latin editions,[j] one of which was the *Sceptical Chymist* itself. They also included another work of Biringuccio,[k] and Otto von Guericke's *Experimenta Nova Magdeburgica* of +1672.[l] But none of the Jesuit scientists of the early +18th century, greatly taken up as they were with astronomical work, had the

[a] Bornet (1), p. 88.   [b] *Ibid.*, pp. 90 ff.
[c] One wonders whether he was a descendant of Chiao Yü,[5] the gunner of the early Ming (*fl. c.* +1345 to +1412), who had been concerned with the writing of an earlier book on gunpowder and artillery, the *Huo Lung Ching*[6] (Fire-Drake Manual). For further information on this important work, the bibliography of which is rather complicated, see Sect. 30.
[d] No. 334 in Bernard-Maître (18). An alternative title, the *Tsê Kho Lu*[7] (Methods for Victory), has sometimes been attributed to the +17th-century edition, but it seems to belong rather to the reprints which were made at the time of the Opium Wars, in 1841 and several subsequent years. See on this Pfister (1), p. 182 (app.), addenda, p. 25* and Pelliot (55, 56); Liu Hsien-Chou (1), p. 84.
[e] In or shortly before +1625 there had been a previous tractate, the *Hsi-Yang Huo Kung Thu Shuo*[8] (Illustrated Treatise on European Gunnery), written by Chang Tao[9] and Sun Hsüeh-Shih,[10] of whom otherwise very little is known. This was not connected with the Jesuit mission, but rather with the detachment of Portuguese gunners who went up to Peking to succour the Ming about that time.
[f] *Huo Kung Chhieh Yao*, ch. 2, pp. 9*b* ff. (pp. 30 ff.).
[g] The full story is given in Section 30.
[h] Much has been written on Boyle's part in the birth of modern chemistry, but here it may suffice to refer to the book of Boas (2).   [i] They arrived in +1687, see Pfister (1), p. 422.
[j] *Chymista Scepticus* (Geneva, +1677), *Introductio ad Historiam Qualitatum Particularium* (Geneva, +1677), *Opera Varia* (Geneva, +1680), and *Experimentorum Novorum Physico-Mechanicorum Continuatio Secunda* (Geneva, +1682). On these see Partington (7), vol. 2, pp. 492, 493, 495, 497. V1112-16.
[k] V3400.   [l] V1768.

[1] 焦勗   [2] 火攻挈要   [3] 武經總要   [4] 洪若翰   [5] 焦玉
[6] 火龍經   [7] 則克錄   [8] 西洋火攻圖說   [9] 張燾   [10] 孫學詩

chemical bent to carry on the work of Johann Schreck and really do something to combine the chemistries of East and West. We hear indeed that two of the Mathematicians-Royal, Joachim Bouvet (Pai Chin[1]) and J. F. Gerbillon (Chang Chhêng[2]), about +1689 'installed in the imperial palace a complete chymical elaboratory, with all the instruments (and vessels) needed for operating in it . . .', but nothing whatever seems to have come of this.[a] In sum, enough has now been said to establish that for various reasons European chemistry of the +17th century failed to exert any influence in China. And perhaps it was because there was so much leeway to make up, quite apart from the growing disorder in the Jesuit mission after +1750, and moreover the paucity of capable men who could understand and interpret it, that nothing at all was done to bring the 'chemical revolution' to China either. Or almost nothing, for there is a strange story to tell about one possible attempt to put the discovery of oxygen into Chinese.

(iii) *A Chinese puzzle—eighth or eighteenth?*

This confronts us indeed with one of the most singular literary puzzles which has presented itself in the whole course of these studies.[b] In 1810 Julius Klaproth published in the *Mémoires* of the Academy of Sciences of St Petersburg a short paper (5) 'on the chemical knowledge of the Chinese in the 8th century'. Introducing this, he said that 'as we have so few exact notions about the state of chemistry among the Ancients, and especially among the asiatick peoples, it seems to me that the following extracts, taken from a Chinese book treating of this science, might offer some interest, for they demonstrate that this people already had, several centuries ago, certain ideas, inexact though they might be, on the effects of oxygen'. He went on to relate that in 1802 he had copied and translated some passages from a Chinese manuscript of sixty-eight pages, one among those which had been brought back from China by the late Mons. Bournon, and (presumably) deposited in one of the libraries of the Russian capital. The title of this 'small collection of chemical and metallurgical experiments' was *Phing Lung Jen*,[3] a phrase which Klaproth translated as 'Confessions of the Peaceful Dragon'.[c] It had been written, he said, according to the MS. itself, by one Maò hhóa[d] in a *ping-shen* year, the first of the Chih-Tê[4] reign-period, i.e. +756. He

---

[a] See Pfister (1), p. 434.   [b] Considered also in a separate paper, Needham (71).

[c] More probable interpretations are suggested on pp. 247–8 below.

[d] Judging from Klaproth's other romanisations, this could hardly be anything else than Mao Hua, since Hs- would be inadmissible here. One could however think of Mao Kua or Mao Khua. Chang Tzu-Kung,[5] in his 1952 translation of Weeks (1), very naturally presumed Mao Hua,[6] though in earlier translations Huang Su-Fêng[7] & Yü Jen-Chün[8] (1936) had adopted Ma Ho,[9] much less likely though retained by Yuan Han-Chhing (1); and Chu Jen-Hung[10] (1937) had suggested Mao[11] as the surname. The Japanese translation (1941) chose Mêng Kao,[12] very improbable; and as Yuan points out, Mao-Hua could be a *hao* or literary name, giving no clue to the family name at all. I should like to take this opportunity of thanking Dr Chang Tzu-Kung again most warmly for much help and many talks during my time in China during the war years (Vol. 1, p. 11), as also Dr Yuan Han-Chhing, whom I knew at Lanchow.

[1] 白晉	[2] 張誠	[3] 平龍認	[4] 至德	[5] 張資珙
[6] 毛華	[7] 黃素封	[8] 俞人駿	[9] 馬和	[10] 朱任宏
[11] 茅	[12] 孟誥			

failed to give the characters for the author's name, but said that he had searched for him in the *Wan Hsing Thung Phu*[1] and the *Wên Hsien Thung Khao* without success. Klaproth then went on to give his partial translations of ch. 3 and ch. 9, the first on the oxygen of the air and its significance for the calces of metals, the second on the metals themselves and some of their derivative compounds, prefacing these by a discussion of the leading ideas in the work.[a]

Klaproth thought he could recognise them as Taoist.

It is easy to see [he wrote] that his [the author's] system is similar to that of the Dáo-chè (*Tao shih*[2]).... In his first chapter the author says: 'All that man can perceive and observe by the senses, and all that he can conceive with his mind and imagination, is composed of two fundamental principles, the Yānn (Yang) and the Ȳne (Yin), which designate the perfect and the imperfect.' This system is represented in the Eight Goúa (*kua*) of Foǔ-hhȳ (Fu Hsi). The Yang is the powerful and the perfected, the Yin is the diametrically opposite. Our author however often diverges from this definition in the course of his work, and one can clearly see that he assumes infinite modifications of these two principles, which manifest themselves in the forms of the world. In this he differs from the system of the *Tao shih*, which explains the difference in forms of visible objects by continual changes in the proportions of Yang and Yin.

Then he plunges in to the supposedly Thang account of oxygen.

Ch. 3. *The Atmosphere*. The Hhiá-chēnn-kí (*hsia shêng chhi*[3])[b] is the *chhi* which rests on the surface of the earth and which rises up to the clouds. When the proportion of Yin, which forms part of its composition, is too great, it is not so perfect (or full) as the *chhi* beyond the clouds. We can feel the *hsia shêng chhi* by the sense of touch, but the elemental fire with which it is mixed makes it invisible to our eyes. There are several methods which purify it and rob it of a part of its Yin. This is done first by things which are modifications of the Yang, such as the metals, sulphur (lieôu-hhouânn, *liu huang*[4]) and charcoal (táne, *than*[5]). When these substances burn they amalgamate with the Yin[c] of the air, and give new combinations of the two fundamental principles.

The Yin of the air (Ký-ȳne, *chhi Yin*[6]) is never pure, but by the aid of fire one can extract it from Tchîne-chě,[d] from saltpetre (Hhò-siaō, *huo hsiao*[7]) and from the stone called Hhé-tânn-

---

[a] The nature of the Chinese language is such that without having the original text before one it is impossible to be sure how far Klaproth understood its true meaning. Accurate recognition of the dictionary equivalents of characters will not always faithfully interpret the true sense, and it is only too easy to introduce ideas which were not originally there.

[b] Literally 'underneath rising-up vapour or *pneuma*', i.e. air. This interpretation is a conjecture of Muccioli (1), for Klaproth gave no characters. Huang Su-Fêng & Yü Jen-Chün supposed *han chen chhi*,[8] impossible on Klaproth's romanisation. Could he have mis-punctuated?

[c] The original has Yang, but this must have been a slip of Klaproth's pen, as Muccioli also saw.

[d] This has puzzled everyone. Klaproth gave no characters, but thought it was a mineral having something to do with grindstones. So perhaps it was (*hsi*) *chen shih*[9] or (*fan*) *ching shih*[10] (RP 76, 100), both meaning magnetite or black iron oxide. It could also have been zinc carbonate, *kan shih*[11] (RP 59), wrongly read as *chhien shih*. And *chün shih*[12] (RP 87) is a synonym for copper sulphate. Yuan Han-Chhing wrote *chhing shih*,[13] lapis lazuli, a complex aluminium-sodium sulphate-silicate, which would not make sense here; cf. Partington (10), p. 426. But he took it to mean some form of calcium carbonate, with what authority we do not know.

[1] 萬姓統譜	[2] 道士	[3] 下升氣	[4] 硫黃	[5] 炭
[6] 氣陰	[7] 火硝	[8] 舍眞氣	[9] 吸針石	[10] 反經石
[11] 乾石	[12] 君石	[13] 青石		

chě.[a] It enters also into the composition of water, where it is so closely bound with the Yang that (its) decomposition becomes very difficult. The elemental fire hides the Yin of the air from our eyes, and we recognise it only by its effects.

For Klaproth this was important, since it showed that the Chinese of the +8th century 'had rather clear ideas about oxygen, which they called *chhi Yin*,[1] or the imperfect (part) of the air'. What else could combine with heated metals, sulphur and carbon, forming new compounds with them? He added that it was 'very interesting that they thought water to be a composite (body), since in Europe it was so long regarded as an element'.

Klaproth's excerpt from ch. 9 was less sensational.

Ch. 9. *The Metals*. There are five principal metals apart from gold; silver, copper, iron, tin and lead.

Gold is the most perfect (Yang), and in general the symbol of the perfection of matter because it contains no Yin whatever; this is why it dominates the four quarters of the world. Silver contains already a little, copper more, and finally lead is the most impure of all the metals. Gold never amalgamates with the Yin of the air, and is always found native. The greatest heat does not change it.

If one purges the silver of its Yin it becomes gold, but as it is always bound to its sulphur, this operation becomes very difficult. Only the silver of Ssī-lôunn-chāne (mountain) in Tiēne-dschoǔ (Thien-chu,[2] India)[b] lends itself to this change. Lao Tzu knew how to change any kind of silver into gold, but did not do it as he himself possessed the golden mountain.[c]

Copper is found native in mountains, or mineralised with the Yin of the air, or with sulphur.[d] When repeatedly melted it loses much of its redness.[e] It is too tightly bound to the Yin to be detached from it. It also readily attracts the Yin of the air, of water, and of alum (Bě-fâne, *pai fan*[3]), the resulting composition being (a kind of) verdigris (Toúnn-sieóu, *thung hsiu*,[4,5] basic copper carbonate).[f]

To get a fine green pigment from copper one must calcine the rust of this metal,[g] and then boil it with white alum in a sufficient amount of water. After it has cooled it will be green,

---

[a] *Hei tan*,[6] *hei than*,[7] is coal, but no remotely likely oxygen-containing substance has been found to fill this bill. Klaproth gave no characters for any of his substances, but commented upon this one as 'a black stone found in marshes', conceivably *hei tan shih*.[8] But no such term, or anything like it, appears in even the greatest Chinese encyclopaedias. Yuan Han-Chhing supposes *hei than shih*,[9] equally meaningless, unless coal, which won't do.

[b] This reference is obscure, though it may well be an allusion to Ceylon.

[c] Of spiritual perfection, no doubt. But of course the Logos of Taoism was always regarded in later times as an alchemist (cf. p. 215).

[d] Ores of copper certainly do include oxides and sulphides. But could this really have been said by anyone in the +8th century?

[e] Perhaps a reference to the making of high-tin bronzes.

[f] The patina and corrosion product of copper is at least as likely to be the basic sulphate (and in sea air the chloride) as the basic carbonate (Partington (10), pp. 330, 333; Mellor (1), p. 378). The term verdigris ('Greek green') also means the acetate, prepared as a pigment and for medicinal use by exposing copper to the fumes of warm vinegar (*PTKM*, ch. 8, (p. 11); *TKKW*, suppl. to ch. 16, Sun & Sun tr., p. 287; cf. Yü Fei-An (1), p. 5). This was called *thung lü*,[10] or better *thung chhing*.[11] Ko Hung already knew (in *PTKM*, *loc. cit.*) that applying it to wood would prevent rotting.

[g] Presumably this means heating to obtain oxides. Sulphates would be formed in the next step.

---

[1] 氣陰   [2] 天竺   [3] 白礬   [4] 銅銹   [5] 銅鏽
[6] 黑丹   [7] 黑炭   [8] 黑澶石   [9] 黑炭石   [10] 銅綠
[11] 銅青

and one must add some natron solution (Guiēne-choùy, *chien shui*[1])[a] which will precipitate the green colour called Siaò-loŭ-chĕ (*hsiao lü sê*[2]).[b] This is used in painting for the colour of plant and bamboo leaves.

To get a blue pigment from copper one must mix three tçán[c] of the rust of red copper with seventeen tçán of sal ammoniac (Naó-chă, *nao sha*[3]) and boil this mixture with pure water. Hhiéne-pânn,[d] who lived in the Han dynasty, was the inventor of this pigment.[e]

If one melts copper with the stone Yânn-chĕ[f] it takes a greenish colour and becomes harder.[g] The utensils made from this copper in the Sung dynasty are much esteemed.[h] It is said that the Eight Kua of Taí-hháo-foú-hhȳ[4] were engraved on a plate of this kind of copper.

Klaproth, who found some difficulty in identifying the substances mentioned in this passage, had less to say about it than about the other. But it merits a word or two. The opening seems to betray a knowledge or half-knowledge of the sulphur–mercury theory of metals (cf. pt. 4), though in the light of what went before a reader might have understood them to be all composed of nitrogen and oxygen; as for 'perfection', it also has a European flavour, though not quite excluded from the classical *Huai Nan Tzu* tradition (cf. p. 24). The explanation of the corrosion of copper is interesting and not ill-phrased, while the making of green and blue pigments from copper salts indicates some practical knowledge. Finally there seems to be a clear mention of brass manufacture.

What is one to make of this curious text? No notice was taken of it for close on eighty years after Klaproth's first presentation in 1807, but then it was disinterred by Duckworth (1) who put the oxygen part into English and taking the date at its face value called for further information. Naturally no one had anything to say. Klaproth was translated into Italian by Guareschi in 1904, and after that wider interest was aroused. Passing mentions were accorded by Moissan[i] and Mellor,[j] while von Lippmann dismissed the whole matter as a late forgery.[k] The most serious discussion was that of Muccioli (1), who concluded that the MS. text was of the +18th rather than the 8th century, Chhing not Thang in date, regarding it as an effort to put into Chinese the

[a] A naturally occurring mixture of the carbonate, sulphate and chloride of sodium, as Klaproth recognised. Cf. pt. 2, p. 181 (the character below is an alternative orthograph), and especially pt. 4.

[b] This must be the carbonate, verditer, the same as *thung lü*,[5] properly so called, or *shih lü*.[6]

[c] Presumably *chhien*,[7] mace, or tenths of an ounce.

[d] Neither Klaproth nor anyone since has been able to pin down this individual.

[e] Perhaps this was the preparation of cupric chloride, unless it was a conversion of the basic to the regular sulphate.

[f] Not certainly identifiable, but one suspects some zinc ore.

[g] This is surely a reference to the making of brass.

[h] How was a Thang writer referring to the Sung? Huang Su-Fêng & Yü Jen-Chün were so upset by this that they transliterated Shang,[8] which was absurd, for brass does not go back that far. Yuan Han-Chhing corrected them but did not offer the obvious escape route that he might have been speaking of the Liu-Sung (+420 to +477). For early brass-making this is not implausible, but one would like some supporting evidence.

[i] (1), vol. 1, pp. 191, 238.

[j] (2), vol. 1, p. 347. He said merely that there could have been no connection between Mao Hua and the definitive later discoveries in Europe.

[k] (1), p. 460. But von Lippmann was so intemperate and so wrong in his general estimate of Chinese alchemy and proto-chemistry that the force of his remarks is now long spent.

| [1] 鹻水 | [2] 小綠色 | [3] 硇砂 | [4] 太昊伏羲 | [5] 銅綠 |
| [6] 石綠 | [7] 錢 | [8] 商 | | |

ideas of the new chemistry, after Priestley and after +1774; the work therefore of one or other of the Jesuits or of their scholarly Chinese friends. This is why we have placed our account of it here. The idea of a late date, in this case the Ming, was also proposed by Chinese chemists and historians of chemistry, notably Huang Su-Fêng (with Fu Wei-Phing[1] and Su Chi-Chhing[2]) in his preface to the 1936 Chinese translation of Weeks' book on the discovery of the chemical elements;[a] but Yuan Han-Chhing was reluctant to accept this, pointing to the general decay of alchemy and chemistry in the Ming period.[b] Yuan Han-Chhing himself inclines to believe in a Thang dating for the document. Some of Muccioli's reasons why 'Mao Hua' could not be pre-Priestley were rather unconvincing, such as the vagueness and sexuality of the Yin-Yang theory, but he was certainly justified in pointing out that the discoveries of Priestley's time would not have been possible without good apparatus of glass, high-temperature equipment, the electric current,[c] and all the other inventions which permitted pneumatic chemistry such as the collection of gases over mercury. But does the 'Mao Hua' MS. pretend to be a document of this kind? The real problem is that its obscurity and *naïveté* could arise either from its speculative (though penetrating) nature, based on only very rough experiments, if it is Thang, or from the difficulties of Jesuits and Chinese scholars in the Chhing attempting to render some idea of the new chemistry into the traditional idiom of that language.

Yuan Han-Chhing, who has considered the matter at length,[d] reviews what 'Mao Hua's' ideas comprised. He was sure that some change occurred in atmospheric air after the combustion of charcoal or sulphur, and that other substances such as salt-petre would give off some 'air' to the air; since it was natural to regard everything as composed of Yin and Yang, it would not have been unnatural to consider atmospheric air as also so composed. Nor would a similar composition of water have been unthinkable; we must beware of reading too much factual background knowledge into the words of the document. Combustion, then, diminished the *Yin chhi* of the air, heating of certain minerals increased it. Here one must bear in mind that even in the +8th century the Chinese had been making the oxides or calces of metals for hundreds of years by heating them in air. Datable about +640, the *Tan Ching Yao Chüeh*[3] (Essentials of the Elixir Manuals), probably by Sun Ssu-Mo,[4] contains a clear process for the making of red mercuric oxide (HgO) by long-continued heating of mercury under conditions of limited admission of air.[e] Similarly the oxides of lead (minium, litharge and massicot) had been made industrially ever since the Warring

---

[a] Cf. the Engl. ed. (1), pp. 35 ff.

[b] (1), p. 229. Cf. p. 208 above. Also if 'Mao Hua' was a pseudonym, why should not the late writer have taken the much more usual course of fathering his book on Ko Hung, Sun Ssu-Mo or Lü Tung-Pin, one or other of the great proto-chemical names of antiquity? Also if the Ming would have been too late for genuineness it would have been much too early for Priestley.

[c] The electrolytic decomposition of water was accomplished in preliminary experiments by G. Beccaria (+1758), P. van Troostwijk & J. R. Deiman (+1789) and G. Pearson (+1797); then definitively with full recognition of the gases by Carlisle & Nicholson in 1800.

[d] (1), pp. 27, 221 ff.

[e] YCCC, ch. 71, p. 12a (tr. Sivin (1), p. 191). Cf. p. 137.

[1] 傅緯平   [2] 蘇繼廎   [3] 丹經要訣   [4] 孫思邈

States period.[a] The fact that nothing at all is said about weighings, such as those which proved the combination of a part of the air with the metal during oxidation, as in the pioneer work of Jean Rey (+1630),[b] may be an argument for the earlier date, since at the end of the eighteenth century the Jesuits or their friends would have been very conscious of this. On the other hand the use of the expression 'elemental fire' is very suspicious, for one of Lavoisier's three great mistakes was the belief that the heat and light evolved in combustion came from an imponderable element, 'caloric', combined with the base of oxygen in oxygen gas (+1786).[c]

The second passage may seem almost more indigenously Chinese than the first, for the alchemical motif is present, and there is no mention of processes which we know the people of medieval China did not use. As for the explanation of the 'rusting' of copper it might again be interpreted much too cleverly by seeing in the reference to the alum, a sulphate, some hint of modern knowledge that the corrosion is generally a basic sulphate. Thus provided one takes the 'Mao Hua' tractate as essentially speculative, inspired guesswork rather than a record of precise experiments, it remains impossible so far to exclude it as a text of the Thang.[d]

Yet there remain many curious, not to say suspicious, circumstances. First, no one has seen the work since Klaproth described it, and until someone can discover it again in one of the great Russian libraries, so that its paper and script can be examined, judgment can only be suspended. Secondly, Thang MSS, apart from the famous buried library of the Tunhuang caves, are extremely rare, though this could of course have been a much later copy, one of a succession handed down through the ages, the kind of thing in fact which Chhü Thai-Su sent to his Jesuit friends to be burnt (p. 224 above). Supposing that one of them had plucked it from the burning? On the other hand the most exhaustive searches, carried on in the Chinese collections and bibliographies of Europe as well as in China, have completely failed to unearth any reference either to Mao Hua or a *Phing Lung Jen*.[e] This in itself is odd because most of the medieval alchemists and iatro-chemists can usually be picked up in two or three collateral references, even if only in the hagiography. As for the title, we have long surmised that *Jen* could have been a mis-reading of *Chih*,[1] which would make the title much more natural. In this case we could take *Phing* as a verb instead of Klaproth's adjective, and translate: 'Records of the Pacification (i.e. Subduing) of the Dragon', that is to say, the formation of the calces of metals (cf. pp. 7–8), an interpretation closely in line with the thinking of alchemists in all the old civilisations. An

---

[a] Cf. Sect. 34.

[b] See Partington (7), vol. 2, pp. 631 ff. Before Lavoisier's definitive solution of the problem in +1777 there had been many demonstrations of the increase of weight on calcination, as by Bayen in +1774 and N. Lemery in +1690 (cf. Partington, *loc. cit.*, vol. 3, pp. 38, 395, 421 ff.).

[c] See Partington (7), vol. 3, pp. 377, 463–5. This point was made to us by Professor Mendoza of the University of Wales at Bangor.

[d] The expression *chhi Yin*, however, 'the Yin of the air', has a very un-Chinese feel about it, and one would rather expect *Yin chhi*. In modern Chinese chemical terminology, oxygen is of course oppositely named *yang*.[2]

[e] What Chhen Kuo-Fu could not find (Yuan, p. 228), it is hardly likely that anyone else will.

[1] 誌   [2] 氧

alternative suggestion is due to Yuan Han-Chhing,[a] who draws attention to a well-known aphorism in *fêng-shui*[1] geomancy:[b] *Shan lung i hsün, phing lung nan jen*,[2] which means: 'The dragon of the mountains (i.e. the lay of the land, or configuration of the earth) is easy to detect and measure; the dragon of the plains is difficult to recognise.' In this case the title would have been right as Klaproth had it, and meta-phorical, signifying that the hidden processes in chemical reactions were as difficult to understand as the 'veins of the earth' in flat country. So that *Phing* would be a noun, and the translation would be: 'Recognition of the Dragon (i.e. the processes) in the Plains (i.e. the baffling chemical phenomena).' Both these possibilities would consort with a date either Thang or Chhing.

Provisionally, for our placing, we have chosen the latter alternative, assuming that the text was the work of one of the Jesuits such as Cibot or Collas, or, even more probably, one of their Chinese friends who had been talking with them about the discovery of oxygen in Europe. But then how can one explain the Thang date on the MS.? Someone could have added it spuriously before the text left China; someone, for that matter, could have insinuated crucial sentences into what was genuinely Thang writing. But against this it must be said that in all the work reported in these volumes we have come across exceedingly few, if any, false datings or claims to antiquity supported by false interpolated passages—where scientific matters were concerned no one in that Confucian culture would have thought it worth while, at least before the very end of the nineteenth century.[c] Spurious ascriptions and attributions of whole books are an entirely different matter, of those there were many, just as in European culture medieval authors liked to father their works on names already venerated,[d] but this kind of thing can almost always be put right by philological and bibliographical criticism, a humanistic technique in which the Chinese themselves excelled earlier than anyone else.[e] Alternatively Klaproth could have made some elementary sinological mistake. But it is difficult to see what it could have been. It is true that he was young at the time. Born in Berlin[f] in +1783, he began the study of Chinese at the age of 14. A voracious polymath in oriental languages (Arabic, Persian, Turkish, Hebrew), he began to publish from 1800 onwards and in 1804 at the age of 21 was called to the St Petersburg Academy of Sciences as 'Adjunctus f.d. orientalis-chen Sprachen u. Literatur.' In the following year he went east with the embassy of Count Golovkin to China, and although this never got beyond the frontier on account of a disagreement about ceremonial, the journey was not unsuccessful for Klaproth

---

[a] (*1*), p. 225.

[b] See Vol. 4, pt. 1, pp. 239ff.

[c] See Vol. 1, p. 43. It is quite inconceivable that any of the Jesuits or their Chinese friends would have put a Thang date—or for that matter an assumed name—on anything which they themselves had written.

[d] Think, apart from anything else, of the Geberian Corpus and the Lullian Corpus. Or in the earlier Arabic world, the Jābirian Corpus.

[e] See Vol. 2, pp. 390ff.

[f] He was the son of Martin H. Klaproth (+1743 to 1817) the celebrated chemist, on whom see Partington (7), vol. 3, pp. 654ff. It was a parallel to the Biots, the chemist-astronomer father J. B. Biot, and the orientalist son E. Biot; cf. Vol. 3, p. 183.

[1] 風水   [2] 山龍易尋平龍難認

since it enabled him to learn Manchu and Mongol by the way, and to collect a great number of books. By 1807 he was a member of the Academy, and then undertook adventurous travels in the south, the Caucasus, Persia and Afghanistan, travels which led before long to his downfall and departure from Russia because his reports were considered to show too clearly the weakness of Russian power in those regions. There is no immediate need to follow his further career in France, his relations with Napoleon, etc., but only to remind ourselves that in the penultimate year of his life, 1834, he published a monograph on the history of the magnetic compass so good that it is still usable at the present day, and greatly helped us in writing Sect. 26(i) in Vol. 4, pt. 1. Young as he was at the time (1802), it is hard to believe either that Klaproth misinterpreted his text, or that he was guilty (as might conceivably be thought possible) of a scholarly hoax designed to attract interest and further his career.[a] But we should be happier if we could find out something definite about the Mons. Bournon who brought the MS. back from China—so far all searches in the French and the sinological literature have proved unavailing. The solution of the puzzle is more likely to come from the Soviet sinologists than anyone else, for they may be able to trace the MS. itself, and they may succeed in finding out who Bournon was and what (from his connections) he is likely to have brought home with him. Perhaps this intriguing document will have the honour of being transferred, in later editions of the present book, from its present place back to the great days of the Thang.

Of one thing we can be reasonably sure: there were Jesuits towards the end of the + 18th century who were very enamoured of the Yin-Yang theory as a general natural philosophy. On 26 June 1789, J. J. M. Amiot, in the course of a letter (7), wrote in these terms:

Here, as in our France, men construct (philosophical) systems, but they fall into desuetude and are soon forgotten when they are not founded on the two great principles Yang and Yin. Yang and Yin, Yin and Yang, when well conceived and well understood, are alone capable of introducing man into the sanctuary of Nature, and unveiling for him even the most secret of her mysteries. Everything happens by the varying combinations of the Yin and Yang, it is by them that all Beings reproduce themselves, and that everything maintains itself in the living state. If the Yang and the Yin were to cease to combine with each other, the whole world would return to chaos. These, you will tell me, are the dreams of the Chinese; we are not necessitated yet to adopt them in preference to those of our European savants. All right, but one day you will adopt them—judging by the productions of your modern Physicists!

Was not Amiot thinking of the great discoveries of Franklin, Galvani and Volta, which introduced the fundamental concept of positive and negative electricity? From + 1780 onwards these terms had been current coin.[b] And could not one of his friends have translated hydrogen and oxygen into terms of Yang and Yin?

---

[a] It is true that this was in a period of great forgeries—after 'Ossian', the *Toparcha Goticus* (Westberg, 1), and a famous Czech deception. But usually these were motivated by nationalist ideals.

[b] See Vol. 2, pp. 278, 467. And, for the course of events in Europe, Cohen's account in Foley (1). Vast developments in electro-chemistry and electro-physiology were to follow.

(iv) *The Kiangnan Arsenal and the sinisation of modern chemistry*

In bidding farewell to the Chinese alchemical tradition after twenty-five centuries it is reasonable to enquire about the date of its last appearance. For all we know, the making of elixirs may still be going on in some remote places, for there have been reports in recent years of obscure Taoist sects being brought to book by agencies of the government of People's China.[a] We have one record of the neighbourhood of 1917 however, mentioned by E. V. Cowdry, the American anatomist who held the chair of this subject at Peking Union Medical College in the twenties of the present century. He wrote:

The [Taoist] priests sometimes attempt to distil mercury. When the distillation has been successfully repeated nine times, the resulting substance will be the 'elixir of life'. One of the priests of the White Cloud Temple is said to have been killed by an explosion during the sixth distillation.[b]

This sounds more like the repeated sublimations of mercuric sulphide during the classical nine cyclical transformations method than any distillation, but perhaps the Taoist was meddling with something else, as his confrères had been warned not to do in the +9th century (cf. p. 78 above). In any case the explosion was the parting shot of a tradition glorious both for its antiquity and its tenacity.

The rest of the story can be briefly told, since in these volumes we do not set out to treat of modern history. During the first half of the nineteenth century Chinese artisans, merchants and minor scholars associating with the European and American traders and missionaries along the eastern seaboard certainly came to know about some of the chemical substances which were by now familiar in Western science, and to some extent used by the Western physicians practising in China. For example, one could think of the halogen elements, oxygen, the purified strong mineral acids,[c] metallic sodium and potassium, etc. Many chemical names seem to have been coined popularly at this time, though not by the literati, who remained relatively uninterested.[d] Fig. 1368 illustrates a scene of this period.[e]

After the middle of the century was reached, things began to move rapidly. The first book in Chinese dealing with modern chemisty came out in 1855; it was the

[a] Also at the outer edges of Chinese culture aurifaction may be going on. While in East Asia in 1971 I read a news item in the *Hongkong Standard* for 16 September entitled 'Housewife cheated by Bomoh'; a Malayan magician, in fact, who had claimed to be able to transmute sea shells into gold, and had got away with $230 on the strength of it.

[b] Cowdry (1), p. 307. When I enjoyed a visit to the Pai Yün Kuan in 1964 no alchemical elaboratory was to be seen, but rich collections of the Taoist alchemical books were in the library. Cf. Fig. 1367.

[c] Sulphuric acid, *huang chhiang shui*,[1] nitric acid, *hsiao chhiang shui*,[2] and hydrochloric acid, *yen chhiang shui*.[3]

[d] The great translator Fryer (1) well appreciated this artisanal and merchant origin. See Yuan Han-Chhing (1), pp. 266ff.

[e] From this time dates what was perhaps the first Chinese–English glossary of scientific (including some chemical) names and terms, the chrestomathy of Bridgman (1) issued at Macao in 1841. Many more, especially in chemistry, were contained in the collection of glossaries published at Fuchow by Doolittle (1) in 1872.

[1] 磺强水　　　　[2] 硝强水　　　　[3] 鹽强水

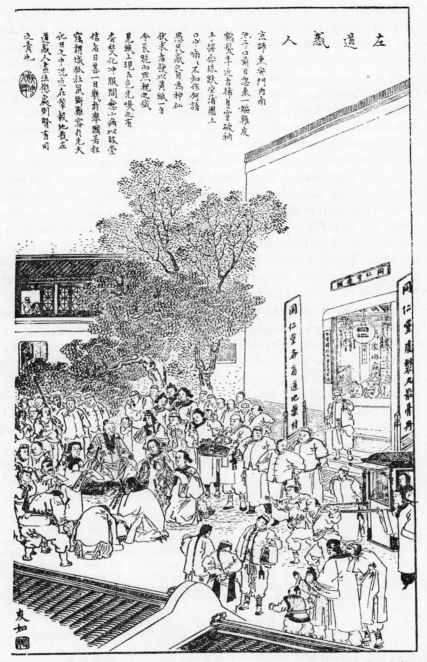

Fig. 1368. A nineteenth-century scene outside the Thung Jen Thang pharmacy (still existing today) in Peking; from an issue of the illustrated magazine *Tien Shih Chai Hua Pao*. Captioned 'A Sinister Humbug', it shows a Taoist woman adept distributing apotropaic cantraps to the believing populace, and the text berates this departure (by no means uncommon) from reliance on orthodox medicines, both traditional and modern. The artist, Wu Yu-Ju, was also the editor of this publication, China's first pictorial, founded in 1884.

Po Wu Hsin Phien[1] (New Treatise on Natural Philosophy and Natural History), written by Benjamin Hobson (Ho Hsin,[2] 1816 to 1873), an English physician working in the missionary hospitals of Canton and Shanghai.[a] We give a specimen page of this with illustrations as Fig. 1369. During this period several other medical missionaries contributed to the chemical literature, notably J. G. Kerr (Chia Yo-Han,[3] 1824 to 1901),[b] whose *Hua-Hsüeh Chhu Chiai*[4] of 1870 (First Steps in Chemistry), done in collaboration with Ho Liao-Jan,[5] was probably the first publication to contain the name of the modern science in its title. Appropriately however for so bureaucratic a culture, the greatest wealth of books on modern chemistry began soon to flow forth from government-sponsored institutions. Not all of these were fully successful. For example, in 1862 a translation bureau called the Thung Wên Kuan[6] was established in Peking.[c] It did not at first engage in scientific publishing, and it was only in the last quarter of the century that this institution brought out translations and adaptations of chemical texts.[d] For example, in 1882 there was the *Hua-Hsüeh Shan Yuan*[7] (Explanation of the Fundamental Principles of Chemistry)[e] by M. A. Billequin (Pi Li-Kan[8]) with Chhêng Lin[9] & Wang Chung-Hsiang.[10] It is interesting that these scholars made an attempt in coining new terms for modern chemical science to incorporate the names of medieval times, even when this meant abandoning the phonetic principle. But when it came to writing the name of manganese 鑽 because *wu ming i*[11] had in times past stood for manganese dioxide (pyrolusite),[f] it was soon obvious that this was an unprofitable side-track, and that the nomenclature of modern chemistry was bound to be constructed on other principles. In effect, manganese to-day is expressed more conveniently by the phonetic character *mêng*.[12]

By far the most significant development at this time was the foundation of the famous Kiangnan Arsenal near Shanghai in 1865. The Chiang-Nan Chi-Chhi Chih-Tsao Chü[13] (Chiangnan Machinery Factory), to give it its proper name, was largely the idea of the far-seeing statesman Ting Jih-Chhang[14] (1823 to 1882),[g] and just how far-sighted it was may be gauged from the fact that two years later a Bureau of Foreign Translations (Fan I Kuan[15]) was set up within it, as well as a School of Foreign Languages (Kuang Fang Yen Kuan[16]).[h] Before the end of the century the

---

[a] For a biography see Wang Chi-Min (1), pp. 14ff. Wang Chi-Min & Wu Lien-Tê (1), pp. 321ff., 358ff., give the Chinese titles of half-a-dozen other valuable books, on medical subjects, written by him.

[b] For a biography see Wang Chi-Min (1), pp. 23ff. Kerr & Ho's book was one which had much influence on the pioneer historian of chemistry in China, Davis's collaborator, Wu Lu-Chhiang, cf. Yuan, p. 33.   [c] Cf. Hummel (2), p. 790 and *sub voce*.

[d] The *Ko Wu Ju Mên*[17] (Introduction to Natural Philosophy) by W. A. P. Martin (Ting Wei-Liang[18]) appeared, however, in 1868. Again see Vol. 1, pp. 48–9.

[e] Cf. Yuan Han-Chhing (1), p. 278. A general list of books on chemistry published during the 19th century in China is given on pp. 289ff.

[f] RP61.   [g] Biography in Hummel (2), p. 721.

[h] The first director was Ying Pao-Shih[19] (1821 to *c*. 1880), the second Fêng Chün-Kuang[20] (1830 to 1878).

[1] 博物新編	[2] 合信	[3] 嘉約翰	[4] 化學初階	[5] 何了然
[6] 同文館	[7] 化學闡原	[8] 畢利干	[9] 承霖	[10] 王鍾祥
[11] 無名異	[12] 錳	[13] 江南機器製造局		[14] 丁日昌
[15] 繙譯館	[16] 廣方言館	[17] 格物入門	[18] 丁韙良	[19] 應寶時
[20] 馮焌光				

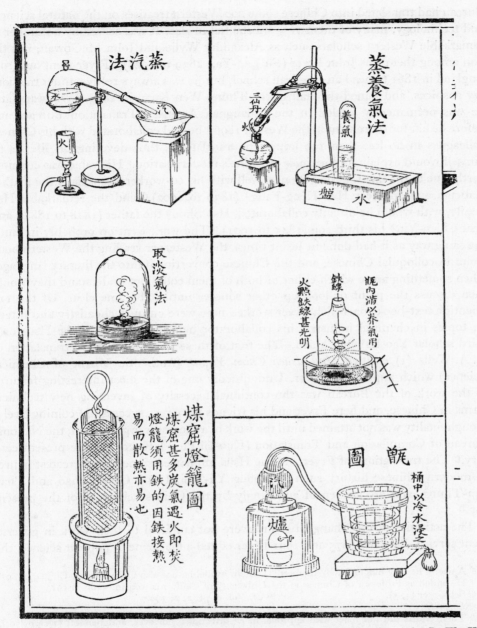

Fig. 1369. A page from the first book in Chinese dealing with modern chemistry, the *Po Wu Hsin Phien*, 1855 (New Treatise on Natural Philosophy and Natural History), written by an English physician, Benjamin Hobson (Ho Hsin, 1816 to 1873), then working in the hospitals of Canton and Shanghai. On the left one sees distillation with a retort, the preparation of nitrogen by heating ammonium dichromate, and Davy's lamp for miners. On the right oxygen is being collected over mercury, and a clean iron wire tipped with sulphur is burning vigorously in it, below, a distillation through a cooled coil is proceeding.

Bureau had translated into Chinese some 200 Western treatises on the natural sciences and technology, many of them very substantial in size. It was served by a number of remarkable Western scholars such as Alexander Wylie and John McGowan, but the lion among them was John Fryer (Fu Lan-Ya,[1] 1839 to 1928).[a] Fryer went out from England in 1861 to head an Anglican school, but he was always restive under missionary auspices, and joined the staff of the Thung Wên Kuan in 1863. Five years later he was permanently settled in the Chiangnan Arsenal's Translation Bureau, and before he left for retirement in the West in 1896[b] he had collaborated with his Chinese colleagues in no less than 129 important translations, thus devoting his life, as he himself would explain, to the cause of China's modernisation.[c] His value was certainly partly due to the fact that he got on so well with his co-workers, especially the mathematician and engineer Hua Hêng-Fang[2] (1830 to 1902),[d] and the remarkable Hsü family, with whom he directly collaborated, Hsü Shou[3] the father (1818 to 1884), and Hsü Chien-Yin[4] his third son (1845 to 1901).[e] The work went on probably in much the same way as it had done in Jesuit times, the Westerner reading the Western book aloud in colloquial Chinese, and the Chinese converting it into the literary language; when something arose which either or both of them could not understand they would then discuss the point or look up other sources until it became clear. Of the 112 scientific text-books on which Fryer worked nine were on pure chemistry and fifteen on topics in chemical industry, his collaborator being either one of the Hsüs or a third scholar Yao Hsüeh-Chhien.[5] The first of these joint works was a translation of D. A. Wells (1) entitled *Hua-Hsüeh Chien Yuan*[6] (Authentic Mirror of Chemical Science) which appeared in 1871. Undoubtedly one of the most interesting features of the work of the Bureau was the continual necessity of inventing new technical terms in Chinese, and here Fryer and his friends were very successful,[f] coining freely, though finality was not attained until the work of a successor organisation, the National Bureau of Compilation and Translation (Kuo-Li Pien I Kuan[7]) in the present century.[g] The translations of Fryer and the Hsüs influenced some of the greatest figures of modern Chinese history such as Khang Yu-Wei, Liang Chhi-Chhao and Than Ssu-Thung, and are spoken of admiringly by the Chinese chemists of the present day.[h]

The activities of the Shanghai group were not confined to their work in government service. Between 1875 and 1891 Fryer edited a magazine of popular science, the

---

[a] A useful literary biography by Bennett (1) is now available. See also Chang Ching-Lu (1), pp. 9ff.
[b] He ended as Professor of Chinese at the University of California, retiring only in 1913.
[c] See Fryer (4, 5).   [d] Cf. Vol. 4, pt. 2, p. 390.
[e] *Loc. cit.*
[f] Their efforts did not meet with approval in all directions, however, and Bennett (1), pp. 30ff., has an interesting account of some of the disputes that went on. Indeed they spread among sinologists in the West, as can be seen from Schlegel (10) criticising Stuart (2). The papers of Fryer (1, 2) and his 'translator's vade mecum' (3) are rare, but worth reading if they can be got hold of.
[g] General usage since 1932, official adoption in 1953.
[h] Cf. Tsêng Chao-Lun (1).

[1] 傅蘭雅   [2] 華蘅芳   [3] 徐壽   [4] 徐建寅   [5] 姚學謙
[6] 化學鑒原   [7] 國立編譯館

Ko Chih Hui Phien[1] (Chinese Scientific and Industrial Magazine); and in 1874 he started the Shanghai Polytechnic Institute and Reading-Rooms (Ko Chih Shu Yuan[2]), to which was attached a publishing house (the Ko Chih Shu Shih[3]) after 1884. For a time this Polytechnic was headed by Wang Thao,[4] the Chinese collaborator of the great sinologist James Legge. Similarly Hsü Chien-Yin edited a collection of modern scientific and chemical writings, the *Ko Chih Tshung Shu*,[5] which appeared between 1897 and 1901. Illustrations from this are shown in Figs. 1370–2.

Meanwhile parallel developments had been proceeding in Japan. Although it is beyond the scope of this work to consider them, the fact that the first book on modern chemistry appeared in Japan some fifteen years earlier than that of Hobson in China is not without interest. Here the great pioneer was Udagawa Yōan[6] with his *Seimi Kaisō*[7] (Treatise on Chemistry), issued in parts between 1837 and 1846.[a] Information on this will be found in Tanaka Minoru (3) and Dōke Tatsumasa (1). As will be seen, no attempt was made to translate the idea of the name of the science into Japanese, only the sound Shê-Mi being reproduced. The Hellenistic 'alchemists', and all the later scholars who strove to explain the root chem-,[b] would have been much surprised to see it in this East Asian guise. For the rest of the story it must suffice to refer the reader to the numerous publications of Japanese scholars.[c]

The language of modern chemistry was an artificial creation from the first, the efforts of Guyton de Morveau (+1782)[d] being standardised in the work of Lavoisier, Berthollet, Fourcroy and others (+1787)—just at the time of the demise of the Jesuit mission in China. Here we cannot follow in any detail the complex processes whereby modern Chinese chemical terminology was formed in the following century; those interested may begin with the recent monograph on the subject by Alleton & Alleton (1). The considerable literature in Chinese includes indispensable dictionaries such as that of Kao Hsien (1).[e] Now modern chemistry provides a perfect example of the age-old dilemma confronting translators from alphabetical languages into Chinese, though naturally the problem had complexities of its own. The Buddhists had faced it long before, from the +2nd century onwards. Should one employ an already existing, in their case Taoist, technical term, and risk a fatal distortion of one's meaning? That was the system of 'explaining by analogy' (*ko i*[8]).[f] Or should one

[a] He was not actually the first Japanese to illustrate chemical apparatus, for as early as 1805 Hashimoto Sōkichi[9] had needed to do this for pharmaceutical purposes in his *Ranka Naigai Sanbō Hōten*[10] (Handbook of the Three Aspects of Dutch Internal and External Medicine), printed in Osaka (Fig. 1373). Information on this and other pioneer Japanese works will be found in the interesting paper of Shimao Eikoh (1).

[b] Cf. pt. 4 below.

[c] E.g. Tanaka Minoru (1, 2, 4); Tsukahara & Tanaka (1); Yamashita (1); Yamazaki (1). The reception of Lavoisier's chemistry in Japan has been studied by Shimao Eikoh (1).

[d] See Partington (7), vol. 3, p. 516.

[e] Published in 1960. And there is Taranzano's scientific dictionary (1) of 1936, perhaps the only one of the kind by a Westerner.

[f] Cf. Vol. 2, p. 409.

[1] 格致彙編 [2] 格致書院 [3] 格致書室 [4] 王韜 [5] 格致叢書
[6] 宇田川榕庵 [7] 舍密開宗 [8] 格義 [9] 橋本宗吉
[10] 蘭科內外三法方典

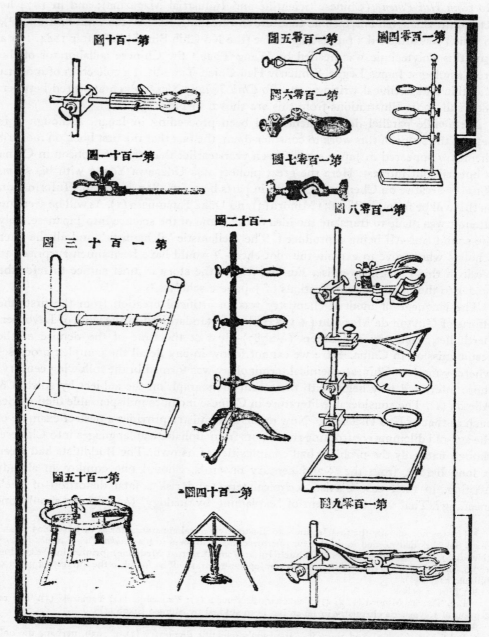

Fig. 1370. Illustration from the *Ko Chih Tshung Shu* (General Treatise on the Natural Sciences), ed. by Hsü Chien-Yin between 1897 and 1901. Hua Hsüeh sect., ch. 1, p. 4b, showing retort stands, clamps, tripods and a Bunsen burner.

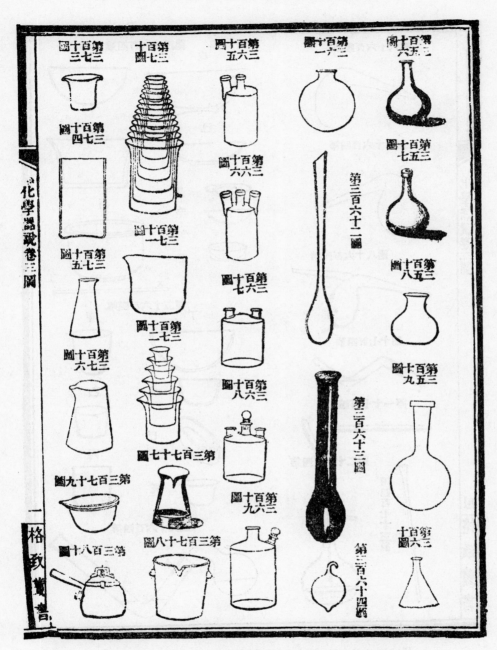

Fig. 1371. Flasks and beakers from the *Ko Chih Tshung Shu*, Hua Hsüeh sect., ch. 3, p. 1a.

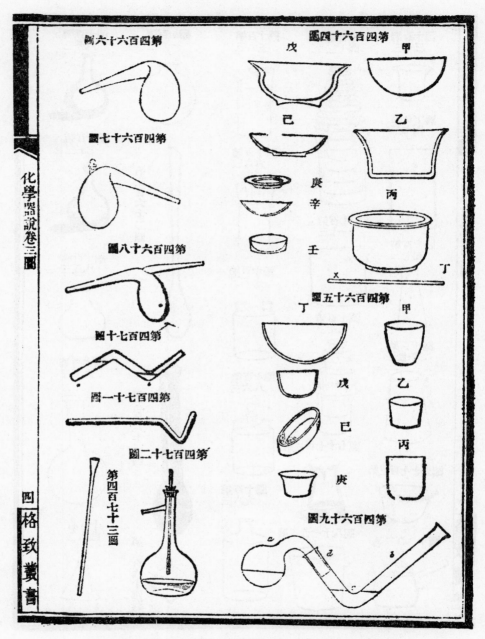

Fig. 1372. Retorts, basins, crucibles, and a wash-bottle from the
Ko Chih Tshung Shu, Hua Hsüeh sect., ch. 3, p. 4*a*.

transliterate the foreign polysyllabic term or name into a string of monosyllables in meaningless juxtaposition, then define the ugly compound resulting? That was the way they ultimately took, with the result that Buddhism retained to the end an indelible quality of foreign-ness within Chinese culture. For chemistry there was of course, as we know, a rich fund of ancient and medieval technical terms for chemical substances (as also indeed for chemical apparatus and operations), but it had grown up on a purely empirical basis, and could not therefore serve for a developed language logically founded on true elements and known atomic combinations. As for the other alternative, strings of meaningless syllables did at one time play a certain part in the Chinese terminology of biochemistry and the medical sciences, but during the past half-century proper translations of the ideas involved have replaced all of them.[a]

The scientific movement of the nineteenth century resolved the dilemma in several ways, partly by adopting a large number of archaic or obsolete characters and giving them precise new technical significances, partly by coining totally new characters according to the classical radical-phonetic system,[b] and partly by joining all these together with a cement of numbers to make polysyllabic formulae as clearly inter- pretable as a term such as calcium perborate[c] is for us. Some hundreds of entirely new ideographs (*hsin tzu*[1]) were thus created, including those needed for the majority of the chemical elements; and hundreds more of old ideographs, traditional but un- used, were pressed into new service as technical terms.[d] For the elements, the metals and semi-metals[e] were based on the semantic radical *chin*,[2] the classical word for metal (such as *pi*[3] for bismuth); the gases on *chhi*[4] (like *lü*,[5] 'the green gas', for chlorine, from *lü*,[6] green); and the earths on *shih*[7] (including *than*[8] for carbon). Apart from the traditionally known metals, mercury retained its unique and very demonstrative ideograph *hung*;[9] and bromine, the only other element liquid at room temperatures, became the only one to bear the normal water radical, *hsiu*[10].[f]

The conviction of the absolute necessity for the retention of the phonetic principle in character-building probably goes back to Hsü Shou and John Fryer, but they did not feel the need for monosyllabism so greatly, leaving oxygen, for example, as *yang chhi*,[11] 'the nourishing gas', instead of reducing it to its present form, *yang*[12] (cf. p. 247 above). Even now, some ambiguities only distinguished tonally still remain, such as *lü suan na*[13] for sodium chlorate and *lü suan na*[14] for sodium aluminate. The Shanghai

---

a For example, the purely phonetic *wei-tha-ming*[15] for vitamin was abandoned for *wei-shêng-su*,[16] 'life-maintaining quintessence'.

b Cf. Vol. 1, pp. 30ff.

c Per- affix is represented by *kao*,[17] 'high', in modern Chinese usage.

d Parallel developments took place in other sciences, but not so successfully. Li Shan-Lan constructed a notation for analytical geometry and infinitesimal calculus based wholly upon Chinese characters, and used it in his translation of the book of Elias Loomis in 1859, as Mei Jung-Chao (1) has shown. Chinese mathematicians preferred, however, to use the international notations.

e A term which goes back to Biringuccio.

f It was an abandoned archaic word for steam or water-vapour.

1 新字	2 金	3 鉍	4 氣	5 氯
6 綠	7 石	8 碳	9 汞	10 溴
11 養氣	12 氧	13 氯酸鈉	14 鋁酸鈉	15 維他命
16 維生素	17 高			

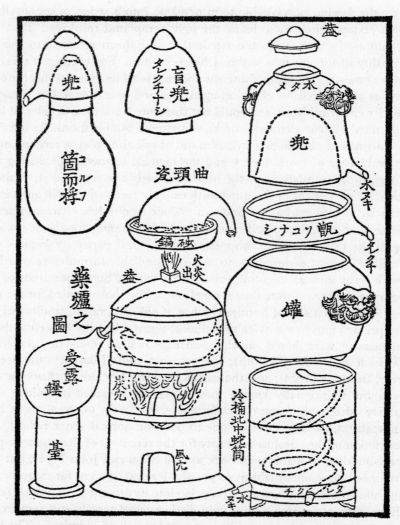

Fig. 1373. Some of the earliest illustrations of chemical apparatus in Japan, a page from the *Ranka Naigai Sanbō Hōten* (Handbook of the Three Aspects of Dutch Internal and External Medicine), published by Hashimoto Sōkichi in 1805. This was the entry of iatro-chemistry in a late phase of the Rangaku period. On the left at the top is an ambix with a helm, annular channel and side-tube; on the right a 'Moor's head' still in which the helm is water-cooled (cf. Vol. 5, pt. 4). On the left at the bottom is a retort protected in the furnace by a sand-bath, and with a receiver in position; on the right a cooling-coil condenser.

group were also still willing to transcribe some organic terms in Buddhist style, e.g. *mi-i-tho-li*[1] for methyl-, instead of the present logical *chia chi*,[2] 'first radical'. Here we have the use of one of the classical cyclical characters,[a] and the application of an old word meaning site or foundation.

So, for example, an amino-acid is *an chi suan*,[3] for *an* was a coinage from the sound of the first syllable of ammonia, while *suan* was the medieval word for sourness, hence adopted throughout in chemistry to denote an acid. To give some further idea of the contemporary use of applied words one could cite *wan*[4] for the single organic bond, as in meth*ane*, and *hsi*[5] for the double bond, as in pin*ene*. Here *wan* was a poetic ancient word for fire, and *hsi* was a disused archaic word meaning fire-coloured. The old word so prominent in alchemy, *huan*,[6] 'cyclical transformation', has found a home in the technical term for reduction *huan yuan*[7] (return to the origin), so that 'reducing sugar' is *huan yuan hsing thang*.[8] Rings are easy to express, as e.g. in *i yang wu yuan*[9] for tetrahydro-furane, the five-membered ring with one heterocyclic oxygen atom. Taking at random any modern chemical term, the sensitive student of Chinese can readily see the subtleties which have gone into its making. For instance, mono-sodium orthophosphate (NaH_2PO_4) is *lin suan erh ching na*,[10] *lin* being an old word meaning glowing, glittering, hence most suitable for phosphorus,[b] and *ching* a coined word for hydrogen, 'the light gas', from *chhing*,[11] light, while *na* came by direct phonetic adoption from natrium. Again, *chhing*,[12] meaning cyan-, was derived from familiar *chhing*,[13] blue-green.

One interesting feature of the development of modern Chinese scientific terminology was the use of the expression *yu chi*[14] for organic, as in *yu chi hua-hsüeh*,[15] organic chemistry. As we know from other Sections, the basic meaning of the word *chi* is machine,[c] though it has also the secondary connotations of motive power, secret process and opportune moment. The character *chi* itself arose in connection with textile technology, from the most ancient of looms, but its semantic significance broadened out through the ages, and a full study of this might throw interesting light on the fundamentals of Chinese scientific thinking. If *yu chi wu*[16] came to mean 'organism' in the modern language, it expressed 'the complex entity which has structure and lower-level components and inter-relations within it', as against the seeming homogeneity, in repetitive array or chaotic flux, of inorganic crystal or earth and sea. It was a little curious that this should have happened, because medieval Chinese philosophy had developed words of great power and content for 'organic pattern', especially *li*[17],[d] but by the nineteenth century this was doubtless felt to be

[a] Cf. Vol. 3, p. 396.
[b] Cf. our study of phosphorescence in Vol. 4, pt. 1, pp. 72ff.
[c] Cf. Vol. 4, pt. 2, pp. 9, 69. Cf. also Sect. 38 in Vol. 6 under *chi*,[18] 'germ', following Vol. 2, pp. 78–9, and Needham & Leslie (1).
[d] Cf. Vol. 2, pp. 472ff.

[1] 迷以脫里 [2] 甲基 [3] 氨基酸 [4] 烷 [5] 烯
[6] 還 [7] 還原 [8] 還原性糖 [9] 一氧伍圜 [10] 磷酸二氫鈉
[11] 輕 [12] 氰 [13] 青 [14] 有機 [15] 有機化學
[16] 有機物 [17] 理 [18] 幾

impossibly old-fashioned, and indeed too much tied up with the psychological, the moral and the spiritual. Yet *wu chi hua-hsüeh*[1] was not a good term for inorganic chemistry since inorganic compounds also have their 'lower-level components', almost *ad infinitum*; but it has become the accepted phrase, and 'organic' vs. 'inorganic', in the light of modern organic philosophy, is perhaps hardly any better.

Thus it is clear that the language has been no fundamental barrier to the development of modern chemical science among the great masses of the Chinese people. A framework of logical rules of nomenclature having been established, chemical papers and books of modern type could be written, and many journals of this kind in Chinese have been appearing through the better part of the past seventy years. From the turn of the century modern chemistry was taught in all the Chinese universities. What will happen in the future depends to some extent on the course taken by projects of alphabetisation, the growth and effects of which it is impossible to foresee. Historically established, however, is the fact that the revivification and adaptation of the ideographic language necessary to include all modern chemistry within its domain was successfully accomplished, so far not only as to permit a vast development of chemical industry, analytical work and university teaching,[a] but also the great achievement of Chinese chemists and biochemists (not without severe international competition) in one of the most striking advances of our own times, the synthesis of active insulin in 1965.[b]

[a] For reviews of all this see Tsêng Chao-Lun (3); Adolph (1, 2).
[b] See Tu Yü-Tshang *et al.* (1); Niu Ching-I *et al.* (1); Kung Yo-Thing *et al.* (1); as also Pastan (1), pp. 391 ff. This was the first synthetic protein ever produced.

[1] 無機化學

BIBLIOGRAPHIES

A CHINESE AND JAPANESE BOOKS BEFORE +1800
B CHINESE AND JAPANESE BOOKS AND JOURNAL ARTICLES SINCE +1800
C BOOKS AND JOURNAL ARTICLES IN WESTERN LANGUAGES

In Bibliographies A and B there are two modifications of the Roman alphabetical sequence: transliterated *Chh-* comes after all other entries under *Ch-*, and transliterated *Hs-* comes after all other entries under *H-*. Thus *Chhen* comes after *Chung* and *Hsi* comes after *Huai*. This system applies only to the first words of the titles. Moreover, where *Chh-* and *Hs-* occur in words used in Bibliography C, i.e. in a Western language context, the normal sequence of the Roman alphabet is observed.

When obsolete or unusual romanisations of Chinese words occur in entries in Bibliography C, they are followed, wherever possible, by the romanisations adopted as standard in the present work. If inserted in the title, these are enclosed in square brackets; if they follow it, in round brackets. When Chinese words or phrases occur romanised according to the Wade–Giles system or related systems, they are assimilated to the system here adopted (cf. Vol. 1, p. 26) without indication of any change. Additional notes are added in round brackets. The reference numbers do not necessarily begin with (1), nor are they necessarily consecutive, because only those references required for this volume of the series are given.

Korean and Vietnamese books and papers are included in Bibliographies A and B. As explained in Vol. 1, pp. 21 ff., reference numbers in italics imply that the work is in one or other of the East Asian languages.

ABBREVIATIONS

See also p. xv

A	Archeion	AJA	American Journ. Archaeology
AA	Artibus Asiae	AJOP	Amer. Journ. Physiol.
AAA	Archaeologia	AJPA	Amer. Journ. Physical Anthropology
AAAA	Archaeology		
A/AIHS	Archives Internationales d'Histoire des Sciences (continuation of Archeion)	AJSC	American Journ. Science and Arts (Silliman's)
		AM	Asia Major
AAN	American Anthropologist	AMA	American Antiquity
AAPWM	Archiv. f. Anat., Physiol., and Wiss. Med. (Joh. Müller's)	AMH	Annals of Medical History
		AMS	American Scholar
ABAW/PH	Abhandlungen d. bayr. Akad. Wiss. München (Phil.-Hist. Klasse)	AMY	Archaeometry (Oxford)
		AN	Anthropos
ACASA	Archives of the Chinese Art Soc. of America	ANATS	Anatolian Studies (British School of Archaeol. Ankara)
ACF	Annuaire du Collège de France	ANS	Annals of Science
ADVC	Advances in Chemistry	ANT	Antaios (Stuttgart)
ADVS	Advancement of Science (British Assoc., London)	ANTJ	Antiquaries Journal
		AP	Aryan Path.
AEM	Anuario de Estudios Medievales (Barcelona)	APH	Actualités Pharmacologiques
		AP/HJ	Historical Journal, National Peiping Academy
AEPHE/SHP	Annuaire de l'Ecole Pratique des Hautes Études (Sect. Sci. Hist. et Philol.)	APAW/PH	Abhandlungen d. preuss. Akad. Wiss. Berlin (Phil.-Hist. Klasse)
AEPHE/SSR	Annuaire de l'Ecole Pratique des Hautes Études (Sect. des Sci. Religieuses)	APHL	Acta Pharmaceutica Helvetica
		APNP	Archives de Physiol. normale et pathologique
AESC	Aesculape (Paris)	AQ	Antiquity
AEST	Annales de l'Est (Fac. des Lettres, Univ. Nancy)	AR	Archiv. f. Religionswissenschaft
		ARB	Annual Review of Biochemistry
AF	Ärztliche Forschung	ARLC/DO	Annual Reports of the Librarian of Congress (Division of Orientalia)
AFG	Archiv. f. Gynäkologie		
AFGR/CINO	Atti della Fondazione Giorgio Ronchi e Contributi dell'Istituto Nazionale di Ottica (Arcetri)	ARMC	Ann. Reports in Medicinal Chemistry
		ARO	Archiv Orientalni (Prague)
AFP	Archivum Fratrum Praedicatorum	ARQ	Art Quarterly
AFRA	Afrasian (student Journal of London Inst. Oriental & African Studies)	ARSI	Annual Reports of the Smithsonian Institution (Washington, D.C.)
		AS/BIHP	Bulletin of the Institute of History and Philology, Academia Sinica
AGMN	Archiv. f. d. Gesch. d. Medizin u. d. Naturwissenschaften (Sudhoff's)	AS/CJA	Chinese Journal of Archaeology, Academia Sinica
AGMW	Abhandlungen z. Geschichte d. Math. Wissenschaft	ASEA	Asiatische Studien; Études Asiatiques
AGNT	Archiv. f. d. Gesch. d. Naturwiss. u. d. Technik (cont. as AGMNT)	ASN/Z	Annales des Sciences Naturelles; Zoologie (Paris)
		ASSF	Acta Societatis Scientiarum Fennicae (Helsingfors)
AGP	Archiv. f. d. Gesch. d. Philosophie		
AGR	Asahigraph	AT	Atlantis
AGWG/PH	Abhdl. d. Gesell. d. Wiss. Z. Göttingen (Phil.-Hist. Kl.)	ATOM	Atomes (Paris)
		AX	Ambix
AHES/AHS	Annales d'Hist. Sociale		
AHOR	Antiquarian Horology	BABEL	Babel; Revue Internationale de la Traduction
AIENZ	Advances in Enzymology		
AIP	Archives Internationales de Physiologie	BCGS	Bull. Chinese Geological Soc.
		BCP	Bulletin Catholique de Pékin

BCS	Bulletin of Chinese Studies (Chhêngtu)		CEM	Chinese Economic Monthly (Shanghai)
BDCG	Ber. d. deutsch. chem. Gesellschaft.		CEN	Centaurus
BDP	Blätter f. deutschen Philosophie		CHA	Chemische Apparatur
BE/AMG	Bibliographie d'Études (Annales du Musée Guimet)		CHEMC	Chemistry in Canada
			CHI	Cambridge History of India
BEC	Bulletin de l'École des Chartes (Paris)		CHIM	Chimica (Italy)
			CHIND	Chemistry and Industry (Journ. Soc. Chem. Ind. London)
BEFED	Bulletin de l'Ecole Française de l'Extrême Orient (Hanoi)		CHJ	Chhing-Hua Hsüeh Pao (Chhing-Hua (Ts'ing-Hua) University Journal of Chinese Studies)
BGSC	Bulletin of the Chinese Geological Survey		
BGTI	Beiträge z. Gesch. d. Technik u. Industrie (continued as Technik Geschichte—see BGTI/TG)		CHJ/T	Chhing-Hua (T'sing-Hua) Journal of Chinese Studies (New Series, publ. Thaiwan)
BGTI/TG	Technik Geschichte		CHWSLT	Chung-Hua Wên-Shih Lun Tshung (Collected Studies in the History of Chinese Literature)
BHMZ	Berg und Hüttenmännische Zeitung		
BIHM	Bulletin of the (Johns Hopkins) Institute of the History of Medicine (cont. as Bulletin of the History of Medicine)		CHYM	Chymia
			CHZ	Chemiker Zeitung
			CIBA/M	Ciba Review (Medical History)
BJ	Biochemical Journal		CIBA/MZ	Ciba Zeitschrift (Medical History)
BJRL	Bull. John Rylands Library (Manchester)		CIBA/S	Ciba Symposia
			CIBA/T	Ciba Review (Textile Technology)
BK	Bunka (Culture), Sendai		CIMC/MR	Chinese Imperial Maritime Customs (Medical Report Series)
BLSOAS	Bulletin of the London School of Oriental and African Studies		
			CIT	Chemie Ingenieur Technik
BM	Bibliotheca Mathematica		CJ	China Journal of Science and Arts
BMFEA	Bulletin of the Museum of Far Eastern Antiquities (Stockholm)		CJFC	Chin Jih Fo Chiao (Buddhism Today), Thaiwan
BMFJ	Bulletin de la Maison Franco–Japonaise (Tokyo)		CLINR	Clinical Radiology
			CLR	Classical Review
BMJ	British Medical Journal		CMJ	Chinese Medical Journal
BNJ	British Numismatic Journ.		CN	Chemical News
BOE	Boethius; Texte und Abhandlungen d. exakte Naturwissenschaften (Frankfurt)		CNRS	Centre National de la Recherche Scientifique
			COCJ	Coin Collectors' Journal
BR	Biological Reviews		COPS	Confines of Psychiatry
BS	Behavioural Science		CP	Classical Philology
BSAA	Bull. Soc. Archéologique d'Alexandrie		CQ	Classical Quarterly
			CR	China Review (Hongkong and Shanghai)
BSAB	Bull. Soc. d'Anthropologie de Bruxelles		
BSCF	Bull. de la Société Chimique de France		CRAS	Comptes Rendus hebdomadaires de l'Acad. des Sciences (Paris)
BSGF	Bull. de la Société Géologique de France		CREC	China Reconstructs
			CRESC	Crescent (Surat)
BSJR	Bureau of Standards Journ. of Research		CRR	Chinese Recorder
			CRRR	Chinese Repository
BSPB	Bull. Soc. Pharm. Bordeaux		CS	Current Science
BUA	Bulletin de l'Université de l'Aurore (Shanghai)		CUNOB	Cunobelin; Yearbook of the British Association of Numismatic Societies
BV	Bharatiya Vidya (Bombay)		
			CUP	Cambridge University Press
			CUQ	Columbia University Quarterly
CA	Chemical Abstracts		CURRA	Current Anthropology
CALM	California Medicine		CVS	Christiania Videnskabsselskabet Skrifter
CBH	Chūgoku Bungaku-hō (Journ. Chinese Literature)		
			CW	Chemische Weekblad
CCJ	Chung-Chi Journal (Chhung-Chi Univ. Coll. Hongkong)		CWR	China Weekly Review
CDA	Chinesisch-Deutschen Almanach (Frankfort a/M)		DAZ	Deutscher Apotheke Zeitung
			DB	The Double Bond

DI	*Die Islam*
DK	*Dōkyō Kenkyū (Researches in the Taoist Religion)*
DMAB	*Abhandlungen u. Berichte d. Deutsches Museum (München)*
DS	*Desalination (International Journ. Water Desalting) (Amsterdam and Jerusalem, Israel)*
DV	*Deutsche Vierteljahrschrift*
DVN	*Dan Viet Nam*
DZZ	*Deutsche Zahnärztlichen Zeit.*
EARLH	*Earlham Review*
EECN	*Electroencephalography and Clinical Neurophysiology*
EG	*Economic Geology*
EHOR	*Eastern Horizon (Hongkong)*
EHR	*Economic History Review*
EI	*Encyclopaedia of Islam*
EMJ	*Engineering and Mining Journal*
END	*Endeavour*
EPJ	*Edinburgh Philosophical Journal (continued as ENPJ)*
ERE	*Encyclopaedia of Religion and Ethics*
ERJB	*Eranos Jahrbuch*
ERYB	*Eranos Yearbook*
ETH	*Ethnos*
EURR	*Europäische Revue (Berlin)*
EXPED	*Expedition (Magazine of Archaeology and Anthropology), Philadelphia*
FCON	*Fortschritte d. chemie d. organischen Naturstoffe*
FER	*Far Eastern Review (London)*
FF	*Forschungen und Fortschritte*
FMNHP/AS	*Field Museum of Natural History (Chicago) Publications; Anthropological Series*
FP	*Federation Proceedings (USA)*
FPNJ	*Folia Psychologica et Neurologica Japonica*
FRS	*Franziskanischen Studien*
GBA	*Gazette des Beaux-Arts*
GBT	*Global Technology*
GEW	*Geloof en Wetenschap*
GJ	*Geographical Journal*
GR	*Geographical Review*
GRM	*Germanisch–Romanische Monatsschrift*
GUJ	*Gutenberg Jahrbuch*
HCA	*Helvetica Chimica Acta*
HE	*Hesperia (Journ. Amer. Sch. Class. Stud. Athens)*
HEJ	*Health Education Journal*
HERM	*Hermes; Zeitschr. f. Klass. Philol.*
HF	*Med Hammare och Fackla (Sweden)*

HHS	*Hua Hsüeh (Chemistry), Ch. Chem. Soc.*
HHSTH	*Hua Hsüeh Thung Hsün (Chemical Correspondent), Chekiang Univ.*
HITC	*Hsüeh I Tsa Chih (Wissen und Wissenschaft), Shanghai*
HJAS	*Harvard Journal of Asiatic Studies*
HMSO	Her Majesty's Stationery Office
HOR	*History of Religion (Chicago)*
HOSC	*History of Science (annual)*
HRASP	*Histoire de l'Acad. Roy. des Sciences, Paris*
HSS	*Hsüeh Ssu (Thought and Learning), Chhêngtu*
HU/BML	*Harvard University Botanical Museum Leaflets*
HUM	*Humanist (RPA, London)*
IA	*Iron Age*
IBK	*Indogaku Bukkyōgaku Kenkyū (Indian and Buddhist Studies)*
IC	*Islamic Culture (Hyderabad)*
ID	*Idan (Medical Discussions), Japan*
IEC/AE	*Industrial and Engineering Chemistry; Analytical Edition*
IEC/I	*Industrial and Engineering Chemistry; Industrial Edition*
IHQ	*Indian Historical Quarterly*
IJE	*Indian Journ. Entomol.*
IJHM	*Indian Journ. History of Medicine*
IJHS	*Indian Journ. History of Science*
IJMR	*Indian Journ. Med. Research*
IMIN	*Industria Mineraria*
IMW	*India Medical World*
INDQ	*Industria y Quimica (Buenos Aires)*
INM	*International Nickel Magazine*
IPEK	*Ipek; Jahrb. f. prähistorische u. ethnographische Kunst (Leipzig)*
IQB	*Iqbal (Lahore), later Iqbal Review (Journ. of the Iqbal Academy or Bazm-i Iqbal)*
IRAQ	*Iraq (British Sch. Archaeol. in Iraq)*
ISIS	*Isis*
ISTC	*I Shih Tsa Chih (Chinese Journal of the History of Medicine)*
IVS	*Ingeniörvidenskabelje Skrifter (Copenhagen)*
JA	*Journal Asiatique*
JAC	*Jahrb. f. Antike u. Christentum*
JACS	*Journ. Amer. Chem. Soc.*
JAHIST	*Journ. Asian History (International)*
JAIMH	*Pratibha; Journ. All-India Instit. of Mental Health*
JALCHS	*Journal of the Alchemical Society (London)*
JAN	*Janus*
JAOS	*Journal of the American Oriental Society*
JAP	*Journ. Applied Physiol.*

JAS	*Journal of Asian Studies* (continuation of *Far Eastern Quarterly, FEQ*)		*Royal Asiatic Society* (North China Branch)
JATBA	*Journal d'Agriculture tropicale et de Botanique appliqué*	*JRAS/P*	*Journ. of the (Royal) Asiatic Soc. of Pakistan*
JBC	*Journ. Biol. Chem.*	*JRIBA*	*Journ. Royal Institute of British Architects*
JBFIGN	*Jahresber. d. Forschungsinstitut f. Gesch. d. Naturwiss.* (Berlin)	*JRSA*	*Journal of the Royal Society of Arts*
JC	*Jimnin Chūgoku (People's China)*, Tokyo	*JS*	*Journal des Sçavans* (1665–1778) and *Journal des Savants* (1816–)
JCE	*Journal of Chemical Education*	*JSA*	*Journal de la Société des Americanistes*
JCP	*Jahrb. f. class. Philologie*		
JCS	*Journal of the Chemical Society*	*JSCI*	*Journ. Soc. Chem. Industry*
JEA	*Journal of Egyptian Archaeology*	*JSHS*	*Japanese Studies in the History of Science* (Tokyo)
JEGP	*Journal of English and Germanic Philology*	*JUB*	*Journ. Univ. Bombay*
JEH	*Journal of Economic History*	*JUS*	*Journ. Unified Science* (continuation of *Erkenntnis*)
JEM	*Journ. Exper. Med.*		
JFI	*Journ. Franklin Institute*	*JWCBRS*	*Journal of the West China Border Research Society*
JGGBB	*Jahrbuch d. Gesellschaft f. d. Gesch. u. Bibliographie des Brauwesens*	*JWCI*	*Journal of the Warburg and Courtauld Institutes*
JGMB	*Journ. Gen. Microbiol.*	*JWH*	*Journal of World History* (UNESCO)
JHI	*Journal of the History of Ideas*		
JHMAS	*Journal of the History of Medicine and Allied Sciences*	*KHS*	*Kho Hsüeh (Science)*
JHS	*Journal of Hellenic Studies*	*KHSC*	*Kho-Hsüeh Shih Chi-Khan* (*Ch. Journ. Hist. of Sci.*)
JI	*Jissen Igaku (Practical Medicine)*	*KHTP*	*Kho Hsüeh Thung Pao* (*Science Correspondent*)
JIM	*Journ. Institute of Metals* (UK)		
JIMA	*Journ. Indian Med. Assoc.*	*KHVL*	*Kungliga Humanistiska Vetenskapsamfundet i Lund Årskerättelse* (*Bull. de la Soc. Roy. de Lettres de Lund*)
JKHRS	*Journ. Kalinga Historical Research Soc.* (Orissa)		
JMBA	*Journ. of the Marine Biological Association* (Plymouth)	*KKD*	*Kiuki Daigaku Sekai Keizai Kenkyūjo Hōkoku* (*Reports of the Institute of World Economics at Kiuki Univ.*)
JNMD	*Journ. Nervous & Mental Diseases*		
JMS	*Journ. Mental Science*		
JNPS	*Journ. Neuropsychiatr.*	*KKTH*	*Khao Ku Thung Hsün* (*Archaeological Correspondent*), cont. as *Khao Ku*
JOP	*Journ. Physiol.*		
JOSHK	*Journal of Oriental Studies* (Hongkong Univ.)	*KKTS*	*Ku Kung Thu Shu Chi Khan* (*Journal of the Imperial Palace Museum and Library*), Thaiwan
JP	*Journal of Philology*		
JPB	*Journ. Pathol. and Bacteriol.*	*KSVA/H*	*Kungl. Svenske Vetenskapsakad. Handlingar*
JPC	*Journ. f. prakt. Chem.*		
JPCH	*Journ. Physical Chem.*	*KVSUA*	*Kungl. Vetenskaps Soc. i Uppsala Årsbok* (*Mem. Roy. Acad. Sci. Uppsala*)
JPH	*Journal de Physique*		
JPHS	*Journ. Pakistan Historical Society*		
JPHST	*Journ. Philos. Studies*	*KW*	*Klinische Wochenschrift*
JPOS	*Journal of the Peking Oriental Society*		
JRAI	*Journal of the Royal Anthropological Institute*	*LA*	*Annalen d. Chemie* (Liebig's)
		LCHIND	*La Chimica e l'Industria* (Milan)
JRAS	*Journal of the Royal Asiatic Society*	*LEC*	*Lettres Édifiantes et Curieuses écrites des Missions Étrangères* (Paris, 1702–1776)
JRAS/B	*Journal of the (Royal) Asiatic Society of Bengal*		
JRAS/BOM	*Journ. Roy. Asiatic Soc., Bombay Branch*	*LH*	*l'Homme; Revue Française d'Anthropologie*
JRAS/KB	*Journal (or Transactions) of the Korea Branch of the Royal Asiatic Society*	*LIN*	*L'Institut* (*Journal Universel des Sciences et des Sociétés Savantes en France et à l'Étranger*)
JRAS/M	*Journal of the Malayan Branch of the Royal Asiatic Society*	*LN*	*La Nature*
JRAS/NCB	*Journal (or Transactions) of the*	*LP*	*La Pensée*

LSYC	Li Shih Yen Chiu (*Journal of Historical Research*), Peking
LSYKK	Li Shih yü Khao Ku (*History and Archaeology; Bulletin of the Shenyang Museum*), Shenyang
LT	Lancet
LYCH	Lychnos (*Annual of the Swedish Hist. of Sci. Society*)
MAAA	Memoirs Amer. Anthropological Association
MAI/NEM	Mémoires de l'Académie des Inscriptions et Belles-Lettres, Paris (*Notices et Extraits des MSS*)
MAIS/SP	Mémoires de l'Acad. Impériale des Sciences, St Pétersbourg
MAS/B	Memoirs of the Asiatic Society of Bengal
MB	Monographiae Biologicae
MBLB	May and Baker Laboratory Bulletin
MBPB	May and Baker Pharmaceutical Bulletin
MCB	Mélanges Chinois et Bouddhiques
MCE	Metallurgical and Chemical Engineering
MCHSAMUC	Mémoires concernant l'Histoire, les Sciences, les Arts, les Mœurs et les Usages, des Chinois, par les Missionnaires de Pékin (Paris 1776–)
MDGNVO	Mitteilungen d. deutsch. Gesellsch. f. Natur. u. Volkskunde Ostasiens
MDP	Mémoires de la Délégation en Perse
MED	Medicus (Karachi)
MEDA	Medica (Paris)
METL	Metallen (Sweden)
MGG	Monatsschrift f. Geburtshilfe u. Gynäkologie
MGGW	Mitteilungen d. geographische Gesellschaft Wien
MGSC	Memoirs of the Chinese Geological Survey
MH	Medical History
MI	Metal Industry
MIE	Mémoires de l'Institut d'Egypte (Cairo)
MIFC	Mémoires de l'Institut Français d'Archéol. Orientale (Cairo)
MIK	Mikrochemie
MIMG	Mining Magazine
MIT	Massachusetts Institute of Technology
MJ	Mining Journal, Railway and Commercial Gazette
MJA	Med. Journ. Australia
MJPGA	Mitteilungen aus Justus Perthes Geogr. Anstalt (Petermann's)
MKDUS/HF	Meddelelser d. Kgl. Danske Videnskabernes Selskab (Hist.-Filol.)
MM	Mining and Metallurgy (New York, contd. as *Mining Engineering*)
MMN	Materia Medica Nordmark
MMVKH	Mitteilungen d. Museum f. Völkerkunde (Hamburg)
MMW	Münchener Medizinische Wochenschrift
MOULA	Memoirs of the Osaka University of Liberal Arts and Education
MP	Il Marco Polo
MPMH	Memoirs of the Peabody Museum of American Archaeology and Ethnology, Harvard University
MRASP	Mémoires de l'Acad. Royale des Sciences (Paris)
MRDTB	Memoirs of the Research Dept. of Tōyō Bunko (Tokyo)
MRS	Mediaeval and Renaissance Studies
MS	Monumenta Serica
MSAF	Mémoires de la Société (Nat.) des Antiquaires de France
MSGVK	Mitt. d. Schlesische Gesellschaft f. Volkskunde
MSIV/MF	Memoire di Mat. e. Fis della Soc. Ital. (Verona)
MSOS	Mitteilungen d. Seminar f. orientalischen Sprachen (Berlin)
MSP	Mining and Scientific Press
MUJ	Museum Journal (Philadelphia)
MUSEON	Le Muséon (Louvain)
N	Nature
NAGE	New Age (New Delhi)
NAR	Nutrition Abstracts and Reviews
NARSU	Nova Acta Reg. Soc. Sci. Upsaliensis
NC	Numismatic Chronicle (and Journ. Roy. Numismatic Soc.)
NCDN	North China Daily News
NCGH	Nihon Chūgoku Gakkai-hō (Bulletin of the Japanese Sinological Society)
NCH	North China Herald
NCR	New China Review
NDI	Niigata Daigaku Igakubu Gakushikai Kaihō (Bulletin of the Medical Graduate Society of Niigata University)
NFR	Nat. Fireworks Review
NHK	Nihon Heibon Keisha (publisher)
NIZ	Nihon Ishigaku Zasshi (Jap. Journ. Hist. Med.)
NN	Nation
NQ	Notes and Queries
NR	Numismatic Review
NRRS	Notes and Records of the Royal Society
NS	New Scientist
NSN	New Statesman and Nation (London)

NU	*The Nucleus*
NUM/SHR	*Studies in the History of Religions* (Supplements to *Numen*)
NW	*Naturwissenschaften*
OAZ	*Ostasiatische Zeitschrift*
ODVS	*Oversigt over det k. Danske Viden-skabernes Selskabs Forhandlinger*
OE	*Oriens Extremus* (Hamburg)
OLZ	*Orientalische Literatur-Zeitung*
ORA	*Oriental Art*
ORCH	*Orientalia Christiana*
ORD	*Ordnance*
ORG	*Organon* (Warsaw)
ORR	*Orientalia* (Rome)
ORS	*Orientalia Suecana*
OSIS	*Osiris*
OUP	Oxford University Press
OUSS	*Ochanomizu University Studies*
OX	*Oxoniensia*
PAAAS	*Proceeding of the British Academy*
PAAQS	*Proceedings of the American Anti-quarian Society*
PAI	*Paideuma*
PAKJS	*Pakistan Journ. Sci.*
PAKPJ	*Pakistan Philos. Journ.*
PAPS	*Proc. Amer. Philos. Soc.*
PCASC	*Proc. Cambridge Antiquarian Soc.*
PEW	*Philosophy East and West* (Univ. Hawaii)
PF	*Psychologische Forschung*
PHI	*Die Pharmazeutische Industrie*
PHREV	*Pharmacological Reviews*
PHY	*Physis* (Florence)
PJ	*Pharmaceut. Journal* (and *Trans. Pharmaceut. Soc.*)
PKAWA	*Proc. Kon. Akad. Wetensch.* *Amsterdam*
PKR	*Peking Review*
PM	*Presse Medicale*
PMG	*Philosophical Magazine*
PMLA	*Publications of the Modern Lan-guage Association of America*
PNHB	*Peking Natural History Bulletin*
POLYJ	*Polytechnisches Journal* (Dingler's)
PPHS	*Proceedings of the Prehistoric Society*
PRGS	*Proceedings of the Royal Geo-graphical Society*
PRIA	*Proceedings of the Royal Irish Academy*
PRPH	*Produits Pharmaceutiques*
PRSA	*Proceedings of the Royal Society* (Series A)
PRSB	*Proceedings of the Royal Society* (Series B)
PRSM	*Proceedings of the Royal Society of Medicine*
PSEBM	*Proc. Soc. Exp. Biol and Med.*
PTRS	*Philosophical Transactions of the Royal Society*
QSGNM	*Quellen u. Studien z. Gesch. d. Naturwiss. u. d. Medizin* (con-tinuation of *Archiv. f. Gesch. d. Math., d. Naturwiss. u. d. Technik, AGMNT,* formerly *Archiv. f. d. Gesch. d. Natur-wiss. u. d. Technik, AGNT*)
QSKMR	*Quellenschriften f. Kunstgeschichte und Kunsttechnik des Mittel-alters u. d. Renaissance* (Vienna)
RA	*Revue Archéologique*
RAA/AMG	*Revue des Arts Asiatiques* (*An-nales du Musée Guimet*)
RAAAS	*Reports, Australasian Assoc. Adv. of Sci.*
RAAO	*Revue d'Assyriologie et d'Archéo-logie Orientale*
RALUM	*Revue de l'Aluminium*
RB	*Revue Biblique*
RBPH	*Revue Belge de Philol. et d'His-toire*
RBS	*Revue Bibliographique de Sinologie*
RDM	*Revue des Mines* (later *Revue Uni-verselle des Mines*)
RGVV	*Religionsgeschichtliche Versuche und Vorarbeiten*
RHR/AMG	*Revue de l'Histoire des Religions* (*Annales du Musée Guimet, Paris*)
RHS	*Revue d'Histoire des Sciences*
RHSID	*Revue d'Histoire de la Sidérurgie* (Nancy)
RIN	*Rivista Italiana di Numismatica*
RKW	*Repertorium f. Kunst. wissen-schaft*
RMY	*Revue de Mycologie*
ROC	*Revue de l'Orient Chrétien*
RP	*Revue Philosophique*
RPA	*Rationalist Press Association* (London)
RPCHG	*Revue de Pathologie comparée et d'Hygiène générale* (Paris)
RPLHA	*Revue de Philol., Litt. et Hist. Ancienne*
RR	*Review of Religion*
RSCI	*Revue Scientifique* (Paris)
RSH	*Revue de Synthèse Historique*
RSI	*Reviews of Scientific Instruments*
RSO	*Rivista di Studi Orientali*
RUB	*Revue de l'Univ. de Bruxelles*
S	*Sinologica* (Basel)
SA	*Sinica* (originally *Chinesische Blätter f. Wissenschaft u. Kunst*)
SAEC	*Supplemento Annuale all'Enciclo-pedia di Chimica*
SAEP	Soc. Anonyme des Études et Pub. (publisher)
SAM	*Scientific American*
SB	*Shizen to Bunka* (*Nature and Culture*)

SBE	Sacred Books of the East series	TAIMME	Transactions of the American Institute of Mining and Metallurgical Engineers
SBK	Seikatsu Bunka Kenkyū (Journ. Econ. Cult.)		
SBM	Svenska Bryggareföreningens Månadsblad	TAPS	Transactions of the American Philosophical Society (cf. MAPS)
SC	Science		
SCI	Scientia	TAS/J	Transactions of the Asiatic Society of Japan
SCIS	Sciences; Revue de la Civilisation Scientifique (Paris)	TBKK	Tōhoku Bunka Kenkyūshitsu Kiyō (Record of the North-Eastern Research Institute of Humanistic Studies), Sendai
SCISA	Scientia Sinica (Peking)		
SCK	Smithsonian Contributions to Knowledge		
SCM	Student Christian Movement (Press)	TCS	Trans. Ceramic Society (formerly Trans. Engl. Cer. Soc., contd as Trans. Brit. Cer. Soc.)
SCON	Studies in Conservation (Journ. Internat. Instit. for the Conservation of Museum objects)		
		TCULT	Technology and Culture
SET	Structure et Evolution des Techniques	TFTC	Tung Fang Tsa Chih (Eastern Miscellany)
		TGAS	Transactions of the Glasgow Archaeological Society
SGZ	Shigaku Zasshi (Historical Journ. of Japan)	TG/T	Tōhō Gakuhō, Tōkyō (Tokyo Journal of Oriental Studies)
SHA	Shukan Asahi		
SHAW/PH	Sitzungsber. d. Heidelberg. Akad. d. Wissensch. (Phil.-Hist. Kl.)	TH	Thien Hsia Monthly (Shanghai)
		THG	Tōhōgaku (Eastern Studies), Tokyo
SHST/T	Studies in the History of Science and Technol. (Tokyo Univ. Inst. Technol.)	TICE	Transactions of the Institute of Chemical Engineers
SI	Studia Islamica (Paris)	TIMM	Transactions of the Institution of Mining and Metallurgy
SIB	Sibrium (Collana di Studi e Documentazioni, Centro di Studi Preistorici e Archeologici Varese)	TJSL	Transactions (and Proceedings) of the Japan Society of London
SILL	Sweden Illustrated	TLTC	Ta Lu Tsa Chih (Continent Magazine), Thaipei
SK	Seminarium Kondakovianum (Recueil d'Études de l'Institut Kondakov)	TMIE	Travaux et Mémoires de l'Inst. d'Ethnologie (Paris)
SM	Scientific Monthly (formerly Popular Science Monthly)	TNS	Transactions of the Newcomen Society
SN	Shirin (Journal of History), Kyoto	TOCS	Transactions of the Oriental Ceramic Society
SNM	Sbornik Nauknych Materialov (Erivan, Armenia)	TP	T'oung Pao (Archives concernant l'Histoire, les Langues, la Géographie, l'Ethnographie et les Arts de l'Asie Orientale), Leiden
SOS	Semitic and Oriental Studies (Univ. of Calif. Publ. in Semitic Philol.)		
SP	Speculum	TQ	Tel Quel (Paris)
SPAW/PH	Sitzungsber. d. preuss. Akad. d. Wissenschaften (Phil.-Hist. Kl.)	TR	Technology Review
		TRAD	Tradition (Zeitschr. f. Firmengeschichte und Unternehmerbiographie)
SPCK	Society for the Promotion of Christian Knowledge		
SPMSE	Sitzungsberichte d. physik. med. Soc. Erlangen	TRSC	Trans. Roy. Soc. Canada
		TS	Tōhō Shūkyō (Journal of East Asian Religions)
SPR	Science Progress		
SSIP	Shanghai Science Institute Publications	TSFFA	Techn. Studies in the Field of the Fine Arts
STM	Studi Medievali	TTT	Theoria to Theory (Cambridge)
SWAW/PH	Sitzungsberichte d. k. Akad. d. Wissenschaften Wien (Phil.-Hist. Klasse), Vienna	TYG	Tōyō Gakuhō (Reports of the Oriental Society of Tokyo)
		TYGK	Tōyōgaku (Oriental Studies), Sendai
		TYKK	Thien Yeh Khao Ku Pao Kao (Archaeological Reports)
TAFA	Transactions of the American Foundrymen's Association		
TAIME	Trans. Amer. Inst. Mining Engineers (continued as TAIMME)	UCC	University of California Chronicle
		UCR	University of Ceylon Review

UNASIA	*United Asia* (India)		ence Materials for History and Archaeology)
UNESC	*Unesco Courier*		
UNESCO	United Nations Educational, Scientific and Cultural Organisation	*WZNHK*	*Wiener Zeitschr. f. Nervenheilkunde*
UUA	*Uppsala Univ. Årsskrift* (*Acta Univ. Upsaliensis*)	*YCHP*	*Yenching Hsüeh Pao* (*Yenching University Journal of Chinese Studies*)
VBA	*Visva-Bharati Annals*	*YJBM*	*Yale Journal of Biology and Medicine*
VBW	*Vorträge d. Bibliothek Warburg*		
VK	*Vijnan Karmee*	*YJSS*	*Yenching Journal of Social Studies*
VKAWA/L	*Verhandelingen d. Koninklijke Akad. v. Wetenschappen te Amsterdam* (Afd. Letterkunde)	*Z*	*Zalmoxis; Revue des Études Religieuses*
VMAWA	*Verslagen en Meded. d. Koninklijke Akad. v. Wetenschappen te Amsterdam*	*ZAC*	*Zeitschr. f. angewandte chemie*
		ZAC/AC	*Angewandte Chemie*
		ZAES	*Zeitschrift f. Aegyptische Sprache u. Altertumskunde*
VVBGP	*Verhandhingen d. Verein z. Beförderung des Gewerbefleisses in Preussen*	*ZASS*	*Zeitschr. f. Assyriologie*
		ZDMG	*Zeitschrift d. deutsch. Morgenländischen Gesellschaft*
WA	*Wissenschaftliche Annalen*	*ZGEB*	*Zeitschr. d. Gesellsch. f. Erdkunde* (Berlin)
WKW	*Wiener klinische Wochenschrift*		
WS	*Wên Shih* (*History of Literature*), Peking	*ZMP*	*Zeitschrift f. Math. u. Physik*
		ZPC	*Zeitschr. f. physiologischen Chemie*
WWTK	*Wên Wu* (formerly *Wên Wu Tshan Khao Tzu Liao, Reference Materials for History and Archaeology)*	*ZS*	*Zeitschr. f. Semitistik*
		ZVSF	*Zeitschr. f. vergl. Sprachforschung*

ADDENDA TO ABBREVIATIONS

ACTAS	*Acta Asiatica* (*Bull. of Eastern Culture*, Tōhō Gakkaì, Tokyo)	*CR/MSU*	*Centennial Review of Arts and Science* (Michigan State University)
BILCA	*Boletim do Instituto Luis de Camões* (Macao)	*ECB*	*Economic Botany*
CFC	*Cahiers Franco-Chinois* (Paris)	*NGM*	*National Geographic Magazine*
CHEM	*Chemistry* (Easton, Pa.)	*POLREC*	*Polar Record*
COMP	*Comprendre* (Soc. Eu. de Culture, Venice)	*PV*	*Pacific Viewpoint* (New Zealand)

A. CHINESE AND JAPANESE BOOKS BEFORE +1800

Each entry gives particulars in the following order:

(a) title, alphabetically arranged, with characters;
(b) alternative title, if any;
(c) translation of title;
(d) cross-reference to closely related book, if any;
(e) dynasty;
(f) date as accurate as possible;
(g) name of author or editor, with characters;
(h) title of other book, if the text of the work now exists only incorporated therein; or, in special cases, references to sinological studies of it;
(i) references to translations, if any, given by the name of the translator in Bibliography C;
(j) notice of any index or concordance to the book if such a work exists;
(k) reference to the number of the book in the *Tao Tsang* catalogue of Wieger (6), if applicable;
(l) reference to the number of the book in the *San Tsang* (Tripiṭaka) catalogues of Nanjio (1) and Takakusu & Watanabe, if applicable.

Words which assist in the translation of titles are added in round brackets.

Alternative titles or explanatory additions to the titles are added in square brackets.

It will be remembered (p. 305 above) that in Chinese indexes words beginning *Chh*- are all listed together after *Ch*-, and *Hs*- after *H*-, but that this applies to initial words of titles only.

Where there are any differences between the entries in these bibliographies and those in Vols. 1–4, the information here given is to be taken as more correct.

An interim list of references to the editions used in the present work, and to the *tshung-shu* collections in which books are available, has been given in Vol. 4, pt. 3, pp. 913 ff., and is available as a separate brochure.

ABBREVIATIONS

C/Han	Former Han.
E/Wei	Eastern Wei.
H/Han	Later Han.
H/Shu	Later Shu (Wu Tai).
H/Thang	Later Thang (Wu Tai).
H/Chin	Later Chin (Wu Tai).
S/Han	Southern Han (Wu Tai).
S/Phing	Southern Phing (Wu Tai).
J/Chin	Jurchen Chin.
L/Sung	Liu Sung.
N/Chou	Northern Chou.
N/Chhi	Northern Chhi.
N/Sung	Northern Sung (before the removal of the capital to Hangchow).
N/Wei	Northern Wei.
S/Chhi	Southern Chhi.
S/Sung	Southern Sung (after the removal of the capital to Hangchow).
W/Wei	Western Wei.

A-Nan Ssu Shih Ching 阿難四事經.
Sūtra on the Four Practices spoken to Ānanda.
India.
Tr. San Kuo, betw. +222 and +230 by Chih-Chhien 支謙.
N/696; TW/493.

A-Phi-Than-Phi Po-Sha Lun 阿毘曇毘婆沙論.
Abhidharma Mahāvibhāsha.
India (this recension not much before +600).
Tr. Hsüan-Chuang, +659 玄奘.
N/1263; TW/1546.

Chang Chen-Jen Chin Shih Ling Sha Lun.
See *Chin Shih Ling Sha Lun.*

Chao Fei-Yen Pieh Chuan 趙飛燕別傳.
[= *Chao Hou I Shih.*]
Another Biography of Chao Fei-Yen [historical novelette].
Sung.
Chhin Shun 秦醇.

Chao Fei-Yen Wai Chuan 趙飛燕外傳.
Unofficial Biography of Chao Fei-Yen (d. −6, celebrated dancing-girl, consort and empress of Han Chhêng Ti).
Ascr. Han, +1st.
Attrib. Ling Hsüan 伶玄.

Chao Hou I Shih 趙后遺事.
A Record of the Affairs of the Empress Chao (−1st century).
See *Chao Fei-Yen Pieh Chuan.*

Chao Hun 招魂.
The Summons of the Soul [ode].
Chou (Chhu), *c.* −240.
Prob. Ching Chhai 景差.
Tr. Hawkes (1), p. 103.

Chen Chhi Huan Yuan Ming 眞氣還元銘.
The Inscription on the Regeneration of the Primary Chhi.
Thang or Sung, must be before the mid +13th century.
Writer unknown.
TT/261.

Chen Chung Chi 枕中記.
[= *Ko Hung Chen Chung Shu.*]
Pillow-Book (of Ko Hung).
Ascr. Chin, *c.* +320, but actually not earlier than the +7th century.
Attrib. Ko Hung 葛洪.
TT/830.

Chen Chung Chi 枕中記.
See *Shê Yang Chen Chung Chi.*

Chen-Chung Hung-Pao Yuan-Pi Shu 枕中鴻寶苑祕書.
The Infinite Treasure of the Garden of Secrets; (Confidential) Pillow-Book (of the Prince of Huai-Nan).
See *Huai-Nan Wang Wan Pi Shu.*
Cf. Kaltenmark (2), p. 32.

Chen Hsi 眞系.
The Legitimate Succession of Perfected, or Realised, (Immortals).
Thang, +805.
Li Po 李渤.
In *YCCC*, ch. 5, pp. 1 *a* ff.

Chen Kao 眞誥.
    Declarations of Perfected, or Realised,
    (Immortals) [visitations and revelations of
    the Taoist pantheon].
    Chin and S/Chhi. Original material from
    +364 to +370, collected from +484 to
    +492 by Thao Hung-Ching (+456 to
    +536), who provided commentary and
    postface by +493 to +498; finished
    +499.
    Original writers unknown.
    Ed. Thao Hung-Ching 陶弘景.
    TT/1004.
Chen Yuan Miao Tao Hsiu Tan Li Yen Chhao
    眞元妙道修丹歷驗抄.
    [= *Hsiu Chen Li Yen Chhao Thu.*]
    A Document concerning the Tried and
    Tested (Methods for Preparing the)
    Restorative Enchymoma of the Mysterious
    Tao of the Primary (Vitalities) [physio-
    logical alchemy].
    Thang or Sung, before +1019.
    Tung Chen Tzu (ps.) 洞眞子.
    In *YCCC*, ch. 72, pp. 17*b* ff.
Chen Yuan Miao Tao Yao Lüeh 眞元妙道要畧.
    Classified Essentials of the Mysterious Tao
    of the True Origin (of Things) [alchemy
    and chemistry].
    Ascr. Chin, +3rd, but probably mostly
    Thang, +8th and +9th, at any rate
    after +7th as it quotes Li Chi.
    Attrib. Chêng Ssu-Yuan 鄭思遠.
    TT/917.
Chêng I Fa Wên (*Thai-Shang*) *Wai Lu I* 正一法
    文太上外籙儀.
    The System of the Outer Certificates, a Thai-
    Shang Scripture.
    Date unknown, but pre-Thang.
    Writer unknown.
    TT/1225.
Chêng Lei Pên Tshao 證類本草.
    See *Ching-Shih Chêng Lei Pei-Chi Pên Tshao*
    and *Chhung-Hsiu Chêng-Ho Ching-Shih*
        Chêng Lei Pei-Yung Pên Tshao
Chêng Tao Pi Shu Shih Chung 證道秘書十種.
    Ten Types of Secret Books on the Verifica-
    tion of the Tao.
    See Fu Chin-Chhüan (6)
Chi Hsiao Hsin Shu 紀效新書.
    A New Treatise on Military and Naval
    Efficiency.
    Ming, c. +1575.
    Chhi Chi-Kuang 戚繼光.
Chi Hsien Chuan 集仙傳.
    Biographies of the Company of the Immortals.
    Sung, c. +1140.
    Tsêng Tshao 曾慥.
Chi I Chi 集異記.
    A Collection of Assorted Stories of Strange
    Events.
    Thang.
    Hsüeh Yung-Jo 薛用弱.
18

Chi Ni Tzu 計倪子.
    [= *Fan Tzu Chi Jan* 范子計然.]
    The Book of Master Chi Ni.
    Chou (Yüeh), −4th century.
    Attrib. Fan Li 范蠡, recording the
    philosophy of his master Chi Jan 計然.
Chi Shêng Fang 濟生方.
    Prescriptions for the Preservation of Health.
    Sung, c. +1267.
    Yen Yung-Ho 嚴用和.
Chi Than Lu 劇談錄.
    Records of Entertaining Conversations.
    Thang, c. +885.
    Khang Phien 康駢 or 骿.
Chi Yün 集韻.
    Complete Dictionary of the Sounds of
    Characters [cf. *Chhieh Yün* and *Kuang
    Yün*].
    Sung, +1037.
    Compiled by Ting Tu 丁度 *et al.*
    Possibly completed in +1067 by Ssuma
    Kuang 司馬光.
Chia-Yu Pên Tshao 嘉祐本草.
    See *Chia-Yu Pu-Chu Shen Nung Pên Tshao.*
Chia-Yu Pu-Chu Shen Nung Pên Tshao 嘉祐補
    註神農本草.
    Supplementary Commentary on the *Pharma-
    copoeia of the Heavenly Husbandman*,
    commissioned in the Chia-Yu reign-
    period.
    Sung, commissioned +1057, finished
    +1060.
    Chang Yü-Hsi 掌禹錫,
    Lin I 林億,
    & Chang Tung 張洞.
Chiang Huai I Jen Lu 江淮異人錄.
    Records of (Twenty-five) Strange Magician-
    Technicians between the Yangtze and the
    Huai River (during the Thang, Wu and
    Nan Thang Dynasties, c. +850 to +950).
    Sung, c. +975.
    Wu Shu 吳淑.
Chiang Wên-Thung Chi 江文通集.
    Literary Collection of Chiang Wên-Thung
    (Chiang Yen).
    S/Chhi, c. +500.
    Chiang Yen 江淹.
Chiao Chhuang Chiu Lu 蕉窗九錄.
    Nine Dissertations from the (Desk at the)
    Banana-Grove Window.
    Ming, c. +1575.
    Hsiang Yuan-Pien 項元汴.
Chien Wu Chi 漸悟集.
    On the Gradual Understanding (of the
    Tao).
    Sung, mid +12th century.
    Ma Yü 馬鈺.
    TT/1128.
Chih Chen Tzu Lung Hu Ta Tan Shih 至眞子
    龍虎大丹詩.
    Song of the Great Dragon-and-Tiger En-
    chymoma of the Perfected-Truth Master.

Chi Chen Tzu Lung Hu Ta Tan Shih (cont.)
Sung, +1026.
Chou Fang (Chih Chen Tzu) 周方.
Presented to the throne by Lu Thien[-Chi]
盧天驥, c. +1115.
TT/266.

Chih-Chhuan Chen-Jen Chiao Chêng Shu 稚川
眞人校證術.
Technical Methods of the Adept (Ko) Chih-
Chhuan (i.e. Ko Hung), with Critical
Annotations [and illustrations of al-
chemical apparatus].
Ascr. Chin, c. +320, but probably later.
Attrib. Ko Hung 葛洪.
TT/895.

Chih Chih Hsiang Shuo San Chhêng Pi Yao 直
指祥說三乘秘要.
See *Wu Chen Phien Chih Chih Hsiang Shuo
San Chhêng Pi Yao*.
Cf. Davis & Chao Yün-Tshung (6).

Chih-Chou hsien-sêng Chin Tan Chih Chih 紙舟
先生金丹直指.
Straightforward Indications about the
Metallous Enchymoma by the Paper-
Boat Teacher.
Sung, prob. +12th.
Chin Yüeh-Yen 金月巖.
TT/239.

Chih Hsüan Phien 指玄篇.
A Pointer to the Mysteries [psycho-physio-
logical alchemy].
Sung, c. +1215.
Pai Yü-Chhan 白玉蟾.
In *Hsiu Chen Shih Shu* (*TT*/260), chs. 1–8.

Chih Kuei Chi 指歸集.
Pointing the Way Home (to Life Eternal); a
Collection.
Sung, c. +1165.
Wu Wu 吳悞.
TT/914.
Cf. Chhen Kuo-Fu (1), vol. 2, pp. 389,
390.

Chih Tao Phien 旨道篇 (or 編).
A Demonstration of the Tao.
Sui or just before, c. +580.
Su Yuan-Ming (or -Lang) 蘇元明(朗)
= Chhing Hsia Tzu 青霞子.
Now extant only in quotations.

Chih Tshao Thu 芝草圖.
See *Thai-Shang Ling-Pao Chih Tshao Thu*.

Chin Hua Chhung Pi Tan Ching Pi Chih 金華
冲碧丹經祕旨.
Confidential Instructions on the Manual of
the Heaven-Piercing Golden Flower
Elixir [with illustrations of alchemical
apparatus].
Sung, +1225.
Phêng Ssu 彭耜 & Mêng Hsü 孟煦
(pref. and ed. Mêng Hsü).
Received from Pai Yü-Chhan 白玉蟾 and
Lan Yuan-Lao 蘭元老.
TT/907.

The authorship of this important work is
obscure. In his preface Mêng Hsü says
that in +1218 he met in the mountains
Phêng Ssu, who transmitted to him a
short work which Phêng himself had received from Pai Yü-Chhan. This is ch. 1
of the present book. Two years later Mêng
met an adept named Lan Yuan-Lao, who
claimed to be an avatar of Pai Yü-Chhan
and transmitted to Mêng a longer text;
this is the part which contains descriptions
of the complicated alchemical apparatus
and appears as ch. 2 of the present work.
The name of the book is taken from that
of the alchemical elaboratory of Lan Yuan-
Lao, which was called Chin Hua Chhung
Pi Tan Shih 金華冲碧丹室.

Chin Hua Tsung Chih 金華宗旨
[= *Thai-I Chin Hua Tsung Chih*, also entitled
Chhang Shêng Shu; former title: *Lü
Tsu Chhuan Shou Tsung Chih*.]
Principles of the (Inner) Radiance of the
Metallous (Enchymoma) [a Taoist *nei tan*
treatise on meditation and sexual tech-
niques, with Buddhist influence].
Ming and Chhing, c. +1403, finalised
+1663, but may have been transmitted
orally from an earlier date. Present title
from +1668.
Writer unknown. Attrib. Lü Yen 呂喦
(Lü Tung-Pin) and his school, late
+8th.
Commentary by Tan Jan-Hui 澹然慧
(1921).
Prefaces by Chang San-Fêng 張三峯
(c. +1410) and several others, some per-
haps apocryphal.
See also *Lü Tsu Shih Hsien-Thien Hsü Wu
Thai-I Chin Hua Tsung Chih*.
Cf. Wilhelm & Jung (1).

Chin Hua Yü I Ta Tan 金華玉液大丹.
The Great Elixir of the Golden Flower (or,
Metallous Radiance) and the Juice of
Jade.
Date unknown, probably Thang.
Writer unknown.
TT/903.

Chin Hua Yü Nü Shuo Tan Ching 金華玉女
說丹經.
Sermon of the Jade Girl of the Golden
Flower about Elixirs and Enchymomas.
Wu Tai or Sung.
Writer unknown.
In *YCCC*, ch. 64, pp. 1 a ff.

Chin I Huan Tan Pai Wên Chüeh 金液還丹百
問訣.
Questions and Answers on Potable Gold
(Metallous Fluid) and Cyclically-
Transformed Elixirs and Enchymomas.
Sung.
Li Kuang-Hsüan 李光玄.
TT/263.

Chin I Huan Tan Yin Chêng Thu 金液還丹印證圖.
Illustrations and Evidential Signs of the Regenerative Enchymoma (constituted by, or elaborated from) the Metallous Fluid.
Sung, prob. +12th, perhaps *c.* +1218, date of preface.
Lung Mei Tzu (ps.) 龍眉子.
TT/148.

Chin Ku Chhi Kuan 今古奇觀.
Strange Tales New and Old.
Ming, *c.* +1620; pr. betw. +1632 and +1644.
Fêng Mêng-Lung 馮夢龍.
Cf. Pelliot (57).

Chin Mu Wan Ling Lun 金木萬靈論.
Essay on the Tens of Thousands of Efficacious (Substances) among Metals and Plants.
Ascr. Chin, *c.* +320. Actually prob. late Sung or Yuan.
Attrib. Ko Hung 葛洪.
TT/933.

Chin Pi Wu Hsiang Lei Tshan Thung Chhi 金碧五相類參同契.
Gold and Caerulean Jade Treatise on the Similarities and Categories of the Five (Substances) and the *Kinship of the Three* [a poem on physiological alchemy].
Ascr. H/Han, *c.* +200.
Attrib. Yin Chhang-Shêng 陰長生.
TT/897.
Cf. Ho Ping-Yü (12).
Not to be confused with the *Tshan Thung Chhi Wu Hsiang Lei Pi Yao*, q.v.

Chin Shih Ling Sha Lun 金石靈砂論.
A Discourse on Metals, Minerals and Cinnabar (by the Adept Chang).
Thang, between +713 and +741.
Chang Yin-Chü 張隱居.
TT/880.

Chin Shih Pu Wu Chiu Shu Chüeh 金石簿五九數訣.
Explanation of the Inventory of Metals and Minerals according to the Numbers Five (Earth) and Nine (Metal) [catalogue of substances with provenances, including some from foreign countries].
Thang, perhaps *c.* +670 (contains a story relating to +664).
Writer unknown.
TT/900.

Chin Shih Wu Hsiang Lei 金石五相類.
[= *Yin Chen Chün Chin Shih Wu Hsiang Lei.*]
The Similarities and Categories of the Five (Substances) among Metals and Minerals (sulphur, realgar, orpiment, mercury and lead) (by the Deified Adept Yin).
Date unknown (ascr. +2nd or +3rd century).

Attrib. Yin Chen-Chün 陰眞君 (Yin Chhang-Shêng).
TT/899.

Chin Tan Chen Chuan 金丹眞傳.
A Record of the Primary (Vitalities, regained by) the Metallous Enchymoma.
Ming, +1615.
Sun Ju-Chung 孫汝忠.

Chin Tan Chêng Li Ta Chhüan 金丹正理大全
Comprehensive Collection of Writings on the True Principles of the Metallous Enchymoma [a florilegium].
Ming, *c.* +1440.
Ed. Han Chhan Tzu 涵蟾子.
Cf. Davis & Chao Yün-Tshung (6).

Chin Tan Chieh Yao 金丹節要.
Important Sections on the Metallous Enchymoma.
Part of *San-Fêng Tan Chüeh* (q.v.).

Chin Tan Chih Chih 金丹直指.
Straightforward Explanation of the Metallous Enchymoma.
Sung, prob. +12th.
Chou Wu-So 周無所.
TT/1058.
Cf. *Chih-Chou hsien-sêng Chin Tan Chih Chih.*
See Chhen Kuo-Fu (1), vol. 2, pp. 447 ff.

Chin Tan Chin Pi Chhien Thung Chüeh 金丹金碧潛通訣.
Oral Instructions explaining the Abscondite Truths of the Gold and Caerulean Jade (Components of the) Metallous Enchymoma.
Date unknown, not earlier than Wu Tai.
Writer unknown.
Incomplete in *YCCC*, ch. 73, pp. 7*a* ff.

Chin Tan Fu 金丹賦.
Rhapsodical Ode on the Metallous Enchymoma.
Sung, +13th.
Writer unknown.
Comm. by Ma Li-Chao 馬涖昭.
TT/258.
Cf. *Nei Tan Fu*, the text of which is very similar.

Chin Tan Lung Hu Ching 金丹龍虎經.
Gold Elixir Dragon and Tiger Manual.
Thang or early Sung.
Writer unknown.
Extant only in quotations, as in *Chu Chia Shen Phin Tan Fa*, q.v.

Chin Tan Pi Yao Tshan Thung Lu 金丹秘要參同錄.
Essentials of the Gold Elixir; a Record of the Concordance (or Kinship) of the Three.
Sung.
Mêng Yao-Fu 孟要甫.
In *Chu Chia Shen Phin Tan Fa*, q.v.

Chin Tan Ssu Pai Tzu 金丹四百字.
The Four-Hundred Word Epitome of the Metallous Enchymoma.

Chin Tan Ssu Pai Tzu (cont.)
Sung, *c.* +1065.
Chang Po-Tuan 張伯端.
In *Hsiu Chen Shih Shu* (*TT*/260), ch. 5,
pp. 1 *a* ff.
TT/1067.
Comms. by Phêng Hao-Ku and Min I-Tê
in *Tao Tsang Hsü Pien* (*Chhu chi*), 21.
Tr. Davis & Chao Yün-Tshung (2).
Chin Tan Ta Chhêng 金丹大成.
Compendium of the Metallous Enchymoma.
Sung, just before +1250.
Hsiao Thing-Chih 蕭廷芝.
In *TTCY* (*mao chi*, 4), and in *TT*/260,
Hsiu Chen Shih Shu, chs. 9–13 incl.
Chin Tan Ta Yao 金丹大要.
[= *Shang Yang Tzu Chin Tan Ta Yao*.]
Main Essentials of the Metallous Enchy-
moma; the true Gold Elixir.
Yuan, +1331 (pref. +1335).
Chhen Chih-Hsü 陳致虛
(Shang Yang Tzu 上陽子).
In *TTCY* (*mao chi*, 1, 2, 3).
TT/1053.
Chin Tan Ta Yao Hsien Phai (*Yuan Liu*) 金丹
大要仙派源流.
[= *Shang Yang Tzu Chin Tan Ta Yao
Hsien Phai*.]
A History of the Schools of Immortals
mentioned in the *Main Essentials of the
Metallous Enchymoma; the true Gold Elixir*.
Yuan, *c.* +1333.
Chhen Chih-Hsü 陳致虛
(Shang Yang Tzu 上陽子).
In *TTCY*, *Chin Tan Ta Yao*, ch. 3, pp.
40 ff.
TT/1056.
Chin Tan Ta Yao Lieh Hsien Chih 金丹大要
列仙誌.
[= *Shang Yang Tzu Chin Tan Ta Yao Lieh
Hsien Chih*.]
Records of the Immortals mentioned in the
*Main Essentials of the Metallous Enchy-
moma; the true Gold Elixir*.
Yuan, *c.* +1333.
Chhen Chih-Hsü 陳致虛
(Shang Yang Tzu 上陽子).
TT/1055.
Chin Tan Ta Yao Pao Chüeh 金丹大藥寶訣.
Precious Instructions on the Great Medi-
cines of the Golden Elixir (Type).
Sung, *c.* +1045.
Tshui Fang 崔昉.
Preface preserved in *Kêng Tao Chi*, ch. 1,
p. 8*b*, but otherwise only extant in
occasional quotations.
Perhaps the same book as the *Wai Tan
Pên Tshao* (q. v.).
Chin Tan Ta Yao Thu 金丹大要圖.
[= *Shang Yang Tzu Chin Tan Ta Yao Thu*.]
Illustrations for the *Main Essentials of the
Metallous Enchymoma; the true Gold Elixir*.

Yuan, +1333.
Chhen Chih-Hsü 陳致虛
(Shang Yang Tzu 上陽子).
Based on drawings and tables of the +10th
century onwards by Phêng Hsiao 彭曉,
Chang Po-Tuan 張伯端 (hence the
name *Tzu Yang Tan Fang Pao Chien
Thu*), Lin Shen-Fêng 林神鳳 and
others.
In *TTCY* (*Chin Tan Ta Yao*, ch. 3,
pp. 26 *a* ff.).
TT/1054.
Cf. Ho Ping-Yü & Needham (2).
Ching Chhu Sui Shih Chi 荊楚歲時記.
Annual Folk Customs of the States of
Ching and Chhu [i.e. of the districts cor-
responding to those ancient States;
Hupei, Hunan and Chiangsi].
Prob. Liang, *c.* +550, but perhaps partly
Sui, *c.* +610.
Tsung Lin 宗懍.
See des Rotours (1), p. cii.
Ching-Shih Chêng Lei Pei-Chi Pên Tshao 經史
證類備急本草.
The Classified and Consolidated Armament-
arium of Pharmaceutical Natural History.
Sung, +1083, repr. +1090.
Thang Shen-Wei 唐慎微.
Ching Shih Thung Yen 警世通言.
Stories to Warn Men.
Ming, *c.* +1640.
Fêng Mêng-Lung 馮夢龍.
Ching Tien Shih Wên 經典釋文.
Textual Criticism of the Classics.
Sui, *c.* +600.
Lu Tê-Ming 陸德明.
Ching Yen Fang 經驗方.
Tried and Tested Prescriptions.
Sung, +1025.
Chang Shêng-Tao 張聲道.
Now extant only in quotations.
Ching Yen Liang Fang 經驗良方.
Valuable Tried and Tested Prescriptions.
Yuan.
Writer unknown.
Chiu Chêng Lu 就正錄.
Drawing near to the Right Way; a Guide
[to physiological alchemy].
Chhing, prefs. +1678, +1697.
Lu Shih-Chhen 陸世忱.
In *Tao Tsang Hsü Pien* (*Chhu chi*), 8.
Chiu Chuan Chhing Chin Ling Sha Tan 九轉青
金靈砂丹.
The Ninefold Cyclically Transformed
Caerulean Golden Numinous Cinnabar
Elixir.
Date unknown.
Writer unknown, but much overlap with
TT/886.
TT/887.
Chiu Chuan Ling Sha Ta Tan 九轉靈砂大
丹.

Chiu Chuan Ling Sha Ta Tan (*cont.*)

The Great Ninefold Cyclically Transformed Numinous Cinnabar Elixir.

Date unknown.

Writer unknown.

TT/886.

Chiu Chuan Ling Sha Ta Tan Tzu Shêng Hsüan Ching 九轉靈砂大丹資聖女經.

Mysterious (or Esoteric) Sagehood-Enhancing Canon of the Great Ninefold Cyclically Transformed Numinous Cinnabar Elixir (or Enchymoma).

Date unknown, probably Thang; the text is in sūtra form.

Writer unknown.

TT/879.

Chiu Chuan Liu Chu Shen Hsien Chiu Tan Ching 九轉流珠神仙九丹經.

Manual of the Nine Elixirs of the Holy Immortals and of the Ninefold Cyclically Transformed Mercury.

Not later than Sung, but contains material from much earlier dates.

Thai-Chhing Chen Jen 太清眞人.

TT/945.

Chiu Huan Chin Tan Erh Chang 九還金丹二章.

Two Chapters on the Ninefold Cyclically Transformed Gold Elixir.

Alternative title of *Ta-Tung Lien Chen Pao Ching, Chin Huan Chin Tan Miao Chüeh* (q.v.).

In *YCCC*, ch. 68, pp. 8 *a* ff.

Chiu Phu 酒譜.

A Treatise on Wine.

Sung, +1020.

Tou Phing 竇苹.

Chiu Shih 酒史.

A History of Wine.

Ming, +16th (but first pr. +1750).

Fêng Shih-Hua 馮時化.

Chiu Thang Shu 舊唐書.

Old History of the Thang Dynasty [+618 to +906].

Wu Tai (H/Chin), +945.

Liu Hsü 劉昫.

Cf. des Rotours (2), p. 64.

For translations of passages see the index of Frankel (1).

Chiu Ting Shen Tan Ching Chüeh

See *Huang Ti Chiu Ting Shen Tan Ching Chüeh.*

Cho Kêng Lu 輟耕錄.

[Sometimes *Nan Tshun Cho Kêng Lu.*]

Talks (at South Village) while the Plough is Resting.

Yuan, +1366.

Thao Tsung-I 陶宗儀.

Chou Hou Pei Chi Fang 肘後備急方.

[= *Chou Hou Tsu Chiu Fang*
or *Chou Hou Pai I Fang*
or *Ko Hsien Ong Chou Hou Pei Chi Fang.*]

Handbook of Medicines for Emergencies.

Chin, *c.* +340.

Ko Hung 葛洪.

Chou Hou Pai I Fang 肘後百一方

See *Chou Hou Pei Chi Fang.*

Chou Hou Tsu Chiu Fang 肘後卒救方

See *Chou Hou Pei Chi Fang.*

Chou I Tshan Thung Chhi 周易參同契.

See also titles under *Tshan Thung Chhi.*

Chou I Tshan Thung Chhi Chieh 周易參同契解.

The *Kinship of the Three and the Book of Changes*, with Explanation.

Text, H/Han, *c.* +140.

Comm., Sung, +1234.

Ed. & comm. Chhen Hsien-Wei 陳顯微.

TT/998.

Chou I Tshan Thung Chhi Chu 周易參同契註.

The *Kinship of the Three and the Book of Changes*, with Commentary.

Text, H/Han, *c.* +140.

Comm. ascr. H/Han, *c.* +160, but probably Sung.

Attrib., ed. and comm. Yin Chhang-Shêng 陰長生.

TT/990.

Chou I Tshan Thung Chhi Chu 周易參同契註.

The *Kinship of the Three and the Book of Changes*, with Commentary.

Text, H/Han, *c.* +140.

Comm. probably Sung.

Ed. and comm. unknown.

TT/991.

Chou I Tshan Thung Chhi Chu 周易參同契註.

The *Kinship of the Three and the Book of Changes*, with Commentary.

Text, H/Han, *c.* +140.

Comm. probably Sung.

Ed. and comm. unknown.

TT/995.

Chou I Tshan Thung Chhi Chu 周易參同契註.

The *Kinship of the Three and the Book of Changes*, with Commentary.

Text, H/Han, *c.* +140.

Comm., Sung, *c.* +1230.

Ed. & comm. Chhu Hua-Ku 儲華谷.

TT/999.

Chou I Tshan Thung Chhi Chu (*TT*/992).

Alternative title for *Tshan Thung Chhi Khao I* (Chu Hsi's) q.v.

Chou I Tshan Thung Chhi Fa Hui 周易參同契發揮.

Elucidations of the *Kinship of the Three and the Book of Changes* [alchemy].

Text, H/Han, *c.* +140.

Comm., Yuan, +1284.

Ed. & comm. Yü Yen 兪琰.

Tr. Wu & Davis (1).

TT/996.

Chou I Tshan Thung Chhi Fên Chang Chu (*Chieh*) 周易參同契分章註 (解).

The *Kinship of the Three and the Book of Changes* divided into (short) chapters, with Commentary and Analysis.

Chou I Tshan Thung Chhi Fên Chang Chu (Chieh)
    (*cont.*)
    Text, Han, *c.* +140.
    Comm., Yuan, *c.* +1330.
    Comm. Chhen Chih-Hsü 陳致虛
        (Shang Yang Tzu 上陽子).
    TTCY pên 93.
Chou I Tshan Thung Chhi Fên Chang Thung
    Chen I 周易參同契分章通眞義.
    The *Kinship of the Three and the Book of*
    Changes divided into (short) chapters for
    the Understanding of its Real Meanings.
    Text, H/Han, *c.* +140.
    Comm., Wu Tai +947.
    Ed. & comm. Phêng Hsiao 彭曉.
    Tr. Wu & Davis (1).
    TT/993.
Chou I Tshan Thung Chhi Shih I 周易參同契
    釋疑.
    Clarification of Doubtful Matters in the
    Kinship of the Three and the Book of
    Changes.
    Yuan, +1284.
    Ed. & comm. Yü Yen 俞琰.
    TT/997.
Chou I Tshan Thung Chhi Su Lüeh 周易參同
    契疏略.
    Brief Explanation of the *Kinship of the Three*
    and the Book of Changes.
    Ming, +1564.
    Ed. & comm. Wang Wên-Lu 王文祿.
Chou I Tshan Thung Chhi Ting Chhi Ko Ming
    Ching Thu 易周參同契鼎器歌明鏡
    圖.
    An Illuminating Chart for the Mnemonic
    Rhymes about Reaction-Vessels in the
    Kinship of the Three and the Book of
    Changes.
    Text, H/Han, *c.* +140 (*Ting Chhi Ko*
    portion only).
    Comm., Wu Tai, +947.
    Ed. & comm. Phêng Hsiao 彭曉.
    TT/994.
Chu Chêng Pien I 諸證辨疑.
    Resolution of Diagnostic Doubts.
    Ming, late +15th.
    Wu Chhiu 吳球.
Chu Chhüan Chi 竹泉集.
    The Bamboo Springs Collection [poems
    and personal testimonies on physiological
    alchemy].
    Ming, +1465.
    Tung Chhung-Li *et al.* 童重理.
    In *Wai Chin Tan* (q.v.), ch. 3.
Chu Chia Shen Phin Tan Fa 諸家神品丹法.
    Methods of the Various Schools for Magical
    Elixir Preparations (an alchemical an-
    thology).
    Sung.
    Mêng Yao-Fu 孟要甫
        (Hsüan Chen Tzu 玄眞子) *et al.*
    TT/911.

Chu Fan Chih 諸蕃志.
    Records of Foreign Peoples (and their Trade).
    Sung, *c.* +1225. (This is Pelliot's dating;
    Hirth & Rockhill favoured between
    +1242 and +1258.)
    Chao Ju-Kua 趙汝适.
    Tr. Hirth & Rockhill (1).
Chu Yeh Thing Tsa Chi 竹葉亭雜記.
    Miscellaneous Records of the Bamboo Leaf
    Pavilion.
    Chhing, begun *c.* +1790 but not finished
    till *c.* 1820.
    Yao Yuan-Chih 姚元之.
Chuan Hsi Wang Mu Wo Ku Fa 傳西王母撮
    固法.
    [= *Thai-Shang Chuan Hsi Wang Mu Wo*
    Ku Fa.]
    A Recording of the Method of Grasping
    the Firmness (taught by) the Mother
    Goddess of the West.
    [Taoist heliotherapy and meditation. 'Grasp-
    ing the firmness' was a technical term for
    a way of clenching the hands during
    meditation.]
    Thang or earlier.
    Writer unknown.
    Fragment in *Hsiu Chen Shih Shu* (*TT*/260),
    ch. 24, p. 1 *a* ff.
    Cf. Maspero (7), p. 376.
Chuang Lou Chi 妝樓記.
    Records of the Ornamental Pavilion.
    Wu Tai or Sung, *c.* +960.
    Chang Mi 張泌.
Chün-Chai Tu Shu Chih 郡齋讀書志.
    Memoir on the Authenticities of Ancient
    Books, by (Chhao) Chün-Chai.
    Sung, +1151.
    Chhao Kung-Wu 晁公武.
Chün-Chai Tu Shu Fu Chih 郡齋讀書附志.
    Supplement to Chün-Chai's (Chhao Kung-
    Wu's) *Memoir on the Authenticities of*
    Ancient Books.
    Sung, *c.* +1200.
    Chao Hsi-Pien 趙希弁.
Chün-Chai Tu Shu Hou Chih 郡齋讀書後志.
    Further Supplement to Chün-Chai's (Chhao
    Kung-Wu's) *Memoir on the Authenticities*
    of Ancient Books.
    Sung, pref. +1151, pr. +1250.
    Chhao Kung-Wu 晁公武, re-compiled by
    Chao Hsi-Pien 趙希弁, from the edi-
    tion of Yao Ying-Chi 姚應績.
Chün Phu 菌譜.
    A Treatise on Fungi.
    Sung, +1245.
    Chhen Jen-Yü 陳仁玉.
Chung Hua Ku Chin Chu 中華古今注.
    Commentary on Things Old and New in
    China.
    Wu Tai (H/Thang), +923 to +926.
    Ma Kao 馬縞.
    See des Rotours (1), p. xcix.

Chung Huang Chen Ching 中黃眞經
  [= *Thai-Chhing Chung Huang Chen Ching* or *Thai Tsang Lun*.]
  True Manual of the Middle (Radiance) of the Yellow (Courts), (central regions of the three parts of the body) [Taoist anatomy and physiology with Buddhist influence].
  Prob. Sung, +12th or +13th.
  Chiu Hsien Chün (ps.) 九仙君.
  Comm. Chung Huang Chen Jen (ps.) 中黃眞人.
  TT/810.
  Completing *TT*/328 and 329 (Wieger).
  Cf. Maspero (7), p. 364.

Chung Lü Chuan Tao Chi 鍾呂傳道集.
  Dialogue between Chungli (Chhüan) and Lü (Tung-Pin) on the Transmission of the Tao (and the Art of Longevity, by Rejuvenation).
  Thang, +8th or +9th.
  Attrib. Chungli Chhüan 鍾離權 and Lü Yen 呂嵒.
  Ed. Shih Chien-Wu 施肩吾.
  In *Hsiu Chen Shih Shu* (*TT*/260), chs.14–16 incl.

Chung Shan Yü Kuei Fu Chhi Ching 中山玉櫃服氣經.
  Manual of the Absorption of the Chhi, found in the Jade Casket on Chung-Shan (Mtn). [Taoist breathing exercises.]
  Thang or Sung, +9th or +10th.
  Attrib. Chang Tao-Ling (Han) 張道陵 or Pi-Yen Chang Tao-chê 碧巖張道者 or Pi-Yen hsien-sêng 碧巖先生.
  Comm. by Huang Yuan-Chün 黃元君.
  In *YCCC*, ch. 60, pp. 1*a* ff.
  Cf. Maspero (7), pp. 204, 215, 353.

Chungli Pa Tuan Chin Fa 鍾離八段錦法.
  The Eight Elegant (Gymnastic) Exercises of Chungli (Chhüan).
  Thang, late +8th.
  Chungli Chhüan 鍾離權.
  In *Hsiu Chen Shih Shu* (*TT*/260), ch. 19.
  Tr. Maspero (7), pp. 418 ff.
  Cf. Notice by Tsêng Tshao in *Lin Chiang Hsien* (*TT*/260, ch. 23, pp. 1*b*, 2*a*) dated +1151. This says that the text was inscribed by Lü Tung-Pin himself on stone and so handed down.

Chhang Chhun Tzu Phan-Hsi Chi 長春子磻溪集.
  Chhiu Chhang-Chhun's Collected (Poems) at Phan-Hsi.
  Sung, *c.* +1200.
  Chhiu Chhu-Chi 邱處機.
  TT/1145.

Chhang Shêng Shu 長生術.
  The Art and Mystery of Longevity and Immortality.
  Alternative title of *Chin Hua Tsung Chih* (q.v.).

Chhen Wai Hsia Chü Chien 塵外遐舉牋.
  Examples of Men who Renounced Official Careers and Shook off the Dust of the World [the eighth and last part (ch. 19) of *Tsun Shêng Pa Chien*, q.v.].
  Ming, +1591.
  Kao Lien 高濂.

Chhi Chü An Lo Chien 起居安樂牋.
  On (Health-giving) Rest and Recreations in a Retired Abode [the third part (Chs. 7, 8) of *Tsun Shêng Pa Chien*, q.v.].
  Ming, +1591.
  Kao Lien 高濂.

Chhi Fan Ling Sha Ko 七返靈砂歌.
  Song of the Sevenfold Cyclically Transformed Numinous Cinnabar (Elixir).
  See *Chhi Fan Tan Sha Chüeh*.

Chhi Fan Ling Sha Lun 七返靈砂論.
  On Numinous Cinnabar Seven Times Cyclically Transformed.
  Alternative title for *Ta-Tung Lien Chen Pao Ching, Hsiu Fu Ling Sha Miao Chüeh* (q.v.).
  In *YCCC*, ch. 69, pp. 1*a* ff.

Chhi Fan Tan Sha Chüeh 七返丹砂訣.
  [= *Wei Po-Yang Chhi Fan Tan Sha Chüeh* or *Chhi Fan Ling Sha Ko*.]
  Explanation of the Sevenfold Cyclically Transformed Cinnabar (Elixir), (of Wei Po-Yang).
  Date unknown (ascr. H/Han).
  Writer unknown (attrib. Wei Po-Yang).
  Comm. by Huang Thung-Chün 黃童君.
  Thang or pre-Thang, before +806.
  TT/881.

Chhi Hsiao Liang Fang 奇效良方.
  Effective Therapeutics.
  Ming, *c.* +1436, pr. +1470.
  Fang Hsien 方賢.

Chhi Kuo Khao 七國考.
  Investigations of the Seven (Warring) States.
  Chhing, *c.* +1660.
  Tung Yüeh 董說.

Chhi Lu 七錄.
  Bibliography of the Seven Classes of Books.
  Liang, +523.
  Juan Hsiao-Hsü 阮孝緒.

Chhi Min Yao Shu 齊民要術.
  Important Arts for the People's Welfare [lit. Equality].
  N/Wei (and E/Wei or W/Wei), between +533 and +544.
  Chia Ssu-Hsieh 賈思勰.
  See des Rotours (1), p.c; Shih Shêng-Han (1).

Chhi Yün Shan Wu Yuan Tzu Hsiu Chen Pien Nan (Tshan Chêng) 棲雲山悟元子修眞辨難參證.
  See *Hsiu Chen Pien Nan (Tshan Chêng)*.

Chhieh Yün 切韻.
  Dictionary of the Sounds of Characters [rhyming dictionary].
  Sui, +601.
  Lu Fa-Yen 陸法言
  See *Kuang Yün*.

Chhien Chin Fang Yen I 千金方衍義.
Dilations upon the *Thousand Golden Remedies*.
Chhing, +1698.
Chang Lu 張璐

Chhien Chin I Fang 千金翼方.
Supplement to the *Thousand Golden Remedies* [i.e. Revised Prescriptions saving lives worth a Thousand Ounces of Gold].
Thang, between +660 and +680.
Sun Ssu-Mo 孫思邈.

Chhien Chin Shih Chih 千金食治.
A Thousand Golden Rules for Nutrition and the Preservation of Health [i.e. Diet and Personal Hygiene saving lives worth a Thousand Ounces of Gold], (included as a chapter in the *Thousand Golden Remedies*).
Thang, +7th (c. +625, certainly before +659).
Sun Ssu-Mo 孫思邈.

Chhien Chin Yao Fang 千金要方.
A Thousand Golden Remedies [i.e. Essential Prescriptions saving lives worth a Thousand Ounces of Gold].
Thang, between +650 and +659.
Sun Ssu-Mo 孫思邈.

Chhien Han Shu 前漢書.
History of the Former Han Dynasty [−206 to +24].
H/Han (begun about +65), c. +100.
Pan Ku 班固, and (after his death in +92) his sister Pan Chao 班昭.
Partial trs. Dubs (2), Pfizmaier (32–34, 37–51), Wylie (2, 3, 10), Swann (1).
Yin-Tê Index, no. 36.

Chhien Hung Chia Kêng Chih Pao Chi Chhêng 鉛汞甲庚至寶集成.
Complete Compendium on the Perfected Treasure of Lead, Mercury, Wood and Metal [with illustrations of alchemical apparatus].
On the translation of this title, cf. Vol. 5, pt. 3.
Has been considered Thang, +808; but perhaps more probably Wu Tai or Sung. Cf. p. 276.
Chao Nai-An 趙耐菴.
TT/912.

Chhien Khun Pi Yün 乾坤秘韞.
The Hidden Casket of Chhien and Khun (kua, i.e. Yang and Yin) Open'd.
Ming, c. +1430.
Chu Chhüan 朱權.
(Ning Hsien Wang 寧獻王, prince of the Ming.)

Chhien Khun Shêng I 乾坤生意.
Principles of the Coming into Being of Chhien and Khun (kua, i.e. Yang and Yin).
Ming, c. +1430.
Chu Chhüan 朱權.

(Ning Hsien Wang 寧獻王, prince of the Ming.)

Chhih Shui Hsüan Chu 赤水玄珠.
The Mysterious Pearl of the Red River [a system of medicine and iatro-chemistry].
Ming, +1596.
Sun I-Khuei 孫一奎.

Chhih Shui Hsüan Chu Chhüan Chi 赤水玄珠全集.
The Mysterious Pearl of the Red River; a Complete (Medical) Collection.
See *Chhih Shui Hsüan Chu*.

Chhih Shui Yin 赤水吟.
Chants of the Red River.
See Fu Chin-Chhüan (1).

Chhih Sung Tzu Chou Hou Yao Chüeh 赤松子肘後藥訣.
Oral Instructions of the Red-Pine Master on Handy (Macrobiotic) Prescriptions.
Pre-Thang.
Writer unknown.
Part of the *Thai-Chhing Ching Thien-Shih Khou Chüeh*.
TT/876.

Chhih Sung Tzu Hsüan Chi 赤松子玄記.
Arcane Memorandum of the Red-Pine Master.
Thang or earlier, before +9th.
Writer unknown.
Quoted in *TT*/928 and elsewhere.

Chhin Hsüan Fu 擒玄賦.
Rhapsodical Ode on Grappling with the Mystery.
Sung, +13th.
Writer unknown.
TT/257.

Chhing Hsiang Tsa Chi 青箱雜記.
Miscellaneous Records on Green Bamboo Tablets.
Sung, c. +1070.
Wu Chhu-Hou 吳處厚.

Chhing Hsiu Miao Lun Chien 清修妙論牋.
Subtile Discourses on the Unsullied Restoration (of the Primary Vitalities) [the first part (chs. 1, 2) of *Tsun Shêng Pa Chien*, q.v.].
Ming, +1591.
Kao Lien 高濂.

Chhing I Lu 清異錄.
Records of the Unworldly and the Strange.
Wu Tai, c. +950.
Thao Ku 陶穀.

Chhing-Ling Chen-Jen Phei Chün (Nei) Chuan 清靈眞人裴君內傳.
Biography of the Chhing-Ling Adept, Master Phei.
L/Sung or S/Chhi, +5th, but with early Thang additions.
Têng Yün Tzu 鄧雲子
(Phei Hsüan-Jen 裴玄仁 was a semi-legendary immortal said to have been born in −178).

Chhing-Ling Chen-Jen Phei Chün (Nei) Chuan
(*cont.*)
In *YCCC*, ch. 105.
Cf. Maspero (7), pp. 386 ff.

Chhing Po Tsa Chih 清波雜志.
Green-Waves Memories.
Sung, +1193.
Chou Hui 周煇.

Chhing Wei Tan Chüeh (or Fa) 清微丹訣(法).
Instructions for Making the Enchymoma in
Calmness and Purity [physiological
alchemy].
Date unknown, perhaps Thang.
Writer unknown.
TT/275.

Chhiu Chhang-Chhun Chhing Thien Ko 邱長春
青天歌.
Chhiu Chhang-Chhun's Song of the Blue
Heavens.
Sung, *c.* +1200.
Chhiu Chhu-Chi 邱處機.
TT/134.

Chhu Chhêng I Shu 褚澄遺書.
Remaining Writings of Chhu Chhêng.
Chhi, *c.* +500, probably greatly remodelled
in Sung.
Chhu Chhêng 褚澄.

Chhü Hsien Shen Yin Shu 臞仙神隱書.
Book of Daily Occupations for Scholars in
Rural Retirement, by the Emaciated
Immortal.
Ming, *c.* +1430.
Chu Chhüan 朱權.
(Ning Hsien Wang 寧獻王, prince of
the Ming.)

Chhu Hsüeh Chi 初學記.
Entry into Learning [encyclopaedia].
Thang, +700.
Hsü Chien 徐堅.

Chhü I Shuo Tsuan 袪疑說纂.
Discussions on the Dispersal of Doubts.
Sung, *c.* +1230.
Chhu Yung 儲泳.

Chhüan-Chen Chi Hsüan Pi Yao 全眞集玄祕要.
Esoteric Essentials of the Mysteries (of the
Tao), according to the Chhüan-Chen
(Perfect Truth) School [the Northern
School of Taoism in Sung and Yuan times].
Yuan, *c.* +1320.
Li Tao-Shun 李道純.
TT/248.

Chhüan-Chen Tso Po Chieh Fa 全眞坐鉢捷法.
Ingenious Method of the Chhüan-Chen
School for Timing Meditation (and other
Exercises) by a (Sinking-) Bowl Clepsydra.
Sung or Yuan.
Writer unknown.
TT/1212.

Chhüan Ching 拳經.
Manual of Boxing.
Chhing, +18th.
Chang Khung-Chao 張孔昭.

Chhun Chhiu Fan Lu 春秋繁露.
String of Pearls on the *Spring and Autumn
Annals.*
C/Han, *c.* −135.
Tung Chung-Shu 董仲舒.
See Wu Khang (1).
Partial trs. Wieger (2); Hughes (1);
d'Hormon (1) (ed.).
Chung-Fa Index no. 4.

Chhun Chhiu Wei Yuan Ming Pao 春秋緯元
命苞.
Apocryphal Treatise on the *Spring and
Autumn Annals*; the Mystical Diagrams
of Cosmic Destiny [astrological-
astronomical].
C/Han, *c.* −1st.
Writer unknown.
In *Ku Wei Shu*, ch. 7.

Chhun Chhiu Wei Yün Tou Shu 春秋緯運斗樞
Apocryphal Treatise on the *Spring and
Autumn Annals*; the Axis of the Turning
of the Ladle (i.e. the Great Bear).
C/Han, −1st or later.
Writer unknown.
In *Ku Wei Shu*, ch. 9, pp. 4 *b* ff. and
YHSF, ch. 55, pp. 22 *a* ff.

Chhun Chu Chi Wên 春渚紀聞.
Record of Things Heard at Spring Island.
Sung, *c.* +1095.
Ho Wei 何薳.

Chhun-yang etc.
See *Shun-yang.*

*Chhung-Hsiu Chêng-Ho Ching-Shih Chêng Lei
Pei-Yung Pên Tshao* 重修政和經史證
類備用本草.
New Revision of the Pharmacopoeia of the
Chêng-Ho reign-period; the Classified
and Consolidated Armamentarium.
(A Combination of the *Chêng-Ho...Chêng
Lei...Pên Tshao* with the *Pên Tshao Yen I.*)
Yuan, +1249; reprinted many times after-
wards, esp. in the Ming, +1468, with at
least seven Ming editions, the last in
+1624 or +1625.
Thang Shen-Wei 唐愼微.
Khou Tsung-Shih 寇宗奭.
Pr. (or ed.) Chang Tshun-Hui 張存惠.

Chhung-Yang Chhüan Chen Chi 重陽全
眞集.
(Wang) Chhung-Yang's [Wang Chê's]
Records of the Perfect Truth (School).
Sung, mid +12th cent.
Wang Chê 王嚞.
TT/1139.

Chhung-Yang Chiao Hua Chi 重陽教化集.
Memorials of (Wang) Chhung-Yang's
[Wang Chê's] Preaching.
Sung, mid +12th cent.
Wang Chê 王嚞.
TT/1140.

Chhung-Yang Chin-Kuan Yü-Suo Chüeh 重陽
金關玉鎖訣.

Chhung-Yang Chin-Kuan Yü-Suo Chüeh (cont.)
(Wang) Chhung-Yang's [Wang's Chê's] Instructions on the Golden Gate and the Lock of Jade.
Sung, mid +12th cent.
Wang Chê 王嚞.
TT/1142.

Chhung-Yang Fên-Li Shih-Hua Chi 重陽分梨十化集.
Writings of (Wang) Chhung-Yang [Wang Chê] (to commemorate the time when he received a daily) Ration of Pears, and the Ten Precepts of his Teacher.
Sung, mid +12th cent.
Wang Chê 王嚞.
TT/1141.

Chhung-Yang Li-Chiao Shih-Wu Lun 重陽立教十五論.
Fifteen Discourses of (Wang) Chhung-Yang [Wang Chê] on the Establishment of his School.
Sung, mid +12th cent.
Wang Chê 王嚞.
TT/1216.

Đai-Viêt Sú-ký Toàn-thú 大越史記全書.
The Complete Book of the History of Great Annam.
Vietnam, c. +1479.
Ngô Si-Liên 吳士連.

Fa Yen 法言.
Admonitory Sayings [in admiration, and imitation, of the *Lun Yü*].
Hsin, +5.
Yang Hsiung 揚雄.
Tr. von Zach (5).

Fa Yuan Chu Lin 法苑珠林.
Forest of Pearls from the Garden of the [Buddhist] Law.
Thang, +668, +688.
Tao-Shih 道世.

Fan Tzu Chi Jan 范子計然.
See *Chi Ni Tzu*.

Fang Hu Wai Shih 方壺外史.
Unofficial History of the Land of the Immortals, Fang-hu. (Contains two *nei tan* commentaries on the *Tshan Thung Chhi*, +1569 and +1573.)
Ming, c. +1590.
Lu Hsi-Hsing 陸西星.
Cf. Liu Tshun-Jen (1, 2).

Fang Yü Chi 方輿記.
General Geography.
Chin, or at least pre-Sung.
Hsü Chiai 徐鍇.

Fei Lu Hui Ta 斐錄彙答.
Questions and Answers on Things Material and Moral.
Ming, +1636.
Kao I-Chih (Alfonso Vagnoni) 高一志.
Bernard-Maître (18), no. 272.

Fên Thu 粉圖.
See *Hu Kang Tzu Fên Thu*.

Fêng Su Thung I 風俗通義.
The Meaning of Popular Traditions and Customs.
H/Han, +175.
Ying Shao 應劭.
Chung-Fa Index, no. 3.

Fo Shuo Fo I Wang Ching 佛說佛醫王經
Buddha Vaidyarāja Sūtra; or *Buddha-prokta Buddha-bhaiṣajyarāja Sūtra* (Sūtra of the Buddha of Healing, spoken by Buddha).
India.
Tr. San Kuo (Wu) +230.
Trs. Liu Yen (Vinayātapa) & Chih-Chhien. 支謙.
N/1327; TW/793.

Fo Tsu Li Tai Thung Tsai 佛祖歷代通載.
General Record of Buddhist and Secular History through the Ages.
Yuan, +1341.
Nien-Chhang (monk) 念常.

Fu Chhi Ching I Lun 服氣精義論.
Dissertation on the Meaning of 'Absorbing the Chhi and the Ching' (for Longevity and Immortality), [Taoist hygienic, respiratory, pharmaceutical, medical and (originally) sexual procedures].
Thang, c. +715.
Ssuma Chhêng-Chên 司馬承貞.
In *YCCC*, ch. 57.
Cf. Maspero (7), pp. 364 ff.

Fu Hung Thu 伏汞圖.
Illustrated Manual on the Subduing of Mercury.
Sui, Thang, J/Chin or possibly Ming.
Shêng Hsüan Tzu 昇女子.
Survives now only in quotations.

Fu Nei Yuan Chhi Ching 服內元氣經.
Manual of Absorbing the Internal Chhi of Primary (Vitality).
Thang, +8th, probably c. +755.
Huan Chen hsien-sêng (Mr Truth-and-Illusion) 幻眞先生.
TT/821, and in *YCCC*, ch. 60, pp. 10b ff.
Cf. Maspero (7), p. 199.

Fu Shih Lun 服石論.
Treatise on the Consumption of Mineral Drugs.
Thang, perhaps Sui.
Writer unknown.
Extant only in excerpts preserved in the *I Hsin Fang* (+982).

Fu Shou Tan Shu 福壽丹書.
A Book of Elixir-Enchymoma Techniques for Happiness and Longevity.
Ming, +1621.
Chêng Chih-Chhiao 鄭之僑 (at least in part).
Partial tr. of the gymnastic material, Dudgeon (1).

Fusō Ryakuki 扶桑畧記.
  Classified Historical Matters concerning the
    Land of Fu-Sang (Japan) [from +898 to
    +1197].
  Japan (Kamakura) +1198.
  Kōen (monk).

Genji Monogatari 源氏物語.
  The Tale of (Prince) Genji.
  Japan, +1021.
  Murasaki Shikibu 紫式部.

Hai Yao Pên Tshao 海藥本草.
  [= *Nan Hai Yao Phu*.]
  Materia Medica of the Countries Beyond
    the Seas.
  Wu Tai (C/Shu), c. +923.
  Li Hsün 李珣.
  Preserved only in numerous quotations in
    Chêng Lei Pên Tshao and later pandects.

Han Fei Tzu 韓非子.
  The Book of Master Han Fei.
  Chou, early −3rd century.
  Han Fei 韓非.
  Tr. Liao Wên-Kuei (1).

Han Kuan I 漢官儀.
  The Civil Service of the Han Dynasty and
    its Regulations.
  H/Han +197.
  Ying Shao 應劭.
  Ed. Chang Tsung-Yuan 張宗源 (+1752
    to 1800).
  Cf. Hummel (2), p. 57.

Han Kung Hsiang Fang 漢宮香方.
  On the Blending of Perfumes in the Palaces
    of the Han.
  H/Han, +1st or +2nd.
  Genuine parts preserved c. +1131 by
    Chang Pang-Chi 張邦基.
  Attrib. Tung Hsia-Chou 薔遐周.
  Comm. by Chêng Hsüan 鄭玄.
  'Restored', c. +1590, by Kao Lien 高濂.

Han Thien Shih Shih Chia 漢天師世家.
  Genealogy of the Family of the Han
    Heavenly Teacher.
  Date uncertain.
  Writers unknown.
  With Pu Appendix, 1918, by Chang Yuan-
    Hsü 張元旭 (the 62nd Taoist Patriarch,
    Thien Shih).
  TT/1442.

Han Wei Tshung-Shu 漢魏叢書.
  Collection of Books of the Han and Wei Dyn-
    asties [first only 38, later increased to
    96].
  Ming, +1592.
  Ed. Thu Lung 屠隆.

Han Wu (Ti) Ku Shih 漢武(帝)故事.
  Tales of (the Emperor) Wu of the Han
    (r. −140 to −87).
  L/Sung and Chhi, late +5th.
  Wang Chien 王儉.

Perhaps based on an earlier work of the
  same kind by Ko Hung 葛洪.
Tr. d'Hormon (1).

Han Wu (Ti) Nei Chuan 漢武(帝)內傳.
  The Inside Story of (Emperor) Wu of the
    Han (r. −140 to −87).
  Material of Chin, L/Sung, Chhi, Liang and
    perhaps Chhen date, +320 to +580,
    probably stabilised about +580.
  Attrib. Pan Ku, Ko Hung, etc.
  Actual writer unknown.
  TT/289.
  Tr. Schipper (1).

Han Wu (Ti) Nei Chuan Fu Lu 漢武(帝)內傳
  附錄.
  See *Han Wu (Ti) Wai Chuan*.

Han Wu (Ti) Wai Chuan 漢武(帝)外傳.
  [= *Han Wu (Ti) Nei Chuan Fu Lu*.]
  Extraordinary Particulars of (Emperor) Wu
    of the Han (and his collaborators), [largely
    biographies of the magician-technicians
    at Han Wu Ti's court].
  Material of partly earlier date collected and
    stabilised in Sui or Thang, early +7th
    century.
  Writers and editor unknown.
  Introductory paragraphs added by Wang
    Yu-Yen 王游嚴 (+746).
  TT/290.
  Cf. Maspero (7), p. 234, and Schipper (1).

Hei Chhien Shui Hu Lun 黑鉛水虎論.
  Discourse on the Black Lead and the Water
    Tiger.
  Alternative title of *Huan Tan Nei Hsiang
    Chin Yo Shih*, q.v.

Ho Chi Chü Fang 和劑局方.
  Standard Formularies of the (Government)
    Pharmacies [based on the *Thai-Phing
    Shêng Hui Fang* and other collections].
  Sung, c. +1109.
  Ed. Chhen Chhêng 陳承, Phei Tsung-
    Yuan 裴宗元, & Chhen Shih-Wên
    陳師文.
  Cf. *SIC*, p. 974.

Honan Chhen Shih Hsiang Phu 河南陳氏香譜.
  See *Hsiang Phu* by Chhen Ching.

Honan Chhêng Shih I Shu 河南程氏遺書.
  Remaining Records of Discourses of the
    Chhêng brothers of Honan [Chhêng I and
    Chhêng Hao, +11th-century Neo-
    Confucian philosophers].
  Sung, +1168, pr. c. +1250.
  Chu Hsi (ed.) 朱熹.
  In *Erh Chhêng Chhüan Shu*, q.v.
  Cf. Graham (1), p. 141.

Honan Chhêng Shih Tshui Yen 河南程氏粹言.
  Authentic Statements of the Chhêng brothers
    of Honan [Chhêng I and Chhêng Hao,
    +11th-century Neo-Confucian philo-
    sophers. In fact more altered and abridged
    than the other sources, which are therefore
    to be preferred.]

Honan Chhêng Shih Tshui Yen (cont.)
Sung, first collected *c.* +1150, supposedly ed. +1166, in its present form by *c.* +1340.
Coll. Hu Yin 胡寅.
Supposed ed. Chang Shih 張栻.
In *Erh Chhêng Chhüan Shu*, q.v., since +1606.
Cf. Graham (1), p. 145.

Honzō-Wamyō 本草和名.
Synonymic Materia Medica with Japanese Equivalents.
Japan, +918.
Fukane no Sukehito 深根輔仁.
Cf. Karow (1).

Hou Han Shu 後漢書.
History of the Later Han Dynasty [+25 to +220].
L/Sung, +450.
Fan Yeh 范曄.
The monograph chapters by Ssuma Piao 司馬彪 (d. +305), with commentary by Liu Chao 劉昭 (*c.* +510), who first incorporated them in the work.
A few chs. tr. Chavannes (6, 16); Pfizmaier (52, 53).
Yin-Tê Index, no. 41.

Hou Tê Lu 厚德錄.
Stories of Eminent Virtue.
Sung, early +12th.
Li Yuan-Kang 李元綱.

Hu Kang Tzu Fên Thu 狐剛子粉圖.
Illustrated Manual of Powders [Salts], by the Fox-Hard Master.
Sui or Thang.
Hu Kang Tzu 狐剛子.
Survives now only in quotations; originally in *TT* but lost. Cf. Vol. 4, pt. 1, p. 308.

Hua Tho Nei Chao Thu 佗佗內照圖.
Hua Tho's Illustrations of Visceral Anatomy.
See *Hsüan Mên Mo Chüeh Nei Chao Thu*.
Cf. Miyashita Saburo (1).

Hua-Yang Thao Yin-Chü Chuan 華陽陶隱居傳.
A Biography of Thao Yin-Chü (Thao Hung-Ching) of Huayang [the great alchemist, naturalist and physician].
Thang.
Chia Sung 賈嵩.
TT/297.

Hua Yen Ching 華嚴經.
Buddha-avataṃsaka Sūtra; The Adornment of Buddha.
India.
Tr. into Chinese, +6th century.
TW/278, 279.

Huai Nan Hung Lieh Chieh 淮南鴻烈解.
See *Huai Nan Tzu*.

Huai Nan Tzu 淮南子.
[= *Huai Han Hung Lieh Chieh* 淮南鴻烈解.]
The Book of (the Prince of) Huai-Nan [compendium of natural philosophy].

C/Han, *c.* −120.
Written by the group of scholars gathered by Liu An (prince of Huai-Nan) 劉安.
Partial trs. Morgan (1); Erkes (1); Hughes (1); Chatley (1); Wieger (2).
Chung-Fa Index, no. 5.
TT/1170.

Huai-Nan (Wang) Wan Pi Shu 淮南 (王) 萬畢術.
[Prob. = *Chen-Chung Hung-Pao Yuan-Pi Shu* and variants.]
The Ten Thousand Infallible Arts of (the Prince of) Huai-Nan [Taoist magical and technical recipes].
C/Han, −2nd century.
No longer a separate book but fragments contained in *TPYL*, ch. 736 and elsewhere
Reconstituted texts by Yeh Tê-Hui in *Kuan Ku Thang So Chu Shu*, and Sun Fêng-I in *Wên Ching Thang Tshung-Shu*.
Attrib. Liu An 劉安.
See Kaltenmark (2), p. 32.
It is probable that the terms *Chen-Chung* 枕中 Confidential Pillow-Book; *Hung-Pao* 鴻寶 Infinite Treasure; *Wan-Pi* 萬畢 Ten Thousand Infallible; and *Yuan-Pi* 苑祕 Garden of Secrets; were originally titles of parts of a *Huai-Nan Wang Shu* 淮南王書 (Writings of the Prince of Huai-Nan) forming the Chung Phien 中篇 (and perhaps also the Wai Shu 外書) of which the present *Huai Nan Tzu* book (q.v.) was the Nei Shu 內書.

Huan Chen hsien-sêng, etc. 幻眞先生.
See *Thai Hsi Ching* and *Fu Nei Yuan Chhi Ching*.

Huan Chin Shu 還金述.
An Account of the Regenerative Metallous Enchymoma.
Thang, probably +9th.
Thao Chih 陶植.
TT/915, also excerpted, in *YCCC*, ch. 70, pp. 13 a ff.

Huan Tan Chou Hou Chüeh 還丹肘後訣.
Oral Instructions on Handy Formulae for Cyclically Transformed Elixirs [with illustrations of alchemical apparatus].
Ascr. Chin, *c.* +320.
Actually Thang, including a memorandum of +875 by Wu Ta-Ling 仵達靈, and the rest probably by other hands within a few years of this date.
Attrib. Ko Hung 葛洪.
TT/908.

Huan Tan Chung Hsien Lun 還丹象仙論.
Pronouncements of the Company of the Immortals on Cyclically Transformed Elixirs.
Sung, +1052.
Yang Tsai 楊在.
TT/230.

Huan Tan Fu Ming Phien 還丹復命篇.
　　Book on the Restoration of Life by the
　　　Cyclically Transformed Elixir.
　　Sung, +12th cent., *c.* +1175.
　　Hsüeh Tao-Kuang 薛道光.
　　TT/1074.
Huan Tan Nei Hsiang Chin Yo Shih 還丹內象
　　金鑰匙.
　　[= *Hei Chhien Shui Hu Lun* and *Hung
　　　Chhien Huo Lung Lun.*]
　　A Golden Key to the Physiological Aspects
　　　of the Regenerative Enchymoma.
　　Wu Tai, *c.* +950.
　　Phêng Hsiao 彭曉.
　　Now but half a chapter in *YCCC*, ch. 70,
　　　pp. 1 *a* ff., though formerly contained in
　　　the *Tao Tsang.*
*Huan Tan Pi Chüeh Yang Chhih-Tzu Shen
　　Fang* 還丹祕訣養赤子神方.
　　The Wondrous Art of Nourishing the
　　　(Divine) Embryo (lit. the Naked Babe) by
　　　the use of the secret Formula of the Re-
　　　generative Enchymoma [physiological
　　　alchemy].
　　Sung, probably late +12th.
　　Hsü Ming-Tao 許明道.
　　TT/229.
Huan Yü Shih Mo 寰宇始末.
　　On the Beginning and End of the World
　　　[the Hebrew-Christian account of crea-
　　　tion, the Four Aristotelian Causes,
　　　Elements, etc.].
　　Ming, +1637.
　　Kao I-Chih (Alfonso Vagnoni) 高一志.
　　Bernard-Maître (18), no. 283.
Huan Yuan Phien 還原篇.
　　Book of the Return to the Origin [poems on
　　　the regaining of the primary vitalities in
　　　physiological alchemy].
　　Sung, *c.* +1140.
　　Shih Thai 石泰.
　　TT/1077. Also in *Hsiu Chen Shih Shu*
　　　(*TT*/260), ch. 2.
Huang Chi Ching Shih Shu 皇極經世書.
　　Book of the Sublime Principle which
　　　governs All Things within the World.
　　Sung, *c.* +1060.
　　Shao Yung 邵雍.
　　TT/1028. Abridged in *Hsing Li Ta Chhüan*
　　　and *Hsing Li Ching I.*
Huang Chi Ho Pi Hsien Ching 皇極闔闢仙經.
　　[= *Yin Chen Jen Tung-Hua Chêng Mo Huang
　　　Chi Ho Pi Chêng Tao Hsien Ching.*]
　　The Height of Perfection (attained by)
　　　Opening and Closing (the Orifices of the
　　　Body); a Manual of the Immortals [phys-
　　　iological alchemy, *nei tan* techniques].
　　Ming or Chhing.
　　Attrib. Yin chen jen (Phêng-Thou)
　　　尹眞人(蓬頭).
　　Ed. Min I-Tê 閔一得, *c.* 1830.
　　In *Tao Tsang Hsü Pien* (*Chhu chi*), 2, from

a MS. preserved at the Blue Goat Temple
　　青羊宮 (Chhêngtu).
Huang Pai Ching 黃白鏡.
　　Mirror of (the Art of) the Yellow and the
　　　White [physiological alchemy].
　　Ming, +1598.
　　Li Wên-Chu 李文燭.
　　Comm. Wang Chhing-Chêng 王清正.
　　In *Wai Chin Tan* coll., ch. 2 (*CTPS, pên*
　　　7).
*Huang-Thien Shang-Chhing Chin Chhüeh Ti
　　Chün Ling Shu Tzu-Wên Shang Ching*
　　皇天上清金闕帝君靈書紫文上經.
　　Exalted Canon of the Imperial Lord of the
　　　Golden Gates, Divinely Written in Purple
　　　Script; a Huang-Thien Shang-Chhing
　　　Scripture.
　　Chin, late +4th, with later revisions.
　　Writer unknown.
　　TT/634.
Huang Thing Chung Ching Ching 黃庭中景經.
　　[= *Thai-Shang Huang Thing Chung Ching
　　　Ching.*]
　　Manual of the Middle Radiance of the
　　　Yellow Courts (central regions of the
　　　three parts of the body) [Taoist anatomy
　　　and physiology].
　　Sui.
　　Li Chhien-Chhêng 李千乘.
　　TT/1382, completing *TT*/398–400.
　　Cf. Maspero (7), pp. 195, 203.
*Huang Thing Nei Ching Wu Tsang Liu Fu Pu
　　Hsieh Thu* 黃庭內景五臟六府補瀉圖
　　Diagrams of the Strengthening and Weaken-
　　　ing of the Five Yin-viscera and the Six
　　　Yang-viscera (in accordance with) the
　　　(*Jade Manual of the*) *Internal Radiance of
　　　the Yellow Courts.*
　　Thang, *c.* +850.
　　Hu An 胡愔.
　　TT/429.
Huang Thing Nei Ching Wu Tsang Liu Fu Thu
　　黃庭內景五臟六府圖.
　　Diagrams of the Five Yin-viscera and the
　　　Six Yang-viscera (discussed in the *Jade
　　　Manual of the*) *Internal Radiance of the
　　　Yellow Courts* [Taoist anatomy and physi-
　　　ology; no illustrations surviving, but much
　　　therapy and pharmacy].
　　Thang, +848.
　　Hu An 胡愔 (title: Thai-pai Shan Chien
　　　Su Nü 太白山見素女.
　　In *Hsiu Chen Shih Shu* (*TT*/260), ch. 54.
　　Illustrations preserved only in Japan, MS. of
　　　before +985.
　　SIC, p. 223; Watanabe Kozo (*1*), pp. 112 ff.
Huang Thing Nei Ching Yü Ching 黃庭內景
　　玉經.
　　[= *Thai-Shang Huang Thing Nei Ching Yü
　　　Ching.*]
　　Jade Manual of the Internal Radiance of the
　　　Yellow Courts (central regions of the

Huang Thing Nei Ching Yü Ching (cont.)
three parts of the body) [Taoist anatomy and physiology]. In 36 *chang*.
L/Sung, Chhi, Liang or Chhen, +5th or +6th. The oldest parts date probably from Chin, about +365.
Writer unknown. Allegedly transmitted by immortals to the Lady Wei (Wei Fu Jen), i.e. Wei Hua-Tshun 魏華存.
TT/328.
Paraphrase by Liu Chhang-Shêng 劉長生 (Sui), *TT*/398.
Comms. by Liang Chhiu Tzu 梁丘子 (Thang), *TT*/399, and Chiang Shen-Hsiu 蔣慎修 (Sung), *TT*/400.
Cf. Maspero (7), p. 239.

Huang Thing Nei Ching Yü Ching Chu 黃庭內景玉經注.
Commentary on (and paraphrased text of) the *Jade Manual of the Internal Radiance of the Yellow Courts*.
Sui.
Liu Chhang-Shêng 劉長生.
TT/398.

Huang Thing Nei Ching (Yü) Ching Chu 黃庭內景(玉)經注.
Commentary on the *Jade Manual of the Internal Radiance of the Yellow Courts*.
Thang, +8th or +9th.
Liang Chhiu Tzu (ps.) 梁丘子.
TT/399, and in *Hsiu Chen Shih Shu* (*TT*/260), chs. 55–57; and in *YCCC*, chs. 11, 12 (where the first 3 *chang* (30 verses) have the otherwise lost commentary of Wu Chhêng Tzu 務成子).
Cf. Maspero (7), pp. 239 ff.

Huang Thing Nei Wai Ching Yü Ching Chieh 黃庭內外景玉經解.
Explanation of the *Jade Manuals of the Internal and External Radiances of the Yellow Courts*.
Sung.
Chiang Shen-Hsiu 蔣慎修.
TT/400.

Huang Thing Wai Ching Yü Ching 黃庭外景玉經.
[= *Thai-Shang Huang Thing Wai Ching Yü Ching*.]
Jade Manual of the External Radiance of the Yellow Courts (central regions of the three parts of the body) [Taoist anatomy and physiology]. In 3 *chüan*.
H/Han, San Kuo or Chin, +2nd or +3rd. Not later than +300.
Writer unknown.
TT/329.
Comms. by Wu Chhêng Tzu 務成子 (early Thang) *YCCC*, ch. 12; Liang Chhiu Tzu 梁丘子 (late Thang), *TT*/260, chs. 58–60; Chiang Shen-Hsiu 蔣慎修 (Sung), *TT*/400.
Cf. Maspero (7), pp. 195 ff., 428 ff.

Huang Thing Wai Ching Yü Ching Chu 黃庭外景玉經註.
Commentary on the *Jade Manual of the External Radiance of the Yellow Courts*.
Sui or early Thang, +7th.
Wu Chhêng Tzu (ps.) 務成子.
In *YCCC*, ch. 12, pp. 30a ff.
Cf. Maspero (7), p. 239.

Huang Thing Wai Ching Yü Ching Chu 黃庭外景玉經註.
Commentary on the *Jade Manual of the External Radiance of the Yellow Courts*.
Thang, +8th or +9th.
Liang Chhiu Tzu (ps.) 梁丘子.
In *Hsiu Chen Shih Shu* (*TT*/260), chs. 58–60.
Cf. Maspero (7), pp. 239 ff.

Huang Ti Chiu Ting Shen Tan Ching Chüeh 黃帝九鼎神丹經訣.
The Yellow Emperor's Canon of the Nine-Vessel Spiritual Elixir, with Explanations.
Early Thang or early Sung, but incorporating as ch. 1 a canonical work probably of the +2nd cent.
Writer unknown.
TT/878. Also, abridged, in *YCCC*, ch. 67, pp. 1a ff.

Huang Ti Nei Ching, Ling Shu 黃帝內經靈樞.
The Yellow Emperor's Manual of Corporeal (Medicine), the Vital Axis [medical physiology and anatomy].
Probably C/Han, c. −1st century.
Writers unknown.
Edited Thang, +762, by Wang Ping 王冰.
Analysis by Huang Wên (1).
Tr. Chamfrault & Ung Kang-Sam (1).
Commentaries by Ma Shih 馬蒔 (Ming) and Chang Chih-Tshung 張志聰 (Chhing) in *TSCC*, *I shu tien*, chs. 67 to 88.

Huang Ti Nei Ching, Ling Shu, Pai Hua Chieh
See Chhen Pi-Liu & Chêng Cho-Jen (1).

Huang Ti Nei Ching, Su Wên 黃帝內經素問.
The Yellow Emperor's Manual of Corporeal (Medicine); Questions (and Answers) about Living Matter [clinical medicine].
Chou, remodelled in Chhin and Han, reaching final form c. −2nd century.
Writers unknown.
Ed. & comm., Thang (+762), Wang Ping 王冰; Sung (c. +1050), Lin I 林億.
Partial trs. Hübotter (1), chs. 4, 5, 10, 11, 21; Veith (1); complete, Chamfrault & Ung Kang-Sam (1).
See Wang & Wu (1), pp. 28 ff.; Huang Wên (1).

Huang Ti Nei Ching Su Wên I Phien 黃帝內經素問遺篇.
The Missing Chapters from the *Questions and Answers of the Yellow Emperor's Manual of Corporeal (Medicine)*.
Ascr. pre-Han.
Sung, preface, +1099.

Huang Ti Nei Ching Su Wên I Phien (cont.)
Ed. (perhaps written by) Liu Wên-Shu
劉溫舒．
Often appended to his *Su Wên Ju Shih Yün
Chhi Ao Lun* (q.v.) 素問入式運氣奧論．

Huang Ti Nei Ching Su Wên, Pai Hua Chieh
See Chou Fêng-Wu, Wang Wan-Chieh &
Hsü Kuo-Chhien (*1*).

*Huang Ti Pa-shih-i Nan Ching Tsuan Thu Chü
Chieh* 黃帝八十一難經纂圖句解．
Diagrams and a Running Commentary for
the *Manual of (Explanations Concerning)
Eighty-one Difficult (Passages) in the Yellow
Emperor's (Manual of Corporeal Medicine)*.
Sung, +1270 (text H/Han, +1st).
Li Kung 李駉．
TT/1012.

Huang Ti Pao Tsang Ching 黃帝寶藏經．
Perhaps an alternative name for *Hsien-
Yuan Pao Tsang (Chhang Wei) Lun*, q.v.

Huang Ti Yin Fu Ching 黃帝陰符經．
See *Yin Fu Ching*.

Huang Ti Yin Fu Ching Chu 黃帝陰符經註．
Commentary on the *Yellow Emperor's Book
on the Harmony of the Seen and the Unseen*.
Sung.
Liu Chhu-Hsüan 劉處玄．
TT/119.

Huang Yeh Fu 黃冶賦．
Rhapsodic Ode on 'Smelting the Yellow'
[alchemy].
Thang, *c.* +840.
Li Tê-Yü 李德裕．
In *Li Wên-Jao Pieh Chi*, ch. 1.

Huang Yeh Lun 黃冶論．
Essay on the 'Smelting of the Yellow'
[alchemy].
Thang, *c.* +830.
Li Tê-Yü 李德裕．
In *Wên Yuan Ying Hua*, ch. 739, p. 15 *a*,
and *Li Wên-Jao Wai Chi*, ch. 4.

Hui Ming Ching 慧命經．
[= *Tsui-Shang I Chhêng Hui Ming Ching*,
also entitled *Hsü Ming Fang*.]
Manual of the (Achievement of) Wisdom
and the (Lengthening of the) Life-Span.
Chhing, +1794.
Liu Hua-Yang 柳華陽．
Cf. Wilhelm & Jung (*1*), editions after 1957.

Hung Chhien Huo Lung Lun 紅鉛火龍論．
Discourse on the Red Lead and the Fire
Dragon.
Alternative title of *Huan Tan Nei Hsiang
Chin Yo Shih*, q.v.

Hung Chhien Ju Hei Chhien Chüeh 紅鉛入黑
鉛訣．
Oral Instructions on the Entry of the Red
Lead into the Black Lead.
Probably Sung, but some of the material
perhaps older.
Compiler unknown.
TT/934.

Huo Kung Chhieh Yao 火攻挈要．
Essentials of Gunnery.
Ming, +1643.
Chiao Hsü 焦勗．
With the collaboration of Thang Jo-Wang
(J. A. Schall von Bell) 湯若望．
Bernard-Maître (18), no. 334.

Huo Lien Ching 火蓮經．
Manual of the Lotus of Fire [physiological
alchemy].
Ming or Chhing.
Attrib. Liu An, 劉安 (Han).
In *Wai Chin Tan*, coll., ch. 1 (*CTPS*, *pên* 6).

Huo Lung Ching 火龍經．
The Fire-Drake (Artillery) Manual.
Ming, +1412.
Chiao Yü 焦玉．
The first part of this book, in three sections,
is attributed fancifully to Chuko Wu-Hou
(i.e. Chuko Liang), and Liu Chi 劉基
(+1311 to +1375) appears as co-editor,
really perhaps co-author.
The second part, also in three sections, is
attributed to Liu Chi alone, but edited,
probably written, by Mao Hsi-Ping
毛希秉 in +1632.
The third part, in two sections, is by Mao
Yuan-I 毛元儀 (*fl.* +1628) and edited
by Chuko Kuang-Jung 諸葛光榮
whose preface is of +1644, Fang Yuan-
Chuang 方元壯 & Chung Fu-Wu 鍾伏武．

Huo Lung Chüeh 火龍訣．
Oral Instructions on the Fiery Dragon
[proto-chemical and physiological alchemy].
Date uncertain, ascr. Yuan, +14th.
Attrib. Shang Yang Tsu Shih 上陽祖師．
In *Wai Chin Tan* (coll.), ch. 3 (*CTPS*, *pên* 8).

Hupei Thung Chih 湖北通志．
Historical Geography of Hupei Province.
Min Kuo, 1921, but based on much older
records.
See Yang Chhêng-Hsi (ed.) (*1*) 楊承禧．

Hsi Chhi Tshung Hua 西溪叢話
(*SKCS* has *Yü* 語).
Western Pool Collected Remarks.
Sung, *c.* +1150.
Yao Khuan 姚寬．

Hsi Chhing Ku Chien 西清古鑑．
Hsi Chhing Catalogue of Ancient Mirrors
(and Bronzes) in the Imperial Collection.
(The collection was housed in the Library
of Western Serenity, a building in the
southern part of the Imperial Palace).
Chhing, +1751.
Liang Shih-Chêng 梁詩正．

Hsi Shan Chhun Hsien Hui Chen Chi 西山羣
仙會眞記．
A True Account of the Proceedings of the Com-
pany of Immortals in the Western Mountains.
Thang, *c.* +800.
Shih Chien-Wu 施肩吾．
TT/243.

Hsi Shang Fu Than 席上腐談.
Old-Fashioned Table Talk.
Yuan, *c.* +1290.
Yü Yen 俞琰.

Hsi Wang Mu Nü Hsiu Chêng Thu Shih Tsê
西王母女修正途十則.
The Ten Rules of the Mother (Goddess)
Queen of the West to Guide Women
(Taoists) along the Right Road of
Restoring (the Primary Vitalities) [phy-
siological alchemy].
Ming or Chhing.
Attrib. Lü Yen 呂喦 (+8th century).
Shen I-Ping *et al.* 沈一炳.
Comm. Min I-Tê 閔一得 (*c.* 1830).
In *Tao Tsang Hsü Pien* (*Chhu chi*), 19.

Hsi-Yang Huo Kung Thu Shuo 西洋火攻圖說.
Illustrated Treatise on European Gunnery.
Ming, before +1625.
Chang Tao 張燾 & Sun Hsüeh-Shih
孫學詩.

Hsi Yo Hua-Shan Chih 西嶽華山誌.
Records of Hua-Shan, the Great Western
Mountain.
Sung, *c.* +1170.
Wang Chhu-I 王處一.
TT/304.

Hsi Yo Tou hsien-sêng Hsiu Chen Chih Nan
西嶽竇先生修眞指南.
Teacher Tou's South-Pointer for the
Regeneration of the Primary (Vitalities),
from the Western Sacred Mountain.
Sung, probably early +13th.
Tou hsien-sêng 竇先生.
In *Hsiu Chen Shih Shu* (*TT*/260), ch. 21,
pp. 1*a* to 6*b*.

Hsi Yu Chi 西遊記.
A Pilgrimage to the West [novel].
Ming, *c.* +1570.
Wu Chhêng-Ên 吳承恩.
Tr. Waley (17).

Hsi Yu Chi.
See *Chhang-Chhun Chen Jen Hsi Yu Chi.*

Hsi Yü Chiu Wên 西域舊聞.
Old Traditions of the Western Countries [a
conflation, with abbreviations, of the
Hsi Yü Wên Chien Lu and the *Shêng Wu
Chi*, q.v.].
Chhing, +1777 and 1842.
Chhun Yuan Chhi-shih-i Lao-jen 椿園七
十一老人 & Wei Yuan 魏源.
Arr. Chêng Kuang-Tsu (1843) 鄭光祖.

Hsi Yü Thu Chi 西域圖記.
Illustrated Record of Western Countries.
Sui, +610.
Phei Chü 裴矩.

Hsi Yü Wên Chien Lu 西域聞見錄.
Things Seen and Heard in the Western
Countries.
Chhing, +1777.
Chhun Yuan Chhi-shih-i Lao-jen
椿園七十一老人.

[The 71-year-old Gentleman of the Cedar
Garden.]
Bretschneider (2), vol. 1, p. 128.

Hsi Yuan Lu 洗冤錄.
The Washing Away of Wrongs (i.e. False
Charges) [treatise on forensic medicine].
Sung, +1247.
Sung Tzhu 宋慈.
Partial tr., H. A. Giles (7).

Hsiang Chhêng 香乘.
Records of Perfumes and Incense [in-
cluding combustion-clocks].
Ming, betw. +1618 and +1641.
Chou Chia-Chou 周嘉胄.

Hsiang Chien 香牋.
Notes on Perfumes and Incense.
Ming, *c.* +1560.
Thu Lung 屠隆.

Huang Kuo 香國.
The Realm of Incense and Perfumes.
Ming.
Mao Chin, 毛晉.

Hsiang Lu 香錄.
[= *Nan Fan Hsiang Lu.*]
A Catalogue of Incense.
Sung, +1151.
Yeh Thing-Kuei 葉廷珪.

Hsiang Phu 香譜.
A Treatise on Aromatics and Incense
[-Clocks].
Sung, *c.* +1073.
Shen Li 沈立.
Now extant only in the form of quotations
in later works.

Hsiang Phu 香譜.
A Treatise on Perfumes and Incense.
Sung, *c.* +1115.
Hung Chhu 洪芻.

Hsiang Phu 香譜.
[= *Hsin Tsuan Hsiang Phu*
or *Honan Chhen shih Hsiang Phu.*]
A Treatise on Perfumes and Aromatic Substances [including incense and combust-
ion-clocks].
Sung, late +12th or +13th; may be as late
as +1330.
Chhen Ching 陳敬.

Hsiang Phu 香譜.
A Treatise on Incense and Perfumes.
Yuan, +1322.
Hsiung Phêng-Lai 熊朋來.

Hsiang Yao Chhao 香藥抄.
Memoir on Aromatic Plants and Incense.
Japan, *c.* +1163.
Kuan-Yu (Kanyu) 觀祐.MS. preserved at
the 滋賀石山寺 Temple. Facsim. re-
prod. in Suppl. to the Japanese Tripiṭaka,
vol. 11.

Hsieh Thien Chi 泄天機.
A Divulgation of the Machinery of Nature
(in the Human Body, permitting the
Formation of the Enchymoma).

Hsieh Thien Chi (cont.)
Chhing, *c.* +1795.
Li Ong (Ni-Wan shih) 李翁 (Mr Ni-Wan).
Written down in 1833 by Min Hsiao-Kên 閔小艮
In *Tao Tsang Hsü Pien* (*Chhu chi*), 4.

Hsien Lo Chi 仙樂集.
(Collected Poems) on the Happiness of the Holy Immortals.
Sung, late +12th cent.
Liu Chhu-Hsüan 劉處玄.
TT/1127.

Hsien-Yuan Huang Ti Shui Ching Yao Fa 軒轅黃帝水經藥法.
(Thirty-two) Medicinal Methods from the Aqueous (Solutions) Manual of Hsien-Yuan the Yellow Emperor.
Date uncertain.
Writer unknown.
TT/922.

Hsien-Yuan Pao Tsang Chhang Wei Lun 軒轅寶藏暢微論.
The Yellow Emperor's Expansive yet Detailed Discourse on the (Contents of the) Precious Treasury (of the Earth) [mineralogy and metallurgy].
Alternative title of *Pao Tsang Lun*, q.v.

Hsien-Yuan Pao Tsang Lun 軒轅寶藏論.
The Yellow Emperor's Discourse on the Contents of the Precious Treasury (of the Earth).
See *Pao Tsang Lun*.

Hsin Hsiu Pên Tshao 新修本草.
The New (lit. Newly Improved) Pharmacopoeia.
Thang, +659.
Ed. Su Ching (= Su Kung) 蘇敬 (蘇恭) and a commission of 22 collaborators under the direction first of Li Chi 李勣 & Yü Chih-Ning 于志寧, then of Chhangsun Wu-Chi 長孫無忌. This work was afterwards commonly but incorrectly known as *Thang Pên Tshao*. It was lost in China, apart from MS. fragments at Tunhuang, but copied by a Japanese in +731 and preserved in Japan though incompletely.

Hsin Lun 新論.
New Discussions.
H/Han, *c.* +10 to +20, presented +25.
Huan Than 桓譚.
Cf. Pokora (9).

Hsin Lun 新論.
New Discourses.
Liang, *c.* +530.
Liu Hsieh 劉勰.

Hsin Thang Shu 新唐書.
New History of the Thang Dynasty [+618 to +906].
Sung, +1061.
Ouyang Hsiu 歐陽修 & Sung Chhi 宋祁.

Cf. des Rotours (2), p. 56.
Partial trs. des Rotours (1, 2); Pfizmaier (66–74). For translations of passages see the index of Frankel (1).
Yin-Tê Index, no. 16.

Hsin Tsuan Hsiang Phu 新纂香譜.
See *Hsiang Phu* by Chhen Ching.

Hsin Wu Tai Shih 新五代史.
New History of the Five Dynasties [+907 to +959].
Sung, *c.* +1070.
Ouyang Hsiu 歐陽修.
For translations of passages see the index of Frankel (1).

Hsin Yü 新語.
New Discourses.
C/Han, *c.* −196.
Lu Chia 陸賈.
Tr. v. Gabain (1).

Hsing Li Ching I 性理精義.
Essential Ideas of the Hsing-Li (Neo-Confucian) School of Philosophers [a condensation of the *Hsing Li Ta Chhüan*, q.v.].
Chhing, +1715.
Li Kuang-Ti 李光地.

Hsing Li Ta Chhüan (*Shu*) 性理大全 (書).
Collected Works of (120) Philosophers of the Hsing-Li (Neo-Confucian) School [*Hsing* = human nature; *Li* = the principle of organisation in all Nature].
Ming, +1415.
Ed. Hu Kuang *et al.* 胡廣.

Hsing Ming Kuei Chih 性命圭旨.
A Pointer to the Meaning of (Human) Nature and the Life-Span [physiological alchemy; the *kuei* is a pun on the two kinds of *thu*, central earth where the enchymoma is formed].
Ascr. Sung, pr. Ming and Chhing, +1615, repr. +1670.
Attrib. Yin Chen Jen 尹眞人.
Written out by Kao Ti 高第.
Prefs. by Yü Yung-Ning *et al.* 余永寧.

Hsing Shih Hêng Yen 醒世恆言.
Stories to Awaken Men.
Ming, *c.* +1640.
Fêng Mêng-Lung 馮夢龍.

Hsiu Chen Chih Nan 修眞指南.
South-Pointer for the Regeneration of the Primary (Vitalities).
See *Hsi Yo Tou hsien-sêng Hsiu Chen Chih Nan*.

Hsiu Chen Li Yen Chhao Thu 修眞歷驗鈔圖.
[= *Chen Yuan Miao Tao Hsiu Tan Li Yen Chhao*.]
Transmitted Diagrams illustrating Tried and Tested (Methods of) Regenerating the Primary Vitalities [physiological alchemy].
Thang or Sung, before +1019.
No writer named but the version in *YCCC*, ch. 72, has Tung Chen Tzu (ps.) 洞眞子.
TT/149.

Hsiu Chen Nei Lien Pi Miao Chu Chüeh 修眞
內煉秘妙諸訣.
Collected Instructions on the Esoteric
Mysteries of Regenerating the Primary
(Vitalities) by Internal Transmutation.
Sung or pre-Sung.
Writer unknown.
Perhaps identical with *Hsiu Chen Pi
Chüeh* (q.v.); now extant only in
quotations.

Hsiu Chen Pi Chüeh 脩眞秘訣.
Esoteric Instructions on the Regeneration of
the Primary (Vitalities).
Sung or pre-Sung, before +1136.
Writer uncertain.
In *Lei Shuo*, ch. 49, pp. 5 *a* ff.

Hsiu Chen Pien Nan (*Tshan Chêng*) 修眞辯難
參證.
[*Chhi Yün Shan Wu Yuan Yzu Hsiu Chen
Pien Nan Tshan Chêng.*]
A Discussion of the Difficulties encountered
in the Regeneration of the Primary
(Vitalities) [physiological alchemy]; with
Supporting Evidence.
Chhing, +1798.
Liu I-Ming 劉一明 (Wu Yuan Tzu
悟元子).
Comm., Min I-Tê 閔一得 (*c.* 1830).
In *Tao Tsang Hsü Pien* (*Chhi chi*), 23.

Hsiu Chen Shih Shu 修眞十書.
A Collection of Ten Tractates and Treat-
ises on the Regeneration of the Primary
(Vitalities) [in fact, many more than
ten].
Sung, *c.* +1250.
Editor unknown.
TT/260.
Cf. Maspero (7), pp. 239, 357.

Hsiu Chen Thai Chi Hun Yuan Thu 修眞太
極混元圖.
Illustrated Treatise on the (Analogy of the)
Regeneration of the Primary (Vitalities)
(with the Cosmogony of) the Supreme
Pole and Primitive Chaos.
Sung, *c.* +1100.
Hsiao Tao-Tshun 蕭道存.
TT/146.

Hsiu Chen Thai Chi Hun Yuan Chih Hsüan Thu
修眞太極混元指玄圖.
Illustrated Treatise Expounding the Mystery
of the (Analogy of the) Regeneration of
the Primary (Vitalities) (with the Cos-
mogony of) the Supreme Pole and
Primitive Chaos.
Thang, *c.* +830.
Chin Chhüan Tzu 金全子.
TT/147.

Hsiu Chen Yen I 修眞演義.
A Popular Exposition of (the Methods of)
Regenerating the Primary (Vitalities)
[Taoist sexual techniques].
Ming, *c.* +1560.

Têng Hsi-Hsien 鄧希賢 (*Tzu Chin
Kuang Yao Ta Hsien* 紫金光耀大仙.
See van Gulik (3, 8).

Hsiu Hsien Pien Huo Lun 修仙辨惑論.
Resolution of Doubts concerning the
Restoration to Immortality.
Sung, *c.* +1220.
Ko Chhang-Kêng 葛長庚
(Pai Yü-Chhan 白玉蟾).
In *TSCC, Shen i tien*, ch. 300, *i wên*, pp.
11 *a* ff.

Hsiu Lien Ta Tan Yao Chih 修錬大丹要旨.
Essential Instructions for the Preparation of
the Great Elixir [with illustrations of
alchemical apparatus].
Probably Sung or later.
Writer unknown.
TT/905.

Hsiu Tan Miao Yung Chih Li Lun 修丹妙用
至理論.
A Discussion of the Marvellous Functions
and Perfect Principles of the Practice of
the Enchymoma.
Late Sung or later.
Writer unknown.
TT/231.
Refers to the Sung adept Hai-Chhan hsien-
sêng 海蟾先生 (Liu Tshao 劉操).

Hsü Chen-Chün Pa-shih-wu Hua Lu 許眞君
八十五化錄.
Record of the Transfiguration of the Adept
Hsü (Hsün) at the Age of Eighty-five.
Chin, +4th cent.
Shih Tshên 施岑.
TT/445.

Hsü Chen-Chün Shih Han Chi 許眞君石函記.
The Adept Hsü (Sun's) Treatise, found in a
Stone Coffer.
Ascr. Chin, +4th cent., perhaps *c.* +370.
Attrib. Hsü Hsün 許遜.
TT/944.
Cf. Davis & Chao Yün-Tshung (6).

Hsü Hsien Chuan 續仙傳.
Further Biographies of the Immortals.
Wu Tai (H/Chou), between +923 and
+936.
Shen Fên 沈汾.
In *YCCC*, ch. 113.

Hsü Ku Chai Chi Suan Fa 續古摘奇算法.
Choice Mathematical Remains Collected to
Preserve the Achievements of Old [magic
squares and other computational examples].
Sung, +1275.
Yang Hui 楊輝.
(In *Yang Hui Suan Fa.*)

Hsü Kuang-Chhi Shou Chi 徐光啓手跡.
Manuscript Remains of Hsü Kuang-Chhi
[facsimile reproductions].
Shanghai, 1962.

Hsü Ming Fang 續命方.
Precepts for Lengthening the Life-span.
Alternative title of *Hui Ming Ching* (q.v.).

Hsü Po Wu Chih 續博物志.
Supplement to the *Record of the Investigation of Things* (cf. *Po Wu Chih*).
Sung, mid +12th century.
Li Shih 李石.

Hsü Shen Hsien Chuan 續神仙傳.
Supplementary Lives of the Hsien (cf. *Shen Hsien Chuan*).
Thang.
Shen Fên 沈汾.

Hsü Shih Shih 續事始.
Supplement to the *Beginnings of All Affairs* (cf. *Shih Shih*).
H/Shu, *c.* +960.
Ma Chien 馬鑑.

Hsü Yen-Chou Shih Hua 許彥周詩話.
Hsü Yen-Chou's Talks on Poetry.
Sung, early +12th, prob. *c.* +1111.
Hsü Yen-Chou 許彥周.

Hsüan Chieh Lu 懸解錄.
See *Hsüan Chieh Lu* 玄解錄.

Hsüan Chieh Lu 玄解錄.
The Mysterious Antidotarium [warnings against elixir poisoning, and remedies for it].
Thang, anonymous preface of +855, prob. first pr. between +847 and +850.
Writer unknown, perhaps Hokan Chi 紇干臮.
The first printed book in any civilisation on a scientific subject.
TT/921, and in *YCCC*, ch. 64, pp. 5*a* ff.

Hsüan Fêng Chhing Hui Lu 玄風慶會錄.
Record of the Auspicious Meeting of the Mysterious Winds [answers given by Chhiu Chhu-Chi (Chhang-Chhun Chen Jen) to Chingiz Khan at their interviews at Samarqand in +1222].
Sung, +1225.
Chhiu Chhu-Chi 邱處機.
TT/173.

Hsüan-Ho Po Ku Thu Lu 宣和博古圖錄.
[= *Po Ku Thu Lu.*]
Hsüan-Ho reign-period Illustrated Record of Ancient Objects [catalogue of the archaeological museum of the emperor Hui Tsung].
Sung, +1111 to +1125.
Wang Fu 王黼 or 戩 *et al.*

Hsüan Kuai Hsü Lu 玄怪續錄.
The *Record of Things Dark and Strange*, continued.
Thang.
Li Fu-Yen 李復言.

Hsüan Mên Mo Chüeh Nei Chao Thu 玄門脈訣內照圖.
[= *Hua Tho Nei Chao Thu.*]
Illustrations of Visceral Anatomy, for the Taoist *Sphygmological Instructions*.
Sung, +1095, repr. +1273 by Sun Huan 孫煥 with the inclusion of Yang Chieh's illustrations.
Attrib. Hua Tho 華佗.
First pub. Shen Chu 沈銖.
Cf. Ma Chi-Hsing (2).

Hsüan Ming Fên Chuan 玄明粉傳.
On the 'Mysterious Bright Powder' (purified sodium sulphate, Glauber's salt).
Thang, *c.* +730.
Liu Hsüan-Chen 劉玄眞.

Hsüan Nü Ching 玄女經.
Canon of the Mysterious Girl [or, the Dark Girl].
Han.
Writer unknown.
Only as fragment in *Shuang Mei Ching An Tshung Shu*, now conflated with *Su Nü Ching*, q.v.
Partial trs., van Gulik (3, 8).

Hsüan Phin Lu 玄品錄.
Record of the (Different) Grades of Immortals.
Yuan.
Chang Thien-Yü 張天雨.
TT/773.
Cf. Chhen Kuo-Fu (1), 1st ed., p. 260.

Hsüan Shih Chih 宣室志.
Records of Hsüan Shih.
Thang, *c.* +860.
Chang Tu 張讀.

Hsüan Shuang Chang Shang Lu 玄霜掌上錄.
Mysterious Frost on the Palm of the Hand; or, Handy Record of the Mysterious Frost [preparation of lead acetate].
Date unknown.
Writer unknown.
TT/938.

I Chen Thang Ching Yen Fang 頤眞堂經驗方.
Tried and Tested Prescriptions of the True-Centenarian Hall (a surgery or pharmacy).
Ming, prob. +15th, *c.* +1450.
Yang shih 楊氏.

I Chi Khao 醫籍考.
Comprehensive Annotated Bibliography of Chinese Medical Literature.
See Taki Mototane (1).

I Chai Ta Fa 醫家大法.
See *I Yin Thang I Chung Ching Kuang Wei Ta Fa.*

I Chien Chih 夷堅志.
Strange Stories fom I-Chien.
Sung, *c.* +1185.
Hung Mai 洪邁.

I Chin Ching 易筋經.
Manual of Exercising the Muscles and Tendons [Buddhist].
Ascr. N/Wei.
Chhing, perhaps +17th.
Attrib. Ta-Mo (Bodhidharma) 達摩
Author unknown.
Reproduced in Wang Tsu-Yuan (1).

I Ching 易經.
The Classic of Changes [Book of Changes].
Chou with C/Han additions.
Compilers unknown.
See Li Ching-Chih (*1, 2*); Wu Shih-Chhang (*1*).
Tr. R. Wilhelm (2); Legge (9); de Harlez (1).
Yin-Tê Index, no. (suppl.) 10.

I Hsin Fang (Ishinhō) 醫心方.
The Heart of Medicine [partly a collection of ancient Chinese and Japanese books].
Japan, +982 (not printed till 1854).
Tamba no Yasuyori 丹波康頼.

I Hsüeh Ju Mên 醫學入門.
Janua Medicinae [a general system of medicine].
Ming, +1575.
Li Chhan 李梴.

I Hsüeh Yuan Liu Lun 醫學源流論.
On the Origins and Progress of Medical Science.
Chhing, +1757.
Hsü Ta-Chhun 徐大椿.
(In *Hsü Ling-Thai I Shu Chhüan Chi*.)

Mên Pi Chih 醫門秘旨.
Confidential Guide to Medicine.
Ming, +1578.
Chang Ssu-Wei 張四維.

I Shan Tsa Tsuan 義山雜纂.
Collected Miscellany of (Li) I-Shan [Li Shang-Yin, epigrams].
Thang, *c.* +850.
Li Shang-Yin 李商隱.
Tr. Bonmarchand (1).

I Shih 逸史.
Leisurely Histories.
Thang.
Lu Shih 盧氏.

I Su Chi 夷俗記.
Records of Barbarian Customs.
Alternative title of *Pei Lu Fêng Su*, q.v.

I Thu Ming Pien 易圖明辨.
Clarification of the Diagrams in the (*Book of*) *Changes* [historical analysis].
Chhing, +1706.
Hu Wei 胡渭.

I Wei Chhien Tso Tu 易緯乾鑿度.
Apocryphal Treatise on the (*Book of*) *Changes*; a Penetration of the Regularities of Chhien (the first *kua*).
C/Han, −1st or +1st century.
Writer unknown.

I Wei Ho Thu Shu 易緯河圖數.
Apocryphal Treatise on the (*Book of*) *Changes*; the Numbers of the Ho Thu (Diagram).
H/Han.
Writer unknown.

I Yin Thang I Chung Ching Kuang Wei Ta Fa 伊尹湯液仲景廣爲大法.
[= *I Chia Ta Fa* or *Kuang Wei Ta Fa*.]
The Great Tradition (of Internal Medicine) going back to I Yin (legendary minister) and his Pharmacal Potions, and to (Chang) Chung-Ching (famous Han physician).
Yuan, +1294.
Wang Hao-Ku 王好古.
ICK, p. 863.

Ishinhō
See *I Hsin Fang*.

Jih Chih Lu 日知錄.
Daily Additions to Knowledge.
Chhing, +1673.
Ku Yen-Wu 顧炎武.

Jih Hua Chu Chia Pên Tshao 日華諸家本草.
The Sun-Rays Master's Pharmaceutical Natural History, collected from Many Authorities.
Wu Tai and Sung, *c.* +972.
Often ascribed by later writers to the Thang, but the correct dating was recognised by Thao Tsung-I in his *Cho Kêng Lu* (+1366) ch. 24, p. 17*b*.
Ta Ming 大明.
(Jih Hua Tzu 日華子 the Sun-Rays Master.)
(Perhaps Thien Ta-Ming 田大明).

Jih Yüeh Hsüan Shu Lun 日月玄樞論.
Discourse on the Mysterious Axis of the Sun and Moon [i.e. Yang and Yin in natural phenomena; the earliest interpretation (or recognition) of the *Chou I Tshan Thung Chhi* (q.v.) as a physiological rather than (or, as well as) a proto-chemical text].
Thang, *c.* +740.
Liu Chih-Ku 劉知古.
Now extant only as quotations in the *Tao Shu* (q.v.), though at one time contained in the *Tao Tsang* separately.

Ju Yao Ching 入藥鏡.
Mirror of the All-Penetrating Medicine (the enchymoma), [rhyming verses].
Wu Tai, *c.* +940.
Tshui Hsi-Fan 崔希範.
TT/132, and in *TTCY* (*hsü chi*, 5).
With commentaries by Wang Tao-Yuan 王道淵 (Yuan); Li Phan-Lung 李攀龍 (Ming) & Phêng Hao-Ku 彭好古 (Ming).
Also in *Hsiu Chen Shih Shu* (*TT*/260), ch. 13, pp. 1*a* ff. with commentary by Hsiao Thing-Chih 蕭廷芝 (Ming).
Also in *Tao Hai Chin Liang*, pp. 35*a* ff., with comm. by Fu Chin-Chhüan 傅金銓 (Chhing).
See also *Thien Yuan Ju Yao Ching*.
Cf. van Gulik (8), pp. 224 ff.

Kan Chhi Shih-liu Chuan Chin Tan 感氣十六轉金丹.
The Sixteen-fold Cyclically Transformed Gold Elixir prepared by the 'Responding

Kan Chhi Shih-liu Chuan Chin Tan (cont.)
to the Chhi' Method [with illustrations of alchemical apparatus].
Sung.
Writer unknown.
TT/904.

Kan Ying Ching 感應經.
On Stimulus and Response (the Resonance of Phenomena in Nature).
Thang, *c.* +640.
Li-Shun-Fêng 李淳風.
See Ho & Needham (2).

Kan Ying Lei Tshung Chih 感應類從志.
Record of the Mutual Resonances of Things according to their Categories.
Chin, *c.* +295.
Chang Hua 張華.
See Ho & Needham (2).

Kao Shih Chuan 高士傳.
Lives of Men of Lofty Attainments.
Chin, *c.* +275.
Huangfu Mi 皇甫謐.

Kêng Hsin Yü Tshê 庚辛玉冊.
Precious Secrets of the Realm of Kêng and Hsin (i.e. all things connected with metals and minerals, symbolised by these two cyclical characters) [on alchemy and pharmaceutics. Kêng-Hsin is also an alchemical synonym for gold].
Ming, +1421.
Chu Chhüan 朱權, (Ning Hsien Wang 寧獻王, prince of the Ming).
Extant only in quotations.

Kêng Tao Chi 庚道集.
Collection of Procedures of the Golden Art (Alchemy).
Sung or Yuan, date unknown but after +1144
Writers unknown.
Compiler, Mêng Hsien chü shih 蒙軒居士.
TT/946.

Khai-Pao Hsin Hsiang-Ting Pên Tshao 開寶新詳定本草.
New and More Detailed Pharmacopoeia of the Khai-Pao reign-period.
Sung, +973.
Liu Han 劉翰, Ma Chih 馬志, and 7 other naturalists, under the direction of Lu To-Hsün 盧多遜.

Khai-Pao Pên Tshao 開寶本草.
See *Khai-Pao Hsin Hsiang-Ting Pên Tshao*.

Khun Yü Ko Chih 坤輿格致.
Investigation of the Earth [Western mining methods based on Agricola's *De Re Metallica*].
Ming, +1639 to 1640, perhaps never printed.
Têng Yü-Han (Johann Schreck) 鄧玉函 & (or) Thang Jo-Wang 湯若望 (John Adam Schall von Bell).

Khung Chi Ko Chih 空際格致.
A Treatise on the Material Composition of the Universe [the Aristotelian Four Elements, etc.].

Ming, +1633.
Kao I-Chih (Alfonso Vagnoni) 高一志.
Bernard-Maître (18), no. 227.

Khung shih Tsa Shuo 孔氏雜說.
Mr Khung's Miscellany.
Sung, *c.* +1082.
Khung Phing-Chung 孔平仲.

Ko Chih Ching Yuan 格致鏡原.
Mirror of Scientific and Technological Origins.
Chhing, +1735.
Chhen Yuan-Lung 陳元龍.

Ko Chih Tshao 格致草.
Scientific Sketches [astronomy and cosmology; part of *Han Yü Thung*, q.v.].
Ming, +1620, pr. +1648.
Hsiung Ming-Yü 熊明遇.

Ko Hsien Ong Chou Hou Pei Chi Fang 葛仙翁肘後備急方.
The Elder-Immortal Ko (Hung's) Handbook of Medicines for Emergencies.
Alt. title of *Chou Hou Pei Chi Fang* (q.v.).
TT/1287.

Ko Hung Chen Chung Shu 葛洪枕中書.
Alt. title of *Chen Chung Chi* (q.v.).

Ko Ku Yao Lun 格古要論.
Handbook of Archaeology, Art and Antiquarianism.
Ming, +1387, enlarged and reissued +1459.
Tshao Chao 曹昭.

Ko Wu Tshu Than 格物麤談.
Simple Discourses on the Investigation of Things.
Sung, *c.* +980.
Attrib. wrongly to Su Tung-Pho 蘇東坡.
Actual writer (Lu) Tsan-Ning (錄)贊寧 (Tung-Pho hsien-sêng). With later additions, some concerning Su Tung-Pho.

Konjaku Monogatari 今昔物語.
Tales of Today and Long Ago (in three collections: Indian, 187 stories and traditions, Chinese, 180, and Japanese, 736).
Japan (Heian), +1107.
Compilers unknown.
Cf. Anon. (103), pp. 97 ff.

Konjaku Monogatarishū 今昔物語集.
See *Konjaku Monogatari*.

Ku Chin I Thung (Ta Chhüan) 古今醫統(大全).
Complete System of Medical Practice, New and Old.
Ming, +1556.
Hsü Chhun-Fu 徐春甫.

Ku Thung Thu Lu 皷銅圖錄.
Illustrated Account of the (Mining), Smelting and Refining of Copper (and other Non-Ferrous Metals).
See Masuda Tsuna (1).

Ku Wei Shu 古微書.
Old Mysterious Books [a collection of the apocryphal Chhan-Wei treatises].
Date uncertain, in part C/Han.
Ed. Sun Chio 孫瑴 (Ming).

Ku Wên Lung Hu Ching Chu Su 古文龍虎經註疏 and *Ku Wên Lung Hu Shang Ching Chu* 古文龍虎上經註.
See *Lung Hu Shang Ching Chu*.

Ku Wên Tshan Thung Chhi Chi Chieh 古文參同契集解.
See *Ku Wên Chou I Tshan Thung Chhi Chu*.

Ku Wên Tshan Thung Chhi Chien Chu Chi Chieh 古文參同契箋註集解.
See *Ku Wên Chou I Tshan Thung Chhi Chu*.

Ku Wên Chou I Tshan Thung Chhi Chu 古文周易參同契註.
Commentary on the Ancient Script Version of the *Kinship of the Three*.
Chhing, +1732.
Ed. and comm. Yuan Jen-Lin 袁仁林.
See Vol. 5, pt. 3.

Ku Wên Tshan Thung Chhi San Hsiang Lei Chi Chieh 古文參同契三相類集解.
See *Ku Wên Chou I Tshan Thung Chhi Chu*.

Kuan Khuei Pien 管窺編.
An Optick Glass (for the Enchymoma).
See Min I-Tê (1).

Kuan Yin Tzu 關尹子.
[= *Wên Shih Chen Ching*.]
The Book of Master Kuan Yin.
Thang, +742 (may be Later Thang or Wu Tai). A work with this title existed in the Han, but the text is lost.
Prob. Thien Thung-Hsiu 田同秀.

Kuang Chhêng Chi 廣成集.
The Kuang-chhêng Collection [Taoist writings of every kind; a florilegium].
Thang, late +9th; or early Wu Tai, before +933.
Tu Kuang-Thing 杜光庭.
TT/611.

Kuang Wei Ta Fa 廣爲大法.
See *I Yin Thang I Chung Ching Kuang Wei Ta Fa*.

Kuang Ya 廣雅.
Enlargement of the *Erh Ya*; *Literary Expositor* [dictionary].
San Kuo (Wei) +230.
Chang I 張揖.

Kuang Yün 廣韻.
Enlargement of the *Chhieh Yün*; *Dictionary of the Sounds of Characters*.
Sung.
(A completion by later Thang and Sung scholars, given its present name in +1011.)
Lu Fa-Yen *et al.* 陸法言.

Kuei Chung Chih Nan 規中指南.
A Compass for the Internal Compasses; or, Orientations concerning the Rules and Measures of the Inner (World) [i.e. the preparation of the enchymoma in the microcosm of man's body].
Sung or Yuan, +13th or +14th.
Chhen Chhung-Su 陳沖素 (Hsü Pai Tzu 盧白子).
TT/240, and in *TTCY* (*shang mao chi*, 5).

Kungyang Chuan 公羊傳.
Master Kungyang's Tradition (or Commentary) on the *Spring and Autumn Annals*.
Chou (with Chhin and Han additions), late −3rd and early −2nd centuries.
Attrib. Kungyang Kao 公羊高 but more probably Kungyang Shou 公羊壽.
See Wu Khang (1); van der Loon (1).

Kuo Shih Pu 國史補.
Emendations to the National Histories.
Thang, *c.* +820.
Li Chao 李肇.

Kuo Yü 國語.
Discourses of the (ancient feudal) States.
Late Chou, Chhin and C/Han, containing much material from ancient written records.
Writers unknown.

Lao Hsüeh An Pi Chi 老學庵筆記.
Notes from the Hall of Learned Old Age.
Sung, *c.* +1190.
Lu Yu 陸游.

Lao Tzu Chung Ching 老子中經.
The Median Canon of Lao Tzu [on physiological micro-cosmography].
Writer unknown.
Pre-Thang.
In *YCCC*, ch. 18.

Lao Tzu Shuo Wu Chhu Ching 老子說五厨經.
Canon of the Five Kitchens [the five viscera] Revealed by Lao Tzu [respiratory techniques].
Thang or pre-Thang.
Writer unknown.
In *YCCC*, ch. 61, pp. 5*b* ff.

Lei Chen Chin Tan 雷震金丹.
Lei Chen's Book of the Metallous Enchymoma.
Ming, after +1420.
Lei Chen (ps. ?) 雷震.
In *Wai Chin Tan*, ch. 5 (*CTPS*, *pên* 10).

Lei Chen Tan Ching 雷震丹經.
Alternative title of *Lei Chen Chin Tan* (q.v.).

Lei Chêng Phu Chi Pên Shih Fang 類證普濟本事方.
Classified Fundamental Prescriptions of Universal Benefit.
Sung, +1253.
Attrib. Hsü Shu-Wei 許叔微 (*fl.* +1132)

Lei Ching Fu I 類經附翼.
Supplement to the Classics Classified; (the Institutes of Medicine).
Ming, +1624.
Chang Chieh-Pin 張介賓.

Lei Kung Phao Chih 雷公炮製.
(Handbook based on the)*Venerable Master Lei's* (*Treatise on*) *the Preparation* (*of Drugs*).
L/Sung, *c.* +470.

Lei Kung Phao Chi (*cont.*)
Lei Hsiao 雷斅.
Ed. Chang Kuang-Tou 張光斗 (Chhing), 1871.

Lei Kung Phao Chih Lun 雷公炮炙論.
The Venerable Master Lei's Treatise on the Decoction and Preparation (of Drugs).
L/Sung, *c.* +470.
Lei Hsiao 雷斅.
Preserved only in quotations in *Chêng Lei Pên Tshao* and elsewhere, and reconstituted by Chang Chi 張驥.
LPC, p. 116.

Lei Kung Phao Chih Yao Hsing (*Fu*) *Chieh* 雷公炮製藥性(賦)解.
(Essays and) Studies on the *Venerable Master Lei's* (*Treatise on*) *the Natures of Drugs and their Preparation*.
First four chapters J/Chin, *c.* +1220.
Li Kao 李杲.
Last six chapters Chhing, *c.* 1650.
Li Chung-Tzu 李中梓.
(Contains many quotations from earlier Lei Kung books, +5th century onwards.)

Lei Kung Yao Tui 雷公藥對.
Answers of the Venerable Master Lei (to Questions) concerning Drugs.
Perhaps L/Sung, at any rate before N/Chhi.
Attrib. Lei Hsiao 雷斅.
Later attrib. a legendary minister of Huang Ti.
Comm. by Hsü Chih-Tshai 徐之才, N/Chhi +565.
Now extant only in quotations.

Lei Shuo 類說.
A Classified Commonplace-Book [a great florilegium of excerpts from Sung and pre-Sung books, many of which are otherwise lost].
Sung, +1136.
Ed. Tsêng Tshao 曾慥.

Li Chi 禮記.
[= *Hsiao Tai Li Chi.*]
Record of Rites [compiled by Tai the Younger].
(Cf. *Ta Tai Li Chi.*)
Ascr. C/Han, *c.* −70/−50, but really H/Han, between +80 and +105, though the earliest pieces included may date from the time of the *Analects* (*c.* −465 to −450).
Attrib. ed. Tai Shêng 戴聖.
Actual ed. Tshao Pao 曹褒.
Trs. Legge (7); Couvreur (3); R. Wilhelm (6).
Yin-Tê Index, no. 27.

Li Hai Chi 蠡海集.
The Beetle and the Sea [title taken from the proverb that the beetle's eye view cannot encompass the wide sea—a biological book].
Ming, late +14th century.
Wang Khuei 王逵.

Li Sao 離騷.
Elegy on Encountering Sorrow [ode].
Chou (Chhu), *c.* −295, perhaps just before −300. Some scholars place it as late as −269.
Chhü Yuan 屈原.
Tr. Hawkes (1).

Li Shih Chen Hsien Thi Tao Thung Chien 歷世眞仙體道通鑑.
Comprehensive Mirror of the Embodiment of the Tao by Adepts and Immortals throughout History.
Prob. Yuan.
Chao Tao-I 趙道一.
TT/293.

Li Tai Ming I Mêng Chhiu 歷代名醫蒙求.
Brief Lives of the Famous Physicians in All Ages.
Sung, +1040.
Chou Shou-Chung 周守忠.

(*Li Tai*) *Shen Hsien* (*Thung*) *Chien* (歷代)神仙(通)鑑.
(Cf. *Shen Hsien Thung Chien.*)
General Survey of the Lives of the Holy Immortals (in all Ages).
Chhing, +1712.
Hsü Tao 徐道 (assisted by Li Li 李理) & Chhêng Yü-Chhi 程毓奇 (assisted by Wang Thai-Su 王太素).

Li Wei Tou Wei I 禮緯斗威儀.
Apocryphal Treatise on the *Record of Rites*; System of the Majesty of the Ladle [the Great Bear].
C/Han, −1st or later.
Writer unknown.

Li Wên-Jao Chi 李文饒集.
Collected Literary Works of Li Tê-Yü (Wên-Jao), (+787 to +849).
Thang, *c.* +855.
Li Tê-Yü 李德裕.

Liang Chhiu Tzu (*Nei* or *Wai*) 梁丘子.
See *Huang Thing Nei Ching* (*Yü*) *Ching Chu* and *Huang Thing Wai Ching* (*Yü*) *Ching Chu.*

Liang Ssu Kung Chi 梁四公記.
Tales of the Four Lords of Liang.
Thang, *c.* +695.
Chang Yüeh 張說.

Liao Yang Tien Wên Ta Pien 寥陽殿問答編.
[= *Yin Chen Jen Liao Yang Tien Wên Ta Pien.*]
Questions and Answers in the (Eastern Cloister of the) Liao-yang Hall (of the White Clouds Temple at Chhing-chhêng Shan in Szechuan) [on physiological alchemy, *nei tan*].
Ming or Chhing.
Attrib. Yin Chen Jen 尹眞人 (Phêng-Thou 蓬頭).
Ed. Min I-Tê 閔一得, *c.* 1830.
In *Tao Tsang Hsü Pien* (*Chhu chi*), 3, from a MS. preserved at the Blue Goat Temple 青羊宮 (Chhêngtu).

Lieh Hsien Chhüan Chuan 列仙全傳.
　　Complete Collection of the Biographies of
　　　the Immortals.
　　Ming, c. +1580.
　　Wang Shih-Chên 王世貞.
　　Collated and corrected by Wang Yün-
　　　Phêng 汪雲鵬.
Lieh Hsien Chuan 列仙傳.
　　Lives of Famous Immortals (cf. *Shen Hsien
　　　Chuan*).
　　Chin, +3rd or +4th century, though
　　　certain parts date from about −35 and
　　　shortly after +167.
　　Attrib. Liu Hsiang 劉向.
　　Tr. Kaltenmark (2).
Lin Chiang Hsien 臨江仙.
　　The Immortal of Lin-chiang.
　　Sung, +1151.
　　Tsêng Tshao 曾慥.
　　In *Hsiu Chen Shih Shu* (*TT*/260), ch. 23,
　　　pp. 1 *a* ff.
*Ling-Pao Chiu Yu Chhang Yeh Chhi Shih Tu
　　Wang Hsüan Chang* 靈寶九幽長夜起
　　尸度亡玄章.
　　Mysterious Cantrap for the Resurrection of
　　　the Body and Salvation from Nothingness
　　　during the Long Night in the Nine Under-
　　　worlds; a Ling-Pao Scripture.
　　Date uncertain.
　　Writer unknown.
　　TT/605.
Ling-Pao Chung Chen Tan Chüeh 靈寶衆眞丹
　　訣.
　　Supplementary Elixir Instructions of the
　　　Company of the Realised Immortals, a
　　　Ling-Pao Scripture.
　　Sung, after +1101.
　　Writer unknown.
　　TT/416.
　　On the term Ling-Pao see Kaltenmark
　　　(4).
Ling-Pao Wu Fu (*Hsü*) 靈寶五符(序).
　　See *Thai-Shang Ling-Pao Wu Fu* (*Ching*).
*Ling-Pao Wu Liang Tu Jen Shang Phin Miao
　　Ching* 靈寶無量度人上品妙
　　經.
　　Wonderful Immeasurable Highly Exalted
　　　Manual of Salvation; a Ling-Pao Scripture.
　　Liu Chhao, perhaps late +5th, probably
　　　finalised in Thang, +7th.
　　Writers unknown.
　　TT/1.
Ling Pi Tan Yao Chien 靈祕丹藥牋.
　　On Numinous and Secret Elixirs and Medi-
　　　cines [the seventh part (chs. 16–18) of
　　　Tsun Shêng Pa Chien, q.v.].
　　Ming, +1591.
　　Kao Lien 高濂.
Ling Piao Lu I 嶺表錄異.
　　Strange Things Noted in the South.
　　Thang, c. +890.
　　Liu Hsün 劉恂.

Ling Sha Ta Tan Pi Chüeh 靈砂大丹祕訣.
　　Secret Doctrine of the Numinous Cinnabar
　　　and the Great Elixir.
　　Sung, after +1101, when the text was
　　　received by Chang Shih-Chung 張侍中.
　　Writer unknown, but edited by a Chhan
　　　abbot Kuei-Yen Chhan-shih 鬼眼
　　　禪師.
　　TT/890.
Ling Shu Ching
　　See *Huang Ti Nei Ching, Ling Shu*.
Ling Wai Tai Ta 嶺外代答.
　　Information on What is Beyond the Passes
　　　(lit. a book in lieu of individual replies to
　　　questions from friends).
　　Sung, +1178.
　　Chou Chhü-Fei 周去非.
Liu Shu Ching Yün 六書精蘊.
　　Collected Essentials of the Six Scripts.
　　Ming, c. +1530.
　　Wei Hsiao 魏校.
Liu Tzu Hsin Lun 劉子新論.
　　See *Hsin Lun*.
Lo-Fou Shan Chih 羅浮山志.
　　History and Topography of the Lo-fou
　　　Mountains (north of Canton).
　　Chhing, +1716 (but based on older
　　　histories).
　　Thao Ching-I 陶敬益.
Lu Hsing Ching 顱囟經.
　　A Tractate on the Fontanelles of the Skull
　　　[anatomical-medical].
　　Late Thang or early Sung, +9th or
　　　+10th.
　　Writer unknown.
Lu Huo Chien Chieh Lu 爐火監戒錄.
　　Warnings against Inadvisable Practices in
　　　the Work of the Stove [alchemical].
　　Sung, c. +1285.
　　Yü Yen 俞琰.
Lu Huo Pên Tshao 爐火本草.
　　Spagyrical Natural History.
　　Possible alternative title of *Wai Tan Pên
　　　Tshao* (q.v.).
Lü Tsu Chhin Yuan Chhun 呂祖沁園春.
　　The (Taoist) Patriarch Lü (Yen's) 'Spring
　　　in the Prince's Gardens' [a brief epi-
　　　grammatic text on physiological alchemy]
　　Thang, +8th (if genuine).
　　Attrib. Lü Yen 呂喦.
　　TT/133.
　　Comm. by Fu Chin-Chhüan 傅金銓
　　　(c. 1822).
　　In *Tao Hai Chin Liang*, p. 45 *a*, and appen-
　　　ded to *Shih Chin Shih* (*Wu Chen Ssu
　　　Chu Phien* ed.).
Lü Tsu Chhuan Shou Tsung Chih 呂祖傳授宗
　　旨.
　　Principles (of Macrobiotics) Transmitted
　　　and Handed Down by the (Taoist)
　　　Patriarch Lü (Yen, Tung-Pin).
　　Orig. title of *Chin Hua Tsung Chih* (q.v.).

Lü Tsu Shih Hsien-Thien Hsü Wu Thai-I Chin Hua Tsung Chih 呂祖師先天虛無太一金華宗旨.

Principles of the (Inner) Radiance of the Metallous (Enchymoma) (explained in terms of the) Undifferentiated Universe, and of all the All-Embracing Potentiality of the Endowment of Primary Vitality, taught by the (Taoist) Patriarch Lü (Yen, Tung-Pin).

Alternative name for *Chin Hua Tsung Chih* (q.v.), but with considerable textual divergences, especially in ch. 1.

Ming and Chhing.

Writers unknown.

Attrib. Lü Yen 呂嵒 (Lü Tung-Pin) and his school, late +8th.

Ed. and comm. Chiang Yuan-Thing 蔣元庭 and Min I-Tê 閔一得, *c.* 1830.

In *TTCY* and in *Tao Tsang Hsü Pien* (*Chhu chi*), 1.

Lü Tsu Shih San Ni I Shih Shuo Shu 呂祖師三尼醫世說述.

A Record of the Lecture by the (Taoist) Patriarch Lü (Yen, Tung-Pin) on the Healing of Humanity by the Three Ni Doctrines (Taoism, Confucianism and Buddhism) [physiological alchemy in mutationist terms].

Chhing, +1664.

Attrib. Lü Yen 呂嵒 (+8th cent.).

Pref. by Thao Thai-Ting 陶太定.

Followed by an appendix by Min I-Tê 閔一得.

In *Tao Tsang Hsü Pien* (*Chhu chi*), 10, 11.

Lun Hêng 論衡.

Discourses Weighed in the Balance.

H/Han, +82 or +83.

Wang Chhung 王充.

Tr. Forke (4); cf. Leslie (3).

Chung-Fa Index, no. 1.

Lung Hu Chhien Hung Shuo 龍虎鉛汞說.

A Discourse on the Dragon and Tiger, (Physiological) Lead and Mercury, (addressed to his younger brother Su Tzu-Yu).

Sung, *c.* +1100.

Su Tung-Pho 蘇東坡.

In *TSCC, Shen i tien*, ch. 300, *i wên*, pp. 6*b* ff.

Lung Hu Huan Tan Chüeh 龍虎還丹訣.

Explanation of the Dragon-and-Tiger Cyclically Transformed Elixir.

Wu Tai, Sung, or later.

Chin Ling Tzu 金陵子.

TT/902.

Lung Hu Huan Tan Chüeh Sung 龍虎還丹訣頌.

A Eulogy of the Instructions for (preparing) the Regenerative Enchymoma of the Dragon and the Tiger (Yang and Yin), [physiological alchemy].

Sung, *c.* +985.

Lin Ta-Ku 林大古
(Ku Shen Tzu 谷神子).

TT/1068.

Lung Hu Shang Ching Chu 龍虎上經註.

Commentary on the *Exalted Dragon-and-Tiger Manual*.

Sung.

Wang Tao 王道.

TT/988, 989.

Cf. Davis & Chao Yün-Tshung (6).

Lung Hu Ta Tan Shih 龍虎大丹詩.

Song of the Great Dragon-and-Tiger Enchymoma.

See *Chih Chen Tzu Lung Hu Ta Tan Shih*.

Lung-Shu Phu-Sa Chuan 龍樹菩薩傳.

Biography of the Bodhisattva Nāgārjuna (+2nd-century Buddhist patriarch).

Prob. Sui or Thang.

Writer unknown.

TW/2047.

Man-Anpō 萬安方.

A Myriad Healing Prescriptions.

Japan, +1315.

Kajiwara Shozen 梶原性全.

Manyōshū 萬葉集.

Anthology of a Myriad Leaves.

Japan (Nara), +759.

Ed. Tachibana no Moroe 橘諸兄.
or Ōtomo no Yakamochi 大伴家持.

Cf. Anon. (103), pp. 14 ff.

Mao Shan Hsien Chê Fu Na Chhi Chüeh 茅山賢者服內氣訣.

Oral Instructions of the Adepts of Mao Shan for Absorbing the Chhi [Taoist breathing exercises for longevity and immortality].

Thang or Sung.

Writer unknown.

In *YCCC*, ch. 58, pp. 3*b* ff.

Cf. Maspero (7), p. 205.

Mao Thing Kho Hua 茅亭客話.

Discourses with Guests in the Thatched Pavilion.

Sung, before +1136.

Huang Hsiu-Fu 黃休復.

Mei-Chhi Shih Chu 梅溪詩注.

(Wang) Mei-Chhi's Commentaries on Poetry.

Short title for *Tung-Pho Shih Chi Chu* (q.v.).

Mêng Chhi Pi Than 夢溪筆談.

Dream Pool Essays.

Sung, +1086; last supplement dated +1091.

Shen Kua 沈括.

Ed. Hu Tao-Ching (*1*); cf. Holzman (*1*).

Miao Chieh Lu 妙解錄.

See *Yen Mên Kung Miao Chieh Lu*.

Miao Fa Lien Hua Ching 妙法蓮華經.

Sūtra on the Lotus of the Wonderful Law

Miao Fa Lien Hua Ching (cont.)
>India.
>Tr. Chin, betw. +397 and +400 by Ku-
>mārajīva (Chiu-Mo-Lo-Shih　鳩摩羅什).
>N/134; TW/262.

Ming I Pieh Lu　名醫別錄.
>Informal (or Additional) Records of
>Famous Physicians (on Materia Medica).
>Ascr. Liang, *c.* +510.
>Attrib. Thao Hung-Ching　陶弘景.
>Now extant only in quotations in the
>pharmaceutical natural histories, and a
>reconstitution by Huang Yü (*1*).
>This work was a disentanglement, made by
>other hands between +523 and +618 or
>+656, of the contributions of Li Tang-
>Chih (*c.* +225) and Wu Phu (*c.* +235)
>and the commentaries of Thao Hung-
>Ching (+492) from the text of the *Shen
>Nung Pên Tshao Ching* itself. In other
>words it was the non-*Pên-Ching* part of
>the *Pên Tshao Ching Chi Chu* (q.v.). It
>may or may not have included some or
>all of Thao Hung-Ching's commentaries.

Ming Shih　明史.
>History of the Ming Dynasty [+1368 to
>+1643].
>Chhing, begun +1646, completed +1736,
>first pr. +1739.
>Chang Thing-Yü　張廷玉 *et al.*

Ming Thang Hsüan Chen Ching Chüeh　明堂玄
　真經訣.
>[=*Shang-Chhing Ming Thang Hsüan Chen
>Ching Chüeh.*]
>Explanation of the Manual of (Recovering
>the) Mysterious Primary (Vitalities of the)
>Cosmic Temple (i.e. the Human Body)
>[respiration and heliotherapy].
>S/Chhi or Liang, late +5th or early +6th
>(but much altered).
>Attrib. to the Mother Goddess of the West,
>Hsi Wang Mu　西王母.
>Writer unknown.
>*TT*/421.
>Cf. Maspero (7), p. 376.

Ming Thang Yuan Chen Ching Chüeh　明堂元
　真經訣.
>See *Ming Thang Hsüan Chen Ching Chüeh.*

Ming Thung Chi　冥通記.
>Record of Communication with the Hidden
>Ones (the Perfected Immortals).
>Liang, +516.
>Chou Tzu-Liang　周子良.
>Ed. Thao Hung-Ching　陶弘景.

Mo Chuang Man Lu　墨莊漫錄.
>Recollections from the Estate of Literary
>Learning.
>Sung, *c.* +1131.
>Chang Pang-Chi　張邦基.

Mo O Hsiao Lu　墨娥小錄.
>A Secretary's Commonplace-Book [popular
>encyclopaedia].

Yuan or Ming, +14th, pr. +1571.
Compiler unknown.

Mo Tzu (incl. *Mo Ching*)　墨子.
>The Book of Master Mo.
>Chou, —4th century.
>Mo Ti (and disciples)　墨翟.
>Tr. Mei Yi-Pao (*1*); Forke (3).
>Yin-Tê Index, no. (suppl.) 21.
>*TT*/1162.

Montoku-Jitsuroku　文德實錄.
>Veritable Records of the Reign of the
>Emperor Montoku [from +851 to
>+858].
>Japan (Heian) +879.
>Fujiwara Mototsune　藤原基經.

Nan Fan Hsiang Lu　南蕃香錄.
>Catalogue of the Incense of the Southern
>Barbarians.
>See *Hsiang Lu.*

Nan Hai Yao Phu　南海藥譜.
>A Treatise on the Materia Medica of the
>South Seas (Indo-China, Malayo-
>Indonesia, the East Indies, etc.).
>Alternative title of *Hai Yao Pên Tshao*,
>q.v. (according to Li Shih-Chen).

Nan Tshun Cho Kêng Lu　南村輟耕錄.
>See *Cho Kêng Lu.*

Nan Yo Ssu Ta Chhan-Shih Li Shih Yuan Wên
　南嶽思大禪師立誓願文.
>Text of the Vows (of Aranyaka Austerities)
>taken by the Great Chhan Master (Hui-)
>Ssu of the Southern Sacred Mountain.
>Chhen, *c.* +565.
>Hui-Ssu　慧思.
>TW/1933, N/1576.

Nei Chin Tan　內金丹.
>[=*Nei Tan Pi Chih* or *Thien Hsien Chih
>Lun Chhang Shêng Tu Shih Nei Lien Chin
>Tan Fa.*]
>The Metallous Enchymoma Within (the
>Body), [physiological alchemy].
>Ming, +1622, part dated +1615.
>Perhaps Chhen Ni-Wan　陳泥丸 (Mr
>Ni-Wan, Chhen), or Wu Chhung-Hsü
>伍沖虛.
>Contains a system of symbols included in
>the text.
>*CTPS*, *pên* 12.

Nei Ching.
>See *Huang Ti Nei Ching, Su Wên* and
>*Huang Ti Nei Ching, Ling Shu.*

Nei Ching Su Wên.
>See *Huang Ti Nei Ching, Su Wên.*

Nei Kung Thu Shuo　內功圖說.
>See Wang Tsu-Yuan (*1*).

Nei Tan Chüeh Fa　內丹訣法.
>See *Huan Tan Nei Hsiang Chin Yo Shih.*

Nei Tan Fu　內丹賦.
>[=*Thao Chen Jen Nai Tan Fu.*]
>Rhapsodical Ode on the Physiological
>Enchymoma.

Nei Tan Fu (*cont.*)
Sung, +13th.
Thao Chih 陶植.
With commentary by an unknown writer.
TT/256.
Cf. *Chin Tan Fu*, the text of which is very similar.

Nei Tan Pi Chih 內丹秘指.
Confidential Directions on the Enchymoma.
Alternative title for *Nei Chin Tan* (q.v.).

Nei Wai Erh Ching Thu 內外二景圖.
Illustrations of Internal and Superficial Anatomy.
Sung, +1118.
Chu Hung 朱肱.
Original text lost, and replaced later; drawings taken from Yang Chieh's *Tshun Chen Huan Chung Thu*.

Nêng Kai Chai Man Lu 能改齋漫錄.
Miscellaneous Records of the Ability-to-Improve-Oneself Studio.
Sung, mid +12th century.
Wu Tshêng 吳曾.

Ni-Wan Li Tsu Shih Nü Tsung Shuang Hsiu Pao Fa 泥丸李祖師女宗雙修寶筏.
See *Nü Tsung Shuang Hsiu Pao Fa*.

Nihon-Koki 日本後記.
Chronicles of Japan, further continued [from +792 to +833].
Japan (Heian), +840.
Fujiwara Otsugu 藤原緒嗣.

Nihon-Koku Ganzai-sho Mokuroku 日本國見在書目錄.
Bibliography of Extant Books in Japan.
Japan (Heian), *c.* +895.
Fujiwara no Sukeyo 藤原佐世.
Cf. Yoshida Mitsukuni (6), p. 196.

Nihon Sankai Meibutsu Zue 日本山海各物圖會.
Illustrations of Japanese Processes and Manufactures (lit., of the Famous Products of Japan).
Japan (Tokugawa), Osaka, +1754.
Hirase Tessai 平瀬徹齋.
Ills. by Hasegawa Mitsunobu 長谷川光 & Chigusa Shinemon 千種屋新右衛門.
Facsim. repr. with introd. notes, Meicho Kankokai, Tokyo, 1969.

Nihon-shoki 日本書記.
See *Nihongi*.

Nihon Ryo-iki 日本靈異記.
Record of Strange and Mysterious Things in Japan.
Japan (Heian), +823.
Writer unknown.

Nihongi 日本記.
[= *Nihon-shoki*.]
Chronicles of Japan [from the earliest times to +696].
Japan (Nara), +720.
Toneri-shinnō (prince), 舍人親王,

Ōno Yasumaro, 大安萬呂,
Ki no Kiyobito *et al.*
Tr. Aston (1).
Cf. Anon. (103), pp. 1 ff.

Nihongi Ryaku 日本記畧.
Classified Matters from the *Chronicles of Japan*.
Japan.

Nittō-Guhō Junrei Gyōki 入唐求法巡禮行記
Record of a Pilgrimage to China in Search of the (Buddhist) Law.
Thang, +838 to +847.
Ennin 圓仁.
Tr. Reischauer (2).

Nü Kung Chih Nan 女功指南.
A Direction-Finder for (Inner) Achievement by Women (Taoists).
[Physiological alchemy, *nei tan* gymnastic techniques, etc.]
See *Nü Tsung Shuang Hsiu Pao Fa*.

Nü Tsung Shuang Hsiu Pao Fa 女宗雙修寶筏.
[= *Ni-Wan Li Tsu Shih Nü Tsung Shuang Hsiu Pao Fa*, or *Nü Kung-Chih Nan*.]
A Precious Raft (of Salvation) for Women (Taoists) Practising the Double Regeneration (of the primary vitalities, for their nature and their life-span, *hsing ming*), [physiological alchemy].
Chhing, *c.* +1795.
Ni-Wan shih 泥丸氏, Li Ong (late +16th), 李翁, Mr Ni-Wan, the Taoist Patriarch Li.
Written down by Thai-Hsü Ong 太虛翁, Shen I-Ping 沈一炳, Ta-Shih (Taoist abbot), *c.* 1820.
In *Tao Tsang Hsü Pien* (*Chhu chi*), 20.
Cf. *Tao Hai Chin Liang*, p. 34*a*, *Shih Chin Shih*, p. 12*a*.

Pai hsien-sêng Chin Tan Huo Hou Thu 白先生金丹火候圖.
Master Pai's Illustrated Tractate on the 'Fire-Times' of the Metallous Enchymoma.
Sung, *c.* +1210.
Pai Yü-Chhan 白玉蟾.
In *Hsiu Chen Shih Shu* (*TT*/260), ch. 1.

Pao Phu Tzu 抱樸 (or 朴) 子.
Book of the Preservation-of-Solidarity Master.
Chin, early +4th century, probably *c.* +320.
Ko Hung 葛洪.
Partial trs. Feifel (1, 2); Wu & Davis (2)
Full tr. Ware (5), *Nei Phien* chs. only.
TT/1171–1173.

Pao Phu Tzu Shen Hsien Chin Shuo Ching 抱朴子神仙金汋經.
The Preservation-of-Solidarity Master's Manual of the Bubbling Gold (Potion) of the Holy Immortals.
Ascr. Chin *c.* +320. Perhaps pre-Thang, more probably Thang.

Pao Phu Tzu Shen Hsien Chin Shuo Ching (*cont.*)
Attrib. Ko Hung 葛洪.
TT/910.
Cf. Ho Ping-Yü (11).

Pao Phu Tzu Yang Shêng Lun 抱朴子養
生論.
The Preservation-of-Solidarity Master's
Essay on Hygiene.
Ascr. Chin *c.* +320.
Attrib. Ko Hung 葛洪.
TT/835.

Pao Shêng Hsin Chien 保生心鑑.
Mental Mirror of the Preservation of Life
[gymnastics and other longevity tech-
niques].
Ming, +1506.
Thieh Fêng chü-shih 鐵峰居士
(The Recluse of Iron Mountain, ps.).
Ed. *c.* +1596 by Hu Wên-Huan 胡文煥.

Pao Shou Thang Ching Yen Fang 保壽堂經
驗方.
Tried and Tested Prescriptions of the Pro-
tection-of-Longevity Hall (a surgery or
pharmacy).
Ming, *c.* +1450.
Liu Sung-shih 劉松石.

Pao Tsang Lun 寶藏論.
[=*Hsien-Yuan Pao Tsang Chhang Wei Lun.*]
(The Yellow Emperor's) Discourse on the
(Contents of the) Precious Treasury (of
the Earth), [mineralogy and metallurgy].
Perhaps in part Thang or pre-Thang; com-
pleted in Wu Tai (S/Han). Tsêng Yuan-
Jung (*1*) notes Chhao Kung-Wu's dating
of it at +918 in his *Chhun Chai Tu Shu
Chih.* Chang Tzu-Kao (*2*), p. 118, also
considers it mainly a Wu Tai work.
Attrib. Chhing Hsia Tzu 青霞子.
If Su Yuan-Ming 蘇元明 and not
another writer of the same pseudonym,
the earliest parts may have been of the
Chin time (+3rd or +4th); cf Yang
Lieh-Yü (*1*).
Now only extant in quotations.
Cf. *Lo-fou Shan Chih,* ch. 4, p. 13 *a.*

Pao Yen Thang Pi Chi 寶顏堂祕笈.
Private Collection of the Pao-Yen Library.
Ming, six collections printed between
+1606 and +1620.
Ed. Chhen Chi-Ju 陳繼儒

Pei Lu Fêng Su 北虜風俗.
[=*I Su Chi.*]
Customs of the Northern Barbarians (i.e.
the Mongols).
Ming, +1594.
Hsiao Ta-Hêng 蕭大亨.

Pei Mêng So Yen 北夢瑣言.
Fragmentary Notes Indited North of
(Lake) Mêng.
Wu Tai (S/Phing), *c.* +950.
Sun Kuang-Hsien 孫光憲.
See des Rotours (4), p. 38.

Pei Shan Chiu Ching 北山酒經.
Northern Mountain Wine Manual.
Sung, +1117.
Chu Hung 朱肱.

Pei Shih 北史.
History of the Northern Dynasties [Nan
Pei Chhao period, +386 to +581].
Thang, *c.* +670.
Li Yen-Shou 李延壽.
For translations of passages see the index of
Frankel (1).

Pên Ching Fêng Yuan 本經逢原.
(Additions to Natural History) aiming at
the Original Perfection of the *Classical
Pharmacopoeia* (*of the Heavenly
Husbandman*).
Chhing, +1695, pr. +1705.
Chang Lu 張璐.
LPC, no. 93.

Pên Tshao Chhiu Chen 本草求眞.
Truth Searched out in Pharmaceutical
Natural History.
Chhing, +1773.
Huang Kung-Hsiu 黃宮繡.

Pên Tshao Ching Chi Chu 本草經集注.
Collected Commentaries on the *Classical
Pharmacopoeia* (*of the Heavenly Husband-
man*).
S/Chhi, +492.
Thao Hung-Ching 陶弘景.
Now extant only in fragmentary form as a
Tunhuang or Turfan MS., apart from
the many quotations in the pharma-
ceutical natural histories, under Thao
Hung-Ching's name.

Pên Tshao Hui 本草滙.
Needles from the Haystack; Selected Essen-
tials of Materia Medica.
Chhing, +1666, pr. +1668.
Kuo Phei-Lan 郭佩蘭.
LPC, no. 84.
Cf. Swingle (4).

Pên Tshao Hui Chien 本草彙箋.
Classified Notes on Pharmaceutical Natural
History.
Chhing, begun +1660, pr. +1666.
Ku Yuan-Chiao 顧元交.
LPC, no. 83.
Cf. Swingle (8).

Pên Tshao Kang Mu 本草綱目.
The Great Pharmacopoeia; or, The Pan-
dects of Natural History (Mineralogy,
Metallurgy, Botany, Zoology etc.),
Arrayed in their Headings and Sub-
headings.
Ming, +1596.
Li Shih-Chen 李時珍.
Paraphrased and abridged tr. Read &
collaborators (2–7) and Read & Pak (1)
with indexes. Tabulation of plants in
Read (1) (with Liu Ju-Chhiang).
Cf. Swingle (7).

Pên Tshao Kang Mu Shih I 本草綱目拾遺.
   Supplementary Amplifications for the
     Pandects of Natural History (of Li Shih-
     Chen).
   Chhing, begun *c.* +1760, first prefaced
     +1765, prolegomena added +1780, last
     date in text 1803.
   Chhing, first pr. 1871.
   Chao Hsüeh-Min 趙學敏.
   LPC, no. 101.
   Cf. Swingle (11).
Pên Tshao Mêng Chhüan 本草蒙筌.
   Enlightenment on Pharmaceutical Natural
     History.
   Ming, +1565.
   Chhen Chia-Mo 陳嘉謨.
Pên Tshao Pei Yao 本草備要.
   Practical Aspects of Materia Medica.
   Chhing, *c.* +1690, second ed. +1694.
   Wang Ang 汪昂.
   LPC, no. 90; ICK, pp. 215 ff.
   Cf. Swingle (4).
Pên Tshao Phin Hui Ching Yao 本草品彙精要.
   Essentials of the Pharmacopoeia Ranked
     according to Nature and Efficacity (Im-
     perially Commissioned).
   Ming, +1505.
   Liu Wên-Thai 劉文泰, Wang Phan 王槃
     & Kao Thing-Ho 高廷和.
Pên Tshao Shih I 本草拾遺.
   A Supplement for the Pharmaceutical
     Natural Histories.
   Thang, *c.* +725.
   Chhen Tshang-Chhi 陳藏器.
   Now extant only in numerous quotations.
Pên Tshao Shu 本草述.
   Explanations of Materia Medica.
   Chhing, before +1665, first pr. +1700.
   Liu Jo-Chin 劉若金.
   LPC, no. 79.
   Cf. Swingle (6).
Pên Tshao Shu Kou Yuan 本草述鉤元.
   Essentials Extracted from the *Explanations
     of Materia Medica*.
   See Yang Shih-Thai (*1*).
Pên Tshao Thu Ching 本草圖經.
   Illustrated Pharmacopoeia; or, Illustrated
     Treatise of Pharmaceutical Natural
     History.
   Sung, +1061.
   Su Sung 蘇頌 *et al.*
   Now preserved only in numerous quota-
     tions in the later pandects of pharma-
     ceutical natural history.
Pên Tshao Thung Hsüan 本草通玄.
   The Mysteries of Materia Medica Un-
     veiled.
   Chhing, begun before +1655, pr. just
     before +1667.
   Li Chung-Tzu 李中梓.
   LPC, no. 75.
   Cf. Swingle (4).

Pên Tshao Tshung Hsin 本草從新.
   New Additions to Pharmaceutical Natural
     History.
   Chhing, +1757.
   Wu I-Lo 吳儀洛.
   LPC, no. 99.
Pên Tshao Yao Hsing 本草藥性.
   The Natures of the Vegetable and Other
     Drugs in the Pharmaceutical Treatises.
   Thang, *c.* +620.
   Chen Li-Yen 甄立言 & (perhaps) Chen
     Chhüan 甄權.
   Now extant only in quotations.
Pên Tshao Yen I 本草衍義.
   Dilations upon Pharmaceutical Natural
     History.
   Sung, pref. +1116, pr. +1119, repr. +1185,
     +1195.
   Khou Tsung-Shih 寇宗奭.
   See also *Thu Ching Yen I Pên Tshao*
     (*TT*/761).
Pên Tshao Yen I Pu I 本草衍義補遺.
   Revision and Amplification of the *Dilations
     upon Pharmaceutical Natural History*.
   Yuan, *c.* +1330.
   Chu Chen-Hêng 朱震亨.
   LPC, no. 47.
   Cf. Swingle (12).
Pên Tshao Yuan Shih 本草原始.
   Objective Natural History of Materia
     Medica; a True-to-Life Study.
   Chhing, begun +1578, pr. +1612.
   Li Chung-Li 李中立.
   LPC, no. 60.
Phan Shan Yü Lu 緐山語錄.
   Record of Discussions at Phan Mountain
     [dialogues of pronouncedly medical
     character on physiological alchemy].
   Sung, prob. early +13th.
   Writer unknown.
   In *Hsiu Chen Shih Shu* (*TT*/260), ch. 53.
Phêng-Lai Shan Hsi Tsao Huan Tan Ko 蓬萊
   山西竈還丹歌.
   Mnemonic Rhymes of the Cyclically
     Transformed Elixir from the Western
     Furnace on Phêng-lai Island.
   Ascr. *c.* −98. Probably Thang.
   Huang Hsüan-Chung 黃玄鐘.
   TT/909.
Phêng Tsu Ching 彭祖經.
   Manual of Phêng Tsu [Taoist sexual tech-
     niques and their natural philosophy].
   Late Chou or C/Han, −4th to −1st.
   Attrib. Phêng Tsu 彭祖.
   Only extant as fragments in *CSHK*
     (Shang Ku Sect.), ch. 16, pp. 5*b* ff.
Phu Chi Fang 普濟方.
   Practical Prescriptions for Everyman.
   Ming, *c.* +1418.
   Chu Hsiao 朱橚 (Chou Ting Wang 周定王,
     prince of the Ming).
   ICK, p. 914.

Pi Yü Chu Sha Han Lin Yü Shu Kuei 碧玉朱砂寒林玉樹匱.
On the Caerulean Jade and Cinnabar Jade-Tree-in-a-Cold-Forest Casing Process.
Sung, early +11th cent.
Chhen Ching-Yuan 陳景元.
TT/891.

Pien Huo Pien 辯惑編.
Disputations on Doubtful Matters.
Yuan, +1348.
Hsieh Ying-Fang 謝應芳.

Pien Tao Lun 辯道論.
On Taoism, True and False.
San Kuo (Wei), c. +230.
Tshao Chih (prince of the Wei), 曹植.
Now extant only in quotations.

Po Wu Chi 博物記.
Notes on the Investigation of Things.
H/Han, c. +190.
Thang Mêng (b) 唐蒙.

Po Wu Chih 博物志.
Records of the Investigation of Things (cf. *Hsü Po Wu Chih*).
Chin, c. +290 (begun about +270).
Chang Hua 張華.

Pu Wu Yao Lan 博物要覽.
The Principal Points about Objects of Art and Nature.
Ming, c. +1560.
Ku Thai 谷泰.

Rokubutsu Shinshi 六物新志.
New Record of Six Things [including the drug mumia]. (In part a translation from Dutch texts.)
Japan, +1786.
Ōtsuki Gentaku 大槻玄澤.

San Chen Chih Yao Yü Chüeh 三眞旨要玉訣.
Precious Instructions concerning the Message of the Three Perfected (Immortals), [i.e. Yang Hsi (*fl.* +370) 楊羲; Hsü Mi (*fl.* +345) 許謐; and Hsü Hui (d. c. +370) 許翽].
Taoist heliotherapy, respiration and meditation.
Chin, c. +365, edited probably in the Thang.
TT/419.
Cf. Maspero (7), p. 376.

San-Fêng Chen Jen Hsüan Than Chhüan Chi 三峯眞人玄譚全集.
Complete Collection of the Mysterious Discourses of the Adept (Chang) San-Fêng [physiological alchemy].
Ming, from c. +1410 (if genuine).
Attrib. Chang San-Fêng 張三峯.
Ed. Min I-Tê (1834) 閔一得.
In *Tao Tsang Hsü Pien (Chhu chi)*, 17.

San-Fêng Tan Chüeh 三峯丹訣 (includes *Chin Tan Chieh Yao* and *Tshai Chen Chi Yao*,

with the *Wu Kên Shu* series of poems, and some inscriptions).
Oral Instructions of (Chang) San-Fêng on the Enchymoma [physiological alchemy].
Ming, from c. +1410 (if genuine).
Attrib. Chang San-Fêng 張三峯.
Ed., with biography, by Fu Chin-Chhüan 傅金銓 (Chi I Tzu 濟一子) c. 1820.

San Phin I Shen Pao Ming Shen Tan Fang 三品頤神保命神丹方.
Efficacious Elixir Prescriptions of Three Grades Inducing the Appropriate Mentality for the Enterprise of Longevity.
Thang, Wu Tai & Sung.
Writers unknown.
YCCC, ch. 78, pp. 1a ff.

San-shih-liu Shui Fa 三十六水法.
Thirty-six Methods for Bringing Solids into Aqueous Solution.
Pre-Thang.
Writer unknown.
TT/923.

San Tshai Thu Hui 三才圖會.
Universal Encyclopaedia.
Ming, +1609.
Wang Chhi 王圻.

San Tung Chu Nang 三洞珠囊.
Bag of Pearls from the Three (Collections that) Penetrate the Mystery [a Taoist florilegium].
Thang, +7th.
Wang Hsüan-Ho (ed.) 王懸河.
TT/1125.
Cf. Maspero (13), p. 77; Schipper (1), p. 11.

San Yen 三言.
See *Hsing Shih Hêng Yen*, *Yü Shih Ming Yen*, *Ching Shih Thung Yen*.

Setsuyō Yoketsu.
See *Shê Yang Yao Chüeh*.

Shan Hai Ching 山海經.
Classic of the Mountains and Rivers.
Chou and C/Han, −8th to −1st.
Writers unknown.
Partial tr. de Rosny (1).
Chung-Fa Index, no. 9.

Shang-Chhing Chi 上清集.
A Literary Collection (inspired by) the Shang-Chhing Scriptures [prose and poems on physiological alchemy].
Sung, c. +1220.
Ko Chhang-Kêng 葛長庚 (Pai Yü-Chhan 白玉蟾).
In *Hsiu Chen Shih Shu TT*/260), chs. 37 to 44

Shang-Chhing Ching 上清經.
[Part of *Thai Shang San-shih-liu Pu Tsun Ching*.]
The Shang-Chhing (Heavenly Purity) Scripture.
Chin, oldest parts date from about +316.
Attrib. Wei Hua-Tshun 魏華存, dictated to Yang Hsi 楊羲.
In *TT*/8.

Shang-Chhing Chiu Chen Chung Ching Nei Chüeh 上清九眞中經內訣.
Confidential Explanation of the Interior Manual of the Nine (Adepts); a Shang-Chhing Scripture.
Ascr. Chin, +4th, probably pre-Thang.
Attrib. Chhih Sung Tzu 赤松子 (Huang Chhu-Phing 黃初平).
TT/901.

Shang Chhing Han Hsiang Chien Chien Thu 上清含象劍鑑圖.
The Image and Sword Mirror Diagram; a Shang-chhing Scripture.
Thang, *c.* +700.
Ssuma Chhêng-Chên 司馬承貞.
TT/428.

Shang-Chhing Hou Shêng Tao Chün Lieh Chi 上清後聖道君列紀.
Annals of the Latter-Day Sage, the Lord of the Tao; a Shang-Chhing Scripture.
Chin, late +4th.
Revealed to Yang Hsi 楊羲.
TT/439.

Shang-Chhing Huang Shu Kuo Tu I 上清黃書過度儀.
The System of the Yellow Book for Attaining Salvation; a Shang-Chhing Scripture [the rituale of the communal Taoist liturgical sexual ceremonies, +2nd to +7th centuries].
Date unknown, but pre-Thang.
Writer unknown.
TT/1276.

Shang-Chhing Ling-Pao Ta Fa 上清靈寶大法.
The Great Liturgies; a Shang-Chhing Ling-Pao Scripture.
Sung, +13th.
Chin Yün-Chung 金允中.
TT/1204, 1205, 1206.

Shang-Chhing Ming Thang Hsüan Chen Ching Chüeh 上清明堂玄眞經訣.
See *Ming Thang Hsüan Chen Ching Chüeh.*

Shang-Chhing San Chen Chih Yao Yü Chüeh 上清三眞旨要玉訣.
See *San Chen Chih Yao Yü Chüeh.*

Shang-Chhing Thai-Shang Pa Su Chen Ching 上清太上八素眞經.
Realisation Canon of the Eight Purifications (or Eightfold Simplicity); a Shang-Chhing Thai-Shang Scripture.
Date uncertain, but pre-Thang.
Writer unknown.
TT/423.

Shang-Chhing Thai-Shang Ti Chün Chiu Chen Chung Ching 上清太上帝君九眞中經.
Ninefold Realised Median Canon of the Imperial Lord; a Shang-Chhing Thai-Shang Scripture.
Compiled from materials probably of Chin period, late +4th.
Writers and editor unknown.
TT/1357.

Shang-Chhing Tung-Chen Chiu Kung Tzu Fang Thu 上清洞眞九宮紫房圖.
Description of the Purple Chambers of the Nine Palaces; a Tung-Chen Scripture of the Shang-Chhing Heavens [parts of the microcosmic body corresponding to stars in the macrocosm].
Sung, probably +12th century.
Writer unknown.
TT/153.

Shang-Chhing Wo Chung Chüeh 上清握中訣.
Explanation of (the Method of) Grasping the Central (Luminary); a Shang-Chhing Scripture [Taoist meditation and helio-therapy].
Date unknown, Liang or perhaps Thang.
Writer unknown.
Based on the procedures of Fan Yu-Chhung 范幼沖 (H/Han).
TT/137.
Cf. Maspero (7), p. 373.

Shang Phin Tan Fa Chieh Tzhu 上品丹法節次.
Expositions of the Techniques for Making the Best Quality Enchymoma [physiological alchemy].
Chhing.
Li Tê-Hsia 李德洽.
Comm. Min I-Tê 閔一德, *c.* 1830.
In *Tao Tsang Hsü Pien (Chhu chi)*, 6.

Shang Shu Ta Chuan 尚書大傳.
Great Commentary on the *Shang Shu* chapters of the *Historical Classic.*
C/Han, *c.* −185.
Fu Shêng 伏勝.
Cf. Wu Khang (1), p. 230.

Shang-Tung Hsin Tan Ching Chüeh 上洞心丹經訣.
An Explanation of the Heart Elixir and Enchymoma Canon; a Shang-Tung Scripture.
Date unknown, perhaps Sung.
Writer unknown.
TT/943.
Cf. Chhen Kuo-Fu (1), vol. 2, pp. 389, 435.

Shang Yang Tzu Chin Tan Ta Yao 上陽子金丹大要.
See *Chin Tan Ta Yao.*

Shang Yang Tzu Chin Tan Ta Yao Hsien Phai (Yuan Liu) 上陽子金丹大要仙派 (源流).
See *Chin Tan Ta Yao Hsien Phai (Yuan Liu).*

Shang Yang Tza Chin Tan Ta Yao Lieh Hsien Chih 上陽子金丹大要列仙誌.
See *Chin Tan Ta Yao Lieh Hsien Chih.*

Shang Yang Tzu Chin Tan Ta Yao Thu 上陽子金丹大要圖.
See *Chin Tan Ta Yao Thu.*

Shao-Hsing Chiao-Ting Ching-Shih Chêng Lei Pei-Chi Pên Tshao 紹興校定經史證類備急本草.

*Shao-Hsing Chiao-Ting Ching-Shih Chêng Lei
    Pei-Chi Pên Tshao (cont.)*|
    The Corrected Classified and Consolidated
    Armamentarium; Pharmacopoeia of the
    Shao-Hsing Reign-Period.
    S/Sung, pres. +1157, pr. +1159, often
    copied and repr. especially in Japan.
    Thang Shen-Wei 唐慎微 ed. Wang Chi-
    Hsien 王繼先 *et al.*
    Cf. Nakao Manzō (*1*, *1*); Swingle (11).
    Illustrations reproduced in facsimile by
    Wada (*1*); Karow (2).
    Facsimile edition of a MS. in the Library of
    Ryokoku University, Kyoto 龍谷大學
    圖書舘
    Ed. with an analytical and historical intro-
    duction, including contents table and in-
    dexes (別册) by Okanishi Tameto 岡西
    爲人 (Shunyōdō, Tokyo, 1971).
Shê Ta Chhêng Lun Shih 攝大乘論釋.
    Mahāyāna-samgraha-bhāshya (Explanatory
    Discourse to assist the Understanding of
    the Great Vehicle).
    India, betw. +300 and +500.
    Tr. Hsüan-Chuang 玄奘, *c.* +650.
    N/1171 (4); TW/1597.
(*Shê Yang*) *Chen Chung Chi* (or *Fang*) (攝養)枕
    中記(方).
    Pillow-Book on Assisting the Nourishment
    (of the Life-Force).
    Thang, early +7th.
    Attrib. Sun Ssu-Mo 孫思邈.
    TT/830, and in YCCC, ch. 33.
Shê Yang Yao Chüeh (*Setsuyō Yoketsu*) 攝養要訣.
    Important Instructions for the Preservation
    of Health conducive to Longevity.
    Japan (Heian), *c.* +820.
    Mononobe Kōsen (imperial physician)
    物部廣泉.
Shen Hsien Chin Shuo Ching 神仙金汋經.
    See *Pao Phu Tzu Shen Hsien Chin Shuo
    Ching.*
Shen Hsien Chuan 神仙傳.
    Lives of the Holy Immortals.
    (Cf. *Lieh Hsien Chuan* and *Hsü Shen Hsien
    Chuan.*)
    Chin, +4th century.
    Attrib. Ko Hung 葛洪.
Shen Hsien Fu Erh Tan Shih Hsing Yao Fa
    神仙服餌丹石行藥法.
    The Methods of the Holy Immortals for
    Ingesting Cinnabar and (Other)
    Minerals, and Using them Medicinally.
    Date unknown.
    Attrib. Ching-Li hsien-sêng 京里先生.
    TT/417.
*Shen Hsien Fu Shih Ling-Chih Chhang-Phu Wan
    Fang* 神仙服食靈芝菖蒲丸方.
    Prescriptions for Making Pills from
    Numinous Mushrooms and Sweet Flag
    (*Calamus*), as taken by the Holy Immortals.
    Date unknown

Writer unknown.
TT/837.
*Shen Hsien Lien Tan Tien Chu San Yuan Pao
    Ching Fa* 神仙鍊丹點鑄三元寶鏡法.
    Methods used by the Holy Immortals to
    Prepare the Elixir, Project it, and Cast
    the Precious Mirrors of the Three Powers
    (or the Three Primary Vitalities), [magical].
    Thang, +902.
    Writer unknown.
    TT/856.
Shen Hsien Thung Chien 神仙通鑑.
    (Cf. (*Li Tai*) *Shen Hsien* (*Thung*) *Chien.*)
    General Survey of the Lives of the Holy
    Immortals.
    Ming, +1640.
    Hsüeh Ta-Hsün 薛大訓.
Shen I Chi 神異記.
    (Probably an alternative title of *Shen I
    Ching*, q.v.)
    Records of the Spiritual and the Strange.
    Chin, *c.* +290.
    Wang Fou 王浮.
Shen I Ching 神異經.
    Book of the Spiritual and the Strange.
    Ascr. Han, but prob. +3rd, +4th or +5th
    century.
    Attrib. Tungfang Shuo 東方朔.
    Probable author, Wang Fou 王浮.
Shen Nung Pên Tshao Ching 神農本草經.
    Classical Pharmacopoeia of the Heavenly
    Husbandman.
    C/Han, based on Chou and Chhin material,
    but not reaching final form before the
    +2nd century.
    Writers unknown.
    Lost as a separate work, but the basis of all
    subsequent compendia of pharmaceutical
    natural history, in which it is constantly
    quoted.
    Reconstituted and annotated by many
    scholars; see Lung Po-Chien (*1*), pp. 2 ff.,
    12 ff.
    Best reconstructions by Mori Tateyuki
    森立之 (1845), Liu Fu 劉復 (1942).
Shen shih Liang Fang 沈氏良方.
    Original title of *Su Shen Liang Fang* (q.v.).
Shen Thien-Shih Fu Chhi Yao Chüeh 申天師
    服氣要訣.
    Important Oral Instructions of the Heavenly
    Teacher (or Patriarch) Shen on the
    Absorption of the Chhi [Taoist breathing
    exercises].
    Thang, *c.* +730.
    Shen Yuan-Chih 申元之.
    Now extant only as a short passage in
    YCCC, ch. 59, pp. 16b ff.
Shêng Chi Tsung Lu 聖濟總錄.
    Imperial Medical Encyclopaedia [issued by
    authority].
    Sung, *c.* +1111 to +1118.
    Ed. by twelve physicians.

Shêng Shih Miao Ching 生尸妙經.

See *Thai-Shang Tung-Hsüan Ling-Pao Mieh Tu* (or *San Yuan*) *Wu Lien Shêng Shih Miao Ching*.

Shêng Shui Yen Than Lu 澠水燕談錄.

Fleeting Gossip by the River Shêng [in Shantung].

Sung, late +11th century (before +1094).

Wang Phi-Chih 王闢之.

Shih Chin Shih 試金石.

On the Testing of (what is meant by) 'Metal' and 'Mineral'.

See Fu Chin-Chhüan (5).

Shih Han Chi 石函記.

See *Hsü Chen Chün Shih Han Chi*.

Shih I Chi 拾遺記.

Memoirs on Neglected Matters.

Chin, *c.* +370.

Wang Chia 王嘉.

Cf. Eichhorn (5).

Shih I Tê Hsiao Fang 世醫得効方.

Efficacious Prescriptions of a Family of Physicians.

Yuan, +1337.

Wei I-Lin 危亦林.

Shih Liao Pên Tshao 食療本草.

Nutritional Therapy; a Pharmaceutical Natural History.

Thang, *c.* +670.

Mêng Shen 孟詵.

Shih Lin Kuang Chi 事林廣記.

Guide through the Forest of Affairs [encyclopaedia].

Sung, between +1100 and +1250; first pr. +1325.

Chhen Yuan-Ching 陳元靚.

(A unique copy of a Ming edition of +1478 is in the Cambridge University Library.)

Shih Ming 釋名.

Explanation of Names [dictionary].

H/Han, *c.* +100.

Liu Hsi 劉熙.

Shih Pien Liang Fang 十便良方.

Excellent Prescriptions of Perfect Convenience.

Sung, +1196.

Kuo Than 郭坦.

Cf. SIC, p. 1119; ICK, p. 813.

Shih Wu Chi Yuan 事物紀原.

Records of the Origins of Affairs and Things.

Sung, *c.* +1085.

Kao Chhêng 高承.

Shih Wu Pên Tshao 食物本草.

Nutritional Natural History.

Ming, +1571 (repr. from a slightly earlier edition).

Attrib. Li Kao 李杲 (J/Chin) or Wang Ying 汪穎 (Ming) in various editions; actual writer Lu Ho 盧和.

The bibliography of this work in its several different forms, together with the questions of authorship and editorship, are complex.

See Lung Po-Chien (1), pp. 104, 105, 106; Wang Yü-Hu (1), 2nd ed. p. 194; Swingle (1, 10).

Shih Yao Erh Ya 石藥爾雅.

The Literary Expositor of Chemical Physic; or, Synonymic Dictionary of Minerals and Drugs.

Thang, +806.

Mei Piao 梅彪.

TT/894.

Shih Yuan 事原.

On the Origins of Things.

Sung.

Chu Hui 朱繪.

Shoku-Nihongi 續日本記.

Chronicles of Japan, continued [from +697 to +791].

Japan (Nara), +797.

Ishikawa Natari 石川,

Fujiwara Tsuginawa 藤原繼繩,

Sugeno Sanemichi 菅野眞道 *et al.*

Shoku-Nihonkoki 續日本後記.

Chronicles of Japan, still further continued [from +834 to +850].

Japan (Heian), +869.

Fujiwara Yoshifusa 藤原良房.

Shou Yü Shen Fang 壽域神方.

Magical Prescriptions of the Land of the Old.

Ming, *c.* +1430.

Chu Chhüan 朱權 (Ning Hsien Wang 寧獻王, prince of the Ming).

Shu Shu Chi I 數術記遺.

Memoir on some Traditions of Mathematical Art.

H/Han, +190, but generally suspected of having been written by its commentator Chen Luan 甄鸞, *c.* +570. Some place the text as late as the Wu Tai period (+10th. cent.), e.g. Hu Shih; and others such as Li Shu-Hua (2) prefer a Thang dating.

Hsü Yo 徐岳.

Shu Yuan Tsa Chi 菽園雜記.

The Bean-Garden Miscellany.

Ming, +1475.

Lu Jung 陸容.

Shuang Mei Ching An Tshung Shu 雙梅景闇叢書.

Double Plum-Tree Collection [of ancient and medieval books and fragments on Taoist sexual techniques].

See Yeh Tê-Hui (1) 葉德輝 in Bib. B.

Shui Yün Lu 水雲錄.

Record of Clouds and Waters [iatro-chemical].

Sung, *c.* +1125.

Yeh Mêng-Tê 葉夢得.

Extant now only in quotations.

Shun Yang Lü Chen-Jen Yao Shih Chih 純陽
呂眞人藥石製.
The Adept Lü Shun-Yang's (i.e. Lü
Tung-Pin's) Book on Preparations of
Drugs and Minerals [in verses].
Late Thang.
Attrib. Lü Tung-Pin 呂洞賓.
TT/896.
Tr. Ho Ping-Yü, Lim & Morsingh (1).

Shuo Wên.
See *Shuo Wên Chieh Tzu.*

Shuo Wên Chieh Tzu 說文解字.
Analytical Dictionary of Characters (lit.
Explanations of Simple Characters and
Analyses of Composite Ones).
H/Han, +121.
Hsü Shen 許慎.

So Sui Lu 瑣碎錄.
Sherds, Orts and Unconsidered Fragments
[iatro-chemical].
Sung, prob. late +11th.
Writer unknown.
Now extant only in quotations. Cf. *Winter's
Tale,* IV, iii, *Timon of Athens,* IV, iii, and
Julius Caesar, IV, i.

Sou Shen Chi 搜神記.
Reports on Spiritual Manifestations.
Chin, c. +348.
Kan Pao 干寶.
Partial tr. Bodde (9).

Sou Shen Hou Chi 搜神後記.
Supplementary Reports on Spiritual
Manifestations.
Chin, late +4th or early +5th century.
Thao Chhien 陶潛.

Ssu Khu Thi Yao Pien Chêng 四庫提要辨證.
See Yü Chia-Hsi (1).

Ssu Shêng Pên Tshao 四聲本草.
Materia Medica Classified according to the
Four Tones (and the Standard Rhymes),
[the entries arranged in the order of the
pronunciation of the first character of
their names].
Thang, c. +775.
Hsiao Ping 蕭炳.

Ssu Shih Thiao Shê Chien 四時調攝牋.
Directions for Harmonising and Strengthen-
ing (the Vitalities) according to the Four
Seasons of the Year [the second part
(chs. 3–6) of *Tsun Shêng Pa Chien,* q.v.].
Ming, +1591.
Kao Lien 高濂.
Partial tr. of the gymnastic material,
Dudgeon (1).

Ssu Shih Tsuan Yao 四時纂要.
Important Rules for the Four Seasons
[agriculture and horticulture, family
hygiene and pharmacy, etc.].
Thang, c. +750.
Han O 韓鄂.

Su Nü Ching 素女經.
Canon of the Immaculate Girl.

Han.
Writer unknown.
Only as fragment in *Shuang Mei Ching An
Tshung Shu,* now containing the *Hsüan Nü
Ching* (q.v.).
Partial trs. van Gulik (3, 8).

Su Nü Miao Lun 素女妙論.
Mysterious Discourses of the Immaculate
Girl.
Ming, c. +1500.
Writer unknown.
Partial tr. van Gulik (3).

Su Shen Liang Fang 蘇沈良方.
Beneficial Prescriptions collected by Su
(Tung-Pho) and Shen (Kua).
Sung, c. +1120. Some of the data go back
as far as +1060. Preface by Lin Ling-Su
林靈素.
Shen Kua 沈括 and Su Tung-Pho
蘇東坡 (posthumous).
The collection was at first called *Shen
shih Liang Fang,* so that most of the
entries are Shen Kua's, but as some certainly stem from Su Tung-Pho, the latter
were probably added by editors at the
beginning of the new century.
Cf. ICK, pp. 737, 732.

Su Wên Ling Shu Ching.
See *Huang Ti Nei Ching, Su Wên* and
Huang Ti Nei Ching, Ling Shu.

Su Wên Nei Ching.
See *Huang Ti Nei Ching, Su Wên.*

Sui Shu 隋書.
History of the Sui Dynasty [+581 to
+617].
Thang, +636 (annals and biographies);
+656 (monographs and bibliography).
Wei Chêng 魏徵 et al.
Partial trs. Pfizmaier (61–65); Balazs (7, 8);
Ware (1).
For translations of passages see the index of
Frankel (1).

Sun Kung Than Phu 孫公談圃.
The Venerable Mr Sung's Conversation
Garden.
Sung, c. +1085.
Sun Shêng 孫升.

Sung Chhao Shih Shih 宋朝事實.
Records of Affairs of the Sung Dynasty.
Yuan, +13th.
Li Yu 李攸.

Sung Shan Thai-Wu hsien-sêng Chhi Ching
嵩山太无先生氣經.
Manual of the (Circulation of the) Chhi,
by Mr Grand-Nothingness of Sung
Mountain.
Thang, +766 to +779.
Prob. Li Fêng-Shih 李奉時 (Thai-Wu
hsien-sêng).
TT/817, and in *YCCC,* ch. 59 (partially),
pp. 7a ff.
Cf. Maspero (7), p. 199.

Sung Shih 宋史.
    History of the Sung Dynasty [+960 to
       +1279].
    Yuan, *c.* +1345.
    Tho-Tho (Toktaga) 脫脫 & Ouyang
       Hsüan 歐陽玄.
    Yin-Tê Index, no. 34.
Szechuan Thung Chih 四川通志.
    General History and Topography of
       Szechuan Province.
    Chhing, +18th century (pr. 1816).
    Ed. Chhang Ming 常明, Yang Fang-
       Tshan 楊芳燦 *et al.*

Ta Chao 大招.
    The Great Summons (of the Soul), [ode].
    Chhu (between Chhin and Han), −206 or
       −205.
    Writer unknown.
    Tr. Hawkes (1), p. 109.
Ta Chih Tu Lun 大智度論.
    Mahā-prajñapāramito-padeśa Śāstra (Com-
       mentary on the Great Sūtra of the Per-
       fection of Wisdom).
    India.
    Attrib. Nāgārjuna, +2nd.
    Mostly prob. of Central Asian origin.
    Tr. Kumārajīva, +406.
    N/1169; TW/1509.
Ta Chün Ku Thung 大鈞皷銅.
    (Illustrated Account of the Mining), Smelt-
       ing and Refining of Copper [and other Non-
       Ferrous Metals], according to the Principles
       of Nature (lit. the Great Potter's Wheel).
    See Masuda Tsuna (1).
Ta Fang Kuang Fo Hua Yen Ching 大方廣佛
    華嚴經.
    Avataṁsaka Sūtra.
    India.
    Tr. Śikshānanda, +699.
    N/88; TW/279.
Ta Huan Tan Chao Chien 大還丹照鑑.
    An Elucidation of the Great Cyclically
       Transformed Elixir [in verses].
    Wu Tai (Shu), +962.
    Writer unknown.
    TT/919.
Ta Huan Tan Chhi Pi Thu 大還丹契祕圖.
    Esoteric Illustrations of the Concordance of
       the Great Regenerative Enchymoma.
    Thang or Sung.
    Writer unknown.
    In *YCCC*, ch. 72, pp. 1 a ff.
    Cf. *Hsiu Chen Li Yen Chhao Thu* and
       Chin I Huan Tan Yin Chêng Thu.
Ta-Kuan Ching-Shih Chêng Lei Pei-Chi Pên
    Tshao 大觀經史證類備急本草.
    The Classified and Consolidated Arma-
       mentarium; Pharmacopoeia of the Ta-
       Kuan reign-period.
    Sung, +1108; repr. +1211, +1214
       (J/Chin), +1302 (Yuan).

Thang Shen-Wei 唐慎微.
    Ed. Ai Shêng 艾晟.
Ta Ming I Thung Chih 大明一統志.
    Comprehensive Geography of the (Chinese)
       Empire (under the Ming dynasty).
    Ming, commissioned +1450, completed
       +1461.
    Ed. Li Hsien 李賢.
Ta Tai Li Chi 大戴禮記.
    Record of Rites [compiled by Tai the Elder]
       (cf. *Hsiao Tai Li Chi*; *Li Chi*).
    Ascr. C/Han, *c.* −70 to −50, but really
       H/Han, between +80 and +105.
    Attrib. ed. Tai Tê 戴德, in fact probably
       ed. Tshao Pao 曹襃.
    See Legge (7).
    Trs. Douglas (1); R. Wilhelm (6).
Ta Tan Chhien Hung Lun 大丹鉛汞論.
    Discourse on the Great Elixir [or Enchym-
       oma] of Lead and Mercury.
    If Thang, +9th, more probably Sung.
    Chin Chu-Pho 金竹坡.
    TT/916.
    Cf. Yoshida Mitsukuni (5), pp. 230–2.
Ta Tan Chi 大丹記.
    Record of the Great Enchymoma.
    Ascr. +2nd cent., but probably Sung,
       +13th.
    Attrib. Wei Po-Yang 魏伯陽.
    TT/892.
Ta Tan Chih Chih 大丹直指.
    Direct Hints on the Great Elixir.
    Sung, *c.* +1200.
    Chhiu Chhu-Chi 邱處機.
    TT/241.
Ta Tan Wên Ta 大丹問答.
    Questions and Answers on the Great Elixir
       (or Enchymoma) [dialogues between
       Chêng Yin and Ko Hung].
    Date unknown, prob. late Sung or Yuan.
    Writer unknown.
    TT/932.
Ta Tan Yao Chüeh Pên Tshao 大丹藥訣
    本草.
    Pharmaceutical Natural History in the form
       of Instructions about Medicines of the
       Great Elixir (Type), [iatro-chemical].
    Possible alternative title of *Wai Tan Pên*
       Tshao (q.v.).
Ta-Tung Lien Chen Pao Ching, Chiu Huan Chin
    Tan Miao Chüeh 大洞錬眞寶經九還
    金丹妙訣.
    Mysterious Teachings on the Ninefold
       Cyclically Transformed Gold Elixir,
       supplementary to the Manual of the
       Making of the Perfected Treasure; a Ta-
       Tung Scripture.
    Thang, +8th, perhaps *c.* +712.
    Chhen Shao-Wei 陳少微.
    TT/884. A sequel to *TT*/883, and in *YCCC*,
       ch. 68, pp. 8 a ff.
    Tr. Sivin (4).

Ta-Tung Lien Chen Pao Ching, Hsiu Fu Ling Sha Miao Chüeh 大洞鍊眞寶經修伏靈砂妙訣.
   Mysterious Teachings on the Alchemical Preparation of Numinous Cinnabar, supplementary to the Manual of the Making of the Perfected Treasure; a Ta-Tung Scripture.
   Thang, +8th, perhaps *c.* +712.
   Chhen Shao-Wei 陳少微.
   TT/883. Alt. title: *Chhi Fan Ling Sha Lun*, as in *YCCC*, ch. 69, pp. 1*a* ff.
   Tr. Sivin (4).

Ta Yu Miao Ching 大有妙經.
   [= *Tung-Chen Thai-Shang Su-Ling Tung-Yuan Ta Yu Miao Ching*.]
   Book of the Great Mystery of Existence [Taoist anatomy and physiology; describes the *shang tan thien*, upper region of vital heat, in the brain].
   Chin, +4th.
   Writer unknown.
   TT/1295.
   Cf. Maspero (7), p. 192.

Tai I Phien 代疑篇.
   On Replacing Doubts by Certainties.
   Ming, +1621.
   Yang Thing-Yün 楊廷筠.
   Preface by Wang Chêng 王徵.

Taketori Monogatari 竹取物語.
   The Tale of the Bamboo-Gatherer.
   Japan (Heian), *c.* +865. Cannot be earlier than *c.* +810 or later than *c.* +955.
   Writer unknown.
   Cf. Matsubara Hisako (1, 2).

Tan Ching Shih Tu 丹經示讀.
   A Guide to the Reading of the Enchymoma Manuals.
   See Fu Chin-Chhüan (3).

Tan Ching Yao Chüeh.
   See *Thai-Chhing Tan Ching Yao Chüeh*.

Tan Fang Ao Lun 丹房奧論.
   Subtle Discourse on the (Alchemical) Elaboratory (of the Human Body, for making the Enchymoma).
   Sung, +1020.
   Chhêng Liao-I 程了一.
   TT/913, and in *TTCY* (*chung mao chi*, 5).

Tan Fang Chien Yuan 丹方鑑源.
   The Mirror of Alchemical Processes (and Reagents); a Source-book.
   Wu Tai (H/Shu), *c.* +938 to +965.
   Tuku Thao 獨孤滔.
   Descr. Fêng Chia-Lo & Collier (1).
   See Ho Ping-Yü & Su Ying-Hui (1).
   TT/918.

Tan Fang Ching Yuan 丹房鏡源.
   The Mirror of the Alchemical Elaboratory; a Source-book.
   Early Thang, not later than +800.
   Writer unknown.

Survives only incorporated in *TT*/912 and in *CLPT*.
   See Ho Ping-Yü & Su Ying-Hui (1).

Tan Fang Hsü Chih 丹房須知.
   Indispensable Knowledge for the Chymical Elaboratory [with illustrations of apparatus].
   Sung, +1163.
   Wu Wu 吳悞.
   TT/893.

Tan Fang Pao Chien Chih Thu 丹房寶鑑之圖.
   [= *Tzu Yang Tan Fang Pao Chien Chih Thu*.]
   Precious Mirror of the Elixir and Enchymoma Laboratory; Tables and Pictures (to illustrate the Principles).
   Sung, *c.* +1075.
   Chang Po-Tuan 張伯端 (Tzu Yang Tzu 紫陽子 or Tzu Yang Chen Jen).
   Incorporated later in *Chin Tan Ta Yao Thu* (q.v.)
   In *Chin Tan Ta Yao* (*TTCY* ed.), ch. 3, pp. 34*a* ff. Also in *Wu Chen Phien* (in *Hsiu Chen Shih Shu*, *TT*/260, ch. 26, pp. 5*a* ff.).
   Cf. Ho Ping-Yü & Needham (2).

Tan I San Chüan 丹擬三卷.
   See Pa Tzu-Yuan (1).

Tan Lun Chüeh Chih Hsin Ching 丹論訣旨心鏡 (*Chien* or *Chao* 鑑, 照 occur as tabu forms in the titles of some versions.)
   Mental Mirror Reflecting the Essentials of Oral Instruction about the Discourses on the Elixir and the Enchymoma.
   Thang, probably +9th.
   Chang Hsüan-Tê 張玄德, criticising the teachings of Ssuma Hsi-I 司馬希夷.
   TT/928, and in *YCCC*, ch. 66, pp. 1*a* ff.
   Tr. Sivin (5).

Tan Thai Hsin Lu 丹臺新錄.
   New Discourse on the Alchemical Laboratory.
   Early Sung or pre-Sung.
   Attrib. Chhing Hsia Tzu 青霞子 or Hsia Yu-Chang 夏有章.
   Extant only in quotations.

Tan-Yang Chen Jen Yü Lu 丹陽眞人玉錄.
   Precious Records of the Adept Tan-Yang.
   Sung, mid +12th cent.
   Ma Yü 馬鈺.
   TT/1044.

Tan-Yang Shen Kuang Tshan 丹陽神光燦.
   Tan Yang (Tzu's Book) on the Resplendent Glow of the Numinous Light.
   Sung, mid +12th cent.
   Ma Yü 馬鈺.
   TT/1136.

Tan Yao Pi Chüeh 丹藥祕訣.
   Confidential Oral Instructions on Elixirs and Drugs.
   Prob. Yuan or early Ming.
   Hu Yen 胡演.
   Now only extant as quotations in the pharmaceutical natural histories.

Tao Fa Hsin Chhuan 道法心傳.
Transmission of (a Lifetime of) Thought on
Taoist Techniques [physiological al-
chemy with special reference to micro-
cosm and macrocosm; many poems and a
long exposition].
Yuan, +1294.
Wang Wei-I 王惟一.
TT/1235, and *TTCY* (*hsia mao chi*, 5).

Tao Fa Hui Yuan 道法會元.
Liturgical and Apotropaic Encyclopaedia of
Taoism.
Thang and Sung.
Writers and compiler unknown.
TT/1203.

Tao Hai Chin Liang 道海津梁.
A Catena (of Words) to Bridge the Ocean
of the Tao.
See Fu Chin-Chhüan (*4*).

Tao Shu 道樞.
Axial Principles of the Tao [doctrinal
treatise, mainly on the techniques of
physiological alchemy].
Sung, early +12th; finished by 1145.
Tsêng Tshao 曾慥.
TT/1005.

Tao Su Fu 擣素賦.
Ode on a Girl of Matchless Beauty [Chao
nü, probably Chao Fei-Yen]; or, Of
What does Spotless Beauty Consist?
C/Han, *c.* −20.
Pan chieh-yü 班婕妤.
In *CSHK*, Chhien Han Sect., ch. 11,
p. 7 a ff.

Tao Tê Ching 道德經.
Canon of the Tao and its Virtue.
Chou, before −300.
Attrib. Li Erh (Lao Tzu) 李耳(老子).
Tr. Waley (*4*); Chhu Ta-Kao (*2*); Lin Yü-
Thang (*1*); Wieger (*7*); Duyvendak (*18*);
and very many others.

Tao Tsang 道藏.
The Taoist Patrology [containing 1464
Taoist works].
All periods, but first collected in the Thang
about +730, then again about +870 and
definitively in +1019. First printed in the
Sung (+1111 to +1117). Also printed in
J/Chin (+1168 to +1191), Yuan (+1244,
+1607). and Ming (+1445, +1598 and
Writers numerous.
Indexes by Wieger (*6*), on which see Pelliot's
review (*58*); and Ong Tu-Chien (Yin-Tê
Index, no. 25).

Tao Tsang Chi Yao 道藏輯要.
Essentials of the Taoist Patrology [con-
taining 287 books, 173 works from the
Taoist Patrology and 114 Taoist works
from other sources].
All periods, pr. 1906 at Erh-hsien-ssu
二仙寺, Chhêngtu.
Writers numerous.

Ed. Ho Lung-Hsiang 賀龍驤 & Phêng
Han-Jan 彭瀚然 (Chhing).

Tao Tsang Hsü Phien Chhu Chi 道藏續篇初集.
First Series of a Supplement to the Taoist
Patrology.
Chhing, early 19th cent.
Edited by Min I-Tê 閔一得.

Tao Yin Yang Shêng Ching 導引養生經.
[= *Thai-Chhing Tao Yin Yang Shêng Ching.*]
Manual of Nourishing the Life-Force (or,
Attaining Longevity and Immortality) by
Gymnastics.
Late Thang, Wu Tai, or early Sung.
Writer unknown.
TT/811, and in *YCCC*, ch. 34.
Cf. Maspero (*7*), pp. 415 ff.

Têng Chen Yin Chüeh 登眞隱訣.
Confidential Instructions for the Ascent to
Perfected (Immortality).
Chin and S/Chhi. Original material from
the neighbourhood of +365 to +366;
commentary (the 'Confidential Instruct-
ions' of the title) by Thao Hung-Ching
(+456 to +536) written between +493
and +498.
Original writer unknown.
Ed. Thao Hung-Ching 陶弘景.
TT/418, but conservation fragmentary.
Cf. Maspero (*7*), pp. 192, 374.

Thai-Chhing Chen Jen Ta Tan 太清眞人大丹.
[Alternative later name of *Thai-Chhing
Tan Ching Yao Chüeh.*]
The Great Elixirs of the Adepts; a Thai-
Chhing Scripture.
Thang, mid +7th (*c.* +640).
Prob. Sun Ssu-Mo 孫思邈.
In *YCCC*, ch. 71.
Tr. Sivin (*1*), pp. 145 ff.

Thai-Chhing Chin I Shen Chhi Ching 太清金
液神氣經.
Manual of the Numinous Chhi of Potable
Gold; a Thai-Chhing Scripture.
Ch. 3 records visitations by the Lady Wei
Hua-Tshun and her companion divinities
mostly paralleling texts in the *Chen Kao*.
They were taken down by Hsü Mi's great-
grandson Hsü Jung-Ti (d. +435), *c.* +430.
Chs 1 and 2 are Thang or Sung, before
+1150. If pre-Thang, cannot be earlier
than +6th.
Writers mainly unknown.
TT/875.

Thai-Chhing Chin I Shen Tan Ching 太清金液
神丹經.
Manual of the Potable Gold (or Metallous
Fluid), and the Magical Elixir (or
Enchymoma); a Thai-Chhing Scripture.
Date unknown, but must be pre-Liang
(Chhen Kuo-Fu (*1*), vol. 2, p. 419). Con-
tains dates between +320 and +330, but
most of the prose is more probably of the
early +5th century.

Thai-Chhing Chin I Shen Tan Ching (cont.)
Preface and main texts of *nei tan* character,
all the rest *wai tan*, including laboratory
instructions.
Writer unknown; chs. variously attributed.
The third chapter, devoted to descriptions
of foreign countries which produced
cinnabar and other chemical substances,
may be of the second half of the +7th
century (see Maspero (14), pp. 95 ff.).
Most were based on Wan Chen's *Nan
Chou I Wu Chih* (+3rd cent.), but not
the one on the Roman Orient (Ta-Chhin)
translated by Maspero. Stein (5) has
pointed out however that the term *Fu-
Lin* for Byzantium occurs as early as
+500 to +520, so the third chapter may
well be of the early +6th century.
TT/873.
Abridged in *YCCC* ch. 65, pp. 1 a ff.
Cf. Ho Ping-Yü (10).

Thai-Chhing Ching Thien-Shih Khou Chüeh
太清經天師口訣.
Oral Instructions from the Heavenly Masters
[Taoist Patriarchs] on the Thai-Chhing
Scriptures.
Date unknown, but must be after the mid
+5th cent. and before Yuan.
Writer unknown.
TT/876.

Thai-Chhing Chung Huang Chen Ching 太清中
黄眞經.
See *Chung Huang Chen Ching*.

Thai-Chhing Shih Pi Chi 太清石壁記.
The Records in the Rock Chamber (lit.
Wall); a Thai-Chhing Scripture.
Liang, early +6th, but includes earlier work
of Chin time as old as the late +3rd,
attributed to Su Yuan-Ming.
Edited by Chhu Tsê hsien-sêng 楚澤先生.
Original writer, Su Yuan-Ming 蘇元明
(Chhing Hsia Tzu 青霞子).
TT/874.
Tr. Ho Ping-Yü (8).
Cf. *Lo-fou Shan Chih*, ch. 4, p. 13 a.

Thai-Chhing Tan Ching Yao Chüeh 太清丹經
要訣.
[= *Thai-Chhing Chen Jen Ta Tan*.]
Essentials of the Elixir Manuals, for Oral
Transmission; a Thai-Chhing Scripture.
Thang, mid +7th (*c.* +640).
Prob. Sun Ssu-Mo 孫思邈.
In *YCCC*, ch. 71.
Tr. Sivin (1), pp. 145 ff.

Thai-Chhing Tao Yin Yang Shêng Ching 太清
導引養生經.
See *Tao Yin Yang Shêng Ching*.

Thai-Chhing Thiao Chhi Ching 太清調氣經.
Manual of the Harmonising of the Chhi; a
Thai-Chhing Scripture [breathing exer-
cises for longevity and immortality].
Thang or Sung, +9th or +10th.

Writer unknown.
TT/813.
Cf. Maspero (7), p. 202.

Thai-Chhing (Wang Lao) (Fu Chhi) Khou Chüeh
(or *Chhuan Fa*) 太清王老服氣口訣
(傳法).
The Venerable Wang's Instructions for
Absorbing the Chhi; a Thai-Chhing
Scripture [Taoist breathing exercises].
Thang or Wu Tai (the name of Wang added
in the +11th).
Writer unknown.
Part due to a woman Taoist, Li I 李液.
TT/815, and in *YCCC*, ch. 62, pp. 1 a ff.
and ch. 59, pp. 10 a ff.
Cf. Maspero (7), p. 209.

Thai-Chhing Yü Pei Tzu 太清玉碑子.
The Jade Stele (Inscription); a Thai-
Chhing Scripture [dialogues between
Chêng Yin and Ko Hung].
Date unknown, prob. late Sung or Yuan.
Writer unknown.
TT/920.
Cf. *Ta Tan Wên Ta* and *Chin Mu Wan Ling
Lun*, which incorporate parallel passages.

*Thai-Chi Chen-Jen Chiu Chuan Huan Tan
Ching Yao Chüeh* 太極眞人九轉還丹
經要訣.
Essential Teachings of the Manual of the
Supreme-Pole Adept on the Ninefold
Cyclically Transformed Elixir.
Date unknown, perhaps Sung on account
of the pseudonym, but the Manual
(*Ching*) itself may be pre-Sui because its
title is in the *Sui Shu* bibliography. Mao
Shan influence is revealed by an account
of five kinds of magic plants or mush-
rooms that grow on Mt Mao, and in-
structions of Lord Mao for ingesting
them.
Writer unknown.
TT/882.
Partial tr. Ho Ping-Yü (9).

Thai-Chi Chen-Jen Tsa Tan Yao Fang 太極眞
人雜丹藥方.
Tractate of the Supreme-Pole Adept on
Miscellaneous Elixir Recipes [with illus-
trations of alchemical apparatus].
Date unknown, but probably Sung on
account of the philosophical significance
of the pseudonym.
Writer unknown.
TT/939.

Thai-Chi Ko Hsien-Ong Chuan 太極葛仙翁傳.
Biography of the Supreme-Pole Elder-
Immortal Ko (Hsüan).
Prob. Ming.
Than Ssu-Hsien 譚嗣先.
TT/447.

Thai Hsi Ching 胎息經.
Manual of Embryonic Respiration.
Thang, +8th, *c.* +755.

Thai Hsi Ching (cont.)
Huan Chen hsien-sêng 幻眞先生
(Mr Truth-and-Illusion).
TT/127, and *YCCC*, ch. 60, pp. 22 *b* ff.
Tr. Balfour (1).
Cf. Maspero (7), p. 211.

Thai Hsi Ching Wei Lun 胎息精微論.
Discourse on Embryonic Respiration and
the Subtlety of the Seminal Essence.
Thang or Sung.
Writer unknown.
In *YCCC*, ch. 58, pp. 1 *a* ff.
Cf. Maspero (7), p. 210.

Thai Hsi Kên Chih Yao Chüeh 胎息根旨要訣.
Instruction on the Essentials of (Under-
standing) Embryonic Respiration [Taoist
respiratory and sexual techniques].
Thang or Sung.
Writer unknown.
In *YCCC*, ch. 58, pp. 4 *b* ff.
Cf. Maspero (7), p. 380.

Thai Hsi Khou Chüeh 胎息口訣.
Oral Explanation of Embryonic Respiration.
Thang or Sung.
Writer unknown.
In *YCCC*, ch. 58, pp. 12 *a* ff.
Cf Maspero (7), p. 198.

Thai Hsi Shui Fa 泰西水法.
Hydraulic Machinery of the West.
Ming, +1612.
Hsiung San-Pa (Sabatino de Ursis) 熊三拔
& Hsü Kuang-Chhi 徐光啓.

Thai Hsüan Pao Tien 太玄寶典.
Precious Records of the Great Mystery [of
attaining longevity and immortality by
physiological alchemy, *nei tan*].
Sung or Yuan, +13th or +14th.
Writer unknown.
TT/1022, and in *TTCY* (*shang mao chi*, 5).

Thai-I Chin Hua Tsung Chih 太一(or 乙)金華
宗旨.
Principles of the (Inner) Radiance of the
Metallous (Enchymoma), (explained in
terms of the) Undifferentiated Universe.
See *Chin Hua Tsung Chih*.

Thai-Ku Chi 太古集.
Collected Works of (Ho) Thai-Ku [Ho Ta-
Thung].
Sung, *c.* +1200.
Ho Ta-Thung 郝大通.
TT/1147.

Thai Ku Thu Tui Ching 太古土兌經.
Most Ancient Canon of the Joy of the Earth;
or, of the Element Earth and the Kua
Tui [mainly on the alchemical sub-
duing of metals and minerals].
Date unknown, perhaps Thang or slightly
earlier.
Attrib. Chang hsien-sêng 張先生.
TT/942.

Thai Pai Ching 太白經.
The Venus Canon.

Thang, *c.* +800.
Shih Chien-Wu 施肩吾.
TT/927.

Thai Phing Ching 太平經.
[= *Thai Phing Chhing Ling Shu*.]
Canon of the Great Peace (and Equality).
Ascr. H/Han, *c.* +150 (first mentioned
+166) but with later additions and inter-
polations.
Part attrib. Yü Chi 于吉.
Perhaps based on the *Thien Kuan Li Pao
Yuan Thai Phing Ching* (*c.* −35) of Kan
Chung-Kho 甘忠可.
TT/1087. Reconstructed text, ed. Wang
Ming (2).
Cf. Yü Ying-Shih (2), p. 84.
According to Hsiung Tê-Chi (1) the parts
which consist of dialogue between a
Heavenly Teacher and a disciple corre-
spond with what the *Pao Phu Tzu*
bibliography lists as *Thai Phing Ching*
and were composed by Hsiang Khai
襄楷.
The other parts would be for the most part
fragments of the *Chia I Ching* 甲乙經,
also mentioned in *Pao Phu Tzu*, and due
to Yü Chi and his disciple Kung Chhung
宮崇 between +125 and +145.

Thai Phing Chhing Ling Shu 太平清領書.
Received Book of the Great Peace and
Purity.
See *Thai Phing Ching*.

Thai-Phing Huan Yü Chi 太平寰宇記.
Thai-Phing reign-period General Descrip-
tion of the World [geographical record].
Sung, +976 to +983.
Yüeh Shih 樂史.

Thai-Phing Hui Min Ho Chi Chü Fang 太平惠
民和劑局方.
Standard Formularies of the (Government)
Great Peace People's Welfare Pharmacies
[based on the *Ho Chi Chü Fang*, etc.].
Sung, +1151.
Ed. Chhen Shih-Wên 陳師文, Phei
Tsung-Yuan 裴完元, and Chhen
Chhêng 陳承.
Cf. Li Thao (1, 6); SIC, p. 973.

Thai-Phing Kuang Chi 太平廣記.
Copious Records collected in the Thai-
Phing reign-period [anecdotes, stories,
mirabilia and memorabilia].
Sung, +978.
Ed. Li Fang 李昉.

Thai-Phing Shêng Hui Fang 太平聖惠方.
Prescriptions Collected by Imperial
Benevolence during the Thai-Phing
reign-period.
Sung, commissioned +982; completed
+992.
Ed. Wang Huai-Yin 王懷隱, Chêng Yen
鄭彥 *et al.*
SIC, p. 921; *Yü Hai*, ch. 63.

Thai-Phing Yü Lan 太平御覽.
 Thai-Phing reign-period Imperial Encyclo-
  paedia (lit. the Emperor's Daily Readings).
 Sung, +983.
 Ed. Li Fang 李昉.
 Some chs. tr. Pfizmaier (84–106).
 Yin-Tê Index, no. 23.
*Thai-Shang Chu Kuo Chiu Min Tsung Chen Pi
 Yao* 太上助國救民總眞祕要.
 Arcane Essentials of the Mainstream of
  Taoism, for the Help of the Nation and
  the Saving of the People; a Thai-Shang
  Scripture [apotropaics and liturgy].
 Sung, +1016.
 Yuan Miao-Tsung 元妙宗.
 TT/1210.
Thai-Shang Chuan Hsi Wang Mu Wo Ku Fa
 太上傳西王母握固法.
 See *Chuan Hsi Wang Mu Wo Ku Fa.*
Thai-Shang Huang Thing Nei (or *Wai* or *Chung*)
 Ching (*Yü*) *Ching* 太上黃庭內(外,中)
 景(玉)經.
 See *Huang Thing,* etc.
Thai-Shang Lao Chün Yang Shêng Chüeh 太上
 老君養生訣.
 Oral Instructions of Lao Tzu on Nourishing
  the Life-Force; a Thai-Shang Scripture
  [Taoist respiratory and gymnastic exer-
  cises].
 Thang.
 Attrib. Hua Tho 華佗 and Wu Phu
  吳普.
 Actual writer unknown.
 TT/814.
Thai-Shang Ling-Pao Chih Tshao Thu 太上靈
 寶芝草圖.
 Illustrations of the Numinous Mushrooms;
  a Thai-Shang Ling-Pao Scripture.
 Sui or pre-Sui.
 Writer unknown.
 TT/1387.
Thai-Shang Ling-Pao Wu Fu (*Ching*) 太上靈
 寶五符(經).
 (Manual of) the Five Categories of For-
  mulae (for achieving Material and
  Celestial Immortality); a Thai-Shang
  Ling-Pao Scripture [liturgical].
 San Kuo, mid +3rd.
 Writers unknown.
 TT/385.
 On the term Ling-Pao see Kaltenmark (4).
Thai-Shang Pa-Ching Ssu-Jui Tzu-Chiang (*Wu-
 Chu*) *Chiang-Shêng Shen Tan Fang* 太上
 八景四蘂紫漿(五珠)降生神丹方.
 Method for making the Eight-Radiances
  Four-Stamens Purple-Fluid (Five-Pearl)
  Incarnate Numinous Elixir; a Thai-
  Shang Scripture.
 Chin, probably late +4th.
 Putatively dictated to Yang Hsi 楊羲.
 In *YCCC*, ch. 68; another version in
  TT/1357.

Thai-Shang Pa Ti Yuan (*Hsüan*) *Pien Ching*
 太上八帝元(玄)變經.
 See *Tung-Shen Pa Ti Yuan* (*Hsüan*) *Pien
  Ching.*
Thai Shang-San-shih-liu pu Tsun Ching 太上
 三十六部尊經.
 The Venerable Scripture in 36 Sections.
 TT/8.
 See *Shang Chhing Ching.*
Thai-Shang Tung Fang Nei Ching Chu 太上洞
 房內經注.
 Esoteric Manual of the Innermost Chamber,
  a Thai-Shang Scripture; with Commen-
  tary.
 Ascr. −1st cent.
 Attrib. Chou Chi-Thung 周季通.
 TT/130.
Thai-Shang Tung-Hsüan Ling-Pao Mieh Tu (or
 San Yuan) *Wu Lien Shêng Shih Miao
 Ching* 太上洞玄靈寶滅度 (or 三元)
 五鍊生尸妙經.
 Marvellous Manual of the Resurrection (or
  Preservation) of the Body, giving Salvation
  from Dispersal, by means of (the Three
  Primary Vitalities and) the Five Trans-
  mutations; a Ling-Pao Thai-Shang Tung-
  Hsüan Scripture.
 Date uncertain.
 Writer unknown.
 TT/366.
Thai-Shang Tung-Hsüan Ling-Pao Shou Tu I
 太上洞玄靈寶授度儀.
 Formulae for the Reception of Salvation; a
  Thai-Shang Tung-Hsüan Ling-Pao
  Scripture [liturgical].
 L/Sung, *c.* +450.
 Lu Hsiu-Ching 陸修靜.
 TT/524.
*Thai-Shang Wei Ling Shen Hua Chiu Chuan
 Tan Sha Fa* 太上衛靈神化九轉丹砂
 法.
 Methods of the Guardian of the Mysteries
  for the Marvellous Thaumaturgical
  Transmutation of Ninefold Cyclically
  Transformed Cinnabar; a Thai-Shang
  Scripture.
 Sung, if not earlier.
 Writer unknown.
 TT/885.
 Tr. Spooner & Wang (1); Sivin (3).
Thai-Shang Yang Shêng Thai Hsi Chhi Ching
 太上養生胎息氣經.
 See *Yang Shêng Thai Hsi Chhi Ching.*
Thai Tsang Lun 胎臟論.
 Discourse on the Foetalisation of the
  Viscera (the Restoration of the Embry-
  onic Condition of Youth and Health).
 Alternative title of *Chung Huang Chen
  Ching* (q.v.).
*Thai-Wei Ling Shu Tzu-Wên Lang-Kan Hua
 Tan Shen Chen Shang Ching* 太微靈書
 紫文琅玕華丹神眞上經.

Thai-Wei Ling Shu Tzu-Wên Lang-Kan Hua Tan Shon Chen Shang Ching (cont.)

Divinely Written Exalted Spiritual Realisation Manual in Purple Script on the Lang-Kan (Gem) Radiant Elixir; a Thai-Wei Scripture.

Chin, late +4th century, possibly altered later.

Dictated to Yang Hsi 楊羲.

TT/252.

Thai-Wu hsien-sêng Fu Chhi Fa 太无先生服氣法.

See *Sung Shan Thai-Wu hsien-sêng Chhi Ching.*

Than hsien-sêng Shui Yün Chi 譚先生水雲集.

Mr Than's Records of Life among the Mountain Clouds and Waterfalls.

Sung, mid +12th cent.

Than Chhu-Tuan 譚處端.

TT/1146.

Thang Hui Yao 唐會要.

History of the Administrative Statutes of the Thang Dynasty.

Sung, +961.

Wang Phu 王溥.

Cf. des Rotours (2), p. 92.

Thang Liu Tien 唐六典.

Institutes of the Thang Dynasty (lit. Administrative Regulations of the Six Ministries of the Thang).

Thang, +738 or +739.

Ed. Li Lin-Fu 李林甫.

Cf. des Rotours (2), p. 99.

Thang Pên Tshao 唐本草.

Pharmacopoeia of the Thang Dynasty.

= *Hsin Hsiu Pên Tshao*, (q.v.).

Thang Yü Lin 唐語林.

Miscellanea of the Thang Dynasty.

Sung, collected c. +1107.

Wang Tang 王讜.

Cf. des Rotours (2), p. 109.

Thao Chen Jen Nei Tan Fu 陶眞人內丹賦.

See *Nei Tan Fu.*

Thi Kho Ko 體殼歌.

Song of the Bodily Husk (and the Deliverance from its Ageing).

Wu Tai or Sung, in any case before +1040

Yen Lo Tzu (ps.) 煙蘿子.

In *Hsiu Chen Shih Shu* (*TT/260*), ch. 18.

Thiao Chhi Ching 調氣經.

See *Thai-Chhing Thiao Chhi Ching.*

Thieh Wei Shan Tshung Than 鐵圍山叢談.

Collected Conversations at Iron-Fence Mountain.

Sung, c. +1115.

Tshai Thao 蔡絛.

Thien-Hsia Chün Kuo Li Ping Shu 天下郡國利病書.

Merits and Drawbacks of all the Countries in the World [geography].

Chhing, +1662.

Ku Yen-Wu 顧炎武.

Thien Hsien Chêng Li Tu 'Fa Tien Ching' 天仙正理讀法點睛.

The Right Pattern of the Celestial Immortals; Thoughts on Reading the *Consecration of the Law.*

See *Fu Chin-Chhüan* (2).

Thien Hsien Chih Lun Chhang Shêng Tu Shih Nei Lien Chin Tan (Chüeh Hsin) Fa 天仙直論長生度世內煉金丹(訣心)法.

(Confidential) Methods for Processing the Metallous Enchymoma; a Plain Discourse on Longevity and Immortality (according to the Principles of the) Celestial Immortals for the Salvation of the World.

Alternative title for *Nei Chin Tan* (q.v.).

Thien Kung Khai Wu 天工開物.

The Exploitation of the Works of Nature.

Ming, +1637.

Sung Ying-Hsing 宋應星.

Tr. Sun Jen I-Tu & Sun Hsüeh-Chuan (1).

Thien-thai Shan Fang Wai Chih 天臺山方外志.

Supplementary Historical Topography of Thien-thai Shan.

Ming.

Chhuan-Têng (monk) 傳燈.

Thien Ti Yin-Yang Ta Lo Fu 天地陰陽大樂賦.

Poetical Essay on the Supreme Joy.

Thang, c. +800.

Pai Hsing-Chien 白行簡.

Thien Yuan Ju Yao Ching 天元入藥鏡.

Mirror of the All-Penetrating Medicine (the Enchymoma; restoring the Endowment) of the Primary Vitalities.

Wu Tai, +940.

Tshui Hsi-Fan 崔希範.

In *Hsiu Chen Shih Shu* (*TT/260*), ch. 21, pp. 6b to 9b; a prose text without commentary, not the same as the *Ju Yao Ching* (q.v.) and ending with a diagram absent from the latter.

Cf. van Gulik (8), pp. 224 ff.

Tho Yo Tzu 橐籥子.

Book of the Bellows-and-Tuyère Master [physiological alchemy in mutationist terms].

Sung or Yuan.

Writer unknown.

TT/1174, and *TTCY* (*hsin mao chi*, 5).

Thou Huang Tsa Lu 投荒雜錄.

Miscellaneous Jottings far from Home.

Thang, c. +835.

Fang Chhien-Li 房千里.

Thu Ching (Pên Tshao) 圖經(本草).

Illustrated Treatise (of Pharmaceutical Natural History). See *Pên Tshao Thu Ching.*

The term *Thu Ching* applied originally to one of the two illustrated parts (the other being a *Yao Thu*) of the *Hsin Hsiu Pên*

Thu Ching (Pên Tshao) (cont.)
> *Tshao* of +659 (q.v.); cf. *Hsin Thang Shu*, ch. 59, p. 21*a* or *TSCCIW*, p. 273. By the middle of the +11th century these had become lost, so Su Sung's *Pên Tshao Thu Ching* was prepared as a replacement. The name *Thu Ching Pên Tshao* was often afterwards applied to Su Sung's work, but (according to the evidence of the *Sung Shih* bibliographies, *SSIW*, pp. 179, 529) wrongly.

Thu Ching Chi-Chu Yen I Pên Tshao 圖經集注衍義本草.
> Illustrations and Collected Commentaries for the *Dilations upon Pharmaceutical Natural History*.
> *TT*/761 (Ong index, no. 767).
> See also *Thu Ching Yen I Pên Tshao*.
> The *Tao Tsang* contains two separately catalogued books, but the *Thu Ching Chi-Chu Yen I Pên Tshao* is in fact the introductory 5 chapters, and the *Thu Ching Yen I Pên Tshao* the remaining 42 chapters of a single work.

Thu Ching Yen I Pên Tshao 圖經衍義本草.
> Illustrations (and Commentary) for the *Dilations upon Pharmaceutical Natural History*. (An abridged conflation of the *Chêng-Ho...Chêng Lei...Pên Tshao* with the *Pên Tshao Yen I*.)
> Sung, c. +1223.
> Thang Shen-Wei 唐愼微, Khou Tsung-Shih 寇宗奭, ed. Hsü Hung 許洪.
> *TT*/761 (Ong index, no. 768).
> See also *Thu Ching Chi-Chu Yen I Pên Tshao*.
> Cf. Chang Tsan-Chhen (*2*); Lung Po-Chien (*1*), nos. 38, 39.

Thu Hsiu Chen Chün Tsao-Hua Chih Nan 土宿眞君造化指南.
> Guide to the Creation, by the Earth's Mansions Immortal.
> See *Tsao-Hua Chih Nan*.

Thu Hsiu Pên Tshao 土宿本草.
> The Earth's Mansions Pharmacopoeia.
> See *Tsao-Hua Chih Nan*.

Thung Hsüan Pi Shu 通玄秘術.
> The Secret Art of Penetrating the Mystery [alchemy].
> Thang, soon after +864.
> Shen Chih-Yen 沈知言.
> *TT*/935.

Thung Su Pien 通俗編.
> Thesaurus of Popular Terms, Ideas and Customs.
> Chhing, +1751.
> Tsê Hao 翟灝.

Thung Ya 通雅.
> Helps to the Understanding of the *Literary Expositor* [general encyclopaedia with much of scientific and technological interest].

Ming and Chhing, finished +1636, pr. +1666.
> Fang I-Chih 方以智.

Thung Yu Chüëh 通幽訣.
> Lectures on the Understanding of the Obscurity (of Nature) [alchemy, proto-chemical and physiological].
> Not earlier than Thang.
> Writer unknown.
> *TT*/906.
> Cf. Chhen Kuo-Fu (*1*), vol. 2, p. 390.

Tien Hai Yü Hêng Chih 滇海虞衡志.
> A Guide to the Region of the Kunming Lake (Yunnan).
> Chhing, c. +1770, pr. +1799.
> Than Tshui 檀萃.

Tien Shu 典術.
> Book of Arts.
> L/Sung.
> Wang Chien-Phing 王建平.

Ting Chhi Ko 鼎器歌.
> Song (or, Mnemonic Rhymes) on the (Alchemical) Reaction-Vessel.
> Han, if indeed originally, as it is now, a chapter of the *Chou I Tshan Thung Chhi* (q.v.).
> It has sometimes circulated separately.
> In *Chou I Tshan Thung Chhi Fên Chang Chu Chieh*, ch. 33 (ch. 3, pp. 7*a* ff.).
> Cf. *Chou I Tshan Thung Chhi Ting Chhi Ko Ming Ching Thu* (*TT*/994).

Ton Isho 頓醫抄.
> Medical Excerpts Urgently Copied.
> Japan, +1304.
> Kajiwara Shozen 梶原性全.

Tongŭi Pogam 東醫寶鑑.
> See *Tung I Pao Chien*.

Tou hsien-sêng Hsiu Chen Chih Nan 寶先生修眞指南.
> See *Hsi Yo Tou hsien-sêng Hsiu Chen Chih Nan*.

Tsao Hua Chhien Chhui 造化鉗鎚.
> The Hammer and Tongs of Creation (i.e. Nature).
> Ming, c. +1430.
> Chu Chhüan 朱權.
> (Ning Hsien Wang 寧獻王, prince of the Ming.)

Tsao-Hua Chih Nan 造化指南.
> [= *Thu Hsiu Pên Tshao*.]
> Guide to the Creation (i.e. Nature).
> Thang, Sung or possibly Ming. A date about +1040 may be the best guess, as there are similarities with the *Wai Tan Pên Tshao* (q.v.).
> Thu Hsiu Chen Chün 土宿眞君 (the Earth's Mansions Immortal).
> Preserved only in quotation, as in *PTKM*.

Tsê Ko Lu 則克錄.
> Methods of Victory.
> Title, in certain editions, of the *Huo Kung Chieh Yao* (q.v.).

Tsêng Kuang Chih Nang Pu 增廣智囊補.
Additions to the *Enlarged Bag of Wisdom* Supplemented.
Ming, c. +1620.
Fêng Mêng-Lung 馮夢龍.

Tshai Chen Chi Yao 採真機要.
Important (Information on the) Means (by which one can) Attain (the Regeneration of the) Primary (Vitalities) [physiological alchemy, poems and commentary].
Part of *San-Fêng Tan Chüeh* (q.v.).

Tshan Thung Chhi 參同契.
The Kinship of the Three; or, The Accordance (of the *Book of Changes*) with the Phenomena of Composite Things [alchemy].
H/Han, +142.
Wei Po-Yang 魏伯陽.

Tshan Thung Chhi.
See also titles under *Chou I Tshan Thung Chhi.*

Tshan Thung Chhi Chang Chü 參同契章句
The *Kinship of the Three* (arranged in) Chapters and Sections.
Chhing, +1717.
Ed. Li Kuang-Ti 李光地.

Tshan Thung Chhi Khao I 參同契考異.
[= *Chou I Tshan Thung Chhi Chu.*]
A Study of the *Kinship of the Three.*
Sung, +1197.
Chu Hsi 朱熹 (originally using pseudonym Tsou Hsin 鄒訢).
TT/992.

Tshan Thung Chhi Shan Yu 參同契闡幽.
Explanation of the Obscurities in the *Kinship of the Three.*
Chhing, pref. +1729, pr. +1735.
Ed and comm. Chu Yuan-Yü 朱元育.
TTCY.

Tshan Thung Chhi Wu Hsiang Lei Pi Yao 參同契五相類祕要
Arcane Essentials of the Similarities and Categories of the Five (Substances) in the *Kinship of the Three* (sulphur, realgar, orpiment, mercury and lead).
Liu Chhao, possibly Thang; prob. between +3rd and +7th cents., must be before the beginning of the +9th cent., though ascr. +2nd.
Writer unknown (attrib. Wei Po-Yang).
Comm. by Lu Thien-Chi 盧天驥, wr. Sung, +1111 to +1117, probably +1114.
TT/898.
Tr. Ho Ping-Yü & Needham (2).

Tshao Mu Tzu 草木子.
The Book of the Fading-like-Grass Master.
Ming, +1378.
Yeh Tzu-Chhi 葉子奇.

Tshê Fu Yuan Kuei 册府元龜.
Collection of Material on the Lives of Emperors and Ministers, (lit. (Lessons of) the Archives, (the True) Scapulimancy);

[a governmental ethical and political encyclopaedia.]
Sung, commissioned +1005, pr. +1013.
Ed. Wang Chhin-Jo 王欽若 & Yang I 楊億.
Cf. des Rotours (2), p. 91.

Tshui Hsü Phien 翠虛篇.
Book of the Emerald Heaven.
Sung, c. +1200.
Chhen Nan 陳楠.
TT/1076.

Tshui Kung Ju Yao Ching Chu (or *Ho*) *Chieh* 崔公入藥鏡註(合)解.
See *Ju Yao Ching* and *Thien Yuan Ju Yao Ching.*

Tshun Chen Huan Chung Thu 存眞環中圖.
Illustrations of the True Form (of the Body) and of the (Tracts of) Circulation (of the Chhi).
Sung, +1113.
Yang Chieh 楊介.
Now partially preserved only in the *Ton-Isho* and the *Man-Anpō* (q.v.). Some of the drawings are in Chu Hung's *Nei Wai Erh Ching Thu*, also in *Hua Tho Nei Chao Thu* and *Kuang Wei Ta Fa* (q.v.).

Tshun Fu Chai Wên Chi 存復齋文集.
Literary Collection of the Preservation-and-Return Studio.
Yuan, +1349.
Chu Tê-Jun 朱德潤.

Tso Chuan 左傳.
Master Tso chhiu's Tradition (or Enlargement) of the *Chhun Chhiu* (*Spring and Autumn Annals*), [dealing with the period −722 to −453].
Late Chou, compiled from ancient written and oral traditions of several States between −430 and −250, but with additions and changes by Confucian scholars of the Chhin and Han, especially Liu Hsin. Greatest of the three commentaries on the *Chhun Chhiu*, the others being the *Kungyang Chuan* and the *Kuliang Chuan*, but unlike them, probably originally itself an independent book of history.
Attrib. Tsochhiu Ming 左邱明.
See Karlgren (8); Maspero (1); Chhi Ssu-Ho (1); Wu Khang (1); Wu Shih-Chhang (1); van der Loon (1), Eberhard, Müller & Henseling (1).
Tr. Couvreur (1); Legge (11); Pfizmaier (1–12).
Index by Fraser & Lockhart (1).

Tso Wang Lun 坐忘論.
Discourse on (Taoist) Meditation.
Thang, c. +715.
Ssuma Chhêng-Chên 司馬承貞.
TT/1024, and in *TTCY* (*shang mao chi*, 5).

Tsui Shang I Chhêng Hui Ming Ching 最上一乘慧命經.
Exalted Single-Vehicle Manual of the Sagacious (Lengthening of the) Life-Span.
See *Hui Ming Ching*.

Tsun Shêng Pa Chien 遵生八牋.
Eight Disquisitions on Putting Oneself in Accord with the Life-Force [a collection of works].
Ming, +1591.
Kao Lien 高濂.
For the separate parts see:
1. *Chhing Hsiu Miao Lun Chien* (chs. 1, 2).
2. *Ssu Shih Thiao Shê Chien* (chs. 3-6).
3. *Chhi Chü An Lo Chien* (chs. 7, 8).
4. *Yen Nien Chhio Ping Chien* (chs. 9, 10).
5. *Yin Chuan Fu Shih Chien* (chs. 11-13).
6. *Yen Hsien Chhing Shang Chien* (chs. 14, 15).
7. *Ling Pi Tan Yao Chien* (chs. 16-18).
8. *Lu Wai Hsia Chü Chien* (ch. 19).

Tsurezuregusa 徙然草.
Gleanings of Leisure Moments [miscellanea, with much on Confucianism, Buddhism and Taoist philosophy].
Japan, c. +1330.
Kenkō hōshi 兼好法師 (Yoshida no Kaneyoshi 吉田兼好).
Cf. Anon. (103), pp. 197 ff.

Tu Hsing Tsa Chih 獨醒雜志.
Miscellaneous Records of the Lone Watcher.
Sung, +1176.
Tsêng Min-Hsing 曾敏行.

Tu I Chih 獨異志.
Things Uniquely Strange.
Thang.
Li Jung 李冗 (*or* 冘).

Tu Jen Ching 度人經.
See *Ling-Pao Wu Liang Tu Jen Shang Phin Miao Ching*.

Tu Shih Fang Yü Chi Yao 讀史方興紀要.
Essentials of Historical Geography.
Chhing, first pr. +1667, greatly enlarged before the author's death in +1692, and pr. c. +1799.
Ku Tsu-Yü 顧祖禹.

Tung-Chen Ling Shu Tzu-Wên Lang-Kan Hua Tan Shang Ching 洞眞靈書紫文琅玕華丹上經.
Divinely Written Exalted Manual in Purple Script on the Lang-Kan (Gem) Radiant Elixir; a Tung-Chen Scripture.
Alternative name of *Thai-Wei Ling Shu Tzu-Wên Lang-Kan Hua Tan Shen Chen Shang Ching* (q.v.).

Tung-Chen Thai-Shang Su-Ling Tung-Yuan Ta Yu Miao Ching 洞眞太上素靈洞元大有妙經.
See *Ta Yu Miao Ching*.

Tung-Chen Thai-Wei Ling Shu Tzu-Wên Shang Ching 洞眞太微靈書紫文上經.
Divinely Written Exalted Canon in Purple Script; a Tung-Chen Thai-Wei Scripture.
See *Thai-Wei Ling Shu Tzu-Wên Lang-Kan Hua Tan Shen Chen Shang Ching*, which it formerly contained.

Tung Hsien Pi Lu 東軒筆錄.
Jottings from the Eastern Side-Hall.
Sung, end +11th.
Wei Thai 魏泰.

Tung-Hsüan Chin Yü Chi 洞玄金玉集.
Collections of Gold and Jade; a Tung-Hsüan Scripture.
Sung, mid +12th cent.
Ma Yü 馬鈺.
TT/1135.

Tung-Hsüan Ling-Pao Chen Ling Wei Yeh Thu 洞玄靈寶眞靈位業圖.
Charts of the Ranks, Positions and Attributes of the Perfected (Immortals); a Tung-Hsüan Ling-Pao Scripture.
Ascr. Liang, early +6th.
Attrib. Thao Hung-Ching 陶弘景.
TT/164.

Tung Hsüan Tzu 洞玄子.
Book of the Mystery-Penetrating Master.
Pre-Thang, perhaps +5th century.
Writer unknown.
In *Shuang Mei Ching An Tshung Shu*.
Tr van Gulik (3).

Tung I Pao Chien 東醫寶鑑.
Precious Mirror of Eastern Medicine [system of medicine].
Korea, commissioned in +1596, presented +1610, printed +1613.
Hǒ Chun 許浚.

Tung-Pho Shih Chi Chu 東坡詩集注.
[= *Mei-Chhi Shih Chu*.]
Collected Commentaries on the Poems of (Su) Tung-Pho.
Sung, c. +1140.
Wang Shih-Phêng 王十朋 (i.e. Wang Mei-Chhi 王梅溪).

Tung Shen Ching 洞神經.
See *Tung Shen Pa Ti Miao Ching Ching* and *Tung Shen Pa Ti Yuan Pien Ching*.

Tung Shen Pa Ti Miao Ching Ching 洞神八帝妙精經.
Mysterious Canon of Revelation of the Eight (Celestial) Emperors; a Tung-Shen Scripture.
Date uncertain, perhaps Thang but more probably earlier.
Writer unknown.
TT/635.

Tung Shen Pa Ti Yuan (Hsüan) Pien Ching 洞神八帝元(玄)變經.
Manual of the Mysterious Transformations of the Eight (Celestial) Emperors; a Tung-Shen Scripture [nomenclature of

Tung Shen Pa Ti Yuan (Hsüan) Pien Ching (cont.)
spiritual beings, invocations, exorcisms, techniques of rapport].
Date uncertain, perhaps Thang but more probably earlier.
Writer unknown.
TT/1187.

Tzu Chin Kuang Yao Ta Hsien Hsiu Chen Yen I 紫金光耀大仙修眞演義.
See *Hsiu Chen Yen I.*

Tzu-Jan Chi 自然集.
Collected (Poems) on the Spontaneity of Nature.
Sung, mid +12th cent.
Ma Yü 馬鈺.
TT/1130.

Tzu-Yang Chen Jen Nei Chuan 紫陽眞人內傳.
Biography of the Adept of the Purple Yang.
H/Han, San Kuo or Chin, before +399.
Writer unknown.
This Tzu-Yang Chen Jen was Chou I-Shan 周義山 (not to be confused with Chang Po-Tuan).
Cf. Maspero (7), p. 201; (13), pp. 78, 103.
TT/300.

Tzu-Yang Chen Jen Wu Chen Phien 紫陽眞人悟眞篇.
See *Wu Chen Phien.*

Tzu Yang Tan Fang Pao Chien Chih Thu 紫陽丹房寶鑑之圖.
See *Tan Fang Pao Chien Chih Thu.*

Wai Chin Tan 外金丹.
Disclosures (of the Nature of) the Metallous Enchymoma [a collection of some thirty tractates on *nei tan* physiological alchemy, ranging in date from Sung to Chhing and of varying authenticity].
Sung to Chhing.
Ed. Fu Chin-Chhüan 傅金銓, c. 1830.
In *CTPS, pên* 6-10 incl.

Wai Kho Chêng Tsung 外科正宗.
An Orthodox Manual of External Medicine.
Ming, +1617.
Chhen Shih-Kung 陳實功.

Wai Kuo Chuan 外國傳.
See *Wu Shih Wai Kuo Chuan.*

Wai Tan Pên Tshao 外丹本草.
Iatrochemical Natural History.
Early Sung, c. +1045.
Tshui Fang 崔昉.
Now extant only in quotations.
Cf. *Chin Tan Ta Yao Pao Chüeh* and *Ta Tan Yao Chüeh Pên Tshao.*

Wai Thai Pi Yao (Fang) 外臺秘要(方).
Important (Medical) Formulae and Prescriptions now revealed by the Governor of a Distant Province.
Thang, +752.
Wang Thao 王燾.
On the title see des Rotours (1), pp. 294,

721. Wang Thao had had access to the books in the Imperial Library as an Academician before his posting as a high official to the provinces.

Wakan Sanzai Zue 和漢三才圖會.
The Chinese and Japanese Universal Encyclopaedia (based on the *San Tshai Thu Hui*).
Japan, +1712.
Terashima Ryōan 寺島良安.

Wamyō-Honzō. See *Honzō-Wamyō.*

Wamyō Ruijūshō 和 (or 倭) 名類聚抄.
General Encyclopaedic Dictionary.
Japan (Heian), +934.
Minamoto no Shitagau 源順.

Wamyōshō 和名抄.
See *Wamyō Ruijushō.*

Wan Hsing Thung Phu 萬姓統譜.
General Dictionary of Biography.
Ming, +1579.
Ling Ti-Chih 凌迪知.

Wan Ping Hui Chhun 萬病回春.
The Restoration of Well-Being from a Myriad Diseases.
Ming, +1587, pr. +1615.
Kung Thing-Hsien 龔廷賢.

Wan Shou Hsien Shu 萬壽仙書.
A Book on the Longevity of the Immortals [longevity techniques, especially gymnastics and respiratory exercises].
Chhing, +18th.
Tshao Wu-Chi 曹無極.
Included in Pa Tzu-Yuan (1).

Wang Hsien Fu 望仙賦.
Contemplating the Immortals; a Hymn of Praise [ode on Wangtzu Chhiao and Chhih Sung Tzu].
C/Han, −14 or −13.
Huan Than 桓譚.
In *CSHK* (Hou Han sect.), ch. 12, p. 7*b*; and several encyclopaedias.

Wang Lao Fu Chhi Khou Chüeh 王老服氣口訣.
See *Thai-Chhing Wang Lao Fu Chhi Khou Chüeh.*

Wang-Wu Chen-Jen Khou Shou Yin Tan Pi Chüeh Ling Phien 王屋眞人口授陰丹秘訣靈篇.
Numinous Record of the Confidential Oral Instructions on the Yin Enchymoma handed down by the Adept of Wang-Wu (Shan).
Thang, perhaps c. +765; certainly between +8th and late +10th.
Probably Liu Shou 劉守.
In *YCCC*, ch. 64, pp. 13*a* ff.

Wang-Wu Chen-Jen Liu Shou I Chen-Jen Khou Chüeh Chin Shang 王屋眞人劉守依眞人口訣進上.
Confidential Oral Instructions of the Adept of Wang-Wu (Shan) presented to the Court by Liu Shou.

Wang-Wu Chen-Jen Liu Shou I Chen-Jen Khou Chüeh Chin Shang (cont.)
Thang, *c.* +785 (after +780); certainly between +8th and late +10th.
Liu Shou 劉守.
In *YCCC*, ch. 64, pp. 14*a* ff.

Wei Lüeh 緯畧.
Compendium of Non-Classical Matters.
Sung, +12th century (end), *c.* +1190.
Kao Ssu-Sun 高似孫.

Wei Po-Yang Chhi Fan Tan Sha Chüeh.
See *Chhi Fan Tan Sha Chüeh.*

Wei Shêng I Chin Ching 衛生易筋經.
See *I Chin Ching.*

Wei Shu 魏書.
History of the (Northern) Wei Dynasty [+386 to +550, including the Eastern Wei successor State].
N/Chhi, +554, revised +572.
Wei Shou 魏收.
See Ware (3).
One ch. tr. Ware (1, 4).
For translations of passages, see the index of Frankel (1).

Wên Shih Chen Ching 文始眞經.
True Classic of the Original Word (of Lao Chün, third person of the Taoist Trinity).
Alternative title of *Kuan Yin Tzu* (q.v.).

Wên Yuan Ying Hua 文苑英華.
The Brightest Flowers in the Garden of Literature [imperially commissioned collection, intended as a continuation of the *Wên Hsüan* (q.v.) and containing therefore compositions written between +500 and +960].
Sung, +987; first pr. +1567.
Ed. Li Fang 李昉, Sung Pai 宋白 *et al.*
Cf des Rotours (2), p. 93.

Wu Chen Phien 悟眞篇.
[= *Tzu-Yang Chen Jen Wu Chen Phien.*]
Poetical Essay on Realising (the Necessity of Regenerating the) Primary (Vitalities) [Taoist physiological alchemy].
Sung, +1075.
Chang Po-Tuan 張伯端.
In, e.g., *Hsiu Chen Shih Shu* (*TT*/260), chs. 26–30 incl.
TT/138. Cf. *TT*/139–43.
Tr. Davis & Chao Yün-Tshung (7).

Wu Chen Phien Chih Chih Hsiang Shuo San Chhêng Pi Yao 悟眞篇直指祥說三乘祕要.
Precise Explanation of the Difficult Essentials of the *Essay on Realising the Necessity of Regenerating the Primary Vitalities,* in accordance with the Three Classes of (Taoist) Scriptures.
Sung, *c.* +1170.
Ong Pao-Kuang 翁葆光.
TT/140.

Wu Chen Phien San Chu 悟眞篇三註.
Three Commentaries on the *Essay on Realising the Necessity of Regenerating the Primary Vitalities* [Taoist physiological alchemy].
Sung and Yuan, completed *c.* +1331.
Hsüeh Tao-Kuang 薛道光 (or Ong Pao-Kuang 翁葆先), Lu Shu 陸墅 & Tai Chhi-Tsung 戴起宗 (or Chhen Chih-Hsü 陳致虛).
TT/139.
Cf. Davis & Chao Yün-Tshung (7).

Wu Chhêng Tzu 務成子.
See *Huang Thing Wai Ching Yü Ching Chu.*

Wu Chhu Ching 五厨經.
See *Lao Tzu Shuo Wu Chhu Ching.*

Wu Hsiang Lei Pi Yao 五相類祕要.
See *Tshan Thung Chhi Wu Hsiang Lei Pi Yao.*

Wu Hsing Ta I 五行大義.
Main Principles of the Five Elements.
Sui, *c.* +600.
Hsiao Chi 蕭吉.

Wu Hsüan Phien 悟玄篇.
Essay on Understanding the Mystery (of the Enchymoma), [Taoist physiological alchemy].
Sung, +1109 or +1169.
Yü Tung-Chen 余洞眞.
TT/1034, and in *TTCY* (*shang mao chi,* 5).

Wu I Chi 武夷集.
The Wu-I Mountains Literary Collection [prose and poems on physiological alchemy].
Sung, *c.* +1220.
Ko Chhang-Kêng 葛長庚 (Pai Yü-Chhan 白玉蟾).
In *Hsiu Chen Shih Shu* (*TT*/260), chs. 45–52.

Wu Kên Shu 無根樹.
The Rootless Tree [poems on physiological alchemy].
Ming, *c.* +1410 (if genuine).
Attrib. Chang San-Fêng 張三峯.
In *San-Fêng Tan Chüeh* (q.v.).

Wu Lei Hsiang Kan Chih 物類相感志.
On the Mutual Responses of Things according to their Categories.
Sung, *c.* +980.
Attrib. wrongly to Su Tung-Pho 蘇東坡.
Actual writer (Lu) Tsan-Ning (monk) 錄贊寧.
See Su Ying-Hui (1, 2).

Wu Li Hsiao Shih 物理小識.
Small Encyclopaedia of the Principles of Things.
Ming and Chhing, finished by +1643, pr. +1664.
Fang I-Chih 方以智.
Cf. Hou Wai-Lu (3, 4).

Wu Lu 吳錄.
   Record of the Kingdom of Wu.
   San Kuo, +3rd century.
   Chang Pho 張勃.
Wu Shang Pi Yao 無上秘要.
   Essentials of the Matchless Books (of
      Taoism), [a florilegium].
   N/Chou, between +561 and +578.
   Compiler unknown.
   TT/1124.
   Cf. Maspero (13), p. 77; Schipper (1), p. 11.
Wu shih Pên Tshao 吳氏本草.
   Mr Wu's Pharmaceutical Natural History.
   San Kuo (Wei), *c.* +235.
   Wu Phu 吳普.
   Extant only in quotations in later literature.
Wu Shih Wai Kuo Chuan 吳時外國傳.
   Records of the Foreign Countries in the
      Time of the State of Wu.
   San Kuo, *c.* +260.
   Khang Thai 康泰.
   Only in fragments in *TPYL* and other
      sources.
Wu Tai Shih Chi.
   See *Hsin Wu Tai Shih.*
Wu Yuan 物原.
   The Origins of Things.
   Ming, +15th.
   Lo Chhi 羅頎.

Yang Hsing Yen Ming Lu 養性延命錄.
   On Delaying Destiny by Nourishing the
      Natural Forces (or, Achieving Longevity
      and Immortality by Regaining the Vitality
      of Youth), [Taoist sexual and respiratory
      techniques].
   Sung, betw. +1013 and +1161 (acc. to
      Maspero), but as it appears in *YCCC* it
      must be earlier than +1020, very prob-
      ably pre-Sung.
   Attrib. Thao Hung-Ching or Sun Ssu-Mo.
   Actual writer unknown.
   TT/831, abridged version in *YCCC*, ch. 32,
      pp. 1 *a* ff.
   Cf. Maspero (7), p. 232.
Yang Hui Suan Fa 楊輝算法.
   Yang Hui's Methods of Computation.
   Sung, +1275.
   Yang Hui 楊輝.
Yang Shêng Shih Chi 養生食忌.
   Nutritional Recommendations and Pro-
      hibitions for Health [appended to *Pao
      Shêng Hsin Chien*, q.v.].
   Ming, *c.* +1506.
   Thieh Fêng Chü-Shih 鐵峰居士.
      (The Recluse of Iron Mountain, ps.).
   Ed. Hu Wên-Huan (*c.* +1596) 胡文煥.
Yang Shêng Tao Yin Fa 養生導引法.
   Methods of Nourishing the Vitality by
      Gymnastics (and Massage), [appended to
      Pao Shêng Hsin Chien, q.v.].
   Ming, *c.* +1506.

Thieh Fêng Chü-Shih 鐵峰居士.
   (The Recluse of Iron Mountain, ps.)
   Ed. Hu Wên-Huan (*c.* +1596) 胡文煥.
Yang Shêng Thai Hsi Chhi Ching 養生胎息氣
   經.
   [= *Thai-Shang Yang Shêng Thai Hsi Chhi
   Ching.*]
   Manual of Nourishing the Life-Force (or,
      Attaining Longevity and Immortality) by
      Embryonic Respiration.
   Late Thang or Sung.
   Writer unknown.
   TT/812.
   Cf. Maspero (7), pp. 358, 365.
Yang Shêng Yen Ming Lu 養生延命錄.
   On Delaying Destiny by Nourishing the
      Natural Forces.
   Alternative title for *Yang Hsing Yen Ming
   Lu* (q.v.).
Yao Chung Chhao 藥種抄.
   Memoir on Several Varieties of Drug Plants.
   Japan, *c.* +1163.
   Kuan-Yu (Kanyu) 觀祐. MS. preserved
      at the 滋賀石山寺 Temple. Facsim.
      reprod. in Suppl. to the Japanese Tripiṭaka,
      vol. 11.
Yao Hsing Lun 藥性論.
   Discourse on the Natures and Properties of
      Drugs.
   Liang (or Thang, if identical with *Pên
   Tshao Yao Hsing*, q.v.).
   Attrib. Thao Hung-Ching 陶弘景.
   Only extant in quotations in books on
      pharmaceutical natural history.
   ICK, p. 169.
Yao Hsing Pên Tshao 藥性本草.
   See *Pên Tshao Yao Hsing.*
Yao Ming Yin Chüeh 藥名隱訣.
   Secret Instructions on the Names of Drugs
      and Chemicals.
   Perhaps an alternative title for the *Thai-
   Chhing Shih Pi Chi* (q.v.).
Yeh Chung Chi 鄴中記.
   Record of Affairs at the Capital of the Later
      Chao Dynasty.
   Chin.
   Lu Hui 陸翽.
   Cf. Hirth (17).
Yen Fan Lu 演繁露.
   Extension of the *String of Pearls* (on the
      Spring and Autumn Annals), [on the
      meaning of many Thang and Sung
      expressions].
   Sung, +1180.
   Chhêng Ta-Chhang 程大昌.
   See des Rotours (1), p. cix.
Yen Hsien Chhing Shang Chien 燕閒清賞牋.
   The Use of Leisure and Innocent Enjoy-
      ments in a Retired Life [the sixth part
      (chs. 14, 15) of *Tsun Shêng Pa Chien*, q.v.].
   Ming, +1591.
   Kao Lien 高濂.

Yen I I Mou Lu 燕翼詒謀錄.
Handing Down Good Plans for Posterity
from the Wings of Yen.
Sung, +1227.
Wang Yung 王栐.

*Yen-Ling hsien-sêng Chi Hsin Chiu Fu Chhi
Ching* 延陵先生集新舊服氣經.
New and Old Manuals of Absorbing the Chhi,
Collected by the Teacher of Yen-Ling.
Thang, early +8th, *c.* +745.
Writer unidentified.
Comm. by Sang Yü Tzu (+9th or +10th)
桑楡子.
TT/818, and (partially) in *YCCC*, ch. 58,
p. 2*a et passim*, ch. 59, pp. 1*a* ff., 18*b* ff.,
ch. 61, pp. 19*a* ff.
Cf. Maspero (7), pp. 220, 222.

Yen Mên Kung Miao Chieh Lu 鴈門公妙解錄.
The Venerable Yen Mên's Record of Marvellous Antidotes [alchemy and elixir
poisoning].
Thang, probably in the neighbourhood of
+847 since the text is substantially
identical with the *Hsüan Chieh Lu*
(q.v.) of this date.
Yen Mên 鴈門 (perhaps a ps. taken
from the pass and fortress on the
Great Wall, cf. Vol. 4, pt. 3, pp. 11,
48 and Fig. 711).
TT/937.

Yen Nien Chhio Ping Chien 延年却病牋.
How to Lengthen one's Years and Ward off
all Diseases [the fourth part (chs. 9, 10)
of *Tsun Shêng Pa Chien*, q.v.].
Ming, +1591.
Kao Lien 高濂.
Partial tr. of the gymnastic material,
Dudgeon (1).

Yen Shou Chhih Shu 延壽赤書.
Red Book on the Promotion of Longevity.
Thang, perhaps Sui.
Phei Yü (or Hsüan) 裴煜 (女).
Extant only in excerpts preserved in the
I Hsin Fang (+982), SIC, p. 465.

Yen Thieh Lun 鹽鐵論.
Discourses on Salt and Iron [record of the
debate of −81 on State control of com-
merce and industry].
C/Han, *c.* −80 to −60.
Huan Khuan 桓寬.
Partial tr. Gale (1); Gale, Boodberg & Lin.

Yin Chen Chün Chin Shih Wu Hsiang Lei 陰真
君金石五相類.
Alternative title of *Chin Shih Wu Hsiang
Lei* (q.v.).

Yin Chen Jen Liao Yang Tien Wên Ta Pien
尹真人寥陽殿問答編.
See *Liao Yang Tien Wên Ta Pien*.

*Yin Chen Jen Tung-Hua Chêng Mo Huang Chi
Ho Pi Chêng Tao Hsien Ching* 尹真人
東華正脈皇極闔闢證道仙經.
See *Huang Chi Ho Pi Hsien Ching*.

Yin Chuan Fu Shih Chien 飲饌服食牋.
Explanations on Diet, Nutrition and
Clothing [the fifth part (chs. 11–13) of
Tsun Shêng Pa Chien, q.v.].
Ming, +1591.
Kao Lien 高濂.

Yin Fu Ching 陰符經.
The Harmony of the Seen and the Unseen.
Thang, *c.* +735 (unless in essence a pre-
served late Warring States document).
Li Chhüan 李筌.
TT/30.
Cf. *TT*/105–124. Also in *TTCY* (*tou chi*, 6).
Tr. Legge (5).
Cf. Maspero (7), p. 222.

Yin Shan Chêng Yao 飲膳正要.
Principles of Correct Diet [on deficiency
diseases, with the aphorism 'many
diseases can be cured by diet alone'].
Yuan, +1330, re-issued by imperial order
in +1456.
Hu Ssu-Hui 忽思慧.
See Lu & Needham (1).

Yin Tan Nei Phien 陰丹內篇.
Esoteric Essay on the Yin Enchymoma.
Appendix to the *Tho Yo Tzu* (q.v.).

*Yin-Yang Chiu Chuan Chhêng Tzu-Chin Tien-
Hua Huan Tan Chüeh* 陰陽九轉成紫
金點化還丹訣.
Secret of the Cyclically Transformed Elixir,
Treated through Nine Yin-Yang Cycles
to Form Purple Gold and Projected to
Bring about Transformation.
Date unknown.
Writer unknown, but someone with Mao
Shan affiliations.
TT/888.

Ying Chhan Tzu Yü Lu 瑩蟾子語錄.
Collected Discourses of the Luminous-
Toad Master.
Yuan, *c.* +1320.
Li Tao-Shun 李道純 (Ying Chhan Tzu
瑩蟾子).
TT/1047.

Ying Yai Shêng Lan 瀛涯勝覽.
Triumphant Visions of the Ocean Shores
[relative to the voyages of Chêng Ho].
Ming, +1451. (Begun +1416 and com-
pleted about +1435.)
Ma Huan 馬歡.
Tr. Mills (11); Groeneveldt (1); Phillips (1);
Duyvendak (10).

Ying Yai Shêng Lan Chi 瀛涯勝覽集.
Abstract of the *Triumphant Visions of the
Ocean Shores* [a refacimento of Ma Huan's
book].
Ming, +1522.
Chang Shêng (*b*) 張昇.
Passages cit. in *TSCC*, *Pien i tien*, chs. 58,
73, 78, 85, 86, 96, 97, 98, 99, 101, 103,
106.
Tr. Rockhill (1).

Yōjōkun 養生訓.
Instructions on Hygiene and the Prolongation of Life.
Japan (Tokugawa), *c.* +1700.
Kaibara Ekiken 貝原益軒 (ed. Sugiyasu Saburō 杉靖三郎).

Yü-Chhing Chin-Ssu Chhing-Hua Pi-Wên Chin-Pao Nei-Lien Tan Chüeh 玉清金笥青華祕文金寶內鍊丹訣.
The Green-and-Elegant Secret Papers in the Jade-Purity Golden Box on the Essentials of the Internal Refining of the Golden Treasure, the Enchymoma.
Sung, late +11th century.
Chang Po-Tuan 張伯端.
TT/237.
Cf. Davis & Chao Yün-Tshung (5).

Yü-Chhing Nei Shu 玉清內書.
Inner Writings of the Jade-Purity (Heaven).
Probably Sung, but present version incomplete, and some of the material may be, or may have been, older.
Compiler unknown.
TT/940.

Yü Fang Chih Yao 玉房指要.
Important Matters of the Jade Chamber.
Pre-Sui, perhaps +4th century.
Writer unknown.
In *I Hsin Fang* (*Ishinhō*) and *Shuang Mei Ching An Tshung Shu*.
Partial trs. van Gulik (3, 8).

Yü Fang Pi Chüeh 玉房祕訣.
Secret Instructions concerning the Jade Chamber.
Pre-Sui, perhaps +4th century.
Writer unknown.
Partial tr. van Gulik (3).
Only as fragment in *Shuang Mei Ching An Tshung Shu* (q.v.).

Yu Huan Chi Wên 游宦紀聞.
Things Seen and Heard on my official Travels.
Sung, +1233.
Chang Shih-Nan 張世南.

Yü Phien 玉篇.
Jade Page Dictionary.
Liang, +543.
Ku Yeh-Wang 顧野王.
Extended and edited in the Thang (+674) by Sun Chhiang 孫強.

Yü Shih Ming Yen 喻世明言.
Stories to Enlighten Men.
Ming, *c.* +1640.
Fêng Mêng-Lung 馮夢龍.

Yü Tung Ta Shen Tan Sha Chen Yao Chüeh 玉洞大神丹砂眞要訣.
True and Essential Teachings about the Great Magical Cinnabar of the Jade Heaven [paraphrase of +8th-century materials].
Thang, not before +8th.
Attrib. Chang Kuo 張果.
TT/889.

Yu-Yang Tsa Tsu 酉陽雜俎.
Miscellany of the Yu-yang Mountain (Cave) [in S.E. Szechuan].
Thang, +863.
Tuan Chhêng-Shih 段成式.
See des Rotours (1), p. civ.

Yuan Chhi Lun 元氣論.
Discourse on the Primary Vitality (and the Cosmogonic Chhi).
Thang, late +8th or perhaps +9th.
Writer unknown.
In *YCCC*, ch. 56.
Cf. Maspero (7), p. 207.

Yuan-Shih Shang Chen Chung Hsien Chi 元始上眞衆仙記.
Record of the Assemblies of the Perfected Immortals; a Yuan-Shih Scripture.
Ascr. Chin, *c.* +320, more probably +5th or +6th.
Attrib. Ko Hung 葛洪.
TT/163.

Yuan Yang Ching 元陽經.
Manual of the Primary Yang (Vitality).
Chin, L/Sung, Chhi or Liang, before +550.
Writer unknown.
Extant only in quotations, in *Yang Hsing Yen Ming Lu*, etc.
Cf. Maspero (7), p. 232.

Yuan Yu 遠遊.
Roaming the Universe; or, The Journey into Remoteness [ode].
C/Han, *c.* −110.
Writer's name unknown, but a Taoist.
Tr. Hawkes (1).

Yüeh Wei Tshao Thang Pi Chi 閱微草堂筆記.
Jottings from the Yüeh-wei Cottage.
Chhing, 1800.
Chi Yün 紀昀.

Yün Chai Kuang Lu 雲齋廣錄.
Extended Records of the Cloudy Studio.
Sung.
Li Hsien-Min 李獻民.

Yün Chhi Yu I 雲溪友議.
Discussions with Friends at Cloudy Pool
Thang, *c.* +870.
Fan Shu 范攄.

Yün Chi Chhi Chhien 雲笈七籤.
The Seven Bamboo Tablets of the Cloudy Satchel [an important collection of Taoist material made by the editor of the first definitive form of the *Tao Tsang* (+1019), and including much material which is not in the Patrology as we now have it].
Sung, *c.* +1022.
Chang Chün-Fang 張君房.
TT/1020.

Yün Hsien Tsa Chi 雲仙雜記.
Miscellaneous Records of the Cloudy Immortals.
Thang or Wu Tai, *c.* +904.
Fêng Chih 馮贄.

Yün Hsien San Lu 雲仙散錄.
　　Scattered Remains on the Cloudy Im-
　　　mortals.
　　Ascr. Thang or Wu Tai, *c.* +904, actually
　　　probably Sung.
　　Attrib. Fêng Chih 馮贄, but probably by
　　　Wang Chih 王銍.

Yün Kuang Chi 雲光集.
　　Collected (Poems) of Light (through the)
　　　Clouds.
　　Sung, *c.* +1170.
　　Wang Chhu-I 王處一.
　　TT/1138.

ADDENDA TO BIBLIOGRAPHY A

Hai Kho Lun 郵客論
　　Guests from Overseas [descriptions of
　　　alchemical exotica] Sung.
　　Li Kuang-Hsüan 李光玄.
　　TT/1033.

Ju Lin Wai Shih 儒林外史.
　　Unofficial History of the World of Learning
　　　[satirical novel on the life of the literati
　　　in the Ming period].
　　Chhing, begun before +1736, completed
　　　+1749.
　　Wu Ching-Tzu 吳敬梓.

Tr. Yang & Yang (1); Tomkinson (2).
　　Cf. Chang Hsin-Tshang (2).

Shih Shuo Hsin Yu 世說新語.
　　New Discourse on the Talk of the Times
　　　[notes of minor incidents from Han to
　　　Chin].
　　Cf. *Hsü Shih Shuo.*
　　L/Sung, +5th.
　　Liu I-Chhing 劉義慶.
　　Commentary by Liu Hsün
　　　劉峻 (Liang).

CONCORDANCE FOR
TAO TSANG BOOKS AND TRACTATES

Wieger nos.		Ong nos.
1	(Ling-Pao Wu Liang) Tu Jen (Shang Phin Miao) Ching	1
8	Thai-Shang San-shih-liu Pu Tsun Ching (contains Shang Chhing Ching)	8
30	(Huang Ti) Yin Fu Ching	31
119	Huang Ti Yin Fu Ching Chu	122
127	Thai Hsi Ching Chu	130
130	Thai-Shang Tung Fang Nei Ching Chu	133
132	(Tshui Kung) Ju Yao Ching Chu Chieh	135
133	Lü Shun-Yang Chen Jen Chhin Yuan Chhun Tan Tzhu Chu Chieh	136
134	(Chhiu Chhang-Chhun) Chhing Thien Ko Chu Shih	137
137	Shang-Chhing Wo Chung Chüeh	140
138	(Tzu-Yang Chen Jen) Wu Chen Phien Chu Su	141
139	Wu Chen Phien San Chu	142
140	(Tzu-Yang Chen Jen) Wu Chen Phien Chih Chih Hsiang Shuo San Chhêng Pi Yao	143
146	Hsiu-Chen Thai Chi Hun Yuan Thu	149
147	Hsiu-Chen Thai Chi Hun Yuan Chih Hsüan Thu	150
148	Chin I Huan Tan Yin Chêng Thu	151
149	Hsiu-Chen Li Yen Chhao Thu	152
153	Shang-Chhing Tung-Chen Chiu Kung Tzu Fang Thu	156
163	Yuan-Shih Shang Chen Chung Hsien Chi	166
164	Tung-Hsüan Ling-Pao Chen Ling Wei Yeh Thu	167
173	Hsüan Fêng Chhing Hui Lu	176
228	Chen I Chin Tan Chüeh	231
229	Huan Tan Pi Chüeh Yang Chhih-Tzu Shen Fang	232
230	Huan Tan Chung Hsien Lun	233
231	Hsiu Tan Miao Yung Chih Li Lun	234
237	Yü-Chhing Chin-Ssu Chhing-Hua Pi-Wên Chin-Pao Nei-Lien Tan Chüeh	240
239	Chih-Chou hsien-sêng Chin Tan Chih Chih	242
240	(Chhen Hsü-Pai hsien-sêng) Kuei Chung Chih Nan	243
241	Ta Tan Chih Chih	244
243	Hsi Shan Chhun Hsien Hui Chen Chi	246
248	Chhüan-Chen Chi Hsüan Pi Yao	251
252	Thai-Wei Ling Shu Tzu-Wên Lang-Kan Hua Tan Shen Chen Shang Ching	255
256	(Thao Chen Jen) Nei Tan Fu	259
257	Chhin Hsüan Fu	260

Wieger nos.		Ong nos.
258	Chin Tan Fu	261
260	Hsiu-Chen Shih Shu	263
261	Chen Chhi Huan Yuan Ming	264
263	Chin I Huan Tan Pai Wên Chüeh	266
266	Chih Chen Tzu Lung Hu Ta Tan Shih	269
275	Chhing Wei Tan Chüeh (or Fa)	278
289	Han Wu (Ti) Nei Chuan	292
290	Han Wu (Ti) Wai Chuan	293
293	Li-Shih Chen Hsien Thi Tao Thung Chien	296
297	Hua-Yang Thao Yin-Chü Chuan	300
300	Tzu-Yang Chen Jen (Chou I-Shan) Nei Chuan	303
301	(San) Mao Shan Chih	304
304	Hsi Yo Hua-Shan Chih	307
328	(Thai-Shang) Huang Thing Nei Ching Yü Ching	331
329	(Thai-Shang) Huang Thing Wai Ching Yü Ching	332
366	(Thai-Shang Tung-Hsüan Ling-Pao Mieh Tu (or San Yuan) Wu Lien) Shêng Shih Miao Ching	369
385	Thai-Shang Ling-Pao Wu Fu (Ching)	388
398	Huang Thing Nei Ching Yü Ching Chu by Liu Chhang-Shêng	401
399	Huang Thing Nei Ching (Yü) Ching Chu by Liang Chhiu Tzu	402
400	Huang Thing Nei Wai Ching Yü Ching Chieh	403
416	Ling-Pao Chung Chen Tan Chüeh	419
417	Shen Hsien Fu Erh Tan Shih Hsing Yao Fa	420
418	Têng Chen Yin Chüeh	421
419	(Shang-Chhing) San Chen Chih Yao Yü Chüeh	422
421	(Shang-Chhing) Ming Thang Yuan Chen Ching Chüeh	424
423	Shang-Chhing Thai-Shang Pa Su Chen Ching	426
428	Shang-Chhing Han Hsiang Chien Chien Thu	431
429	Huang Thing Nei Ching Wu Tsang Liu Fu Pu Hsieh Thu	432
439	Shang-Chhing Hou Shêng Tao Chün Lieh Chi	442
445	(Hsi Shan) Hsü Chen-Chün (Hsü Hsün) Pa-shih-wu Hua Lu	448
447	Thai-Chi Ko Hsien-Ong (or Kung) (Ko Hsüan) Chuan	450
524	Thai-Shang Tung-Hsüan Ling-Pao Shou Tu I	528
605	Ling-Pao Chiu Yu Chhang Yeh Chhi Shih Tu Wang Hsüan Chang	610
611	Kuang Chhêng Chi	616

Wieger nos.		Ong nos.
634	*Huang-Thien Shang-Chhing Chin Chhüeh Ti Chün Ling Shu Tzu-Wên Shang Ching*	639
635	*Tung Shen Pa Ti Miao Ching Ching*	640
761	*Thu Ching (Chi Chu) Yen I Pên Tshao*	{767 {768
773	*Hsüan Phin Lu*	780
810	*(Thai-Chhing) Chung Huang Chen Ching*	816
811	*(Thai-Chhing) Tao Yin Yang-Shêng Ching*	817
812	*(Thai-Shang) Yang-Shêng Thai-Hsi Chhi Ching*	818
813	*Thai-Chhing Thiao Chhi Ching*	819
814	*Thai-Shang Lao Chün Yang-Shêng Chüeh*	820
815	*Thai-Chhing (Wang Lao) Fu Chhi Khou Chüeh (or Chhuan Fa)*	821
817	*Sung Shan Thai-Wu hsien-sêng Chhi Ching*	823
818	*(Yen-Ling hsien-sêng Chi) Hsin Chiu Fu Chhi Ching*	824
821	*(Huan-Chen hsien-sêng) Fu Nei Yuan Chhi Chüeh*	827
830	*Chen Chung Chi*	836
(830)	*(Shê Yang) Chen Chung Chi (or Fang)*	(836)
831	*Yang Hsing Yen Ming Lu*	837
835	*(Pao Phu Tzu) Yang Shêng Lun*	841
838	*Shang-Chhing Ching Chen Tan Pi Chüeh*	844
856	*Shen Hsien Lien Tan Tien Chu San Yuan Pao Chao Fa*	862
873	*Thai-Chhing (or Shang-Chhing) Chin I Shen Tan Ching*	879
874	*Thai-Chhing Shih Pi Chi*	880
875	*Thai-Chhing Chin I Shen Chhi Ching*	881
876	*(Thai-Chhing Ching) Thien-Shih Khou Chüeh*	882
878	*(Huang Ti) Chiu Ting Shen Tan Ching Chüeh*	884
879	*Chiu Chuan Ling Sha Ta Tan Tzu Shêng Hsüan Ching*	885
880	*(Chang Chen Jen) Chin Shih Ling Sha Lun*	886
881	*(Wei Po-Yang) Chhi Fan Tan Sha Chüeh*	887
882	*(Thai-Chi Chen Jen) Chiu Chuan Huan Tan Ching Yao Chüeh*	888
883	*(Ta-Tung Lien Chen Pao Ching) Hsiu Fu Ling Sha Miao Chüeh*	889
884	*(Ta-Tung Lien Chen Pao Ching) Chiu Huan Chin Tan Miao Chüeh*	890
885	*(Thai-Shang Wei Ling Shen Hua) Chiu Chuan Tan Sha Fa*	891
886	*Chiu Chuan Ling Sha Ta Tan*	892
887	*Chiu Chuan Chhing Chin Ling Sha Tan*	893
888	*Yin-Yang Chiu Chuan Chhêng Tzu Chin Tien Hua Huan Tan Chüeh*	894
889	*Yü-Tung Ta Shen Tan Sha Chen Yao Chüeh*	895
890	*Ling Sha Ta Tan Pi Chüeh*	896
891	*Pi Yü Chu Sha Han Lin Yü Shu Kuei*	897
892	*Ta Tan Chi*	898
893	*Tan Fang Hsü Chih*	899
894	*Shih Yao Erh Ya*	900
895	*(Chih-Chhuan Chen Jen) Chiao Chêng Shu*	901
896	*Shun-Yang Lü Chen Jen Yao Shih Chih*	902
897	*Chin Pi Wu Hsiang Lei Tshan Thung Chhi*	903
898	*Tshan Thung Chhi Wu Hsiang Lei Pi Yao*	904
899	*(Yin Chen Chün) Chin Shih Wu Hsiang Lei*	905
900	*Chin Shih Pu Wu Chiu Shu Chüeh*	906
901	*Shang-Chhing Chiu Chen Chung Ching Nei Chüeh*	907
902	*Lung Hu Huan Tan Chüeh*	908
903	*Chin Hua Yü I Ta Tan*	909
904	*Kan Chhi Shih-liu Chuan Chin Tan*	910
905	*Hsiu Lien Ta Tan Yao Chih (or Chüeh)*	911
906	*Thung Yu Chüeh*	912
907	*Chin Hua Chhung Pi Tan Ching Pi Chih*	913
908	*Huan Tan Chou Hou Chüeh*	914
909	*Phêng-Lai Shan Hsi Tsao Huan Tan Ko*	915
910	*(Pao Phu Tzu) Shen Hsien Chin Shuo Ching*	916
911	*Chu Chia Shen Phin Tan Fa*	917
912	*Chhien Hung Chia Kêng Chih Pao Chi Chhêng*	918
913	*Tan Fang Ao Lun*	919
914	*Chih Kuei Chi*	920
915	*Huan Chin Shu*	921
916	*Ta Tan Chhien Hung Lun*	922
917	*Chen Yuan Miao Tao Yao Lüeh*	923
918	*Tan Fang Chien Yuan*	924
919	*Ta Huan Tan Chao Chien*	925
920	*Thai-Chhing Yü Pei Tzu*	926
921	*Hsüan Chieh Lu*	927
922	*(Hsien-Yuan Huang Ti) Shui Ching Yao Fa*	928
923	*San-shih-liu Shui Fa*	929
927	*Thai Pai Ching*	933
928 932	*Tan Lun Chüeh Chih Hsin Chien Ta Tan Wên Ta*	934 938
933	*Chin Mu Wan Ling Lun*	939
934	*Hung Chhien Ju Hei Chhien Chüeh*	940
935	*Thung Hsüan Pi Shu*	941
937	*Yen Mên Kung Miao Chieh Lu (=921)*	943
938	*Hsüan Shuang Chang Shang Lu*	944
939	*Thai-Chi Chen Jen Tsa Tan Yao Fang*	945
940	*Yü Chhing Nei Shu*	946
942	*Thai-Ku Thu Tui Ching*	948
943	*Shang-Tung Hsin Tan Ching Chüeh*	949
944	*(Hsü Chen-Chün) Shih Han Chi*	950

Wieger nos.		Ong nos.	Wieger nos.		Ong nos.
945	*Chiu Chuan Liu (or Ling) Chu Shen Hsien Chiu Tan Ching*	951	1077	*Huan Yuan Phien*	1083
946	*Kêng Tao Chi*	952	1087	*Thai Phing Ching*	1093
988	*(Ku Wên) Lung Hu Ching Chu Su*	994	1124	*Wu Shang Pi Yao*	1130
989	*(Ku Wên) Lung Hu Shang Ching Chu*	995	1125	*San Tung Chu Nang*	1131
			1127	*Hsien Lo Chi*	1133
990	*Chou I Tshan Thung Chhi (Chu)* comm. by Yin Chhang-Shêng	996	1128	*Chien Wu Chi*	1134
			1130	*Tzu-Jan Chi*	1136
991	*Chou I Tshan Thung Chhi Chu* comm. anon.	997	1135	*Tung-Hsüan Chin Yü Chi*	1141
			1136	*Tan-Yang Shen Kuang Tshan*	1142
992	*Tshan Thung Chhi Khao I (or Chou I Tshan Thung Chhi Chu)* comm. by Chu Hsi	998	1138	*Yün Kuang Chi*	1144
			1139	*(Wang) Chhung-Yang Chhüan Chen Chi*	1145
993	*Chou I Tshan Thung Chhi Fên Chang Thung Chen I* comm. by Phêng Hsiao	999	1140	*(Wang) Chhung-Yang Chiao Hua Chi*	1146
994	*Chou I Tshan Thung Chhi Ting Chhi Ko Ming Ching Thu* comm. by Phêng Hsiao	1000	1141	*(Wang) Chhung-Yang Fên-Li Shih-Hua Chi*	1147
			1142	*(Wang) Chhung-Yang (Chen Jen) Chin-Kuan (or Chhüeh) Yü-So Chüeh*	1148
995	*Chou I Tshan Thung Chhi Chu* comm. anon.	1001	1145	*Chhang-Chhun Tzu Phan-Chhi Chi*	1151
996	*Chou I Tshan Thung Chhi Fa Hui* comm. by Yü Yen	1002	1146	*Than hsien-sêng Shui Yün Chi*	1152
			1147	*Thai-Ku Chi*	1153
997	*Chou I Tshan Thung Chhi Shih I* comm. by Yü Yen	1003	1162	*Mo Tzu*	1168
			1170	*Huai Nan (Tzu) Hung Lieh Chieh*	1176
998	*Chou I Tshan Thung Chhi Chieh* comm. by Chhen Hsien-Wei	1004	1171	*Pao Phu Tzu, Nei Phien*	1177
			1172	*Pao Phu Tzu, Pieh Chih*	1178
999	*Chou I Tshan Thung Chhi Chu* comm. by Chhu Hua-Ku	1005	1173	*Pao Phu Tzu, Wai Phien*	1179
			1174	*Tho Yo Tzu*	1180
1004	*Chen Kao*	1010	1187	*Tung Shen Pa Ti Yuan (Hsüan) Pien Ching*	1193
1005	*Tao Shu*	1011			
1020	*Yün Chi Chhi Chhien*	1026	1204		1211
1022	*Thai Hsüan Pao Tien*	1028	1205	*Shang-Chhing Ling-Pao Ta Fa*	1212
1024	*Tso Wang Lun*	1030	1206		1213
1028	*Huang Chi Ching Shih (Shu)*	1034	1212	*Chhüan-Chen Tso Po Chieh Fa*	1219
1034	*Wu Hsüan Phien*	1040	1216	*(Wang) Chhung-Yang Li-Chiao Shih-Wu Lun*	1223
1044	*Tan-Yang Chen Jen Yü Lu*	1050			
1047	*Ying Chhan Tzu Yü Lu*	1053	1225	*Chêng I Fa Wên (Thai-Shang) Wai Lu I*	1233
1053	*(Shang Yang Tzu) Chin Tan Ta Yao*	1059	1235	*Tao Fa Hsin Chhuan*	1243
1054	*(Shang Yang Tzu) Chin Tan Ta Yao Thu*	1060	1273	*Shang-Chhing Ching Pi Chüeh*	1281
			1276	*Shang-Chhing Huang Shu Kuo Tu I*	1284
1055	*(Shang Yang Tzu) Chin Tan Ta Yao Lieh Hsien Chih*	1061	1287	*Ko Hsien-Ong (Ko Hung) Chou Hou Pei Chi Fang*	1295
1056	*(Sheng Yang Tzu) Chin Tan Ta Yao Hsien Phai (Yuan Liu)*	1062	1295	*(Tung-Chen Thai-Shang Su-Ling Tung-Yuan) Ta Yu Miao Ching*	1303
1058	*Chin Tan Chih Chih*	1064	1357	*Shang-Chhing Thai-Shang Ti Chün Chiu Chen Chung Ching*	1365
1067	*Chin Tan Ssu Pai Tzu (Chu)*	1073	1382	*Huang Thing Chung Ching Ching*	1390
1068	*Lung Hu Huan Tan Chüeh Sung*	1074	1405	*Thai-Shang Lao Chün Thai Su Ching* (see index s.v. *Thai Su Chuan*)	1413
1074	*Huan Tan Fu Ming Phien*	1080			
1076	*Tshui Hsü Phien*	1082	1442	*Han Thien Shih Shih Chia*	1451

B. CHINESE AND JAPANESE BOOKS AND JOURNAL ARTICLES SINCE +1800

Achiwa Gorō (1) 阿知波五郎.
Rangaku-ki no Shizen Ryū-nō-setsu Kenkyū
蘭學期の自然良能說研究.
A Study of the Theory of Nature-Healing in the Period of Dutch Learning in Japan.
ID, 1965 No. 31, 2223.

Akitsuki Kanei (1) 秋月觀瑛.
Kōrō Kannen no Shiroku 黃老觀念の
系譜.
On the Genealogy of the Huang-Lao Concept (in Taoism).
THG, 1955, **10**, 69.

Andō Kōsei (1) 安藤更生.
Kanshin 鑑眞.
Life of Chien-Chen (+688 to +763), [outstanding Buddhist missionary to Japan, skilled also in medicine and architecture].
Bijutsu Shuppansha, Tokyo 1958, repr. 1963.
Abstr. *RBS*, 1964, **4**, no. 889.

Andō Kōsei (2) 安藤更生.
Nihon no Miira 日本のミイラ.
Mummification in Japan.
Mainichi Shimbunsha, Tokyo, 1961.
Abstr. *RBS*, 1968, **7**, no. 575.

Anon. (10).
Tunhuang Pi Hua Chi 敦煌壁畫集.
Album of Coloured Reproductions of the fresco-paintings at the Tunhuang cave-temples.
Peking, 1957.

Anon. (11).
Changsha Fa Chüeh Pao-Kao 長沙發掘
報告.
Report on the Excavations (of Tombs of the Chhu State, of the Warring States period, and of the Han Dynasties) at Chhangsha.
Acad. Sinica Archaeol. Inst., Kho-Hsüeh, Peking, 1957.

Anon. (17).
Shou-hsien Tshai Hou Mu Chhu Thu I Wu 壽縣蔡侯墓出土遺物.
Objects Excavated from the Tomb of the Duke of Tshai at Shou-hsien.
Acad. Sinica. Archaeol. Inst., Peking, 1956.

Anon. (27).
Shang-Tshun-Ling Kuo Kuo Mu Ti
上村嶺虢國墓地.
The Cemetery (and Princely Tombs) of the State of (Northern) Kuo at Shang-tshun-ling (near Shen-hsien in the San-mên Gorge Dam Area of the Yellow River).

Institute of Archaeology, Academia Sinica, Peking, 1959 (Field Expedition Reports, Ting Series, no. 10), (Yellow River Excavations Report no. 3).

Anon. (28).
Yünnan Chin-Ning Shih-Chai Shan Ku Mu Chhün Fa-Chüeh Pao-Kao 雲南晉寧
石寨山古墓羣發掘報告.
Report on the Excavation of a Group of Tombs (of the Tien Culture) at Shih-chai Shan near Chin-ning in Yunnan.
2 vols.
Yunnan Provincial Museum.
Wên-Wu, Peking, 1959.

Anon. (57).
Chung Yao Chih 中藥志.
Repertorium of Chinese Materia Medica (Drug Plants and their Parts, Animals and Minerals).
4 vols.
Jen-min Wei-shêng, Peking, 1961.

Anon. (73) (Anhui Medical College Physiotherapy Dept.).
Chung I An-Mo Hsüeh Chien Phien 中醫
按摩學簡編.
Introduction to the Massage Techniques in Chinese Medicine.
Jen-min Wei-shêng, Peking, 1960, repr. 1963

Anon. (74) (National Physical Education Council).
Thai Chi Chhüan Yün Tung 太極拳運動.
The Chinese Boxing Movements [instructions for the exercises].
Jen-min Thi-yü, Peking, 1962.

Anon. (77).
Chhi Kung Liao Fa Chiang I 氣功療法講義.
Lectures on Respiratory Physiotherapy.
Kho-hsüeh Chi-shu, Shanghai, 1958.

Anon. (78).
Chung-Kuo Chih Chhien chih Ting Liang Fên-Hsi 中國制錢之定量分析.
Analyses of Chinese Coins (of different Dynasties).
KHS, 1921, **6** (no. 11), 1173.
Table reprinted in Wang Chin (2), p. 88.

Anon. (100).
Shao-Hsing Chiu Niang Tsao 紹興酒釀造.
Methods of Fermentation (and Distillation) of Wine used at Shao-hsing (Chekiang).
Chhing Kung Yeh, Peking, 1958.

Anon. (101).
Chung-Kuo Ming Tshai Phu 中國名菜譜.
Famous Dishes of Chinese Cookery.
12 vols.
Chhing Kung Yeh, Peking, 1965.

Anon. (*103*).
> *Nihon Miira no Kenkyū* 日本ミイラの
> 研究.
> Researches on Mummies (and Self-Mummi-
> fication) in Japan.
> Heibonsha, Tokyo, 1971.

Anon. (*104*).
> *Chhangsha Ma Wang Tui i Hao Han Mu
> Fa-Chüeh Chien-Pao* 長沙馬王堆一號
> 漢墓發掘簡報.
> Preliminary Report on the Excavation of Han
> Tomb No. 1 at Ma-wang-tui (Hayagriva
> Hill) near Chhangsha [the Lady of Tai,
> c. −180].
> Wên Wu, Peking, 1972.

Anon. (*105*).
> *Kōkogaku-shō no Shin-Hakken; Nisen-yonen
> mae no Kinue Orimono Sono-hoka* 考古
> 學上の新發見；二千余年まえの緝
> 繪織物その他.
> A New Discovery in Archaeology; Painted
> Silks, Textiles and other Things more than
> Two Thousand Years old.
> *JC*, 1972 (no. 9), 68, with colour-plates.

Anon. (*106*).
> *Wên-Hua Ta Ko-Ming Chhi Chien Chhu Thu
> Wên Wu* 文化大革命期間出土
> 文物.
> Cultural Relics Unearthed during the period
> of the Great Cultural Revolution (1965–71),
> vol. 1 [album].
> Wên Wu, Peking, 1972.

Anon. (*109*).
> *Chung-Kuo Kao Têng Chih-Wu Thu Chien*
> 中國高等植物圖鑑.
> *Iconographia Cormophytorum Sinicorum*
> (Flora of Chinese Higher Plants).
> 2 vols. Kho-Hsüeh, Peking, 1972 (for Nat.
> Inst. of Botany).

Anon. (*110*).
> *Chhang Yung Chung Tshao Yao Thu Phu*
> 常用中草藥圖譜.
> Illustrated Flora of the Most Commonly
> Used Drug Plants in Chinese Medicine.
> Jen-min Wei-shêng, Peking, 1970.

Anon. (*111*).
> *Man-chhêng Han Mu Fa-Chüeh Chi Yao*
> 滿城漢墓發掘紀要.
> The Essential Findings of the Excavations of
> the (Two) Han Tombs at Man-chhêng
> (Hopei), [Liu Shêng, Prince Ching of
> Chung-shan, and his consort Tou Wan].
> *KKTH*, 1972, (no. 1), 8.

Anon. (*112*).
> *Man-chhêng Han Mu 'Chin Lou Yü-I' ti
> Chhing-Li ho Fu-Yuan* 滿城漢墓「金縷
> 玉衣」的清理和復原.
> On the Origin and Detailed Structure of
> the Jade Body-cases Sewn with Gold
> Thread found in the Han Tombs at Man-
> chhêng.
> *KKTH*, 1972, (no. 2), 39.

Anon. (*113*).
> *Shih Than Chi-nan Wu-ying Shan Chhu-
> Thu-ti Hsi Han, Lo Wu, Tsa Chi, Yen Huan
> Thao Yung* 試談濟南无影山出土的
> 西漢樂舞雜技宴歡陶俑.
> A Discourse on the Early Han pottery models
> of musicians, dancers, acrobats and mis-
> cellaneous artists performing at a banquet,
> discovered in a Tomb at Wu-ying Shan
> (Shadowless Hill) near Chinan (in Shan-
> tung province).
> *WWTK*, 1972, (no. 5), 19.

Anon. (*115*).
> *Tzhu-Hang Ta Shih Chuan* 慈航大師傳.
> A Biography of the Great Buddhist Teacher,
> Tzhu-Hang (d., self-mummified, 1954).
> Thaipei, 1959 (Kan Lu Tshung Shu ser.
> no. 11).

Aoki Masaru (*1*) 青水正兒.
> *Chūka Meibutsu Kō* 中華名物考.
> Studies on Things of Renown in (Ancient
> and Medieval) China, [including aro-
> matics, incense and spices].
> Shunjūsha, Tokyo, 1959.
> Abstr. *RBS*, 1965, *5*, no. 836.

Asahina Yasuhiko (*1*) (ed.) 朝比奈泰彦 with
> 16 collaborators.
> *Shōsōin Yakubutsu* 正倉院藥物.
> The Shōsōin Medicinals; a Report on
> Scientific Researches.
> With an English abstract by Obata Shige-
> yoshi.
> Shokubutsu Bunken Kankō-kai, Osaka, 1955.

Chan Jan-Hui (*1*) 湛然慧.
> Alternative orthography of Tan Jan-Hui (*1*).

Chang Chhang-Shao (*1*) 張昌紹.
> *Hsien-tai-ti Chung Yao Yen-Chiu* 現代的
> 中藥研究.
> Modern Researches on Chinese Drugs.
> Kho-hsüeh Chi-shu, Shanghai, 1956.

Chang Chhi-Yün (*2*) (ed.) 張其昀.
> *Chung-Hua Min Kuo Ti-Thu Chi* 中華民
> 國地圖集.
> Atlas of the Chinese Republic (5 vols.):
> vol. 1 Thaiwan, vol. 2, Central Asia,
> vol. 3 North China, vol. 4 South China,
> vol. 5 General maps.
> National Defence College
> National Geographical } Thaipei, 1962–3.
> Institute

Chang Ching-Lu (*1*).
> *Chung-Kuo Chin-Tai Chhu-Pan Shih-Liao
> Chhu Phien* 中國近代出版史料初編.
> Materials for a History of Modern Book-
> Publishing in China, Pt. 1.

Chang Hsin-Chhêng (*1*) 張心澂.
> *Wei Shu Thung Khao* 偽書通考.
> A Complete Investigation of the (Ancient
> and Medieval) Books of Doubtful
> Authenticity.
> 2 vols., Com. Press, 1939, repr. 1957.

Chang Hsing-Yün (1) 章杏雲.
　　Yin Shih Pien 飲食辯.
　　A Discussion of Foods and Beverages.
　　1814, repr. 1824.
　　Cf. Dudgeon (2).
Chang Hsüan (1) 張瑄.
　　*Chung Wên Chhang Yung San Chhien Tzu
　　　Hsing I Shih* 中文常用三千字形義釋.
　　Etymologies of Three Thousand Chinese
　　　Characters in Common Use.
　　Hongkong Univ. Press, 1968.
Chang Hung-Chao (1) 章鴻釗.
　　Shih Ya 石雅.
　　Lapidarium Sinicum; a Study of the Rocks,
　　　Fossils and Minerals as known in
　　　Chinese Literature.
　　Chinese Geol. Survey, Peiping: 1st ed. 1921,
　　　2nd ed. 1927.
　　MGSC (ser. B), no. 2, 1–432 (with Engl.
　　　summary).
　　Crit. P. Demiéville, *BEFEO*, 1924, **24**,
　　　276.
Chang Hung-Chao (3) 章鴻釗.
　　Chung-Kuo Yung Hsin ti Chhi-Yuan
　　　中國用鋅的起源.
　　Origins and Development of Zinc Tech-
　　　nology in China.
　　KHS, 1923, **8** (no. 3), 233, repr. in Wang
　　　Chin (2), p. 21.
　　Cf. Chang Hung-Chao (2).
Chang Hung-Chao (6) 章鴻釗.
　　*Tsai Shu Chung-Kuo Yung Hsin ti Chhi-
　　　Yuan* 再述中國用鋅的起源.
　　Further Remarks on the Origins and
　　　Development of Zinc Technology in
　　　China.
　　KHS, 1925, **9** (no. 9), 1116, repr. in Wang
　　　Chin (2), p. 29.
　　Cf. Chang Hung-Chao (3).
Chang Hung-Chao (8) 章鴻釗.
　　*Lo shih 'Chung-Kuo I-Lan' Chüan Chin
　　　Shih I Chêng* 洛氏「中國伊蘭」卷金
　　　石譯證.
　　Metals and Minerals as Treated in Laufer's
　　　'Sino-Iranica', translated with Commen-
　　　taries.
　　MGSC 1925 (Ser. B), no. 3, 1–119.
　　With English preface by Ong Wên-Hao.
Chang Tzu-Kao (1) 張子高.
　　Kho-Hsüeh Fa Ta Lüeh Shih 科學發達
　　　略史.
　　A Classified History of the Natural Sciences.
　　Com. Press, Shanghai, 1923, repr. 1936.
Chang Tzu-Kao (2) 張子高.
　　Chung-Kuo Hua-Hsüeh Shih Kao (*Ku-Tai
　　　chih Pu*) 中國化學史稿(古代之部).
　　A Draft History of Chemistry in China
　　　(Section on Antiquity).
　　Kho-Hsüeh, Peking, 1964.
Chang Tzu-Kao (3) 張子高.
　　Lien Tan Shu Fa-Shêng yü Fa-Chan
　　　鍊丹術發生與發展.

On the Origin and Development of Chinese
　　Alchemy.
　　CHJ, 1960, **7** (no. 2), 35.
Chang Tzu-Kao (4) 張子高.
　　*Tshung Tu Hsi Thung Chhi Than Tao 'Wu'
　　　Tzu Pên I* 從鍍錫銅器談到「鋈」字
　　　本義.
　　Tin-Plated Bronzes and the Possible
　　　Original Meaning of the Character *wu*.
　　AS/CJA, 1958 (no. 3), 73.
Chang Tzu-Kao (5) 張子高.
　　*Chao Hsüeh-Min 'Pên Tshao Kang Mu
　　　Shih I' Chu Shu Nien-Tai, Chien-Lun
　　　Wo-Kuo Shou-Tzhu Yung Chhiang-Shiu
　　　Kho Thung Pan Shih* 趙學敏「本草綱
　　　目拾遺」著述年代兼論我國首次用强
　　　水刻銅版事.
　　On the Date of Publication of Chao Hsüeh-
　　　Min's *Supplement to the Great Pharma-
　　　copoeia*, and the Earliest Use of Acids for
　　　Etching Copper Plates in China.
　　KHSC, 1962, **1** (no. 4), 106.
Chang Tzu-Kao (6) 張子高.
　　*Lun Wo Kuo Niang Chiu Chhi-Yuan ti
　　　Shih-Tai Wên-Thi* 論我國釀酒起源的
　　　時代問題.
　　On the Question of the Origin of Wine in
　　　China.
　　CHJ, 1960, **17** (7), no. 2, 31.
Chang Tzu-Kung (1) 張資珙.
　　*Lüeh Lun Chung-Kuo ti Nieh Chih Pai-
　　　Thung ho tha tsai Li-Shih shang yü Ou-
　　　Ya Ko Kuo ti Kuan-Hsi* 略論中國的
　　　鎳質白銅和他在歷史上與歐亞各國
　　　的關係.
　　On Chinese Nickel and Paktong, and on
　　　their Role in the Historical Relations
　　　between Asia and Europe.
　　KHS, 1957, **33** (no. 2), 91.
Chang Tzu-Kung (2) 張資珙.
　　Yuan Su Fa-Hsien Shih 元素發現史.
　　The Discovery of the Chemical Elements
　　　(a translation of Weeks (1), with some
　　　40% of original material added).
　　Shanghai, 1941.
Chang Wên-Yuan (1).
　　Thai Chi Chhüan Chhang Shih Wên Ta
　　　太極拳常識問答.
　　Explanation of the Standard Principles of
　　　Chinese Boxing.
　　Jen-min Thi-yü, Peking, 1962.
Chao Pi-Chhen (1) 趙避塵.
　　Hsing Ming Fa Chüeh Ming Chih 性命法
　　　訣明指.
　　A Clear Explanation of the Oral Instruct-
　　　ions concerning the Techniques of the
　　　Nature and the Life-Span.
　　Chhi-shan-mei, Thaipei, Thaiwan, 1963.
　　Tr. Lu Khuan-Yü (4).
Chi Yün (1).
　　Yüeh Wei Tshao Thang Pi Chi 閱微草堂
　　　筆記.

Chi Yün (1) (cont.)
　Jottings from the Yüeh-wei Cottage.
　1800.
Chia Tsu-Chang & Chia Tsu-Shan (1)　賈祖璋
　賈祖珊.
　Chung-Kuo Chih-Wu Thu Chien　中國植物
　圖鑑.
　Illustrated Dictionary of Chinese Flora
　[arranged on the Engler system; 2602
　entries].
　Chung-hua, Peking, 1936, repr. 1955, 1958.
Chia Yo-Han (Kerr, J. G.)　嘉約翰 & Ho
　Liao-Jan　何了然 (1).
　Hua Hsüeh Chhu Chiai　化學初階.
　First Steps in Chemistry.
　Canton, 1870.
Chiang Thien-Shu (1)　蔣天樞.
　'Chhu Tzhu Hsin Chu' Tao Lun　「楚辭新
　注」導論.
　A Critique of the New Commentary on the
　Odes of Chhu.
　CHWSLT, 1962, 1, 81.
　Abstr. RBS, 1969, 8, no. 558.
Chiang Wei-Chhiao (1) [Yin Shih Tzu]　蔣維喬.
　Yin Shih Tzu Ching Tso Fa　因是子靜坐
　法.
　Yin Shih Tzu's Methods of Meditation
　[Taoist].
　Shih-yung, Hongkong, 1914, repr. 1960,
　1969.
　With Buddhist addendum, Hsü Phien
　續編.
　Cf. Lu Khuan-Yü (1), pp. 167, 193.
Chi ang Wei-Chhiao (2) [Yin Shih Tzu]　蔣維喬
　Ching Tso Fa Chi Yao　靜坐法輯要.
　The Important Essentials of Meditation
　Practice.
　Repr. Thaiwan Yin Ching Chhu, Thaipei,
　1962.
[Chiang Wei-Chhiao] (3) Yin Shih Tzu　蔣維喬.
　Hu Hsi Hsi Ching Yang Shêng Fa　呼吸習
　靜養生法.
　Methods of Nourishing the Life-Force by
　Respiratory Physiotherapy and Medita-
　tion Technique.
　Repr. Thai-Phing, Hongkong, 1963.
[Chiang Wei-Chhiao] (4) Yin Shih Tzu　蔣維喬.
　Yin Shih Tzu Ching Tso Wei Shêng Shih
　Yen Than　因是子靜坐衛生實驗談.
　Talks on the Preservation of Health by
　Experiments in Meditation.
　Printed together with the Hsü Phien of (1).
　Tzu-Yu, Thaichung, Thaiwan｝
　　　　　　　　　　　Hongkong｝ 1957.
　Cf. Lu Khuan-Yü, (1), pp. 157, 160, 193.
Chiang Wei-Chhiao (5)　蔣維喬.
　Chung-Kuo-ti Hu Hsi Hsi Ching Yang
　Shêng Fa (Chhi Kung Fang Chih Fa)
　中國的呼吸習靜養生法(氣功防
　治法).
　The Chinese Methods of Prolongevity by
　Respiratory and Meditational Technique

　(Hygiene and Health due to the Circula-
　tion of the Chhi).
　Wei-Shêng, Shanghai, 1956, repr. 1957.
Chiang Wei-Chhiao 蔣維喬 & Liu Kuei-Chen
　(1).
　Chung I Than Chhi Kung Liao Fa　中醫談
　氣功療法.
　Respiratory Physiotherapy in Chinese
　Medicine.
　Thai-Phing, Hongkong, 1964.
Chieh Hsi-Kung (1)　解希恭.
　Thai-yuan Tung-thai-pao Chhu Thu ti Han
　Tai Thung Chhi　太原東太堡出土的
　漢代銅器.
　Bronze Objects of Han Date Excavated at
　Tung-thai-pao Village near Thaiyuan
　(Shansi), [including five unicorn-foot
　horse-hoof gold pieces, about 140 gms. wt.,
　with almost illegible inscriptions].
　WWTK, 1962 (no. 4/5), no. 138-9, 66
　(71), ill. p. 11.
　Abstr. RBS, 1969, 8, no. 360 (p. 196).
Chikashige Masumi (1) = (1)　近重眞澄.
　Tōyō Renkinjutsu; Kagakujō yori mitaru
　Tōyōjōdai no Bunka
　東洋錬金術；化學上より見たる東
　洋上代の文化.
　East Asian Alchemy; the Culture of
　East Asia in Early Times seen from the
　Chemical Point of View.
　Rokakuho Uchida, Tokyo, 1929, repr.
　1936.
　Based partly on (4) and on papers in
　SN 1918, 3 (no. 2) and 1919, 4 (no. 2).
Chikashige Masumi (2)　近重眞澄.
　Tōyō Kodōki no Kagaku-teki Kenkyū
　東洋古銅器の化學的研究.
　A Chemical Investigation of Ancient
　Chinese Bronze [and Brass] Vessels.
　SN, 1918, 3 (no. 2), 177.
Chikashige Masumi (3)　近重眞澄.
　Kagaku yori mitaru Tōyōjōdai no Bunka
　化學より觀たる東洋上代の文化.
　The Culture of Ancient East Asia seen
　from the Viewpoint of Chemistry.
　SN, 1919, 4 (no. 2), 169.
Chikashige Masumi (4)　近重眞澄.
　Tōyō Kodai Bunka no Kagakukan
　東洋古代文化の化學觀.
　A Chemical View of Ancient East Asian
　Culture.
　Pr. pr. Tokyo, 1920.
Chojiya Heibei (1).
　Shoseki Seirenho　硝石製煉法.
　The Manufacture of Saltpetre.
　Yedo, 1863.
Chŏn Sangun (2)　全相運.
　Han'guk kwahak kisul sa
　韓國科學技術史.
　A Brief History of Science and Technology
　in Korea.
　World Science Co. Seoul, 1966.

Chou Fêng-Wu 周鳳梧, Wang Wan-Chieh
    王萬杰 & Hsü Kuo-Chhien 徐國仟 (1).
    Huang Ti Nei Ching Su Wên, Pai Hua Chieh
    黄帝內經素問白話解.
    The *Yellow Emperor's Manual of Corporeal*
    (*Medicine*); *Questions* (*and Answers*) about
    Living Matter; done into Colloquial
    Language.
    Jen-min Wei-shêng, Peking, 1963.
Chou Shao-Hsien (1) 周紹賢.
    Tao Chia yü Shen Hsien 道家與神仙.
    The Holy Immortals of Taoism; the
    Development of a Religion.
    Chung-Hua, Thaipei (Thaiwan), 1970.
Chu Chi-Hai (1) 朱季海.
    '*Chhu Tzhu*' *Chieh Ku Shih I* 「楚辭」解故
    識遺.
    Commentary on Parts of the *Odes of Chhu*
    (especially *Li Sao* and *Chiu Pien*), [with
    special attention to botanical identifi-
    cations].
    CHWSLT, 1962, **2**, 77.
    Abstr. *RBS*, 1969, **8**, no. 557.
Chu Lien (1) 朱璉.
    Hsin Chen Chiu Hsüeh 新針灸學.
    New Treatise on Acupuncture and Moxi-
    bustion.
    Jen-min Wei-shêng, Peking, 1954.
Chhang Pi-Tê (1) 昌彼得.
    Shuo Fu Khao 說郛考.
    A Study of the *Shuo Fu* Florilegium.
    Chinese Planning Commission for East
    Asian Studies, Thaipei (Thaiwan), 1962.
Chhen Ching (1) 陳經.
    Chhiu Ku Ching Shê Chin Shih Thu 求古
    精舍金石圖.
    Illustrations of Antiques in Bronze and
    Stone from the Spirit-of-Searching-Out-
    Antiquity Cottage.
Chhen Kung-Jou (3) 陳公柔.
    Pai-Sha Thang Mu Chien Pao 白沙唐墓
    簡報.
    Preliminary Report on (the Excavation of) a
    Thang Tomb at the Pai-sha (Reservoir),
    (in Yü-hsien, Honan).
    KKTH, 1955 (no. 1), 22.
Chhen Kuo-Fu (1) 陳國符.
    '*Tao Tsang*' *Yuan Liu Khao* 「道藏」源流考.
    A Study on the Evolution of the Taoist
    Patrology.
    1st ed. Chung-Hua, Shanghai, 1949.
    2nd ed. in 2 vols., Chung-Hua, Peking, 1963.
Chhen Mêng-Chia (4) 陳夢家.
    Yin Hsü Pu Tzhu Tsung Shu 殷虛卜辭綜
    述.
    A study of the Characters on the Shang
    Oracle-Bones.
    Kho-Hsüeh, Peking, 1956.
Chhen Pang-Hsien (1) 陳邦賢.
    Chung Kuo I-Hsüeh Shih 中國醫學史.
    History of Chinese Medicine.
    Com. Press, Shanghai, 1937, 1957.

Chhen Phan (7) 陳槃.
    Chan-Kuo Chhin Han Chien Fang-Shih
    Khao Lun 戰國秦漢間方士考論.
    Investigations on the Magicians of the
    Warring States, Chhin and Han periods.
    AS/BIHP, 1948, **17**, 7.
Chhen Pi-Liu 陳璧琉 & Chêng Cho-Jen (1)
    鄭卓人.
    Ling Shu Ching, Pai Hua Chieh 靈樞經
    白話解.
    The *Yellow Emperor's Manual of Corporeal*
    (*Medicine*); *the Vital Axis*; done into
    Colloquial Language.
    Jen-min Wei-shêng, Peking, 1963.
Chhen Thao (1).
    Chhi Kung Kho-Hsüeh Chhang Shih 氣功
    科學常識.
    A General Introduction to the Science of
    Respiratory Physiotherapy.
    Kho-hsüeh Chi-shu, Shanghai, 1958.
Chhen Wên-Hsi (1) 陳文熙.
    Lu-kan-shih 'Tutty' Thou-shih Thang-Thi
    爐甘石 Tutty 鍮石鑪銻.
    A Study of the Designations of Zinc Ores,
    lu-kan-shih, tutty and brass.
    HITC 1933, **12**, 839; 1934, **13**, 401.
Chhen Yin-Kho (3) 陳寅恪.
    Thien Shih Tao yü Pin-Hai Ti-Yü chih
    Kuan-Hsi 天師道與濱海地域之關係
    On the Taoist Church and its Relation to
    the Coastal Regions of China (*c.* +126 to
    +536).
    AS/BIHP, 1934 (no. 3/4), 439.
Chhen Yuan (4) 陳垣.
    Shih Hui Chü Li 史諱舉例.
    On the Tabu Changes of Personal Names in
    History; Some Examples.
    Chung-Hua, Peking, 1962, repr. 1963.

Dohi Keizō (1) 土肥慶藏.
    Shōsōin Yakushi no Shiteki Kōsatsu 正倉
    院藥種の史的考察.
    Historical Investigation of the Drugs pre-
    served in the Imperial Treasury at
    Nara.
    In *Zoku Shōsōin Shiron* 續正倉院史論.
    1932, No. 15, Neiyaku 寧藥.
    1st pagination, p. 133.
Dōno Tsurumatsu (1) 道野鶴松.
    Kodai no Shina ni okeru Kagakushisō toku
    ni Genzoshisō ni tsuite 古代の支那に
    於ける化學思想特に元素思想に就
    いて.
    On Ancient Chemical Ideas in China, with
    Special Reference to the Idea of Ele-
    ments [comparison with the Four
    Aristotelian Elements and the Spagyrical
    Tria Prima].
    TG/T, 1931, **1**, 159.

Fan Hsing-Chun (6) 范行準.
    Chung Hua I-Hsüeh Shih 中華醫學史.

Fan Hsing-Chun (6) (cont.)
    Chinese Medical History.
    ISTC, 1947, **1** (no. 1), 37, (no. 2), 21;
      1948, **1** (no. 3/4), 17.
Fan Hsing-Chun (12) 范行準.
    Liang Han San Kuo Nan Pei Chhao Sui
      Thang I Fang Chien Lu 兩漢三國南北
    朝隋唐醫方簡錄.
    A Brief Bibliography of (Lost) Books on
      Medicine and Pharmacy written during
      the Han, Three Kingdoms, Northern and
      Southern Dynasties and Sui and Thang
      Periods.
    CHWSLT, 1965, **6**, 295.
Fêng Chhêng-Chün (1) 馮承鈞.
    Chung-Kuo Nan-Yang Chiao-Thung Shih
    中國南洋交通史.
    History of the Contacts of China with the
      South Sea Regions.
    Com. Press, Shanghai, 1937, repr. Thai-
      Phing, Hongkong, 1963.
Fêng Chia-Shêng (1) 馮家昇.
    Huo-Yao ti Fa-Hsien chi chhi Chhuan Pu
    火藥的發現及其傳佈.
    The Discovery of Gunpowder and its
      Diffusion.
    AP/HJ, 1947, **5**, 29.
Fêng Chia-Shêng (2) 馮家昇.
    Hui Chiao Kuo wei Huo-Yao yu Chung-Kuo
      Chhuan Ju Ou-Chou ti Chhiao Liang 回
    教國爲火藥由中國傳入歐州的橋梁.
    The Muslims as the Transmitters of Gun-
      powder from China to Europe.
    AP/HJ, 1949, 1.
Fêng Chia-Shêng (3) 馮家昇.
    Tu Hsi-Yang ti Chi Chung Huo-Chhi Shih
      Hou 讀西洋的幾種火器史後.
    Notes on reading some of the Western
      Histories of Firearms.
    AP/HJ, 1947, **5**, 279.
Fêng Chia-Shêng (4) 馮家昇.
    Huo-Yao ti Yu Lai chi chhi Chhuan Ju
      Ou-chou ti Ching Kuo 火藥的由來及
    其傳入歐洲的經過.
    On the Origin of Gunpowder and its Trans-
      mission to Europe.
    Essay in Li Kuang-Pi & Chhien Chün-Yeh
      (q.v.), p. 33.
    Peking, 1955.
Fêng Chia-Shêng (5) 馮家昇.
    Lien Tan Shu ti Chhêng Chhang chi chhi
      Hsi Chhuan 煉丹術的成長及其西傳.
    Achievements of (ancient Chinese) Al-
      chemy and its Transmission to the West.
    Essay in Li Kuang-Pi & Chhien Chün-Yeh
      (q.v.), p. 120.
    Peking 1955.
Fêng Chia-Shêng (6) 馮家昇.
    Huo-Yao ti Fa-Ming ho Hsi Chhuan 火藥
    的發明和西傳.
    The Discovery of Gunpowder and its
      Transmission to the West.

Hua-Tung, Shanghai, 1954.
    Revised ed. Jen-Min, Shanghai, 1962.
Fu Chhin-Chia (1) 傅勤家.
    Chung-Kuo Tao Chiao Shih 中國道教史.
    A History of Taoism in China.
    Com. Press, Shanghai, 1937.
Fu Chin-Chhüan (1) 傅金銓.
    Chhih Shui Yin 赤水吟.
    Chants of the Red River [physiological
      alchemy].
    1823.
    In *CTPS*, *pên* 4.
Fu Chin-Chhüan (2) 傅金銓.
    Thien Hsien Chêng Li Tu 'Fa Tien Ching'
    天仙正理讀法點睛.
    The Right Pattern of the Celestial Im-
      mortals; Thoughts on Reading the
      Consecration of the Law [physiological
      alchemy. *Tien ching* refers to the cere-
      mony of painting in the pupils of the eyes
      in an image or other representation].
    1820.
    In *CTPS*, *pên* 5.
Fu Chin-Chhüan (3) 傅金銓.
    Tan Ching Shih Tu 丹經示讀.
    A Guide to the Reading of the Enchymoma
      Manuals [dialogue of pupil and teacher on
      physiological alchemy]
    c. 1825.
    In *CTPS*, *pên* 11.
Fu Chin-Chhüan (4) 傅金銓.
    Tao Hai Chin Liang 道海津梁.
    A Catena (of Words) to Bridge the Ocean
      of the Tao [mutationism, Taoist–Buddh-
      ist–Confucian syncretism, and physio-
      logical alchemy].
    1822.
    In *CTPS*, *pên* 11.
Fu Chin-Chhüan (5) 傅金銓.
    Shih Chin Shih 試金石.
    On the Testing of (what is meant by)
      'Metal' and 'Mineral'.
    c. 1820.
    In *Wu Chen Phien Ssu Chu* ed.
Fu Chin-Chhüan (6) (ed.) 傅金銓.
    Chêng Tao Pi Shu Shih Chung 證道秘書十種
    Ten Types of Secret Books on the Verifica-
      tion of the Tao.
    Early 19th.
Fu Lan-Ya (Fryer, John) 傅蘭雅 & Hsü Shou
    徐壽 (1) (tr.).
    Hua-Hsüeh Chien Yuan 化學鑑原.
    Authentic Mirror of Chemical Science
      (translation of Wells, 1).
    Chiangnan Arsenal Transl. Bureau, Shang-
      hai, 1871.
Fukui Kōjun (1) 福井康順.
    Tōyō Shisō no Kenkyū 東洋思想の研究.
    Studies in the History of East Asian
      Philosophy.
    Risōsha, Tokyo, 1956.
    Abstr. *RBS*, 1959, **2**, no. 564.

Fukunaga Mitsuji (1) 福永光司.
Hōzensetsu no Keisei 封禪說の形成.
The Evolution of the Theory of the Fêng and Shan Sacrifices (in Chhin and Han Times).
TS, 1954, **1** (no. 6), 28, (no. 7), 45.

Harada Yoshito 原田淑人 & Tazawa Kingo (1) 田澤金吾.
Rakurō Gokan-en Ō Ku no Fumbo 樂浪五官掾王旰の墳墓.
Lo-Lang; a Report on the Excavation of Wang Hsü's Tomb in the Lo-Lang Province (an ancient Chinese Colony in Korea).
Tokyo Univ. Tokyo, 1930.

Hasegawa Usaburo (1) 長谷川卯三郎.
Shin Igaku Zen 新醫學禪.
New Applications of Zen Buddhist Techniques in Medicine.
So Gensha, Tokyo, 1970. (In the Hara-o-tsukuruzen Series.)

Hiraoka Teikichi (2) 平岡楨吉.
'*Enanji' ni arawareta Ki no Kenkyū* 「淮南子」に現われた氣の研究.
Studies on the Meaning and the Conception of 'chhi' in the *Huai Nan Tzu* book.
Kan Gi Bunka Gakkai, Tokyo, 1961.
Abstr. *RBS*, 1968, **7**, no. 620.

Ho Han-Nan (1) 何漢南.
Sian Shih Hsi-yao-shih Tshun Thang Mu Chhing-Li Chi 西安市西窰實村唐墓清理記.
A Summary Account of the Thang Tomb at Hsi-yao-shih Village near Sian [the tomb which yielded early Arabic coins].
Cf. Hsia Nai (3).
KKTH, 1965, no. 8 (no. 108), pp. 383, 388.

Ho Hsin (1) (Hobson, Benjamin) 合信.
Po Wu Hsin Phien 博物新編.
New Treatise on Natural Philosophy and Natural History [the first book on modern chemistry in Chinese].
Shanghai, 1855.

Ho Ping-Yü 何丙郁 & Chhen Thieh-Fan (1) 陳鐵凡.
Lun 'Shun Yang Lü Chen Jen Yao Shih Chih' ti Chu Chhing Shih-Tai 論「純陽呂眞人藥石製」的著成時代.
On the Dating of the 'Manipulations of Drugs and Minerals, by the Adept Lü Shun-Yang', a Taoist Pharmaceutical and Alchemical Manual.
JOSHK, 1971, **9**, 181–228.

Ho-Ping-Yü 何丙郁 & Su Ying-Hui 蘇瑩輝 (1).
'*Tan Fang Ching Yuan' Khao* 「丹房鏡源」考.
On the *Mirror of the Alchemical Elaboratory*, (a Thang Manual of Practical Experimentation).
JOSHK, 1970, **8** (no. 1), 1, 23.

Hori Ichirō (1) 堀一郎.
Yudono-san Kei no Sokushimbutsu (Miira) to

sono Haikei 湯殿山系の即身佛(ミイラ)とその背景.
The Preserved Buddhas (Mummies) at the Temples on Yudono Mountain.
TBKK, 1961, no. 35 (no. 3).
Repr. in Hori Ichirō (2), p. 191.

Hori Ichirō (2) 堀一郎.
Shūkyō Shūzoku no Seikatsu Kisei 宗敎習俗の生活規制.
Life and Customs of the Religious Sects (in Buddhism).
Miraisha, Tokyo, 1963.

Hou Pao-Chang (1) 侯寶璋.
Chung-Kuo Chieh-Phou Shih 中國解剖史
A History of Anatomy in China.
ISTC, 1957, **8** (no. 1), 64.

Hou Wai-Lu (3) 侯外廬.
Fang I-Chih—Chung-Kuo ti Pai Kho Chhüan Shu Phai Ta Chê-Hsüeh Chia 方以智—中國的百科全書派大哲學家.
Fang I-Chih—China's Great Encyclopaedist Philosopher.
LSYC, 1957 (no. 6), 1; 1957 (no. 7), 1.

Hou Wai-Lu (4) 侯外廬.
Shih-liu Shih-Chi Chung-Kuo ti Chin-Pu ti Chê-Hsüeh Ssu-Chhao Kai-Shu 十六世紀中國的進步的哲學思潮槪述.
Progressive Philosophical Thinking in +16th-century China.
LSYC, 1959 (no. 10), 39.

Hou Wai-Lu 侯外廬, Chao Chi-Pin 趙紀彬, Tu Kuo-Hsiang 杜國庠 & Chhiu Han-Shêng (1) 邱漢生.
Chung-Kuo Ssu-Hsiang Thung Shih 中國思想通史.
General History of Chinese Thought.
5 vols.
Jen-Min, Peking, 1957.

Hu Shih (7) 胡適.
Lun Hsüeh Chin Chu, ti-i Chi 論學近著第一集.
Recent Studies on Literature (first series).

Hu Yao-Chen (1).
Chhi Kung Chien Shen Fa 氣功健身法.
Respiratory Exercises and the Strengthening of the Body.
Thai-Phing, Hongkong, 1963.

Huang Chu-Hsün (1) 黃著勳.
Chung-Kuo Khuang Chhan 中國鑛產.
The Mineral Wealth and Productivity of China.
2nd ed., Com. Press, Shanghai, 1930.

Hung Huan-Chhun (1) 洪煥椿.
Shih chih Shih-san Shih-Chi Chung-Kuo Kho-Hsüeh-ti Chu-Yao Chhêng-Chiu. 十至十三世紀中國科學的主要成就.
The Principal Scientific (and Technological) Achievements in China from the +10th to the +13th centuries (inclusive), [the Sung period].
LSYC, 1959, **5** (no. 3), 27.

Hung Yeh (2) 洪業.
   Tsai Shuo 'Hsi Ching Tsa Chi' 再說「西
   京雜記」.
   Further Notes on the *Miscellaneous Records
   of the Western Capital* [with a study of
   the dates of Ko Hung].
   AS/BIHP, 1963, **34** (no. 2), 397.]
Hsia Nai (2) 夏鼐.
   Khao-Ku-Hsüeh Lun Wên Chi 考古學論
   文集.
   Collected Papers on Archaeological Subjects.
   Academia Sinica, Peking, 1961.
Hsia Nai (3) 夏鼐.
   *Sian Thang Mu Chhu Thu A-la-pa Chin
   Pi* 西安唐墓出土阿拉伯金幣.
   Arab Gold Coins unearthed from a Thang
   Dynasty Tomb (at Hsi-yao-thou Village)
   near Sian, Shensi (gold dīnārs of the
   Umayyad Caliphs 'Abd al-Malik, +702,
   'Umar ibn 'Abd al-'Azīz, +718, and
   Marwān II, +746).
   Cf. Ho Han-Nan (1).
   KKTH, 1965, no. 8 (no. 108), 420, with
   figs 1–6 on pl. 1.
Hsiang Ta (3) 向達.
   Thang Tai Chhang-An yü Hsi Yü Wên Ming
   唐代長安與西域文明.
   Western Cultures at the Chinese Capital
   (Chhang-an) during the Thang Dynasty.
   YCHP Monograph series, no. 2, Peiping,
   1933.
Hsieh Sung-Mu (1) 謝誦穆.
   Chung-Kuo Li-Tai I-Hsüeh Wei Shu Khao
   中國歷代醫學偽書考.
   A Study of the Authenticity of (Ancient
   and Medieval) Chinese Medical Books.
   ISTC, 1947, **1** (no. 1), 53.
Hsiung Tê-Chi (1) 熊德基.
   '*Thai Phing Ching*' *ti Tso-Chê ho Ssu-
   Hsiang chi chhi yü Huang Chin ho Thien
   Shih Tao ti Kuan-Hsi* 「太平經」的作
   者和思想及其與黃巾和天師道的
   關係.
   The Authorship and Ideology of the *Canon
   of the Great Peace*; and its Relation with
   the Yellow Turbans (Rebellion) and the
   Taoist Church (Tao of the Heavenly
   Teacher).
   LSYC 1962 (no. 4), 8.
   Abstr. *RBS*, 1969, **8**, no. 737.
Hsü Chien-Yin (1) 徐建寅.
   Ko Chih Tshung Shu 格致叢書.
   A General Treatise on the Natural Sciences.
   Shanghai, 1901.
Hsü Chih-I (1) 徐致一.
   Wu Chia Thai Chi Chhüan 吳家太極拳.
   Chinese Boxing Calisthenics according to
   the Wu Tradition.
   Hsin-Wên, Hongkong, 1969.
Hsü Chung-Shu (7) 徐中舒.
   Chin Wên Chia Tzhu Shih Li 金文嘏辭
   釋例.

Terms and Forms of the Prayers for Bles-
   sings in the Bronze Inscriptions.
   AS/BIHP, 1936, (no. 4), 15.
Hsü Chung-Shu (8) 徐中舒.
   Chhen Hou Ssu Chhi Khao Shih 陳侯四
   器考釋.
   Researches on Four Bronze Vessels of the
   Marquis Chhen [i.e. Prince Wei of Chhi
   State, r. −378 to −342].
   AS/BIHP, 1934, (no. 3/4), 499.
Hsü Ti-Shan (1) 許地山.
   Tao Chiao Shih 道教史.
   History of Taoism.
   Com. Press, Shanghai, 1934.
Hsü Ti-Shan (2) 許地山.
   Tao Chia Ssu-Hsiang yü Tao Chiao 道家
   思想與道教.
   Taoist Philosophy and Taoist Religion.
   YCHP, 1927, **2**, 249.
Hsüeh Yü (1) 薛愚.
   Tao-Chia Hsien Yao chih Hua-Hsüeh Kuan
   道家仙藥之化學觀.
   A Look at the Chemical Reactions in-
   volved in the Elixir-making of the Taoists.
   HSS, 1942, **1** (no. 5), 126.
Huang Lan-Sun (1) (ed.) 黃蘭孫.
   *Chung-Kuo Yao-Wu-ti Kho-Hsüeh Yen-
   Chiu* 中國藥物的科學研究.
   Scientific Researches on Chinese Materia
   Medica.
   Chhien-Chhing Thang, Shanghai, 1952.

Imai Usaburō (1) 今井宇三郎.
   '*Goshinpen*' *no Seisho to Shisō* 悟眞篇の
   成書と思想.
   The *Poetical Essay on Realising the...
   Primary Vitalities* [by Chang Po-Tuan,
   +1075]; its System of Thought and how
   it came to be written.
   TS, 1962, **19**, 1.
   Abstr. *RBS*, 1969, **8**, no. 799.
Ishihara Akira (1) 石原明.
   Gozōnyūtai no Igi ni tsuite 五臟入胎の意
   義について.
   The Buddhist Meaning of the Visceral
   Models (in the Sakyamuni Statue at the
   Seiryōji Temple).
   NIZ, 1956, **7** (nos. 1–3), 5.
Ishihara Akira (2) 石原明.
   Indo Kaibōgaku no Seiritsu to sono Ryūden
   印度解剖學の成立とその流傳.
   On the Introduction of Indian Anatomical
   Knowledge (to China and Japan).
   NIZ, 1956, **7** (nos. 1–3), 64.
Ishii Masako (1) 石井昌子.
   Kōhon '*Chen Kao*' 稿本「眞誥.」
   Draft of an Edition of the *Declarations of
   Perfected Immortals*, (with Notes on
   Variant Readings).
   In several volumes.
   Toyoshima Shobō, Tokyo, (for Dōkyō
   Kankōkai 道教刊行會), 1966–.

Ishii Masako (2) 石井昌子.
    'Chen Kao' no Seiritsu o Meguru Shiryō-teki Kentō; '*Têng Chen Yin Chüeh*', '*Chen Ling Wei Yeh Thu*' *oyobi* '*Wu Shang Pi Yao*' *tono Kankei wo Chūshin-ni* 「眞誥」の成立をみぐる資料的檢討;「登眞隱訣」,「眞靈位業圖」及び「無上秘要」との關係を中心に.
    Documents for the Study of the Formation of the *Declarations of Perfected Immortals*....
    DK, 1968, **3**, 79–195 (with French summary on p. iv).

Ishii Masako (3) 石井昌子.
    'Chen Kao' no Seiritsu ni Kansuru Kōsatsu 「眞誥」の成立に關する一考察.
    A Study of the Formation of the *Declarations of Perfected Immortals*.
    DK, 1965, **1**, 215 (French summary, p. x).

Ishii Masako (4) 石井昌子.
    Thao Hung-Ching Denkikō 陶弘景傳記考.
    A Biography of Thao Hung-Ching.
    DK, 1971, **4**, 29–113 (with French summary, p. iv).

Ishijima Yasutaka (1) 石島快隆.
    Hōbokushi Insho Kō 抱朴子引書考.
    A Study of the Books quoted in the *Pao Phu Tzu* and its Bibliography.
    BK, 1956, **20**, 877.
    Abstr. *RBS*, 1959, **2**, no. 565.

Itō Kenkichi (1) 伊藤堅吉.
    Sei no Mihotoke 性のみほとけ.
    Sexual Buddhas (Japanese Tantric images etc.).
    Zufushinsha, Tokyo, 1965.

Itō Mitsutōshi (1) 伊藤光遠.
    Yang Shêng Nei Kung Pi Chüeh 養生內功祕訣.
    Confidential Instructions on Nourishing the Life Force by Gymnastics (and other physiological techniques).
    Tr. from the Japanese by Tuan Chu-Chün 段竹君.
    Thaipei (Thaiwan), 1966.

Jen Ying-Chhiu (1) 任應秋.
    Thung Su Chung-Kuo I-Hsüeh Shih Hua 通俗中國醫學史話.
    Popular Talks on the History of Medicine.
    Chungking, 1957.

Jung Kêng (3) 容庚.
    Chin Wên Pien 金文編.
    Bronze Forms of Characters.
    Peking, 1925, repr. 1959.

Kao Chih-Hsi (1) 高至喜.
    Niu Têng 牛鐙.
    An 'Ox Lamp' (bronze vessel of Chhien Han date, probably for sublimation, with the boiler below formed in the shape of an ox, and the rising tubes a continuation of its horns).
    WWTK, 1959 (no. 7), 66.

Kao Hsien (1) *et al.* 高銛.
    Hua-Hsüeh Yao Phin Tzhu-Tien 化學藥品辭典.
    Dictionary of Chemistry and Pharmacy (based on T. C. Gregory (1), with the supplement by A. Rose & E. Rose).
    Shanghai Sci & Tech. Pub., Shanghai, 1960.

Kawabata Otakeshi 川端男剪 & Yoneda Yūtarō 米田祐太郎 (1)
    Tōsei Biyaku-kō 東西媚藥考.
    Die Liebestränke in Europa und Orient.
    Bunkiūsha, Tokyo.

Kawakubo Teirō (1) 川久保悌郎.
    Shindai Manshū ni okeru Shōka no Zokusei ni tsuite 清代滿洲における燒鍋の簇生について.
    On the (Kao-liang) Spirits Distilleries in Manchuria in the Chhing Period and their Economic Role in Rural Colonisation.
    Art. in *Wada Hakase Koki Kinen Tōyōshi Ronsō* (Wada Festschrift) 和田博士古稀記念東洋史論叢, Kōdansha, Tokyo, 1961, p. 303.
    Abstr. *RBS*, 1968, **7**, no. 758.

Khung Chhing-Lai *et al.* (1) 孔慶萊 (13 collaborators).
    Chih-Wu-Hsüeh Ta Tzhu Tien 植物學大辭典.
    General Dictionary of Chinese Flora.
    Com. Press, Shanghai and Hongkong, 1918, repr. 1933 and often subsequently.

Kimiya Yasuhiko (1) 木宮泰彦.
    Nikka Bunka Kōryūshi 日華文化交流史.
    A History of Cultural Relations between Japan and China.
    Fuzambō, Tokyo, 1955.
    Abstr *RBS*, 1959, **2**, no. 37.

Kobayashi Katsuhito (1) 小林勝人.
    Yō Shu Gakuha no hitobite 楊朱學派の人々.
    On the Disciples and Representatives of the (Hedonist) School of Yang Chu.
    TYGK, 1961, **5**, 29.
    Abstr. *RBS*, 1968, **7**, no. 606.

Koyanagi Shikita (1) 小柳司氣太.
    Tao Chiao Kai Shuo 道敎概說.
    A Brief Survey of Taoism.
    Tr. Chhen Pin-Ho 陳斌和.
    Com. Press, Shanghai, 1926.
    Repr. Com. Press, Thaipei, 1966.

Kuo Mo-Jo (8) 郭沫若.
    Chhu Thu Wên Wu Erh San Shih 出土文物二三事.
    One or two Points about Cultural Relics recently Excavated (including Japanese coin inscriptions).
    WWTK, 1972 (no. 3), 2.

Kuo Pao-Chün (1) 郭寶鈞.
    Hsün-hsien Hsin-tshun Ku Tshan Mu chih Chhing Li 濬縣辛村古殘墓之清理.

Kuo Pao-Chün (1) (cont.)
Preliminary Report on the Excavations at
the Ancient Cemetery of Hsin-tshun
village, Hsün-hsien (Honan).
TYKK, 1936, 1, 167.
Kuo Pao-Chün (2) 郭寶鈞.
Hsün-hsien Hsin-tshun 濬縣辛村.
(Archaeological Discoveries at) Hsin-tshun
Village in Hsün-hsien (Honan).
Inst. of Archaeology, Academia Sinica,
Peking, 1964 (Field Expedition Reports,
I series, no. 13).
Kurihara Keisuke (1) 栗原圭介.
Gusai no Gireiteki Igi 虞祭の「儀禮」的
意義.
The Meaning and Practice of the Yü
Sacrifice, as seen in the *Personal
Conduct Ritual*.
NCGH, 1961, 13, 19.
Abstr. *RBS*, 1968, 7, no. 615.
Kuroda Genji (1) 黑田源次.
Ki 氣.
On the Concept of Chhi (*pneuma*; in
ancient Chinese thought).
TS, 1954 (no. 4/5), 1; 1955 (no. 7), 16.

Lai Chia-Tu (1) 賴家度.
'*Thien Kung Khai Wu*' *chi chhi Chu chê*;
Sung Ying-Hsing 「天工開物」及其著者
朱應星.
The *Exploitation of the Works of Nature* and
its Author; Sung Ying-Hsing.
Essay in Li Kuang-Pi & Chhien Chün-
Yeh (q.v.), p. 338.
Peking, 1955.
Lai Tou-Yen (1) 賴斗岩.
I Shih Sui Chin 醫史碎錦.
Medico-historical Gleanings.
ISTC, 1948, 2 (nos 3/4), 41.
Lao Kan (6) 勞榦.
*Chung-Kuo Tan-Sha chih Ying-Yung chi
chhi Thui-Yen* 中國丹砂之應用及其
推演.
The Utilisation of Cinnabar in China and
its Historical Implications.
AS/BIHP, 1936, 7 (no. 4), 519.
Li Chhiao-Phing (1) 李喬苹.
Chung-Kuo Hua-Hsüeh Shih 中國化學
史.
History of Chemistry in China.
Com. Press, Chhangsha, 1940, 2nd (en-
larged) ed. Thaipei, 1955.
Li Kuang-Pi 李光璧 & Chhien Chün-Yeh (1)
錢君曄.
*Chung-Kuo Kho-Hsüeh Chi-Shu Fa-Ming ho
Kho-Hsüeh Chi-Shu Jen Wu Lun Chi*
中國科學技術發明和科學技術人物
論集.
Essays on Chinese Discoveries and Inven-
tions in Science and Technology, and on
the Men who made them.
San-lien Shu-tien, Peking, 1955.

Li Nien (4) 李儼.
Chung Suan Shih Lun Tshung 中算史論
叢.
Gesammelte Abhandlungen ü. die Ge-
schichte d. chinesischen Mathematik.
3 vols. 1933–5; 4th vol. (in 2 parts), 1947.
Com. Press, Shanghai.
Li Nien (21) 李儼.
Chung Suan Shih Lun Tshung (second
series) 中算史論叢.
Collected Essays on the History of Chinese
Mathematics—vol. 1, 1954; vol. 2,
1954; vol. 3, 1955; vol. 4, 1955; vol. 5,
1955.
Kho-Hsüeh, Peking.
Li Shu-Hua (3) 李書華.
Li Shu-Hua Yu Chi 李書華遊記.
Travel Diaries of Li Shu-Hua [recording
visits to temples and other notable places
around Huang Shan, Fang Shan, Thien-
thai Shan, Yen-tang Shan etc. in 1935
and 1936].
Chhuan-chi Wên-hsüeh, Thaipei, 1969.
Li Shu-Huan (1) 李叔還.
Tao Chiao Yao I Wên Ta Chi Chhêng 道教
要義問答集成.
A Catechism of the Most Important Ideas
and Doctrines of the Taoist Religion.
Pr. Kao-hsiung and Thaipei, Thaiwan, 1970.
Distributed by the Chhing Sung Kuan
(Caerulean Pine-tree Taoist Abbey),
Chhing Shan (Castle Peak), N.T.
Hongkong.
Liang Chin (1) 梁津.
Chou Tai Ho Chin Chhêng Fên Khao 周代
合金成分考.
A Study of the Analysis of Alloys of the
Chou period.
KHS, 1925, 9 (no. 3), 1261; repr. in Wang
Chin (2), p. 52.
Lin Thien-Wei (1) 林天蔚.
Sung-Tai Hsiang-Yao Mou-I Shih Kao
宋代香藥貿易史稿.
A History of the Perfume and Drug
Trade during the Sung Dynasty.
Chung-kuo Hsüeh-shê, Hongkong, 1960.
Ling Shun-Shêng (6) 凌純聲.
Chung-Kuo Chiu chih Chhi Yuan 中國酒
之起源.
On the Origin of Wine in China.
AS/BIHP, 1958, 29, 883 (Chao Yuan-Jen
Presentation Volume).
Liu Kuei-Chen (1).
Chhi Kung Liao Fa Shih Chien 氣功療法
實踐.
The Practice of Respiratory Physiotherapy.
Hopei Jen-min, Paoting, 1957.
Also published as: *Shih Yen Chhi Kung Liao
Fa* 試驗氣功療法.
Experimental Tests of Respiratory Physio-
therapy.
Thai-Phing, Hongkong, 1965.

Liu Po (*1*) 劉波.
  Mo-Ku chi chhi Tsai-Phei 蘑菇及其栽培.
  Mushrooms, Toadstools, and their Culti-
    vation.
  Kho-Hsüeh, Peking, 1959, repr, 1960, 2nd
    ed. enlarged, 1964.
Liu Shih-Chi (*1*) 劉仕驥.
  Chung-Kuo Tsang Su Sou Chhi 中國葬俗
    搜奇.
  A Study of the Curiosities of Chinese Burial
    Customs.
  Shanghai Shu-chü, Hongkong, 1957.
Liu Shou-Shan *et al.* (*1*) 劉壽山.
  *Chung Yao Yen-Chiu Wên-Hsien Tsê-Yao
    1820–1961* 中藥研究文獻摘要.
  A Selection of the Most Important Findings
    in the Literature on Chinese Drugs from
    1820 to 1961.
  Kho-Hsüeh, Peking, 1963.
Liu Wên-Tien (*2*) 劉文典.
  Huai Nan Hung Lieh Chi Chieh 淮南鴻
    烈集解.
  Collected Commentaries on the *Huai Nan
    Tzu* Book.
  Com. Press, Shanghai, 1923, 1926.
Liu Yu-Liang (*1*) 劉友樑.
  Khuang Wu Yao yü Tan Yao 礦物藥與
    丹藥.
  The Compounding of Mineral and In-
    organic Drugs in Chinese Medicine.
  Sci. & Tech. Press, Shanghai, 1962.
Lo Hsiang-Lin (*3*) 羅香林.
  *Thang Tai Kuang-chou Kuang-Hsiao Ssu
    yü Chung-Yin Chiao-Thung chih Kuan-
    Hsi* 唐代廣州光孝寺與中印交通之
    關係.
  The Kuang-Hsiao Temple at Canton during
    the Thang period, with reference to Sino-
    Indian Relations.
  Chung-kuo Hsüeh-shê, Hongkong, 1960.
Lo Tsung-Chen (*1*) 羅宗真.
  *Chiangsu I-Hsing Chin Mu Fa-Chüeh Pao-
    Kao* 江蘇宜興晉墓發掘報告
    (with a postscript by Hsia Nai 夏鼐).
  Report of an Excavation of a Chin Tomb at
    I-hsing in Chiangsu [that of Chou Chhu,
    d. +297, which yielded the belt-orna-
    ments containing aluminium; see p. 192).
  AS/CJA, 1957 (no. 4), no. 18, 83.
  Cf. Shen Shih-Ying (*1*); Yang Kên (*1*).
Lo Tsung-Chen (*2*) 羅宗真.
  Rejoinder to Shen Shih-Ying (*1*).
  KKTH, 1963 (no. 3), 165.
Lu Khuei-Shêng (*1*) (ed.) 陸奎生.
  Chung Yao Kho-Hsüeh Ta Tzhu-Tien
    中藥科學大辭典.
  Dictionary of Scientific Studies of Chinese
    Drugs.
  Shanghai Pub. Co., Hongkong, 1957.
Lung Po-Chien (*1*) 龍伯堅.
  Hsien Tshun Pên-Tshao Shu Lu 現存本草
    書錄.

Bibliographical Study of Extant Pharma-
    copoeias and Treatises on Natural History
    (from all periods).
  Jên-min Wei-shêng, Peking, 1957.
Ma Chi-Hsing (*2*) 馬繼興.
  Sung-Tai-ti Jen Thi Chieh Phou Thu
    宋代的人體解剖圖.
  On the Anatomical Illustrations of the Sung
    Period.
  ISTC, 1957, **8** (no. 2), 125.
Maeno Naoaki (*1*) 前野直彬.
  Meikai Yūkō 冥界游行.
  On the Journey into Hell [critique of
    Duyvendak (20) continued; a study of the
    growth of Chinese conceptions of hell].
  CBH, 1961, **14**, 38; **15**, 33.
  Abstr. *RBS*, 1968, **7**, no. 636.
Mao Phan-Lin (*1*) 茆泮林.
  Ed. & comm. *Huai Nan Wan Pi Shu* (q.v.).
  In *Lung Chhi Ching Shih Tshung-Shu* 龍溪
    精舍叢書.
  Collection from the Dragon Pool Studio.
  Ed. Chêng Kuo-Hsün 鄭國勳 (1917).
  c. 1821.
Masuda Tsuna (*1*) 增田綱謹. Master-
    Craftsman to the Sumitomo Family.
  Kodō Zuroku 皷銅圖錄.
  Illustrated Account of the (Mining,) Smelting
    and Refining of Copper (and other Non-
    Ferrous Metals).
  Kyoto, 1801.
  Tr. in *CRRR*, 1840, **9**, 86.
Masutomi Kazunosuke (*1*) 益富壽之助.
  *Shōsōin Yakubutsu o Chūshin to suru Kodai
    Sekiyaku no Kenkyū* 正倉院藥物を中
    心とする古代石藥の研究.
  A study of Ancient Mineral Drugs based on
    the chemicals preserved in the Shōsōin
    (Treasury, at Nara).
  Nihon Kōbutsu shumi no Kai, Kyoto, 1957.
Matsuda Hisao (*1*) 松田壽男.
  Jūen to Ninjin to Chōbi 戎鹽と人參と貂皮.
  On Turkestan salt, Ginseng and Sable Furs.
  SGZ, 1957, **66**, 49.
Mei Jung-Chao (*1*) 梅榮照.
  *Wo Kuo ti-i pên Wei-chi-fên Hsüeh ti i-pên;
    'Tai Wei Chi Shih Chi' Chhu Pan I Pai
    Chou Nien* 我國第一本微積分學的
    譯本;「代微積拾級」出版一百周年.
  The Centenary of the First Translation into
    Chinese of a book on Analytical Geo-
    metry and Calculus; (Li Shan-Lan's
    translation of Elias Loomis).
  KHSC, 1960, **3**, 59.
  Abstr. *RBS*, 1968, **7**, no. 747.
Mêng Nai-Chhang (*1*) 孟乃昌.
  *Kuan-yü Chung-Kuo Lien-Tan-Shu Chung
    Hsiao-Suan-ti Ying Yung* 關於中國煉
    丹術中硝酸的應用.
  On the (Possible) Applications of Nitric
    Acid in (Mediaeval) Chinese Alchemy.
  KHSC, 1966, **9**, 24.

Michihata Ryōshū (*1*) 道端良秀.
　Chūgoku Bukkyō no Kishin 中國佛教の
　　鬼神.
　The 'Gods and Spirits' in Chinese Buddhism.
　IBK, 1962, **10**, 486.
　Abstr. *RBS*, 1969, **8**, no. 700.
Mikami, Yoshio (*16*) 三上義夫.
　*Shina no Muki Sanrui ni kansuru Chishiki
　　no Hajime* 支那の無機酸類に關する
　　知識の始め.
　Le Premier Savoir des Acides Inorganiques
　　en Chine.
　JI, 1931, **1** (no. 1), 95.
Miki Sakae (*1*) 三木榮.
　Chōsen Igakushi oyobi Shippeishi 朝鮮醫
　　學史及疾病史.
　A History of Korean Medicine and of
　　Diseases in Korea.
　Sakai, Osaka, 1962.
Miki Sakae (*2*) 三木榮.
　Taikei Sekai Igakushi; Shoshi Teki Kenkyū
　　體系世界醫學史；書誌的研究.
　A Systematic History of World Medicine;
　　Bibliographical Researches.
　Tokyo, 1972.
Min I-Tê (*1*) 閔一得.
　Kuan Khuei Pien 管窺編.
　An Optick Glass (for the Enchymoma).
　c. 1830.
　In *Tao Tsang Hsü Pien* (*Chhu chi*), 7.
Miyagawa Torao *et al.* (*1*) 宮川寅雄.
　*Chhangsha Kanbo no Kiseki; Yomigaeru Tai
　　Hou Fu Jen no Sekai* 長沙漢墓の奇
　　跡；よみがえる軑侯夫人の世界
　Marvellous Relics from a Han Tomb; the
　　World of the Resurrected Lady of Tai.
　SHA 1972 (増刊) no. 9–10.
　Other important picture also in *AGR* 1972,
　　25 Aug.
Miyashita Saburō (*1*) 宮下三郎.
　*Kanyakū, Shūseki no Yakushi-gaku teki
　　Kenkyū* 漢藥；秋石の藥史學的研究.
　A Historical-Pharmaceutical Study of the
　　Chinese Drug 'Autumn Mineral' (*chhiu
　　shih*).
　Priv. pr. Osaka, 1969.
Miyashita Saburō (*2*) 宮下三郎.
　*Senroku-jūichi-ren ni Chin Katsu ga Seizō
　　shita Sei-horumonzai ni tsuite* 一〇六一
　　年に沈括が製造した性ホルモン劑に
　　ついて.
　On the Preparation of 'Autumn Mineral'
　　[Steroid Sex Hormones from Urine] by
　　Shen Kua in +1061.
　NIZ, 1965, **11** (no. 2), 1.
Mizuno Seiichi (*3*) 水野清一.
　Indai Seidō Bunka no Kenkyū 殷代靑銅
　　文化の研究.
　Researches on the Bronze Culture of the
　　Shang (Yin) Period.
　Kyoto, 1953.
Morita Kōmon (*1*) 森田幸門.

Josetsu 序說.
　Introduction to the Special Number of
　　Nihon Ishigaku Zasshi (*Journ. Jap. Soc.
　　Hist. of Med.*) on the Model Human
　　Viscera in the Cavity of the Statue of
　　Sakyamuni (Buddha) at the Seiriyōji
　　Temple at Saga (near Kyoto).
　NIZ, 1956, **7** (nos. 1–3), 1.
Murakami Yoshimi (*3*) 村上嘉實.
　Chūgoku no Sennin; Hōbokushi no Shisō
　　中國の仙人；抱朴子の思想.
　On the Immortals of Chinese (Taoism); a Study
　　of the Thought of Pao Phu Tzu (Ko Hung).
　Heirakuji Shoten, Tokyo, 1956; repr. Se-u
　　Sōshō, Kyoto, 1957.
　Abstr. *RBS*, 1959, **2**, nos. 566, 567.
Nakajima Satoshi (*1*) 中島敏.
　Shina ni okeru Shisshiki Shūdōhō no Kigen
　　支那に於ける濕式收銅法の起源.
　The Origins and Development of the Wet
　　Method for Copper Production in China.
　In Miscellany of Oriental Studies presented
　　to Prof. Katō Gen'ichi, Tokyo 1950,
　　加藤博士還曆記念東洋史集說.
　Also *TYG*, 1945, **27** (no. 3).
Nakao, Manzō (*1*) 中尾万三.
　Shokuryō-honsō no Kōsatsu 「食療本草」の
　　考察.
　A Study of the [Tunhuang MS. of the] *Shih
　　Liao Pên Tshao* (Nutritional Therapy;
　　a Pharmaceutical Natural History), [by
　　Mêng Shen, c. +670].
　SSIP, 1930, **1** (no. 3).
Nakaseko Rokuro (*1*).
　Sekai Kwagakushi 世界化學史.
　General History of Chemistry.
　Kaniya Shoten, Kyoto, 1927.
　Rev. M. Muccioli, *A*, 1928, **9**, 379.

Ōbuchi Ninji (*1*) 大淵忍爾.
　Dōkyō-shi no Kenkyū 道教史の研究.
　Researches on the History of Taoism and
　　the Taoist Church.
　Okayama, 1964.
Ogata Kōan (*1*) 緒方洪庵.
　Byōgaku Tsūron 病學通論.
　Survey of Pathology (after Christopher
　　Hufeland's theories).
　Tokyo, 1849.
Ogata Kōan (*2*) 緒方洪庵.
　Hushi Keiken Ikun 扶氏經驗遺訓.
　Mr Hu's (Christopher Hufeland's) Well-tested
　　Advice to Posterity [medical macrobiotics].
　Tokyo, 1857.
Ogata Tamotsu (*1*) 小片保.
　*Waga Kuni Sokushinbutsu seiritsu ni Kansuru
　　Shomondai* 我國即身佛成立に關す
　　る諸問題.
　The Self-Mummified Buddhas of Japan, and
　　Several (Anatomical) Questions concern-
　　ing them.
　NDI (Spec. No.), 1962 (no. 15), 16, with 8 pls.

Okanishi Tameto (2) 岡西爲人.
  Sung I-chhien I Chi Khao 宋以前醫籍考.
  Comprehensive Annotated Bibliography of
    Chinese Medical Literature in and before
    the Sung Period.
  Jên-min Wei-shêng, Peking, 1958.
Okanishi Tameto (4) 岡西爲人.
  Tan Fang chih Yen-Chiu 丹方之研究.
  Index to the 'Tan' Prescriptions in Chinese
    Medical Works.
  In *Huang Han I-Hsüeh Tshung-Shu*, 1936,
    vol. 11.
Okanishi Tameto (5) 岡西爲人.
  Chhung Chi 'Hsin Hsiu Pên Tshao' 重輯
    「新修本草」.
  Newly Reconstituted Version of the *New
    and Improved Pharmacopoeia* (of +659).
  National Pharmaceutical Research Insti-
    tute, Thaipei, 1964.
Ong Tu-Chien (1) 翁獨健.
  'Tao Tsang' Tzu Mu Yin Tê 道藏子目
    引得.
  An Index to the Taoist Patrology.
  Harvard-Yenching, Peiping, 1935.
Ong Wên-Hao (1) 翁文灝.
  Chung-Kuo Kung Chhan Chih Lüeh 中國
    鑛產誌畧.
  The Mineral Resources of China (Metals
    and Non-Metals except Coal).
  MGSC (Ser. B), 1919, no. 1, 1–270.
  With English contents-table.
Ōya Shin'ichi (1) 大矢眞一.
  Nihon no Sangyō Gijutsu 日本の產業技
    術.
  Industrial Arts and Technology in (Old)
    Japan.
  Sanseido, Tokyo, 1970.

Pa Tzu-Yuan (1) (ed.) 巴子園.
  Tan I San Chüan 丹擬三卷.
  Three Books of Draft Memoranda on
    Elixirs and Enchymomas.
  1801.
Phan Wei (1) 潘霨.
  Wei Shêng Yao Shu 衛生要術.
  Essential Techniques for the Preservation
    of Health [based on earlier material on
    breathing exercises, physical culture and
    massage etc. collected by Hsü Ming-
    Fêng 徐鳴峰].
  1848, repr. 1857.
Pi Li-Kan (Billequin, M. A.) 畢利干,
    Chhêng Lin 承霖 & Wang Chung-
    Hsiang 王鍾祥 (1).
  Hua-Hsüeh Shan Yuan 化學闡原.
  Explanation of the Fundamental Principles
    of Chemistry.
  Thung Wên Kuan, Peking, 1882.

Richie, Donald ドナルド・リチー & Itō
    Kenkichi (1) = (1) 伊藤堅吉.
  Danjozō 男女像.

Images of the Male and Female Sexes
    [= The Erotic Gods].
  Zufushinsha, Tokyo, 1967.

Sanaka Sō (1) 佐中壯.
  *Tō Inkyo Shōden; Sono Senjutsu o tsujite
    mita Honzōgaku to Senyaku to no Kankei*
    陶隱居小傳；その撰述を通じて見
    た本草學と仙藥との關係.
  A Biography of Thao Hung-Ching; his
    Knowledge of Botany and Medicines of
    Immortality.
  Art. in *Wada Hakase Koki Kinen Tōyōshi
    Ronsō* (Wada Festschrift) 和田博士古
    稀記念東洋史論叢. Kōdansha, Tokyo,
    1961, p. 447.
  Abstr. *RBS*, 1968, **7**, no. 756.
Sawa Ryūken (1) 左和隆研.
  Nihon Mikkyō, sono Tenkai to Bijutsu
    日本密敎その展開と美術.
  Esoteric (Tantric) Buddhism in Japan; its
    Development and (Influence on the) Arts.
  NHK, Tokyo, 1966, repr. 1971.
Shen Shih-Ying (1) 沈時英.
  *Kuan-yü Chiangsu I-Hsing Hsi Chin Chou
    Chhu Mu Chhu-Thu Tai-Shih Chhêng-
    Fên Wên-Thi* 關于江蘇宜興西晉周處
    墓出土帶飾成分問題.
  Notes on the Chemical Composition of the
    Belt Ornaments from the Western
    Chin Period (+265 to +316) found in
    the Tomb of Chou Chhu at I-hsing in
    Chiangsu.
  KKTH, 1962 (no. 9), 503.
  Eng. tr. by N. Sivin (unpub.).
  Cf. Lo Tsung-Chen (1); Yang Kên (1).
Shih Shu-Chhing (2) 史樹青.
  Ku-Tai Kho-Chi Shih Wu Ssu Khao 古代
    科技事物四考.
  Four Notes on Ancient Scientific Tech-
    nology; (*a*) Ceramic objects for medical
    heat-treatment; (*b*) Mercury silvering of
    bronze mirrors; (*c*) Cardan Suspension
    perfume burners; (*d*) Dyeing stoves.
  WWTK, 1962 (no. 3), 47.
Shima Kunio (1) 島邦男.
  Inkyo Bokuji Kenkyū 殷墟卜辭研究.
  Researches on the Shang Oracle-Bones and
    their Inscriptions.
  Chūgokugaku Kenkyūkai, Hirosaki, 1958.
  Abstr. in *RBS*, 1964, **4**, no. 520.
Shinoda Osamu (1) 篠田統.
  Daki Hyō Shōkō 曖氣樽小考.
  A Brief Study of the 'Daki' [*Nuan Chhi*]
    Temperature Stabiliser (used in breweries
    for the saccharification vats, cooling them
    in summer and warming them in winter).
  MOULA, ser. B, 1963 (no. 12), 217.
Shinoda Osamu (2) 篠田統.
  Chūsei no Sake 中世の酒.
  Wine-Making in Medieval (China and
    Japan).

Shinoda Osamu (2) (cont.)
　Art. in Yabuuchi Kiyoshi (25), p. 321.
Su Fên　蘇芬, Chu Chia-Hsüan 朱稼軒 et
　al. (1).
　　Tzhu-Hang Fa Shih Phu-Sa Ssu Pu Hsiu
　　航慈法師菩薩四不朽.
　　The Self-Mummification of the Abbot and
　　Bodhisattva, Tzhu-Hang (d. 1954).
　　CJFC, 1959 (no. 27), pp. 15, 21, etc.
Su Ying-Hui (1)　蘇瑩輝.
　　Lun 'Wu Lei Hsiang Kan Chih' chih Tso-
　　Chhêng Shih-Tai　論「物類相感志」之作
　　成時代.
　　On the Time of Completion of the Mutual
　　Responses of Things according to their
　　Categories.
　　TLTC 1970, 40 (no. 10).
Su Ying-Hui (2)　蘇瑩輝.
　　'Wu Lei Hsiang Kan Chih' Fên Chüan Yen-
　　Ko Khao-Lüeh 「物類相感志」分卷沿
　　革考略.
　　A Study of the Transmission of the Mutual
　　Responses of Things according to their
　　Categories and the Vicissitudes in the
　　Numbering of its Chapters.
　　KKTS, 1970, 1 (no. 2), 23.
Sun Fêng-I (1)　孫馮翼.
　　Ed. & comm. Huai Nan Wan Pi Shu (q.v.).
　　In Wên Ching Thang Tshung-Shu 問經堂
　　叢書.
　　Collection from the Hall of Questioning the
　　Classics.
　　+1797 to 1802.
Sun Tso-Yün (1)　孫作雲.
　　Shuo Yü Jen 說羽人.
　　On the Feathered and Winged Immortals
　　(of early Taoism).
　　LSYKK, 1948.

Takeuchi Yoshio (1)　武內義雄.
　　Shinsen Setsu　神僊說.
　　The Holy Immortals (a study of ancient
　　Taoism).
　　Tokyo, 1935.
Taki Mototane (1)　多紀元胤.
　　I Chi Khao (Iseki-kō)　醫籍考.
　　Comprehensive Annotated Bibliography of
　　Chinese Medical Literature (Lost or
　　Still Existing).
　　c. +1825, pr. 1831.
　　Repr. Tokyo, 1933, and Chinese-Western
　　Medical Research Society, Shanghai, 1936,
　　with introdn. by Wang Chi-Min.
Takizawa Bakin (1)　瀧澤馬琴.
　　Kinsei-setsu Bishōnen-roku 近世說美少
　　年錄.
　　Modern Stories of Youth and Beauty.
　　Japan (Yedo), c. 1820.
Tan Jan-Hui (1) (ed.)　澹然慧.
　　'Chhang Shêng Shu', 'Hsü Ming Fang' Ho
　　Khan 「長生術」「續命方」合刋.
　　A Joint Edition of the Art and Mystery of

Longevity and Immortality and the Pre-
　cepts for Lengthening the Life-Span. [The
　former work is that previously entitled
　Thai-I Chin Hua Tsung Chih (q.v.) and
　the latter is that previously entitled
　Tsui-Shang I Chhêng Hui Ming Ching
　(q.v.).]
　Peiping, 1921 (the edition used by Wilhelm
　& Jung, 1).
Thang Yung-Thung 湯用彤 & Thang I-Chieh
　(1)　湯一介.
　　Khou Chhien-Chih ti Chu-Tso yü Ssu-
　　Hsiang　寇謙之的著作與思想.
　　On the Doctrines and Writings of (the
　　Taoist reformer) Khou Chhien-Chih (in
　　the Northern Wei period).
　　LSYC, 1961, 8 (no. 5), 64.
　　Abstr. RBS, 1968, 7, no. 659.
Ting Hsü-Hsien (1)　丁緒賢.
　　Hua Hsüeh Shih Thung Khao 化學史
　　通考.
　　A General Account of the History of
　　Chemistry.
　　2 vols., Com. Press, Shanghai, 1936, repr.
　　1951.
Ting Wei-Liang (Martin, W. A. P.) (1)
　丁韙良.
　　Ko Wu Ju Mên　格物入門.
　　An Introduction to Natural Philosophy.
　　Thung Wên Kuan, Peking, 1868.
Ting Wên-Chiang (1)　丁文江.
　　Biography of Sung Ying-Hsing 宋應星
　　(author of the Exploitation of the Works of
　　Nature).
　　In the Hsi Yung Hsüan Tshung-Shu 喜詠
　　軒叢書, ed. Thao Hsiang 陶湘.
　　Peiping, 1929.
Tōdō Kyōshun (1)　藤堂恭俊.
　　Shina Jōdokyō ni okeru Zuichiku Yōgo
　　Setsu no seiritsu Katei ni tsuite シナ淨土
　　教における隋逐擁護說の成立過程
　　について.
　　On the Origin of the Invocation to the
　　25 Bodhisattvas for Protection against
　　severe Judgments; a Practice of
　　the Chinese Pure Land (Amidist)
　　School.
　　Art. in Tsukamoto Hakase Shōju Kinen
　　Bukkyōshigaku Ronshū (Tsukamoto
　　Festschrift) 塚本博士頌壽記念佛教
　　史學論集, Kyoto, 1961, p. 502.
　　Abstr. RBS, 1968, 7, no. 664.
Tokiwa, Daijō (1)　常盤大定.
　　Dōkyō Gaisetsu 道教概說.
　　Outline of Taoism.
　　TYG, 1920, 10 (no. 3), 305.
Tokiwa, Daijō (2)　常盤大定.
　　Dōkyō Hattatsu-shi Gaisetsu 道教發達史
　　概說.
　　General Sketch of the Development of
　　Taoism.
　　TYG, 1921, 11 (no. 2), 243.

Tsêng Chao-Lun (1) 曾昭掄
The Translations of the Chiangnan Arsenal Bureau.
TFTC, 1951, **38** (no. 1), 56.

Tsêng Chao-Lun (2) 曾昭掄.
Chung Wai Hua-Hsüeh Fa-Chan Kai Shu 中外化學發展概述.
Chinese and Western Chemical Discoveries; an Outline.
TFTC, 1953, **40** (no. 18), 33.

Tsêng Chao-Lun (3) 曾昭掄.
Erh-shih Nien Lai Chung-Kuo Hua-Hsüeh chih Chin-Chan 二十年來中國化學之進展
Advances in Chemistry in China during the past Twenty Years.
KHS, 1936, **19** (no. 10), 1514.

Tsêng Hsi-Shu (1) 曾熙署.
Ssu Thi Ta Tzu Tien 四體大字典.
Dictionary of the Four Scripts.
Shanghai, 1929.

Tsêng Yuan-Jung (1) 曾遠榮.
Chung-Kuo Yung Hsin chih Chhi-Yuan 中國用鋅之起源.
Origins and Development of Zinc Technology in China [with a dating of the *Pao Tsang Lun*].
Letter of Oct. 1925 to Wang Chin.
Pr. in Wang Chin (2), p. 92.

Tshai Lung-Yün (1) 蔡龍雲.
Ssu Lu Hua Chhüan 四路華拳.
Chinese Boxing Calisthenics on the Four Directions System.
Jen-min Thi-yu, Peking, 1959, repr. 1964.

Tshao Yuan-Yü (1) 曹元宇.
Chung-Kuo Ku-Tai Chin-Tan-Chia ti Shê-Pei ho Fang-Fa 中國古代金丹家的設備和方法.
Apparatus and Methods of the Ancient Chinese Alchemists.
KHS, 1933, **17** (no. 1), 31.
Reprinted in Wang Chin (2), p. 67.
Engl. précis by Barnes (1).
Engl. abstr. by H. D. C[ollier], *ISIS*, 1935, **23**, 570.

Tshao Yuan-Yü (2) 曹元宇.
Chung-Kuo Tso Chiu Hua-Hsüeh Shih-Liao 中國作酒化學史料.
Materials for the History of Fermentation (Wine-making) Chemistry in China.
HITC, 1922, **6** (no. 6), 1.

Tshao Yuan-Yü (3) 曹元宇.
Kuan-yü Thang-Tai mei-yu Chêng-Liu Chiu ti Wên-thi 關于唐代沒有蒸餾酒的問題.
On the Question of whether Distilled Alcoholic Liquors were known in the Thang Period.
KHSC, 1963, no. 6, 24.

Tsuda Sōkichi (2) 津田左右吉.
Shinsen Shisō ni kansuru ni-san no Kōsatsu 神仙思想に關する二三の考察.
Some Considerations and Researches on the Holy Immortals (and the Immortality Cult in Ancient Taoism).
In *Man-Sen Chiri Rekishi Kenkyū Hōkoku* 滿鮮地理歷史研究報告 (Research Reports on the Historical Geography of Manchuria and Korea), 1924, no. 10, 235

Tsumaki, Naoyoshi (1) 妻木直良.
Dōkyō no Kenkyū 道敎の研究.
Studies in Taoism.
TYG, 1911, **1** (no. 1), 1; (no. 2), 20; 1912, **2** (no. 1), 58.

Tuan Wên-Chieh (1) (ed.) 段文杰.
Yü Lin Khu 楡林窟.
The Frescoes of Yü-lin-khu [i.e. Wan-fo-hsia, a series of cave-temples in Kansu].
Tunhuang Research Institute, Chung-kuo Ku-tien I-shu, Peking, 1957.

Tzu Chhi (1) 梓溪.
Chhing Thung Chhi Ming Tzhu Chieh-Shuo 青銅器名詞解說.
An Explanation of the Terminology of (Ancient) Bronze Vessels.
WWTK, 1958 (no. 1), 1; (no. 2), 55; (no. 3), 1; (no. 4), 1; (no. 5), 1; (no. 6), 1; (no. 7), 68.

Udagawa Yōan (1) 宇田川榕庵.
Seimi Kaisō 舍密開宗.
Treatise on Chemistry [largely a translation of W. Henry (1), but with added material from other books, and some experiments of his own].
Tokyo, 1837–46.
Cf. Tanaka Minoru (3).

Umehara Sueji (3) 梅原末治.
Senoku Seishō Shinshūhen 泉屋清賞;新收編.
New Acquisitions of the Sumitomo Collection of Ancient Bronzes (Kyoto); a Catalogue.
Kyoto, 1961.
With English contents-table.

Wada Hisanori (1) 和田久德.
'*Namban Kōroku*' to '*Shohanshi*' to no Kankei 「南蕃香錄」と「諸蕃志」との關係.
On the *Records of Perfumes and Incense of the Southern Barbarians* [by Yeh Thing-Kuei, c. +1150] and the *Records of Foreign Peoples* [by Chao Ju-Kua, c. +1250, for whom it was an important source].
OUSS, 1962, **15**, 133.
Abstr. *RBS*, 1969, **8**, no. 183.

Wang Chi-Liang 王季梁 & Chi Jen-Jung (1) 紀絪容.
Chung-Kuo Hua-Hsüeh Chieh chih Kuo-Chhü yü Wei-Lai 中國化學界之過去與未來.
The Past and Future of Chemistry [and Chemical Industry] in China.
HHSTH, 1942, 3.

Wang Chi-Wu (*1*)　王輯五.
Chung-Kuo Jih-Pên Chiao Thung Shih
中國日本交通史.
A History of the Relations and Connections
between China and Japan.
Com. Pr., Thaiwan, 1965 (*Chung-Kuo
Wên-Hua Shih Tshung-Shu* ser.).

Wang Chin (*1*)　王璡.
Chung-Kuo chih Kho-Hsüeh Ssu-Hsiang
中國之科學思想.
On (the History of) Scientific Thought in
China.
Art. in *Kho-Hsüeh Thung Lun*　(科學通
論).
Sci. Soc. of China, Shanghai, 1934.
Orig. pub. *KHS*, 1922, **7** (no. 10), 1022.

Wang Chin (*2*) (ed.)　王璡.
*Chung-Kuo Ku-Tai Chin-Shu Hua-Hsüeh chi
Chin Tan Shu* 中國古代金屬化學及
金丹術.
Alchemy and the Development of Metal-
lurgical Chemistry in Ancient and Medi-
eval China [collective work].
Chung-kuo Kho-hsüeh Thu-shu I-chhi
Kung-ssu, Shanghai, 1955.

Wang Chin (*3*)　王璡.
*Chung-Kuo Ku-Tai Chin Shu Yuan Chih
chih Hua-Hsüeh*　中國古代金屬原質
之化學.
The Chemistry of Metallurgical Operations
in Ancient and Medieval China [smelt-
ing and alloying].
KHS, 1919, **5** (no. 6), 555; repr. in Wang
Chin (*2*), p. 1.

Wang Chin (*4*)　王璡.
*Chung-Kuo Ku-Tai Chin Shu Hua Ho Wu
chih Hua Hsüeh*　中國古代金屬化合物
之化學.
The Chemistry of Compounds containing
Metal Elements in Ancient and Medieval
China.
KHS, 1920, **5** (no. 7), 672; repr. in Wang
Chin (*2*), p. 10.

Wang Chin (*5*)　王璡.
*Wu Shu Chhien Hua-Hsüeh Chhêng-Fên chi
Ku-Tai Ying-Yung Chhien Hsi Hsin La
Khao*　五銖錢化學成分及古代應用
鉛錫鋅鑞考.
An Investigation of the Ancient Tech-
nology of Lead, Tin, Zinc and *la*, together
with Chemical Analyses of the Five-Shu
Coins [of the Han and subsequent
periods].
KHS, 1923, **8** (no. 9), 839; repr. in Wang
Chin (*2*), p. 39.

Wang Chin (*6*)　王璡.
Chung-Kuo Thung Ho Chin nei chih Nieh
中國銅合金內之鎳.
On the Chinese Copper Alloys containing
Nickel [paktong] etc.
KHS, 1929, **13**, 1418; abstr. in Wang Chin
(*2*), p. 91.

Wang Chin (*7*)　王璡.
*Chung-Kuo Ku-Tai Chiu-Ching Fa-Chiao
Yeh chih i pan*　中國古代酒精發酵業
之一斑.
A Brief Study of the Alcoholic Fermentation
Industry in Ancient (and Medieval)
China.
KHS, 1921, **6** (no. 3), 270.

Wang Chin (*8*)　王璡.
*Chung-Kuo Ku-Tai Thao Yeh chih Kho-
Hsüeh Kuan*　中國古代陶業之科學
觀.
Scientific Aspects of the Ceramics Industry
in Ancient China.
KHS, 1921, **6** (no. 9), 869.

Wang Chin (*9*)　王璡.
*Chung-Kuo Huang Thung Yeh chih Chhüan
Shêng Shih-Chhi*　中國黃銅業之全盛
時期.
On the Date of Full Development of the
Chinese Brass Industry.
KHS, 1925, **10**, 495.

Wang Chin (*10*)　王璡.
*I-Hsing Thao Yeh Yuan Liao chih Kho-
Hsüeh Kuan*　宜興陶業原料之科學觀
Scientific Aspects of the Raw Materials of
the I-hsing Ceramics Industry.
KHS, 1932, **16** (no. 2), 163.

Wang Chin (*11*)　王璡.
*Chung-Kuo Ku-Tai Hua-Hsüeh ti Chhêng-
Chiu*　中國古代化學的成就.
Achievements of Chemical Science in
Ancient and Medieval China.
KHTP, 1951, **2** (no. 11), 1142.

Wang Chin (*12*)　王璡.
Ko Hung i-chhien chih Chin Tan Shih Lüeh
葛洪以前之金丹史略.
A Historical Survey of Alchemy before Ko
Hung (*c.* +300).
HITC, 1935, **14**, 145, 283.

Wang Hsien-Chhien (*3*)　王先謙.
Shih Ming Su Chêng Pu　釋名疏證補.
Revised and Annotated Edition of the
[Han] *Explanation of Names* [dictionary].
Peking, 1895.

Wang Khuei-Kho (*1*) (tr.)　王奎克.
*San-shih-liu Shui-Fa—Chung-Kuo Ku Tai
Kuan-yü Shui Jung I ti I Chung Tsao
Chhi Lien Tan Wên Hsien*　三十六水
法—中國古代關於水溶液的一種早
期煉丹文獻.
The *Thirty-six Methods of Bringing Solids
into Aqueous Solution*—an Early Chinese
Alchemical Contribution to the Problem
of Dissolving (Mineral Substances), [a
partial translation of Tshao Thien-Chhin,
Ho Ping-Yü & Needham, J. (*1*).]
KHSC, 1963, **5**, 67.

Wang Khuei-Kho (*2*)　王奎克.
*Chung-Kuo Lien-Tan-Shu Chung ti Chin-I
ho Hua-Chhih*　中國煉丹術中的金液
和華池.

Wang Khuei-Kho (2) (cont.)
'Potable Gold' and Solvents (for Mineral Substances) in (Medieval) Chinese Alchemy.
KHSC, 1964, **7**, 53.
Wang Ming (2) 王明.
'*Thai Phing Ching*' Ho Chiao 「太平經」合校.
A Reconstructed Edition of the *Canon of the Great Peace (and Equality)*.
Chung-Hua, Peking and Shanghai, 1960.
Wang Ming (3) 王明.
'*Chou I Tshan Thung Chhi*' Khao Chêng 「周易參同契」考證.
A Critical Study of the *Kinship of the Three*.
AS/BIHP, 1948, **19**, 325.
Wang Ming (4) 王明.
'*Huang Thing Ching*' Khao 「黃庭經」考.
A Study on the *Manuals of the Yellow Courts*.
AS/BIHP, 1948, **20**A.
Wang Ming (5) 王明.
'*Thai Phing Ching*' Mu Lu Khao 「太平經」目錄考.
A Study of the Contents Tables of the *Canon of the Great Peace (and Equality)*.
WS, 1965, no. 4, 19.
Wang Tsu-Yuan (1) 王祖源.
Nei Kung Thu Shuo 內功圖說.
Illustrations and Explanations of Gymnastic Exercises [based on an earlier presentation by Phan Wei (q.v.) using still older material from Hsü Ming-Fêng].
1881.
Modern reprs. Jen-min Wei-shêng, Peking, 1956; Thai-Phing, Hongkong, 1962.
Wang Yeh-Chhiu, 王治秋 Wang Chung-Shu 王仲殊 & Hsia Nai (1) 夏鼐.
Bunka Dai-Kakumei-Kikan Shutsudo Bumbutsu Tenran 文化大革命期間出土文物展覽.
Articles to accompany the Exhibition of Cultural Relics Excavated (in Ten Provinces of China) during the Period of the Great Cultural Revolution.
JC, 1971 (no. 10), 31, with colour-plates.
Watanabe Kōzō (1) 渡邊幸三.
Genzai suru Chūgoku Kinsei made no Gozō Rokufu Zu no Gaisetsu 現存する中國近世までの五藏六府圖の概說
General Remarks on (the History of) Dissection and Anatomical Illustration in China.
NIZ, 1956, **7** (nos. 1–3), 88.
Watanabe Kōzō (2) 渡邊幸三.
Seiryōji Shaka Tainai Gozō no Kaibō-gakuteki Kenkyū 清凉寺釋迦胎內五藏の解剖學的研究.
An Anatomical Study (of Traditional

Chinese Medicine) in relation to the Visceral Models in the Sakyamuni Statue at the Seiryōji Temple (at Saga, near Kyoto).
NIZ, 1956, **7** (nos. 1–3), 30.
Wei Yuan (1) 魏源.
Shêng Wu Chi 聖武記.
Records of the Warrior Sages [a history of the military operations of the Chhing emperors].
1842.
Wên I-To (3) 聞一多.
Shen Hua yü Shih 神話與詩.
Religion and Poetry (in Ancient Times), [contains a study of the Taoist immortality cult and a theory of its origins].
Peking, 1956 (posthumous).
Wu Chhêng-Lo (2) 吳承洛.
Chung-Kuo Tu Liang Hêng Shih 中國度量衡史.
History of Chinese Metrology [weights and measures].
Com. Press, Shanghai, 1937; 2nd ed. Shanghai, 1957.
Wu Shih-Chhang (1) 吳世昌.
Mi Tsung Su Hsiang Shuo Lüeh 密宗塑像說略.
A Brief Discussion of Tantric (Buddhist) Images.
AP/HJ, 1935, **1**.
Wu Tê-To (1) 吳德鐸.
Thang Sung Wên-Hsien chung Kuan-yü Chêng-Liu Chiu yü Chêng-Liu Chhi Wên-thi 唐宋文獻中關于蒸餾酒與蒸餾器問題.
On the Question of Liquor Distillation and Stills in the Literature of the Thang and Sung Periods.
KHSC, 1966, no. 9, 53.

Yabuuchi Kiyoshi (11) (ed.) 藪內清.
'*Tenkō Kaibutsu*' no Kenkyū 「天工開物」の研究.
A Study of the *Thien Kung Khai Wu* (Exploitation of the Works of Nature, +1637).
Japanese translation of the text, with annotative essays by several hands.
Kōseisha, Tokyo, 1953.
Rev. Yang Lien-Shêng, *HJAS*, 1954, **17**, 307.
English translation of the text (sparingly annotated). See Sun & Sun (1).
Chinese translations of the eleven essays:
(a) '*Thien Kung Khai Wu*' chih Yen-Chiu by Su Hsiang-Yü 蘇薌雨 et al.
Tshung-Shu Wei Yuan Hui, Thaiwan, and Chi-Shêng, Hongkong, 1956
(b) '*Thien Kung Khai Wu*' Yen-Chiu Wên Chi by Chang Hsiung 章熊 & Wu Chieh 吳傑.
Com. Press, Peking, 1961.

Yabuuchi Kiyoshi (25) (ed.) 藪內清.
  Chūgoku Chūsei Kagaku Gijutsushi no
    Kenkyū 中國中世科學技術史の研究.
  Studies in the History of Science and Tech-
    nology in Medieval China [a collective
    work].
  Kadokawa Shoten (for the Jimbun Kagaku
    Kenkyusō), Tokyo, 1963.
Yamada Keiji (1) 山田慶兒.
  'Butsurui sōkan shi' no seiritsu 「物類相感
    志」の成立.
  The Organisation of the Book Wu Lei
    Hsiang Kan Chih (Mutual Responses of
    Things according to their Categories).
  SBK, 1965, 13, 305.
Yamada Keiji (2) 山田慶兒.
  Chūsei no Shizen-kan 中世の自然觀.
  The Naturalism of the (Chinese) Middle
    Ages [with special reference to Taoism,
    alchemy, magic and apotropaics].
  Art. in Yabuuchi (25), pp. 55-110.
Yamada Kentarō (1) 山田憲太郎.
  Tōzai Kōyakushi 東西香藥史.
  A History of Perfumes, Incense, Aromatics
    and Spices in East and West.
  Tokyo, 1958.
Yamada Kentarō (2) 山田憲太郎.
  Kōryō no Rekishi 香料の歷史.
  History of Perfumes, Incense and Aro-
    matics.
  Tokyo, 1964 (Kiino Kuniya Shinshō, ser. B,
    14).
Yamada Kentarō (3) 山田憲太郎.
  Ogawa Kō-ryō Jihō 小川香料時報.
  News from the Ogawa Company; A History
    of the Incense, Spice and Perfume
    Industry (in Japan).
  Ogawa & Co. (pr. pr.), Osaka, 1948.
Yamada Kentarō (4) 山田憲太郎.
  Tōa Kō-ryōshi 東亞香料史.
  A History of Incense, Aromatics and Per-
    fumes in East Asia.
  Toyōten, Tokyo, 1942.
Yamada Kentarō (5) 山田憲太郎.
  Chūgoku no Ansoku-kō to Seiyō no Benzoin
    to no Genryū. 中國の安息香と西洋
    のベンゾインとの源.
  A Study of the Introduction of an-hsi
    hsiang (gum guggul, bdellium) into China,
    and that of gum benzoin into Europe.
  SB, 1951 (no. 2), 1-36.
Yamada Kentarō (6) 山田憲太郎.
  Chūsei no Chūgokujin to Arabiajin ga Shitte
    ita Ryūnō no Sanshutsuchi toku ni
    Baritsu (no) Kuni ni tsuite 中世の中國
    人とアラビア人が知つていた龍腦
    の產出地とくに婆律國について
  On the knowledge which the Medieval
    Chinese and Arabs possessed of Baros
    camphor (from Dryobalanops aromatica)
    and its Place of Production, Borneo.
  NYGDR, 1966 (no. 5), 1.
Yamada Kentarō (7) 山田憲太郎.
  Ryūnō-kō (Sono Shōhinshi-teki Kōsatsu)
    龍腦考(その商品史的考察)
  A Study of Borneo or Baros camphor
    (from Dryobalanops aromatica), and the
    History of the Trade in it.
  NYGDR, 1967 (no. 10), 19.
Yamada Kentarō (8) 山田憲太郎.
  Chin sunawachi Kō 沈すなわ香.
  On the 'Sinking Aromatic' (garroo wood,
    Aquilaria agallocha).
  NYGDR, 1970, 7 (no. 1), 1.
Yang Chhêng-Hsi (1) et al. 楊承禧.
  Hu-Pei Thung Chih 湖北通志.
  Historical Geography of Hupei Province.
  1921.
Yang Kên (1) 楊根.
  Chin Tai Lü Thung ho-chin-ti Chien-Ting
    chi chhi Yeh-Lien Chi-Shu-ti Chhu-Pu
    Than-Thao 晉代鋁銅合金的鑑定及
    其冶煉技術的初步探討.
  An Aluminium-Copper Alloy of the Chin
    Dynasty (+265 to +420); its Determina-
    tion and a Preliminary Study of the
    Metallurgical Technology (which it
    Implies).
  AS/CJA, 1959 (no. 4), no. 26, 91.
  Eng. tr. by D. Bryan (unpub.) for the
    Aluminium Development Association,
    1962.
  Cf. Lo Tsung-Chen (1); Shen Shih-
    Ying (1).
Yang Lieh-Yü (1) 楊烈宇.
  Chung-Kuo Ku-Tai Lao-Tung Jen-Min tsai
    Chin-Shu chi Ho-Chin Ying-Yung-shang-
    ti Chhêng-Chiu 中國古代勞動人民在
    金屬及合金應用上的成就.
  Ancient Chinese Achievements in Practical
    Metal and Alloy Technology.
  KHTP, 1955, 5 (no. 10), 77.
Yang Lien-Shêng (2) 楊聯陞.
  Tao Chiao chih Tzu-Po yü Fo Chiao chih
    Tzu-Phu 道教之自搏與佛教之自撲
  Penitential Self-Flagellation, Violent Pros-
    tration and similar practices in Taoist and
    Buddhist Religion.
  Art. in Tsukamoto Hakase Shōju Kinen
    Bukkyōshigaku Ronshū (Tsukamoto
    Festschrift), 塚本博士頌壽記念佛教
    史學論集. Kyoto, 1961, p. 962.
  Abstr. RBS, 1968, 7, no. 642.
  Also AS/BIHP, 1962, 34, 275; abstr. in
    RBS, 1969, 8, no. 740.
Yang Ming-Chao (1) 楊明照.
  Critical Notes on the Pao Phu Tzu
    book.
  BCS, 1944, 4.
Yang Po-Chün (1) 楊伯峻.
  Lüeh Than Wo-Kuo Shih-Chi shang kuan-
    yü Shih Thi Fang-Fu-ti Chi-Tsai ho
    Ma-wang-tui I-hao Han-Mu Mu-Chu
    Wén-Thi 略談我國史籍上關於尸體

Yang Po-Chun (*1*) (*cont.*)
防腐的記載和馬王堆一號漢墓墓主
問題.
A Brief Discussion of Some Historical Texts
concerning the Preservation of Human
Bodies in an Incorrupt State, especially in
connection with the Han Burial in Tomb
no. 1 at Ma-wang-tui.
WWTK, 1972 (no. 9), 36.
Yang Shih-Thai (*1*) 楊時泰.
Pên Tshao Shu Kou Yuan 本草述鉤元.
Essentials Extracted from the *Explanations
of Materia Medica*.
Pref. 1833, first pr. 1842.
LPC, no. 108.
Yeh Tê-Hui (*1*) (ed.) 葉德輝.
Shuang Mei Ching An Tshung Shu 雙梅景
闇叢書.
Double Plum-Tree Collection [of ancient
and medieval books and fragments on
Taoist sexual techniques].
Contains *Su Nü Ching* (incl. *Hsüan Nü
Ching*), *Tung Hsüan Tzu*, *Yü Fang Chih
Yao*, *Yü Fang Pi Chüeh*, *Thien Ti Yin
Yang Ta Lo Fu*, etc. (*qq.v.*).
Chhangsha, 1903 and 1914.
Yeh Tê-Hui (*2*) 葉德輝.
Ed. & comm. *Huai Nan Wan Pi Shu* (q.v.).
In *Kuan Ku Thang So Chu Shu* 觀古堂
所著書.
Writings from the Hall of Pondering
Antiquity.
Chhangsha, 1919.
Yen Tun-Chieh (*20*) 嚴敦傑.
*Chung-Kuo Ku-Tai Tzu-Jan Kho-Hsüeh
ti Fa-Chan chi chhi Chhêng-Chiu* 中國
古代自然科學的發展及其成就.
The Development and Achievements of the
Chinese Natural Sciences (down to 1840).
KHSC, 1969, **1** (no. 3), 6.
Yen Tun-Chieh (*21*) 嚴敦傑.
Hsü Kuang-Chhi 徐光啟.
A Biography of Hsü Kuang-Chhi.
Art. in *Chung-Kuo Ku-Tai Kho-Hsüeh Chia*,
ed. Li Nien (*27*), 2nd ed. p. 181.
Yin Shih Tzu 丙是子.
See Chiang Wei-Chhiao.
Yoshida Mitsukuni (*2*) 吉田光邦
'*Tenkō Kaibutsu' no Seiren Shuzō Gijutsu*
「天工開物」の 製錬鑄造技術.
Metallurgy in the *Thien Kung Khai Wu*
(Exploitation of the Works of Nature,
+1637).
Art. in Yabuuchi Kiyoshi (*11*), p. 137.
Yoshida Mitsukuni (*5*) 吉田光邦.
Chūsei no Kagaku (Rentan-jitsu) to Senjitsu
中世の 化學(煉丹術)と 仙術.
Chemistry and Alchemy in Medieval China.
Art. in Yabuuchi (*25*), p. 200.
Yoshida Mitsukuni (*6*) 吉田光邦.
Renkinjutsu 錬金術.
(An Introduction to the History of)

Alchemy (and Early Chemistry in China
and Japan).
Chūō Kōronsha, Tokyo, 1963.
Yoshida Mitsukuni (*7*) 吉田光邦.
Chūgoku Kagaku-gijutsu-shi Ronshū 中國科
學技術史論集.
Collected Essays on the History of Science
and Technology in China.
Tokyo, 1972.
Yoshioka Yoshitoyo (*1*) 吉岡義豐.
Dōkyō Keiten Shiron 道教經典史論.
Studies on the History of the Canonical
Taoist Literature.
Dōkyō Kankōkai, Tokyo, 1955, repr. 1966.
Abstr. *RBS*, 1957, **1**, no. 415.
Yoshioka Yoshitoyo (*2*) 吉岡義豐.
*Sho Tō ni okeru Butsu-Dō Ronsō no ichi
Shiryō, 'Dōkyō Gisū' no seiritsu ni tsuite*
初唐における 佛道論爭の 一資料
「道教義樞」の 成立について.
The *Tao Chiao I Shu* (Basic Principles of
Taoism, by Mêng An-Phai 孟安排,
c. +660) and its Background; a Contri-
bution to the Study of the Polemics
between Buddhism and Taoism at the
Beginning of the Thang Period.
IBK, 1956, **4**, 58.
Cf. *RBS*, 1959, **2**, no. 590.
Yoshioka Yoshitoyo (*3*) 吉岡義豐
Eisei e no Nagai Dōkyō 永生への 願い道
敎.
Taoism; the Quest for Material Immortality
and its Origins.
Tankōsha, Tokyo, 1972 (Sekai no Shūkyō,
no. 9).
Yü Chia-Hsi (*1*) 余嘉錫.
Ssu Khu Thi Yao Pien Chêng 四庫提要
辨證.
A Critical Study of the Annotations in the
'Analytical Catalogue of the *Complete
Library of the Four Caetgories* (of
Literature)'.
1937.
Yü Fei-An (*1*) 于非闇.
Chung-Kuo Hua Yen-Sê-ti Yen-Chiu 中國
畫顏色的研究.
A Study of the Pigments Used by Chinese
Painters.
Chao-hua Mei-shu, Peking, 1955, 1957.
Yü Yün-Hsiu (*1*) 余雲岫.
Ku-tai Chi-Ping Ming Hou Su I 古代疾
病名候疏義.
Explanations of the Nomenclature of
Diseases in Ancient Times.
Jen-min Wei-shêng, Shanghai, 1953.
Rev. Nguyen Tran-Huan, *RHS*, 1956, **9**, 275.
Yuan Han-Chhing (*1*) 袁翰青.
Chung-Kuo Hua-Hsüeh Shih Lun Wên Chi
中國化學史論文集.
Collected Papers in the History of Chemistry
in China.
San-Lien, Peking, 1956.

Addenda to Bibliography B on pages 477–8.

C. BOOKS AND JOURNAL ARTICLES IN WESTERN LANGUAGES

ABBOTT, B. C. & BALLENTINE, D. (1).'The "Red Tide" Alga, a toxin from *Gymnodinium veneficum*. *JMBA*, 1957, **36**, 169.

ABEGG, E., JENNY, J. J. & BING, M. (1). 'Yoga'. *CIBA/M*, 1949, **7** (no. 74), 2578. *CIBA/MZ*, 1948, **10**, (no. 121), 4122.
 Includes: 'Die Anfänge des Yoga' and 'Der klassische Yoga' by E. Abegg; 'Der Kundalinī-Yoga' by J. J. Jenny; and 'Über medizinisches und psychologisches in Yoga' by M. Bing & J. J. Jenny.

ABICH, M. (1). 'Note sur la Formation de l'Hydrochlorate d'Ammoniaque à la Suite des Éruptions Volcaniques et surtout de celles du Vésuve.' *BSGF*, 1836, **7**, 98.

ABRAHAMS, H. J. (1). Introduction to the Facsimile Reprint of the 1530 Edition of the English Translation of H. Brunschwyk's *Vertuose Boke of Distillacyon*. Johnson, New York and London, 1971 (Sources of Science Ser., no. 79).

ABRAHAMSOHN, J. A. G. (1). 'Berättelse om *Kien* [*chien*], elt Nativt Alkali Minerale från China...' *KSVA/H*, 1772, **33**, 170. Cf. von Engeström (2).

ABRAMI, M., WALLICH, R. & BERNAL, P. (1). 'Hypertension Artérielle Volontaire.' *PM*, 1936, **44** (no. 17), 1 (26 Feb.).

ACHELIS, J. D. (1). 'Über den Begriff Alchemie in der Paracelsischen Philosophie.' *BDP*, 1929–30, **3**, 99.

ADAMS, F. D. (1). *The Birth and Development of the Geological Sciences*. Baillière, Tindall & Cox, London, 1938; repr. Dover, New York, 1954.

ADNAN ADIVAR (1). 'On the *Tanksuq-nāmah-i Ilkhān dar Funūn-i 'Ulūm-i Khiṭāi*.' *ISIS*, 1940 (appeared 1947), **32**, 44.

ADOLPH, W. H. (1). 'The Beginnings of Chemical Research in China.' *PNHB*, 1950, **18** (no. 3), 145.

ADOLPH, W. H. (2). 'Observations on the Early Development of Chemical Education in China.' *JCE*, 1927, **4**, 1233, 1488.

ADOLPH, W. H. (3). 'The Beginnings of Chemistry in China.' *SM*, 1922, **14**, 441. Abstr. *MCE*, 1922, **26**, 914.

AGASSI, J. (1). 'Towards an Historiography of Science.' Mouton, 's-Gravenhage, 1963. (History and Theory; Studies in the Philosophy of History, Beiheft no. 2.)

AHMAD, M. & DATTA, B. B. (1). 'A Persian Translation of the +11th-Century Arabic Alchemical Treatise '*Ain al-Ṣan'ah wa 'Aun al-Ṣana'ah* (Essence of the Art and Aid to the Workers) [by 'Abd al-Malik al-Ṣāliḥī al-Khwārizmī al-Kathī, +1034].' *MAS/B*, 1927, **8**, 417. Cf. Stapleton & Azo (1).

AIGREMONT, Dr [ps. S. Schultze] (1). *Volkserotik und Pflanzenwelt; eine Darstellung alter wie moderner erotischer und sexuelle Gebräuche, Vergleiche, Benennungen, Sprichwörter, Redewendungen, Rätsel, Volkslieder, erotischer Zaubers und Aberglaubens, sexuelle Heilkunde die sich auf Pflanzen beziehen*. 2 vols. Trensinger, Halle, 1908. Re-issued as 2 vols. bound in one, Bläschke, Darmstadt, n.d. (1972).

AIKIN, A. & AIKIN, C. R. (1). *A Dictionary of Chemistry and Mineralogy*. 2 vols. Phillips, London, 1807.

AINSLIE, W. (1). *Materia Indica; or, some Account of those Articles which are employed by the Hindoos and other Eastern Nations in their Medicine, Arts and Agriculture; comprising also Formulae, with Practical Observations, Names of Diseases in various Eastern Languages, and a copious List of Oriental Books immediately connected with General Science, etc. etc.* 2 vols. Longman, Rees, Orme, Brown & Green, London, 1826.

AITCHISON, L. (1). *A History of Metals*. 2 vols. McDonald & Evans, London, 1960.

ALEXANDER, GUSTAV (1). *Herrengrunder Kupfergefässe*. Vienna, 1927.

ALEXANDER, W. & STREET, A. (1). *Metals in the Service of Man*. Pelican Books, London, 1956 (revised edition).

ALI, M. T., STAPLETON, H. E. & HUSAIN, M. H. (1). 'Three Arabic Treatises on Alchemy by Muḥammad ibn Umail [al-Ṣādiq al-Tamīnī] (d. c. +960); the *Kitāb al-Mā' al-Waraqī wa'l Arḍ al-Najmīyah* (Book of the Silvery Water and the Starry Earth), the *Risālat al-Shams Ila'l Hilāli* (Epistle of the Sun to the Crescent Moon), and the *al-Qaṣīdat al-Nūnīyah* (Poem rhyming in Nūn) —edition of the texts by M.T.A.; with an Excursus (with relevant Appendices) on the Date, Writings and Place in Alchemical History of Ibn Umail, an Edition (with glossary) of an early mediaeval Latin rendering of the first half of the *Mā' al-Waraqī*, and a Descriptive Index, chiefly of the alchemical authorities quoted by Ibn Umail [Senior Zadith Filius Hamuel], by H.E.S. & M.H.H.' *MAS/B*, 1933, **12** (no. 1), 1–213.

ALLEN, E. (ed.) (1). *Sex and Internal Secretions; a Survey of Recent Research*. Williams & Wilkins, Baltimore, 1932.

ALLEN, H. WARNER (1). *A History of Wine; Great Vintage Wines from the Homeric Age to the Present Day*. Faber & Faber, London, 1961.

ALLETON, V. & ALLETON, J. C. (1). *Terminologie de la Chimie en Chinois Moderne*. Mouton, Paris and The Hague, 1966. (Centre de Documentation Chinois de la Maison des Sciences de l'Homme, and VIe Section de l'École Pratique des Hautes Études, etc.; Matériaux pour l'Étude de l'Extrême-Orient Moderne et Contemporain; Études Linguistiques, no. 1.)

AMIOT, J. J. M. (7). 'Extrait d'une Lettre...' *MCHSAMUC*, 1791, **15**, v.

AMIOT, J. J. M. (9). 'Extrait d'une Lettre sur la Secte des Tao-sée [Tao shih].' *MCHSAMUC*, 1791, **15**, 208–59.

ANAND, B. K. & CHHINA, G. S. (1). 'Investigations of Yogis claiming to stop their Heart Beats.' *IJMR*, 1961, **49**, 90.

ANAND, B. K., CHHINA, G. S. & BALDEV SINGH (1). 'Studies on Shri Ramanand Yogi during his Stay in an Air-tight Box.' *IJMR*, 1961, **49**, 82.

ANAND, MULK RAJ (1). *Kama-Kala; Some Notes on the Philosophical Basis of Hindu Erotic Sculpture*. Nagel; Geneva, Paris, New York and Karlsruhe, 1958.

ANAND, MULK RAJ & KRAMRISCH, S. (1). *Homage to Khajuraho*. With a brief historical note by A. Cunningham. Marg, Bombay, n.d. (c. 1960).

ANDERSON, J. G. (8). 'The Goldsmith in Ancient China.' *BMFEA*, 1935, **7**, 1.

ANDŌ KŌSEI (1). 'Des Momies au Japon et de leur Culte.' *LH*, 1968, 8 (no. 2), 5.

ANIANE, M. (1). 'Notes sur l'Alchimie, "Yoga" Cosmologique de la Chrétienté Mediévale'; art. in *Yoga, Science de l'Homme Intégrale*. Cahiers du Sud, Loga, Paris, 1953.

ANON. (83). 'Préparation de l'Albumine d'Oeuf en Chine.' *TP*, 1897 (1e sér.), **8**, 452.

ANON. (84). *Beytrag zur Geschichte der höhern Chemie*. 1785. Cf. Ferguson (1), vol. 1, p. 111.

ANON. (85). *Aurora Consurgens* (first half of the +14th cent.). In ANON. (86). *Artis Auriferae*. Germ. tr. 'Aufsteigung der Morgenröthe' in Morgenstern (1).

ANON. (86). *Artis Auriferae, quam Chemiam vocant, Volumina Duo, quae continent 'Turbam Philosophorum', aliosą antiquiss. auctores, quae versa pagina indicat; Accessit noviter Volumen Tertium...* Waldkirch, Basel, 1610. One of the chief collections of standard alchemical authors' (Ferguson (1), vol. 1, p. 51).

ANON. (87). *Musaeum Hermeticum Reformatum et Amplificatum* (twenty-two chemical tracts). à Sande, Frankfurt, 1678 (the original edition, much smaller, containing only ten tracts, had appeared at Frankfurt, in 1625; see Ferguson (1), vol. 2, p. 119). Tr. Waite (8).

ANON. (88). *Probierbüchlein, auff Golt, Silber, Kupffer und Bley, Auch allerley Metall, wie man die Zunutz arbeyten und Probieren Soll. c.* 1515 or some years earlier; first extant pr. ed., Knappe, Magdeburg, 1524. Cf. Partington (7), vol. 2, p. 66. Tr. Sisco & Smith (2).

ANON. (89). (in Swedish) *METL*, 1960 (no. 3), 95.

ANON. (90). 'Les Chinois de la Dynastie Tsin [Chin] Connaissaient-ils déjà l'Alliage Aluminium–Cuivre?' *RALUM*, 1961, 108. Eng. tr. 'Did the Ancient Chinese discover the First Aluminium–Copper Alloy?' *GBT*, 1961, 41.

ANON. [initialled Y.M.] (91). 'Surprenante Découverte; un Alliage Aluminium–Cuivre réalisé en Chine à l'Époque Tsin [Chin].' *LN*, 1961 (no. 3316), 333.

ANON. (92). *British Encyclopaedia of Medical Practice; Pharmacopoeia Supplement* [proprietary medicines]. 2nd ed. Butterworth, London, 1967.

ANON. (93). *Gehes Codex d. pharmakologische und organotherapeutische Spezial-präparate...* [proprietary medicines]. 7th ed. Schwarzeck, Dresden, 1937.

ANON. (94). *Loan Exhibition of the Arts of the Sung Dynasty* (Catalogue). Arts Council of Great Britain and Oriental Ceramic Society, London, 1960.

ANON. (95). Annual Reports, Messrs Schimmel & Co., Distillers, Miltitz, near Leipzig, 1893 to 1896.

ANON. (96). Annual Report, Messrs Schimmel & Co., Distillers, Miltitz, near Leipzig, 1911.

ANON. (97). *Decennial Reports on Trade etc. in China and Korea* (Statistical Series, no. 6), 1882–1891. Inspectorate-General of Customs, Shanghai, 1893.

ANON. (98). 'Saltpetre Production in China.' *CEM*, 1925, **2** (no. 8), 8.

ANON. (99). 'Alkali Lands in North China [and the sodium carbonate (*chien*) produced there].' *JSCI*, 1894, **13**, 910.

ANON. (100). *A Guide to Peiping [Peking] and its Environs*. Catholic University (Fu-Jen) Press, for Peking Bookshop (Vetch), Peking, 1946.

ANON. (101) (ed.). *De Alchemia: In hoc Volumine de Alchemia continentur haec: Geber Arabis, philosophi solertissimi rerumque naturalium, praecipue metallicarum peritissimi...* (4 books); *Speculum Alchemiae* (Roger Baeon); *Correctorium Alchemiae* (Richard Anglici); *Rosarius Minor; Liber Secretorum Alchemicae* (Calid = Khalid); *Tabula Smaragdina* (with commentary of Hortulanus)...etc. Petreius, Nuremberg, 1541. Cf. Ferguson (1), vol. 1, p. 18.

ANON. (103). *Introduction to Classical Japanese Literature.* Kokusai Bunka Shinkokai (Soc. for Internat. Cultural Relations), Tokyo, 1948.

ANON. (104). *Of a Degradation of Gold made by an Anti-Elixir; a Strange Chymical Narative.* Herringman, London, 1678. 2nd ed. *An Historical Account of a Degradation of Gold made by an Anti-Elixir; a Strange Chymical Narrative. By the Hon. Robert Boyle, Esq.* Montagu, London, 1739.

ANON. (105). 'Some Observations concerning Japan, made by an Ingenious Person that hath many years resided in that Country...' *PTRS*, 1669, **4** (no. 49), 983.

ANON. (113). 'A 2100-year-old Tomb Excavated; the Contents Well Preserved.' *PKR*, 1972, no. 32 (11 Aug.), 10. *EHOR*, 1972, **11** (no. 4), 16 (with colour-plates). [The Lady of Tai (d. *c.* −186), incorrupted body, with rich tomb furnishings.] The article also distributed as an offprint at showings of the relevant colour film, e.g. in Hongkong, Sept. 1972.

ANON. (114). 'A 2100-year-old Tomb Excavated.' *CREC*, 1972, **21** (no. 9), 20 (with colour-plates). [The Lady of Tai, see previous entry.]

ANON. (115). *Antiquities Unearthed during the Great Proletarian Cultural Revolution.* n.d. [Foreign Languages Press, Peking, 1972]. With colour-plates. Arranged according to provinces of origin.

ANON. (116). *Historical Relics Unearthed in New China* (album). Foreign Languages Press, Peking, 1972.

ANTENORID, J. (1). 'Die Kenntnisse der Chinesen in der Chemie.' *CHZ*, 1902, **26** (no. 55), 627.

ANTZE, G. (1). 'Metallarbeiten aus Peru.' *MMVKH*, 1930, **15**, 1.

APOLLONIUS OF TYANA. *See* Conybeare (1); Jones (1).

ARDAILLON, E. (1). *Les Mines de Laurion dans l'Antiquité.* Inaug. Diss. Paris. Fontemoing, Paris, 1897.

ARLINGTON, L. C. & LEWISOHN, W. (1). *In Search of Old Peking.* Vetch, Peiping, 1935.

ARMSTRONG, E. F. (1). 'Alcohol through the Ages.' *CHIND*, 1933, **52** (no. 12), 251, (no. 13), 279. (Jubilee Memorial Lecture of the Society of Chemical Industry.)

ARNOLD, P. (1). *Histoire des Rose-Croix et les Origines de la Franc-Maçonnerie.* Paris, 1955.

AROUX, E. (1). *Dante, Hérétique, Révolutionnaire et Socialiste; Révélations d'un Catholique sur le Moyen Age.* 1854.

AROUX, E. (2). *Les Mystères de la Chevalerie et de l'Amour Platonique au Moyen Age.* 1858.

ARSENDAUX, H. & RIVET, P. (1). 'L'Orfèvrerie du Chiriqui et de Colombie'. *JSA*, 1923, **15**, 1.

ASCHHEIM, S. (1). 'Weitere Untersuchungen über Hormone und Schwangerschaft; das Vorkmmen der Hormone im Harn der Schwangeren.' *AFG*, 1927, **132**, 179.

ASCHHEIM, S. & ZONDEK, B. (1). 'Hypophysenvorderlappen Hormon und Ovarialhormon im Harn von Schwangeren.' *KW*, 1927, **6**, 1322.

ASHBEE, C. R. (1). *The Treatises of Benvenuto Cellini on Goldsmithing and Sculpture; made into English from the Italian of the Marcian Codex...* Essex House Press, London, 1898.

ASHMOLE, ELIAS (1). *Theatrum Chemicum Britannicum; Containing Severall Poeticall Pieces of our Famous English Philosophers, who have written the Hermetique Mysteries in their owne Ancient Language, Faithfully Collected into one Volume, with Annotations thereon by E. A. Esq.* London, 1652. Facsim. repr. ed. A. G. Debus, Johnson, New York and London, 1967 (Sources of Science ser. no. 39).

ASTON, W. G. (tr.) (1). '*Nihongi*', *Chronicles of Japan from the Earliest Times to +697.* Kegan Paul, London, 1896; repr. Allen & Unwin, London, 1956.

ATKINSON, R. W. (2). '[The Chemical Industries of Japan; I,] Notes on the Manufacture of *oshiroi* (White Lead).' *TAS/J*, 1878, **6**, 277.

ATKINSON, R. W. (3). 'The Chemical Industries of Japan; II, *Ame* [dextrin and maltose].' *TAS/J*, 1879, **7**, 313.

[ATWOOD, MARY ANNE] (1) (Mary Anne South, Mrs Atwood). *A Suggestive Enquiry into the Hermetic Mystery; with a Dissertation on the more Celebrated of the Alchemical Philosophers, being an Attempt towards the Recovery of the Ancient Experiment of Nature.* Trelawney Saunders, London, 1850. Repr. with introduction by W. L. Wilmhurst, Tait, Belfast, 1918, repr. 1920. *Hermetic Philosophy and Alchemy; a Suggestive Enquiry.* Repr. New York, 1960.

AVALON, A. (ps.). *See* Woodroffe, Sir J.

AYRES, LEW (1). *Altars of the East.* New York, 1956.

BACON, J. R. (1). *The Voyage of the Argonauts.* London, 1925.

BACON, ROGER
    Compendium Studii Philosophiae, +1271. *See* Brewer (1).
    De Mirabili Potestatis Artis et Naturae et de Nullitate Magiae, bef. +1250. *See* de Tournus (1); T. M [oufet]? (1); Tenney Davis (16).
    De Retardatione Accidentium Senectutis etc., +1236 to +1245. *See* R. Browne (1); Little & Withington (1).
    De Secretis operibus Artis et Naturae et de Nullitate Magiae, bef. +1250. *See* Brewer (1).
    Opus Majus, +1266. *See* Bridges (1); Burke (1); Jebb (1).
    Opus Minus, +1266 or +1267. *See* Brewer (1).

Opus Tertium, +1267. *See* Little (1); Brewer (1).

Sanioris Medicinae etc., pr. +1603. *See* Bacon (1).

Secretum Secretorum (ed.), c. +1255, introd. c. +1275. *See* Steele (1).

BACON, ROGER (1). *Sanioris Medicinae Magistri D. Rogeri Baconis Angli De Arte Chymiae Scripta.* Schönvetter, Frankfurt, 1603. Cf. Ferguson (1), vol. 1, p. 63.

BAGCHI, B. K. & WENGER, M. A. (1). 'Electrophysiological Correlates of some Yogi Exercises.' *EECN*, 1957, **7** (suppl.), 132.

BAIKIE, J. (1). 'The Creed [of Ancient Egypt].' *ERE*, vol. iv, p. 243.

BAILEY, CYRIL (1). *Epicurus; the Extant Remains.* Oxford, 1926.

BAILEY, SIR HAROLD (1). 'A Half-Century of Irano-Indian Studies.' *JRAS*, 1972 (no. 2), 99.

BAILEY, K. C. (1). *The Elder Pliny's Chapters on Chemical Subjects.* 2 vols. Arnold, London, 1929 and 1932.

BAIN, H. FOSTER (1). *Ores and Industry in the Far East; the Influence of Key Mineral Resources on the Development of Oriental Civilisation.* With a chapter on Petroleum by W. B. Heroy. Council on Foreign Relations, New York, 1933.

BAIRD, M. M., DOUGLAS, C. G., HALDANE, J. B. S. & PRIESTLEY, J. G. (1). 'Ammonium Chloride Acidosis.' *JOP*, 1923, **57**, xli.

BALAZS, E. (= S.) (1). 'La Crise Sociale et la Philosophie Politique à la Fin des Han.' *TP*, 1949, **39**, 83.

BANKS, M. S. & MERRICK, J. M. (1). 'Further Analyses of Chinese Blue-and-White [Porcelain and Pottery].' *AMY*, 1967, **10**, 101.

BARNES, W. H. (1). 'The Apparatus, Preparations and Methods of the Ancient Chinese Alchemists.' *JCE*, 1934, **11**, 655. 'Diagrams of Chinese Alchemical Apparatus' (an abridged translation of Tshao Yuan-Yü, 1). *JCE*, 1936, **13**, 453.

BARNES, W. H. (2). 'Possible References to Chinese Alchemy in the −4th or −3rd Century.' *CJ*, 1935, **23**, 75.

BARNES, W. H. (3). 'Chinese Influence on Western Alchemy.' *N*, 1935, **135**, 824.

BARNES, W. H. & YUAN, H. B. (1). 'Thao the Recluse (+452 to +536); Chinese Alchemist.' *AX*, 1946, **2**, 138. Mainly a translation of a short biographical paper by Tshao Yuan-Yü.

LA BARRE, W. (1). 'Twenty Years of Peyote Studies.' *CURRA*, 1960, **1**, 45.

LA BARRE, W. (2). *The Peyote Cult.* Yale Univ. Press, New Haven, Conn., repr. Shoestring Press, Hamden, Conn. 1960. (Yale Univ. Publications in Anthropology, no. 19.)

BARTHOLD, W. (2). *Turkestan down to the Mongol Invasions.* 2nd ed. London, 1958.

BARTHOLINUS, THOMAS (1). *De Nivis Usu Medico Observationes Variae.* Copenhagen, 1661.

BARTON, G. A. (1). '[The "Abode of the Blest" in] Semitic [including Babylonian, Jewish and ancient Egyptian, Belief].' *ERE*, ii, 706.

DE BARY, W. T. (3) (ed.). *Self and Society in Ming Thought.* Columbia Univ. Press, New York and London, 1970.

BASU, B. N. (1) (tr.). *The 'Kāmasūtra' of Vātsyāyana* [prob. +4th century]. Rev. by S. L. Ghosh. Pref. by P. C. Bagchi. Med. Book Co., Calcutta, 1951 (10th ed.).

BAUDIN, L. (1). 'L'Empire Socialiste des Inka [Incas].' *TMIE*, 1928, no. 5.

BAUER, W. (3). 'The Encyclopaedia in China.' *JWH*, 1966, **9**, 665.

BAUER, W. (4). *China und die Hoffnung auf Glück; Paradiese, Utopien, Idealvorstellungen.* Hanser, München, 1971.

BAUMÉ, A. (1). *Éléments de Pharmacie.* 1777.

BAWDEN, F. C. & PIRIE, N. W. (1). 'The Isolation and Some Properties of Liquid Crystalline Substances from Solanaceous Plants infected with Three Strains of Tobacco Mosaic Virus.' *PRSB*, 1937, **123**, 274.

BAWDEN, F. C. & PIRIE, N. W. (2). 'Some Factors affecting the Activation of Virus Preparations made from Tobacco Leaves infected with a Tobacco Necrosis Virus.' *JGMB*, 1950, **4**, 464.

BAYES, W. (1). *The Triple Aspect of Chronic Disease, having especial reference to the Treatment of Intractable Disorders affecting the Nervous and Muscular System.* Churchill, London, 1854.

BAYLISS, W. M. (1). *Principles of General Physiology.* 4th ed. Longmans Green, London, 1924.

BEAL, S. (2) (tr.). *Si Yu Ki [Hsi Yü Chi], Buddhist Records of the Western World*, transl. from the Chinese of Hiuen Tsiang [Hsüan-Chuang]. 2 vols. Trübner, London, 1881, 1884, 2nd ed. 1906. Repr. in 4 vols. with new title; *Chinese Accounts of India, translated from the Chinese of Hiuen Tsiang.* Susil Gupta, Calcutta, 1957.

BEAUVOIS, E. (1). 'La Fontaine de Jouvence et le Jourdain dans les Traditions des Antilles et de la Floride.' *MUSEON*, 1884, **3**, 404.

BEBEY, F. (1). 'The Vibrant Intensity of Traditional African Music.' *UNESC*, 1972, **25** (no. 10), 15. (On p. 19, a photograph of a relief of Ouroboros in Dahomey.)

BEDINI, S. A. (5). 'The Scent of Time; a Study of the Use of Fire and Incense for Time Measurement in Oriental Countries.' *TAPS*, 1963 (N.S.), **53**, pt. 5, 1–51. Rev. G. J. Whitrow, *A/AIHS*, 1964, **17**, 184.

BEDINI, S. A. (6). 'Holy Smoke; Oriental Fire Clocks.' *NS*, 1964, **21** (no. 380), 537.

VAN BEEK, G. W. (1). 'The Rise and Fall of Arabia Felix.' *SAM*, 1969, **221** (no. 6), 36.

BEER, G. (1) (ed. & tr.). 'Das Buch Henoch [Enoch]' in *Die Apokryphen und Pseudepigraphien des alten Testaments*, ed. E. Kautzsch, 2 vols. Mohr (Siebeck), Tübingen, Leipzig ¦and Freiburg i/B, 1900, vol. 2 (Pseudepigraphien), pp. 217 ff.

LE BEGUE, JEAN (1). *Tabula de Vocabulis Synonymis et Equivocis Colorum* and *Experimenta de Coloribus* (MS. BM. 6741 of +1431). Eng. tr. Merrifield (1), vol. 1, pp. 1–321.

BEH, Y. T. *See* Kung, S. C., Chao, S. W., Bei, Y. T. & Chang, C. (1).

BEHANAN, KOVOOR T. (1). *Yoga; a Scientific Evaluation*. Secker & Warburg, London, 1937. Paperback repr. Dover, New York and Constable, London. n.d. (*c.* 1960).

BEHMEN, JACOB. *See* Boehme, Jacob.

BELL, SAM HANNA (1). *Erin's Orange Lily*. Dobson, London, 1956.

BELPAIRE, B. (3). 'Note sur un Traité Taoiste.' *MUSEON*, 1946, **59**, 655.

BENDALL, C. (1) (ed.). *Subhāṣita-saṃgraha*. Istas, Louvain, 1905. (Muséon Ser. nos. 4 and 5.)

BENEDETTI-PICHLER, A. A. (1). 'Micro-chemical Analysis of Pigments used in the Fossae of the Incisions of Chinese Oracle-Bones.' *IEC/AE*, 1937, **9**, 149. Abstr. *CA*, 1938, **31**, 3350.

BENFEY, O. T. (1). 'Dimensional Analysis of Chemical Laws and Theories.' *JCE*, 1957, **34**, 286.

BENFEY, O. T. (2) (ed.). *Classics in the Theory of Chemical Combination*. Dover, New York, 1963. (Classics of Science, no. 1.)

BENFEY, O. T. & FIKES, L. (1). 'The Chemical Prehistory of the Tetrahedron, Octahedron, Icosahedron and Hexagon.' *ADVC*, 1966, **61**, 111. (Kekulé Centennial Volume.)

BENNETT, A. A. (1). *John Fryer; the Introduction of Western Science and Technology into Nineteenth-Century China*. Harvard Univ. Press, Cambridge, Mass. 1967. (Harvard East Asian Monographs, no. 24.)

BENSON, H., WALLACE, R. K., DAHL, E. C. & COOKE, D. F. (1). 'Decreased Drug Abuse with Transcendental Meditation; a Study of 1862 Subjects.' In 'Hearings before the Select Committee on Crime of the House of Representatives (92nd Congress)', U.S. Govt. Washington, D.C. 1971, p. 681 (Serial no. 92-1).

BENTHAM, G. & HOOKER, J. D. (1). *Handbook of the British Flora; a Description of the Flowering Plants and Ferns indigenous to, or naturalised in, the British Isles*. 6th ed. 2 vols. (1 vol. text, 1 vol. dra wings). Reeve, London, 1892. repr. 1920.

BENVENISTE, E. (1). 'Le Terme *obryza* et la Métallurgie de l'Or.' *RPLHA*, 1953, **27**, 122.

BENVENISTE, E. (2). *Textes Sogdiens* (facsimile reproduction, transliteration, and translation with glossary). Paris, 1940. Rev. W. B. Hemming, *BLSOAS*, **11**.

BERENDES, J. (1). *Die Pharmacie bei den alten Culturvölkern; historisch-kritische Studien*. 2 vols. Tausch & Grosse, Halle, 1891.

BERGMAN, FOLKE (1). *Archaeological Researches in Sinkiang*. Reports of the Sino-Swedish [scientific] Expedition [to Northwest China]. 1939, vol. 7 (pt. 1).

BERGMAN, TORBERN (1). *Opuscula Physica et Chemica, pleraque antea seorsim edita, jam ab Auctore collecta, revisa et aucta*. 3 vols. Edman, Upsala, 1779–83. Eng. tr. by E. Cullen, *Physical and Chemical Essays*, 2 vols. London, 1784, 1788; the 3rd vol. Edinburgh, 1791.

BERGSØE, P. (1). 'The Metallurgy and Technology of Gold and Platinum among the Pre-Columbian Indians.' *IVS*, 1937, no. A44, 1–45. Prelim. pub. *N*, 1936, **137**, 29.

BERGSØE, P. (2). 'The Gilding Process and the Metallurgy of Copper and Lead among the Pre-Columbian Indians.' *IVS*, 1938, no. A46. Prelim. pub. 'Gilding of Copper among the Pre-Columbian Indians.' *N*, 1938, **141**, 829.

BERKELEY, GEORGE, BP. (1). *Siris; Philosophical Reflections and Enquiries concerning the Virtues of Tar-Water*. London, 1744.

BERNAL, J. D. (1). *Science in History*. Watts, London, 1954. (Beard Lectures at Ruskin College, Oxford.) Repr. 4 vols. Penguin, London, 1969.

BERNAL, J. D. (2). *The Extension of Man; a History of Physics before 1900*. Weidenfeld & Nicolson, London, 1972. (Lectures at Birkbeck College, London, posthumously published.)

BERNARD, THEOS (1). *Haṭhayoga; the Report of a Personal Experience*. Columbia Univ. Press, New York, 1944; Rider, London, 1950. Repr. 1968.

BERNARD-MAÎTRE, H. (3). 'Un Correspondant de Bernard de Jussieu en China; le Père le Chéron d'Incarville, missionaire français de Pékin, d'après de nombreux documents inédits.' *A/AIHS*, 1949, **28**, 333, 692.

BERNARD-MAÎTRE, H. (4). 'Notes on the Introduction of the Natural Sciences into the Chinese Empire.' *YJSS*, 1941, **3**, 220.

BERNARD-MAÎTRE, H. (9). 'Deux Chinois du 18e siècle à l'École des Physiocrates Français.' *BUA*, 1949 (3e sér.), **10**, 151.

BERNARD-MAÎTRE, H. (17). 'La Première Académie des Lincei et la Chine.' *MP*, 1941, 65.

BERNARD-MAÎTRE, H. (18). 'Les Adaptations Chinoises d'Ouvrages Européens; Bibliographie chronologique depuis la venue des Portugais à Canton jusqu'à la Mission française de Pékin (+1514 à +1688).' *MS*, 1945, **10**, 1-57, 309-88.

BERNAREGGI, E. (1). 'Nummi Pelliculati' (silver-clad copper coins of the Roman Republic). *RIN*, 1965, **67** (5th. ser., **13**), 5.

BERNOULLI, R. (1). 'Seelische Entwicklung im Spiegel der Alchemie u. verwandte Disciplinen.' *ERJB*, 1935, **3**, 231-87. Eng. tr. 'Spiritual Development as reflected in Alchemy and related Disciplines.' *ERYB*, 1960, **4**, 305. Repr. 1970.

BERNTHSEN, A. *See* Sudborough, J. J. (1).

BERRIMAN, A. E. (2). 'A Sumerian Weight-Standard in Chinese Metrology during the Former Han Dynasty (−206 to −23).' *RAAO*, 1958, **52**, 203.

BERRIMAN, A. E. (3). 'A New Approach to the Study of Ancient Metrology.' *RAAO*, 1955, **49**, 193.

BERTHELOT, M. (1). *Les Origines de l'Alchimie*. Steinheil, Paris, 1885. Repr. Libr. Sci. et Arts, Paris, 1938.

BERTHELOT, M. (2). *Introduction à l'Étude de la Chimie des Anciens et du Moyen-Age*. First published at the beginning of vol. 1 of the *Collection des Anciens Alchimistes Grecs* (see Berthelot & Ruelle), 1888. Repr. sep. Libr. Sci. et Arts, Paris, 1938. The 'Avant-propos' is contained only in Berthelot & Ruelle; there being a special Preface in Berthelot (2).

BERTHELOT, M. (3). Review of de Mély (1), *Lapidaires Chinois*. *JS*, 1896, 573.

BERTHELOT, M. (9). Les Compositons Incendiaires dans l'Antiquité et Moyen Ages.' *RDM*, 1891, **106**, 786.

BERTHELOT, M. (10). *La Chimie au Moyen Age;* vol. 1, *Essai sur la Transmission de la Science Antique au Moyen Age* (Latin texts). Impr. Nat. Paris, 1893. Photo. repr. Zeller, Osnabrück; Philo, Amsterdam, 1967. Rev. W. P[agel], *AX*, 1967, **14**, 203.

BERTHELOT, M. (12). 'Archéologie et Histoire des Sciences; avec Publication nouvelle du Papyrus Grec chimique de Leyde, et Impression originale du *Liber de Septuaginta* de Geber.' *MRASP*, 1906, **49**, 1-377. Sep. pub. Philo, Amsterdam, 1968.

BERTHELOT, M. [P. E. M.]. *See* Tenney L. Davis' biography (obituary), with portrait. *JCE*, 1934, **11** (585) and Boutaric (1).

BERTHELOT, M. & DUVAL, R. (1). *La Chimie au Moyen Age*; vol. 2, *l'Alchimie Syriaque*. Impr. Nat. Paris, 1893. Photo. repr. Zeller, Osnabrück; Philo, Amsterdam, 1967. Rev. W. P[agel], *AX*, 1967, **14**, 203.

BERTHELOT, M. & HOUDAS, M. O. (1). *La Chimie au Moyen Age*; vol. 3, *l'Alchimie Arabe*. Impr. Nat. Paris, 1893. Photo repr. Zeller, Osnabrück; Philo, Amsterdam, 1967. Rev. W. P[agel], *AX*, 1967, **14**, 203.

BERTHELOT, M. & RUELLE, C. E. (1). *Collection des Anciens Alchimistes Grecs*. 3 vols. Steinheil, Paris, 1888. Photo. repr. Zeller, Osnabrück, 1967.

BERTHOLD, A. A. (1). 'Transplantation der Hoden.' *AAPWM*, 1849, **16**, 42. Engl. tr. by D. P. Quiring. *BIHM*, 1944, **16**, 399.

BERTRAND, G. (1). Papers on laccase. *CRAS*, 1894, **118**, 1215; 1896, **122**, 1215; *BSCF*, 1894, **11**, 717; 1896, **15**, 793.

BERTUCCIOLI, G. (2). 'A Note on Two Ming Manuscripts of the *Pên Tshao Phin Hui Ching Yao*.' *JOSHK*, 1956, **2**, 63. Abstr. *RBS*, 1959, **2**, 228.

BETTENDORF, G. & INSLER, V. (1) (ed.). *The Clinical Application of Human Gonadotrophins*. Thieme, Stuttgart, 1970.

BEURDELEY, M. (1) (ed.). *The Clouds and the Rain; the Art of Love in China*. With contributions by K. Schipper on Taoism and sexuality, Chang Fu-Jui on literature and poetry, and J. Pimpaneau on perversions. Office du Livre, Fribourg and Hammond & Hammond, London, 1969.

BEVAN, E. R. (1). 'India in Early Greek and Latin Literature.' *CHI*, Cambridge, 1935, vol. 1, ch. 16, p. 391.

BEVAN, E. R. (2). *Stoics and Sceptics*. Oxford, 1913.

BEVAN, E. R. (3). *Later Greek Religion*. Oxford, 1927.

BEZOLD, C. (3). *Die 'Schatzhöhle'; aus dem Syrische Texte dreier unedirten Handschriften in's Deutsche übersetzt und mit Anmerkungen versehen...nebst einer Arabischen Version nach den Handschriften zu Rom, Paris und Oxford*. 2 vols. Hinrichs, Leipzig, 1883, 1888.

BHAGVAT, K. & RICHTER, D. (1). 'Animal Phenolases and Adrenaline.' *BJ*, 1938, **32**, 1397.

BHAGVAT SINGHJI, H. H. (Maharajah of Gondal) (1). *A Short History of Aryan Medical Science*. Gondal, Kathiawar, 1927.

BHATTACHARYA, B. (1) (ed.). *Guhya-samāja Tantra, or Tathāgata-guhyaka*. Orient. Instit., Baroda, 1931. (Gaekwad Orient. Ser. no. 53.)

BHATTACHARYA, B. (2). *Introduction to Buddhist Esoterism*. Oxford, 1932.

BHISHAGRATNA, (KAVIRAJ) KUNJA LAL SHARMA (1) (tr.). *An English Translation of the 'Sushruta Samhita',* *based on the original Sanskrit Text.* 3 vols. with an index volume, pr. pr. Calcutta, 1907–18. Re-issued, Chowkhamba Sanskrit Series Office, Varanasi, 1963. Rev. M. D. Grmek, *A/AIHS*, 1965, **18**, 130.

BIDEZ, J. (1). *'l'Épître sur la Chrysopée' de Michel Psellus* [with Italian translation]; [also] *Opuscules et Extraits sur l'Alchimie, la Météorologie et la Démonologie...* (Pt. VI of *Catalogue des Manuscrits Alchimiques Grecques*). Lamertin, for Union Académique Internationale, Brussels, 1928.

BIDEZ, J. (2). *Vie de Porphyre le Philosophe Neo-Platonicien avec les Fragments des Traités περὶ ἀγαλμάτων et 'De Regressu Animae'.* 2 pts. Univ. Gand, Leipzig, 1913. (Receuil des Trav. pub. Fac. Philos. Lettres, Univ. Gand.)

BIDEZ, J. & CUMONT, F. (1). *Les Mages Hellenisés; Zoroastre, Ostanès et Hytaspe d'après la Tradition Grecque.* 2 vols. Belles Lettres, Paris, 1938.

BIDEZ, J., CUMONT, F., DELATTE, A. HEIBERG, J. L., LAGERCRANTZ, O., KENYON, F., RUSKA, J. & DE FALCO, V. (1) (ed.). *Catalogue des Manuscrits Alchimiques Grecs.* 8 vols. Lamertin, Brussels, 1924–32 (for the Union Académique Internationale).

BIDEZ, J., CUMONT, F., DELATTE, A., SARTON, G., KENYON, F. & DE FALCO, V. (1) (ed.). *Catalogue des Manuscrits Alchimiques Latins.* 2 vols. Union Acad. Int., Brussels, 1939–51.

BIOT, E. (1) (tr.). *Le Tcheou-Li ou Rites des Tcheou* [Chou]. 3 vols. Imp. Nat., Paris, 1851. (Photo-grapically reproduced, Wêntienko, Peiping, 1930.)

BIOT, E. (17). 'Notice sur Quelques Procédés Industriels connus en Chine au XVIe siècle.' *JA*, 1835 (2e sér.), **16**, 130.

BIOT, E. (22). 'Mémoires sur Divers Minéraux Chinois appartenant à la Collection du Jardin du Roi.' *JA*, 1839 (3e sér.), **8**, 206.

BIRKENMAIER, A. (1). 'Simeon von Köln oder Roger Bacon?' *FRS*, 1924, **2**, 307.

AL-BĪRŪNĪ, ABŪ AL-RAIḤĀN MUḤAMMAD IBN-AḤMAD. *Ta'rīkh al-Hind* (History of India). *See* Sachau (1).

BISCHOF, K. G. (1). *Elements of Chemical and Physical Geology,* tr. B. H. Paul & J. Drummond from the 1st German edn. (3 vols., Marcus, Bonn, 1847-54), Harrison, London, 1854 (for the Cavendish Society). 2nd German ed. 3 vols. Marcus, Bonn, 1863, with supplementary volume, 1871.

BLACK, J. DAVIDSON (1). 'The Prehistoric Kansu Race.' *MGSC* (Ser. A.), 1925, no. 5.

BLAKNEY, R. B. (1). *The Way of Life; Lao Tzu—a new Translation of the 'Tao Tê Ching'.* Mentor, New York, 1955.

BLANCO-FREIJEIRO, A. & LUZÓN, J. M. (1). 'Pre-Roman Silver Miners at Rio Tinto.' *AQ*, 1969, **43**, 124.

DE BLANCOURT, HAUDICQUER (1). *L'Art de la Verrerie...* Paris, 1697. Eng. tr. *The Art of Glass...with an Appendix containing Exact Instructions for making Glass Eyes of all Colours.* London, 1699.

BLAU, J. L. (1). *The Christian Interpretation of the Cabala in the Renaissance.* Columbia Univ. Press, New York, 1944. (Inaug. Diss. Columbia, 1944.)

BLOCHMANN, H. F. (1) (tr.). *The 'Ā'īn-i Akbarī' (Administration of the Mogul Emperor Akbar) of Abū'l Faẓl 'Allāmī.* Rouse, Calcutta, 1873. (Bibliotheca Indica, N.S., nos. 149, 158, 163, 194, 227, 247 and 287.)

BLOFELD, J. (3). *The Wheel of Life; the Autobiography of a Western Buddhist.* Rider, London, 1959.

BLOOM, ANDRÉ [METROPOLITAN ANTHONY] (1). 'Contemplation et Ascèse; Contribution Orthodoxe', art. in *Technique et Contemplation.* Études Carmelitaines, Paris, 1948, p. 49.

BLOOM, ANDRÉ [METROPOLITAN ANTHONY] (2). 'l'Hésychasme, Yoga Chrétien?', art. in *Yoga*, ed. J. Masui, Paris, 1953.

BLOOMFIELD, M. (1) (tr.). *Hymns of the Atharva-veda, together with Extracts from the Ritual Books and the Commentaries.* Oxford, 1897 (*SBE*, no. 42). Repr. Motilal Banarsidass, Delhi, 1964.

BLUNDELL, J. W. F. (2). *Medicina Mechanica.* London.

BOAS, G. (1). *Essays on Primitivism and Related Ideas in the Middle Ages.* Johns Hopkins Univ. Press, Baltimore, 1948.

BOAS, MARIE (2). *Robert Boyle and Seventeenth-Century Chemistry.* Cambridge, 1958.

BOAS, MARIE & HALL, A. R. (2). 'Newton's Chemical Experiments.' *A/AIHS*, 1958, **37**, 113.

BOCHARTUS, S. (1). *Opera Omnia, hoc est Phaleg, Canaan, et Hierozoicon.* Boutesteyn & Luchtmans, Leiden and van de Water, Utrecht, 1692. [The first two books are on the geography of the Bible and the third on the animals mentioned in it.]

BOCTHOR, E. (1). *Dictionnaire Français–Arabe,* enl. and ed. A. Caussin de Perceval. Didot, Paris, 1828–9. 3rd ed. Didot, Paris, 1864.

BODDE, D. (5). 'Types of Chinese Categorical Thinking.' *JAOS*, 1939, **59**, 200.

BODDE, D. (9). 'Some Chinese Tales of the Supernatural; Kan Pao and his *Sou Shen Chi*.' *HJAS*, 1942, **6**, 338.

BODDE, D. (10). 'Again Some Chinese Tales of the Supernatural; Further Remarks on Kan Pao and his *Sou Shen Chi*.' *JAOS*, 1942, **62**, 305.

BOECKH, A. (1) (ed.). *Corpus Inscriptionum Graecorum.* 4 vols. Berlin, 1828–77.

BOEHME, JACOB (1). *The Works of Jacob Behmen, the Teutonic Theosopher... To which is prefixed, the Life of the Author, with Figures illustrating his Principles, left by the Rev. W. Law.* Richardson, 4 vols. London, 1764–81. *See* Ferguson (1), vol. I, p. 111. Based partly upon: *Idea Chemiae Böhmianae Adeptae; das ist, ein Kurtzer Abriss der Bereitung deß Steins der Weisen, nach Anleitung deß Jacobi Böhm...* Amsterdam, 1680, 1690; and: *Jacob Böhms kurtze und deutliche Beschreibung des Steins der Weisen, nach seiner Materia, aus welcher er gemachet, nach seiner Zeichen und Farbe, welche im Werck erscheinen, nach seiner Kraft und Würckung, und wie lange Zeit darzu erfordert wird, und was insgemein bey dem Werck in acht zu nehmen...* Amsterdam, 1747.

BOEHME, JACOB (2). *The Epistles of Jacob Behmen, aliter Teutonicus Philosophus, translated out of the German Language.* London, 1649.

BOERHAAVE, H. (1). *Elementa Chemiae, quae anniversario labore docuit, in publicis, privatisque, Scholis.* 2 vols. Severinus and Imhoff, Leiden, 1732. Eng. tr. by P. Shaw: *A New Method of Chemistry, including the History, Theory and Practice of the Art.* 2 vols. Longman, London, 1741, 1753.

BOERHAAVE, HERMANN. *See* Lindeboom (1).

BOERSCHMANN, E. (11). 'Peking, eine Weltstadt der Baukunst.' *AT*, 1931 (no. 2), 74.

BOLL, F. (6). 'Studien zu Claudius Ptolemäus.' *JCP*, 1894, **21** (Suppl.), 155.

BOLLE, K. W. (1). *The Persistence of Religion; an Essay on Tantrism and Sri Aurobindo's Philosophy.* Brill, Leiden, 1965. (Supplements to *Numen*, no. 8.)

BONI, B. (3). 'Oro e Formiche Erodotee.' *CHIM*, 1950 (no. 3).

BONMARCHAND, G. (1) (tr.). 'Les Notes de Li Yi-Chan [Li I-shan], (Yi-Chan Tsa Tsouan [*I-Shan Tsa Tsuan*]), traduit du Chinois; Étude de Littérature Comparée.' *BMFJ*, 1955 (N.S.) **4** (no. 3), 1–84.

BONNER, C. (1). 'Studies in Magical Amulets, chiefly Graeco-Egyptian.' Ann Arbor, Michigan, 1950. (Univ. Michigan Studies in Humanities Ser., no. 49.)

BONNIN, A. (1). *Tutenag and Paktong; with Notes on other Alloys in Domestic Use during the Eighteenth Century.* Oxford, 1924.

BONUS, PETRUS, of Ferrara (1). *M. Petri Boni Lombardi Ferrariensis Physici et Chemici Excellentiss. Introductio in Artem Chemiae Integra, ab ipso authore inscripta Margarita Preciosa Novella; composita ante annos plus minus ducentos septuaginta, Nune multis mendis sublatis, comodiore, quam antehâc, forma edita, et indice revum ad calcem adornata.* Foillet, Montbeliard, 1602. 1st ed. Lacinius ed. Aldus, Venice, 1546. Tr. Waite (7). *See* Leicester (1), p. 86. Cf. Ferguson (1), vol. I, p. 115.

BORNET, P. (2). 'Au Service de la Chine; Schall et Verbiest, maîtres-fondeurs, I. les Canons.' *BCP*, 1946 (no. 389), 160.

BORNET, P. (3) (tr.). 'Relation Historique' [de Johann Adam Schall von Bell, S.J.]; Texte Latin avec Traduction française.' Hautes Études, Tientsin, 1942 (part of *Lettres et Mémoires d'Adam Schall S.J.* ed H. Bernard[-Maître]).

BORRICHIUS, O. (1). *De Ortu et Progressu Chemiae.* Copenhagen, 1668.

BOSE, D. M., SEN, S.-N., SUBBARAYAPPA, B. V. et al. (1). *A Concise History of Science in India.* Baptist Mission Press, Calcutta, for the Indian National Science Academy, New Delhi, 1971.

BOSON, G. (1). 'Alcuni Nomi di Pietri nelle Inscrizioni Assiro-Babilonesi.' *RSO*, 1914, **6**, 969.

BOSON, G. (2). 'I Metalli e le Pietri nelle Inscrizioni Assiro-Babilonesi.' *RSO*, 1917, **7**, 379.

BOSON, G. (3). *Les Métaux et les Pierres dans les Inscriptions Assyro-Babyloniennes.* Munich, 1914.

BOSTOCKE, R. (1). *The Difference between the Ancient Physicke, first taught by the godly Forefathers, insisting in unity, peace and concord, and the Latter Physicke...* London, 1585. Cf. Debus (12).

BOUCHÉ-LECLERCQ, A. (1). *L'Astrologie Grecque.* Leroux, Paris, 1899.

BOURKE, J. G. (1). 'Primitive Distillation among the Tarascoes.' *AAN*, 1893, **6**, 65.

BOURKE, J. G. (2). 'Distillation by Early American Indians.' *AAN*, 1894, **7**, 297.

BOURNE, F. S. A. (2). *The Lo-fou Mountains; an Excursion.* Kelly & Walsh, Shanghai, 1895.

BOUTARIC, A. (1). *Marcellin Berthelot (1827 à 1907).* Payot, Paris, 1927.

BOVILL, E. W. (1). 'Musk and Amber[gris].' *NQ*, 1954.

BOWERS, J. Z. & CARUBBA, R. W. (1). 'The Doctoral Thesis of Engelbert Kaempfer: "On Tropical Diseases, Oriental Medicine and Exotic Natural Phenomena".' *JHMAS*, 1970, **25**, 270.

BOYLE, ROBERT (1). *The Sceptical Chymist; or, Chymico-Physical Doubts and Paradoxes, touching the Experiments whereby Vulgar Spagyrists are wont to endeavour to evince their Salt, Sulphur and Mercury to be the True Principles of Things.* Crooke, London, 1661.

BOYLE, ROBERT (4). 'A New Frigoric Experiment.' *PTRS*, 1666, **1**, 255.

BOYLE, ROBERT (5). *New Experiments and Observations touching Cold.* London, 1665. Repr. 1772.

BOYLE, ROBERT. *See* Anon. (104).

BRADLEY, J. E. S. & BARNES, A. C. (1). *Chinese–English Glossary of Mineral Names.* Consultants' Bureau, New York, 1963.

BRASAVOLA, A. (1). *Examen Omnium Sinplicium Medicamentorum.* Rome, 1536.

BRELICH, H. (1). 'Chinese Methods of Mining Quicksilver.' *TIMM*, 1905, **14**, 483.

BRELICH, H. (2). 'Chinese Methods of Mining Quicksilver.' *MJ*, 1905 (27 May), 578, 595.

BRETSCHNEIDER, E. (1). *Botanicon Sinicum; Notes on Chinese Botany from Native and Western Sources,* 3 vols.
  Vol. 1 (Pt. 1, no special sub-title) contains
      ch. 1. Contribution towards a History of the Development of Botanical Knowledge among Eastern
      Asiatic Nations.
      ch. 2. On the Scientific Determination of the Plants Mentioned in Chinese Books.
      ch. 3. Alphabetical List of Chinese Works, with Index of Chinese Authors.
      app. Celebrated Mountains of China (list)
      Trübner, London, 1882 (printed in Japan); also pub. *JRAS/NCB*, 1881 (n.s.), **16**, 18–230 (in
      smaller format).
  Vol. 2, Pt. II, *The Botany of the Chinese Classics,* with Annotations, Appendixes and Indexes by
      E. Faber, contains
      Corrigenda and Addenda to Pt. I
      ch. 1. Plants mentioned in the *Erh Ya.*
      ch. 2. Plants mentioned in the *Shih Ching,* the *Shu Ching,* the *Li Chi,* the *Chou Li* and other Chinese
      classical works.
      Kelly & Walsh, Shanghai etc. 1892; also pub. *JRAS/NCB*, 1893 (n.s.), **25**, 1–468.
  Vol. 3, Pt. III, *Botanical Investigations into the Materia Medica of the Ancient Chinese,* contains
      ch. 1. Medicinal Plants of the *Shen Nung Pên Tshao Ching* and the [*Ming I*] *Pieh Lu*
      with indexes of geographical names, Chinese plant names and Latin generic names.
      Kelly & Walsh, Shanghai etc., 1895; also pub. *JRAS/NCB*, 1895 (n.s.), **29**, 1–623.
BRETSCHNEIDER, E. (2). *Mediaeval Researches from Eastern Asiatic Sources; Fragments towards the Know-
  ledge of the Geography and History of Central and Western Asia from the +13th to the +17th
  century.* 2 vols. Trübner, London, 1888. New ed. Routledge & Kegan Paul, 1937. Photo-reprint, 1967.
BREUER, H. & KASSAU, E. (1). *Eine einfache Methode zur Isolierung von Steroiden aus biologischen Medien
  durch Mikrosublimation.* Proc. 1st International Congress of Endocrinology, Copenhagen, 1960,
  Session XI (*d*), no. 561.
BREUER, H. & NOCKE, L. (1). 'Stoffwechsel der Oestrogene in der menschlichen Leber'; art. in VIter
  Symposium d. Deutschen Gesellschaft f. Endokrinologie, *Moderne Entwicklungen auf dem Gesta-
  gengebiet Hormone in der Veterinärmedizin.* Kiel, 1959, p. 410.
BREWER, J. S. (1) (ed.). *Fr. Rogeri Bacon Opera quaedam hactenus inedita.* Longman, Green, Longman &
  Roberts, London, 1859 (Rolls Series, no. 15). Contains *Opus Tertium* (*c.* +1268), part of *Opus
  Minus* (*c.* +1267), part of *Compendium Studii Philosophiae* (+1272), and the *Epistola de Secretis
  Operibus Artis et Naturae et de Nullitate Magiae* (*c.* +1270).
BRIDGES, J. H. (1) (ed.). *The 'Opus Maius'* [*c.* +1266] *of Roger Bacon.* 3 vols. Oxford, 1897–1900.
BRIDGMAN, E. C. (1). *A Chinese Chrestomathy, in the Canton Dialect.* S. Wells Williams, Macao, 1841.
BRIDGMAN, E. C. & WILLIAMS, S. WELLS (1). 'Mineralogy, Botany, Zoology and Medicine' [sections of
  a Chinese Chrestomathy], in Bridgman (1), pp. 429, 436, 460 and 497.
BRIGHTMAN, F. E. (1). *Liturgies, Eastern and Western.* Oxford, 1896.
BROMEHEAD, C. E. N. (2). 'Aetites, or the Eagle-Stone.' *AQ*, 1947, **21**, 16.
BROOKS, CHANDLER McC., GILBERT, J. L., LEVEY, H. A. & CURTIS, D. R. (1). *Humors, Hormones and
  Neurosecretions; the Origins and Development of Man's present Knowlege of the Humoral Control of
  Body Function.* New York State Univ. N.Y. 1962.
BROOKS, E. W. (1). 'A Syriac Fragment [a chronicle extending from +754 to +813].' *ZDMG*, 1900, **54**,
  195.
BROOKS, G. (1). *Recherches sur le Latex de l'Arbre à Laque d'Indochine; le Laccol et ses Derivés.* Jouve,
  Paris, 1932.
BROOKS, G. (2). 'La Laque Végétale d'Indochine.' *LN*, 1937 (no. 3011), 359.
BROOMHALL, M. (1). *Islam in China.* Morgan & Scott, London, 1910.
BROSSE, T. (1). *Études instrumentales des Techniques du Yoga; Expérimentation psychosomatique...* with
  an Introduction 'La Nature du Yoga dans sa Tradition' by J. Filliozat, École Française d'Extrême-
  Orient, Paris, 1963 (Monograph series, no. 52).
BROUGH, J. (1). 'Soma and *Amanita muscaria.*' *BLSOAS*, 1971, **34**, 331.
BROWN-SÉQUARD, C. E. (1). 'Du Rôle physiologique d'un thérapeutique d'un Suc extrait de Test-
  icules d'Animaux, d'après nombre de faits observés chez l'Homme.' *APNP*, 1889, **21**, 651.
BROWNE, C. A. (1). 'Rhetorical and Religious Aspects of Greek Alchemy; including a Commentary and
  Translation of the Poem of the Philosopher Archelaos upon the Sacred Art.' *AX*, 1938, **2**, 129; 1948,
  3, 15.
BROWNE, E. G. (1). *Arabian Medicine.* Cambridge, 1921. Repr. 1962. (French tr. H. J. P. Renaud;
  Larose, Paris, 1933.)
BROWNE, RICHARD (1) (tr.). *The Cure of Old Age and the Preservation of Youth* (tr. of Roger Bacon's
  De Retardatione Accidentium Senectutis...). London, 1683.
BROWNE, SIR THOMAS (1). *Religio Medici.* 1642.

BRUCK, R. (1) (tr.). 'Der Traktat des Meisters Antonio von Pisa.' *RKW*, 1902, **25**, 240. A + 14th-century treatise on glass-making.

BRUNET, P. & MIELI, A. (1). *L'Histoire des Sciences (Antiquité)*. Payot, Paris, 1935. Rev. G. Sarton, *ISIS*, 1935, **24**, 444.

BRUNSCHWYK, H. (1). '*Liber de arte Distillandi de Compositis': Das Buch der waren Kunst zu distillieren die Composita und Simplicia; und das Buch 'Thesaurus Pauperum', Ein schatz der armen genannt Micarium die brösamlin gefallen von den büchern d'Artzny und durch Experiment von mir Jheronimo Brunschwick uff geclubt und geoffenbart zu trost denen die es begehren*. Grüninger, Strassburg, 1512. (This is the so-called 'Large Book of Distillation'.) Eng. tr. *The Vertuose Boke of the Distillacyon*..., Andrewe, or Treveris, London, 1527, 1528 and 1530. The last reproduced in facsimile, with an introduction by H. J. Abrahams, Johnson, New York and London, 1971 (Sources of Science Ser., no. 79).

BRUNSCHWYK, H. (2). '*Liber de arte distillandi de simplicibus' oder Buch der rechten Kunst zu distillieren die eintzigen Dinge*. Grüninger, Strassburg, 1500. (The so-called 'Small Book of Distillation'.)

BRUNTON, T. LAUDER (1). *A Textbook of Pharmacology, Therapeutics and Materia Medica*. Adpated to the United States Pharmacopoeia by F. H. Williams. Macmillan, London, 1888.

BRYANT, P. L. (1). 'Chinese Camphor and Camphor Oil.' *CJ*, 1925, **3**, 228.

BUCH, M. (1). 'Die Wotjäken, eine ethnologische Studie.' *ASSF*, 1883, **12**, 465.

BUCK, J. LOSSING (1). *Land Utilisation in China; a Study of 16,786 Farms in 168 Localities, and 38,256 Farm Families in Twenty-two Provinces in China, 1929 to 1933*. Univ. of Nanking, Nanking and Commercial Press, Shanghai, 1937. (Report in the International Research Series of the Institute of Pacific Relations.)

BUCKLAND, A. W. (1). 'Ethnological Hints afforded by the Stimulants in Use among Savages and among the Ancients.' *JRAI*, 1879, **8**, 239.

BUDGE, E. A. WALLIS (4) (tr.). *The Book of the Dead; the Papyrus of Ani in the British Museum*. Brit. Mus., London, 1895.

BUDGE, E. A. WALLIS (5). *First Steps in [the Ancient] Egyptian [Language and Literature]; a Book for Beginners*. Kegan Paul, Trench & Trübner, London, 1923.

BUDGE, E. A. WALLIS (6) (tr.). *Syrian Anatomy, Pathology and Therapeutics; or, 'The Book of Medicines' —the Syriac Text, edited from a Rare Manuscript, with an English Translation*... 2 vols. Oxford, 1913.

BUDGE, E. A. WALLIS (7) (tr.). *The 'Book of the Cave of Treasures'; a History of the Patriarchs and the Kings and their Successors from the Creation to the Crucifixion of Christ, translated from the Syriac text of BM Add. MS. 25875*. Religious Tract Soc. London, 1927.

BUHOT, J. (1). *Arts de la Chine*. Editions du Chène, Paris, 1951.

BÜLFFINGER, G. B. (1). *Specimen Doctrinae Veterum Sinarum Moralis et Politicae; tanquam Exemplum Philosophiae Gentium ad Rem Publicam applicatae; Excerptum Libellis Sinicae Genti Classicis, Confucii sive Dicta sive Facta Complexis*. Frankfurt a/M, 1724.

BULLING, A. (14). 'Archaeological Excavations in China, 1949 to 1971.' *EXPED*, 1972, **14** (no. 4), 2; **15** (no. 1), 22.

BURCKHARDT, T. (1). *Alchemie*. Walter, Freiburg i/B, 1960. Eng. tr. by W. Stoddart: *Alchemy; Science of the Cosmos, Science of the Soul*. Stuart & Watkins, London, 1967.

BURKE, R. B. (1) (tr.). *The 'Opus Majus' of Roger Bacon*. 2 vols. Philadelphia and London, 1928.

BURKILL, I. H. (1). *A Dictionary of the Economic Products of the Malay Peninsula* (with contributions by W. Birtwhistle, F. W. Foxworthy, J. B. Scrivener & J. G. Watson). 2 vols. Crown Agents for the Colonies, London, 1935.

BURKITT, F. C. (1). *The Religion of the Manichees*. Cambridge, 1925.

BURKITT, F. C. (2). *Church and Gnosis*. Cambridge, 1932.

BURNAM, J. M. (1). *A Classical Technology edited from Codex Lucensis 490*. Boston, 1920.

BURNES, A. (1). *Travels into Bokhara*... 3 vols. Murray, London, 1834.

BURTON, A. (1). *Rush-bearing; an Account of the Old Customs of Strewing Rushes, Carrying Rushes to Church, the Rush-cart; Garlands in Churches, Morris-Dancers, the Wakes, and the Rush*. Brook & Chrystal, Manchester, 1891.

BUSHELL, S. W. (2). *Chinese Art*. 2 vols. For Victoria and Albert Museum, HMSO, London, 1909; 2nd ed. 1914.

CABANÈS, A. (1). *Remèdes d'Autrefois*. 2nd ed. Maloine, Paris, 1910.

CALEY, E. R. (1). 'The Leyden Papyrus X; an English Translation with Brief Notes.' *JCE*, 1926, **3**, 1149.

CALEY, E. R. (2). 'The Stockholm Papyrus; an English Translation with Brief Notes.' *JCE*, 1927, **4**, 979.

CALEY, E. R. (3). 'On the Prehistoric Use of Arsenical Copper in the Aegean Region.' *HE*, 1949, **8** (Suppl.), 60 (Commemorative Studies in Honour of Theodore Leslie Shear).

CALEY, E. R. (4). 'The Earliest Use of Nickel Alloys in Coinage.' *NR*, 1943, **1**, 17.

CALEY, E. R. (5). 'Ancient Greek Pigments.' *JCE*, 1946, **23**, 314.

CALEY, E. R. (6). 'Investigations on the Origin and Manufacture of Orichalcum', art. in *Archaeological Chemistry*, ed. M. Levey. Pennsylvania University Press, Philadelphia, Pennsylvania, 1967, p. 59.

CALEY, E. R. & RICHARDS, J. C. (1). *Theophrastus on the Stones*. Columbus, Ohio, 1956.

CALLOWAY, D. H. (1). 'Gas in the Alimentary Canal.' Ch. 137 in *Handbook of Physiology*, sect. 6, 'Alimentary Canal', vol. 5, 'Bile; Digestion; Ruminal Physiology'. Ed. C. F. Code & W. Heidel. Williams & Wilkins, for the American Physiological Society, Washington, D.C. 1968.

CALMET, AUGUSTIN (1). *Dissertations upon the Appearances of Angels, Daemons and Ghosts, and concerning the Vampires of Hungary, Bohemia, Moravia and Silesia*. Cooper, London, 1759, tr. from the French ed. of 1745. Repr. with little change, under the title: *The Phantom World, or the Philosophy of Spirits, Apparitions, etc....*, ed. H. Christmas, 2 vols. London, 1850.

CAMMANN, S. VAN R. (4). 'Archaeological Evidence for Chinese Contacts with India during the Han Dynasty.' *S*, 1956, **5**, 1; abstr. *RBS*, 1959, **2**, no. 320.

CAMMANN, S. VAN R. (5). 'The "Bactrian Nickel Theory".' *AJA*, 1958, **62**, 409. (Commentary on Chêng & Schwitter, 1.)

CAMMANN, S. VAN R. (7). 'The Evolution of Magic Squares in China.' *JAOS*, 1960, **80**, 116.

CAMMANN, S. VAN R. (8). 'Old Chinese Magic Squares.' *S*, 1962, **7**, 14. Abstr. L. Lanciotti, *RBS*, 1969, **8**, no. 837.

CAMMANN, S. VAN R. (9). 'The Magic Square of Three in Old Chinese Philosophy and Religion.' *HOR*, 1961, **1** (no. 1), 37. Crit. J. Needham, *RBS*, 1968, **7**, no. 581.

CAMMANN, S. VAN R. (10). 'A Suggested Origin of the Tibetan Maṇḍala Paintings.' *ARQ*, 1950, **13**, 107.

CAMMANN, S. VAN R. (11). 'On the Renewed Attempt to Revive the "Bactrian Nickel Theory".' *AJA*, 1962, **66**, 92 (rejoinder to Chêng & Schwitter, 2).

CAMMANN, S. VAN R. (12). 'Islamic and Indian Magic Squares.' *HOR*, 1968, **8**, 181, 271.

CAMMANN, S. VAN R. (13). Art. 'Magic Squares' in *EB* 1957 ed., vol. XIV, p. 573.

CAMPBELL, D. (1). *Arabian Medicine and its Influence on the Middle Ages*. 2 vols. (the second a bibliography of Latin MSS translations from Arabic). Kegan Paul, London, 1926.

CARATINI, R. (1). 'Quadrature du Cercle et Quadrature des Lunules en Mésopotamie.' *RAAO*, 1957, **51**, 11.

CARBONELLI, G. (1). *Sulle Fonti Storiche della Chimica e dell'Alchimia in Italia*. Rome, 1925.

CARDEW, S. (1). 'Mining in China in 1952.' *MJ*, 1953, **240**, 390.

CARLID, G. & NORDSTRÖM, J. (1). *Torbern Bergman's Foreign Correspondence* (with brief biography by H. Olsson). Almqvist & Wiksell, Stockholm, 1965.

CARLSON, C. S. (1). 'Extractive and Azeotropic Distillation.' Art. in *Distillation*, ed. A. Weissberger (*Technique of Organic Chemistry*, vol. 4), p. 317. Interscience, New York, 1951.

CARR, A. (1). *The Reptiles*. Time-Life International, Holland, 1963.

CARTER, G. F. (1). 'The Preparation of Ancient Coins for Accurate X-Ray Fluorescence Analysis.' *AMY*, 1964, **7**, 106.

CARTER, T. F. (1). *The Invention of Printing in China and its Spread Westward*. Columbia Univ. Press, New York, 1925, revised ed. 1931. 2nd ed. revised by L. Carrington Goodrich. Ronald, New York, 1955.

CARY, G. (1). *The Medieval Alexander*. Ed. D. J. A. Ross. Cambridge, 1956. (A study of the origins and versions of the Alexander-Romance; important for medieval ideas on flying-machine and diving-bell or bathyscaphe.)

CASAL, U. A. (1). 'The Yamabushi.' *MDGNVO*, 1965, **46**, 1.

CASAL, U. A. (2). 'Incense.' *TAS/J*, 1954 (3rd ser.), **3**, 46.

CASARTELLI, L. C. (1). '[The State of the Dead in] Iranian [and Persian Belief].' *ERE*, vol. XI, p. 847.

CASE, R. E. (1). 'Nickel-containing Coins of Bactria, −235 to −170.' *COCJ*, 1934, **102**, 117.

CASSIANUS, JOHANNES. *Conlationes*, ed. Petschenig. Cf. E. C. S. Gibson tr. (1).

CASSIUS, ANDREAS (the younger) (1). *De Extremo illo et Perfectissimo Naturae Opificio ac Principe Terraenorum Sidere Auro de admiranda ejus Natura...Cogitata Nobilioribus Experimentis Illustrata*. Hamburg, 1685. Cf. Partington (7), vol. 2, p. 371; Ferguson (1), vol. 1, p. 148.

CEDRENUS, GEORGIUS (1). *Historiōn Archomenē* (c. +1059), ed. Bekker (in *Corp. Script. Hist. Byz.* series).

CENNINI, CENNINO (1). *Il Libro dell'Arte*. MS on dyeing and painting, 1437. Eng. trs. C. J. Herringham, Allen & Unwin, London, 1897; D. V. Thompson, Yale Univ. Press. New Haven, Conn. 1933.

CERNY, J. (1). *Egyptian Religion*.

CHADWICK, H. (1) (tr.). *Origen 'Contra Celsum'; Translated with an Introduction and Notes*. Cambridge, 1953.

CHAMBERLAIN, B. H. (1). *Things Japanese*. Murray, London, 2nd ed. 1891; 3rd ed. 1898.

CHAMPOLLION, J. F. (1). '*L'Égypte sous les Pharaons; ou Recherches sur la Geographie, la Religion, la Langue, les Écritures, et l'Histoire de l'Égypte avant l'invasion de Cambyse*. De Bure, Paris, 1814.

CHAMPOLLION, J. F. (2). *Grammaire Égyptien en Écriture Hieroglyphique*. Didot, Paris, 1841.

CHAMPOLLION, J. F. (3). *Dictionnaire Égyptien en Écriture Hieroglyphique.* Didot, Paris, 1841.

CHANG, C. *See* Kung, S. C., Chao, S. W., Pei, Y. T. & Chang, C. (1).

CHANG CHUNG-YUAN (1). 'An Introduction to Taoist Yoga.' *RR*, 1956, **20**, 131.

CHANG HUNG-CHAO (1). *Lapidarium Sinicum; a Study of the Rocks, Fossils and Minerals as known in Chinese Literature* (in Chinese with English summary). Chinese Geological Survey, Peiping, 1927. *MGSC* (ser. B), no. 2.

CHANG HUNG-CHAO (2). 'The Beginning of the Use of Zinc in China.' *BCGS*, 1922, **2** (no. 1/2), 17. Cf. Chang Hung-Chao (3).

CHANG HUNG-CHAO (3). 'New Researches on the Beginning of the Use of Zinc in China.' *BCGS*, 1925, **4** (no. 1), 125. Cf. Chang Hung-Chao (6).

CHANG HUNG-CHAO (4). 'The Origins of the Western Lake at Hangchow.' *BCGS*, 1924, **3** (no. 1), 26. Cf. Chang Hung-Chao (5).

CHANG HSIEN-FÊNG (1). 'A Communist Grows in Struggle.' *CREC*, 1969, **18** (no. 4), 17.

CHANG KUANG-YU & CHANG CHÊNG-YU (1). *Peking Opera Make-up; an Album of Cut-outs.* Foreign Languages Press, Peking, 1959.

CHANG TZU-KUNG (1). 'Taoist Thought and the Development of Science; a Missing Chapter in the History of Science and Culture-Relations.' Unpub. MS., 1945. Now in *MBPB*, 1972, **21** (no. 1), **7** (no. 2), 20.

CHARLES, J. A. (1). 'Early Arsenical Bronzes—a Metallurgical View.' *AJA*, 1967, **71**, 21. A discussion arising from the data in Renfrew (1).

CHARLES, J. A. (2). 'The First Sheffield Plate.' *AQ*, 1968, **42**, 278. With an appendix on the dating of the Minoan bronze dagger with silver-capped copper rivet-heads, by F. H. Stubbings.

CHARLES, J. A. (3). 'Heterogeneity in Metals.' *AMY*, 1973, **15**, 105.

CHARLES, R. H. (1) (tr.). *The 'Book of Enoch', or 'I Enoch', translated from the Editor's Ethiopic Text, and edited with the Introduction, Notes and Indexes of the First Edition, wholly recast, enlarged and re-written, together with a Reprint from the Editor's Text of the Greek Fragments.* Oxford, 1912 (first ed., Oxford, 1893).

CHARLES, R. H. (2) (ed.). *The Ethiopic Version of the 'Book of Enoch', edited from 23 MSS, together with the Fragmentary Greek and Latin Versions.* Oxford, 1906.

CHARLES, R. H. (3). *A Critical History of the Doctrine of a Future Life in Israel, in Judaism, and in Christianity; or, Hebrew, Jewish and Christian Eschatology from pre-Prophetic Times till the Close of the New Testament Canon.* Black, London, 1899. Repr. 1913 (Jowett Lectures, 1898–9).

CHARLES, R. H. (4) 'Gehenna', art. in Hastings, *Dictionary of the Bible*, Clark, Edinburgh, 1899, vol. 2, p. 119.

CHARLES, R. H. (5) (ed.). *The Apocrypha and Pseudepigrapha of the Old Testament in English; with Introductions, and Critical and Explanatory Notes, to the Several Books...* 2 vols. Oxford, 1913 (1 Enoch is in vol. 2).

CHATLEY, H. (1). MS. translation of the astronomical chapter (ch. 3, Thien Wên) of *Huai Nan Tzu.* Unpublished. (Cf. note in *O*, 1952, **72**, 84.)

CHATLEY, H. (37). 'Alchemy in China.' *JALCHS*, 1913, **2**, 33.

CHATTERJI, S. K. (1). 'India and China; Ancient Contacts—What India received from China.' *JRAS/B*, 1959 (n.s.), **1**, 89.

CHATTOPADHYAYA, D. (1). 'Needham on Tantrism and Taoism.' *NAGE*, 1957, **6** (no. 12), 43; 1958, **7** (no. 1), 32.

CHATTOPADHYAYA, D. (2). 'The Material Basis of Idealism.' *NAGE*, 1958, **7** (no. 8), 30.

CHATTOPADHYAYA, D. (3) 'Brahman and Maya.' *ENQ*, 1959, **1** (no. 1), 25.

CHATTOPADHYAYA, D. (4). *Lokāyata, a Study in Ancient Indian Materialism.* People's Publishing House, New Delhi, 1959.

CHAVANNES, E. (14). *Documents sur les Tou-Kiue [Thu-Chüeh] (Turcs) Occidentaux, receuillis et commentés par E. C....* Imp. Acad. Sci., St Petersburg, 1903. Repr. Paris, with the inclusion of the 'Notes Additionelles', n. d.

CHAVANNES, E. (17). 'Notes Additionelles sur les Tou-Kiue [Thu-Chüeh] (Turcs) Occidentaux.' *TP*, 1904, **5**, 1–110, with index and errata for Chavannes (14).

CHAVANNES, E. (19). 'Inscriptions et Pièces de Chancellerie Chinoises de l'Époque Mongole.' *TP*, 1904, **5**, 357–447; 1905, **6**, 1–42; 1908, **9**, 297–428.

CHAVANNES, E. & PELLIOT, P. (1). 'Un Traité Manichéen retrouvé en Chine, traduit et annoté.' *JA*, 1911 (10e sér), **18**, 499; 1913 (11e sér), **1**, 99, 261.

CH'ÊN, JEROME. *See* Chhen Chih-Jang.

CHÊNG, C. F. & SCHWITTER, C. M. (1). 'Nickel in Ancient Bronzes.' *AJA*, 1957, **61**, 351. With an appendix on chemical analysis by X-ray fluorescence by K. G. Carroll.

CHÊNG, C. F. & SCHWITTER, C. M. (2). 'Bactrian Nickel and [the] Chinese [Square] Bamboos.' *AJA*, 1962, **66**, 87 (reply to Cammann, 5).

CHÊNG MAN-CHHING & SMITH, R. W. (1). *Thai-Chi; the 'Supreme Ultimate' Exercise for Health, Sport and Self-Defence*. Weatherhill, Tokyo, 1966.

CHÊNG TÊ-KHUN (2) (tr.). 'Travels of the Emperor Mu.' *JRAS/NCB*, 1933, **64**, 142; 1934, **65**, 128.

CHÊNG TÊ-KHUN (7). 'Yin Yang, Wu Hsing and Han Art.' *HJAS*, 1957, **20**, 162.

CHÊNG TÊ-KHUN (9). *Archaeology in China*.
Vol. 1, *Prehistoric China*. Heffer, Cambridge, 1959.
Vol. 2, *Shang China*. Heffer, Cambridge, 1960.
Vol. 3, *Chou China*, Heffer, Cambridge, and Univ. Press, Toronto, 1963.
Vol. 4, *Han China* (in the press).

CHENG WOU-CHAN. See Shêng Wu-Shan.

CHEO, S. W. *See* Kung, S. C., Chao, S. W., Pei, Y. T. & Chang, C. (1).

CHEYNE, T. K. (1). *The Origin and Religious Content of the Psalter*. Kegan Paul, London, 1891. (Bampton Lectures.)

CHHEN CHIH-JANG (1). *Mao and the Chinese Revolution*. Oxford, 1965. With 37 Poems by Mao Tsê-Tung, translated by Michael Bullock & Chhen Chih-Jang.

CHHEN SHOU-YI (3). *Chinese Literature; a Historical Introduction*. Ronald, New York, 1961.

CHHU TA-KAO (2) (tr.). *Tao Tê Ching, a new translation*. Buddhist Lodge, London, 1937.

CHIKASHIGE, MASUMI (1). *Alchemy and other Chemical Achievements of the Ancient Orient; the Civilisation of Japan and China in Early Times as seen from the Chemical (and Metallurgical) Point of View*. Rokakuho Uchida, Tokyo, 1936. Rev. Tenney L. Davis, *JACS*, 1937, **59**, 952. Cf. Chinese résumé of Chakashige's lectures by Chhen Mêng-Yen, *KHS*, 1920, **5** (no. 3), 262.

CHIU YAN TSZ. See Yang Tzu-Chiu (1).

CHOISY, M. (1). *La Métaphysique des Yogas*. Ed. Mont. Blanc, Geneva, 1948. With an introduction by P. Masson-Oursel.

CHŎN SANGŬN (1). *Science and Technology in Korea; Traditional Instruments and Techniques*. M.I.T. Press, Cambridge, Mass. 1972.

CHOU I-LIANG (1). 'Tantrism in China.' *HJAS*, 1945, **8**, 241.

CHOULANT, L. (1). *History and Bibliography of Anatomic Illustration*. Schuman, New York, 1945, tr. from the German (Weigel, Leipzig, 1852) by M. Frank, with essays by F. H. Garrison, M. Frank, E. C. Streeter & Charles Singer, and a Bibliography of M. Frank by J. C. Bay.

CHOU YI-LIANG. See Chou I-Liang.

CHU HSI-THAO (1). 'The Use of Amalgam as Filling Material in Dentistry in Ancient China.' *CMJ*, 1958, **76**, 553.

CHWOLSON, D. (1). *Die Ssabier und der Ssabismus*. 2 vols. Imp. Acad. Sci., St Petersburg, 1856. (On the culture and religion of the Sabians, Ṣābi, of Harrān, 'pagans' till the +10th century, a people important for the transmission of the Hermetica, and for the history of alchemy, Harrān being a cross-roads of influences from the East and West of the Old World.)

[CIBOT, P. M.] (3). 'Notice du Cong-Fou [*Kung fu*], des Bonzes Tao-sée [*Tao Shih*].' *MCHSAMUC*, 1779, **4**, 441. Often ascribed, as by Dudgeon (1) and others, to J. J. M. Amiot, but considered Cibot's by Pfister (1), p. 896.

[CIBOT, P. M.] (5). 'Notices sur différens Objets; (1) Vin, Eau-de-Vie et Vinaigre de Chine, (2) Raisins secs de Hami, (3) Notices du Royaume de Hami, (4) Rémèdes [*pao-hsing shih, khu chiu*], (5) Teinture chinoise, (6) Abricotier [selection, care of seedlings, and grafting], (7) Armoise.' *MCHSAMUC*, 1780, **5**, 467–518.

CIBOT, P. M. (11). (posthumous). 'Notice sur le Cinabre, le Vif-Argent et le *Ling sha*.' *MCHSAMUC*, 1786, **11**, 304.

CIBOT, P. M. (12) (posthumous). 'Notice sur le Borax.' *MCHSAMUC*, 1786, **11**, 343.

CIBOT, P. M. (13) (posthumous). 'Diverses Remarques sur les Arts-Pratiques en Chine; Ouvrages de Fer, Art de peindre sur les Glaces et sur les Pierres.' *MCHSAMUC*, 1786, **11**, 361.

CIBOT, P. M. (14). 'Notice sur le Lieou-li [*Liu-li*], ou Tuiles Vernissées.' *MCHSAMUC*, 1787, **13**, 396.

[CIBOT, P. M.] (16). 'Notice du Ché-hiang [*Shê hsiang*, musk and the musk deer].' *MCHSAMUC*, 1779, **4**, 493.

[CIBOT, P. M.] (17). 'Quelques Compositions et Recettes pratiquées chez les Chinois ou consignées dans leurs Livres, et que l'Auteur a crues utiles ou inconnues en Europe [on felt, wax, conservation of oranges, bronzing of copper, etc. etc.].' *MCHSAMUC*, 1779, **4**, 484.

CLAPHAM, A. R., TUTIN, T. G. & WARBURG, E. F. (1). *Flora of the British Isles*. 2nd ed. Cambridge, 1962.

CLARK, A. J. (1). *Applied Pharmacology*. 7th ed. Churchill, London, 1942.

CLARK, E. (1). 'Notes on the Progress of Mining in China.' *TAIME*, 1891, **19**, 571. (Contains an account (pp. 587 ff.) of the recovery of silver from argentiferous lead ore, and cupellation by traditional methods, at the mines of Yen-tang Shan.)

CLARK, R. T. RUNDLE (1). *Myth and Symbol in Ancient Egypt*. Thames & Hudson, London, 1959.

CLARK, W. G. & DEL GIUDICE, J. (1) (ed.). *Principles of Psychopharmacology*. Academic Press, New York and London, 1971.

CLARKE, J. & GEIKIE, A. (1). *Physical Science in the Time of Nero, being a Translation of the 'Quaestiones Naturales' of Seneca, with notes by Sir Archibald Geikie*. Macmillan, London, 1910.

CLAUDER, GABRIEL (1). *Inventum Cinnabarinum, hoc est Dissertatio Cinnabari Nativa Hungarica, longa circulatione in majorem efficaciam fixata et exalta*. Jena, 1684.

CLEAVES, F. W. (1). 'The Sino-Mongolian Inscription of +1240 [edict of the empress Törgene, wife of Ogatai Khan (+1186 to +1241) on the cutting of the blocks for the Yuan edition of the *Tao Tsang*].' *HJAS*, 1960, **23**, 62.

CLINE, W. (1). *Mining and Metallurgy in Negro Africa*. Banta, Menasha, Wisconsin, 1937 (mimeographed). (General Studies in Anthropology, no. 5, Iron.)

CLOW, A. & CLOW, NAN L. (1). *The Chemical Revolution; a Contribution to Social Technology*. Batchworth, London, 1952.

CLOW, A. & CLOW, NAN L. (2). 'Vitriol in the Industrial Revolution.' *EHR*, 1945, **15**, 44.

CLULEE, N. H. (1). 'John Dee's Mathematics and the Grading of Compound Qualities.' *AX*, 1971, **18**, 178.

CLYMER, R. SWINBURNE (1). *Alchemy and the Alchemists; giving the Secret of the Philosopher's Stone, the Elixir of Youth, and the Universal Solvent; Also showing that the True Alchemists did not seek to transmute base metals into Gold, but sought the Highest Initiation or the Development of the Spiritual Nature in Man...* 4 vols. Philosophical Publishing Co. Allentown, Pennsylvania, 1907. The first two contain the text of Hitchcock (1), but 'considerably re-written and with much additional information, mis-information and miscellaneous nonsense interpolated' (Cohen, 1).

COGHLAN, H. H. (1). 'Metal Implements and Weapons [in Early Times before the Fall of the Ancient Empires].' Art. in *History of Technology*, ed. C. Singer, E. J. Holmyard & A. R. Hall. Oxford, 1954, vol. 1, p. 600.

COGHLAN, H. H. (3). 'Etruscan and Spanish Swords of Iron.' *SIB*, 1957, **3**, 167.

COGHLAN, H. H. (4). 'A Note upon Iron as a Material for the Celtic Sword.' *SIB*, 1957, **3**, 129.

COGHLAN, H. H. (5). *Notes on Prehistoric and Early Iron in the Old World; including a Metallographic and Metallurgical Examination of specimens selected by the Pitt Rivers Museum, and contributions by I. M. Allen*. Oxford, 1956. (Pitt Rivers Museum Occasional Papers on Technology, no. 8.)

COGHLAN, H. H. (6). 'The Prehistorical Working of Bronze and Arsenical Copper.' *SIB*, 1960, **5**, 145.

COHAUSEN, J. H. (1). *Lebensverlängerung bis auf 115 Jahre durch den Hauch junger Mädchen*. Orig. title: *Der wieder lebende Hermippus, oder curieuse physikalisch-medizinische Abhandlung von der seltener Art, sein Leben durch das Anhauchen Junger- Mägdchen bis auf 115 Jahr zu verlängern, aus einem römischen Denkmal genommen, nun aber mit medicinischen Gründen befestiget, und durch Beweise und Exempel, wie auch mit einer wunderbaren Erfindung aus der philosophischen Scheidekunst erläutert und bestätiget von J. H. C....* Alten Knaben, (Stuttgart?), 1753. Latin ed. Andreae, Frankfurt, 1742. Reprinted in *Der Schatzgräber in den literarischen und bildlichen Seltenheiten, Sonderbarkeiten, etc., hauptsächlich des deutschen Mittelalters*, ed. J. Scheible, vol. 2. Scheible, Stuttgart and Leipzig, 1847. Eng. tr. by J. Campbell, *Hermippus Redivivus; or, the Sage's Triumph over Old Age and the Grave, wherein a Method is laid down for prolonging the Life and Vigour of Man*. London, 1748, repr. 1749, 3rd ed. London, 1771. Cf. Ferguson (1), vol. 1, pp. 168 ff.; Paal (1).

COHEN, I. BERNARD (1). 'Ethan Allen Hitchcock; Soldier–Humanitarian–Scholar; Discoverer of the "True Subject" of the Hermetic Art.' *PAAQS*, 1952, 29.

COLEBY, L. J. M. (1). *The Chemical Studies of P. J. Macquer*. Allen & Unwin, London, 1938.

COLLAS, J. P. L. (3) (posthumous). 'Sur un Sel appellé par les Chinois *Kièn*.' *MCHSAMUC*, 1786, **11**, 315.

COLLAS, J. P. L. (4) (posthumous). 'Extrait d'une Lettre de Feu M. Collas, Missionnaire à Péking, 1e Sur la Chaux Noire de Chine, 2e Sur une Matière appellée Lieou-li [*Liu-li*], qui approche du Verre, 3e Sur une Espèce de Mottes à Brûler.' *MCHSAMUC*, 1786, **11**, 321.

COLLAS, J. P. L. (5) (posthumous). 'Sur le Hoang-fan [*Huang fan*] ou vitriol, le Nao-cha [*Nao sha*] ou Sel ammoniac, et le Hoang-pé-mou [*Huang po mu*].' *MCHSAMUC*, 1786, **11**, 329.

COLLAS, J. P. L. (6) (posthumous). 'Notice sur le Charbon de Terre.' *MCHSAMUC*, 1786, **11**, 334.

COLLAS, J. P. L. (7) (posthumous). 'Notice sur le Cuivre blanc de Chine, sur le Minium et l'Amadou.' *MCHSAMUC*, 1786, **11**, 347.

COLLAS, J. P. L. (8) (posthumous). 'Notice sur un Papier doré sans Or.' *MCHSAMUC*, 1786, **11**, 351.

COLLAS, J. P. L. (9) (posthumous). 'Sur la Quintessence Minérale de M. le Comte de la Garaye.' *MCHSAMUC*, 1786, **11**, 298.

COLLIER, H. B. (1). 'Alchemy in Ancient China.' *CHEMC*, 1952, 41 (101).

COLLINS, W. F. (1). *Mineral Enterprise in China*. Revised edition, Tientsin Press, Tientsin, 1922. With an appendix chapter on 'Mining Legislation and Development' by Ting Wên-Chiang (V. K. Ting), and a memorandum on 'Mining Taxation' by G. G. S. Lindsey.

CONDAMIN, J. & PICON, M. (1). 'The Influence of Corrosion and Diffusion on the Percentage of Silver in Roman Denarii.' *AMY*, 1964, **7**, 98.

DE CONDORCET, A. N. (1). *Esquisse d'un Tableau Historique des Progrès de l'Esprit Humain.* Paris, 1795. Eng. tr. by J. Barraclough, *Sketch for a Historical Picture of the Human Mind.* London, 1955.

CONNELL, K. H. (1). *Irish Peasant Society; Four Historical Essays.* Oxford, 1968. ('Illicit Distillation', pp. 1–50.)

CONRADY, A. (1). 'Indischer Einfluss in China in 4-jahrh. v. Chr.' *ZDMG*, 1906, **60**, 335.

CONRADY, A. (3). 'Zu *Lao-Tze*, cap. 6' (The valley spirit). *AM*, 1932, **7**, 150.

CONRING, H. (1). *De Hermetica Aegyptiorum Vetere et Paracelsicorum Nova Medicina.* Muller & Richter, Helmstadt, 1648.

CONYBEARE, F. C. (1) (tr.). *Philostratus [of Lemnos); the 'Life of Apollonius of Tyana'.* 2 vols. Heinemann, London, 1912, repr. 1948. (Loeb Classics series.)

CONZE, E. (8). 'Buddhism and Gnosis.' *NUM/SHR*, 1967, **12**, 651 (in *Le Origini dello Gnosticismo*).

COOPER, W. C. & SIVIN, N. (1). 'Man as a Medicine; Pharmacological and Ritual Aspects of Traditional Therapy using Drugs derived from the Human Body.' Art. in Nakayama & Sivin (1), p. 203.

CORBIN, H. (1). 'De la Gnose antique à la Gnose Ismaelienne.' Art. in *Atti dello Convegno di Scienze Morali, Storiche e Filologiche*—'Oriente ed Occidente nel Medio Evo'. Acc. Naz. dei Lincei, Rome, 1956 (Atti dei Convegni Alessandro Volta, no. 12), p. 105.

CORDIER, H. (1). *Histoire Générale de la Chine.* 4 vols. Geuthner, Paris, 1920.

CORDIER, H. (13). 'La Suppression de la Compagnie de Jésus et de la Mission de Péking' (1774). *TP* 1916, **17**, 271, 561.

CORDIER, L. (1). 'Observations sur la Lettre de Mons. Abel Rémusat...sur l'Existence de deux Volcans brûlans dans la Tartarie Centrale.' *JA*, 1824, **5**, 47.

CORDIER, V. (1). *Die chemischen Zeichensprache Einst und Jetzt.* Leykam, Graz, 1928.

CORNARO, LUIGI (1). *Discorsi della Vita Sobria.* 1558. Milan, 1627. Eng. tr. by J. Burdell, *The Discourses and Letters of Luigi Cornaro on a Sober and Temperate Life.* New York, 1842. Also nine English translations before 1825, incl. Dublin, 1740.

CORNER, G. W. (1). *The Hormones in Human Reproduction.* Univ. Press, Princeton, N.J. 1946.

CORNFORD, F. M. (2). *The Laws of Motion in Ancient Thought.* Inaug. Lect. Cambridge, 1931.

CORNFORD, F. M. (7). *Plato's Cosmology; the 'Timaeus' translated, with a running commentary.* Routledge & Kegan Paul, London, 1937, repr. 1956.

COVARRUBIAS, M. (2). *The Eagle, the Jaguar, and the Serpent; Indian Art of the Americas—North America (Alaska, Canada, the United States).* Knopf, New York, 1954.

COWDRY, E. V. (1). 'Taoist Ideas of Human Anatomy.' *AMH*, 1925, **3**, 301.

COWIE, A. T. & FOLLEY, S. J. (1). 'Physiology of the Gonadotrophins and the Lactogenic Hormone.' Art. in *The Hormones...*, ed. G. Pincus, K. V. Thimann & E. B. Astwood. Acad. Press, New York, 1948–64, vol. 3, p. 309.

COYAJI, J. C. (2). 'Some Shahnamah Legends and their Chinese Parallels.' *JRAS/B*, 1928 (n.s.), **24**, 177.

COYAJI, J. C. (3). '*Bahram Yasht;* Analogues and Origins.' *JRAS/B*, 1928 (n.s.), **24**, 203.

COYAJI, J. C. (4). 'Astronomy and Astrology in the *Bahram Yasht.*' *JRAS/B*, 1928 (n.s.), **24**, 223.

COYAJI, J. C. (5). 'The *Shahnamah* and the *Fêng Shen Yen I.*' *JRAS/B*, 1930 (n.s.), **26**, 491.

COYAJI, J. C. (6). 'The *Sraosha Yasht* and its Place in the History of Mysticism.' *JRAS/B*, 1932 (n.s.), **28**, 225.

CRAIG, SIR JOHN (1). 'Isaac Newton and the Counterfeiters.' *NRRS*, 1963, **18**, 136.

CRAIG, SIR JOHN (2). 'The Royal Society and the Mint.' *NRRS*, 1964, **19**, 156.

CRAIGIE, W. A. (1). '[The State of the Dead in] Teutonic [Scandinavian, Belief].' *ERE*, vol. xi, p. 851.

CRAVEN, J. B. (1). *Count Michael Maier, Doctor of Philosophy and of Medicine, Alchemist, Rosicrucian, Mystic (+1568 to +1622); his Life and Writings.* Peace, Kirkwall, 1910.

CRAWLEY, A. E. (1). *Dress, Drinks and Drums; Further Studies of Savages and Sex*, ed. T. Besterman. Methuen, London, 1931.

CREEL, H. G. (7). 'What is Taoism?' *JAOS*, 1956, **76**, 139.

CREEL, H. G. (11). *What is Taoism?, and other Studies in Chinese Cultural History.* Univ. Chicago Press, Chicago, 1970.

CRESSEY, G. B. (1). *China's Geographic Foundations; a Survey of the Land and its People.* McGraw-Hill, New York, 1934.

CROCKET, R., SANDISON, R. A. & WALK, A. (1) (ed.). *Hallucinogenic Drugs and their Psychotherapeutic Use.* Lewis, London, 1961. (Proceedings of a Quarterly Meeting of the Royal Medico-Psychological Association.) Contributions by A. Cerletti and others.

CROFFUT, W. A. (1). *Fifty Years in Camp and Field; the Diary of Major-General Ethan Allen Hitchcock, U.S. Army.* Putnam, New York and London, 1909. 'A biography including copious extracts from the diaries but relatively little from the correspondence' (Cohen, 1).

CROLL, OSWALD (1). *Basilica Chymica.* Frankfurt, 1609.
CRONSTEDT, A. F. (1). *An Essay towards a System of Mineralogy.* London, 1770, 2nd ed. 1788 (greatly enlarged and improved by J. H. de Magellan). Tr. by G. von Engeström fom the Swedish *Försök till Mineralogie eller Mineral-Rikets Upställning.* Stockholm, 1758.
CROSLAND, M. P. (1). *Historical Studies in the Language of Chemistry.* Heinemann, London, 1962.
CUMONT, F. (4). *L'Égypte des Astrologues.* Fondation Égyptologique de la Reine Elisabeth, Brussels, 1937.
CUMONT, F. (5) (ed.). *Catalogus Codic. Astrolog. Graecorum.* 12 vols. Lamertin, Brussels, 1929–.
CUMONT, F. (6) (ed.). *Textes et Monuments Figurés relatifs aux Mystères de Mithra.* 2 vols. Lamertin, Brussels, 1899.
CUMONT, F. (7). 'Masque de Jupiter sur un Aigle Éployé [et perché sur le Corps d'un Ouroboros]; Bronze du Musée de Bruxelles.' Art. in *Festschrift f. Otto Benndorf.* Hölder, Vienna, 1898, p. 291.
CUMONT, F. (8). '*La Cosmogonie Manichéenne d'après Théodore bar Khōni* [Bp. of Khalkar in Mesopotamia, c. +600]. Lamertin, Brussels, 1908 (Recherches sur le Manichéisme, no. 1). Cf. Kugener & Cumont, (1, 2).
CUMONT, F. (9). 'La Roue à Puiser les Âmes du Manichéisme.' *RHR/AMG,* 1915, **72**, 384.
CUNNINGHAM, A. (1). 'Coins of Alexander's Successors in the East.' *NC,* 1873 (n.s.), **13**, 186.
CURWEN, M. D. (1) (ed.). *Chemistry and Commerce.* 4 vols. Newnes, London, 1935.
CURZON, G. N. (1). *Persia and the Persian Question.* London, 1892.
CYRIAX, E. F. (1). 'Concerning the Early Literature on Ling's Medical Gymnastics.' *JAN,* 1926, **30**, 225.

DALLY, N. (1). *Cinésiologie, ou Science du Mouvement dans ses Rapports avec l'Éducation, l'Hygiène et la Thérapie; Études Historiques, Théoriques et Pratiques.* Librairie Centrale des Sciences, Paris, 1857.
DALMAN, G. (1). *Arbeit und Sitte in Palästina.*
  Vol. 1 *Jahreslauf und Tageslauf* (in two parts).
  Vol. 2 *Der Ackerbau.*
  Vol. 3 *Von der Ernte zum Mehl* (*Ernten, Dreschen, Worfeln, Sieben, Verwahren, Mahlen*).
  Vol. 4 *Brot, Öl und Wein.*
  Vol. 5. *Webstoff, Spinnen, Weben, Kleidung.*
  Bertelsmann, Gütensloh, 1928– . (Schriften d. deutschen Palästina-Institut, nos. 3, 5, 6, 7, 8; Beiträge z. Forderung christlicher Theologie, ser. 2, Sammlung Wissenschaftlichen Monographien, nos. 14, 17, 27, 29, 33, 36.)
  Vol. 6 *Zeltleben, Vieh- und Milch-wirtschaft, Jagd, Fischfang.*
  Vol 7. *Das Haus, Hühnerzucht, Taubenzucht, Bienenzucht.*
  Olms, Hildesheim, 1964. (Schriften d. deutschen Palästina-Institut, nos. 9, 10; Beiträge z. Forderung Christlicher Theologie. ser. 2, Sammlung Wissenschaftlichen Monographien, nos. 41, 48.)
DANA, E. S. (1). *A Textbook of Mineralogy, with an Extended Treatise on Crystallography and Physical Mineralogy.* 4th ed. rev. & enlarged by W. E. Ford. Wiley, New York, 1949.
DARMSTÄDTER, E. (1) (tr.). *Die Alchemie des Geber* [containing *Summa Perfectionis, Liber de Investigatione Perfectionis, Liber de Inventione Veritatis, Liber Fornacum,* and *Testamentum Geberis,* in German translation]. Springer, Berlin, 1922. Rev. J. Ruska, *ISIS,* 1923, **5**, 451.
DAS, M. N. & GASTAUT, H. (1). 'Variations de l'Activité électrique du Cerveau, du Coeur et des Muscles Squelettiques au cours de la Méditation et de l'Extase Yogique. Art. in *Conditionnement et Reactivité en Électro-encéphalographie,* ed. Fischgold & Gastaut (1). Masson, Paris, 1957, pp. 211 ff.
DASGUPTA, S. N. (3). *A Study of Patañjali.* University Press, Calcutta, 1920.
DASGUPTA, S. N. (4). *Yoga as Philosophy and Religion.* London, 1924.
DAUBRÉE, A. (1). 'La Génération des Minéraux dans la Pratique des Mineurs du Moyen Age d'après le "Bergbüchlein".' *JS,* 1890, 379, 441.
DAUMAS, M. (5). 'La Naissance et le Developpement de la Chimie en Chine.' *SET,* 1949, **6**, 11.
DAVENPORT, JOHN (1), ed. A. H. Walton. *Aphrodisiacs and Love Stimulants, with other chapters on the Secrets of Venus; being the two books by John Davenport entitled 'Aphrodisiacs and Anti-Aphrodisiacs'* [London, pr. pr. 1869, but not issued till 1873] *and 'Curiositates Eroticae Physiologiae; or, Tabooed Subjects Freely Treated'* [London, pr. pr. 1875]; *now for the first time edited, with Introduction and Notes* [and the omission of the essays 'On Generation' and 'On Death' from the second work] *by A.H.W....* Lyle Stuart, New York, 1966.
DAVID, SIR PERCIVAL (3). *Chinese Connoisseurship; the 'Ko Ku Yao Lun'* [+1388], (*Essential Criteria of Antiquities)—a Translation made and edited by Sir P. D....with a Facsimile of the Chinese Text.* Faber & Faber, London, 1971.
DAVIDSON, J. W. (1). *The Island of Formosa, past and present.* Macmillan, London, 1903.
DAVIES, D. (1). 'A Shangri-La in Ecuador.' *NS,* 1973, **57**, 236. On super-centenarians, especially in the Vilcabamba Valley in the Andes.

DAVIES, H. W., HALDANE, J. B. S. & KENNAWAY, E. L. (1). 'Experiments on the Regulation of the Blood's Alkalinity.' *JOP*, 1920, **54**, 32.

DAVIS, TENNEY L. (1). 'Count Michael Maier's Use of the Symbolism of Alchemy.' *JCE*, 1938, **15**, 403.

DAVIS, TENNEY L. (2). 'The Dualistic Cosmogony of Huai Nan Tzu and its Relations to the Background of Chinese and of European Alchemy.' *ISIS*, 1936, **25**, 327.

DAVIS, TENNEY L. (3). 'The Problem of the Origins of Alchemy.' *SM*, 1936, **43**, 551.

DAVIS, TENNEY L. (4). 'The Chinese Beginnings of Alchemy.' *END*, 1943, **2**, 154.

DAVIS, TENNEY L. (5). 'Pictorial Representations of Alchemical Theory.' *ISIS*, 1938, **28**, 73.

DAVIS, TENNEY L. (6). 'The Identity of Chinese and European Alchemical Theory.' *JUS*, 1929, **9**, 7. This paper has not been traceable by us. The reference is given in precise form by Davis & Chhen Kuo-Fu (2), but the journal in question seems to have ceased publication after the end of vol. 8.

DAVIS, TENNEY L. (7). 'Ko Hung (Pao Phu Tzu), Chinese Alchemist of the +4th Century.' *JCE*, 1934, **11**, 517.

DAVIS, TENNEY L. (8). 'The "Mirror of Alchemy" [*Speculum Alchemiae*] of Roger Bacon, translated into English.' *JCE*, 1931, **8**, 1945.

DAVIS, TENNEY L. (9). 'The Emerald Table of Hermes Trismegistus; Three Latin Versions Current among Later Alchemists.' *JCE*, 1926, **3**, 863.

DAVIS, TENNEY L. (10). 'Early Chinese Rockets.' *TR*, 1948, **51**, 101, 120, 122.

DAVIS, TENNEY L. (11). 'Early Pyrotechnics; I, Fire for the Wars of China, II, Evolution of the Gun, III, Early Warfare in Ancient China.' *ORD*, 1948, **33**, 52, 180, 396.

DAVIS, TENNEY L. (12). 'Huang Ti, Legendary Founder of Alchemy.' *JCE*, 1934, **11**, (635).

DAVIS, TENNEY L. (13). 'Liu An, Prince of Huai-Nan.' *JCE*, 1935, **12**, (1).

DAVIS, TENNEY L. (14). 'Wei Po-Yang, Father of Alchemy.' *JCE*, 1935, **12**, (51).

DAVIS, TENNEY L. (15). 'The Cultural Relationships of Explosives.' *NFR*, 1944, **1**, 11.

DAVIS, TENNEY L. (16) (tr.). *Roger Bacon's Letter concerning the Marvellous Power of Art and Nature, and concerning the Nullity of Magic...with Notes and an Account of Bacon's Life and Work.* Chem. Pub. Co., Easton, Pa. 1923. Cf. T. M[oufet] (1659).

DAVIS, TENNEY L. See Wu Lu-Chhiang & Davis.

DAVIS, TENNEY L. & CHAO YÜN-TSHUNG (1). 'An Alchemical Poem by Kao Hsiang-Hsien [+14th cent.]'. *ISIS*, 1939, **30**, 236.

DAVIS, TENNEY L. & CHAO YÜN-TSHUNG (2). 'The Four-hundred Word *Chin Tan* of Chang Po-Tuan [+11th cent.].' *PAAAS*, 1940, **73**, 371.

DAVIS, TENNEY L. & CHAO YÜN-TSHUNG (3). 'Three Alchemical Poems by Chang Po-Tuan.' *PAAAS*, 1940, **73**, 377.

DAVIS, TENNEY L. & CHAO YÜN-TSHUNG (4). 'Shih Hsing-Lin, disciple of Chang Po-Tuan [+11th cent.] and Hsieh Tao-Kuang, disciple of Shih Hsing-Lin.' *PAAAS*, 1940, **73**, 381.

DAVIS, TENNEY L. & CHAO YÜN-TSHUNG (5). 'The Secret Papers in the Jade Box of Chhing-Hua.' *PAAAS*, 1940, **73**, 385.

DAVIS, TENNEY L. & CHAO YÜN-TSHUNG (6). 'A Fifteenth-century Chinese Encyclopaedia of Alchemy.' *PAAAS*, 1940, **73**, 391.

DAVIS, TENNEY L. & CHAO YÜN-TSHUNG (7). 'Chang Po-Tuan of Thien-Thai; his *Wu Chen Phien* (Essay on the Understanding of the Truth); a Contribution to the Study of Chinese Alchemy.' *PAAAS*, 1939, **73**, 97.

DAVIS, TENNEY L. & CHAO YÜN-TSHUNG (8). 'Chang Po-Tuan, Chinese Alchemist of the +11th Century.' *JCE*, 1939 **16**, 53.

DAVIS, TENNEY L. & CHAO YÜN-TSHUNG (9). 'Chao Hsüeh-Min's Outline of Pyrotechnics [*Huo Hsi Lüeh*]; a Contribution to the History of Fireworks.' *PAAAS*, 1943, **75**, 95.

DAVIS, TENNEY, L. & CHHEN KUO-FU (1) (tr.). 'The Inner Chapters of *Pao Phu Tzu*.' *PAAAS*, 1941, **74**, 297. [Transl. of chs. 8 and 11; précis of the remainder.]

DAVIS, TENNEY L. & CHHEN KUO-FU (2). 'Shang Yang Tzu, Taoist writer and commentator on Alchemy.' *HJAS*, 1942, **7**, 126.

DAVIS, TENNEY L. & NAKASEKO ROKURO (1). 'The Tomb of Jofuku [Hsü Fu] or Joshi [Hsü Shih]; the Earliest Alchemist of Historical Record.' *AX*, 1937, **1**, 109, ill. *JCE*, 1947, **24**, (415).

DAVIS, TENENY L. & NAKASEKO ROKURO (2). 'The Jofuku [Hsü Fu] Shrine at Shingu; a Monument of Earliest Alchemy.' *NU*, 1937, **15** (no. 3), 60. 67.

DAVIS, TENNEY L. & WARE, J. R. (1). 'Early Chinese Military Pyrotechnics.' *JCE*, 1947, **24**, 522.

DAVIS, TENNEY L. & WU LU-CHHIANG (1). 'Ko Hung on the Yellow and the White.' *JCE*, 1936, **13**, 215.

DAVIS, TENNEY L. & WU LU-CHHIANG (2). 'Ko Hung on the Gold Medicine.' *JCE*, 1936, **13**, 103.

DAVIS, TENNEY L. & WU LU-CHHIANG (3). 'Thao Hung-Chhing.' *JCE*, 1932, **9**, 859.

DAVIS, TENNEY L. & WU LU-CHHIANG (4). 'Chinese Alchemy.' *SM*, 1930, **31**, 225. Chinese tr. by Chhen Kuo-Fu in *HHS*, 1936, **3**, 771.

DAVIS, TENNEY L. & WU LU-CHHIANG (5). 'The Advice of Wei Po-Yang to the Worker in Alchemy.' *NU*, 1931, **8**, 115, 117. Repr. *DB*, 1935, **8**, 13.

DAVIS, TENNEY L. & WU LU-CHHIANG (6). 'The Pill of Immortality.' *TR*, 1931, **33**, 383.

DAWKINS, J. M. (1). *Zinc and Spelter; Notes on the Early History of Zinc from Babylon to the +18th Century, compiled for the Curious.* Zinc Development Association, London, 1950. Repr. 1956.

DEANE, D. V. (1). 'The Selection of Metals for Modern Coinages.' *CUNOB*, 1969, no. 15. 29.

DEBUS, A. G. (1). *The Chemical Dream of the Renaissance.* Heffer, Cambridge,1968. (Churchill College Overseas Fellowship Lectures, no. 3.)

DEBUS, A. G. (2). Introduction to the facsimile edition of Elias Ashmole's *Theatrum Chemicum Britannicum* (1652). Johnson, New York and London, 1967. (Sources of Science ser., no. 39.)

DEBUS, A. G. (3). 'Alchemy and the Historian of Science.' (An essay-review of C. H. Josten's *Elias Ashmole.*) *HOSC*, 1967, **6**, 128.

DEBUS, A. G. (4). 'The Significance of the History of Early Chemistry.' *JWH*, 1965, **9**, 39.

DEBUS, A. G. (5). 'Robert Fludd and the Circulation of the Blood.' *JHMAS*, 1961, **16**, 374.

DEBUS, A. G. (6). 'Robert Fludd and the Use of Gilbert's *De Magnete* in the Weapon-Salve Controversy.' *JHMAS*, 1964, **19**, 389.

DEBUS, A. G. (7). 'Renaissance Chemistry and the Work of Robert Fludd.' *AX*, 1967, **14**, 42.

DEBUS, A. G. (8). 'The Sun in the Universe of Robert Fludd.' Art. in *Le Soleil à la Renaissance; Sciences et Mythes*, Colloque International, April 1963. Brussels, 1965, p. 261.

DEBUS, A. G. (9). 'The Aerial Nitre in the +16th and early +17th Centuries.' Communication to the Xth International Congress of the History of Science, Ithaca, N.Y. 1962. In *Communications*, p. 835.

DEBUS, A. G. (10). 'The Paracelsian Aerial Nitre.' *ISIS*, 1964, **55**, 43.

DEBUS, A. G. 11). 'Mathematics and Nature in the Chemical Texts of the Renaissance.' *AX*, 1968, **15**, 1.

DEBUS, A. G. (12). 'An Elizabethan History of Medical Chemistry' [R. Bostocke's *Difference between the Auncient Phisicke...and the Latter Phisicke*, +1585]. *ANS*, 1962, **18**, 1.

DEBUS, A. G. (13). 'Solution Analyses Prior to Robert Boyle.' *CHYM*, 1962, **8**, 41.

DEBUS, A. G. (14). 'Fire Analysis and the Elements in the Sixteenth and Seventeenth Centuries.' *ANS*, 1967, **23**, 127.

DEBUS, A. G. (15). 'Sir Thomas Browne and the Study of Colour Indicators.' *AX*, 1962, **10**, 29.

DEBUS, A. G. (16). 'Palissy, Plat, and English Agricultural Chemistry in the Sixteenth and Seventeenth Centuries.' *A/AIHS*, 1968, **21** (nos. 82–3), 67.

DEBUS, A. G. (17). 'Gabriel Plattes and his Chemical Theory of the Formation of the Earth's Crust.' *AX*, 1961, **9**, 162.

DEBUS, A. G. (18). *The English Paracelsians.* Oldbourne, London, 1965; Watts, New York, 1966. Rev. W. Pagel, *HOSC*, 1966, **5**, 100.

DEBUS, A. G. (19). 'The Paracelsian Compromise in Elizabethan England.' *AX*, 1960, **8**, 71.

DEBUS, A. G. (20) (ed.). *Science, Medicine and Society in the Renaissance; Essays to honour Walter Pagel.* 2 vols. Science History Pubs (Neale Watson), New York, 1972.

DEBUS, A. G. (21). 'The Medico-Chemical World of the Paracelsians.' Art. in *Changing Perspectives in the History of Science*, ed. M. Teich & R. Young (1), p. 85.

DEDEKIND, A. (1). *Ein Beitrag zur Purpurkunde.* 1898.

DEGERING, H. (1). 'Ein Alkoholrezept aus dem 8. Jahrhundert.' [The earliest version of the *Mappae Clavicula*, now considered *c*. +820.] *SPAW/PH*, 1917, **36**, 503.

DELZA, S. (1). *Body and Mind in Harmony; Thai Chi Chhüan (Wu Style), an Ancient Chinese Way of Exercise.* McKay, New York, 1961.

DEMIÉVILLE, P. (2). Review of Chang Hung-Chao (1), *Lapidarium Sinicum.* *BEFEO*, 1924, **24**, 276.

DEMIÉVILLE, P. (8). 'Momies d'Extrême-Orient.' *JS*, 1965, 144.

DENIEL, P. L. (1). *Les Boissons Alcooliques Sino-Vietnamiennes.* Inaug. Diss. Bordeaux, 1954. (Printed Dong-nam-a, Saigon).

DENNELL, R. (1). 'The Hardening of Insect Cuticles.' *BR*, 1958, **33**, 178.

DEONNA, W. (2). 'Le Trésor des Fins d'Annecy.' *RA*, 1920 (5ᵉ sér), **11**, 112.

DEONNA, W. (3). 'Ouroboros.' *AA*, 1952, **15**, 163.

DEVASTHALI, G. V. (1). *The Religion and Mythology of the Brāhmaṇas with particular reference to the 'Satapatha-brāhmaṇa'.* Univ. of Poona, Poona, 1965. (Bhau Vishnu Ashtekar Vedic Research series, no. 1.)

DEVÉRIA, G. (1). 'Origine de l'Islamisme en Chine; deux Légendes Mussulmanes Chinoises; Pélérinages de Ma Fou-Tch'ou.' In *Volume Centenaire de l'Ecole des Langues Orientales Vivantes, 1795–1895.* Leroux, Paris, 1895, p. 305.

DEY, K. L. (1). *Indigenous Drugs of India.* 2nd ed. Thacker & Spink, Calcutta, 1896.

DEYSSON, G. (1). 'Hallucinogenic Mushrooms and Psilocybine.' *PRPH*, 1960, **15**, 27.

DIELS, H. (1). *Antike Technik*. Teubner, Leipzig and Berlin, 1914; enlarged 2nd ed. 1920 (rev. B. Laufer, *AAN*, 1917, **19**, 71). Photolitho reproducton, Zeller, Osnabrück, 1965.

DIELS, H. (3). 'Die Entdeckung des Alkohols.' *APAW/PH*, 1913, no. 3, 1–35.

DIELS, H. (4). 'Etymologica' (incl. 2. χυμεία). *ZVSF*, 1916 (NF), **47**, 193.

DIELS, H. (5). *Fragmente der Vorsokratiker*. 7th ed., ed. W. Kranz. 3 vols.

DIERGART, P. (1) (ed.). *Beiträge aus der Geschichte der Chemie dem Gedächtnis v. Georg W. A. Kahlbaum...* Deuticke, Leipzig and Vienna, 1909.

DIHLE, A. (2). 'Neues zur Thomas-Tradition.' *JAC*, 1963, **6**, 54.

DILLENBERGER, J. (1). *Protestant Thought and Natural Science; a Historical Interpretation*. Collins, London, 1961.

DIMIER, L. (1). *L'Art d'Enluminure*. Paris, 1927.

DINDORF, W. *See* John Malala and Syncellos, Georgius.

DIVERS, E. (1). 'The Manufacture of Calomel in Japan.' *JSCI*, 1894, **13**, 108. Errata, p. 473.

DIXON, H. B. F. (1). 'The Chemistry of the Pituitary Hormones.' Art. in *The Hormones...* ed. G. Pincus, K. V. Thimann & E. B. Astwood. Academic Press, New York, 1948–64, vol. 5, p. 1.

DOBBS, B. J. (1). 'Studies in the Natural Philosophy of Sir Kenelm Digby.' *AX*, 1971, **18**, 1.

DODWELL, C. R. (1) (ed. & tr.). *Theophilus [Presbyter]; De Diversis Artibus (The Various Arts)* [probably by Roger of Helmarshausen, c. +1130]. Nelson, London, 1961.

DOHI KEIZO (1). 'Medicine in Ancient Japan; A Study of Some Drugs preserved in the Imperial Treasure House at Nara.' In *Zoku Shōsōin Shiron*, 1932, no. 15, Neiyaku. 1st pagination, p. 113.

DONDAINE, A. (1). 'La Hierarchie Cathare en Italie.' *AFP*, 1950, **20**, 234.

DOOLITTLE, J. (1). *A Vocabulary and Handbook of the Chinese Language*. 2 vols. Rozario & Marcal, Fuchow, 1872.

DORESSE, J. (1). *Les Livres Secrets des Gnostiques d'Égypte*. Plon, Paris, 1958–.
  Vol. 1. *Introduction aux Écrits Gnostiques Coptes découverts à Khénoboskion*.
  Vol. 2. '*L'Évangile selon Thomas*', ou '*Les Paroles Secrètes de Jésus*'.
  Vol. 3. '*Le Livre Secret de Jean*'; '*l'Hypostase des Archontes*' ou '*Livre de Nōréa*'.
  Vol. 4. '*Le Livre Sacré du Grand Esprit Invisible*' ou '*Évangile des Égyptiens*'; '*l'Épître d'Eugnoste le Bienheureux*'; '*La Sagesse de Jésus*'.
  Vol. 5 '*L'Évangile selon Philippe*.'

DORFMAN, R. I. & SHIPLEY, R. A. (1). *Androgens; their Biochemistry, Physiology and Clinical Significance*. Wiley, New York and Chapman & Hall, London, 1956.

DOUGLAS, R. K. (2). *Chinese Stories*. Blackwood, Edinburgh and London, 1883. (Collection of translations previously published in *Blackwood's Magazine*.)

DOUTHWAITE, A. W. (1). 'Analyses of Chinese Inorganic Drugs.' *CMJ*, 1890, **3**, 53.

DOZY, R. P. A. & ENGELMANN, W. H. (1). *Glossaire des Mots Espagnols et Portugais dérivés de l'Arabe*. 2nd ed. Brill, Leiden, 1869.

DOZY, R. P. A. & DE GOEJE, M. J. (2). *Nouveaux Documents pour l'Étude de la Religion des Ḥarrāniens*. Actes du 6e Congr. Internat. des Orientalistes, Leiden, 1883. 1885, vol. 2, pp. 281ff., 341 ff.

DRAKE, N. F. (1). 'The Coal Fields of North-East China.' *TAIME*, 1901, **31**, 492, 1008.

DRAKE, N. F. (2). 'The Coal Fields around Tsê-Chou, Shansi.' *TAIME*, 1900, **30**, 261.

DRONKE, P. (1). 'L'Amor che Move il Sole e l'Altre Stelle.' *STM*, 1965 (3ᵃ ser.), **6**, 389.

DRONKE, P. (2). 'New Approaches to the School of Chartres.' *AEM*, 1969, **6**, 117.

DRUCE, G. C. (1). 'The Ant-Lion.' *ANTJ*, 1923, **3**, 347.

DU, Y., JIANG, R. & TSOU, C. (1). See Tu Yü-Tshang, Chiang Jung-Chhing & Tsou Chhêng-Lu (1).

DUBLER, C. E. (1). *La 'Materia Medica' de Dioscorides; Transmission Medieval y Renacentista*. 5 vols. Barcelona, 1955.

DUBS, H. H. (4). 'An Ancient Chinese Stock of Gold [Wang Mang's Treasury].' *JEH*, 1942, **2**, 36.

DUBS, H. H. (5). 'The Beginnings of Alchemy.' *ISIS*, 1947, **38**, 62.

DUBS, H. H. (34). 'The Origin of Alchemy.' *AX*, 1961, **9**, 23. Crit. abstr. J. Needham, *RBS*, 1968, **7**, no. 755.

DUCKWORTH, C. W. (1). 'The Discovery of Oxygen.' *CN*, 1886, **53**, 250.

DUDGEON, J. (1). 'Kung-Fu, or Medical Gymnastics.' *JPOS*, 1895, **3** (no. 4), 341–565.

DUDGEON, J. (2). 'The Beverages of the Chinese' (on tea and wine). *JPOS*, 1895, **3**, 275.

DUDGEON, J. (4). '[Glossary of Chinese] Photographic Terms', in Doolittle (1), vol. 2, p. 518.

DUHR, J. (1). *Un Jésuite en Chine, Adam Schall*. Desclée de Brouwer, Paris, 1936. Engl. adaptation by R. Attwater, *Adam Schall, a Jesuit at the Court of China, 1592 to 1666*. Geoffrey Chapman, London, 1963. Not very reliable sinologically.

DUNCAN, A. M. (1). 'The Functions of Affinity Tables and Lavoisier's List of Elements.' *AX*, 1970, **17**, 28.

DUNCAN, A. M. (2). 'Some Theoretical Aspects of Eighteenth-Century Tables of Affinity.' *ANS*, 1962, **18**, 177, 217.

DUNCAN, E. H. (1). 'Jonson's "Alchemist" and the Literature of Alchemy.' *PMLA*, 1946, **61**, 699.

DUNLOP, D. M. (5). 'Sources of Silver and Gold in Islam according to al-Hamdānī (+10th century).' *SI*, 1957, **8**, 29.

DUNLOP, D. M. (6). *Arab Civilisation to A.D. 1500*. Longman, London and Librairie du Liban, Beirut, 1971.

DUNLOP, D. M. (7). *Arabic Science in the West*. Pakistan Historical Soc., Karachi, 1966. (Pakistan Historical Society Pubs. no. 35.)

DUNLOP, D. M. (8). 'Theodoretus-Adhrīṭūs.' Communication to the 26th International Congress of Orientalists, New Delhi, 1964. Summaries of papers, p. 328.

DÜNTZER, H. (1). *Life of Goethe*. 2 vols. Macmillan, London, 1883.

DÜRING, H. I. (1). 'Aristotle's Chemical Treatise, *Meteorologica* Bk. IV, with Introduction and Commentary.' *GHA*, 1944 (no. 2), 1–112. Sep. pub., Elander, Goteborg, 1944.

DURRANT, P. J. (1). *General and Inorganic Chemistry*. 2nd ed. repr. Longmans Green, London, 1956.

DUVEEN, D. I. & WILLEMART, A. (1). 'Some +17th-Century Chemists and Alchemists of Lorraine.' *CHYM*, 1949, **2**, 111.

DUYVENDAK, J. J. L. (18) (tr.). '*Tao Tê Ching*', the Book of the Way and its Virtue. Murray, London, 1954 (Wisdom of the East Series). Crit. revs. P. Demiéville, *TP*, 1954, **43**, 95; D. Bodde, *JAOS*, 1954, **74**, 211.

DUYVENDAK, J. J. L. (20). 'A Chinese *Divina Commedia*.' *TP*, 1952, **41**, 255. (Also sep. pub. Brill, Leiden, 1952.)

DYSON, G. M. (1). 'Antimony in Pharmacy and Chemistry; I, History and Occurrence of the Element; II, The Metal and its Inorganic Compounds; III, The Organic Antimony Compounds in Therapy. *PJ*, 1928, **121** (4th ser. **67**), 397, 520.

EBELING, E. (1). 'Mittelassyrische Rezepte zur Bereitung (Herstellung) von wohlriechenden Salben.' *ORR*, 1948 (n.s.) **17**, 129, 299; 1949, **18**, 404; 1950, **19**, 265.

ECKERMANN, J. P. (1). *Gespräche mit Goethe*. 3 vols. Vols. 1 and 2, Leipzig, 1836. Vol. 3, Magdeburg, 1848. Eng. tr. 2 vols. by J. Oxenford, London, 1850. Abridged ed. *Conversations of Goethe with Eckermann*, Dent, London, 1930. Ed. J. K. Moorhead, with introduction by Havelock Ellis.

EDKINS, J. (17). 'Phases in the Development of Taoism.' *JRAS/NCB*, 1855 (1st ser.), **5**, 83.

EDKINS, J. (18). 'Distillation in China.' *CR*, 1877, **6**, 211.

EFRON, D. H., HOLMSTEDT, BO & KLINE, N. S. (1) (ed.). *The Ethno-pharmacological Search for Psychoactive Drugs*. Washington, D.C. 1967. (Public Health Service Pub. no. 1645.) Proceedings of a Symposium, San Francisco, 1967.

EGERTON, F. N. (1). 'The Longevity of the Patriarchs; a Topic in the History of Demography.' *JHI*, 1966, **27**, 575.

EGGELING, J. (1) (tr.). The '*Satapatha-brāhmaṇa*' according to the Text of the Mādhyandina School. 5 vols. Oxford, 1882–1900 (*SBE*, nos. 12, 26, 41, 43, 44). Vol. 1 repr. Motilal Banarsidass, Delhi, 1963.

EGLOFF, G. & LOWRY, C. D. (1). 'Distillation as an Alchemical Art.' *JCE*, 1930, **7**, 2063.

EICHHOLZ, D. E. (1). 'Aristotle's Theory of the Formation of Metals and Minerals.' *CQ*, 1949, **43**, 141.

EICHHOLZ, D. E. (2). *Theophrastus '*De Lapidibus*'*. Oxford, 1964.

EICHHORN, W. (6). 'Bemerkung z. Einführung des Zölibats für Taoisten.' *RSO*, 1955, **30**, 297.

EICHHORN, W. (11) (tr.). The *Fei-Yen Wai Chuan*, with some notes on the *Fei-Yen Pieh Chuan*. Art. in *Eduard Erkes in Memoriam 1891–1958*, ed. J. Schubert. Leipzig, 1962. Abstr. *TP*, 1963, **50**, 285.

EISLER, R. (4). 'l'Origine Babylonienne de l'Alchimie; à propos de la Découverte Récente de Récettes Chimiques sur Tablettes Cunéiformes.' *RSH*, 1926, **41**, 5. Also *CHZ*, 1926 (nos. 83 and 86); *ZASS*, 1926, 1.

ELIADE, MIRCEA (1). *Le Mythe de l'Eternel Retour; Archétypes et Répétition*. Gallimard, Paris, 1949. Eng. tr. by W. R. Trask, *The Myth of the Eternal Return*. Routledge & Kegan Paul, London, 1955.

ELIADE, MIRCEA (4). 'Metallurgy, Magic and Alchemy.' *Z*, 1938, **1**, 85.

ELIADE, MIRCEA (5). *Forgerons et Alchimistes*. Flammarion, Paris, 1956. Eng. tr. S. Corrin, *The Forge and the Crucible*. Harper, New York, 1962. Rev. G. H[eym], *AX*, 1957, **6**, 109.

ELIADE, MIRCEA (6). *Le Yoga, Immortalité et Liberté*. Payot, Paris, 1954. Eng. tr. by W. R. Trask. Pantheon, New York, 1958.

ELIADE, MIRCEA (7). *Imgaes and Symbols; Studies in Religious Symbolism*. Tr. from the French (Gallimard, Paris, 1952) by P. Mairet. Harvill, London, 1961.

ELIADE, M. (8). 'The Forge and the Crucible: a Postscript.' *HOR*, 1968, **8**, 74–88.

ELLINGER, T. U. H. (1). *Hippocrates on Intercourse and Pregnancy; an English Translation of '*On Semen*' and '*On the Development of the Child*'*. With introd. and notes by A. F. Guttmacher. Schuman, New York, 1952.

ELLIS, G. W. (1). 'A Vacuum Distillation Apparatus.' *CHIND*, 1934, **12**, 77 (*JSCI*, **53**).

ELLIS, W. (1). *History of Madagascar*. 2 vols. Fisher, London and Paris, 1838.

VON ENGESTRÖM, G. (1). 'Pak-fong, a White Chinese Metal' (in Swedish). *KSVA/H*, 1776, **37**, 35.

VON ENGESTRÖM, G. (2). 'Försök på Förnt omtalle Salt eller *Kien* [*chien*].' *KSVA/H*, 1772, **33**, 172. Cf. Abrahamsohn (1).

D'ENTRECOLLES, F. X. (1). *Lettre au Père Duhalde* (on alchemy and various Chinese discoveries in the arts and sciences, porcelain, artificial pearls and magnetic phenomena) dated 4 Nov. 1734. *LEC*, 1781, vol. 22, pp. 91 ff.

D'ENTRECOLLES, F. X. (2). *Lettre au Père Duhalde* (on botanical subjects, fruits and trees, including the persimmon and the lichi; on medicinal preparations isolated from human urine; on the use of the magnet in medicine; on the feathery substance of willow seeds; on camphor and its sublimation; and on remedies for night-blindness) dated 8 Oct. 1736. *LEC*, 1781, vol. 22, pp. 193 ff.

ST EPHRAIM OF SYRIA [d. +373]. *Discourses to Hypatius* [against the Theology of Mani, Marcion and Bardaisan]. See Mitchell, C. W. (1).

EPHRAIM, F. (1). *A Textbook of Inorganic Chemistry*. Eng. tr. P. C. L. Thorne. Gurney & Jackson, London, 1926.

ERCKER, L. (1). *Beschreibung Allefürnemsten Mineralischem Ertzt und Berckwercks Arten*... Prague, 1574. 2nd ed. Frankfurt, 1580. Eng. tr. by Sir John Pettus, as *Fleta Minor, or, the Laws of Art and Nature, in Knowing, Judging, Assaying, Fining, Refining and Inlarging the Bodies of confin'd Metals*... Dawks, London, 1683. See Sisco & Smith (1); Partington (7), vol. 2, pp. 104 ff.

ERKES, E. (1) (tr.). 'Das Weltbild d. *Huai Nan Tzu*.' (Transl. of ch. 4.) *OAZ*, 1918, **5**, 27.

ERMAN, A. & GRAPOW, H. (1). *Wörterbuch d. Aegyptische Sprache*. 7 vols. (With *Belegstellen*, 5 vols. as supplement.) Hinrichs, Leipzig, 1926–.

ERMAN, A. & GRAPOW, H. (2). *Aegyptisches Handwörterbuch*. Reuther & Reichard, Berlin, 1921.

ERMAN, A. & RANKE, H. (1). *Aegypten und aegyptisches Leben in Altertum*. Tübingen, 1923.

ESSIG, E. O. (1). *A College Entomology*. Macmillan, New York, 1942.

ESTIENNE, H. (1) (Henricus Stephanus). *Thesaurus Graecae Linguae*. Geneva, 1572; re-ed. Hase, de Sinner & Fix, 8 vols. Didot, Paris, 1831–65.

ETHÉ, H. (1) (tr.). *Zakarīya ibn Muḥ. ibn Maḥmūd al-Qazwīnī's Kosmographie; Die Wunder der Schöpfung* [c. +1275]. Fues (Reisland), Leipzig, 1868. With notes by H. L. Fleischer. Part I only; no more published.

EUGSTER, C. H. (1). 'Brève Revue d'Ensemble sur la Chimie de la Muscarine.' *RMY*, 1959, **24**, 1.

EUONYMUS PHILIATER. See Gesner, Conrad.

EVOLA, J. (G. C. E.) (1). *La Tradizione Ermetica*. Bari, 1931. 2nd ed. 1948.

EVOLA, J. (G. C. E.) (2). *Lo Yoga della Potenza, saggio sui Tantra*. Bocca, Milan, 1949. Orig. pub. as *l'Uomo come Potenza*.

EVOLA, J. (G. C. E.) (3). *Metafisica del Sesso*. Atanòr, Rome, 1958.

EWING, A. H. (1). *The Hindu Conception of the Functions of Breath; a Study in early Indian Psychophysics*. Inaug. Diss. Johns Hopkins University. Baltimore, 1901; and in *JAOS*, 1901, **22** (no. 2).

FABRE, M. (1). *Pékin, ses Palais, ses Temples, et ses Environs*. Librairie Française, Tientsin, 1937.

FABRICIUS, J. A. (1). *Bibliotheca Graeca*... Edition of G. C. Harles, 12 vols. Bohn, Hamburg, 1808.

FABRICIUS, J. A. (2). *Codex Pseudepigraphicus Veteris Testamenti, Collectus, Castigatus, Testimoniisque Censuris et Animadversionibus Illustratus*. 3 vols. Felginer & Bohn, Hamburg, 1722–41.

FABRICIUS, J. A. (3). *Codex Apocryphus Novi Testamenti, Collectus, Castigatus, Testimoniisque Censuris et Animadversionibus Illustratus*. 3 vols. in 4. Schiller & Kisner, Hamburg, 1703–19.

FARABEE, W. C. (1). 'A Golden Hoard from Ecuador.' *MUJ*, 1912.

FARABEE, W. C. (2). 'The Use of Metals in Prehistoric America.' *MUJ*, 1921.

FARNWORTH, M., SMITH, C. S. & RODDA, J. L. (1). 'Metallographic Examination of a Sample of Metallic Zinc from Ancient Athens.' *HE*, 1949, **8** (Suppl.) 126. (Commemorative Studies in Honour of Theodore Leslie Shear.)

FEDCHINA, V. N. (1). 'The +13th-century Chinese Traveller, [Chhiu] Chhang-Chhun' (in Russian), in *Iz Istorii Nauki i Tekhniki Kitaya* (Essays in the History of Science and Technology in China), p. 172. Acad. Sci. Moscow, 1955.

FEHL, N. E. (1). 'Notes on the Lü Hsing [chapter of the *Shu Ching*]; proposing a Documentary Theory.' *CCJ*, 1969, **9** (no. 1), 10.

FEIFEL, E. (1) (tr.). '*Pao Phu Tzu (Nei Phien)*, chs. 1–3.' *MS*, 1941, **6**, 113.

FEIFEL, E. (2) (tr.). '*Pao Phu Tzu (Nei Phien)*, ch. 4.' *MS*, 1944, **9**, 1.

FEIFEL, E. (3) (tr.). '*Pao Phu Tzu (Nei Phien)*, ch. 11, Translated and Annotated.' *MS*, 1946, **11** (no. 1), 1.

FEISENBERGER, H. A. (1). 'The [Personal] Libraries of Newton, Hooke and Boyle.' *NRRS*, 1966, **21**, 42.

FÊNG CHIA-LO & COLLIER, H. B. (1). 'A Sung-Dynasty Alchemical Treatise; the "Outline of Alchemical Preparations" [*Tan Fang Chien Yuan*], by Tuku Thao [+10th cent.].' *JWCBRS*, 1937, **9**, 199.

FÊNG HAN-CHI (H. Y. Fêng) & SHRYOCK, J. K. (2). 'The Black Magic in China known as *Ku*.' *JAOS*, 1935, **65**, 1. Sep. pub. Amer. Oriental Soc. Offprint Ser. no. 5.

FERCHL, F. & SÜSSENGUTH, A. (1). *A Pictorial History of Chemistry*. Heinemann, London, 1939.

FERDY, H. (1). *Zur Verhütung der Conception*. 1900.

FERGUSON, JOHN (1). *Bibliotheca Chemica; a Catalogue of the Alchemical, Chemical and Pharmaceutical Books in the Collection of the late James Young of Kelly and Durris*... 2 vols. Maclehose, Glasgow, 1906.

FERGUSON, JOHN (2). *Bibliographical Notes on Histories of Inventions and Books of Secrets*. 2 vols. Glasgow, 1898; repr. Holland Press, London, 1959. (Papers collected from *TGAS*.)

FERGUSON, JOHN (3). 'The "Marrow of Alchemy" [1654–5].' *JALCHS*, 1915, **3**, 106.

FERRAND, G. (1). *Relations de Voyages et Textes Géographiques Arabes, Persans et Turcs relatifs à l'Extrême Orient, du 8ᵉ au 18ᵉ siècles, traduits, revus et annotés etc*. 2 vols. Leroux, Paris, 1913.

FERRAND, G. (2) (tr.). *Voyage du marchand Sulaymān en Inde et en Chine redigé en +851; suivi de remarques par Abū Zayd Haṣan (vers +916)*. Bossard, Paris, 1922.

FESTER, G. (1). *Die Entwicklung der chemischen Technik, bis zu den Anfängen der Grossindustrie*. Berlin, 1923. Repr. Sändig, Wiesbaden, 1969.

FESTUGIÈRE, A. J. (1). *La Révélation d'Hermès Trismégiste*, I. *L'Astrologie et les Sciences Occultes*. Gabalda, Paris, 1944. Rev. J. Filliozat, *JA*, 1944, **234**, 349.

FESTUGIÈRE, A. J. (2). 'L'Hermétisme.' *KHVL*, 1948, no. 1, 1–58.

FIERZ-DAVID, H. E. (1). *Die Entwicklungsgeschichte der Chemie*. Birkhauser, Basel, 1945. (Wissenschaft und Kultur ser., no. 2.) Crit. E. J. Holmyard, *N*, 1946, **158**, 643.

FIESER, L. F. & FIESER, M. (1). *Organic Chemistry*. Reinhold, New York; Chapman & Hall, London, 1956.

FIGUIER, L. (1). *l'Alchimie et les Alchimistes; ou, Essai Historique et Critique sur la Philosophie Hermétique*. Lecou, Paris, 1854. 2nd ed. Hachette, Paris, 1856. 3rd ed. 1860.

FIGUROVSKY, N. A. (1). 'Chemistry in Ancient China, and its Influence on the Progress of Chemical Knowledge in other Countries' (in Russian). Art. in *Iz Istorii Nauki i Tekhniki Kitaya*. Moscow, 1955, p. 110.

FILLIOZAT, J. (1). *La Doctrine Classique de la Médécine Indienne*. Imp. Nat., CNRS and Geuthner, Paris, 1949.

FILLIOZAT, J. (2). 'Les Origines d'une Technique Mystique Indienne.' *RP*, 1946, **136**, 208.

FILLIOZAT, J. (3). 'Taoisme et Yoga.' *DVN*, 1949, **3**, 1.

FILLIOZAT, J. (5). Review of Festugière (1). *JA*, 1944, **234**, 349.

FILLIOZAT, J. (6). 'La Doctrine des Brahmanes d'après St Hippolyte.' *JA*, 1945, **234**, 451; *RHR/AMG*, 1945, **128**, 59.

FILLIOZAT, J. (7). 'L'Inde et les Échanges Scientifiques dans l'Antiquité.' *JWH*, 1953, **1**, 353.

FILLIOZAT, J. (10). 'Al-Bīrūnī et l'Alchimie Indienne.' Art. in *Al-Bīrūnī Commemoration Volume*. Iran Society, Calcutta, 1958, p. 101.

FILLIOZAT, J. (11). Review of P. C. Ray (1) revised edition. *ISIS*, 1958, **49**, 362.

FILLIOZAT, J. (13). 'Les Limites des Pouvoirs Humains dans l'Inde.' Art. in *Les Limites de l'Humain*. Études Carmelitaines, Paris, 1953, p. 23.

FISCHER, OTTO (1). *Die Kunst Indiens, Chinas und Japans*. Propylaea, Berlin, 1928.

FISCHGOLD, H. & GASTAUT, H. (1) (ed.). *Conditionnement et Reactivité en Électro-encephalographie*. Masson, Paris, 1957. For the Féderation Internationale d'Électro-encéphalographie et de Neuro-physiologie Clinique, Report of 5th Colloquium, Marseilles, 1955 (*Electro-encephalography and Clinical Neurophysiology*, Supplement no. 6).

FLIGHT, W. (1). 'On the Chemical Compositon of a Bactrian Coin.' *NC*, 1868 (n.s.), **8**, 305.

FLIGHT, W. (2). 'Contributions to our Knowledge of the Composition of Alloys and Metal-Work, for the most part Ancient.' *JCS*, 1882, **41**, 134.

FLORKIN, M. (1). *A History of Biochemistry. Pt. I, Proto-Biochemistry; Pt. II, From Proto-Biochemistry to Biochemistry*. Vol. 30 of *Comprehensive Biochemistry*, ed. M. Florkin & E. H. Stotz. Elsevier, Amsterdam, London and New York, 1972.

FLUDD, ROBERT (3). *Tractatus Theologo-Philosophicus, in Libros Tres distributus; quorum I, De Vita, II, De Morte, III, De Resurrectione; Cui inseruntur nonnulla Sapientiae Veteris...Fragmenta;...collecta Fratribusq a Cruce Rosea dictis dedicata à Rudolfo Otreb Brittano*. Oppenheim, 1617.

FLÜGEL, G. (1) (ed. & tr.). *The 'Fihrist al-'Ulūm' (Index of the Sciences)* [by Abū'l-Faraj ibn abū-Ya'qūb al-Nadīm]. 2 vols. Leipzig, 1871–2.

FLÜGEL, G. (2) (tr.). *Lexicon Bibliographicum et Encyclopaedicum, a Mustafa ben Abdallah Katib Jelebi dicto et nomine Haji Khalfa celebrato compositum*...(the *Kashf al-Ẓunūn* (Discovery of the Thoughts of Muṣṭafā ibn 'Abdallāh Haji Khalfa, or Ḥajji Khalīfa, +17th-century Turkish (bibliographer). 7 vols. Bentley (for the Or. Tr. Fund Gt. Br. & Ireland), London and Leipzig, 1835–58.

FOHNAHN, A. (1). 'New Chemical Terminology in Chinese.' *JAN*, 1927, **31**, 395.

FOLEY, M. G. (1) (tr.). *Luigi Galvani: 'Commentary on the Effects of Electricity on Muscular Motion'*, *translated into English*...[from *De Viribus Electricitatis in Motu Musculari Commentarius*, Bologna, +1791]; *with Notes and a Critical Introduction by I. B. Cohen, together with a Facsimile*...*and a Bibliography of the Editions and Translations of Galvani's Book prepared by J. F. Fulton & M. E. Stanton*. Burndy Library, Norwalk, Conn. U.S.A. 1954. (Burndy Library Publications, no.10.)

FORBES, R. J. (3). *Metallurgy in Antiquity; a Notebook for Archaeologists and Technologists*. Brill, Leiden, 1950 (in press since 1942). Rev. V. G. Childe, *A/AIHS*, 1951, **4**, 829.

[FORBES, R. J.] (4*a*). *Histoire des Bitumes, des Époques les plus Reculées jusqu'à l'an 1800*. Shell, Leiden, n.d.

FORBES, R. J. (4*b*). *Bitumen and Petroleum in Antiquity*. Brill, Leiden, 1936.

FORBES, R. J. (7). 'Extracting, Smelting and Alloying [in Early Times before the Fall of the Ancient Empires].' Art. in A *History of Technology*, ed. C. Singer, E. J. Holmyard & A. R. Hall. vol. 1, p.572. Oxford, 1954.

FORBES, R. J. (8). 'Metallurgy [in the Mediterranean Civilisations and the Middle Ages].' In *A History of Technology*, ed. C. Singer *et al.* vol. 2, p. 41. Oxford, 1956.

FORBES, R. J. (9). *A Short History of the Art of Distillation*. Brill, Leiden, 1948.

FORBES, R. J. (10). *Studies in Ancient Technology*. Vol. 1, *Bitumen and Petroleum in Antiquity; The Origin of Alchemy; Water Supply*. Brill, Leiden, 1955. (Crit. Lynn White, *ISIS*, 1957, **48**, 77.)

FORBES, R. J. (16). 'Chemical, Culinary and Cosmetic Arts' [in early times to the Fall of the Ancient Empires]. Art. in *A History of Technology*, ed. C. Singer *et al.* Vol. 1, p. 238. Oxford, 1954.

FORBES, R. J. (20). *Studies in Early Petroleum History*. Brill, Leiden, 1958.

FORBES, R. J. (21). *More Studies in Early Petroleum History*. Brill, Leiden, 1959.

FORBES, R. J. (26). 'Was Newton an Alchemist?' *CHYM*, 1949, **2**, 27.

FORBES, R. J. (27). *Studies in Ancient Technology*. Vol. 7, *Ancient Geology; Ancient Mining and Quarrying*; *Ancient Mining Techniques*. Brill, Leiden, 1963.

FORBES, R. J. (28). *Studies in Ancient Technology*. Vol. 8, *Synopsis of Early Metallurgy; Physico-Chemical Archaeological Techniques; Tools and Methods; Evolution of the Smith (Social and Sacred Status); Gold; Silver and Lead; Zinc and Brass*. Brill, Leiden, 1964. A revised version of Forbes (3).

FORBES, R. J. (29). *Studies in Ancient Technology*. Vol. 9, *Copper; Tin; Bronze; Antimony; Arsenic; Early Story of Iron*. Brill, Leiden, 1964. A revised version of Forbes (3).

FORBES, R. J. (30). *La Destillation à travers les Ages*. Soc. Belge pour l'Étude du Pétrole, Brussels, 1947.

FORBES, R. J. (31). 'On the Origin of Alchemy.' *CHYM*, 1953, **4**, 1.

FORBES, R. J. (32). Art. 'Chemie' in *Real-Lexikon f. Antike und Christentum*, ed. T. Klauser, 1950–3, vol. 2, p. 1061.

FORBES, T. R. (1). 'A[rnold] A[dolf] Berthold [1803–61] and the First Endocrine Experiment; some Speculations as to its Origin.' *BIHM*, 1949, **23**, 263.

FORKE, A. (3) (tr.). *Me Ti [Mo Ti] des Sozialethikers und seiner Schüler philosophische Werke*. Berlin, 1922. (*MSOS*, Beibände, **23–25**).

FORKE, A. (4) (tr.). '*Lun-Hêng*', *Philosophical Essays of Wang Chhung*. Vol. 1, 1907. Kelly & Walsh, Shanghai; Luzac, London; Harrassowitz, Leipzig. Vol. 2, 1911 (with the addition of Reimer, Berlin). (*MSOS*, Beibände, **10** and **14**.) Photolitho Re-issue, Paragon, New York, 1962. Crit. P. Pelliot, *JA*, 1912 (10e sér.), **20**, 156.

FORKE, A. (9). *Geschichte d. neueren chinesischen Philosophie* (i.e. from the beginning of the Sung to modern times). De Gruyter, Hamburg, 1938. (Hansische Univ. Abhdl. a. d. Geb. d. Auslands-kunde, no. 46 (ser. B, no. 25).)

FORKE, A. (12). *Geschichte d. mittelälterlichen chinesischen Philosophie* (i.e. from the beginning of the Former Han to the end of the Wu Tai). De Gruyter, Hamburg, 1934. (Hamburg. Univ. Abhdl. a. d. Geb. d. Auslandskunde, no. 41 (ser. B, no. 21).)

FORKE, A. (13). *Geschichte d. alten chinesischen Philosophie* (i.e. from antiquity to the beginning of the Former Han). De Gruyter, Hamburg, 1927. (Hamburg. Univ. Abhdl. a. d. Geb. d. Auslands-kunde, no. 25 (ser. B, no. 14).)

FORKE, A. (15). 'On Some Implements mentioned by Wang Chhung' (1. Fans, 2. Chopsticks, 3. Burning Glasses and Moon Mirrors). Appendix III to Forke (4).

FORKE, A. (20). 'Ko Hung der Philosoph und Alchymist.' *AGP*, 1932, **41**, 115. Largely incorporated in (12), pp. 204 ff.

FÖRSTER, E. (1). *Roger Bacon's 'De Retardandis Senectutis Accidentibus et de Sensibus Conservandis' und Arnald von Villanova's 'De Conservanda Juventutis et Retardanda Senectute'*. Inaug. Diss. Leipzig, 1924.

FOWLER, A. M. (1). 'A Note on ἄμβροτος.' *CP*, 1942, **37**, 77.

FRÄNGER, W. (1). *The Millennium of Hieronymus Bosch*. Faber, London, 1952.

FRANCKE, A. H. (1). 'Two Ant Stories from the Territory of the Ancient Kingdom of Western Tibet; a Contribution to the Question of Gold-Digging Ants.' *AM*, 1924, **1**, 67.

FRANK, B. (1). '*Kata-imi* et *Kata-tagae;* Étude sur les Interdits de Direction à l'Époque Heian.' *BMFJ*, 1958 (n.s.), **5** (no. 2–4), 1–246.

FRANKE, H. (17). 'Das chinesische Wort für "Mumie" [mummy].' *OR*, 1957, **10**, 253.

FRANKE, H. (18). 'Some Sinological Remarks on Rashīd al-Dīn's "History of China".' *OR*, 1951, **4**, 21.

FRANKE, W. (4). *An Introduction to the Sources of Ming History*. Univ. Malaya Press, Kuala Lumpur and Singapore, 1968.

FRANKFORT, H. (4). *Ancient Egyptian Religion; an Interpretation*. Harper & Row, New York, 1948. Paperback ed. 1961.

FRANTZ, A. (1). 'Zink und Messing im Alterthum.' *BHMZ*, 1881, **40**, 231, 251, 337, 377, 387.

FRASER, SIR J. G. (1). *The Golden Bough*. 3-vol. ed. Macmillan, London, 1900; superseded by 12-vol. ed. (here used), Macmillan, London, 1913–20. Abridged 1-vol. ed. Macmillan, London, 1923.

FRENCH, J. (1). *Art of Distillation*. 4th ed. London, 1667.

FRENCH, P. J. (1). *John Dee; the World of an Elizabethan Magus*. Routledge & Kegan Paul, London, 1971.

FREUDENBERG, K., FRIEDRICH, K. & BUMANN, I. (1). 'Über Cellulose und Stärke [incl. description of a molecular still].' *LA*, 1932, **494**, 41 (57).

FREUND, IDA (1). *The Study of Chemical Composition; an Account of its Method and Historical Development, with illustrative quotations*. Cambridge, 1904. Repr. Dover, New York, 1968, with a foreword by L. E. Strong and a brief biography by O. T. Benfey.

FRIEDERICHSEN, M. (1). 'Morphologie des Tien-schan [Thien Shan].' *ZGEB*, 1899, **34**, 1–62, 193–271. Sep. pub. Pormetter, Berlin, 1900.

FRIEDLÄNDER, P. (1). 'Über den Farbstoff des antiken Purpurs aus *Murex brandaris*.' *BDCG*, 1909, **42**, pt. 1, 765.

FRIEND, J. NEWTON (1). *Iron in Antiquity*. Griffin, London, 1926.

FRIEND, J. NEWTON (2). *Man and the Chemical Elements*. London, 1927.

FRIEND, J. NEWTON & THORNEYCROFT, W. E. (1). 'The Silver Content of Specimens of Ancient and Mediaeval Lead.' *JIM*, 1929, **41**, 105.

FRITZE, M. (1) (tr.). *Pancatantra*. Leipzig, 1884.

FRODSHAM, J. D. (1) (tr.). *The Poems of Li Ho* (*+791 to +817*). Oxford, 1970.

FROST, D. V. (1). 'Arsenicals in Biology; Retrospect and Prospect.' *FP*, 1967, **26** (no. 1), 194.

FRYER, J. (1). *An Account of the Department for the Translation of Foreign Books of the Kiangnan Arsenal*. *NCH*, 28 Jan. 1880, and offprinted.

FRYER, J. (2). 'Scientific Terminology; Present Discrepancies and Means of Securing Uniformity.' *CRR*, 1872, **4**, 26, and sep. pub.

FRYER, J. (3). *The Translator's Vade-Mecum*. Shanghai, 1888.

FRYER, J. (4). 'Western Knowledge and the Chinese.' *JRAS/NCB*, 1886, **21**, 9.

FRYER, J. (5). 'Our Relations with the Reform Movement.' Unpublished essay, 1909. See Bennett (1), p. 151.

FUCHS, K. W. C. (1). *Die vulkanische Erscheinungen der Erde*. Winter, Leipzig and Heidelberg, 1865.

FUCHS, W. (7). 'Ein Gesandschaftsbericht ü. Fu-Lin in chinesischer Wiedergabe aus den Jahren +1314 bis +1320.' *OE*, 1959, **6**, 123.

FÜCK, J. W. (1). 'The Arabic Literature on Alchemy according to al-Nadīm (+987); a Translation of the Tenth Discourse of the Book of the Catalogue (*al-Fihrist*), with Introduction and Commentary.' *AX*, 1951, **4**, 81.

DE LA FUENTE, J. (1). *Yalalag; una Villa Zapoteca Serrana*. Museo Nac. de Antropol. Mexico City, 1949. (Ser. Científica, no. 1.)

FYFE, A. (1). 'An Analysis of Tutenag or the White Copper of China.' *EPJ*, 1822, **7**, 69.

GADD, C. J. (1). 'The Ḥarrān Inscriptions of Nabonidus [of Babylon, –555 to –539].' *ANATS*, 1958, **8**, 35.

GADOLIN, J. (1). *Observationes de Cupro Albo Chinensium Pe-Tong vel Pack-Tong*. *NARSU*, 1827, **9** 137.

GALLAGHER, L. J. (1) (tr.). *China in the 16th Century; the Journals of Matthew Ricci, 1583–1610*. Random House, New York, 1953. (A complete translation, preceded by inadequate bibliographical details, of Nicholas Trigault's *De Christiana Expeditione apud Sinas* (1615). Based on an earlier publication: *The China that Was; China as discovered by the Jesuits at the close of the 16th Century: from the Latin of Nicholas Trigault*. Milwaukee, 1942.) Identifications of Chinese names in Yang Lien-Shêng (4). Crit. J. R. Ware, *ISIS*, 1954, **45**, 395.

GANZENMÜLLER, W. (1). *Beiträge zur Geschichte der Technologie und der Alchemie*. Verlag Chemie, Weinheim, 1956. Rev. W. Pagel, *ISIS*, 1958, **49**, 84.

GANZENMÜLLER, W. (2). *Die Alchemie im Mittelalter*. Bonifacius, Paderborn, 1938. Repr. Olms, Hildesheim, 1967. French, tr. by Petit-Dutailles, Paris, n.d. (*c.* 1940).

GANZENMÜLLER, W. (3). '*Liber Florum Geberti;* alchemistischen Öfen und Geräte in einer Handschrift des 15. Jahrhunderts.' *QSGNM*, 1942, **8**, 273. Repr. in (1), p. 272.

GANZENMÜLLER, W. (4). 'Zukunftsaufgaben der Geschichte der Alchemie.' *CHYM*, 1953, **4**, 31.

GANZENMÜLLER, W. (5). 'Paracelsus und die Alchemie des Mittelalters.' *ZAC/AC*, 1941, **54**, 427.

VON GARBE, R. K. (3) (tr.). *Die Indischen Mineralien, ihre Namen und die ihnen zugeschriebenen Kräfte; Narahari's 'Rāja-nighaṇṭu' [King of Dictionaries], varga XIII, Sanskrit und Deutsch, mit kritischen und erläuternden Anmerkungen herausgegeben...* Hirzel, Leipzig, 1882.

GARBERS, K. (1) (tr.). *'Kitāb Kimiya al-Itr wa'l-Tas'idat'; Buch über die Chemie des Parfüms und die Destillationen von Ya'qub ibn Ishaq al-Kindī; ein Beitrag zur Geschichte der arabischen Parfümchemie und Drogenkunde aus dem 9tr Jahrh. A.D., übersetzt...* Brockhaus, Leipzig, 1948. (Abhdl. f.d. Kunde des Morgenlandes, no. 30.) Rev. A. Mazaheri, *A/AIHS*, 1951, **4** (no. 15), 521.

GARNER, SIR HARRY (1). 'The Composition of Chinese Bronzes.' *ORA*, 1960, **6** (no. 4), 3.

GARNER, SIR HARRY (2). *Chinese and Japanese Cloisonné Enamels.* Faber & Faber, London, 1962.

GARNER, SIR HARRY (3). 'The Origins of "Famille Rose" [polychrome decoration of Chinese Porcelain].' *TOCS*, 1969.

GEBER (ps. of a Latin alchemist *c*. +1290). *The Works of Geber, the most famous Arabian Prince and Philosopher, faithfully Englished by R. R., a Lover of Chymistry [Richard Russell].* James, London, 1678. Repr. and ed. E. J. Holmyard. Dent, London, 1928.

GEERTS, A. J. C. (1). *Les Produits de la Nature Japonaise et Chinoise, Comprenant la Dénomination, l'Histoire et les Applications aux Arts, à l'Industrie, à l'Economie, à la Médécine, etc. des Substances qui dérivent des Trois Régnes de la Nature et qui sont employées par les Japonais et les Chinois: Partie Inorganique et Minéralogique...* [only part published]. 2 vols. Levy, Yokohama; Nijhoff, 's Gravenhage, 1878, 1883. (A paraphrase and commentary on the mineralogical chapters of the *Pên Tshao Kang Mu*, based on Ono Ranzan's commentary in Japanese.)

GEERTS, A. J. C. (2). 'Useful Minerals and Metallurgy of the Japanese; [Introduction and] A, Iron.' *TAS/J*, 1875, **3**, 1, 6.

GEERTS, A. J. C. (3). 'Useful Minerals and Metallurgy of the Japanese; [B], Copper.' *TAS/J*, 1875, **3**, 26.

GEERTS, A. J. C. (4). 'Useful Minerals and Metallurgy of the Japanese; C, Lead and Silver.' *TAS/J*, 1875, **3**, 85.

GEERTS, A. J. C. (5). 'Useful Minerals and Metallurgy of the Japanese; D, Quicksilver.' *TAS/J*, 1876, **4**, 34.

GEERTS, A. J. C. (6). 'Useful Minerals and Metallurgy of the Japanese; E, Gold' (with twelve excellent pictures on thin paper of gold mining, smelting and cupellation from a traditional Japanese mining book). *TAS/J*, 1876, **4**, 89.

GEERTS, A. J. C. (7). 'Useful Minerals and Metallurgy of the Japanese; F, Arsenic' (reproducing the picture from *Thien Kung Khai Wu*). *TAS/J*, 1877, **5**, 25.

GEHES CODEX. See Anon. (93).

GEISLER, K. W. (1). 'Zur Geschichte d. Spirituserzeugung.' *BGTI*, 1926, **16**, 94.

GELBART, N. R. (1). 'The Intellectual Development of Walter Charleton.' *AX*, 1971, **18**, 149.

GELLHORN, E. & KIELY, W. F. (1). 'Mystical States of Consciousness; Neurophysiological and Clinical Aspects.' *JNMD*, 1972, **154**, 399.

GENZMER, F. (1). 'Ein germanisches Gedicht aus der Hallstattzeit.' *GRM*, 1936, **24**, 14.

GEOGHEGAN, D. (1). 'Some Indications of Newton's Attitude towards Alchemy.' *AX*, 1957, **6**, 102.

GEORGII, A. (1). *Kinésithérapie, ou Traitement des Maladies par le Mouvement selon la Méthode de Ling... suivi d'un Abrégé des Applications de Ling à l'Éducation Physique.* Baillière, Paris, 1847.

GERNET, J. (3). *Le Monde Chinois.* Colin, Paris, 1972. (Coll. Destins du Monde.)

GESNER, CONRAD (1). *De Remediis secretis, Liber Physicus, Medicus et partiam Chymicus et Oeconomicus in vinorum diversi apparatu, Medicis & Pharmacopoiis omnibus praecipi necessarius nunc primum in lucem editus.* Zürich, 1552, 1557; second book edited by C. Wolff, Zürich, 1569; Frankfurt, 1578.

GESNER, CONRAD (2). *Thesaurus Euonymus Philiatri, Ein köstlicher Schatz....* Zürich, 1555. Eng. tr. Daye, London, 1559, 1565. French tr. Lyon, 1557.

GESSMANN, G. W. (1). *Die Geheimsymbole der Chemie und Medizin des Mittelalters; eine Zusammenstellung der von den Mystikern und Alchymisten gebrauchten geheimen Zeichenschrift, nebst einen Kurzgefassten geheimwissenschaftlichen Lexikon.* Pr. pr. Graz, 1899, then Mickl, München, 1900.

GETTENS, R. J., FITZHUGH, E. W., BENE, I. V. & CHASE, W. T. (1). *The Freer Chinese Bronzes. Vol. 2. Technical Studies.* Smithsonian Institution, Washington, D.C. 1969 (Freer Gallery of Art Oriental Studies, no. 7). See also Pope, Gettens, Cahill & Barnard (1).

GHOSH, HARINATH (1). 'Observations on the Solubility *in vitro* and *in vivo* of Sulphide of Mercury, and also on its Assimilation, probable Pharmacological Action and Therapeutic Utility.' *IMW*, 1 Apr. 1931.

GIBB, H. A. R. (1). 'The Embassy of Hārūn al-Rashīd to Chhang-An.' *BLSOAS*, 1922, **2**, 619.

GIBB, H. A. R. (4). *The Arab Conquests in Central Asia.* Roy. Asiat. Soc., London, 1923. (Royal Asiatic Society, James G. Forlong Fund Pubs. no. 2.)

GIBB, H. A. R. (5). 'Chinese Records of the Arabs in Central Asia.' *BLSOAS*, 1922, **2**, 613.

GIBBS, F. W. (1). 'Invention in Chemical Industries [+1500 to +1700].' Art. in *A History of Technology*, ed. C. Singer *et al.* Vol. 3, p. 676. Oxford, 1957.

GIBSON, E. C. S. (1) (tr.). *Johannes Cassianus' 'Conlationes'* in 'Select Library of Nicene and Post-Nicene Fathers of the Christian Church'. Parker, Oxford, 1894, vol. 11, pp. 382 ff.

GICHNER, L. E. (1). *Erotic Aspects of Hindu Sculpture*. Pr. pr., U.S.A. (no place of publication stated), 1949.

GICHNER, L. E. (2). *Erotic Aspects of Chinese Culture*. Pr. pr., U.S.A. (no place of publication stated), c. 1957.

GIDE, C. (1). *Les Colonies Communistes et Coopératives*. Paris, 1928. Eng. tr. by E. F. Row. *Communist and Cooperative Colonies*. Harrap, London, 1930.

GILDEMEISTER, E. & HOFFMANN, F. (1). *The Volatile Oils*. Tr. E. Kremers. 2nd ed., 3 vols. Longmans Green, London, 1916. (Written under the auspices of Schimmel & Co., Distillers, Miltitz near Leipzig.)

GILDEMEISTER, J. (1). 'Alchymie.' *ZDMG*, 1876, **30**, 534.

GILES, H. A. (2). *Chinese–English Dictionary*. Quaritch, London, 1892, 2nd ed. 1912.

GILES, H. A. (7) (tr.). 'The *Hsi Yüan Lu* or "Instructions to Coroners"; (Translated from the Chinese).' *PRSM*, 1924, **17**, 59.

GILES, H. A. (14). *A Glossary of Reference on Subjects connected with the Far East*. 3rd ed. Kelly & Walsh, Shanghai, 1900.

GILES, L. (6). *A Gallery of Chinese Immortals ('hsien'), selected biographies translated from Chinese sources (Lieh Hsien Chuan, Shen Hsien Chuan, etc.)*. Murray, London, 1948.

GILES, L. (7). 'Wizardry in Ancient China.' *AP*, 1942, **13**, 484.

GILES, L. (14). 'A Thang Manuscript of the *Sou Shen Chi*.' *NCR*, 1921, **3**, 378, 460.

GILLAN, H. (1). *Observations on the State of Medicine, Surgery and Chemistry in China (+1794)*, ed. J. L. Cranmer-Byng (2). Longmans, London, 1962.

GLAISTER, JOHN (1). *A Textbook of Medical Jurisprudence, Toxicology and Public Health*. Livingstone, Edinburgh, 1902. 5th ed. by J. Glaister the elder and J. Glaister the younger, Edinburgh, 1931. 6th ed. title changed to *Medical Jurisprudence and Toxicology*, J. Glaister the younger, Edinburgh, 1938. 7th ed. Edinburgh, 1942, 9th ed. Edinburgh, 1950. 10th to 12th eds. (same title), Edinburgh, 1957 to 1966 by J. Glaister the younger & E. Rentoul.

GLISSON, FRANCIS (1). *Tractatus de Natura Substantiae Energetica, seu de Vita Naturae ejusque Tribus Primus Facultatibus; I, Perceptiva; II, Appetitiva; III, Motiva, Naturalibus*. Flesher, Brome & Hooke, London, 1672. Cf. Pagel (16, 17); Temkin (4).

GLOB, P. V. (1). *Iron-Age Man Preserved*. Faber & Faber, London; Cornell Univ. Press, Ithaca, N.Y. 1969. Tr. R. Bruce-Mitford from the Danish *Mosefolket; Jernalderens Mennesker bevaret i 2000 År*.

GLOVER, A. S. B. (1) (tr.). 'The Visions of Zosimus', in Jung (3).

GMELIN, J. G. (1). *Reise durch Russland*. 3 vols. Berlin, 1830.

GOAR, P. J. See Syncellos, Georgius.

GODWIN, WM. (1). *An Enquiry concerning Political Justice, and its Influence on General Virtue and Happiness*. London, 1793.

GOH THEAN-CHYE. See Ho Ping-Yü, Ko Thien-Chi *et al.*

GOLDBRUNNER, J. (1). *Individuation; a study of the Depth Psychology of Carl Gustav Jung*. Tr. from Germ. by S. Godman. Hollis & Carter, London, 1955.

GOLTZ, D. (1). *Studien zur Geschichte der Mineralnamen in Pharmazie, Chemie und Medizin von den Anfängen bis Paracelsus*. 1971. (Sudhoffs Archiv. Beiheft, no. 14.)

GONDAL, MAHARAJAH OF. See Bhagvat Singhji.

GOODFIELD, J. & TOULMIN, S. (1). 'The Qaṭṭāra; a Primitive Distillation and Extraction Apparatus still in Use.' *ISIS*, 1964, **55**, 339.

GOODMAN, L. S. & GILMAN, A. (ed.) (1). *The Pharmacological Basis of Therapeutics*. Macmillan, New York, 1965.

GOODRICH, L. CARRINGTON (1). *Short History of the Chinese People*. Harper, New York, 1943.

GOODWIN, B. (1). 'Science and Alchemy', art. in *The Rules of the Game*... ed. T. Shanin (1), p. 360.

GOOSENS, R. (1). Un Texte Grec relatif à l'aśvamedha' [in the Life of Apollonius of Tyana by Philostratos]. *JA*, 1930, **217**, 280.

GÖTZE, A. (1). 'Die "Schatzhöhle"; Überlieferung und Quelle.' *SHAW/PH*, 1922, no. 4.

GOULD, S. J. (1). 'History *versus* Prophecy; Discussion with J. W. Harrington.' *AJSC*, 1970, **268**, 187. With reply by J. W. Harrington, p. 189.

GOWLAND, W. (1). 'Copper and its Alloys in Prehistoric Times.' *JRAI*, 1906, **36**, 11.

GOWLAND, W. (2). 'The Metals in Antiquity.' *JRAI*, 1912, **42**, 235. (Huxley Memorial Lecture 1912.)

GOWLAND, W. (3). 'The Early Metallurgy of Silver and Lead.' Pt. I, 'Lead' (no more published). *AAA* 1901, **57**, 359.

GOWLAND, W. (4). 'The Art of Casting Bronze in Japan.' *JRSA*, 1895, **43**. Repr. *ARSI*, 1895, 609.

GOWLAND, W. (5). 'The Early Metallurgy of Copper, Tin and Iron in Europe as illustrated by ancient Remains, and primitive Processes surviving in Japan.' *AAA*, 1899, **56**, 267.

GOWLAND, W. (6). 'Metals and Metal-Working in Old Japan.' *TJSL*, 1915, **13**, 20.

GOWLAND, W. (7). 'Silver in Roman and earlier Times.' Pt. I, 'Prehistoric and Protohistoric Times' (no more published). *AAA*, 1920, **69**, 121.

GOWLAND, W. (8). 'Remains of a Roman Silver Refinery at Silchester' (comparisons with Japanese technique). *AAA*, 1903, **57**, 113.

GOWLAND, W. (9). *The Metallurgy of the Non-Ferrous Metals*. Griffin, London, 1914. (Copper, Lead, Gold, Silver, Platinum, Mercury, Zinc, Cadmium, Tin, Nickel, Cobalt, Antimony, Arsenic, Bismuth, Aluminium.)

GOWLAND, W. (10). 'Copper and its Alloys in Early Times.' *JIM*, 1912, **7**, 42.

GOWLAND, W. (11). 'A Japanese Pseudo-Speiss (*Shirome*), and its Relation to the Purity of Japanese Copper and the Presence of Arsenic in Japanese Bronze.' *JSCI*, 1894, **13**, 463.

GOWLAND, W. (12). 'Japanese Metallurgy; I, Gold and Silver and their Alloys.' *JCSI*, 1896, **15**, 404. No more published.

GRACE, V. R. (1). *Amphoras and the Ancient Wine Trade*. Amer. School of Classical Studies, Athens and Princeton, N.J., 1961.

GRADY, M. C. (1). 'Préparation Electrolytique du Rouge au Japan.' *TP*, 1897 (1ᵉ sér.), **8**, 456.

GRAHAM, A. C. (5). '"Being" in Western Philosophy compared with *shih/fei* and *yu/wu* in Chinese Philosophy.' With an appendix on 'The Supposed Vagueness of Chinese'. *AM*, 1959, **7**, 79.

GRAHAM, A. C. (6) (tr.). *The Book of Lieh Tzu*. Murray, London, 1960.

GRAHAM, A. C. (7). 'Chuang Tzu's "Essay on Making Things Equal".' Communication to the First International Conference of Taoist Studies, Villa Serbelloni, Bellagio, 1968.

GRAHAM, D. C. (4). 'Notes on the Han Dynasty Grave Collection in the West China Union University Museum of Archaeology [at Chhêngtu].' *JWCBRS*, 1937, **9**, 213.

GRANET, M. (5). *La Pensée Chinoise*. Albin Michel, Paris, 1934. (Evol. de l'Hum. series, no. 25 bis.)

GRANT, R. McQ. (1). *Gnosticism; an Anthology*. Collins, London, 1961.

GRASSMANN, H. (1). 'Der Campherbaum.' *MDGNVO*, 1895, **6**, 277.

GRAY, B. (1). 'Arts of the Sung Dynasty.' *TOCS*, 1960, 13.

GRAY, J. H. (1). '[The "Abode of the Blest" in] Persian [Iranian, Thought].' *ERE*, vol. ii, p. 702.

GRAY, J. H. (1). *China: a History of the Laws, Manners and Customs of the People*. Ed. W. G. Gregor. 2 vols. Macmillan, London, 1878.

GRAY, W. D. (1). *The Relation of Fungi to Human Affairs*. Holt, New York, 1959.

GREEN, F. H. K. (1). 'The Clinical Evaluation of Remedies.' *LT*, 1954, 1085.

GREEN, R. M. (1) (tr.). *Galen's Hygiene; 'De Sanitate Tuenda'*. Springfield, Ill. 1951.

GREENAWAY, F. (1). 'Studies in the Early History of Analytical Chemistry.' Inaug. Diss. London, 1957.

GREENAWAY, F. (2). *The Historical Continuity of the Tradition of Assaying*. Proc. Xth Int. Congr. Hist. of Sci., Ithaca, N.Y., 1962, vol. 2, p. 819.

GREENAWAY, F. (3). 'The Early Development of Analytical Chemistry.' *END*, 1962, **21**, 91.

GREENAWAY, F. (4). *John Dalton and the Atom*. Heinemann, London, 1966.

GREENAWAY, F. (5). 'Johann Rudolph Glauber and the Beginnings of Industrial Chemistry.' *END*, 1970, **29**, 67.

GREGORY, E. (1). *Metallurgy*. Blackie, London and Glasgow, 1943.

GREGORY, J. C. (1). *A Short History of Atomism*. Black, London, 1931.

GREGORY, J. C. (2). 'The Animate and Mechanical Models of Reality.' *JPHST*, 1927, **2**, 301. Abridged in 'The Animate Model of Physical Process'. *SPR*, 1925.

GREGORY, J. C. (3). 'Chemistry and Alchemy in the Natural Philosophy of Sir Francis Bacon (+1561 to +1626).' *AX*, 1938, **2**, 93.

GREGORY, J. C. (4). 'An Aspect of the History of Atomism.' *SPR*, 1927, **22**, 293.

GREGORY, T. C. (1). *Condensed Chemical Dictionary*. 1950. Continuation by A. Rose & E. Rose. Chinese tr. by Kao Hsien (1).

GRIERSON, SIR G. A. (1). *Bihar Peasant Life*. Patna, 1888; reprinted Bihar Govt., Patna, 1926.

GRIERSON, P. (2). 'The Roman Law of Counterfeiting.' Art. in *Essays in Roman Coinage*, Mattingley Presentation Volume, Oxford, 1965, p. 240.

GRIFFITH, E. F. (1). *Modern Marriage*. Methuen, London, 1946.

GRIFFITH, F. LL. & THOMPSON, H. (1). *'The Demotic Magical Papyrus of London and Leiden [+3rd Cent.]*. 3 vols. Grevel, London, 1904–9.

GRIFFITH, R. T. H. (1) (tr.). *The Hymns of the 'Atharva-veda'*. 2 vols. Lazarus, Benares, 1896. Repr. Chowkhamba Sanskrit Series Office, Varanasi, 1968.

GRIFFITHS, J. GWYN (1) (tr.). *Plutarch's 'De Iside et Osiride'*. University of Wales Press, Cardiff, 1970.

GRINSPOON, L. (1). 'Marihuana.' *SAM*, 1969, **221** (no. 6), 17.

GRMEK, M. D. (2). 'On Ageing and Old Age; Basic Problems and Historical Aspects of Gerontology and Geriatrics.' *MB*, 1958, **5** (no. 2).

DE GROOT, J. J. M. (2). *The Religious System of China.* Brill, Leiden, 1892.
  Vol. 1, Funeral rites and ideas of resurrection.
    2 and 3, Graves, tombs, and *fêng-shui*.
    4, The soul, and nature-spirits.
    5, Demonology and sorcery.
    6, The animistic priesthood (*wu*).

GRÖSCHEL-STEWART, U. (1). 'Plazentahormone.' *MMN*, 1970, **22**, 469.

GROSIER, J. B. G. A. (1). *De la Chine; ou, Description Générale de cet Empire*, etc. 7 vols. Pillet & Bertrand, Paris, 1818–20.

GRUMAN, G. J. (1). 'A History of Ideas about the Prolongation of Life; the Evolution of Prolongevity Hypotheses to 1800.' Inaug. Diss., Harvard University, 1965. *TAPS*, 1966 (n.s.), **56** (no. 9), 1–102.

GRUMAN, G. J. (2). 'An Introduction to the Literature on the History of Gerontology.' *BIHM*, 1957, **31**, 78.

GUARESCHI, S. (1). Tr. of Klaproth (5). *SAEC*, 1904, **20**, 449.

GUERLAC, H. (1). 'The Poets' Nitre.' *ISIS*, 1954, **45**, 243.

GUICHARD, F. (1). 'Properties of saponins of *Gleditschia*.' *BSPB*, 1936, **74**, 168.

VAN GULIK, R. H. (3). '*Pi Hsi Thu Khao*'; *Erotic Colour-Prints of the Ming Period, with an Essay on Chinese Sex Life from the Han to the Chhing Dynasty (−206 to +1644)*. 3 vols. in case. Privately printed. Tokyo, 1951 (50 copies only, distributed to the most important Libraries of the world). Crit. W. L. Hsü, *MN*, 1952, **8**, 455; H. Franke, *ZDMG*, 1955 (NF) **30**, 380.

VAN GULIK, R. H. (4). 'The Mango "Trick" in China; an essay on Taoist Magic.' *TAS/J*, 1952 (3rd ser.), **3**, 1.

VAN GULIK, R. H. (8). *Sexual Life in Ancient China; a Preliminary Survey of Chinese Sex and Society from c. −1500 to +1644*. Brill, Leiden, 1961. Rev. R. A. Stein, *JA*, 1962, **250**, 640.

GUNAWARDANA, R. A. LESLIE H. (1). 'Ceylon and Malaysia; a Study of Professor S. Paranavitana's Research on the Relations between the Two Regions.' *UCR*, 1967, **25**, 1–64.

GUNDEL, W. (4). Art. 'Alchemie' in *Real-Lexikon f. Antike und Christentum*, ed. T. Klauser, 1950–3, vol. 1, p. 239.

GUNDEL, W. & GUNDEL, H. G. (1). *Astrologumena; das astrologische Literatur in der Antike und ihre Geschichte.* Steiner, Wiesbaden, 1966. (*AGMW* Beiheft, no. 6, pp. 1–382.)

GUNTHER, R. T. (3) (ed.). *The Greek Herbal of Dioscorides, illustrated by a Byzantine in +512, englished by John Goodyer in +1655, edited and first printed, 1933*. Pr. pr. Oxford, 1934, photolitho repr. Hafner, New York, 1959.

GUNTHER, R. T. (4). *Early Science in Cambridge.* Pr. pr. Oxford, 1937.

GUPPY, H. B. (1). 'Samshu-brewing in North China.' *JRAS/NCB*, 1884, **18**, 63.

GURE, D. (1). 'Jades of the Sung Group.' *TOCS*, 1960, 39.

GUTZLAFF, C. (1). 'On the Mines of the Chinese Empire.' *JRAS/NCB*, 1847, 43.

GYLLENSVÅRD, BO (1). *Chinese Gold and Silver [-Work] in the Carl Kempe Collection.* Stockholm, 1953; Smithsonian Institution, Washington, D.C., 1954.

GYLLENSVÅRD, BO (2). 'Thang Gold and Silver.' *BMFEA*, 1957, **29**, 1–230.

HACKIN, J. & HACKIN, J. R. (1). *Recherches archéologiques à Begram, 1937*. Mémoires de la Délégation Archéologique Française en Afghanistan, vol. 9. Paris, 1939.

HACKIN, J., HACKIN, J. R., CARL, J. & HAMELIN, P. (with the collaboration of J. Auboyer, V. Elisséeff, O. Kurz & P. Stern) (1). *Nouvelles Recherches archéologiques à Begram (ancienne Kāpiśi), 1939–1940*. Mémoires de la Délégation Archéologique Française en Afghanistan, vol. 11. Paris, 1954. (Rev. P. S. Rawson, *JRAS*, 1957, 139.)

HADD, H. E. & BLICKENSTAFF, R. T. (1). *Conjugates of Steroid Hormones.* Academic Press, New York and London, 1969.

HADI HASAN (1). *A History of Persian Navigation.* Methuen, London, 1928.

HAJI KHALFA (or Ḥajji Khalfa). See Flügel (2).

HALBAN, J. (1). 'Über den Einfluss der Ovarien auf die Entwicklung der Genitales.' *MGG*, 1900, **12**, 496.

HALDANE, J. B. S. (2). 'Experiments on the Regulation of the Blood's Alkalinity.' *JOP*, 1921, **55**, 265.

HALDANE, J. B. S. (3). 'Über Halluzinationen infolge von Änderungen des Kohlensäuredrucks.' *PF*, 1924, **5**, 356.

HALDANE, J. B. S. See also Baird, Douglas, Haldane & Priestley (1); Davies, Haldane & Kennaway (1).

HALDANE, J. B. S., LINDER, G. C., HILTON, R. & FRASER, F. R. (1). 'The Arterial Blood in Ammonium Chloride Acidosis.' *JOP*, 1928, **65**, 412.

HALDANE, J. B. S., WIGGLESWORTH, V. B. & WOODROW, C. E. (1). 'Effect of Reaction Changes on Human Inorganic Metabolism.' *PRSB*, 1924, **96**, 1.

HALDANE, J. B. S., WIGGLESWORTH, V. B. & WOODROW, C. E. (2). 'Effect of Reaction Changes on Human Carbohydrate and Oxygen Metabolism.' *PRSB*, 1924, **96**, 15.

HALEN, G. E. (1). *De Chemo Scientiarum Auctore*. Upsala, 1694.

HALES, STEPHEN (2). *Philosophical Experiments; containing Useful and Necessary Instructions for such as undertake Long Voyages at Sea, shewing how Sea Water may be made Fresh and Wholsome*. London, 1739.

HALL, E. T. (1). 'Surface Enrichment of Buried [Noble] Metal [Alloys].' *AMY*, 1961, **4**, 62.

HALL, E. T. & ROBERTS, G. (1). 'Analysis of the Moulsford Torc.' *AMY*, 1962, **5**, 28.

HALL, F. W. (1). '[The "Abode of the Blest" in] Greek and Roman [Culture].' *ERE*, vol. ii, p. 696.

HALL, H. R. (1). 'Death and the Disposal of the Dead [in Ancient Egypt].' Art. in *ERE*, vol. iv, p. 458.

HALL, MANLY P. (1). *The Secret Teachings of All Ages*. San Francisco, 1928.

HALLEUX, R. (1). 'Fécondité des Mines et Sexualité des Pierres dans l'Antiquité Gréco-Romaine.' *RBPH*, 1970, **48**, 16.

HALOUN, G. (2). Translations of *Kuan Tzu* and other ancient texts made with the present writer, unpub. MSS.

HAMARNEH, SAMI, K. & SONNEDECKER, G. (1). *A Pharmaceutical View of Albucasis (al-Zahrāwī) in Moorish Spain*. Brill, Leiden, 1963.

HAMMER-JENSEN, I. (1). 'Deux Papyrus à Contenu d'Ordre Chimique.' *ODVS*, 1916 (no. 4), 279.

HAMMER-JENSEN, I. (2). 'Die ältesten Alchemie.' *MKDVS/HF*, 1921, **4** (no. 2), 1–159.

HANBURY, DANIEL (1). *Science Papers, chiefly Pharmacological and Botanical*. Macmillan, London, 1876.

HANBURY, DANIEL (2). 'Notes on Chinese Materia Medica.' *PJ*, 1861, **2**, 15, 109, 553; 1862, **3**, 6, 204, 260, 315, 420. German tr. by W. C. Martius (without Chinese characters), *Beiträge z. Materia Medica Chinas*. Kranzbühler, Speyer, 1863. Revised version, with additional notes, references and map, in Hanbury (1), pp. 211 ff.

HANBURY, DANIEL (6). 'Note on Chinese Sal Ammoniac.' *PJ*, 1865, **6**, 514. Repr. in Hanbury (1), p. 276.

HANBURY, DANIEL (7). 'A Peculiar Camphor from China [Ngai Camphor from *Blumea balsamifera*]. *PJ*, 1874, **4**, 709. Repr. in Hanbury (1), pp. 393 ff.

HANBURY, DANIEL (8). 'Some Notes on the Manufactures of Grasse and Cannes [and Enfleurage].' *PJ*, 1857, **17**, 161. Repr. in Hanbury (1), pp. 150 ff.

HANBURY, DANIEL (9). 'On Otto of Rose.' *PJ*, 1859, **18**, 504. Repr. in Hanbury (1), pp. 164 ff.

HANSFORD, S. H. (1). *Chinese Jade Carving*. Lund Humphries, London, 1950.

HANSFORD, S. H. (2) (ed.). *The Seligman Collection of Oriental Art; Vol. 1, Chinese, Central Asian and Luristan Bronzes and Chinese Jades and Sculptures*. Arts Council G. B., London, 1955.

HANSON, D. (1). *The Constitution of Binary Alloys*. McGraw-Hill, New York, 1958.

HARADA, YOSHITO & TAZAWA, KINGO (1). *Lo-Lang; a Report on the Excavation of Wang Hsü's Tomb in the Lo-Lang Province, an ancient Chinese Colony in Korea*. Tokyo University, Tokyo, 1930.

HARBORD, F. W. & HALL, J. W. (1). *The Metallurgy of Steel*. 2 vols. 7th ed. Griffin, London, 1923.

HARDING, M. ESTHER (1). *Psychic Energy; its Source and Goal*. With a foreword by C. G. Jung. Pantheon, New York, 1947. (Bollingen series, no. 10.)

VON HARLESS, G. C. A. (1). *Jakob Böhme und die Alchymisten; ein Beitrag zum Verständnis J. B.'s...* Berlin, 1870. 2nd ed. Hinrichs, Leipzig, 1882.

HARRINGTON, J. W. (1). 'The First "First Principles of Geology".' *AJSC*, 1967, **265**, 449.

HARRINGTON, J. W. (2). 'The Prenatal Roots of Geology; a Study in the History of Ideas.' *AJSC*, 1969, **267**, 592.

HARRINGTON, J. W. (3). 'The Ontology of Geological Reasoning; with a Rationale for evaluating Historical Contributions.' *AJSC*, 1970, **269**, 295.

HARRIS, C. (1). '[The State of the Dead in] Christian [Thought].' *ERE*, vol. xi, p. 833.

HARRISON, F. C. (1). 'The Miraculous Micro-Organism' (*B. prodigiosus* as the causative agent of 'bleeding hosts'). *TRSC*, 1924, **18**, 1.

HARRISSON, T. (8). 'The *palang*; its History and Proto-history in West Borneo and the Philippines.' *JRAS/M*, 1964, **37**, 162.

HARTLEY, SIR HAROLD (1). 'John Dalton, F.R.S. (1766 to 1844) and the Atomic Theory; a Lecture to commemorate his Bicentenary.' *PRSA*, 1967, **300**, 291.

HARTNER, W. (12). *Oriens-Occidens; ausgewählte Schriften zur Wissenschafts- und Kultur-geschichte (Festschrift zum 60. Geburtstag)*. Olms, Hildesheim, 1968. (Collectanea, no. 3.)

HARTNER, W. (13). 'Notes on Picatrix.' *ISIS*, 1965, **56**, 438. Repr. in (12), p. 415.

HASCHMI, M. Y. (1). 'The Beginnings of Arab Alchemy.' *AX*, 1961, **9**, 155.

HASCHMI, M. Y. (2). '*The Propagation of Rays*'; the Oldest Arabic Manuscript about Optics (*the Burning-Mirror*), [a text written by] *Yaʻkub ibn Ishaq al-Kindī*, Arab Philosopher and Scholar of the +9th Century. Photocopy, Arabic text and Commentary. Aleppo, 1967.

HASCHMI, M. Y. (3). 'Sur l'Histoire de l'Alcool.' Résumés des Communications, XIIth International Congress of the History of Science, Paris, 1968, p. 91.

HASCHMI, M. Y. (4). 'Die Anfänge der arabischen Alchemie.' Actes du XIe Congrès International d'Histoire des Sciences, Warsaw, 1965, p. 290.

HASCHMI, M. Y. (5). 'Ion Exchange in Arabic Alchemy.' Proc. Xth Internat. Congr. Hist. of Sci., Ithaca, N.Y. 1962, p. 541. Summaries of Communications, p. 56.

HASCHMI, M. Y. (6). 'Die Geschichte der arabischen Alchemie.' DMAB, 1967, **35**, 60.

HATCHETT, C. (1). 'Experiments and Observations on the Various Alloys, on the Specific Gravity, and on the Comparative Wear of Gold...' PTRS, 1803, **93**, 43.

HAUSHERR, I. (1). 'La Méthode d'Oraison Hésychaste.' ORCH, 1927, **9**, (no. 2), 102.

HÄUSSLER, E. P. (1). 'Über das Vorkommen von a-Follikelhormon (3-oxy-17 Keto-1, 3, 5-oestratriën) im Hengsturin.' HCA, 1934, **17**, 531.

HAWKES, D. (1) (tr.). 'Chhu Tzhu'; the Songs of the South—an Ancient Chinese Anthology. Oxford, 1959. Rev. J. Needham, NSN (18 Jul. 1959).

HAWKES, D. (2). 'The Quest of the Goddess.' AM, 1967, **13**, 71.

HAWTHORNE, J. G. & SMITH, C. S. (1) (tr.). 'On Divers Arts'; the Treatise of Theophilus [Presbyter], translated from the Mediaeval Latin with Introduction and Notes...[probably by Roger of Helmars-hausen, c. +1130]. Univ. of Chicago Press, Chicago, 1963.

HAY, M. (1). Failure in the Far East; Why and How the Breach between the Western World and China First Began (on the dismantling of the Jesuit Mission in China in the late +18th century). Spearman, London; Scaldis, Wetteren (Belgium), 1956.

HAYS, E. E. & STEELMAN, S. L. (1). 'The Chemistry of the Anterior Pituitary Hormones.' Art. in The Hormones..., ed. G. Pincus, K. V. Thimann & E. B. Astwood. Academic Press, New York, 1948–64. vol. 3, p. 201.

HEDBLOM, C. A. (1). 'Disease Incidence in China [16,000 cases].' CMJ, 1917, **31**, 271.

HEDFORS, H. (1) (ed. & tr.). The 'Compositiones ad Tingenda Musiva'... Uppsala, 1932.

HEDIN, SVEN A., BERGMAN, F. et al. (1). History of the Expedition in Asia, 1927/1935. 4 vols. Reports of the Sino-Swedish [Scientific] Expedition [to NW China]. 1936. Nos. 23, 24, 25, 26.

HEIM, R. (1). 'Old and New Investigations on Hallucinogenic Mushrooms from Mexico.' APH, 1959, **12**, 171.

HEIM, R. (2). Champignons Toxiques et Hallucinogènes. Boubée, Paris, 1963.

HEIM, R. & HOFMANN, A. (1). 'Psilocybine.' CRAS, 1958, **247**, 557.

HEIM, R., WASSON, R. G. et al. (1). Les Champignons Hallucinogènes du Mexique; Études Ethnologiques, Taxonomiques, Biologiques, Physiologiques et Chimiques. Mus. Nat. d'Hist. Nat. Paris, 1958.

VON HEINE-GELDERN, R. (4). 'Die asiatische Herkunft d. südamerikanische Metalltechnik.' PAI, 1954, **5**, 347.

HEMNETER, E. (1). 'The Influence of the Caste-System on Indian Trades and Crafts.' CIBA/T, 1937, **1** (no. 2), 46.

H[EMSLEY], W. B. (1). 'Camphor.' N, 1896, **54**, 116.

HENDERSON, G. & HURVITZ, L. (1). 'The Buddha of Seiryō-ji [Temple at Saga, Kyoto]; New Finds and New Theory.' AA, 1956, **19**, 5.

HENDY, M. F. & CHARLES, J. A. (1). 'The Production Techniques, Silver Content and Circulation History of the +12th-Century Byzantine Trachy.' AMY, 1970, **12**, 13.

HENROTTE, J. G. (1). 'Yoga et Biologie.' ATOM, 1969, **24** (no. 265), 283.

HENRY, W. (1). Elements of Experimental Chemistry. London, 1810. German. tr. by F. Wolff, Berlin, 1812. Another by J. B. Trommsdorf.

HERMANN, A. (1). 'Das Buch Kmj.t und die Chemie.' ZAES, 1954, **79**, 99.

HERMANN, P. (1). Een constelijk Distileerboec inhoudende de rechte ende waerachtige conste der distilatiën om alderhande wateren der cruyden, bloemen ende wortelen ende voorts alle andere dinge te leeren distileren opt alder constelijcste, alsoo dat dies gelyke noyt en is gheprint geweest in geen derley sprake... Antwerp, 1552.

HERMANNS, M. (1). Die Nomaden von Tibet. Vienna, 1949. Rev. W. Eberhard, AN, 1950, **45**, 942.

HERRINGHAM, C. J. (1). The 'Libro dell'Arte' of Cennino Cennini [+1437]. Allen & Unwin, London, 1897.

HERRMANN, A. (2). Die Alten Seidenstrassen zw. China u. Syrien; Beitr. z. alten Geographie Asiens, I (with excellent maps). Berlin, 1910. (Quellen u. Forschungen z. alten Gesch. u. Geographie, no. 21; photographically reproduced, Tientsin, 1941).

HERRMANN, A. (3). 'Die Alten Verkehrswege zw. Indien u. Süd-China nach Ptolemäus.' ZGEB, 1913, 771.

HERRMANN, A. (5). 'Die Seidenstrassen vom alten China nach dem Romischen Reich.' MGGW, 1915, **58**, 472.

HERRMANN, A. (6). *Die Verkehrswege zw. China, Indien und Rom um etwa 100 nach Chr.* Leipzig, 1922 (Veröffentlichungen d. Forschungs-instituts f. vergleich. Religionsgeschichte a.d. Univ. Leipzig, no. 7.)

HERTZ, W. (1). 'Die Sage vom Giftmädchen.' *ABAW/PH*, 1893, **20**, no. 1. Repr. in *Gesammelte Abhandlungen*, ed. v. F. von der Leyen, 1905, pp. 156–277.

D'HERVEY ST DENYS, M. J. L. (3). *Trois Nouvelles Chinoises, traduites pour la première fois.* Leroux, Paris, 1885. 2nd ed. Dentu, Paris, 1889.

HEYM, G. (1). 'The *Aurea Catena Homeri* [by Anton Joseph Kirchweger, +1723].' *AX*, 1937, **1**, 78. Cf. Ferguson (1), vol. 1, p. 470.

HEYM, G. (2). 'Al-Rāzī and Alchemy.' *AX*, 1938, **1**, 184.

HICKMAN, K. C. D. (1). 'A Vacuum Technique for the Chemist' (molecular distillation). *JFI*, 1932, **213**, 119.

HICKMAN, K. C. D. (2). 'Apparatus and Methods [for Molecular Distillation].' *IEC/I*, 1937, **29**, 968.

HICKMAN, K. C. D. (3). 'Surface Behaviour in the Pot Still.' *IEC/I*, 1952, **44**, 1892.

HICKMAN, K. C. D. & SANFORD, C. R. (1). 'The Purification, Properties and Uses of Certain High-Boiling Organic Liquids.' *JPCH*, 1930, **34**, 637.

HICKMAN, K. C. D. & SANFORD, C. R. (2). 'Molecular stills.' *RSI*, 1930, **1**, 140.

HICKMAN, K. C. D. & TREVOY, D. J. (1). 'A Comparison of High Vacuum Stills and Tensimeters.' *IEC/I*, 1952, **44**, 1903.

HICKMAN, K. C. D. & WEYERTS, W. (1). 'The Vacuum Fractionation of Phlegmatic Liquids.' *JACS*, 1930, **52**, 4714.

HIGHMORE, NATHANIEL (1). *The History of Generation, examining the several Opinions of divers Authors, especially that of Sir Kenelm Digby, in his Discourse of Bodies.* Martin, London, 1651.

HILGENFELD, A. (1). *Die Ketzergeschichte des Urchristenthums.* Fues (Reisland), Leipzig, 1884.

HILLEBRANDT, A. (1). *Vedische Mythologie.* Breslau, 1891–1902.

HILTON-SIMPSON, M. W. (1). *Arab Medicine and Surgery; a Study of the Healing Art in Algeria.* Oxford, 1922.

HIORDTHAL, T. See Hjortdahl, T.

HIORNS, A. H. (1). *Metal-Colouring and Bronzing.* Macmillan, London and New York, 1892. 2nd ed. 1902.

HIORNS, A. H. (2). *Mixed Metals or Metallic Alloys.* 3rd ed. Macmillan, London and New York, 1912.

HIORNS, A. H. (3). *Principles of Metallurgy.* 2nd ed. Macmillan, London, 1914.

HIRTH, F. (2) (tr.). 'The Story of Chang Chhien, China's Pioneer in West Asia.' *JAOS*, 1917, **37**, 89. (Translation of ch. 123 of the *Shih Chi*, containing Chang Chhien's Report; from §18–52 inclusive and 101 to 103. §98 runs on to §104, 99 and 100 being a separate interpolation. Also tr. of ch. 111 containing the biogr. of Chang Chhien.)

HIRTH, F. (7). *Chinesische Studien.* Hirth, München and Leipzig, 1890.

HIRTH, F. (9). *Über fremde Einflüsse in der chinesischen Kunst.* G. Hirth, München and Leipzig. 1896.

HIRTH, F. (11). 'Die Länder des Islam nach Chinesischen Quellen.' *TP*, 1894, **5** (Suppl.). (Translation of, and notes on, the relevant parts of the *Chu Fan Chih* of Chao Ju-Kua; subsequently incorporated in Hirth & Rockhill.)

HIRTH, F. (25). 'Ancient Porcelain; a study in Chinese Mediaeval Industry and Trade.' G. Hirth, Leipzig and Munich; Kelly & Walsh, Shanghai, Hongkong, Yokohama and Singapore, 1888.

HIRTH, F. & ROCKHILL, W. W. (1) (tr.). *Chau Ju-Kua; His work on the Chinese and Arab Trade in the 12th and 13th centuries, entitled 'Chu-Fan-Chi'.* Imp. Acad. Sci, St Petersburg, 1911. (Crit. G. Vacca *RSO*, 1913, **6**, 209; P. Pelliot, *TP*, 1912, **13**, 446; E. Schaer, *AGNT*, 1913, **6**, 329; O. Franke, *OAZ*, 1913, **2**, 98; A. Vissière, *JA*, 1914 (11e sér.), **3**, 196.)

HISCOX, G. D. (1) (ed.). *The Twentieth Century Book of Recipes, Formulas and Processes; containing nearly 10,000 selected scientific, chemical, technical and household recipes, formulas and processes for use in the laboratory, the office, the workshop and in the home.* Lockwood, London; Henley, New York, 1907. Lexicographically arranged. 4th ed., Lockwood, London; Henley, New York, 1914. Retitled *Henley's Twentieth Century Formulas, Recipes and Processes; containing 10,000 selected household and workshop formulas, recipes, proceeses and money-saving methods for the practical use of manufacturers, mechanics, housekeepers and home workers.* Spine title unchanged; index of contents added and 2 entries omitted.

HITCHCOCK, E. A. (1). *Remarks upon Alchemy and the Alchemists, indicating a Method of discovering the True Nature of Hermetic Philosophy; and showing that the Search after the Philosopher's Stone had not for its Object the Discovery of an Agent for the Transmutation of Metals—Being also an attempt to rescue from undeserved opprobrium the reputation of a class of extraordinary thinkers in past ages.* Crosby & Nichols, Boston, 1857. 2nd ed. 1865 or 1866. See also Clymer (1); Croffut (1).

HITCHCOCK, E. A. (2). *Remarks upon Alchymists, and the supposed Object of their Pursuit; showing that the Philosopher's Stone is a mere Symbol, signifying Something which could not be expressed openly without incurring the Danger of an Auto-da-Fé. By an Officer of the United States Army.* Pr. pr. Herald, Carlisle, Pennsylvania, 1855. This pamphlet was the first form of publication of the material enlarged in Hitchcock (1).

HJORTDAHL, T. (1). 'Chinesische Alchemie', art. in Kahlbaum Festschrift (1909), ed. Diergart (1): *Beiträge aus der Geschichte der Chemie*, pp. 215–24. Comm. by E. Chavannes, *TP*, 1909 (2e sér.), **10**, 389.

HJORTDAHL, T. (2). 'Fremstilling af Kemiens Historie' (in Norwegian). *CVS*, 1905, **1** (no. 7).

HO JU (1). *Poèmes de Mao Tsê-Tung* (French translation). Foreign Languages Press, Peking, 1960. 2nd ed., enlarged, 1961.

HO PENG YOKE. See Ho Ping-Yü.

HO PING-YÜ (5). 'The Alchemical Work of Sun Ssu-Mo.' Communication to the American Chemical Society's Symposium on Ancient and Archaeological Chemistry, at the 142nd Meeting, Atlantic City, 1962.

HO PING-YÜ (7). 'Astronomical Data in the Annamese *Ðai Việt Sú-Ký Toàn-thú*; an early Annamese Historical Text.' *JAOS*, 1964, **84**, 127.

HO PING-YÜ (8). 'Draft translation of the *Thai-Chhing Shih Pi Chi* (Records in the Rock Chamber); an alchemical book (*TT*/874) of the Liang period (early +6th Century, but including earlier work as old as the late +3rd).' Unpublished.

HO PING-YÜ (9). Précis and part draft translation of the *Thai Chi Chen-Jen Chiu Chuan Huan Tan Ching Yao Chüeh* (Essential Teachings of the Manual of the Supreme-Pole Adept on the Ninefold Cyclically Transformed Elixir); an alchemical book (*TT*/882) of uncertain date, perhaps Sung but containing much earlier metarial.' Unpublished.

HO PING-YÜ (10). 'Précis and part draft translation of the *Thai-Chhing Chin I Shen Tan Ching* (Manual of the Potable Gold and Magical Elixir; a Thai-Chhing Scripture); an alchemical book (*TT*/873) of unknown date and authorship but prior to +1022 when it was incorporated in the *Yün Chi Chhi Chhien*.' Unpublished.

HO PING-YÜ (11). 'Notes on the *Pao Phu Tzu Shen Hsien Chin Shuo Ching* (The Preservation-of-Solidarity Master's Manual of the Bubbling Gold (Potion) of the Holy Immortals); an alchemical book (*TT*/910) attributed to Ko Hung (c. +320).' Unpublished.

HO PING-YÜ (12). 'Notes on the *Chin Pi Wu Hsiang Lei Tshan Thung Chhi* (Gold and Caerulean Jade Treatise on the Similarities and Categories of the Five (Substances) and the *Kinship of the Three*); an alchemical book (*TT*/897) attributed to Yin Chhang-Shêng (H/Han, c. +200), but probably of somewhat later date.' Unpublished.

HO PING-YÜ (13). 'Alchemy in Ming China (+1368 to +1644).' Communication to the XIIth International Congress of the History of Science, Paris, 1968. Abstract Vol. p. 174. Communications, Vol. 3A, p. 119.

HO PING-YÜ (14). 'Taoism in Sung and Yuan China.' Communication to the First International Conference of Taoist Studies, Villa Serbelloni, Bellagio, 1968.

HO PING-YÜ (15). 'The Alchemy of Stones and Minerals in the Chinese Pharmacopoeias.' *CCJ*, 1968, **7**, 155.

HO PING-YÜ (16). 'The System of the *Book of Changes* and Chinese Science.' *JSHS*, 1972, No. 11, 23.

HO PING-YÜ & CHHEN THIEH-FAN (1) = (1). 'On the Dating of the *Shun-Yang Lü Chen-Jen Yao Shih Chih*, a Taoist Pharmaceutical and Alchemical Manual.' *JOSHK*, 1971, **9**, 181 (229).

HO PING-YÜ, KO THIEN-CHI & LIM, BEDA (1). 'Lu Yu (+1125 to 1209), Poet-Alchemist.' *AM*, 1972, 163.

HO PING-YÜ, KO THIEN-CHI & PARKER, D. (1). 'Pai Chü-I's Poems on Immortality.' *HJAS*, 1974, **34**, 163.

HO PING-YÜ & LIM, BEDA (1). 'Tshui Fang, a Forgotten +11th-Century Alchemist [with assembly of citations, mostly from *Pên Tshao Kang Mu*, probably transmitted by *Kêng Hsin Yü Tshê*].' *JSHS*, 1972, No. 11, 103.

HO PING-YÜ, LIM, BEDA & MORSINGH, FRANCIS (1) (tr.). 'Elixir Plants: the *Shun-Yang Lü Chen-Jen Yao Shih Chih* (Pharmaceutical Manual of the Adept Lü Shun-Yang)' [in verses]. Art. in Nakayama & Sivin (1), p. 153.

HO PING-YÜ & NEEDHAM, JOSEPH (1). 'Ancient Chinese Observations of Solar Haloes and Parhelia.' *W*, 1959, **14**, 124.

HO PING-YÜ & NEEDHAM, JOSEPH (2). 'Theories of Categories in Early Mediaeval Chinese Alchemy' (with transl. of the *Tshan Thung Chhi Wu Hsiang Lei Pi Yao*, c. +6th to +8th cent.). *JWCI*, 1959, **22**, 173.

HO PING-YÜ & NEEDHAM, JOSEPH (3). 'The Laboratory Equipment of the Early Mediaeval Chinese Alchemists.' *AX*, 1959, **7**, 57.

HO PING-YÜ & NEEDHAM, JOSEPH (4). 'Elixir Poisoning in Mediaeval China.' *JAN*, 1959, **48**, 221.

HO PING-YÜ & NEEDHAM, JOSEPH. See Tshao Thien-Chhin, Ho Ping-Yü & Needham, J.

HOEFER, F. (1). *Histoire de la Chimie*. 2 vols. Paris, 1842–3. 2nd ed. 2 vols. Paris, 1866–9.

HOENIG, J. (1). 'Medical Research on Yoga.' *COPS*, 1968, **11**, 69.

HOERNES, M. (1). *Natur- und Ur-geschichte der Menschen*. Vienna and Leipzig, 1909.

HOERNLE, A. F. R. (1) (ed. & tr.). *The Bower Manuscript; Facsimile Leaves, Nagari Transcript, Romanised Transliteration and English Translation with Notes*. 2 vols. Govt. Printing office, Calcutta, 1893– 1912. (Archaeol. Survey of India, New Imperial Series, no. 22.) Mainly pharmacological text of late +4th cent. but with some chemistry also.

HOFF, H. H., GUILLEMIN, L. & GUILLEMIN, R. (1) (tr. and ed.). *The 'Cahier Rouge' of Claude Bernard*. Schenkman, Cambridge, Mass. 1967.

HOFFER, A. & OSMOND, H. (1). *The Hallucinogens*. Academic Press, New York and London, 1968. With a chapter by T. Weckowicz.

HOFFMANN, G. (1). Art. 'Chemie' in A. Ladenburg (ed.), *Handwörterbuch der Chemie*, Trewendt, Breslau, 1884, vol. 2, p. 516. This work forms Division 2, Part 3 of W. Förster (ed.), *Encyklopaedie der Naturwissenschaften* (same publisher).

HOFMANN, K. B. (1). 'Zur Geschichte des Zinkes bei den Alten.' *BHMZ*, 1882, **41**, 492, 503.

HOLGEN, H. J. (1). 'Iets over de Chineesche Alchemie.' *CW*, 1917, **24**, 400.

HOLGEN, H. J. (2). 'Iets uit de Geschiedenis van de Chineesche Mineralogie en Chemische Technologie.' *CW*, 1917, **24**, 468.

HOLLOWAY, M. (1). *Heavens on Earth; Utopian Communities in America, +1680 to 1880*. Turnstile, London, 1951. 2nd ed. Dover, New York, 1966.

HOLMYARD, E. J. (1). *Alchemy*. Penguin, London, 1957.

HOLMYARD, E. J. (2). 'Jābir ibn Ḥayyān [including a bibliography of the Jābirian corpus].' *PRSM*, 1923, **16** (Hist. Med. Sect.), 46.

HOLMYARD, E. J. (3). 'Some Chemists of Islam.' *SPR*, 1923, **18**, 66.

HOLMYARD, E. J. (4). 'Arabic Chemistry [and Cupellation].' *N*, 1922, **109**, 778.

HOLMYARD, E. J. (5). '*Kitāb al-'Ilm al-Muktasab fī Zirāʿat al-Dhahab*' (*Book of Knowledge acquired concerning the Cultivation of Gold), by Abū'l Qāsim Muḥammad ibn Aḥmad al-Irāqī* [d. c. +1300]; the Arabic text edited with a translation and introduction. Geuthner, Paris, 1923.

HOLMYARD, E. J. (7). 'A Critical Examination of Berthelot's Work on Arabic Chemistry.' *ISIS*, 1924, **6**, 479.

HOLMYARD, E. J. (8). 'The Identity of Geber.' *N*, 1923, **111**, 191.

HOLMYARD, E. J. (9). 'Chemistry in Mediaeval Islam.' *CHIND*, 1923, **42**, 387. *SCI*, 1926, 287.

H[OLMYARD], E. J. (10). 'The Accuracy of Weighing in the +8th Century.' *N*, 1925, **115**, 963.

HOLMYARD, E. J. (11). 'Maslama al-Majrīṭī and the *Rutbat al-Ḥakīm* [(The Sage's Step)].' *ISIS*, 1924, **6**, 293.

HOLMYARD, E. J. (12) (ed.). *The 'Ordinall of Alchimy' by Thomas Norton of Bristoll* (c. +1440; facsimile reproduction from the *Theatrum Chemicum Brittannicum* (+1652) with annotations by Elias Ashmole). Arnold, London, 1928.

HOLMYARD, E. J. (13). 'The Emerald Table.' *N*, 1923, **112**, 525.

HOLMYARD, E. J. (14). 'Alchemy in China.' *AP*, 1932, **3**, 745.

HOLMYARD, E. J. (15). 'Aidamir al-Jildakī [+14th-century alchemist].' *IRAQ*, 1937, **4**, 47.

HOLMYARD, E. J. (16). 'The Present Position of the Geber Problem.' *SPR*, 1925, **19**, 415.

HOLMYARD, E. J. (17). 'An Essay on Jābir ibn Ḥayyān.' Art. in *Studien z. Gesch. d. Chemie; Festgabe f. E. O. von Lippmann zum 70. Geburtstage*, ed. J. Ruska (37). Springer, Berlin, 1927, p. 28.

HOLMYARD, E. J. & MANDEVILLE, D. C. (1). '*Avicennae De Congelatione et Conglutinatione Lapidum*', being Sections of the '*Kitāb al-Shifā*'; the Latin and Arabic texts edited with an English translation of the latter and with critical notes. Geuthner, Paris, 1927. Rev. G. Sarton, *ISIS*, 1928, **11**, 134.

HOLTORF, G. W. (1). *Hongkong—World of Contrasts*. Books for Asia, Hongkong, 1970.

HOMANN, R. (1). *Die wichtigsten Körpergottheiten im 'Huang Thing Ching'* (Inaug. Diss. Tübingen). Kümmerle, Göppingen, 1971. (Göppinger Akademische Beiträge, no. 27.)

HOMBERG, W. (1). Chemical identification of a carved realgar cup brought from China by the ambassador of Siam. *HRASP*, 1703, 51.

HOMMEL, R. P. (1). *China at Work; an illustrated Record of the Primitive Industries of China's Masses, whose Life is Toil, and thus an Account of Chinese Civilisation*. Bucks County Historical Society, Doylestown, Pa.; John Day, New York, 1937.

HOMMEL, W. (1). 'The Origin of Zinc Smelting.' *EMJ*, 1912, **93**, 1185.

HOMMEL, W. (2). 'Über indisches und chinesisches Zink.' *ZAC*, 1912, **25**, 97.

HOMMEL, W. (3). 'Chinesisches Zink.' *CHZ*, 1912, **36**, 905, 918.

HOOVER, H. C. & HOOVER, L. H. (1) (tr.). *Georgius Agricola 'De Re Metallica' translated from the 1st Latin edition of 1556, with biographical introduction, annotations and appendices upon the development of mining methods, metallurgical processes, geology, mineralogy and mining law from the earliest times to the 16th century*. 1st ed. Mining Magazine, London, 1912; 2nd ed. Dover, New York, 1950.

HOOYKAAS, R. (1). 'The Experimental Origin of Chemical Atomic and Molecular Theory before Boyle.' *CHYM*, 1949, **2**, 65.

HOOYKAAS, R. (2). 'The Discrimination between "Natural" and "Artificial" Substances and the Development of Corpuscular Theory.' *A/AIHS*, 1947, **1**, 640.

HOPFNER, T. (1). *Griechisch-Aegyptischer Offenbarungszauber.* 2 vols. photolitho script. (Studien z. Palaeogr. u. Papyruskunde, ed. C. Wessely, nos. 21, 23.)

HOPKINS, A. J. (1). *Alchemy, Child of Greek Philosophy.* Columbia Univ. Press, New York, 1934. Rev. D. W. Singer, *A*, 1936, **18**, 94; W. J. Wilson, *ISIS*, 1935, **24**, 174.

HOPKINS, A. J. (2). 'A Defence of Egyptian Alchemy.' *ISIS*, 1938, **28**, 424.

HOPKINS, A. J. (3). 'Bronzing Methods in the Alchemical Leiden Papyri.' *CN*, 1902, **85**, 49.

HOPKINS, A. J. (4). 'Transmutation by Colour; a Study of the Earliest Alchemy.' Art. in *Studien z. Gesch. d. Chemie* (von Lippmann Festschrift), ed. J. Ruska. Springer, Berlin, 1927, p. 9.

HOPKINS, E. W. (3). 'Soma.' Art. in *ERE*, vol. xi, p. 685.

HOPKINS, E. W. (4). 'The Fountain of Youth.' *JAOS*, 1905, **26**, 1–67.

HOPKINS, L. C. (17). 'The Dragon Terrestial and the Dragon Celestial; I, A Study of the *Lung* (terrestrial).' *JRAS*, 1931, 791.

HOPKINS, L. C. (18). 'The Dragon Terrestial and the Dragon Celestial; II, A Study of the *Chhen* (celestial).' *JRAS*, 1932, 91.

HOPKINS, L. C. (25). 'Metamorphic Stylisation and the Sabotage of Significance; a Study in Ancient and Modern Chinese Writing.' *JRAS*, 1925, 451.

HOPKINS, L. C. (26). 'Where the Rainbow Ends.' *JRAS*, 1931, 603.

HORI ICHIRO (1). 'Self-Mummified Buddhas in Japan; an Aspect of the Shugen-dō ('Mountain Asceticism') Cult.' *HOR*, 1961, **1** (no. 2), 222.

HORI ICHIRO (2). *Folk Religion in Japan; Continuity and Change*, ed. J. M. Kitagawa & A. L. Miller. Univ. of Tokyo Press, Tokyo and Univ. of Chicago Press, Chicago, 1968. (Haskell Lectures on the History of Religions, new series, no. 1.)

D'HORME, E. & DUSSAUD, R. (1). *Les Religions de Babylonie et d'Assyrie, des Hittites et des Hourrites, des Phéniciens et des Syriens.* Presses Univ. de France, Paris, 1945. (Mana, Introd. à l'Histoire des Religions, no. 1, pt. 2.)

D'HORMON, A. *et al.* (1) (ed. & tr.). '*Han Wu Ti Ku Shih;* Histoire Anecdotique et Fabuleuse de l'Empereur Wou [Wu] des Han' in *Lectures Chinoises.* École Franco-Chinoise, Peiping, 1945 (no. 1), p. 28.

D'HORMON, A. (2) (ed.). *Lectures Chinoises.* École Franco-Chinoise, Peiping, 1945–. No. 1 contains text and tr. of the *Han Wu Ti Ku Shih*, p. 28.

HOURANI, G. F. (1). *Arab Seafaring in the Indian Ocean in Ancient and Early Mediaeval Times.* Princeton Univ. Press, Princeton, N.J. 1951. (Princeton Oriental Studies, no. 13.)

HOWARD-WHITE, F. B. (1). *Nickel, an Historical Review.* Methuen, London, 1963.

HOWELL, E. B. (1) (tr.). '*Chin Ku Chhi Kuan;* story no. XIII, the Persecution of Shen Lien.' *CJ*, 1925, **3**, 10.

HOWELL, E. B. (2) (tr.). *The Inconstancy of Madam Chuang, and other Stories from the Chinese . . .* (from the *Chin Ku Chhi Kuan, c.* +1635). Laurie, London, n.d. (1925).

HRISTOV, H., STOJKOV, G. & MIJATER, K. (1). *The Rila Monastery [in Bulgaria]; History, Architecture, Frescoes, Wood-Carvings.* Bulgarian Acad. of Sci., Sofia, 1959. (Studies in Bulgaria's Architectural Heritage, no. 6.)

HSIA NAI (6). 'Archaeological Work during the Cultural Revolution.' *CREC*, 1971, **20** (no. 10), 31.

HSIA NAI, KU YEN-WEN, LAN HSIN-WÊN *et al.* (1). *New Archaeological Finds in China.* Foreign Languages Press, Peking, 1972. With colour-plates, and Chinese characters in footnotes.

HSIAO WÊN (1). 'China's New Discoveries of Ancient Treasures.' *UNESC*, 1972, **25** (no. 12), 12.

HTIN AUNG, MAUNG (1). *Folk Elements in Burmese Buddhism.* Oxford, 1962. Rev. P. M. R[attansi], *AX*, 1962, **10**, 142.

HUANG TZU-CHHING (1). 'Über die alte chinesische Alchemie und Chemie.' *WA*, 1957, **6**, 721.

HUANG TZU-CHHING (2). 'The Origin and Development of Chinese Alchemy.' Unpub. MS. of a lecture in the Physiological Institute of Chhinghua University, *c.* 1942 (dated 1944). A preliminary form of Huang Tzu-Chhing (1) but with some material which was omitted from the German version, though that was considerably enlarged.

HUANG TZU-CHHING & CHAO YÜN-TSHUNG (1) (tr.). 'The Preparation of Ferments and Wines [as described in the *Chhi Min Yao Shu* of] Chia Ssu-Hsieh of the Later Wei Dynasty [*c.* +540]; with an introduction by T. L. Davis.' *HJAS*, 1945, **9**, 24. Corrigenda by Yang Lien-Shêng, 1946, **10**, 186.

HUANG WÊN (1). '*Nei Ching*, the Chinese Canon of Medicine.' *CMJ*, 1950, **68**, 17 (originally M.D. Thesis, Cambridge, 1947).

HUANG WÊN (2). *Poems of Mao Tsê-Tung, translated and annotated.* Eastern Horizon Press, Hongkong, 1966.

HUARD, P. & HUANG KUANG-MING (M. WONG) (1). 'La Notion de Cercle et la Science Chinoise.' *A/AIHS*, 1956, **9**, 111. (Mainly physiological and medical.)

HUARD, P. & HUANG KUANG-MING (M. WONG) (2). *La Médecine Chinoise au Cours des Siècles.* Dacosta, Paris, 1959.

HUARD, P. & HUANG KUANG-MING (M. WONG) (3). 'Évolution de la Matière Médicale Chinoise.' *JAN*, 1958, **47**. Sep. pub. Brill, Leiden, 1958.

HUARD, P. & HUANG KUANG-MING (M. WONG) (5). 'Les Enquêtes Françaises sur la Science et la Technologie Chinoises au 18e Siècle.' *BEFEO*, 1966, **53**, 137–226.

HUARD, P. & HUANG KUANG-MING (M. WONG) (7). *Soins et Techniques du Corps en Chine, au Japon et en Inde; Ouvrage précédé d'une Étude des Conceptions et des Techniques de l'Éducation Physique, des Sports et de la Kinésithérapie en Occident dépuis l'Antiquité jusquà l'Époque contemporaine.* Berg International, Paris, 1971.

HUARD, P., SONOLET, J. & HUANG KUANG-MING (M. WONG) (1). 'Mesmer en Chine; Trois Lettres Médicales [MSS] du R. P. Amiot; rédigées à Pékin, de +1783 à +1790. *RSH*, 1960, **81**, 61.

HUBER, E. (1). 'Die mongolischen Destillierapparate.' *CHA*, 1928, **15**, 145.

HUBER, E. (2). *Der Kampf um den Alkohol im Wandel der Kulteren.* Trowitsch, Berlin, 1930.

HUBER, E. (3). *Bier und Bierbereitung bei den Völkern der Urzeit*,
Vol. 1. *Babylonien und Ägypten.*
Vol. 2. *Die Völker unter babylonischen Kultureinfluss; Auftreten des gehopften Bieres.*
Vol. 3. *Der ferne Osten und Äthiopien.*
Gesellschaft f. d. Geschichte und Bibliographie des Brauwesens, Institut f. Gärungsgewerbe, Berlin, 1926–8.

HUBICKI, W. (1). 'The Religious Background of the Development of Alchemy and Chemistry at the Turn of the +16th and +17th Centuries.' Communication to the XIIth Internat. Congr. Hist. of Sci. Paris, 1968. Résumés, p. 102. Actes, vol. 3A, p. 81.

HUFELAND, C. (1). *Makrobiotik; oder die Kunst das menschliche Leben zu verlängern.* Berlin, 1823. *The Art of Prolonging Life.* 2 vols. Tr. from the first German ed. London, 1797. Hebrew tr. Lemberg (Lwów), 1831.

HUGHES, A. W. MCKENNY (1). 'Insect Infestation of Churches.' *JRIBA*, 1954.

HUGHES, E. R. (1). *Chinese Philosophy in Classical Times.* Dent, London, 1942. (Everyman Library, no. 973.)

HUGHES, M. J. & ODDY, W. A. (1). 'A Reappraisal of the Specific Gravity Method for the Analysis of Gold Alloys.' *AMY*, 1970, **12**, 1.

HUMBERT, J. P. L. (1). *Guide de la Conversation Arabe.* Paris, Bonn and Geneva, 1838.

VON HUMBOLDT, ALEXANDER (1). *Cosmos; a Sketch of a Physical Description of the Universe.* 5 vols. Tr. E. Cotté, B. H. Paul & W. S. Dallas. Bohn, London, 1849–58.

VON HUMBOLDT, ALEXANDER (3). *Examen Critique de l'Histoire de la Géographie du Nouveau Continent, et des Progrès de l'Astronomie Nautique au 15e et 16e Siècles.* 2 vols. Paris, 1837.

VON HUMBOLDT, ALEXANDER (4). *Fragmens de Géologie et de Climatologie Asiatique.* 2 vols. Gide, de la Forest & Delaunay, Paris, 1831.

HUMMEL, A. W. (6). 'Astronomy and Geography in the Seventeenth Century [in China].' (On Hsiung Ming-Yü's work.) *ARLC/DO*, 1938, 226.

HUNGER, H., STEGMÜLLER, O., ERBSE, H. *et al.* (1). *Geschichte der Textüberlieferung der antiken und mittelälterlichen Literatur.* 2 vols. Vol. 1, *Antiken Literatur.* Atlantis, Zürich, 1964. See Ineichen, Schindler, Bodmer *et al.* (1).

HUSAIN, YUSUF (1) (ed. & tr.). '*Ḥauḍ al-Ḥayāt* [=*Baḥr al-Ḥayāt* (The Ocean, or Water, of Life)], la Version Arabe de l'*Amritkunḍa* [text and French précis transl.].' *JA*, 1928, **213**, 291.

HUTTEN, E. H. (1). 'Culture, One and Indivisible.' *HUM*, 1971, **86** (no. 5), 137.

HUZZAYIN, S. A. (1). *Arabia and the Far East; their commercial and cultural relations in Graeco-Roman and Irano-Arabian times.* Soc. Royale de Géogr. Cairo, 1942.

ICHIDA, MIKINOSUKE (1). 'The Hu Chi, mainly Iranian Girls, found in China during the Thang Period.' *MRDTB*, 1961, **20**, 35.

IDELER, J. L. (1) (ed.). *Physici et Medici Graeci Minores.* 2 vols. Reimer, Berlin, 1841.

IHDE, A. J. (1). 'Alchemy in Reverse; Robert Boyle on the Degradation of Gold.' *CHYM*, 1964, **9**, 47. Abstr. in Proc. Xth Internat. Congr. Hist. of Sci., Ithaca, N.Y., 1962, p. 907.

ILG, A. (1). 'Theophilus Presbyter *Schedula Diversarum Artium*; I, Revidierter Text, Übersetzung und Appendix.' *QSKMR*, 1874, **7**, 1–374.

IMBAULT-HUART, C. (1). 'La Légende du premier Pape des Taoistes, et l'Histoire de la Famille Pontificale des Tchang [Chang], d'après des Documents Chinois, traduits pour la première fois.' *JA*, 1884 (8e sér.), **4**, 389. Sep. pub. Impr. Nat. Paris, 1885.

IMBAULT-HUART, C. (2). 'Miscellanées Chinois.' *JA*, 1881 (7e sér.), **18**, 255, 534.

INEICHEN, G., SCHINDLER, A., BODMER, D. *et al.* (1). *Geschichte der Textüberlieferung der antiken und mittelälterlichen Literatur.* 2 vols. Vol. 2, *Mittelälterlichen Literatur.* Atlantis, Zürich, 1964. See Hunger, Stegmüller, Erbse *et al.* (1).

INTORCETTA, P., HERDTRICH, C., [DE] ROUGEMONT, F. & COUPLET, P. (1) (tr.). '*Confucius Sinarum Philosophus, sive Scientia Sinensis, latine exposita*'...; *Adjecta est: Tabula Chronologica Monarchiae Sinicae juxta cyclos annorum LX, ab anno post Christum primo, usque ad annum praesentis Saeculi 1683* [by P. Couplet, pr. 1686]. Horthemels, Paris, 1687. Rev. in *PTRS*, 1687, **16** (no. 189), 376.

IYENGAR, B. K. S. (1). *Light on Yoga ('Yoga Bīpika')*. Allen & Unwin, London, 2nd ed. 1968, 2nd imp. 1970.

IYER, K. C. VIRARAGHAVA (1). 'The Study of Alchemy [in Tamilnad, South India].' Art. in *Acarya [P.C.] Ray Commemoration Volume*, ed. H. N. Datta, Meghned Saha, J. C. Ghosh *et al.* Calcutta, 1932, p. 460.

JACKSON, R. D. & VAN BAVEL, C. H. M. (1). 'Solar distillation of water from Soil and Plant materials; a simple Desert Survival technique.' *S*, 1965, **149**, 1377.

JACOB, E. F. (1). 'John of Roquetaillade.' *BJRL*, 1956, **39**, 75.

JACOBI, HERMANN (3). '[The "Abode of the Blest" in] Hinduism.' *ERE*, vol. ii, p. 698.

JACOBI, JOLANDE (1). *The Psychology of C. G. Jung; an Introduction with Illustrations.* Tr. from Germ. by R. Manheim. Routledge & Kegan Paul, London, 1942. 6th ed. (revised), 1962.

JACQUES, D. H. (1). *Physical Perfection.* New York, 1859.

JAGNAUX, R. (1). *Histoire de la Chimie.* 2 vols. Baudry, Paris, 1891.

JAHN, K. & FRANKE, H. (1). *Die China-Geschichte des Rašīd ad-Dīn [Rashīd al-Dīn]; Übersetzung, Kommentar, Facsimiletafeln.* Böhlaus, Vienna, 1971. (Österreiche Akademie der Wissenschaften, Phil.-Hist. Kl., Denkschriften, no. 105; Veröffentl. d. Kommission für Gesch. Mittelasiens, no. 1.) This is the Chinese section of the *Jāmiʿ al-Tawārīkh*, finished in +1304, the whole by +1316. See Meredith-Owens (1).

JAMES, MONTAGUE R. (1) (ed. & tr.). *The Apocryphal New Testament; being the Apocryphal Gospels, Acts, Epistles and Apocalypses, with other Narratives and Fragments, newly translated by....* Oxford, 1924, repr. 1926 and subsequently.

JAMES, WILLIAM (1). *Varieties of Religious Experience; a Study in Human Nature.* Longmans Green, London, 1904. (Gifford Lectures, 1901–2.)

JAMSHED BAKHT, HAKIM, S. & MAHDIHASSAN, S. (1). 'Calcined Metals or *kushtas*; a Class of Alchemical Preparations used in Unani-Ayurvedic Medicine.' *MED*, 1962, **24**, 117.

JAMSHED BAKHT, HAKIM, S. & MAHDIHASSAN, S. (2). 'Essences [(*araqiath*)]; a Class of Alchemical Preparations [used in Unani-Ayurvedic Medicine].' *MED*, 1962, **24**, 257.

JANSE, O. R. T. (6). 'Rapport Préliminaire d'une Mission archéologique en Indochine.' *RAA/AMG*, 1935, **9**, 144, 209; 1936, **10**, 42.

JEBB, S. (1). *Fratris Rogeri Bacon Ordinis Minorum 'Opus Majus' ad Clementum Quartum Pontificem Romanum* [r. +1265 to +1268] *ex MS. Codice Dublinensi, cum aliis quibusdam collato, nunc primum edidit...* Bowyer, London, 1733.

JEFFERYS, W. H. & MAXWELL, J. L. (1). *The Diseases of China, including Formosa and Korea.* Bale & Danielsson, London, 1910. 2nd ed., re-written by Maxwell alone. ABC Press, Shanghai, 1929.

JENYNS, R. SOAME (3). *Archaic [Chinese] Jades in the British Museum.* Brit. Mus. Trustees, London, 1951.

JOACHIM, H. H. (1). 'Aristotle's Conception of Chemical Combination.' *JP*, 1904, **29**, 72.

JOHN OF ANTIOCH (fl. +610) (1). *Historias Chronikēs apo Adam.* See Valesius, Henricus (1).

JOHN MALALA (prob. = Joh. Scholasticus, Patriarch of Byzantium, d. +577). *Chronographia*, ed. W. Dindorf. Weber, Bonn, 1831 (in *Corp. Script. Hist. Byz.* series).

JOHNSON, A. CHANDRAHASAN & JOHNSON, SATYABAMA (1). 'A Demonstration of Oesophageal Reflux using Live Snakes.' *CLINR*, 1969, **20**, 107.

JOHNSON, C. (1) (ed.) (tr.). '*De Necessariis Observantiis Scaccarii Dialogus (Dialogus de Scaccario)*', '*Discourse on the Exchequer*', by Richard Fitznigel, Bishop of London and Treasurer of England [c. +1180], *text and translation, with introduction.* London, 1950.

JOHNSON, OBED S. (1). *A Study of Chinese Alchemy.* Commercial Press, Shanghai, 1928. Ch. tr. by Huang Su-Fêng: *Chung-Kuo Ku-Tai Lien-Tan Shu.* Com. Press, Shanghai, 1936. Rev. B. Laufer, *ISIS*, 1929, **12**, 330; H. Chatley, *JRAS/NCB*, 1928, *NCDN*, 9 May 1928. Cf. Waley (14).

JOHNSON, R. P. (1). 'Note on some Manuscripts of the *Mappae Clavicula*.' *SP*, 1935, **10**, 72.

JOHNSON, R. P. (2). '*Compositiones Variae*'... *an Introductory Study.* Urbana, Ill. 1939. (Illinois Studies in Language and Literature, vol. 23, no. 3.)

JONAS, H. (1). *The Gnostic Religion.* Beacon, Boston, 1958.

JONES, B. E. (1). *The Freemason's Guide and Compendium.* London, 1950.

JONES, C. P. (1) (tr.). *Philostratus' 'Life of Apollonius'*, with an introduction by G. W. Bowersock, Penguin, London, 1970.

DE JONG, H. M. E. (1). *Michael Maier's 'Atalanta Fugiens'; Sources of an Alchemical Book of Emblems.* Brill, Leiden, 1969. (Janus Supplements, no. 8.)

JOPE, E. M. (3). 'The Tinning of Iron Spurs; a Continuous Practice from the +10th to the +17th Century.' *OX*, 1956, **21**, 35.

JOSEPH, L. (1). 'Gymnastics from the Middle Ages to the Eighteenth Century.' *CIBA/S*, 1949, **10**, 1030.

JOSTEN, C. H. (1). 'The Text of John Dastin's "Letter to Pope John XXII".' *AX*, 1951, **4**, 34.

JOURDAIN, M. & JENYNS, R. SOAME (1). *Chinese Export Art.* London, 1950.

JOYCE, C. R. B. & CURRY, S. H. (1) (ed.). *The Botany and Chemistry of* Cannabis. Williams & Wilkins, Baltimore, 1970. Rev. *SAM*, 1971, **224** (no. 3), 238.

JUAN WEI-CHOU. See Wei Chou-Yuan.

JULIEN, STANISLAS (1) (tr.). *Voyages des Pélerins Bouddhistes.* Impr. Imp., Paris, 1853–8. 3 vols. (Vol. 1 contains Hui Li's Life of Hsüan-Chuang; Vols. 2 and 3 contain Hsüan-Chuang's *Hsi Yu Chi.*)

JULIEN, STANISLAS (11). 'Substance anaesthésique employée en Chine dans le Commencement du 3e Siècle de notre ére pour paralyser momentanément la Sensibilité.' *CRAS*, 1849, **28**, 195.

JULIEN, STANISLAS & CHAMPION, P. (1). *Industries Anciennes et Modernes de l'Empire Chinois, d'après des Notices traduites du Chinois.*...(paraphrased précis accounts based largely on *Thien Kung Khai Wu*; and eye-witness descriptions from a visit in 1867). Lacroix, Paris, 1869.

JULIUS AFRICANUS. *Kestoi.* See Thevenot, D. (1).

JUNG, C. G. (1). *Psychologie und Alchemie.* Rascher, Zürich, 1944. 2nd ed. revised, 1952. Eng. tr. R. F. C. Hull [& B. Hannah], *Psychology and Alchemy.* Routledge & Kegan Paul, London, 1953 (Collected Works, vol. 12). Rev. W. Pagel, *ISIS*, 1948, **39**, 44; G. H[eym], *AX*, 1948, **3**, 64.

JUNG, C. G. (2). 'Synchronicity; an Acausal Connecting Principle' [on extra-sensory perception]; essay in the collection *The Structure and Dynamics of the Psyche.* Routledge & Kegan Paul, London, 1960 (Collected Works, vol. 8). Rev. C. Allen, *N*, 1961, **191**, 1235.

JUNG, C. G. (3). *Alchemical Studies.* Eng. tr. from the Germ., R. F. C. Hull. Routledge & Kegan Paul, London, 1968 (Collected Works, vol. 13). Contains the 'European commentary' on the *Thai-I Chin Hua Tsung Chih,* pp. 1–55, and the 'Interpretation of the Visions of Zosimos', pp. 57–108.

JUNG, C. G. (4). *Aion; Researches into the Phenomenology of the Self.* Eng. tr. from the Germ., R. F. C. Hull. Routledge & Kegan Paul, London, 1959 (Collected Works, vol. 9, pt. 2).

JUNG, C. G. (5). *Paracelsica.* Rascher, Zürich and Leipzig, 1942. Eng. tr. from the Germ., R. F. C. Hull.

JUNG, C. G. (6). *Psychology and Religion; West and East.* Eng. tr. from the Germ., R. F. C. Hull. Routledge & Kegan Paul, London, 1958 (63 corr.) (Collected Works, vol. 11). Contains the essay 'Transformation Symbolism in the Mass'.

JUNG, C. G. (7). *Memories, Dreams and Reflections.* Recorded by A. Jaffé, tr. R. & C. Winston. New York and London, 1963.

JUNG, C. G. (8). *Mysterium Conjunctionis; an Enquiry into the Separation and Synthesis of Psychic Opposites in Alchemy.* Eng. tr. from the Germ., R. F. C. Hull. Routledge & Kegan Paul, London, 1963 (Collected Works, vol. 14). Orig. ed. *Mysterium Conjunctionis; Untersuchung ü. die Trennung u. Zusammensetzung der seelische Gegensätze in der Alchemie,* 2 vols. Rascher, Zürich, 1955, 1956 (Psychol. Abhandlungen, ed. C. G. J., nos. 10, 11).

JUNG, C. G. (9). 'Die Erlösungsvorstellungen in der Alchemie.' *ERJB*, 1936, 13–111.

JUNG, C. G. (10). *The Integration of the Personality.* Eng. tr. S. Dell. Farrar & Rinehart, New York and Toronto, 1939, Kegan Paul, Trench & Trübner, London, 1940, repr. 1941. Ch. 5, 'The Idea of Redemption in Alchemy' is the translation of Jung (9).

JUNG, C. G. (11). 'Über Synchronizität.' *ERJB*, 1952, **20**, 271.

JUNG, C. G. (12). *Analytical Psychology; its Theory and Practice.* Routledge, London, 1968.

JUNG, C. G. (13). *The Archetypes and the Collective Unconscious.* Eng. tr. by R. F. C. Hull. Routledge & Kegan Paul, London, 1959 (Collected Works, vol. 9, pt. 1).

JUNG, C. G. (14). 'Einige Bemerkungen zu den Visionen des Zosimos.' *ERJB*, 1938. Revised and expanded as 'Die Visionen des Zosimos' in *Von der Wurzeln des Bewusstseins; Studien ü. d. Archetypus.* In *Psychologische Abhandlungen.* Zürich, 1954, vol. 9.

JUNG, C. G. & PAULI, W. (1). *The Interpretation of Nature and the Psyche.*
(*a*) 'Synchronicity; an Acausal Connecting Principle', by C. G. Jung.
(*b*) 'The Influence of Archetypal Ideas on the Scientific Theories of Kepler', by W. Pauli.
Tr. R. F. C. Hull. Routledge & Kegan Paul, London, 1955.
Orig. pub. in German as *Naturerklärung und Psyche,* Rascher, Zürich, 1952 (Studien aus dem C. G. Jung Institut, no. 4).

KAHLBAUM, G. W. A. See Diergart, P. (Kahlbaum Festschrift).

KAHLE, P. (7). 'Chinese Porcelain in the Lands of Islam.' *TOCS*, 1942, 27. Reprinted in Kahle (3), p. 326, with Supplement, p. 351 (originally published in *WA*, 1953, **2**, 179 and *JPHS*, 1953, **1**, 1).

KAHLE, P. (8). 'Islamische Quellen über chinesischen Porzellan.' *ZDMG*, 1934, **88**, 1, *OAZ*, 1934, **19** (N.F.), 69.

KALTENMARK, M. (2) (tr.). *Le 'Lie Sien Tchouan' [Lieh Hsien Chuan]; Biographies Légendaires des Immortels Taoistes de l'Antiquité.* Centre d'Etudes Sinologiques Franco-Chinois (Univ. Paris), Peking, 1953. Crit. P. Demiéville, *TP*, 1954, **43**, 104.

KALTENMARK, M. (4). 'Ling Pao; Note sur un Terme du Taoisme Religieux', in *Mélanges publiés par l'Inst. des Htes. Etudes Chin.* Paris, 1960, vol. 2, p. 559 (Bib. de l'Inst. des Htes. Et. Chin. vol. 14).

KANGRO, H. (1). *Joachim Jungius' [+1587 to +1657] Experimente und Gedanken zur Begründung der Chemie als Wissenschaft; ein Beitrag zur Geistesgeschichte des 17. Jahrhunderts.* Steiner, Wiesbaden, 1968 (Boethius; *Texte und Abhandlungen z. Gesch. d. exakten Naturwissenschaften*, no. 7). Rev. R. Hooykaas, *A/AIHS*, 1970, **23**, 299.

KAO LEI-SSU (1) (Aloysius Ko, S.J.). 'Remarques sur un Écrit de M. P[auw] intitulé "Recherches sur les Égyptiens et les Chinois" (1775).' *MCHSAMUC*, 1777, **2**, 365–574 (in some editions, 2nd pagination, 1–174).

KAO, Y. L. (1). 'Chemical Analysis of some old Chinese Coins.' *JWCBRS*, 1935, **7**, 124.

KAPFERER, R. (1). 'Der Blutkreislauf im altchinesischen Lehrbuch *Huang Ti Nei Ching*.' *MMW*, 1939 (no. 18), 718.

KARIMOV, U. I. (1) (tr.). *Neizvestnoe Sovrineniye al-Rāzī 'Kniga Taishnvi Taishi' (A Hitherto Unknown Work of al-Rāzī, 'Book of the Secret of Scerets').* Acad. Sci. Uzbek SSR, Tashkent, 1957. Rev. N. A. Figurovsky, tr. P. L. Wyvill, *AX*, 1962, **10**, 146.

KARLGREN, B. (18). 'Early Chinese Mirror Inscriptions.' *BMFEA*, 1934, **6**, 1.

KAROW, O. (2) (ed.). *Die Illustrationen des Arzneibuches der Periode Shao-Hsing* (Shao-Hsing Pên Tshao Hua Thu) *vom Jahre +1159, ausgewählt und eingeleitet.* Farbenfabriken Bayer Aktiengesellschaft (Pharmazeutisch-Wissenschaftliche Abteilung), Leverkusen, 1956. Album selected from the *Shao-Hsing Chiao-Ting Pên Tshao Chieh-Thi* published by Wada Toshihiko, Tokyo, 1933.

KASAMATSU, A. & HIRAI, T. (1). 'An Electro-encephalographic Study of Zen Meditation (*zazen*).' *FPNJ*, 1966, **20** (no. 4), 315.

KASSAU, E. (1). 'Charakterisierung einiger Steroidhormone durch Mikrosublimation.' *DAZ*, 1960, **100**, 1102.

KAZANCHIAN, T. (1). *Laboratornaja Technika i Apparatura v Srednevekovoj Armenii po drevnim Armjanskim Alchimicheskim Rukopisjam* (in Armenian with Russian summary). *SNM*, 1949, **2**, 1–28.

KEFERSTEIN, C. (1). *Mineralogia Polyglotta.* Anton, Halle, 1849.

KEILIN, D. & MANN, T. (1). 'Laccase, a blue Copper-Protein Oxidase from the Latex of *Rhus succedanea*.' *N*, 1939, **143**, 23.

KEILIN, D. & MANN, T. (2). 'Some Properties of Laccase from the Latex of Lacquer-Trees.' *N*, 1940, **145**, 304.

KEITH, A. BERRIEDALE (5). *The Religion and Philosophy of the Vedas and Upanishads.* 2 vols. Harvard Univ. Press, Cambridge (Mass.), 1925. (Harvard Oriental Series, nos. 31, 32.)

KEITH, A. BERRIEDALE (7). '[The State of the Dead in] Hindu [Belief].' *ERE*, vol. xi, p. 843.

KELLING, R. (1). *Das chinesische Wohnhaus; mit einem II Teil über das frühchinesische Haus unter Verwendung von Ergebnissen aus Übungen von Conrady im Ostasiatischen Seminar der Universität Leipzig,* von Rudolf Keller und Bruno Schindler. Deutsche Gesellsch. für Nat. u. Völkerkunde Ostasiens, Tokyo, 1935 (*MDGNVO*, Supplementband no. 13). Crit. P. Pelliot, *TP*, 1936, **32**, 372.

KENNEDY, J. (1). 'Buddhist Gnosticism, the System of Basilides.' *JRAS*, 1902, 377.

KENNEDY, J. (2). 'The Gospels of the Infancy, the *Lalita Vistara*, and the *Vishnu Purana*; or, the Transmission of Religious Ideas between India and the West.' *JRAS*, 1917, 209, 469.

KENT, A. (1). 'Sugar of Lead.' *MBLB*, 1961, **4** (no. 6), 85.

KERNEIZ, C. (1). *Les 'Asanas', Gymnastique immobile du Hathayoga.* Tallandier, Paris, 1946.

KERNEIZ, C. (2). *Le Yoga.* Tallandier, Paris, 1956. 2nd ed. 1960.

KERR, J. G. (1). '[Glossary of Chinese] Chemical Terms', in Doolittle (1), vol. 2, p. 542.

KEUP, W. (1) (ed.). *The Origin and Mechanisms of Hallucinations.* Plenum, New York and London, 1970.

KEYNES, J. M. (Lord Keynes) (1) (posthumous). 'Newton the Man.' Essay in *Newton Tercentenary Celebrations* (July 1946). Royal Society, London, 1947, p. 27. Reprinted in *Essays in Biography*.

KHORY, RUSTOMJEE NASERWANJEE & KATRAK, NANABHAI NAVROSJI (1). *Materia Medica of India and their Therapeutics.* Times of India, Bombay, 1903.

KHUNRATH, HEINRICH (1). *Amphitheatrum Sapientiae Aeternae Solius Verae, Christiano-Kabalisticum, Divino-Magicum, necnon Physico-Chymicum, Tetriunum, Catholicon...* Prague, 1598; Magdeburg, 1602; Frankfurt, 1608, and many other editions.

KIDDER, J. E. (1). *Japan before Buddhism.* Praeger, New York; Thames & Hudson, London, 1959.

KINCH, E. (1). 'Contributions to the Agricultural Chemistry of Japan.' *TAS/J*, 1880, **8**, 369.

KING, C. W. (1). *The Natural History of Precious Stones and of the Precious Metals*. Bell & Daldy, London, 1867.

KING, C. W. (2). *The Natural History of Gems or Decorative Stones*. Bell & Daldy, London, 1867.

KING, C. W. (3). *The Gnostics and their Remains*. 2nd ed. Nutt, London, 1887.

KING, C. W. (4). *Handbook of Engraved Gems*. 2nd ed. Bell, London, 1885.

KLAPROTH, J. (5). 'Sur les Connaissances Chimiques des Chinois dans le 8ème Siècle.' *MAIS/SP*, 1810, 2, 476. Ital. tr., S. Guareschi, *SAEC*, 1904, 20, 449.

KLAPROTH, J. (6). *Mémoires relatifs à l'Asie...* 3 vols. Dondey Dupré, Paris, 1826.

KLAPROTH, M. H. (1). *Analytical Essays towards Promoting the Chemical Knowledge of Mineral Substances*. 2 vols. Cadell & Davies, London, 1801.

KNAUER, E. (1). 'Die Ovarientransplantation.' *AFG*, 1900, 60, 322.

KNOX, R. A. (1). *Enthusiasm; a Chapter in the History of Religion, with special reference to the +17th and +18th Centuries*. Oxford, 1950.

KO, ALOYSIUS. See Kao Lei-Ssu.

KOBERT, R. (1). 'Chronische Bleivergiftung in klassischen Altertume.' Art. in Kahlbaum Festschrift (1909), ed. Diergart (1), pp. 103–19.

KOPP, H. (1). *Geschichte d. Chemie*. 4 vols. 1843–7.

KOPP, H. (2). *Beiträge zur Geschichte der Chemie*. Vieweg, Braunschweig, 1869.

KRAMRISCH, S. (1). *The Art of India; Traditions of Indian Sculpture, Painting and Architecture*. Phaidon, London, 1954.

KRAUS, P. (1). 'Der Zusammenbruch der Dschābir-Legende; II, Dschābir ibn Ḥajjān und die Isma'ilijja.' *JBFIGN*, 1930, 3, 23. Cf. Ruska (1).

KRAUS, P. (2). 'Jābir ibn Ḥayyān; Contributions à l'Histoire des Idées Scientifiques dans l'Islam; I, Le Corpus des Écrits Jābiriens.' *MIE*, 1943, 44, 1–214. Rev. M. Meyerhof, *ISIS*, 1944, 35, 213.

KRAUS, P. (3). 'Jābir ibn Ḥayyān; Contributions à l'Histoire des Idées Scientifiques dans l'Islam; II, Jābir et la Science Grecque.' *MIE*, 1942, 45, 1–406. Rev. M. Meyerhof, *ISIS*, 1944, 35, 213.

KRAUS, P. (4) (ed.). *Jābir ibn Ḥayyān; Essai sur l'Histoire des Idées Scientifiques dans l'Islam*. Vol. 1. *Textes Choisis*. Maisonneuve, Paris and El-Kandgi, Cairo, 1935. No more appeared.

KRAUS, P. (5). *L'Épître de Beruni sur al-Rāzī (Risālat al-Bīrūnī fī Fihrist Kutub Muḥammad ibn Zakarīyā al-Rāzī) [c. +1036]*. Paris, 1936.

KRAUS, P. & PINES, S. (1). 'Al-Rāzī.' Art. in *EI*, vol. iii, pp. 1134 ff.

KREBS, M. (1). *Der menschlichen Harn als Heilmittel; Geschichte, Grundlagen, Entwicklung, Praxis*. Marquardt, Stuttgart, 1942.

KRENKOW, F. (2). 'The Oldest Western Accounts of Chinese Porcelain.' *IC*, 1933, 7, 464.

KROLL, J. (1). *Die Lehren des Hermes Trismegistos*. Aschendorff, Münster i.W., 1914. (Beiträge z. Gesch. d. Philosophie des Mittelalters, vol. 12, no. 2.)

KROLL, W. (1). 'Bolos und Demokritos.' *HERM*, 1934, 69, 228.

KRÜNITZ, J. G. (1). *Ökonomisch-Technologische Enzyklopädie*. Berlin, 1773–81.

KUBO NORITADA (1). 'The Introduction of Taoism to Japan.' In *Religious Studies in Japan*, no. 11 (no. 105), 457. See Soymié (5), p. 281 (10).

KUBO NORITADA (2). 'The Transmission of Taoism to Japan, with particular reference to the *san shih* (three corpses theory).' *Proc. IXth Internat. Congress of the History of Religions*, Tokyo, 1958, p. 335.

KUGENER, M. A. & CUMONT, F. (1). *Extrait de la CXXIII ème 'Homélie' de Sévère d'Antioch*. Lamertin, Brussels, 1912. (Recherches sur le Manichéisme, no. 2.)

KUGENER, M. A. & CUMONT, F. (2). *L'Inscription Manichéenne de Salone [Dalmatia]*. (A tombstone or consecration memorial of the Manichaean Virgin Bassa.) Lamertin, Brussels, 1912. (Recherches sur le Manichéisme, no. 3.)

KÜHN, F. (3). *Die Dreizehnstöckige Pagode* (Stories translated from the Chinese). Steiniger, Berlin, 1940.

KÜHNEL, P. (1). *Chinesische Novellen*. Müller, München, 1914.

KUNCKEL, J. (1). '*Ars Vitraria Experimentalis*', oder *Vollkommene Glasmacher-Kunst...* Frankfurt and Leipzig; Amsterdam and Danzig, 1679. 2nd ed. Frankfurt and Leipzig, 1689. 3rd ed. Nuremberg, 1743, 1756. French tr. by the Baron d'Holbach, Paris, 1752.

KUNCKEL, J. (2). '*Collegium Physico-Chemicum Experimentale*', oder *Laboratorium Chymicum; in welchem deutlich und gründlich von den wahren Principiis in der Natur und denen gewürckten Dingen so wohl über als in der Erden, als Vegetabilien, Animalien, Mineralien, Metallen..., nebst der Transmutation und Verbesserung der Metallen gehandelt wird...* Heyl, Hamburg and Leipzig, 1716.

KUNG, S. C., CHAO, S. W., PEI, Y. T. & CHANG, C. (1). 'Some Mummies Found in West China.' *JWCBRS*, 1939, 11, 105.

KUNG YO-THING, TU YÜ-TSHANG, HUANG WEI-TÊ, CHHEN CHHANG-CHHING & seventeen other collaborators (1). 'Total Synthesis of Crystalline Insulin.' *SCISA*, 1966, 15, 544.

LACAZE-DUTHIERS, H. (1). 'Tyrian purple.' *ASN/Z*, 1859 (4e sér.), **12**, 5.

LACH, D. F. (5). *Asia in the Making of Europe.* 2 vols. Univ. Chicago Press, Chicago and London, 1965–.

LACH, D. F. (6). 'The Sinophilism of Christian Wolff (+1679 to +1754).' *JHI*, 1953, **14**, 561.

LAGERCRANTZ, O. (1). *Papyrus Graecus Holmiensis.* Almquist & Wiksells, Upsala, 1913. (The first publication of the +3rd-cent. technical and chemical Stockholm papyrus.) Cf. Caley (2).

LAGERCRANTZ, O. (2). 'Über das Wort Chemie.' *KVSUA*, 1937–8, 25.

LAMOTTE, E. (1) (tr.). *Le Traité de la Grande Vertu de Sagesse de Nāgārjuna (Mahāprajñāpāramitā-śāstra).* 3 vols. Muséon, Louvain, 1944 (Bibl. Muséon, no. 18). Rev. P. Demiéville, *JA*, 1950, **238**, 375.

LANDUR, N. (1). 'Compte Rendu de la Séance de l'Académie des Sciences [de France] du 24 Août 1868.' *LIN* (1e section), 1868, **36** (no. 1808), 273. Contains an account of a communication by M. Chevreul on the history of alchemy, tracing it to the *Timaeus;* with a critical paragraph by Landur himself maintaining that in his view much (though not all) of ancient and mediaeval alchemy was disguised moral and mystical philosophy.

LANE, E. W. (1). *An Account of the Manners and Customs of the Modern Egyptians (1833 to 1835).* Ward Lock, London, 3rd ed. 1842; repr. 1890.

LANGE, E. F. (1). 'Alchemy and the Sixteenth-Century Metallurgists.' *AX*, 1966, **13**, 92.

LATTIMORE, O. & LATTIMORE, E. (1) (ed.). *Silks, Spices and Empire; Asia seen through the Eyes of its Discoverers.* Delacorte, New York, 1968. (Great Explorers Series, no. 3.)

LAUBRY, C. & BROSSE, T. (1). 'Documents recueillis aux Indes sur les "Yoguis" par l'enregistrement simultané du pouls, de la respiration et de l'electrocardiogramme.' *PM*, 1936, **44** (no. 83), 1601 (14 Oct.). Rev. J. Filliozat, *JA*, 1937, 521.

LAUBRY, C. & BROSSE, T. (2). 'Interférence de l'Activité Corticale sur le Système Végétatif Neuro-vasculaire.' *PM*, 1935, **43** (no. 84). (19 Oct.)

LAUFER, B. (1). *Sino-Iranica; Chinese Contributions to the History of Civilisation in Ancient Iran.* *FMNHP/AS*, 1919, **15**, no. 3 (Pub. no. 201). Rev. and crit. Chang Hung-Chao, *MGSC*, 1925 (ser. B), no. 5.

LAUFER, B. (8). *Jade; a Study in Chinese Archaeology and Religion. FMNHP/AS*, 1912, **10**, 1–370. Repub. in book form, Perkins, Westwood & Hawley, South Pasadena, 1946. Rev. P. Pelliot, *TP*, 1912, **13**, 434.

LAUFER, B. (10). 'The Beginnings of Porcelain in China.' *FMNHP/AS*, 1917, **15**, no. 2 (Pub. no. 192), (includes description of +2nd-century cast-iron funerary cooking-stove).

LAUFER, B. (12). 'The Diamond; a study in Chinese and Hellenistic Folk-Lore.' *FMNHP/AS*, 1915, **15**, no. 1 (Pub. no. 184).

LAUFER, B. (13). 'Notes on Turquois in the East.' *FMNHP/AS*, 1913, **13**, no. 1 (Pub. no. 169).

LAUFER, B. (15). 'Chinese Clay Figures, Pt. I; Prolegomena on the History of Defensive Armor.' *FMNHP/AS*, 1914, **13**, no. 2 (Pub. no. 177).

LAUFER, B. (17). 'Historical Jottings on Amber in Asia.' *MAAA*, 1906, **1**, 211.

LAUFER, B. (24). 'The Early History of Felt.' *AAN*, 1930, **32**, 1.

LAUFER, B. (28). 'Christian Art in China.' *MSOS*, 1910, **13**, 100.

LAUFER, B. (40). 'Sex Transformation and Hermaphrodites in Ancient China.' *AJPA*, 1920, **3**, 259.

LAUFER, B. (41). 'Die Sage von der goldgrabenden Ameisen.' *TP*, 1908, **9**, 429.

LAUFER, B. (42). *Tobacco and its Use in Asia.* Field Mus. Nat. Hist., Chicago, 1924. (Anthropology Leaflet, no. 18.)

LEADBEATER, C. W. (1). *The Chakras, a Monograph.* London, n.d.

LECLERC, L. (1) (tr.). 'Le Traité des Simples par Ibn al-Beithar.' *MAI/NEM*, 1877, **23**, 25; 1883, **26**.

LECOMTE, LOUIS (1). *Nouveaux Mémoires sur l'État présent de la Chine.* Anisson, Paris, 1696. (Eng. tr. *Memoirs and Observations Topographical, Physical, Mathematical, Mechanical, Natural, Civil and Ecclesiastical, made in a late journey through the Empire of China, and published in several letters, particularly upon the Chinese Pottery and Varnishing, the Silk and other Manufactures, the Pearl Fishing, the History of Plants and Animals, etc.* translated from the Paris edition, etc., 2nd ed. London, 1698. Germ. tr. Frankfurt, 1699–1700. Dutch tr. 's Graavenhage, 1698.)

VON LECOQ, A. (1). *Buried Treasures of Chinese Turkestan; an Account of the Activities and Adventures of the 2nd and 3rd German Turfan Expeditions.* Allen & Unwin, London, 1928. Eng. tr. by A. Barwell of *Auf Hellas Spuren in Ost-turkestan.* Berlin, 1926.

VON LECOQ, A. (2). *Von Land und Leuten in Ost-Turkestan...* Hinrichs, Leipzig, 1928.

LEDERER, E. (1). 'Odeurs et Parfums des Animaux.' *FCON*, 1950, **6**, 87.

LEDERER, E. & LEDERER, M. (1). *Chromatography; a Review of Principles and Applications.* Elsevier, Amsterdam and London, 1957.

LEEDS, E. T. (1). 'Zinc Coins in Mediaeval China.' *NC*, 1955 (6th ser.), **14**, 177.

LEEMANS, C. (1) (ed. & tr.). *Papyri Graeci Musei Antiquarii Publici Lugduni Batavi...* Leiden, 1885. (Contains the first publication of the +3rd-cent. chemical papyrus Leiden X.) Cf. Caley (1).

VAN LEERSUM, E. C. (1). *Préparation du Calomel chez les anciens Hindous*. Art. in Kahlbaum Festschrift (1909), ed. Diergart (1), pp. 120–6.

LEFÉVRE, NICOLAS (1). *Traicté de la Chymie*. Paris, 1660, 2nd ed. 1674. Eng. tr. *A Compleat Body of Chymistry*. Pulleyn & Wright, London, 1664. repr. 1670.

LEICESTER, H. M. (1). *The Historical Background of Chemistry*. Wiley, New York, 1965.

LEICESTER, H. M. & KLICKSTEIN, H. S. (1). 'Tenney Lombard Davis and the History of Chemistry.' *CHYM*, 1950, **3**, 1.

LEICESTER, H. M. & KLICKSTEIN, H. S. (2) (ed.). *A Source-Book in Chemistry, +1400 to 1900*. McGraw-Hill, New York, 1952.

LEISEGANG, H. (1). *Der Heilige Geist; das Wesen und Werden der mystisch-intuitiven Erkenntnis in der Philosophie und Religion der Griechen*. Teubner, Leipzig and Berlin, 1919; photolitho reprint, Wissenschaftliche Buchgesellschaft, Darmstadt, 1967. This constitutes vol. 1 of Leisegang (2).

LEISEGANG, H. (2). '*Pneuma Hagion*'; *der Ursprung des Geistbegriffs der synoptischen Evangelien aus d. griechischen Mystik*. Hinrichs, Leipzig, 1922. (Veröffentlichungen des Forschungsinstituts f. vergl. Religionsgeschichte an d. Univ. Leipzig, no. 4.) This constitutes vol. 2 of Leisegang (1).

LEISEGANG, H. (3). *Die Gnosis*. 3rd ed. Kröner, Stuttgart, 1941. (Kröners Taschenausgabe, no. 32.) French tr.: '*La Gnose*'. Paris, 1951.

LEISEGANG, H. (4). 'The Mystery of the Serpent.' *ERYB*, 1955, 218.

LENZ, H. O. (1). *Mineralogie der alten Griechen und Römer deutsch in Auszügen aus deren Schriften*. Thienemann, Gotha, 1861. Photo reprint, Sändig, Wiesbaden, 1966.

LEPESME, P. (1). 'Les Coléoptères des Denrées alimentaires et des Produits industriels entreposés.' Art. in *Encyclopédie Entomologique*, vol. xxii, pp. 1–335. Lechevalier, Paris, 1944.

LESSIUS, L. (1). *Hygiasticon; seu Vera Ratio Valetudinis Bonae et Vitae…ad extremam Senectute Conservandae*. Antwerp, 1614. Eng. tr. Cambridge, 1634; and two subsequent translations.

LEVEY, M. (1). 'Evidences of Ancient Distillation, Sublimation and Extraction in Mesopotamia.' *CEN*, 1955, **4**, 23.

LEVEY, M. (2). *Chemistry and Chemical Technology in Ancient Mesopotamia*. Elsevier, Amsterdam and London, 1959.

LEVEY, M. (3). 'The Earliest Stages in the Evolution of the Still.' *ISIS*, 1960, **51**, 31.

LEVEY, M. (4). 'Babylonian Chemistry; a Study of Arabic and −2nd Millennium Perfumery.' *OSIS*, 1956, **12**, 376.

LEVEY, M. (5). 'Some Chemical Apparatus of Ancient Mesopotamia.' *JCE*, 1955, **32**, 180.

LEVEY, M. (6). 'Mediaeval Arabic Toxicology; the "Book of Poisons" of Ibn Waḥshīya [+10th cent.] and its Relation to Early Indian and Greek Texts.' *TAPS*, 1966, **56** (no. 7), 1–130.

LEVEY, M. (7). 'Some Objective Factors in Babylonian Medicine in the Light of New Evidence.' *BIHM*, 1961, **35**, 61.

LEVEY, M. (8). 'Chemistry in the *Kitāb al-Sumum* (Book of Poisons) by Ibn al-Waḥshīya [al-Nabaṭī, fl. +912].' *CHYM*, 1964, **9**, 33.

LEVEY, M. (9). 'Chemical Aspects of Medieval Arabic Minting in a Treatise by Manṣūr ibn Ba'ra [c: +1230].' *JSHS*, 1971, Suppl. no. 1.

LEVEY, M. & AL-KHALEDY, NOURY (1). *The Medical Formulary [Aqrābādhīn] of [Muḥ. ibn ʿAlī ibn ʿUmar] al-Samarqandī [c. +1210], and the Relation of Early Arabic Simples to those found in the indigenous Medicine of the Near East and India*. Univ. Pennsylvania Press, Philadelphia, 1967.

LÉVI, S. (2). 'Ceylan et la Chine.' *JA*, 1900 (9ᵉ sér.), **15**, 411. Part of Lévi (1).

LÉVI, S. (4). 'On a Tantric Fragment from Kucha.' *IHQ*, 1936, **12**, 204.

LÉVI, S. (6). *Le Népal; Étude Historique d'un Royaume Hindou*. 3 vols. Paris, 1905–8. (Annales du Musée Guimet, Bib. d'Études, nos. 17–19.)

LÉVI, S. (8). 'Un Nouveau Document sur le Bouddhisme de Basse Époque dans l'Inde.' *BLSOAS*, 1931, **6**, 417. (Nāgārjuna and gold refining.)

LÉVI, S. (9). 'Notes Chinoises sur l'Inde; V, Quelques Documents sur le Bouddhisme Indien dans l'Asie Centrale, pt. 1.' *BEFEO*, 1905, **5**, 253.

LÉVI, S. (10). 'Vajrabodhi à Ceylan.' *JA*, 1900, (9ᵉ sér.) **15**, 418. Part of Lévi (1).

LEVOL, A. (1). 'Analyse d'un Échantillon de Cuivre Blanc de la Chine.' *RCA*, 1862, **4**, 24.

LEVY, ISIDORE (1). 'Sarapis; V, la Statue Mystérieuse.' *RHR/AMG*, 1911, **63**, 124.

LEWIS, BERNARD (1). *The Arabs in History*. London.

LEWIS, M. D. S. (1). *Antique Paste Jewellery*. Faber, London, 1970. Rev. G. B. Hughes, *JRSA*, 1972, **120**, 263.

LEWIS, NORMAN (1). *A Dragon Apparent; Travels in Indo-China*. Cape, London, 1951.

LI CHHIAO-PHING (1) = (1). *The Chemical Arts of Old China* (tr. from the 1st, unrevised, edition, Chhangsha, 1940, but with additional material). J. Chem. Ed., Easton, Pa. 1948. Revs. W. Willetts, *ORA*, 1949, **2**, 126; J. R. Partington, *ISIS*, 1949, **40**, 280; Li Cho-Hao, *JCE*, 1949, **26**, 574. The Thaipei ed. of 1955 (Chinese text) was again revised and enlarged.

Lɪ Cʜo-Hᴀo (1). 'Les Hormones de l'Adénohypophyse.' *SCIS*, 1971, nos. 74–5, 69.

Lɪ Cʜo-Hᴀo & Eᴠᴀɴs, H. M. (1). 'Chemistry of the Anterior Pituitary Hormones.' Art. in *The Hormones*.... Ed. G. Pincus, K. V. Thimann & E. B. Astwood. Academic Press, New York, 1948–64, vol. 1, p. 633.

Lɪ Hᴜɪ-Lɪɴ (1). *The Garden Flowers of China*. Ronald, New York, 1959. (Chronica Botanica series, no. 19.)

Lɪ Kᴜo-Cʜʜɪɴ & Wᴀɴɢ Cʜʜᴜɴɢ-Yᴜ (1). *Tungsten, its History, Geology, Ore-Dressing, Metallurgy, Chemistry, Analysis, Applications and Economics.* Amer. Chem. Soc., New York, 1943 (Amer. Chem. Soc. Monographs, no. 94). 3rd ed. 1955 (A. C. S. Monographs, no. 130).

Lɪᴀɴɢ, H. Y. (1). 'The Wah Chang [Hua-Chhang, Antimony] Mines.' *MSP*, 1915, **111**, 53. (The initials are given in the original as H. T. Liang, but this is believed to be a misprint.)

Lɪᴀɴɢ, H. Y. (2). 'The Shui-khou Shan [Lead and Zinc] Mine in Hunan.' *MSP*, 1915, **110**, 914.

Lɪᴀɴɢ Po-Cʜʜɪᴀɴɢ (1). 'Überblick ü. d. seltenste chinesische Lehrbuch d. Medizin *Huang Ti Nei Ching*.' *AGMN*, 1933, **26**, 121.

Lɪʙᴀᴠɪᴜs, Aɴᴅʀᴇᴀs (1). *Alchemia. Andr. Libavii, Med. D[oct.], Poet. Physici Rotemburg. Operâ e Dispersis passim Optimorum Autorum, Veterum et Recentium exemplis potissimum, tum etian praeceptis quibusdam operosè collecta, adhibitâ; ratione et experientia, quanta potuit esse, methodo accuratâ explicata, et In Integrum Corpus Redacta*... Saur & Kopff, Frankfurt, 1597. Germ. tr. by F. Rex *et al.* Verlag Chemie, Weinheim, 1964.

Lɪʙᴀᴠɪᴜs, Aɴᴅʀᴇᴀs (2). *Singularium Pars Prima: in qua de abstrusioribus difficilioribusque nonnullis in Philosophia, Medicina, Chymia etc. Quaestionibus; utpote de Metallorum, Succinique Natura, de Carne fossili, ut credita est, de gestatione cacodaemonum, Veneno, aliisque rarioribus, quae versa indicat pagina, plurimis accuratè disseritur.* Frankfurt, 1599. Part II also 1599. Parts III and IV, 1601.

Lɪᴄʜᴛ, S. (1). 'The History [of Therapeutic Exercise].' Art. in *Therapeutic Exercise*, ed. S. Licht, Licht, New Haven, Conn. 1958, p. 380. (Physical Medicine Library, no. 3.)

Lɪᴇʙᴇɴ, F. (1). *Geschichte d. physiologische Chemie*. Deuticke, Leipzig and Vienna, 1935.

Lɪɴ Yü-Tʜᴀɴɢ (1) (tr.). *The Wisdom of Laotse [and Chuang Tzu] translated, edited and with an introduction and notes.* Random House, New York, 1948.

Lɪɴ Yü-Tʜᴀɴɢ (7). *Imperial Peking; Seven Centuries of China* (with an essay on the Art of Peking, by P. C. Swann). Elek, London, 1961.

Lɪɴ Yü-Tʜᴀɴɢ (8). *The Wisdom of China*. Joseph, London (limited edition) 1944; (general circulation edition) 1949.

Lɪɴᴅʙᴇʀɢ, D. C. & Sᴛᴇɴᴇᴄᴋ, N. H. (1). 'The Sense of Vision and the Origins of Modern Science', art. in *Science, Medicine and Society in the Renaissance* (Pagel Presentation Volume), ed. Debus (20), vol. 1, p. 29.

Lɪɴᴅᴇʙooᴍ, G. A. (1). *Hermann Boerhaave; the Man and his Work.* Methuen, London, 1968.

Lɪɴɢ, P. H. (1). *Gymnastikens Allmänna Grunder*...(in Swedish). Leffler & Sebell, Upsala and Stockholm, 1st part, 1834, 2nd part, 1840 (based on observations and practice from 1813 onwards). Germ. tr.: *P. H. Ling's Schriften über Leibesübungen* (with posthumous additions), by H. F. Massmann, Heinrichshofen, Magdeburg, 1847. Cf. Cyriax (1).

Lɪɴᴋ, Aʀᴛʜᴜʀ E. (1). 'The Taoist Antecedents in Tao-An's [+312 to +385] Prajñā Ontology.' Communication to the First International Conference of Taoist Studies, Villa Serbelloni, Bellagio, 1968.

ᴠoɴ Lɪᴘᴘᴍᴀɴɴ, E. O. (1). *Entstehung und Ausbreitung der Alchemie, mit einem Anhange, Zur älteren Geschichte der Metalle; ein Beitrag zur Kulturgeschichte.* 3 vols. Vol. 1, Springer, Berlin, 1919. Vol. 2, Springer, Berlin, 1931. Vol. 3, Verlag Chemie, Weinheim, 1954 (posthumous, finished in 1940, ed. R. von Lippmann).

ᴠoɴ Lɪᴘᴘᴍᴀɴɴ, E. O. (3). *Abhandlungen und Vorträge zur Geschichte d. Naturwissenschaften.* 2 vols. Vol. 1, Veit, Leipzig, 1906. Vol. 2, Veit, Leipzig, 1913.

ᴠoɴ Lɪᴘᴘᴍᴀɴɴ, E. O. (4). *Geschichte des Zuckers, seiner Darstellung und Verwendung, seit den ältesten Zeiten bis zum Beginne der Rübenzuckerfabrikation; ein Beitrag zur Kulturgeschichte.* Hesse, Leipzig, 1890.

ᴠoɴ Lɪᴘᴘᴍᴀɴɴ, E. O. (5). 'Chemisches bei Marco Polo.' *ZAC*, **21**, 1778. Repr. in (3), vol. 2, p. 258.

ᴠoɴ Lɪᴘᴘᴍᴀɴɴ, E. O. (6). 'Die spezifische Gewichtsbestimmung bei Archimedes.' Repr. in (3), vol. 2, p. 168.

ᴠoɴ Lɪᴘᴘᴍᴀɴɴ, E. O. (7). 'Zur Geschichte d. Saccharometers u. d. Senkspindel.' Repr. in (3), vol. 2, pp. 171, 177, 183.

ᴠoɴ Lɪᴘᴘᴍᴀɴɴ, E. O. (8). 'Zur Geschichte der Kältemischungen.' Address to the General Meeting of the Verein Deutscher Chemiker, 1898. Repr. in (3). vol. 1, p. 110.

ᴠoɴ Lɪᴘᴘᴍᴀɴɴ, E. O. (9). *Beiträge z. Geschichte d. Naturwissenschaften u. d. Technik.* 2 vols. Vol. 1, Springer, Berlin, 1925. Vol. 2, Verlag Chemie, Weinheim, 1953 (posthumous, ed. R. von Lippmann). Both vols. photographically reproduced, Sändig, Niederwalluf, 1971.

VON LIPPMANN, E. O. (10). 'J. Ruska's Neue Untersuchungen ü. die Anfänge der Arabischen Alchemie.' *CHZ*, 1925, 2, 27.

VON LIPPMANN, E. O. (11). 'Some Remarks on Hermes and Hermetica.' *AX*, 1938, **2**, 21.

VON LIPPMANN, E. O. (12). 'Chemisches u. Alchemisches aus Aristoteles.' *AGNT*, 1910, **2**, 233–300.

VON LIPPMANN, E. O. (13). 'Beiträge zur Geschichte des Alkohols.' *CHZ*, 1913, **37**, 1313, 1348, 1358, 1419, 1428. Repr. in (9), vol. 1, p. 60.

VON LIPPMANN, E. O. (14). 'Neue Beiträge zur Geschichte dez Alkohols.' *CHZ*, 1917, **41**, 865, 883, 911. Repr. in (9), vol. 1, p. 107.

VON LIPPMANN, E. O. (15). 'Zur Geschichte des Alkohols.' *CHZ*, 1920, **44**, 625. Repr. in (9), vol. 1, p. 123.

VON LIPPMANN, E. O. (16). 'Kleine Beiträge zur Geschichte d. Chemie.' *CHZ*, 1933, **57**, 433. 1. Zur Geschichte des Alkohols. 2. Der Essig des Hannibal. 3. Künstliche Perlen und Edelsteine. 4. Chinesische Ursprung der Alchemie.

VON LIPPMANN, E. O. (17). 'Zur Geschichte des Alkohols und seines Namens.' *ZAC*, 1912, **25**, 1179, 2061.

VON LIPPMANN, E. O. (18). 'Einige Bemerkungen zur Geschichte der Destillation und des Alkohols.' *ZAC*, 1912, **25**, 1680.

VON LIPPMANN, E. O. (19). 'Zur Geschichte des Wasserbades vom Altertum bis ins 13. Jahrhundert.' Art. in Kahlbaum Festschrift (1909), ed. Diergart (1), pp. 143–57.

VON LIPPMANN, E. O. (20). *Urzeugung und Lebenskraft; Zur Geschichte dieser Problem von den ältesten Zeiten an bis zu den Anfängen des 20. Jahrhunderts.* Springer, Berlin, 1933.

VON LIPPMANN, E. O. Biography, see Partington (19).

VON LIPPMANN, E. O. & SUDHOFF, K. (1). 'Thaddäus Florentinus (Taddeo Alderotti) über den Weingeist.' *AGMW*, 1914, **7**, 379. (Latin text, and comm. only.)

LIPSIUS, A. & BONNET, M. (1). *Acta Apostolorum Apocrypha.* 2 vols. in 3 parts. Mendelssohn, Leipzig, 1891–1903.

LITTLE, A. G. (1) (ed.). *Part of the 'Opus Tertium'* [c. +1268] of Roger Bacon. Aberdeen, 1912.

LITTLE, A. G. & WITHINGTON, E. (1) (ed.). *Roger Bacon's 'De Retardatione Accidentium Senectutis', cum aliis Opusculis de Rebus Medicinalibus.* Oxford, 1928. (Pubs. Brit. Soc. Franciscan Studies, no. 14.) Also printed as Fasc. 9 of Steele (1). Cf. the Engl. tr. of the *De Retardatione* by R. Browne, London, 1683.

LIU MAO-TSAI (1). *Kutscha und seine Beziehungen zu China vom 2 Jahrhundert v. bis zum 6 Jh. n. Chr.* 2 vols. Harrassowitz, Wiesbaden, 1969. (Asiatische Forschungen [Bonn], no. 27.)

LIU MAU-TSAI. See Liu Mao-Tsai.

LIU PÊN-LI, HSING SHU-CHIEH, LI CHHÊNG-CHHIU & CHANG TAO-CHUNG (1). 'True Hermaphroditism; a Case Report.' *CMJ*, 1959, **78**, 449.

LIU TSHUN-JEN (1). 'Lu Hsi-Hsing and his Commentaries on the *Tshan Thung Chhi.*' *CHJ/T*, 1968, (n.s.) **7**, (no. 1), 71.

LIU TSHUN-JEN (2). 'Lu Hsi-Hsing [+1520 to c. +1601]; a Confucian Scholar, Taoist Priest and Buddhist Devotee of the +16th Century.' *ASEA*, 1965, **18–19**, 115.

LIU TSHUN-JEN (3). 'Taoist Self-Cultivation in Ming Thought.' Art. in *Self and Society in Ming Thought*, ed. W. T. de Bary. Columbia Univ. Press, New York, 1970, p. 291.

LIU TS'UN-YAN. See Liu Tshun-Jen.

LLOYD, G. E. R. (1). *Polarity and Analogy; Two Types of Argumentation in Greek Thought.* Cambridge, 1971.

LLOYD, SETON (2). 'Sultantepe, II.' *ANATS*, 1954, **4**, 101.

LLOYD, SETON & BRICE, W. (1), with a note by C. J. Gadd. 'Ḥarrān.' *ANATS*, 1951, **1**, 77–111.

LLOYD, SETON & GÖKÇE, NURI (1), with notes by R. D. Barnett. 'Sultantepe, I.' *ANATS*, 1953, **3**, 27.

LO, L. C. (1) (tr.). 'Liu Hua-Yang; *Hui Ming Ching*, Das Buch von Bewusstsein und Leben.' In *Chinesische Blätter*, vol. 3, no. 1, ed. R. Wilhelm.

LOEHR, G. (1). 'Missionary Artists at the Manchu Court.' *TOCS*, 1962, **34**, 51.

LOEWE, M. (5). 'The Case of Witchcraft in −91; its Historical Setting and Effect on Han Dynastic History' (*ku* poisoning). *AM*, 1970, **15**, 159.

LOEWE, M. (6). 'Khuang Hêng and the Reform of Religious Practices (−31).' *AM*, 1971, **17**, 1.

LOEWE, M. (7). 'Spices and Silk; Aspects of World Trade in the First Seven Centuries of the Christian Era.' *JRAS*, 1971, 166.

LOEWENSTEIN, P. J. (1). *Swastika and Yin-Yang.* China Society Occasional Papers (n. s.), China Society, London, 1942.

VON LÖHNEYSS, G. E. (1). *Bericht vom Bergwerck, wie man diselben bawen und in güten Wolstande bringen sol, sampt allen dazu gehörigen Arbeiten, Ordnung und Rechtlichen Processen beschrieben durch G.E.L.* Zellerfeld, 1617. 2nd ed. Leipzig, 1690.

LONICERUS, ADAM (1). *Kräuterbuch.* Frankfort, 1578.

LORGNA, A. M. (1). 'Nuove Sperienze intorno alla Dolcificazione dell'Acqua del Mare.' *MSIV/MF*, 1786, **3**, 375. 'Appendice alla Memoria intorno alla Dolcificazione dell'Acqua del Mare.' *MSIV/MF*, 1790, **5**, 8.

LOTHROP, S. (1). 'Coclé; an Archaeological Study of Central Panama.' *MPMH*, 1937, **7**.

LOUIS, H. (1). 'A Chinese System of Gold Milling.' *EMJ*, 1891, 640.

LOUIS, H. (2). 'A Chinese System of Gold Mining.' *EMJ*, 1892, 629.

LOVEJOY, A. O. & BOAS, G. (1). *A Documentary History of Primitivism and Related Ideas*. Vol. 1. *Primitivism and Related Ideas in Antiquity*. Johns Hopkins Univ. Press, Baltimore, 1935.

LOWRY, T. M. (1). *Historical Introduction to Chemistry*. Macmillan, London, 1936.

LU GWEI-DJEN (1). 'China's Greatest Naturalist; a Brief Biography of Li Shih-Chen.' *PHY*, 1966, **8**, 383. Abridgment in Proc. XIth Internat. Congress of the History of Science, Warsaw, 1965, Summaries, vol. 2, p. 364; Actes, vol. 5, p. 50.

LU GWEI-DJEN (2). 'The Inner Elixir (*Nei Tan*); Chinese Physiological Alchemy.' Art. in *Changing Perspectives in the History of Science*, ed. M. Teich & R. Young. Heinemann, London, 1973, p. 68.

LU GWEI-DJEN & NEEDHAM, JOSEPH (1). 'A Contribution to the History of Chinese Dietetics.' *ISIS*, 1951, **42**, 13 (submitted 1939, lost by enemy action; again submitted 1942 and 1948).

LU GWEI-DJEN & NEEDHAM, JOSEPH (3). 'Mediaeval Preparations of Urinary Steroid Hormones.' *MH*, 1964, **8**, 101. Prelim. pub. *N*, 1963, **200**, 1047. Abridged account, *END*, 1968, **27** (no. 102), 130.

LU GWEI-DJEN & NEEDHAM, JOSEPH (4). 'Records of Diseases in Ancient China', art. in *Diseases in Antiquity*, ed. D. R. Brothwell & A. T. Sandison. Thomas, Springfield, Ill. 1967, p. 222.

LU GWEI-DJEN, NEEDHAM, JOSEPH & NEEDHAM, D. M. (1). 'The Coming of Ardent Water.' *AX*, 1972, **19**, 69.

LU KHUAN-YÜ (1). *The Secrets of Chinese Meditation; Self-Cultivation by Mind Control as taught in the Chhan, Mahāyāna and Taoist Schools in China*. Rider, London, 1964.

LU KHUAN-YÜ (2). *Chhan and Zen Teaching* (Series Two). Rider, London, 1961.

LU KHUAN-YÜ (3). *Chhan and Zen Teaching* (Series Three). Rider, London, 1962.

LU KHUAN-YÜ (4) (tr.). *Taoist Yoga; Alchemy and Immortality—a Translation, with Introduction and Notes, of 'The Secrets of Cultivating Essential Nature and Eternal Life' (Hsing Ming Fa Chüeh Ming Chih) by the Taoist Master Chao Pi-Chhen, b. 1860*. Rider, London, 1970.

LUCAS, A. (1). *Ancient Egyptian Materials and Industries*. Arnold, London (3rd ed.), 1948.

LUCAS, A. (2). 'Silver in Ancient Times.' *JEA*, 1928, **14**, 315.

LUCAS, A. (3). 'The Occurrence of Natron in Ancient Egypt.' *JEA*, 1932, **18**, 62.

LUCAS, A. (4). 'The Use of Natron in Mummification.' *JEA*, 1932, **18**, 125.

LÜDY-TENGER, F. (1). *Alchemistische und chemische Zeichen*. Berlin, 1928. Repr. Lisbing, Würzburg, 1972.

LUK, CHARLES. See LU KHUAN-YÜ.

LUMHOLTZ, C. S. (1). *Unknown Mexico; a Record of Five Years' Exploration among the Tribes of the Western Sierra Madre; in the Tierra Caliente of Tepic and Jalisco; and among the Tarascos of Michoacan*, 2 vols. Macmillan, London, 1903.

LUTHER, MARTIN (1). *Werke*. Weimarer Ausgabe.

MACALISTER, R. A. S. (2). *The Excavation of [Tel] Gezer, 1902–05 and 1907–09*. 3 vols. Murray, London, 1912.

McAULIFFE, L. (1). *La Thérapeutique Physique d'Autrefois*. Paris, 1904.

McCLURE, C. M. (1). 'Cardiac Arrest through Volition.' *CALM*, 1959, **90**, 440.

McCONNELL, R. G. (1). *Report on Gold Values in the Klondike High-Level Gravels*. Canadian Geol. Survey Reports, 1907, 34.

McCULLOCH, J. A. (2). '[The State of the Dead in] Primitive and Savage [Cultures].' *ERE*, vol. xi, p. 817.

McCULLOCH, J. A. (3). '[The "Abode of the Blest" in] Primitive and Savage [Cultures].' *ERE*, vol. ii, p. 680.

McCULLOCH, J. A. (4). '[The "Abode of the Blest" in] Celtic [Legend].' *ERE*, vol. ii, p. 688.

McCULLOCH, J. A. (5). '[The "Abode of the Blest" in] Japanese [Thought].' *ERE*, vol. ii, p. 700.

McCULLOCH, J. A. (6). '[The "Abode of the Blest" in] Slavonic [Lore and Legend].' *ERE*, vol. ii, p. 706.

McCULLOCH, J. A. (7). '[The "Abode of the Blest" in] Teutonic [Scandinavian, Belief].' *ERE*, vol. ii, p. 707.

McCULLOCH, J. A. (8). 'Incense.' Art. in *ERE*, vol. vii, p. 201.

McCULLOCH, J. A. (9). 'Eschatology.' Art. in *ERE*, vol. v, p. 373.

McCULLOCH, J. A. (10). 'Vampires.' *ERE*, vol. xii, p. 589.

McDONALD, D. (1). *A History of Platinum*. London, 1960.

MacDONELL, A. A. (1). 'Vedic Religion.' *ERE*, vol. xii, p. 601.

McGovern, W. M. (1). *Early Empires of Central Asia*. Univ. of North Carolina Press, Chapel Hill, 1939.

McGowan, D. J. (2). 'The Movement Cure in China' (Taoist medical gymnastics). *CIMC/MR*, 1885 (no. 29), 42.

McGuire, J. E. (1). 'Transmutation and Immutability; Newton's Doctrine of Physical Qualities.' *AX*, 1967, **14**, 69.

McGuire, J. E. (2). 'Force, Active Principles, and Newton's Invisible Realm.' *AX*, 1968, **15**, 154.

McGuire, J. E. & Rattansi, P. M. (1). 'Newton and the "Pipes of Pan".' *NRRS*, 1966, **21**, 108.

McKenzie, R. Tait (1). *Exercise in Education and Medicine*. Saunders, Philadelphia and London, 1923.

McKie, D. (1). 'Some Notes on Newton's Chemical Philosophy, written upon the Occasion of the Tercentenary of his Birth.' *PMG*, 1942 (7th ser.), **33**, 847.

McKie, D. (2). 'Some Early Chemical Symbols.' *AX*, 1937, **1**, 75.

McLachlan, H. (1). *Newton; the Theological Manuscripts*. Liverpool, 1950.

Macquer, P. J. (1). *Élémens de la Théorie et de la Pratique de la Chimie*. 2 vols, Paris, 1775. (The first editions, uncombined, had been in 1749 and 1751 respectively, but this contained accounts of the new discoveries.) Eng. trs. London, 1775, Edinburgh, 1777. Cf. Coleby (1).

Madan, M. (1) (tr.). *A New and Literal Translation of Juvenal and Persius, with Copious Explanatory Notes by which these difficult Satirists are rendered easy and familiar to the Reader*. 2 vols. Becket, London, 1789.

Maenchen-Helfen, O. (4). *Reise ins asiatische Tuwa*. Berlin, 1931.

de Magalhaens, Gabriel (1). *Nouvelle Relation de la Chine*. Barbin, Paris, 1688 (a work written in 1668). Eng. tr. *A New History of China, containing a Description of the Most Considerable Particulars of that Vast Empire*. Newborough, London, 1688.

Magendie, F. (1). *Mémoire sur la Déglutition de l'Air atmosphérique*. Paris, 1813.

Mahdihassan, S. (2). 'Cultural Words of Chinese Origin' [*firoza* (Pers) = turquoise, *yashb* (Ar) = jade, *chamcha* (Pers) = spoon, *top* (Pers, Tk, Hind) = cannon, *silafchi* (Tk) = metal basin]. *BV*, 1950, **11**, 31.

Mahdihassan, S. (3). 'Ten Cultural Words of Chinese Origin' [*huqqa* (Tk), *qaliyan* (Tk) = tobacco-pipe, *sunduq* (Ar) = box, *piali* (Pers), *findjan* (Ar) = cup, *jaushan* (Ar) = armlet, *safa* (Ar) = turban, *qasai, qasab* (Hind) = butcher, *kah-kashan* (Pers) = Milky Way, *tugra* (Tk) = seal]. *JUB*, 1949, **18**, 110.

Mahdihassan, S. (5). 'The Chinese Origin of the Words Porcelain and Polish.' *JUB*, 1948, **17**, 89.

Mahdihassan, S. (6). 'Carboy as a Chinese Word.' *CS*, 1948, **17**, 301.

Mahdihassan, S. (7). 'The First Illustrations of Stick-Lac and their probable origin.' *PKAWA*, 1947, **50**, 793.

Mahdihassan, S. (8). 'The Earliest Reference to Lac in Chinese Literature.' *CS*, 1950, **19**, 289.

Mahdihassan, S. (9). 'The Chinese Origin of Three Cognate Words: Chemistry, Elixir, and Genii.' *JUB*, 1951, **20**, 107.

Mahdihassan, S. (11). 'Alchemy in its Proper Setting, with Jinn, Sufi, and Suffa as Loan-Words from the Chinese.' *IQB*, 1959, **7** (no. 3), 1.

Mahdihassan, S. (12). 'Alchemy and its Connection with Astrology, Pharmacy, Magic and Metallurgy.' *JAN*, 1957, **46**, 81.

Mahdihassan, S. (13). 'The Chinese Origin of Alchemy.' *UNASIA*, 1953, **5** (no. 4), 241.

Mahdihassan, S. (14). 'The Chinese Origin of the Word Chemistry.' *CS*, 1946, **15**, 136. 'Another Probable Origin of the Word Chemistry from the Chinese.' *CS*, 1946, **15**, 234.

Mahdihassan, S. (15). 'Alchemy in the Light of its Names in Arabic, Sanskrit and Greek.' *JAN*, 1961, **49**, 79.

Mahdihassan, S. (16). 'Alchemy a Child of Chinese Dualism as illustrated by its Symbolism.' *IQB*, 1959, **8**, 15.

Mahdihassan, S. (17). 'On Alchemy, Kimiya and Iksir.' *PAKPJ*, 1959, **3**, 67.

Mahdihassan, S. (18). 'The Genesis of Alchemy.' *IJHM*, 1960, **5** (no. 2), 41.

Mahdihassan, S. (19). 'Landmarks in the History of Alchemy.' *SCI*, 1963, **57**, 1.

Mahdihassan, S. (20). 'Kimiya and Iksir; Notes on the Two Fundamental Concepts of Alchemy.' *MBLB*, 1962, **5** (no. 3), 38. *MBPB*, 1963, **12** (no. 5), 56.

Mahdihassan, S. (21). 'The Early History of Alchemy.' *JUB*, 1960, **29**, 173.

Mahdihassan, S. (22). 'Alchemy; its Three Important Terms and their Significance.' *MJA*, 1961, 227.

Mahdihassan, S. (23). 'Der Chino-Arabische Ursprung des Wortes Chemikalie.' *PHI*, 1961, **23**, 515.

Mahdihassan, S. (24). 'Das Hermetische Siegel in China.' *PHI*, 1960, **22**, 92.

Mahdihassan, S. (25). 'Elixir; its Significance and Origin.' *JRAS/P*, 1961, **6**, 39.

Mahdihassan, S. (26). 'Ouroboros as the Earliest Symbol of Greek Alchemy.' *IQB*, 1961, **9**, 1.

Mahdihassan, S. (27). 'The Probable Origin of Kekulé's Symbol of the Benzene Ring.' *SCI*, 1960, **54**, 1.

MAHDIHASSAN, S., (28). 'Alchemy in the Light of Jung's Psychology and of Dualism.' *PAKPJ*, 1962, **5**, 95.

MAHDIHASSAN, S. (29). 'Dualistic Symbolism; Alchemical and Masonic.' *IQB*, 1963, 55.

MAHDIHASSAN, S. (30). 'The Significance of Ouroboros in Alchemy and Primitive Symbolism.' *IQB*, 1963, 18.

MAHDIHASSAN, S. (31). 'Alchemy and its Chinese Origin as revealed by its Etymology, Doctrines and Symbols.' *IQB*, 1966, 22.

MAHDIHASSAN, S. (32). 'Stages in the Development of Practical Alchemy.' *JRAS/P*, 1968, **13**, 329.

MAHDIHASSAN, S. (33). 'Creation, its Nature and Imitation in Alchemy.' *IQB*, 1968, 80.

MAHDIHASSAN, S. (34). 'A Positive Conception of the Divinity emanating from a Study of Alchemy.' *IQB*, 1969, **10**, 77.

MAHDIHASSAN, S. (35). '*Kursi* or throne; a Chinese word in the *Koran*.' *JRAS/BOM*, 1953, **28**, 19.

MAHDIHASSAN, S. (36). '*Khazana*, a Chinese word in the *Koran*, and the associated word "Godown".' *JRAS/BOM*, 1953, **28**, 22.

MAHDIHASSAN, S. (37). 'A Cultural Word of Chinese Origin; *ta'un* meaning Plague in Arabic.' *JUB*, 1953, **22**, 97. *CRESC*, 1950, 31.

MAHDIHASSAN, S. (38). 'Cultural Words of Chinese Origin; *qaba, aba, diba, kimkhwab* (kincob).' *JKHRS*, 1950, **5**, 203.

MAHDIHASSAN, S. (39). 'The Chinese Origin of the Words Kimiya, Sufi, Dervish and Qalander, in the Light of Mysticism.' *JUB*, 1956, **25**, 124.

MAHDIHASSAN, S. (40). 'Chemistry a Product of Chinese Culture.' *PAKJS*, 1957, **9**, 26.

MAHDIHASSAN, S. (41). 'Lemnian Tablets of Chinese Origin.' *IQB*, 1960, **9**, 49.

MAHDIHASSAN, S. (42). 'Über einige Symbole der Alchemie.' *PHI*, 1962, **24**, 41.

MAHDIHASSAN, S. (43). 'Symbolism in Alchemy; Islamic and other.' *IC*, 1962, **36** (no. 1), 20.

MAHDIHASSAN, S. (44). 'The Philosopher's Stone in its Original Conception.' *JRAS/P*, 1962, **7** (no. 2), 263.

MAHDIHASSAN, S. (45). 'Alchemie im Spiegel hellenistisch-buddhistische Kunst d. 2. Jahrhunderts.' *PHI*, 1965, **27**, 726.

MAHDIHASSAN, S. (46). 'The Nature and Role of Two Souls in Alchemy.' *JRAS/P*, 1965, **10**, 67.

MAHDIHASSAN, S. (47). 'Kekulé's Dream of the Ouroboros, and the Significance of this Symbol.' *SCI*, 1961, **55**, 187.

MAHDIHASSAN, S. (48). 'The Natural History of Lac as known to the Chinese; Li Shih-Chen's Contribution to our Knowledge of Lac.' *IJE*, 1954, **16**, 309.

MAHDIHASSAN, S. (49). 'Chinese Words in the Holy Koran; *qirtas* (paper) and its Synonym *kagaz*.' *JUB*, 1955, **24**, 148.

MAHDIHASSAN, S. (50). 'Cultural Words of Chinese Origin; *kutcherry* (government office), *tusser* (silk).' Art. in Karmarker Commemoration Volume, Poona, 1947–8, p. 97.

MAHDIHASSAN, S. (51). 'Union of Opposites; a Basic Theory in Alchemy and its Interpretation.' Art. in *Beiträge z. alten Geschichte und deren Nachleben*, Festschrift f. Franz Altheim, ed. R. Stiehl & H. E. Stier, vol. 2, p. 251. De Gruyter, Berlin, 1970.

MAHDIHASSAN, S. (52). 'The Genesis of the Four Elements, Air, Water, Earth and Fire.' Art. in Gulam Yazdani Commemoration Volume, Hyderabad, Andhra, 1966, p. 251.

MAHDIHASSAN, S. (53). 'Die frühen Bezeichnungen des Alchemisten, seiner Kunst und seiner Wunderdroge.' *PHI*, 1967, .

MAHDIHASSAN, S. (54). 'The *Soma* of the Aryans and the *Chih* of the Chinese.' *MBPB*, 1972, **21** (no. 3), 30.

MAHDIHASSAN, S. (55). 'Colloidal Gold as an Alchemical Preparation.' *JAN*, 1972, **58**, 112.

MAHLER, J. G. (1). *The Westerners among the Figurines of the Thang Dynasty of China*. Ist. Ital. per il Med. ed Estremo Or., Rome, 1959. (Ser. Orientale Rom, no. 20.)

MAHN, C. A. F. (1). *Etymologische Untersuchung auf dem Gebiete der Romanischen Sprachen*. Dümmler, Berlin, 1858, repr. 1863.

MAIER, MICHAEL (1). *Atalanta Fugiens*, 1618. Cf. Tenney Davis (1); J. Read (1); de Jong (1).

MALHOTRA, J. C. (1). 'Yoga and Psychiatry; a Review.' *JNPS*, 1963, **4**, 375.

MANUEL, F. E. (1). *Isaac Newton, Historian*. Cambridge, 1963.

MANUEL, F. E. (2). *The Eighteenth Century Confronts the Gods*. Harvard Univ. Press, Cambridge, Mass. 1959.

MAQSOOD ALI, S. ASAD & MAHDIHASSAN, S. (4). 'Bazaar Medicines of Karachi; [IV], Inorganic Drugs.' *MED*, 1961, **23**, 125.

DE LA MARCHE, LECOY (1). 'L'Art d'Enluminer; Traité Italien du XVe Siecle' (*De Arte Illuminandi*, Latin text with introduction). *MSAF*, 1888, **47** (5e sér.), **7**, 248.

MARÉCHAL, J. R. (3). *Reflections upon Prehistoric Metallurgy; a Research based upon Scientific Methods*. Brimberg, Aachen (for Junker, Lammersdorf), 1963. French and German editions appeared in 1962.

MARSHALL, SIR JOHN (1). *Taxila; An Illustrated Account of Archaeological Excavations carried out at Taxila under the orders of the Government of India between the years 1913 and 1934.* 3 vols. Cambridge, 1951.

MARTIN, W. A. P. (2). *The Lore of Cathay.* Revell, New York and Chicago, 1901.

MARTIN, W. A. P. (3). *Hanlin Papers.* 2 vols. Vol. 1. Trübner, London; Harper, New York, 1880; Vol. 2. Kelly & Walsh, Shanghai, 1894.

MARTIN, W. A. P. (8). 'Alchemy in China.' A paper read before the Amer. Or. Soc. 1868; abstract in *JAOS*, 1871, **9**, xlvi. *CR*, 1879, **7**, 242. Repr. in (3), vol. 1, p. 221; (2), pp. 44 ff.

MARTIN, W. A. P. (9). *A Cycle of Cathay.* Oliphant, Anderson & Ferrier, Edinburgh and London; Revell New York, 1900.

MARTINDALE, W. (1). *The Extra Pharmacopoeia; incorporating Squire's 'Companion to the Pharmacopoeia'.* 1st edn. 1883. 25th edn., ed. R. G. Todd, Pharmaceutical Press, London, 1967.

MARX, E. (2). Japanese peppermint oil still. *MDGNVO*, 1896, **6**, 355.

MARYON, H. (3). 'Soldering and Welding in the Bronze and Early Iron Ages.' *TSFFA*, 1936, **5** (no. 2).

MARYON, H. (4). 'Prehistoric Soldering and Welding' (a précis of Maryon, 3). *AQ*, 1937, **11**, 208.

MARYON, H. (5). 'Technical Methods of the Irish Smiths.' *PRIA*, 1938, **44**c, no. 7.

MARYON, H. (6). *Metalworking and Enamelling; a Practical Treatise.* 3rd ed. London, 1954.

MASON, G. H. (1). *The Costume of China.* Miller, London, 1800.

MASON, H. S. (1). 'Comparative Biochemistry of the Phenolase Complex.' *AIENZ*, 1955, **16**, 105.

MASON, S. F. (2). 'The Scientific Revolution and the Protestant Reformation; I, Calvin and Servetus in relation to the New Astronomy and the Theory of the Circulation of the Blood.' *ANS*, 1953, **9** (no. 1).

MASON, S. F. (3). 'The Scientific Revolution and the Protestant Reformation; II, Lutheranism in relation to Iatro-chemistry and the German Nature-philosophy.' *ANS*, 1953, **9** (no. 2).

MASPERO, G. (2). *Histoire ancienne des Peuples d'Orient.* Paris, 1875.

MASPERO, H. (7). 'Procédés de 'nourrir le principe vital' dans la Religion Taoiste Ancienne.' *JA*, 1937, **229**, 177 and 353.

MASPERO, H. (9). 'Notes sur la Logique de Mo-Tseu [Mo Tzu] et de son École.' *TP*, 1928, **25**, 1.

MASPERO, H. (13). *Le Taoisme.* In *Mélanges Posthumes sur les Religions et l'Histoire de la Chine*, vol. 2, ed. P. Demiéville, SAEP, Paris, 1950. (Publ. du Mus. Guimet, Biblioth. de Diffusion, no 58.) Rev. J. J. L. Duyvendak, *TP*, 1951, **40**, 372.

MASPERO, H. (14). *Études Historiques.* In *Mélanges Posthumes sur les Religions et l'Histoire de la Chine*, vol. 3, ed. P. Demiéville. Civilisations du Sud, Paris, 1950. [Publ. du Mus. Guimet, Biblioth. de Diffusion, no. 59.) Rev. J. J. L. Duyvendak, *TP*, 1951, **40**, 366.

MASPERO, H. (19). 'Communautés et Moines Bouddhistes Chinois au 2e et 3e Siècles.' *BEFEO*, 1910, **10**, 222.

MASPERO, H. (20)., 'Les Origines de la Communauté Bouddhiste de Loyang.' *JA*, 1934, **225**, 87.

MASPERO, H. (22). 'Un Texte Taoiste sur l'Orient Romain.' *MIFC*, 1937, **17**, 377 (*Mélanges G. Maspero*, vol. 2). Reprinted in Maspero (14), pp. 95 ff.

MASPERO, H. (31). Review of R. F. Johnston's *Buddhist China* (London, 1913). *BEFEO*, 1914, **14** (no. 9), 74.

MASPERO, H. (32). *Le Taoïsme et les Religions Chinoises.* (Collected posthumous papers, partly from (12) and (13) reprinted, partly from elsewhere, with a preface by M. Kaltenmark.) Gallimard, Paris, 1971. (Bibliothèque des Histoires, no. 3.)

MASSÉ, H. (1). *Le Livre des Merveilles du Monde.* Chêne, Paris, 1944. (Album of colour-plates from al-Qazwīnī's Cosmography, *c.* +1275, with introduction, taken from Bib. Nat. Suppl. Pers. MS. 332.)

MASSIGNON, L. (3). 'The Qarmatians.' *EI*, vol. ii, pt. 2, p. 767.

MASSIGNON, L. (4). 'Inventaire de la Littérature Hermétique Arabe.' App. iii in Festugière (1), 1944. (On the role of the Sabians of Ḥarrān, who adopted the Hermetica as their Scriptures.)

MASSIGNON, L. (5). 'The Idea of the Spirit in Islam.' *ERYB*, 1969, **6**, 319 (*The Mystic Vision*, ed. J. Campbell). Tr. from the German in *ERJB*, 1945, **13**, 1.

MASSON, L. (1). 'La Fontaine de Jouvence.' *AESC*, 1937, **27**, 244; 1938, **28**, 16.

MASSON-OURSEL, P. (4). *Le Yoga.* Presses Univ. de France, Paris, 1954. (Que Sais-je? ser. no. 643.)

AL-MASʿŪDĪ. See de Meynard & de Courteille.

MATCHETT, J. R. & LEVINE, J. (1). 'A Molecular Still designed for Small Charges.' *IEC/AE*, 1943, **15**, 296.

MATHIEU, F. F. (1). *La Géologie et les Richesses Minières de la Chine.* Impr. Comm. et Industr., la Louvière, n.d. (1924), paginated 283–529, with 4 maps (from Pub. de l'Assoc. des Ingénieurs de l'École des Mines de Mons).

MATSUBARA, HISAKO (1) (tr.). *Die Geschichte von Bambus-sammler und dem Mädchen Kaguya* [the *Taketori Monogatari*, *c.* +866], with illustrations by Mastubara Naoko. Langewiesche-Brandt, Ebenhausen bei München, 1968.

MATSUBARA, HISAKO (2). 'Dies-seitigkeit und Transzendenz im *Taketori Monogatari*.' Inaug. Diss. Ruhr Universität, Bochum, 1970.

MATTHAEI, C. F. (1) (tr.). *Nemesius Emesenus 'De Natura Hominis' Graece et Latine (c. +400)*. Halae Magdeburgicae, 1802.

MATTIOLI, PIERANDREA (2). *Commentarii in libros sex Pedacii Dioscoridis Anazarbei de materia medica*.... Valgrisi, Venice, 1554, repr. 1565.

MAUL, J. P. (1). 'Experiments in Chinese Alchemy.' Inaug. Diss., Massachusetts Institute of Technology, 1967.

MAURIZIO, A. (1). *Geschichte der gegorenen Getränke*. Berlin and Leipzig, 1933.

MAXWELL, J. PRESTON (1). 'Osteomalacia and Diet.' *NAR*, 1934, **4** (no. 1), 1.

MAXWELL, J. PRESTON, HU, C. H. & TURNBULL, H. M. (1). 'Foetal Rickets [in China].' *JPB*, 1932, **35**, 419.

MAYERS, W. F. (1). *Chinese Reader's Manual*. Presbyterian Press, Shanghai, 1874; reprinted 1924.

MAZZEO, J. A. (1). 'Notes on John Donne's Alchemical Imagery.' *ISIS*, 1957, **48**, 103.

MEAD, G. R. S. (1). *Thrice-Greatest Hermes; Studies in Hellenistic Theosophy and Gnosis—Being a Translation of the Extant Sermons and Fragments of the Trismegistic Literature, with Prolegomena, Commentaries and Notes*. 3 vols. Theosophical Pub. Soc., London and Benares, 1906.

MEAD, G. R. S. (2) (tr.). '*Pistis Sophia'; a [Christian] Gnostic Miscellany; being for the most part Extracts from the 'Books of the Saviour', to which are added Excerpts from a Cognate Literature*. 2nd ed. Watkins, London, 1921.

MECHOULAM, R. & GAONI, Y. (1). 'Recent Advances in the Chemistry of Hashish.' *FCON*, 1967, **25**, 175.

MEHREN, A. F. M. (1) (ed. & tr.). *Manuel de la Cosmographie du Moyen-Âge, traduit de l'Arabe; 'Nokhbet ed-Dahr fi Adjaib-il-birr wal-Bahr [Nukhbat al-Dahr fi 'Ajāib al-Birr wa'l Bahr]' de Shems ed-Din Abou-Abdallah Mohammed de Damas* [Shams al-Dīn Abū 'Abd-Allāh al-Anṣarī al-Ṣūfī al-Dimashqī; *The Choice of the Times and the Marvels of Land and Sea, c. +1310*]... St Petersburg, 1866 (text), Copenhagen, 1874 (translation).

MEILE, P. (1). 'Apollonius de Tyane et les Rites Védiques.' *JA*, 1945, **234**, 451.

MEISSNER, B. (1). *Babylonien und Assyrien*. Winter, Heidelberg, 1920, Leipzig, 1925.

MELLANBY, J. (1). 'Diphtheria Antitoxin.' *PRSB*, 1908, **80**, 399.

MELLOR, J. W. (1). *Modern Inorganic Chemistry*. Longmans Green, London, 1916; often reprinted.

MELLOR, J. W. (2). *Comprehensive Treatise on Inorganic and Theoretical Chemistry*. 15 vols. Longmans Green, London, 1923.

MELLOR, J. W. (3). 'The Chemistry of the Chinese Copper-red Glazes.' *TCS*, 1936, **35**.

DE MÉLY, F. (1) (with the collaboration of M. H. Courel). *Les Lapidaires Chinois*. Vol. 1 of *Les Lapidaires de l'Antiquité et du Moyen Age*. Leroux, Paris, 1896. (Contains facsimile reproduction of the mineralogical section of *Wakan Sanzai Zue*, chs. 59, 60, and 61.) Crit. rev. M. Berthelot, *JS*, 1896, 573).

DE MÉLY, F. (6). 'L'Alchimie chez les Chinois et l'Alchimie Grecque.' *JA*, 1895 (9e sér.), **6**, 314.

DE MENASCE, P. J. (2). 'The Cosmic Noria (Zodiac) in Parsi Thought.' *AN*, 1940, **35–6**, 451.

MEREDITH-OWENS, G. M. (1). 'Some Remarks on the Miniatures in the [Royal Asiatic] Society's *Jāmi' al-Tawārīkh* (MS. A27 of +1314) [by Rashīd al-Dīn, finished +1316]. *JRAS*, 1970 (no. 2, Wheeler Presentation Volume), 195. Includes a brief account of the section on the History of China; cf. Jahn & Franke (1).

MERRIFIELD, M. P. (1). *Original Treatises dating from the +12th to the +18th Centuries on the Arts of Painting in Oil, Miniature, and the Preparation of Colour and Artificial Gems*. 2 vols. London, 1847, London, 1849.

MERRIFIELD, M. P. (2). *A Treatise on Painting [Cennino Cennini's], translated from Tambroni's Italian text of 1821*. London, 1844.

MERSENNE, MARIN (3). *La Verité des Sciences, contre les Sceptiques on Pyrrhoniens*. Paris, 1625. Facsimile repr. Frommann, Stuttgart and Bad Cannstatt, 1969. Rev. W. Pagel, *AX*, 1970, **17**, 64.

MERZ, J. T. (1). *A History of European Thought in the Nineteenth Century*. 2 vols. Blackwood, Edinburgh and London, 1896.

METCHNIKOV, E. (= I. I.) (1). *The Nature of Man; Studies in Optimistic Philosophy*. Tr. P. C. Mitchell, New York, 1903, London, 1908; rev. ed. by C. M. Beadnell, London, 1938.

METTLER, CECILIA C. (1). *A History of Medicine*. Blakiston, Toronto, 1947.

METZGER, H. (1). *Newton, Stahl, Boerhaave et la Doctrine Chimique*. Alcan, Paris, 1930.

DE MEURON, M. (1). 'Yoga et Médecine; propos du Dr J. G. Henrotte recueillis par...' *MEDA*, 1968 (no. 69), 2.

MEYER, A. W. (1). *The Rise of Embryology*. Stanford Univ. Press, Palo Alto, Calif. 1939.

VON MEYER, ERNST (1). *A History of Chemistry, from earliest Times to the Present Day; being also an Introduction to the Study of the Science*. 2nd ed., tr. from the 2nd Germ. ed. by G. McGowan. Macmillan, London, 1898.

MEYER, H. H. & GOTTLIEB, R. (1). *Die experimentelle Pharmakologie als Grundlage der Arzneibehandlung.* 9th ed. Urban & Schwarzenberg, Berlin and Vienna, 1936.

MEYER, P. (1). *Alexandre le Grand dans la Litterature Française du Moyen Age.* 2 vols. Paris, 1886.

MEYER, R. M. (1). *Goethe.* 3 vols. Hofmann, Berlin, 1905.

MEYER-STEINEG, T. & SUDHOFF, K. (1). *Illustrierte Geschichte der Medizin.* 5th ed. revised and enlarged, ed. R. Herrlinger & F. Kudlien. Fischer, Stuttgart, 1965.

MEYERHOF, M. (3). 'On the Transmission of Greek and Indian Science to the Arabs.' *IC*, 1937, **11**, 17.

MEYERHOF, M. & SOBKHY, G. P. (1) (ed. & tr.). *The Abridged Version of the 'Book of Simple Drugs' of Aḥmad ibn Muḥammad al-Ghāfiqī of Andalusia by Gregorius Abu'l-Faraj (Bar Hebraeus).* Govt. Press, Cairo, 1938. (Egyptian University Faculty of Med. Pubs. no. 4.)

DE MEYNARD, C. BARBIER (3). '"L'Alchimiste", Comédie en Dialecte Turc Azeri [Azerbaidjani].' *JA*, 1886 (8ᵉ sér.), **7**, 1.

DE MEYNARD, C. BARBIER & DE COURTEILLE, P. (1) (tr.). *Les Prairies d'Or* (the *Murūj al-Dhahab* of al-Mas'ūdī, +947). 9 vols. Paris, 1861–77.

MIALL, L. C. (1). *The Early Naturalists, their Lives and Work* (+1530 to +1789). Macmillan, London, 1912.

MICHELL, H. (1). *The Economics of Ancient Greece.* Cambridge, 1940. 2nd ed. 1957.

MICHELL, H. (2). 'Oreichalcos.' *CLR*, 1955, **69** (n.s. **5**), 21.

MIELI, A. (1). *La Science Arabe, et son Rôle dans l'Evolution Scientifique Mondiale.* Brill, Leiden, 1938. Repr. 1966, with additional bibliography and analytic index by A. Mazaheri.

MIELI, A. (3). *Pagine di Storia della Chimica.* Rome, 1922.

MIGNE, J. P. (1) (ed.). *Dictionnaire des Apocryphes; ou, Collection de tous les Livres Apocryphes relatifs à l'Ancien et au Nouveau Testament, pour la plupart, traduits en Français pour la première fois sur les textes originaux; et enrichie de préfaces, dissertations critiques, notes historiques, bibliographiques, géographiques et theologiques...* 2 vols. Migne, Paris, 1856. Vols. 23 and 24 of his *Troisième et Dernière Encyclopédie Théologique,* 60 vols.

MILES, L. M. & FÊNG, C. T. (1). 'Osteomalacia in Shansi.' *JEM*, 1925, **41**, 137.

MILES, W. (1). 'Oxygen-consumption during Three Yoga-type Breathing Patterns.' *JAP*, 1964, **19**, 75.

MILLER, J. INNES (1). *The Spice Trade of the Roman Empire, −29 to +641.* Oxford, 1969.

MILLS, J. V. (11). *Ma Huan['s] 'Ying Yai Shêng Lan', 'The Overall Survey of the Ocean's Shores' [1433]; translated from the Chinese text edited by Fêng Chhêng-Chün, with Introduction, Notes and Appendices...* Cambridge, 1970. (Hakluyt Society Extra Series, no. 42.)

MINGANA, A. (1) (tr.). *An Encyclopaedia of the Philosophical and Natural Sciences, as taught in Baghdad about +817; or, the 'Book of Treasures' by Job of Edessa: the Syriac Text Edited and Translated...* Cambridge, 1935.

MITCHELL, C. W. (1). *St Ephraim's Prose Refutations of Mani, Marcion and Bardaisan;...from the Palimpsest MS. Brit. Mus. Add. 14623...*Vol. 1. *The Discourses addressed to Hypatius.* Vol. 2. *The Discourse called 'Of Domnus', and Six other Writings.* Williams & Norgate, London, 1912–21. (Text and Translation Society Series.)

MITRA, RAJENDRALALA (1). 'Spirituous Drinks in Ancient India.' *JRAS/B*, 1873, **42**, 1–23.

MIYASHITA SABURŌ (1). 'A Link in the Westward Transmission of Chinese Anatomy in the Later Middle Ages.' *ISIS*, 1968, **58**, 486.

MIYUKI MOKUSEN (1). 'Taoist Zen Presented in the *Hui Ming Ching*.' Communication to the First International Conference of Taoist Studies, Villa Serbelloni, Bellagio, 1968.

MIYUKI MOKUSEN (2). 'The "Secret of the Golden Flower", Studies and [a New] Translation.' Inaug. Diss., Jung Institute, Zürich, 1967.

MODEL, J. G. (1). *Versuche und Gedanken über ein natürliches oder gewachsenes Salmiak.* Leipzig, 1758.

MODI, J. J. (1). 'Haoma.' Art. in *ERE*, vol. vi, p. 506.

MOISSAN, H. (1). *Traité de Chimie Minérale.* 5 vols. Masson, Paris, 1904.

MONTAGU, B. (1) (ed.). *The Works of Lord Bacon.* 16 vols. in 17 parts. Pickering, London, 1825–34.

MONTELL, G. (2). 'Distilling in Mongolia.' *ETH*, 1937 (no. 5), **2**, 321.

DE MONTFAUCON, B. (1). *L'Antiquité Expliquée et Representée en Figures.* 5 vols. with 5-vol. supplement. Paris, 1719. Eng. tr. by D. Humphreys, *Antiquity Explained, and Represented in Sculptures, by the Learned Father Montfaucon.* 5 vols. Tonson & Watts, London, 1721–2.

MONTGOMERY, J. W. (1). 'Cross, Constellation and Crucible; Lutheran Astrology and Alchemy in the Age of the Reformation.' *AX*, 1964, **11**, 65.

MOODY, E. A. & CLAGETT, MARSHALL (1) (ed. and tr.). *The Mediaeval Science of Weights ('Scientia de Ponderibus'); Treatises ascribed to Euclid, Archimedes, Thabit ibn Qurra, Jordanus de Nemore, and Blasius of Parma.* Univ. of Wisconsin Press, Madison, Wis., 1952. Revs. E. J. Dijksterhuis, *A/AIHS*, 1953, **6**, 504; O. Neugebauer, *SP*, 1953, **28**, 596.

MOORE-BENNETT, A. J. (1). 'The Mineral Areas of Western China.' *FER*, 1915, 225.

MORAN, S. F. (1). 'The Gilding of Ancient Bronze Statues in Japan.' *AA*, 1969, **30**, 55.

DE MORANT, G. SOULIÉ (2). *L'Acuponcture Chinoise.* 4 vols.
   I. *l'Énergie (Points, Méridiens, Circulation).*
   II. *Le Maniement de l'Energie.*
   III. *Les Points et leurs Symptômes.*
   IV. *Les Maladies et leurs Traitements.*
   Mercure de France, Paris, 1939–. Re-issued as 5 vols. in one, with 1 vol. of plates, Maloine, Paris, 1972.
MORERY, L. (1). *Grand Dictionnaire Historique; ou le Mélange Curieux de l'Histoire Sacrée et Profane...* 1688, Supplement 1689. Later editions revised by J. Leclerc. 9th ed. Amsterdam and The Hague, 1702. Eng. tr. revised by Jeremy Collier, London, 1701.
MORET, A. (1). 'Mysteries, Egyptian.' *ERE*, vol. ix, pp. 74–5.
MORET, A. (2). *Kings and Gods in Egypt.* London, 1912.
MORET, A. (3). 'Du Caractère Religieux de la Royauté Pharaonique.' *BE/AMG*, 1902, **15**, 1–344.
MORFILL, W. R. & CHARLES, R. H. (1) (tr.). *The 'Book of the Secrets of Enoch'* [2 Enoch], *translated from the Slavonic...* Oxford, 1896.
MORGENSTERN, P. (1) (ed.). '*Turba Philosophorum*'; *Das ist, Das Buch von der güldenen Kunst, neben andern Authoribus, welche mit einander 36 Bücher in sich haben. Darinn die besten vrältesten Philosophi zusamen getragen, welche tractiren alle einhellig von der Universal Medicin, in zwey Bücher abgetheilt, unnd mit Schönen Figuren gezieret. Jetzundt newlich zu Nutz und Dienst allen waren Kunstliebenden der Natur (so der Lateinischen Sprach unerfahren) mit besondern Fleiß, mühe unnd Arbeit trewlich an tag geben...* König, Basel, 1613. 2nd ed. Krauss, Vienna, 1750. Cf. Ferguson (1), vol. 2, pp. 106 ff.
MORRIS, IVAN I. (1). *The World of the Shining Prince; Court Life in Ancient Japan* [in the Heian Period, +782 to +1167, here particularly referring to Late Heian, +967 to +1068]. Oxford, 1964.
MORRISON, P. & MORRISON, E. (1). 'High Vacuum.' *SAM*, 1950, **182** (no. 5), 20.
MORTIER, F. (1). 'Les Procédés Taoistes en Chine pour la Prolongation de la Vie Humaine.' *BSAB*, 1930, **45**, 118.
MORTON, A. A. (1). *Laboratory Technique in Organic Chemistry.* McGraw-Hill, New York and London, 1938.
MOSS, A. A. (1). 'Niello.' *SCON*, 1955, **1**, 49.
M[OUFET], T[HOMAS] (of Caius, d. +1605)? (1). '*Letter* [of Roger Bacon] *concerning the Marvellous Power of Art and Nature.* London, 1659. (Tr. of *De Mirabili Potestate Artis et Naturae, et de Nullitate Magiae.*) French. tr. of the same work by J. Girard de Tournus, Lyons, 1557, Billaine, Paris, 1628. Cf. Ferguson (1), vol. 1, pp. 52, 63-4, 318, vol. 2, pp. 114, 438.
MOULE, A. C. & PELLIOT, P. (1) (tr. & annot.). *Marco Polo (+1254 to +1325); The Description of the World.* 2 vols. Routledge, London, 1938. Further notes by P. Pelliot (posthumously pub.). 2 vols. Impr. Nat. Paris, 1960.
MUCCIOLI, M. (1). 'Intorno ad una Memoria di Giulio Klaproth sulle "Conoscenze Chimiche dei Cinesi nell 8 Secolo".' *A*, 1926, **7**, 382.
MUELLER, K. (1). 'Die Golemsage und die sprechenden Statuen.' *MSGVK*, 1918, **20**, 1–40.
MUIR, J. (1). *Original Sanskrit Texts.* 5 vols. London, 1858–72.
MUIR, M. M. PATTISON (1). *The Story of Alchemy and the Beginnings of Chemistry.* Hodder & Stoughton, London, 1902. 2nd ed. 1913.
MUIRHEAD, W. (1). '[Glossary of Chinese] Mineralogical and Geological Terms. In Doolittle (1), vol. 2, p. 256.
MUKAND SINGH, THAKUR (1). *Ilajul Awham (On the Treatment of Superstitions).* Jagat, Aligarh, 1893 (in Urdu).
MUKERJI, KAVIRAJ B. (1). *Rasa-jala-nidhi; or, Ocean of Indian Alchemy.* 2 vols. Calcutta, 1927.
MULTHAUF, R. P. (1). 'John of Rupescissa and the Origin of Medical Chemistry.' *ISIS*, 1954, **45**, 359.
MULTHAUF, R. P. (2). 'The Significance of Distillation in Renaissance Medical Chemistry.' *BIHM*, 1956, **30**, 329.
MULTHAUF, R. P. (3). 'Medical Chemistry and "the Paracelsians".' *BIHM*, 1954, **28**, 101.
MULTHAUF, R. P. (5). *The Origins of Chemistry.* Oldbourne, London, 1967.
MULTHAUF, R. P. (6). 'The Relationship between Technology and Natural Philosophy, c. +1250 to +1650, as illustrated by the Technology of the Mineral Acids.' Inaug. Diss., Univ. California, 1953.
MULTHAUF, R. P. (7). 'The Beginnings of Mineralogical Chemistry.' *ISIS*, 1958, **49**, 50.
MULTHAUF, R. P. (8). 'Sal Ammoniac; a Case History in Industrialisation.' *TCULT*, 1965, **6**, 569.
MUS, P. (1). 'La Notion de Temps Réversible dans la Mythologie Bouddhique.' *AEPHE/SSR*, 1939, 1.

AL-NADĪM, ABŪ'L-FARAJ IBN ABŪ YA'QŪB. See Flügel, G. (1).
NADKARNI, A. D. (1). *Indian Materia Medica.* 2 vols. Popular, Bombay, 1954.

NAGEL, A. (1). 'Die Chinesischen Küchengott.' *AR*, 1908, **11**, 23.

NAKAYAMA SHIGERU & SIVIN, N. (1) (ed.). *Chinese Science; Explorations of an Ancient Tradition*. M.I.T. Press, Cambridge, Mass., 1973. (M.I.T. East Asian Science Ser. no. 2.)

NANJIO, B. (1). *A Catalogue of the Chinese Translations of the Buddhist Tripiṭaka*. Oxford, 1883. (See Ross, E. D, 3.)

NARDI, S. (1) (ed.). *Taddeo Alderotti's Consilia Medicinalia'*, *c.* +*1280*. Turin, 1937.

NASR, SEYYED HOSSEIN. See Said Husain Nasr.

NAU, F. (2). 'The translation of the *Tabula Smaragdina* by Hugo of Santalla (mid +12th century).' *ROC*, 1907 (2e sér.), **2**, 105.

NEAL, J. B. (1). 'Analyses of Chinese Inorganic Drugs.' *CMJ*, 1889, **2**, 116; 1891, **5**, 193,

NEBBIA, G. & NEBBIA-MENOZZI, G. (1). 'A Short History of Water Desalination.' Art. from *Acqua Dolce dal Mare*. IIª Inchiesta Internazionale, Milan, Fed. delle Associazioni Sci. e Tecniche, 1966, pp. 129–172.

NEBBIA, G. & NEBBIA-MENOZZI, G. (2). 'Early Experiments on Water Desalination by Freezing.' *DS*, 1968, **5**, 49.

NEEDHAM, DOROTHY M. (1). *Machina Carnis; the Biochemistry of Muscle Contraction in its Historical Development*. Cambridge, 1971.

NEEDHAM, JOSEPH (2). *A History of Embryology*. Cambridge, 1934. 2nd ed., revised with the assistance of A. Hughes. Cambridge, 1959; Abelard-Schuman, New York, 1959.

NEEDHAM, JOSEPH (25). 'Science and Technology in China's Far South-East.' *N*, 1946, **157**, 175. Reprinted in Needham & Needham (1).

NEEDHAM, JOSEPH (27). 'Limiting Factors in the Advancement of Science as observed in the History of Embryology.' *YJBM*, 1935, **8**, 1. (Carmalt Memorial Lecture of the Beaumont Medical Club of Yale University.)

NEEDHAM, JOSEPH (30). 'Prospection Géobotanique en Chine Médiévale.' *JATBA*, 1954, **1**, 143.

NEEDHAM, JOSEPH (31). 'Remarks on the History of Iron and Steel Technology in China (with French translation; 'Remarques relatives à l'Histoire de la Sidérurgie Chinoise'). In *Actes du Colloque International 'Le Fer à travers les Ages'*, pp. 93, 103. Nancy, Oct. 1955. (*AEST*, 1956, Mémoire no. 16.)

NEEDHAM, JOSEPH (32). *The Development of Iron and Steel Technology in China*. Newcomen Soc. London, 1958. (Second Biennial Dickinson Memorial Lecture, Newcomen Society.) Précis in *TNS*, 1960, **30**, 141; rev. L. C. Goodrich, *ISIS*, 1960, **51**, 108. Repr. Heffer, Cambridge, 1964. French tr. (unrevised, with some illustrations omitted and others added by the editors), *RHSID*, 1961, **2**, 187, 235; 1962, **3**, 1, 62.

NEEDHAM, JOSEPH (34). 'The Translation of Old Chinese Scientific and Technical Texts.' Art. in *Aspects of Translation*, ed. A. H. Smith, Secker & Warburg, London, 1958. p. 65. (Studies in Communication, no. 22.) Also in *BABEL*, 1958, **4** (no. 1), 8.

NEEDHAM JOSEPH (36). *Human Law and the Laws of Nature in China and the West*. Oxford Univ. Press, London, 1951. (Hobhouse Memorial Lectures at Bedford College, London, no. 20.) Abridgement of (37).

NEEDHAM, JOSEPH (37). 'Natural Law in China and Europe.' *JHI*, 1951, **12**, 3 & 194 (corrigenda, 628).

NEEDHAM, JOSEPH (45). 'Poverties and Triumphs of the Chinese Scientific Tradition.' Art. in *Scientific Change; Historical Studies in the Intellectual, Social and Technical Conditions for Scientific Discovery and Technical Invention from Antiquity to the Present*, ed. A. C. Crombie, p. 117. Heinemann, London, 1963. With discussion by W. Hartner, P. Huard, Huang Kuang-Ming, B. L. van der Waerden and S. E. Toulmin (Symposium on the History of Science, Oxford, 1961). Also, in modified form: 'Glories and Defects...' in *Neue Beiträge z. Geschichte d. alten Welt*, vol. 1, *Alter Orient und Griechenland*, ed. E. C. Welskopf, Akad. Verl. Berlin, 1964. French tr. (of paper only) by M. Charlot, 'Grandeurs et Faiblesses de la Tradition Scientifique Chinoise', *LP*, 1963, no. 111. Abridged version; 'Science and Society in China and the West', *SPR*, 1964, **52**, 50.

NEEDHAM, JOSEPH (47). 'Science and China's Influence on the West.' Art. in *The Legacy of China*, e R. N. Dawson. Oxford, 1964, p. 234.

NEEDHAM, JOSEPH (48). 'The Prenatal History of the Steam-Engine.' (Newcomen Centenary Lecture.) *TNS*, 1963, **35**, 3–58.

NEEDHAM, JOSEPH (50). 'Human Law and the Laws of Nature.' Art. in *Technology, Science and Art; Common Ground*. Hatfield Coll. of Technol., Hatfield, 1961, p. 3. A lecture based upon (36) and (37), revised from Vol. 2, pp. 518 ff. Repr. in *Social and Economic Change* (Essays in Honour of Prof. D. P. Mukerji), ed. B. Singh & V. B. Singh. Allied Pubs. Bombay, Delhi etc., 1967, p. 1.

NEEDHAM, JOSEPH (55). 'Time and Knowledge in China and the West.' Art. in *The Voices of Time; a Cooperative Survey of Man's Views of Time as expressed by the Sciences and the Humanities*, ed. J. T. Fraser. Braziller, New York, 1966, p. 92.

NEEDHAM, JOSEPH (56). *Time and Eastern Man.* (Henry Myers Lecture, Royal Anthropological Institute, 1964.) Royal Anthropological Institute, London, 1965.

NEEDHAM, JOSEPH (58). 'The Chinese Contribution to Science and Technology.' Art. in *Reflections on our Age* (Lectures delivered at the Opening Session of UNESCO at the Sorbonne, Paris, 1946), ed. D. Hardman & S. Spender. Wingate, London, 1948, p. 211. Tr. from the French *Conférences de l'Unesco.* Fontaine, Paris, 1947, p. 203.

NEEDHAM, JOSEPH (59). 'The Roles of Europe and China in the Evolution of Oecumenical Science.' *JAHIST*, 1966, **1**, 1. As Presidential Address to Section X, British Association, Leeds, 1967, in *ADVS*, 1967, **24**, 83.

NEEDHAM, JOSEPH (60). 'Chinese Priorities in Cast Iron Metallurgy.' *TCULT*, 1964, **5**, 398.

NEEDHAM, JOSEPH (64). *Clerks and Craftsmen in China and the West* (Collected Lectures and Addresses). Cambridge, 1970.

NEEDHAM, JOSEPH (65). *The Grand Titration; Science and Society in China and the West.* (Collected Addresses.) Allen & Unwin, London, 1969.

NEEDHAM, JOSEPH (67). *Order and Life* (Terry Lectures). Yale Univ. Press, New Haven, Conn.; Cambridge, 1936. Paperback edition (with new foreword), M.I.T. Press, Cambridge, Mass. 1968. Italian tr. by M. Aloisi, *Ordine e Vita*, Einaudi, Turin, 1946 (Biblioteca di Cultura Scientifica, no. 14).

NEEDHAM, JOSEPH (68). 'Do the Rivers Pay Court to the Sea? The Unity of Science in East and West.' *TTT*, 1971, **5** (no. 2), 68.

NEEDHAM, JOSEPH (70). 'The Refiner's Fire; the Enigma of Alchemy in East and West.' Ruddock, for Birkbeck College, London, 1971 (Bernal Lecture). French tr. (with some additions and differences), 'Artisans et Alchimistes en Chine et dans le Monde Hellénistique.' *LP*, 1970, no. 152, 3 (Rapkine Lecture, Institut Pasteur, Paris).

NEEDHAM, JOSEPH (71). 'A Chinese Puzzle—Eighth or Eighteenth?', art. in *Science, Medicine and Society in the Renaissance* (Pagel Presentation Volume), ed. Debus (20), vol. 2, p. 251.

NEEDHAM, JOSEPH & LU GWEI-DJEN (1). 'Hygiene and Preventive Medicine in Ancient China.' *JHMAS*, 1962, **17**, 429; abridged in *HEJ*, 1959, **17**, 170.

NEEDHAM, JOSEPH & LU GWEI-DJEN (3). 'Proto-Endocrinology in Mediaeval China.' *JSHS*, 1966, **5**, 150.

NEEDHAM, JOSEPH & NEEDHAM, DOROTHY M. (1) (ed.). *Science Outpost.* Pilot Press, London, 1948.

NEEDHAM, JOSEPH & ROBINSON, K. (1). 'Ondes et Particules dans la Pensée Scientifique Chinoise.' *SCIS*, 1960, **1** (no. 4), 65.

NEEDHAM, JOSEPH, WANG LING & PRICE, D. J. DE S. (1). *Heavenly Clockwork; the Great Astronomical Clocks of Mediaeval China.* Cambridge, 1960. (Antiquarian Horological Society Monographs, no. 1.) Prelim. pub. *AHOR*, 1956, **1**, 153.

NEEF, H. (1). *Die im 'Tao Tsang' enthaltenen Kommentare zu 'Tao-Tê-Ching' Kap. VI.* Inaug. Diss. Bonn, 1938.

NEOGI, P. (1). *Copper in Ancient India.* Sarat Chandra Roy (Anglo-Sanskrit Press), Calcutta, 1918. (Indian Assoc. for the Cultivation of Science, Special Pubs. no. 1.)

NEOGI, P. & ADHIKARI, B. B. (1). 'Chemical Examination of Ayurvedic Metallic Preparations; I, *Shata-puta lauha* and *Shahashra-puta lauha* (Iron roasted a hundred or a thousand times).' *JRAS/B*, 1910 (n.s.), **6**, 385.

NERI, ANTONIO (1). *L'Arte Vetraria distinta in libri sette*... Giunti, Florence, 1612. 2nd ed. Rabbuiati, Florence, 1661, Batti, Venice, 1663. Latin tr. *De Arte Vitraria Libri Septem, et in eosdem Christoph. Merretti...Observationes et Notae.* Amsterdam, 1668. German tr. by F. Geissler, Frankfurt and Leipzig, 1678. English tr. by C. Merrett, London, 1662. Cf. Ferguson (1), vol. 2, pp. 134 ff.

NEUBAUER, C. & VOGEL, H. (1). *Handbuch d. Analyse d. Harns.* 1860, and later editions, including a revision by A. Huppert, 1910.

NEUBURGER, A. (1). *The Technical Arts and Sciences of the Ancients.* Methuen, London, 1930. Tr. by H. L. Brose from *Die Technik d. Altertums.* Voigtländer, Leipzig, 1919. (With a drastically abbreviated index and the total omission of the bibliographies appended to each chapter, the general bibliography, and the table of sources of the illustrations).

NEUBURGER, M. (1). 'Théophile de Bordeu (1722 bis 1776) als Vorläufer d. Lehre von der inneren Sekretion.' *WKW*, 1911 (pt. 2), 1367.

NEUMANN, B. (1). 'Messing.' *ZAC*, 1902, **15**, 511.

NEUMANN, B. & KOTYGA, G. (1) (with the assistance of M. Rupprecht & H. Hoffmann). 'Antike Gläser.' *ZAC*, 1925, **38**, 776, 857; 1927, **40**, 963; 1928, **41**, 203; 1929, **42**, 835.

NEWALL, L. C. (1). 'Newton's Work in Alchemy and Chemistry.' Art. in *Sir Isaac Newton, 1727 to 1927*, Hist. Sci. Soc. London, 1928, pp. 203–55.

NGUYEN DANG TÂM (1). 'Sur les Bokétonosides, Saponosides du Boket ou *Gleditschia fera* Merr. (*australis* Hemsl.; *sinensis* Lam.).' *CRAS*, 1967, **264**, 121.

Niu Ching-I, Kung Yo-Thing, Huang Wei-Tê, Ko Liu-Chün & eight other collaborators (1). 'Synthesis of Crystalline Insulin from its Natural A-Chain and the Synthetic B-Chain.' *SCISA*, 1966, **15**, 231.

Noble, S. B. (1). 'The Magical Appearance of Double-Entry Book-keeping' (derivation from the mathematics of magic squares). Unpublished MS., priv. comm.

Nock, A. D. & Festugière, A. J. (1). *Corpus Hermeticum* [Texts and French translation]. Belles lettres, Paris, 1945–54.
Vol. 1, Texts I to XII; text established by Nock, tr. Festugière.
Vol. 2, Texts XIII to XVIII, Asclepius; text established by Nock, tr. Festugière.
Vol. 3, Fragments from Stobaeus I to XXII; text estab. and tr. Festugière.
Vol. 4, Fragments from Stobaeus XXIII to XXIX (text estab. and tr. Festugière) and Miscellaneous Fragments (text estab. Nock, tr. Festugière).
(Coll. Universités de France, Assoc. G. Budé.)

Noel, Francis (2). *Philosophia Sinica; Tribus Tractatibus primo Cognitionem primi Entis, secundo Ceremonias erga Defunctos, tertio Ethicam juxta Sinarum mentem complectens.* Univ. Press, Prague, 1711. Cf. Pinot (2), p. 116.

Noel, Francis (3) (tr.). *Sinensis Imperii Libri Classici Sex; nimirum: Adultorum Schola* [*Ta Hsüeh*], *Immutabile Medium* [*Chung Yung*], *Liber Sententiarum* [*Lun Yü*], *Mencius* [*Mêng Tzu*], *Filialis Observantia* [*Hsiao Ching*], *Parvulorum Schola* [*San Tzu Ching*], *e Sinico Idiomate in Latinum traducti....* Univ. Press, Prague, 1711. French tr. by Pluquet, *Les Livres Classiques de l'Empire de la Chine, précédés d'observations sur l'Origine, la Nature et les Effets de la Philosophie Morale et Politique dans cet Empire.* 7 vols. De Bure & Barrois, Didot, Paris, 1783–86. The first three books had been contained in Intorcetta *et al.* (1) *Confucius Sinarum Philosophus...*, the last three were now for the first time translated.

Noel, Francis (5) (tr.). MS. translation of the *Tao Tê Ching*, sent to Europe between +1690 and +1702. Present location unknown. See Pfister (1), p. 418.

Nordhoff, C. (1). *The Communistic Societies of the United States, from Personal Visit and Observation.* Harper, New York, 1875. 2nd ed. Dover, New York, 1966, with an introduction by M. Holloway.

Norin, E. (1). 'Tzu Chin Shan, an Alkali-Syenite Area in Western Shansi; Preliminary Notes.' *BGSC*, 1921, no. 3, 45–70.

Norpoth, L. (1). 'Paracelsus—a Mannerist?', art. in *Science, Medicine and Society in the Renaissance* (Pagel Presentation Volume), ed. Debus (20), vol. 1, p. 127.

Norton, T. (1). *The Ordinall of Alchimy* (c. 1+440). See Holmyard (12).

Noyes, J. H. [of Oneida] (1). *A History of American Socialisms.* Lippincott, Philadelphia, 1870. 2nd ed. Dover, New York, 1966, with an introduction by M. Holloway.

O'Flaherty, W. D. (1). 'The Submarine Mare in the Mythology of Siva.' *JRAS*, 1971, 9.

O'Leary, De Lacy (1). *How Greek Science passed to the Arabs.* Routledge & Kegan Paul, London, 1948.

Oakley, K. P. (2). 'The Date of the "Red Lady" of Paviland.' *AQ*, 1968, **42**, 306.

Ōbuchi, Ninji (1). 'How the *Tao Tsang* Took Shape.' Contribution to the First International Conference of Taoist Studies, Villa Serbelloni, Bellagio, 1958.

Oesterley, W. O. E. & Robinson, T. H. (1). *Hebrew Religion; its Origin and Development.* SPCK, London, 2nd ed. 1937, repr. 1966.

Ogden, W. S. (1). 'The Roman Mint and Early Britain.' *BNJ*, 1908, **5**, 1–50.

Ohsawa, G. See Sakurazawa, Nyoiti (1).

d'Ollone, H., Vissière, A., Blochet, E. *et al.* (1). *Recherches sur les Mussulmans Chinois.* Leroux, Paris, 1911. (Mission d'Ollone, 1906–1909: cf. d'Ollone, H.: *In Forbidden China*, tr. B. Miall, London, 1912.)

Olschki, L. (4). *Guillaume Boucher; a French Artist at the Court of the Khans.* Johns Hopkins Univ. Press, Baltimore, 1946 (rev. H. Franke, *OR*, 1950, **3**, 135).

Olschki, L. (7). *The Myth of Felt.* Univ. of California Press, Los Angeles, Calif., 1949.

Ong Wên-Hao (1). 'Les Provinces Métallogéniques de la Chine.' *BGSC*, 1920, no. 2, 37–59.

Ong Wên-Hao (2). 'On Historical Records of Earthquakes in Kansu.' *BGSC*, 1921, no. 3, 27–44.

Oppert, G. (2). 'Mitteilungen zur chemisch-technischen Terminologie im alten Indien; (1) Über die Metalle, besonders das Messing, (2) der Indische Ursprung der Kadmia (Calaminaris) und der Tutia.' Art. in Kahlbaum Festschrift (1909), ed. Diergart (1), pp. 127–42.

Orschall, J. C. (1). '*Sol sine Veste'; Oder dreyssig Experimenta dem Gold seinen Purpur auszuziehen...* Augsburg, 1684. Cf. Partington (7), vol. 2, p. 371; Ferguson (1), vol. 2, pp. 156 ff.

da Orta, Garcia (1). *Coloquios dos Simples e Drogas he cousas medicinais da India compostos pello Doutor Garcia da Orta.* de Endem, Goa, 1563. Latin epitome by Charles de l'Escluze, Plantin, Antwerp, 1567. Eng. tr. *Colloquies on the Simples and Drugs of India* with the annotations of the Conde de Ficalho, 1895, by Sir Clements Markham. Sotheran, London, 1913.

OSMOND, H. (1) 'Ololiuqui; the Ancient Aztec Narcotic.' *JMS*, 1955, **101**, 526.

OSMOND, H. (2). 'Hallucinogenic Drugs in Psychiatric Research.' *MBLB*, 1964, **6** (no. 1), 2.

OST, H. (1). *Lehrbuch der chemischen Technologie*. 11th ed. Jänecke, Leipzig, 1920.

OTA, K. (1). 'The Manufacture of Sugar in Japan.' *TAS/J*, 1880, **8**, 462.

OU YUN-JOEI. See Wu Yün-Jui in Roi & Wu (1).

OUSELEY, SIR WILLIAM (1) (tr.). *The 'Oriental Geography 'of Ebn Haukal, an Arabian Traveller of the Tenth Century* [Abū al-Qāsim Muḥammad Ibn Ḥawqal, *fl.* +943 to +977]. London, 1800. (This translation, done from a Persian MS., is in fact an abridgement of the *Kitāb al-Masālik wa'l-Mamālik*, 'Book of the Roads and the Countries', of Ibn Ḥawqal's contemporary, Abū Ishāq Ibrāhīm ibn Muḥammad al-Fārisī al-Iṣṭakhrī.)

PAAL, H. (1). *Johann Heinrich Cohausen, +1665 bis +1750; Leben und Schriften eines bedeutenden Arztes aus der Blütezeit des Hochstiftes Münster, mit kulturhistorischen Betrachtungen*. Fischer, Jena, 1931. (Arbeiten z. Kenntnis d. Gesch. d. Medizin im Rheinland und Westfalen, no. 6.)

PAGEL, W. (1). 'Religious Motives in the Medical Biology of the Seventeenth Century.' *BIHM*, 1935, **3**, 97.

PAGEL, W. (2). 'The Religious and Philosophical Aspects of van Helmont's Science and Medicine.' *BIHM*, Suppl. no. 2, 1944.

PAGEL, W. (10). *Paracelsus; an Introduction to Philosophical Medicine in the Era of the Renaissance*. Karger, Basel and New York, 1958. Rev. D. G[eoghegan], *AX*, 1959, **7**, 169.

PAGEL, W. (11). 'Jung's Views on Alchemy.' *ISIS*, 1948, **39**, 44.

PAGEL, W. (12). 'Paracelsus; Traditionalism and Mediaeval Sources.' Art. in *Medicine, Science and Culture*, O. Temkin Presentation Volume, ed. L. G. Stevenson & R. P. Multhauf. Johns Hopkins Press. Baltimore, Md. 1968, p. 51.

PAGEL, W. (13). 'The Prime Matter of Paracelsus.' *AX*, 1961, **9**, 117.

PAGEL, W. (14). 'The "Wild Spirit" (Gas) of John-Baptist van Helmont (+1579 to +1644), and Paracelsus.' *AX*, 1962, **10**, 2.

PAGEL, W. (15). 'Chemistry at the Cross-Roads; the Ideas of Joachim Jungius.' *AX*, 1969, **16**, 100. (Essay-review of Kangro, 1.)

PAGEL, W. (16). 'Harvey and Glisson on Irritability, with a Note on van Helmont.' *BIHM*, 1967, **41**, 497.

PAGEL, W. (17). 'The Reaction to Aristotle in Seventeenth-Century Biological Thought.' Art. in Singer Commemoration Volume, *Science, Medicine and History*, ed. E. A. Underwood. Oxford, 1953, vol. 1, p. 489.

PAGEL, W. (18). 'Paracelsus and the Neo-Platonic and Gnostic Tradition.' *AX*, 1960, **8**, 125.

PALÉOLOGUE, M. G. (1). *L'Art Chinois*. Quantin, Paris, 1887.

PALLAS, P. S. (1). *Sammlungen historischen Nachrichten ü. d. mongolischen Völkerschaften*. St Petersburg, 1776. Fleischer, Frankfurt and Leipzig, 1779.

PALMER, A. H. (1). 'The Preparation of a Crystalline Globulin from the Albumin fraction of Cow's Milk.' *JBC*, 1934, **104**, 359.

[PALMGREN, N.] (1). 'Exhibition of Early Chinese Bronzes arranged on the Occasion of the 13th International Congress of the History of Art.' *BMFEA*, 1934, **6**, 81.

PÁLOS, S. (2). *Atem und Meditation; Moderne chinesische Atemtherapie als Vorschule der Meditation—Theorie, Praxis, Originaltexte*. Barth, Weilheim, 1968.

PARANAVITANA, S. (4). *Ceylon and Malaysia*. Lake House, Colombo, 1966.

DE PAREDES, J. (1). *Recopilacion de Leyes de los Reynos de las Indias*. Madrid, 1681.

PARENNIN, D. (1). 'Lettre à Mons. [J. J.] Dortous de Mairan, de l'Académie Royale des Sciences (on demonstrations to Chinese scholars of freezing-point depression, fulminate explosions and chemical precipitation, without explanations but as a guarantee of theological veracity; on causes of the alleged backwardness of Chinese astronomy, including imperial displeasure at ominous celestial phenomena; on the pretended origin of the Chinese from the ancient Egyptians; on famines and scarcities in China; and on the aurora borealis)'. *LEC*, 1781, vol. 22, pp. 132 ff., dated 28 Sep. 1735.

PARKES, S. (1). *Chemical Essays, principally relating to the Arts and Manufactures of the British Dominions*. 5 vols. Baldwin, Cradock & Joy, London, 1815.

PARTINGTON, J. R. (1). *Origins and Development of Applied Chemistry*. Longmans Green, London, 1935.

PARTINGTON, J. R. (2). 'The Origins of the Atomic Theory.' *ANS*, 1939, **4**, 245.

PARTINGTON, J. R. (3). 'Albertus Magnus on Alchemy.' *AX*, 1937, **1**, 3.

PARTINGTON, J. R. (4). *A Short History of Chemistry*. Macmillan, London, 1937, 3rd ed. 1957.

PARTINGTON, J. R. (5). *A History of Greek Fire and Gunpowder*. Heffer, Cambridge, 1960.

PARTINGTON, J. R. (6). 'The Origins of the Planetary Symbols for Metals.' *AX*, 1937, **1**, 61.

PARTINGTON, J. R. (7). *A History of Chemistry*.

Vol. 1, pt. 1. *Theoretical Background* [Greek, Persian and Jewish].

Vol. 2. +*1500* to +*1700*.

Vol. 3. +*1700* to *1800*.

Vol. 4. *1800 to the Present Time.*

Macmillan, London, 1961– . Rev. W. Pagel, *MH*, 1971, **15**, 406.

PARTINGTON, J. R. (8). 'Chinese Alchemy.'

  (*a*) *N*, 1927, **119**, 11.

  (*b*) *N*, 1927, **120**, 878; comment on B. E. Read (11).

  (*c*) *N*, 1931, **128**, 1074; dissent from von Lippmann (1).

PARTINGTON, J. R. (9). 'The Relationship between Chinese and Arabic Alchemy.' *N*, 1928, **120**, 158.

PARTINGTON, J. R. (10). *General and Inorganic Chemistry...* 2nd ed. Macmillan, London, 1951.

PARTINGTON, J. R. (11). 'Trithemius and Alchemy.' *AX*, 1938, **2**, 53.

PARTINGTON, J. R. (12). 'The Discovery of Mosaic Gold.' *ISIS*, 1934, **21**, 203.

PARTINGTON, J. R. (13). 'Bygone Chemical Technology.' *CHIND*, 1923 (n.s.), **42** (no. 26), 636.

PARTINGTON, J. R. (14). 'The Kerotakis Apparatus.' *N*, 1947, **159**, 784.

PARTINGTON, J. R. (15). 'Chemistry in the Ancient World.' Art. in *Science, Medicine and History*, Singer Presentation Volume, ed. E. A. Underwood. Oxford, 1953, vol. 1, p. 35. Repr. with slight changes, 1959, 241.

PARTINGTON, J. R. (16). 'An Ancient Chinese Treatise on Alchemy [the *Tshan Thung Chhi* of Wei Po-Yang].' *N*, 1935, **136**, 287.

PARTINGTON, J. R. (17). 'The Chemistry of al-Rāzī.' *AX*, 1938, **1**, 192.

PARTINGTON, J. R. (18). 'Chemical Arts in the Mount Athos Manual of Christian Iconography [prob. +13th cent., MSS of +16th to +18th centuries].' *ISIS*, 1934, **22**, 136.

PARTINGTON, J. R. (19). 'E. O. von Lippmann [biography].' *OSIS*, 1937, **3**, 5.

PASSOW, H., ROTHSTEIN, A. & CLARKSON, T. W. (1). 'The General Pharmacology of the Heavy Metals.' *PHREV*, 1961, **13**, 185

PASTAN, I. (1). 'Biochemistry of the Nitrogen-containing Hormones.' *ARB*, 1966, **35** (pt. 1), 367,

DE PAUW, C. (1). *Recherches Philosophiques sur les Égyptiens et les Chinois...* (vols. IV and V of *Oeuvres Philosophiques*), Cailler, Geneva, 1774. 2nd ed. Bastien, Paris, Rep. An. III (1795). Crit. Kao Lei-Ssu [Aloysius Ko, S.J.], *MCHSAMUC*, 1777, **2**, 365, (2nd pagination) 1–174.

PECK, E. S. (1). 'John Francis Vigani, first Professor of Chemistry in the University of Cambridge, +1703 to +1712, and his Cabinet of Materia Medica in the Library of Queens' College.' *PCASC*, 1934, **34**, 34.

PELLIOT, P. (1). Critical Notes on the Earliest Reference to Tea. *TP*, 1922, **21**, 436.

PELLIOT, P. (3). 'Notes sur Quelques Artistes des Six Dynasties et des Thang.' *TP*, 1923, **22**, 214. (On the Bodhidharma legend and the founding of Shao-lin Ssu on Sung Shan, pp. 248 ff., 252 ff.)

PELLIOT, P. (8). 'Autour d'une Traduction Sanskrite du *Tao-tŏ-king* [*Tao Tê Ching*].' *TP*, 1912, **13**, 350.

PELLIOT, P. (10). 'Les Mongols et la Papauté.'

  Pt. 1 'La Lettre du Grand Khan Güyük à Innocent IV [+1246].'

  Pt. 2*a* 'Le Nestorien Siméon Rabban-Ata.'

  Pt. 2*b* 'Ascelin [Azelino of Lombardy, a Dominican, leader of the first diplomatic mission to the Mongols, +1245 to +1248].'

  Pt. 2*c* 'André de Longjumeau [Dominican envoy, +1245 to +1247].'

  ROC, 1922, **23** (sér. 3, **3**), 3–30; 1924, **24** (sér. 3, **4**), 225–335; 1931, **28** (sér. 3, **8**), 3–84.

PELLIOT, P. (47). *Notes on Marco Polo; Ouvrage Posthume.* 2 vols. Impr. Nat. and Maisonneuve, Paris, 1959.

PELLIOT, P. (54). 'Le Nom Persan du Cinabre dans les Langues "Altaiques".' *TP*, 1925, **24**, 253.

PELLIOT, P. (55). 'Henri Bosmans, S.J.' *TP*, 1928, **26**, 190.

PELLIOT, P. (56). Review of Cordier (12), *l'Imprimerie Sino-Européenne en Chine. BEFEO*, 1903, **3**, 108.

PELLIOT, P. (57). 'Le *Kin Kou K'i Kouan* [*Chin Ku Chhi Kuan*, Strange Tales New and Old, *c.* +1635]' (review of E. B. Howell, 2). *TP*, 1925, **24**, 54.

PELLIOT, P. (58). Critique of L. Wieger's *Taoisme. JA*, 1912 (10ᵉ sér.), **20**, 141.

PELSENEER, J. (3). 'La Réforme et l'Origine de la Science Moderne.' *RUB*, 1954, **5**, 406.

PELSENEER, J. (4). 'L'Origine Protestante de la Science Moderne.' *LYCH*, 1947, 246. Repr. *GEW*, **47**.

PELSENEER, J. (5). 'La Réforme et le Progrès des Sciences en Belgique au 16ᵉ Siècle.' Art. in *Science, Medicine and Hisory*, Charles Singer Presentation Volume, ed. E. A. Underwood, Oxford, 1953, vol. 1, p. 280.

PENZER, N. M. (2). *Poison-Damsels; and other Essays in Folklore and Anthropology.* Pr. pr. Sawyer, London, 1952.

PERCY, J. (1). *Metallurgy; Fuel, Fire-Clays, Copper, Zinc and Brass.* Murray, London, 1861.

PERCY, J. (2). *Metallurgy; Iron and Steel.* Murray, London, 1864.

PERCY, J. (3). *Metallurgy; Introduction, Refractories, Fuel.* Murray, London, 1875.

PERCY, J. (4). *Metallurgy; Silver and Gold.* Murray, London, 1880.

PEREIRA, J. (1). *Elements of Materia Medica and Therapeutics.* 2 vols. Longman, Brown, Green & Longman, London, 1842.

PERKIN, W. H. & KIPPING, F. S. (1). *Organic Chemistry,* rev. ed. Chambers, London and Edinburgh, 1917.

PERRY, E. S. & HECKER, J. C. (1). 'Distillation under High Vacuum.' Art. in *Distillation,* ed. A. Weissberger (*Technique of Organic Chemistry,* vol. 4), p. 495. Interscience, New York, 1951.

PERTOLD, O. (1). 'The Liturgical Use of *mahuḍa* liquor among the Bhīls.' *ARO,* 1931, **3,** 406.

PETERSON, E. (1). 'La Libération d'Adam de l'Ἀνάγκη.' *RB,* 1948, **55,** 199.

PETRIE, W. M. FLINDERS (5). 'Egyptian Religion.' Art. in *ERE,* vol. v, p. 236.

PETTUS, SIR JOHN (1). *Fleta Minor; the Laws of Art and Nature, in Knowing, Judging, Assaying, Fining,
Refining and Inlarging the Bodies of confin'd Metals*... The first part is a translation of Ercker (1),
the second contains: *Essays on Metallic Words, as a Dictionary to many Pleasing Discources.* Dawkes,
London, 1683; reissued 1686. See Sisco & Smith (1); Partington (7), vol. 2, pp. 104 ff.

PETTUS, SIR JOHN (2). *Fodinae Regales; or, the History, Laws and Places of the Chief Mines and Mineral
Works in England, Wales and the English Pale in Ireland; as also of the Mint and Mony; with a
Clavis explaining some difficult Words relating to Mines, Etc.* London, 1670. See Partington (7),
vol. 2, p. 106.

PFISTER, R. (1). 'Teinture et Alchimie dans l'Orient Hellénistique.' *SK,* 1935, **7,** 1–59.

PFIZMAIER, A. (95) (tr.). 'Beiträge z. Geschichte d. Edelsteine u. des Goldes.' *SWAW/PH,* 1867, **58,**
181, 194, 211, 217, 218, 223, 237. (Tr. chs. 807 (coral), 808 (amber), 809 (gems), 810, 811 (gold),
813 (in part), *Thai-Phing Yü Lan.*)

PHARRIS, B. B., WYNGARDEN, L. J. & GUTKNECHT, G. D. (1). Art. in *Gonadotrophins, 1968,* ed. E.
Rosenberg, p. 121.

PHILALETHA, EIRENAEUS (or IRENAEUS PHILOPONUS). Probably pseudonym of George Starkey (*c.* +1622
to +1665, *q.v.*). See Ferguson (1), vol. 2, pp. 194, 403.

PHILALETHES, EUGENIUS. See Vaughan, Thomas (+1621 to +1665), and Ferguson (1), vol. 2, p. 197.

PHILIPPE, M. (1). 'Die Braukunst der alten Babylonier im Vergleich zu den heutigen Braumethoden.'
In Huber, E. (3), *Bier und Bierbereitung bei den Völkern d. Urzeit,* vol. 1, p. 29.

PHILIPPE, M. (2). 'Die Braukunst der alten Ägypter im Lichte heutiger Brautechnik.' In Huber, E. (3),
Bier und Bierbereitung bei den Völkern d. Urzeit, vol. 1, p. 55.

PHILLIPPS, T. (SIR THOMAS) (1). 'Letter...communicating a Transcript of a MS. Treatise on the
Preparation of Pigments, and on Various Processes of the Decorative Arts practised in the Middle
Ages, written in the +12th Century and entitled *Mappae Clavicula.*' *AAA,* 1847, **32,** 183.

PHILOSTRATUS OF LEMNOS. See Conybeare (1); Jones (1).

PIANKOFF, A. & RAMBOVA, N. (1). *Egyptian Mythological Papyri.* 2 vols. Pantheon, New York, 1957
(Bollingen Series, no. 40).

PINCHES, T. G. (1). 'Tammuz.' *ERE,* vol. xii, p. 187. 'Heroes and Hero-Gods (Babylonian).' *ERE,*
vol. vi, p. 642.

PINCUS, G., THIMANN, K. V. & ASTWOOD, E. B. (1) (ed.). *The Hormones; Physiology, Chemistry and
Applications.* 5 vols. Academic Press, New York, 1948–64.

PINOT, V. (2). *Documents Inédits relatifs à la Connaissance de la Chine en France de 1685 à 1740.* Geuthner,
Paris, 1932.

PITTS, F. N. (1). 'The Biochemistry of Anxiety.' *SAM,* 1969, **220** (no. 2), 69.

PIZZIMENTI, D. (1) (ed. & tr.). *Democritus 'De Arte Magna' sive 'De Rebus Naturalibus', necnon Synesii et
Pelagii et Stephani Alexandrini et Michaelis Pselli in eundem Commentaria.* Padua, 1572, 1573,
Cologne, 1572, 1574 (cf. Ferguson (1), vol. 1, p. 205). Repr. J. D. Tauber: *Democritus Abderyta
Graecus 'De Rebus Sacris Naturalibus et Mysticis', cum Notis Synesii et Pelagii*... Nuremberg, 1717.

PLESSNER, M. (1). 'Picatrix' Book on Magic and its Place in the History of Spanish Civilisation.' Communication to the IXth International Congress of the History of Science, Barcelona and Madrid,
1959. Abstract in *Guiones de las Communicaciones,* p. 78. A longer German version appears in the
subsequent *Actes* of the Congress, p. 312.

PLESSNER, M. (2). 'Hermes Trismegistus and Arab Science.' *SI,* 1954, **2,** 45.

PLESSNER, M. (3). 'Neue Materialen z. Geschichte d. *Tabula Smaragdina.*' *DI,* 1927, **16,** 77. (A critique
of Ruska, 8.)

PLESSNER, M. (4). 'Jābir ibn Ḥayyān und die Zeit der Entstehung der arabischen Jābir-schriften.'
ZDMG, 1965, **115,** 23.

PLESSNER, M. (5). 'The Place of the *Turba Philosophorum* in the Development of Alchemy.' *ISIS,* 1954,
45, 331.

PLESSNER, M. (6). 'Vorsokratischen Philosophie und Griechischer Alchemie in Arabisch-Lateinische
Traktat; *Turba Philosophorum.*' *BOE,* **4** (in the press).

PLESSNER, M. (7). 'The *Turba Philosophorum;* a Preliminary Report on Three Cambridge Manuscripts.'
AX, 1959, **7,** 159. (These MSS are longer than that used by Ruska (6) in his translation, but the
authenticity of the additional parts has not yet been established.)

PLESSNER, M. (8). 'Geber and Jābir ibn Ḥayyān; an Authentic +16th-Century Quotation from Jābir.' *AX*, 1969, **16**, 113.

PLOSS, E. E., ROOSEN-RUNGE, H., SCHIPPERGES, H. & BUNTZ, H. (1). *Alchimia; Ideologie und Technologie.* Moos, München, 1970.

POISSON, A. (1). *Théories et Symboles des Alchimistes, le Grand Oeuvre; suivi d'un Essai sur la Bibliographie Alchimique du XIXe Siècle.* Paris, 1891. Repr. 1972.

POISSONNIER, P. J. (1). *Appareil Distillatoire présenté au Ministre de la Marine.* Paris, 1779.

POKORA, T. (4). 'An Important Crossroad of Chinese Thought' (Huan Than, the first coming of Buddhism; and Yogistic trends in ancient Taoism). *ARO*, 1961, **29**, 64.

POLLARD, A. W. (2) (ed.). *The Travels of Sir John Mandeville; with Three Narratives in illustration of it— The Voyage of Johannes de Plano Carpini, the Journal of Friar William de Rubruquis, the Journal of Friar Odoric.* Macmillan, London, 1900. Repr. Dover, New York; Constable, London, 1964.

POMET, P. (1). *Histoire Générale des Drogues.* Paris, 1694. Eng. tr. *A Compleat History of Druggs.* 2 vols. London, 1735.

DE PONCINS, GONTRAN (1). *From a Chinese City.* New York, 1957.

POPE, J. A., GETTENS, R. J., CAHILL, J. & BARNARD, N. (1). *The Freer Chinese Bronzes.* Vol. 1, Catalogue. Smithsonian Institution, Washington, D.C. 1967. (Freer Gallery of Art Oriental Studies, no. 7; Smithsonian Publication, no. 4706.) See also Gettens, Fitzhugh, Bene & Chase (1).

POPE-HENNESSY, U. (1). *Early Chinese Jades.* Benn, London, 1923.

PORKERT, MANFRED (1). *The Theoretical Foundations of Chinese Medicine.* M.I.T. Press, Cambridge, Mass. 1973. (M.I.T. East Asian Science and Technology Series, no. 3.)

PORKERT, MANFRED (2). 'Untersuchungen einiger philosophisch-wissenschaftlicher Grundbegriffe und Beziehungen in Chinesischen.' *ZDMG*, 1961, **110**, 422.

PORKERT, MANFRED (3). 'Wissenschaftliches Denken im alten China—das System der energetischen Beziehungen.' *ANT*, 1961, **2**, 532.

DELLA PORTA, G. B. (3). *De distillatione libri IX; Quibus certa methodo, multiplici artificii: penitioribus naturae arcanis detectis cuius libet mixti, in propria elementa resolutio perfectur et docetur.* Rome and Strassburg, 1609.

POSTLETHWAYT, MALACHY (1). *The Universal Dictionary of Trade and Commerce; translated from the French of Mons. [Jacques] Savary [des Bruslons], with large additions.* 2 vols. London, 1751–5. 4th ed. London, 1774.

POTT, A. F. (1). 'Chemie oder Chymie?' *ZDMG*, 1876, **30**, 6.

POTTIER, E. (1). 'Observations sur les Couches profondes de l'Acropole [& Nécropole] à Suse.' *MDP*, 1912, **13**, 1, and pl. xxxvii, 8.

POUGH, F. H. (1). 'The Birth and Death of a Volcano [Parícutin in Mexico].' *END*, 1951, **10**, 50.

[VON PRANTL, K.] (1). 'Die Keime d. Alchemie bei den Alten.' *DV*, 1856 (no. 1), no. 73, 135.

PREISENDANZ, K. (1). 'Ostanes.' Art. in Pauly–Wissowa, *Real-Encyklop. d. class. Altertumswiss.* Vol. xviii, pt. 2, cols. 1609 ff.

PREISENDANZ, K. (2). 'Ein altes Ewigkeitsymbol als Signet und Druckermarke.' *GUJ*, 1935, 143.

PREISENDANZ, K. (3). 'Aus der Geschichte des Uroboros; Brauch und Sinnbild.' Art. in E. Fehrle Festschrift, Karlsruhe, 1940, p. 194.

PREUSCHEN, E. (1). 'Die Apocryphen Gnostichen Adamschriften aus dem Armenischen übersetzt und untersucht.' Art. in Festschrift f. Bernhard Stade, sep. pub. Ricker (Töpelmann), Giessen, 1900.

PRYOR, M. G. M. (1). 'On the Hardening of the Ootheca of *Blatta orientalis* (and the cuticle of insects in general).' *PRSB*, 1940, **128**, 378, 393.

PRZYŁUSKI, J. (1). 'Les Unipédes.' *MCB*, 1933, **2**, 307.

PRZYŁUSKI, J. (2). (*a*) 'Une Cosmogonie Commune à l'Iran et à l'Inde.' *JA*, 1937, **229**, 481. (*b*) 'La Théorie des Eléments.' *SCI*, 1933.

PUECH, H. C. (1). *Le Manichéisme; son Fondateur, sa Doctrine.* Civilisations du Sud, SAEP, Paris, 1949. (Musée Guimet, Bibliothèque de Diffusion, no. 56.)

PUECH, H. C. (2). 'Catharisme Médiéval et Bogomilisme.' Art. in *Atti dello Convegno di Scienze Morali, Storiche e Filologiche*—'Oriente ed Occidente nel Medio Evo'. Accad. Naz. di Lincei, Rome, 1956 (Atti dei Convegni Alessandro Volta, no. 12), p. 56.

PUECH, H. C. (3). 'The Concept of Redemption in Manichaeism.' *ERYB*, 1969, **6**, 247 (*The Mystic Vision*, ed. J. Campbell). Tr. from the German in *ERJB*, 1936, **4**, 1.

PUFF VON SCHRICK, MICHAEL. See von Schrick.

PULLEYBLANK, E. G. (11). 'The Consonantal System of Old Chinese.' *AM*, 1964, **9**, 206.

PULSIFER, W. H. (1). *Notes for a History of Lead; and an Enquiry into the Development of the Manufacture of White Lead and Lead Oxides.* New York, 1888.

PUMPELLY, R. (1). 'Geological Researches in China, Mongolia and Japan, during the years 1862 to 1865.' *SCK*, 1866, **202**, 77.

PUMPELLY, R. (2). 'An Account of Geological Researches in China, Mongolia and Japan during the Years 1862 to 1865.' *ARSI*, 1866, **15**, 36.

PURKINJE, J. E. (PURKYNĚ). See Teich (1).

DU PUY-SANIÈRES, G. (1). 'La Modification Volontaire du Rhythme Respiratoire et les Phenomènes qui s'y rattachent.' *RPCHG*, 1937 (no. 486).

QUIRING, H. (1). *Geschichte des Goldes; die goldenen Zeitalter in ihrer kulturellen und wirtschaftlichen Bedeutung.* Enke, Stuttgart, 1948.

QUISPEL, G. (1). 'Gnostic Man; the Doctrine of Basilides.' *ERYB*, 1969, **6**, 210 (*The Mystic Vision*, ed. J. Campbell). Tr. from the German in *ERJB*, 1948, **16**, 1.

RAMAMURTHI, B. (1). 'Yoga; an Explanation and Probable Neurophysiology.' *JIMA*, 1967, **48**, 167.

RANKING, G. S. A. (1). 'The Life and Works of Rhazes.' (Biography and Bibliography of al-Rāzī.) Proc. XVIIth Internat. Congress of Medicine, London, 1913. Sect. 23, pp. 237–68.

RAO, GUNDU H. V., KRISHNASWAMY, M., NARASIMHAIYA, R. L., HOENIG, J. & GOVINDASWAMY, M. V. (1). 'Some Experiments on a Yogi in Controlled States.' *JAIMH*, 1958, **1**, 99.

RAO, SHANKAR (1). 'The Metabolic Cost of the (Yogi) Head-stand Posture.' *JAP*, 1962, **17**, 117.

RAO, SHANKAR (2). 'Oxygen-consumption during Yoga-type Breathing at Altitudes of 520 m. and 3800 m.' *IJMR*, 1968, **56**, 701.

RATLEDGE, C. (1). 'Cooling Cells for Smashing.' *NS*, 1964, **22**, 693.

RATTANSI, P. M. (1). 'The Literary Attack on Science in the Late Seventeenth and Eighteenth Centuries.' Inaug. Diss. London, 1961.

RATTANSI, P. M. (2). 'The Intellectual Origins of the Royal Society.' *NRRS*, 1968, **23**, 129.

RATTANSI, P. M. (3). 'Newton's Alchemical Studies', art. in *Science, Medicine and Society in the Renaissance* (Pagel Presentation Volume), ed. Debus (20), vol. 2, p. 167.

RATTANSI, P. M. (4). 'Some Evaluations of Reason in +16th- and +17th-Century Natural Philosophy', art. in *Changing Perspectives in the History of Science*, ed. M. Teich & R. Young. Heinemann, London, 1973, p. 148.

RATZEL, F. (1). *History of Mankind.* Tr. A. J. Butler, with introduction by E. B. Tylor. 3 vols. London, 1896–8.

RAWSON, P. S. (1). *Tantra.* (Catalogue of an Exhibition of Indian Religious Art, Hayward Gallery, London, 1971.) Arts Council of Great Britain, London, 1971.

RAY, P. (1). 'The Theory of Chemical Combination in Ancient Indian Philosophies.' *IJHS*, 1966, **1**, 1.

RAY, P. C. (1). *A History of Hindu Chemistry, from the Earliest Times to the middle of the 16th cent. A.D., with Sanskrit Texts, Variants, Translation and Illustrations.* 2 vols. Chuckerverty & Chatterjee, Calcutta, 1902, 1904, repr. 1925. New enlarged and revised edition in one volume, ed. P. Ray, retitled *History of Chemistry in Ancient and Medieval India*, Indian Chemical Society, Calcutta, 1956. Revs. J. Filliozat, *ISIS*, 1958, **49**, 362; A. Rahman, *VK*, 1957, 18.

RAY, P. C. See Tenney L. Davis' biography (obituary), with portrait. *JCE*, 1934, **11** (535).

RAY, T. (1) (tr.). *The 'Ananga Ranga'* [written by Kalyana Malla, for Lad Khan, a son of Ahmed Khan Lodi, *c.* +1500], pref. by G. Bose. Med. Book Co. Calcutta, 1951 (3rd ed.).

RAZDAN, R. K. (1). 'The Hallucinogens.' *ARMC*, 1970 (1971), **6**.

RAZOOK. See Razuq.

RAZUQ, FARAJ RAZUQ (1). 'Studies on the Works of al-Ṭughrā'ī.' Inaug. Diss., London, 1963.

READ, BERNARD E. (with LIU JU-CHHIANG) (1). *Chinese Medicinal Plants from the 'Pên Tshao Kang Mu'*, *A.D. 1596...a Botanical, Chemical and Pharmacological Reference List.* (Publication of the Peking Nat. Hist. Bull.) French Bookstore, Peiping, 1936 (chs. 12–37 of *PTKM*). Rev. W. T. Swingle, *ARLC/DO*, 1937, 191. Originally published as *Flora Sinensis*, Ser. A, vol. 1, *Plantae Medicinalis Sinensis*, 2nd ed., *Bibliography of Chinese Medicinal Plants from the Pên Tshao Kang Mu, A.D. 1596*, by B. E. Read & Liu Ju-Chhiang. Dept. of Pharmacol. Peking Union Med. Coll. & Peking Lab. of Nat. Hist. Peking, 1927. First ed. Peking Union Med. Coll. 1923.

READ, BERNARD E. (2) (with LI YÜ-THIEN). *Chinese Materia Medica; Animal Drugs.*

		Serial nos.	Corresp. with chaps. of *Pên Tshao Kang Mu*
Pt. I	Domestic Animals	322–349	50
II	Wild Animals	350–387	51*A* & *B*
III	Rodentia	388–399	51*B*
IV	Monkeys and Supernatural Beings	400–407	51*B*
V	Man as a Medicine	408–444	52

PNHB, **5** (no. 4), 37–80; **6** (no. 1), 1–102. (Sep. issued, French Bookstore, Peiping, 1931.)

	Serial nos.	Corresp. with chaps. of *Pên Tshao Kang Mu*

READ, BERNARD E. (3) (with LI YÜ-THIEN). *Chinese Materia Medica; Avian Drugs.*

    Pt. VI Birds — 245–321 — 47, 48, 49
    PNHB, 1932, **6** (no. 4), 1–101. (Sep. issued, French Bookstore, Peiping, 1932.)

READ, BERNARD E. (4) (with LI YÜ-THIEN). *Chinese Materia Medica; Dragon and Snake Drugs.*

    Pt. VII Reptiles — 102–127 — 43
    PNHB, 1934, **8** (no. 4), 297–357. (Sep. issued, French Bookstore, Peiping, 1934.)

READ, BERNARD E. (5) (with YU CHING-MEI). *Chinese Materia Medica; Turtle and Shellfish Drugs.*

    Pt. VIII Reptiles and Invertebrates — 199–244 — 45, 46
    PNHB, (Suppl.) 1939, 1–136. (Sep. issued, French Bookstore, Peiping, 1937.)

READ, BERNARD E. (6) (with YU CHING-MEI). *Chinese Materia Medica; Fish Drugs.*

    Pt. IX Fishes (incl. some amphibia, octopoda and crustacea) — 128–199 — 44
    PNHB (Suppl.), 1939. (Sep. issued, French Bookstore, Peiping, n.d. prob. 1939.)

READ, BERNARD E. (7) (with YU CHING-MEI). *Chinese Materia Medica; Insect Drugs.*

    Pt. X Insects (incl. arachnidae etc.) — 1–101 — 39, 40, 41, 42
    PNHB (Suppl.), 1941. (Sep. issued, Lynn, Peiping, 1941.)

READ, BERNARD E. (10). 'Contributions to Natural History from the Cultural Contacts of East and West.' *PNHB*, 1929, **4** (no. 1), 57.

READ, BERNARD E. (11). 'Chinese Alchemy.' *N*, 1927, **120**, 877.

READ, BERNARD E. (12). 'Inner Mongolia; China's Northern Flowery Kingdom.' (This title is a reference to the abundance of wild flowers on the northern steppes, but the article also contains an account of the saltpetre industry and other things noteworthy at Hochien in S.W. Hopei.) *PJ*, 1926, **61**, 570.

READ, BERNARD E. & LI, C. O. (1). 'Chinese Inorganic Materia Medica.' *CMJ*, 1925, **39**, 23.

READ, BERNARD E. & PAK, C. (PAK KYEBYŎNG) (1). *A Compendium of Minerals and Stones used in Chinese Medicine, from the 'Pên Tshao Kang Mu'. PNHB*, 1928, **3** (no. 2), i–vii, 1–120. Revised and enlarged, issued separately, French Bookstore, Peiping, 1936 (2nd ed.). Serial nos. 1–135, corresp. with chaps. of *Pên Tshao Kang Mu*, 8, 9, 10, 11.

READ, J. (1). *Prelude to Chemistry; an Outline of Alchemy, its Literature and Relationships.* Bell, London, 1936.

READ, J. (2). 'A Musical Alchemist [Count Michael Maier].' Abstract of Lecture, Royal Institution, London, 22 Nov. 1935.

READ, J. (3). *Through Alchemy to Chemistry.* London, 1957.

READ, T. T. (1). 'The Mineral Production and Resources of China' (metallurgical notes on tours in China, with analyses by C. F. Wang, C. H. Wang & F. N. Lu). *TAIMME*, 1912, **43**, 1–53.

READ, T. T. (2). 'Chinese Iron castings.' *CWR*, 1931 (16 May).

READ, T. T. (3). 'Metallurgical Fallacies in Archaeological Literature.' *AJA*, 1934, **38**, 382.

READ, T. T. (4). 'The Early Casting of Iron; a Stage in Iron Age Civilisation.' *GR*, 1934, **24**, 544.

READ, T. T. (5). 'Iron, Men and Governments.' *CUQ*, 1935, **27**, 141.

READ, T. T. (6). 'Early Chinese Metallurgy.' *MI*, 1936 (6 March), p. 308.

READ, T. T. (7). 'The Largest and the Oldest Iron Castings.' *IA*, 1936, **136** (no. 18, 30 Apr.), 18 (the lion of Tshang-chou, +954, the largest).

READ, T. T. (8). 'China's Civilisation Simultaneous, not Osmotic' (letter). *AMS*, 1937, **6**, 249.

READ, T. T. (9). 'Ancient Chinese Castings.' *TAFA*, 1937 (Preprint no. 37–29 of June), 30.

READ, T. T. (10). 'Chinese Iron—A Puzzle.' *HJAS*, 1937, **2**, 398.

READ, T. T. (11). Letter on 'Pure Iron—Ancient and Modern'. *MM*, 1940 (June), p. 294.

READ, T. T. (12). 'The Earliest Industrial Use of Coal.' *TNS*, 1939, **20**, 119.

READ, T. T. (13). 'Primitive Iron-Smelting in China.' *IA*, 1921, **108**, 451.

REDGROVE, H. STANLEY (1). 'The Phallic Element in Alchemical Tradition.' *JALCHS*, 1915, **3**, 65. Discussion, pp. 88 ff.

REGEL, A. (1). 'Reisen in Central-Asien, 1876–9.' *MJPGA*, 1879, **25**, 376, 408. 'Turfan.' *MJPGA*,
    1880, **26**, 205. 'Meine Expedition nach Turfan.' *MJPGA*, 1881, **27**, 380. Eng. tr. *PRGS*, 1881, 340.
REID, J. S. (1). '[The State of the Dead in] Greek [Thought].' *ERE*, vol. xi, p. 838.
REID, J. S. (2). '[The State of the Dead in] Roman [Culture].' *ERE*, vol. xi, p. 839.
REINAUD, J. T. & FAVÉ, I. (1). *Du Feu Grégeois, des Feux de Guerre, et des Origines de la Poudre à Canon,
    d'après des Textes Nouveaux*. Dumaine, Paris, 1845. Crit. rev. by D[efrémer]y, *JA*, 1846 (4e sér.),
    7, 572; E. Chevreul, *JS*, 1847, 87, 140, 209.
REINAUD, J. T. & FAVÉ, I. (2). 'Du Feu Grégeois, des Feux de Guerre, et des Origines de la Poudre à
    Canon chez les Arabes, les Persans et les Chinois.' *JA*, 1849 (4e sér.), **14**, 257.
REINAUD, J. T. & FAVÉ, I. (3). Controverse à propos du Feu Grégeois; Réponse aux Objections de
    M. Ludovic Lalanne.' *BEC*, 1847 (2e sér.), **3**, 427.
REITZENSTEIN, R. (1). *Die Hellenistischen Mysterienreligionen, nach ihren Grundgedanken und Wirkungen*.
    Leipzig, 1910. 3rd, enlarged and revised ed. Teubner, Berlin and Leipzig, 1927.
REITZENSTEIN, R. (2). *Das iranische Erlösungsmysterium; religionsgeschichtliche Untersuchungen*. Marcus &
    Weber, Bonn, 1921.
REITZENSTEIN, R. (3). '*Poimandres*'; *Studien zur griechisch-ägyptischen und frühchristlichen Literatur*.
    Teubner, Leipzig, 1904.
REITZENSTEIN, R. (4). *Hellenistische Wundererzählungen*.
  Pt. I   *Die Aretalogie* [Thaumaturgical Fabulists]; *Ursprung, Begriff, Umbildung ins Weltliche*.
  Pt. II *Die sogenannte Hymnus der Seele in den Thomas-Akten*.
  Teubner, Leipzig, 1906.
RÉMUSAT, J. P. A. (7) (tr.). *Histoire de la Ville de Khotan, tirée des Annales de la Chine et traduite du
    Chinois; suivie de Recherches sur la Substance Minérale appelée par les Chinois Pierre de Iu [Jade] et
    sur le Jaspe des Anciens*. [Tr. of *TSCC*, *Pien i tien*, ch. 55.] Doublet, Paris, 1820. Crit. rev. J.
    Klaproth (6), vol. 2, p. 281.
RÉMUSAT, J. P. A. (9). 'Notice sur l'Encyclopédie Japonoise et sur Quelques Ouvrages du Même Genre'
    (mostly on the *Wakan Sanzai Zue*). *MAI/NEM*, 1827, **11**, 123. Botanical lists, with Linnaean Latin
    identifications, pp. 269–305; list of metals, p. 231, precious stones, p. 232; ores, minerals and
    chemical substances, pp. 233–5.
RÉMUSAT, J. P. A. (10). 'Lettre de Mons. A. R....à Mons. L. Cordier...sur l'Existence de deux
    Volcans brûlans dans la Tartarie Centrale [a translation of passages from *Wakan Sanzai Zue*].' *JA*,
    1824, **5**, 44. Repr. in (11), vol. 1, p. 209.
RÉMUSAT, J. P. A. (11). *Mélanges Asiatiques; ou, Choix de Morceaux de Critique et de Mémoires relatifs
    aux Réligions, aux Sciences, aux Coutumes, à l'Histoire et à la Géographie des Nations Orientales*.
    2 vols. Dondey-Dupré, Paris, 1825–6.
RÉMUSAT, J. P. A. (12). *Nouveaux Mélanges Asiatiques; ou, Recueil de Morceaux de Critique et de Mémoires
    relatifs aux Religions, aux Sciences, aux Coutumes, à l'Histoire et à la Géographie des Nations Orientales*.
    2 vols. Schubart & Heideloff and Dondey-Dupré, Paris, 1829.
RÉMUSAT, J. P. A. (13). *Mélanges Posthumes d'Histoire et de Littérature Orientales*. Imp. Roy., Paris,
    1843.
RENAULD, E. (1) (tr.). *Michel Psellus' 'Chronographie', ou Histoire d'un Siècle de Byzance, +976
    à +1077; Texte établi et traduit*... 2 vols. Paris, 1938. (Collection Byzantine Budé, nos.
    1, 2.)
RENFREW, C. (1). 'Cycladic Metallurgy and the Aegean Early Bronze Age.' *AJA*, 1967, **71**, 1. See
    Charles (1).
RENOU, L. (1). *Anthologie Sanskrite*. Payot, Paris, 1947.
RENOU, L. & FILLIOZAT, J. (1). *L'Inde Classique; Manuel des Études Indiennes*. Vol. 1, with the collabora-
    tion of P. Meile, A. M. Esnoul & L. Silburn. Payot, Paris, 1947. Vol. 2, with the collaboration of
    P. Demiéville, O. Lacombe & P. Meile. École Française d'Extrême Orient, Hanoi; Impr. Nationale,
    Paris, 1953.
RETI, LADISLAO (6). *Van Helmont, Boyle, and the Alkahest*. Clark Memorial Library, Univ. of California,
    Los Angeles, 1969. (In *Some Aspects of Seventeenth-Century Medicine and Science*, Clark Library
    Seminar, no. 27, 1968.)
RETI, LADISLAO (7). 'Le Arte Chimiche di Leonardo da Vinci.' *LCHIND*, 1952, **34**, 655, 721.
RETI, LADISLAO (8). 'Taddeo Alderotti and the Early History of Fractional Distillation' (in Spanish). MS.
    of a Lecture in Buenos Aires, 1960.
RETI, LADISLAO (10). 'Historia del Atanor desde Leonardo da Vinci hasta "l'Encyclopédie" de Diderot.'
    INDQ, 1952, **14** (no. 10), 1.
RETI, LADISLAO (11). 'How Old is Hydrochloric Acid?' *CHYM*, 1965, **10**, 11.
REUVENS, C. J. C. (1). *Lettres à Mons. Letronne...sur les Papyrus Bilingues et Grecs et sur Quelques Autres
    Monumens Gréco-Égyptiens du Musée d'Antiquités de l'Université de Leide*. Luchtmans, Leiden,
    1830. Pagination separate for each of the three letters.

REX, FRIEDEMANN, ATTERER, M., DEICHGRÄBER, K. & RUMPF, K. (1). *Die 'Alchemie' des Andreas Libavius, ein Lehrbuch der Chemie aus dem Jahre 1597, zum ersten mal in deutscher Übersetzung... herausgegeben...* Verlag Chemie, Weinheim, 1964.

REY, ABEL (1). *La Science dans l'Antiquité.* Vol. 1 *La Science Orientale avant les Grecs*, 1930, 2nd ed. 1942; Vol. 2 *La Jeunesse de la Science Grecque*, 1933; Vol. 3 *La Maturité de la Pensée Scientifique en Grèce*, 1939; Vol. 4 *L'Apogée de la Science Technique Grecque (Les Sciences de la Nature et de l'Homme, les Mathematiques, d'Hippocrate à Platon)*, 1946. Albin Michel, Paris. (Evol. de l'Hum. Ser. Complementaire.)

RHENANUS, JOH. (1). *Harmoniae Imperscrutabilis Chymico-Philosophicae Decades duae.* Frankfurt, 1625. See Ferguson (1), vol. 2, p. 264.

RIAD, H. (1). 'Quatre Tombeaux de la Nécropole ouest d'Alexandrie.' (Report of the −2nd-century sāqīya fresco at Wardian.) *BSAA*, 1967, **42**, 89. Prelim. pub., with cover colour photograph, *AAAA*, 1964, **17** (no. 3).

RIBÉREAU-GAYON, J. & PEYNARD, E. (1). *Analyse et Contrôle des Vins...* 2nd ed. Paris, 1958.

RICE, D. S. (1). 'Mediaeval Ḥarrān; Studies on its Topography and Monuments.' *ANATS*, 1952, **2**, 36–83.

RICE, TAMARA T. (1). *The Scythians.* 3rd ed. London, 1961.

RICE, TAMARA T. (2). *Ancient Arts of Central Asia.* Thames & Hudson, London, 1965.

RICHET, C. (1) (ed.). *Dictionnaire de Physiologie.* 6 vols. Alcan, Paris, 1895–1904.

RICHIE, D. & ITO KENKICHI (1) = (1). *The Erotic Gods; Phallicism in Japan* (English and Japanese text and captions). Zufushinsha, Tokyo, 1967.

VON RICHTHOFEN, F. (2). *China; Ergebnisse eigener Reisen und darauf gegründeter Studien.* 5 vols. and Atlas. Reimer, Berlin, 1877–1911.
 Vol. 1 Einleitender Teil
 Vol. 2 Das nördliche China
 Vol. 3 Das südliche China (ed. E. Tiessen)
 Vol. 4 Palaeontologischer Teil (with contributions by W. Dames et al.)
 Vol. 5 Abschliessende palaeontologischer Bearbeitung der Sammlung... (by F. French).
 (Teggart Bibliography says 5 vols. +2 Atlas Vols.)

VON RICHTHOFEN, F. (6). *Letters on Different Provinces of China.* 6 parts, Shanghai, 1871–2.

RICKARD, T. A. (2). *Man and Metals.* Fr. tr. by F. V. Laparra, *L'Homme et les Métaux.* Gallimard, Paris, 1938. Rev. L. Febvre, *AHES/AHS*, 1940, **2**, 243.

RICKARD, T. A. (3). *The Story of the Gold-Digging Ants.* UCC, 1930.

RICKETT, W. A. (1) (tr.). *The 'Kuan Tzu' Book.* Hongkong Univ. Press, Hong Kong, 1965. Rev. T. Pokora, *ARO*, 1967, **35**, 169.

RIDDELL, W. H. (2). Earliest representations of dragon and tiger. *AQ*, 1945, **19**, 27.

RIECKERT, H. (1). 'Plethysmographische Untersuchungen bei Konzentrations- und Meditations-Übungen.' *AF*, 1967, **21**, 61.

RIEGEL, BEISWANGER & LANZL (1). Molecular stills. *IEC/A*, 1943, **15**, 417.

RIETHE, P. (1). 'Amalgamfüllung Anno Domini 1528' [A MS. of therapy and pharmacy drawn from the practice of Johannes Stocker, d. +1513]. *DZZ*, 1966, **21**, 301.

RITTER, H. (1) (ed.). *Pseudo-al-Majrīṭī 'Das Ziel des Weisen' [Ghāyat al-Ḥakīm].* Teubner, Leipzig, 1933. (Studien d. Bibliothek Warburg, no. 12.)

RITTER, H. (2) '*Picatrix*, ein arabisches Handbuch hellenistischer Magie.' *VBW*, 1923, **1**, 94. A much enlarged and revised form of this lecture appears as the introduction to vol. 2 of Ritter & Plessner (1), pp. xx ff.

RITTER, H. (3) (tr.). *Al-Ghazzālī's* (al-Ṭusī, +1058 to +1112) '*Das Elixir der Glückseligkeit' [Kīmiyā al-Sa'āda].* Diederichs, Jena, 1923. (Religiöse Stimmen der Völkers; die Religion der Islam, no. 3.)

RITTER, H. & PLESSNER, M. (1). '*Picatrix'; das 'Ziel des Weisen' [Ghāyat al Ḥakīm] von Pseudo-Majrīṭī* 2 vols. Vol. 1, Arabic text, ed. H. Ritter. Teubner, Leipzig and Berlin, 1933 (Studien der Bibliothek Warburg, no. 12). Vol. 2, German translation, with English summary (pp. lix–lxxv), by H. Ritter & M. Plessner. Warburg Inst. London, 1962 (Studies of the Warburg Institute, no. 27). Crit. rev. W. Hartner, *DI*, 1966, **41**, 175, repr. Hartner (12), p. 429.

RITTER, K. (1). *Die Erdkunde im Verhaltnis z. Natur und z. Gesch. d. Menschen; oder, Allgemeine Vergleichende Geographie.* Reimer, Berlin, 1822–59. 19 vols., the first on Africa, all the rest on Asia. Indexes after vols. 5, 13, 16 and 17.

RITTER, K. (2). *Die Erdkunde von Asien.* 5 vols. Reimer, Berlin, 1837 (part of Ritter, 1).

RIVET, P. (2). 'Le Travail de l'Or en Colombie.' *IPEK*, 1926, **2**, 128.

RIVET, P. (3). 'L'Orfèvrerie Colombienne; Technique, Aire du Dispersion, Origines.' Communication to the XXIst International Congress of Americanists, The Hague, 1924.

RIVET, P. & ARSENDAUX, H. (1). 'La Métallurgie en Amérique pre-Colombienne.' *TMIE*, 1946, no. 39.

ROBERTS [-AUSTEN], W. C. (1). 'Alloys used for Coinage' (Cantor Lectures). *JRSA*, 1884, **32**, 804, 835, 881.

ROBERTS-AUSTEN, W. C. (2). 'Alloys' (Cantor Lectures). *JRSA*, 1888, **36**, 1111, 1125, 1137.

ROBERTSON, T. BRAILSFORD & RAY, L. A. (1). 'An Apparatus for the Continuous Extraction of Solids at the Boiling Temperature of the Solvent.' *RAAAS*, 1924, **17**, 264.

ROBINSON, B. W. (1). 'Royal Asiatic Society MS. no. 178; an unrecorded Persian Painter.' *JRAS*, 1970 (no. 2, Wheeler Presentation Volume), 203. ('Abd al Karīm, active *c*. +1475, who illustrated some East Asian subjects.)

ROBINSON, G. R. & DEAKERS, T. W. (1). 'Apparatus for sublimation of anthracene.' *JCE*, 1932, **9**, 1717.

DE ROCHAS D'AIGLUN, A. (1). *La Science des Philosophes et l'Art des Thaumaturges dans l'Antiquité.* Dorbon, Paris. 1st ed. n.d. (1882), 2nd ed. 1912.

DE ROCHEMONTEIX, C. (1). *Joseph Amiot et les Derniers Survivants de la Mission Française à Pékin (1750 à 1795); Nombreux Documents inédits, avec Carte.* Picard, Paris, 1915.

ROCKHILL, W. W. (1). 'Notes on the Relations and Trade of China with the Eastern Archipelago and the Coast of the Indian Ocean during the +15th Century.' *TP*, 1914, **15**, 419; 1915, **16**, 61, 236, 374, 435, 604.

ROCKHILL, W. W. (5) (tr. & ed.). *The Journey of William of Rubruck to the Eastern Parts of the World (+1253 to +1255) as narrated by himself; with Two Accounts of the earlier Journey of John of Pian de Carpine.* Hakluyt Soc., London, 1900 (second series, no. 4).

RODWELL, G. F. (1). *The Birth of Chemistry.* London, 1874.

RODWELL, J. M. (1). *Aethiopic and Coptic Liturgies and Prayers.* Pr. pr. betw. 1870 and 1886.

ROGERS, R. W. (1). '[The State of the Dead in] Babylonian [and Assyrian Culture].' *ERE*, vol. xi, p. 828.

ROI, J. (1). *Traité des Plantes Médicinales Chinoises.* Lechevalier, Paris, 1955. (Encyclopédie Biologique ser. no. 47.) No Chinese characters, but a photocopy of those required is obtainable from Dr Claude Michon, 8 bis, Rue Desilles, Nancy, Meurthe & Moselle, France.

ROI, J. & WU YÜN-JUI (OU YUN-JOEI) (1). 'Le Taoisme et les Plantes d'Immortalité.' *BUA*, 1941 (3e sér.), **2**, 535.

ROLANDI, G. & SCACCIATI, G. (1). 'Ottone e Zinco presso gli Antichi' (Brass and Zinc in the Ancient World). *IMIN*, 1956, **7** (no. 11), 759.

ROLFINCK, WERNER (1). *Chimia in Artis Formam Redacta.* Geneva, 1661, 1671, Jena, 1662, and later editions.

ROLLESTON, SIR HUMPHREY (1). *The Endocrine Organs in Health and Disease, with an Historical Review.* London, 1936.

RÖLLIG, W. (1). 'Das Bier im alten Mesopotamien.' *JGGBB*, 1970 (for 1971), 9–104.

RONCHI, V. (5). 'Scritti di Ottica; Tito Lucrezio Caro, Leonardo da Vinci, G. Rucellai, G. Fracastoro, G. Cardano, D. Barbaro, F. Maurolico, G. B. della Porta, G. Galilei, F. Sizi, E. Torricelli, F. M. Grimaldi, G. B. Amici [a review].' *AFGR/CINO*, 1969, **24** (no. 3), 1.

RONCHI, V. (6). 'Philosophy, Science and Technology.' *AFGR/CINO*, 1969, **24** (no. 2), 168.

RONCHI, V. (7). 'A New History of the Optical Microscope.' *IJHS*, 1966, **1**, 46.

RONCHI, V. (8). 'The New History of Optical Microscopy.' *ORG*, 1968, **5**, 191.

RORET, N. E. (1) (ed.). *Manuel de l'Orfévre*, part of *Encyclopédie Roret* (or *Manuels Roret*). Roret, Paris, 1825– . Berthelot (1, 2) used the ed. of 1832.

ROSCOE, H. E. & SCHORLEMMER, C. (1). *A Treatise on Chemistry.* Macmillan, London, 1923.

ROSENBERG, E. (1) (ed.). *Gonadotrophins, 1968.* 1969.

ROSENBERG, M. (1). *Geschichte der Goldschmiedekunst auf technische Grundlage.* Frankfurt-am-Main.
  Vol. 1 *Einführung*, 1910.
  Vol. 2 *Niello*, 1908.
  Vol. 3 (in 3 parts) *Zellenschmelz*, 1921, 1922, 1925.
  Re-issued in one vol., 1972.

VON ROSENROTH, K. & VAN HELMONT, F. M. (1) (actually anon.). *Kabbala Denudata, seu Doctrina Hebraeorum Transcendentalis et Metaphysica*, etc. Lichtenthaler, Sulzbach, 1677.

ROSENTHAL, F. (1) (tr.). *The 'Muqaddimah' [of Ibn Khaldun]; an Introduction to History.* Bollingen, New York, 1958. Abridgement by N. J. Dawood, London, 1967.

ROSS, E. D. (3). *Alphabetical List of the Titles of Works in the Chinese Buddhist Tripitaka.* Indian Govt. Calcutta, 1910. (See Nanjio, B.)

ROSSETTI, GABRIELE (1). *Disquisitions on the Anti-Papal spirit which produced the Reformation; its Secret Influence on the Literature of Europe in General and of Italy in Particular.* Tr. C. Ward from the Italian. 2 vols. Smith & Elder, London, 1834.

ROSSI, P. (1). *Francesco Bacone; dalla Magia alla Scienza.* Laterza, Bari, 1957. Eng. tr. by Sacha Rabinovitch, *Francis Bacon; from Magic to Science.* Routledge & Kegan Paul, London, 1968.

ROTH, H. LING (1). *Oriental Silverwork, Malay and Chinese; a Handbook for Connoisseurs, Collectors, Students and Silversmiths.* Truslove & Hanson, London, 1910, repr. Univ. Malaya Press, Kuala Lumpur, 1966.

ROTH, MATHIAS (1). *The Prevention and Cure of Many Chronic Diseases by Movements.* London, 1851.

ROTHSCHUH, K. E. (1). *Physiologie; der Wandel ihrer Konzepte, Probleme und Methoden vom 16. bis 19. Jahrhundert.* Alber, Freiburg and München, 1968. (Orbis Academicus, Bd. 2, no. 15.)

DES ROTOURS, R. (3). 'Quelques Notes sur l'Anthropophagie en Chine.' *TP*, 1963, **50**, 386. 'Encore Quelques Notes...' *TP*, 1968, **54**, 1.

ROUSSELLE, E. (1). 'Der lebendige Taoismus im heutigen China.' *SA*, 1933, **8**, 122.

ROUSSELLE, E. (2). 'Yin und Yang vor ihrem Auftreten in der Philosophie.' *SA*, 1933, **8**, 41.

ROUSSELLE, E. (3). 'Das Primat des Weibes im alten China.' *SA*, 1941, **16**, 130.

ROUSSELLE, E. (4a). 'Seelische Führung im lebenden Taoismus.' *ERJB*, 1933, **1** (a reprint of (6), with (5) intercalated). Eng. tr., 'Spiritual Guidance in Contemporary Taoism.' *ERYB*, 1961, **4**, 59 ('Spiritual Disciplines', ed. J. Campbell). Includes footnotes but no Chinese characters.

ROUSSELLE, E. (4b). *Zur Seelischen Führung im Taoismus; Ausgewählte Aufsätze.* Wissenschaftl. Buchgesellsch., Darmstadt, 1962. (A collection of three reprinted articles (7), (5) and (6), including footnotes, and superscript references to Chinese characters, but omitting the characters themselves.)

ROUSSELLE, E. (5). '*Ne Ging Tu* [*Nei Ching Thu*], "Die Tafel des inneren Gewebes"; ein Taoistisches Meditationsbild mit Beschriftung.' *SA*, 1933, **8**, 207.

ROUSSELLE, E. (6). 'Seelische Führung im lebenden Taoismus.' *CDA*, 1934, 21.

ROUSSELLE, E. (7). 'Die Achse des Lebens.' *CDA*, 1933, 25.

ROUSSELLE, E. (8). 'Dragon and Mare; Figures of Primordial Chinese Mythology' (personifications and symbols of Yang and Yin, and the *kua* Chhien and Khun), *ERYB*, 1969, **6**, 103 (*The Mystic Vision*, ed. J. Campbell). Tr. from the German in *ERJB*, 1934, **2**, 1.

RUDDY, J. (1). 'The Big Bang at Sudbury.' *INM*, 1971 (no. 4), 22.

RUDELSBERGER, H. (1). *Chinesische Novellen aus dem Urtext übertragen.* Insel Verlag, Leipzig, 1914. 2nd ed., with two tales omitted, Schroll, Vienna, 1924.

RUFUS, W. C. (2). 'Astronomy in Korea.' *JRAS/KB*, 1936, **26**, 1. Sep. pub. as *Korean Astronomy.* Literary Department, Chosen Christian College, Seoul (Eng. Pub. no. 3), 1936.

RUHLAND, MARTIN (RULAND) (1). *Lexicon Alchemiae, sive Dictionarium Alchemisticum, cum obscuriorum Verborum et rerum Hermeticarum, tum Theophrast-Paracelsicarum Phrasium, Planam Explicationem Continens.* Palthenius, Frankfurt, 1612; 2nd ed. Frankfurt, 1661. Photolitho repr., Olms, Hildesheim, 1964. Cf. Ferguson (1), vol. 2, p. 303.

RULAND, M. See Ruhland, Martin.

RUSH, H. P. (1) Biography of A. A. Berthold. *AMH*, 1929, **1**, 208.

RUSKA, J. For bibliography see Winderlich (1).

RUSKA, J. (1). 'Die Mineralogie in d. arabischen Litteratur.' *ISIS*, 1913, **1**, 341.

RUSKA, J. (2). 'Der Zusammenbruch der Dschābir-Legende; I, die bisherigen Versuche das Dschābirproblem zu lösen.' *JBFIGN*, 1930, **3**, 9. Cf. Kraus (1).

RUSKA, J. (3). 'Die Siebzig Bücher des Ğābir ibn Ḥajjān.' Art. in *Studien z. Gesch. d. Chemie; Festgabe f. E. O. von Lippmann zum 70. Geburtstage*, ed. J. Ruska. Springer, Berlin, 1927, p. 38.

RUSKA, J. (4). *Arabische Alchemisten.* Vol. 1, *Chālid* [Khālid] *ibn Jazīd ibn Muʿāwija* [Muʿawiya]. Winter, Heidelberg, 1924 (Heidelberger Akten d. von Portheim Stiftung, no. 6). Rev. von Lippmann (10); *ISIS*, 1925, **7**, 183. Repr. with Ruska (5), Sändig, Wiesbaden, 1967.

RUSKA, J. (5). *Arabische Alchemisten.* Vol. 2, *Ğaʿfar* [Jaʿfar] *al-Ṣādiq, der sechste Imām.* Winter, Heidelberg, 1924 (Heidelberger Akten d. von Portheim Stiftung, no. 10). Rev. von Lippmann (10). Repr. with Ruska (4), Sändig, Wiesbaden, 1967.

RUSKA, J. (6). '*Turba Philosophorum*; ein Beitrag z. Gesch. d. Alchemie.' *QSGNM*, 1931, **1**, 1–368.

RUSKA, J. (7). 'Chinesisch-arabische technische Rezepte aus der Zeit der Karolinger.' *CHZ*, 1931, **55**, 297.

RUSKA, J. (8). '*Tabula Smaragdina*'; ein Beitrag z. Gesch. d. Hermetischen Literatur. Winter, Heidelberg, 1926 (Heidelberger Akten d. von Portheim Stiftung, no. 16).

RUSKA, J. (9). 'Studien zu Muḥammad ibn ʿUmail al-Tamīnī's *Kitāb al-Māʾal al-Waraqī waʾl-Ard al-Najmīyah.*' *ISIS*, 1936, **24**, 310.

RUSKA, J. (10). 'Der Urtext der *Tabula Chemica.*' *A*, 1934, **16**, 273.

RUSKA, J. (11). 'Neue Beiträge z. Gesch. d. Chemie (1. Die Namen der Goldmacherkunst, 2. Die Zeichen der griechischen Alchemie, 3. Griechischen Zeichen in Syrischer Überlieferung, 4. Ü. d. Ursprung der neueren chemischen Zeichen, 5. Kataloge der Decknamen, 6. Die metallurgischen Künste). *QSGNM*, 1942, **8**, 305.

RUSKA, J. (12). 'Über das Schriftenverzeichniss des Ğābir ibn Ḥajjān [Jābir ibn Ḥayyān] und die Unechtheit einiger ihm zugeschriebenen Abhandlungen.' *AGMN*, 1923, **15**, 53.

RUSKA, J. (13). 'Sal Ammoniacus, Nušādir und Salmiak.' *SHAW/PH*, 1923 (no. 5), 1–23.

RUSKA, J. (14). 'Übersetzung und Bearbeitungen von al-Rāzī's Buch "Geheimnis der Geheimnisse" [*Kitāb Sirr al-Asrār*].' *QSGNM*, 1935, **4**, 153–238; 1937, **6**, 1–246.

RUSKA, J. (15). 'Die Alchemie al-Rāzī's.' *DI*, 1935, **22**, 281.

RUSKA, J. (16). 'Al-Bīrūnī als Quelle für das Leben und die Schriften al-Rāzī's.' *ISIS*, 1923, **5**, 26.

RUSKA, J. (17). 'Ein neuer Beitrag zur Geschichte des Alkohols.' *DI*, 1913, **4**, 320.

RUSKA, J. (18). 'Über die von Abulqāsim al-Zuhrāwī beschriebene Apparatur zur Destillation des Rosenwassers.' *CHA*, 1937, **24**, 313.

RUSKA, J. (19) (tr.). *Das Steinbuch des Aristoteles; mit literargeschichtlichen Untersuchungen nach der arabischen Handschrift der Bibliothèque Nationale herausgegeben und übersetzt.* Winter, Heidelberg, 1912. (This early +9th-century text, the earliest of the Arabic lapidaries and widely known later as (Lat.) *Lapidarium Aristotelis*, must be termed Pseudo-Aristotle; it was written by some Syrian who knew both Greek and Eastern traditions, and was translated from Syriac into Arabic by Luka bar Serapion, or Lūqā ibn Sarāfyūn.)

RUSKA, J. (20). 'Über Nachahmung von Edelsteinen.' *QSGNM*, 1933, **3**, 316.

RUSKA, J. (21). *Das 'Buch der Alaune und Salze'; ein Grundwerk der spät-lateinischen Alchemie* [Spanish origin, +11th cent.]. Verlag Chemie, Berlin, 1935.

RUSKA, J. (22). 'Wem verdankt Man die erste Darstellung des Weingeists?' *DI*, 1913, **4**, 162.

RUSKA, J. (23). 'Weinbau und Wein in den arabischen Bearbeitungen der Geoponika.' *AGNT*, 1913, **6**, 305.

RUSKA, J. (24). *Das Steinbuch aus der 'Kosmographie' des Zakariya ibn Maḥmūd al-Qazwīnī* [c. +1250] *übersetzt und mit Anmerkungen versehen...* Schmersow (Zahn & Baendel), Kirchhain N-L, 1897. (Beilage zum Jahresbericht 1895–6 der prov. Oberrealschule Heidelberg.)

RUSKA, J. (25). 'Der Urtext d. *Tabula Smaragdina*.' *OLZ*, 1925, **28**, 349.

RUSKA, J. (26). 'Die Alchemie des Avicenna.' *ISIS*, 1934, **21**, 14.

RUSKA, J. (27). 'Über die dem Avicenna zugeschriebenen alchemistischen Abhandlungen.' *FF*, 1934, **10**, 293.

RUSKA, J. (28). 'Alchemie in Spanien.' *ZAC/AC*, 1933, **46**, 337; *CHZ*, 1933, **57**, 523.

RUSKA, J. (29). 'Al-Rāzī (Rhazes) als Chemiker.' *ZAC*, 1922, **35**, 719.

RUSKA, J. (30). 'Über die Anfänge der wissenschaftlichen Chemie.' *FF*, 1937, **13**.

RUSKA, J. (31). 'Die Aufklärung des Jābir-Problems.' *FF*, 1930, **6**, 265.

RUSKA, J. (32). 'Über die Quellen von Jābir's Chemische Wissen.' *A*, 1926, **7**, 267.

RUSKA, J. (33). 'Über die Quellen des [Geber's] *Liber Claritatis*.' *A*, 1934, **16**, 145.

RUSKA, J. (34). 'Studien zu den chemisch-technischen Rezeptsammlungen des *Liber Sacerdotum* [one of the texts related to *Mappae Clavicula*, etc.].' *QSGNM*, 1936, **5**, 275 (83–125).

RUSKA, J. (35). 'The History and Present Status of the Jābir Problem.' *JCE*, 1929, **6**, 1266 (tr. R. E. Oesper); *IC*, 1937, **11**, 303.

RUSKA, J. (36). 'Alchemy in Islam.' *IC*, 1937, **11**, 30.

RUSKA, J. (37) (ed.). *Studien z. Geschichte d. Chemie; Festgabe E. O. von Lippmann zum 70. Geburtstage...* Springer, Berlin, 1927.

RUSKA, J. (38). 'Das Giftbuch des Ǵābir ibn Ḥajjān.' *OLZ*, 1928, **31**, 453.

RUSKA, J. (39). 'Der Salmiak in der Geschichte der Alchemie.' *ZAC*, 1928, **41**, 1321; *FF*, 1928, **4**, 232.

RUSKA, J. (40). 'Studien zu Severus [or Jacob] bar Shakko's "Buch der Dialoge".' *ZASS*, 1897, **12**, 8, 145.

RUSKA, J. & GARBERS, K. (1). 'Vorschriften z. Herstellung von scharfen Wässern bei Jābir und Rāzī.' *DI*, 1939, **25**, 1.

RUSKA, J. & WIEDEMANN, E. (1). 'Beiträge z. Geschichte d. Naturwissenschaften, LXVII; Alchemistische Decknamen. *SPMSE*, 1924, **56**, 17. Repr. in Wiedemann (23), vol. 2, p. 596.

RUSSELL, E. S. (1). *Form and Function; a Contribution to the History of Animal Morphology.* Murray, London, 1916.

RUSSELL, E. S. (2). *The Interpretation of Development and Heredity; a Study in Biological Method.* Clarendon Press, Oxford, 1930.

RUSSELL, RICHARD (1) (tr.). *The Works of Geber, the Most Famous Arabian Prince and Philosopher...* [containing *De Investigatione, Summa Perfectionis, De Inventione* and *Liber Fornacum*]. James, London, 1678. Repr., with an introduction by E. J. Holmyard, Dent, London, 1928.

RYCAUT, SIR PAUL (1). *The Present State of the Greek Church.* Starkey, London, 1679.

SACHAU, E. (1) (tr.). *Alberuni's India.* 2 vols. London, 1888; repr. 1910.

DE SACY, A. I. SILVESTRE (1). 'Le "Livre du Secret de la Création", par le Sage Bélinous [Balīnās; Apollonius of Tyana, attrib.].' *MAI/NEM*, 1799, **4**, 107–58.

DE SACY, A. I. SILVESTRE (2). *Chrestomathie Arabe; ou, Extraits de Divers Écrivains Arabes, tant en Prose qu'en Vers...* 3 vols. Impr. Imp. Paris, 1806. 2nd ed. Impr. Roy. Paris, 1826–7.

SAEKI, P. Y. (1). *The Nestorian Monument in China*. With an introductory note by Lord William Gascoyne-Cecil and a pref. by Rev. Prof. A. H. Sayce. SPCK, London, 1916.

SAEKI, P. Y. (2). *The Nestorian Documents and Relics in China*. Maruzen, for the Toho Bunkwa Gakuin, Tokyo, 1937, second (enlarged) edn. Tokyo, 1951.

SAGE, B. M. (1). 'De l'Emploi du Zinc en Chine pour la Monnaie.' *JPH*, 1804, **59**, 216. Eng. tr. in Leeds (1) from *PMG*, 1805, **21**, 242.

SAHLIN, C. (1). 'Cementkopper, en historiske Översikt.' *HF*, 1938, **9**, 100. Résumé in Lindroth (1) and *SILL*, 1954.

SAID HUSAIN NASR (1). *Science and Civilisation in Islam* (with a preface by Giorgio di Santillana). Harvard University Press, Cambridge, Mass. 1968.

SAID HUSAIN NASR (2). *The Encounter of Man and Nature; the Spiritual Crisis of Modern Man*. Allen & Unwin, London, 1968.

SAID HUSAIN NASR (3). *An Introduction to Islamic Cosmological Doctrines*. Cambridge, Mass. 1964.

SAKURAZAWA, NYOITI [OHSAWA, G.] (1). *La Philosophie de la Médecine d'Extrême-Orient; le Livre du Jugement Suprême*. Vrin, Paris, 1967.

SALAZARO, D. (1). *L'Arte della Miniatura nel Secolo XIV, Codice della Biblioteca Nazionale di Napoli...* Naples, 1877. The MS. Anonymus, *De Arte Illuminandi* (so entitled in the Neapolitan Library Catalogue, for it has no title itself). Cf. Partington (12).

SALMONY, A. (1). *Carved Jade of Ancient China*. Gillick, Berkeley, Calif., 1938.

SALMONY, A. (2). 'The Human Pair in China and South Russia.' *GBA*, 1943 (6e sér.), **24**, 321.

SALMONY, A. (4). *Chinese Jade through* [i.e. until the end of] *the* [Northern] *Wei Dynasty*. Ronald, New York, 1963.

SALMONY, A. (5). *Archaic Chinese Jades from the Edward and Louise B. Sonnenschein Collection*. Chicago Art Institute, Chicago, 1952.

SAMBURSKY, S. (1). *The Physical World of the Greeks*. Tr. from the Hebrew edition by M. Dagut. Routledge & Kegan Paul, London, 1956.

SAMBURSKY, S. (2). *The Physics of the Stoics*. Routledge & Kegan Paul, London, 1959.

SAMBURSKY, S. (3). *The Physical World of Late Antiquity*. Routledge & Kegan Paul, London, 1962. Rev. G. J. Whitrow, *A/AIHS*, 1964, **17**, 178.

SANDYS, J. E. (1). *A History of Classical Scholarship*. 3 vols. Cambridge, 1908. Repr. New York, 1964.

DI SANTILLANA, G. (2). *The Origins of Scientific Thought*. University of Chicago Press, Chicago, 1961.

DI SANTILLANA, G. & VON DECHEND, H. (1). *Hamlet's Mill; an Essay on Myth and the Frame of Time*. Gambit, Boston, 1969.

SARLET, H., FAIDHERBE, J. & FRENCK, G. 'Mise en evidence chez différents Arthropodes d'un Inhibiteur de la D-acidaminoxydase.' *AIP*, 1950, **58**, 356.

SARTON, GEORGE (1). *Introduction to the History of Science*. Vol. 1, 1927; Vol. 2, 1931 (2 parts); Vol. 3, 1947 (2 parts). Williams & Wilkins, Baltimore, (Carnegie Institution Pub. no. 376.)

SARTON, GEORGE (13). Review of W. Scott's 'Hermetica' (1). *ISIS*, 1926, **8**, 342.

SARWAR, G. & MAHDIHASSAN, S. (1). 'The Word *Kimiya* as used by Firdousi.' *IQB*, 1961, **9**, 21.

SASO, M. R. (1). 'The Taoists who did not Die.' *AFRA*, 1970, no. 3, 13.

SASO, M. R. (2). *Taoism and the Rite of Cosmic Renewal*. Washington State University Press, Seattle, 1972.

SASO, M. R. (3). 'The Classification of Taoist Sects and Ranks observed in Hsinchu and other parts of Northern Thaiwan.' *AS/BIE* 1971, **30** (vol. 2 of the Presentation Volume for Ling Shun-Shêng).

SASO, M. R. (4). 'Lu Shan, Ling Shan (Lung-hu Shan) and Mao Shan; Taoist Fraternities and Rivalries in Northern Thaiwan.' Unpubl. MS. 1973.

SASTRI, S. S. SURYANARAYANA (1). *The 'Sāṃkhya Kārikā' of Iśvarakrsna*. University Press, Madras, 1930.

SATYANARAYANAMURTHI, G. G. & SHASTRY, B. P. (1). 'A Preliminary Scientific Investigation into some of the unusual physiological manifestations acquired as a result of Yogic Practices in India.' *WZNHK*, 1958, **15**, 239.

SAURBIER, B. (1). *Geschichte der Leibesübungen*. Frankfurt, 1961.

SAUVAGET, J. (2) (tr.). *Relation de la Chine et de l'Inde, redigée en +857 (Akhbār al-Ṣīn wa'l-Hind)*. Belles Lettres, Paris, 1948. (Budé Association, Arab Series.)

SAVILLE, M. H. (1). *The Antiquities of Manabi, Ecuador*. 2 vols. New York, 1907, 1910.

SAVILLE, M. H. (2). *Indian Notes*. New York, 1920.

SCHAEFER, H. (1). *Die Mysterien des Osiris in Abydos*. Leipzig, 1901.

SCHAEFER, H. W. (1). *Die Alchemie; ihr ägyptisch-griechischer Ursprung und ihre weitere historische Entwicklung*. Programm-Nummer 260, Flensburg, 1887; phot. reprod. Sändig, Wiesbaden, 1967.

SCHAFER, E. H. (1). 'Ritual Exposure [Nudity, etc.] in Ancient China.' *HJAS*, 1951, **14**, 130.

SCHAFER, E. H. (2). 'Iranian Merchants in Thang Dynasty Tales.' *SOS*, 1951, **11**, 403.

SCHAFER, E. H. (5). 'Notes on Mica in Medieval China.' *TP*, 1955, **43**, 265.

SCHAFER, E. H. (6). 'Orpiment and Realgar in Chinese Technology and Tradition.' *JAOS*, 1955, **75**, 73.

SCHAFER, E. H. (8). 'Rosewood, Dragon's-Blood, and Lac.' *JAOS*, 1957, **77**, 129.

SCHAFER, E. H. (9). 'The Early History of Lead Pigments and Cosmetics in China.' *TP*, 1956, **44**, 413.

SCHAFER, E. H. (13). *The Golden Peaches of Samarkand; a Study of Thang Exotics*. Univ. of Calif. Press, Berkeley and Los Angeles, 1963. Rev. J. Chmielewski, *OLZ*, 1966, **61**, 497.

SCHAFER, E. H. (16). *The Vermilion Bird; Thang Images of the South*. Univ. of Calif. Press, Berkeley and Los Angeles, 1967. Rev. D. Holzman, *TP*, 1969, **55**, 157.

SCHAFER, E. H. (17). 'The Idea of Created Nature in Thang Literature' (on the phrases *tsao wu chê* and *tsao hua chê*). *PEW*, 1965, **15**, 153.

SCHAFER, E. H. & WALLACKER, B. E. (1). 'Local Tribute Products of the Thang Dynasty.' *JOSHK*, 1957, **4**, 213.

SCHEFER, C. (2). 'Notice sur les Relations des Peuples Mussulmans avec les Chinois dépuis l'Extension de l'Islamisme jusqu'à la fin du 15e Siècle.' In *Volume Centenaire de l'École des Langues Orientales Vivantes, 1795–1895*. Leroux, Paris, 1895, pp. 1–43.

SCHELENZ, H. (1). *Geschichte der Pharmazie*. Berlin, 1904; photographic reprint, Olms, Hildesheim, 1962.

SCHELENZ, H. (2). *Zur Geschichte der pharmazeutisch-chemischen Destilliergeräte*. Miltitz, 1911. Reproduced photographically, Olms, Hildesheim, 1964. (Publication supported by Schimmel & Co., essential oil distillers, Miltitz.)

SCHIERN, F. (1). *Über den Ursprung der Sage von den goldgrabenden Ameisen*. Copenhagen and Leipzig, 1873.

SCHIPPER, K. M. (1) (tr.). *L'Empereur Wou des Han dans la Légende Taoiste; le 'Han Wou-Ti Nei-Tchouan [Han Wu Ti Nei Chuan]'*. Maisonneuve, Paris, 1965. (Pub. de l'École Française d'Extrême Orient, no. 58.)

SCHIPPER, K. M. (2). 'Priest and Liturgy; the Live Tradition of Chinese Religion.' MS. of a Lecture at Cambridge University, 1967.

SCHIPPER, K. M. (3). 'Taoism; the Liturgical Tradition.' Communication to the First International Conference of Taoist Studies, Villa Serbelloni, Bellagio, 1968.

SCHIPPER, K. M. (4). 'Remarks on the Functions of "Inspector of Merits" [in Taoist ecclesiastical organisation; with a description of the Ordination ceremony in Thaiwan Chêng-I Taoism].' Communication to the Second International Conference of Taoist Studies, Chino (Tateshina), Japan, 1972.

SCHLEGEL, G. (10). 'Scientific Confectionery' (a criticism of modern chemical terminology in Chinese). *TP*, 1894 (1e sér.), **5**, 147.

SCHLEGEL, G. (11). 'Le Tchien [*Chien*] en Chine.' *TP*, 1897 (1e sér.), **8**, 455.

SCHLEIFER, J. (1). 'Zum Syrischen Medizinbuch; II, Der therapeutische Teil.' *RSO*, 1939, **18**, 341. (For Pt I see *ZS*, 1938 (n.s.), **4**, 70.)

SCHMAUDERER, E. (1). 'Kenntnisse ü. das Ultramarin bis zur ersten künstlichen Darstellung um 1827.' *BGTI/TG*, 1969, **36**, 147.

SCHMAUDERER, E. (2). 'Künstliches Ultramarin im Spiegel von Preisaufgaben und der Entwicklung der Mineralanalyse im 19. Jahrhundert.' *BGTI/TG*, 1969, **36**, 314.

SCHMAUDERER, E. (3). 'Die Entwicklung der Ultramarin-fabrikation im 19. Jahrhundert.' *TRAD*, 1969, **3–4**, 127.

SCHMAUDERER, E. (4). 'J. R. Glaubers Einfluss auf die Frühformen der chemischen Technik.' *CIT*, 1970, **42**, 687.

SCHMAUDERER, E. (5). 'Glaubers Alkahest; ein Beispiel für die Fruchtbarkeit alchemischer Denkansätze im 17. Jahrhundert.'; in the press.

SCHMIDT, C. (1) (ed.). *Koptisch-Gnostische Schriften* [including *Pistis Sophia*]. Hinrichs, Leipzig, 1905 (Griech. Christliche Schriftsteller, vol. 13). 2nd ed. Akad. Verlag, Berlin, 1954.

SCHMIDT, R. (1) (tr.). *Das 'Kāmasūtram' des Vātsyāyana; die indische Ars Amatoria nebst dem vollständigen Kommentare (Jayamangalā) des Yasodhara—aus dem Sanskrit übersetzt und herausgegeben...* Berlin, 1912. 7th ed. Barsdorf, Berlin, 1922.

SCHMIDT, R. (2). *Beitäage z. Indischen Erotik; das Liebesleben des Sanskritvolkes, nach den Quellen dargestellt von R. S...* 2nd ed. Barsdorf, Berlin, 1911. Reissued under the imprint of Linser, in the same year.

SCHMIDT, R. (3) (tr.). *The 'Rati Rahasyam' of Kokkoka* [said to be +11th cent. under Rājā Bhōja]. Med. Book Co. Calcutta, 1949. (Bound with Tatojaya (1), *q.v.*)

SCHMIDT, W. A. (1). *Die Griechischen Papyruskunden der K. Bibliothek Berlin; III, Die Purpurfärberei und der Purpurhandel in Altertum*. Berlin, 1842.

SCHMIEDER, K. C. (1). *Geschichte der Alchemie*. Halle, 1832.

SCHRIMPF, R. (1). 'Bibliographie Sommaire des Ouvrages publiés en Chine durant la Période 1950–60 sur l'Histoire du Développement des Sciences et des Techniques Chinoises.' *BEFEO*, 1963, **51**, 615. Includes chemistry and chemical industry.

SCHNEIDER, W. (1). 'Über den Ursprung des Wortes "Chemie".' *PHI*, 1959, **21**, 79.

SCHNEIDER, W. (2). 'Kekule und die organische Strukturchemie.' *PHI*, 1958, **20**, 379.

SCHOLEM, G. (3). *Jewish Gnosticism, Merkabah* [apocalyptic or Messianic] *Mysticism, and the Talmudic Tradition*. New York, 1960.

SCHOLEM, G. (4). 'Zur Geschichte der Anfänge der Christlichen Kabbala.' Art. in L. Baeck Presentation Volume, London, 1954.

SCHOTT, W. (2). 'Ueber ein chinesisches Mengwerk, nebst einem Anhang linguistischer Verbesserungen zu zwei Bänden der Erdkunde Ritters' [the *Yeh Huo Pien* of Shen Tê-Fu (Ming)]. *APAW/PH*, 1880, no. 3.

SCHRAMM, M. (1). 'Aristotelianism; Basis of, and Obstacle to, Scientific Progress in the Middle Ages— Some Remarks on A. C. Crombie's "From Augustine to Galileo".' *HOSC*, 1963, **2**, 91; 1965, **4**, 70.

VON SCHRICK, MICHAEL PUFF (1). *Hienach volget ein nüczliche Materi von manigerley ausgeprânten Wasser, wie Man die nüczen und pruchen sol zu Gesuntheyt der Menschen; Ûn das Puchlein hat Meiyster Michel Schrick, Doctor der Erczney durch lijebe und gepet willen erberen Personen ausz den Pûchern zu sammen colligiert un beschrieben*. Augsburg, 1478, 1479, 1483, etc.

SCHUBARTH, DR (1). 'Ueber das chinesisches Weisskupfer und die vom Vereine angestellten Versuche dasselbe darzustellen.' *VVBGP*, 1824, **3**, 134. (The Verein in question was the Verein z. Beförderung des Gewerbefleisses in Preussen.)

SCHULTES, R. E. (1). *A Contribution to our Knowledge of* Rivea corymbosa, *the narcotic Ololiuqui of the Aztecs*. Botanical Museum, Harvard Univ. Cambridge, Mass. 1941.

SCHULTZE, S. See Aigremont, Dr.

SCHURHAMMER, G. (2). 'Die Yamabushis nach gedrückten und ungedrückten Berichten d. 16. und 17. Jahrhunderts.' *MDGNVO*, 1965, **46**, 47.

SCOTT, HUGH (1). *The Golden Age of Chinese Art; the Lively Thang Dynasty*. Tuttle, Rutland, Vt. and Tokyo, 1966.

SCOTT, W. (1) (ed.). *Hermetica*. 4 vols. Oxford, 1924–36.
   Vol 1, Introduction, Texts and Translation, 1924.
   Vol 2, Notes on the Corpus Hermeticum, 1925.
   Vol 3, Commentary; Latin Asclepius and the Hermetic Excerpts of Stobaeus, 1926 (posthumous ed. A. S. Ferguson).
   Vol 4, Testimonia, Addenda, and Indexes (posthumous, with A. S. Ferguson).
   Repr. Dawson, London, 1968. Rev. G. Sarton, *ISIS*, 1926, **8**, 342.

SÉBILLOT, P. (1). *Les Travaux Publics et les Mines dans les Traditions et les Superstitions de tous les Peuples*. Paris, 1894.

SEGAL, J. B. (1). 'Pagan Syrian Monuments in the Vilayet of Urfa [Edessa].' *ANATS*, 1953, **3**, 97.

SEGAL, J. B. (2). 'The Ṣābian Mysteries; the Planet Cult of Ancient Ḥarrān.' Art. in *Vanished Civilisations*, ed. E. Bacon. 1963.

SEIDEL, A. (1). 'A Taoist Immortal of the Ming Dynasty; Chang San-Fêng.' Art. in *Self and Society in Ming Thought*, ed. W. T. de Bary. Columbia Univ. Press, New York, 1970, p. 483.

SEIDEL, A. (2). *La Divinisation de Lao Tseu [Lao Tzu] dans le Taoisme des Han*. École Française de l'Extrême Orient, Paris, 1969. (Pub. de l'Éc. Fr. de l'Extr. Or., no. 71.) A Japanese version is in *DK*, 1968, **3**, 5–77, with French summary, p. ii.

SELIMKHANOV, I. R. (1). 'Spectral Analysis of Metal Articles from Archaeological Monuments of the Caucasus.' *PPHS*, 1962, **38**, 68.

SELYE, H. (1). *Textbook of Endocrinology*. Univ. Press and *Acta Endocrinologica*, Montreal, 1947.

SEN, SATIRANJAN (1). 'Two Medical Texts in Chinese Translation.' *VBA*, 1945, **1**, 70.

SENCOURT, ROBERT (1). *Outflying Philosophy; a Literary Study of the Religious Element in the Poems and Letters of John Donne and in the Works of Sir Thomas Browne and Henry Vaughan the Silurist, together with an Account of the Interest of these Writers in Scholastic Philosophy, in Platonism and in Hermetic Physic; with also some Notes on Witchcraft*. Simpkin, Marshall, Hamilton & Kent, London, n.d. (1923).

SENGUPTA, KAVIRAJ N. N. (1). *The Ayurvedic System of Medicine*. 2 vols. Calcutta, 1925.

SERRUYS, H. (1) (tr.). '*Pei Lu Fêng Su*; Les Coutumes des Esclaves Septentrionaux [Hsiao Ta-Hêng's book on the Mongols, +1594].' *MS*, 1945, **10**, 117–208.

SEVERINUS, PETRUS (1). *Idea Medicinae Philosophicae*, 1571. 3rd ed. The Hague, 1660.

SEWTER, E. R. A. (1) (tr.). *Fourteen Byzantine Rulers; the 'Chronographia' of Michael Psellus* [+1063, the last part by +1078]. Routledge & Kegan Paul, London; Yale Univ. Press, New Haven, Conn., 1953. 2nd revised ed. Penguin, Baltimore, and London, 1966.

SEYBOLD, C. F. (1). Review of J. Lippert's 'Ibn al-Qifṭī's *Ta'rīkh al-Ḥukamā*', auf Grund der Vorarbeiten Aug. Müller (Dieter, Leipzig, 1903).' *ZDMG*, 1903, **57**, 805.

SEYYED HOSSEIN NASR. See Said Husain Nasr.

SEZGIN, F. (1). 'Das Problem des Jābir ibn Ḥayyān im Lichte neu gefundener Handschriften.' *ZDMG*, 1964, **114**, 255.

SHANIN, T. (1) (ed.). *The Rules of the Game; Cross-Disciplinary Essays on Models in Scholarly Thought.* Tavistock, London, 1972.

SHAPIRO, J. (1). 'Freezing-out, a Safe Technique for Concentration of Dilute Solutions.' *S*, 1961, **133**, 2063.

SHASTRI, KAVIRAJ KALIDAS (1). *Catalogue of the Rasashala Aushadhashram Gondal* (Ayurvedic Pharmaceutical Works of Gondal), [founded by the Maharajah of Gondal, H. H. Bhagvat Singhji]. 22nd ed. Gondal, Kathiawar, 1936. 40th ed. 1952.

SHAW, THOMAS (1). *Travels or Observations relating to sereral parts of Barbary and the Levant.* Oxford, 1738; London, 1757; Edinburgh, 1808. *Voyages dans la Régence d'Alger.* Paris, 1830.

SHEA, D. & FRAZER, A. (1) (tr.). *The 'Dabistan', or School of Manners* [by Mobed Shah, +17th Cent.], *translated from the original Persian, with notes and illustrations...* 2 vols. Paris, 1843.

SHEAR, T. L. (1). 'The Campaign of 1939 [excavating the ancient Athenian agora].' *HE*, 1940, **9**, 261.

SHÊN TSUNG-HAN (1). *Agricultural Resources of China.* Cornell Univ. Press, Ithaca, N.Y., 1951.

SHÊNG WU-SHAN (1). *Érotologie de la Chine; Tradition Chinoise de l'Érotisme.* Pauvert, Paris, 1963. (Bibliothèque Internationale d'Érotologie, no. 11.) Germ. tr. *Die Erotik in China*, ed. Lo Duca. Desch, Basel, 1966. (Welt des Eros, no. 5.)

SHEPPARD, H. J. (1). 'Gnosticism and Alchemy.' *AX*, 1957, **6**, 86. 'The Origin of the Gnostic-Alchemical Relationship.' *SCI*, 1962, **56**, 1.

SHEPPARD, H. J. (2). 'Egg Symbolism in Alchemy.' *AX*, 1958, **6**, 140.

SHEPPARD, H. J. (3). 'A Survey of Alchemical and Hermetic Symbolism.' *AX*, 1960, **8**, 35.

SHEPPARD, H. J. (4). 'Ouroboros and the Unity of Matter in Alchemy; a Study in Origins.' *AX*, 1962, **10**, 83. 'Serpent Symbolism in Alchemy.' *SCI*, 1966, **60**, 1.

SHEPPARD, H. J. (5). 'The Redemption Theme and Hellenistic Alchemy.' *AX*, 1959, **7**, 42.

SHEPPARD, H. J. (6). 'Alchemy; Origin or Origins?' *AX*, 1970, **17**, 69.

SHEPPARD, H. J. (7). 'Egg Symbolism in the History of the Sciences.' *SCI*, 1960, **54**, 1.

SHEPPARD, H. J. (8). 'Colour Symbolism in the Alchemical *Opus*.' *SCI*, 1964, **58**, 1.

SHERLOCK, T. P. (1). 'The Chemical Work of Paracelsus.' *AX*, 1948, **3**, 33.

SHIH YU-CHUNG (1). 'Some Chinese Rebel Ideologies.' *TP*, 1956, **44**, 150.

SHIMAO EIKOH (1). 'The Reception of Lavoisier's Chemistry in Japan.' *ISIS*, 1972, **63**, 311.

SHIRAI, MITSUTARŌ (1). 'A Brief History of Botany in Old Japan.' Art. in *Scientific Japan, Past and Present*, ed. Shinjo Shinzo. Kyoto, 1926. (Commemoration Volume of the 3rd Pan-Pacific Science Congress.)

SIGERIST, HENRY E. (1). *A History of Medicine.* 2 vols. Oxford, 1951. Vol. 1, *Primitive and Archaic Medicine.* Vol. 2, *Early Greek, Hindu and Persian Medicine.* Rev. (vol. 2), J. Filliozat, *JAOS*, 1926, **82**, 575.

SIGERIST, HENRY E. (2). *Landmarks in the History of Hygiene.* London, 1956.

SIGGEL, A. (1). *Die Indischen Bücher aus dem 'Paradies d. Weisheit über d. Medizin' des 'Alī Ibn Sahl Rabban al-Ṭabarī.* Steiner, Wiesbaden, 1950. (Akad. d. Wiss. u. d. Lit. in Mainz; Abhdl. d. geistes- und sozial-wissenschaftlichen Klasse, no. 14.) Crit. O. Temkin, *BIHM*, 1953, **27**, 489.

SIGGEL, A. (2). *Arabisch-Deutsches Wörterbuch der Stoffe aus den drei Natur-reichen die in arabischen alchemistischen Handschriften vorkommen; nebst Anhang, Verzeichnis chemischer Geräte.* Akad. Verlag. Berlin, 1950. (Deutsche Akad. der Wissenchaften zu Berlin; Institut f. Orientforschung, Veröffentl. no. 1.)

SIGGEL, A. (3). *Decknamen in der arabischen Alchemistischen Literatur.* Akad. Verlag. Berlin, 1951. (Deutsche Akad. der Wissenschaften zu Berlin; Institut f. Orientforschung, Veröffentl. no. 5.) Rev. M. Plessner, *OR*, **7**, 368.

SIGGEL, A. (4). 'Das Sendschreiben "Das Licht über das Verfahren des Hermes der Hermesse dem, der es begehrt".' (*Qabas al-Qabīs fī Tadbīr Harmas al-Harāmis*, early +13th cent.) *DI*, 1937, **24**, 287.

SIGGEL, A. (5) (tr.). '*Das Buch der Gifte*' [*Kitāb al-Sumūm wa daf 'maḍārrihā*] des *Jābir ibn Ḥayyān* [Kr/2145]; *Arabische Text in Faksimile...übers. u. erläutert...* Steiner, Wiesbaden, 1958. (Veröffentl. d. Orientalischen Komm. d. Akad. d. Wiss. u. d. Lit. no. 12.) Cf Kraus (2), pp. 156 ff.

SIGGEL, A. (6). 'Gynäkologie, Embryologie und Frauenhygiene aus dem "Paradies der Weisheit [*Firdaws al-Ḥikma*] über die Medizin" des Abū Ḥasan 'Alī ibn Sahl Rabban al-Ṭabarī [d. c. +860], nach der Ausgabe von Dr. M. Zubair al-Ṣiddīqī (Sonne, Berlin-Charlottenberg, 1928).' *QSGNM*, 1942, **8**, 217.

SILBERER, H. (1). *Probleme der Mystik und ihrer Symbolik.* Vienna, 1914. Eng. tr. S. E. Jelliffe, *Problems of Mysticism, and its Symbolism.* Moffat & Yard, New York, 1917.

SINGER, C. (1). *A Short History of Biology*. Oxford, 1931.

SINGER, C. (3). 'The Scientific Views and Visions of St. Hildegard.' Art. in Singer (13), vol. 1, p. 1. Cf. Singer (16), a parallel account.

SINGER, C. (4). *From Magic to Science; Essays on the Scientific Twilight*. Benn, London, 1928.

SINGER, C. (8). *The Earliest Chemical Industry; an Essay in the Historical Relations of Economics and Technology, illustrated from the Alum Trade*. Folio Society, London, 1948.

SINGER, C. (13) (ed.). *Studies in the History and Method of Science*. Oxford, vol. 1, 1917; vol. 2, 1921. Photolitho reproduction, Dawson, London, 1955.

SINGER, C. (16). 'The Visions of Hildegard of Bingen.' Art. in Singer (4), p. 199.

SINGER, C. (23). 'Alchemy' (art. in *Oxford Classical Dictionary*). Oxford.

SINGER, CHARLES (25). *A Short History of Anatomy and Physiology from the Greeks to Harvey*. Dover, New York, 1957. Revised from *The Evolution of Anatomy*. Kegan Paul, Trench, & Trubner, London, 1925.

SINGER, D. W. (1). *Giordano Bruno; His Life and Thought, with an annotated Translation of his Work 'On the Infinite Universe and Worlds'*. Schuman, New York, 1950.

SINGER, D. W. (2). 'The Alchemical Writings attributed to Roger Bacon.' *SP*, 1932, **7**, 80.

SINGER, D. W. (3). 'The Alchemical Testament attributed to Raymund Lull.' *A*, 1928, **9**, 43. (On the pseudepigraphic nature of the Lullian corpus.)

SINGER, D. W. (4). 'l'Alchimie.' Communiction to the IVth International Congress of the History of Medicine, Brussels, 1923. Sep. pub. de Vlijt, Antwerp, 1927.

SINGER, D. W., ANDERSON, A. & ADDIS, R. (1). *Catalogue of Latin and Vernacular Alchemical Manuscripts in Great Britain and Ireland before the 16th Century*. 3 vols. Lamertin, Brussels, 1928–31 (for the Union Académique Internationale).

SINGLETON, C. S. (1) (ed.). *Art, Science and History in the Renaissance*. Johns Hopkins, Baltimore, 1968.

SISCO, A. G. & SMITH, C. S. (1) (tr.). *Lazarus Ercker's Treatise on Ores and Assaying, translated from the German edition of +1580*. Univ. Chicago Press, Chicago, 1951.

SISCO, A. G. & SMITH, C. S. (2). '*Bergwerk- und Probier-büchlein*'; *a Translation from the German of the 'Berg-büchlein', a Sixteenth-Century Book on Mining Geology, by A. G. Sisco, and of the 'Probier-büchlein', a Sixteenth-Century Work on Assaying, by A. G. Sisco & C. S. Smith; with technical annotations and historical notes*. Amer. Institute of Mining and Metallurgical Engineers, New York, 1949.

SIVIN, N. (1). 'Preliminary Studies in Chinese Alchemy; the *Tan Ching Yao Chüeh* attributed to Sun Ssu-Mo (+581? to after +674).' Inaug. Diss., Harvard University, 1965. Published as: *Chinese Alchemy; Preliminary Studies*. Harvard Univ. Press, Cambridge, Mass. 1968. (Harvard Monographs in the History of Science, no. 1.) Ch. 1 sep. pub. *JSHS*, 1967, **6**, 60. Revs. J. Needham, *JAS*, 1969, 850; Ho Ping-Yü, *HJAS*, 1969, **29**, 297; M. Eliade, *HOR*, 1970, **10**, 178.

SIVIN, N. (1a). 'On the Reconstruction of Chinese Alchemy.' *JSHS*, 1967, **6**, 60 (essentially ch. 1 of Sivin, 1).

SIVIN, N. (2). 'Quality and Quantity in Chinese Alchemy.' Priv. circ. 1966; expanded as: 'Reflections on Theory and Practice in Chinese Alchemy.' Contribution to the First International Conference of Taoist Studies, Villa Serbelloni, Bellagio, 1968.

SIVIN, N. (3). Draft Translation of *Thai-Shang Wei Ling Shen Hua Chiu Chuan Tan Sha Fa* (*TT*/885). Unpublished MS., copy deposited in Harvard-Yenching Library for circulation.

SIVIN, N. (4). Critical Editions and Draft Translations of the Writings of Chhen Shao-Wei (*TT*/883 and 884, and *YCCC*, chs. 68–9). Unpublished MS.

SIVIN, N. (5). Critical Edition and Draft Translation of *Tan Lun Chüeh Chih Hsin Ching* (*TT*/928 and *YCCC*, ch. 66). Unpublished MS.

SIVIN, N. (6). 'William Lewis as a Chemist.' *CHYM*, 1962, **8**, 63.

SIVIN, N. (7). 'On the *Pao Phu Tzu* (*Nei Phien*) and the Life of Ko Hung (+283 to +343).' *ISIS*, 1969. **60**, 388.

SIVIN, N. (8). 'Chinese Concepts of Time.' *EARLH*, 1966, **1**, 82.

SIVIN, N. (9). *Cosmos and Computation in Chinese Mathematical Astronomy*. Brill, Leiden, 1969. Reprinted from *TP*, 1969, **55**.

SIVIN, N. (10). 'Chinese Alchemy as a Science.' Contrib. to '*Nothing Concealed*' (Wu Yin Lu); *Essays in Honour of Liu (Aisin-Gioro) Yü-Yün*, ed. F. Wakeman, Chinese Materials and Research Aids Service Centre, Thaipei, Thaiwan, 1970, p. 35.

SKRINE, C. P. (1). 'The Highlands of Persian Baluchistan.' *GJ*, 1931, **78**, 321.

DE SLANE, BARON MCGUCKIN (2) (tr.). *Ibn Khallikan's Dictionary* (translation of Ibn Khallikān's *Kitāb Wafayāt al-A'yān*, a collection of 865 biographies, +1278). 4 vols. Paris, 1842–71.

SMEATON, W. A. (1). 'Guyton de Morveau and Chemical Affinity.' *AX*, 1963, **11**, 55.

SMITH, ALEXANDER (1). *Introduction to Inorganic Chemistry*. Bell, London, 1912.

SMITH, C. S. (4). 'Matter versus Materials; a Historical View.' *SC*, 1968, **162**, 637.

SMITH, C. S. (5). 'A Historical View of One Area of Applied Science—Metallurgy.' Art. in *Applied Science and Technological Progress*. A Report to the Committee on Science and Astronautics of the United States House of Representatives by the National Academy of Sciences, Washington, D.C. 1967.

SMITH, C. S. (6). 'Art, Technology and Science; Notes on their Historical Interaction.' *TCULT*, 1970, **11**, 493.

SMITH, C. S. (7). 'Metallurgical Footnotes to the History of Art.' *PAPS*, 1972, **116**, 97. (Penrose Memorial Lecture, Amer. Philos. Soc.)

SMITH, C. S. & GNUDI, M. T. (1) (tr. & ed.). *Biringuccio's 'De La Pirotechnia' of +1540, translated with an introduction and notes*. Amer. Inst. of Mining and Metallurgical Engineers, New York, 1942, repr. 1943. Reissued, with new introductory material. Basic Books, New York, 1959.

SMITH, F. PORTER (1). *Contributions towards the Materia Medica and Natural History of China, for the use of Medical Missionaries and Native Medical Students*. Amer. Presbyt. Miss. Press, Shanghai; Trübner, London, 1871.

SMITH, F. PORTER (2). 'Chinese Chemical Manufactures.' *JRAS/NCB*, 1870 (n.s.), **6**, 139.

SMITH, R. W. (1). 'Secrets of Shao-Lin Temple Boxing.' Tuttle, Rutland, Vt. and Tokyo, 1964.

SMITH, T. (1) (tr.). *The 'Recognitiones' of Pseudo-Clement of Rome* [c. +220]. In Ante-Nicene Christian Library, ed. A. Roberts & J. Donaldson, Clark, Edinburgh, 1867. vol. 3, p. 297.

SMITH, T., PETERSON, P. & DONALDSON, J. (1) (tr.). *The Pseudo-Clementine Homilies* [c. +190, attrib. Clement of Rome, *fl.* +96]. In Ante-Nicene Christian Library, ed. A. Roberts & J. Donaldson, Clark, Edinburgh, 1867. vol. 17.

SMITHELLS, C. J. (1). 'A New Alloy of High Density.' *N*, 1937, **139**, 490.

SMYTHE, J. A. (1). *Lead; its Occurrence in Nature, the Modes of its Extraction, its Properties and Uses, with Some Account of its Principal Compounds*. London and New York, 1923.

SNAPPER, I. (1). *Chinese Lessons to Western Medicine; a Contribution to Geographical Medicine from the Clinics of Peiping Union Medical College*. Interscience, New York, 1941.

SNELLGROVE, D. (1). *Buddhist Himalaya; Travels and Studies in Quest of the Origins and Nature of Tibetan Religion*. Oxford, 1957.

SNELLGROVE, D. (2). *The 'Hevajra Tantra', a Critical Study*. Oxford, 1959.

SODANO, A. R. (1) (ed. & tr.). *Porfirio* [Porphyry of Tyre]; *Lettera ad Anebo* (Greek text and Italian tr.). Arte Tip., Naples, 1958.

SOLLERS, P. (1). 'Traduction et Presentation de quelques Poèmes de Mao Tsê-Tung.' *TQ*, 1970, no. 40, 38.

SOLLMANN, T. (1). *A Textbook of Pharmacology and some Allied Sciences*. Saunders, 1st ed. Philadelphia and London, 1901. 8th ed., extensively revised and enlarged, Saunders, Philadelphia and London, 1957.

SOLOMON, D. (1) (ed.). *LSD, the Consciousness-Expanding Drug*. Putnam, New York, 1964. Rev. W. H. McGlothlin, *NN*, 1964, **199** (no. 15), 360.

SOYMIÉ, M. (4). 'Le Lo-feou Chan (Lo-fou Shan]; Étude de Géographie Religieuse.' *BEFEO*, 1956, **48**, 1–139.

SOYMIÉ, M. (5). 'Bibliographie du Taoisme; Études dans les Langues Occidentales' (pt. 2). *DK*, 1971, **4**, 290–225 (1–66); with Japanese introduction, p. 288 (3).

SOYMIÉ, M. (6). 'Histoire et Philologie de la Chine Médiévale et Moderne; Rapport sur les Conférences' (on the date of *Pao Phu Tzu*). *AEPHE/SHP*, 1971, 759.

SOYMIÉ, M. & LITSCH, F. (1). 'Bibliographie du Taoisme; Études dans les Langues Occidentales' (pt. 1). *DK*, 1968, **3**, 318–247 (1–72); with Japanese introduction, p. 316 (3).

SPEISER, E. A. (1). *Excavations at Tepe Gawra*. 2 vols. Philadelphia, 1935.

SPENCER, J. E. (3). 'Salt in China.' *GR*, 1935, **25**, 353.

SPENGLER, O. (1). *The Decline of the West*, tr. from the German, *Die Untergang des Abendlandes*, by C. F. Atkinson. 2 vols. Vol. 1, *Form and Actuality;* vol. 2, *Perspectives of World History*. Allen & Unwin, London, 1926, 1928.

SPERBER, D. (1). 'New Light on the Problem of Demonetisation in the Roman Empire.' *NC*, 1970 (7th ser.), **10**, 112.

SPETER, M. (1). 'Zur Geschichte der Wasserbad-destillation; das "Berchile" Abul Kasims.' *APHL*, 1930, **5** (no. 8), 116.

SPIZEL, THEOPHILUS (1). *De Re Literaria Sinensium Commentarius...* Leiden, 1660 (frontispiece, 1661).

SPOONER, R. C. (1). 'Chang Tao-Ling, the first Taoist Pope.' *JCE*, 1938, **15**, 503.

SPOONER, R. C. (2). 'Chinese Alchemy.' *JWCBRS*, 1940 (A), **12**, 82.

SPOONER, R. C. & WANG, C. H. (1). 'The Divine Nine-Turn Tan-Sha Method, a Chinese Alchemical Recipe.' *ISIS*, 1948, **28**, 235.

VAN DER SPRENKEL, O. (1). 'Chronology, Dynastic Legitimacy, and Chinese Historiography' (mimeographed). Paper contributed to the Study Conference at the London School of Oriental Studies

1956, but not included with the rest in *Historians of China and Japan*, ed. W. G. Beasley & E. G. Pulleybank, 1961.

SQUIRE, S. (1) (tr.). *Plutarch 'De Iside et Osiride', translated into English* (sep. pagination, text and tr.). Cambridge, 1744.

STADLER, H. (1) (ed.). *Albertus Magnus 'De Animalibus, libri XXVI.'* 2 vols. Münster i./W., 1916–21.

STANLEY, R. C. (1). *Nickel, Past and Present.* Proc. IInd Empire Mining and Metallurgical Congress, 1928, pt. 5, Non-Ferrous Metallurgy, 1–34.

STANNUS, H. S. (1). 'Notes on Some Tribes of British Central Africa [esp. the Anyanja of Nyasaland]. *JRAI*, 1912, **40**, 285.

STAPLETON, H. E. (1). 'Sal-Ammoniac; a Study in Primitive Chemistry.' *MAS/B*, 1905, **1**, 25.

STAPLETON, H. E. (2). 'The Probable Sources of the Numbers on which Jābirian Alchemy was based.' *A/AIHS*, 1953, **6**, 44.

STAPLETON, H. E. (3). 'The Gnomon as a possible link between one type of Mesopotamian *Ziggurat* and the Magic Square Numbers on which Jābirian Alchemy was based.' *AX*, 1957, **6**, 1.

STAPLETON, H. E. (4). 'The Antiquity of Alchemy.' *AX*, 1953, **5**, 1. The Summary also printed in *A/AIHS*, 1951, **4** (no. 14), 35.

STAPLETON, H. E. (5). 'Ancient and Modern Aspects of Pythagoreanism; I, The Babylonian Sources of Pythagoras' Mathematical Knowledge; II, The Part Played by the Human Hand with its Five Fingers in the Development of Mathematics; III, Sumerian Music as a possible intermediate Source of the Emphasis on Harmony that characterises the −6th-century Teaching of both Pythagoras and Confucius; IV, The Belief of Pythagoras in the Immaterial, and its Co-existence with Natural Phenomena.' *OSIS*, 1958, **13**, 12.

STAPLETON, H. E. & AZO, R. F. (1). 'Alchemical Equipment in the +11th Century.' *MAS/B*, 1905, **1**, 47. (Account of the '*Ainu al-San'ah wa-l 'Aunu al-Sana'ah* (Essence of the Art and Aid to the Workers) by Abū-l Ḥakīm al-Sālihī al-Kāthī, +1034.) Cf. Ahmad & Datta (1).

STAPLETON, H. E. & AZO, R. F. (2). 'An Alchemical Compilation of the +13th Century.' *MAS/B*, 1910, **3**, 57. (A florilegium of extracts gathered by an alchemical copyist travelling in Asia Minor and Mesopotamia about +1283.)

STAPLETON, H. E., AZO, R. F. & HUSAIN, M. H. (1). 'Chemistry in Iraq and Persia in the +10th Century.' *MAS/B*, 1927, **8**, 315–417. (Study of the *Madkhal al-Ta'līmī* and the *Kitāb al-Asrār* of al-Rāzī (d. +925), the relation of Arabic alchemy with the Sabians of Ḥarrān, and the role of influences from Hellenistic culture, China and India upon it.) Revs. G. Sarton, *ISIS*, 1928, **11**, 129; J. R. Partington, *N*, 1927, **120**, 243.

STAPLETON, H. E., AZO, R. F., HUSAIN, M. H. & LEWIS, G. L. (1). 'Two Alchemical Treatises attributed to Avicenna.' *AX*, 1962, **10**, 41.

STAPLETON, H. E. & HUSAIN, H. (1) (tr.). 'Summary of the Cairo Arabic MS. of the "Treatise of Warning (*Risālat al-Ḥaḍar*)" of Agathodaimon, his Discourse to his Disciples when he was about to die.' Published as Appendix B in Stapleton (4), pp. 40 ff.

STAPLETON, H. E., LEWIS, G. L. & TAYLOR, F. SHERWOOD (1). 'The Sayings of Hermes as quoted in the *Mā al-Waraqī* of Ibn Umail' (c. +950). *AX*, 1949, **3**, 69.

STARKEY, G. [Eirenaeus Philaletha] (1). *Secrets Reveal'd; or, an Open Entrance to the Shut-Palace of the King; Containing the Greatest Treasure in Chymistry, Never yet so plainly Discovered. Composed by a most famous English-man styling himself Anonymus, or Eyrenaeus Philaletha Cosmopolita, who by Inspiration and Reading attained to the Philosophers Stone at his Age of Twenty-three Years, A.D. 1645...* Godbid for Cooper, London, 1669. Eng. tr. of first Latin ed. *Introitus Apertus...* Jansson & Weyerstraet, Amsterdam, 1667. See Ferguson (1), vol. 2, p. 192.

STARKEY, G. [Eirenaeus Philaletha] (2). *Arcanum Liquoris Immortalis, Ignis-Aquae Seu Alkehest.* London, 1683, Hamburg, 1688. Eng. tr. 1684.

STAUDENMEIER, LUDWIG (1). *Die Magie als experimentelle Wissenschaft.* Leipzig, 1912.

STEELE, J. (1) (tr.). *The 'I Li', or Book of Etiquette and Ceremonial.* 2 vols. London, 1917.

STEELE, R. (1) (ed.). *Opera Hactenus Inedita Rogeri Baconi.* 9 fascicles in 3 vols. Oxford, 1914–.

STEELE, R. (2). 'Practical Chemistry in the +12th Century; Rasis *De Aluminibus et Salibus*, the [text of the] Latin translation by Gerard of Cremona, [with an English précis].' *ISIS*, 1929, **12**, 10.

STEELE, R. (3) (tr.). *The Discovery of Secrets* [a Jābirian Corpus text]. Luzac (for the Geber Society), London, 1892.

STEELE, R. & SINGER, D. W. (1). 'The Emerald Table [*Tabula Smaragdina*].' *PRSM*, 1928, **21**, 41.

STEIN, O. (1). 'References to Alchemy in Buddhist Scriptures.' *BLSOAS*, 1933, **7**, 263.

STEIN, R. A. (5). 'Remarques sur les Mouvements du Taoisme Politico-Religieux au 2e Siècle ap. J. C.' *TP*, 1963, **50**, 1–78. Japanese version revised by the author, with French summary of the alterations. *DK*, 1967, **2**.

STEIN, R. A. (6). 'Spéculations Mystiques et Thèmes relatifs aux "Cuisines" [*chhu*] du Taoisme.' *ACF*, 1972, **72**, 489.

STEINGASS, F. J. (1). *A Comprehensive Persian–English Dictionary*. Routledge & Kegan Paul, London, 1892, repr. 1957.

STEININGER, H. (1). *Hauch- und Körper-seele, und der Dämon, bei 'Kuan Yin Tzu'*. Harrassowitz, Leipzig, 1953. (Sammlung orientalistischer Arbeiten, no. 20.)

STEINSCHNEIDER, M. (1). 'Die Europäischen Übersetzungen aus dem Arabischen bis mitte d. 17. Jahrhunderts. A. Schriften bekannter Übersetzer; B, Übersetzungen von Werken bekannter Autoren deren Übersetzer unbekannt oder unsicher sind.' *SWAW /PH*, 1904, **149** (no. 4), 1–84; 1905, **151** (no. 1), 1–108. Also sep. issued. Repr. Graz, 1956.

STEINSCHNEIDER, M. (2). 'Über die Mondstationen (Naxatra) und das Buch Arcandam.' *ZDMG*, 1864, **18**, 118. 'Zur Geschichte d. Übersetzungen ans dem Indischen in Arabische und ihres Einflusses auf die Arabische Literatur, insbesondere über die Mondstationen (Naxatra) und daraufbezüglicher Loosbücher.' *ZDMG*, 1870, **24**, 325; 1871, **25**, 378. (The last of the three papers has an index for all three.)

STEINSCHNEIDER, M. (3). 'Euklid bei den Arabern.' *ZMP*, 1886, **31** (Hist. Lit. Abt.), 82.

STEINSCHNEIDER, M. (4) *Gesammelte Schriften*, ed. H. Malter & A. Marx. Poppelauer, Berlin, 1925.

STENRING, K. (1) (tr.). *The Book of Formation, 'Sefer Yetzirah', by R. Akiba ben Joseph...* With introd. by A. E. Waite, Rider, London, 1923.

STEPHANIDES, M. K. (1). *Symbolai eis tēn Historikē tōn Physikōn Epistēmōn kai Idiōs tēs Chymeias* (in Greek). Athens, 1914. See Zacharias (1).

STEPHANIDES, M. K. (2). Study of Aristotle's views on chemical affinity and reaction. *RSCI*, 1924, **62**, 626.

STEPHANIDES, M. K. (3). *Psammourgikē kai Chymeia* (Ψαμμουργικὴ καὶ Χυμεία) [in Greek]. Mytilene, 1909.

STEPHANIDES, M. K. (4). 'Chymeutische Miszellen.' *AGNWT*, 1912, **3**, 180.

STEPHANUS OF ALEXANDRIA. *Megalēs kai Hieras Technēs* [*Chymeia*]. Not in the *Corpus Alchem. Gr.* (Berthelot & Ruelle) but in Ideler (1), vol. 2.

STEPHANUS, HENRICUS. See Estienne, H. (1).

STILLMAN, J. M. (1). *The Story of Alchemy and Early Chemistry*. Constable, London and New York, 1924. Repr. Dover, New York, 1960.

STRASSMEIER, J. N. (1). *Inschriften von Nabuchodonosor* [−6th cent.]. Leipzig, 1889.

STRASSMEIER, J. N. (2). *Inschriften von Nabonidus* [r. −555 to −538]. Leipzig. 1889.

STRAUSS, BETTINA (1). 'Das Giftbuch des Shānāq; eine literaturgeschichtliche Untersuchung.' *QSGNM*, 1934, **4**, 89–152.

VON STRAUSS-&-TORNEY, V. (1). 'Bezeichnung der Farben Blau und Grün in Chinesischen Alterthum.' *ZDMG*, 1879, **33**, 502.

STRICKMANN, M. (1). 'Notes on Mushroom Cults in Ancient China.' Rijksuniversiteit Gent (Gand), 1966. (Paper to the 4e Journée des Orientalistes Belges, Brussels, 1966.)

STRICKMANN, M. (2). 'On the Alchemy of Thao Hung-Ching.' Unpub. MS. Revised version contributed to the 2nd International Conference of Taoist Studies, Tateshina, Japan, 1972.

STRICKMANN, M. (3). 'Taoism in the Lettered Society of the Six Dynasties.' Contribution to the 2nd International Conference of Taoist Studies, Chino (Tateshina), Japan, 1972.

STROTHMANN, R. (1). 'Gnosis Texte der Ismailiten; Arabische Handschrift Ambrosiana H 75.' *AGWG/PH*, 1943 (3rd ser.), no. 28.

STRZODA, W. (1). *Die gelben Orangen der Prinzessin Dschau, aus dem chinesischen Urtext*. Hyperion Verlag, München, 1922.

STUART, G. A. (1). *Chinese Materia Medica; Vegetable Kingdom, extensively revised from Dr F. Porter Smith's work*. Amer. Presbyt. Mission Press, Shanghai, 1911. An expansion of Smith, F.P. (1).

STUART, G. A. (2). 'Chemical Nomenclature.' *CRR*, 1891; 1894, **25**, 88; 1901, **32**, 305.

STUHLMANN, C. C. (1). 'Chinese Soda.' *JPOS*, 1895, **3**, 566.

SUBBARAYAPPA, B. V. (1). 'The Indian Doctrine of Five Elements.' *IJHS*, 1966, **1**, 60.

SUDBOROUGH, J. J. (1). *A Textbook of Organic Chemistry; translated from the German of A. Bernthsen, edited and revised*. Blackie, London, 1906.

SUDHOFF, K. (1). 'Eine alchemistische Schrift des 13. Jahrhunderts betitelt *Speculum Alkimie Minus*, eines bisher unbekannten Mönches Simeon von Köln.' *AGNT*, 1922, **9**, 53.

SUDHOFF, K. (2). 'Alkoholrezept aus dem 8. Jahrhundert?' [The earliest version of the *Mappae Clavicula*, now considered c. +820.] *NW*, 1917, **16**, 681.

SUDHOFF, K. (3). 'Weiteres zur Geschichte der Destillationstechnik.' *AGNT*, 1915, **5**, 282.

SUDHOFF, K. (4). 'Eine Herstellungsanweisung für "Aurum Potabile" und "Quinta Essentia" von dem herzogliche Leibarzt Albini di Moncalieri (14ter Jahrh.).' *AGNT*, 1915, **5**, 198.

SÜHEYL ÜNVER, A. (1). *Tanksuknamei Ilhan der Fünunu Ulumu Hatai Mukaddinesi* (Turkish tr.) T. C. Istanbul Universitesi Tib Tarihi Enstitusu Adet 14. Istanbul, 1939.

SÜHEYL ÜNVER, A. (2). *Wang Shu-ho eseri hakkinda* (Turkish with Eng. summary). Tib. Fak. Mecmuasi. Yil 7, Sayr 2, Umumi no. 28. Istanbul, 1944.

AL-SUHRAWARDY, ALLAMA SIR ABDULLAH AL-MAMUN (1) (ed.). *The Sayings of Muḥammad* [ḥadith]. With foreword by M. K. (Mahatma) Gandhi. Murray, London, 1941. (Wisdom of the East series.)

SUIDAS (1). *Lexicon Graece et Latine...(c. +1000)*, ed. Aemilius Portus & Ludolph Kuster, 3 vols. Cambridge, 1705.

SULLIVAN, M. (8). 'Kendi' (drinking vessels, Skr. *kundika*, with neck and side-spout). *ACASA*, 1957, **11**, 40.

SUN JEN I-TU & SUN HSÜEH-CHUAN (1) (tr.). '*Thien Kung Khai Wu*', *Chinese Technology in the Seventeenth Century, by Sung Ying-Hsing*. Pennsylvania State Univ. Press; University Park & London, Penn. 1966.

SUTER, H. (1). *Die Mathematiker und Astronomen der Araber und ihre Werke*. Teubner, Leipzig, 1900. (Abhdl. z. Gesch. d. Math. Wiss. mit Einschluss ihrer Anwendungen, no. 10; supplement to *ZMP*, **45**.) Additions and corrections in *AGMW*, 1902, no. 14.

SUZUKI SHIGEAKI (1). 'Milk and Milk Products in the Ancient World.' *JSHS*, 1965, **4**, 135.

SWEETSER, WM. (1). *Human Life*. New York, 1867.

SWINGLE, W. T. (12). 'Notes on Chinese Accessions; chiefly Medicine, Materia Medica and Horticulture.' *ARLC/DO*, 1928/1929, 311. (On the *Pên Tshao Yen I Pu I*, the *Yeh Tshai Phu*, etc.; including translations by M. J. Hagerty.)

SYNCELLOS, GEORGIOS (1). *Chronographia* (c. +800), ed. W. Dindorf. Weber, Bonn, 1829 (in *Corp. Script. Hist. Byz.* series). Ed. P. J. Goar, Paris, 1652.

TANAKA, M. (1). *The Development of Chemistry in Modern Japan*. Proc. XIIth Internat. Congr. Hist. of Sci., Paris, 1968. Abstracts & Summaries, p. 232; Actes, vol. 6, p. 107.

TANAKA, M. (2). 'A Note to the History of Chemistry in Modern Japan, [with a Select List of the most important Contributions of Japanese Scientists to Modern Chemistry].' *SHST/T*, Special Issue for the XIIth Internat. Congress of the Hist. of Sci., Paris, 1968.

TANAKA, M. (3). 'Einige Probleme der Vorgeschichte der Chemie in Japan; Einführung und Aufnahme der modernen Materienbegriffe.' *JSHS*, 1967, **6**, 96.

TANAKA, M. (4). 'Ein Hundert Jahre der Chemie in Japan.' *JSHS*, 1964, **3**, 89.

TARANZANO, C. (1). *Vocabulaire des Sciences Mathématiques, Physiques et Naturelles*. 2 vols. Hsien-hsien, 1936.

TARN, W. W. (1). *The Greeks in Bactria and India*. Cambridge, 1951.

TASLIMI, MANUCHECHR (1). 'An Examination of the *Nihāyat al-Ṭalab* (The End of the Search) [by 'Izz al-Dīn Aidamur ibn 'Ali ibn Aidamur al-Jildakī, c. +1342] and the Determination of its Place and Value in the History of Islamic Chemistry.' Inaug. Diss. London, 1954.

TATARINOV, A. (2). 'Bemerkungen ü. d. Anwendung schmerzstillender Mittel bei den Operationen, und die Hydropathie, in China.' Art. in *Arbeiten d. k. Russischen Gesandschaft in Peking über China, sein Volk, seine Religion, seine Institutionen, socialen Verhältnisse, etc.*, ed. C. Abel & F. A. Mecklenburg. Heinicke, Berlin, 1858. Vol. 2, p. 467.

TATOJAYA, YATODHARMA (1) (tr.). *The 'Kokkokam' of Ativira Rama Pandian* [a Tamil prince at Madura, late +16th cent.]. Med. Book Co., Calcutta, 1949. Bound with R. Schmidt (3).

TAYLOR, F. SHERWOOD (2). 'A Survey of Greek Alchemy.' *JHS*, 1930, **50**, 109.

TAYLOR, F. SHERWOOD (3). *The Alchemists*. Heinemann, London, 1951.

TAYLOR, F. SHERWOOD (4). *A History of Industrial Chemistry*. Heinemann, London, 1957.

TAYLOR, F. SHERWOOD (5). 'The Evolution of the Still.' *ANS*, 1945, **5**, 185.

TAYLOR, F. SHERWOOD (6). 'The Idea of the Quintessence.' Art. in *Science, Medicine and History* (Charles Singer Presentation Volume), ed. E. A. Underwood, Oxford, 1953. Vol. 1, p. 247.

TAYLOR, F. SHERWOOD (7). 'The Origins of Greek Alchemy.' *AX*, 1937, **1**, 30.

TAYLOR, F. SHERWOOD (8) (tr. and comm.). 'The Visions of Zosimos [of Panopolis].' *AX*, 1937, **1**, 88.

TAYLOR, F. SHERWOOD (9) (tr. and comm.). 'The Alchemical Works of Stephanos of Alexandria.' *AX*, 1937, **1**, 116; 1938, **2**, 38.

TAYLOR, F. SHERWOOD (10). 'An Alchemical Work of Sir Isaac Newton.' *AX*, 1956, **5**, 59.

TAYLOR, F. SHERWOOD (11). 'Symbols in Greek Alchemical Writings.' *AX*, 1937, **1**, 64.

TAYLOR, F. SHERWOOD & SINGER, CHARLES (1). 'Pre-scientific Industrial Chemistry [in the Mediterranean Civilisations and the Middle Ages].' Art. in *A History of Technology*, ed. C. Singer *et al.* Oxford, 1956. Vol. 2, p. 347.

TAYLOR, J. V. (1). *The Primal Vision; Christian Presence amid African Religion*. SCM Press, London, 1963.

TEGENGREN, F. R. (1). 'The Iron Ores and Iron Industry of China; including a summary of the Iron situation of the Circum-Pacific Region.' *MGSC*, 1921 (Ser. A), no. 2, pt. I, pp. 1–180, with Chinese abridgement of 120 pp. 1923 (Ser. A), no. 2, pt. II, pp. 181–457, with Chinese abridgement of 190 pp. The section on the Iron Industry starts from p. 297: 'General Survey; Historical Sketch' [based mainly on Chang Hung-Chao (*1*)], pp. 297–314; 'Account of the Industry [traditional] in different Provinces', pp. 315–64; 'The Modern Industry', pp. 365–404; 'Circum-Pacific Region', pp. 405-end.

TEGENGREN, F. R. (2). 'The Hsi-khuang Shan Antimony Mining Fields in Hsin-hua District, Hunan.' *BGSC*, 1921, no. 3, 1–25.

TEGENGREN, F. R. (3). 'The Quicksilver Deposits of China.' *BGSC*, 1920, no. 2, 1–36.

TEGGART, F. J. (1). *Rome and China; a Study of Correlations in Historical Events.* Univ. of California Press, Berkeley, Calif. 1939.

TEICH, MIKULÁŠ (1) (ed.). *J. E. Purkyně, 'Opera Selecta'.* Prague, 1948.

TEICH, MIKULÁŠ (2). 'From "Enchyme" to "Cyto-Skeleton"; the Development of Ideas on the Chemical Organisation of Living Matter.' Art. in *Changing Perspectives in the History of Science...*, ed. M. Teich & R. Young, Heinemann, London, 1973, p. 439.

TEICH, MIKULÁŠ & YOUNG, R. (1) (ed.). *Changing Perspectives in the History of Science...* Heinemann, London, 1973.

TEMKIN, O. (3). 'Medicine and Graeco-Arabic Alchemy.' *BIHM*, 1955, **29**, 134.

TEMKIN, O. (4). 'The Classical Roots of Glisson's Doctrine of Irritation.' *BIHM*, 1964, **38**, 297.

TEMPLE, SIR WM. (3). 'On Health and Long Life.' In *Works*, 1770 ed. vol. 3, p. 266.

TESTE, A. (1). *Homoeopathic Materia Medica, arranged Systematically and Practically.* Eng. tr. from the French, by C. J. Hempel. Rademacher & Shelk, Philadelphia, 1854.

TESTI, G. (1). *Dizionario di Alchimia e di Chimica Antiquaria.* Mediterranea, Rome, 1950. Rev. F. S[herwood] T[aylor], *AX*, 1953, **5**, 55.

THACKRAY, A. (1). '"Matter in a Nut-shell"; Newton's "Opticks" and Eighteenth-Century Chemistry.' *AX*, 1968, **15**, 29.

THELWALL, S. & HOLMES, P. (1) (tr.). *The Writings of Tertullian* [c. +200]. In Ante-Nicene Christian Library, ed. A. Roberts & J. Donaldson. Clark, Edinburgh, 1867, vols. 11, 15 and 18.

THEOBALD, W. (1). 'Der Herstelling der Bronzefarbe in Vergangenheit und Gegenwart.' *POLYJ*, 1913, **328**, 163.

THEOPHANES (+758 to +818) (1). *Chronographia*, ed. Classen (in *Corp. Script. Hist. Byz.* series).

[THEVENOT, D.] (1) (ed.). *Scriptores Graeci Mathematici, Veterum Mathematicorum Athenaei, Bitonis, Apollodori, Heronis et aliorum Opera Gr. et Lat. pleraque nunc primum edita* [including the *Kestoi* of Julius Africanus]. Paris, 1693.

THOMAS, E. J. (2). '[The State of the Dead in] Buddhist [Belief].' *ERE*, vol. xi, p. 829.

THOMAS, SIR HENRY (1). 'The Society of Chymical Physitians; an Echo of the Great Plague of London, +1665.' Art. in Singer Presentation Volume, *Science, Medicine and History*, ed. E. A. Underwood. 2 vols. Oxford, 1953. Vol. 2, p. 56.

THOMPSON, D. V. (1). *The Materials of Mediaeval Painting.* London, 1936.

THOMPSON, D. V. (2) (tr.). *The 'Libro dell' Arte' of Cennino Cennini* [+1437]. Yale Univ. Press, New Haven, Conn. 1933.

THOMPSON, NANCY (1). 'The Evolution of the Thang Lion-and-Grapevine Mirror.' *AA*, 1967, **29**. Sep. pub. Ascona, 1968; with an addendum on the Jen Shou Mirrors by A. C. Soper.

THOMPSON, R. CAMPBELL (5). *On the Chemistry of the Ancient Assyrians* (mimeographed, with plates of Assyrian cuneiform tablets, romanised transcriptions and translations). Luzac, London, 1925.

THOMS, W. J. (1). *Human Longevity; its Facts and Fictions.* London, 1873.

THOMSEN, V. (1). 'Ein Blatt in türkische "Runen"-schrift aus Turfan.' *SPAW/PH*, 1910, 296. Followed by F. C. Andreas: 'Zwei Soghdische Exkurse zu V. Thomsen's "Ein Blatt...".' 307.

THOMSON, JOHN (2). '[Glossary of Chinese Terms for] Photographic Chemicals and Apparatus.' In Doolittle (1), vol. 2, p. 319.

THOMSON, T. (1). *A History of Chemistry.* 2 vols. Colburn & Bentley, London, 1830.

THORNDIKE, LYNN (1). *A History of Magic and Experimental Science.* 8 vols. Columbia Univ. Press, New York:
      Vols. 1 & 2 (The First Thirteen Centuries), 1923, repr. 1947;
      Vols. 3 and 4 (Fourteenth and Fifteenth Centuries), 1934;
      Vols. 5 and 6, (Sixteenth Century), 1941;
      Vols. 7 and 8 (Seventeenth Century), 1958.
      Rev. W. Pagel, *BIHM*, 1959, **33**, 84.

THORNDIKE, LYNN (6). 'The *cursus philosophicus* before Descartes.' *A/AIHS*, 1951, **4** (**30**), 16.

THORPE, SIR EDWARD (1). *History of Chemistry.* 2 vols. in one. Watts, London, 1921.

THURSTON, H. (1). *The Physical Phenomena of Mysticism*, ed. J. H. Crehan. Burns & Oates, London, 1952. French tr. by M. Weill, *Les Phenomènes Physiques du Mysticisme aux Frontières de la Science.* Gallimard, Paris, 1961.

TIEFENSEE, F. (1). *Wegweiser durch die chinesischen Höflichkeits-Formen.* Deutschen Gesellsch. f. Natur- u. Völkerkunde Ostasiens, Tokyo, 1924 (*MDGNVO*, **18**), and Behrend, Berlin, 1924.

TIMKOVSKY, G. (1). *Travels of the Russian Mission through Mongolia to China, and Residence in Peking in the Years 1820–1*, with corrections and notes by J. von Klaproth. Longmans, Rees, Orme, Brown & Green, London, 1827.

TIMMINS, S. (1). 'Nickel German Silver Manufacture', art. in *The Resources, Products and Industrial History of Birmingham and the Midland Hardware District*, ed. S. Timmins. London, 1866, p. 671.

TOBLER, A. J. (1). *Excavations at Tepe Gawra*. 2 vols. Philadelphia, 1950.

TOLL, C. (1). *Al-Hamdānī, 'Kitāb al-Jauharatain' etc., 'Die beiden Edelmetalle Gold und Silber'*, herausgegeben u. übersetzt... University Press, Uppsala, 1968 (*UUA*, Studia Semitica, no. 1).

TOLL, C. (2). 'Minting Technique according to Arabic Literary Sources.' *ORS*, 1970, **19–20**, 125.

TORGASHEV, B. P. (1). *The Mineral Industry of the Far East*. Chali, Shanghai, 1930.

DE TOURNUS, J. GIRARD (1) (tr.). *Roger Bachon de l'Admirable Pouvoir et Puissance de l'Art et de Nature, ou est traicté de la pierre Philosophale*. Lyons, 1557, Billaine, Paris, 1628. Tr. of *De Mirabili Potestate Artis et Naturae, et de Nullitate Magiae*.

TRIGAULT, NICHOLAS (1). *De Christiana Expeditione apud Sinas*. Vienna, 1615; Augsburg, 1615. Fr. tr.: *Histoire de l'Expédition Chrétienne au Royaume de la Chine, entrepris par les PP. de la Compagnie de Jésus, comprise en cinq livres...tirée des Commentaires du P. Matthieu Riccius, etc.* Lyon, 1616; Lille, 1617; Paris, 1618. Eng. tr. (partial): *A Discourse of the Kingdome of China, taken out of Ricius and Trigautius*, In *Purchas his Pilgrimes*. London, 1625, vol. 3, p. 380. Eng. tr. (full): see Gallagher (1). Trigault's book was based on Ricci's *I Commentarj della Cina* which it follows very closely, even verbally, by chapter and paragraph, introducing some changes and amplifications, however. Ricci's book remained unprinted until 1911, when it was edited by Venturi (1) with Ricci's letters; it has since been more elaborately and sumptuously edited alone by d'Elia (2).

TSHAO THIEN-CHHIN, HO PING-YÜ & NEEDHAM, JOSEPH (1). 'An Early Mediaeval Chinese Alchemical Text on Aqueous Solutions' (the *San-shih-liu Shui Fa*, early +6th century). *AX*, 1959, **7**, 122. Chinese tr. by Wang Khuei-Kho (*1*), *KHSC*, 1963, no. 5, 67.

TSO, E. (1). 'Incidence of Rickets in Peking; Efficacy of Treatment with Cod-liver Oil.' *CMJ*, 1924, **38**, 112.

TSUKAHARA, T. & TANAKA, M. (1). 'Edward Divers; his Work and Contribution to the Foundation of [Modern] Chemistry in Japan.' *SHST/T*, 1965, 4.

TU YÜ-TSHANG, CHIANG JUNG-CHHING & TSOU CHHÊNG-LU (1). 'Conditions for the Successful Resynthesis of Insulin from its Glycyl and Phenylalanyl Chains.' *SCISA*, 1965, **14**, 229.

TUCCI, G. (4). 'Animadversiones Indicae; VI, A Sanskrit Biography of the Siddhas, and some Questions connected with Nāgārjuna.' *JRAS/B*, 1930, **26**, 138.

TUCCI, G. (5). *Teoria e Practica del Maṇḍala*. Rome, 1949. Eng. tr. London, 1961.

ULSTADT, PHILIP (1). *Coelum Philosophorum seu de Secretis Naturae Liber*. Strassburg, 1526 and many subsequent eds.

UNDERWOOD, A. J. V. (1). 'The Historical Development of Distilling Plant.' *TICE*, 1935, **13**, 34.

URDANG, G. (1). 'How Chemicals entered the Official Pharmacopoeias.' *A/AIHS*, 1954, **7**, 303.

URE, A. (1). A *Dictionary of Arts, Manufactures and Mines*. 1st ed., 2 vols, London, 1839. 5th ed. 3 vols. ed. R. Hunt, Longman, Green, Longman & Roberts, London, 1860.

VACCA, G. (2). 'Nota Cinesi.' *RSO*, 1915, **6**, 131. (1) A silkworm legend from the *Sou Shen Chi*. (2) The fall of a meteorite described in *Mêng Chhi Pi Than*. (3) Invention of movable type printing (*Mêng Chhi Pi Than*). (4) A problem of the mathematician I-Hsing (chess permutations and combinations) in *Mêng Chhi Pi Than*. (5) An alchemist of the +11th century (*Mêng Chhi Pi Than*).

VAILLANT, A. (1) (tr.). *Le Livre des Secrets d'Hénoch; Texte Slave et Traduction Française*. Inst. d'Études Slaves, Paris, 1952. (Textes Publiés par l'Inst. d'Ét. Slaves, no. 4.)

VALESIUS, HENRICUS (1). *Polybii, Diodori Siculi, Nicolai Damasceni, Dionysii Halicar[nassi], Appiani, Alexand[ri] Dionis[ii] et Joannis Antiocheni, Excerpta et Collectaneis Constantini Augusti [VII] Porphyrogenetae...nunc primum Graece edidit, Latine vertit, Notisque illustravit*. Du Puis, Paris, 1634.

DE LA VALLÉE-POUSSIN, L. (9). '[The "Abode of the Blest" in] Buddhist [Belief].' *ERE*, vol. ii, p. 686.

VANDERMONDE, J. F. (1). 'Eaux, Feu (et Cautères), Terres etc., Métaux, Minéraux et Sels, du *Pên Ts'ao Kang Mou*.' MS., accompanied by 80 (now 72) specimens of inorganic substances collected and studied at Macao or on Poulo Condor Island in +1732, then presented to Bernard de Jussieu, who deposited them in the Musée d'Histoire Naturelle at Paris. The samples were analysed for E. Biot (22) by Alexandre Brongniart (in 1835 to 1840), and the MS. text (which had been acquired from the de Jussieu family by the Museum in 1857) printed in excerpt form by de Mély (1), pp. 156–248. Between 1840 and 1895 the collection was lost, but found again by Lacroix, and the MS. text, not catalogued at the time of acquisition, was also lost, but found again by Deniker; both in time for the work of de Mély.

VARENIUS, BERNARD (1). *Descriptio Regni Japoniae et Siam; item de Japoniorum Religione et Siamensium; de Diversis Omnium Gentium Religionibus...* Hayes, Cambridge, 1673.

VARENIUS, BERNARD (2). *Geographiae Generalis, in qua Affectiones Generales Telluris explicantur summa cura quam plurimus in locis Emendata, et XXXIII Schematibus Novis, aere incisis, una cum Tabb. aliquot quae desiderabantur Aucta et Illustrata, ab Isaaco Newton, Math. Prof. Lucasiano apud Cantabrigiensis.* Hayes, Cambridge, 1672. 2nd ed. (*Auctior et Emendatior*), 1681.

VÄTH, A. (1) (with the collaboration of L. van Hée). *Johann Adam Schall von Bell, S. J., Missionar in China, Kaiserlicher Astronom und Ratgeber am Hofe von Peking; ein Lebens- und Zeit-bild.* Bachem, Köln, 1933. (Veröffentlichungen des Rheinischen Museums in Köln, no. 2.) Crit. P. Pelliot, *TP*, 1934, **31**, 178.

VAUGHAN, T. [Eugenius Philalethes] (1), (attrib.), *A Brief Natural History, intermixed with a Variety of Philosophical Discourses, and Observations upon the Burning of Mount Aetna; with Refutations of such vulgar Errours as our Modern Authors have omitted.* Smelt, London, 1669. See Ferguson (1), vol. 2, p. 197; Waite (4), p. 492.

VAUGHAN, T. (Eugenius Philalethes] (2). *Magia Adamica; or, the Antiquitie of Magic, and the Descent thereof from Adam downwards proved; Whereunto is added, A Perfect and True Discoverie of the True Coelum Terrae, or the Magician's Heavenly Chaos, and First Matter of All Things.* London, 1650. Repr. in Waite (5). Germ. ed. Amsterdam, 1704. See Ferguson (1), vol. 2, p. 196.

DE VAUX, B. CARRA (5). '*L'Abrégé des Merveilles*' (*Mukhtaṣaru'l-'Ajā'ib*) *traduit de l'Arabe*...(A work attributed to al-Mas'ūdī.) Klincksieck, Paris, 1898.

VAVILOV, S. I. (1). 'Newton and the Atomic Theory.' Essay in *Newton Tercentenary Celebrations Volume* (July 1946). Royal Society, London, 1947, p. 43.

DE VEER, GERARD (1). 'The Third Voyage Northward to the Kingdoms of Cathaia, and China, Anno 1596.' In *Purchas his Pilgrimes*, 1625 ed., vol. 3, pt. 2, bk. iii. p. 482; ed. of McLehose, Glasgow, 1906, vol. 13, p. 91.

VEI CHOW JUAN. See Wei Chou-Yuan.

VELER, C. D. & DOISY, E. A. (1). 'Extraction of Ovarian Hormone from Urine.' *PSEBM*, 1928, **25**, 806.

VON VELTHEIM, COUNT (1). *Von den goldgrabenden Ameisen und Greiffen der Alten; eine Vermuthung.* Helmstadt, 1799.

VERHAEREN, H. (1). *L'Ancienne Bibliothèque du Pé-T'ang.* Lazaristes Press, Peking, 1940.

DI VILLA, E. M. (1). *The Examination of Mines in China.* North China Daily Mail, Tientsin, 1919.

DE VILLARD, UGO MONNERET (2). *Le Leggende Orientali sui Magi Evangelici.* Vatican City, 1952. (Studie Testi, no. 163.)

DE VISSER, M. W. (2). *The Dragon in China and Japan.* Müller, Amsterdam, 1913. Orig. in *VKAWA/L*, 1912 (n. r.), **13** (no. 2.).

V[OGT], E. (1). 'The Red Colour Used in [Palaeolithic and Neolithic] Graves.' *CIBA/T*, 1947, **5** (no. 54), 1968.

VOSSIUS, G. J. (1). *Etymologicon Linguae Latinae.* Martin & Allestry, London, 1662; also Amsterdam, 1695, etc.

WADDELL, L. A. (4). '[The State of the Dead in] Tibetan [Religion].' *ERE*, vol. xi, p. 853.

WAITE, A. E. (1). *Lives of Alchemystical Philosophers, based on Materials collected in 1815 and supplemented by recent Researches; with a Philosophical Demonstration of the True Principles of the Magnum Opus or Great Work of Alchemical Re-construction, and some Account of the Spiritual Chemistry...; to Which is added, a Bibliography of Alchemy and Hermetic Philosophy.* Redway, London, 1888. Based on: [Barrett, Francis], (attrib.). *The Lives of Alchemystical Philosophers; with a Critical Catalogue of Books in Occult Chemistry, and a Selection of the most Celebrated Treatises on the Theory and Practice of the Hermetic Art.* Lackington & Allen, London, 1814, with title-page slightly changed, 1815. See Ferguson (1), vol. 2, p. 41. The historical material in both these works is now totally unreliable and outdated; two-thirds of it concerns the 17th century and later periods, even as enlarged and re-written by Waite. The catalogue is 'about the least critical compilation of the kind extant'.

WAITE, A. E. (2). *The Secret Tradition in Alchemy; its Development and Records.* Kegan Paul, Trench & Trübner, London; Knopf, New York, 1926.

WAITE, A. E. (3). *The Hidden Church of the Holy Graal* [Grail]; *its Legends and Symbolism considered in their Affinity with certain Mysteries of Initiation and Other Traces of a Secret Tradition in Christian Times.* Rebman, London, 1909.

WAITE, A. E. (4) (ed.). *The Works of Thomas Vaughan; Eugenius Philalethes*...Theosophical Society, London, 1919.

WAITE, A. E. (5) (ed.). *The Magical Writings of Thomas Vaughan (Eugenius Philalethes); a verbatim reprint of his first four treatises;* '*Anthroposophia Theomagica*', '*Anima Magica Abscondita*', '*Magia Adamica*' *and the* '*Coelum Terrae*'. Redway, London, 1888.

WAITE, A. E. (6) (tr.). *The Hermetic and Alchemical Writings of Aureolus Philippus Theophrastus Bombast of Hohenheim, called Paracelsus the Great*...2 vols. Elliott, London, 1894. A translation of the Latin Works, Geneva, 1658.

WAITE, A. E. (7) (tr.). *The 'New Pearl of Great Price', a Treatise concerning the Treasure and most precious Stone of the Philosophers* [by P. Bonus of Ferrara, *c.* +1330]. Elliott, London, 1894. Tr. from the Aldine edition (1546).

WAITE, A. E. (8) (tr.). *The Hermetic Museum Restored and Enlarged; most faithfully instructing all Disciples of the Sopho-Spagyric Art how that Greatest and Truest Medicine of the Philosophers' Stone may be found and held; containing Twenty-two most celebrated Chemical Tracts.* 2 vols. Elliott, London, 1893, later repr. A translation of Anon. (87).

WAITE, A. E. (9). *The Brotherhood of the Rosy Cross; being Records of the House of the Holy Spirit in its Inward and Outward History.* Rider, London, 1924.

WAITE, A. E. (10). *The Real History of the Rosicrucians.* London, 1887.

WAITE, A. E. (11) (tr.). *The 'Triumphal Chariot of Antimony', by Basilius Valentinus, with the Commentary of Theodore Kerckringius.* London, 1893. A translation of the Latin *Currus Triumphalis Antimonii*, Amsterdam, 1685.

WAITE, A. E. (12). *The Holy Kabbalah; a Study of the Secret Tradition in Israel as unfolded by Sons of the Doctrine for the Benefit and Consolation of the Elect dispersed through the Lands and Ages of the Greater Exile.* Williams & Norgate, London, 1929.

WAITE, A. E. (13) (tr.). *The 'Turba Philosophorum', or, 'Assembly of the Sages'; called also the 'Book of Truth in the Art' and the Third Pythagorical Synod; an Ancient Alchemical Treatise translated from the Latin, [together with] the Chief Readings of the Shorter Codex, Parallels from the Greek Alchemists, and Explanations of Obscure Terms.* Redway, London, 1896.

WAITE, A. E. (14). 'The Canon of Criticism in respect of Alchemical Literature.' *JALCHS*, 1913, **1**, 17. His reply to the discussion, p. 32.

WAITE, A. E. (15). 'The Beginnings of Alchemy.' *JALCHS*, 1915, **3**, 90. Discussion, pp. 101 ff.

WAITE, A. E. See also Stenring (1).

WAKEMAN, F. (1) (ed.). *Wu Yin Lu, 'Nothing Concealed'; Essays in Honour of Liu (Aisin-Gioro) Yü-Yün.* Chinese Materials and Research Aids Service Centre, Thaipei, Thaiwan, 1970.

WALAAS, O. (1) (ed.). *The Molecular Basis of Some Aspects of Mental Activity.* 2 vols. Academic Press, London and New York, 1966–7.

WALDEN, P. (1). *Mass, Zahl und Gewicht in der Chemie der Vergangenheit; ein Kapitel aus der Vorgeschichte des Sogenannten quantitative Zeitalters der Chemie.* Enke, Stuttgart, 1931. Repr. Liebing, Würzburg, 1970. (Samml. chem. u. chem. techn. Vorträge, N.F. no. 8.)

WALDEN, P. (2). 'Zur Entwicklungsgeschichte d. chemischen Zeichen.' Art. in *Studien z. Gesch. d. Chemie* (von Lippmann Festschrift), ed. J. Ruska. Springer, Berlin, 1927, p. 80.

WALDEN, P. (3). *Geschichte der Chemie.* Universitätsdruckerei, Bonn, 1947. 2nd ed. Athenäum, Bonn, 1950.

WALDEN, P. (4). 'Paracelsus und seine Bedeutung für die Chemie.' *ZAC/AC*, 1941, **54**, 421.

WALEY, A. (1) (tr.). *The Book of Songs.* Allen & Unwin, London, 1937.

WALEY, A. (4) (tr.). *The Way and its Power; a Study of the 'Tao Tê Ching' and its Place in Chinese Thought.* Allen & Unwin, London, 1934. Crit. Wu Ching-Hsiang, *TH*, 1935, **1**, 225.

WALEY, A. (10) (tr.). *The Travels of an Alchemist; the Journey of the Taoist [Chhiu] Chhang-Chhun from China to the Hindu-Kush at the summons of Chingiz Khan, recorded by his disciple Li Chih-Chhang.* Routledge, London, 1931. (Broadway Travellers Series.) Crit. P. Pelliot, *TP*, 1931, **28**, 413.

WALEY, A. (14). 'Notes on Chinese Alchemy, supplementary to Johnson's "Study of Chinese Alchemy".' *BLSOAS*, 1930, **6**, 1. Revs. P. Pelliot, *TP*, 1931, **28**, 233; Tenney L. Davis, *ISIS*, 1932, **17**, 440.

WALEY, A. (17). *Monkey, by Wu Chhêng-Ên.* Allen & Unwin, London, 1942.

WALEY, A. (23). *The Nine Songs; a study of Shamanism in Ancient China* [the *Chiu Ko* attributed traditionally to Chhü Yuan]. Allen & Unwin, London, 1955.

WALEY, A. (24). 'References to Alchemy in Buddhist Scriptures.' *BLSOAS*, 1932, **6**, 1102.

WALEY, A. (27) (tr.). *The Tale of Genji.* 6 vols. Allen & Unwin, London; Houghton Mifflin, New York, 1925–33.
  Vol. 1   *The Tale of Genji.*
  Vol. 2   *The Sacred Tree.*
  Vol. 3   *A Wreath of Cloud.*
  Vol. 4   *Blue Trousers.*
  Vol. 5   *The Lady of the Boat.*
  Vol. 6   *The Bridge of Dreams.*

WALKER, D. P. (1). 'The Survival of the "Ancient Theology" in France, and the French Jesuit Missionaries in China in the late Seventeenth Century.' MS. of Lecture at the Cambridge History of Science Symposium, Oct. 1969. Pr. in Walker (2) pp. 194 ff.

WALKER, D. P. (2). *The Ancient Theology; Studies in Christian Platonism from the +15th to the +18th Century*. Duckworth, London, 1972.

WALKER, D. P. (3). 'Francis Bacon and *Spiritus*', art. in *Science, Medicine and Society in the Renaissance* (Pagel Presentation Volume), ed. Debus (20), vol. 2, p. 121.

WALKER, W. B. (1). 'Luigi Cornaro; a Renaissance Writer on Personal Hygiene.' *BIHM*, 1954, **28**, 525.

WALLACE, R. K. (1). 'Physiological Effects of Transcendental Meditation.' *SC*, 1970, **167**, 1751.

WALLACE, R. K. & BENSON, H. (1). 'The Physiology of Meditation.' *SAM*, 1972, **226** (no. 2), 84.

WALLACE, R. K., BENSON, H. & WILSON, A. F. (1). 'A Wakeful Hypometabolic Physiological State.' *AJOP*, 1971, **221**, 795.

WALLACKER, B. E. (1) (tr.). *The 'Huai Nan Tzu' Book*, [*Ch.*] *11; Behaviour, Culture and the Cosmos*. Amer. Oriental Soc., New Haven, Conn. 1962. (Amer. Oriental Series, no. 48.)

VAN DE WALLE, B. (1). 'Le Thème de la Satire des Métiers dans la Littérature Egyptienne.' *CEG*, 1947, **43**, 50.

WALLESER, M. (3). 'The Life of Nāgārjuna from Tibetan and Chinese Sources.' *AM* (Hirth Anniversary Volume), **1**, 1.

WALSHE, W. G. (1). '[Communion with the Dead in] Chinese [Thought and Liturgy].' *ERE*, vol. iii, p. 728.

WALTON, A. HULL. See Davenport, John.

WANG, CHHUNG-YU (1). *Bibliography of the Mineral Wealth and Geology of China*. Griffin, London, 1912.

WANG, CHHUNG-YU (2). *Antimony; its History, Chemistry, Mineralogy, Geology, Metallurgy, Uses, Preparations, Analysis, Production and Valuation; with Complete Bibliographies*. Griffin, London, 1909.

WANG CHHUNG-YU (3). *Antimony; its Geology, Metallurgy, Industrial Uses and Economics*. Griffin, London, 1952. ('3rd edition' of Wang Chhung-Yu(2), but it omits the chapters on the history, chemistry, mineralogy and analysis of antimony, while improving those that are retained.)

WANG CHI-MIN & WU LIEN-TÊ (1). *History of Chinese Medicine*. Nat. Quarantine Service, Shanghai, 1932, 2nd ed. 1936.

WANG CHIUNG-MING (1). 'The Bronze Culture of Ancient Yunnan.' *PKR*, 1960 (no. 2), 18. Reprinted in mimeographed form, Collet's Chinese Bookshop, London, 1960.

WANG LING (1). 'On the Invention and Use of Gunpowder and Firearms in China.' *ISIS*, 1947, **37**, 160.

WARE, J. R. (1). 'The *Wei Shu* and the *Sui Shu* on Taoism.' *JAOS*, 1933, **53**, 215. Corrections and emendations in *JAOS*, 1934, **54**, 290. Emendations by H. Maspero, *JA*, 1935, **226**, 313.

WARE, J. R. (5) (tr.). *Alchemy, Medicine and Religion in the China of +320; the 'Nei Phien' of Ko Hung* ('*Pao Phu Tzu*'). M.I.T. Press, Cambridge, Mass. and London, 1966. Revs. Ho Ping-Yü, *JAS*, 1967, **27**, 144; J. Needham, *TCULT*, 1969, **10**, 90.

WARREN, W. F. (1). *The Earliest Cosmologies; the Universe as pictured in Thought by the Ancient Hebrews, Babylonians, Egyptians, Greeks, Iranians and Indo-Aryans—a Guidebook for Beginners in the Study of Ancient Literatures and Religions*. Eaton & Mains, New York; Jennings & Graham, Cincinnati, 1909.

WASHBURN, E. W. (1). 'Molecular Stills.' *BSJR*, 1929, **2** (no. 3), 476. Part of a collective work by E. W. Washburn, J. H. Bruun & M. M. Hicks: *Apparatus and Methods for the Separation, Identification and Determination of the Chemical Constituents of Petroleum*, p. 467.

WASITZKY, A. (1). 'Ein einfacher Mikro-extraktionsapparat nach dem Soxhlet-Prinzip.' *MIK*, 1932, **11**, 1.

WASSON, R. G. (1). 'The Hallucinogenic Fungi of Mexico; an Enquiry into the Origins of the Religious Idea among Primitive Peoples.' *HU/BML*, 1961, **19**, no. 7. (Ann. Lecture, Mycol. Soc. of America.)

WASSON, R. G. (2). '*Ling Chih* [the Numinous Mushroom]; Some Observations on the Origins of a Chinese Conception.' Unpub. MS. Memorandum, 1962.

WASSON, R. G. (3). *Soma; Divine Mushroom of Immortality*. Harcourt, Brace & World, New York; Mouton, The Hague, 1968. (Ethno-Mycological Studies, no. 1.) With extensive contributions by W. D. O'Flaherty. Rev. F. B. J. Kuiper, *IIJ*, 1970, **12** (no. 4), 279; followed by comments by R. G. Wasson, 286.

WASSON, R. G. (4). 'Soma and the Fly-Agaric; Mr Wasson's Rejoinder to Prof. Brough.' Bot. Mus. Harvard Univ. Cambridge, Mass. 1972. (Ethno-Mycological Studies, no. 2.)

WASSON, R. G. & INGALLS, D. H. H. (1). 'The Soma of the *Rig Veda*; what was it?' (Summary of his argument by Wasson, followed by critical remarks by Ingalls.) *JAOS*, 1971, **91** (no. 2). Separately issued as: *R. Gordon Wasson on Soma and Daniel H. H. Ingalls' Response*. Amer. Oriental Soc. New Haven, Conn. 1971. (Essays of the Amer. Orient. Soc. no. 7.)

WASSON, R. G. & WASSON, V. P. (1). *Mushrooms, Russia and History*. 2 vols. Pantheon, New York, 1957.

WATERMANN, H. I. & ELSBACH, E. B. (1). 'Molecular stills.' *CW*, 1929, **26**, 469.

WATSON, BURTON (1) (tr.). '*Records of the Grand Historian of China*', *translated from the 'Shih Chi' of Ssuma Chhien*. 2 vols. Columbia University Press, New York, 1961.

WATSON, R., Bp of Llandaff (1). *Chemical Essays*. 2 vols. Cambridge, 1781; vol. 3, 1782; vol. 4, 1786; vol. 5, 1787. 2nd ed. 3 vols. Dublin, 1783. 5th ed. 5 vols., Evans, London, 1789. 3rd ed. Evans, London, 1788. 6th ed. London, 1793–6.

WATSON, WM. (4). *Ancient Chinese Bronzes*. Faber & Faber, London, 1962.

WATTS, A. W. (2). *Nature, Man and Woman; a New Approach to Sexual Experience*. Thames & Hudson, London; Pantheon, New York, 1958.

WAYMAN, A. (1). 'Female Energy and Symbolism in the Buddhist Tantras.' *HOR*, 1962, **2**, 73.

WESBTER, C. (1). 'English Medical Reformers of the Puritan Revolution; a Background to the "Society of Chymical Physitians".' *AX*, 1967, **14**, 16.

WEEKS, M. E. (1). *The Discovery of the Elements; Collected Reprints of a series of articles published in the* Journal of Chemical Education; *with Illustrations collected by F. B. Dains*. Mack, Easton, Pa. 1933. Chinese tr. *Yuan Su Fa-Hsien Shih* by Chang Tzu-Kung, with additional material. Shanghai, 1941.

WEI CHOU-YUAN (VEI CHOU JUAN) (1). 'The Mineral Resources of China.' *EG*, 1946, **41**, 399–474

VON WEIGEL, C. E. (1). *Observationes Chemicae et Mineralogicae*. Pt. 1, Göttingen, 1771; pt. 2, Gryphiae, 1773.

WEISS, H. B. & CARRUTHERS, R. H. (1). *Insect Enemies of Books* (63 pp. with extensive bibliography). New York Public Library, New York, 1937.

WELCH, HOLMES, H. (1). *The Parting of the Way; Lao Tzu and the Taoist Movement*. Beacon Press, Boston, Mass. 1957.

WELCH, HOLMES H. (2). 'The Chang Thien Shih ["Taoist Pope"] and Taoism in China.' *JOSHK*, 1958, **4**, 188.

WELCH, HOLMES H. (3). 'The Bellagio Conference on Taoist Studies.' *HOR*, 1970, **9**, 107.

WELLMANN, M. (1). 'Die Stein- u. Gemmen-Bücher d. Antike.' *QSGNM*, 1935, **4**, 86.

WELLMANN, M. (2). 'Die Φυσικά des Bolos Democritos und der Magier Anaxilaos aus Larissa.' *APAW/PH*, 1928 (no. 7).

WELLMANN, M. (3). 'Die "Georgika" des [Bolus] Demokritos.' *APAW/PH*, 1921 (no. 4), 1–.

WELLS, D. A. (1). *Principles and Applications of Chemistry*. Ivison, Blakeman & Taylor, New York and Chicago, 1858. Chinese tr. by J. Fryer & Hsü Shou, Shanghai, 1871.

WELTON, J. (1). *A Manual of Logic*. London, 1896.

WENDTNER, K. (1). 'Assaying in the Metallurgical Books of the +16th Century.' Inaug. Diss. London, 1952.

WENGER, M. A. & BAGCHI, B. K. (1). 'Studies of Autonomic Functions in Practitioners of Yoga in India.' *BS*, 1961, **6**, 312.

WENGER, M. A., BAGCHI, B. K. & ANAND, B. K. (1). 'Experiments in India on the "Voluntary" Control of the Heart and Pulse.' *CIRC*, 1961, **24**, 1319.

WENSINCK, A. J. (2). *A Handbook of Early Muhammadan Tradition, Alphabetically Arranged*. Brill, Leiden, 1927.

WENSINCK, A. J. (3). 'The Etymology of the Arabic Word *djinn*.' *VMAWA*, 1920, 506.

WERTHEIMER, E. (1). Art. 'Arsenic' in *Dictionnaire de Physiologie*, ed. C. Richet, vol. i. Paris.

WERTIME, T. A. (1). 'Man's First Encounters with Metallurgy.' *SC*, 1964, **146**, 1257.

WEST, M. (1). 'Notes on the Importance of Alchemy to Modern Science in the Writings of Francis Bacon and Robert Boyle.' *AX*, 1961, **9**, 102.

WEST, M. L. (1). *Early Greek Philosophy and the Orient*. Oxford, 1971.

WESTBERG, F. (1). *Die Fragmente des* Toparcha Goticus (*Anonymus Tauricus*, '*Zapisk gotskogo toparcha*'); *Nachdruck der Ausgabe St. Petersburg, 1901, mit einem wissenchafts-geschichtlichen Vorwort in englischer Sprache von Ihor Ševčenko* (*Washington*). Zentralantiquariat der D. D. R., Leipzig, 1971. (Subsidia Byzantina, no. 18.)

WESTBROOK, J. H. (1). 'Historical Sketch [of Intermetallic Compounds].' Xerocopy of art. without indication of place or date of pub., comm. by the author, General Electric Co., Schenectady, N.Y.

WESTERBLAD, C. A. (1). *Pehr Henrik Ling; en Lefnadsteckning och några Sympunkter* [in Swedish]. Norstedt, Stockholm, 1904. *Ling, the Founder of the Swedish Gymnastics*. London, 1909.

WESTERBLAD, C. A. (2). *Ling; Tidshistoriska Undersökningar* [in Swedish]. Norstedt, Stockholm. Vol. 1, *Den Lingska Gymnastiken i dess Upphofsmans Dagar*, 1913. Vol. 2, *Personlig och allmän Karakteristik samt Litterär Analys*, 1916.

WESTFALL, R. S. (1). 'Newton and the Hermetic Tradition', art. in *Science, Medicine and Society in the Renaissance* (Pagel Presentation Volume), ed. Debus (20), vol. 2. p. 183.

WEULE, K. (1). *Chemische Technologie der Naturvölker*. Stuttgart, 1922.

WEYNANTS-RONDAY, M. (1). *Les Statues Vivantes; Introduction à l'Étude des Statues Égyptiennes*... Fond. Egyptol. Reine Elis:, Brussels, 1926.

WHELER, A. S. (1). 'Antimony Production in Hunan Province.' *TIMM*, 1916, **25**, 366.

WHELER, A. S. & LI, S. Y. (1). 'The Shui-ko-shan [Shui-khou Shan] Zinc and Lead Mine in Hunan [Province].' *MIMG*, 1917, **16**, 91.

WHITE, J. H. (1). *The History of the Phlogiston Theory*. Arnold, London, 1932.

WHITE, LYNN (14). *Machina ex Deo; Essays in the Dynamism of Western Culture*. M.I.T. Press, Cambridge, Mass. 1968.

WHITE, LYNN (15). 'Mediaeval Borrowings from Further Asia.' *MRS*, 1971, **5**, 1.

WHITE, W. C., Bp. of Honan (3). *Bronze Culture of Ancient China; an archaeological Study of Bronze Objects from Northern Honan dating from about −1400 to −771*. Univ. of Toronto Press, Toronto, 1956 (Royal Ontario Museum Studies, no. 5).

WHITFORD, J. (1). 'Preservation of bodies after arsenic poisoning.' *BMJ*, 1884, pt. 1, 504.

WHITLA, W. (1). *Elements of Pharmacy, Materia Medica and Therapeutics*. Renshaw, London, 1903.

WHITNEY, W. D. & LANMAN, C. R. (1) (tr.). *Atharva-veda Saṃhitā*. 2 vols. Harvard Univ. Press, Cambridge, Mass. 1905. (Harvard Oriental Series, nos. 7, 8.)

WIBERG, A. (1). 'Till Frågan om Destilleringsförfarandets Genesis; en Etnologisk-Historisk Studie' [in Swedish]. *SBM*, 1937 (nos. 2–3), 67, 105.

WIDENGREN, GEO. (1). 'The King and the Tree of Life in Ancient Near Eastern Religion.' *UUA*, 1951, **4**, 21.

WIEDEMANN, E. (7). 'Beiträge z. Gesch. d. Naturwiss.; VI, Zur Mechanik und Technik bei d. Arabern.' *SPMSE*, 1906, **38**, 1. Repr. in (23), vol. 1, p. 173.

WIEDEMANN, E. (11). 'Beiträge z. Gesch. d. Naturwiss.; XV, Über die Bestimmung der Zusammensetzung von Legierungen.' *SPMSE*, 1908, **40**, 105. Repr. in (23), vol. 1, p. 464.

WIEDEMANN, E. (14). 'Beiträge z. Gesch. d. Naturwiss.; XXV, Über Stahl und Eisen bei d. muslimischen Völkern.' *SPMSE*, 1911, **43**, 114. Repr. in (23), vol. 1, p. 731.

WIEDEMANN, E. (15). 'Beiträge z. Gesch. d. Naturwiss.; XXIV, Zur Chemie bei den Arabern' (including a translation of the chemical section of the *Mafātīḥ al-ʿUlūm* by Abū ʿAbdallah al-Khwārizmī al-Kātib, *c.* +976). *SPMSE*, 1911, **43**, 72. Repr. in (23), vol. 1, p. 689.

WIEDEMANN, E. (21). 'Zur Alchemie bei den Arabern.' *JPC*, 1907, **184** (N.F. **76**), 105.

WIEDEMANN, E. (22). 'Über chemische Apparate bei den Arabern.' Art. in the Kahlbaum Gedächtnisschrift: *Beiträge aus d. Gesch. d. Chemie*... ed. P. Diergart (1), 1909, p. 234.

WIEDEMANN, E. (23). *Aufsätze zur arabischen Wissenschaftsgeschichte* (a reprint of his 79 contributions in the series 'Beiträge z. Geschichte d. Naturwissenschaften' in *SPMSE*), ed. W. Fischer, with full indexes. 2 vols. Olm, Hildesheim and New York, 1970.

WIEDEMANN, E. (24). 'Beiträge z. Gesch. d. Naturwiss.; I, Beiträge z. Geschichte der Chemie bei den Arabern.' *SPMSE*, 1902, **34**, 45. Repr. in (23), vol. 1, p. 1.

WIEDEMANN, E. (25). 'Beiträge z. Gesch. d. Naturwiss.; LXIII, Zur Geschichte der Alchemie.' *SPMSE*, 1921, **53**, 97. Repr. in (23), vol. 2, p. 545.

WIEDEMANN, E. (26). 'Beiträge z. Mineralogie u.s.w. bei den Arabern.' Art. in *Studien z. Gesch. d. Chemie* (von Lippmann Festschrift), ed. J. Ruska. Springer, Berlin, 1927, p. 48.

WIEDEMANN, E. (27). 'Beitrage z. Gesch. d. Naturwiss.; II, 1. Einleitung, 2. Ü. elektrische Erscheinungen, 3. Ü. Magnetismus, 4. Optische Beobachtungen, 5. Ü. einige physikalische usf. Eigenschaften des Goldes, 6. Zur Geschichte d. Chemie (a) Die Darstellung der Schwefelsäure durch Erhitzen von Vitriolen, die Wärme-entwicklung beim Mischen derselben mit Wasser, und ü. arabische chemische Bezeichnungen, (b) Astrologie and Alchemie, (c) Anschauungen der Araber ü. die Metallverwandlung und die Bedeutung des Wortes al-Kimiya.' *SPMSE*, 1904, **36**, 309. Repr. in (23), vol. 1, p. 15.

WIEDEMANN, E. (28). 'Beiträge z. Gesch. d. Naturwiss.; XL, Über Verfälschungen von Drogen usw. nach Ibn Bassām und Nabarāwī.' *SPMSE*, 1914, **46**, 172. Repr. in (23), vol. 2, p. 102.

WIEDEMANN, E. (29). 'Zur Chemie d. Araber.' *ZDMG*, 1878, **32**, 575.

WIEDEMANN, E. (30). 'Al-Kīmīyā.' Art. in *Encyclopaedia of Islam*, vol. ii, p. 1010.

WIEDEMANN, E. (31). 'Beiträge zur Gesch. der Naturwiss.; LVII, Definition verschiedener Wissenschaften und über diese verfasste Werke.' *SPMSE*, 1919, **50–51**, 1. Repr. in (23), vol. 2, p. 431.

WIEDEMANN, E. (32). *Zur Alchemie bei den Arabern*. Mencke, Erlangen, 1922. (Abhandlungen zur Gesch. d. Naturwiss. u. d. Med., no. 5.) Translation of the entry on alchemy in Haji Khalfa's Bibliography and of excerpts from al-Jildakī, with a biographical glossary of Arabic alchemists.

WIEDEMANN, E. (33). 'Beiträge zur Gesch. der Naturwiss.; V, Auszüge aus arabischen Enzyklopädien und anderes.' *SPMSE*, 1905, **37**, 392. Repr. in (23), vol. 1, p. 109.

WIEGER, L. (2). *Textes Philosophiques*. (Ch and Fr.) Mission Press, Hsien-hsien, 1930.

WIEGER, L. (3). *La Chine à travers les Ages; Précis, Index Biographique et Index Bibliographique*. Mission Press, Hsien-hsien, 1924. Eng. tr. E. T. C. Werner.

WIEGER, L. (6) *Taoisme*. Vol. 1. *Bibliographie Générale*: (1) Le Canon (Patrologie); (2) Les Index Officiels et Privés. Mission Press. Hsien-hsien, 1911. Crit. P. Pelliot, *JA*, 1912 (10ᵉ Sér.) **20**, 141.

WIEGER, L. (7). *Taoisme*. Vol. 2. *Les Pères du Système Taoiste* (tr. selections of Lao Tzu, Chuang Tzu, Lieh Tzu). Mission Press, Hsien-hsien, 1913.

WIEGLEB, J. C. (1). *Historisch-kritische Untersuchung der Alchemie, oder den eingebildeten Goldmacher-kunst; von ihrem Ursprunge sowohl als Fortgange, und was nun von ihr zu halten sey.* Hoffmanns Wittwe und Erben, Weimar, 1777. 2nd ed. 1793. Photolitho repr. of the original ed., Zentral-Antiquariat D.D.R. Leipzig, 1965. Cf. Ferguson (1), vol. 2, p. 546.

WIGGLESWORTH, V. B. (1). 'The Insect Cuticle.' *BR*, 1948, **23**, 408.

WILHELM, HELLMUT (6). 'Eine Chou-Inschrift über Atemtechnik.' *MS*, 1948, **13**, 385.

WILHELM, RICHARD & JUNG, C. G. (1). *The Secret of the Golden Flower; a Chinese Book of Life* (including a partial translation of the *Thai-I Chin Hua Tsung Chih* by R. W. with notes, and a 'European commentary' by C. G. J.).

    Eng. ed. tr. C. F. Baynes, (with C. G. J.'s memorial address for R. W.). Kegan Paul, London and New York, 1931. From the Germ. ed. *Das Geheimnis d. goldenen Blute; ein chinesisches Lebensbuch.* Munich, 1929.

    Abbreviated preliminary version: '*Tschang Scheng Shu* [*Chhang Shêng Shu*]; die Kunst das mensch-lichen Leben zu verlängern.' *EURR*, 1929, **5**, 530.

    Revised Germ. ed. with new foreword by C. G. J., Rascher, Zürich, 1938. Repr. twice, 1944.

    New Germ. ed. entirely reset, with new foreword by Salome Wilhelm, and the partial translation of a Buddhist but related text, the *Hui Ming Ching*, from R. W.'s posthumous papers, Zürich, 1957.

    New Eng. ed. including all the new material, tr. C. F. Baynes. Harcourt, New York and Routledge, London, 1962, repr. 1965, 1967, 1969. Her revised tr. of the 'European commentary' alone had appeared in an anthology: *Psyche und Symbol*, ed. V. S. de Laszlo. Anchor, New York, 1958. Also tr. R. F. C. Hull for C. G. J.'s *Collected Works*, vol. 13, pp. 1–55, i.e. Jung (3).

WILLETTS, W. Y. (1). *Chinese Art*. 2 vols. Penguin, London, 1958.

WILLETTS, W. Y. (3). *Foundations of Chinese Art; from Neolithic Pottery to Modern Architecture.* Thames & Hudson, London, 1965. Revised, abridged and re-written version of (1), with many illustrations in colour.

WILLIAMSON, G. C. (1). *The Book of 'Famille Rose'* [polychrome decoration of Chinese Porcelain]. London, 1927.

WILSON, R. McLACHLAN (1). *The Gnostic Problem; a Study of the Relations between Hellenistic Judaism and the Gnostic Heresy.* Mowbray, London, 1958.

WILSON, R. McLACHLAN (2). *Gnosis and the New Testament.* Blackwell, Oxford, 1968.

WILSON, R. McLACHLAN (3) (ed. & tr.). *New Testament Apocrypha* (ed. E. Hennecke & W. Schnee-melcher). 2 vols. Lutterworth, London, 1965.

WILSON, W. (1) (tr.). *The Writings of Clement of Alexandria* (b. c. +150] (including *Stromata*, c. +200). In Ante-Nicene Christian Library, ed. A. Roberts & J. Donaldson. Clark, Edinburgh, 1867, vols 4 and 12.

WILSON, W. J. (1). 'The Origin and Development of Graeco-Egyptian Alchemy.' *CIBA/S*, 1941, **3**, 926.

WILSON, W. J. (2) (ed.). 'Alchemy in China.' *CIBA/S*, 1940, **2** (no. 7), 594.

WILSON, W. J. (2a). 'The Background of Chinese Alchemy.' *CIBA/S*, 1940, **2** (no. 7), 595.

WILSON, W. J. (2b). 'Leading Ideas of Early Chinese Alchemy.' *CIBA/S*, 1940, **2** (no. 7), 600.

WILSON, W. J. (2c). 'Biographies of Early Chinese Alchemists.' *CIBA/S*, 1940, **2** (no. 7), 605.

WILSON, W. J. (2d). 'Later Developments of Chinese Alchemy.' *CIBA/S*, 1940, **2** (no. 7), 610.

WILSON, W. J. (2e). 'The Relation of Chinese Alchemy to that of other Countries.' *CIBA/S*, 1940, **2** (no. 7), 618.

WILSON, W. J. (3). 'An Alchemical Manuscript by Arnaldus [de Lishout] de Bruxella [written from +1473 to +1490].' *OSIS*, 1936, **2**, 220.

WINDAUS, A. (1). 'Über d. Entgiftung der Saponine durch Cholesterin.' *BDCG*, 1909, **42**, 238.

WINDAUS, A. (2). 'Über d. quantitative Bestimmung des Cholesterins und der Cholesterinester in einigen normalen und pathologischen Nieren.' *ZPC*, 1910, **65**, 110.

WINDERLICH, R. (1) (ed.). *Julius Ruska und die Geschichte d. Alchemie, mit einem Völlstandigen Verzeichnis seiner Schriften; Festgabe zu seinem 70. Geburtstage...* Ebering, Berlin, 1937. (Abhdl. z. Gesch. d. Med. u. d. Naturwiss., no. 19).

WINDERLICH, R. (2). 'Verschüttete und wieder aufgegrabene Quellen der Alchemie des Abendlandes' (a biography of J. Ruska and an account of his work). Art. in Winderlich (1), the Ruska Presentation Volume.

WINKLER, H. A. (1). *Siegel und Charaktere in der Mohammedanische Zauberei.* De Gruyter, Berlin and Leipzig, 1930. (*DI* Beiheft, no. 7.)

WISE, T. A. (1). *Commentary on the Hindu System of Medicine.* Thacker, Ostell & Lepage, Calcutta; Smith Elder, London, 1845.

WISE, T. A. (2). *Review of the History of Medicine [among the Asiatic Nations].* 2 vols. Churchill, London, 1867.

WOLF, A. (1) (with the co-operation of F. Dannemann & A. Armitage). *A History of Science, Technology, and Philosophy in the 16th and 17th Centuries.* Allen & Unwin, London, 1935; 2nd ed., revised by D. McKie, London, 1950. Rev. G. Sarton, *ISIS*, 1935, **24**, 164.

WOLF, A. (2). *A History of Science, Technology and Philosophy in the 18th Century.* Allen & Unwin, London, 1938; 2nd ed. revised by D. McKie, London, 1952.

WOLF, JOH. CHRISTOPH (1). *Manichaeismus ante Manichaeos, et in Christianismo Redivivus; Tractatus Historico-Philosophicus...* Liebezeit & Stromer, Hamburg, 1707. Repr. Zentralantiquariat D. D. R., Leipzig, 1970.

WOLF, T. (1). *Viajes Cientificos.* 3 vols. Guayaquil, Ecuador, 1879.

WOLFF, CHRISTIAN (1). 'Rede über die Sittenlehre der Sineser', pub. as *Oratio de Sinarum Philosophia Practica* [that morality is independent of revelation]. Frankfurt a/M, 1726. The lecture given in July 1721 on handing over the office of Pro-Rector, for which Christian Wolff was expelled from Halle and from his professorship there. See Lach (6). The German version did not appear until 1740 in vol. 6 of Wolff's *Kleine Philosophische Schriften*, Halle.

WOLTERS, O. W. (1). 'The "Po-Ssu" Pine-Trees.' *BLSOAS*, 1960, **23**, 323.

WONG K. CHIMIN. See Wang Chi-Min.

WONG, M. or MING. See Huang Kuang-Ming, Huard & Huang Kuang-Ming.

WONG MAN. See Huang Wên.

WONG WÊN-HAO. See Ong Wên-Hao.

WOOD, I. F. (1). '[The State of the Dead in] Hebrew [Thought].' *ERE*, vol. xi, p. 841.

WOOD, I. F. (2). '[The State of the Dead in] Muhammadan [Muslim, Thought].' *ERE*, vol. xi, p. 849.

WOOD, R. W. (1). 'The Purple Gold of Tut'ankhamēn.' *JEA*, 1934, **20**, 62.

WOODCROFT, B. (1) (tr.). *The 'Pneumatics' of Heron of Alexandria.* Whittingham, London, 1851.

WOODROFFE, SIR J. G. (ps. A. Avalon) (1). *Śakti and Śakta; Essays and Addresses on the Śakta Tantra-śāstra.* 3rd ed. Ganesh, Madras; Luzac, London, 1929.

WOODROFFE, SIR J. G. (ps. A. Avalon) (2). *The Serpent Power* [Kuṇḍalinī Yoga], *being the Śat-cakra-nirūpana* [i.e. ch. 6 of Pūrnānanda's *Tattva-chintāmaṇi*] *and 'Pādukā-panchaka', two works on Laya Yoga...* Ganesh, Madras; Luzac, London, 1931.

WOODROFFE, SIR J. G. (ps. A. Avalon) (3) (tr.). *The Tantra of the Great Liberation, 'Mahā-nirvāna Tantra', a translation from the Sanskrit.* London, 1913. Ganesh, Madras, 1929 (text only).

WOODS, J. H. (1). *The Yoga System of Patañjali; or, the Ancient Hindu Doctrine of Concentration of Mind...* Harvard Univ. Press, Cambridge, Mass. 1914. (Harvard Oriental Series, no. 17.)

WOODWARD, J. & BURNETT, G. (1). *A Treatise on Heraldry, British and Foreign...* 2 vols. Johnston, Edinburgh and London, 1892.

WOOLLEY, C. L. (4). 'Excavations at Ur, 1926–7, Part II.' *ANTJ*, 1928, **8**, 1 (24), pl. viii, 2.

WOULFE, P. (1). 'Experiments to show the Nature of *Aurum Mosaicum*.' *PTRS*, 1771, **61**, 114.

WRIGHT, SAMSON, (1). *Applied Physiology.* 7th ed. Oxford, 1942.

WU KHANG (1). *Les Trois Politiques du Tchounn Tsieou [Chhun Chhiu] interpretées par Tong Tchong-Chou [Tung Chung-Shu] d'après les principes de l'école de Kong-Yang [Kungyang].* Leroux, Paris, 1932. (Includes tr. of ch. 121 of *Shih Chi*, the biography of Tung Chung-Shu.)

WU LU-CHHIANG. See Tenney L. Davis' biography (obituary). *JCE*, 1936, **13**, 218.

WU LU-CHHIANG & DAVIS, T. L. (1) (tr.). 'An Ancient Chinese Treatise on Alchemy entitled *Tshan Thung Chhi*, written by Wei Po-Yang about +142...' *ISIS*, 1932, **18**, 210. Critique by J. R. Partington, *N*, 1935, **136**, 287.

WU LU-CHHIANG & DAVIS, T. L. (2) (tr.). 'An Ancient Chinese Alchemical Classic; Ko Hung on the Gold Medicine, and on the Yellow and the White; being the 4th and 16th chapters of *Pao Phu Tzu...*' *PAAAS*, 1935, **70**, 221.

WU YANG-TSANG (1). 'Silver Mining and Smelting in Mongolia.' *TAIME*, 1903, **33**, 755. With a discussion by B. S. Lyman, pp. 1038 ff. (Contains an account of the recovery of silver from argentiferous lead ore, and cupellation by traditional methods, at the mines of Ku-shan-tzu and Yen-tung Shan in Jehol province. The discussion adds a comparison with traditional Japanese methods observed at Hosokura). Abridged version in *EMJ*, 1903, **75**, 147.

WULFF, H. E. (1). *The Traditional [Arts and] Crafts of Persia; their Development, Technology and Influence on Eastern and Western Civilisations.* M.I.T. Press, Cambridge, Mass. 1966. Inaug. Diss. Univ. of New South Wales, 1964.

WUNDERLICH, E. (1). 'Die Bedeutung der roten Farbe im Kultus der Griechern und Römer.' *RGVV* 1925, **20**, 1.

YABUUCHI KIYOSHI (9). 'Astronomical Tables in China, from the Han to the Thang Dynasties.' Eng. art. in Yabuuchi Kiyoshi (25) (ed.), *Chūgoku Chūsei Kagaku Gijutsushi no Kenkyū* (Studies in the History of Science and Technology in Mediaeval China). Jimbun Kagaku Kenkyusō, Tokyo, 1963.

YAMADA KENTARO (1). *A Short History of Ambergris [and its Trading] by the Arabs and the Chinese in the Indian Ocean.* Kinki University, 1955, 1956. (Reports of the Institute of World Economics, *KKD*, nos. 8 and 11.)

YAMADA KENTARO (2). *A Study of the Introduction of 'An-hsi-hsiang' into China and of Gum Benzoin into Europe.* Kinki University, 1954, 1955. (Reports of the Institute of World Economics, *KKD*, nos. 5 and 7.)

YAMASHITA, A. (1). 'Wilhelm Nagayoshi Nagai [Nakai Nakayoshi], Discoverer of Ephredrin; his Contributions to the Foundation of Organic Chemistry in Japan.' *SHST/T*, 1965, 11.

YAMAZAKI, T. (1). 'The Characteristic Development of Chemical Technology in Modern Japan, chiefly in the Years between the two World Wars.' *SHST/T*, 1965, 7.

YAN TSZ CHIU. See Yang Tzu-Chiu (1).

YANG LIEN-SHÊNG (8). 'Notes on Maspero's "Les Documents Chinois de la Troisième Expédition de Sir Aurel Stein en Asie Centrale".' *HJAS*, 1955, **18**, 142.

YANG TZU-CHIU (1). 'Chemical Industry in Kuangtung Province.' *JRAS/NCB*, 1919, **50**, 133.

YATES, FRANCES A. (1). *Giordano Bruno and the Hermetic Tradition.* Routledge & Kegan Paul, London, 1964. Rev. W. P[agel], *AX*, 1964, **12**, 72.

YATES, FRANCES A. (2). 'The Hermetic Tradition in Renaissance Science.' Art. in *Art, Science and History in the Renaissance*, ed. C. S. Singleton. Johns Hopkins Univ. Press, Baltimore, 1968, p. 255.

YATES, FRANCES A. (3). *The Rosicrucian Enlightenment.* Routledge & Kegan Paul, London, 1972.

YEN CHI (1). 'Ancient Arab Coins in North-West China.' *AQ*, 1966, **40**, 223.

YETTS, W. P. (4). 'Taoist Tales; III, Chhin Shih Huang's Ti's Expeditions to Japan.' *NCR*, 1920, **2**, 290.

YOUNG, S. & GARNER, SIR HARRY M. (1). 'An Analysis of Chinese Blue-and-White [Porcelain]', with 'The Use of Imported and Native Cobalt in Chinese Blue-and-White [Porcelain].' *ORA*, 1956 (n. s.), **2** (no. 2).

YOUNG, W. C. (1) (ed.). *Sex and Internal Secretions.* 2 vols. Williams & Wilkins, Baltimore, 1961.

YŪ YING-SHIH (2). 'Life and Immortality in the Mind of Han China.' *HJAS*, 1965, **25**, 80.

YUAN WEI-CHOU. See Wei Chou-Yuan.

YULE, SIR HENRY (1) (ed.). *The Book of Ser Marco Polo the Venetian, concerning the Kingdoms and Marvels of the East, translated and edited, with Notes, by H. Y....*, 1st ed. 1871, repr. 1875. 2 vols. ed. H. Cordier. Murray, London, 1903 (reprinted 1921). 3rd ed. also issued Scribners, New York, 1929. With a third volume, *Notes and Addenda to Sir Henry Yule's Edition of Ser Marco Polo*, by H. Cordier. Murray, London, 1920.

YULE, SIR HENRY (2). *Cathay and the Way Thither; being a Collection of Mediaeval Notices of China.* 2 vols. Hakluyt Society Pubs. (2nd ser.) London, 1913–15. (1st ed. 1866.) Revised by H. Cordier, 4 vols. Vol. 1 (no. 38), *Introduction; Preliminary Essay on the Intercourse between China and the Western Nations previous to the Discovery of the Cape Route.* Vol. 2 (no. 33), *Odoric of Pordenone.* Vol. 3 (no. 37), *John of Monte Corvino and others.* Vol. 4 (no. 41), *Ibn Baṭṭuṭa and Benedict of Goes.* (Photo-litho reprint, Peiping, 1942.)

YULE, H. & BURNELL, A. C. (1). *Hobson-Jobson; being a Glossary of Anglo-Indian Colloquial Words and Phrases....* Murray, London, 1886.

YULE & CORDIER. See Yule (1).

ZACHARIAS, P. D. (1). 'Chymeutike, the real Hellenic Chemistry.' *AX*, 1956, **5**, 116. Based on Stephanides (1), which it expounds.

ZIMMER, H. (1). *Myths and Symbols in Indian Art and Civilisation*, ed. J. Campbell. Pantheon (Bollingen), Washington, D.C., 1947.

ZIMMER, H. (3). 'On the Significance of the Indian Tantric Yoga.' *ERYB*, 1961, **4**, 3, tr. from German in *ERJB*, 1933, **1**.

ZIMMER, H. (4). 'The Indian World Mother.' *ERYB*, 1969, **6**, 70 (*The Mystic Vision*, ed. J. Campbell). Tr. from the German in *ERJB*, 1938, **6**, 1.

ZIMMERN, H. (1). 'Assyrische Chemische-Technische Rezepte; insbesondere f. Herstellung farbiger glasierter Ziegel, im Umschrift und Übersetzung.' *ZASS*, 1925, **36** (N.F. **2**), 177.

ZIMMERN, H. (2). 'Babylonian and Assyrian [Religion].' *ERE*, vol. ii, p. 309.

ZONDEK, B. & ASCHHEIM, S. (1). 'Hypophysenvorderlappen und Ovarium; Beziehungen der endokrinen Drüsen zur Ovarialfunktion.' *AFG*, 1927, **130**, 1.

ZURETTI, C. O. (1). *Alchemistica Signa; Glossary of Greek Alchemical Symbols.* Vol. 8 of Bidez, Cumont, Delatte, Heiberg *et al.* (1).

ZURETTI, C. O. (2). *Anonymus 'De Arte Metallica seu de Metallorum Conversione in Aurum et Argentum'* [early +14th cent. Byzantine]. Vol. 7 of Bidez, Cumont, Delatte, Heiberg *et al.* (1), 1926.

ADDENDA TO BIBLIOGRAPHY C

ANDRIEU, M. (1). *'Immixtio et Consecratio'; la Consécration par Contact dans les Documents Liturgiques du Moyen Âge.* Picard, Paris, 1924. (Univ. de Strasbourg, Bibl. de l'Inst. de Droit Canonique, no. 2.)

ANON. (117). 'The Study of a Body Two Thousand Years Old [the Lady of Tai, d. *c.* −166].' *CREC*, 1973, **22** (no. 10), 32. (Cf. pt. 2, p. 304.)

ARDAILLON, E. (1). *Les Mines de Laurion dans l'Antiquité* (Inaug. Diss., Paris). Thorin, Paris, 1897. Repr. in Kounas (1). (Cf. pt. 2, pp. 41, 218.)

BAILEY, SIR HAROLD (2). 'Trends in Iranian Studies.' *MRDTB*, 1971, **29**, 1. (Cf. pt. 2, p. 116 (g), another statement on the etymology of *soma-haoma*.)

BAILEY, SIR HAROLD (3). 'The Range of the Colour *zar-* in Khotan Saka Texts.' Art. in *Mémorial Jean de Menasce*, 1974, ed. Gignoux & Tafazzoli, p. 369. (On colour names in relation to *soma-haoma*.)

BENFEY, O. T. (3). 'How not to Die from the Elixir of Life; or, were the Chinese first again? Kidney Stones *vs.* Mercury Poisoning; Eat Spinach with your Tuna; Popeye to the Rescue; and Another Research Problem for *Chemistry*'s Readers.' *CHEM*, 1974, **47** (no. 7), 2 (Editorial).

BEYER, S. (1). *The Cult of Tārā; Magic and Ritual in Tibet.* Univ. California Press, Berkeley and Los Angeles, 1973. (Includes material on alchemy in Tibet.)

BOECKH, AUGUST (1). *A Dissertation on the Silver Mines of Laurion in Attica* (1815). Repr. in Kounas (1). (Originally part of Boeckh's *The Public Economy of Athens*, tr. from the German by G. C. Lewis, 2nd ed., Parker, London, 1842, pp. 615–74.)

BOXER, C. R. (5). 'A Note on the Interaction of Portuguese and Chinese Medicine at Macao and Peking in the +16th to +18th Centuries.' *BILCA*, 1974, **8**, 33. (Contains biographical material on J. F. Vandermonde, cf. pt. 2, pp. 160–1. He was appointed medical officer to the city of Macao in 1723 and returned to France in 1732.)

BULLING, A. (15). 'The "Guide of the Souls" Picture in the Western Han Tomb in Ma Wang Tui near Chhangsha.' *ORA*, 1974, **20**, 1.

CAPON, E. & McQUITTY, W. (1). *Princes of Jade.* Cardinal (Nelson), London, 1973. (Cf. pt. 2, p. 303, the tombs and jade body cases of Liu Shêng and Tou Wan; also the tomb of the Lady of Tai, pp. 8 ff., 162 ff.)

CASTANEDA, CARLOS (1). *The Teachings of Don Juan; a Yaqui Way of Knowledge.* Ballantine Books, New York, 1974, orig. pub. California Univ. Press, 1968.

CASTANEDA, CARLOS (2). *A Separate Reality; Further Conversations with Don Juan.* Simon & Schuster (Touchstone), New York, 1971; Pocket Books, New York, 1973.

CASTANEDA, CARLOS (3). *Journey to Ixtlan; the Lessons of Don Juan.* Simon & Schuster, New York, 1972; Pocket Books, New York, 1974.

CHANG, H. C. (CHANG HSIN-CHANG). *See* Chang Hsin-Tshang.

CHANG, HSIN-TSHANG (1). *Allegory and Courtesy in [Edmund] Spenser; a Chinese View.* Edinburgh Univ. Press, Edinburgh, 1956. Rev. Liu Jung-Ju (J. J. Y. Liu), *JRAS*, 1956, 87.

CHANG HSIN-TSHANG (2) (tr. and comm.). *Chinese Literature; Popular Fiction and Drama.* University Press, Edinburgh, 1973.

CHŎN SANGUN (1). *Science and Technology in Korea; Traditional Instruments and Techniques.* M.I.T. Press, Cambridge, Mass., 1974. (M.I.T. East Asian Science Series, no. 4.) Based largely on Chŏn Sangun (2).

DAY, JOAN (1). *Bristol Brass; the History of the Industry.* David & Charles, Newton Abbot, 1973.

DECAISNE, J. (1). 'Note sur les Deux Espèces de Nerprun [buckthorn] qui fournissent le Vert de Chine.' *CRAS*, 1857, **44**. Repr. in Rondot, Persoz & Michel (1), pp. 139 ff.

DIGBY, SIR KENELM (1). *A Choice Collection of Rare Chymical Secrets and Experiments in Philosophy. As also Rare and unheard-of Medicines, Menstruums, and Alkahests; with the True Secret of Volatilising the fixt Salt of Tartar. Collected and Experimented by the Honourable and truly Learned Sir Kenelm Digby, Kt., Chancellour to Her Majesty the Queen-Mother. Hitherto kept Secret since his Decease but now Published for the good and benefit of the Publick, by George Hartman, London: Printed for the Publisher, and are to be Sold by the Book-Sellars of London, and at his own House in Hewes Court in Black-Fryers.* London, 1682. *See* Dobbs (3).

DOBBS, B. J. (2). 'Studies in the Natural Philosophy of Sir Kenelm Digby; II, Digby and Alchemy.' *AX*, 1973, **20**, 143.

DOBBS, B. J. (3). 'Studies in the Natural Philosophy of Sir Kenelm Digby; III, Digby's Experimental Alchemy—the Book of Secrets.' *AX*, 1974, **21**, 1.

DŌKE TATSUMASA (1). 'Udagawa Yōan; a Pioneer Scientist of Early Nineteenth-Century Feudal Japan.' *JSHS*, 1973, **12**, 99.

EICHHORN, W. (12). 'Die Wiedereinrichtung der Staatsreligion im Anfang der Sung-Zeit.' *MS*, 1964, **23**, 205–63.

EVANS, R. J. W. (1). *Rudolf II and his World; a Study in Intellectual History*, +*1576 to* +*1162*. Oxford, 1973.

FARRAND, W. R. (1). 'Frozen Mammoths and Modern Geology.' *SC*, 1961, **133**, 729. (Cf. pt. 2, p. 304.)

FUKUI KŌJUN (1). 'A Study of the *Chou I Tshan Thung Chhi*.' *ACTAS*, 1974, no. 27, 19.

GERSHEVITCH, I. (2). 'An Iranianist's View of the Soma Controversy.' Art. in *Mémorial Jean de Menasce*, 1974, ed. Gignoux & Tafazzoli, p. 45.

GIBERT, L. (1). *Dictionnaire Historique et Géographique de la Mandchourie*. Missions Étrangères, Hong-kong, 1934.

GIGNOUX, P. & TAFAZZOLI, A. (1) (ed.). *Mémorial Jean de Menasce*. Imp. Orientaliste, Louvain, 1974. (Pub. Fondation Culturelle Iranienne, no. 185.)

GRANET, M. (1). *Danses et légendes de la Chine Ancienne*. 2 vols. Alcan, Paris, 1926.

GRANT, EDWARD (1) (ed. and tr.). *A Source-Book of Mediaeval Science*. Harvard University Press, Cambridge, Mass., 1973. (Contains translations of essential passages on alchemy from Ibn Sīnā (Avicenna), Petrus Bonus, Albertus Magnus, Thomas Aquinas and Albert of Saxony.)

HARNER, M. J. (1). *Hallucinogens and Shamanism*. Oxford, 1973. Rev. *SAM*, 1973, **229** (no. 4), 129. (Cf. pt. 2, p. 116.)

HEFFERN, R. (1). *Secrets of the Mind-Altering Plants of Mexico*. Pyramid, New York, 1974. (Cf. pt. 2, p. 116.)

HUNG HSIN (1). 'A Treasure Lake [the Hungtsê Hu, in the Lower Huai River Valley in Chiangsu province].' *CREC*, 1974, **23** (no. 8), 46. (Cf. pt. 2, p. 121 on *Ganoderma lucidum*, now cultivated for pharmaceutical use in China; nephritis, neurasthenia, general tonic properties, no hallucinogens.)

HUXLEY, ALDOUS (1). *The Doors of Perception; and, Heaven and Hell*. Penguin, London, 1959; many times reprinted. The first orig. pub. Chatto & Windus, London, 1954, the second, Chatto & Windus, London, 1956. (Cf. pt. 2, p. 116.)

HYDE, MARGARET O. (1). *Mind Drugs*. McGraw-Hill, New York, 1972; 2nd ed., revised and enlarged, Simon & Schuster (Pocket Books), New York, 1973. (Cf. pt. 2, pp. 116, 150.)

JEON SANG-WOON. *See* Chŏn Sangun.

KERR, G. H. (1). *Okinawa; the History of an Island People*. Tokyo, 1958. (States on p. 40 that the voyages of exploration which discovered the Liu-Chhiu Islands in +607 and later were motivated by an interest in Phêng-Lai and the other isles of the holy immortals (cf. p. 18) on the part of the Sui emperor, Yang Ti (cf. p. 132). But *Sui Shu*, ch. 81, p. 10*a*, does not bear this out. It records that the first sighting was reported in +605, by a sea-captain, Ho Man, and that the first successful reconnaissance was made in +607 by a military officer, Chu Kuan, who was accompanied by the sailor. This was followed by an expeditionary force under Chhen Lêng in +609.)

KIMURA EIICHI (1). 'Taoism and Chinese Thought.' *ACTAS*, 1974, no. 27, 1.

KOUNAS, DIONYSIOS A. (1) (ed.). *Studies on the Ancient Silver Mines at Laurion*. Coronado Press, Lawrence, Kansas, 1972. (Reprints by litho-offset Boeckh (1), Ardaillon (1) and Xenophon 'On the Revenues' (*c.* −355), tr. H. G. Dakyns; with an introduction by Kounas.)

LONGO, V. G. (1). *Neuropharmacology and Behaviour*. Freeman, New York, 1973. Rev. *SAM*, 1973, **229** (no. 4), 129. (Cf. pt. 2. p. 116.)

MCINTYRE, L., SEIDLER, N. & SEIDLER, R. (1). 'The Lost Empire of the Incas.' *NGM*, 1973, **144** (no. 6), 729–87. (Includes an account of the perfect preservation by freezing, over five centuries, of the bodies of men and boys sacrificed at shrines from 17,000 to 20,000 ft altitude in the Andean range of mountains.) (Cf. pt. 2, p. 304.)

DEL MAR, ALEXANDER (1). *A History of the Precious Metals from the Earliest Times to the Present*. Begun, 1858; laid aside, 1862; completed, 1879. Bell, London, 1880; 2nd ed., revised, 1902. Repr. Burt Franklin, New York, 1968.

MATHER, R. B. (1). 'The Fine Art of Conversation; the Yen Yü Phien of the *Shih Shuo Hsin Yü*.' *JAOS*, 1971, **91**, 222. (The story of Ho Yen and the Five-Mineral Powder (p. 45 above) is translated and discussed on p. 232.)

MICHEL, A. F. (1). 'Sur la Matière Colorante des Nerpruns [buckthorns] Indigènes.' In Rondot, Persoz & Michel (1), pp. 183ff.

MISH, J. L. (1). 'Creating an Image of Europe for China; [Giulio] Aleni's [+1582 to +1649] *Hsi Fang Ta Wên* (Questions and Answers concerning the Western World), with Introduction, Translation and Notes.' *MS*, 1964, **23**, 1–87. Abstr. *RBS*, 1973, **10**, no. 866. (Contains discussions of geography, astronomy and geomancy in China and the West, with mention of alchemy on p. 76. This interview with the Jesuit was edited by Chiang Tê-Ching, later Minister of Rites under the last Ming emperor, in +1637.)

MIYAKAWA HISAYUKI (1). 'The Legate Kao Phien [d. +887] and a Taoist Magician, Lü Yung-Chih, in the Time of Huang Chhao's Rebellion [+875 to +884].' *ACTAS*, 1974, no. 27, 75. (The Taoist entourage of the general who suppressed it, including several alchemists, Chuko Yin, Tshai Thien and Shenthu shêng = Pieh-Chia.) (Cf. pp. 173–4.)

MURAKAMI YOSHIMI (1). 'The Affirmation of Desire in Taoism.' *ACTAS*, 1974, no. 27, 57.

NEEDHAM, JOSEPH (43). 'The Past in China's Present.' *CR/MSU*, 1960, **4**, 145 and 281; repr. with some omissions, *PV*, 1963, **4**, 115. French tr.: *Du Passé Culturel, Social et Philosophique Chinois dans ses Rapports avec la Chine Contemporaine*, by G. M. Merkle-Hunziker. *COMP*, 1960, no. 21–2, 261; 1962, no. 23–4,113; repr. in *CFC*, 1960, no. 8, 26; 1962, no. 15–16, 1.

PANTHEO, GIOVANNI AGOSTINO (1). *Voarchadumia contra Alchimiam; Ars Distincta ab Archimia & Sophia; cum Additionibus, Proportionibus, Numeris et Figuris opportunis*...[an assayer's counterblast to aurifactive claims]. Venice, 1530; 2nd ed. Paris, 1550. (The opening word, meaning gold thoroughly refined, is said to be formed from a 'Chaldaean' word and a Hebrew phrase.) Ferguson (1), vol. 2, pp. 166–7. (Cf. pt. 2. p. 32.)

PEARSON, R. J. (1). *The Archaeology of the Ryukyu [Liu-Chhiu] Islands; a Regional Chronology from −3000 to the Historic Period*. Univ. Hawaii Press, Honolulu, 1969.

PERSOZ, J. (1). 'Sur une Matière Colorante Verte qui vient de Chine' (*lü kao*, from buckthorn bark and twigs). *CRAS*, 1852, **35**, 558. Repr. in Rondot, Persoz & Michel (1), pp. 129ff.

PERSOZ, J. (2). 'Des Propriétés Chimiques et Tinctoriales du Vert de Chine' (*lü kao, Rhamnus* spp.). In Rondot, Persoz & Michel (1), pp. 151ff.

PERSOZ, J. (3). 'Sur la Teinture en Jaune avec le *hoang-tchi [huang-chih]*' (*Gardenia* spp.). In Rondot, Persoz & Michel (1), pp. 199ff., cf. p. 86ff.

PORTER, W. N. (1) (tr.). *The Miscellany of a Japanese Priest, being a Translation of the 'Tsurezuregusa'* [+1338], *by Kenkō [Hōshi], Yoshida [no Kaneyoshi]*. With an introduction by Ichikawa Sanki. Clarendon (Milford), London, 1914. Repr. Tuttle, Rutland, Vt. and Tokyo, 1974.

RENONDEAU, G. (1). *Le 'Shugendō'; Histoire, Doctrine et Rites des Yamabushi*. Paris, 1965. (Cahiers de la Société Asiatique, no. 18.) (Cf. pt. 2, p. 299.)

RONDOT, N. (3). 'Notice du Vert de Chine et de la Teinture en Vert chez les Chinois...' (*Rhamnus* spp.). In Rondot, Persoz & Michel (1), pp. 1ff.

RONDOT, N., PERSOZ, J. & MICHEL, A. F. (1). *Notice du Vert de Chine et de la Teinture en Vert chez les Chinois, suivie d'une Étude des Propriétés Chimiques et Tinctoriales du 'lo-kao [lü kao]' par Mons. J. P...., et de Recherches sur la Matière Colorante des Nerpruns Indigènes par Mons. A. F. M....*, Lahure (for the Chambre de Commerce de Lyon), Paris, 1858.

RUDENKO, S. I. (2). *Frozen Tombs of Siberia; the Pazyryk [perma-frost] Burials of Iron-Age Horsemen*. Dent, London, 1970. (With an explanatory introduction by the translator, M. W. Thompson.) (Cf. pt. 2, p. 304.)

RUDOLPH, R. C. (8). 'Two Recently Discovered Han Tombs.' *AAAA*, 1973, **26** (no. 2), 106.

SANFORD, J. H. (1). 'Japan's "Laughing Mushrooms".' *ECB*, 1972, **26**, 174. (Cf. pt. 2, p. 121.)

SCHULTES, R. E. & HOFMANN, A. (1). *The Botany and Chemistry of Hallucinogens*. Thomas, New York, 1973. Rev. *SAM*, 1973, **229** (no. 4), 129. (Cf. pt. 2, p. 116.)

SIGISMUND, R. (1). *Die Aromata in ihrer Bedeutung für Religion, Sitten, Gebräuche, Handel und Geographie des Alterthums bis zu den ersten Jahrhunderten unserer Zeitrechnung*. Winter, Leipzig, 1884; photo-litho reprint, Zentralantiquariat d. D.D.R., Leipzig, 1974. (Cf. pt. 2, pp. 136ff.)

SMITH, C. S. (8). 'An Examination of the Arsenic-Rich Coating on a Bronze Bull from Horoztepe.' Art. in *Application of Science...Works of Art*, 1973, ed. W. J. Young (1), p. 96.

SMITH, C. S. & HAWTHORNE, J. G. (1) (ed. & tr.). '*Mappae Clavicula*; A Little Key to the World of Medieval Techniques.' *TAPS*, 1974 (N.S.), **64** (pt. 4), 1–128. (Annotated translation based on a collation of the Sélestat and Phillipps-Corning MSS, with reproductions of both.)

SOLOMON, D. (2) (ed.). *The Marijuana Papers*. New York, 1966; 2nd ed., revised, Granada (Panther Books), 1969; repr. 1970, 1972. (Cf. pt. 2, p. 150.)

TAMBURELLO, A. (1). '"Taoismus" in Japan' (tr. from Ital. by G. Glaesser). *ANT*, 1970, **12**, 125. (Cf. pt. 2, p. 300.)

TÊNG SSU-YÜ & BIGGERSTAFF, K. (1). *An Annotated Bibliography of Selected Chinese Reference Works*. Harvard-Yenching Instit. Peiping, 1936. (Yenching Journ. Chin. Studies, monograph no. 12.)

TOMKINSON, L. (2) (tr.). 'Tales of Ming Scholars (the *Ju Lin Wai Shih*).' Unpub. MS. in the East Asian History of Science Library, Cambridge.

TSUNODA RYUSAKU & GOODRICH, L. CARRINGTON (1). 'Japan in the Chinese Dynastic Histories.' Perkins, South Pasadena, 1951. (Perkins Asiatic Monographs, no. 2.)

VERESHCHAGIN, N. K. (1). 'The Mammoth (Woolly Elephant) "Cemeteries" of North-East Siberia.' *POLREC*, 1974, **17** (no. 106), 3. (Cf. pt. 2, p. 304.)

VERHAEREN, H. (2) (ed.). *Catalogue de la Bibliothèque du Pé-T'ang [the Pei Thang Jesuit Library] in Peking*. Lazaristes Press, Peking, 1949. Photographically reproduced, Belleslettres (Cathasia). Paris, 1969.

WAGNER, R. G. (1). 'Lebens-stil und Drogen im chinesischen Mittelalter.' *TP*, 1973, **59**, 79–178. (An exhaustive study of the tonic Han Shih San, a powder of four inorganic substances (Ca, Mg, Si) and nine plant substances containing alkaloids.)

WALLACKER, B. E. (2). 'Liu An, Second Prince of Huai-Nan (*c.* −180 to −122).' *JAOS*, 1972, **92**, 36.

WELCH, HOLMES H. (4). *The Practice of Chinese Buddhism, 1900 to 1950*. Harvard Univ. Press, Cambridge, Mass., 1967.

WÊN PIEN (1). 'The World's Oldest Painting on Silk; Visions of Heaven, Earth and the Underworld in a Two-Thousand-Year-Old Chinese Tomb [the Lady of Tai, d. *c.* −166].' *UNESC*, 1974, **27** (no. 4), 18.

WILHELM, RICHARD (2) (tr.). '*I Ging*' [*I Ching*]; *Das Buch der Wandlungen*. 2 vols. (3 books, pagination of 1 and 2 continuous in first volume). Diederichs, Jena, 1924. (Eng. tr. C. F. Baynes (2 vols.). Bollingen-Pantheon, New York, 1950.) (See Vol. 2, p. 308.)

YANG, GLADYS & YANG HSIEN-YI (1) (tr.). *The Scholars* ('*Ju Liu Wai Shih*'). 1957. (With an appendix on official rank and on the examination system.)

YASUDA YURI (1) (tr.). *Old Tales of Japan*. With illustrations by Sakakura Yoshinobu and Mitsui Eiichi. Tuttle, Tokyo and Rutland, Vt., 1947, repr. 1953. Revised ed. 1956, many times reprinted. (Contains (pp. 133 ff.) a very abridged version in English of the *Taketori Monogatari* (+9th cent.), 'The Luminous Princess'.)

YATES, ROBIN (1). Translations of the alchemical poems of Lu Kuei-Mêng (unpub. MS.).

YOUNG, W. J. (1) (ed.). *The Application of Science in the Examination of Works of Art*. Museum of Fine Arts, Boston, 1973.

YÜ YING-SHIH (1). *Trade and Expansion in Han China; a Study in the Structure of Sino-Barbarian Economic Relations*. Univ. Calif. Press, Berkeley and Los Angeles, 1967.

GENERAL INDEX

by Muriel Moyle

Notes

(1) Articles (such as 'the', 'al-', etc.) occurring at the beginning of an entry, and prefixes (such as 'de', 'van', etc.) are ignored in the alphabetical sequence. Saints appear among all letters of the alphabet according to their proper names. Styles such as Mr, Dr, if occurring in book titles or phrases, are ignored; if with proper names, printed following them.

(2) The various parts of hyphenated words are treated as separate words in the alphabetical sequence. It should be remembered that, in accordance with the conventions adopted, some Chinese proper names are written as separate syllables while others are written as one word.

(3) In the arrangement of Chinese words, Chh- and Hs- follow normal alphabetical sequence, and *ü* is treated as equivalent to *u*.

(4) References to footnotes are not given except for certain special subjects with which the text does not deal. They are indicated by brackets containing the superscript letter of the footnote.

(5) Explanatory words in brackets indicating fields of work are added for Chinese scientific and technological persons (and occasionally for some of other cultures), but not for political or military figures (except kings and princes).

Abhidharma Mahāvibhāshā, 164

Abscesses, 47

Academy of Sciences of St Petersburg, *Mémoires*, 242

Account of the Cyclically (-Transformed) Metallous Enchymoma. See *Huan Chin Shu*

Acetate cations, 102

Acetic acid, 98, 99, 137

Action contrary to Nature, 52

Acupuncture, 177

Additions to the Enlarged Bag of Wisdom. See *Tsêng Kuang Chih Nang Pu*

The Adept Hsü's Treatise found in a Stone Coffer. See *Hsü Chen Chün Shih Han Chi*

The Adept Ko (Hung)'s Prescriptions for Emergencies. See *Ko Hsien Ong Chou Hou Pei Chi Fang*

The Adept Lü Shun-Yang's (Book) on Preparations of Drugs and Minerals. See *Shun-Yang Lü Chen-Jen Yao Shih Chih*

Agassi, Joseph, xxx

Agricola, Georgius (+16th-century metallurgist), 235, 236, 239 (e)

Ai Ti (Chin emperor, r+362 to +365), 112

Air. *See* Atmosphere

'*Ajā'ib al-Makhlūqāt* (Marvels of Creation), 174 (h)

Albertus Magnus (*c*. +1200 to +1280), 12 (i), 137

Alchemia (Libavius), 236

Alchemical apparatus. *See* Apparatus

Alchemical ceremonies, 69, 72, 106, 196

Alchemical experiments. *See* Experiments

Alchemical formulae, 198

Alchemical furnaces, xxxiv, 30, 36, 71, 133, 199

Alchemical inscription on a stele in Szechuan, 195–6

Alchemical laboratories, 118, 184–5, 218, 229; of Lao Tzu, 215

Alchemical *mantram*, 159

Alchemical reactions and reagents, 66, 181

Alchemical substances, 4, 66

decrease in number used, 182, 208, 217–19, 231, 238

Alchemical synonyms and metaphors, 69–70, 104, 152–7

Alchemical texts, xxxiv, 113, 117, 200

before the time of Ko Hung, 105

dating of, 152

destruction of, 208, 224–5, 247

style of, 74, 99–100, 104, 130, 138, 152, 171, 182

Alchemical theories, 143–51, 182

Alchemical transmutation, 39, 68, 81, 100–1, 111, 197

The Alchemist (Ben Jonson), 213–15

Alchemists

links between, 11–12, 76–81, 111–12, 119

of Muslim Spain, 124 (a)

secretiveness of, 38, 74, 87, 99, 100, 104, 128, 130, 152, 189

women, 38–9, 42, 169–71, 191–2, 205, 209

Alchemy

Arabic. *See* Arabic alchemy

Buddhism and, 165–6

Burmese. *See* Burmese alchemy

class-distinctions and, 208

decline of, 182, 246

and the desire for wealth, 194

Alchemy (*cont.*)
  dragon-slaying metaphor in, 8
  first extant treatise on, 43, 50
  Hellenistic. *See* Hellenistic alchemy
  Indian. *See* Indian alchemy
  inhibiting factors in the spread of, 74–5
  in Japan. *See* Japan
  Libavius' definition of, 236
  and magic, 225
  on rhe threshold of the chemical age, 229
  origins of, 7
  physiological. See *Nei tan*
  practical laboratory. See *Wai tan*
  proto-chemical, 216, 219, 247
  and sex, 214
  'spiritual', 100
  Taoist influence on, 106
  Western. *See* Western alchemy
Aleurites. See Thung-tree
Alexandrians. *See* Hellenistic culture
Alfonso X (King of Castile, El Sabio, r. +1252
    onwards, d. +1284), 191 (b)
Alkaline salt, 84
Alkalis, 160, 232, 234
Alkanet, 126 (d)
Alleton & Alleton (1), 255
Alloys and alloying, 189, 219, 233
  arsenic, 199
  beryllium, 190
  of coinage, 28, 37
  copper, 199
  copper–arsenic, 102, 129
  gold-containing, 49
  gold-like, 37, 99, 159, 186, 193, 196, 207 (c), 233
  lead, 199
  male or female, 97 (jj)
  mercury, 199
  silver-like, 193, 207 (c)
  zinc and lead, 136
Alluvial clays, 84
Almadén, mercury mines of, 3 (f), 6
Altair (asterism), 70 (c)
Altruism. *See* Ethics
Aludel, 159, 196, 198, 199
Alum, 78, 83, 84, 96 (d), 128, 129–30, 133, 159,
    173, 201, 237, 239, 244, 247
  yellow, 181
Amalgamation process, 4, 66, 67–8, 84, 197
Amalgams, 71, 123, 219
  gold, 68, 73, 89, 109, 194
  lead, 67, 73, 172, 209
  silver, cinnabar, lead and mercury, 141
  tin, 128
Amber, 97 (cc)
Ambix, 199, 259
Amerindians, 3 (e)
Amethyst, 89
Amino-acid (*an chi suan*), 260–1
Amiot, J. J. M. (Jesuit, +1774), 222, 249
Ammonium chloride, 69 (d), 78, 196, 233
Amulets. *See* Talismans
An Chhi shêng (legendary immortal, perhaps a

Taoist adept in or before the Han), 11, 31,
    32, 43, 76, 77
An Lu-Shan rebellion, 114, 141 (d), 182
Analytical methods, 28
Anatomy, 177
Anawrahta (Burmese king, +1065), 166
Ancestor-worship, 221, 223
Animal genera
  Laccifer, 126 (h)
  Lakshadia, 126 (h)
Animals
  experiments on, 51, 97 (ff), 105
  lists of, 12, 14
  in the sun and the moon, 4, 21
  synonyms relating to, 104–5
Annam, 75, 139
Anthology of a Myriad Leaves. See *Manyōshū*
Antibiotics, 47
Antimonic acid, 236
Antimony, 233
Aphrodisiacs, 129 (c), 1790
Apotropaics, 44, 114, 175, 177
Apparatus
  alchemical, xxxiv, 66, 73, 110, 130, 133, 159,
    160, 198–9, 210
  chemical, 246, 255 (a); illustrations, 256–9
Appetite, inhibition of, 146
Apsaras, 174 (h)
Aqua fortis (nitric acid), 237, 240
Aqua regia (mixture of nitric and hydrochloric
    acids), 237, 239, 240
Aqueous reactions, xxxiv
Aqueous solutions, 25 (g), 102, 110, 118
Arabic alchemy, 45, 124 (a), 166, 195
Arabic culture, xxxiv
  failure to produce modern science, xxvi–xxviii
Arabs, xxv, 212, 220
Aragonite, 136
Arcane Essentials of the Similarities and Cate-
    gories of the Five (Substances) in the
    Kinship of the Three. See *Tshan Thung Chhi
    Wu Hsiang Lei Pi Yao*
Archimedean screw, 235
Argentan, 233
Argentifaction, 1, 26, 75, 81, 83, 88, 89, 100, 101,
    103, 129, 170, 173–4, 186, 188, 192, 193,
    195, 209, 212, 215, 219, 227, 244
  and Confucian morality, 168
Ari monks (Burma), 166
Aristotle, 235
Aristotelian proto-chemists, 172
Arsenate cations, 102
Arsenic, 102, 131, 135, 137, 232, 233
  alloys of. *See* Alloys and alloying
  compounds, 102, 129
  as elixir ingredient, 133
  tonic medication with, 146
Arsenic disulphide. *See* Realgar
Arsenic trioxide. *See* Arsenious oxide *and*
    Arsenolite
Arsenic trisulphide. *See* Orpiment
Arsenious oxide, 98, 133, 144, 181, 198

Arsenolite (*yü shih*), 96 (d), 129, 131, 133
Arthaśāstra, 165
Artillery tactics, 241
Artisans. *See* Technicians
Asaṅga, 164
Asbestos, 176 (h)
Ascending to the heavens, 41, 46
Asceticism, 176, 205, 214
Assaying, 206 (g), 211
Astrology, 106
Astronomy, xx, xxiii, xxx, xxxi, 218, 220, 221, 223
    Ptolemaic, xxv
Aśvaghoṣa (Mahāyānist monk), 43 (d)
Atmosphere, 243, 246
Atomic theory, 221
Atomism, 149
Attar of roses, 170
Aurifaction, xx, xxxiii, xxxiv, 1, 12, 26–7, 28–9,
    31, 32, 35, 36–7, 38, 39, 43, 44, 48, 49, 57,
    81, 82–3, 84, 88, 89, 100, 180, 184, 186–8,
    190, 192, 193–4, 195, 197, 198, 207, 209,
    212, 219, 220
    arsenolite used in, 129
    Burmese, 166
    Confucian morality and, 168
    and immortality, 33, 57, 75
    Indian, 161, 162, 164–5
    hardly known in Japan, 174–5, 180
    methods, 101–4
    in modern China, 250 (a)
    poverty and, 1, 2
    and purification of oneself, 100
    on the threshold of the chemical age, 229
    Western. *See* Western aurifaction
    without intent to deceive, 189
Aurifiction, xx, xxxiii, xxxiv, 26, 28, 36, 38, 57,
    215
    government prohibition of, 26, 29 (a)
    hardly known in Japan, 174–5, 180
Auspicious days. *See* Lucky and unlucky days
Authentic Mirror of Chemical Science. See
    Hua-Hsüeh Chien Yuan
'Autumn mineral'. See *Chhiu shih*
Avataṃsaka Sūtra, 164
Axis of the Tao. See *Tao Shu*
Ayurvedic medicine, 46 (c), 48, 49
Azedarach. *See* Plant genera, *Melia*
Azurite, 15, 25, 78
    aqueous solutions of, 102

Bacon, Francis (+1560 to +1626), 108, 225 (d)
Bacon, Roger (+1214 to +1292), xxv
Bailey, K. C., 16
Baking-powder, 232
Bamboo tubes, used in laboratory equipment, 143
Banner of painted silk, 21–3
Barbarian invasions, 190
Barbier de Meynard (3), 215
Bardach's test, 46 (c)
Barnes & Yuan (1), 123
Basilica Chymica (Croll), 236
Bead amulets, 3

Becher, J. J. (chemist, +1635 to +1682), 218
Beneficial Prescriptions collected by Su Tung-Pho
    and Shen Kua. See *Su Shen Liang Fang*
Bengali Muslims, xxviii
Bergman, Torbern (chemist, +1735 to +1784),
    233
Berthelot & Duval, 124
Berthollet, C. L. (chemist, +1748 to +1822), 255
Bhasma (herbo-metallic preparations), 48, 49
Bibliography of Extant Books in Japan. See
    Nihon-koku Ganzai-sho Mokuroku
Billequin, M. A. (French engineer in China,
    1880), 252
Biochemistry, xxx, 147, 260, 262
Bio-geochemical prospecting, 232
'Biography of Liu Hung', 168, 169
Biography of the Supreme-Pole Elder-Immortal
    Ko. See *Thai-Chi Ko Hsien-Ong Chuan*
Biology, xxxi
Biringuccio, Vanuccio (metallurgist, +1480 to
    +1539), 241
al-Bīrūnī, Abū al-Raihān (geographer, +11th
    century), xxviii, 161, 164
Bismuth (*pi*), 260
Black, Joseph (chemist, +1728 to +1799), 221
Black, as auspicious colour, 5
'Black chalk', 233
'Black lead juice', 78
Bleaching, 232
Blood, as elixir ingredient, 9, 87
Bodhisattvas, 164
Body-weight, loss of, 120 (b)
Boils, 78
Bolus of Mendes (Graeco-Egyptian naturalist,
    early −2nd century), 48
Bombs, 184
Bomoh (Malay word for magician or alchemist),
    250 (a)
Bone diseases, 130
Bone-marrow, poisoning of, 144 (c)
Book-burning, 208, 224–5, 247
Book of Changes. See *I Ching*
Book of Daily Occupations for Scholars in Rural
    Retirement, by the Emaciated Immortal.
    See *Chhü Hsien Shen Yin Shu*
Book of the Emerald Heaven. See *Tshui Hsü
    Phien*
Book of the Firing-Process. See *Huo Chi*
Book of the Golden Hall Master. See *Chin Lou
    Tzu*
Book of the Infinite Treasure in the Secret
    Garden. See *Hung Pao Yuan Pi Shu*
Book of Lao Tzu's Conversion of Foreigners.
    See *Lao Tzu Hua Hu Ching*
Book of Master Chi Ni. See *Chi Ni Tzu*
Book of Master Chuang. See *Chuang Tzu*
Book of Master Han Fei. See *Han Fei Tzu*
Book of Master Kuan. See *Kuan Tzu*
Book of Master Lieh. See *Lieh Tzu*
Book of Master Mo. See *Mo Tzu*
Book of the Preservation-of-Solidarity Master.
    See *Pao Phu Tzu*

Book of the Preservation-of-Solidarity Master: Esoteric Chapters. See *Pao Phu Tzu (Nei Phien)*

Book of the Preservation-of-Solidarity Master: Exoteric Chapters. See *Pao Phu Tzu (Wai Phien)*

Book of the Realm of Kêng and Hsin. See *Kêng Hsin Ching*

Book on the Restoration of Life by the Cyclically-Transformed Elixir. See *Huan Tan Fu Ming Phien*

Book of the Return to the Source. See *Huan Yuan Phien*

Borax, 196, 198, 232
Borneo camphor, 170
Botanical garden, Khaifêng, 210
Botany, xxxi, 211, 221
Bournon, Mons., 242, 249
Bouvet, Joachim (Jesuit Mathematician-Royal of the King of France, +1685), 242
Boxer Rebellion, 117
Boxing (*thai chi chhüan*), 209
Boyle, Robert (chemist, +1627 to +1691), 172, 218, 221, 225 (d), 235, 241
Boym, Michel (Jesuit botanist), 234
'Brahmin' books, 160
Brahmin pharmaceutics. See *Po-lo-mên Yao Fang*
Brasavola, Antonio (physician and pharmacist, +1500 to +1555), 238
Brass, 136, 187 (g)
  considered as auspicious, 31 (a)
  manufacture of, 245
Breathing exercises. See Respiratory exercises
Bridgman, E. C., 47
Brief Explanation of the *Kinship of the Three*. See *Chou I Tshan Thung Chhi Shu Lüeh*
'Bright window dust' (motes in sunbeams), 73, 149–50
Brine, and preservation of dried meat, 104
Bromine (*hsiu*), 259
Bronze, 129, 219
  changed into gold, 161, 164
  high-tin, 244 (e)
Bronzing, 103
'Bronzing dips', 147
Brownian motion, 151
Buddha images, gilding of, 175
Buddhism and Buddhists, 114 (a), 116, 120, 139, 149, 175, 176, 208, 219, 224, 255
  and Indian alchemy, 160–7
  practice of self-mutilation, 165
  Shingon, 177
  Taoist influence in, 191
  Theravadin, 166
Buddhist chemical technical terms, 255, 259, 261
Buddhist monks, 139, 160, 174, 193–4
  persecution of, 165
'Bumping' during boiling, 160 (a)
Bureau of Foreign Translations, 252
Bureaucratic feudalism, xxvi
Burial customs, 2–3

Burmese alchemy, 166–7
Burning-mirror, 71

Calamine, 234
Calcareous spar, 14
Calcite, 137
Calcium acetate, 236
Calcium carbonate, 133, 243 (d)
  stalactitic, 14
  trigonal, 14
Calcium polysulphides, 103
Calcium sulphate, 88, 89 (b), 131
  hydrated, 14
Calendrical science, 177
Calomel, 123–8, 131, 133, 198
  in India, 124
  sublimation of, 137
  technical terms. See Technical terms
'Caloric', 247
Camphor, 170, 234
Cannon casting, 237, 240–1
Canon of the Great Peace and Equality. See *Thai Phing Ching*
Canon of the Mysterious Girl. See *Hsüan Nü Ching*
Canon of the Tao and its Virtue. See *Tao Tê Ching*
Canton, 252
Capitalism, xxviii, 214
'Capitalists' chapter of the *Shih Chi*, 6
Carapaces, 210
Carbon, 79, 159, 244
  reducing power of, 17, 243
  technical term, 260
'Casing' processes, 196, 207 (g)
Cassia. See Plant genera, *Cinnamomum*
Categories, theories of, 69, 74, 145, 146, 201
Caustic lye, 137
Cavendish, Henry (chemist, +1731 to 1810), 221
'Caves of cinnabar', 6
Celibacy, 205, 214
Cement, 233
Cementation, 49
Ceramics. See Porcelain
Cereal method for becoming an immortal, 37
Cereal wines, distillation of spirits from, 231
Cereals, abstaining from, 9 (c), 25, 29, 112, 118, 120, 175
Ceremonies. See Alchemical ceremonies
Ceruse, 15, 16, 17, 126, 127
Cerussa, 16
Cerussite, 16
Ceylon, 244 (b)
Chandragupta Maurya (Indian king, −320), 164–5
Chang Chao-Chhi (Later Han adept), 43
Chang Chen-Jen Chin Shih Ling-Sha Lun (A Discourse on Metals, Minerals and Cinnabar by the Adept Chang), 143
Chang Chêng-Chhang (Yuan and Ming alchemist, c. +1391), 209
Chang Chhang (Han Confucian, −1st century), 35

Chang Chi-Hsien (Taoist at the court of Hui Tsung, +1105), 190
Chang Chih-Fang (Thang general), 174
Chang Chün-Fang (Taoist editor and bibliographer, c. +1020), 115, 142, 196
Chang Chung (Chin adept, fl. +307), 112
Chang family (hereditary Taoist patriarchs), 44
Chang Hsü-Pai (Taoist at the court of Hui Tsung, +1118), 191
Chang Hsüan (Ming novelist, +17th century), 213
Chang Hsüan-Tê (Thang alchemical writer), 172, 173
Chang Kai-Tha (alchemist), 96 (f)
Chang Khai (magician), 105
Chang Kuo (Thang alchemist, c. +685 to +756), 141, 142
Chang Kuo-Hsiang (Ming compiler, +1607), 44, 117
Chang Liang (adviser to Han Kao Tsu and aspirant to hsien-ship, d. −187), 20, 44
Chang Ling. See Chang Tao-Ling
Chang Mu (Sung physiological alchemist), 205
Chang Pang-Chi (Sung writer, +1131), 185
Chang Po-Tuan (Taoist physiological alchemist, d. +1082), 200–1, 203, 208
Chang San-Fêng (Ming physiological alchemist, c. +1391), 209
Chang San-Fêng's Instructions about Enchymomas. See San-Fêng Tan Chüeh
Chang Tao-Ling (Taoist theocrat, fl. c. +156), 20, 43–4, 96 (c), 118, 190, 209, 212
Chang Thien-Yü (Yuan Taoist), 43
Chang Tzu-Chhiao (court scholar, −1st century), 13
Chang Tzu-Ho's elixir method, 92
Chang Tzu-Kao (2), 37, 79; (3), 79; (5), 239
Chang Wu-Mêng (Sung alchemist, +10th century), 195, 196, 200
Chang Yao (Professor of Macrobiotics in Northern Wei, c. +400), 118
Chang Yen-Phien (Ming Taoist), 212
Chang Yin-Chü (Thang alchemist, c. +713 to c. +742), 143–4
Chang Yü-Chhu (Ming editor, +15th century), 44, 116
Chang Yung (Governor of I-chou in Sung, +995), 192, 228 (e), 230
Chang Yung-Tê (Sung alchemist, +991), 188
Change and decay, control of. See Putrefaction, preservation from
Chao (Prince of Yen State, r. −311 to −278), 13
Chao Chi. See Hui Tsung
Chao Chieh-Hao (Han woman alchemist, c. −95), 39
Chao Hsüan-Lang (putative founder of Taoism and ancestral patron of the Sung emperors), 183
Chao Hsüeh-Min (Chhing scientific encyclopaedist, c. +1760), 239
Chao Kuang-Hsin (Chin alchemist, d. c. +345), 112–13

Chao Kuei-Chen (Thang alchemist and Taoist, d. +846), 166
Chao Ming (magician), 105
Chao Nai-An (Thang alchemical compiler, c. +808), 152, 158–9, 171
Chao Shêng (Han alchemist), 43
Chao Tao-I (Taoist hagiographer, probably Yuan), 19, 20, 40, 43
Chao Yeh (Han writer, +1st century), 5
Chao Yu-Chhin (Sung alchemist), 206
Charcoal, 17, 101, 102, 138, 243, 246
Charlatans, 106, 212–13, 215
Charms. See Talismans
Chem- (etymol. root), 255
Chemical affinity, 68, 145
Chemical change, 44, 103, 111, 138, 197
Chemical elements, 260
  discovery of, 246
Chemical industry, 177, 231, 262
Chemical names, 250
Chemical papyri, 137
Chemical reactants, cyclical heating and cooling of, 60
Chemical reactions, 62 (d), 197
  products of, 210
Chemical substances, 130, 155, 234
  division into male and female, 166
  familiar in Western science, 250
  identification of, 138, 157–8
  industrial, 219
  lethal effects of, 135
  lists of, 12, 14, 152
  preparation of, from minerals, 195
  symbols for, never developed by Chinese alchemists, 152
  technical terms. See Technical terms
Chemical technology, 15, 229, 230, 236
Chemical terminology in China, 255, 259–62
  See also Technical terms
Chemical theory, and sexual phenomena, 40
Chemistry, xxxi, 79, 81, 85, 167, 207, 216, 218, 236, 242, 246
  empirical, 229, 231
  inhibiting factors on the spread of, in China, 74–5
  military, 240
  pneumatic, 246
  Western. See Western chemistry
Chemistry (modern), 220–1, 246, 252, 261
  first book in Chinese dealing with, 250, 252, 255
  first book in Japanese dealing with, 255
  Jesuits' failure to bring modern chemistry to China, 221–2, 241–2
  terminology. See Technical terms
Chen Chung Chi (Records of the Pillow-Book), 110, 135, 146
Chen Chung Hung Pao (Infinite Treasure Pillow-Book), 38
Chen jen (perfected immortal, adept, lit. 'true man'), 18 (a)
Chen Kao (Declarations of Perfected Immortals), 120

Chen Tsung (Sung emperor, r. +998 to +1022), 115, 183–5, 186, 196

Chen yen, school of Buddhism. *See* Shingon

Chen Yuan Miao Tao Yao Lüeh (Classified Essentials of the Mysterious Tao of the True Origin (of Things)), 78, 159

Chêng Ao (Thang and Wu Tai elixir-seeker, +866 to +939), 180

Chêng Chhiao (Sung bibliographer, *c.* +1150), 130, 158

Chêng i ('Perfect Unity'), 113

Chêng Lei Pên Tshao (Classified Pharmaceutical Natural History), 158

Chêng Po-Chhiao (Warring States adept), 13

Chêng Tao Pi Shu Shih Chung (Ten Types of Secret Books on the Verification of the Tao), 219

Chêng-Thung Tao Tsang (Taoist Patrology of the Chêng-Thung reign period), 44, 116, 117, 216

Chêng Yang Tzu (Thang or Sung adept), 200

Chêng Yin (Chin alchemist, teacher of Ko Hung), 77, 78, 79, 80, 82, 121

du Chesne, Joseph (iatrochemist, +1521 to +1609), 124

Chessmen, magnetised, 32, 105

Chha nü, 56

Chhang-an (Sian), 112, 169

Chhang Chen Tzu. *See* Than Chhu-Tuan

Chhang Chhun Tzu. *See* Chhiu Chhu-Chi

Chhang-Chhun Tzu Phan-hsi Chi (Chhiu Chhang-Chhun's Collected (Poems) at Phan-hsi), 205

Chhang O (legendary, wife of Yi the Archer), 4, 21

Chhang Shêng Tzu (Later Han adept), 43

Chhang Shêng Tzu. *See* Liu Chhu-Hsüan

Mr Chhen's elixir method, 92

Chhen Chih-Hsü (Yuan Taoist writer and physiological alchemist, +14th century), 51, 54, 57, 206, 208

Chhen Ching-Yuan (Taoist writer, *fl.* early +11th century), 196

Chhen Chung-Liang (Prefect of Fêng-hsiang, +11th century), 193–4

Chhen Hsi-Liang (Sung scholar and official, d. *c.* +1067), 193

Chhen Hsien-Wei (Sung commentator, +1234), 54, 64

Chhen Hsin-Chia (Ming Minister of War, +1642), 240

Chhen Kuan-Wu. *See* Chhen Chih-Hsü

Chhen Kuo-Fu (1), 57, 79, 81, 198, 247 (e)

Chhen Nan (Taoist alchemist, +12th century), 203

Chhen Pao-Chih (Chêng-I hsien-sêng, adept, +472 to +549), 112

Chhen Phing (adviser of Han Kao Tsu and aspirant to *hsien*-ship, d. −178), 20

Chhen Shao-Wei (Thang alchemical writer, *fl. c.* +712), 121, 142, 143

Chhen Thuan (Taoist philosopher and mutationist, d. +989), 194, 200

Chhen Tshang-Chhi (pharmaceutical naturalist, +739), 31 (a)

Chhen Tsun-Kuei (historian of astronomy), xx

Chhen Yao-Tso (Sung civil official, +1008), 115

Chhen Yü (Thang alchemist, +17th century), 121

Chhen Yün-Shêng (Wu Tai alchemist, *fl.* +904 to +943), 173

Chhêng Liao-I (Sung writer on physiological alchemy, *c.* +1020), 197

Chhêng Lin & Wang Chung-Hsiang, 252

Chhêng Thang (legendary emperor), 196

Chhêng Ti (Han emperor, r. −32 to −7), 36

Chhêng Tsu (Ming emperor), 116, 209

Chhêng-tu, 117

Chhêng Wang (Chou king), 3

Chhêng Wei (Han alchemist), 38–9, 40, 100

Chhêngkung Hsing (Taoist teacher of Khou Chhien-Chih, *c.* +400), 118

Chhi (vapour or emanation), 149, 150
circulation of, 196
of primary vitality, 154–5

Chhi (State), 13, 31, 34, 47

Chhi Fan Ling Sha Lun (Discourse on Sevenfold Cyclically-Transformed Numinous Cinnabar), 142

Chhi Kuo Khao (Investigations of the Seven (Warring) States), 125

Mr Chhi-Li's elixir method, 88, 92

Chhi Lu (*Hsien Tao Lu*) (Taoist Section of the Bibliography of the Seven Classes of Books), 113

Chhi Lun (Sung civil official, +1008), 115

Chhi Min Yao Shu (Important Arts for the People's Welfare), 126

Chhien Chin I Fang (Supplement to the *Thousand Golden Remedies*), 135

Chhien Chin Yao Fang (Thousand Golden Remedies), 178 (d)

Chhien Han Shu (History of the Former Han Dynasty), 13, 20, 23, 24, 25, 26–7, 28, 35, 36, 37, 113

Chhien hung ('lead and mercury'), 78

Chhien Hung Chia Kêng Chih Pao Chi Chhêng (Complete Compendium on the Perfect Treasure of Lead, Mercury, Wood and Metal), 152, 158, 160

Chhien Khun Pi Yün (The Hidden Casket of Yin and Yang Opened), 211

Chhien Khun Shêng I (Principles of the Coming into Being of Yin and Yang), 211

Chhien Yang Tzu. *See* Yü Mu-Shun

Chhih Fu (legendary immortal), 9

Chhih shih chih. See Red bole clay

Chhih Sung Tzu (the Red Pine Master), 9, 10, 20, 88, 105 (b)
his elixir method, 91

Chhin dynasty, 4, 5, 44
tombs, 16

Chhin (State), 125

Chhin Chhêng-Tsu (Director-General of Medical Services and Professor of Medicine in the Liu Sung, *fl.* +420 to +470), 45

Chhin Shih Huang Ti (First Emperor, d. −210), 2, 4, 8, 11, 13, 17–19, 27, 28, 33, 34–5, 37, 49, 52, 197, 225, 229

Chhing (Szechuanese widow and owner of mercury mines, c. −245 to −210), 6

Chhing Hsia Tzu (philosophical name of more than one alchemical writer), 130, 159, 180, 211

Chhing Hsiang Tsa Chi (Miscellaneous Records on Green Bamboo Tablets), 186

Chhing Hsü Tzu. *See* Chao Nai-An

Chhing Lin Tzu (Master Chhing-Lin, Han or San Kuo alchemist), 95, 103

Chhiu ('searching for'), 44

Chhiu Chhang-Chhun Chhing Thien Ko (Chhiu Chhang-Chhun's Song of the Blue Heavens), 204

Chhiu Chhang-Chhun's Collected (Poems) at Phan-hsi. See *Chhang-Chhun Tzu Phan-hsi Chi*

Chhiu Chhu-Chi (fourth of the Seven Masters of the Northern School of Taoism, +1148 to +1227), 204–5

Chhiu shih ('autumn mineral', a purified mixture of urinary steroid hormones), 78

Chhü Hsien Shen Yin Shu (Book of Daily Occupations for Scholars in Rural Retirement, by the Emaciated Immortal), 211

Chhu Hua-Ku (alchemical commentator, c. +1230), 54

Chhü I Shuo Tsuan (Discussions on the Dispersal of Doubts), 206

Chhü Thai-Su (Christian scientific scholar, +1605), 224, 247

Chhu-Tsê hsien-sêng (Chhi and Liang alchemist, c. +5th century), 130

Chhu Tzhu (Elegies of Chhu (State)), 23

Chhu Yung (Sung scholar, c. +1230), 206–8

Chhüan-Lo. *See* Kwŏllŭk

Chhun Chu Chi Wên (Record of Things Heard at Spring Island), 193

Chhung Fang (Sung adept, c. early +11th century), 200

Chhung-Hsü Kuan (Temple of the Vast Inane, a Taoist abbey in the Lo-fou Shan mtns., associated with Ko Hung), Fig. 1353 a, caption

Chhung Shang (Warring States adept), 13

Chhung tao (repeated transmutation, a technique for prolonging life), 14

Chhung-Yang Chhüan Chen Chi (Wang Chhung-Yang's Records of the Perfect-Truth School), 203

Chhung-Yang Chiao Hua Chi (Memorials of Wang Chhung-Yang's Preaching), 204

Chhung-Yang Chin-Kuan Yü-So Chüeh (Wang Chhung-Yang's Instructions on the Golden Gate and the Lock of Jade), 204

Chhung-Yang Fên-Li Shih-Hua Chi (Writings of Wang Chhung-Yang (to commemorate the time when he received a daily) Ration of Pears, and the Ten Precepts of his Teacher), 204

Chhung-Yang Li-Chiao Shih-Wu Lun (Fifteen Discourses of Wang Chhung-Yang on the Establishment of his School), 204

Chhung Yang Tzu. *See* Wang Chung-Fu

Chi Chhiu Tzu (Master Chi-Chhiu, Han or San Kuo alchemist), 88, 95
his elixir method, 92

Chi Han (Chin botanist, and Governor of Kuang-chou, +306), 80

Chi I Tzu. *See* Chin-Chhüan

Chi Ni Tzu. *See* Fan Li

Chi Ni Tzu (The Book of Master Chi Ni), 11, 14, 15, 16, 17, 127

Chi Yen (Chi Ni Tzu, naturalist philosopher, late −4th or early −3rd century), 6, 14

Chia I (Han Confucian statesman, *fl.* −175), 28

Chia Sung (Thang writer), 120

Chiang Fu (Thang adept), 147

Chiang Huai I Jen Lu (Records of (Twenty-five) Strange Magician-Technicians between the Yangtze and the Huai River, during the Thang, Wu and Southern Thang Dynasties), 169

Chiang Shu-Mou (Chhin aspirant to *hsien*-ship, −3rd century), 19

Chiang Yen (Liu Sung poet, +444 to +505), 52

Chiangnan Arsenal. *See* Kiangnan Arsenal

Chiangsi (province), 169

Chiangsu (province), 168 (a)

Chiao-chih (Annam), 81

Chiao Hsü (Ming gunner, +1643), 241

Chiao Kan (Han writer, c. −40), 6

Chiao-Kuang Chen Jen (Northern Chou adept, +6th century), 112

Chien (naturally occurring soda, a mixture of salts), 232, 234

Chien-Chen (Kanshin, Buddhist monk, +688 to +763), 177

Chien Wu Chi (On the Gradual Understanding (of the Tao)), 204

Chih- Chhuan Chen Jen Chiao Chêng Shu (Technical Methods of the Adept (Ko) Chih-Chhuan, with Critical Annotations), 110

Chih Fa-Lin (Indian monastic traveller, +664), 139

Chih Hsüan Phien (On the Demonstration of the Mystery), 194

Chih I Tzu. *See* Chao Nai-An

Chih Kuei Chi (Pointing the Way Home to Life Eternal), 198–9

Chih-Tshung (Chisō, Buddhist monk, +562), 177

Chihli I (Warring States adept), 7

Chikashige Masumi (1), 84

'Child's-play method of making gold', 84 (f), 95

Chile, 232

Chin (dynasty), 113, 198, 199

Chin (gold or metal, an elixir ingredient), 73, 87

Chin Chu-Pho (Thang alchemical writer), 172

Chin hua, 96 (y)

Chin Hua Chhung-Pi Tan Ching Pi Chih (Confidential Hints on the Manual of the Heaven-Piercing Golden Flower Elixir), 199, **200**, 203, 208

Chin Hua Yü Nü Shao Tan Ching (Elixir Manual according to the Teaching of the Jade Girl on the Golden Flower), 197

Chin i. See Potable gold

Chin I Ching (Manual of Potable Gold), 96 (bb)

Chin i hua shen tan (elixir), 179 (a)

Chin I Tan Ching (Manual of the Potable Gold Elixir), 78

Chin Ku Chhi Kuan (Strange Tales New and Old), 175, 212, 213, 230

Chin kung, 56

Teacher Chin Lou (Chin alchemist), 95

Chin Lou Tzu. *See* Hsiao I

Chin Lou Tzu (Book of the Golden Hall Master), 18

Chin Mu Wan Ling Lun (Essay on the Tens of Thousands of Efficacious (Substances) among Metals and Plants), 109

Chin Pi Ching (Gold and Caerulean Jade Manual), 150

Chin Pi Wu Hsiang Lei Tshan Thung Chhi (Golden Jade Treatise on the Similarities and Categories of the Five (Substances) and the *Kinship of the Three*), 51, 150

Chin Shih Pu Wu Chiu Shu Chüeh (Explanation of the Inventory of Metals and Minerals according to the Numbers Five and Nine), 138, 140

Chin Shu (History of the Chin Dynasty), 76, 79, 97 (cc), 112

Chin Tan Ssu-Pai Tzu (Four Hundred Word Epitome of the Metallous Enchymoma), 200

Chin Tan Ta Yao (Main Essentials of the Metallous Enchymoma; the True Gold Elixir), 206, 208

Chin Tan Ta Yao Thu (Illustrations for the *Main Essentials* . . .), 206, 208

Chinese culture
attempted 'debunking' of, 228
and the failure to develop modern science, xxv–xxvi, xxvii, 219

Chinese and Japanese Universal Encyclopaedia. See *Wakan Sanzai Zue*

Chinese language, and the development of modern chemical science, 261–2

Ching-chou, tribute products from, 4

Ching Hsiang (Prince of Chhu, r. −294 to −261), 7

Ching-tê-chen, centre of the porcelain industry, 230 (b)

Ching Ti (Han emperor, r. −156 to −141), 26

Ching Tsung (Thang emperor, r. +825 to +826), 151, 182

Chingiz Khan, 205

Chinhŭng (King of Silla (Korea), r. +540 to +575), 167

Chiu Chuan Liu Chu Shen Hsien Chiu Tan Ching (Manual of the Nine Elixirs of the Holy Immortals and of the Ninefold Cyclically-Transformed Mercury), 85

Chiu Huang Pên Tshao (Natural History of Emergency Food Plants), 209–10

Chiu Ling Tzu. *See* Huang Hua

Chiu Thang Shu (Old History of the Thang Dynasty), 132, 169

Chiu Ting Tan Ching (Manual of the Nine Reaction-Vessels Elixir), 78

Chiu Yuan Tzu (Later Han adept), 43

Chŏng Yutha (Korean Pharmacognostic Master, in Japan, +554), 177

Chou (period), 5, 6, 44, 46
burial customs, 3

Chou (Shang High King), 17

Chou Chi-Thung (Han adept and writer, c. −60), 40

Chou I Tshan Thung Chhi (The Kinship of the Three; or, the Accordance of the *Book of Changes* with the Phenomena of Composite Things), 51

Chou I Tshan Thung Chhi Chieh (The *Kinship of the Three* with Explanations), 54, 64

Chou I Tshan Thung Chhi Chu (The *Kinship of the Three* with Commentary by Yin Chhang-Shêng), 53

Chou I Tshan Thung Chhi Chu. Alternative title for *Tshan Thung Chhi Khao I* (Chu Hsi's) q.v.

Chou I Tshan Thung Chhi Fa Hui (Elucidation of the *Kinship of the Three*), 54, 65

Chou I Tshan Thung Chhi Fên Chang Chu (Commentary on the *Kinship of the Three* divided into short chapters), 54, 206

Chou I Tshan Thung Chhi Fên Chang Thung Chen I (The *Kinship of the Three* divided into (short) chapters for the Understanding of its Real Meanings), 53

Chou I Tshan Thung Chhi Khao I (Investigations and Criticisms of the *Kinship of the Three*), 53

Chou I Tshan Thung Chhi Shih (Clarification of Doubtful Matters in the *Kinship of the Three*), 54

Chou I Tshan Thung Chhi Shu Lüeh (Brief Explanation of the *Kinship of the Three*), 54

Chou I Tshan Thung Chhi Ting Chhi Ko Ming Ching Thu (An Illuminating Chart for the Mnemonic Rhymes about the Reaction-Vessels in the *Kinship of the Three* and the *Book of Changes*), 53, 163

Chou I Tshan Thung Chhi Tshê Su (Penetrating Explanation of the *Kinship of the Three*), 54–5

Chou-khou-tien, remains of palaeolithic man at, 3

Chou Tan (Imperial Physician in Northern Wei, c. +400), 118

Chou Ting Wang. *See* Chu Hsiao

Chou Tzu-Liang (Taoist adept, +515), 121

Christianity, 230
and Confucianism, 221, 223
and modern science, xxvi, xxvii, xxviii

Chromatography, 48 (c)

Chronicles of Japan. See *Nihongi*

Chronicles of Japan, continued. See *Shoku-Nihongi*

Chronicles of Japan, still further continued. See *Shoku-Nihonkoki*

Chrysanthemum, as elixir ingredient, 113

Chrysopoia (the making of the perfect metal, gold), 57

Chu Chhüan (Ning Hsien Wang, Prince of the Ming, naturalist, metallurgist and alchemist, +1390 to +1448), 210–12

Chu Chia Shen Phin Tan Fa (Methods of the Various Schools for Magical Elixir Preparations), 109–10, 137, 159, 197

Chu Chu (legendary Taoist alchemist), 96 (v), 127 his elixir method, 92

Chu Chü (physiological alchemist, d. +1242), 202

Chu Hsi (Neo-Confucian philosopher, +1130 to +1200), 51, 53, 56, 183

Chu Hsiao (Chou Ting Wang, Prince of the Ming, naturalist and botanist, c. +1380 to +1425), 209

Chu hsien-sêng (Mr Chu, Han alchemist, +2nd century), 76, 77

Chu I-Chhien (Sung Taoist, +1008), 115

Chu Ju-Tzu (Chin adept, d. c. +345), 113

Chu Phing-Man, 7–8

Chü-shêng (plant), 72, 97 (cc)

Chu Yuan-Yü (Chhing alchemical commentator, +1669), 54, 71

Chuang Chou (Warring States philosopher, −4th century), xxiii, 111

Chuang Tzu (The Book of Master Chuang), 7–8

Chüeh Tung Tzu. *See* Li Hsiu

Chung Hua Ku Chin Chu (Commentary on Things Old and New in China), 125

Chymical Elaboratory Practice. See *Tan Fang Hsü Chih*

Cibot, P. M. (Jesuit, +1774), 221, 222, 230–1, 232, 233, 234, 248

Cicada

of jade, placed in the mouth of the dead, 3

metamorphosis like (i.e. *shih chieh* of corpse-free immortals), 112

Cinnabar, 3, 6, 19, 57, 66, 68, 76, 79, 81, 85, 86–7, 127, 133, 171, 201, 217–18, 219, 231

antiquity of, 4

as an elixir ingredient, 87, 89, 98, 112–13, 131, 149, 173; taken alone, 142, 144, 196

grades of, 142

ingestion of, 78, 111, 119, 142

medicinal use of, 46, 150–1, 231

mercury and, 25, 104, 111, 142–3

use for red ink, 6

technical terms. *See* Technical terms

transformation into gold, 29, 30, 102, 104, 198

'Cinnabar water', 24

Cinnamon, 9, 11

Circulation of the *chhi* of primary vitality, 154

Citric acid, 49

Civil Service of the Han Dynasty and its Regulations. See *Han Kuan I*

Clamshell calcite, 136

Clarification of Doubtful Matters in the *Kinship*

of the Three. See *Chou I Tshan Thung Chhi Shih I*

Class-distinctions, and alchemy, 208

Classic of the Mountains and Rivers. See *Shan Hai Ching*

Classified Essentials of the Mysterious Tao of the True Origin (of Things). See *Chen Yuan Miao Tao Yao Lüeh*

Classified Historical Matters about Fu-Sang (Japan). See *Fusō Ryakuki*

Classified Matters from the Chronicles of Japan. See *Nihongi Ryaku*

Classified Pharmaceutical Natural History. See *Chêng Lei Pên Tshao*

Clays, 12, 14, 15, 84, 131, 133

coloured, ingestion of, 12

Clepsydra, xxxiii

Clockwork, xxx, xxxiii, 115–16

'Cloud-and-frost elixir', 20

Coal, recognition of plant origin of, 97 (cc)

industry in China, 233–4

Coffins, 5 (k)

of the Lady of Tai, 21–3

Coinage

counterfeit, 28

debasement of, 26, 28

minting of, 26, 27, 28

silver and tin, 28

in Wang Mang's time, 37

Coining Edict. *See* Edicts

Coke, manufacture of, 234

'Cold elixir' (*han tan*), 83

Collas, J. P. L. (Jesuit, +1774), 221, 222, 230–1, 232, 233, 234, 248

Collected Commentaries on the Classical Pharmacopoeia of the Heavenly Husbandman. See *Pên Tshao Ching Chi Chu*

Collected Miscellany of Li Shang-Yin. See *I-Shan Tso Tsuan*

Collected (Poems) on the Happiness of the Holy Immortals. See *Hsien Lo Chi*

Collected (Poems) of Light (through the Clouds). See *Yün Kuang Chi*

Collected (Poems) on the Spontaneity of Nature. See *Tzu-Jan Chi*

Collected Works of (Ho) Thai-Ku. See *Thai-Ku Chi*

Collection of Procedures of the Golden Art. See *Kêng Tao Chi*

Collection of Ten Books on the Regeneration of the Primary Vitalities. See *Hsiu Chen Shih Shu*

Collections of Gold and Jade; a Tung-Hsüan Scripture. See *Tung-Hsüan Chin Yü Chi*

College of All Sages, 185 (a)

Colloidal gold. *See* Gold

Colour-changes of indicators, 9 (e)

Colours

appearing in chemical operations, 9 (e), 196

social valuation of, 5

symbolic, 67

Combustion, 246–7

Commentary on the Great Sūtra of the Perfection of Wisdom. See *Mahā-prajñāpāramito-padeśa Śāstra*
Commentary on the *Kinship of the Three* divided into short chapters. See *Chou I Tshan Thung Chhi Fên Chang Chu*
Commentary on Things Old and New in China. See *Chung Hua Ku Chin Chu*
Compassion, in Buddhist thought, 165
Complete Book of the History of Annam. See *Đai-Viêt Sú-ký Toàn-thú*
Complete Collection of the Biographies of the Immortals. See *Lieh Hsien Chhüan Chuan*
Complete Compendium on the Perfect Treasure of Lead, Mercury, Wood and Metal. See *Chhien Hung Chia Kêng Chih Pao Chi Chhêng*
Comprehensive Mirror of the Embodiment of the Tao by Adepts and Immortals throughout History. See *Li-Shih Chen Hsien Thi Tao Thung Chien*
Condensation, reflux, 196
Condensation-aggregation (*kuei*), 196
Condenser (cooling coil), 259
Conference on the principles of administration (−81), 34
Confidential Hints on the Manual of the Heaven-Piercing Golden Flower Elixir. See *Chin Hua Chhung-Pi Tan Ching Pi Chih*
Confucian classics, 229
 'discovery' of ancient-script version of, 55
 first translation into a Western language, 223
Confucian morality, 168, 194
Confucianism, 108, 165, 190, 209, 225, 248
 and Christianity, 221, 223
 and the cult of immortality, 34–5, 37, 53, 107
 disdain for artisanal technology, 219
Confucius, veneration of, 221
Confucius Sinarum Philosophus, 223
Conjunctio oppositorum. See Contraries, marriage of
Conjuring tricks, 171, 227
Consecration by contact, 38 (b)
Contagion, 38 (b)
Contraries, marriage of, 69, 145, 148 (g), 149, 166 (d)
'Controlling', of chemicals, 158 (c). See also 'Killing' and 'Subduing'
Cooling systems, 199, 259
Copernicus, xxvii
Copper, 88, 101, 140, 147, 233, 244
 corrosion of, 244 (f), 245, 247
 as elixir ingredient, 78, 87
 'gilding' of, by zinc–mercury amalgam, 31 (b)
 green flame of, 140
 mining. See Mines and mining
 pigments from, 244–5
 resources, nationalisation of, 28
 rusting of, 244–5
 salts of, 163
 wet precipitation of, from solutions by iron, 25 (h), 104, 190 (b), 207 (e)

Copper acetate, 136
Copper anions, 102
Copper–arsenic alloys. See Alloys and alloying
Copper carbonate, 15, 78, 102, 136, 244
Copper sulphate, 102, 143, 196, 243 (d)
Copper vessels, 87
Copperas, 89, 233, 237–8, 239
Cornelian Law against coining and aurifiction (−81), 29 (a)
Corpse-free (*shih chieh*) immortals, 21, 112
Corrosion, of copper, 245, 247
Corrosive sublimate, 123, 128, 233
 technical terms. See Technical terms
Cosmetics, 16, 17, 125–7
Cosmogonic enumeration order of the Five Elements, 181 (b)
Counter-current flow, xx
Counterfeit things, 104
Couplet, Philippe (Jesuit, +1624 to +1692), 223
Cover-names, 82 (d), 104, 148 (g), 152 ff.
Cowdry, E. V. (American anatomist), 250
Crane birds, 21, 23 (e)
Cranmer-Byng, J. L., 234
Creation, Hebrew-Christian account of, 235
 ex nihilo, 210 (i)
Criminals, used in elixir experiments, 118
Croll, Oswald (chemist, *c.* +1580 to +1609), 236
Cronstedt, A. F. (chemist, +1702 to 1765), 233
Crow in the sun, 21
'Crow's-beak gold', 188
Cryptogamic hallucinogens, 8 (g)
Crystallisation, 25, 120 (a), 232
Cupellation, xxxiii, 36, 39 (a), 73, 84, 140, 189
Cupric chloride, 245 (e)
Cupro-nickel, 39 (b), 136, 193, 233, 234
Cyanides, 88 ff., 98, 147
Cyclical change, xxiii
Cyclical charts, 62–5
Cyclical signs, 59, 60
Cyclical transformations method, 14 (a), 126, 198, 250, 261
Cyclically-transformed elixir. See Elixirs of longevity and immortality
Cyprus, 49

Đai-Viêt Sú-ký Toàn-thú (The Complete Book of the History of Annam), 75
Daigo (Japanese emperor, r. +897 to +930), 179
Dalton, John (chemist, +1766 to 1844), 218, 221
Davis, Tenney L. (1), 124
Davis, Tenney L. & Chao Yün-Tshung, 200
Davis, Tenney L. & Chhen Kuo-Fu (1), 81
Davis, Tenney L. & Wu Lu-Chhiang (3), 123
De Aluminibus et Salibus, 124 (a)
De Re Literaria Sinensium (Spizel), 7 (d)
De Re Metallica (Agricola), 235
De Simplicium Medicamentorum Temperamentis (Galen), 16 (d)
Dead bodies, preservation of. See Putrefaction, preservation from
Death. See Temporary death
Debility (*lêng*), 129

Declarations of Perfected Immortals. See *Chen Kao*

Decline of the West (Spengler), xxiii

Deflagrating mixture, 138

The Deified Adept Yin (Chhang-Shêng's Book) on the Similarities and Categories of the Five (Substances) among Metals and Minerals. See *Yin Chen Chün Chin Shih Wu Hsiang Lei*

Demonifuge properties of mirrors, 106, Fig. 1362*a* caption

Dew-mirrors, 31 (a)

Dhammazedi (Burmese king, r. +1460 onwards), 167

Dietary regimens, and the attainment of longevity and material immortality, 1, 8, 9, 20, 114

Digestion methods (*shui fa*), 78 (d)

Dilations upon Pharmaceutical Natural History. See *Pên Tshao Yen I*

Diminution in number of substances used in elixirs, 182, 208, 217–19, 231

Diocletian decree against aurification (+296), 29 (a)

Dioscorides (+2nd-century pharmacist and physician), 16, 238

Diplōsis. See Doubling

Direct Hints on the Great Elixir. See *Ta Tan Chih Chih*

Discourse of Nāgārjuna on Eye (Diseases). See *Lung-Shu Yen Lun*

Discourse on the Cooling Regimen Powder. See *Han Shih San Lun*

Discourse on the Great Elixir of Lead and Mercury. See *Ta Tan Chhien Hung Lun*

Discourse on Metals, Minerals and Cinnabar by the Adept Chang. See *Chang Chen Jen Chin Shih Ling-Sha Lun*

Discourse on the Precious Treasury of the Earth. See *Pao Tsang Lun*

Discourse on the Sevenfold Cyclically-Transformed Cinnabar. See *Chhi Fan Ling Sha Lun*

Discourse of the Yellow Emperor on the Contents of the Precious Treasury of the Earth. See *Hsien-Yuan Pao Tsang Lun*

Discourses on Salt and Iron. See *Yen Thieh Lun*

Discourses weighed in the Balance. See *Lun Hêng*

Discussion on (the Use of) Mineral Drugs. See *Lun Shih*

Discussions on the Dispersal of Doubts. See *Chhü I Shuo Tsuan*

Discussions with Friends at Cloudy Pool. See *Yün Chhi Yu I*

Diseases, 45, 46, 47, 130, 163, 231

Disinfection, 46 (c)

Disputations on Doubtful Matters. See *Pien Huo Phien*

Distillation, 25, 26 (a), 68 (b), 103, 220, 231, 250
  apparatus for, 199, 229 (b), 253
  descensory, 143, 199
  of essential vegetable oils, 170

'Diverus and Lazarus', 23 (a)

Divination, 106, 177

Divine retribution, 106

Diviner's board, 173

Dodder, 168 (g)

Dog's bile, 84 (j), 88

Dominicans, 221

Dosage, 83, 112, 119, 135

'The Dose alone is what makes something a Poison or not' (Paracelsus), 135 (d)

Double-hours, 59, 73

Doubling (*diplōsis*, the 'debasement' of precious metals), 136, 213

Dragon, as symbol, 66, 248

Dragon-and-Tiger Book. See *Lung Hu Ching*

Dragon-slaying, 7–8

'Dragon sprout' (*lung ya*), 147

'Dragon's fat' (*lung kao*). See *Lung kao*

Dream Pool Essays. See *Mêng Chhi Pi Than*

Dreams, 183

Drowned persons, floating of, 71 (d)

Drugs, 82, 177
  of immortality. *See* Elixirs of longevity and immortality
  mineral, 47–8, 135, 163, 168
  plant. *See* Plant drugs
  rare, 116
  in the Shōsōin, Nara, 178
  sulpha-, 47

'Dual-cultivation', 200 (a)

Dualism, Persian, 72

Dubs, H. H., 4, 37

Duckworth, C. W. (1), 245

Dung, 16 (d), 78

Dutch vinegar process, 15, 16, 17, 124

Dye-plants, 126

'Dyeing', 136

Dyers, 83

Dyer's bugloss, 126 (d)

Early Han (dynasty, −206 to +24), 4, 6, 17, 36, 44, 46

Earth, sphericity of, 218

Earthworms, excreta of, 133 (d)

Eating and drinking from vessels of alchemical gold or silver, as way to immortality, 31, 49, 212

'Edible elixir' (*erh tan*), 83

Edicts
  against Buddhist monks and nuns, 165–6
  anti-coining (−144), 26, 29 (a), 33, 35 (d), 46
  ordering the burning of Taoist books (+1281), 116

Egyptians, 16, 100 (a), 136, 165

The Eight Immortals, 148

The Eight Venerable Adepts. *See* Pa Kung

Electrical science, xxv, 246, 249

Electro-chemical series of the elements, 145

Electrum spear-head, 49

'Elegant Girl by the Riverside' (*ho shang chha nü*), 68, 104, 157

Elegies of Chhu (State). See *Chhu Tzhu*

Elémens de Chymie (Macquer), 230–1

Elements. *See* Five-Element theory
'Elixir flower' (*tan hua*), 83
Elixir ingredients, 9, 11, 76, 112, 144–5, 147
Elixir Manual according to the Teaching of the Jade Girl on the Golden Flower. See *Chin Hua Yü Nü Shuo Tan Ching*
'Elixir-mother' (*tan mu*), 206–7
Elixir poisoning, 74, 78, 120, 133, 135, 138, 144, 146, 147, 151, 168, 171, 180, 182, 185, 194, 208, 209, 212
    means of counteracting, 119, 168 (g), 208
    warnings against, 168–9, 172, 173, 185
Elixir Techniques of the Min Shan Mountains. See *Min Shan Tan Fa*
Elixirs of longevity and immortality, xxv, xxx, xxxiv, 1, 2, 4, 9, 12, 13, 20, 24, 25, 26, 31, 33, 36, 40, 43, 44, 46, 51, 68, 71, 78, 81–106, 114, 160, 166, 173, 177, 180, 191, 196, 197, 199, 201, 209, 229
    arsenical and mercurial compounds in, 131
    'cloud-and-frost', 20
    and colours, 98
    'cyclically-transformed', 20, 73, 74, 82, 83, 109, 131, 140, 194, 196
    decrease in number of substances used for making. *See* Alchemical substances
    dosage, 135
    expeditions in search of. *See* Voyages in search of . . .
    experiments with, 51, 97 (ff), 105, 118
    'fire-times', 61–2, 63, 64, 65, 171, 196, 199
    glittering (possibly phosphorescent), 148
    ingestion of, xxxiv, 75–6, 104, 111–13, 179
    in Japan, 175, 177, 178–80
    knowledge of, transmitted to the West, 223
    lists of, 34, 90 ff.
    major and minor, 82
    from mineral and metallic substances, 3, 9, 45, 47–8, 72–3, 74, 75, 76, 78, 83–4, 86–7, 88–9, 98–9, 123, 129, 130, 131, 133, 138, 142, 144, 146, 148–9, 152, 185, 207
    movement away from, 185–6
    named after adepts, 88
    overdose, 112
    poisonous. *See* Elixir poisoning
    possibly still being made in modern China, 250
    preparative methods for, 83–6, 88, 98–9, 109, 130–1, 133, 135; erroneous or dangerous, 78
    and sexual hyperactivity, 214
    technical terms. *See* Technical terms
    Wei Po-Yang on, 72
    See also Cinnabar, as an elixir ingredient; Gold, as an elixir ingredient
Elucidation of the Great Cyclically-Transformed Elixir. See *Ta Huan Tan Chao Chien*
Elucidation of the *Kinship of the Three*. See *Chou I Tshan Thung Chhi Fa Hui*
Emblica. See Plant genera, *Phyllanthus*
Embryology, 70
'Embryonic respiration', 154
En-no-Shōkaku (perhaps the founder of the *shugendō* cult, b. +634), 175

Enchymoma, xxxv, 154
Endocrinology, xxx
Ennin (Japanese monk, *c.* +840), 165–6
d'Entrecolles, F. X. (Jesuit, +1734), 230
Epigrams, 168 (c), 195–6
Epsom salt, 158
Equality of the forms of human experience, xxvi–xxvii
Erh (cakes, meat dumplings, or bait), 97 (dd)
Erh I Shih Lu (Veritable Records of Heaven and Earth), 126
Ersatz (imitation) productions, 137
Esoteric Manual of the Innermost Chamber, with Commentary. See *Thai-Shang Tung Fang Nei Ching Chu*
Essay on the Tens of Thousands of Efficacious (Substances) among Metals and Plants. See *Chin Mu Wan Ling Lun*
Essential Hints on the Preparation of Powerful Elixirs. See *Hsiu Lien Ta Tan Yao Chih*
Essential Teachings of the Manual of the Supreme-Pole Adept on the Ninefold Cyclically-Transformed Elixir. See *Thai Chi Chen-Jen Chiu Chuan Huan Tan Ching Yao Chüeh*
Essentials of the Elixir Manuals. See *Tan Ching Yao Chüeh*
Essentials of the Elixir Manuals for Oral Transmission; a Thai-Chhing Scripture. See *Thai-Chhing Tan Ching Yao Chüeh*
Essentials of Gunnery. See *Huo Kung Chhieh Yao*
Essentials of the Magnificence of the Three Heavens. See *San Tung Chhiung Kang*
Essentials of the Pharmacopoeia Ranked according to Nature and Efficacity. See *Pên Tshao Phin Hui Ching Yao*
Etching
    of glass, 233 (h)
    of metal plates, 239
Ethics, xxvii
    and alchemy, 1 (f), 101, 106, 194
    Confucian, 168, 194
    inapplicability to non-human natural things, 17 (b)
    scientific, xxviii–xxix, 227
Eucharist, 9 (e), 38 (b)
Euclidean geometry, xxv
European alchemy. *See* Western alchemy
'Evil eye', 83 (a)
Exalted Manual of the Dragon and Tiger. See *Lung Hu Shang Ching*
Examen Omnium Simplicium Medicamentorum . . . (Brasavola), 238
Excreta
    of earthworm, 133 (d)
    of man, 84 (j)
    of sparrow, 136 (f)
Executions, of miscreants, 35 (e)
Exorcism, 6, 165, 166
Expansion-dispersion (*shen*), 196
Expeditions. *See* Voyages

Experimenta Nova Magdeburgica (von Guericke), 241

Experiments and experimentation
alchemical, 82, 100, 107, 108, 117, 138, 147, 148, 182, 197, 206, 230
chemical and metallurgical, 193
of Chu Chhüan, 210
lack of theories of modern type, 219
on man and animals, 51, 97 (ff), 105, 118

'Explaining by analogy' (*ko i*), 255

Explanation of the Dragon-and-Tiger Cyclically-Transformed Elixir. See *Lung Hu Huan Tan Chüeh*

Explanation of the Fundamental Principles of Chemistry. See *Hua-Hsüeh Shan Yuan*

Explanation of the Inventory of Metals and Minerals according to the Numbers Five and Nine. See *Chin Shih Pu Wu Chiu Shu Chüeh*

Explanation of Names. See *Shih Ming*

Explanation of the Obscurities in the *Kinship of the Three*. See *Tshan Thung Chhi Chhan Yu*

Explanation of the Yellow Emperor's Manual of the Nine-Vessel Spiritual Elixir. See *Huang Ti Chiu Ting Shen Tan Ching Chüeh*

Explanatory Discourse to assist the Understanding of the Great Vehicle. See *Shê Ta Chhêng Lun Shih*

Exploitation of the Works of Nature. See *Thien Kung Khai Wu*

Explosives, xxx, 78–9, 138–9, 158–9, 184, 226, 237–8, 241

Eye diseases, 163

Face-powder, or face-cream, 4, 16, 17, 126

Face-tattooing, 28

Faeces, human, 84 (j)

Fahrenheit, G. D. (physicist, +1689 to +1736), 225 (d)

Fairy islands. *See* Magical islands

Fan Chung-Yen (Sung scholar and high official, +989 to +1052), 182, 192

Fan I Kuan. *See* Bureau of Foreign Translations

Fan Li (Warring States administrator, merchant, philosopher and perhaps alchemist, −5th century), 9, 11, 14

Fan Shu (Thang scholar and poet, *c.* +870), 167–8, 169

Fan Tzu Chi Jan. See *Chi Ni Tzu*

Fang-Chang (island of the immortals), 13

Fang hsiang shih exorcist, 23 (d)

Fang Hui (legendary immortal), 9

Fang I-Chih (Chhing scientific encyclopaedist, +1664), 218, 236, 239

Fang-shih (adepts, magicians), 17, 106

Fanyang mountain, 43

'Father Cinnamon', 9 (e)

Feathers, as substance used in chemical preparations, 210
in cloaks worn by Taoists and goddesses, 176 (h)

Feifel, E. (1), 79, 81, 88

Fên yeh system (astrological geography), 190 (c)

Fêng Chia-Lo & Collier (1), 181

Fêng Chia-Shêng, 159

Fêng Chün-Ta (the Blue Ox Master, expert on sexual techniques, +1st or +2nd century), 40, 48

Fêng Mêng-Lung (Ming novelist, d. +1646), 212–13, 214, 215

Fêng Su Thung I (The Meaning of Popular Traditions and Customs), 27

Fêng Tê-Chih (Sung Taoist, +1008), 115

Fêng Wei-Liang (Thang alchemist), 121

Ferghana, 18

Fermentation, 25, 38 (b)
industries, 231
technology, 211

Ferric oxide, 15
hydrated, 14

Ferric sulphate, 233

Ferrous sulphate. *See* Copperas

Feudalism, xxvi, xxviii

Fifteen Discourses of Wang Chhung-Yang on the Establishment of his School. See *Chhung-Yang Li-Chiao Shih-Wu Lun*

Filliozat, J. (10), 163

Filtration, 232

Fire-Drake Manual. See *Huo Lung Ching*

Fire (element), 157 (i), 172–3, 217 (f)

'Fire-times' (*huo hou*), 61–2, 63, 64, 65, 171, 196, 199

First printed book on a scientific subject in any civilisation, 167–9

First Series of a Supplement to the Taoist Patrology. See *Tao Tsang Hsü Phien Chhu Chi*

First Steps in Chemistry. See *Hua-Hsüeh Chhu Chiai*

Fish, 137

Fishing-rod, reel of, 12, 104 (g)

Five Dynasties, 180, 182

Five-Element theory, xxv, 5, 12, 52, 58, 60, 67, 72, 74, 143, 157, 171–2, 181
enumeration orders, 181 (b)

Five Minerals, 86, 96 (d), 199
therapeutic use of, 46, 47

'Five Yellow (Substances) pill', 181

'Fixed elixir' (*fu tan*), 83

Fleeting Gossip beside the River Shêng. See *Shêng Shui Yen Than Lu*

Flight ecstasy, 4, 8, 175

Flora Cochinchinensis (Loureiro), 222–3

'Flowing-pearl elixir', 40

Fo-Thu-Têng (missionary monk and thaumaturgist, *fl.* +310), 112

Folk cosmology, 19

Folklore, 4, 23
Japanese, 176

de Fontaney, Jean (Jesuit Mathematician-Royal of the King of France, +1685), 241

Forest of Symbols of the *Book of Changes*. See *I Lin*

Forgeries, literary, 249 (a)

Former Han (period). *See* Early Han (dynasty)

du Four, Vital (chemical practitioner, d. +1327), 237

The Four Ancillaries (*ssu fu*), divisions of Taoist literature, 113

Four Aristotelian causes, 235

Four Books of Shen Nung. See *Shen Nung Ssu Ching*

Four Element theories, 172, 234, 235

Four Hundred Word Epitome of the Metallous Enchymoma. See *Chin Tan Ssu-Pai Tzu*

The Four Magical (Things), 201

Four seasons, 72

de Fourcroy, A. F. (chemist, +1755 to 1809), 221, 255

Fox, 189

Franciscans, 221

Franklin, Benjamin (+1706 to +1790), 249

Freezing of water, 225–6

Freezing-out process, 147 (f)

Fréret, Nicolas (scholar and sinologist, +1688 to +1749), 224

Frescoes
Koguryǒ tombs, 167
painting of, 136
Tunhuang cave-temples, 150, 165 (d)

Fruit preservation, 211

Fryer, John (translator, 1839 to 1928), 254–5, 259

Fu. See Talismans

Fu (Han Dowager-Empress), 82

Fu Chien (Shih Tsu, emperor of the Early Chhin dynasty, r. +357 to +385), 112

Fu Chin-Chhuan (authority on Taoist physiological alchemy, early +19th century), 54, 153, 218–19

Fu-Hsi (the organiser god), 21

Fu-Hsi system (of the *Book of Changes*), 63, 65

Fu Huo Fan Fa (Method of Subduing Alum (or Vitriol) by Fire), 159

Fu phên tzu (plant), 88, 98

Fu-sang (legendary tree), 19, 21

Fu Shih Hsien Ching (Manual of Longevity and Immortality produced by Diet and Drugs), 118

Fuchs Expedition, 70 (a)

Fujiwara no Sukeyo (Japanese bibliographer, *c.* 895), 178

Fujiwara no Tadahira (Japanese Regent, d. +949), 179

Fukui Kōjun (1), 57

Fulminating gold, 226 (b), 236

Fumes, produced by manufacture of copperas, 239

Furnace, Spirit, 29, 31, 240 (f)

Furnaces. *See* Alchemical furnaces

Fusō Ryakuki (Classified Historical Matters about Fu-Sang), 176 (b)

Galen, 16 (d), 230

Galileo, xxv, 108, 220, 221, 236

Galvani, A. (physicist, +1737 to +1798), 249

de la Garaye, Comte, 233

Gases
collection of, over mercury, 246

nature of, 221
technical terms, 260

Gaubil, Antoine (Jesuit historian of Chinese astronomy, +1732), xx

Geber (ps. of a Latin alchemist, *c.* +1290), 123

Geberian Corpus (late +13th and early +14th centuries), 237, 248 (d)

Gem-mining, 2

General Catalogue of Precious Writings. See *Pao Wên Thung Lu*

General Dictionary of Biography. See *Wan Hsing Thung Phu*

General Encyclopaedic Dictionary. See *Wamyō Ruijūshō*

General Survey of the Lives of the Holy Immortals. See *Shen Hsien Thung Chien*

Generation, maternal and paternal contributions to, 70

Geography, 177, 212

Geomancy, 248

Geometry, xxv

Gerbillon, J. F. (Jesuit Mathematician-Royal of the King of France, +1685), 242

Geriatric medicine, 211

'Germ', 261 (c)

'German silver', 233

Ghosts and Spirits, 36–7, 106, Fig. 1362a, caption

Gilding, 103, 123, 207 (a)

Gillan, Hugh (physician to Lord Macartney's embassy, +1793), 234

Gilt paper, 233

Glass-making, 103 (f), 236 (h)

Glauber's salt, 158, 232

Glaze frits, 233

Gleanings of Leisure Moments. See *Tsurezureguas*

Glucose, 98

Glucosides, cyanogenetic, 88 (h)

God of the Stove. *See* Furnace Spirit

Gold, 196, 219, 244
'artificial' or 'counterfeit', 26, 27, 28, 33, 35, 99, 137, 159, 175, 186–7, 189, 198; colour of, 102; detection of, 140, 189; prohibition of, 26, 29 (a); superior to natural gold, 1–2, 100; the contrary doctrine, 189
changed into silver, 161
coinage, 37
colloidal, 49–50, 98, 99
colour, 31 (b), 189
curative powers of, 188–9
definitions of, 36
dissolving of, 98–9, 237
-faking. *See* Aurifiction
and immortality, 1, 31 (b), 34
ingestion of, 2, 33, 34–5, 40, 49, 83, 99, 144, 151
as an elixir ingredient, 1, 89, 100, 111, 118, 133, 135
in Japan, 174–5
lists of, 159
-making. *See* Aurifaction
projection for, 199

Gold (*cont.*)
purple, 75, 173, 194
technical terms. *See* Technical terms
toxicity of, 144
vessels. *See* Vessels of gold
Gold and Caerulean Jade Manual. See *Chin Pi Ching*
Gold amalgam. *See* Amalgams
'Gold can be made, and men can find salvation', xxxiv, 27
Gold dust (*chin hsieh*), 188
'Gold elixir', 133
Gold leaf, 99, 233
Gold-mines, in Japan, 174 (f)
Gold paint, 233
'Gold paste', 4, 102
Gold vessels. *See* Vessels of gold
Gold tortoise, 188
Golden Elixir of the Inner World. See *Nei Chin Tan*
'Golden Flower' (*chin hua*), 66, 67, 78
Golden Jade Treatise on the Similarities and Categories of the Five (Substances) and the *Kinship of the Three*. See *Chin Pi Wu Hsiang Lei Tshan Thung Chhi*
Goldsmiths. *See* Technicians
Golovkin, Count, his embassy to China, 1805, 248–9
Graeco-Egyptian papyri, 28
Graham, George (clockmaker), xxx
'Grand-Purity elixir' (*Thai-Chhing tan*), 85–6
'Grand-Purity Golden Potion magical elixir' (*Thai-Chhing chin i shen tan*), 43
Graphite, ingestion of, 78
Graves. *See* Burial customs
Great Bear, 64
The Great Elixirs of the Adepts . . . See *Thai-Chhing Chen Jen Ta Tan*
Great Pharmacopoeia. See *Pên Tshao Kang Mu*
'Great Unity Three Messengers elixir', 133
Greek alchemical manuscripts, 152
Greeks, xxx, 3 (f), 16, 234
Green-and-Elegant Secret Papers in the Jade-Purity Golden Box on the Essentials of the Internal Refining of the Enchymoma, the Golden Treasure. See *Yü-Chhing Chin-Ssu Chhing-Hua Pi-Wên Chin-Pao Nei-Lien Tan Chüeh*
'Grey tin', 103 (d)
Grosier, J. B. G. A. (sinologist, 1819), 230
Growth and transformation of ores and metals in the earth, capable of acceleration by the alchemist in the laboratory, 24, 142 (f)
Guanine, 137
von Guericke, Otto (physicist, +1602 to +1686), 241
Guareschi, S., 245
Guide to the Creation, by the Earth's Mansions Immortal. See *Thu Hsiu Chen Chün Tsao-Hua Chih Nan*
Gunpowder, xxx, 79 (a), 139, 159, 184, 226, 237, 238, 241

Gymnastic exercises, 20
Gypsum, 14, 89 (b), 131

Haematite, 3, 87
brown, 9 (c), 14, 96 (s), 131, 146
red, 15
Haemopoiesis, 144 (c)
Hai Chhan Tzu. *See* Liu Tshao
Hai Chhiung Tzu. *See* Pai Yü-Chhan
Hailstorms, 226
Hainan, 232
Hair, used in chemical preparations, 210
Hallucinogens, 8 (g)
Halogen elements, 250
The Hammer and Tongs of Creation. See *Tsao-Hua Chhien Chhui*
Han (period), 2, 5, 16, 17, 72, 231
Han Chhung (Later Han alchemist, +2nd century), 40
Han Chung (Chhin adept, −3rd century), 18
Han Chung (alchemist, before +300, perhaps identical with the preceding), 88, 93
Han Fei Tzu (The Book of Master Han Fei), 7, 227
Han Hsü Tzu. *See* Chu Chhüan
Han Kuan I (The Civil Service of the Han Dynasty and its Regulations), 17
Han Shih San Lun (Discourse on the Cooling Regimen Powder), 45
Han Wu Ti. *See* Wu Ti
Han Wu Ti Nei Chuan (Inside Story of the Emperor Wu of the Han), 34
Handbook of the Secret Teaching concerning Elixirs. See *Tan Lun Chüeh Chih Hsin Chien*
Handbook of the Three Aspects of Dutch Internal and External Medicine. See *Ranka Naigai Sanbō Hōten*
'Handy elixir', 88
Handy Formulae for Cyclically-Transformed Elixirs. See *Huan Tan Chou Hou Chüeh*
Hangchow, 81 (b), 234
Hanuman, 215 (b)
Harvey, William, xxx
Hashimoto Sōkichi (Japanese pharmacist and chemist, 1805), 255 (a), 259
Hataka (liquid which will turn bronze into gold), 164
The Heart of Medicine. See *I Hsin Fang*
Heating processes
technical terms, 143
Heating times. *See* 'Fire-times'
Heliotherapy, 12
Helix. See Plant genera, *Hedra*
Hellenistic culture, 166
proto-chemistry in, 45, 57, 102–3, 136, 255
Hellenistic papyri. *See* Chemical papyri
Hellenistic proto-chemical Corpus, 29, 36, 195
Hemp-seed oil, 104
Henckel, J. F. (chemist, +1679 to +1744), 147 (e)
Herbo-metallic preparations, 48–9
Herbs of immortality, 32, 33 (e), 44, 48, 49, 229

Herodotus, xxiv (a)
Hexagrams, 60–1, 66
Hhé-tânn-chê, 243–4
The Hidden Casket of Yin and Yang Opened.
  See *Chhien Khun Pi Yün*
Hides, used in chemical preparations, 210
Hierogamy, 37 (d)
High-temperature equipment, 246
Hinamitachi (Japanese student, *c.* +602), 177
Hipparchus (*c.* −140), xxx
Historical Classic. See *Shu Ching*
Historical Records. See *Shih Chi*
History of the Chin Dynasty. See *Chin Shu*
History of the Former Han Dynasty. See *Chhien Han Shu*
History of the Ming Dynasty. See *Ming Shih*
History of the Northern Dynasties. See *Pei Shih*
History of the (Northern) Wei Dynasty. See *Wei Shu*
History of the Southern Dynasties. See *Nan Shih*
History of the Sui Dynasty. See *Sui Shu*
History of the Sung Dynasty. See *Sung Shih*
History of the Three Kingdoms period. See *San Kuo Chih*
Ho chhê (lead), 56, 127–8
Hŏ Chun (Korean medical writer, +1610), 167
Ho Chün-Shih. See *Ma Ming-Shêng*
Ho Fa-Chhêng (Sung alchemist, +10th century), 173–4
Ho Liao-Jan (chemical writer, 1870), 252
Ho Lu (King of Wu State, d. *c.* −495), 5
Ho Phing-Shu. See *Ho Yen*
Ho Shang Chang Jen (the Old Man leaning on a staff by the Riverside; legendary or semi-legendary pre-Chhin alchemical adept), 11, 77
Ho shang chha nü. See 'Elegant girl by the river-side'
Ho Ta-Thung (Sixth of the Seven Masters of the Northern School of Taoism, +1140 to +1212), 205
Ho Wei (Sung Scholar and writer, *c.* +1095), 193
Ho Yen (Later Han and San Kuo scholar and courtier, +190 to +249), 45, 48
Hobson, Benjamin (physician and chemist, 1816 to 1873), 252, 255
Hokan Chi (Thang Governor of Chiangsi, +847), 167–9
Holan Hsi-Chen (Sung Taoist, d. +1010), 184, 194
Honey, 137, 196
  as elixir ingredient, 87
  as ingredient in an explosive mixture, 78–9, 159
Hooke, Robert, xxx
Hōraisan, 18 (b)
Hormones, xxx, 78 (c), 179 (a), 220
Horticulture, 211
Hou Khai (alchemical adept in the Liang, d. +573), 112
Hsi (double bond), 261
Hsi Tsung (Later Chou emperor, +956), 194
Hsi Tsung (Thang emperor, +874 to +888), 114

Hsi Wang Mu (goddess), 4, 34
Hsi-Yang Huo Kung Thu Shuo (Illustrated Treatise on European Gunnery), 241 (e)
Hsi Yo Hua-shan Chih (Records of Hua-shan, the Great Western Mountain), 205
Hsi Yu Chi (Pilgrimage to the West), 215
Hsi Yuan Wên Chien Lu (Things seen and heard in the Western Garden), 213
Hsia Fu (Later Han alchemist), 40
Hsiang lei (the similarities of categories of substances), 146
Hsiao I. See *Liang Yuan Ti*
Hsiao Shih (alchemical immortal), 4, 125, 127
Hsiao Wên Ti (Northern Wei emperor, r. +471 to +500), 119
Hsiao Yen. See *Wu Ti* (Liang emperor)
Hsieh Chih-Chien (Later Han alchemical adept), 43
Hsieh Ying-Fang (sceptical writer in the Yuan period, +1348), 209
Hsien. See Immortals
Hsien Lo Chi (Collected (Poems) on the Happiness of the Holy Immortals), 204
Hsien-mên (= shaman ?), 13 (b)
Hsien Tsung (Thang emperor, r. +806 to 820), 151, 182
Hsien-Yuan Pao Tsang Lun (Discourse of the Yellow Emperor on the Contents of the Precious Treasury of the Earth), 130 (c), 180 (b), 211
Hsienmên Kao (Warring States adept), 13, 32, 34
  his elixir method, 88, 91
Hsienyuan Chi (late Thang thaumaturgist), 211 (b)
Hsin (dynasty, +9 to +23), 37
Hsin Hsiu Pên Tshao (Newly Reorganised Pharmacopoeia), 16, 31 (a)
Hsin Lun (New Discussions), 26, 38, 82
Hsin Thang Shu (New History of the Thang Dynasty), 25, 141, 146, 151
Hsin Wu Tai Shih (New History of the Five Dynasties), 180
Hsiu Chen Shih Shu (A Collection of Ten Books on the Regeneration of the Primary Vitalities), 203
Hsiu Lien Ta Tan Yao Chih (Essential Hints on the Preparation of Powerful Elixirs), 198
Hsiung Ming-Yü (Ming scientific encyclopaedist, *c.* +1620), 218
Hsü Chen Chün Pa-shih-wu Hua Lu (Record of the Transfiguration of the Adept Hsü (Hsün) at (the Age of) Eighty-Five), 111
Hsü Chen Chün Shih Han Chi (The Adept Hsü (Hsün's) Treatise found in a Stone Coffer), 111
Hsü Chhêng-Po. See *Hsü Chien*
Hsü Chien (Northern Wei alchemist, *c.* +490), 119
Hsü Chien-Yin (translator of scientific books, 1845 to 1901), 254, 255, 256
Hsü Fu (alchemical adept sent out by Chhin Shih Huang Ti to find the islands of the immortals in the Eastern Ocean, *fl.* −219), 17–18, 19, 27, 34, 175 (b)

Hsü Hsüan (Taoist bibliographer, +989), 115
Hsü Huang-Min (Taoist alchemist, late +4th century), 121
Hsü Hui (Taoist alchemist, *fl. c.* +360), 121
Hsü Kuang-Chhi (+17th-century scientific scholar and high official), 235, 237, 238, 239
Hsü Ling-Fu (Thang alchemist), 121
Hsü Mai (Taoist alchemist, *fl.* +340 to +365), 76, 77, 121
Hsü Mi (Taoist alchemist and leader of the Mao Shan school, d. +373), 121
Hsü Po Wu Chih (Supplement to the *Record of the Investigation of Things*), 126
Hsü Shih Shih (Supplement to the *Beginnings of All Affairs*), 126
Hsü Shou (translator of scientific books, 1818 to 1884), 254, 259
Hsü Sun (Taoist alchemist and writer, *c.* +290 to +374), 110–11
Hsü Tao-Mo (Thang alchemist, +7th century), 121
Hsü Thung Chih (Supplement to the *Historical Collections*), 151
Hsü Ti-Shan (*1*), 23
Hsü Tshung-Shih (alchemist and commentator, +2nd century), 50, 55, 77
Hsü Yen-Chou (Sung writer, *c.* +1110), 150, 191, 195
Hsüan (Prince of the State of Chhi, −4th century), 13
Hsüan Chen Tzu. *See* Mêng Yao-Fu
Hsüan Chieh Lu (Mysterious Antidotarium), 168–9, 171
Hsüan-Chuang (Hsüan-Tsang, Buddhist monastic pilgrim, *c.* +640), 162, 163, 164, 165, 215 (b)
Hsüan Fêng Chhing Hui Lu (Record of the Auspicious Meeting of the Mysterious Winds), 204
Hsüan huang (elixir component), 83, 84 (j), 85
Hsüan kao, 84 (j)
Hsüan ming lung kao ('mysterious bright dragon's fat'), 88–9, 98
Hsüan Nü (Mysterious Girl, Taoist goddess), 85, 153, 197
Hsüan Nü Ching (Canon of the Mysterious Girl), 177 (b)
Hsüan Phin Lu (Record of the (Different) Grades of Immortals), 40
Hsüan Shuang Chang Shang Lu (Mysterious Frost on the Palm of the Hand), 218
Hsüan Ti (Early Han emperor, r. −73 to −48), 35
Hsüan Tsung (Thang emperor, r. +713 to +755), 114, 140, 141, 145, 146
Hsüan Tsung (Thang emperor, r. +847 to +859), 182
Hsüan-Tu Kuan (Taoist abbey), 114
Hsüan Tu Pao Tsang (Precious Patrology of the Mysterious Capital), 116, 205
Hsüan-Ying (Buddhist monk, +649), 114 (a)
Hsüeh Chi-Chhang (Thang alchemist, +8th century), 121

Hsüeh Tzu-Hsien (Taoist alchemist in Sung, +12th century), 203
Hsüeh Yü (*1*), 95
Htin Aung (1), 166
Hu fên (lead carbonate), 68
Hu Su (Sung scholar and civil servant), 192
Hua chhih (bath of vinegar and saltpetre, or vinegar itself, or any solvent bath), 98
Hua-Chhing Kung (Taoist abbey), 141
Hua chin shih ('metal-dissolving mineral', saltpetre), 89
Hua Hêng-Fang (mathematician and engineer, 1830 to 1902), 254
Hua-Hsüeh Chhu Chiai (First Steps in Chemistry), 252
Hua-Hsüeh Chien Yuan (Authentic Mirror of Chemical Science), 254
Hua-Hsüeh Shan Yuan (Explanation of the Fundamental Principles of Chemistry), 252
Hua Ling-Ssu (Prefect of Lu-chiang in San Kuo or Chin), 100
Hua-Yang Thao Yin-Chü Chuan (The Life of Thao Yin-Chü (i.e. Thao Hung-Ching) of Huayang), 120
Huai Nan Hung Lieh Chieh. See Huai Nan Tzu
Huai Nan Pien Hua Shu (The (Prince of) Huai-Nan's Book of Change and Transformation), 25
Huai Nan Tzu. See Liu An
Huai Nan Tzu (The Book of (the Prince of) Huai-Nan), 5, 14, 17, 23, 24, 25, 60, 245
Huai Nan Wan Pi Ching (Manual of the Ten Thousand Infallible (Arts) of the Prince of Huai-Nan), 24–5
Huai Nan Wan Pi Shu (The Ten Thousand Infallible Arts of the Prince of Huai-Nan), 5, 14, 25
Huan Chin Shu (Account of the Cyclically (-Transformed) Metallous Enchymoma,) 172, 173
Huan Khuan (Han official and writer, *c.* −80), 34
Huan tan (cyclically-transformed elixir). *See* Elixirs of longevity and immortality
Huan Tan Chou Hou Chüeh (Handy Formulae for Cyclically-Transformed Elixirs), 110, 171–2
Huan Tan Fu Ming Phien (Book on the Restoration of Life by the Cyclically-Transformed Elixir), 203
Huan Than (Han official and writer, *c.* +10), 26, 38, 39, 82
Huan Ti (Later Han emperor, r. +147 to +168), 40
Huan Yu Shih Mo (On the Beginning and End of the World), 235
Huan yuan (reduction), 261
Huan Yuan Phien (Book of the Return to the Source), 203
Huang, Arcadius, 224 (b)
Huang Chhao (+9th-century leader of an uprising), 114, 182

Huang ching ('deer-bamboo', *Polygonatum falcatum*), 112

Huang Chui (Han alchemist, *c.* −121), 31

Huang fan (yellow iron alum, or ferric sulphate), 233

Huang Hsüan-Chung (alchemical official during the time of Han Wu Ti), 33

Huang Hua (Chiu Ling Tzu, alchemical adept, −2nd century), 20

Huang Lan encyclopaedia, 4

Huang Pai Ching (Mirror of Alchemy), 216, 217, 219

Huang Shih Kung (the Old Gentleman of the Yellow Stone), 20, 40

Huang Ti, 31

Huang Ti Chiu Ting Ching (the Yellow Emperor's Manual of the Nine Vessels), 43

Huang Ti Chiu Ting Shen Tan Ching (The Yellow Emperor's Manual of the Nine-Vessel Magical Elixir), 50 (b), 83, 96 (a)

Huang Ti Chiu Ting Shen Tan Ching Chüeh (Explanation of the Yellow Emperor's Manual of the Nine-Vessel Spiritual Elixir), 50 (b), 84

Huang Ti Nei Ching (The Yellow Emperor's Manual (of Corporeal Medicine)), 47

Huang Ti Yin Fu Ching (The Yellow Emperor's Book on the Harmony of the Seen and the Unseen), 204

Huang yeh-jen (Hermit Huang, +4th-century alchemist in Lo-fou Shan, pupil of Ko Hung), 121, Fig. 1353 *a*, caption

Huangfu Mi (eminent medical writer, especially on acupuncture, and biographer of great men and immortals, +215 to +282), 11

Hui Tsung (Sung emperor, r. +1101 to +1125), 115, 145, 190–1

Hun 'souls', xxxiv, 46, 48, 70, 71, 73

Hun Thien theory (astronomy), 107 (b)

Hung Pao Yuan Pi Shu (Book of the Infinite Treasure in the Secret Garden), 14, 24

Hunger, recipes for the prevention of, 146

'Hungry ghosts', 166 (a)

Huo Chi (Book of the Firing-Process), 71–2

Huo hou. See 'Fire-times'

Huo Kung Chhieh Yao (Essentials of Gunnery), 241

Huo Lung Ching (Fire-Drake Manual), 241 (c)

Hydraulic Machinery of the West. See *Thai Hsi Shui Fa*

Hydro-mechanical clockwork, 115–16

Hydrochloric acid, 237, 239

Hydrocyanic acid, 88, 98

Hydrofluoric acid, 233 (h)

Hydrogen, 261
  and oxygen, 249

Hygiene, 110, 135 (g)

Hypnosis, 186, 189

I (jet), 97 (cc)

I Chhieh Tao Ching Yin I (Titles of all the Taoist Canons and their Meanings), 114

I Ching (Book of Changes), xxv, 51, 52, 53, 61, 65, 66, 71, 74, 128, 143, 145, 181, 201, 217

I Chou Shu (Lost Books of Chou), 3

I Hsin Fang (The Heart of Medicine), 167, 177

I-Hsing (+8th-century Tantric monk, mathematician and astronomer), xxx

I Lin (Forest of Symbols of the *Book of Changes*), 6

I-Shan Tsa Tsuan (Collected Miscellany of Li Shang-Yin), 168 (c)

I Yin (minister of the semi-legendary emperor, Chhêng Thang), 196

I Yü Thu Chih (Illustrated Record of Strange Countries), 212

Iatro-chemical Natural History. See *Wai Tan Pên Tshao*

Iatro-chemistry, xx, xxv, xxxiii, xxxv, 193, 210, 220, 259

Ibn Abū Uṣaybi'a (+1203 to +1270), 225 (d)

Ibn al-Haitham, xxviii

Ibn Khurdādhbih (*c.* +885), 174 (h)

Ice cream, 225 (d)

Ideographs, as technical terms, 260

Illuminating Chart for the Mnemonic Rhyme about the Reaction-Vessels in the *Kinship of the Three* and the *Book of Changes*. See *Chou I Tshan Thung Chhi Ting Chhi Ko Ming Ching Thu*

Illustrated Record of Strange Countries. See *I Yü Thu Chih*

Illustrated Treatise on European Gunnery. See *Hsi-Yang Huo Kung Thu Shuo*

Illustrations for the *Main Essentials of the Metallous Enchymoma; the true Gold Elixir.* See *Chin Tan Ta Yao Thu*

Imitations. See *Ersatz* productions

Immortality of the soul, 223–4. *See also* Material immortality

Immortals, 8, 9–11, 13–14, 19, 25, 33, 36, 46, 110, 227
  corpse-free, 21
  and Japanese culture, 175
  journey in search of, 119
  and magic, 106

Imperial Academy, 185

Imperial Academy of Medicine, 212

'Imperial Baldachin elixir', 171

Imperial Commentary on the Canon of the Virtue of the Tao. See *Yü Chieh Tao Tê Ching*

Imperial Medical College, 163

Imperial patronage of alchemists and Taoists, 141, 146–7, 148, 151, 190–1, 209

Imperial University, 192

Imperial Workshops, 35–6, 82, 189

Important Arts for the People's Welfare. See *Chhi Min Yao Shu*

Important Instructions for the Preservation of Health conducive to Longevity. See *Shê Yang Yao Chüeh*

Important (Medical) Formulae and Prescriptions . . . See *Wai Thai Pi Yao*

Important Rules for the Four Seasons. See *Ssu Shih Tsuan Yao*

Incarnation, xxviii
Incense, xxxiv, 233
India
    calomel in, 124
    contacts with China. *See* Transmissions and
        contacts
    'projection' in, 100 (a)
Indian alchemist, at the Chinese court (+7th
    century), 237
Indian alchemy, 160–5
Indian atomism, 160 (d)
Indian Buddhist monks. *See* Buddhist monks
Indian chemistry, 162
Indian medicine, 46 (c), 48, 49
Indian thaumaturgical devotees, 49
Indigo, 136
Indus Valley, 16
Industrial chemicals, 219
Infinite Treasure Pillow-Book. See *Chen Chung
    Hung Pao*
Ink
    red, 6
    white lead, 17
Inner Book of Wei Po-Yang. See *Wei Po-Yang
    Nei Ching*
Inorganic substances, 44, 45, 182, 210
    aqueous solutions of, 102, 139 (g)
    elixir preparations made from, 45–6, 149
    ingestion of, 44–5, 47–8
    therapeutic use of, 46, 47, 48, 49, 211, 212, 230
    toxicity of, 135
Inside Story of the Emperor Wu of the Han. See
    Han Wu Ti Nei Chuan
Inside Story of the Spiritual Lord Mao. See
    Mao Chün Nei Chuan
'Instantly Successful elixir' (*li chhêng tan*), 87
Insulin, 262
Intorcetta, Prosper (Jesuit, +1625 to +1696),
    223
Investigation of the Earth. See *Khun Yü Ko Chih*
Investigation of the Seven (Warring) States. See
    Chhi Kuo Khao
Investigations and Criticisms of the *Kinship of
    the Three*. See *Chou I Tshan Thung Chhi
    Khao I*
Ion exhange, 25 (h), 104, 187 (g), 207 (e)
Iōsis. See 'Purpling'
Iron, 244
    'gold' made from, 187–8
    transformed to look like copper, 104
Iron and steel industry, 81, 233, 234
Iron alum, 89
    yellow, 233
Iron compounds, 3
Iron ore, 15
Iron oxide, 243 (d)
Iron rust, 78
Iron vessels, casting of, 234
'Irrigation' processes, 207 (g)
Isaiah, xxiv (a)
Ishinhō. See *I Hsin Fang*
Isinglass, 26

Islamic science, xxvi–xxviii
Ivy, tincture of, 147

Jābirian Corpus (+9th and +10th centuries),
    124 (a), 187 (g), 248 (d)
Jade
    amulets, 3
    aqueous solution of, 118
    artificial white, 136
    ingestion of, 9, 25, 43, 87, 102
    transmutation of, 111
Jade Girls. See Yü Nü
The Jade-Tablet Master. . . See *Thai-Chhing
    Yü Pei Tzu*
Jansenists, 221
Japan
    alchemy in, 174–80
    chemical apparatus, 259
    chemical technology, 15, 174
    gold in, 174–5
    medicine, 177
    medieval specimen of calomel in, 128
    metallurgical technology, 174
    modern chemistry in, 255
    Taoism in, 175–6
    tomb of Jofuku, 18
Japanese process for calomel, 125 (b)
Jen Kuang (legendary immortal), 9
Jen Tsung (Sung emperor, r. +1022 to +1063),
    184
Jesuit library, Peking, 235
Jesuit mission, failure of, 221–36, 242, 255
Jesuits, 218, 238, 240, 241, 246, 247, 248, 249
    and alchemy, 224–5, 227, 230
    failure to transmit modern chemistry to China,
        220–3, 234
    transmissions to the West, 221, 223, 230–2
    suppression of, 222–3
Jesus, biographies of, in Taoist collections, 191 (a)
Jet, 96 (cc)
Jih Hua Chu Chia Pên Tshao (The Sun-Rays
    Master's Pharmaceutical Natural History . . .),
    124
Jofuku. *See* Hsü Fu
Jonson, Ben (+1572 to +1637), 213–15
Jottings from the Eastern Side-Hall. See *Tung
    Hsien Pi Lu*
Ju hsiang (aromatic nuts of *Pistacia* spp.), 198
Ju Lin Wai Shih (Unofficial History of the World
    of Learning), 215
Juan Hsiao-Hsü (Liang bibliographer, +523),
    113
Jui Tsung (Thang emperor, r. +710 to +712),
    140
Jujube-dates, 31, 137
Jurchen Tartars, 116

Kai Kung (alchemical adept, –2nd century), 11,
    20, 77
Kalinite, 133
Kan Chan (Chin alchemist, +4th century),
    111

Kan Chhi Shih-liu Chuan Chin Tan (The Sixteen-fold Cyclically-Transformed Gold Elixir prepared by the 'Responding to the Chhi' Method), 198

Kan Pao (Chin writer on strange phenomena, *c.* +348), 80

Kan Shih (Han expert in Taoist sexual techniques, +1st or +2nd century), 40, 105

Kansu salt, 84, 101

Kao Hsien (*1*), 225

Kao Lei-Ssu (Chinese scholar and Jesuit, +1734 to *c.* +1790), 228–9, 240

Kao Li-Shih (powerful Thang eunuch, *d.* +763), 141

Kao Shih Chuan (Lives of Men of Lofty Attainments), 11

Kao Thing-Ho (physician and pharmaceutical naturalist, +1505), 216

Kao Tsu (Early Han emperor, r. −206 to −194), 20, 23

Kao Tsung (Thang emperor, r. +650 to +684), 132, 140, 183

Kautilya (prime minister of Chandragupta Maurya, *c.* −300), 164–5

Kên hsüeh ('hard snow'), 128

Teacher Kêng (Thang woman alchemist at court, +9th century), 169–71, 191

Kêng Hsin Ching (Book of the Realm of Kêng and Hsin), 43

Kêng Hsin Yü Tshê (Precious Secrets (lit. Jade Pages) of the Realm of Kêng and Hsin), 210–11

Kêng Tao Chi (Collection of Procedures of the Golden Art), 182, 197

Kepler, Johannes, xxvii, 220, 221, 236

Kerr, J. G. (medical missionary, 1824 to 1901), 252

Khaifêng, botanical garden at, 210

Khang (Prince of Chiao-Tung, −113), 32

Khang Fêng Tzu (Master Khang-Fêng, Chin or pre-Chin alchemist), 91

The Khang-Hsi (Chhing) emperor (r. +1662 to +1723), 221, 225

Khang Yu-Wei, 254

al-Khāzinī, xxviii

Khou Chhien-Chih (Taoist religious leader in Northern Wei, d. +448), 118

Khou Tsung-Shih (pharmaceutical naturalist, +1116), 188–9

Khua-Fu the Boaster (mythological being), 196

'Khua Miao Shu Tan Kho Thi Chin' (story about alchemy in the *Chin Ku Chhi Kuan*), 213

Khubilai Khan (Mongol, later Yuan, emperor, r. +1206 to +1294, on the Chinese throne from +1280 onwards), 44, 116, 209

Khun-Lun Shan mountains, 24, 36

Khun Yü Ko Chih (Investigation of the Earth), 235

Khung Chi Ko Chih (Treatise on the Material Composition of the Universe), 235

al-Khwārizmī al-Kathī (+1034), 124 (a)

Kiangnan Arsenal, 252, 254

'Killing' of metals or other substances, 8 (a). *See also* 'Subduing' and 'Controlling'

Kim Kagi (Korean alchemist, +9th century), 167

al-Kindī, 45

Kinetic theory of matter, 151

'Kingdoms of Women', 174 (h)

Kinkazan (island), 174 (f)

Kinsei-setsu Bishōnen-roku (Modern Stones of Youth and Beauty), 175

Kinship of the Three. See *Tshan Thung Chhi*

The Kinship of the Three; or, the Accordance of the *Book of Changes* with the Phenomena of Composite Things. See *Chou I Tshan Thung Chhi*

The *Kinship of the Three* divided into (short) chapters for the Understanding of its Real Meanings. See *Chou I Tshan Thung Chhi Fên Chang Thung Chen I*

The Kinship of the Three with Commentary. See *Chou I Tshan Thung Chhi Chu*

The *Kinship of the Three* with Explanations. See *Chou I Tshan Thung Chhi Chieh*

Kitāb al-Asrār, 124 (a)

Kitāb al-Masālik wa'l-Mamālik, 174 (h)

Kitāb al-Naqd (Book of Testing, or, of Coinage), 187 (g)

Kitchen God. *See* Furnace Spirit

Klaproth, Julius (orientalist, b. +1783), 242–9

Klaproth, Martin (chemist, +1743 to 1817), 248 (f)

Knox, Ronald, 108

Ko, Aloysius. *See* Kao Lei-Ssu

Ko Chih Hui Phien (Chinese Scientific and Industrial Magazine), 255

Ko Chih Shu Shih (publishing house), 255

Ko Chih Tshao (Scientific Sketches), 218

Ko Chih Tshung Shu (General Treatise on the Natural Sciences), 255, 256–8

Ko Hsien Ong Chou Hou Pei Chi Fang (The Adept Ko (Hung)'s Prescriptions for Emergencies), 110

Ko Hsüan (alchemist, Ko Hung's great-uncle, +164 to +244), 76, 77, 121

Ko Hung (scholar and eminent Taoist alchemist, +283 to +343), 1, 19, 36, 38, 46, 49, 50, 52, 75–113, 119, 121, 125, 127, 128, 138, 145, 161, 167, 171, 176, 182, 199, 219

Ko i, 255

Koguryŏ, tomb frescoes (Korea), 167

Kokonor, Lake, 232

Konjaku Monogatari (Tales of Today and Long Ago), 176 (a)

Korea, 167, 177

Kou Chien (King of Yüeh, r. −496 to −470), 14

Ku Sung Tzu (Thang or Wu Tai alchemist), 150

Ku Yung (Han Confucian scholar, *c.* −32 to −7), 36

Kuan Tzu (Book of Master Kuan), 229

Kuang Chhêng Tzu (mythological hermit), 8

Kuang Fang Yen Kuan. *See* School of Foreign Languages

Kuang Ning Tzu. *See* Ho Ta-Thung

Kuang Ya dictionary, 5
Kuang Yün dictionary, 5
Kuangsi (province), 232
Kuangtung (province), 81
Kuei (ancient term for a demon, and later the Neo-Confucian term for condensation-aggregation processes), 196
Kuei Fu ('Father Cinnamon'), 9 (e)
Kuei Ku Tzu (Warring States philosopher), 19
Kumārajīva (translator, +406), 161, 162, 163
Kume no sennin (Japanese immortal with magical powers, +9th century), 175–6
Kunckel, Johann (glassmaker, alchemist and industrial chemist, +1630 to +1703), 103 (f)
Kuo Chhung-Chen (Thang alchemist, +7th century), 121
Kuo Hsü-Chou (alchemical friend of Pai Chü-I, +818), 148
Kuo-Li Pien I Kuan. *See* National Bureau of Compilation and Translation
Kuo Shih-Chün (Sung or pre-Sung physician), 12 (e)
Kushta (herbo-metallic preparations), 48, 49
Kwanroku. *See* Kwŏllŭk
Kwŏllŭk (Korean monk who brought many sciences to Japan, +602), 177

La (probably an alloy of zinc, tin and lead), 136
Labelled inorganic drug specimens, Sian find, 144 (f)
Laboratories. *See* Alchemical laboratories
Laboratory equipment. *See* Alchemical apparatus
Lacquer, 104
  as elixir ingredient, 87
Lady of the Moon, 4
Lady of Tai (d. *c.* −166), 21–3
Lake salt, 133
Lamotte, E. (1), 162, 163
Lan Kung-Chhi (alchemical adept, +3rd or +4th century), 111
Lan Yuan-Lao (Sung expert in alchemica apparatus and techniques, *c.* +1210), 199, 202
Lao Hsüeh An Pi Chi (Notes from the Hall of Learned Old Age), 190
Lao Kan (6), 5, 6, 44, 67
Lao Tzu, 1–2, 117, 177, 215, 244
  as patronal ancestor of the Thang emperors, 183
  worship of, 40, 52, 141
Lao Tzu Hua Hu Ching (Book of Lao Tzu's Conversion of Foreigners), 116
Lapis lazuli, 131, 243 (d)
Later Han (dynasty, +25 to +220), 5, 7, 40, 44, 45, 46, 75
Lavoisier, A. L. (chemist, +1743 to +1794), 218, 221, 222, 247, 255
Lawrence, D. H., 151 (a)
Leaching, 49
Lead, 66–7, 71, 100, 101, 103, 133, 144, 155, 171, 173, 201, 217–18, 219, 234, 244
  black, 78

  as elixir ingredient, 87, 133, 135, 173, 196
  ingestion of, 78
  medical use of, 15–16
  metallic, 14, 15, 17, 67, 73
  as a prime source of the elixir, 74
  red, 6
  salts of, 3, 16
  solution in mercury, 66
  technical terms. *See* Technical terms
  and tin, 100
  toxicity of, 144
  white. *See* White lead
Lead acetate, 15, 16, 17, 124, 171
Lead amalgam, 67, 73, 172, 209
Lead carbonate, 14, 15, 16, 67, 68 (e), 84, 124, 127, 136
  face-powder, 4, 17, 126
  medical uses of, 16
  oldest name for, 16
'Lead elixir', 135
Lead oxides, 3 (f), 6, 14, 67, 68 (e), 87, 103, 246
Lecture on a Secret Method for Yellow Gold by Master Lingyang. See *Ling Yang Tzu Shuo Huang Chin Pi Fa*
Legends. *See* Folklore
Lei Hsiao (physician, pharmacist and alchemist, +5th century), 119
Lei Kung Phao Chih Lun (The Venerable Master Lei's Treatise on the Decoction and Preparation of Drugs), 119
Leicester, H. M. (1), 137
Leonite, 89
'Letters from Heaven', 183–4
Lévi, S. (8), 164
Lexicography, alchemical, 151 ff.
Lexicon Alchemiae (Ruhland), 152
Li (organic pattern), 261
Li Chüeh (Sung physiological alchemist), 205
Li Chung-Fu (Han adept, supposed teacher of Tso Tzhu), 76, 77
Li Han-Kuang (Thang alchemist, +8th century), 121
Li Hsiu (Han alchemical adept, −2nd century), 20
Li Hsün (Earlier Shu merchant of drugs and perfumes, pharmaceutical naturalist and chemical practitioner, *fl.* +919 to +925), 180 (e)
Li I-Shan. *See* Li Shang-Yin
Li Kên (Taoist alchemist), 100
Li Kuo-Chih (alchemist, +5th century), 121
Li Lin-Fu (eminent Thang minister and owner of water-mills, d. +752), 141
Li Pa-Pai (Han adept), 19
Li Pai (poet, +701 to +762), 140, 148, 150
Li Phu-Wên (Taoist teacher of Khou Chhien-Chih, *c.* +400), 118
Li Pi (Sung alchemical writer, *c.* +1020), 185
Li Shang-Yin (Thang epigrammatist, +813 to *c.* +858), 168 (c)
Li Shao-Chün (Han aurifactive alchemist, −133), 29–31, 33–4, 40, 137

Li Shao-Yün (Sung woman alchemist, *c.* +1111), 191–2

Li Shêng (founder of the Southern Thang dynasty), 180

Li Shih-Chen (pharmaceutical naturalist, +1518 to +1593), 97 (cc), 125, 127–8, 158, 210, 216, 220, 240

Li Shih Chen Hsien Thi Tao Thung Chien (Comprehensive Mirror of the Embodiment of the Tao by Adepts and Immortals throughout History), 88

Li shih Lien Tan Fa (Mr Li's Techniques for Preparing Elixirs), 150

Li Shun (leader of an uprising, +995), 192

Li Shun-Hsing (alchemical adept, +6th century), 112

Li Tao-Yin (aurifactive alchemist, +9th century), 180

Mr Li's Techniques for Preparing Elixirs. See *Li shih Lien Tan Fa*

Li Thien-Ching (Ming high official, *c.* +1639), 235

Li Wên (alchemist), 91

Li Wên-Chu (Ming alchemical writer, +1598), 216, 217, 219

Liang (period), 34, 199

Liang Chhi-Chhao, 254

Liang I Tzu (Master Liang-I, Chin or pre-Chin alchemist, if a historical character), 94, 99

Liang Ling-Tsan (horological engineer, *c.* +720), xxx

Liang Shen (Chin Taoist, d. +318), 111

Liang Yuan Ti. *See* Yuan Ti

Libavius, Andreas (chemist, *c.* +1540 to +1616), 137, 218, 236

von Liebig, Justus (founder of organic chemistry, 1803 to 1873), 221

Lieh Hsien Chhüan Chuan (Complete Collection of the Biographies of the Immortals), 10, 41, 42, 122, 134

Lieh Hsien Chuan (Lives of Famous Immortals), 9, 11, 12, 19, 88, 96 (n), 96 (q), 104, 127

Lieh Tzu (The Book of Master Lieh), 229

Lieh Yü-Khou (Warring States philosopher, –4th century), 8

Lien ('effecting chemical transformation in', or 'preparing by heating'), 44

Life of the Bodhisattva Nāgārjuna. See *Lung-Shu Phu-Sa Chuan*

Life of Thao Yin-Chü of Huayang. See *Hua-Yang Thao Yin-Chü Chuan*

Lin Ling-Su (Taoist at the court of Hui Tsung, +1118), 191, 193

Ling Sha Ta Tan Pi Chüeh (Secret Doctrine of the Numinous Cinnabar and the Great Elixir), 198

Ling Yang Tzu Shuo Huang Chin Pi Fa (Lecture on a Secret Method for Yellow Gold by Master Lingyang), 12

Lingyang Tzu-Ming, 12, 104, 153, 156, 157

Lingyang Tzu-Ming Ching (Manual (of Physiological Alchemy) by Lingyang Tzu-Ming), 12

von Lippmann, E. O. (1), 245

Litharge, 14, 15, 67, 246

Litmus, 9 (e)

Liturgical practices, 3 (i), 108, 110, 120

Mr Liu's elixir method, 92

Liu An (Prince of Huai-Nan, –178 to –122), 13, 14, 23–4, 25, 26, 27, 29, 35, 106

Liu An-Min (Marquis of Yang-chhêng, –60), 35

Liu Chêng (Later Han alchemical adept), 43

Liu Chhu-Hsüan (third of the Seven Masters of the Northern School of Taoism, +1147 to +1203), 204

Liu Chih-Chhang (Sung aurifactive alchemist, +1125), 190

Liu Hsiang (scholar and alchemist, *c.* –60), 9, 13, 14, 24, 27, 35–6, 82

Liu Hsiao-Sun (Sui scholar and writer, late +6th century), 126

Liu Hung (Han alchemist, *fl.* +122), 168

Liu I-Chhing (scholar and writer, +5th century), 45

Liu i ni ('six-and-one paste'), 133

Liu Jung (Han adept, –2nd century), 20

Liu Khuan (Later Han adept, traditionally +121 to +186), 40

Liu Phing (son or daughter of Liu An, patron of aurifactors), 26

Liu Pi-Chiang (Han jurist, –164 to –85), 14 (b)

Liu Tao-Ho (Thang alchemist, *c.* +650), 121, 132, 140

Liu Tao-Kung (Later Han alchemist), 43

Liu Tê (Han scholar, b. *c.* –126), 14, 24

Liu Thai-Pin (aspirant to *hsien*-ship, –3rd century), 19

Liu Tshao (founder of the Southern School of Taoism, early +11th century), 147–8, 200

Liu Tshun-Jen (1), 57

Liu Wên-Thai (physician and pharmaceutical naturalist, +1505), 216

Liu Yuan (Ming adept, *c.* +1391), 91, 209

Lives of the Holy Immortals. See *Shen Hsien Chuan*

Lives of Famous Immortals. See *Lieh Hsien Chuan*

Lives of Men of Lofty Attainments. See *Kao Shih Chuan*

Lo Chhen Kung (Chhin adept, *fl.* –3rd century), 11, 77

Lo-Chhêng, Marquis of (–113), 32

Lo Chhung (Taoist adept, +5th century), 119

Lo-fou Shan mountains, 81, Fig. 1353 *a*, caption

Lo Hsia Kung (disciple of Mao Shih Kung, –3rd century), 11

Lo shu (magic square), 6–7

Logos, in Taoism, 244 (c). *See also* Lao Tzu

London Pharmacopoeia (+1618), 124

Longevity. *See* Material immortality and longevity

Lost Books of Chou. See *I Chou Shu*

Lotus, 207

Lou Ching (adventurer, and military and civil official, perhaps patron of aurifactors, *c.* –210), 56

Loureiro, João (Jesuit botanist, +1774), 222–3, 234

Lü fan (green vitriol or copperas), 89

Lu Hsi-Hsing (alchemical commentator, *c.* +1570), 54

Lu Hsiu-Ching (Taoist bibliographer, +437), 113, 121

Lu Huo Chien Chieh Lu (Warnings against Inadvisable Practices in the Work of the Stove), 193, 194

Lu Huo Pên Tshao (Pharmaceutical Natural History of the Stove and Furnace), 220

Lu I-Chhung (Liang alchemist, +6th century) 121

Teacher Lu Li (Chin or pre-Chin alchemist), 95

Lu Shan (mountain), 148

Lü shih Chhun Chhiu (Master Lü's Spring and Autumn Annals), 24

Lu Thang Tzu. *See* Chao Shêng

Lu Thien-Chi (Education Commissioner and alchemical commentator, *c.* +1111), 124, 145, 197

Lü Tung-Pin (Thang adept and alchemist, probably +8th century), 147–8, 198, 200, 203

Lü Tzu-Hua (Han alchemist, +2nd century), 76, 77

Lu Yu (Sung scholar, poet and writer, +1125 to +1209), 190, 197 (e)

Luan Ta (Han court magician, −113), 32–3, 105

Lüchhiu Fang-Yuan (Thang alchemist, d. +902), 121

Lunacy, feigned, 186

Lucky and unlucky days, 106, 176

Lullian Corpus (+14th century), 236 (h), 248 (d)

Lun Hêng (Discourses weighed in the Balance), 107

Lun Shih (Discussion on (the Use of) Mineral Drugs), 47

Lucidity, of writing, 74, 79(e), 87, 133

Lunar mansions (*hsiu*), 62

Lung Chhuan Lüeh Chih (Classical Records from Dragon River), 193–4

Lung Chhuan Pieh Chih (Further Records from Dragon River), 188 (b)

Lung Hu Ching (Dragon-and Tiger Manual), 144

Lung Hu Huan Tan Chüeh (Explanation of the Dragon-and-Tiger Cyclically-Transformed Elixir), 88

Lung Hu Shang Ching (Exalted Manual of the Dragon and Tiger), 56

Lung kao ('dragon fat', perhaps mercury), 84 (j), 87, 88, 98

Lung-Mêng. *See* Nāgārjuna

Lung-Mu. *See* Nāgārjuna

Lung Po. *See* Lung Shu

Lung-Shêng. *See* Nāgārjuna

Lung-Shu. *See* Nāgārjuna

Lung Shu (Later Han adept), if historical, 40

Lung-Shu Phu-Sa Chuan (Life of the Bodhisattva Nāgārjuna), 162–3

Lung-Shu Phu-Sa Ho Hsiang Fa (Methods of the Bodhisattva Nāgārjuna for Compounding Perfumes, or Incense), 163

Lung-Shu Phu-Sa Yang Shêng Fang (Macrobiotic Prescriptions of the Bodhisattva Nāgārjuna), 165

Lung-Shu Phu-Sa Yao Fang (Pharmaceutics of the Bodhisattva Nāgārjuna), 163

Lung-Shu Yen Lun (Discourse of Nāgārjuna on Eye (Diseases)), 163

Lung-Yü (daughter of Mu Kung, Duke of Chhin, −7th century), 4, 125–7

Lustration, 196

Lute (sealing-compound), 133

Ma-chi Shan mountains, 82

Ma Kao (encyclopaedic writer, *c.* +925), 125

Ma-Ku (goddess), 39

Ma Lang (Chin alchemist, +4th century), 12

Ma Lao (Chin alchemist, +4th century), 121

Ma Ming. *See* Aśvaghoṣa

Ma Ming-Shêng (Later Han adept, +2nd century), 43, 76, 77

Ma-wang Tui Tomb No. 1, body of the Lady of Tai found in, 21

Ma Yü (first of the Seven Masters of the Northern School of Taoism, +1123 to +1183), 204

Macartney Embassy, 234

McGowan, John, 254

Macquer, P. J. (chemist, +1718 to +1784), 230–1, 233

Macrobiogens, xxxiv

Macrobiotic Prescriptions of the Bodhisattva Nāgārjuna. *See Lung-Shu Phu-Sa Yang Shêng Fang*

Macrobiotics, xx, xxxiii, xxxiv

Mādhyamika logic, 161

Magic, 3, 16, 25, 105, 110, 114
    Confucian criticism of, 190, 225
    and *hsien*-ship, 13, 29, 106
    natural, 177
    Taoist, 106, 108, 120

Magic daggers, 97 (jj)

Magic mushrooms, 9, 18, 40, 167

Magic squares, 6, 166

'Magical elixir'. See *Shen tan*

Magical islands, 13, 18–19, 33

Magical technicians. *See* Technicians

Magnesium silicate, hydrated, 15

Magnesium silicon oxide. *See* Talc

Magnesium sulphate, 158

Magnetic compass, 218, 249

Magnetic phenomena, xxv

Magnetic pole, 218

Magnetisation, 230

Magnetised chessmen, 32, 105

Magnetite, 86, 87, 89, 98, 99, 131, 243 (d)

Mahā-prajñāpāramito-padeśa Śāstra (Commentary on the Great Sūtra of the Perfection of Widsom), 161

Mahāyāna-saṃgraha-bhāshya, 164

Mahdihassan, S., 48, 49

Main Essentials of the Metallous Enchymoma; the True Gold Elixir. See *Chin Tan Ta Yao*

de Mairan, J. J. Dortous (physical chemist, +1678 to +1771), 225

Maithuna, 166 (a)

Malabar nightshade. *See* Plant genera, *Basella*

Malachite, 15, 26, 78, 86, 87, 104, 136, 144, 159

Malic acid, 49, 160 (a)

Mallow. *See* Plant genera, *Malva*

'Mallow' anthocyanin, 126

Manchu alchemical books, 55 (c)

Manchu invasion, 236, 237, 240, 241

Manchu translation of New Testament, 223 (c)

Manganese, 252

Mangu Khan (Mongol emperor, r. +1251 to +1259), 44, 116

Manichaeism, 72, 115 (a)

Manual labour, 219

Manual of the Grand-Purity Elixir. See *Thai-Chhing Tan Ching*

Manual of Longevity and Immortality produced by Diet and Drugs. See *Fu Shih Hsien Ching*

Manual of the Nine Elixirs of the Holy Immortals and of the Ninefold Cyclically-Transformed Mercury. See *Chiu Chuan Liu Chu Shen Hsien Chiu Tan Ching*

Manual of the Nine Reaction-Vessels Elixir. See *Chiu Ting Tan Ching*

Manual (of Physiological Alchemy) by Lingyang Tzu-Ming. See *Lingyang Tzu-Ming Ching*

Manual of Potable Gold. See *Chin I Ching*

Manual of the Potable Gold and the Magical Elixir. See *Thai-Chhing Chin I Shen Tan Ching*

Manual of the Potable Gold Elixir. See *Chin I Tan Ching*

Manual of the Ten Thousand Infallible (Arts) of the Prince of Huai-Nan. See *Huai Nan Wan Pi Ching*

Manual of the Thirty-Six Aqueous Solutions. See *San-shih-liu Shui Ching*

'Manuals of the Immortals', 123, 130

Manyōshū (Anthology of a Myriad Leaves), 175

Mao Chün Nei Chuan (Inside Story of the Spiritual Lord Mao), 39–40

Mao Chung (youngest brother of the Taoist hierarch, Mao Ying, —1st century), 39 (b)

Maò hhóa (Mao Hua ?), 242

Mao Hsi Kung (Han alchemist), 77

'Mao Hua' tractate, 242 (d), 246, 247

Mao Ku (second brother of the Taoist hierarch, Mao Ying, —1st century), 39 (b)

Mao Po-Tao (Later Han alchemist), 43

Mao-Shan mountains, 76, 115, 186

Mao Shan School of Taoism, 39 (b), 41, 108 (b), 120

Mao Shih Kung (pupil of An Chhi shêng), 11

Mao Ying (Taoist alchemist, one of the Three Lords Mao, *c.* —48), 39, 41

Maps, 4, 218

Marriage of Contraries. *See* Contraries, marriage of

Mary the Jewess (Alexandrian proto-chemist and inventor of laboratory apparatus, +2nd century), 191

Maslama al-Majrītī (d. +1007), 137

Maspero, Henri (13), 8

Massicot, 246

Master Mo's Elixir Methods. See *Mo Tzu Tan Fa*

Master Tsochhiu's Enlargement of the Spring and Autumn Annals. See *Tso Chuan*

Material immortality and longevity, xxx, xxxiv, 1, 2, 4, 8, 9, 20, 29, 31, 34, 35, 36, 40, 46, 51, 52, 57, 72, 75, 99, 105, 166, 193, 195, 200, 205, 212, 220, 224, 227

drinking from golden vessels and, 1, 31, 33, 49, 212

elixirs of. *See* Elixirs of longevity and immortality

techniques for attaining. *See* Techniques

Wang Chhung's views on, 107

Mathematics, xx, xxxi, 220, 221, 223

Matter and spirit, 149

Mattioli, P. A. (botanist, pharmacist and chemist, +1501 to +1577), 238

Maximilian, Emperor, 31 (c)

The Meaning of Popular Traditions and Customs. See *Fêng Su Thung I*

Mechanical clock, xxxiii

Mechanical invention, 191

Medicine, xxxi, xxxiii, 19, 45–7, 119, 123, 160, 186, 220, 232, 233

alchemy and, 130–1, 145, 167, 211, 212

cinnabar used in, 231

dosage, 135

geriatric, 211

in India, 46 (c), 48, 49

in Japan, 177

lead used in, 15–16

prescriptions, 135, 144, 171

psychosomatic, xxxi

terminology, 260

veterinary, 211

Medicine of immortality. *See* Elixirs of longevity and immortality

Meditation techniques, xxxv, 8

Mei Piao (Thang alchemical lexicographer, +806), 129, 138, 152–3, 156, 171, 182

Mellor, J. W., 245

Mémoires concernant les Chinois . . ., 231

Memorials of Wang Chhung-Yang's Preaching. See *Chuung-Yang Chiao Hua Chi*

Mêng Chan-Jan (Thang alchemist, +8th century), 121

Mêng Chhang (Later Shu emperor, +938 to +965), 180

Mêng Chhi Pi Than (Dream Pool Essays), 187

Mêng Fa Shih (Taoist bibliographer, +5th century), 113

Mêng Hsü (Sung alchemist, expert in apparatus and techniques, +1225), 199, 202–3, 219

Mêng Nai-Chhang (1), 89, 99

Mêng Shen (Thang proto-chemical naturalist, +621 to +718), 121, 132, 140, 145

Mêng Yao-Fu (alchemical compiler in Sung), 110, 197
Mercuric anions, 102
Mercuric chloride. *See* Corrosive sublimate
Mercuric oxide, red, 246
Mercuric sulphide, 21, 46 (c), 68, 74, 85, 102, 131, 179
 cyclical transformations of, 126, 250
Mercurous chloride. *See* Calomel
Mercury, 3–5, 46, 47, 48, 57, 58, 66, 67, 68, 71, 73, 88–9, 98, 123–4, 144, 159, 171, 173, 199, 201, 217–18, 219
 antiquity of, 4–5
 association with gold, 31 (b)
 aurifaction and, 102, 190
 calcination of, 137
 cinnabar and, 111, 195
 converted into silver, 192
 cyclical transformations of, 14 (a), 126
 as an elixir ingredient, 87, 133, 135–6, 138, 173, 196
 ingestion of, 9, 40, 78, 89
 isolation of, from plant tissues, 147 (e), 207, 231–2
 metallic, 12, 231
 mines, 6
 as prime source of the elixir, 74
 projection, 174
 small doses of, 119
 sulphur and, 14 (a), 74, 78, 79, 198, 238
 technical terms. *See* Technical terms
 toxicity of, 144
 transformed into 'gold', 198
 volatility of, 68 (b), 148 (g)
Mercury, chlorides of, 124–5
 first preparation of, in China, 128, 129
 sublimation of, 124, 127
Meru, Mt, 24 (f)
Metacinnabarite, 160 (a)
Metal (element), 73, 74, 87
Metal-workers, 219, 220
Metallurgical aurifictors. *See* Technicians
Metallurgy, 45, 186, 189, 210, 211, 219, 236, 240
 in Japan, 174
 in Korea, 167
 powder-, 190
 Western. *See* Western metallurgy
Metals, xxx, 210, 243–4, 260
 auspicious, 31 (a)
 calces of, 243, 246, 247
 growth and transmutation of, in the earth, 24, 142 (f)
 'healing' of, by herbal drugs, 49
 oxides of, 246
 sulphur-mercury theory of, 238, 245
 in therapy, 48
Metaphors. *See* Alchemical synonyms and metaphors
Methane, 21, 261
Methods of the Bodhisattva Nāgārjuna for Compounding Perfumes, or Incense. See *Lung-Shu Phu-Sa Ho Hsiang Fa*

Methods for Preparing Five Numinous Elixirs. See *Wu Ling Tan Fa*
Methods of the Various Schools for Magical Elixir Preparations. See *Chu Chia Shen Phin Tan Fa*
Mezuzah, 153 (a)
Mezzabarba, Carlo (Latin Patriarch of Alexandria and papal legate to China, *fl.* +1710 to +1725), 222 (a)
Mica, 87, 118, 131
 ingestion of, 9, 40, 120
Dr Michael. *See* Yang Thing-Yün
Mikami Yoshio, xx; (16), 239
Military chemistry, 240
Military Encyclopaedia. See *Wu Ching Tsung Yao*
Military inventions, 184
Millwrights, 177
Min I-Tê (Taoist editor, early 19th century), 218
Min Shan Tan Fa (Elixir Techniques of the Min Shan Mountains), 96 (f)
Mineral acids, 160, 237–42, 250
 first knowledge of, 139, 158
Mineral remedies, 46–7, 48–9, 144 (f), 211, 212, 230–1, 233
Mineralogical prospecting, 104 (d)
Mineralogy, 211
Minerals
 aqueous solutions of, 118
 growth and transmutation of, in the earth, xxix, 24, 196, 226
 identification of, 143
 ingestion of, 9, 12, 47–8
 lists of, 12, 14
 rare, 116
Ming (dynasty), 55, 209–12, 219, 227, 237, 240, 246
 alchemical works, 216–19
 bronzes, 219
 novels and plays, 212
Ming chhuang chen ('bright window dust'), 73, 149, 150
Ming Shih (History of the Ming Dynasty), 209
Mining, 210, 211, 236
 copper, 28
 gold, 174 (f)
 mercury, 6
 placer gold, 2 (b)
Minium. *See* Lead oxides
'Minor Cyclically-transformed elixir', 133
Minting of coinage, 26, 27, 28
Mirabilite, 158
Mirror of the Alchemical Elaboratory: a Source-book. See *Tan Fang Ching Yuan*
Mirror of Alchemical Processes (and Reagents): a Source-book. See *Tan Fang Chien Yuan*
Mirror of Alchemy. See *Huang Pai Ching*
Mirrors
 burning-, 71
 as demonifuges, 106, Fig. 1362a, caption
 dew-, 31 (a)
 moon-, 71

Mirza Fath-ali Akhunzadé (Turkish playwright, 1851), 215

Miscellaneous Records on Green Bamboo Tablets. See *Chhing Hsiang Tsa Chi*

Miscellaneous Records of the Lone Watcher. See *Tu Hsing Tsa Chih*

Missionaries
  Jesuit. *See* Jesuits
  medical, 252

Miu Shih-Têng (Han encyclopaedist, +220), 4

Mnemonic Rhymes of the Cyclically-Transformed Elixir from …Phêng-lai Island. See *Phêng-lai Shan Hsi Tsao Huan Tan Ko*

Mo Chuang Man Lu (Recollections from the Estate of Literary Learning), 185

Mo O Hsiao Lu (Secretary's Commonplace Book), 208, 218

Mo Shu Shang Hsia Ching (Pulse Manual in Two Chapters), 47

Mo Ti (Mo Tzu, Warring States philosopher, −4th century), 92

Mo Tzu (The Book of Master Mo), 16, 127

Mo Tzu Tan Fa (Master Mo's Elixir Methods), 96 (u)

Modern Stories of Youth and Beauty. See *Kinsei-setsu Bishōnen-roku*

Mohists, 96 (u)

Moissan, H. (chemist, 1852 to 1907), 245

Money. *See* Coinage

Mongols, 44, 116, 208

Monkey, 215

Mono-sodium orthophosphate (*lin suan erh ching na*), 261

Mononobe Kōsen (Japanese imperial physician, *c.* +820), 179

Monotheism, xxvi

Montoku (Japanese emperor, r. +851 to +858), 176 (c), 179

Montoku Jitsuroku (Veritable Records of the Reign of the Emperor Montoku), 176 (c)

Moon, influence on tides, 106

'Moon-lover'. *See* Chandragupta Maurya

Moon-mirror, 71

Morality. *See* Ethics

Mordants, 232

de Morveau, Guyton (chemist, +1737 to 1816), 255

'Mosaic gold', 99 (f), 103, 196

Motes, in sunbeams, 73, 149–50

Motion of bodies, 149, 150

'Mountain pomegranate' (*shan shih liu*). *See* Plant genera, *Berberis*, *Rhododendron* and *Rosa*

Mountains
  and the carrying out of alchemical processes, 36
  and the attainment of *hsien*-ship, 106

Movable-type printing, 187

Mu Kung (Duke of Chhin, r. −658 to −620), 4, 125, 127

Mu Tsung (Thang emperor, r. +821 to +824), 151, 182

Mu Wang (Chou High King, −10th century), 4

Muccioli, M. (1), 245, 246

Muḥammad, the Prophet, biographies of, in Taoist collections, 191 (a)

Mulberry, 18, 19
  leaves, 147
  wood, ingestion of, 25, 78

Mullah Ibrahim Khalil the Alchemist, 215

Mummification. *See* Self-mummification

'Muriate of quicksilver', 124

Mushrooms. *See* Magic mushrooms

Music, magical effects of, 12 (i), 125 (e)

Musical gamut, 59

Musk, 133

Mutationist inversion, xx

Mutual Production Order, 58, 67

Myrobalans, 179, 189 (b)

Mysterious Antidotarium. See *Hsüan Chieh Lu*

Mysterious Frost on the Palm of the Hand. See *Hsüan Shuang Chang Shang Lu*

Mysterious Girl. *See* Hsüan Nü

'Mysterious liquid' (*hsüan shui*, synonym for mercury), 87, 89

Mysterious Teachings on the Ninefold Cyclically-Transformed Gold Elixir . . . See *Ta-Tung Lien Chen Pao Ching, Chiu Huan Chin Tan Miao Chüeh*

Mysterious Teachings on the Processing of Numinous Cinnabar . . . See *Ta-Tung Lien Chen Pao Ching, Hsiu Fu Ling-Sha Miao Chüeh*

Mysterium conjunctionis. *See* Contraries, marriage of

Mystical religion, and the development of natural science, xxvi, 107

Nafs (chthonic spirit, somewhat analogous to *pho*), 48

'Nāgārjuna', identity of, 161–4

Nāgārjuna (Buddhist philosopher and patriarch, probably +2nd century, perhaps also alchemist), 161–4

Nāgārjuna's Discussions on Ophthalmology. See *Yen Kho Lung-Mu Lun*

Nan Chi Tzu. *See* Liu Jung

Nan Fang Tshao Mu Chuang (Records of the Plants and Trees of the Southern Regions), 80

Nan Shih (History of the Southern Dynasties), 119

Nan Tsung. *See* Taoism, Southern School

Nanpo Tzu-Khuei, 8

Nao sha (sal ammoniac), 233

Nao shui (aqua regia), 239

Nara
  Shōsōin Treasury, 84
  Tōdaiji temple, 175
  Tōshōdaiji temple, 178

Nārāyaṇasvāmin (Indian chemical practitioner in China, *c.* +649), 160

Nasr, Said Husain, xxvi

National Bureau of Compilation and Translation, 254

National Dispensaries, 188

Natrium, 261

Natron, 232, 245

Natural calamities, and the unethical behaviour of rulers, 106

Natural change, slow and spontaneous, of ores and minerals in the earth, capable of acceleration in the alchemical laboratory, 24

Natural History of Emergency Food Plants. See *Chiu Huang Pên Tshao*

Natural knowledge, evolutionary development of, xxiii

Natural science, and mystical religion, 107

Nature
  regularities in, 74, 107 (c)
  and the scientific revolution, xxvii, xxviii
  uniqueness and unpredictability in, 107 (c)

Nei Chin Tan (Metallous Enchymoma within the Body; or, Golden Elixir of the Inner World), 153–5

Nei tan (physiological alchemy), xx, xxxiv–xxxv, 12, 57, 74, 150, 153, 154–5, 160, 168, 171, 172, 173, 181, 182, 196, 197, 198, 200–1, 203, 205, 206, 208, 209, 216–17, 218–19
  in Korea, 167
  terminology, 217

Neo-Confucianism, 196, 219
  and modern science, xxvi

Neolithic age, 3

New Discourse on the Alchemical Laboratory. See *Tan Thai Hsin Lu*

New Discourses on the Talk of the Times. See *Shih Shuo Hsin Yü*

New Discussions. See *Hsin Lun*

New History of the Five Dynasties. See *Hsin Wu Tai Shih*

New History of the Thang Dynasty. See *Hsin Thang Shu*

New Treatise on Natural Philosophy and Natural History. See *Po Wu Hsin Phien*

Newly Reorganised Pharmacopoeia. See *Hsin Hsiu Pên Tshao*

Newton, Isaac, xxv, xxx

Ngô Si-Liên (Annamese historian, c. +1479), 75

Ni Hung-Pao (Ming Minister of Agriculture, c. +1640), 236

Nickel, 233, 236

Niello, 31 (a)

Night-shining jewel, 176 (h)

Nihon-koku Ganzai-sho Mokuroku (Bibliography of Books Extant in Japan), 178

Nihon Ryo-iki (Record of Strange and Mysterious Things in Japan), 175

Nihongi (Chronicles of Japan), 177 (c)

Nihongi Ryaku (Classified Matters from the Chronicles of Japan), 176 (b)

Nimmyō (Japanese emperor, r. +833 to +850), 179 (a)

'Nine flower elixir' (*chiu hua tan*), 113

'Nine-vessel elixir of the Yellow Emperor' (*Huang Ti chiu ting tan*), 130–1

'Ninefold Radiance elixir' (*chiu kuang tan*), 86

Ning Hsien Wang. *See* Chu Chhüan

Nitrate, 79, 241

Nitre, 129, 157–8, 232, 245
  ingestion of, 9, 78
  technical terms. *See* Technical terms

Nitric acid, 89, 102, 232, 237, 238–9

Nitrogen, 245

Nitron. *See* Natron

Nitto-Guhō Junrei Giyōki (Record of a Pilgrimage to China in Search of the Buddhist Law), 165

Niu Wên-Hou (disciple of the adept Yin Thung, +4th century), 112

Nobel, Alfred, 226 (b)

Nomenclature. *See* Technical terms

Northern School of Taoism. *See* Taoism

Notes from the Hall of Learned Old Age. See *Lao Hsüeh An Pi Chi*

Notes on the Nourishing and Prolonging of Life. See *Yang Shêng Yen Ming Lu*

Novels and plays, 212–15

Nü Yü (legendary immortal), 8

Nutritional science, 132 (d), 140 (b)

Ob Ugrians, 3 (a)

'Observations on the State of Medicine, Surgery and Chemistry in China' (Gillan), 234

Ōbuchi, Ninji (1), 113

Ochre, 15

Ode on a Girl of Matchless Beauty. See *Tao Su Fu*

'Oil of bricks', 237 (d)

Okada Masayuki, 165

Okanishi, Tameto (1), 132

Old Bolsheviks, 223

Old History of the Thang Dynasty. See *Chiu Thang Shu*

Oleum sulphuris, 238

Oleum vitrioli, 238

On the Beginning and End of the World. See *Huan Yu Shih Mo*

On the Cold Forest Jade-Cinnabar Casing Process. See *Pi Yü Chu Sha Han Lin Yü Shu Kuei*

On the Demonstration of the Mystery. See *Chih Hsüan Phien*

On the Gradual Understanding (of the Tao). See *Chien Wu Chi*

On reading the *Kinship of the Three*. See *Tu Tshan Thung Chhi*

On Replacing Doubts by Certainties. See *Tai I Phien*

Ono Imoko (Japanese ambassador to China, +607), 177

Ophthalmology, 163

Oracle-bones, 3

Oral Explanation of the *Kinship of the Three*. See *Tshan Thung Chhi Khou I*

Oral instruction, and alchemy, 82, 87, 104, 155, 196

Order of Mutual Production. *See* Mutual Production Order

Ore indications ('subtle sprouts', *ling miao*), 210

Organic acids, 147
Organic philosophy, 261
Organic substances, 181, 182, 186, 196
Orpiment, 78, 87, 99, 131, 135, 137, 144, 159, 181, 198, 201
Orrery, xxx
Ox bile, 101, 102
Oxalic acid, 49
Oxidation, 247
Oxygen, 242, 243-4, 245, 247, 250
    European discovery of, 248
    and hydrogen, 249
    technical terms. *See* Technical terms
Oyster-shells, 84, 133

Pa Hsien. *See* The Eight Immortals
Pa Kung (the Eight Venerable Adepts, at the court of the Prince of Huai-Nan), 23
Paekche, 177
Pagel, Walter, xxx
Pai Chü-I (poet, +722 to +846), 148-9, 163
Pai Yü-Chhan (Sung alchemist, *fl.* +1209 to +1224), 199, 203
Pai-Yün Kuan. *See* Temples
Paint, and painting, 245
    white lead, 17
Paktong (cupro-nickel), 39 (b), 136, 193, 233, 234
Palaeolithic age, 3
Pan chieh-yü (Court lady, *c.* −20), 17
Pan Yungphung (Korean Pharmacognostic Master, +554), 177
Pao Ching (scholar of alchemical interests and father-in-law of Ko Hung, perhaps identical with the radical philosopher Pao Ching-Yen), 76, 77, 80, 121, 142
Pao Ku (daughter of Pao Ching and wife of Ko Hung, also skilled in alchemy and medicine, especially acupuncture), 39 (d), 76
Pao-Kuang Tao-Jen (Yuan Taoist physician), 163
Pao Phu Tzu. *See* Ko Hung
Pao Phu Tzu (Book of the Preservation-of-Solidarity Master), 1, 5, 7, 27 (c), 39, 78, 99, 102, 105, 107, 109, 110, 113, 127-8, 198
Pao Phu Tzu (*Nei Phien*) (Book of the Preservation-of-Solidarity Master: Esoteric Chapters), 81-2, 83, 88, 109
Pao Phu Tzu Shen Hsien Chin Shuo Ching (The Preservation-of-Solidarity Master's Manual of the Bubbling Gold (Potion) of the Holy Immortals), 109
Pao Phu Tzu (*Wai Phien*) (Book of the Preservation-of-Solidarity Master: Exoteric Chapters), 80, 81
Pao Phu Tzu Yang Shêng Lun (The Preservation-of-Solidarity Master's Essay on Hygiene), 109
Pao sha lung ya (elixir), 19
Pao Tsang Lun (Discourse on the Precious Treasury of the Earth), 159, 180, 181
Pao Wên Thung Lu (General Catalogue of Precious Writings), 115
Paper, 234

Paracelsus (Theophrastus von Hohenheim, +1493 to +1541), xxv, xxx, 108, 123-4, 135, 137, 185, 220, 235 (h), 236
Paraguay, Jesuit communities in, 223 (a)
Parennin, Dominique (Jesuit, +1665 to +1741), 225-7
Partington, J. R. (7), 233; (8), 33
de Pauw, C. (Dutch philosopher, +1774), 227-9, 240
Peach, gum, 19
    wood, Fig. 1362*a*, caption
Pearls, 157
    artificial, 137, 230
    ingestion of, 35, 99
Pei Chi Tzu. *See* Yin Hêng
Pei Shih (History of the Northern Dynasties), 119, 131
Pei Tsung. *See* Taoism, Northern School of
Peking, 233
    translation bureau, 252
Peking Union Medical College, 250
Pelliot, P., 168
Pên Ching. See *Shen Nung Pên Tshao Ching*
Pên tshao tradition, 16
Pên Tshao Ching Chi Chu (Collected Commentaries on the Classical Pharmacopoeia (of the Heavenly Husbandman)), 123, 167
Pên Tshao Kang Mu (Great Pharmacopoeia), 15, 127, 158, 216
Pên Tshao Kang Mu Shih I (Supplementary Amplifications for the *Great Pharmacopoeia*), 239
Pên Tshao Phin Hui Ching Yao (Essentials of the Pharmacopoeia Ranked according to Nature and Efficacity), 216, 232 (b)
Pên Tshao Shih I (A Supplement for the Pharmaceutical Natural Histories), 31
Pên Tshao Yen I (Dilations upon Pharmaceutical Natural History), 188
Penetrating Explanation of the *Kinship of the Three*. See *Chou I Tshan Thung Chhi Tshê Su*
'Perfect Unity' (*chêng i*), 113
'Perfection', and 'imperfection', 243, 244, 245
Perfume oils, 48
Periodic Table, 31 (b)
Permeation, penetration and rest as opposed to ceaseless motion, 149
Persia, 136, 139
Persian dualism, 72
Persian lilac. *See* Plant genera, *Melia*
Personal deity, xxvi, 210 (i)
Pessimism, xxiv
    Wang Chhung's, 107
Pestle and mortar, 198
Phan Shih-Chêng (Thang alchemist, *fl.* +650 to +680), 121, 132, 140
Pharmaceutical Natural History of the Stove and Furnace. See *Lu Huo Pên Tshao*
Pharmaceutics of the Bodhisattva Nāgārjuna. See *Lung-Shu Phu-Sa Yao Fang*
Pharmaco-sexual techniques, 167
Pharmacological literature, 5

Pharmacopoeia of the Heavenly Husbandman. See *Shen Nung Pên Tshao Ching*

Pharmacy, 119, 178, 210, 219–20, 251

Phêng Hsiao (alchemical commentator, +947), 50, 51, 53, 63

Phêng-Lai (magical island of the immortals), 13, 18, 31, 33, 35, 36, 176 (h)

Phêng-lai Shan Hsi Tsao Huan Tan Ko (Mnemonic Rhymes of the Cyclically-Transformed Elixir from the Western Furnace on Phêng-Lai Island), 33–4

Phêng Ssu (Taoist alchemist, +13th century), 199

his collaboration with Mêng Hsü, 203, 208

Phêng Tsu (legendary immortal of great age, the Chinese Methuselah, expert in sexual techniques), 9

Philosophers' stone, 38, 100

Philosophical names, styles or pseudonyms

Chêng Yang Tzu, 200

Chhang Chen Tzu (Than Chhu-Tuan)

Chhang Chhun Tzu (Chhiu Chhu-Chi)

Chhang Shêng Tzu, 43

Chhang Shêng Tzu (Liu Chhu-Hsüan)

Chhien Yang Tzu (Yü Mu-Shun)

Chhih Sung Tzu, 9, 10, 20, 88, 91, 105 (b)

Chhing Hsia Tzu (Su Yuan-Ming and at least one other), 130, 159, 180, 211

Chhing Hsü Tzu (Chao Nai-An)

Chhing Lin Tzu (Chhing Lin), 95, 103

Chhüan Yang Tzu (Yü Yen)

Chhung Yang Tzu (Wang Chung-Fu)

Chi Chhiu Tzu (Chi Chhiu), 88, 95

Chi I Tzu (Fu Chin-Chhüan)

Chi Ni Tzu (Fan Li)

Chih I Tzu (Chao Nai-An)

Chin Lou Tzu (Hsiao I and at least one other), 95

Chiu Ling Tzu (Huang Hua)

Chüeh Tung Tzu (Li Hsiu)

Hai Chhan Tzu (Liu Tshao)

Hai Chhiung Tzu (Pai Yü-Chhan)

Han Fei Tzu (Han Fei)

Han Hsü Tzu (Chu Chhüan)

Hsien Mên Tzu (Hsienmên Kao)

Hsüan Chen Tzu (Mêng Yao-Fu)

Huai Nan Tzu (Liu An)

Khang Fêng Tzu (Khang Fêng), 91

Ku Sung Tzu, 150

Kuang Chhêng Tzu, 8

Kuang Ning Tzu (Ho Ta-Thung)

Liang I Tzu, 94, 99

Ling Yang Tzu (Lingyang Tzu-Ming)

Lu Thang Tzu (Chao Shêng)

Nan Chi Tzu (Liu Jung)

Pao Phu Tzu (Ko Hung)

Pei Chi Tzu (Yin Hêng)

Shang Yang Tzu (Chhen Chih-Hsü)

Shen Yin Tzu, 114

Shun Yang Tzu (Lü Tung-Pin)

Tan Yang Tzu (Ma Yü)

Thien Jan Tzu (Ho Ta-Thung)

Tshao I Tzu (Lou Ching)

Tshui Hsüan Tzu (Shih Thai)

Tshui Wên Tzu (Tshui Wên), 88, 91

Wu Chhêng Tzu, 88, 91, 95

Ying Chhan Tzu (Li Tao-Shun)

Yü Pei Tzu, 79, 195 (b)

Yü Yang Tzu (Wang Chhu-I)

Yuan Yang Tzu, 56

Yün Ya Tzu (Wei Po-Yang)

Phing Lung Jen, 242, 247

dating of, 248–9

suggested translations of the title, 247–8

Pho 'souls', xxxiv, 46, 48, 70, 71, 73

Phonetic principle in character-building, 259

Phosphorescence, 148 (d), 261 (b)

Phosphorus (*lin*), 261

Phu, tribal people of, 3

Physic gardens, 179

Physicians

Taoist, 212

Western, in China, 250

Physics, xxxi, 70, 220

Physiological alchemy. See *Nei tan*

Pi Shêng (chemical worker, and inventor of movable-type printing, c. +1000 to +1050), 187

Pi Yü Chu Sha Han Lin Yü Shu Kuei (On the Cold Forest Jade-Cinnabar Casing Process), 196

Piao and *li* (patefact and subdite, technical terms in medicine), 146

Pien Huo Phien (Disputations on Doubtful Matters), 209

Pigments, 244–5

organic, 232

Pilgrimage to the West. See *Hsi Yu Chi*

Pimihu (Pimiko, Japanese sorceress and ruler, c. +2nd century), Fig. 1363, caption

Pinchbeck, 189

Pine leaves, as food, 132, 146

Pine resin, 97 (cc)

Pine-seed elixirs, 9, 180

Pinene, 261

Ping shih (common salt, gypsum, or leonite), 89, 98

Piston-bellows, 133 (c)

Placer mining, 2 (b)

Plant accumulators, 207 (f), 232

Plant acids, in cementation processes, 49

Plant drugs, 33, 46, 112, 130, 139, 168

combination with metallic preparations, 48–9

idea that a herbal drug might heal a diseased metal, 49

in Japan, 178, 179

Plant genera

Ailanthus, 96 (aa), 99

Aleurites, 26, 233

Amanita, 11 (c)

Anchusa, 126 (d)

Aristolochia, 159

Asparagus, 11, 112 (b)

Atractylis (*Atractylodes*), 11, 40, 113, 177

Basella, 126 (f)

Plant genera (*cont.*)
  Berberis, 126 (j)
  Caesalpinia, 136 (k)
  Carthamus, 126 (d)
  Cinnamomum, 11
  Cuscuta, 15, 168 (g)
  Eucommia, 199 (c)
  Fomes, 11
  Gleditschia, 138 (a), 159, 232
  Gyrophora, 11
  Hedera, 147
  Hibiscus, 19 (c)
  Ixeris, 96 (cc)
  Lecanora, 9 (e)
  Lindera, 18 (f)
  Lithospermum, 126 (d)
  Lycium, 179 (b)
  Malva, 9 (e)
  Melia, 136
  Morus, 147
  Mulgedium, 96 (cc)
  Nelumbo, 207 (f)
  Pachyma, 11, 97 (cc)
  Paulownia, 24 (e)
  Pentaglottis, 126 (d)
  Phyllanthus, 189 (b)
  Pistacia, 198 (f)
  Polygonatum, 112 (b)
  Poria, 11
  Portulaca, 147, 207 (f)
  Quisqualis, 12 (e)
  Rhododendron, 126 (j)
  Rosa, 126 (j)
  Rubus, 11, 84 (j), 88, 89, 98, 168 (g)
  Sesamum, 96 (cc)
  Terminalia, 179 (c)
  Zizyphus, 31, 137
Plants, 196, 210
  containing chemical substances, 147
  containing metals, 207, 232
  counteracting the poison of elixirs, 119, 168 (g), 208
  as elixir ingredients. *See* Elixirs of longevity and immortality
  of immortality. *See* Herbs of immortality
  lists of, 12, 14
  safe to use for food, 210
'Plating' (*tu*), 130, 187, 189
Pleistocene age, 3
Pliny (+77), 16
Pneuma, 149
Po Hai (the Yellow Sea), 13
Po-lo-mên Yao Fang (Brahmin Pharmaceutics), 160
Po Wu Hsin Phien (New Treatise on Natural Philosophy and Natural History), 252
Poetical Essay on Realising the Necessity of Regenerating the Primary Vitalities. *See* *Wu Chen Phien*
Poetry, and ideas of alchemy and immortality, 148–50
Pointing the Way Home to Life Eternal. *See* *Chih Kuei Chi*

de Poirot, Louis (Jesuit, 1814), 223
Poisoning, 74, 78, 120, 135, 144, 146, 147, 151, 168–9, 231
Poisonous effects overcome when substances are employed together, 144
Poisonous substances, degrees of tolerance to, 135
Polar-equatorial system (astronomy), xxx
Pollen, as food, 146
Polo, Marco, 174 (h)
Pŏpchang (Paekche Buddhist priest, +685), 177
Porcelain industry and techniques, 230, 234, 237
'Potable gold', 1, 40, 49, 82, 83, 99, 178–9
  ingredients of, 88–9
Potash alum, 15, 86, 87, 133, 139 (f)
  aqueous solutions of, 102
Potassium, 250
  sulphates of, 45 (i), 138
Potassium aluminium sulphate, 15, 133
Potassium anions, 102
Potassium auricyanide, 99
Potassium flame test, xxix, 139
Potassium nitrate, 14, 139, 158, 238
Poverty, and aurifaction, 1, 2, 193, 194
Pre-established harmony, 107 (c)
Precession of the equinoxes, xxx
Precious Golden Badges, 188, 190
Precious Mirror of Eastern Medicine. *See* *Tongŭi Pogam*
Precious Patrology of the Mysterious Capital. *See* *Hsüan Tu Pao Tsang*
Precious Patrology of the Mysterious Capital (i.e. the Taoist Church) collected in the Great Chin Dynasty. *See* *Ta Chin Hsüan Tu Pao Tsang*
Precious Records of the Adept Tan-Yang. *See* *Tan-Yang Chen Jen Yü Lu*
Precious Secrets of the Realm of Kêng and Hsin. *See* *Kêng Hsin Yü Tshê*
Precipitation of copper from solutions by ionic exchange, 25 (h), 104, 207 (e)
Pregnant women, and elixirs, 171 (e)
Prescriptions, medical, 135, 144, 171
Preservation of dead bodies. *See* Putrefaction, inhibition of
The Preservation-of-Solidarity Master's Essay on Hygiene. *See* *Pao Phu Tzu Yang Shêng Lun*
The Preservation-of-Solidarity Master's Manual of the Bubbling Gold (Potion) of the Holy Immortals. *See* *Pao Phu Tzu Shen Hsien Chin Shuo Ching*
Priestley, Joseph (chemist, +1733 to 1804), 218, 221, 246
Prima materia, 163, 166
The Prince of Huai-Nan's Book of Change and Transformation. *See* *Huai Nan Pien Hua Shu*
Principles of the Coming into Being of Yin and Yang. *See* *Chhien Khun Shêng I*
Printing, 114 (c)
  first printed book on a scientific subject, 167–9
  movable-type, 187
  of the *Tao Tsang*, 115–16

Private wealth, as aim of aurifaction, 2, 168, 214

Professorship of Alchemy, establishment of, 117–18

Projection (*tien*), 38, 39, 84, 85, 88, 89, 100–1, 136, 161, 162, 174, 199

Prospecting
  bio-geochemical, 232
  mineralogical, 104 (d)

Proto-biochemistry, xxxiv

Proto-chemistry, xxxiii, 57, 103, 117, 136, 160, 172, 182, 197, 200, 203, 210, 211, 216, 223

Proto-gunpowder, 79, 159

Proto-scientific theories, 115, 120, 143, 171

Proverbs, 27, 56 (c), 197

Pseudo-Democritus (Hellenistic proto-chemist, probably +1st century), 48

Pseudonyms. *See* Philosophical names . . .

Psimithion, 16

Psycho-physiological alchemy. See *Nei tan*

Ptolemaic planetary astronomy, xxv

Pu Chhêng (magician-technician, perhaps another name for Shangchhêng Kung), 106

Puliang I (disciple of the legendary adept Khuang Chhêng Tzu), 8

Pulmonary abscess, 47

Pulse Manual in Two Chapters. See *Mo Shu Shang Hsia Ching*

Purity, of traditional chemical products, 128 (j)

'Purple of Cassius', 50 (a)

'Purple girl' (*tzu nü*, amethyst), 89

'Purple powder' (*tzu hsüeh*), 188–9

'Purple sheen gold', 173

'Purple snow' (*tzu hsüeh*), 179

'Purple travelling elixir' (*tzu yu tan*), 138

'Purpling' (*iōsis*), 173, 194

Purslane, 147 (e)
  containing mercury, 207 (f)

Putrefaction, inhibition of, xxx, xxxiv, 21, 75 (e), 104

al-Qazwīnī, Muḥammad ibn Maḥmūd (c. +1275), 174 (h)

Quantitative analysis, 9 (e)

Quantitative measurements, 69, 191

Quartz, 137, 199
  use in elixirs, 78, 131, 178

Quercetanus. *See* du Chesne

Questions about Jade. See *Yü Tshê Ching*

Questions and Answers on the Great Elixir. See *Ta Tan Wên Ta*

Quicksilver. *See* Mercury

Rabbit in the moon, 4, 21

Rainbow, xxix

Rangaku period, 259

Rangoon creeper. *See* Plant genera, *Quisqualis*

Ranka Naigai Sanbō Hōten (Handbook of the Three Aspects of Dutch Internal and External Medicine), 255(a), 260

Rasa-vaiśeshika Sūtra, 163

Rasaratnākara, 161–2

Rasarṇava Tantra, 124

Raspberry. *See* Plant genera, *Rubus*

Rationalism, 106–7

Ravelins, 241

Ray, P. C. (1), 161, 162, 163–4

al-Razī, Ibn Zakarīyā (+865 to c. +920), 45, 124 (a), 174 (h)

Reaction-vessels, 71, 73, 74, 133, 196, 199

Reactions between substances of the same category, 68–9, 145

Realgar, 78, 83, 86, 87, 88, 89, 98, 99, 101, 109, 112, 131, 133, 135, 137, 144, 146, 159, 181, 198, 201
  aqueous solutions of, 102
  toxicity of, 144

de Reaumur, R. A. F. (physicist, chemist, comparative biochemist and metallurgist, +1683 to +1757), 225 (d)

Recherches Philosophiques sur les Egyptiens et les Chinois (de Pauw), 228

'Recluse' and 'worldly' environments of alchemical activities, significance of, 36, 39, 82–3, 200

Recollections from the Estate of Literary Learning. See *Mo Chuang Man Lu*

Record of the Assemblies of Perfected Immortals. See *Yuan-Shih Shang Chen Chung Hsien Chi*

Record of the Auspicious Meeting of the Mysterious Winds. See *Hsüan Fêng Chhing Hui Lu*

Record of the (Different) Grades of Immortals. See *Hsüan Phin Lu*

Record of the Great Elixir. See *Ta Tan Chi*

Record of a Pilgrimage to China in Search of the Buddhist Law. See *Nitto-Guhō Junrei Giyōki*

Record of Strange and Mysterious Things in Japan. See *Nihon Ryo-iki*

Record of Things Heard at Spring Island. See *Chhun Chu Chi Wên*

Record of the Transfiguration of the Adept Hsü (Hsün) at (the Age of) Eighty-Five. See *Hsü Chen Chün Pa-shih-wu Hua Lu*

Records of Hua-shan, the Great Western Mountain. See *Hsi Yo Hua-shan Chih*

Records of the Origins of Affairs and Things. See *Shih Wu Chi Yuan*

Records of the Pillow-Book. See *Chen Chung Chi*

Records of the Plants and Trees of the Southern Regions. See *Nan Fang Tshao Mu Chuang*

Records in the Rock Chamber; a Thai-Chhing Scripture. See *Thai-Chhing Shih Pi Chi*

Records of (Twenty-five) Strange Magician-Technicians . . . See *Chiang Huai I Jen Lu*

Records of the Western Countries in the Time of the Great Thang Dynasty. See *Ta Thang Hsi Yü Chi*

Red
  as the auspicious colour, 5–7, 31 (b)
  and symbolic revivification of the dead, 2–3

Red bole clay (*chhih shih chih*), 14, 84, 131, 133

'Red dust' (the illusory nature of worldly things), 149 (b)

Red Flag, 6 (a)

'Red flowery elixir', 43
Red ink, 6
'Red Lady of Paviland' skeleton, 3 (a)
Red lead. *See* Lead oxides
Red Pine Master. *See* Chhih Sung Tzu
Red Regions (*tan thien*, centres of vital heat in the human body, three, each consisting of nine components), 151
'Red salt', 103 (c)
Red sappan pigment, 136
'Red snow' (*hung hsüeh*), 179
Redoubts, 241
Reduction (*huan yuan*), 261
Reel of the fishing-rod, 12
'Refined elixir' (*lien tan*), 83
Reflux condensation, 196
Regularity and repetitiveness in Nature, *vs.* uniqueness and unpredictability, 107 (c)
Religious experience, 108
  and natural science, xxvii, 107
Renou & Filliozat (1), 163
Reports on Spiritual Manifestations. See *Sou Shen Chi*
Respiratory exercises, 8, 12, 23 (e), 82, 112, 141, 176, 198, 199, 201, 205
Retribution, 106, 194
Revivification, 137
Revolts and uprisings
  of An Lu-Shan, 114, 141 (d), 182
  Boxer, 117
  of Huang Chhao, 114, 182
  of Shih Ssu-Ming, 114, 182
  of Wang Hsiao-Po and Li Shun, 192
Rey, Jean (chemist, +1582 to after +1645), 247
Rhinoceros horn, 133
Ricci, Matteo (leader of the Jesuit mission, +1552 to +1610), 218, 221, 224, 227, 229, 238
Rites Controversy, 221–3
'River chariot' (*ho chhê*, lead), 66, 67
Rock-salt, 78, 87, 142–3, 233
Rocket weapons, 184
Rolfinck, Werner (chemist, +1599 to +1673), 232 (a)
Roman Empire, amount of gold held by, 37
Romans, 3 (f), 29 (a)
Rouge powders and creams, 126, 127
Royal Society, 152
Rudolf II (Holy Roman Emperor, r. +1576 to +1612), 191 (b)
Rūḥ (ouranic spirit, somewhat analogous to *hun*), 48
Ruhland, Martin (alchemical lexicographer, +1612), 152
Russia, 249
Rust. *See* Corrosion

A Sack of Pearls from the Three Heavens. See *San Tung Chu Nang*
Sacrifices, 37, 108
  fêng and *shan*, 31, 33
  to the furnace spirit, 29, 31, 240 (f)

Sadvaha (Indian king), 162
Safflower (*hung hua*). *See* Plant genera, *Carthamus*
Saga (Japanese emperor, r. +809 to +823), 178
Śakti goddesses, 166 (a)
Sal ammoniac, 69, 146, 159, 173, 195, 198, 210 (f), 233, 239, 245
Saliva, 57, 155
  of foxes, 189
Salt, 78, 89, 102, 131, 137, 181, 234
  and preservation of dried meat, 104
Saltpetre, 14, 79, 98, 102, 109, 137–8, 139, 146, 159, 196, 198, 234, 237, 239, 240, 243, 246
  cooling of water by the addition of, 225
  ingestion of, 9, 78, 89, 144
  technical term, 89
San-Fêng Tan Chüeh (Chang San-Fêng's Instructions about Enchymomas), 209
San Kuo Chih (History of the Three Kingdoms period), 96 (w), 97 (cc)
San Mao Chün. *See* The Three Lords Mao
San-shih-liu Shui Ching (Manual of the Thirty-Six Aqueous Solutions), 105
San-shih-liu Shui Fa (Thirty-Six Methods for Bringing Solids into Aqueous Solutions), 105, 120
San tung. See the Three Heavens
San Tung Chhiung Kang (Essentials of the Magnificence of the Three Heavens), 114
San Tung Chu Nang (A Sack of Pearls from the Three Heavens), 112, 114
San Yen. See the Three Collections
Śāṇaka, identity of, 164
Sanaka Sō (1), 123
Sandalwood oil, 48
di Santillana, Giorgio, xxvii
Saponins, 138 (a), 232
Sappan. *See* Plant genera, *Caesalpinia*
Sātavāhana (Indian king, +2nd century), 162
de Saussure, Léopold (historian of astronomy, 1866 to 1925), xx
Sceptical Chymist (Boyle), 172, 235, 241
Sceptical rationalism, 106–7
  combined with empiricism at the Renaissance, 107 (e)
'Scepticism' and 'enthusiasm', marriage of, at the Renaissance, 107 (e), 108
Schall von Bell, John Adam (Jesuit, +1640), 221, 235–6, 237, 239, 240–1
Scheele, C. W. (chemist, +1742 to +1786), 221
Schipper, K. (1), 34
Schlegel (11), 232
Scholasticism, 172
School of Foreign Languages (Peking), 252, 254
School of Naturalists (Yin-Yang Chia), 5, 12–19
Shreck, Johann (Jesuit, +1576 to +1630), 228 (c), 235, 236, 242
Science
  continuity and universality of, xxiii–xxv xxvi–xxvii
  Islamic. *See* Islamic science
  modern, xxiv–xxxi, 108, 152, 217, 220

Science (*cont.*)
and morality, xxviii–xxix, 227
philosophy of, 149
religion and, xxiii, xxvii, 107
research in, 107 (e)
women and, 191
Scientific revolution
inhibition of, in China, xxv, 219
in the West, 75
Scientific Sketches. See *Ko Chih Tshao*
Scroll painting, 136
Scyths, 229
Secrecy. *See* Alchemists, secretiveness of
Secret Art of Penetrating the Mystery. See *Thung Hsüan Pi Shu*
Secret Doctrine of the Numinous Cinnabar and the Great Elixir. See *Ling Sha Ta Tan Pi Chüeh*
Secret Instructions concerning the Jade Chamber. See *Yü Fang Pi Chüeh*
Secretary's Commonplace Book. See *Mo O Hsiao Lu*
Seimi Kaisō (Treatise on Chemistry), 255
Selections from the *Tao Tsang*. See *Tao Tsang Chi Yao*
Selenite, 88, 131
Self-cultivation (*hsiu ming*), 205, 219
Self-mummification, 176
Self-mutilation, Buddhist, 165
Self-purification, 167 (b)
Semen, 155
Sesame oil, 137
Setsuyō Yoketsu. See *Shê Yang Yao Chüeh*
Seven Bamboo Tablets of the Cloudy Satchel. See *Yün Chi Chhi Chhien*
The Seven Divisions (*pu*, of Taoist literature), 113
'Seven-fold cyclically-transformed cinnabar', 138
The Seven Perfect-Truth Masters (of the Northern School of Taoism), 204–5
Sex
and Taoist ideas in Japan, 176
and Taoism in China, 37 (d), 205
and alchemy, 57, 69, 145, 201, 214,
of metals and minerals, 88 (c), 95, 97, 166 (d)
Tantric liturgical, 166 (a)
Sex hormones. *See* Hormones
Sexual techniques, 39, 55 (a), 56 (c), 57, 70–1, 82, 114, 201, 205
Shamanism, 8, 13 (b), 17, 50
Shan Hai Ching (classic of the Mountains and Rivers), 23, 24, 229
Shan Tao-Khai (Taoist adept, *fl. c.* +310), 112
Shang (period), 5, 17
Shang-Chhing (class of Taoist scriptures), 76
Shang-Chhing Han Hsiang Chien Chien Thu (The Sword and Mirror Diagram embodying the Image; a Shang-Chhing Scripture), 141
Shang Yang Tzu. *See* Chhen Chih-Hsü
Shang Yang Tzu Chin Tan Ta Yao. See *Chin Tan Ta Yao*
Shangchhêng Kung. *See* Pu Chhêng
Shanghai, 252

Shanghai Polytechnic Institute and Reading-Rooms, 255
Shao lien ('transmutation by heating', an alchemical phrase but perhaps also a reference to the Buddhist practice of self-mutilation), 165
Shao Ong (Early Han court thaumaturgist, −121), 31, 33
Shê Ta Chhêng Lun Shih (Explanatory Discourse to assist the Understanding of the Great Vehicle), 164
Shê Yang Yao Chüeh (Important Instructions for the Preservation of Health conducive to Longevity), 179
Shen (ancient term for a god, later the spirit in man; and in Neo-Confucian philosophy, a term for expansion-dispersion processes), 196
Shen Chih-Yen (Thang alchemical writer, *c.* +864), 171
Shen-Chou (magical island of the immortals), 18
Shen Hsien Chuan (Lives of the Holy Immortals), 52, 75, 101, 110
Shen Hsien Thung Chien (General Survey of the Lives of the Holy Immortals), 30, 39, 51
Shen I-Ping (Taoist, +1708 to +1786), 218
Shen Kua (+11th-century astronomer, engineer and high official), 187
Shen Nung Pên Tshao Ching (Pharmacopoeia of the Heavenly Husbandman), 5, 9 (e), 11 (b), 46–7, 105
Shen Nung Ssu Ching (Four Books of Shen Nung), 105
Shen tan ('magical elixir'), 83
Shen Tsung (Ming emperor, r. +1573 to +1620), 44, 117
Shen Wan-San (Ming alchemist, +15th century), 209
Shen Yin Tzu (Thang Taoist editor), 114
Shen Yü-Hsia (Ming woman alchemist, +15th century), 209
Shêng Shui Yen Than Lu (Fleeting Gossip beside the River Shêng), 187
Shenthu Pieh-Chia (Taoist alchemist, *c.* +885), 174
Shenthu Ssu-Ma, 174
Shih Chi (Historical Records), 6, 12, 17, 20, 29, 47
Shih chung huang tzu (brown haematite), 9 (c)
Shih Ho (aurifactive alchemist, +4th century), 111
Shih hsien-sêng tan fa (Teacher Shih's elixir method), 91
Shih Khuan-Shu (Han alchemist, *c.* −121), 31–2
Shih Ming (Explanation of Names Dictionary), 16, 126
Shih Shuo Hsin Yü (New Discourses on the Talk of the Times), 45
Shih Ssu-Ming's rebellion (+8th century), 114, 182
Shih Thai (Sung physiological alchemist, d. +1158), 203
Shih Tshên (Taoist biographer, +4th century), 111

Shih Tsu (Chhien Chin emperor). *See* Fu Chien

Shih Tsung (Northern Chou emperor, r. +557 to +561), 165

Shih Tsung (Jurchen Chin emperor, r. +1161 to +1190), 116

Shih Tsung (Ming emperor, r. +1522 to +1567), 212

Shih Tzu-Hsin (Secretary in a Han Prime Minister's office, and would-be aurifactor), 82–3

Shih Wu Chi Yuan (Records of the Origins of Affairs and Things), 126

Shih Yao Erh Ya (Synonymic Dictionary of Minerals and Drugs), 88, 89, 129, 138, 151–3, 156

Shimao, Eikoh (1), 255

Shingon Buddhism, 177

Shingū, tomb of Jofuku (Hsü Fu) at, 18

Shintoism, 175–6

al-Shīrāzī, Quṭb al-Dīn (+1300), xxix

Shoku-Nihongi (Chronicles of Japan, continued), 174 (f)

Shoku-Nihonkoki (Chronicles of Japan, still further continued), 179 (a)

Shōsōin Treasury, Nara, 84, 128 (j), 178

Shu Chi-Chen (Chin alchemist, +4th century) 121

Shu Ching (Historical Classic), 4

Shugendō (mountain asceticism) cult, 175

Shui yin. *See* Mercury

Shui yin fên (quicksilver powder, i.e. calomel), 124

Shui-yin shuang tan (calomel), 128

Shun Yang Tzu. *See* Lü Tung-Pin

Shun-Yang Lü Chen-Jen Yao Shih Chih (The Adept Lü Shun-Yang's (Book) on Preparations of Drugs and Minerals), 19, 147–8

Shunyü I (physician, −216 to c. −147), 46–7

Shunyü Shu-Thung (Han administrative officer and alchemist, c. +150), 50, 55, 77

Shuo Wên dictionary, 5

Si-vúóng (Annamese king, +2nd century), 75

Siberian tribes, 174

Śikshānanda (translator, +699), 164

Siliceous clays, 15

Silver, 1, 155, 163, 217, 219, 233, 244
  black lead and, 78
  blackened by the vapours of sulphur, 31 (a)
  changed into 'gold', 161, 244
  in coinage, 37
  ingestion of, 2, 78, 99, 135, 144
  making of. *See* Argentifaction
  projection for, 199
  transformed to look like gold, 104
  vessels of, 212

Silver chloride, 236, 239

Silvering, 123, 207 (a)

Similarities and Categories of the Five (Substances). See *Wu Hsiang Lei*

Similarities and Categories of the Five (Substances) in the *Kinship of the Three*. See *Tshan Thung Chhi Wu Hsiang Lei*

Similarities and Categories of (Substances formed

by) the Five Elements. See *Wu Hsing Hsiang Lei*

Similia cum similibus agunt, 145, 201

Sinophilism, movement against, 228

Sivin, Nathan, xx–xxi, xxiv, xxv, xxix, xxxi; (1), 102, 132, 133

Sixteen-fold Cyclically-Transformed Gold Elixir prepared by the 'Responding to the Chhi' Method. See *Kan Chhi Shih-liu Chuan Chin Tan*

Slave-girls, 192

Small Encyclopaedia of the Principles of Things. See *Wu Li Hsiao Shih*

Smelting, 84, 102, 139, 232, 236

'Snow-silver', 170

Soap, 137
  -making, 232

Soap-bean pods, 138, 159, 232

Soapstone, 15, 84

Social and economic factors, in the break-through of the scientific revolution, xxviii

Soda. *See* Natron, Nitre

Sodium, 233, 250

Sodium aluminate (*lü suan na*), 259

Sodium borate, 232

Sodium carbonate, 158, 232

Sodium chlorate (*lü suan na*), 259

Sodium chloride, 196, 232

Sodium nitrate, 158, 232

Sodium sulphate, 131, 143, 145, 158, 232

'Soft elixir' (*jou tan*), 83

Solar plexus, 151 (a)

Solders, 232

Solubilisation processes, 160 (a)

'Song of Potable Gold', 150

Sŏnjo (Korean king, r. +1568 to +1608), 167

'Sophic lead', 155, 217 (c)

'Sophic mercury', 201 (b), 217 (c)

'Sophic sulphur', 201 (b)

Sores, 78

Sorites, 29 (f), 31 (b)

Sou Shen Chi (Reports on Spiritual Manifestations), 80

'Soul-recalling elixir', 99

Southern School of Taoism. *See* Taoism

Soxhlet continuous extraction process, 170

Spagyrical. *See* Alchemical

Spanish colonialism, 223 (a)

Spengler, Oswald, xxiii–xxiv

Spirits and ghosts. *See* Ghosts and Spirits

Spizel, Theophilus (sinologist, +1660), 7 (d)

Ssu fu. *See* Four Ancillaries

Ssu Ming (Director of Destinies, a Taoist divinity), 29 (c)

Ssu Shih Tsuan Yao (Important Rules for the Four Seasons), 178 (d)

Ssuma Chhêng-Chên (Taoist alchemist, c. +639 to c. +727), 121, 140–1

Ssuma Chhien (Early Han Historiographer-Royal), 2, 29, 33, 46

Stalactites, used in elixirs, 131, 179

Stalagmites, used in elixirs, 131

Stannic sulphide (mosaic gold), 99 (f), 233 (g)
    earliest known preparation of, 103
Staphylococcal infections, 47
State support of research, 29, 31, 35, 82, 117,
    185, 189, 252
Stein, O. (1), 162, 164
Stick lac (*tzu kung*), 126
Stills
    for distilling mercury, 198
    'Moor's head', 259
Stone beads, 3
Stove platform, 198
Strange Tales New and Old. See *Chin Ku Chhi
    Kuan*
Strickmann, M. (2), 123
Study of the Kinship of the Three. See *Tshan
    Thung Chhi Khao I*
Styles. *See* Philosophical names . . .
Styria (Austria), peasants of, 135
Su Chhê. *See* Su Ying-Pin
Su Lin (Han adept, d. c. −60), 40
Su Lo (Hsin magician-technician, +10), 37
Su Shen Liang Fang (Beneficial Prescriptions
    collected by Su Tung-Pho & Shen Kua), 193
Su Shih. *See* Su Tung-Pho
Su Sung (high official, astronomer and engineer,
    +1020 to +1101), xxx
Su Tung-Pho (poet, +1036 to +1101), 193–4
Su Tzu-Chan. *See* Su Tung-Pho
Su Tzu-Yu. *See* Su Ying-Pin
Su Ying-Pin (scholar and writer, brother of Su
    Tung-Pho, +1039 to +1112), 182, 193, 194
Su Yuan-Lang. *See* Su Yuan-Ming
Su Yuan-Ming (alchemist, physician and writer),
    +3rd and +4th centuries), 130, 145
'Subduing', of chemicals, 137 (m), 138, 158 (c)
    See also 'Killing' and 'Controlling'
Sublimation process, 131, 137
    of chlorides of mercury, 124
    of mercury and sulphur, 74, 231
    purification by repeated, 85
Subtle Discourse on the Alchemical Elaboratory.
    See *Tan Fang Ao Lun*
'Subtle sprouts' (*ling miao*), 210
Succinic acid, 236
Sufis, xxvi
Dr Sui (Archiater-Royal or Physician to the
    Prince of Chhi, d. c. −160), 47
Sui Shu (History of the Sui Dynasty, biblio-
    graphy), 12, 24, 25, 52, 160, 163, 199
Suicide, Buddhist and political, 165 (d)
Sulpha-drugs, 47
Sulphate cations, 102
Sulphide films, 147
Sulphur, 14, 67, 68, 74, 87, 89, 98, 131, 133, 135,
    137–8, 144, 159, 173, 181, 196, 201, 243, 244
    and attainment of immortality, 146
    combustion of, 246
    heated with realgar, saltpetre and honey, 78–9
    polymeric, 137
Sulphur and mercury, 74, 78, 79
    cyclical transformation of, 14 (a), 126, 198

Sulphur–mercury theory of metals, 238, 245
Sulphuric acid, 237–8, 239–40
Sulphydryl groups, 147
Sun Kung Than Phu (The Venerable Mr Sun's
    Conversation Garden), 193
Sun Pu-Erh (woman adept, seventh and last of
    the Seven Masters of the Northern School of
    Taoism, +1119 to +1183), 205
The Sun-Rays Master's Pharmaceutical Natural
    History . . . See *Jih Hua Chu Chia Pên Tshao*
Sun Shêng (Sung scholar and writer, c. +1085),
    193
Sun-spots, xxix
Sun Ssu-Mo (alchemist, physician and writer,
    c. +581 to after +672), 46, 121, 132–8, 145,
    159, 197, 219, 220, 246
    his style, 133
Sun Thai-Chhung (Thang alchemist, +744), 148
Sun Wên-Thao (Liang alchemist, +6th century),
    121
Sun Yu-Yo (Taoist alchemist, d. +489), 121
Sung (dynasty), 190
    decline of alchemy in, 182
    alchemical compendia in, 196–9
Sung-shan Mountains, 119, 132, 146
Sung Shih (History of the Sung Dynasty), 79,
    158, 163, 183, 194
Sung Tê-Fang (Taoist editor, +1237), 116, 205
Sung Wu-Chi (Warring States adept), 13
Sung Ying-Hsing (encyclopaedist of technology,
    +1637), 126, 219 (d)
Supplement for the Pharmaceutical Natural
    Histories. See *Pên Tshao Shih I*
Supplement to the Beginnings of All Things. See
    Hsü Shih Shih
Supplement to the *Record of the Investigation of
    Things*. See *Hsü Po Wu Chih*
Supplement to the Thousand Golden Remedies.
    See *Chhien Chin I Fang*
Supplementary Amplifications for the Great
    Pharmacopoeia. See *Pên Tshao Kang Mu
    Shih I*
Supplementary Taoist Patrology of the Wan-Li
    reign-period. See *Wan-Li Hsü Tao Tsang*
Surface-enrichment techniques, 49, 129, 147,
    207 (b), 230
The Sword and Mirror Diagram embodying the
    Image. . . See *Shang-Chhing Han Hsiang
    Chien Chien Thu*
Symbolic animals, 153 (e), 153 (f), 156 (a), 156 (e)
Symbolic colours, 67
Symbolic correlations, xxv, 58, 96 (d), 181
Symbolic notation in Chinese alchemy, 152–5
Sympathetic magic, 3
Synonymic Dictionary of Minerals and Drugs.
    See *Shih Yao Erh Ya*
Synonyms. *See* Alchemical synonyms and meta-
    phors
Szechuan (province), 173, 174, 192

Ta Chih Tu Lun (translation of the *Mahā-
    prajñāpāramito-padeśa Śāstra*), 161, 162

Ta Chin Hsüan Tu Pao Tsang (Precious Patrology of the Mysterious Capital (i.e. the Taoist Church) collected in the Great Chin Dynasty), 116

Ta Fang Kuang Fo Hua Yen Ching (translation of the *Avataṁsaka Sutra*), 164

Ta Huan Tan Chao Chien (An Elucidation of the Great Cyclically-Transformed Elixir), 180, 181

Ta Sung Thien Kung Pao Tsang (Treasures of the Heavenly Palace; the Great Sung Patrology), 115

Ta Tan Chi (Record of the Great Elixir), 51

Ta Tan Chhien Hung Lun (Discourse on the Great Elixir of Lead and Mercury), 172

Ta Tan Chih Chih (Direct Hints on the Great Elixir), 205

Ta Tan Wên Ta (Questions and Answers on the Great Elixir), 79

Ta Thang Hsi Yü Chi (Records of the Western Countries in the Time of the Great Thang Dynasty), 162

Ta-Tung Lien Chen Pao Ching, Chiu Huan Chin Tan Miao Chüeh (Mysterious Teachings on the Ninefold Cyclically-Transformed Gold Elixir, according to the Precious Manual of the Re-casting of the Primary (Vitalities); a Ta-Tung Scripture), 141–2, 143

Ta-Tung Lien Chen Pao Ching, Hsiu Fu Ling-Sha Miao Chüeh (Mysterious Teachings on the Processing of Numinous Cinnabar, according to the Precious Manual of the Re-casting of the Primary (Vitalities); a Ta-Tung Scripture), 141, 143

Ta-Yuan. *See* Ferghana

Taboos, 106

Tabula Smaragdina, 193

Tai Hou chhi tzu. See Lady of Tai

Tai I Phien (On Replacing Doubts by Certainties), 228 (c)

Takada Chitsugi (Japanese pharmaceutical adept, c. +850), 179

Taketori Monogatari (Tale of the Bamboo-Gatherer), 176

Takizawa Bakin (Japanese novelist, c. 1820), 175

Talc, 15, 133

Tale of the Bamboo-Gatherer. See *Taketori Monogatari*

Tales of Today and Long Ago. See *Konjaku Monogatari*

Talismans (*fu*), 3, 7, 106, 114, 153 (a), 176, 188, 190, Fig. 1362a, caption

Tamba no Yasuyori (Japanese physician and compiler of medical texts, +982), 177

Tamjing (Korean Buddhist priest, +610), 177

Tan (cinnabar, or elixir), 4, 86–7, 89, 99

Tan Ching Yao Chüeh (Essentials of the Elixir Manuals), 128, 133, 135, 137, 138, 246

Tan-Chou (magical island of the immortals), 19

Tan Fang Ao Lun (Subtle Discourse on the Alchemical Elaboratory), 197

Tan Fang Ching Yuan (The Mirror of the Alchemical Elaboratory; a Source-book), 158

Tan Fang Chien Yuan (Mirror of Alchemical Processes (and Reagents); a Source-book), 158, 180, 181, 211

Tan Fang Hsü Chih (Indispensable Knowledge for the Chymical Elaboratory), 198

Tan Lun Chüeh Chih Hsin Chien (Handbook of the Secret Teaching concerning Elixirs), 172, 173

Tan sha. See Cinnabar

Tan Thai Hsin Lu (New Discourse on the Alchemical Laboratory), 211

Tan thien. See Red Regions

Tan-Yang Chen Jen Yü Lu (Precious Records of the Adept Tan-Yang), 204

Tan-Yang Shen Kuang Tshan (Tan Yang (Tzu's Book) on the Resplendent Glow of the Numinous Light), 204

Tan Yang Tzu. *See* Ma Yü

Tanaka Minoru (3), 255

Tancke, Joachim (anatomist, alchemist and chemical practitioner, +1537 to +1609), 236 (h). *See* 'Valentine, Basil'

Tang Shan mountains, 127

Tanning, 210 (g)

Tantrism, 36 (f), 161, 166, 176–7

Tao kuei (a 'large knife-point' of a substance, rough unit of measurement), 72 (f)

the Tao of Nature, 59–60, 108–9, 210 (i)

Tao Shu (Axis of the Tao), 55

Tao Su Fu (Ode on a Girl of Matchless Beauty), 17 (f)

Tao Tê Ching (Canon of the Tao and its Virtue), 52, 116, 194, 217, 222 (e)

Tao Tsang (Taoist Patrology), 40, 44, 51, 53, 54, 78, 79, 105, 109, 110, 111, 112, 113–17, 120, 128, 132, 142, 145, 146, 158, 172, 196, 197, 203, 204, 205
   burning of, 208
   first definitive edition of (+1019), 115, 183, 185
   printing of, 115, 190, 216

Tao Tsang Chi Yao (Selections from the *Tao Tsang*), 54, 117

Tao Tsang Hsü Phien Chhu Chi (First Series of a Supplement to the Taoist Patrology), 218

Tao Wu Ti (Thopa Kuei, emperor of the Northern Wei Dynasty, r. +377 to +409), 117

Taoism and Taoists, xxiii, xxxiv, 8, 43, 52, 55, 56, 75, 118, 169, 184, 219, 224, 243
   and attainment of immortality, 4, 21, 23
   and Buddhism, 191
   and the development of natural science, 107–8
   and elixir-making, 166, 250
   and experimentation, 107 (e)
   in Japan, 175–6
   liturgies, 108, 166, 190
   and magic, 106, 108, 120
   manual work and, 208, 219–20
   and modern science, xxvi
   myths and legends, 23

Taoism and Taoists (*cont.*)
    Northern School (Pei Tsung), 200, 203–8
    poverty of, 2
    red regarded as 'auspicious' by, 6
    secret cover-names used by, 104
    Southern School (Nan Tsung), 200–3, 208, 218
    supported by Sung emperors, 182, 190–1
    suspected of subversive political ideas, 208
    women and, 39
Taoist alchemists, 4, 26, 28–9, 36, 76, 106, 212
Taoist ceremonies, 69, 72, 106, 196
Taoist Church, 120, 161
Taoist Patrology. See *Tao Tsang*
Taoist Patrology of the Chêng-Thung reign period. See *Chêng-Thung Tao Tsang*
The Taoist Patrology. compiled in honour of Royal Longevity. See *Wan Shou Tao Tsang*
Taoist Section of the Bibliography of the Seven Classes of Books. See *Chhi Lu (Hsien Tao Lu)*
Taoist writings, 105, 113–17
    destruction of, 44, 114, 182
    search for, 115, 182
Tartars, 229
Tchîne-chĕ, 243
Tea-drinking, 68 (g)
Technical knowledge, and alchemy, 219–20
Technical Methods of the Adept (Ko) Chih-Chhuan, with Critical Annotations. See *Chih-Chhuan Chen Jen Chiao Chêng Shu*
Technical terms, 206
    alchemy, 152
    ambix, 199
    amethyst, 89
    amino-acid, 261
    biochemistry, 259
    bismuth, 259
    bromine, 259
    calomel, 124, 125, 127, 128–9, 133
    carbon, 259
    categories, 145, 146
    chemical substances, 14–16, 232 (c), 259, 261
    chemistry (modern), 252, 254, 255, 261
    cinnabar, 66, 87, 150, 153, 156, 157
    common salt, 89
    copperas, 89, 239
    corrosive sublimate, 124–5
    'doubling', 136
    elixirs, 20, 45, 83–5, 86–7, 88
    gases, 259
    gold, 87, 104
    heating processes, 143
    hydrogen, 261
    jet, 96 (cc)
    lead, 66, 67, 68, 127–8; metallic, 157
    medical sciences, 259
    mercury, 4, 5, 66, 84 (j), 87, 88, 104, 153, 156, 157, 259
    metal, 259
    minium, 68 (e)
    mono-sodium orthophosphate, 261
    natron, or nitron, 232, 245

    nitre, 157
    nitric acid, 239
    oxygen, 247 (d), 259
    pharmacy, 146
    phosphorus, 261
    plant drugs, 47
    projection, 38 (b), 101
    realgar, 89
    reduction, 261
    sal ammoniac, 239
    saltpetre, 89
    silver, 104
    sodium aluminate, 259
    sodium chlorate, 259
    tetrahydro-furane, 261
    vinegar, 157
    vitamin, 259 (a)
Technicians, 28, 35–6, 38, 107, 219–20
Techniques for attaining longevity and material immortality, 13, 14, 29, 75, 205
    dietary. *See* Dietary regimens
    gymnastic, 20
    meditational, xxxv, 8
    pharmaco-sexual, 167
    respiratory. *See* Respiratory exercises
    sexual. *See* Sexual techniques
Technology, and morality, xxviii
Temkin, O. (3), 45
Templars, 223
Temples
    Arising of the Yang, Terrace of the. *See* Chhu-Yang Thai
    Buddhist, renamed by Taoists, 191
    Chhien-fo-tung (Mo-kao-khu, Buddhist), 150, 165 (d)
    Chhu-Yang Thai (Taoist), Ko Ling, Hangchow, 81 (b)
    Chhung-Hsü Kuan (Taoist), Lo-fou Shan, 81 (a), Fig. 1353*a*, caption
    Erh-Hsien Ssu (Taoist), Chhêng-tu, 117 (b)
    Eternal Heaven. *See* Thien-Chhang Kuan
    Golden Glory Island Shrine. *See* Kinkazan Jingu
    Grand Unity. *See* Thai-I Kuan
    Great Eastern. *See* Tōdaiji
    Great Loving-Kindness. *See* Ta-Tzhu Ssu
    Hsüan-Tu Kuan (Taoist), 114
    Khai-Yuan Ssu (Buddhist), Tingchow, built by Wang Chieh, 187
    Kinkazan Jingu (Shinto), 174 (f)
    Mysterious Capital. *See* Hsüan-Tu Kuan
    Opening of Origins. *See* Khai-Yuan Ssu
    Pai-Yün Kuan (Taoist), Peking, 117, 250
    Pure Yang. *See* Shun-Yang Kung
    Shun-Yang Kung (Taoist), Thaiyuan, in honour of Lü Tung-Pin, 148
    Ta-Tzhu Ssu (Buddhist), 192
    Taoist, 195
    Thai-I Kuan (Taoist), 140
    Thang Temple of Monks from the Four Directions. *See* Tōshōdaiji
    Thien-Chhang Kuan (Taoist), 116

Temples (*cont.*)
  Tōdaiji (Buddhist), Nara, 175
  Tōshōdaiji (Buddhist), Nara, 178
  Tunhuang cave-temples. *See* Chhien-fo-tung
  Two Immortals. *See* Erh-Hsien Ssu
  Vast Inane. *See* Chhung-Hsü Kuan
  White Clouds. *See* Pai-Yün Kuan
Temporary death, and resurrection, 51, 166
Ten-day week, 19
The Ten Thousand Infallible Arts of the Prince
  of Huai-nan. See *Huai Nan Wan Pi Shu*
Ten Types of Secret Books on the Verification of
  the Tao. See *Chêng Tao Pi Shu Shih Chung*
Tendai wuyaku (drug-plant), 18 (b)
Têng Yo (Chin Governor of Canton, +4th
  century), 81
Têng Yü (Liang adept, d. +515), 120
Tequesquite, 232
Terebinthus. See Plant genera, *Pistacia*
Terrentius. *See* Schreck, Johann
Tetrahydro-furane (*i yang wu yuan*), 261
Textile technology, 261
Thai-Chhing Chen Jen, 85 (b)
Thai-Chhing Chen Jen Ta Tan (The Great Elixirs
  of the Adepts; a Thai-Chhing Scripture),
  132 (g)
Thai-Chhing Chin I Shen Tan Ching (Manual of
  the Potable Gold (or Metallous Fluid) and the
  Magical Elixir (or Enchymoma); a Thai-
  Chhing Scripture), 96 (c), 99
Thai-Chhing Shih Pi Chi (The Records of the
  Rock Chamber; a Thai-Chhing Scripture),
  96 (s), 99, 130, 131, 138, 198
Thai-Chhing Tan Ching (Manual of the Grand-
  Purity Elixir), 78, 86, 96 (c)
Thai-Chhing Tan Ching Yao Chüeh (Essentials
  of the Elixir Manuals for Oral Transmission;
  a Thai-Chhing Scripture), 132, 197, 198
Thai-Chhing Yü Pei Tzu (The Jade-Tablet
  Master, a Thai-Chhing Scripture), 79
*Thai Chi Chen-Jen Chiu Chuan Huan Tan Ching
  Yao Chüeh* (Essential Teachings of the
  Manual of the Supreme-Pole Adept on the
  Ninefold Cyclically-Transformed Elixir), 199
Thai Chi Chen-Jen Tsa Tan Yao Fang (Tractate
  of the Supreme-Pole Adept on Miscellaneous
  Elixir Recipes), 199
Thai-Chi Ko Hsien-Ong Chuan (Biography of the
  Supreme-Pole Elder-Immortal Ko), 76
Thai-Ho Chen Jen. *See* Yin Kuei
Thai Hsi Shui Fa (Hydraulic Machinery of the
  West), 234–5
Thai Hsüan Nü (legendary woman alchemist), 42
Thai-I hsün shuo chung shih (probably realgar),
  89
Thai-I Kuan (temple), 140
'Thai-I Soul-Recalling elixir', 92
Thai-Ku Chi (Collected Works of (Ho) Thai-Ku),
  205
Thai-pai Shan mountains, 141
Thai Phing Ching (Canon of the Great Peace and
  Equality), 6 (a), 75

Thai-Phing Yü Lan (Thai-Phing reign-period
  Imperial Encyclopaedia), 15, 26, 39
Thai Shan, Mt, 33
Thai-Shang Tung Fang Nei Ching Chu (Esoteric
  Manual of the Innermost Chamber, with
  Commentary), 40
Thai Tsu (Later Liang emperor, r. +907 to
  +914), 180
Thai Tsung (Thang emperor, r. +627 to +649),
  132
Thai Tsung (Sung emperor, r. +976 to +997),
  115, 182, 194
Thai Wu Ti (Northern Wei emperor, r. +424 to
  +452), 118
Thai Yang Nü. *See* Thai Hsüan Nü
Thai-Yang Tao, 6 (a)
Thai Yin Nü. *See* Thai Hsüan Nü
Than Chhu-Tuan (second of the Seven Masters
  of the Northern School of Taoism, +1123
  to +1185), 204
Than hsien-sêng Shui Yün Chi (Mr Than's
  Records of Life among the Mountain Clouds
  and Waterfalls), 204
Than Ssu-Hsien (Taoist biographer), 76
Than Ssu-Thung, 254
Thang (period), 79, 114, 146, 160, 199, 201, 231,
  246
  alchemy, 198
  bibliography, 52
  Indian-Chinese chemical contacts, 139–40
  literature and poetry, 148–50
  manuscripts, 247
Thang Kung-Fang (Han immortal), 19
Thang Shen-Wei (Sung pharmaceutical natura-
  list, +1080), 31 (a)
Thao Chhu-Ching. *See* Thao Tan
Thao Chih (Thang writer on physiological
  alchemy), 172, 173
Thao Chung-Wên (Ming high official, +16th
  century), 212
Thao Hung-Ching (pharmaceutical naturalist,
  alchemist and physician, +456 to +536),
  xxix, 39 (b), 46, 108 (b), 110, 113, 119–20,
  121, 122, 123, 127, 128, 129, 136, 140, 145,
  219, 220
Thao Tan (Chin adept), 112
Thaumaturgists, 49, 160, 175
Theophrastus (d. −287), 16
Theories of categories. *See* Categories
Theories of modern type, intrinsic to the scientific
  revolution, 219
Theravadin Buddhism, 166
Thien (technical term, to add to or to augment in
  'doubling'), 136
Thien Jan Tzu. *See* Ho Ta-Thung
Thien Kung Khai Wu (Exploitation of the Works
  of Nature), 219 (d)
Thien Liang-I (Sui or Thang alchemist), 121
Thien ming sha (probably potash alum), 139(f)
Thien Shih Tao (Taoist church), 44
Things seen and heard in the Western Garden.
  See *Hsi Yuan Wên Chien Lu*

Thirty-Six Methods for Bringing Solids into Aqueous Solutions. See *San-shih-liu Shui Fa*

Thölde, Johann (industrial chemist, *fl.* +1580 to +1612), 123. *See* 'Valentine, Basil'

Thopa Kuei. *See* Tao Wu Ti

The Thousand Golden Remedies. See *Chhien Chin Yao Fang*

Thread chromatography, 48 (c)

The Three Classes of Drugs (princely, ministerial, and adjutant), 144, 146

The Three Collections (*San Yen*, of stories by Fêng Mêng-Lung, d. +1646), 212–13

Three Commentaries on the 'Poetical Essay on the Understanding of the Truth'. See *Wu Chen Phien San Chu*

The 'Three Corpses' (*san shih*, in the body), 176

The Three Heavens (*san tung*, divisions of Taoist literature), 113

Three Kingdoms (period), 45, 96 (w), 97 (cc), 113

The Three Lords Mao (San Mao Chün), 39 (b)

'Three Messengers elixir', 131

The Three Primary Vitalities, xx, 52

The Three 'Red Regions'. *See* Red Regions

The Three Yellow (substances), 201

Thu Hsiu Chen Chün Tsao-Hua Chih Nan (Guide to (lit. South-pointing Compass for) the Creation, by the Earth's Mansions Immortal), 210

Thu Hsiu Pên Tshao. See *Thu Hsiu Chen Chün Tsao-Hua Chih Nan*

Thu Po (the Earth Lord), 23

Thunder and lightning, 226

Thung Chih Lüeh (bibliography in the *Thung Chih*), 130, 135, 158, 211 (c)

Thung Hsüan Pi Shu (Secret Art of Penetrating the Mystery), 171

Thung-tree wood, 26

Thung Wên Kuan (translation bureau in Peking), 252

Tibet, 161, 166, 167 (c), 232

Tides, 106

Tien, or *Tien hua* (projection), 38 (b), 39, 84, 85, 88–9, 100, 101, 136, 161–2, 174, 199

Tiger, as symbol, 66

Tiles, 174, 228 (d), 233

Time, xxix, xxxiv

Time-controlling substances, xxxiv

Time-keeping, xxx, xxxiii

Tin, 16, 100, 234, 244
 amalgam, 128

Tin acetate, 236

Tin-foil, 234

Tin sulphide. *See* Stannic sulphide

Ting (nail, or ingot), 102

Ting I (Chin alchemist, +4th century), 76, 77, 121, 142

Ting Jih-Chhang (statesman, 1823 to 1882), 252

Titles of all the Taoist Canons and their Meanings. See *I Chhieh Tao Ching Yin I*

Toad in the moon, 21

Tōdaiji temple, Nara, 175

Tomb-guardian figures, 23 (d)

Tombs, 3
 Chhin, 16
 of Chhin Shih Huang Ti, 4–5
 Han, 16
 of Ho Lu, King of Wu, 5
 of Jofuku, 18
 Koguryŏ, 167

Tompion, Thomas, xxx

Tongŭi Pogam (Precious Mirror of Eastern Medicine), 167

Tonics, 178

Toothache, 130

Tōshōdaiji temple, Nara, 178

Tou (Empress Dowager of the Han, −133), 29

Tou Tzu-Ming. *See* Lingyang Tzu-Ming

de Tournon, Charles Maillard (Latin Patriarch of Antioch and papal legate in China, *d.* +1710), 222 (a)

Tractate of the Supreme-Pole Adept on Miscellaneous Elixir Recipes. See *Thai Chi Chen-Jen Tsa Tan Yao Fang*

Translations
 from alphabetical languages into Chinese, 255, 259
 into Chinese, of Western treatises on science and technology, 254

Transliteration, 259, 261

Transmissions and contacts, 237
 Chinese–Japanese, 177
 from East to West, 221, 223, 230–1, 234
 Indian–Chinese, 139–40, 164
 from West to East, 136, 234, 250

Transmutation. *See* Alchemical transmutation

'Travelling Canteen', 29 (f)

Treasures of the Heavenly Palace; the Great Sung Patrology. See *Ta Sung Thien Kung Pao Tsang*

Treatise on Chemistry. See *Seimi Kaisō*

Treatise on the Material Composition of the Universe. See *Khung Chi Ko Chih*

Trebuchets, 184

Trees
 bearing 'pure fruit-maidens', 166
 of the sun and the moon, 19

Tribal peoples
 Amerindians, 3 (e)
 Jurchen Tartars, 116
 Manchus, 229
 Ob Ugrians, 3 (a)
 of Phu, 3
 Scyths, 229
 Siberian, 174

Trigault, Nicholas (Jesuit, +1621), 227, 235, 236

Trigonal calcium carbonate, 14

Tripiṭaka (Buddhist Patrology), 160

Trithemius (Abbot Johannes, +1462 to +1516), 31 (c)

True and Essential Teachings about the Great Magical Cinnabar of the Jade Cavern. See *Yü Tung Ta Shen Tan-Sha Chen Yao Chüeh*

Tsao-hua chê (the Author, or Foundation, of Change), 210 (i)

Tsao-Hua Chhien Chhui (The Hammer and Tongs of Creation), 211

Tsao-wu chê (the Author of Things, or Nature), 210 (i)

Tsêng Kuang Chih Nang Pu (Additions to the Enlarged Bag of Wisdom), 213

Tsêng Kung-Liang (military encyclopaedist, +1044), 184

Tsêng Min-Hsing (Sung scholar and writer, +1176), 189

Tsêng Tshao (Sung Taoist writer), 55

Tshai Nü (The Chosen Girl; Taoist goddess), 92

Tshai Thien (Thang alchemist, c. +885), 173

Tshai Yung (Later Han scholar), 40

Tshan Thung Chhi (Kinship of the Three), 5, 17, 50, 51, 52, 53, 57, 87, 105, 130, 133, 148, 150, 173, 182, 183, 199, 200, 206, 208, 217, 218–19
earliest extant edition, 53
ku wên (ancient script) text, 55
literary style of, 56, 74
origins of the text, 56
various interpretations of, 56 ff.

Tshan Thung Chhi Chhan Yu (Explanation of the Obscurities in the *Kinship of the Three*), 54, 71

Tshan Thung Chhi Khao I (A Study of the *Kinship of the Three*), 53

Tshan Thung Chhi Khou I (Oral Explanation of the *Kinship of the Three*), 55

Tshan Thung Chhi Wu Hsiang Lei (The Similarities and Categories of the Five (Substances) in the *Kinship of the Three*), 124, 144, 145, 197

Tshan Thung Chhi Wu Hsiang Lei Pi Yao (Arcane Essentials of the Similarities and Categories of the Five (Substances) in the *Kinship of the Three*), 51, 145

Tshao I Tzu. *See* Lou Ching

Tshao Tshan (Early Han minister and alchemist, d. −190), 11, 20, 77

Tshao Tshao (+155 to +220, founder of the Wei State in the Three Kingdoms period), 106

Tshao Yuan-Yü (1), 67

Tshui Fang (Sung iatro-chemist and naturalist, c. +1045), 185, 210, 220

Tshui Hao (Northern Wei alchemist, c. +440), 118

Tshui Hsü Phien (Book of the Emerald Heaven), 203

Tshui Hsüan Tzu. *See* Shih Thai

Tshui Tsê (Northern Wei alchemist, c. +450), 118

Tshui Wên Tzu (Master Tshui-Wên, Han, San Kuo or Chin alchemist), 88, 91

Tso Chuan (Master Tsochhiu's Enlargement of the Spring and Autumn Annals), 6

Tso Tzhu (magician and alchemist in Han, +155 to +220), 76, 77, 105

Tso Yuan-Fang. *See* Tso Tzhu

Tso Yuan-Tsê (Thang alchemist), 121

Tsou Yen (Naturalist philosopher, c. −350 to −270), 2, 5, 6, 7, 12–13, 14, 28

Tsung Hsiao-Tzu (Thang alchemist and diviner, c. +880), 173

Tsurezuregusa (Gleanings of Leisure Moments), 176 (a)

Tu ('poison' or 'active principle'), 135 (d)

Tu Chung (semi-legendary adept), 199 (f)

Tu Hsing Tsa Chih (Miscellaneous Records of the Lone Watcher), 189

Tu Kuang-Thing (Thang and Wu Tai alchemist, +850 to +933), 121

Tu Tshan Thung Chhi (On reading the *Kinship of the Three*), 55

Tuan Chhao-Yung (Ming alchemist, +16th century), 212

Tucci, G. (4), 163

Tuku Thao (Wu Tai alchemical writer, c. +950), 158, 180, 211

Tung Fêng (Han alchemist and physician in Annam, +2nd century), 75

Tung Hsien Pi Lu (Jottings from the Eastern Side-Hall), 192

Tung-Hsüan Chin Yü Chi (Collections of Gold and Jade; a Tung-Hsüan Scripture), 204

Tung Mi (Official of the Board of Rites, interested in dietary and pharmaceutical macrobiotics, c. +400), 117

Tung oil, 233

Tung Yüeh (Chhing historian, +1660), 125

Tungkuo Yen-Nien (Han adept, +1st or +2nd century), 40

Tunhuang cave-temples, 150, 165 (d)

Tunhuang MSS, 146, 247

'Turkestan' salt, 133

Tzu-Jan Chi (Collected (Poems) on the Spontaneity of Nature), 204

Tzu-Yang Chen Jen. *See* Chang Po-Tuan (d. +1082), Chou Chi-Thung (c. −60)

Udagawa Yōan (chemist, +1798 to 1846), 255

Udyāna, 139

Unani medicine, 49

Underworld, 21, 23

Unity and continuity of human culture, xxiii, xxx

Unofficial History of the World of Learning. See *Ju Lin Wai Shih*

Ur, 49

Urinary hormones, 78 (c), 179 (a)

Urine, xxx, 181; elixirs from, 36, 78

de Ursis, Sabatino (Jesuit, +1612), 235

Vagnoni, Alfonso (Jesuit, +1633), 235

'Valentine, Basil', ps. (chemist, c. 1600), 123, 137, 237

Vamacāra, 36 (f), 166

Vaporisation, 103

Vauquelin, L. N. (chemist, +1763 to 1829), 221

Vega (asterism), 70 (c)

Veneers, 129

The Venerable Li's elixir method, 92

The Venerable Master Lei's Treatise on the Decoction and Preparation of Drugs. See *Lei Kung Phao Chih Lun*

The Venerable Mr Sun's Conversation Garden. See *Sun Kung Than Phu*

The Venerable Yen Mên's Explanations of the Mysteries. See *Yen Mên Kung Miao Chieh Lu*

Verdigris, 136, 244

Verditer, 245 (b)

Veritable Records of Heaven and Earth. See *Erh I Shih Lu*

Veritable Records of the Reign of the Emperor Montoku. See *Montoku Jitsuroku*

Vermilion, 74, 125, 126, 127
paint used in tomb burials, 3

Vessels of 'gold', 1, 31, 33, 49, 140, 212

Vessels of 'silver', 212

Veterinary medicine, 211

Vieta, Franciscus (mathematician, +1540 to +1603), 221

Vietnam, alchemy in, 167

Vinegar, 87, 88, 89, 98, 99, 102, 109, 131, 135, 181, 218
making of, 231

Vinegar process. *See* Dutch vinegar process

Virgil, 3 (f)

Vitalis of Furno. *See* du Four, Vital

Vitamins, 147 (e)
technical term, 259 (a)

Vitriol, 97 (cc), 128, 135, 143, 159
blue, 101
green, 89, 237, 238, 239
white, 233

Vitruvius (engineer and builder, *c.* −27), 16

Volatility, 26 (a), 68 (b)

Volta, Alessandro (physicist, +1745 to 1827), 249

Volumetric analysis, 9 (e)

Voyages in search of the elixir of immortality, 13, 17–19, 31, 33, 34–5

Vyāḍi (semi-legendary Indian alchemist), 164

Wai tan practical laboratory alchemy), xx, xxxv, 57, 66, 74, 100, 153, 155, 171, 182, 196, 198, 203, 206, 208, 216, 220

Wai Tan Pên Tshao (Iatro-chemical Natural History), 185, 210, 220

Wai Thai Pi Yao (Important (Medical) Formulae and Prescriptions now revealed by the Governor of a Distant Province), 163

Wakan Sanzai Zue (Chinese and Japanese Universal Encyclopaedia), 174 (e)

Waley, A., 215; (4), 72; (24), 160, 164

Wamyō Ruijūshō (General Encyclopaeic Dictionary), 179 (d)

Wan (technical term for single inter-carbon bond), 261

Wan Hsing Thung Phu (General Dictionary of Biography), 243

Wan-Li Hsü Tao Tsang (Supplementary Taoist Patrology of the Wan-Li reign-period), 44, 117, 216

Wan Shou Tao Tsang (The Taoist Patrology compiled in honour of Royal Longevity), 115–16

Wang Chê. *See* Wang Chung-Fu

Wang Chêng (scholar interested in physics and engineering, collaborator of Johann Schreck, +1620), 228 (c)

Wang Chhang (Han alchemist, +1st century), 43, 44

Wang Chhin-Jo (Taoist scholar and editor, +1008), 115

Wang Chhu-I (fifth of the Seven Masters of the Northern School of Taoism, +1142 to +1198), 205

Wang Chhung (sceptical philosopher, +27 to +97), 106

Wang Chhung-Yang's Instructions on the Golden Gate and the Lock of Jade. See *Chhung-Yang Chin-Kuan Yü-So Chüeh*

Wang Chhung-Yang's Records of the Perfect-Truth School. See *Chhung-Yang Chhüan Chen Chi*

Wang Chieh (aurifictive metallurgical technologist, *fl. c.* +980 to *c.* +1020), 186–90, 192, 207 (c), 219

Wang Chin (Physician-in-Attendance of the Imperial Academy of Medicine, in the Ming, +16th century), 212

Wang Chün's elixir method, 92

Wang Chung-Fu (founder of the Northern School of Taoism, +1113 to +1170), 148, 203, 208

Wang Chung-Kao (Han alchemist, −2nd century), 23

Wang Fu (Taoist commentator), 55

Wang fu jen (d. −121, consort of Han Wu Ti), 31

Wang Hsi-Chih (calligrapher, +321 to +379), 76

Wang Hsi-I (Thang recluse, +8th century), 146

Wang Hsiao-Po (leader of an uprising, +995), 192

Wang Hsing (Han alchemist, +1st or +2nd century), 40, 49

Wang Hsüan-Ho (Thang Taoist bibliographer), 112

Wang Hsüan-I (Taoist mountain recluse, +750), 141

Wang I (commentator, *fl.* +115 to +135), 12

Wang Jo-Na (Thang and Wu Tai Taoist, *fl.* +875 to +920), 180

Wang Jung (Thang and Wu Tai alchemist, +873 to +921), 180

Wang Khuei-Ko (2), 88, 89, 98

Wang Kuei (Thang alchemist, +7th century), 121

Wang Mang (Hsin emperor, r. +9 to +23), 29, 37–8, 39

Wang Ming (3), 57

Wang Pao (court scholar, −1st century), 13

Wang Phan (pharmaceutical naturalist, +1505), 216

Wang Phi-Chih (Sung writer, *c.* +1090), 187

Wang Ssu-Chen (Han alchemist, *fl.* +180), 76, 77

Wang Tao-I (disciple of Yin Thung, +4th century), 112

Wang Than (Han adept, −198 to −123), 199 (g)

Wang Thao (Thang medical writer, +752), 163

Wang Thao (19th-century collaborator of James Legge), 255
Wang Tzu-Hsi (Taoist at the court of Hui Tsung, +1111), 190–1
Wang Tzu-Yuan. *See* Wang Yen
Wang Wei-Hsüan (Han alchemist, +2nd century), 40
Wang Wên (Ming ophthalmologist), 163
Wang Wên-Lu (Ming commentator, +1564), 54
Wang-wu Mountains, 118
Wang Yang (Han scholar-official suspected of alchemy), 27
Wang Yen (Later Shu emperor, r. +918 to +925), 174
Wang Yen (Taoist bibliographer, d. +604), 112, 114
Wang Yü-Chhêng (Taoist bibliographer, *fl.* +989), 115
Wang Yuan-Chih (Thang alchemist, +7th century), 121, 132
Wangyu Rungtha (Korean Professor of Medicine, +554), 177
Wāqwāq (Arabic name for Japan), Fig. 1363, caption
Ware, J. R. (5), 79, 80, 81, 82, 84, 88, 128
Warnings against Inadvisable Practices in the Work of the Stove. *See Lu Huo Chien Chieh Lu*
Warring States (period), 6, 7, 8, 46, 48, 246–7
Water
  composition of, 244, 246
  electrolytic decomposition of, 246 (c)
  freezing of, 225–6
Water (element), 73, 74, 172–3
Water-jackets, 199
Water-mills, 141 (d)
Water-raising devices, 235
Water-tanks, 235
Wax, 78
Weeks, M. E., 246
Wei State (Three Kingdoms period), 45
Wei (Prince of the State of Chhi, −4th century), 13
Wei Ching-Chao (Thang alchemist, +8th century), 121
Wei Fa-Chao (Thang alchemist, +7th century), 121
Wei-hsi (jet), 96 (cc)
Wei Po-Yang (Later Han alchemist and writer on alchemical theory, mid +2nd century), 17, 50–75, 76, 77, 173, 182, 191, 201
Wei Po-Yang Nei Ching (The Inner Book of Wei Po-Yang), 52, 105
Wei Shang (Chin or pre-Chin thaumaturgist), 105
Wei Shu (History of the (Northern) Wei Dynasty), 117
Wei Shu-Chhing (Han adept, +1st or +2nd century), 40
Wei Thai (Sung scholar and writer, end of +11th century), 192
Wei Wên-Hsiu (Northern Wei alchemist, *c.* +450), 118

Weighing of reactants, 69 (c), 191
Wells, D. A. (1), 254
Wên Hsien Thung Khao (Comprehensive Study of (the History of) Civilisation), 243
Wên Hsüan Ti (Northern Chhi emperor, r. +550 to +560), 131
Wên Ti (Han emperor, r. −179 to −156), 27
Wên Wang system (of the *Book of Changes*), 63
Wesley, John, 108
Western (European) alchemy, xxxiii, xxxv, 57, 75, 100, 136, 172, 195, 214, 228
  and moral intentions, 101
  style of writing, 181
Western chemistry, 123–4, 172, 218, 233, 234, 237
  failure to exert any influence in China till 19th century, 221–2, 241–2
Western culture, xxv
Western embryological theories, 70 (h)
Western metallurgy, 236
Western misconceptions about China, 228–9
Western proto-chemistry, 136, 172
Western science, xxvii, xxviii, 107 (e), 250
'Wet copper method', 207 (e)
White (colour), 5, 127 (a)
White lead, 15, 68, 103, 133, 136
  ink, 17
  most ancient reference to artificially made, 16, 124
  paint, 17
'White-out', 70 (a)
Wild raspberry. *See* Plant genera, *Rubus*
Wine, 104, 137, 234
  consecration of, 38 (b)
  as elixir ingredient, 87, 99
Wo Chhüan (legendary immortal), 9
Women
  alchemists, 38, 39, 42, 169–71, 191–2, 205, 209
  Kingdoms of. *See* Kingdoms
  pregnancy and elixirs. *See* Pregnant women
  Taoists, 39
  scientists, 191
Wood (element), 172–3
Woolley, C. L., 49
World-views, xxiv, xxix
  Chinese, xxxiv
  religious, xxvii
  scientific, xxviii
'Worldly' *vs.* 'recluse' environments, 36, 39, 82–3, 200
Writings of Wang Chhung-Yang . . . and the Ten Precepts of his Teacher. See *Chhung-Yang Fên-Li Shih-Hua Chi*
Wu (shamans, magicians), 17
Wu Chen Phien (Poetical Essay on Realising the Necessity of Regenerating the Primary Vitalities), 200, 203, 206, 219
Wu Chen Phien San Chu (Three Commentaries on the *Poetical Essay on Realising the Necessity of Regenerating the Primary Vitalities*), 206
Wu Chhêng-Ên (Ming novelist, *c.* +1570), 215

Wu Chhêng Tzu (Han, San Kuo or Chin alchemist), 88, 91
  his aurifaction method, 95
Wu Chhu-Hou (Sung writer, c. +1070), 186
Wu Ching Tsung Yao (Collection of the Most Important Military Techniques), 159, 184, 241
Wu Ching-Tzu (Chhing novelist, fl. +1736 to +1749), 215
Wu Hsiang Lei (Similarities and Categories of the Five (Substances)), 51
Wu Hsing Hsiang Lei (Similarities and Categories of (Substances formed by) the Five Elements), 53
Wu Li Hsiao Shih (Small Encyclopaedia of the Principles of Things), 218, 239
Wu Ling Tan Fa (Methods for Preparing Five Numinous Elixirs), 96 (e)
Wu Lu-Chhiang & Tenney L. Davis (1), 54, 56–7, 58, 60, 67, 73, 103; (2), 81
Wu Mêng (Chin alchemist, d. +374, one of the leaders of the Taoist Church), 76, 121, 142
Wu shih san (medicinal powder of the five minerals), 45
Wu Shu (scholar and writer, c. +975), 169
Wu Ta-Wên (Government Secretary at Chhêngtu, studied alchemy without success), 39, 100
Wu Tai. See Five dynasties
Wu Tai Shih Chi. See Hsin Wu Tai Shih
Wu-thai Shan, mountains, 139
Wu Ti (Early Han emperor, r. −140 to −86), 11, 13, 23, 27, 29, 31–4, 39, 106, 199
Wu Ti (Liang emperor, r. +502 to +549), 119–20
Wu Ti (Northern Chou emperor, r. +561 to +578), 112, 114
Wu Tsê Thien (Thang empress, r. +685 to +705), 140
Wu Tshêng (scholar and writer, +12th century), 150
Wu Tsung (Thang emperor, r. +841 to +847), 151, 165, 166, 182
Wu Wu (writer on alchemical apparatus, +1163), 198
Wu Yüeh Chhun Chhiu (Spring and Autumn Annals of the States of Wu and Yüeh), 5
Wu Yün (Thang alchemist, +8th century), 140
Wylie, Alexander, 54, 254

Xenophon (d. c. −350), 16

Yamabushi ('mountain monk') practices, 176
Yang Chiai (Sung scholar-official interested in alchemy), 228 (d), 230
Yang Hsi (Taoist religious writer, +330 to +387), 76, 77, 121
Yang Shen (Ming scholar), 55
Yang Shêng-An. See Yang Shen
Yang Shêng Yen Ming Lu (Notes on the Nourishing and Prolonging of Life), 120
Yang Tê-Wang (Chinese scholar and Jesuit), 228 (a)

Yang Thing-Yün (Christian apologist, +1621), 228 (c)
Yang Ti (Sui emperor, r. +604 to +617), 132. See also Kerr (1) in Addendum to Bibliography C.
Yao (legendary emperor), 9
Yao Hsüeh-Chhien (translator, 19th century), 254
Yeh Mêng-Tê (Sung scholar interested in iatrochemistry, +1077 to +1148), 204 (l)
Yeh Tê-Hui (editor), 25
Yellow (colour), 173 (d)
  associated with the Earth element, 58, 67
  as imperial colour, 5 (l)
the Yellow Emperor. See Huang Ti
The Yellow Emperor's Book on the Harmony of the Seen and the Unseen. See Huang Ti Yin Fu Ching
The Yellow Emperor's Manual (of Corporeal Medicine). See Huang Ti Nei Ching
The Yellow Emperor's Manual of the Nine-Vessel Magical Elixir. See Huang Ti Chiu Ting Shen Tan Ching
The Yellow Emperor's Manual of the Nine Vessels. See Huang Ti Chiu Ting Ching
Yellow River, 32
'Yellow silver'. See Brass
'Yellow Sprout' (huang ya, metallic lead, sulphur, or litharge), 66, 67, 68
Yen (State), 13, 31, 34
Yen Kho Lung-Mu Lun (Nāgārjuna's Discussions on Ophthalmology), 163
Yen Mên Kung Miao Chieh Lu (The Venerable Yen Mên's Explanations of the Mysteries), 171
Yen Shih-Ku (Thang commentator, +641), 27, 32 (b)
Yen Thieh Lun (Discourses on Salt and Iron), 34
Yi the Archer, 4
Yi Chunmin (Korean Taoist, +16th century), 167
Yi Su-gwang (Korean Taoist, +14th century), 167
Yin and Yang 'souls', 48, 70
Yin and Yang theory, xxv, 12, 13, 25, 40, 47, 48, 52, 53, 59, 64, 69, 74, 243, 244, 246, 249
  and alchemical theory, 143, 144–5, 155, 157
  and Persian dualism, 72
Yin Chen-Chün Chin Shih Wu Hsiang Lei (The Deified Adept Yin (Chhang-Shêng's Book) on the Similarities and Categories of the Five (Substances) among Metals and Minerals), 88, 89, 145–6, 157
Yin Chen Jen. See Yin Chhang-Shêng
Yin Chhang-Shêng (Later Han alchemist, fl. +120 to +210), 43, 49, 51, 53, 75–6, 77, 195
Yin Hêng (Han adept, −2nd century), 20
Yin Kuei (Chin alchemist, c. +290 to +307), 101, 111–12
Yin Kung-Tu. See Yin Kuei
Yin Ling-Chien. See Yin Thung
Yin Thung (Chin adept, d. +388), 112
Ying Chhan Tzu. See Li Tao-Shun
Ying-Chou (island of the immortals), 13

Ying hua, 207

Ying I-Chieh (Thang alchemist), 121

Ying Shao (Later Han scholar and writer, d. +195), 17 (c), 27

Ying Tsung (Ming emperor, +1459), 209

Yogacāra, xxxv, 166. *See also* Nei tan

Yogasara, 163 (c)

Yoginis, 174 (h)

Yoshida Mitsukuni, (5), 68, 72, 79, 105, 131

Yōshyō sennin (Japanese thaumaturgist, *fl.* +901), 176

Yü the Great (semi-legendary emperor), 16

Yü Chang (Sui Taoist, d. +614), 112

Yü-Chhing Chin-Ssu Chhing-Hua Pi-Wên Chin-Pao Nei-Lien Tan Chüeh (The Green-and-Elegant-Secret Papers in the Jade-Purity Golden Box on the Essentials of the Internal Refining of the Enchymoma, the Golden Treasure), 200

Yu chi (organic chemistry), 26

Yü Chieh Tao Tê Ching (Imperial Commentary on the *Canon of the Tao and its Virtue*), 190

Yü Fan (Taoist commentator, *c.* +230), 52

Yü Fang Pi Chüeh (Secret Instructions concerning the Jade Chamber), 177 (b)

Yü Hsi (astronomer, *fl.* +307 to +338), xxx

Yu-kan-tzu (fruit). *See* Plant genera, *Phyllanthus*

Yü Kung (Tribute of Yü). Chapter of the *Shu Ching*, *q.v.*

Yü Mu-Chhun (Chhing commentator, 1841), 54

Yü Nü (Jade Girls; Taoist angels), 197, 207

Yü Pei Tzu, 79, 195 (f)

Yü shêng (disciple of Wei Po-Yang, +2nd century), 51, 77

Yü Tshê Ching (Questions about Jade), 97 (cc)

Yü Tung Ta Shen Tan-Sha Chen Yao Chüeh (True and Essential Teachings about the Great Magical Cinnabar of the Jade Cavern), 141, 142

Yü Yang Tzu. *See* Wang Chhu-I

Yü Yen (Yuan Taoist commentator, late +13th century), 51, 54, 56, 65, 69, 193

Yuan (dynasty), 219

Yuan Han-Chhing (*1*), 50, 67, 68, 79, 110, 197, 198, 246, 248

Yuan Shih Shang Chen Chung Hsien Chi (Record of the Assemblies of Perfected Immortals; a Yuan-Shih Scripture), 110

Yuan Ti (Liang emperor, r. +552 to +555), 18

Yuan Yang Tzu (alchemical commentator), 56

Yüeh Hsia Kung (Chhin alchemist), 77

Yüeh Tzu-Chhang (Chin or pre-Chin alchemist), 91

Yün Chi Chhi Chhien (Seven Bamboo Tablets of the Cloudy Satchel), 44, 115, 123, 133, 142, 182, 185, 196–7

Yün Chhi Yu I (Discussions with Friends at Cloudy Pool), 167

Yün Kuang Chi (Collected (Poems) of Light (through the Clouds)), 205

Yun Kunphyŏng (Korean alchemist, +16th century), 167

Yün Ya Tzu. *See* Wei Po-Yang, 56

Yün Yang Tao Jen. *See* Chu Yuan-Yü

Yunnan (province), 233

Zinc, 187 (g), 233, 234, 236, 245 (f)

Zinc carbonate (zinc-bloom), 234, 243 (d)

Zinc sulphate, 233

Zoology, 221

Zoroastrianism, 72

Zosimos of Panopolis (+4th century), 109

ADDENDA TO BIBLIOGRAPHY B

An Chih-Min (*1*) 安志敏.
　Chhangsha Hsin Fa-hsien-ti Han Po Hua
　　Shih-Than 長沙新發現的西漢帛畫
　　試探.
　A Tentative Interpretation of the Western
　　(Former) Han Silk Painting recently
　　discovered at Chhangsha (Ma-wang-tui,
　　No. 1 Tomb).
　KKTH, 1973, no. 1 (no. 124), 43.
An Chih-Min (*2*) 安志敏.
　Chin Pan yü Chin Ping; *Chhu Han Chin*
　　Pi chi chhi yu Kuan Wên-Thi 金版與
　　金餅; 楚漢金幣及其有關問題.
　The 'chin pan' and 'chin ping' currency;
　　a study of the Gold Coins of the Chhu
　　State and the Han dynasty, and Some
　　Related Problems.
　AS/CJA, 1973, No. 39 (no. 2), 61.
Anon. (*116*)
　Hsien-yang Shih Chin Nien Fa-hsien-ti
　　I-Phi Chhin Han I Wu 咸陽市近年
　　發現的一批秦漢遺物.
　Recent Finds of the Chhin and Han
　　Dynasties at the City of Hsienyang.
　KKTH, 1973 (no. 3), 167 and pl. 10, fig. 2.
Anon. (*117*).
　Kuan-yü Chhangsha Ma-Wang-Tui
　　I-hao Han Mu ti Tso Than Chi-Yao
　　關於長沙馬王堆一號漢墓的座談
　　記要.
　Report of an Editorial Round-Table
　　Discussion on the Han Tomb Ma-
　　wang-tui No. 1 at Chhangsha.
　KKTH, 1972, no. 5 (no. 122), 37.
Anon. (*118*).
　Hsi Han Po Hua 西漢帛畫.
　A Silk Painting of the Western (Former)
　　Han, [the T-shaped Banner in the
　　Tomb of the Lady of Tai, *c.*–166],
　　(album, with explanatory introduction).
　Wên-Wu, Peking, 1972.
Anon. (*124*).
　Tshung Sian Nan Chiao Chhu-thu-ti
　　I-Yao Wên-Wu Khan Thang-Tai I-Yao-
　　ti Fa-Chan 從西安南郊出土的醫
　　藥文物看唐代醫藥的發展.
　The Development of Pharmaceutical
　　Chemistry in the Thang Period as seen
　　from the specimens of (inorganic) drugs
　　recovered from the Hoard in the
　　Southern Suburbs of Sian [the silver
　　boxes with the labelled chemicals,
　　c.+756].
　WWTK, 1972 (no. 6), 52.

Anon. (*125*).
　Chhangsha Ma-Wang-Tui I-hao Han Mu
　　Fa-Chüeh Pao-Kao yü Wên-Tzu Thu
　　Lu 長沙馬王堆一號漢墓發掘
　　報告與文資圖錄.
　The Excavation of Han Tomb No. 1 at
　　Ma-wang-tui (Hayagriva Hill) near
　　Chhangsha; Illustrated (Full) Report
　　including the Literary Evidence.
　2 Vols. (text and album).
　Wên-Wu, Peking, 1973, for the Hunan
　　Provincial Museum and the Archaeo-
　　logical Institute of Academia Sinica.
　Rev. *EHOR*, 1974, **13** (no. 3), 66.
Anon. (*126*).
　Ma-Wang-Tui-I hao Han Mu Nü Shih
　　Yen-Chiu-ti Chi-ko Wên-Thi 馬王堆
　　一號漢墓女屍研究的幾個問題.
Anon. (*126*)
　Some problems raised by the Researches
　　into the (Preservation of the) Female
　　Corpse in the Han Tomb No. 1 at Ma-
　　wang-tui [the Lady of Tai; anatomy,
　　pathology and chemistry].
　WWTK, 1973, no. 7 (no. 206), 74.
　Preceded by a one-page press statement on
　　the same subject by the Hsinhua News
　　Agency.

Chhen Shun-Hua (*1*) 陳舜華.
　Ma-Wang-Tui Han Mu 馬王堆漢墓.
　On the Han Tomb at Ma-wang-tui [the
　　tomb of the Lady of Tai, with particular
　　reference to the causes of the conservation
　　of her body from decay].
　Chung-hua, Hongkong, 1973.
Chhin Ling-Yün (*1*) 秦嶺雲.
　Pei-ching Fa-Hai Ssu Ming Tai Pi Hua
　　北京法海寺明代壁畫.
　The Ming Frescoes of the Fa-Hai Ssu
　　Temple at Peking.
　Chung-Kuo Ku Tien I Shu, Peking,
　　1958.
　Cf. Fig. 1312 in Vol. 5. pt. 2.

Hashimoto Sōkichi (*1*) 橋本宗吉..
　Ranka Naigai Sanbō Hōten 蘭科內外三
　　法方典.
　Handbook of the Three Aspects of Dutch
　　Internal and External Medicine.
　Naniwa (Osaka), 1805.
　Cf. Shimao Eikoh (*1*).

I Ping (1) 一冰.
Thang-Tai Yeh Yin Shu Chhu Than
唐代冶銀術初探.
A Preliminary Study of Silver Smelting
during the Thang Period.
WWTK, 1972 (no. 6), 40.

Kao Yao-Thing (1) 高耀亭.
Ma-Wang-Tui I-hao Han Mu Sui Tsang
Phin chung Kung Shih Yung ti Shou
Lei 馬王堆一號漢墓隨葬品中
供食用的獸類.
On the kinds of Animals used for the
Sacrificial Food Offerings in the Han
Tomb No. 1 at Ma-wang-tui, at the
burial (of the Lady of Tai).
WWTK, 1973, no. 9 (no. 208), 76.

Kêng Chien-Thing (1) 耿鑒庭.
Sian Nan Chiao Thang-Tai Chiao Tsang
li ti I-Yao Wên-Wu 西安南郊唐代
窖藏裡的醫藥文物.
Cultural Objects of Pharmaceutical
Interest found in a Hoard in the Southern
Suburbs of Sian [the silver boxes with
labelled inorganic chemicals, c.+756].
WWTK, 1972 (no. 6), 56.

Ku Thieh-Fu (1) 顧鐵符.
Shih Lun Chhangsha Han Mu li Pao-
Tshun Thiao Chien 試論長沙漢墓的
保存條件.
A Tentative Discussion of the Various
Preservative Conditions in the Tomb at
Chhangsha (of the Lady of Tai).
KKTH, 1972, no. 6 (no. 123), 53.

Li Thao (14) 李濤.
Chung-Kuo Tui-Yü Chin-Tai Chi Chung
Chi Chhu I-Hsüeh-ti Kung-Hsien 中國
對於近代幾種基礎醫學的貢獻.
Chinese Contributions to some of the
Fundamental Principles of Modern
Medicine.
ISTC, 1955, 7 (no. 2), 110.

Ma Yung (1) 馬雍.
Lun Chhangsha Ma-Wang-Tui I-hao
Han Mu Chhu Thu Po Hua Ti Ming
Chhêng Ho Tso Yung 論長沙馬王
堆一號漢墓出土帛畫的名稱和作用.
A Discussion of the Name and Function
of the Silk Painting Unearthed from the
Han Tomb No. 1 at Ma-wang-tui,
Chhangsha.
KKTH, 1973 (no. 2), 118.

Miyashita Saburō (3) 宮下三郎.
Tonkōbon 'Chō Chūkei Gozōron' Kōyakuchū
敦煌本張仲景「五藏論」校譯注.
Critical Edition and Annotated Japanese
Translation of three Tunhuang MSS of
Chang Chung-Ching's Wu Tsang Lun

(Treatise on the Five Viscera) [S 5614,
P 2755, 2378].
TG/K, 1964, 35, 289.
Abstr. RBS, 1973, 10, no. 897.
Rev. P. Demiéville, TP, 1970, 56, 18.

Nakamura Mokuko (1) 中村木公.
Meika Chōjujitsu Rekidan 名家長壽實
歷談.
A Discussion of Some Celebrated Cases
of Great Longevity.
Tokyo, 1907.

Shih Shu-Chhing (4) 史樹青.
Wo Kuo Ku-Tai-ti Chin Tsho Kung I
我國古代的金錯工藝.
The Art of Inlaying (Gold and Silver) in
Ancient China.
WWTK, 1973, no. 6 (no. 205), 66.

Shih Wei (1) 史爲.
Chhangsha Ma-Wang-Tui I-hao Han Mu-ti
Kuan Kuo Chih-Tu 長沙馬王堆
一號漢墓的棺槨制度.
On the Design and Construction of the
Inner and Outer Coffins in the Ma-
wang-tui Tomb No. 1 at Chhangsha.
KKTH, 1972, no 6 (no. 123), 48.

Sun Tso-Yün (2) 孫作雲.
Ma-Wang-Tui I-hao Han Mu Chhi Kuan
Hua Khao Shih 馬王堆一號漢墓
漆棺畫考釋.
The Mythological Paintings on the Lacquer
Coffins of the Han Tomb at Ma-wang-
tui, No. 1 (Chhangsha).
KKTH, 1973, no. 4 (no. 127), 247.

Sun Tso-Yun (3) 孫作雲.
Chhangsha Ma-Wang-Tui I-hao Han Mu
Chhu Thu Hua Fan Khao Shih 長沙
馬王堆一號漢墓出土畫幡考釋.
Studies on the Silk Painting Unearthed
from the Han Tomb Ma-wang-tui No. 1
at Chhangsha.
KKTH, 1973, no. 1 (no. 124), 54.

Yü Hsing-Wu (1) 于省吾.
Kuan-yü Chhangsha Ma-Wang-Tui I-hao
Han Mu Nei Kuan Kuan Shih Chieh
Shuo 關于長沙馬王堆一號漢墓
內棺棺飾的解說.
An Interpretation of the Decorations on
the Inner Coffin of Han Tomb
Ma-wang-tui No. 1 at Chhangsha.
KKTH, 1973 (no. 2), 126.

Yü Ying-Shih (1) 余英時.
Fang I-Chih Wan Chieh Khao 方以智晚
節考.
Fang I-Chih; his last Years and his Death.
Inst. of Adv. Chinese Studies and Research,
New Asia College, Chinese University,
Hongkong, 1972.

夏	HSIA kingdom (legendary?)	c. −2000 to c. −1520
商	SHANG (YIN) kingdom	c. −1520 to c. −1030
周	CHOU dynasty (Feudal Age) { Early Chou period	c. −1030 to −722
	Chhun Chhiu period 春秋	−722 to −480
	Warring States (Chan Kuo) period 戰國	−480 to −221
First Unification 秦	CHHIN dynasty	−221 to −207
漢 HAN dynasty {	Chhien Han (Earlier or Western)	−202 to +9
	Hsin interregnum	+9 to +23
	Hou Han (Later or Eastern)	+25 to +220
三國	SAN KUO (Three Kingdoms period)	+221 to +265
First Partition 蜀	SHU (HAN) +221 to +264	
魏	WEI +220 to +265	
吳	WU +222 to +280	
Second Unification 晉	CHIN dynasty : Western	+265 to +317
	Eastern	+317 to +420
劉宋	(Liu) SUNG dynasty	+420 to +479
Second Partition	Northern and Southern Dynasties (Nan Pei chhao)	
齊	CHHI dynasty	+479 to +502
梁	LIANG dynasty	+502 to +557
陳	CHHEN dynasty	+557 to +589
魏 {	Northern (Thopa) WEI dynasty	+386 to +535
	Western (Thopa) WEI dynasty	+535 to +556
	Eastern (Thopa) WEI dynasty	+534 to +550
北齊	Northern CHHI dynasty	+550 to +577
北周	Northern CHOU (Hsienpi) dynasty	+557 to +581
Third Unification 隋	SUI dynasty	+581 to +618
唐	THANG dynasty	+618 to +906
Third Partition 五代	WU TAI (Five Dynasty period) (Later Liang, Later Thang (Turkic), Later Chin (Turkic), Later Han (Turkic) and Later Chou)	+907 to +960
遼	LIAO (Chhitan Tartar) dynasty	+907 to +1124
	West LIAO dynasty (Qarā-Khiṭāi)	+1124 to +1211
西夏	Hsi Hsia (Tangut Tibetan) state	+986 to +1227
Fourth Unification 宋	Northern SUNG dynasty	+960 to +1126
宋	Southern SUNG dynasty	+1127 to +1279
金	CHIN (Jurchen Tartar) dynasty	+1115 to +1234
元	YUAN (Mongol) dynasty	+1260 to +1368
明	MING dynasty	+1368 to +1644
清	CHHING (Manchu) dynasty	+1644 to +1911
民國	Republic	+1912

N.B. When no modifying term in brackets is given, the dynasty was purely Chinese. Where the overlapping of dynasties and independent states becomes particularly confused, the tables of Wieger (1) will be found useful. For such periods, especially the Second and Third Partitions, the best guide is Eberhard (9). During the Eastern Chin period there were no less than eighteen independent States (Hunnish, Tibetan, Hsienpi, Turkic, etc.) in the north. The term 'Liu chhao' (Six Dynasties) is often used by historians of literature. It refers to the south and covers the period from the beginning of the +3rd to the end of the +6th centuries, including (San Kuo) Wu, Chin, (Liu) Sung, Chhi, Liang and Chhen. For all details of reigns and rulers see Moule & Yetts (1).

SUMMARY OF THE CONTENTS OF VOLUME 5

CHEMISTRY AND CHEMICAL TECHNOLOGY

Part 2, Spagyrical Discovery and Invention:

Magisteries of Gold and Immortality

33 Alchemy and Chemistry

Introduction; the historical literature

Primary sources
Secondary sources

Concepts, terminology and definitions

Aurifiction and aurifaction in the West
  The theory of *chrysopoia*
  The persistence of the aurifactive dream
The artisans' cupel and the enigma of aurifactive
  philosophy
Gold and silver in ancient China
Cupellation and cementation in ancient China
Aurifaction in the *Pao Phu Tzu* book
The drug of deathlessness; macrobiotics and immortality-
  theory in East and West
  Hellenistic metaphor and Chinese reality
  Ideas about the after-life in East and West
  The *hun* and *pho* souls
  Material immortality; the *hsien* and the celestial
    bureaucracy
Macrobiotics and the origin of alchemy in ancient China
The missing element; liturgy and the origins of Chinese
  alchemy
  Incense, prototypal reactant
  Fumigation, expellant and inductant
Nomenclature of chemical substances

The metallurgical-chemical background; identifications of alchemical processes

The availability of metallic elements
Golden uniform-substrate alloys
  The origin of the brasses
  The origins of zinc
  Other golden alloys
Arsenical copper
Silvery uniform-substrate alloys
  Paktong ('Tanyang copper', cupro-nickel)
  Chinese nickel in Greek Bactria?
  Other silvery alloys
Amalgams
The treatment of metal and alloy surfaces
  Superficial enrichment; the addition of a layer of
    precious metal (gilding and silvering)
  Superficial enrichment; the withdrawal of a layer of
    base metal (cementation)
  The deposition of coloured surface-films ('tingeing',
    bronzing, pickling, dipping)
  'Purple sheen gold' and *shakudō*
Violet alloys, 'purple of Cassius', ruby glass, mosaic gold,
  and the *panacea antimonialis*
Thang lists of 'golds' and 'silvers', artificial and genuine

The physiological background; verifications of
the efficacy of elixirs

Initial exhilaration
Terminal incorruptibility

Part 3, Spagyrical Discovery and Invention:

Historical Survey, from Cinnabar Elixirs to Synthetic Insulin

The historical development of alchemy and early
  chemistry

The origins of alchemy in Chou, Chhin and Early Han;
  its relation with Taoism
  The School of Naturalists and the First Emperor
  Aurifiction and aurifaction in the Han
  The three roots of elixir alchemy
Wei Po-Yang; the beginnings of alchemical literature in
  the Later Han (+2nd cent.)
Ko Hung, systematiser of Chinese alchemy (c. +300), and
  his times
  Fathers and masters
  The *Pao Phu Tzu* book and its elixirs
  Character and contemporaries
Alchemy in the Taoist Patrology (*Tao Tsang*)
The golden age of alchemy; from the end of Chin (+400)
  to late Thang (+800)
  The Imperial Elaboratory of the Northern Wei and the
    Taoist Church at Mao Shan
  Alchemy in the Sui re-unification
  Chemical theory and spagyrical poetry under the
    Thang
  Chemical lexicography and classification in the Thang
  Buddhist echoes of Indian alchemy

The silver age of alchemy; from the late Thang (+800) to
  the end of the Sung (+1300)
  The first scientific printed book, and the court alchemist
    Mistress Kêng
  From proto-chemistry to proto-physiology
  Alchemy in Japan
  Handbooks of the Wu Tai
  Theocratic mystification, and the laboratory in the
    National Academy
  The emperor's artificial gold factory under Metallurgist
    Wang Chieh
  Social aspects, conventional attitudes and gnomic
    inscriptions
  Alchemical compendia and books with illustrations
  The Northern and Southern Schools of Taoism
Alchemy in its decline; Yuan, Ming and Chhing
  The Emaciated Immortal, Prince of the Ming
  Ben Jonson in China
  Chinese alchemy in the age of Libavius and Becher
  The legacy of the Chinese alchemical tradition
The coming of modern chemistry
  The failure of the Jesuit mission
  Mineral acids and gunpowder
  A Chinese puzzle—eighth or eighteenth?
  The Kiangnan Arsenal and the sinisation of modern
    chemistry

Part 4, Spagyrical Discovery and Invention:

Apparatus and Theory

Laboratory apparatus and equipment

The laboratory bench
The stoves *lu* and *tsao*
The reaction-vessels *ting* (tripod, container, cauldron) and *kuei* (box, casing, container, aludel)
The sealed reaction-vessels *shen shih* (aludel, lit. magical reaction-chamber) and *yao fu* (chemical pyx)
Steaming apparatus, water-baths, cooling jackets, condenser tubes and temperature stabilisers
Sublimation apparatus
Distillation and extraction apparatus
  Destillatio per descensum
  The distillation of sea-water
  East Asian types of still
  The stills of the Chinese alchemists
  The evolution of the still
  The geographical distribution of still types
The coming of Ardent Water
  The Salernitan quintessence
  Ming naturalists and Thang 'burnt-wine'
  Liang 'frozen-out wine'
  From icy mountain to torrid still
  Oils in stills; the rose and the flame-thrower
Laboratory instruments and accessory equipment

Reactions in aqueous medium

The formation and use of a mineral acid
'Nitre' and *hsiao*; the recognition and separation of soluble salts
Saltpetre and copperas as limiting factors in East and West
The precipitation of metallic copper from its salts by iron
The role of bacterial enzyme actions
Geodes and fertility potions
Stabilised lacquer latex and perpetual youth

The theoretical background of elixir alchemy [with Nathan Sivin]

Introduction
  Areas of uncertainty
  Alchemical ideas and Taoist revelations
The spectrum of alchemy
The role of time
  The organic development of minerals and metals
  Planetary correspondences, the First Law of Chinese physics, and inductive causation
  Time as the essential parameter of mineral growth
  The subterranean evolution of the natural elixir
The alchemist as accelerator of cosmic process
  Emphasis on process in theoretical alchemy
  Prototypal two-element processes
  Correspondences in duration
  Fire phasing
Cosmic correspondences embodied in apparatus
  Arrangements for microcosmic circulation
  Spatially oriented systems
  Chaos and the egg
Proto-chemical anticipations
  Numerology and gravimetry
  Theories of categories

Comparative survey

China and the Hellenistic world
  Parallelisms of dating
  The first occurrence of the term 'chemistry'
  The origins of the root 'chem-'
  Parallelisms of content
  Parallelisms of symbol
China and the Arabic world
  Arabic alchemy in rise and decline
  The meeting of the streams
  Material influences
  Theoretical influences
  The name and concept of 'elixir'
Macrobiotics in the Western world

Part 5, Spagyrical Discovery and Invention:

Physiological Alchemy

The outer and the inner macrobiogens; the elixir and the enchymoma

Esoteric traditions in European alchemy
Chinese physiological alchemy; the theory of the enchymoma (*nei tan*) and the three primary vitalities
  The quest for material immortality
  Rejuvenation by the union of opposites; an *in vivo* reaction
The *Hsiu Chen* books and the *Huang Thing* canons
The historical development of physiological alchemy
The techniques of macrobiogenesis
  Respiration control, aerophagy, salivary deglutition and the circulation of the *chhi*
  Gymnastics, massage and physiotherapeutic exercise
  Meditation and mental concentration
  Phototherapeutic procedures
  Sexuality and the role of theories of generation
The borderline between proto-chemical (*wai tan*) and physiological (*nei tan*) alchemy

Late enchymoma literature of Ming and Chhing
  The 'Secret of the Golden Flower' unveil'd
Chinese physiological alchemy (*nei tan*) and the Indian Yoga, Tantric and Hathayoga systems
  Originalities and influences; similarities and differences
Conclusions; *nei tan* as proto-biochemistry

The enchymoma in the test-tube; medieval preparations of urinary steroid and protein hormones

Introduction
The sexual organs in Chinese medicine
Proto-endocrinology in Chinese medical theory
The empirical background
The main iatro-chemical preparations
Comments and variant processes
The history of the technique

DATE DUE

12 Sept 96			
GAYLORD			PRINTED IN U.S.A.